# Handbook of Mathematical Geosciences

B. S. Daya Sagar · Qiuming Cheng
Frits Agterberg
Editors

# Handbook of Mathematical Geosciences

## Fifty Years of IAMG

*Editors*
B. S. Daya Sagar
Systems Science and Informatics Unit
Indian Statistical Institute–Bangalore Centre
Bengaluru
India

Frits Agterberg
Geological Survey of Canada
Ottawa, ON
Canada

Qiuming Cheng
State Key Lab of Geological Processes and Mineral Resources
China University of Geosciences
Beijing
China

https://doi.org/10.1007/978-3-319-78999-6

*Dedicated to*
*Daniel F. Merriam and Richard A. Reyment*
*(Fathers of the IAMG)*

# Foreword

The International Association for Mathematical Geosciences (IAMG) was founded during the 23rd International Geological Congress in Prague, August 1968. Within the Earth Sciences, the IAMG has played a prominent role during the past 50 years by living up to its mandate to promote, worldwide, the advancement of mathematics, statistics, and informatics in the geosciences. Under its auspices there have been and continue to be important developments in applications of mathematics, statistics and computer science in the Earth Sciences. To give two examples: IAMG members Georges Matheron and Jean Serra developed geostatistics and mathematical morphology resulting in methods that are now widely applied in other branches of science and engineering; John Aitchison invented methods to circumvent the problem of spurious correlations that often arise in compositional data analysis of petrological and geochemical data. IAMG members later followed up on developing this topic now used in other fields of science and in the social sciences as well. During the first 30 years of its existence, IAMG stood as the abbreviation of International Association for Mathematical Geology, but its current name was adopted to widen its scope and provide a home to scientists who are not only geologists but who perform research in other fields of science and engineering. From the beginning, prominent mathematical statisticians including John Tukey, Geoffrey Watson, and Franklin Graybill played a prominent part within the IAMG by providing advice and collaborating in research projects.

In addition to organizing or co-sponsoring international conferences, workshops, and lecture series, the IAMG established three successful scientific journals: Mathematical Geosciences, Computers & Geosciences, and Natural Resources Research (formerly: Nonrenewable Resources). In total, five types of IAMG awards were created to honor William Christian Krumbein (1902–1979), Andrew Borisovich Vistelius (1915–1995), John Cedric Griffiths (1912–1992), Felix Chayes (1916–1993), and Georges Matheron (1930–2000), who were pioneers in mathematical geology. The book in front of us "Handbook of Mathematical Geosciences: Fifty Years of IAMG" published to celebrate the Golden Anniversary of the IAMG contains 45 chapters prepared by IAMG award winners, founding members, and distinguished lecturers. It covers new theoretical developments, applications, reviews of subfields of the mathematical geosciences, and historical information on the IAMG, especially in its early years.

Bill Krumbein, as a geologist, first started using a digital computer in 1958, and gradually more mathematical geologists began working with digital computers in the 1960s. This involved the development of computer programs written in FORTRAN or ALGOL to use existing statistical techniques such as analysis of variance, multiple regression, multivariate statistical techniques, and time series analysis that had been developed during the first half of the twentieth century. Also, new methods including trend surface analysis and geostatistical ore reserve estimation techniques were developed specifically for solving geoscience problems. Dan Merriam established the “Kansas Geological Survey Computer Contributions.” In this series, 50 computer programs were published between 1966 and 1970. During this time period, Dick Reyment worked closely with Dan to establish the IAMG.

Computers brought about further important changes that were rapidly adopted by mathematical geologists including geographic information systems (GIS), exploratory data analysis, the fast Fourier transform, mathematical morphology, fractals, and nonlinear models. Even more recently, our world has entered the “Big Data” era, with the production of data with unprecedented speed and in large quantities. The new knowledge obtained through digital analysis and the novel methods of data mining are greatly benefitting human decision making. People’s life, working, and thinking are being subjected to drastic changes. “Big Data” resulted in the emergence of “Data Science” which, to some extent, is affecting all fields of science both in how scientific research is being conducted using digital data and by facilitating the use of scientific methods to study the digital data.

Nowadays, geosciences and geological research are mainly characterized by the following words: “Systematic,” “Comprehensive,” “Quantitative,” “Three-dimensional,” “New-model,” “Green,” “Intelligent,” and “Beneficial to People.” In this regard, Mathematical Geosciences and the IAMG play an increasingly important role, prompting the advancement of the geosciences in the future. Earth science and geological studies are data-intensive. If we want to solve geological problems and use the results in a meaningful way, we have to obtain and work with many different kinds of data obtained by using sound geological concepts and methods borrowed from physics, chemistry, and remote sensing. Geoscience experts in the latter fields of science make invaluable contributions to our understanding of the Earth and the geological processes that took place millions of year ago. In all these endeavors, mathematics plays a significant role. This is where the IAMG is exceedingly helpful. Geology is characterized by the four “Deeps”: Its data and processes are deep in the Earth, deep under the sea, deep in outer space, and deep in time. It is not easy to obtain comprehensive geological data sets in practice. Data collection can be very expensive. Much attention is to be paid to costs and benefits.

Earth scientists should always do their best to define target populations from which truly representative samples are to be drawn. Geological samples almost never fully comprise the entire population of study because of differences in space and time. There is no "overall data completeness" or "comprehensive data" in geological science and practice. Other methods of data collection have to be developed and used in order to make the random samples fit the target populations as closely as possible so that information loss because of spatial restrictions is minimized.

The ultimate purpose of Earth Science is to promote progress and development of human society: The products of the Earth's evolution over millions of years are to be used to our advantage, and we have to guard against the negative effects of the different types of disasters that can be associated with geological processes. Geological data have particular characteristic features that reflect time and cause of origin, spatial environments, and genesis. They can manifest different outcomes reflecting spatial and temporal conditions. When faced with geological data, one should not only know the "What?" but also the "Why?" and the "How?" for the data: What they truly mean and how they are to be used. One should not only establish "correlations" but also "causality" and spatiotemporal relations. Geology differs from most other areas in the Big Data era in that the focus is on the "What?" only and on correlations without causality and the "Why."

The laws of physics and chemistry have not changed through geologic time. This fact underlies the principle of actualism already understood by geologists in the nineteenth century. Some early geologists already surmised that the ice ages of which the effects can be clearly seen on the surface of the Earth were caused by minor systematic fluctuations in amount of radiation received from the sun. A full explanation of this periodicity was provided in the theory of Milankovitch. This theory currently is used to estimate ages of stage boundaries in the geologic timescale during the past 65 million years with a precision that is better than precisions provided by geochronological dating methods.

The age of the Earth is about 4.5 billion years, and it is in its middle age. Taking 90 years as expectation of human age, for example, this means that one year in our life is approximately equivalent to 50 million years in the past of the Earth. Thus, the factor of difference is about 4,500,000,000/90 = 50,000,000. The following examples illustrate the change of perspective needed to understand geological processes. Earthquakes with a magnitude greater than 8.0 earthquakes on the Gutenberg–Richter scale occur about once a year. Consequently, about 50 million such earthquakes probably have occurred over the last 50 million years. The speed of tectonic plates is of the order of 1–10 cm/year. Thus, plates have moved 500–5000 km per 50 million years. It explains why oceans are opening and closing over geologic time.

Early in the nineteenth century, it became known that most coal deposits originated during the Carboniferous. More recently, Earth scientists have developed theories about the genesis of ore and hydrocarbon deposits that help to make new discoveries. Recognition of importance of bio-factors has aided in the understanding of various geological processes including ore and hydrocarbon formation, as well as distribution of pollutants in the ecosystem. Increasingly, mathematics and statistics are fruitfully employed in the discovery process as abundantly exemplified in many of the chapters in this Handbook. All of the preceding considerations illustrate the complexity and particularities of geological data as well as their usefulness and importance. Fully comprehensive geological data collection, their effective computer-based treatment, rational analysis, and translation into digital knowledge, all depend on the guidance provided by powerful theory based on mathematics with applications of efficient methods.

Initially, most IAMG members were located within the USA or Europe. These regions continue to have relatively many members, but China and other Asian countries now also constitute a large regional group. In 1990, a workshop was organized at the China University of Geosciences in Wuhan at which the participants included Richard McCammon, IAMG President at the time as well as four future IAMG Presidents. Now, the IAMG's China Section holds annual meetings attended by hundreds of mathematical geoscientists. Increasingly, it became felt that mathematical geoscience is making an indispensable contribution in China to aid in the prediction of occurrences of mineral resources, especially in the non-traditional regions such as deep Earth and in covered regions and the assessment of hazards such as earthquakes and landslides. As society develops from its industrialization to post-industrialization stage, environmental and ecological applications become increasingly important to establish and reduce the effects of regional patterns of pollution. Other anticipated areas of applications are urban space utilization and agricultural products under the new concepts of green and low-carbon development.

Beijing, China

Pengda Zhao
Academician of the Chinese Academy of Sciences,
China University of Geosciences

Ottawa, Canada
February 2018

Frits Agterberg
Geological Survey of Canada

# Preface

The International Association of Mathematical Geosciences (IAMG) was formed in 1968, and the year 2018 is marked as its Golden Anniversary. The "Handbook of Mathematical Geosciences: Fifty Years of IAMG" released during the IAMG Conference held at Olomouc and Prague (Czech Republic), September 2–8, 2018, motivates readers including professional geomathematicians, and undergraduate and postgraduate students to learn about the fifty years of contributions by award-winning mathematical geoscientists. This book that showcases the success of the IAMG celebrating its fifty years of existence is a compilation of 45 chapters. Compiled by academics, scientists, and engineers who are the recipients of IAMG's accolades such as the Krumbein Medal/Chayes Prize/Vistelius Award/Griffiths Award/Matheron Lectureship/Distinguished Lectureship/Honorary Membership as well as IAMG Founding Members, this Handbook covers 45 chapters on topics such as mathematical geosciences, mathematical morphology, geostatistics, fractals and multifractals, spatial statistics, multipoint geostatistics, compositional data analysis, informatics, geocomputation, numerical methods, and chaos theory in the geosciences categorized broadly into theory, general applications, exploration and resource estimation, reviews, and reminiscences. Unique features of this book include the following:

- Contributions by award-winning mathematical geoscientists of interest to academics/researchers/students engaged in applications of mathematics, statistics, computers, and informatics.
- A unique fusion of geology, hydrology, mining engineering, geoengineering, and applications of quantitative techniques and methodology in the aforementioned fields.
- Historical perspectives on how the IAMG evolved during the past fifty years.
- Past, present, and future demands for mathematical geosciences in academics, industry, and the professions.
- Pathbreaking mathematical frameworks/approaches/methodologies/algorithms to deal with varied aspects usually encountered by geoscientists.

The first ten chapters are categorized as theoretical, followed by seven chapters (from 11 to 17) in the general applications part. Chapters 18–26 and 27–35 are, respectively, categorized as exploration and resources estimation, and reviews. The last ten chapters (from 36 to 45) are categorized as reminiscences. What follows includes a brief summary for each of the chapters of the Handbook.

Chapter 1 by Dubrule reviews relationships between Bayesian methods, geostatistics, and ensemble Kalman filtering which are well discussed and reviewed. The author rightly mentions that (i) inversion techniques are not discussed and (ii) fast-growing machine learning algorithms are challenging the geostatistical and Bayesian formalisms.

In Chap. 2, Baddeley compares and contrasts various statistical methods–such as logistic regression, Poisson point process models, maximum entropy, monotone regression, nonparametric survey estimates, recursive partitioning, and receiver operating characteristic curves–for predicting the occurrence of mineral deposits.

Chapter 3 by Schaeben is concerned with testing joint conditional independence of categorical random variables with a newly proposed standard likelihood ratio test. How it resolves limitations obvious with "omnibus" and "new omnibus" tests is explained with a strong theoretical basis invoking the Hammersley–Clifford theorem.

The sample space approach for modeling compositional data is explained in Chap. 4 by Egozcue and Pawlowsky-Glahn. Interestingly, perturbation between elements and its opposite, i.e., difference perturbation, appear to be Matheron–Serra's morphological dilations and erosions or Minkowski additions and subtractions. Repeated perturbations and their inverted versions (difference perturbations) seem to be multiscale morphological dilations and erosions.

Possible methods required to refocus and streamline expert geological judgment inputs along with analytical methods are reviewed by Kaufman in Chap. 5.

Remotely sensed satellite data acquisition via various sensing mechanisms pose challenges particularly in developing filters meant for feature extraction or retrieval. Many developed filters yield promising results, but could not be generalized due to varied complexities involved in sensing mechanisms leading to the acquisition of different types of satellite images. For instance, filters that work fine for satellite images acquired via optical sensing mechanisms would not yield appropriate results for those images acquired via microwave sensing mechanisms. Besides, satellite images now available are with a large number of channels at high spatial/temporal/spectral resolutions making the ability to map features with high degree of precision more challenging. However, due to availability of filters that cannot be generalized for images acquired by different mechanisms, there is a need for the development of filters with strong theoretical basis. Cressie contributes rich content in Chap. 6, and the ideas provided in this chapter are of fundamental importance.

Deutsch in his Chap. 7 provides convincing arguments/discussions that are logical and powerful on why the ensemble of realizations needs to be considered instead of one single realization for proper planning, decision making, and uncertainty assessment.

In the past forty years, how criteria and arguments are employed in comparing binary coefficients in multivariate statistical analysis is reviewed in Chap. 8 by Hohn.

Armstrong, Mondaini, and Camargo provide a sociological study based on Google retrievals in Chap. 9. How research in geosciences diffuses within academia and into industry is studied in this chapter, whereby the research idea employed is plurigaussian simulation invented in France. This study is someway related to "scientometrics." The obvious choice to carry out this type of study is complex network based analysis, small-world network analysis (due to Duncan Watts and Steven Strogatz). Such ideas in social network analysis were predominantly developed by Barabasi and his group.

In the first part of Chap. 10, Cheng gave an excellent overview chronologically on how mathematical geosciences or geomathematics evolved in the last fifty years by also providing (i) historical connections between the mathematics and the geosciences, and (ii) a new definition of mathematical geosciences. An introduction to fractal density and singularity analysis and related subjects to solve geological problems discussing geological principles with case studies related to earthquakes is provided in the second part of this chapter. Cheng demonstrated the application of his original concept of fractal density and the local singularity to model the clustering frequency of earthquakes of the Pacific subduction zones. Much stronger singularity is discovered via the clustering frequency of earthquakes in the colder and older western boundaries of Pacific plates than that of the hotter and younger eastern boundaries of the Pacific plates.

Use of electrofacies in reservoir characterization is provided with demonstration on a giant clastic oil reservoir, the Amal field of Libya, in Chap. 11 by Davis.

In Chap. 12, morphological medians and weighted morphological medians are employed by Serra in a new elegant approach demonstrated on shoreline extrapolations. Quench stripe generation, based on these novel two types of medians provides the main basis in predicting the locations of the shorelines.

A comprehensive review of geostatistical methods to analyze remote-sensing data is presented in Chap. 13 by Militino, Ugarte, and P′erez-Goya. This review highlights the importance of geostatistics in processing and analysis of remotely sensed satellite data available in multiple spatial/temporal/spectral resolutions acquired via a host of different sensing mechanisms.

Chapter 14 by Goovaerts contains an interesting first application of space–time geostatistics to assess lead levels recorded in drinking water of public distribution system in Flint, Michigan.

Statistical Parametric Mapping (SPM)—popular in medical imaging science to evaluate differences between individual pairs of images or average images—applied on examples drawn from environmental and geoscience contexts is reviewed in Chap. 15 by McKenna. Extending the application of SPM to the hundreds of channels of hyperspectral remotely sensed satellite data would provide new insights into remote-sensing scientists.

In the interesting Chap. 16, Buccianti shows how compositional data analysis has a role in dealing with water chemistry. The author puts Illya Prigogine's ideas and concepts (including dissipative structures, dynamical systems, open and closed systems that respectively draw energy from external sources and from within, self-organized criticality, universal power laws, time irreversibility) into a new perspective. It reminds the reader of the popular book on Chaos: Man's New Dialogue with Nature by Illya Prigogine and Isabelle Stengers.

Chapter 17 by Grunsky, Drew, and Smith is the outcome of a major project concerned with soil geochemical analyses in the USA via principal component analysis and compositional framework approach. The material is presented with many maps, tabular data, and supplementary information.

Work carried out across three decades by Dowd and his group on the quantification of uncertainty in mineral/energy/environmental applications via various approaches is reviewed with a focus on mineral and energy resources, and environmental applications in Chap. 18.

Olea in Chap. 19 explains uncertainty, geostatistics, and kriging methods on the basis of a coal seam example. Three ad hoc methods, namely distance analysis, kriging, and stochastic simulation, are employed for evaluation of their usage for predicting changes in uncertainty due to changes in spatially correlated samples. Also included is a demonstration of the efficacy of these methods on real data for the Anderson coal bed. It is inferred that the stochastic simulation-based approach outperforms distance and kriging-based methods.

The topic in relation to predicting molybdenum deposit growth as a function of cutoff grade via a nonlinear model constructed by using data from several deposits is addressed in Chap. 20 by Schuenemeyer, Drew, and Bliss. Predicting molybdenum deposit growth cutoff grades is decided on the basis of a prior model derived by plotting cutoff grade as a function of deposit grade.

Chapter 21 by Pan provides a discussion with focus on several aspects of mineral resources, mineral resource estimation, and associated features with more information on how/why details provided in this chapter are of fundamental importance.

Mineral resource assessment problems and involved three types of errors are discussed in Chap. 22 by Singer. Also presented in this chapter are possible ways to avoid these errors. The chapter is written in a way that can be understood by non-mathematicians or non-statisticians.

In Chap. 23 by Bonham-Carter and Grunsky, two exploratory multivariate methods, namely proximity regression and residual principal component analysis, are applied to analyze geochemical survey data. The first method is useful in making predictions of spatial proximity to geological features, whereas the second method is a recommended way for partitioning geochemical elements into clusters.

Chapter 24 by Doveton is concerned with an approach to compositional data analysis that is significantly different from the Aitchison/Pawlowsky-Glahn/Egozcue approach to CoData problem-solving.

Two parts of Chap. 25 by Soares and Azevedo, respectively, provide the (i) state of the art in recent geostatistical seismic inversion methods and their applications to evaluate reservoir properties, and (ii) seismic inversion-based methodology to assess uncertainty and risks at early stage of exploration.

In Chap. 26, Agterberg provides rich information-related studies to understand the differences in the degree of heterogeneities in the spatial distribution of metal deposits between the regional level and global level. It is interesting to see that de Wijs' work formed the basis for this new version of the model that provides a framework for explaining difference between regional and worldwide distributions. The de Wijs model has also been used elsewhere in the iterated bisection process to compute multifractal spectra that provide a host of dimensions such as topological dimension, capacity dimension, and information dimension. A host of such dimensions is of immense use to understand not only spatial but also temporal distribution patterns.

Chapter 27 by Caers provides views on why philosophical principles are required to be translated into workable practices.

Various approaches involving spatial statistics, geological variables, geometry and topology of geological objects to develop coherent Earth models are well documented as an excellent review in Chap. 28 by Caumon.

Origins of kriging, its success, and its new application domains across the last five decades, and the role of IAMG journals popularizing this technique by publishing in English are explained in Chap. 29 by Chilès and Desassis.

Recent advances in Multiple-Point Statistics (MPS)—that is important and significant in handling complex and realistic phenomena of relevance to the Earth sciences—are thoroughly reviewed in Chap. 30 by Tahmasebi.

Mariethoz provides interesting views on the conceptual differences between the concurrent approaches of Minimum Point Statistics and Covariance-Based Geostatistics in Chap. 31 with an illustrated example.

Srivastava provides information on the origin of Multiple-Point Statistics (MPS) algorithms along with many personal reminiscences in Chap. 32.

Chapter 33 by van den Boogaart and Tolosana-Delgado contains useful new proposals. This chapter provides state of the art and mathematical building blocks for solutions in predictive geometallurgy—i.e., the understanding of geometallurgy. The chapter further explores possible links between geometallurgical problems and relevant techniques taken from mathematical geosciences. From the insights provided into this chapter, the next generation of mathematical geoscientists and experts in geoinformatics would surely benefit.

Chapter 34 by Ma provides possible links between mathematical geosciences and Data Science. Many learning techniques such as artificial intelligence, active learning, machine learning and intelligence, and deep learning approaches together now play a much bigger role in pattern discovery from massive data sets—predictive geosciences. The journey from toy models developed by nonlinear physicists to predictive models has posed several newer challenges. Data Science would bring under one umbrella the powerful theories, algorithms available under different names in different disciplines.

Daya Sagar reviews potential applications of nonlinear mathematical morphological transformations to deal with a host of challenges encountered in geosciences and Geographical Information Science (GISci) with a large number of excellent case studies shown illustratively in Chap. 35.

Many recollections by IAMG members from the old days are provided in Chap. 36 by Cubitt and Henley, with contributions provided by T. Victor (Vic) Loudon, EHT (Tim) Whitten, John Gower, Daniel (Dan) Merriam, Thomas (Tom) Jones, and Hannes Thiergärtner. Also provided in this chapter is information on those pioneering scientists who were instrumental in forming and shaping the IAMG. The chapter is immensely useful for young generation mathematical geoscientists in order to know and appreciate the hard work of peers and scientists of earlier generations.

How the applications of forward and inverse models in particular in Earth science-related problems evolved over a period of 70 years is lucidly explained in simplest possible language by Whitten in Chap. 37. Besides this, how other approaches in particular applications of scaling theories or fractal geometry and theory of chaos, in other words nonlinear approaches—that have already shown significant success in modeling and characterization of various phenomena and processes of relevance to the Earth sciences—can be foreseen in the next 50 years to give a scope for further research.

Václav Němec's professional and personal reminiscences are chronologically provided in Chap. 38 by Němec, along with details on the IAMG's formation and personal early development.

Chap. 39 by Henley provides a rounded view of the life and works, and a glimpse of the legacy of Andrey Vistelius, first President of the IAMG.

Many theoretical sound techniques, algorithms, and software tools developed have shown promising results in certain application-specific domains but with limited utility in terms of generalization. Thiergärtner's interesting and genuine views, opinions, and recommendations in Chap. 40 are thought provoking.

Application of kriging, inverse distance methods, and the variogram in multivariate data analysis, spatial estimation, and in texture-based classification are shown with simple illustrations by Carr in Chap. 41.

Full in Chap. 42 provides a review of the development and applications of a linear unmixing method fairly extensively used by geologists during the past 50 years.

Chapter 43 on Pearce Element Ratios provides insight into the evolution of melts in volcanic systems along with many personal memories and (from the point of view of compositional data analysis) a somewhat antiquated method of approach. An excellent review with extensive Skaergaard applications is provided in this chapter by Nicholls.

Myers in Chap. 44 gives a helpful set of reflections by a mathematician who adopted geostatistics as a principal field of research and has made many important contributions to the field along with personal reminiscences on IAMG and the *Journal of Mathematical Geosciences*.

Agterberg in his Chap. 45 provides a holistic view on the beginnings of IAMG and about the academics/scientists/engineers who were instrumental in shaping the IAMG and making it a most successful association promoting worldwide the advancement of mathematics, statistics, and informatics in the geosciences. This chapter enlightens and motivates the young generation mathematical geoscientists.

Bangalore, India — B. S. Daya Sagar
Beijing, China — Qiuming Cheng
Ottawa, Canada — Frits Agterberg

# Contents

## Part V Reminiscences

# Editors and Contributors

## About the Editors

**B. S. Daya Sagar** is a Full Professor of the Systems Science and Informatics Unit (SSIU) at the Indian Statistical Institute. He received his M.Sc. and Ph.D. degrees in Geoengineering and Remote Sensing from the Faculty of Engineering, Andhra University, Visakhapatnam, India, in 1991 and 1994, respectively. He is also first Head of the SSIU. Earlier, he worked in the College of Engineering, Andhra University, Centre for Remote Imaging, Sensing and Processing (CRISP), and the National University of Singapore in various positions during 1992–2001. He served as Associate Professor and Researcher in the Faculty of Engineering and Technology (FET), Multimedia University, Malaysia, during 2001–2007. Since 2017, he has been a Visiting Professor at the University of Trento, Trento, Italy. His research interests include mathematical morphology, GISci, digital image processing, fractals and multifractals, their applications in extraction, analyses, and modeling of geophysical patterns. He has published over 85 papers in journals and has authored and/or guest edited 11 books and/or special theme issues for journals. He recently authored a book entitled *Mathematical Morphology in Geomorphology and GISci*, CRC Press: Boca Raton, 2013, p. 546. He recently co-edited two special issues on "Filtering and Segmentation with Mathematical Morphology" for *IEEE Journal of Selected Topics in Signal Processing* (v. 6, no. 7, p. 737–886, 2012), and "Applied Earth Observation and Remote Sensing in India" for *IEEE Journal of Selected*

*Topics in Applied Earth Observation and Remote Sensing* (v. 10, no. 12, p. 5149–5328, 2017). He is an elected Fellow of Royal Geographical Society (1999), Indian Geophysical Union (2011), and was a Member of New York Academy of Sciences during 1995–1996. He received the Dr. Balakrishna Memorial Award from Andhra Pradesh Academy of Sciences in 1995, the Krishnan Gold Medal from Indian Geophysical Union in 2002, and the "Georges Matheron Award-2011 (with Lecturership)" of the International Association for Mathematical Geosciences. He is the Founding Chairman of Bangalore Section IEEE GRSS Chapter. He is on the Editorial Boards of Computers and Geosciences, and Frontiers: Environmental Informatics.

**Qiuming Cheng** did his Ph.D. degree in Earth Science under supervision of Dr. Frits Agterberg at the University of Ottawa in 1994. He spent a year at the Geological Survey of Canada as a PDF under the supervision of Dr. Graeme Bonham-Carter and soon became a Faculty Member at York University, Toronto, Canada, in 1995 with cross-appointments in the Department of Earth and Space Science and Engineering and the Department of Geography. He was promoted to associate professor in 1997 and full professor in 2002. He was awarded a Changjiang Scholar Professorship in China by the China's Ministry of Education where he has set up and leads the State Key Lab of Geological Processes and Mineral Resources (GPMR) located on both campuses of China University of Geosciences in Beijing and Wuhan. Currently, he holds a Thousand Talent National Special Professorship of China, serving as the Founding Director of the GPMR laboratory. He has specialized in mathematical geoscience with research focus on nonlinear mathematical modeling of Earth processes and geoinformatics techniques for prediction of mineral resources. He has authored and co-authored more than 300 research articles. He has been awarded several prestigious awards including the Krumbein Medal, the highest award by the International Association for Mathematical Geosciences (IAMG). He was an elected President of the International Association for Mathematical Geosciences (IAMG) during 2012–16. He is the President of International

Union of Geological Sciences (IUGS) for the period between 2016 and 2020. He is an international leader in the application of nonlinear mathematics and geoinformatics to the analysis, modeling, and prediction of a wide range of geological processes and mineral resources quantitative assessment. His primary research interest involves the interdisciplinary study of nonlinear properties of the Earth's systems, as well as quantitative assessment and prediction of natural resources and environmental impacts. His research on fractal density and local singularity analysis theory and geomathematical models has made major impacts in several geoscientific disciplines, including those concerned with ocean ridge heat flow, magmatic flare-up during continent crustal growth and formation of supercontinents, earthquakes, floods, hydrothermal mineralization, and prediction of deeply buried mineral deposits.

**Frits Agterberg** is a Dutch-born Canadian Mathematical Geologist who served at the Geological Survey of Canada in Ottawa. He attended Utrecht University in the Netherlands from 1954 to 1961. With other founding members, he was instrumental in establishing the International Association for Mathematical Geosciences (IAMG) in 1968. He received the IAMG's William Christian Krumbein Medal in 1978, and he was IAMG Distinguished Lecturer in 2004. In 2017, he was conferred with the Honorary Membership of the IAMG. He has authored or co-authored over 250 scientific papers and 5 books. He has served the IAMG in many ways, including being its President from 2004 to 2008. After defending his doctoral thesis on structural geology of the Italian Alps at Utrecht University and a one-year fellowship at the University of Wisconsin in Madison, he became "petrological statistician" in his first job at the Geological Survey of Canada (GSC) in 1962. He was asked to create the GSC Geomathematics Section in 1971. He retired from the GSC in 1996 but still has an office at their Ottawa headquarters. In 1968, he became associated with the University of Ottawa where he taught a "statistics in geology" course for 25 years and has supervised six geomathematical Ph.D. students. From 1978 to 1989, he directed the Quantitative Stratigraphy Project of the International Geological Correlation Program. From 1981 to 2001, he was a Correspondent of the Royal Netherlands Academy of Arts and Sciences. During the past 20 years, primarily in collaboration with Qiuming Cheng, his colleagues, and students at the China University of Geosciences in Wuhan and Beijing and at York University, Toronto, he has worked on applications of multifractals to study the spatial distribution of metals in rocks and orebodies.

## Contributors

**Frits Agterberg** Geological Survey of Canada, Ottawa, ON, Canada

**M. Armstrong** Escola de Matemática Aplicada, Fundação Getulio Vargas, Rio de Janeiro, Brazil; MINES Paristech, PSL Research University, CERNA – Centre for Industrial Economy, i3, CNRS UMR 9217, Paris, France

**Leonardo Azevedo** CERENA, Instituto Superior Técnico, Universidade de Lisboa, Lisbon, Portugal

**Adrian Baddeley** Department of Mathematics and Statistics, Curtin University, Perth, WA, Australia

**James D. Bliss** Southwest Statistical Consulting, LLC, Cortez, CO, USA

**G. F. Bonham-Carter** Merrickville, ON, Canada

**Antonella Buccianti** Department of Earth Sciences, University of Florence, Florence, Italy; CNR-IGG, Unit of Florence, Florence, Italy

**Jef Caers** Stanford University, Stanford, USA

**S. Camargo** Escola de Matemática Aplicada, Fundação Getulio Vargas, Rio de Janeiro, Brazil

**James R. Carr** Department of Geological Sciences and Engineering, University of Nevada, Reno, Reno, NV, USA

**Guillaume Caumon** GeoRessources-ENSG, Université de Lorraine – CNRS–CREGU, Vandoeuvre-lès-Nancy, France

**Qiuming Cheng** State Key Lab of Geological Processes and Mineral Resources, China University of Geosciences, Beijing, China

**Jean-Paul Chilès** Centre of Geosciences, Mines ParisTech, Fontainebleau, France

**Noel Cressie** Distinguished Professor, National Institute for Applied Statistics Research Australia (NIASRA), School of Mathematics and Applied Statistics, University of Wollongong, Wollongong, Australia

**John Cubitt** Holt, Wrexham, UK

**John C. Davis** Heinemann Oil GmbH, Baldwin City, KS, USA

**B. S. Daya Sagar** Systems Science and Informatics Unit, Indian Statistical Institute-Bangalore Centre, Bengaluru, India

**Nicolas Desassis** Centre of Geosciences, Mines ParisTech, Fontainebleau, France

**Clayton V. Deutsch** University of Alberta, Edmonton, Canada

**John H. Doveton** Kansas Geological Survey, Lawrence, KS, USA

**Peter Dowd, FREng, FTSE** The University of Adelaide, Adelaide, Australia

**L. J. Drew** United States Geological Survey, Reston, VA, USA

**Lawrence J. Drew** Southwest Statistical Consulting, LLC, Cortez, CO, USA

**Olivier Dubrule** Imperial College London, London, UK

**Juan José Egozcue** Department of Civil and Environmental Engineering, Universidad Politécnica de Cataluña, Barcelona, Spain

**William E. Full** GXStat, LLC, Wichita, KS, USA

**Pierre Goovaerts** BioMedware, Inc, Jerome, MI, USA

**E. C. Grunsky** Department of Earth and Environmental Sciences, University of Waterloo, Waterloo, ON, Canada; China University of Geosciences, Beijing, China

**Stephen Henley** Resources Computing International Limited, Matlock, Derbyshire, UK

**Michael E. Hohn** West Virginia Geological and Economic Survey, Morgantown, USA

**G. M. Kaufman** Management Emeritus, E62-437, Sloan School of Management MIT, Cambridge, MA, USA

**Xiaogang Ma** Department of Computer Science, University of Idaho, Moscow, ID, USA

**Gregoire Mariethoz** Institute of Earth Surface Dynamics (IDYST), University of Lausanne, Lausanne, Switzerland

**Sean A. McKenna** IBM Research, Dublin, Ireland

**A. F. Militino** Department of Statistics and O.R., Public University of Navarra (Spain), Pamplona, Spain; InaMat (Institute for Advanced Materials), Pamplona, Spain

**R. Mohan Srivastava** TriStar Gold Inc., Toronto, ON, Canada

**A. Mondaini** Department of Physics, UERJ, Rio de Janeiro, Brazil

**Donald E. Myers** Department of Mathematics, University of Arizona, Tucson, AZ, USA

**J. Nicholls** Department of Geoscience, University of Calgary, Calgary, AB, Canada

**Václav Němec** Praha 10 - Strašnice, Czech Republic

**Ricardo A. Olea** U.S. Geological Survey, Reston, VA, USA

**Guocheng Pan** China Hanking Holdings, Shenyang, Liaoning, People's Republic of China

**Vera Pawlowsky-Glahn** Department of Computer Science, Applied Mathematics and Statistics, University of Girona, Girona, Spain

**U. Pérez-Goya** Department of Statistics and O.R., Public University of Navarra (Spain), Pamplona, Spain

**Jean Serra** Ecole des Mines de Paris, Paris, France

**Helmut Schaeben** Geophysics and Geoinformatics, TU Bergakademie Freiberg, Freiberg, Germany

**John H. Schuenemeyer** Southwest Statistical Consulting, LLC, Cortez, CO, USA

**Donald A. Singer** Cupertino, CA, USA

**D. B. Smith** United States Geological Survey, Denver, CO, USA

**Amílcar Soares** CERENA, Instituto Superior Técnico, Universidade de Lisboa, Lisbon, Portugal

**Pejman Tahmasebi** Department of Petroleum Engineering, University of Wyoming, Laramie, WY, USA

**Hannes Thiergärtner** Department of Geosciences, Free University of Berlin, Berlin, Germany

**R. Tolosana-Delgado** Helmholtz Institute Freiberg for Resource Technology, Freiberg, Germany

**M. D. Ugarte** Department of Statistics and O.R., Public University of Navarra (Spain), Pamplona, Spain; InaMat (Institute for Advanced Materials), Pamplona, Spain

**K. G. van den Boogaart** Helmholtz Institute Freiberg for Resource Technology, Freiberg, Germany

**E. H. Timothy Whitten** Riverside, Widecombe-in-the-Moor, Devon, UK

# Part I
# Theory

# Chapter 1
# Kriging, Splines, Conditional Simulation, Bayesian Inversion and Ensemble Kalman Filtering

**Olivier Dubrule**

**Abstract** This chapter discusses, from a theoretical point of view, how the geostatistical approach relates to other commonly-used models for inversion or data assimilation in the petroleum industry. The formal relationship between point Kriging and splines or radial basis functions is first presented. The generalizations of Kriging to the estimation of average values or values affected by measurement errors are also addressed. Two algorithms are often used for conditional simulation: the "rough plus smooth" approach consists of adding a smooth correction to a non-conditional simulation, whilst sequential Gaussian simulation allows the point-by-point construction of the realizations. As with Kriging, conditional simulation can be applied to average values or to data affected by measurement errors. Geostatistical inversion generates high-resolution realizations of vertical impedance traces constrained by seismic amplitudes. If the relationship between impedance and amplitude data is linearized, geostatistical inversion is a particular case of Bayesian inversion. Because of the non-linearity of production data vis-à-vis the variables of the earth model, their assimilation is harder than that of seismic data. Ensemble Kalman filtering, if considered from a geostatistical viewpoint, consists of using a large number—or ensemble—of realizations to calculate empirical covariances between the dynamic data and the parameters of the geostatistical model. These covariances are then used in the equations for interpolating the mismatch between simulated and new production data using a coKriging-like formalism. Interestingly, most of these techniques can be expressed using the same generic equation by which an initial model not honouring some newly arrived data is made conditional to these data by adding a (co-)Kriged interpolation of the data mismatches to the initial model. In spite of their similar equations, Bayesian inversion, geostatistics and ensemble Kalman filtering have a different approach to the inference of the covariance models used by these equations.

**Keywords** Dual Kriging • Radial basis functions • Geostatistical inversion
Energy-based methods • Prediction error filter

---

O. Dubrule (✉)
Imperial College London, London, UK
e-mail: o.dubrule@imperial.ac.uk



B. S. Daya Sagar et al. (eds.), *Handbook of Mathematical Geosciences*,
https://doi.org/10.1007/978-3-319-78999-6_1

## 1.1 Introduction

Fifty years ago, when geostatistics was pioneered by Matheron (1971), its main applications were Kriging and the change of support for mining applications. At the time, geostatistics was presented as a new discipline, without much reference to its relationships with other mathematical interpolation and modeling techniques. This has now changed as the relationships between geostatistics and such techniques as splines, regularization, Bayesian inversion, or ensemble Kalman filtering have become clearer. This convergence is fascinating and has led to many significant developments allowing the integration of multi-disciplinary data into 3-D geostatistical earth models.

This chapter discusses approaches for generating 2-D or 3-D subsurface models constrained by geological (wells), seismic or dynamic data. In spite of the wealth of data available, the uncertainty on the 3-D earth model remains high in most cases. Approaches that are designed to generate one unique "deterministic" model often pick the smoothest one. This is not realistic in situations where the Earth Model is used for flow simulation, as the results are biased if the model heterogeneities are not representative of that of the actual reservoir. More generally, non-linear operations, such as the application of cut-offs, may give biased results when applied to deterministic smooth models such as those produced by Kriging.

The multi-realization approach is now routinely applied to subsurface parameters inversion. *Looking at the mean provides much less information than looking at a movie of realizations. ...By construction, each of the realizations captures the essential random fluctuations of the actual field from which the data were extracted* (Tarantola 2005). This is a fundamental change. The traditional inversion approach could be formulated as "How to find an estimate of the spatial parameters which is as close as possible to the first guess values of these parameters and which provides, through forward modeling, an output which is as close as possible to the available data" (modified from Evensen 2007). These first guess values are usually a smooth (Kriging-like) spatial model of these parameters. Now the question has changed to "Find the probability density function (pdf) of 3-D models constrained by all the existing data, and provide techniques for sampling realizations from this pdf".

This chapter, written from a geostatistical perspective, discusses the convergence between the existing techniques.

Deterministic approaches such as Kriging, splines, regularization- or energy-based methods generate a single model of the subsurface, which usually minimizes or maximizes an optimisation criterion. These approaches are closely related and their formal relationships are discussed.

Geostatistical simulation is then revisited, and two key simulation algorithms are discussed; The first one is sequential Gaussian simulation and the second one is the "rough plus smooth" combination of an unconditional simulation plus a smooth correction term. These two algorithms have helped bridge the gap between geostatistics and inversion.

Two successful approaches are then discussed for integrating seismic and dynamic data into the earth model. Rather than using an approach merely based on statistical correlations between data and model parameters, it is assumed that there exists a deterministic relationship (or forward model) between model parameters and data, possibly including a random error.

The first approach, geostatistical inversion, produces reservoir-scale models of acoustic or elastic parameters constrained by single- or multi-offset seismic amplitude data. The value of using sequential Gaussian simulation to calculate seismically-constrained realizations is discussed. In situations where the forward model is linear, geostatistical inversion can be formulated as a particular case of Bayesian seismic inversion.

The second approach, ensemble Kalman filtering, consists of sequentially updating an "ensemble" of geostatistical realizations using dynamic data as they are acquired in time. The key idea here is to statistically derive the covariance terms of the equation used in Bayesian inversion from an ensemble of realizations rather than from a theoretical covariance model. The formal relationship between ensemble Kalman filtering and co-Kriging is discussed.

Most of the above techniques can be shown to use the same kind of formalism, where the mismatch between newly arrived data and the current model is interpolated and used to update this model.

One of the conclusions of this chapter is that the equations of Bayes, geostatistics or ensemble Kalman filtering are closely related. However, this relationship is mostly formal as the three techniques differ in their approach to the covariances used in the equations. Geostatisticians first fit a model to the data, whilst Bayesians start from a model based on general "prior" information. Only later in the process do they introduce the well data. And ensemble Kalman filtering directly uses the experimental covariances calculated from the realizations of the ensemble.

The topic of joint inversion of seismic and dynamic data is not discussed here, in spite of the interesting on-going developments in 4-D seismic data inversion. This is because the objective of this chapter is to address formal relationships between the different formalisms rather than discuss specific applications.

## 1.2 Deterministic Aspects of Geostatistics

### *1.2.1 Simple Stationary Kriging*

The basic model used by geostatistics is that of stationary random functions of order 2: a spatial property $z(\mathbf{x})$ at location $\mathbf{x}$ is represented by a random function $Z(\mathbf{x})$, which is assumed to follow a trend $m(\mathbf{x})$ and a stationary covariance $C(\mathbf{h})$

$$m(\mathbf{x}) = E(Z(\mathbf{x})) \tag{1.1a}$$

$$C(\mathbf{h}) = E(Z(\mathbf{x})Z(\mathbf{x}+\mathbf{h})) - E(Z(\mathbf{x}))E(Z(\mathbf{x}+\mathbf{h})) \tag{1.1b}$$

At each unsampled location $\mathbf{x}$, the value of $Z(\mathbf{x})$ is estimated by a linear combination $Z_k(\mathbf{x})$ of the values $Z_i = Z(\mathbf{x}_i)$ at the $n$ data points $(\mathbf{x}_i)_{i=1,\dots,n}$. Kriging is the best linear unbiased estimator, in the sense that it is unbiased and that it minimizes the estimation variance. If the trend $m(\mathbf{x})$ is known at each location $\mathbf{x}$, the simple Kriging (Chilès and Delfiner 2012, p. 151) system of equations is obtained

$$Z_k(\mathbf{x}) - m(\mathbf{x}) = \sum_{i=1}^{n} \lambda_i (Z_i - m(\mathbf{x}_i)) \tag{1.2a}$$

$$\text{with } \sum_{i=1}^{n} \lambda_i C(\mathbf{x_i} - \mathbf{x_j}) = C(\mathbf{x} - \mathbf{x_j}) \text{ for } j \in (1, \dots, n) \tag{1.2b}$$

### 1.2.2 Kriging with Intrinsic Random Functions of Order k

Matheron (1973) generalized the above model to that of Intrinsic Random Functions of Order k (IRF-k), where the definition of the variogram as a generalized covariance of order zero and of generalized covariances of order k leads to a model based on the stationarity of generalized increments of order k.

With k-IRFs, the model only considers linear combinations of $Z(\mathbf{x})$ that filter polynomials of order k (such polynomials being likely to represent a trend). Simple Kriging is not applicable any more. For instance, if k = 1 in two dimensions, and if $K(\mathbf{h})$ designates the generalized covariance of order k (GC-k), the kriging system becomes

$$Z_k(\mathbf{x}) = \sum_{i=1}^{n} \lambda_i Z_i \tag{1.3a}$$

$$\begin{aligned} &\text{with } \sum_{i=1}^{n} \lambda_i K(\mathbf{x_i} - \mathbf{x_j}) + \mu_0 + \mu_1 x_{j1} + \mu_2 x_{j2} = K(\mathbf{x} - \mathbf{x_j}) \text{for } j \in (1, \dots, n) \\ &\text{and } \sum_{i=1}^{n} \lambda_i = 1 \quad \sum_{i=1}^{n} \lambda_i x_{i1} = x_1 \quad \sum_{i=1}^{n} \lambda_i x_{i2} = x_2 \end{aligned} \tag{1.3b}$$

where the coordinates of each point $\mathbf{x}$ of the plane are written as $\mathbf{x} = (x_1, x_2)$.

### 1.2.3 Kriging Extensions

The goal here is not to discuss the details of Kriging, as there are plenty of excellent textbooks for this (Chilès and Delfiner 2012, p. 150). However, two features of Kriging deserve to be discussed, as they facilitate the understanding of the relationship between Kriging, splines and Bayesian approaches.

#### 1.2.3.1 Generalization of Kriging to the Interpolation of Average Values

Kriging is a linear interpolator. The data used by Kriging do not have to be point values, but they can be any linear function of the parameters of interest; Hansen et al. (2006) call these "volume support data". In particular, Kriging can be used to estimate the average value of a parameter $Z(v_{\mathbf{x}})$ at a location x by a linear combination volume of support data $Z(v_{\mathbf{x}_i})$ (Chilès and Delfiner 2012, p. 198)

$$Z_k(v_{\mathbf{x}}) = \sum_{i=1}^{n} \lambda_i Z(v_{\mathbf{x}_i}) \tag{1.4}$$

This property of Kriging, extensively used in mining applications, is of significant interest in the context of linear inversion of volume support data (Hansen et al. 2006). The Kriging equations associated with Eq. 1.4 are not given here, as they are a bit heavy, but conceptually simple thanks to the linear property of Kriging.

#### 1.2.3.2 Error CoKriging

Error coKriging (Dubrule 2003) is a generalization of Kriging to the situation where measurements $Y_i$ of the parameter $Z_i$ at data points $\mathbf{x_i}$ are affected by an unbiased random error

$$Y_i = Z_i + \varepsilon_i \text{ with } E(\varepsilon_i) = 0 \text{ and } Var(\varepsilon_i) = C_{\varepsilon_i} \tag{1.5}$$

In this situation, error coKriging allows the estimation of $Z(\mathbf{x})$ at any unsampled location $\mathbf{x}$ from a linear combination of values $Y_i$ (the random measurement error attached to each data can be zero or not) (Dubrule 2003; Hansen et al. 2006; Chilès and Delfiner 2012, p. 216)

$$Z_k(\mathbf{x}) = \sum_{i=1}^{n} \lambda_i Y_i \tag{1.6}$$

#### 1.2.3.3 Dual Kriging

If a global neighborhood is used, that is if all the available data are used to estimate $Z(\mathbf{x})$ at every single location $\mathbf{x}$, the Kriging equations (Eq. 1.3) can be inverted to obtain the dual Kriging system (for interpolation in the case of Kriging and smoothing in the case of error coKriging). For example, in two dimensions for a k-IRF of order 1

$$z_k(\mathbf{x}) = z_k(x_1, x_2) = a_0 + a_1 x_1 + a_2 x_2 + \sum_{i=1}^{n} b_i K(\mathbf{x} - \mathbf{x_i}) \tag{1.7}$$

where the conditions on the coefficients $(a_0, a_1, a_2, b_1, \ldots, b_n)$ are different for Kriging and error coKriging (Dubrule 1983)

$$\text{Kriging:} \quad \sum_{i=1}^{n} b_i = \sum_{i=1}^{n} b_i x_{i1} = \sum_{i=1}^{n} b_i x_{i2} = 0 \quad \text{and} \quad z_k(x_{i1}, x_{i2}) = z_i \tag{1.8}$$

$$\text{Error coKriging:} \sum_{i=1}^{n} b_i = \sum_{i=1}^{n} b_i x_{i1} = \sum_{i=1}^{n} b_i x_{i2} = 0 \quad \text{and} \quad z_k(x_{i1}, x_{i2}) + b_i C_{\varepsilon_i} = y_i \tag{1.9}$$

### 1.2.4 *Kriging and Splines*

#### 1.2.4.1 Interpolating Splines

Splines are a popular method for deterministic interpolation and approximation (Micula and Micula 1999). In 2-D, interpolating splines calculate a function honouring the data and minimizing an energy functional. Harmonic splines minimize the stretching energy of a membrane while biharmonic splines minimize the bending energy of an elastic plate. The biharmonic spline function can be written using a similar expression as Eq. 1.7 (Duchon 1975), but with a specific model for the generalized covariance function

$$K(\mathbf{x} - \mathbf{x_i}) = \left((x_1 - x_{i1})^2 + (x_2 - x_{i2})^2\right) Log\left(\sqrt{(x_1 - x_{i1})^2 + (x_2 - x_{i2})^2}\right) \tag{1.10}$$

Splines and Kriging are a particular case of a more general class of interpolators, called radial basis functions (Billings et al. 2002a, b). With splines, the polynomial in Eq. 1.7 belongs to the kernel of the operator $T$ that is minimized by the spline function ($T$ is the gradient for harmonic splines and the laplacian for biharmonic splines), whilst the function $K(\mathbf{h})$ is the Green function associated with the operator $T'T$, where $T'$ is the transposed operator of $T$ (Matheron 1981a)

$$T'TK(\mathbf{h}) = \delta \tag{1.11}$$

where $\delta$ is the Dirac Function. Choosing the energy functional minimized by splines is equivalent to fixing the degree of the trend function and the generalized covariance model for Kriging. For harmonic splines, these are respectively a constant and the De Wijs variogram in *Logh* (Chilès and Delfiner 2012, p. 94).

The consequence of Eq. 1.11 on the spectral density of the generalized covariance $K(\mathbf{h})$ is straightforward. For example, the spectral densities associated with the harmonic and biharmonic splines are power laws, representing fractal models. Szeliski and Terzopoulos (1989) and Micula and Micula (1999) discuss this relationship between Splines and fractals.

### 1.2.4.2 Smoothing Splines

Smoothing splines are used in situations where measurements at data points are affected by a random error (Eq. 1.5). In two dimensions, they compute a function $f(x_1, x_2)$ minimizing the sum of a spline energy functional plus a weighted distance to the $n$ data

$$\|Tf\|^2 + \theta \sum_{i=1}^{n} \frac{(f(x_{i1}, x_{i2}) - y_i)^2}{C_{\varepsilon_i}} \tag{1.12}$$

The smoothing biharmonic spline function has the same expression as that of Kriging and error coKriging (Eq. 1.7) but with the following relationships

$$\sum_{i=1}^{n} b_i = \sum_{i=1}^{n} b_i x_{i1} = \sum_{i=1}^{n} b_i x_{i2} = 0 \quad \text{and} \quad f(x_{1i}, x_{i2}) + b_i \frac{C_{\varepsilon_i}}{\theta} = y_i \tag{1.13}$$

Smoothing biharmonic splines are identical to error Cokriging as long as the generalized covariance used by error Cokriging is the function $\theta K(\mathbf{x} - \mathbf{x_i})$, where $K(\mathbf{x} - \mathbf{x_i})$ is given by Eq. 1.10 (Matheron 1981a; Dubrule 2003). This is a general relationship between smoothing splines and coKriging, which are formally equivalent if the generalized covariance $K(\mathbf{h})$ is that satisfying Eq. 1.11, with the coefficient of $K(\mathbf{h})$ equal to the smoothing parameter $\theta$ of Eq. 1.12.

#### 1.2.4.3 Kriging and Regularization—The Discrete Case

The discrete case is the situation where interpolation is performed at the nodes of a regular grid and each data point is located at one of the nodes of this grid. If $p$ is the total number of grid nodes, the number $n$ of data points is such that $n < p$.

In the discrete case, Matheron (1981b) also demonstrated the equivalence between splines and Kriging, and between smoothing splines and error coKriging. Both the Kriged and spline values $z_u$ minimize

$$\sum_{u,v=1}^{p} z_u B_{uv} z_v + \sum_{i=1}^{n} \frac{(z_i - y_i)^2}{C_{\varepsilon_i}} \tag{1.14}$$

where the $u$ and $v$ indices designate all the $p$ grid points where the interpolation takes place, whilst $i$ indices designate the $n$ data points. The minimization of Eq. 1.14 is performed according to the unknown values $z_u$ at all grid nodes (including those unknown values $z_i$ where a data point with measured value $y_i$ is present). The first term of Eq. 1.14 can be interpreted as a quadratic energy function traditionally used in inverse problems. In the regularization context, the choice of this quadratic form is driven by smoothing considerations, often using Briggs' finite difference Laplacian (or spline) "roughening" operator (Briggs 1974; Bolondi et al. 1976). Seen from the geostatistical perspective, $B_{uv}$ is the inverse of the covariance matrix in the stationary case and a pseudo-inverse of the generalized covariance matrix in the k-IRF case (Matheron 1981b). Equation 1.14 confirms the clear relationship between the inverse of the (generalized) covariance and the spline differential operator.

Kriging can thus be formalized in the frame of energy-based estimation techniques such as splines. This comes from the relationship between the inverse of the covariance function and the roughening filter implicit in the quadratic regularization term. It will be shown below that the regularization term can also be regarded, in the Bayesian inversion context, as an expression of the prior knowledge about the variable under study.

### *1.2.5 Kriging and Bayesian Inversion*

#### 1.2.5.1 Bayesian Linear Inversion

Here it may be useful to recall the general expression of the posterior mean and covariance in the case of Bayesian linear inversion of a multigaussian function. A very good reference for this is Tarantola (2005).

In the discrete case, consider a stationary multigaussian random vector $z$ of dimension $p$ containing the grid values $z_u$ over a two or three-dimensional regular grid of size $p$. Assume also that a vector $y$ contains the $n$ data $y_i$. It is assumed again that the data are affected by an error vector $\varepsilon$ of dimension $n$, and also that these data are a linear function of the $p$ values of $z$ over the grid

$$y = Fz + \varepsilon \tag{1.15}$$

where the vector $\varepsilon$ has mean zero and covariance matrix $C_\varepsilon$ and $F$ is a matrix of dimension $n \times p$. In the multigaussian case, thanks to the Bayes formula relating the posterior pdf $f_{post}(z)$ to the prior pdf $f_{prio}(z)$ and the likelihood function $g(y|z)$, the prior mean vector $m$ (dimension $p$) and covariance matrix $C$ (dimension $p \times p$) of $z$ are updated using the information brought by the data vector $y$

$$f_{post}(z) \propto f_{prio}(z) g(y/z) \propto \exp\left[(z-m)' C^{-1}(z-m)\right] \times \exp\left[(y-Fz)' C_\varepsilon^{-1}(y-Fz)\right] \tag{1.16}$$

$f_{post}(z)$ is a multigaussian function with the mean vector

$$m_{post} = m + C F' \left(FCF' + C_\varepsilon\right)^{-1}(y - F m) \tag{1.17}$$

and the covariance matrix

$$C_{post} = C - CF' \left(FCF' + C_\varepsilon\right)^{-1} FC \tag{1.18}$$

#### 1.2.5.2 Kriging and Bayesian Inversion

Equation 1.17 can also be written

$$m_{post} = m + \Lambda(y - Fm) = (I - \Lambda F)m + \Lambda y \tag{1.19}$$

with

$$\Lambda = C F' \left(FCF' + C_\varepsilon\right)^{-1} \tag{1.20}$$

In can be checked that $\Lambda$ is also the $p \times n$ matrix giving at each line $u$ the $n$ simple Kriging (or error coKriging) weights associated with the Kriging of the value $z_u$ at node $u$. Comparing the first part of Eq. 1.19 with Eq. 1.2 shows that, in the multigaussian case, $m_{post}$ is equal to simple Kriging and that the matrix $C_{post}$ contains the variances and covariances of simple Kriging at each node u of the regular grid.

### 1.2.6 Energy-Based Versus Probabilistic Estimates

The minimization of Eq. 1.14 leads to either Kriging or splines if the (inverse of) the covariance (Kriging) and the differential operator (splines) are properly chosen. Minimizing the expression in Eq. 1.14 is equivalent to maximizing

$$\exp\left(-\left(\sum_{u,v=1}^{p} z_u B_{uv} z_v + \sum_{i=1}^{n} \frac{(z_i - y_i)^2}{C_{\varepsilon_i}}\right)\right) \tag{1.21}$$

This is also the expression (up to a multiplicative constant) of the conditional multivariate distribution in the multigaussian case, as given by Bayes theorem (Eq. 1.16), in the case where $m = 0$, where the matrix $C_\varepsilon$ is diagonal and where the data are point values. The first term represents the prior pdf and the second the likelihood function. Kriging which is equal to the mean of the posterior pdf, also maximizes this pdf in the multigaussian case.

Expression (1.21) relates the world of energy functionals (such as splines) with that of probability functions (such as Kriging). More generally regularization and maximum a posteriori Bayesian estimates are identical if the prior covariance used in Bayesian inversion is properly chosen. The equivalence between an energy function and a probability distribution is also used in statistical mechanics, as the probability of a particular configuration is inversely related to its energy. Suppose that the vector $z$ minimizes an energy functional $E(z)$. Using the results of Geman and Geman (1984), Szeliski and Terzopoulos (1989) associate a probability to this energy through the Boltzmann (or Gibbs) distribution $p(z)$ defined as

$$p(z) = \frac{1}{Z}\exp\left(-\frac{E(z)}{T}\right) \tag{1.22}$$

where $Z$ and $T$ are positive constants. If Bayes' theorem is applied to the above prior pdf $p(z)$ and the posterior pdf is maximized, the formalism of splines is obtained.

### 1.2.7 Conclusion on Kriging

Three different ways of calculating a Kriging interpolator have been discussed

- using the basic approach where Kriging is calculated at each location as a linear combination of the data (Eq. 1.2)
- using Eq. 1.7, where the expression of dual Kriging, or more generally of radial basis functions, is used

- minimizing Eq. 1.14 in the discrete case, where Kriging values are calculated on a discrete grid by a minimization incorporating a regularization and a distance to the data term.

Kriging, although derived using a probabilistic formalism, is still a deterministic technique, in the sense that one unique or "best" model is produced, In most cases, Kriging provides a representation that is very smooth. As a result the application of non- linear operators to Kriged models will provide biased results (Dubrule 2003). This is one of the reasons for the success of conditional simulation.

## 1.3 Stochastic Aspects of Geostatistics: Conditional Simulation

With conditional simulation, the approach is stochastic. A large number of realizations are generated, which match the data (if the simulation is conditional) and share the first (mean) and second order (stationary covariance or generalized covariance) moments of the modeled random function. The main benefit of conditional simulation is that it produces realizations that behave away from the well data the same way as the well data themselves (Dubrule 2003). This is not true with Kriging, which produces a model that is smoother away from the wells than it is at the wells.

Conditional simulation can also be regarded as a technique for generating realizations of the conditional multigaussian pdf fully characterized by Eqs. 1.17 and 1.18. In other words, the realizations "vibrate" around their Kriging mean with a variance at each location equal to the Kriging variance.

A number of conditional simulation algorithms have been developed (Chilès and Delfiner 2012, p. 478). Among them, two are routinely used in the petroleum industry and are particularly interesting in relation with the inversion of seismic and production data.

### *1.3.1 Method 1: "Smooth Plus Rough" or "Rough Plus Smooth" Algorithm*

$Z(\mathbf{x})$ can be simply written as the sum of Kriging plus the Kriging error

$$Z(\mathbf{x}) = Z_k(\mathbf{x}) + (Z(\mathbf{x}) - Z_k(\mathbf{x})) \tag{1.23}$$

The "smooth plus rough" (Oliver 1996) simulation method writes a conditional simulation $Z_{cs}(\mathbf{x})$ as the sum of Kriging plus a simulation of the Kriging error. A non-conditional simulation $Z_{ncs}(\mathbf{x})$ of $Z(\mathbf{x})$ is generated first, which honors the mean and the covariance of $Z(\mathbf{x})$, then the conditional simulation $Z_{cs}(\mathbf{x})$ is calculated as

$$Z_{cs}(\mathbf{x}) = Z_k(\mathbf{x}) + (Z_{ncs}(\mathbf{x}) - Z_{ncsk}(\mathbf{x})) \tag{1.24}$$

where $Z_{ncsk}(\mathbf{x})$ designates Kriging of $Z_{ncs}(\mathbf{x})$ using as data the values $Z_{ncs}(\mathbf{x}_i)$ of the non-conditional simulation at the conditioning data locations. Thus to the smooth term $Z_k(\mathbf{x})$ is added the rough term $(Z_{ncs}(\mathbf{x}) - Z_{ncsk}(\mathbf{x}))$. Chilès and Delfiner (2012, p. 495) show that $Z_{cs}(\mathbf{x})$ honors the data and has the same (generalized) covariance as $Z_{ncs}(\mathbf{x})$ (and hence as $Z(\mathbf{x})$).

Equation 1.24 can also be expressed in the form of a "rough plus smooth" equation

$$Z_{cs}(\mathbf{x}) = Z_{ncs}(\mathbf{x}) + (Z_k(\mathbf{x}) - Z_{ncsk}(\mathbf{x})) \tag{1.25}$$

Using Eq. 1.17, Eq. 1.25 can be written in the discrete case, assuming that the data are average values of the gridded values and are affected by a measurement error. At location $u$ of the discrete grid

$$z_{ucs} = z_{uncs} + CF' \left(FCF' + C_\varepsilon\right)^{-1} (y - Fz_{uncs}) \tag{1.26}$$

Equation 1.26 shows that conditional simulation is obtained by adding to a non-conditional simulation a Kriging of the mismatch $(y - Fz_{uncs})$ between the data and the unconditional simulation at the data location. This formalism will appear to be quite general and will facilitate the understanding of the relationship between conditional simulation and Kalman Filtering.

### *1.3.2 Method 2: Sequential Gaussian Simulation (SGS)*

SGS (Deutsch and Journel 1998) is probably the most popular and flexible conditional simulation technique used in applications. SGS works under the multi-gaussian assumption and sequentially draws random locations within the simulated grid. At each new random location, the value is first Kriged from the previously simulated values and the well data. Then, a random value is sampled from the Gaussian pdf with mean equal to the Kriged value and variance equal to the Kriging variance (SGS uses the property that, in the multivariate normal case, univariate conditional distributions are also Gaussian). Then the sampled value is merged with the rest of the dataset, and a new random location is chosen within the simulated grid. The grid points where a data point is present are treated the same way as grid points with no data if the error $\varepsilon$ affecting the data is different from zero. If all the data are exact, then the grid nodes with data points are left unchanged. The result is a Gaussian realization constrained by the data values and satisfying the input statistics (mean and covariance function).

The main difference between "rough plus smooth" and SGS is that SGS works sequentially, grid point by grid point. The sequential nature of SGS is well suited to the geostatistical inversion of seismic data. Indeed, at each grid node, the sequential approach can make sure that the sampled value is compatible with both the previously generated points and the seismic data at the same location, thus combining the advantage of single trace inversion with that of spatial coupling. This will be discussed in Sect. 1.4.

### *1.3.3 Spectrum and Conditional Simulation*

Since the frequency spectrum is the Fourier transform of the covariance (Chilès and Delfiner 2012, p. 66), the spectrum of a conditional simulation is the same as that of the data. Conditional simulation addresses the following statement from Claerbout (2002) about seismic data interpolation: *Of all the assumptions we could make to fill empty bins, one that people usually find easiest to agree with is that the spectrum should be the same in the empty-bin regions as where bins are filled.*

Claerbout (2002) also defines the Prediction Error Filter (PEF) as the linear operator $T$ that transforms the data into a white noise. In other words, $T'T$ is the inverse of the covariance. Based on Eq. 1.11, this also means that $T$ is the spline operator associated with the covariance of the data. Claerbout (2002) shows that unconditional simulations can be generated by applying $T^{-1}$ to a white noise. This is the same technique as that used by Oliver (1988) and Oliver (1995) who applies what he calls the square root of the covariance function to a white noise.

## 1.4 Geostatistical Inversion of Seismic Data

### *1.4.1 Deterministic Seismic Inversion*

Until the mid-nineties or so, most seismic inversion studies were deterministic, in the sense that they generated a single "best" model, usually at the same resolution as the seismic data. Often, regularization-based or Bayesian methods were used, which led to the generation of one "maximum posterior" or "optimal for a given norm (often L2)" 3-D acoustic impedance model (Tarantola 2005).

If the seismic inversion problem is linearized as with Fatti et al.'s (1994) model, the reflection coefficient $r(\theta)$ at seismic time $t$ for a seismic block of offset $\theta$ can be written

$$r(\theta) = a_1(\theta)\frac{\partial LogI_p(t)}{\partial t} + a_2(\theta)\frac{\partial LogI_s(t)}{\partial t} \tag{1.27a}$$

$$\text{and} \quad y(\theta) = w(\theta)^* r(\theta) + \varepsilon(\theta) \tag{1.27b}$$

where $I_p(t)$ and $I_s(t)$ are the compressive and shear impedances at time $t$, $a_1(\theta)$ and $a_2(\theta)$ are offset-related parameters, $w(\theta)$ is the seismic wavelet for offset $\theta$ and $\varepsilon(\theta)$ is noise. This model is linear in the logarithm of $I_p(t)$ and $I_s(t)$. Thus, as long as the logarithms of impedances are inverted, the seismic amplitudes can we written as in Eq. 1.15 as a linear function of the logarithms of impedances, and the posterior mean obtained by multigaussian Bayesian seismic inversion (Eqs. 1.17 and 1.18) is identical to Kriging. The solution can also be regarded as a regularization-based solution, where the norm controlling the smoothness is derived from the inverse of the covariance.

At the time when only deterministic inversion was used, geostatisticians often treated seismic data as "soft" information, making use only of statistical correlations between seismic and reservoir parameters in order to constrain the earth models. This "soft" approach to seismic data allowed the development of some interesting interpolation techniques such as external drift or collocated coKriging (Dubrule 2003). However it also led to reservoir models not fully compatible with the seismic data as, if a seismic forward model such as that of Eq. 1.27 was applied to them, the actual seismic data was not recovered.

The above approaches proved sufficient until the late eighties or so, as seismic data were used at rather large scale. Thanks to the development of 3-D earth modeling at the reservoir scale in the early nineties, it became necessary to work with models at higher resolution than seismic data, and hence to quantify the uncertainty attached to these models. Then the availability of 4D seismic data also called for new technology to better constrain the earth models. Geostatistical inversion, described below, was developed with these issues in mind.

### *1.4.2 Geostatistical Inversion (GI)*

The original GI algorithm (Bortoli et al. 1992; Haas and Dubrule 1994) used SGS to simulate high-resolution acoustic impedance traces constrained by seismic data. SGS starts by picking a random cell within a regular two-dimensional grid. At this cell, a large number of possible acoustic impedance vertical traces are generated by SGS, then the trace that best matches the actual seismic trace at this location is selected. Then SGS moves to another random location of the two-dimensional grid, etc. until the whole model is filled with high-resolution impedance traces. Initially SGS appeared to be well suited to this application, as it allowed the use of any kind of forward model—linear or not—relating the acoustic impedance trace generated by SGS to the seismic amplitude trace at the same location. The acoustic impedance vertical traces simulated by SGS typically have higher frequency content than the seismic amplitudes, which makes them non-unique. This uncertainty can be quantified by generating multiple conditional simulations. Unfortunately the use of SGS proved to take too much computer time for large seismic datasets.

By revisiting the above GI algorithm in a Bayesian framework and in the linear context of Fatti's model (Eq. 1.27), authors such as Buland and Omre (2003) or Escobar et al. (2006) not only clarified the GI formalism but also provided a straightforward conditional simulation algorithm based on Eqs. 1.17 and 1.18 which was more efficient then SGS for sampling acoustic impedance traces compatible with seismic amplitudes. Whilst Bayesian inversion provided an expression of the posterior mean and covariance of the impedances multiGaussian pdf, GI allowed the sampling of reservoir-scale impedance realizations from this pdf.

As convincingly shown by Francis (2006a, b) or Escobar et al. (2006), cut-off operations such as those used to translate acoustic impedance into facies can be applied to GI realizations, thus avoiding statistical bias if these cut-offs were applied to Kriging.

## 1.5 Kalman Filtering and Ensemble Kalman Filtering

### *1.5.1 Kalman Filtering (KF)*

Suppose that a Gaussian random vector $Z_{t-1}$ has evolved until time $(t-1)$ and that $Z_{t-1}$ is an unbiased estimate of the unknown true state vector $z_{t-1}$ at time $(t-1)$

$$Z_{t-1} = z_{t-1} + R_{t-1} \text{ with } E(R_{t-1}) = 0 \text{ and } Var(Z_{t-1}) = Var(R_{t-1}) = C_{t-1} \quad (1.28)$$

If the model error is neglected, the forward model relating the true state vector at time $(t-1)$ with the state vector at time $t$ is assumed to be a linear function $L_t$

$$z_1 = L_t z_{t-1} \quad (1.29)$$

At time step $t$, the unknown true state of the system has evolved according to Eq. 1.29 and a vector $d_t$ of $n$ new data may also be available. Assume that these data are linear functions of the state vector $z_t$, and can be expressed as in Eq. 1.15

$$d_t = F_t z_t + \varepsilon_t \quad (1.30)$$

where the error vector $\varepsilon_t$ has mean zero and covariance matrix $C_{\varepsilon_t}$.

KF (Kalman 1960) aims to combine the information provided about $z_t$ by the forward model $L_t$ applied to the estimate $Z_{t-1}$ (Eq. 1.29) with the information provided by the data $d_t$ (Eq. 1.30). Bayes can be used for this, $L_t Z_{t-1}$ playing the role of the prior distribution. It is easy to verify that the covariance of the random vector $L_t Z_{t-1}$ is $C_t = L_t C_{t-1} L_t^{'}$. Hence, under Gaussian assumptions the best estimate is (from Eq. 1.19)

$$Z_t = L_t Z_{t-1} + \Lambda_t (d_t - F_t L_t Z_{t-1}) \tag{1.31}$$

where the kriging weights matrix $\Lambda_t$ (as in Eq. 1.20) is now called the Kalman gain

$$\Lambda_t = C_t F_t^{'} \left( F_t C_t F_t^{'} + C_{\varepsilon_t} \right)^{-1} \tag{1.32}$$

$Z_{t-1}$ as defined in Eq. 1.28 can represent any kind of unbiased estimate based on all the information available at time $(t-1)$. Kriging and conditional simulation are both unbiased estimates of $z_t$, only their variance is different and is of course minimum if $Z_{t-1}$ is Kriging and larger if $Z_{t-1}$ is simulation. Chilès and Delfiner (2012) (p. 497) show that the variance of the difference between a random function and its conditional simulation is twice the Kriging variance. In case $Z_{t-1}$ is simulation, Eq. 1.31 looks like the "rough plus smooth" method (Eq. 1.26) with $L_t Z_{t-1}$ playing the role of the non conditional simulation. Equation 1.31 makes the estimate $L_t Z_{t-1}$ conditional to the new data $d_t$ by adding an interpolation of the mismatch between $F_t L_t Z_{t-1}$ and the data.

In standard geostatistical applications, the observations are often spatial and hence assimilated simultaneously, while KF processes information sequentially, time step after time step. Tarantola (2005) (in Appendix 6.18) shows that, if in a linear least-squares problem the dataset can be divided into subsets with zero covariance between them, then solving one global inverse problem is equivalent to solving a series of smaller problems using the posterior state and covariance matrix of each partial problem as prior information for the next. Oliver et al. (2008) also show (in Chap. 11) that, under the same assumptions as Tarantola (2005), the step by step computation of KF provides (in the multigaussian case) the same result as would be obtained by integrating all the data in one single step. In the case where $L_t$ is the identity function, these two results also imply that simple Kriging would provide the same result if data were incorporated sequentially into the Kriging system, or in one single batch (under the assumptions that each batch of data has zero covariance with the others).

### *1.5.2 Constraining Reservoir Models by Production Data*

Fluid flow models are strongly non-linear, and linear approximations such as those already discussed for seismic modeling or KF cannot be used.

A distinction must be made between "history-matching", where a single reservoir model is modified until the flow simulation matches the production data, and "constraining reservoir models by production data", where reservoir model realizations compatible with production data are generated. Here the discussion focuses on the second objective rather than the first one. Some techniques to address this objective are based on rigorous approaches such as Markov-Chain Monte Carlo (MCMC) or Genetic Algorithms (GA) (Oliver et al. 2008). But these are very time-consuming and often unpractical. Ensemble Kalman filtering appears to be a more practical approach for incorporating production data into the reservoir model.

### *1.5.3 Ensemble Kalman Filtering (EnKF) Versus Conditional Simulation*

EnKF (Evensen 2007; Oliver et al. 2008) starts with an ensemble of initial realizations that are not constrained by production data. Typically the state vector $z_t$ at time $t$ contains permeabilities, porosities, saturations, pressures, and thermodynamic variables at the simulator grid nodes followed by a vector of predicted production data at each well $i$ at time $t$.

The following notation is used for a given state vector $z_t$

$$z_t = \left(z_{ut}^1, z_{ut}^2, \ldots z_{ut}^k, q_{it}^{1*}, \ldots . q_{it}^{l*}\right) \tag{1.33}$$

It is assumed that there are $k$ gridded variables in $z_t$, that the simulator grid is composed of $p$ cells $u$, and that there are $n$ wells $i$ each with $l$ new production data at time $t$. The total size of the state vector $z_t$ is $kp + nl$. The predicted data vector $d_t^*$ is the vector of size $nl$

$$d_t^* = \left(q_{1t}^{1*}, \ldots . q_{nt}^{1*}, q_{1t}^{2*}, \ldots . q_{nt}^{2*}, \ldots, q_{1t}^{l*}, \ldots . q_{nt}^{l*}\right) \tag{1.34}$$

The relation between state vector and predicted data is

$$d_t^* = Pz_t \text{ with } P = \left(O_{nlxkp}, I_{nlxnl}\right) \tag{1.35}$$

$P$ is a $nl \times (kp + nl)$ matrix. The function $f_t$, which represents the flow simulator, is non-linear. If the model errors are neglected

$$z_t = f_t(z_{t-1}) \tag{1.36}$$

does not modify the rock properties (unless they are affected by changes in pressure and saturation), but replaces the pressure, saturation, and simulated data with new values at time $t$.

The problem is now to calculate the best estimate of the state vector $z_t$ combining the information provided by the flow simulation forward model $f_t(z_{t-1})$ and that provided by the new data $d_t$.

If $f_t$ is a linear function, this is the standard KF domain of application and Eq. 1.31 applies, $L_t$ playing the role of $f_t$. But now $f_t$ is non-linear. It would still be convenient to update the state vector through a generalization of Eq. 1.31

$$z_t = f_t(z_{t-1}) + \Lambda_t(d_t - Pf_t(z_{t-1})) \tag{1.37}$$

where the Kalman gain $\Lambda_t$ is obtained using Eq. 1.32. Assuming that there is no error associated with the data, Eq. 1.32 can be simplified into

$$\Lambda_t = C_t P' \left(PC_t P'\right)^{-1} \tag{1.38}$$

Equation 1.38 requires the knowledge of the covariance $C_t$ of $f_t(z_{t-1})$, in other words the covariance of the image of the state vector after application of the flow simulation model $f_t$. $f_t$ is non-linear and this covariance cannot be simply calculated —as in the linear case—from the covariance at the previous step. EnKF addresses this issue by statistically deriving this covariance using the information from the multiple realizations, typically about a hundred of them. This is the key idea behind EnKF.

There are of course a number of issues resulting from the fact that the covariances are calculated from a finite number of realizations of the ensemble. The first one is spurious correlation, because the ensemble members are not independent except in the starting ensemble. The second one is that if the number of realizations in the ensemble is not large enough, then the covariances are poorly estimated. Standard geostatistics addresses this by fitting mathematical models to the experimental covariances, in order to smooth the spurious correlations.

### 1.5.4 Ensemble Kalman Filtering and Its Relationship with CoKriging

In Eq. 1.37, focus now on the rock properties in the state vector. $f_t(z_{t-1})$ leaves the rock properties unchanged, as only the time-dependent state vectors in the simulator grid are calculated by one time-step of the flow simulator, whilst $\Lambda_t(d_t - Pf_t(z_{t-1}))$ is a linear combination of the differences between observed and predicted production data at each well. Thus EnKF interpolates between the wells by calculating a linear combination of these differences across the field, then adds these interpolated difference to the rock properties model. Is it possible to reformulate EnKF as a well by well geostatistical approach?

The term $\Lambda_t$ in Eq. 1.37 is the Kalman gain as given by Eq. 1.38. In the case where there is no error affecting the data, Eqs. 1.37 and 1.38 can be written

$$z_t - f_t(z_{t-1}) = C_t P' \left(P C_t P'\right)^{-1} (d_t - P f_t(z_{t-1})) \tag{1.39}$$

The left-hand side is the update calculated by EnKF for the property of interest as the time step evolves from $(t-1)$ to $t$. The Kalman gain coefficients of the right-hand side are nothing else than the simple coKriging weights (see for instance Chilès and Delfiner 2012, p. 303).

Thus, each estimate of a 3-D spatial parameter such as porosity or permeability at time $(t-1)$ is updated at time $t$ by a linear combination of all the inconsistencies generated by this parameter at the data points. Since, in the case of flow simulations, many parameters are involved in the production profiles prediction, all the individual parameters' 3-D models must be corrected in a consistent way, which is why multivariate coKriging—and not univariate Kriging applies here.

## 1.6 Beyond the Formal Relationship Between Geostatistics and Bayes

### *1.6.1 Two Identical Formalisms but Different Assumptions*

The above developments show that techniques such as conditional simulation, Bayesian inversion, geostatistical inversion and ensemble Kalman filtering follow a similar mathematical formalism.

However, their philosophy of application differs in the way the covariance is approached. This can be understood by looking again as Bayes rule as presented in Eq. 1.16

$$f_{post}(z) \propto f_{prio}(z) g(y/z) \tag{1.40}$$

With geostatistics, the experimental (generalized) covariance calculated on the data $y$ is fitted by a model which becomes the covariance of the unconditional distribution $f_{prio}(z)$. Then the data $y$ are used a second time through the simulation conditioning process of Eq. 1.26.

With Bayes, the covariance model associated with $f_{prio}(z)$ is a prior based on local or analog knowledge, but not on the data themselves (Tarantola 2005). This prior is transformed into a posterior covariance through the conditioning process of Eq. 1.40.

With geostatistics, the aim of conditional simulation is to generate realizations that match the data and satisfy the input covariance; the SGS and rough plus smooth algorithms work only if the data themselves satisfy this input covariance. But the random function $Z_{cs}(\mathbf{x})$ of Eqs. 1.24 and 1.25 is not an ergodic or even a stationary random function; its variance at each location $\mathbf{x}$ is equal to the Kriging variance and changes with $\mathbf{x}$, as it is zero at the data points. In other words, the covariance of the random function $Z(\mathbf{x})$ is different from that of $Z_{cs}(\mathbf{x})$ conditionally to the data

(Chilès and Delfiner 2012, p. 497). But the covariance calculated on a single conditional realization does not "see" any difference between the grid cells associated with data points and those not associated with data points. It is only as the realizations change, leaving the data unchanged, that the covariance across realizations appears non-stationary and hence non-ergodic.

On the other hand, Bayes combines a prior covariance—usually different from that of the data—with a data-based likelihood, resulting into a posterior pdf that sits somewhere between the prior and the likelihood. Bayes *updates* prior covariances based on new data whilst conditional simulation *anchors* the realizations against the hard data (Escobar, personal communication).

### 1.6.2 Model Falsifiability

Tarantola (2006) challenges the geostatistical and Bayes formalisms if models are to be falsifiable or have a scientific meaning: *I suggest that the setting, in principle, for an inverse problem should be as follows: use all available prior information to sequentially create models of the system, potentially an infinite number of them. For each model, solve the forward modeling problem, compare the predictions to the actual observations and use some criterion to decide if the fit is acceptable or unacceptable, given the uncertainties in the observations and, perhaps, in the physical theory being used. The unacceptable models have been falsified, and must be dropped. The collection of all the models that have not been falsified represent the solution of the inverse problem.* Thus, Tarantola (2006) offers to keep all the prior realizations that are compatible with the data. Thus the data are used to validate or reject the prior realizations, rather than update the prior pdf into the posterior.

### 1.6.3 Looking Ahead: Machine Learning and Falsifiability

The fast growth in machine learning algorithms (Goodfellow et al. 2016) is challenging the geostatistical and Bayesian formalisms in situations where data are plenty. Thanks to this large number of data, the approach used to falsify a convolutional neural network model (for instance) relating input parameters to data is often to test whether the convolutional model works as well on a training (or calibration) dataset as on a test dataset not used for training. The prior model itself is completely data-driven, which contradicts Tarantola (2006) but the validation step is along the lines of his above recommendations! This topic is likely to generate interesting discussions in the future.

## 1.7 Conclusion

The objective of this chapter was to discuss the convergence observed over the last fifty years between geostatistics and other modelling and inversion techniques.

A formal convergence exists between the main techniques used to constrain reservoir models by multi-disciplinary data. Kriging, splines, conditional simulation, geostatistical inversion and ensemble Kalman filtering can be interpreted using either the geostatistical formalism or Bayes.

Most of these techniques amount to the same approach where an initial model is updated by using a linear combination of the mismatches between the new data and their prediction from the initial model (Eqs. 1.19, 1.26, 1.31 and 1.39).

However the methods above have a different philosophy towards the inference of the covariances used in these calculations. Bayes uses the data to update a prior pdf which is independent of the data. Geostatistics generate realizations of conditional simulations that reproduce the modeled covariance—or the spectrum—of the data. EnKF does not model a covariance but directly uses the empirical covariances derived from the ensemble realizations and their flow simulations.

**Acknowledgements** The author would like to thank Imperial College London, and Total for seconding him there as a Visiting Professor. Igor Escobar and Danila Kuznetsov are thanked for the fruitful discussions held at the Total Geoscience Research Centre in Aberdeen (UK). The author will also never forget the passionate and illuminating discussions with Albert Tarantola at the Café Beaubourg in Paris!

## References

Billings SD, Beaston RK, Newsam GN (2002a) Interpolation of geophysical data using continuous global surfaces. Geophysics 67(6):1810–1822

Billings SD, Newsam GN, Beaston RK (2002b) Smooth fitting of geophysical data using continuous global surfaces. Geophysics 67(6):1823–1834

Bolondi G, Rocca F, Zanoletti S (1976) Automatic contouring of faulted subsurfaces. Geophysics 41(6):1377–1393

Bortoli LJ, Alabert F, Haas, A, Journel AG (1992) Constraining stochastic images to seismic data. In: Soares A (ed) Proceedings of the international geostatistics congress, Troia 1992. Kluwer Publishers, Dordrecht, Netherlands

Briggs IC (1974) Machine contouring using minimum curvature. Geophysics 39(1):39–48

Buland A, Omre H (2003) Bayesian linearized AVO inversion. Geophysics 68(1):185–198

Chilès JP, Delfiner P (2012) Geostatistics. Modeling spatial uncertainty. Wiley series in probability and statistics, 2nd edn. Wiley

Claerbout J (2002) Image estimation by example: geophysical soundings image construction: multidimensional autoregression. http://sepwww.stanford.edu/sep/prof

Deutsch CV, Journel AG (1998) GSLIB, geostatistical software library and user's guide, 2nd edn. Oxford University Press

Dubrule O (1983) Two methods with different objectives: splines and Kriging. Math Geol 15(2): 245–257

Dubrule O (2003) Geostatistics for seismic data integration in 3-D earth models, SEG/EAGE Course Notes. SEG Publications

Duchon J (1975) Fonctions Splines du Type Plaque Mince en Dimension 2, Séminaire d'Analyse Numérique, No. 231, U.S.M.G, Grenoble, France

Escobar I, Williamson P, Cherett A, Doyen PM, Bornard R, Moyen R, Crozat T (2006) Fast geostatistical inversion in a stratigraphic grid, SEG Annual Meeting, New Orleans

Evensen G (2007) Data assimilation. The ensemble Kalman filter. Springer

Fatti JL, Smith GC, Vail PJ, Strauss PJ, Levitt PR (1994) Detection of gas in sandstone reservoirs using AVO analysis: a 3-D seismic case history using the geostack technique. Geophysics 59:1362–1376

Francis A (2006a) Understanding stochastic inversion part 1. First Break 24:69–77

Francis A (2006b) Understanding stochastic inversion part 2. First Break 24:79–84

Geman D, Geman S (1984) Stochastic relaxation, Gibbs distribution and the Bayesian restoration of images. IEEE Trans Pattern Anal Mach Intell PAMI-6(6):721–741

Goodfellow I, Bengio Y, Courville A (2016) Deep learning. MIT Press

Haas A, Dubrule O (1994) Geostatistical inversion—a sequential method of stochastic reservoir modeling constrained by seismic data. First Break 12(11):561–569

Hansen TM, Journel AG, Tarantola A, Mosegaard K (2006) Linear inverse Gaussian theory and geostatistics. Geophysics 71(6):101–111

Kalman RE (1960) A new approach to linear filter and prediction problems. J Basic Eng 82:35–45

Matheron G (1971) The theory of regionalized variables and its applications. http://cg.ensmp.fr/bibliotheque/public/MATHERON_Ouvrage_00167.pdf

Matheron G (1973) The intrinsic random functions and their applications. Adv Appl Probab 5:439–468

Matheron G (1981a) Splines and Kriging, their formal equivalence. Syracuse University, Geol Contrib 8:77–95

Matheron G (1981b) Splines et Krigeage: Le Cas Fini. http://cg.ensmp.fr/bibliotheque/public/MATHERON_Rapport_00228.pdf

Micula G, Micula S (1999) Handbook of splines. Springer Science & Business Media

Oliver DS (1988) Calculation of the inverse of the covariance. Math Geol 30(7):911–933

Oliver DS (1995) Moving averages for Gaussian simulation in two and three dimensions. Math Geol 27(8):939–960

Oliver DS (1996) On conditional simulation to inaccurate data. Math Geol 28(6):811–817

Oliver D, Reynolds AC, Liu N (2008) Inverse theory for petroleum reservoir characterization and history-matching. Cambridge University Press

Szeliski R, Terzopoulos D (1989) From splines to fractals. Comput Graph 23(3):51–60

Tarantola A (2005) inverse problem theory and model parameter estimation. SIAM, Philadelphia

Tarantola A (2006) Popper, Bayes and the inverse problem. Nat Phys 2:492–494

# Chapter 2
# A Statistical Commentary on Mineral Prospectivity Analysis

**Adrian Baddeley**

**Abstract** We compare and contrast several statistical methods for predicting the occurrence of mineral deposits on a regional scale. Methods include logistic regression, Poisson point process modelling, maximum entropy, monotone regression, nonparametric curve estimation, recursive partitioning, and ROC (Receiver Operating Characteristic) curves. We discuss the use and interpretation of these methods, the relationships between them, their strengths and weaknesses from a statistical standpoint, and fallacies about them. Potential improvements and extensions include models with a flexible functional form; techniques which take account of sampling effort, deposit endowment and spatial association between deposits; conditional simulation and prediction; and diagnostics for validating the analysis.

## 2.1 Introduction

The pioneering work of Agterberg (1974) developed a statistical strategy for predicting the likely occurrence of mineral deposits. In essence, the observed association between known deposits and other known geostructural or geochemical information is used to predict the spatially-varying abundance of unknown deposits. The association between predictors and deposits is modelled by logistic regression.

This general approach to prospectivity analysis has been extended and adopted across a wide range of applications, for predicting mineral deposits (Chung and Agterberg 1980; Bonham-Carter 1995), archaeological finds (Scholtz 1981; Kvamme 1983), landslides (Chung and Fabbri 1999; Gorsevski et al. 2006), animal and plant species (Franklin 2009) and other features which can be treated as points at the scale of interest. Extensions and modifications include logistic regression for sampled data, maximum entropy, and weights-of-evidence modelling.

However, the scientific literature contains many conflicting statements about the interpretation of these methods. For example, there are different understandings of

A. Baddeley (✉)
Department of Mathematics and Statistics, Curtin University, GPO Box U1987,
Perth, WA 6845, Australia
e-mail: adrian.baddeley@curtin.edu.au



B. S. Daya Sagar et al. (eds.), *Handbook of Mathematical Geosciences*,
https://doi.org/10.1007/978-3-319-78999-6_2

the fundamental scope and validity of logistic regression, about the degree of flexibility inherent in the assumptions, and about the interpretation of the results. This is a concern, because misunderstanding of a statistical technique poses the obvious risk that it may be mis-applied, its results misinterpreted, or its performance incorrectly evaluated.

In statistical science the understanding of these techniques has also changed dramatically over the last four decades. The modern synthesis of statistical modelling permits a new and deeper appreciation of prospectivity methods. New tools from statistical science may enable exploration geologists to perform a more searching analysis of their survey data.

Accordingly, this article offers a commentary and critique of prospectivity analysis from the standpoint of modern statistical methodology. We begin by examining the fundamentals of logistic regression, explaining the interpretation of the results, and discussing its strengths and weaknesses. We explain the close relationship between logistic regression, point process modelling, and maximum entropy methods. We canvas some alternative methods which are less well known, including monotone regression, nonparametric regression, recursive partitioning models, and ROC curves. (The popular weights-of-evidence method is not discussed here, but will be treated in detail in another article.) New tools include robust estimation, model selection and variable selection, conditional prediction and model diagnostics. Several unanswered questions in prospectivity analysis are identified as topics for future research in statistical methodology.

## 2.2 Example Data

For the sake of demonstration and discussion, we shall use a vastly oversimplified example. The Murchison geological survey data shown in Fig. 2.1 record the spatial locations of gold deposits and associated geological features in the Murchison area of Western Australia. They are extracted from a regional survey (scale 1:500,000) of the Murchison area made by the Geological Survey of Western Australia (Watkins and Hickman 1990). The features shown in the Figure are the known locations of gold deposits, the known or inferred locations of geological faults, and greenstone outcrop. The study region is contained in a $330 \times 400$ km rectangle. At this scale, gold deposits are point-like, i.e. their spatial extent is negligible. These data were previously analysed in Foxall and Baddeley (2002), Brown et al. (2002); see also Groves et al. (2000), Knox-Robinson and Groves (1997). Data were kindly provided by Dr. Carl Knox-Robinson, and permission granted by Dr. Tim Griffin, Geological Survey of Western Australia and by Dr. Knox-Robinson.

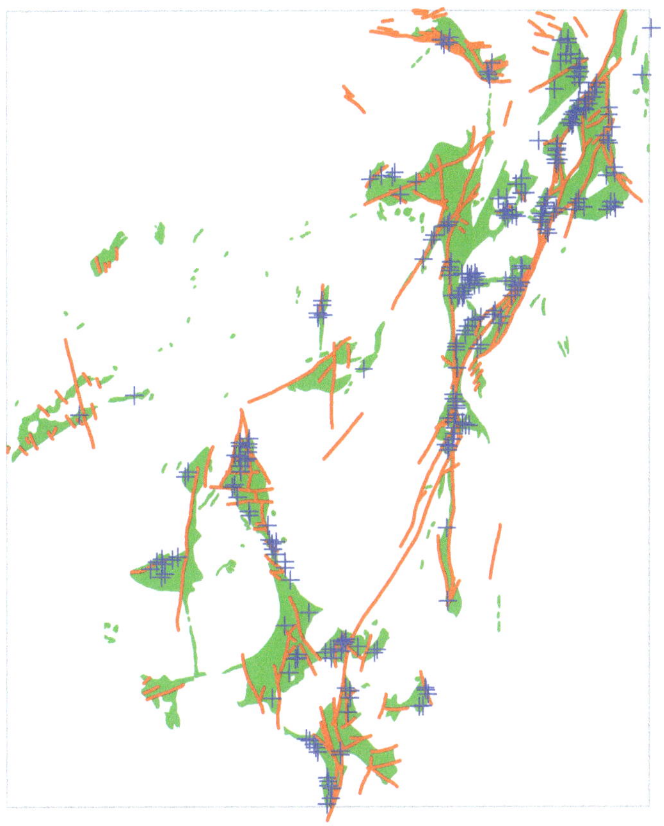

**Fig. 2.1** Murchison geological survey data. Known gold deposits (blue crosses), major geological faults (red lines) and greenstone outcrop (green shading) in a survey region about 200 by 300 km across

Evidently, both the geological fault pattern and the greenstone outcrop have "predictive" value for gold prospectivity, because gold deposits are strongly associated with proximity to both features. For the purposes of analysis in this article, we require predictors to be spatial variables. A predictor $Z$ should be a function $Z(u)$ defined at any spatial location $u$. For a map of rock type such as the greenstone outcrop, the simplest choice for the predictor value $Z(u)$ at location $u$ is the "indicator" equal to 1 if the location $u$ falls inside the greenstone, and 0 if it falls outside. For a map of linear features such as geological faults, a common choice for the predictor value $Z(u)$ is the distance from $u$ to the nearest fault. Figure 2.2 shows contours of this distance function for the Murchison data.

It is important to note that our choice of spatial predictor $Z(u)$ will affect the results of the analysis: the results would usually be different if we replace the distance function in Fig. 2.2 by the squared distance or the square root of distance, etc. Several other choices of spatial predictor derived from geological fault data are canvassed in Berman and Turner (1992). Likewise for the greenstone outcrop we could have chosen another predictor, such as the distance function of the greenstone. The choice of predictor can be revisited after the analysis, as discussed in Sect. 2.4.6.

Fig. 2.2 Contours of distance to the nearest fault in the Murchison survey

## 2.3 Logistic Regression

Here we recapitulate and re-examine some details of the logistic regression technique, for the purposes of discussion.

### 2.3.1 *Basics of Logistic Regression*

Logistic regression is a general statistical technique for modelling the relationship between a binary response variable and a numerical explanatory variable (Berkson 1955; McCullagh and Nelder 1989; Dobson and Barnett 2008; Hosmer and Lemeshow 2000). The use of logistic regression to predict the presence/absence of point events was pioneered in geology by Agterberg (1974, 1980), apparently on the suggestion of the statistician Tukey (1972): see Agterberg (2001). The study region is divided into pixels; in each pixel the presence or absence of any deposits is recorded; then logistic regression is used to predict the probability of the presence of a deposit as a function of predictor variables. This was later independently rediscovered in archaeology (Scholtz 1981; Hasenstab 1983; Kvamme 1983, 1995) and is now a standard technique in GIS applications (Bonham-Carter 1995) including spatial ecology (Franklin 2009).

The study region is divided into pixels of equal area. For each pixel, we record whether mineral deposits are present or absent. We then fit the *logistic regression* relationship

$$\log \frac{p}{1-p} = \alpha + \beta z \tag{2.1}$$

where $p$ is the probability of presence of a deposit (or deposits) in a given pixel, and $z$ is the corresponding value of the predictor variable.

Here $\alpha$ and $\beta$ are model parameters which are estimated from the data. Some writers state that the interpretation of $\alpha$ and $\beta$ is "obscure" (Wheatley and Gillings 2002, p. 175), perhaps because of the unfamiliar form of the left hand side of (2.1). The quantity $p/(1-p)$ is the *odds* of presence against absence, that is, the probability $p$ of presence of a deposit, divided by the probability $1-p$ of absence. The left hand side of (2.1) is the logarithm of the odds of presence. (In this paper 'log' always refers to the natural logarithm, with base $e$.) The logistic regression relationship (2.1) states that the *log odds* of presence is a linear function of the predictor $z$. The straight line has slope $\beta$ and intercept $\alpha$. The transformation $\log(p/(1-p))$ ensures that

$$p = \frac{e^{\alpha+\beta z}}{1+e^{\alpha+\beta z}} \tag{2.2}$$

is a well-defined probability value (between 0 and 1) for any possible values of $\alpha, \beta$ and $z$. The log odds is the "canonical" choice of transformation in order to satisfy some desirable statistical properties (McCullagh and Nelder 1989), and arises naturally in many applications. Bookmakers often quote gambling odds that are equally spaced on a logarithmic scale, such as the sequence 2:1, 4:1, 8:1, 16:1. Since logistic regression is widely used in medical and public health research, standard statistical textbooks contain many useful ways to interpret and explain these quantities (Hosmer and Lemeshow 2000).

Once the parameters $\alpha, \beta$ have been estimated from data (as detailed in Sect. 2.3.3), the predicted probabilities $p_j$ can be computed using (2.2) and displayed as colours or greyscales in a pixel image, as shown in Fig. 2.3. Qualitative interpretation of the map seems to be adequate for many purposes, while many writers recommend using only the *sign* of the slope parameter $\beta$ (Gorsevski et al. 2006, pp. 405–407). However, much more can be done with the fitted logistic regression, as we discuss below.

The general appearance of Fig. 2.3 is very similar to the contour plot of distance to nearest fault in Fig. 2.2. This is a foreseeable consequence of the simple model (2.1) which implies that contours of probability are contours of distance to nearest fault. This is not true of more complicated models involving several predictors.

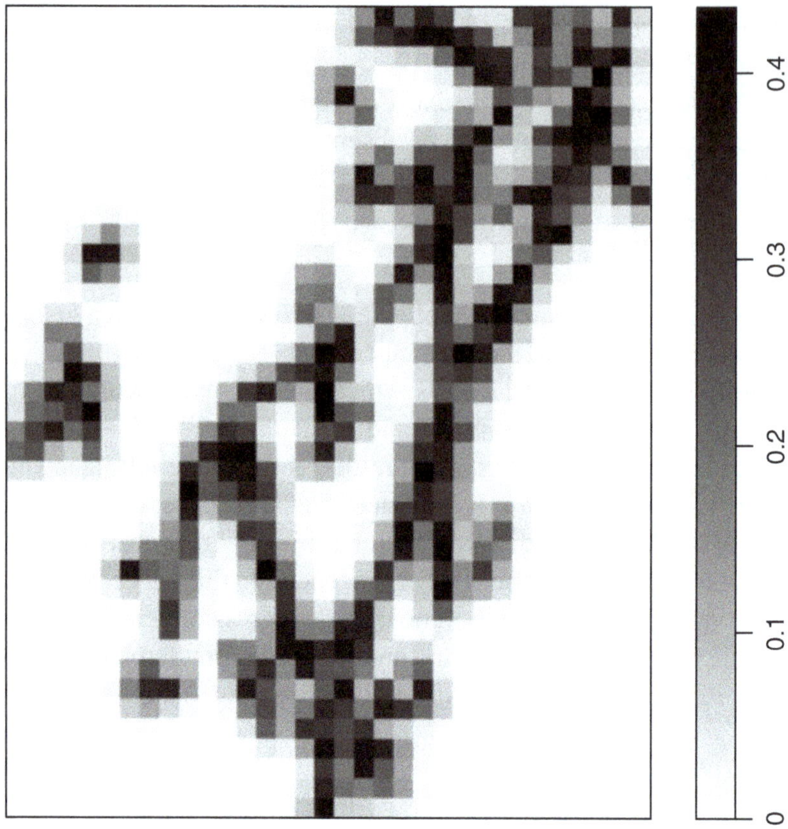


**Fig. 2.3** Fitted probability of a gold deposit in each 10-km-square pixel in the Murchison survey, estimated by logistic regression. Pixel values are probabilities (between 0 and 1)

### 2.3.2 *Flexibility and Validity*

Some writers describe logistic regression as a 'nonparametric' technique (Kvamme 2006, p. 24), which would suggest that it is able to detect and respond to any kind of relationship, not specified in advance, between the predictor $z$ and the presence probability $p$. On the contrary, logistic regression is a parametric model of a very simple kind. The relationship $z$ and $p$ is rigidly defined by Eqs. (2.1) and (2.2): *the relationship is linear* on the scale of the log odds. The position of the line is determined by the two parameters $\alpha$ and $\beta$. Logistic regression could be *false* for a particular application: that is, the model assumptions could be incorrect.

Logistic regression is an example of a "generalised linear model" (McCullagh and Nelder 1989; Dobson and Barnett 2008), essentially a linear regression of the transformed probabilities against the predictor. In the analysis of the Murchison data shown here, if we replace the distance function $Z(u)$ by its square $Z(u)^2$, or square root $\sqrt{Z(u)}$, etc. in the logistic regression, we obtain a different model, which is incompatible with the original model. If the log odds are a linear function of squared distance, then they are *not* a linear function of distance. Consequently, the choice of predictor variable is very important, and it involves an implicit *model assumption* about the relationship between presence probability $p$ and predictor $z$. Even the sign of the fitted slope parameter $\beta$ could be misleading if the predictor was chosen incorrectly.

Such freedom as does exist in the logistic regression model is the freedom to choose the predictor or predictors $Z(u)$. Once the predictor is chosen, the model becomes rigid. If there is concern about the form of relationship between $p$ and $Z$, one simple strategy is to fit a polynomial, instead of linear, relationship between the log odds and the predictor variable.

Statistical science has developed an armory of techniques for "validating" a regression analysis (Harrell 2001; Hosmer and Lemeshow 2000). These include diagnostics for checking the validity of the logistic regression relationship (2.1), measures of sensitivity of the fitted model to the data, techniques for selecting the most important variables and the most informative models, and measures of goodness-of-fit. As far as the author is aware, these techniques are rarely used in geoscience. This presents the risk of failing to detect situations where logistic regression analysis is not appropriate. Model validation is a kind of "due diligence" for data analysts.

A weakness of all parametric modelling is that, because of its "low degrees of freedom", the model predictions at a given location are heavily influenced by the entire dataset, including data observed under very different conditions. In the Murchison example, the predicted probability of presence of a gold deposit declines dramatically between distances 0, 1 and 2 km from the nearest fault. This is not necessarily a reflection of the observed frequency of occurrence of gold deposits at these distances: rather, it is a consequence of the large negative value of the estimated slope parameter $\beta$, which arises because of the scarcity of gold deposits at much larger distances.

Extension of the logistic regression technique to account for characteristics of the mineral deposits, such as total endowment of gold, would be problematic because it would effectively require a model for the probability distribution of the endowment (and this might also be spatially-varying). However, it is straightforward to apply logistic regression to different subsets of the deposits, for example to predict the occurrence of deposits with endowment exceeding a specified threshold.

The logistic regression technique described here assumes that the relationship (2.1) holds throughout the study region, with the same parameter values $\alpha$, $\beta$ throughout. This assumption can be avoided using geographically-weighted logistic regression (Lloyd 2011) or local likelihood estimation (Loader 1999; Baddeley 2017) which allow the parameters to be spatially-varying.

### 2.3.3 *Fitting Procedure and Implicit Assumptions*

For the discussion it will be important to know a few details about the procedure that is used to fit the logistic regression relationship.

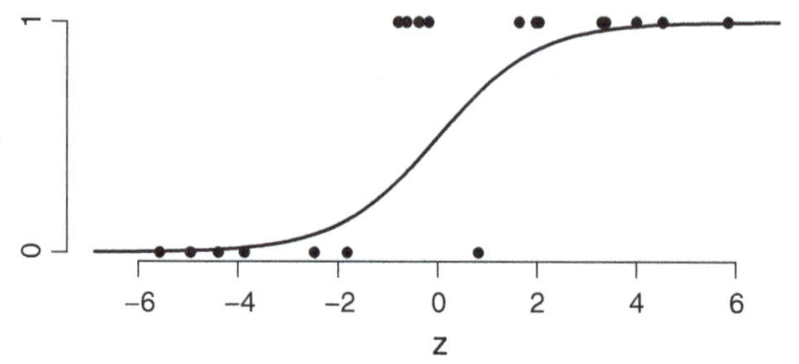


**Fig. 2.4** Illustrative example of binary responses $y$ (filled circles) and fitted probabilities (solid curve) plotted against predictor value $z$

Suppose there are $N$ pixels, with covariate values $z_1, \ldots, z_N$ respectively, and pixel presence/absence indicators $y_1, \ldots, y_N$ respectively, where $y_j = 1$ if the $j$-th pixel contains a mineral deposit, and $y_j = 0$ if not. The goal is to fit a relationship of the form (2.2). This is not a simple matter of curve-fitting, because the data $(z_j, y_j)$ do not lie "along" or "near" the curve in any sense. See Fig. 2.4. Instead, it is necessary to specify a measure of closeness or agreement between the curve and the observed data: the model is fitted by choosing the parameter values $\alpha, \beta$ which make this agreement as close as possible.

The classical fitting method is *maximum likelihood*. Given the data $y_1, \ldots, y_N$ and $z_1, \ldots, z_N$, define the *likelihood* $L(\alpha, \beta)$ to be the theoretical probability of obtaining the observed pattern of outcomes $(y_1, \ldots, y_N)$, as a function of the unknown parameter values $\alpha$ and $\beta$. The likelihood is a measure of agreement between the logistic regression curve and the observed data.

To find the likelihood, first consider a single pixel $j$ where $j = 1, 2, \ldots, N$. The probability of obtaining a presence ($y_j = 1$) in this pixel is

$$p_j = \frac{e^{\alpha+\beta z_j}}{1 + e^{\alpha+\beta z_j}} \tag{2.3}$$

and the probability of an absence ($y_j = 0$) is $1 - p_j$. The likelihood for pixel $j$ is the probability of obtaining the *observed* outcome $y_j$,

$$L_j = \mathbb{P}\{Y_j = y_j\} = \begin{cases} p_j & \text{if } y_j = 1 \\ 1 - p_j & \text{if } y_j = 0 \end{cases}$$

or more compactly

$$L_j = p_j^{y_j}(1 - p_j)^{1-y_j} = \left(\frac{p_j}{1 - p_j}\right)^{y_j} (1 - p_j)$$

which is a function $L_j = L_j(\alpha, \beta)$ of the unknown values of the parameters. Then the full likelihood is the predicted probability of the entire observed pattern of presences and absences $(y_1, \ldots, y_N)$,

$$L = L_1 L_2 \ldots L_N, \tag{2.4}$$

and is a function $L = L(\alpha, \beta)$ of the unknown parameter values $\alpha, \beta$. Equation (2.4) assumes that the outcomes in different pixels are statistically independent of each other, because the likelihood is obtained by multiplying likelihood contributions from each pixel. That is, the logistic regression technique, as it is commonly applied to presence/absence data, makes two assumptions:

1. the probability of presence $p$ is related to the predictor variable $z$ by a logistic regression relationship (2.1);
2. presence/absence outcomes in different pixels are statistically independent of each other.

The (parametric) *maximum likelihood* fitting rule is to choose the values of the parameters $\alpha, \beta$ which maximise the likelihood $L(\alpha, \beta)$. This is a standard procedure in classical statistics, carrying with it many useful additional tools such as standard errors, confidence intervals, and significance tests (Hogg and Craig 1970; Freedman et al. 2007).

Ignoring some pathological cases (e.g. where no deposits are observed), the likelihood is maximised by setting its partial derivatives to zero. Equivalently we may work with the derivatives of $\log L$. This yields the *score equations* for logistic regression

$$\sum_{j=1}^{N} p_j = \sum_{j=1}^{N} y_j \tag{2.5}$$

$$\sum_{j=1}^{N} p_j z_j = \sum_{j=1}^{N} y_j z_j \tag{2.6}$$

obtained by setting $\partial \log L / \partial \alpha = 0$ and $\partial \log L / \partial \beta = 0$ respectively. Typically the score equations have a unique solution in $(\alpha, \beta)$, giving the maximum likelihood estimates $\widehat{\alpha}, \widehat{\beta}$ of the parameters. There are no explicit formulae for $\widehat{\alpha}, \widehat{\beta}$ and the score equations must be solved numerically.

The score equations (2.5)–(2.6) have a commonsense interpretation in their own right. In (2.5) the right hand side is the observed number of deposits, while the left hand side is the expected (mean) number of deposits according to the model. In (2.6) the right hand side is the sum of the predictor values at the observed deposits, while the left hand side is the expected (mean) value of this sum according to the model. In this case maximum likelihood is equivalent to the "method of moments" in which parameters are estimated by equating the observed value of a statistic to its theoretical mean value.

Logistic regression is a simple two-parameter model, equivalent to linear regression on a transformed scale. The parameters are estimated using the entire dataset, as shown by Eq. (2.4) or (2.5)–(2.6). Consequently, the presence probability predicted by logistic regression, for a pixel with predictor value $z$, is influenced by data where the predictor value is very different from $z$, as discussed above.

It is not obligatory to use maximum likelihood estimation to fit the logistic regression model. Although maximum likelihood is theoretically optimal if the logistic regression model is true, it may fail if the model is false ("non-robust to mis-specification") and it is sensitive to anomalies in the data ("non-robust against outliers"). Robustness against outliers can be improved using *penalised likelihood* in which the likelihood $L$ is multiplied by a term $b(\alpha, \beta)$ which penalises large parameter values.

## 2.3.4 Pixel Size and Model Consistency

### Dependence on Pixel Size

The results of a logistic regression analysis clearly depend on the size of the pixels used. Table 2.1 shows estimates of the parameters $\alpha$ and $\beta$ in the logistic regression of gold deposits against distance from the nearest fault, in the Murchison data, obtained using different pixel grid sizes. Estimates of the slope parameter $\beta$ are roughly consistent between different grids. The estimate of the intercept parameter $\alpha$ becomes lower (more negative) as the pixels become smaller, so that the predicted presence probabilities also become smaller: this is intuitively reasonable, since a smaller pixel must have a smaller chance of containing a deposit.

The score equations help to explain Table 2.1. If the pixel grid is subdivided into a finer grid, the right-hand sides of (2.5) and (2.6) are unchanged, so the left-hand sides must also be unchanged. Since the number of pixels $N$ has been increased by the subdivision, the predicted probabilities $p_j$ must decrease by the same proportion $f$, the ratio of pixel areas in the two grids. Using $\log(p/(1-p)) \approx \log p$ for small $p$, the estimate of $\alpha$ must decrease by approximately $\log f$.

In order to make the results approximately consistent between different pixel sizes, the logistic regression (2.1) could be modified to

$$\log \frac{p}{1-p} = \log A + \alpha + \beta z \tag{2.7}$$

**Table 2.1** Fitted logistic regression parameters for Murchison data

| Pixel size (km) | $\alpha$ | $\beta$ |
|---|---|---|
| 10 | −0.260 | −0.243 |
| 5 | −1.321 | −0.282 |
| 2 | −2.947 | −0.261 |
| 1 | −4.303 | −0.266 |
| 0.5 | −5.681 | −0.270 |

**Table 2.2** Fitted logistic regression parameters for Murchison data, adjusted for pixel area

| Pixel size (km) | $\alpha$ | $\beta$ |
|---|---|---|
| 10 | −4.844 | −0.243 |
| 5 | −4.532 | −0.282 |
| 2 | −4.332 | −0.261 |
| 1 | −4.302 | −0.266 |
| 0.5 | −4.293 | −0.270 |

where $A$ is the pixel area used. In the language of statistical modelling, the constant $\log A$ plays the role of an *offset* in the model formula. The resulting, adjusted estimates for the parameters $\alpha, \beta$ from the Murchison data are shown in Table 2.2, and they are indeed approximately consistent across different pixel sizes. They could have been obtained from the results in Table 2.1 by subtracting $\log A$ from the estimates of $\alpha$.

For reasons explained below, slightly better consistency is achieved by replacing logistic regression (2.1) by *complementary log–log regression*

$$\log(-\log(1 - p)) = \log A + \alpha + \beta z. \tag{2.8}$$

### Large Pixels

Large pixel sizes are preferred by some researchers. A common justification is that predictions are desired for large spatial regions, for example, the probability that the entire exploration lease contains at least one deposit. Some researchers also feel that small pixel sizes are inappropriate because they lead to tiny probability values, which may be considered physically unrealistic.

However, large pixels are not needed in order to predict the probability of a deposit in a large spatial region $R$. Suppose that a logistic regression model has been fitted using a fine grid of pixels. If the region $R$ is decomposed into pixels, the probability $p(R)$ of presence of at least one deposit in $R$ satisfies

$$1 - p(R) = \prod_{j \in R} (1 - p_j), \tag{2.9}$$

where $\prod_{j \in R}$ denotes the product over all pixels in $R$. The left hand side is the probability that there are no deposits in $R$. On the right hand side, $(1 - p_j)$ is the probability that there are no deposits in pixel $j$, and since pixel outcomes are assumed to be independent, these pixel absence probabilities should be multiplied together. Hence, $p(R)$ can be calculated using presence probabilities for a fine pixel grid.

Moreover, the use of large pixels in logistic regression causes difficulties, related to the aggregation of points into geographical areas (Elliott et al. 2000; Waller and Gotway 2004; Wakefield 2007, 2004). The most important of these is the statistical bias due to aggregation ('ecological bias', Wakefield (2004, 2007) or 'aggregation bias', Dean and Balshaw (1997), Alt et al. (2001)). The 'ecological fallacy' (Robinson 1950) is the incorrect belief that a model fitted to aggregated data will apply equally to the original un-aggregated data. The 'modifiable area unit problem' (Openshaw 1984) or 'change-of-support' (Gotway and Young 2002; Banerjee and Gelfand 2002; Cressie 1996) is the problem of reconciling models that were fitted using different pixel sizes or aggregation levels.

Our analysis in Baddeley et al. (2010) shows that aggregation bias is highly dependent on the smoothness of the predictor as a function of spatial location. The distance-to-nearest-fault predictor in the Murchison example, and indeed the distance transform of any spatial feature, is a Lipschitz-continuous function of spatial location, which leads to relatively small aggregation bias. This is illustrated by Table 2.2. However, a predictor which indicates a classification, such as rock type, may have very substantial bias due to aggregation, persisting even at small pixel sizes (Baddeley et al. 2010).

Strictly speaking it can be *impossible* to reconcile two spatial logistic regression models fitted to the same spatial point pattern data using different pixel grids. Two such models are often logically incompatible (Baddeley et al. 2010), because the product rule (2.9) is incompatible with the logistic relation (2.1). It may help to recall that the pixels are artificial. A logistic regression model, using pixels of a particular size, makes an implicit assumption about the spatial random process of points in continuous space. For different pixel sizes, the corresponding assumptions are different, and generally incompatible. There is no random process in continuous space which satisfies a logistic regression model when it is discretised on *every* pixel grid. Two research teams who apply spatial logistic regression to the same data, but using different pixel sizes, may obtain results that cannot be reconciled exactly. This incompatibility can be eliminated by using complementary log–log regression (2.8) instead of logistic regression.

### Small Pixels

Mathematical theory suggests that pixels should be as small as possible, in order to reduce the unwanted effects of aggregation (Baddeley et al. 2010). However, if this is taken literally, several practical problems arise. Small pixel size implies a large number of pixels. Software for logistic regression may suffer from numerical overflow. In a fine pixellation, the overwhelming majority of pixels do not contain a data point, so the overwhelming majority of response values $y_j$ are zero. This may cause numerical instability and algorithm failure. The standard algorithm for fitting logistic regression, Iteratively-Reweighted Least Squares (McCullagh and Nelder 1989), relies on second-order Taylor approximation of the log likelihood: the algorithm itself may fail when it encounters a numerically singular matrix, or the associated statistical tools may behave incorrectly due to the *Hauck-Donner effect* (Hauck and Donner 1977).

One valid strategy for avoiding these problems is to take only a random sample of the absence-pixels (the pixels with $y_j = 0$), and to apply logistic regression to the subsampled data, using an additional offset to adjust for the sampling (Baddeley et al. 2015, Sect. 9.10).

A more natural and comprehensive solution is described in the next section.

## 2.4 Poisson Point Process Models

Pixels are artificial, so it is reasonable to ask whether logistic regression for pixel data has a well-defined meaning in continuous space, without reference to the pixel grid and pixel size. The appropriate meaning is that of the Poisson point process, studied below.

### 2.4.1 *Logistic Regression with Infinitesimal Pixels*

Logistic regressions fitted using different pixel sizes may be logically incompatible, except when the pixel size is very small. Accordingly, the only consistent interpretation of logistic regression is obtained by making the pixels *infinitesimal*.

Infinitesimal pixel size is a mathematical rather than a physical concept; it is comparable to the use of infinitesimal increments $dx$ in differential and integral calculus. The practical user will not be required to "construct" infinitesimal pixels; they will exist only in the mathematical theory. Real physical measurements will be expressed as integrals over these infinitesimal pixels.

The presence probability $p$ in an infinitesimal pixel will be infinitesimal. A more tangible quantity is the *intensity* or *rate* $\lambda$, loosely defined as the expected number of deposit points per unit area. In a pixel of very small area $A$, at most one deposit point will be present, so the expected number of points is equal to the probability of presence, and we have $\lambda \approx p/A$.

Logistic regression with infinitesimal pixels can be derived heuristically by letting the pixel size tend to zero. A rigorous argument is laid out in Baddeley et al. (2010), Warton and Shepherd (2010a, b). Assume that, for a small enough pixel size, logistic regression holds in the adjusted form (2.7), and that pixel outcomes are independent. Since $p$ is small, $\log(p/(1-p)) \approx \log p$, so that the logistic regression implies

$$\log p = \log A + \alpha + \beta z$$

or equivalently

$$\log \lambda = \alpha + \beta z.$$

This gives a consistent limit as pixel area tends to zero. In the limit, the intensity $\lambda(u)$ at a spatial location $u$ is a loglinear function of the predictor,

$$\lambda(u) = \exp(\alpha + \beta Z(u)) \tag{2.10}$$

where $Z(u)$ is the predictor value at location $u$.

Contrary to the claim that logistic regression is a flexible "nonparametric" model, we conclude that logistic regression is tantamount to assuming a loglinear (exponential) relationship between the density of deposits per unit area $\lambda$ and the predictor variable $Z$.

### 2.4.2 *Poisson Point Process*

Logistic regression, as commonly applied to presence/absence data, implicitly assumes that pixel outcomes are independent of each other. If independence holds for sufficiently small pixel size then, invoking the classical Poisson limit theorem, the random number of deposits falling in any spatial region $R$ must follow a Poisson distribution.

**Definition 1** A random variable $K$ taking nonnegative integer values has a Poisson distribution with mean $\mu$ if

$$\mathbb{P}\{K = k\} = e^{-\mu} \frac{\mu^k}{k!} \tag{2.11}$$

for any $k = 0, 1, 2, \ldots$.

Consequently (Warton and Shepherd 2010a, b; Baddeley et al. 2010; Renner et al. 2015)

**Theorem 1** *If logistic regression holds in the adjusted form* (2.7) *for sufficiently small pixels, then the random spatial pattern of deposit points must follow a* Poisson point process *with intensity of the form* (2.10).

**Definition 2** The spatial Poisson point process with intensity function $\lambda(u)$, $u \in \mathbb{R}^2$ is characterised by the following properties:

(**PP1**) Poisson counts: the number $n(\mathbf{X} \cap B)$ of points falling in any region $B$ has a Poisson distribution;

(**PP2**) intensity: the number $n(\mathbf{X} \cap B)$ of points falling in a region $B$ has expected value

$$\mu(B) = \mathbb{E}[n(\mathbf{X} \cap B)] = \int_B \lambda(u)\,\mathrm{d}u; \tag{2.12}$$

(**PP3**) independence: if $B_1, B_2, \ldots$ are disjoint regions of space then $n(\mathbf{X} \cap B_1)$, $n(\mathbf{X} \cap B_2), \ldots$ are independent random variables;

(**PP4**) conditional property: given that $n(\mathbf{X} \cap B) = n$, the $n$ points are independent and identically distributed, with common probability density

$$f(u) = \frac{\lambda(u)}{I} \tag{2.13}$$

where $I = \int_B \lambda(u)\,\mathrm{d}u$.

The intensity function $\lambda(u)$ completely determines the Poisson point process model. It encapsulates both the abundance of points (by Eq. (2.12)) and the spatial distribution of individual point locations (by Eq. (2.13)). Values of intensity have dimension length$^{-2}$.

The properties listed above can be used directly to simulate random realisations of the Poisson process. See Daley and Vere-Jones (2003, 2008) for an authoritative treatise on point processes, or (Baddeley et al. 2015, Chaps. 5, 9) for an introduction, and Kutoyants (1998), Møller and Waagepetersen (2004) for further details of statistical theory for point processes.

Theorem 1 establishes a logically consistent, physical meaning in continuous space for the logistic regression model fitted to pixel presence/absence data. Whereas logistic regression models can be somewhat difficult to interpret in practical terms, the infinitesimal-pixel limit of logistic regression is a very simple model, a Poisson point process whose intensity $\lambda(u)$ depends exponentially (log-linearly) on the predictor $Z(u)$ through (2.10). This model is well-studied, and permits highly detailed predictions to be made about various quantities, such as the expected number of points in a target region (using **PP2**), the probability of exactly $k$ points in a target region (using **PP1**), and the probability distribution of distance from a fixed starting location to the nearest random point.

The conclusion of Theorem 1 remains true in the more general case where the pixel outcomes are weakly dependent on each other (Baddeley et al. 2010, Theorem 3).

From a statistical perspective, the Poisson point process is the fundamental model, while logistic regression is a practical technique for fitting this model approximately on a discretised grid. The connection between them is not a surprise: indeed it is strongly suggested by the standard 'infinitesimal' description of the Poisson point process (Breiman 1968). It is inconceivable that Tukey (1972) was unaware of this connection.

### *2.4.3 Fitting a Poisson Point Process Model*

#### Fitting Procedures

We emphasise the distinction between a statistical model and the procedure used to fit the model. The statistical model is a description of both the systematic tendencies and the random variability in the observations, and allows us to make predictions. The model must first be fitted to the observed data. The fitting procedure is not uniquely determined by the model (unless we choose to follow a rule such as maximum likelihood) and there may be several possible choices of procedure, each with its own merits.

The Poisson point process, with loglinear intensity (2.10), has been identified as the relevant model for spatial point pattern data in continuous space. We shall now mention several possible fitting procedures for this model.

First we consider maximum likelihood. Suppose that the observed deposit locations are $x_1, \ldots, x_n$ in study region $W$. Then the log likelihood of the Poisson point process with intensity function $\lambda(u)$ is

$$\log L = \sum_{i=1} \log \lambda(x_i) + \int_W (1 - \lambda(u))\, du. \tag{2.14}$$

This can be derived either from the characteristic properties (**PP1**)–(**PP4**) of the Poisson process, or by taking the limit of the logistic regression likelihood (2.4), with appropriate rescaling, as pixel size tends to zero. See Baddeley et al. (2010), Warton and Shepherd (2010a, b), Baddeley et al. (2015, Sect. 9.7).

For the loglinear intensity model (2.10), the score equations are obtained by setting the partial derivatives of (2.14) to zero, giving

$$\int_W \lambda(u)\, du = n \tag{2.15}$$

$$\int_W Z(u)\lambda(u)\, du = \sum_{i=1}^{n} Z(x_i) \tag{2.16}$$

and these are also the infinitesimal-pixel limits of the logistic regression score equations (2.5)–(2.6). The score equations have the same "method-of-moments" interpretation as in the discrete case: namely the left hand side of each equation is the theoretical mean value, under the model, of the statistic that is evaluated for the observed data on the right hand side.

The main practical challenge in fitting the model is the fact that Eqs. (2.14) or (2.15)–(2.16) involve an integral over the study region. Unless this integral can be simplified using calculus, it must be approximated numerically.

An important case where the integral *can* be simplified is where $Z(u)$ takes only the values 0 and 1. This predictor might represent a particular rock type such as the greenstone in the Murchison example. If this is the only predictor, then the integrals in (2.14)–(2.16) can be evaluated exactly, given only the area of the greenstone and non-greenstone regions, because the integrands are constant in each region. Then the model can be fitted exactly. This case is a rare exception.

### Pixel Regression

The simplest approximation of an integral is the midpoint rule, using the sum of values of the integrand at a regular grid of sample points. This leads to the logistic regression technique of Sect. 2.3. The observed spatial locations $x_1, \ldots, x_n$ of the deposits are discretised into pixel presence-absence indicators $y_1, \ldots, y_N$. The predictor $Z$ is evaluated at the pixel centres $c_j$ to give predictor values $z_j = Z(c_j)$, and logistic regression of $y$ against $z$ is performed.

Procedures of this type are well-established in statistical science. Lewis (1972) and Tukey's former student Brillinger (Brillinger 1978; Brillinger and Segundo 1979; Brillinger and Preisler 1986) showed that the likelihood of a general point process in one-dimensional time, or a Poisson point process in higher dimensions, can be usefully approximated by the likelihood of logistic regression for the discretised process. Asymptotic equivalence was established in Besag et al. (1982).

This makes it practicable to fit spatial Poisson point process models of general form to point pattern data (Berman and Turner 1992; Clyde and Strauss 1991; Baddeley and Turner 2000, 2005) by enlisting efficient and reliable software already developed for generalized linear models. Approximation of a stochastic process by a generalized linear model is now commonplace in applied statistics (Lindsey 1992, 1995, 1997; Lindsey and Mersch 1992).

Complementary log–log regression is more appropriate than logistic regression in this context. A Poisson random variable $K$ with mean $\mu$ has probability $\mathbb{P}\{K = 0\} = e^{-\mu}$ of taking the value zero, by (2.11), and has probability $p = 1 - e^{-\mu}$ of taking a positive value. In a Poisson point process with intensity function $\lambda(u)$, the presence probability of at least one deposit in a given region $B$ is therefore

$$p(B) = 1 - e^{-\mu(B)} = 1 - \exp(-\int_B \lambda(u)\,\mathrm{d}u).$$

Inverting this relationship, the expected number of points in $B$ is

$$\int_B \lambda(u)\,\mathrm{d}u = -\log(1 - p(B))$$

If $B$ is a small pixel of area $A$, and the intensity has the loglinear form (2.10), then the relationship between presence probability and the predictor variable is

$$\log(-\log(1 - p)) = \log A + \alpha + \beta z,$$

which follows the complementary log–log regression relationship (2.8) rather than the logistic regression (2.1). However, the discrepancy is small in many cases, and the logistic function $\log(p/(1 - p))$ has slightly better numerical and computational properties, because it is the theoretically "canonical" link function (McCullagh and Nelder 1989).

**Berman-Turner Device**

In numerical analysis, an integral can often be approximated more accurately using a *quadrature rule*, based on a small number of well-chosen sample points, rather than a dense grid of sample points. Berman and Turner (1992) applied this principle to the Poisson point process likelihood (2.14) and developed an efficient fitting procedure based on a relatively small number of sample points.

In the Berman-Turner scheme, the sample points $u_1, \ldots, u_m$ consist of the observed deposit locations $x_1, \ldots, x_n$ together with a complementary set of "dummy" points $u_{n+1}, \ldots, u_m$. The integral of any function $f$ is approximated using the quadrature rule

$$\int_W f(u)\,\mathrm{d}u \approx \sum_k w_k f(u_k),$$

where $w_1, \ldots, w_m$ are numerical weights chosen appropriately. For example, the weights $w_k$ may be the areas of the tiles of the Dirichlet-Voronoï tessellation (Okabe et al. 1992) of $W$ associated with the quadrature points $u_1, \ldots, u_m$. The Poisson process log likelihood (2.14) is then approximated by

$$\begin{aligned}\log \mathrm{BTL} &= \sum_{i=1} \log \lambda(x_i) + \sum_k w_k(1 - \lambda(u_k)) \\ &= \sum_k (I_k \log \lambda(u_k) + w_k(1 - \lambda(u_k)))\end{aligned} \tag{2.17}$$

where $I_k = 1$ if the quadrature point $u_k$ is a data point, and $I_k = 0$ if it is a dummy point. The approximate log likelihood (2.17) has the same form as the (weighted) log likelihood of a Poisson regression model, and can be fitted reliably using existing statistical software (Berman and Turner 1992). The Berman-Turner technique is the main algorithm for point process modelling in the software package `spatstat` (Baddeley et al. 2015, Chap. 9).

If the predictor variables are smooth functions of spatial location, then the Berman-Turner device is extremely efficient, because of the properties of numerical quadrature (Berman and Turner 1992; Baddeley and Turner 2000). This applies, for example, to the distance function of the geological faults in the Murchison example. The approximation is less accurate when the predictor is discontinuous, such as an indicator of rock type.

**Conditional Logistic Regression**

An alternative fitting method involves placing the "dummy" sample points at random. This is the equivalent of the procedure, already described for pixel presence/absence data, of randomly selecting a subset of the pixels where no deposit is present.

Suppose the dummy point pattern is randomly generated according to a Poisson point process with known intensity $\delta > 0$. Combine the two point patterns, data $\mathbf{x}$ and dummy $\mathbf{d}$, into a single pattern $\mathbf{v} = \mathbf{x} \cup \mathbf{d}$; this is a realisation of a random point process with intensity $\kappa(u) = \lambda(u) + \delta$. Given $\mathbf{v} = \{v_1, \ldots, v_J\}$, that is, given only the locations of the combined pattern of data and dummy points, let $s_1, \ldots, s_J$ be indicators such that $s_j = 1$ if the point $v_j$ is a data point, and $s_j = 0$ if it is a dummy point. The probability $q_j = \mathbb{P}\{s_j = 1\}$ that a given random point $v_j$ is actually a data point, equals

$$q_j = \frac{\lambda(v_j)}{\lambda(v_j) + \delta},$$

the ratio of the intensity of $\mathbf{x}$ to the intensity of $\mathbf{v}$. Hence

$$\log \frac{q_j}{1 - q_j} = \log \lambda(v_j) - \log \delta = \alpha + \beta Z(v_j) - \log \delta. \tag{2.18}$$

The data/dummy status $s_j$ of each point $v_j$ is independent of other points. It follows that the conditional likelihood of the data/dummy status of the points of $\mathbf{v}$, given their locations, is the likelihood of logistic regression in the form (2.18). The Poisson point process model with loglinear intensity (2.10) could be fitted by logistic regression of $s_j$ on $z_j = Z(v_j)$ given $\mathbf{v}$.

This technique relies on the independence properties of the Poisson point process, and is a counterpart of the well-known relationship between logistic regression and loglinear Poisson models in contingency tables (Dobson and Barnett 2008; McCullagh and Nelder 1989).

Several versions of this technique have been used for point pattern data in continuous space (Diggle and Rowlingson 1994; Baddeley et al. 2014). By using random sample points, the technique avoids bias which may occur in numerical quadrature, while potentially increasing variability due to random sampling. The variance contribution due to randomisation can be estimated, and appears to be acceptable in many cases (Baddeley et al. 2014).

**Maximum Entropy**

The principle of maximum entropy (Dutta 1966) is often used in ecology, for example, to study the influence of habitat variables on the spatial distribution of animals or plants (Dudík et al. 2007; Elith et al. 2011; Phillips et al. 2006). Conceptually this method considers all possible spatial distributions, and finds the spatial distribution which maximises a quantity called entropy, subject to constraints implied by the observed data. The constraints are equivalent to the score equations (2.15)–(2.16) or (2.5)–(2.6). The maximum entropy solution is a probability distribution which is a *loglinear* function of the predictors. It was shown in Renner and Warton (2013) that this solution is equivalent to fitting a loglinear Poisson point process, or equivalent to logistic regression on a fine pixel grid. An analogy could be drawn with the stretching of a string: a string may take on any shape, but if we demand that the string be stretched as tight as possible, it will take up a straight line. Thus, this analysis principle is equivalent to fitting a Poisson point process model with loglinear intensity.

### *2.4.4 Murchison Example*

Here we give a worked example of Poisson point process modelling for the Murchison data of Fig. 2.1. The gold deposit locations are assumed to follow a Poisson process with intensity $\lambda(u)$ assumed to be a loglinear function of distance to the nearest fault,

$$\lambda(u) = \exp(\alpha + \beta d(u)), \tag{2.19}$$

where $\alpha, \beta$ are parameters and $d(u)$ is the distance from location $u$ to the nearest geological fault. Contours of $d(u)$ are shown in Fig. 2.2. The model (2.19) corresponds

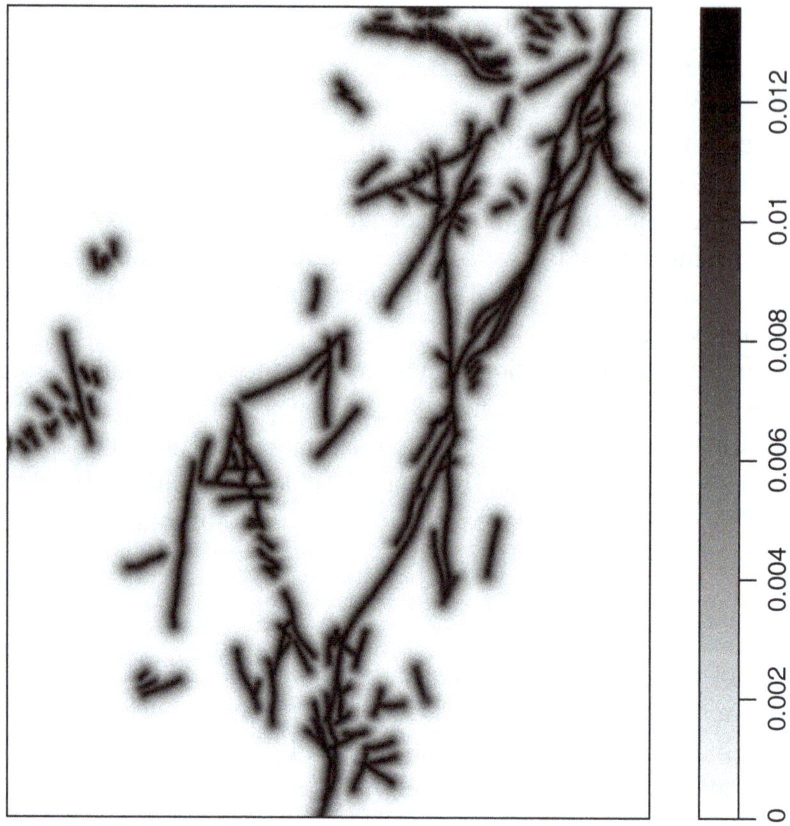


**Fig. 2.5** Fitted intensity of gold deposits in the Murchison survey according to the loglinear Poisson point process model. Pixel values are intensities (number of deposits per square kilometre).

to logistic regression of pixel presence/absence indicators against distance to nearest fault.

We used the Berman-Turner device as implemented in the `spatstat` package (Baddeley et al. 2015) in the function `ppm`. The fitted parameters are $\widehat{\alpha} = -4.34$ and $\widehat{\beta} = -0.26$ km$^{-1}$. These values are quite similar to the estimates in Table 2.2, as expected. The fitted intensity function,

$$\lambda(u) = \exp(-4.34 - 0.26\, d(u)), \tag{2.20}$$

is displayed as a greyscale image in Fig. 2.5. Note that the spatial resolution of Fig. 2.5 is finer than the spacing of sample points used to fit the model; indeed $\lambda(u)$ can be evaluated at any location $u$ in continuous space, using (2.20).

The fitted intensity relationship (2.20) can be interpreted directly. The estimated intensity of gold deposits in the immediate vicinity of a geological fault is about $\exp(-4.34) = 0.013$ deposits per square kilometre or 1.3 deposits per 100 km$^2$. This intensity decreases by a *factor* of $\exp(-0.26) = 0.77$ for every additional kilometre away from a fault. At a distance of 10 km, the intensity has fallen by a factor of $\exp(10 \times (-0.26)) = 0.074$ to $\exp(-4.34 + 10 \times (-0.26)) = 0.001$ deposits per square kilometre or 0.1 deposits per 100 km$^2$. Figure 2.6 shows the effect of the

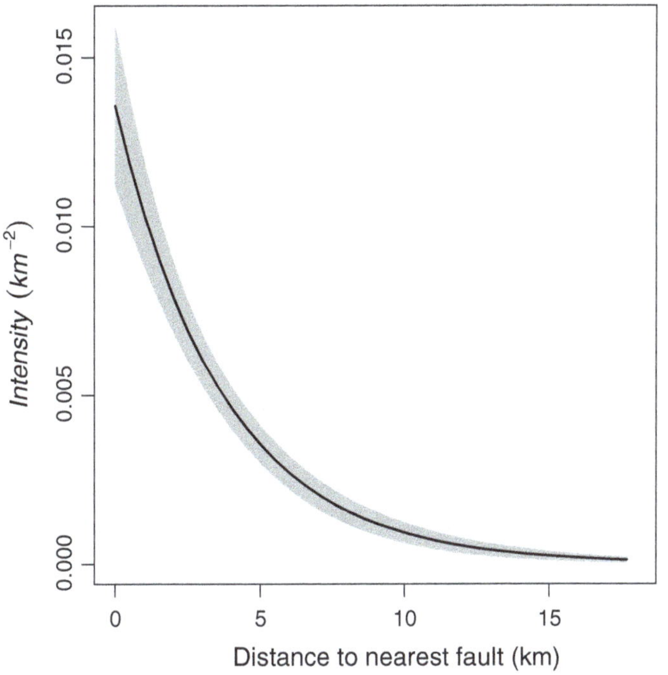


**Fig. 2.6** Fitted intensity of Murchison gold deposits as a function of distance to the nearest fault, assuming it is a loglinear function of distance. Solid line: maximum likelihood estimate. Shading: pointwise 95% confidence interval

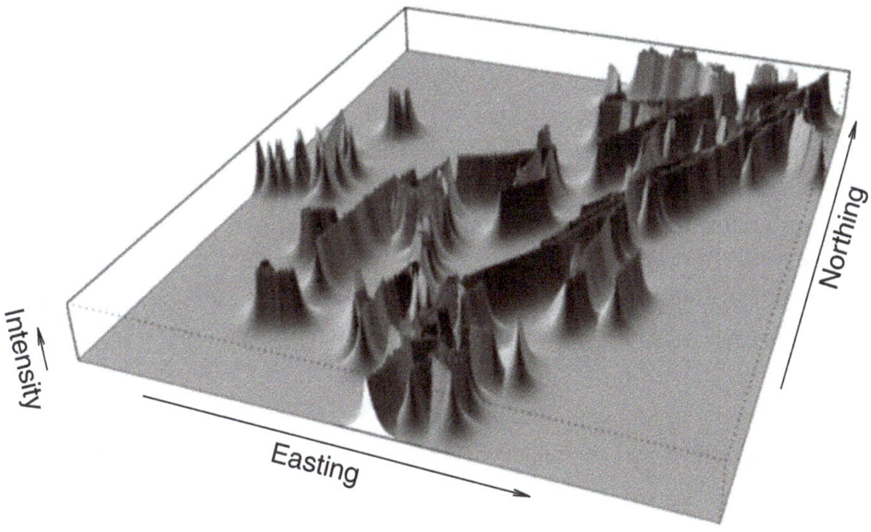


**Fig. 2.7** Perspective view of fitted intensity surface of loglinear Poisson point process model of Murchison gold deposits against distance from nearest fault

distance covariate on the intensity function, according to the fitted loglinear Poisson model.

Figure 2.7 shows a perspective view of the fitted intensity function, treated as a surface in three dimensions. Note that, fortuitously, the southern edge of the perspective plot in Fig. 2.7 shows the shape of the fitted intensity curve in Fig. 2.6.

We caution again that this analysis has not fitted a highly flexible model in which the abundance of gold deposits depends, in some unspecified way, on the distance to the nearest fault. Rather, the very specific loglinear relationship (2.19) has been fitted. The flexible part of this analysis is the freedom to choose another predictor variable or variables to replace the distance function $d(u)$. Once the predictors are chosen, the analysis becomes rigidly parametric.

### 2.4.5 *Statistical Inference*

The Poisson point process model with loglinear intensity (2.10) belongs to the class of "exponential family" models (McCullagh and Nelder 1989). Statistical inference has been studied in detail for this class (Barndorff-Nielsen 1978) and for the Poisson process in particular (Kutoyants 1998; Rathbun and Cressie 1994).

A full set of standard tools is available for statistical inference. These include standard errors and confidence intervals for the parameter estimates, hypothesis tests (likelihood ratio test, score test), and variable selection and model selection (analysis of deviance, Akaike information criterion). See Baddeley et al. (2015, Chap. 9) for a full implementation.

Table 2.3 shows the estimated standard errors and 95% confidence intervals for the parameters in the loglinear model fitted to the Murchison data. These are asymptotic standard errors based on the Fisher information matrix.

Analysis of variance, or in this case, analysis of deviance (McCullagh and Nelder 1989; Hosmer and Lemeshow 2000; Dobson and Barnett 2008) supports a formal hypothesis test of statistical significance for the dependence on a predictor variable. For example the likelihood ratio test of the null hypothesis $\beta = 0$ against the alternative $\beta \neq 0$ indicates very strong evidence that gold deposit abundance is dependent on the distance to the nearest fault.

Recently-developed tools for model selection in point process models include Sufficient Dimension Reduction (Guan and Wang 2010).

**Table 2.3** Standard errors and confidence intervals for parameters in loglinear Poisson model of Murchison data

| | Estimate | SE | 95% CI |
|---|---|---|---|
| $\alpha$ | −4.30 | 0.09 | [−4.47, −4.13] |
| $\beta$ | −0.27 | 0.02 | [−0.31, −0.23] |

### 2.4.6 Diagnostics

A fitted model is not like a fitted shoe. A shoe must approximately match the shape of the wearer's foot before we call it fitted. On the contrary, statistical software "fits" a model to data on the assumption that the model is true, and does not check that the model describes the data at all.

Diagnostic quantities and diagnostic plots for a fitted model should be used to check the model assumptions. For linear regression and linear models, diagnostics are highly developed in statistical theory and applied statistical practice (Atkinson 1985). For logistic regression in a general context, diagnostics are also available (Landwehr et al. 1984; Dobson and Barnett 2008) and these extend to the "exponential family" class of models, at least in theory.

Diagnostics for the Poisson point process model, corresponding to the well-known diagnostics for logistic regression, were developed in Baddeley et al. (2013a, b). Two of these are shown here for the Murchison data.

The *influence* measure $e_i$ is the effect on the fitted log likelihood of deleting the $i$th deposit point $x_i$ (Baddeley et al. 2013a). Figure 2.8 shows circles of diameter proportional to $e_i$ centred at the deposit locations $x_i$. The geological fault pattern is also shown. In this Figure, large circles represent observations which had a large effect on the resulting fitted model. There is one very large circle at middle left of the Figure, and we notice that there are no geological faults near this deposit. That is,

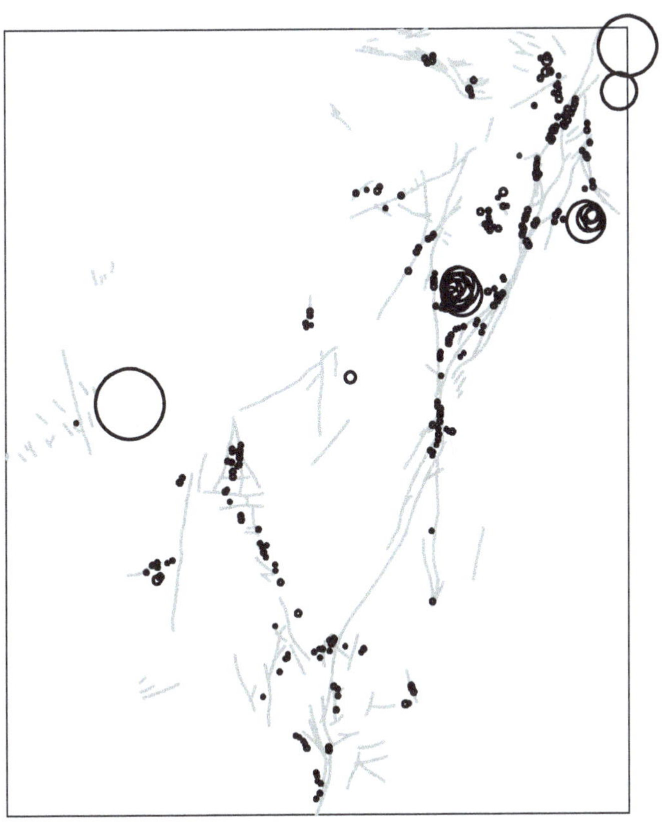

**Fig. 2.8** Influence diagnostic for the loglinear Poisson model of gold deposits against distance to nearest fault. Circle diameters are proportional to the influence of each deposit. Grey lines are geological faults

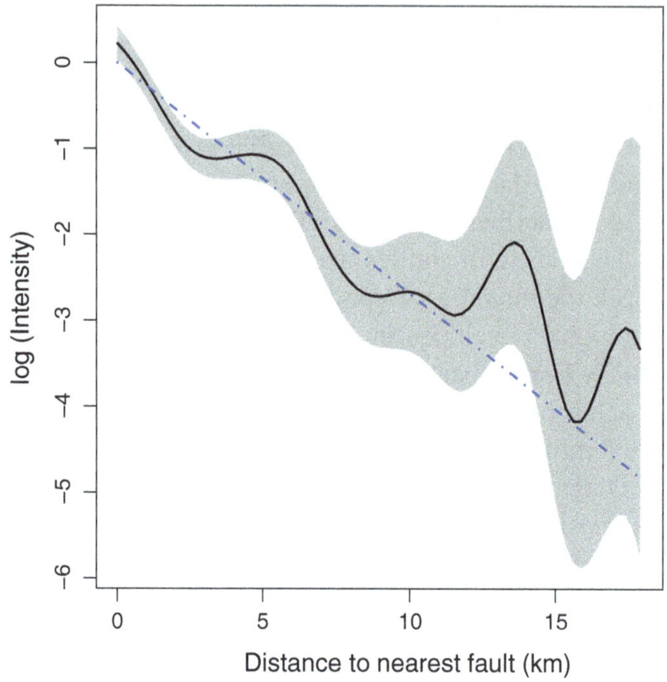


**Fig. 2.9** Partial residual plot for loglinear Poisson model of Murchison gold deposits as a function of distance to the nearest fault. Solid line: smoothed partial residual. Shading: pointwise 95% confidence interval. Dot-dash line: fitted model

the influence diagnostic identifies this deposit as anomalous, perhaps an "outlier", with respect to the fitted model in which deposits are most likely to occur close to a geological fault. This is an entirely data-driven diagnostic, and tells us only that this observation is anomalous with respect to the model. It is unable to tell us whether the deposit is truly anomalous in geological terms, or whether the survey perhaps failed to detect an existing geological fault near this location.

Strategies for dealing with anomalous data include outlier detection and removal, and robust model-fitting which is resistant to the effects of outliers. Robust parameter estimation for Poisson point process models was developed in Assunção and Guttorp (1999).

Figure 2.9 shows a *partial residual* plot (Baddeley et al. 2013b) for the Murchison gold deposits against distance to nearest fault. Assuming that the loglinear model (2.19) is approximately true, say $\log \lambda(u) = \alpha + \beta d(u) + H(d(u))$ where the error $H(d)$ is small, this procedure forms an estimate $\widehat{H}(d)$ of the error term, adds it to the fitted linear term, and plots $\widehat{\alpha} + \widehat{\beta} d + \widehat{H}(d)$ against values of distance $d$. If the model is correct, this plot should be a straight line. Departures from the straight line can be interpreted as suggesting the correct form of dependence. Figure 2.9 suggests there is a minor departure from the loglinear model.

An alternative way to explore non-linearity is to fit a polynomial or spline function in place of the linear function on the right hand side of (2.1) or (2.19). In order to avoid over-fitting and numerical instability, the model should be fitted by *penalised* maximum likelihood, in which the log likelihood (2.14) is augmented by a penalty term that discourages extreme values of the parameters which might produce a wildly-oscillating polynomial. Figure 2.10 shows a penalised maximum

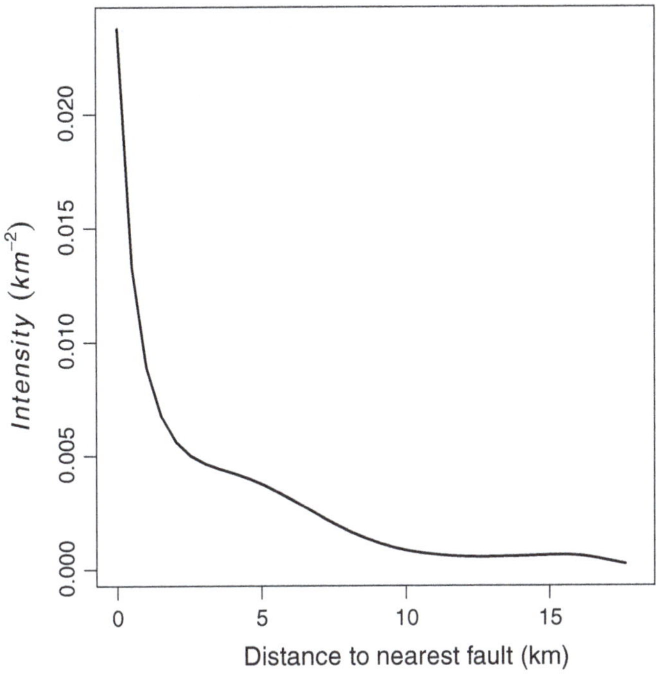


**Fig. 2.10** Fitted intensity of Murchison gold deposits assuming the log intensity is a fifth-order B-spline function of distance to nearest fault

likelihood fit of a model in which the log intensity is a fifth-order B-spline function of distance to the nearest fault. The model was fitted in the `spatstat` package using code for Generalised Additive Models (GAM) (Hastie and Tibshirani 1990). This fit also suggests minor departure from the loglinear model.

### 2.4.7 *Rationale for Prediction*

Up to this point, our commentary on prospectivity analysis applies equally well to the analysis of archaeological finds, plant species distribution, etc., using logistic regression and related tools. However, the key goal of prospectivity analysis is the *prediction of previously-unknown deposits*, and this sets it apart from other applications. This prediction problem deserves more attention from the statistical community, and we shall identify several topics for research.

The rationale for predicting "new" mineral deposits is clearest when we extrapolate from a fully-explored region to an unexplored region. We might extrapolate from a previous, fully-explored mining lease to a newly-granted exploration lease which is geologically analogous. We fit a model to the fully-explored region, obtaining estimates of the model parameters $\alpha, \beta$, which we believe can be extrapolated to the unexplored region. Applying the fitted model relationship to the predictor variables for the new region, we obtain explicit predictions for the mineral deposits in the new region. These predictions may include expected numbers of deposits, probability of no deposits, probability distribution of distance to the nearest deposit, and so on. These predictions are valid calculations even if the geological structure in the two regions was formed at the same epoch, because of the assumption of independence between deposits. Essentially the fully-explored region is used to obtain estimates of the parameters of the "laws" which apply to both regions, and these laws are then applied to the new region.

The statistical reasoning is far more complicated when we wish to predict hitherto-undiscovered mineral deposits from known deposits *in the same region*. It would be futile to assume that the region has been fully explored, since this would imply there are no deposits remaining to be discovered. Instead our statistical model must now recognise two categories of deposits, known and unknown. The methods described above can be re-deployed if we assume that the true spatial pattern of *all* deposits (whether known or unknown) is a Poisson point process with intensity function $\kappa(u)$, say, and that a deposit existing at a location $u$ will be detected with probability $P(u)$, independently of other deposits. Then, by the "thinning" property of the Poisson process, the pattern of *detected* deposits is also a Poisson process, with intensity $\lambda(u) = P(u)\kappa(u)$; the pattern of *undetected* deposits is a Poisson process with intensity $\xi(u) = (1 - P(u))\kappa(u)$; and the detected and undetected deposits are independent of each other. Fitting a Poisson point process model to the observed mineral deposits allows us to estimate $\lambda(u)$ only. If the detection probability $P(u)$ is known, then it becomes feasible to back-calculate $\kappa(u) = \lambda(u)/P(u)$ and

$$\xi(u) = \frac{1 - P(u)}{P(u)} \lambda(u). \tag{2.21}$$

It is then possible to make predictions or conditional simulations of the undetected deposits. The independence property of the Poisson process implies that the prediction or conditional simulation depends only on the fitted model parameters, and does not otherwise depend on the observed deposits. The conditional simulation is a realisation of the Poisson process of the assumed loglinear form with the parameter values fitted from the data: the simulated realisation is independent of the observed deposits, given the fitted model parameters.

This argument is an instance of the prediction approach to survey sampling inference (Royall 1988). The difficulty is that the detection probability $P(u)$ will depend on the detection method, the spatially-varying amount of survey effort, and other factors. If $P(u)$ can be estimated from data, perhaps by comparing the results of successive surveys of the same region, then the form of (2.21) suggests that the appropriate model is a logistic regression for $P(u)$ on explanatory variables. If no information is available about $P(u)$, we could make the simplifying assumption that $P(u) \equiv P$ is constant; then $\xi(u)$ is a constant multiple of $\lambda(u)$, so that at least the relative prospectivity of different locations $u$ can be assessed from a plot of $\lambda(u)$.

Other, non-Poisson point processes can also serve as models of mineral deposits (Baddeley et al. 2015, Chaps. 12 and 13) and support prediction and conditional simulation. In such models, the presence of a point affects the probability of presence of a point at nearby locations. In this case the conditional simulation does depend on the observed deposit locations (Møller and Waagepetersen 2004; Baddeley et al. 2015).

A more realistic model of the detection process would envisage that the discovery of a new deposit will encourage the exploration geologist to survey the surrounding areas more intensively, increasing the detection probability in these surrounding areas. This destroys the independence structure: the pattern of observed deposits is no longer a Poisson point process, and is spatially clustered. Non-Poisson point process models would be needed to describe the spatial pattern of observed deposits, and even if the spatial pattern of all deposits is assumed to be Poisson, the pattern of undiscovered deposits is both non-Poisson and dependent on the observed deposits. A full analysis of the prediction problem would require the deployment of Missing Data principles (Little and Rubin 2002).

In prospectivity analysis it may or may not be desirable to fit any explicit relationship between deposit abundance and predictors such as distance to the nearest fault. Often the objective is simply to select a distance threshold, so as to delimit the area which is considered highly prospective (high predicted intensity) for the mineral. The ROC curve (Sect. 2.7) is more relevant to this exercise. However, if credible models can be fitted, they contain much more valuable predictive information.

## 2.5 Monotone Regression

The remainder of this article describes three alternative analysis techniques, genuinely different from logistic regression, which do not seem to be widely used in prospectivity analysis. These techniques are genuinely "non-parametric" in the sense that they assume only that the intensity or rate of mineral deposits $\lambda(u)$ is a function of the predictor variable $Z(u)$ at the same location $u$,

$$\lambda(u) = \rho(Z(u)) \tag{2.22}$$

where $\rho(z)$ is a function to be estimated. We do not assume that $\rho(z)$ has any particular functional form.

The assumption (2.22) is encountered frequently. In geological applications where the points are the locations of mineral deposits, $\rho$ is an index of the prospectivity (Bonham-Carter 1995) or predicted frequency of deposits as a function of geological and geochemical covariates $z$. In ecological applications where the points are the locations of individual organisms, $\rho$ is a "resource selection function" (Manly et al. 1993) reflecting preference for particular environmental conditions $z$.

In *monotone regression*, we assume that $\rho(z)$ is a monotone function of $z$, either monotone increasing (non-decreasing):

$$z_1 < z_2 \quad \text{implies} \quad \rho(z_1) \leq \rho(z_2)$$

or monotone decreasing (non-increasing):

$$z_1 < z_2 \quad \text{implies} \quad \rho(z_1) \geq \rho(z_2).$$

Sager (1982) considered the log-likelihood of the Poisson point process with intensity (2.22),

$$\log L = \sum_i \log \rho(Z(x_i)) - \int_W \rho(Z(u))\, du, \tag{2.23}$$

and showed that the log-likelihood can be maximised over the class of all monotone functions $\rho$. The optimal function $\widehat{\rho}(z)$ is the *nonparametric maximum likelihood* estimate of $\rho(z)$ under the monotonicity constraint, or simply the *monotone regression* estimate.

To simplify the discussion, assume that $\rho(z)$ is monotone decreasing, and that the values of $Z(u)$ are real numbers greater than or equal to zero. Sager (1982) showed that the monotone regression estimate $\widehat{\rho}(z)$ is piecewise constant, with jumps occurring only at the observed values $z_i = Z(x_i)$ of the predictor at the deposit point locations. For any $z$ let

$$A(z) = |\{u \in W : Z(u) \le z\}| \tag{2.24}$$

be the area of the subset of the survey region where the covariate value is less than or equal to $z$. Also let $N(z) = \sum_i \mathbf{1}\{z_i \le z\}$ be the number of data points for which the covariate value is less than or equal to $z$. In the Murchison example, $A(z)$ is the area lying closer than $z$ kilometres from the nearest fault, and $N(z)$ is the number of deposits lying in this region. Then the monotone regression estimate $\widehat{\rho}(z)$ is the maximum of simple functions

$$\widehat{\rho}(z) = \max_i \rho_i(z) \tag{2.25}$$

where

$$\rho_i(z) = \begin{cases} \frac{N(z_i)}{A(z_i)} & \text{if } z < z_i \\ 0 & \text{if } z \ge z_i. \end{cases} \tag{2.26}$$

The monotone regression estimate $\widehat{\rho}(z)$ can be computed rapidly using the Pool Adjacent Violators algorithm (Barlow et al. 1972) or the following Maximum Upper Sets algorithm (Sager 1982):

1. Sort the data values as $z_1 \le z_2 \le \cdots \le z_n$.
2. Initialise $K = 0$ and $z_K = 0$. These represent the largest data value whose status has been resolved so far.
3. For each $i > K$ consider the interval $[z_K, z_i]$, evaluate the empirical intensity $\rho_{i,K} = (N(z_i) - N(z_K)/(A(z_i) - A(z_K))$, and find the value $i^*$ which maximises $\rho_{i,K}$.
4. For all $z$ lying in the interval $[z_K, z_{i^*}]$, set $\rho(z) = \rho_{i^*,K}$.
5. If $i^* = n$, exit. Otherwise, set $K = i^*$ and go to step 3.

**Fig. 2.11** Fitted intensity of gold deposits as a function of distance to the nearest fault, using monotone regression

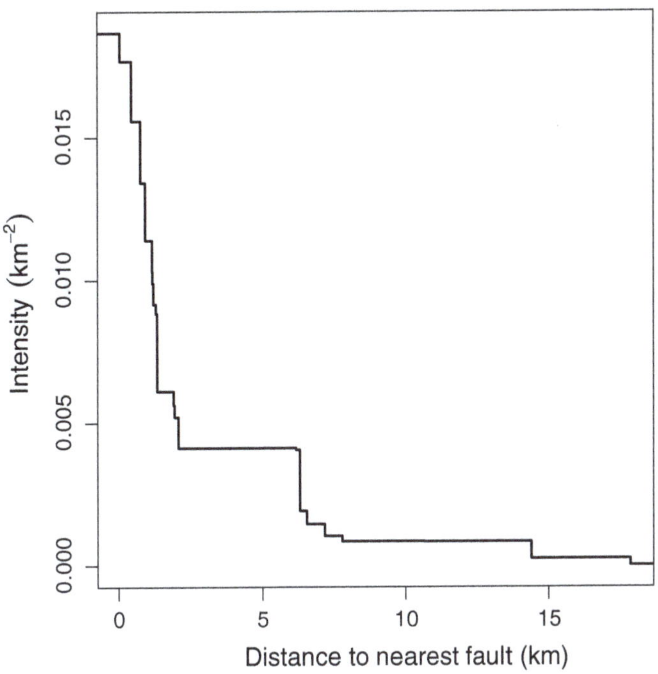


Figure 2.11 shows the monotone regression estimate of the intensity of gold deposits as a function of distance to nearest fault in the Murchison data. The curve has the same overall shape as the exponential curve implied by the loglinear Poisson point process model or logistic regression (Fig. 2.6), except for a prominent plateau between $z = 2$ and $z = 6$ km.

Note also that the monotone regression estimate of intensity at small distances $z$ is higher (Fig. 2.11) than in the loglinear Poisson model (Fig. 2.6). This is expected, since the estimate $\widehat{\rho}(z)$ depends primarily on the part of the survey where $Z(u) \leq z$. This is more satisfactory than the behaviour of the loglinear Poisson model for which the fitted curve depends on the entire dataset. If, for example, we were to restrict the study area to the region lying at most 20 km from the nearest fault, the estimates of the parameters $\alpha, \beta$ in the loglinear Poisson model could change markedly, while the monotone regression in Fig. 2.11 would be unchanged.

Figure 2.12 shows a perspective plot of the predicted intensity implied by the monotone regression. Compared with Fig. 2.7, this shows qualitatively the same effect of a dense concentration close to the geological faults, but with a different profile (again fortuitously displayed at the southern edge of the plot).

Sager (1982) shows that this method extends to multiple predictor variables. The author believes it can also be extended to allow points to have weights determined by the mineral endowment of the deposit, or a similar characteristic.

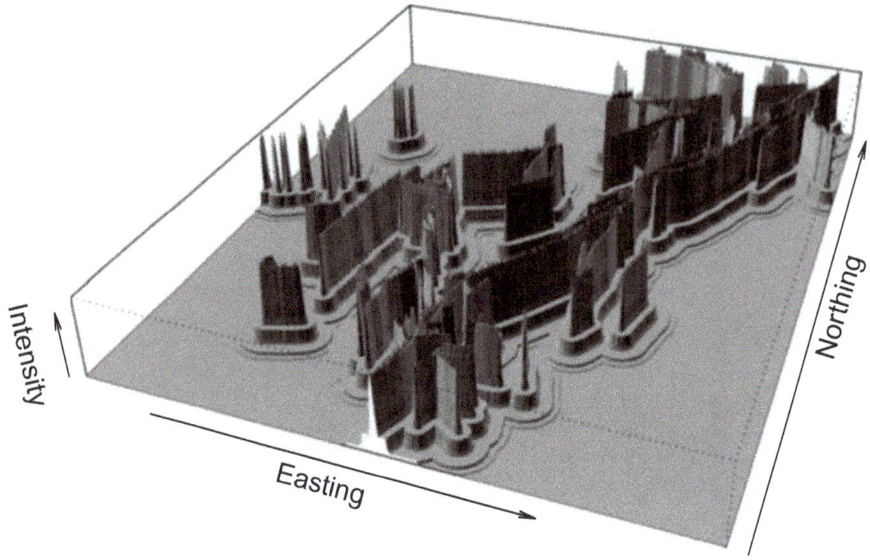


**Fig. 2.12** Perspective view of fitted intensity using monotone regression

## 2.6 Nonparametric Curve Estimation

A second alternative to logistic regression is nonparametric curve estimation, in which we assume that the intensity is a smooth function of the predictor, $\lambda(u) = \rho(Z(u))$, and estimate the function $\rho(z)$ by nonparametric smoothing. This was developed in Baddeley et al. (2012), Guan (2008).

Assume that Eq. (2.22) holds, and that $\rho(z)$ is a continuous function of $z$, and that $Z(u)$ is at least a continuous function of location $u$, without further constraints. Nonparametric estimation of $\rho$ is closely connected to estimation of a probability density from biased sample data (Jones 1991; El Barmi and Simonoff 2000) and to the estimation of relative densities (Handcock and Morris 1999). Under the smoothness assumptions, $\rho$ is proportional to the ratio of two probability densities, the numerator being the density of covariate values at the points of the point process, while the denominator is the density of covariate values at random locations in space. Kernel smoothing can be used to estimate the function $\rho$ as a relative density (Baddeley et al. 2012; Guan 2008).

Define the *spatial distribution function* (Lahiri 1999; Lahiri et al. 1999) as the cumulative distribution function of the covariate value $Z(U)$ at a random point $U$ uniformly distributed in $W$:

$$G(z) = \frac{1}{|W|} \int_W \mathbf{1}\{Z(u) \le z\} \, \mathrm{d}u. \tag{2.27}$$

Here we use the 'indicator' notation: $\mathbf{1}\{\dots\}$ equals 1 if the statement '...' is true, and 0 if the statement is false. Equivalently $G(z) = A(z)/A(\infty) = A(z)/|W|$ where $A(z)$, defined in (2.24), is the area of the set of all locations in $W$ where the covariate value is less than or equal to $z$. In practice $G(z)$ would often be estimated by evaluating the covariate at a fine grid of pixel locations, and forming the cumulative distribution function

$$G(z) = \frac{\#\{\text{pixels } u : Z(u) \le z\}}{\#\text{pixels}}. \tag{2.28}$$

Three estimators of $\rho$ proposed in Baddeley et al. (2012) are

$$\hat{\rho}_{\mathrm{R}}(z) = \frac{1}{|W|G'(z)} \sum_i \kappa(Z(x_i) - z) \tag{2.29}$$

$$\hat{\rho}_{\mathrm{W}}(z) = \sum_i \frac{1}{|W|G'(Z(x_i))} \kappa(Z(x_i) - z) \tag{2.30}$$

$$\hat{\rho}_T(z) = \frac{1}{|W|} \sum_i \kappa(G(Z(x_i)) - G(z)) \tag{2.31}$$

where $x_1, \dots, x_n$ are the data points, $Z(x_i)$ are the observed values of the covariate $Z$ at the data points, $|W|$ is the area of the observation window $W$, and $\kappa$ is a *one-dimensional* smoothing kernel—smoothing is conducted on the observed values $Z(x_i)$ rather than in the window $W$. The derivative $G'(z)$ is usually approximated by differentiating a smoothed estimate of $G$. The estimators (2.29)–(2.31) were developed in Baddeley et al. (2012) by adapting estimators from kernel smoothing (Jones 1991; El Barmi and Simonoff 2000). An estimator similar to (2.29) was proposed in Guan (2008).

Figure 2.13 shows the fitted estimate of intensity for the Murchison gold deposits as a function of distance to the nearest fault. The plot shows the ratio estimator $\hat{\rho}_R(z)$ against $z$, equation (2.29), together with the pointwise 95% confidence interval for $\rho(z)$ based on asymptotic theory assuming a Poisson process (Baddeley et al. 2012). Tickmarks on the horizontal axis show the observed distance values $z_i = Z(x_i)$ at the deposits.

The overall shape of Fig. 2.13 is consistent with Figs. 2.6 and 2.11. A plateau of intensity is visible between $z = 2.5$ and $z = 5.5$ km, consistent with the plateau seen in Fig. 2.11. The peak of intensity in Fig. 2.13 occurs at about $z = 1$ km, rather than at distance $z = 0$, but this may be an artefact of the smoothing procedure, as it is not seen in the other two estimates (2.30) and (2.31).

Figure 2.14 shows a perspective view of the predicted intensity using nonparametric curve estimation. This is quite similar to the surface obtained by monotone regression, shown in Fig. 2.12.

The nonparametric curve estimate has the attractive property that $\hat{\rho}(z)$ depends only on the survey information from locations where the predictor value is approximately equal to $z$. In the Murchison example, the estimated intensity $\hat{\rho}(z)$ of gold deposits at a distance $z$ from the nearest fault, is estimated using only the deposits and non-deposit locations which lie approximately $z$ km from the nearest fault. Although smoothing artefacts may be present, this property means that the nonparametric

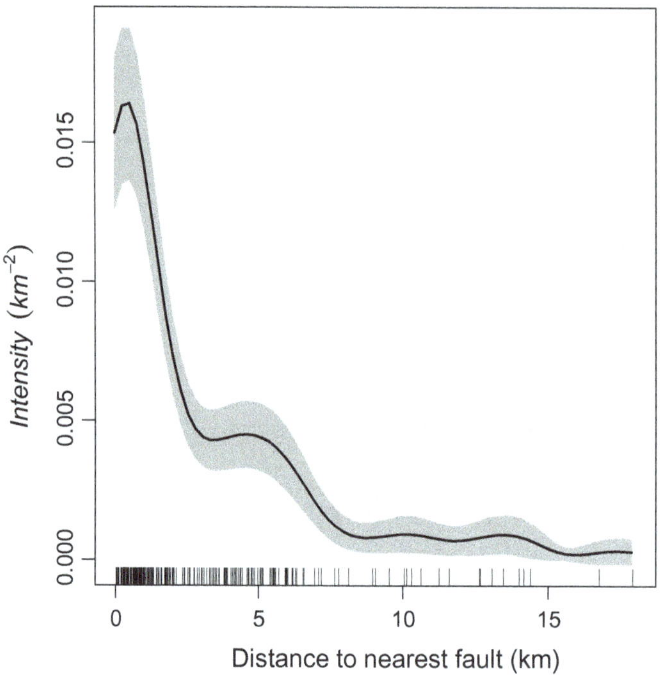


**Fig. 2.13** Fitted intensity of gold deposits as a function of distance to the nearest fault, using kernel-based estimator

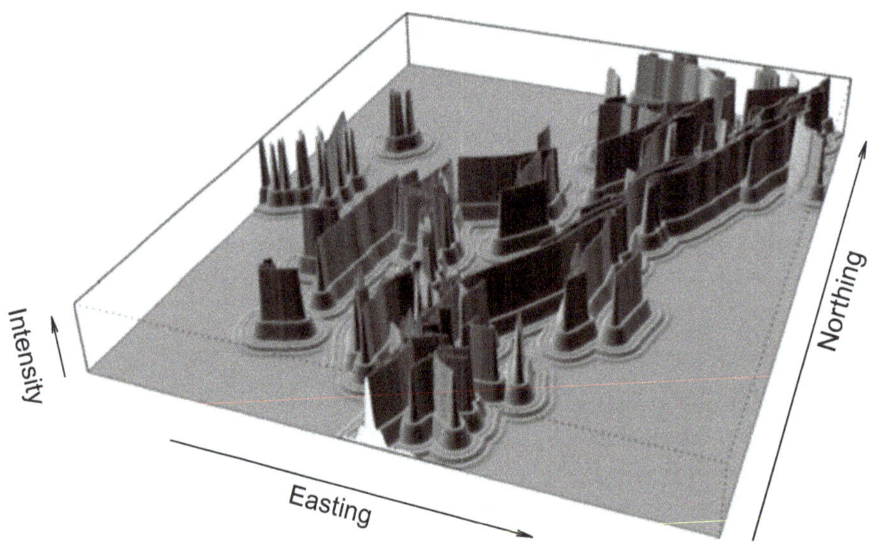


**Fig. 2.14** Perspective view of fitted intensity using nonparametric curve estimate of $\rho$

curve estimate can be treated as an estimate of the true relationship between intensity and predictor.

The estimators (2.29)–(2.31) can be modified to incorporate numerical weights, for example, representing the endowment of each deposit. Then $\rho(z)$ has the interpretation of the expected *total endowment* per unit area, of deposits at a distance $z$ from the nearest fault.

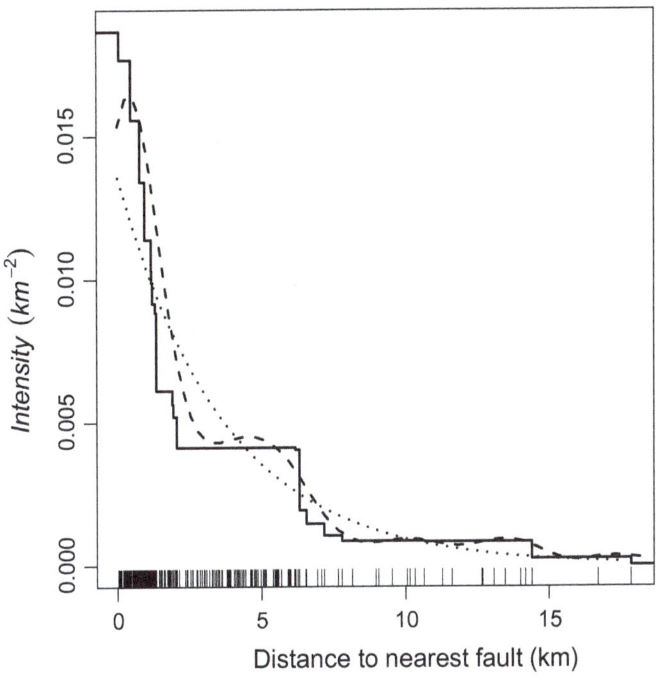


**Fig. 2.15** Fitted intensity of Murchison gold deposits as a function of distance to the nearest fault, estimated by three different methods: monotone regression (solid lines), kernel smoothing (dashed lines) and logistic regression (dotted lines)

Figure 2.15 shows the three estimates of $\rho(z)$ together. Logistic regression, log-linear Poisson point process modelling, and maximum entropy methods effectively assume that prospectivity is an exponential function $\rho(z) = AB^z = \exp(\alpha + \beta z)$, while monotone regression assumes $\rho(z)$ is a decreasing function of $z$, and kernel estimation assumes $\rho(z)$ is a smooth function of $z$ without further restriction.

This analysis assumes that the intensity at a location $u$ depends *only* on the covariate value $Z(u)$. To validate the assumption (2.22) we can compare the predicted intensity $\hat{\rho}(Z(u))$ assuming (2.22) with a (spatial) kernel estimate $\hat{\lambda}(u)$ which does not assume (2.22). If the assumption is not true, $\widehat{\rho}(z)$ is still meaningful: it is effectively an estimate of the average intensity $\lambda(u)$ over all locations $u$ where $Z(u) = z$.

## 2.7 ROC Curves

Suppose we agree that the ultimate goal of prospectivity analysis is to decide which parts of an exploration area are most likely to contain a valuable deposit. Then the essential task is to *classify* different parts of the exploration area into areas of high and low prospectivity, rather than necessarily needing to model the degree of prospectivity at every location.

The *Receiver Operating Characteristic (ROC)* curve (Krzanowski and Hand 2009) is a summary of the performance of a classifier. It is often applied to medical diagnostic tests (Nam and D'Agostino 2002) when the test is based on thresholding a quantitative assay. Suppose for example that a medical test returns a "positive" result (predicting a high risk of disease) if the patient's blood cholesterol level exceeds a threshold $t$. For a given choice of threshold $t$, the "true positive rate" $\mathrm{TP}(t)$ is the fraction of patients with the disease who return a correct, positive test result. The "false positive rate" $\mathrm{FP}(t)$ is the fraction of patients who do not have the disease but who return an incorrect, positive test result. The ROC curve is a plot of true positive rate $\mathrm{TP}(t)$ against false positive rate $\mathrm{FP}(t)$ for all thresholds $t$. A good classifier has a large true positive rate in comparison to the false positive rate, so the ROC curve of a good classifier will lie well above the diagonal line on the graph.

The same technique can be applied to prospectivity analysis (Rakshit et al. 2017), taking the mineral deposits as the "disease", and using either a spatial predictor $Z(u)$ or a fitted model intensity $\lambda(u)$ to classify pixels into high or low prospectivity classes. Suppose that $Z(u)$ is a real-valued spatial predictor. Calculate the empirical cumulative distribution function of $Z$ at the observed deposit locations,

$$\widehat{F}_{\mathbf{x}}(t) = \frac{1}{n(\mathbf{x})} \sum_i \mathbf{1}\{Z(x_i) \leq t\}$$

and the empirical "spatial distribution function" of $Z(u)$ over all locations $u$ in $W$,

$$F_W(t) = \frac{1}{|W|} \int_W \mathbf{1}\{Z(u) \leq t\} \, \mathrm{d}u.$$

Then the ROC plot is a graph of $1 - \widehat{F}_{\mathbf{x}}(t)$ against $1 - F_W(t)$ for all $t$. Equivalently it is a plot of $R_+(p) = 1 - F_{\mathbf{x}}(F_W^{-1}(1-p))$ against $p$. Applied statisticians would recognise this as a form of the classical P–P plot.

The formulae above assume that larger values of $Z$ are more prospective. If smaller values of $Z$ are more prospective, then the appropriate ROC plot is a graph of $F_{\mathbf{x}}(t)$ against $F_w(t)$ for all $t$, or equivalently a graph of $R_-(p) = F_{\mathbf{x}}(F_W^{-1}(p))$ against $p$. This is the P–P plot of $F_{\mathbf{x}}$ against $F_W$.

Figure 2.16 shows the ROC curve for the Murchison gold deposits against distance from nearest fault, assuming smaller distances are more prospective. The horizontal axis shows the fraction of area in the survey region which lies less than $t$ km away from a fault, and the vertical axis shows the fraction of deposits which lie less than $t$ km from a fault. For example, we may read off the plot that 60% of all known deposits lie in a region occupying 10% of the survey area defined by a distance threshold. This has a useful practical interpretation. The threshold itself is not shown on the ROC plot but could be obtained from the spatial cumulative distribution function. The ROC curve depends on the choice of the study region (Jiménez-Valverde 2012).

The interpretation of ROC curves in spatial analysis is controversial. Some writers suggest (Fielding and Bell 1997) that the ROC can be used to evaluate the goodness-of-fit of a species distribution model, or equivalently the goodness-of-fit of a prospectivity analysis. Others disagree (Lobo et al. 2007, p. 146) and argue that the ROC is an indicator of predictive power—the ability to segregate pixels reliably into two classes of high and low prospectivity.

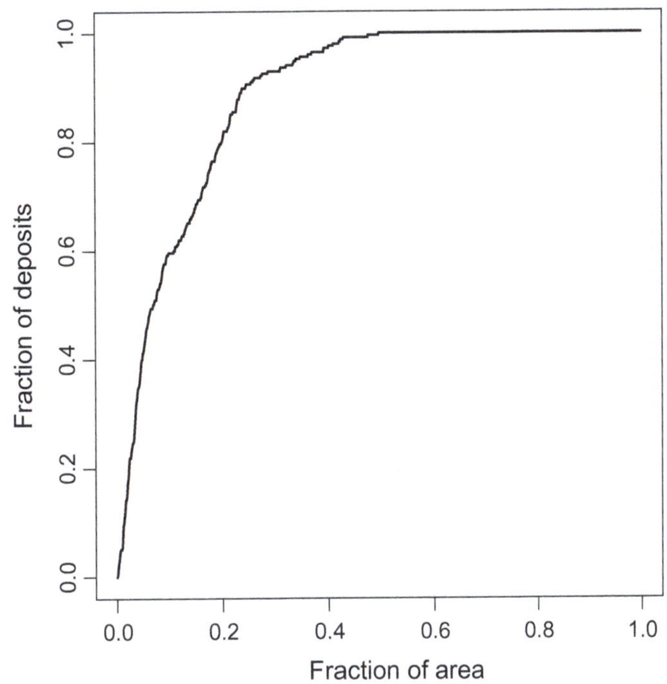


**Fig. 2.16** Empirical ROC curve for gold deposits against distance from the nearest fault in the Murchison survey rectangle

The ROC can be based either on a real-valued predictor variable $Z(u)$ or on a fitted model intensity $\lambda(u)$. In the latter case it is tempting to regard the ROC as a summary of the predictive power of the fitted model (Lobo et al. 2007; Austin 2007; Thuiller et al. 2003). However, if the model is logistic regression, or a loglinear Poisson point process, or if the intensity is estimated using monotone regression, then the fitted intensity $\lambda(u)$ is a monotone function of the predictor $Z(u)$. Thresholding $\lambda(u)$ is equivalent to thresholding $Z(u)$, so that the ROC curves derived from any of these models are identical. For example, the ROC cannot be used to compare the predictive power of logistic regression against that of monotone regression. It would be more appropriate to regard the ROC as a summary of the inherent predictive power of the predictor variable $Z(u)$ itself (Rakshit et al. 2017).

The ROC does have a connection with the other techniques described in this article. Suppose that the point process intensity $\lambda(u)$ depends on the predictor $Z(u)$ through a function $\rho(z)$ as in (2.22). Then we show in (Rakshit et al. 2017) that the slope of the ROC curve is closely related to $\rho$. If large values of the predictor are more prospective, the slope of the ROC curve is

$$\frac{\mathrm{d}}{\mathrm{d}p}R_{+}(p) = \frac{\mathrm{d}}{\mathrm{d}p}\left[1 - F_{\mathbf{x}}(F_W^{-1}(1-p))\right] = \frac{1}{\kappa}\rho(F_W^{-1}(1-p))$$

while if small values of the predictor are prospective, the slope is

$$\frac{\mathrm{d}}{\mathrm{d}p}R_{-}(p) = \frac{\mathrm{d}}{\mathrm{d}p}\left[F_{\mathbf{x}}(F_W^{-1}(p))\right] = \frac{1}{\kappa}\rho(F_W^{-1}(p))$$

where $\kappa$ is the average intensity over the study region. Analysis using the ROC curve is not fundamentally different from fitting a point process model or pixel presence/absence regression model, but may be a more practically useful presentation of the same information.

## 2.8 Recursive Partitioning

Classification and Regression Tree (CART) (Breiman et al. 1984) or Recursive Partitioning methods offer another alternative approach. Given one or many predictor variables, these methods predict the response by thresholding the predictors. The result is a prediction rule, organised as a logical tree, in which each fork of the tree is a threshold operation on one of the predictors. This kind of rule would appear to be well-suited to the practical needs of prospectivity analysis.

For a single predictor variable, the result of recursive partitioning is a piecewise-constant function $\widehat{\rho}(z)$ which is not constrained to be monotone. Figure 2.17 shows the estimated intensity of the Murchison gold deposits as a function of distance to the nearest fault only, using recursive partitioning. Any number of predictor variables can be included in the analysis.

### Software and Data

All analyses in this chapter were performed using the `spatstat` library (Baddeley et al. 2015) which is a contributed extension package for the R statistical software system (R Development Core Team 2011). Both R and `spatstat` can be downloaded from https://cran.r-project.org. The Murchison data are included in

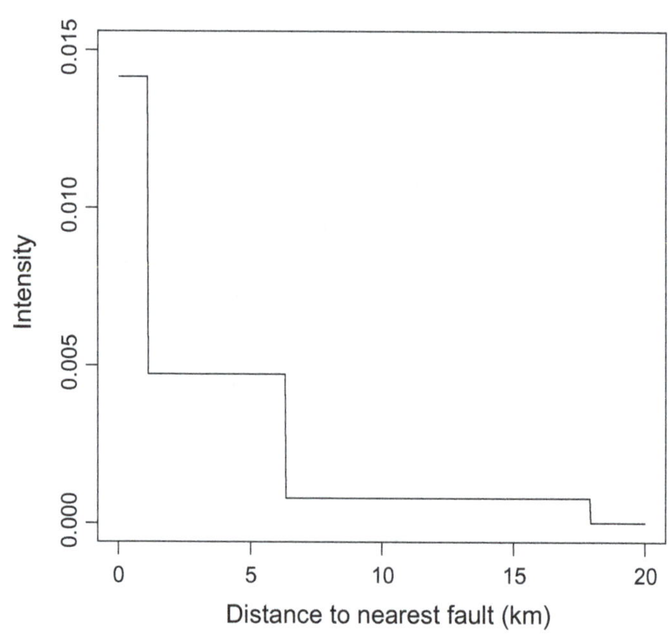


**Fig. 2.17** Intensity of Murchison gold deposits as a function of distance from the nearest fault, estimated by recursive partitioning

`spatstat`. Software scripts for the analyses in this chapter are available at www.spatstat.org.

**Acknowledgements** This research would not have been possible without enthusiastic support and collaboration of the Centre for Exploration Targeting (CET) at the University of Western Australia, in particular Professor Eun-Jung Holden. The research was funded by the Australian Research Council under a Discovery Outstanding Researcher Award.
This article includes summaries of the results of joint research with the Perth Spatial Point Processes Group (Ya-Mei Chang, Andrew Hardegen, Thomas Lawrence, Gopalan Nair, Suman Rakshit and Yong Song) and with collaborators Ege Rubak, Rob Foxall and Rolf Turner. Valuable advice was given by Professors Allan Trench and Mike Dentith.

## References

Agterberg F (1974) Automatic contouring of geological maps to detect target areas for mineral exploration. J Int Assoc Math Geol 6:373–395

Agterberg F (2001) Appreciation of contributions by Georges Matheron and John Tukey to a mineral-resources research project. Nat Resource Res 10:287–295

Alt J, King G, Signorino C (2001) Aggregation among binary, count and duration models: estimating the same quantities from different levels of data. Polit Anal 9:21–44

Assunção R, Guttorp P (1999) Robustness for inhomogeneous Poisson point processes. Ann Inst Stat Math 51:657–678

Atkinson A (1985) Plots, transformations and regression. No. 1 in Oxford statistical science series. Oxford University Press, Clarendon

Austin M (2007) Species distribution models and ecological theory: a critical assessment and some possible new approaches. Ecol Modell 200:1–19

Baddeley A (2017) Local composite likelihood for spatial point processes. Spat Stat (in press). https://doi.org/10.1016/j.spasta.2017.03.001

Baddeley A, Turner R (2000) Practical maximum pseudolikelihood for spatial point patterns (with discussion). Austr New Zealand J Stat 42(3):283–322

Baddeley A, Turner R (2005) Spatstat: an R package for analyzing spatial point patterns. J Stat Softw 12(6):1–42. www.jstatsoft.org. ISSN 1548-7660

Baddeley A, Berman M, Fisher N, Hardegen A, Milne R, Schuhmacher D, Turner R (2010) Spatial logistic regression and change-of-support for Poisson point processes. Electron J Stat 4:1151–1201. https://doi.org/10.1214/10-EJS581

Baddeley A, Chang Y, Song Y, Turner R (2012) Nonparametric estimation of the dependence of a spatial point process on a spatial covariate. Stat Interface 5:221–236

Baddeley A, Chang Y, Song Y (2013a) Leverage and influence diagnostics for spatial point processes. Scand J Stat 40:86–104. https://doi.org/10.1111/j.1467-9469.2011.00786.x

Baddeley A, Chang YM, Song Y, Turner R (2013b) Residual diagnostics for covariate effects in spatial point process models. J Comput Graph Stat 22:886–905

Baddeley A, Coeurjolly JF, Rubak E, Waagepetersen R (2014) Logistic regression for spatial Gibbs point processes. Biometrika 101(2):377–392

Baddeley A, Rubak E, Turner R (2015) Spatial point patterns: methodology and applications with R. Chapman and Hall/CRC, London

Banerjee S, Gelfand A (2002) Prediction, interpolation and regression for spatially misaligned data. Sanhkya A 64:227–245

Barlow R, Bartholomew D, Bremner J, Brunk H (1972) Statistical inference under order restrictions. Wiley, New York

Barndorff-Nielsen O (1978) Information and exponential families in statistical theory. Wiley, Chichester, New York, Brisbane, Toronto

Berkson J (1955) Maximum likelihood and minimum $\chi^2$ estimates of the logistic function. J Am Stat Assoc 50:130–162

Berman M, Turner T (1992) Approximating point process likelihoods with GLIM. Appl Stat 41:31–38

Besag J, Milne R, Zachary S (1982) Point process limits of lattice processes. J Appl Probab 19:210–216

Bonham-Carter G (1995) Geographic information systems for geoscientists: modelling with GIS. No. 13 in computer methods in the geosciences. Pergamon Press, Elsevier, Kidlington, Oxford, UK

Breiman L (1968) Probability. Addison-Wesley, Reading

Breiman L, Friedman J, Stone C, Olshen R (1984) Classification and regression trees. Chapman and Hall/CRC

Brillinger D (1978) Comparative aspects of the study of ordinary time series and of point processes. In: Krishnaiah P (ed) Developments in statistics. Academic Press, New York, London, pp 33–133

Brillinger D, Preisler H (1986) Two examples of quantal data analysis: a) multivariate point process, b) pure death process in an experimental design. In: Proceedings, XIII international biometric conference. International Biometric Society, Seattle, pp 94–113

Brillinger D, Segundo J (1979) Empirical examination of the threshold model of neuron firing. Biol Cybern 35:213–220

Brown W, Gedeon T, Baddeley A, Groves D (2002) Bivariate J-function and other graphical statistical methods help select the best predictor variables as inputs for a neural network method of mineral prospectivity mapping. In: Bayer U, Burger H, Skala W (eds) IAMG 2002: 8th annual conference of the international association for mathematical geology, vol 1. International Association of Mathematical Geology, pp 257–268

Chung C, Agterberg F (1980) Regression models for estimating mineral resources from geological map data. Math Geol 12:473–488

Chung C, Fabbri A (1999) Probabilistic prediction models for landslide hazard mapping. Photogramm Eng Remote Sens 62(12):1389–1399

Clyde M, Strauss D (1991) Logistic regression for spatial pair-potential models. In: Possolo A (ed) Spatial statistics and imaging. Lecture notes—monograph series, vol 20, chap. II. Institute of Mathematical Statistics, Hayward, CA, pp 14–30. ISBN 0-940600-27-7

Cressie N (1996) Change of support and the modifiable area unit problem. Geogr Syst 3:159–180

Daley D, Vere-Jones D (2003) An introduction to the theory of point processes. Volume I: Elementary theory and methods, 2nd edn. Springer, New York

Daley D, Vere-Jones D (2008) An introduction to the theory of point processes. Volume II: General theory and structure, 2nd edn. Springer, New York

Dean C, Balshaw R (1997) Efficiency lost by analyzing counts rather than exact times in Poisson and overdispersed Poisson regression. J Am Stat Assoc 92:1387–1398

Diggle P, Rowlingson B (1994) A conditional approach to point process modelling of elevated risk. J R Stat Soc Ser A 157(3):433–440

Dobson A, Barnett A (2008) An introduction to generalized linear models, 3rd edn. CRC Press

Dudík M, Phillips SJ, Schapire R (2007) Maximum entropy density estimation with generalized regularization and an application to species distribution modeling. J Mach Learn Res 8:1217–1260

Dutta M (1966) On maximum (information-theoretic) entropy estimation. Sankhya: Indian J Stat Series A 28:319–328

El Barmi H, Simonoff J (2000) Transformation based density estimation for weighted distributions. J Nonparametr Stat 12:861–878

Elith J, Phillips SJ, Hastie T, Dudík M, Chee YE, Yates CJ (2011) A statistical explanation of MaxEnt for ecologists. Divers Distrib 17:43–57

Elliott P, Wakefield J, Best N, Briggs D (eds) (2000) Spatial epidemiology: methods and applications. Oxford University Press, Oxford

Fielding A, Bell J (1997) A review of methods for the assessment of prediction errors in conservation presence/absence models. Environ Conserv 24:38–49

Foxall R, Baddeley A (2002) Nonparametric measures of association between a spatial point process and a random set, with geological applications. Appl Stat 51(2):165–182

Franklin J (2009) Mapping species distributions: spatial inference and prediction. Cambridge University Press, Cambridge, UK

Freedman D, Pisani R, Purves R (2007) Statistics, 4th edn. Norton, New York

Gorsevski P, Gessler P, Folz R, Elliott W (2006) Spatial prediction of landslide hazard using logistic regression and ROC analysis. Trans GIS 10:395–415

Gotway C, Young L (2002) Combining incompatible spatial data. J Am Stat Assoc 97:632–648

Groves D, Goldfarb R, Knox-Robinson C, Ojala J, Gardoll S, Yun G, Holyland P (2000) Late-kinematic timing of orogenic gold deposits and significance for computer-based exploration techniques with emphasis on the Yilgarn Block, Western Australia. Ore Geol Rev 17:1–38

Guan Y (2008) On consistent nonparametric intensity estimation for inhomogeneous spatial point processes. J Am Stat Assoc 103:1238–1247

Guan Y, Wang H (2010) Sufficient dimension reduction for spatial point processes directed by Gaussian random fields. J R Stat Soc Ser B 72:367–387

Handcock M, Morris M (1999) Relative distribution methods in the social sciences. Springer, New York

Harrell F (2001) Regression modeling strategies. Springer, New York

Hasenstab R (1983) A preliminary cultural resource sensitivity analysis for the proposed flood control facilities construction in the Passaic River basin of New Jersey. Technical report, Soil Systems, Inc, New York, submitted to the Passaic River Basin Special Studies Branch, Department of the Army, USA. New York District Army Corps of Engineers

Hastie T, Tibshirani R (1990) Generalized additive models. Chapman and Hall/CRC

Hauck W Jr, Donner A (1977) Wald's test as applied to hypotheses in logit analysis. J Am Stat Assoc 72(360, Part 1):851–853

Hogg R, Craig A (1970) Introduction to mathematical statistics, 3rd edn. Macmillan

Hosmer D, Lemeshow S (2000) Applied logistic regression, 2nd edn. Wiley

Jiménez-Valverde A (2012) Insights into the area under the receiver operating characteristic curve (AUC) as a discrimination measure in species distribution modelling. Glob Ecol Biogeogr 21:498–507

Jones M (1991) Kernel density estimation for length-biased data. Biometrika 78:511–519

Knox-Robinson C, Groves D (1997) Gold prospectivity mapping using a geographic information system (GIS), with examples from the Yilgarn Block of Western Australia. Chronique de la Recherche Minière 529:127–138

Krzanowski W, Hand D (2009) ROC curves for continuous data. Chapman and Hall/CRC Press, London/Boca Raton

Kutoyants Y (1998) Statistical inference for spatial Poisson processes. No. 134 in lecture notes in statistics. Springer, New York

Kvamme K (1983) Computer processing techniques for regional modeling of archaeological site locations. Adv Comput Archaeol 1:26–52

Kvamme K (1995) A view from across the water: the North American experience in archeological GIS. In: Lock G, Stančič Z (eds) Archaeology and geographical information systems: a European perspective, pp 1–14. CRC Press

Kvamme K (2006) There and back again: revisiting archeological locational modeling. In: Mehrer M, KLWescott (eds) GIS and archaeological site modelling, pp 3–40. CRC Press

Lahiri S (1999) Asymptotic distribution of the empirical spatial cumulative distribution function predictor and prediction bands based on a subsampling method. Probab Theory Rel Fields 114:55–84

Lahiri S, Kaiser M, Cressie N, Hsu N (1999) Prediction of spatial cumulative distribution functions using subsampling. J Am Stat Assoc 94:86–97

Landwehr J, Pregibon D, Shoemaker A (1984) Graphical methods for assessing logistic regression models. J Am Stat Assoc 79:61–83

Lewis P (1972) Recent results in the statistical analysis of univariate point processes. In: Lewis P (ed) Stochastic point processes. Wiley, New York, pp 1–54

Lindsey J (1992) The analysis of stochastic processes using GLIM. Springer, Berlin

Lindsey J (1995) Modelling frequency and count data. Oxford University Press

Lindsey J (1997) Applying generalized linear models. Springer

Lindsey J, Mersch G (1992) Fitting and comparing probability distributions with log linear models. Comput Stat Data Anal 13:373–384

Little R, Rubin D (2002) Statistical analysis with missing data, 2nd edn. Wiley

Lloyd C (2011) Local models for spatial analysis. CRC Press, Boca Raton, FL

Loader C (1999) Local regression and likelihood. Springer, New York

Lobo J, Jiménez-Valverde A, Real R (2007) AUC: a misleading measure of the performance of predictive distribution models. Glob Ecol Biogeogr 17(2):145–151

Manly B, McDonald L, Thomas D (1993) Resource selection by animals: statistical design and analysis for field studies. Chapman and Hall, London

McCullagh P, Nelder J (1989) Generalized linear models, 2nd edn. Chapman and Hall

Møller J, Waagepetersen R (2004) Statistical inference and simulation for spatial point processes. Chapman and Hall/CRC, Boca Raton, FL

Nam BH, D'Agostino R (2002) Discrimination index, the area under the ROC curve. In: Huber-Carol C, Balakrishnan N, Nikulin M, Mesbah M (eds) Goodness-of-fit tests and model validity. Birkhäuser, Basel, pp 267–279

Okabe A, Boots B, Sugihara K (1992) Spatial tessellations: concepts and applications of Voronoi diagrams. Wiley series in probability and mathematical statistics. Wiley, Chichester, England, New York

Openshaw S (1984) The modifiable area unit problem. Geo Books, Norwich

Phillips SJ, Anderson RP, Schapire RE (2006) Maximum entropy modeling of species geographic distributions. Ecol Modell 190:231–259

R Development Core Team (2011) R: a language and environment for statistical computing. R Foundation for Statistical Computing, Vienna, Austria. http://www.R-project.org/. ISBN 3-900051-07-0

Rakshit S, Rubak E, Nair G, Baddeley A (2017) Meaning of the ROC curve for a spatial point pattern (manuscript in preparation)

Rathbun S, Cressie N (1994) Asymptotic properties of estimators of the parameters of spatial inhomogeneous Poisson point processes. Adv Appl Probab 26:122–154

Renner I, Warton D (2013) Equivalence of MAXENT and Poisson point process models for species distribution modeling in ecology. Biometrics 69:274–281

Renner I, Elith J, Baddeley A, Fithian W, Hastie T, Phillips S, Popovic G, Warton D (2015) Point process models for presence-only analysis. Methods Ecol Evol 6(4):366–379. https://doi.org/10.1111/2041-210X.12352

Robinson W (1950) Ecological correlations and the behavior of individuals. Am Sociol Rev 15:351–357

Royall R (1988) The prediction approach to sampling theory. In: Krishnaiah P, Rao C (eds) Handbook of statistics, vol 6 (Sampling). Elsevier, Amsterdam, pp 399–413

Sager T (1982) Nonparametric maximum likelihood estimation of spatial patterns. Ann Stat 10:1125–1136

Scholtz S (1981) Location choice models in Sparta. In: III RL, Ottinger J, Scholtz S, Limp W, Watkins B, Jones RD (eds) Settlement predictions in sparta: a locational analysis and cultural resource assessment on the uplands of Calhoun County, Arkansas, no. 14 in Arkansas archaeological survey research series. Arkansas Archaeological Survey, Fayetteville, Arkansas, USA, pp 207–222

Thuiller W, Vayreda J, Pino J, Sabate S, Lavorel S, Gracia C (2003) Large-scale environmental correlates of forest tree distributions in Catalonia (NE Spain). Glob Ecol Biogeogr 12:313–325
Tukey J (1972) Discussion of paper by F.P. Agterberg and S.C. Robinson. Bull Int Stat Inst 44(1):596; Proceedings, 38th Congress. International Statistical Institute
Wakefield J (2004) A critique of statistical aspects of ecological studies in spatial epidemiology. Environ Ecol Stat 11:31–54
Wakefield J (2007) Disease mapping and spatial regression with count data. Biostatistics 8:158–183
Waller L, Gotway C (2004) Applied spatial statistics for public health data. Wiley
Warton D, Shepherd L (2010a) Correction note—Poisson point process models solve the "pseudo-absence problem" for presence-only data in ecology. Ann Appl Stat 4:2203–2204
Warton D, Shepherd L (2010b) Poisson point process models solve the "pseudo-absence problem" for presence-only data in ecology. Ann Appl Stat 4:1383–1402
Watkins K, Hickman A (1990) Geological evolution and mineralization of the Murchison Province, Western Australia. Bulletin 137, Geological Survey of Western Australia, published by Department of Mines, Western Australia, 1990. Available online from Department of Industry and Resources, State Government of Western Australia. www.doir.wa.gov.au
Wheatley D, Gillings M (2002) Spatial technology and archaeology: the archaeological applications of GIS. Taylor and Francis

# Chapter 3
# Testing Joint Conditional Independence of Categorical Random Variables with a Standard Log-Likelihood Ratio Test

**Helmut Schaeben**

**Abstract** While tests for pairwise conditional independence of random variables have been devised, testing joint conditional independence of several random variables seems to be a challenge in general. Restriction to categorical random variables implies in particular that their common distribution may initially be thought of as contingency table, and then in terms of a log-linear model. Thus, Hammersley–Clifford theorem applies, and provides insight in the factorization of the log-linear model corresponding to assumptions of independence or conditional independence. Such assumptions simplify the full joint log-linear model, and in turn any conditional distribution. If the joint log-linear model corresponding to the assumption of joint conditional independence given the conditioning variable is not sufficiently large to explain some data according to a standard log-likelihood test, its null–hypothesis of joint conditional independence may be rejected with respect to some significance level. Enlarging the log-linear model by some product terms of variables and running the log-likelihood test on different models may provide insight which variables are lacking conditional independence. Since the joint distribution determines any conditional distribution, the series of tests eventually provides insight which variables and product terms a proper logistic regression model should comprise.

## 3.1 Introduction

Conditional independence is a probabilistic approach to causality (Suppes 1970; Dawid 1979, 2004, 2007; Spohn 1980, 1994; Pearl 2009; Chalak and White 2012) while for instance correlation is obviously not as it is a symmetric relationship. Features of conditional independence are

H. Schaeben (✉)
Geophysics and Geoinformatics, TU Bergakademie Freiberg, Freiberg, Germany
e-mail: schaeben@tu-freiberg.de





B. S. Daya Sagar et al. (eds.), *Handbook of Mathematical Geosciences*,
https://doi.org/10.1007/978-3-319-78999-6_3

- Conditionally independent random variables are conditionally uncorrelated.
- Conditionally independent random variables may be significantly correlated or not.
- Independence does not imply conditional independence and vice versa.
- Pairwise conditional independence does not imply joint conditional independence.

Statistical tests for pairwise conditional independence of random variables have been devised, e.g., Bergsma (2004), Su and White (2007), Su and White (2008), Song (2009), Bergsma (2010), Huang (2010), Zhang et al. (2011), Bouezmarni et al. (2012), Györfi and Walk (2012), Doran et al. (2014), Ramsey (2014), Huang et al. (2016), testing joint conditional independence of several random variables seems to be a challenge in general. For the special case of dichotomous variables, the "omnibus test" (Bonham-Carter 1994) and the "new omnibus test" (Agterberg and Cheng 2002) have been suggested.

Weak conditional independence of random variables was introduced in Wong and Butz (1999), and elaborated on in Butz and Sanscartier (2002). Extended conditional independence has recently been introduced in Constantinou and Dawid (2015). The definition of weak conditional independence given in Cheng (2015) refers to conditional independent random events, and rephrases conditional independence in terms of ratios of conditional probabilities rather than conditional probabilities to avoid the distinction of conditional independence given a conditioning event or its complement. This definition becomes irrelevant when proceeding from elementary probability of events to probability of random variables, and to the general definition of conditionally independent random variables.

Conditional independence is an issue in a Bayesian approach to estimate posterior (conditional) probabilities of a dichotomous random target variable in terms of weights-of-evidence (Good 1950, 1960, 1985). In turn, conditional independence is the major mathematical assumption of potential modeling with weights of evidence, cf. (Bonham-Carter et al. 1989; Agterberg and Cheng 2002; Schaeben 2014b), e.g., applied to prospectivity modeling of mineral deposits. The method requires a training dataset laid out in regular cells (pixels, voxels) of equal physical size representing the support of probabilities. The sum of posterior probabilities over all cells equals the sum of the target variable over all cells. Deviations indicate a violation of the assumption of conditional independence, and are used as statistic of a test (Agterberg and Cheng 2002) which involves a normality assumption. Funny enough, ArcSDM calculates so-called normalized probabilities, i.e., posterior probabilities rescaled so that the overall measure of conditional independence is satisfied (ESRI 2018); of course, the trick does not fix any problem. Violation of the assumption of conditional independence does not only corrupt the posterior (conditional) probabilities estimated with weights of evidence, but also their ranks, cf. (Schaeben 2014b), which is worse. Thus, the method of weights-of-evidence requires the mathematical modeling assumption of conditional independence to yield reasonable predictions. However, conditional independence is an issue with respect to logistic regression, too.

## 3.2 From Contingency Tables to Log-Linear Models

A comprehensive exposure of log-linear models is Christensen (1997). Let $\mathbf{Z}$ be a random vector of categorical random variables $\mathsf{Z}_\ell, \ell = 0, \dots, m$, i.e., $\mathbf{Z} = (\mathsf{Z}_0, \mathsf{Z}_1, \dots, \mathsf{Z}_m)^\mathsf{T}$. It is completely characterized by its distribution

$$p_\kappa = P_\mathbf{Z}(s_\kappa) = P(\mathbf{Z} = s_\kappa) = P\left((\mathsf{Z}_0, \dots, \mathsf{Z}_m) = (s_{k_0}, \dots, s_{k_m})\right)$$

with the multi-index $\kappa = (k_0, \dots, k_m)$, where $s_{k_\ell}$ with $k_\ell = 1, \dots, K_\ell$ denotes all possible categories of the categorical random variable $\mathsf{Z}_\ell, \ell = 0, \dots, m$. Since it is assumed that there is a total of $K_\ell$ different categories with $P_{\mathsf{Z}_\ell}(s_{k_\ell}) > 0$, there is a total of $\prod_{\ell=0}^m K_\ell$ different categorical states for $\mathbf{Z} = \bigotimes_{\ell=0}^m \mathsf{Z}_\ell$.

The distribution of a categorical random vector may initially be thought of as being provided by contingency tables. More conveniently, the distribution of a categorical random vector $\mathbf{Z}$ can generally be written in terms of a log-linear model as

$$\log p_\kappa = \sum_\kappa w_\kappa f_\mathbf{Z}^\kappa(z)$$

with

$$\begin{aligned} w_\kappa &= \log p_\kappa, \\ f_\mathbf{Z}^\kappa(z) &= \mathbf{I}_{\{s_\kappa\}}(z) = \mathbf{I}_{\{s_{k_0}, \dots, s_{k_m}\}}(z_0, \dots, z_m). \end{aligned}$$

## 3.3 Independence, Conditional Independence of Random Variables

If the random variables $\mathsf{Z}_\ell, \ell = 1, \dots, m$, are independent, then the joint probability of any subset of random variables $\mathsf{Z}_\ell$ can be factorized into the product of the individual probabilities, i.e.,

$$P_{\bigotimes_{\ell \in M} \mathsf{Z}_\ell} = \bigotimes_{\ell \in M} P_{\mathsf{Z}_\ell}.$$

where $M$ denotes any non-empty subset of the set $\{1, \dots, m\}$. In particular

$$P_\mathbf{Z} = P_{\bigotimes_{\ell=1}^m \mathsf{Z}_\ell} = \bigotimes_{\ell=1}^m P_{\mathsf{Z}_\ell}.$$

If the random variables $\mathsf{Z}_\ell, \ell = 1, \dots, m$, are conditionally independent given $\mathsf{Z}_0$, then the joint conditional probability of any subset of random variables $\mathsf{Z}_\ell$ given $\mathsf{Z}_0$ can be factorized into the product of the individual conditional probabilities, i.e.,

$$P_{\bigotimes_{\ell \in M} Z_\ell | Z_0} = \bigotimes_{\ell \in M} P_{Z_\ell | Z_0}, \tag{3.1}$$

and in particular

$$P_{\bigotimes_{\ell=1}^{m} Z_\ell | Z_0} = \bigotimes_{\ell=1}^{m} P_{Z_\ell | Z_0}.$$

## 3.4 Logistic Regression, and Its Special Case of Weights-of-Evidence

Conditional expectation of a dichotomous random target variable $Z_0$ given a $m$–variate random predictor vector $\mathbf{Z} = (Z_1, \ldots, Z_m)^\mathsf{T}$ is equal to a conditional probability, i.e.,

$$\mathrm{E}(Z_0 \mid \mathbf{Z}) = P(Z_0 = 1 \mid \mathbf{Z}).$$

Then the ordinary logistic regression model (without interaction terms) neglecting the error term yields

$$\operatorname{logit} P(Z_0 = 1 \mid \mathbf{Z}) = \beta_0 + \boldsymbol{\beta}^\mathsf{T} \mathbf{Z}, \quad \beta_0 \in \mathbb{R}, \boldsymbol{\beta} \in \mathbb{R}^m.$$

Omitting the error term it can be rewritten in terms of a probability as

$$P\left(Z_0 = 1 \mid \mathbf{Z}\right) = \Lambda\left(\beta_0 + \boldsymbol{\beta}^\mathsf{T} \mathbf{Z}\right),$$

where $\Lambda$ denotes the logistic function. The logistic regression model with interaction terms reads in terms of a logit transformed probability

$$\operatorname{logit} P(Z_0 = 1 \mid \mathbf{Z}) = \beta_0 + \sum_{\ell} \beta_\ell Z_\ell + \sum_{\ell_i, \ldots, \ell_j} \beta_{\ell_i, \ldots, \ell_j} Z_{\ell_i} \ldots Z_{\ell_j}), \tag{3.2}$$

and in terms of a probability

$$P\left(Z_0 = 1 \mid \mathbf{Z}\right) = \Lambda\left(\beta_0 + \sum_{\ell} \beta_\ell Z_\ell + \sum_{\ell_i, \ldots, \ell_j} \beta_{\ell_i, \ldots, \ell_j} Z_{\ell_i} \ldots Z_{\ell_j})\right).$$

If all predictor variables are dichotomous variables and conditionally independent given the target variable then the parameters of the ordinary logistic regression model simplify to

$$\beta_0 = \operatorname{logit} P(Z_0 = 1) + W^{(0)}, \quad \beta_\ell = C_\ell, \ \ell = 1, \ldots, m,$$

with contrasts

$$C_\ell = W_\ell^{(1)} - W_\ell^{(0)}, \;\; \ell = 1, \ldots, m,$$

defined as differences of weights of evidence

$$W_\ell^{(1)} = \ln \frac{P(Z_\ell = 1 \mid Z_0 = 1)}{P(Z_\ell = 1 \mid Z_0 = 0)}, \quad W_\ell^{(0)} = \ln \frac{P(Z_\ell = 0 \mid Z_0 = 1)}{P(Z_\ell = 0 \mid Z_0 = 0)},$$

and with $W^{(0)} = \sum_{\ell=1}^{m} W_\ell^{(0)}$ provided all conditional probabilities are different from 0 (Schaeben 2014b). Obviously the model parameters become independent of one another, and can be estimated by mere counting. This special case of a logistic regression model is usually referred to as the method of "weights-of-evidence". In turn, the canonical generalization of Bayesian weights-of-evidence is logistic regression.

That weights of evidence $W_\ell$ agree with the logistic regression parameters $\beta_\ell$ in case of joint conditional independence becomes obvious when recalling

$$\begin{aligned} C_\ell &= W_\ell^{(1)} - W_\ell^{(0)} \\ &= \ln \frac{P(Z_\ell = 1 \mid Z_0 = 1)}{P(Z_\ell = 1 \mid Z_0 = 0)} - \ln \frac{P(Z_\ell = 0 \mid Z_0 = 1)}{P(Z_\ell = 0 \mid Z_0 = 0)} \\ &= \ln \left( \frac{O(Z_0 = 1 \mid Z_\ell = 1)}{O(Z_0 = 1 \mid Z_\ell = 0)} \right) = \beta_\ell, \end{aligned}$$

which is the log odds ratio, the usual interpretation of $\beta_\ell$ (Hosmer and Lemeshow 2000).

If $\mathbf{Z}$ comprises $m$ dichotomous predictor variables $Z_\ell, \ell = 1, \ldots, m$, there are $2^m$ possible different realizations $z_k, k = 1, \ldots, 2^m$, of $\mathbf{Z}$. Then

$$\begin{aligned} \sum_{i=1}^{n} \widehat{P}\big(Z_0 = 1 \mid \mathbf{Z} = z(i)\big) &= \sum_{k=1}^{2^m} \widehat{P}(Z_0 = 1 \mid \mathbf{Z} = z_k)\, H(\mathbf{Z} = z_k) \\ &= \sum_{k=1}^{2^m} \widehat{P}(Z_0 = 1 \mid \mathbf{Z} = z_k)\, n\, \widehat{P}(\mathbf{Z} = z_k) \\ &= n\widehat{P}(Z_0 = 1) = \sum_{i=1}^{n} z_0(i), \end{aligned}$$

where the last equation is an application of the formula of total probability. It is a constitutive equation to estimate the parameters of a logistic regression model and holds always for fitted logistic regression models. With respect to weights-of-evidence, the test statistic of the so-called "new omnibus test" of conditional independence (Agterberg and Cheng 2002) is

$$t = \sum_{i=1}^{n} \left( \widehat{P}\left(Z_0 = 1 \mid \mathbf{Z} = z\,(i)\right) - z_0(i) \right)$$

and should not be too large for conditional independence to be reasonably assumed.

## 3.5 Hammersley–Clifford Theorem

Rephrasing the proper statement (Lauritzen 1996) casually, the Hammersley–Clifford Theorem states that a probability distribution with a positive density satisfies one of the Markov properties with respect to an undirected graph $G$ if and only if its density can be factorized over the cliques of the graph. Since the distribution of a categorical random vector can be represented in terms of a log-linear model, Hammersley–Clifford theorem applies. Given $(m+1)$ random variables $Z_0, \ldots, Z_m$, there is a total of $\binom{m+1}{\ell+1}$ different product terms each involving $(\ell+1)$ variables, $\ell = 0, \ldots, m$, summing to a total of $\sum_{\ell=0}^{m} \binom{m+1}{\ell+1} = 2^{m+1} - 1$ different terms. Thus there is a total of $(m+1)$ single variable terms, and a total of $2^{m+1} - (m+2)$ multi variable terms.

The full log-linear model encompasses all terms and reads

$$\log p_\kappa = \sum_{\ell=0}^{m} \sum_{\alpha \in C_{\ell+1}^{m+1}} \sum_{\kappa(\alpha)} \phi_{\kappa(\alpha)} \, \mathbf{I}_{s_{\kappa(\alpha)}}(z_{\kappa(\alpha)}) \tag{3.3}$$

where $\alpha \in C_{\ell+1}^{m+1}$ denotes an $(\ell+1)$-combination of the set $\{1, \ldots, m+1\} \subset \mathbb{N}$, and $\kappa(\alpha) = (k_{i_1}, \ldots, k_{i_{\ell+1}})$ denotes a multi-index with $(\ell+1)$ entries $k_{i_\ell} = 1, \ldots, K_{i_\ell}$, for $\ell = 0, \ldots, m$. The random vector $\mathbf{Z}_{\kappa(\alpha)}$ is the product of any tuple of $(\ell+1)$ components of $\mathbf{Z}$, the total number of which is $\binom{m+1}{\ell+1}$.

Assumptions of independence or conditional independence simplify the distribution of $\mathbf{Z}$, i.e., its full log-linear model, considerably. Assuming independence for all its components $Z_\ell, \ell = 0, \ldots, m$, the log-linear model simplifies according to Eq. (3.1) to

$$\log p_\kappa = \sum_{\ell=0}^{m} \log p_{k_\ell} = \sum_{\ell=0}^{m} \sum_{k_\ell=1}^{K_\ell} \phi_{k_\ell} \, \mathbf{I}_{\{s_{k_\ell}\}}(z_\ell), \tag{3.4}$$

where $\phi_{k_\ell} = \log p_{k_\ell}$.

Assuming joint conditional independence of all components $Z_\ell, \ell = 1, \ldots, m$, given $Z_0$, the log-linear model, Eq. (3.3), simplifies according to Eq. (3.1) to

$$\log p_\kappa = \sum_{\ell=0}^{m} \sum_{k_\ell=1}^{K_\ell} \phi_{k_\ell} \, \mathbf{I}_{\{s_{k_\ell}\}}(z_\ell) + \sum_{\ell=1}^{m} \sum_{\alpha \in \{0,\ell\}} \sum_{\kappa(\alpha)} \phi_{\kappa(\alpha)} \, \mathbf{I}_{\{s_{\kappa(\alpha)}\}}(z_{\kappa(\alpha)}). \tag{3.5}$$

Thus the latter model, Eq. (3.5), assuming conditional independence differs from the model for independence, Eq. (3.4), in the additional product terms $\mathsf{Z}_0 \otimes \mathsf{Z}_\ell, \ell = 1, \ldots, m$.

Any violation of joint conditional independence given $\mathsf{Z}_0$ results in additional cliques of the graph and in additional product terms. Assuming that conditional independence given $\mathsf{Z}_0$ does not hold for a particular subset $\mathsf{Z}_{\ell_1}, \ldots, Z_{\ell_k}$ of variables $\mathsf{Z}_\ell$ results in an enlarging of the log-linear model of Eq. (3.5) by additional terms referring to $\mathsf{Z}_0 \otimes \bigotimes_{\ell=1}^{k} \bigotimes_{\ell_i \in C_\ell^k} \mathsf{Z}_{\ell_i}$ and $\bigotimes_{\ell=1}^{k} \bigotimes_{\ell_i \in C_\ell^k} \mathsf{Z}_{\ell_i}$, respectively.

## 3.6 Testing Joint Conditional Independence of Categorical Random Variables

The statistic of the likelihood ratio test (Neyman and Pearson 1933; Casella and Berger 2001) is the ratio of the maximized likelihood of a restricted model and the maximized likelihood of the full model. The assumption of the likelihood ratio test concerns the choice of the model family of distributions.

The null-hypothesis is that a given log-linear model is sufficiently large to represent the joint distribution. If the random variables are categorical, the full log-linear model is always sufficiently large as was explicitly shown above. More interesting are tests whether a smaller log-linear model is sufficiently large. Testing the null-hypothesis whether a log-linear model encompassing one-variable and two-variable terms, all of which involve $\mathsf{Z}_0$, is sufficiently large provides a test of conditional independence of all $\mathsf{Z}_\ell, \ell = 1, \ldots, m$, given $\mathsf{Z}_0$ because this log-linear model is sufficiently large in case of conditional independence given $\mathsf{Z}_0$. Thus, a reasonable rejection of the initial null-hypothesis implies a reasonable rejection of the assumption of conditional independence given $\mathsf{Z}_0$.

## 3.7 Conditional Distribution, Logistic Regression

Since the joint distribution implies all marginal and conditional distribution, respectively, the conditional distribution

$$P_{\mathsf{Z}_0 | \bigotimes_{\ell=1}^{m} \mathsf{Z}_\ell} = \frac{P_{\bigotimes_{\ell=0}^{m} \mathsf{Z}_\ell}}{P_{\bigotimes_{\ell=1}^{m} \mathsf{Z}_\ell}} \tag{3.6}$$

is explicitly given here by

$$\frac{P_{\bigotimes_{\ell=0}^{m} \mathsf{Z}_\ell}(s_{k_0}, \ldots, s_{k_\ell})}{P_{\bigotimes_{\ell=1}^{m} \mathsf{Z}_\ell}(s_{k_1}, \ldots, s_{k_\ell})} = \frac{P_{\bigotimes_{\ell=0}^{m} \mathsf{Z}_\ell}(s_{k_0}, \ldots, s_{k_\ell})}{\sum_{k_0=1}^{K_0} P_{\bigotimes_{\ell=0}^{m} \mathsf{Z}_\ell}(s_{k_0}, s_{k_1}, \ldots, s_{k_\ell})}.$$

Assuming independence, Eq. (3.6) immediately reveals

$$P_{Z_0|\bigotimes_{\ell=1}^m Z_\ell} = P_{Z_0}.$$

Assuming conditional independence of all $Z_\ell, \ell = 1, \ldots, m$, given $Z_0$ and further that $Z_0$ is dichotomous, then

$$P_{Z_0|\bigotimes_{\ell=1}^m Z_\ell}(1 \mid s_{k_1}, \ldots, s_{k_\ell}) = \frac{P_{\bigotimes_{\ell=0}^m Z_\ell}(1, s_{k_1}, \ldots, s_{k_\ell})}{\sum_{i=0}^1 P_{\bigotimes_{\ell=0}^m Z_\ell}(i, s_{k_1}, \ldots, s_{k_\ell})} \tag{3.7}$$

with

$$P_{\bigotimes_{\ell=0}^m Z_\ell}(1, s_{k_1}, \ldots, s_{k_\ell}) = \exp\left(\phi_1 + \sum_{\ell=1}^m \phi_{k_\ell} + \sum_{\ell=1}^m \sum_{k_\ell=1}^{K_\ell} \phi_{1,k_\ell}\right)$$

and

$$\sum_{i=0}^1 P_{\bigotimes_{\ell=0}^m Z_\ell}(i, s_{k_1}, \ldots, s_{k_\ell}) = \sum_{i=0}^1 \exp\left(\phi_i + \sum_{\ell=1}^m \phi_{k_\ell} + \sum_{\ell=1}^m \sum_{k_\ell=1}^{K_\ell} \phi_{i,k_\ell}\right)$$

Thus,

$$\begin{aligned}
&\frac{P_{\bigotimes_{\ell=0}^m Z_\ell}(1, s_{k_1}, \ldots, s_{k_\ell})}{\sum_{s=0}^1 P_{\bigotimes_{\ell=0}^m Z_\ell}(s, s_{k_1}, \ldots, s_{k_\ell})} \\
&= \frac{\exp\left(\phi_1 \mathbf{I}_{\{1\}}(1) + \sum_{\ell=1}^m \phi_{1,k_\ell} \mathbf{I}_{\{1,s_{k_\ell}\}}(1, Z_\ell)\right)}{\sum_{s=0}^1 \exp\left(\phi_s \mathbf{I}_{\{s\}}(1) + \sum_{\ell=1}^m \phi_{k_\ell} \mathbf{I}_{\{s,s_{k_\ell}\}}(1, Z_\ell)\right)} \\
&= \frac{\exp\left(\phi_1 + \sum_{\ell=1}^m \phi_{1,s_{k_\ell}} \mathbf{I}_{\{s_{k_\ell}\}}(Z_\ell)\right)}{1 + \exp\left(\phi_1 + \sum_{\ell=1}^m \phi_{1,k_\ell} \mathbf{I}_{\{s_{k_\ell}\}}(Z_\ell)\right)} \\
&= \Lambda\left(\phi_1 + \sum_{\ell=1}^m \phi_{1,k_\ell} \mathbf{I}_{\{s_{k_\ell}\}}(Z_\ell)\right).
\end{aligned}$$

Finally,

$$P_{Z_0|\bigotimes_{\ell=1}^m Z_\ell} = \Lambda\left(\beta_0 + \sum_{\ell=1}^m \beta_\ell Z_\ell\right),$$

which is obviously logistic regression

$$\operatorname{logit} P_{Z_0|\bigotimes_{\ell=1}^m Z_\ell} = \beta_0 + \sum_{\ell=1}^m \beta_\ell Z_\ell. \tag{3.8}$$

It should be noted that additional product terms in the joint probability $P_{\bigotimes_{\ell=0}^{m} Z_\ell}$ on the right hand side of Eq. (3.7) of the form $\bigotimes_{\ell=1}^{k} \bigotimes_{\ell_i \in C_\ell^k} Z_{\ell_i}$ including $Z_\ell, \ell = 1, \ldots, m$, only, i.e., not including $Z_0$, would not effect the form of the conditional probability, Eq. (3.8). Additional product terms of the form $Z_0 \otimes \bigotimes_{\ell=1}^{k} \bigotimes_{\ell_i \in C_\ell^k} Z_{\ell_i}$, i.e., including $Z_0$, result in a logistic regression model with interaction terms, Eq. (3.2).

Ordinary logistic regression is optimum, if the joint probability of the (dichotomous) target variable and the predictor variables is of log-linear form and all predictor variables are jointly conditionally independent given the target variable; in particular, it is optimum if the predictor variables are categorical and jointly conditionally independent given the target variable (Schaeben 2014a). Logistic regression with interaction terms is optimum, if the joint probability of the (dichotomous) target variable and the predictor variables is of log-linear form and the interaction terms correspond to lacking conditionally independence given the target variable; for categorical predictor variables, interaction terms can compensate for any lack of conditional independence exactly. Logistic regression with interaction terms is optimum in case of lacking conditional independence (Schaeben 2014a).

## 3.8 Practical Applications

The practical application of the log-likelihood ratio test of joint conditional independence generally includes the following steps

- test the null-hypothesis that the full log-linear model is sufficiently large to represent the joint probability of all predictor variables and the target variables;
- if the first null-hypothesis is not reasonably rejected, test the null-hypotheses that smaller log-linear models are sufficiently large; in particular;
- test the null hypothesis that the log-linear model without any interaction term is sufficiently large;
- if the final null-hypothesis is rejected, then the predictor variables must not be assumed to be jointly conditionally independent given the target variable.

### *3.8.1 Practical Application with Fabricated Indicator Data*

#### 3.8.1.1 The Data Set BRY

The data set BRY is derived from the https://en.wikipedia.org/wiki/Conditional_independence. Initially it comprises three random events $B$, $R$, $Y$, denoting the subsets of the set of all 49 pixels which are blue, red or yellow with given probabilities $P(B) = \frac{18}{49} = 0.367, P(R) = \frac{16}{49} = 0.326, P(Y) = \frac{12}{49} = 0.244$. The random events

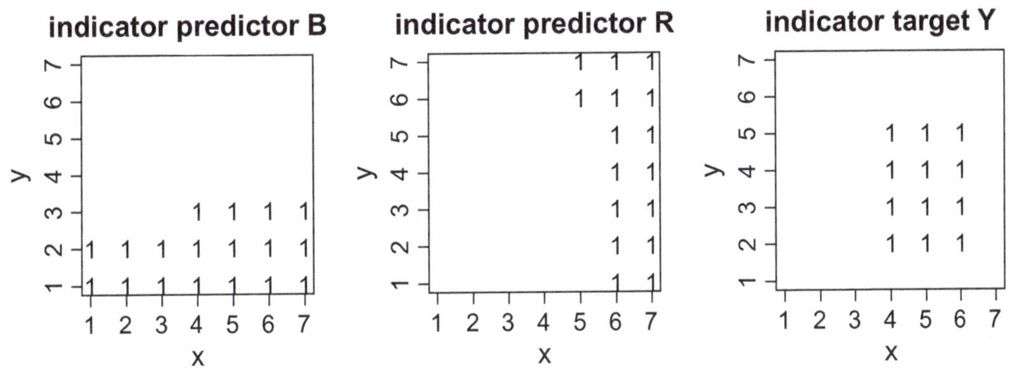


**Fig. 3.1** Map images of random events $B, R, Y$.

$B, R, Y$ are distinguished from their corresponding random indicator variables $\mathsf{B}, \mathsf{R}, \mathsf{Y}$ defined as usually, e.g.,

$$\mathsf{B}(\omega) = \mathbf{I}_B(\omega), \omega \in \Omega,$$

where $\mathbf{I}$ denotes the indicator variable. They are assigned to pixels of a $7 \times 7$ digital map image, Fig. 3.1.

It should be noted that in this example any spatial references are solely owed to the purpose of visualization as map images, and that the test itself does not take any spatial references or spatially induced dependences into account.

Checking independence according to its definition in reference to random events, the figures

$$P(B \cap R) = 0.122, \quad P(B)\, P(R) = 0.119$$

indicate that the random events $B$ and $R$ are not independent. However, the deviation is small.

Next, conditional independence is checked in terms of its definition referring to random events. Since conditional independence of the random events $B$ and $R$ given $Y$ does not imply conditional independence of the random events $B$ and $R$ given the complement $\complement Y$, two checks are required. The results are

$$P(B \cap R \mid Y) = \frac{1}{6} = P(B \mid Y)\, P(R \mid Y)$$

$$P(B \cap R \mid \complement Y) = \frac{4}{37} \neq \left(\frac{12}{37}\right)^2 = P(B \mid \complement Y)\, P(R \mid \complement Y),$$

and indicate that the random events $B$ and $R$ are conditionally independent given the random event $Y$, but that they are not conditionally independent given the complement $\complement Y$. It should be noted that the deviation of the joint conditional probability and the product of the two individual conditional probabilities in terms of their ratio is 1.027. In fact, the events $B$ and $R$ are conditionally independent given either $Y$ or $\complement Y$ if one white pixel, e.g. pixel (1,7) with $\mathsf{B} = \mathsf{R} = \mathsf{Y} = 0$, is omitted.

Generalizing the view to random variables B, R, Y and their unique joint realization as shown in Fig. 3.1, Pearson's $\chi^2$ test with Yates' continuity correction of the null-hypothesis of independence of the random variables B and R given the data returns a $p$-value of 1 indicating that the null-hypothesis cannot reasonably be rejected.

The likelihood ratio test is applied with respect to the log-linear distribution corresponding to the null-hypothesis of conditional independence and results in a $p$-value of 0.996 indicating that the null-hypothesis cannot reasonably be rejected.

Thus, given the data the tests suggest to infer that the random variables B and R are independent and conditionally independent given the random variable Y.

#### 3.8.1.2 The Data Set SCCI

The next data set SCCI comprises three random events $B_1, B_2, T$ with given probabilities $P(B_1) = P(B_2) = P(T) = \frac{7}{49} = \frac{7}{49} = 0.142$. They are assigned to pixels of a $7 \times 7$ digital map image, Fig. 3.2.

Checking independence according to its definition for random events, the figures

$$P(B_1 \cap B_2) = 0.102, \quad P(B_1)\, P(B_2) = 0.020$$

indicate that the random events $B_1$ and $B_2$ are not independent.

Next, conditional independence is checked in terms of its definition referring to random events. Since conditional independence of the random events $B_1$ and $B_2$ given $T$ does not imply conditional independence of the random events $B_1$ and $B_2$ given $\complement T$, two checks are required. The results are

$$\begin{aligned} P(B_1 \cap B_2 \mid T) = 0.714 \neq \quad & 0.734 = P(B_1 \mid T)\, P(B_2 \mid T) \\ P(B_1 \cap B_2 \mid \complement T) = 0 \neq & \; 0.0005 = P(B_1 \mid \complement T)\, P(B_2 \mid \complement T), \end{aligned}$$

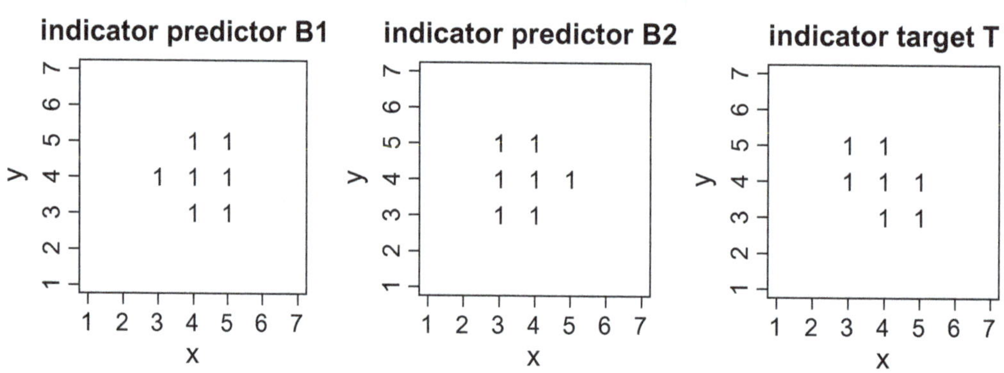


**Fig. 3.2** Map images of random events $B_1, B_2, T$ with $P(B_1) = P(B_2) = P(T) = \frac{7}{49} = \frac{7}{49} = 0.142$.

and indicate that the random events $B_1$ and $B_2$ are neither conditionally independent given the random event $T$ nor given the complement $\complement T$.

Testing the null-hypothesis of independence of the random variables $\mathsf{B}_1$ and $\mathsf{B}_2$ with Pearson's $\chi^2$ test with Yates' continuity correction given the data returns a $p$-value of practically equal to 0 indicating that the null-hypothesis should be rejected. The likelihood ratio test is applied with respect to the log-linear distribution corresponding to the null-hypothesis of conditional independence and results in a $p$-value of 0.825 indicating that the null-hypothesis cannot reasonably be rejected.

Thus, given the data the tests imply that the random variables $\mathsf{B}_1$ and $\mathsf{B}_2$ are not independent but conditionally independent given the random variable $\mathsf{T}$.

## 3.9 Discussion and Conclusions

Since pairwise conditional independence does not imply joint conditional independence, the $\chi^2$-test (Bonham-Carter 1994) of independence given $\mathsf{Z}_0 = 1$ does not apply to checking the modeling assumption of weights-of-evidence. The disadvantage of both the "omnibus" test (Bonham-Carter 1994) and the "new omnibus" test (Agterberg and Cheng 2002) is twofold. First, it involves an assumption of normal distribution which itself should be subject to a test. Second, weights-of-evidence has to be applied to calculate the test statistic which is the sum of all predicted conditional probabilities within the training data set. If the test actually suggests rejection of the null-hypothesis of conditional independence, the user learns that the application of weights-of-evidence was not mathematically authorized to predict the conditional probabilities. The standard likelihood ratio test suggested here resolves both shortcomings.

**Acknowledgements** The author would like to thank Prof. Juanjo Egozcue, UPC Barcelona, Spain, and Prof. K. Gerald van den Boogaart, HIF, Germany, for emphatic and instructive discussions of conditional independence.

## References

Agterberg FP, Cheng Q (2002) Conditional independence test for weights-of-evidence modeling. Nat Resour Res 11:249–255

Bergsma, WP (2004) Testing conditional independence for continuous random variables. Eurandom Report Vol. 2004048, Eindhoven

Bergsma WP (2010) Nonparametric testing of conditional independence by means of the partial copula, SSRN: https://ssrn.com/abstract=1702981, https://doi.org/10.2139/ssrn.1702981

Bonham-Carter GF (1994) Geographic information systems for geoscientists: modeling with GIS. Elsevier, Pergamon

Bonham-Carter GF, Agterberg FP, Wright DF (1989) Weights of evidence modeling: a new approach to mapping mineral potential. In: Agterberg FP, Bonham-Carter GF (eds.) Statistical Applications in the Earth Sciences, Geological Survey of Canada, Paper 89-9, pp. 171–183

Bouezmarni T, Rombouts JVK, Taamouti A (2012) Nonparametric copula-based test for conditional independence with applications to granger causality. J Bus Econ Stat 30:275–287

Butz CJ, Sanscartier MJ (2002) Properties of weak conditional independence. In: Alpigini JJ, Peters JF, Skowron A, Zhong N (eds) Rough Sets and Current Trends in Computing: third international conference, RSCTC 2002 Malvern, PA, USA, 14–16 October, 2002, vol 2475. Proceedings Lecture Notes in Computer Science. Springer, Berlin, pp 349–356

Casella G, Berger RL (2001) Statistical inference, 2nd edn. Duxburry Thomson Learning

Chalak K, White H (2012) Causality, conditional independence, and graphical separation in settable systems. Neural Comput 24:1611–1668

Cheng Q (2015) BoostWofE: A new sequential weights of evidence model reducing the effect of conditional dependency. Math Geosci 47:591–621

Christensen R (1997) Log–Linear models and logistic regression, 2nd edn. Springer, Berlin

Constantinou P, Dawid AP (2015) Extended conditional independence and applications in causal inference, arXiv:1512.00245

Dawid AP (1979) Conditional independence in statistical theory. J R Stat Soc B 41:1–31

Dawid AP (2004) Probability, causality and the empirical world: A Bayes-de Finetti-Popper-Borel synthesis. Stat Sci 19:44–57

Dawid AP (2007) Fundamentals of statistical causality. Research Report 279, Department of Statistical Science, University College London

Doran G, Muandet K, Zhang K, Schölkopf B (2014) A permutation-based kernel conditional independence test. Proceedings of UAI

ESRI: ArcSDM3.1 User Guide Spatial Data Modeller 3 Extension for ArcMap 9.1 file: ///C|/arcgis/ArcSDM/Documentation/sdmrspns.htm(9of16)2/22/200612:05:17PM)

Good IJ (1950) Probability and the weighing of evidence. Griffin, London

Good IJ (1960) Weight of evidence, corroboration, explanatory power, information and the utility of experiments. J R Stat Soc B 22:319–331

Good IJ (1985) Weight of evidence: A brief survey. Bayesian Stat 2:249–270

Györfi L, Walk H (2012) Strongly consistent nonparametric tests of conditional independence. Stat Probab Lett 82:1145–1150

Hosmer DW, Lemeshow S (2000) Applied logistic regression, 2nd edn. Wiley, New Jersey

Huang T-M (2010) Testing conditional independence using maximal nonlinear conditional correlation. Ann Stat 38:2047–2091

Huang M, Sun Y, White H (2016) A flexible nonparametric test for conditional independence. Econom. Theory 32:1434–1482

Lauritzen SL (1996) Graphical models. Clarendon Press, Oxford

Neyman J, Pearson ES (1933) On the problem of the most efficient tests of statistical hypotheses. Philos Trans R Soc Lond A 231:289–337

Pearl J (2009) Causality: models, reasoning, and inference, 2nd edn. Cambridge University Press, New York

Ramsey JD (2014) A scalable conditional independence test for nonlinear, non Gaussian data, arXiv:1401.5031

Schaeben H (2014a) A mathematical view of weights-of-evidence, conditional independence, and logistic regression in terms of Markov random fields. Math Geosci 46:691–709

Schaeben H (2014b) Potential modeling: conditional independence matters. Int J Geomath 5:99–116

Song K (2009) Testing conditional independence via Rosenblatt transforms. Ann Statist 37:4011–4045

Spohn W (1980) Stochastic independence, causal independence, and shieldability. J Philos Log 9:73–99

Spohn W (1994) On the properties of conditional independence. In: Suppes P, Humphreys P (eds) Scientific Philosopher, vol 1. Probability and Probabilistic Causality. Kluwer, Dordrecht, pp 173–194

Su L, White H (2007) A consistent characteristic function-based test for conditional independence. J Econom 141:807–834

Su L, White H (2008) A nonparametric Hellinger metric test for conditional independence. Econo Theory 24:829–864

Suppes P (1970) A probabilistic theory of causality. North-Holland, Amsterdam

Wong SKM, Butz CJ (1999) Contextual weak independence in Bayesian networks. In: Fifteenth conference on uncertainty in artificial intelligence, pp. 670–679

Zhang K, Peters J, Janzing D, Schölkopf B (2011) Kernel-based conditional independence test and application in causal discovery. In: Cozman FG, Pfeffer A (eds) Proceedings of the 27th Conference on Uncertainty in Artificial Intelligence (UAI 2011), Barcelona, Spain, July 14–17, 2011. AUAI Press, Corvallis, OR, USA, pp 804–813

# Chapter 4
# Modelling Compositional Data. The Sample Space Approach

**Juan José Egozcue and Vera Pawlowsky-Glahn**

**Abstract** Compositions describe parts of a whole and carry relative information. Compositional data appear in all fields of science, and their analysis requires paying attention to the appropriate sample space. The log-ratio approach proposes the simplex, endowed with the Aitchison geometry, as an appropriate representation of the sample space. The main characteristics of the Aitchison geometry are presented, which open the door to statistical analysis addressed to extract the relative, not absolute, information. As a consequence, compositions can be represented in Cartesian coordinates by using an isometric log-ratio transformation. Standard statistical techniques can be used with these coordinates.

**Keywords** Compositional data analysis · Aitchison geometry
Simplex · Variation matrix · Biplot · Balance dendrogram · ilr · clr

**AMS Subjet classifications** 62-07 · 62-02

## 4.1 Introduction

The difficulties when dealing with compositional data have been known for more than a century. Indirectly, Pearson (1897) described some of these problems and coined the term spurious correlation. They are easily illustrated using the early characterizations of compositional data, which relay on the constant sum constraint (CSC). For instance, Chayes (1960, 1962) and Connor and Mosimann (1969) based

---

J. J. Egozcue (✉)
Department of Civil and Environmental Engineering, Universidad Politécnica de Cataluña, Barcelona, Spain
e-mail: juan.jose.egozcue@upc.edu

V. Pawlowsky-Glahn
Department of Computer Science, Applied Mathematics and Statistics, University of Girona, Girona, Spain
e-mail: vera.pawlowsky@udg.edu



B. S. Daya Sagar et al. (eds.), *Handbook of Mathematical Geosciences*,
https://doi.org/10.1007/978-3-319-78999-6_4

their analysis on the fact that a vector of proportions $\mathbf{x} = (x_1, x_2, \ldots, x_D)$ satisfies the CSC,

$$\sum_{i=1}^{D} x_i = \kappa > 0\,, \quad x_i > 0\,, \quad i = 1, 2, \ldots, D\,. \tag{4.1}$$

It defines the $\kappa$-simplex of $D$ components or parts. Here the simplex is denoted $\mathbb{S}^D$, with no reference to the positive constant $\kappa$. Data fulfilling the CSC were called *constrained or closed data*. In the eighties, promoted by J. Aitchison, this kind of data were recognized as *compositional data* (Aitchison and Shen 1980; Aitchison 1982, 1986). In the last reference, additional conditions were added to the original CSC characterisation, leading to the formulation of some principles for compositional data analysis. They were the starting point on which the log-ratio approach to compositional data is based. These principles have been reformulated several times in order to depurate and to clarify them for users (Aitchison and Egozcue 2005; Egozcue 2009; Egozcue and Pawlowsky-Glahn 2011a; Pawlowsky-Glahn et al. 2015). Nonetheless, they have been contested from different points of view (e.g. Scealy and Welsh 2014), arguing that they match the conditions for the application of log-ratio methods. But not all data satisfying the CSC (4.1), for instance admitting that some parts can be zero, are automatically adequate for a log-ratio analysis. In the last decade, in which the log-ratio approach has shown to be useful in a large number of applications, it also became clear that it can be rigorously applied to problems in which the CSC is not fulfilled, or where the components do not represent proportions. The key point for this change of the paradigm represented by the CSC, is the conception of compositions as equivalence classes of vectors which positive components are proportional (Barceló-Vidal et al. 2001; Martín-Fernández et al. 2003; Pawlowsky-Glahn et al. 2015; Barceló-Vidal and Martín-Fernández 2016), and the related idea that the simplex is just a representation of the sample space of compositions. This fact is a direct consequence of the scale invariance of compositions (Aitchison 1986) but, up to now, its implications have not been completely recognised.

This contribution aims at a reformulation of the principles of compositional data analysis in their log-ratio version, presenting them as a practical and natural need in many situations of data analysis. Section 4.2 discusses scale invariance and compositional equivalence and Sect. 4.3 presents the simplex as an appropriate sample space for compositional data. Perturbation, the group operation between compositions, is shown to be a natural operation in Sect. 4.4. The Aitchison distance and the requirements on it are discussed in Sect. 4.5. The consequence of the previous sections is the Euclidean space structure of the simplex, which has been termed Aitchison geometry (Pawlowsky-Glahn and Egozcue 2001). The Aichison geometry has been shown to be useful for the modelling and analysis of compositions, centring the interest in the relative information contained in the data. Some of these elements are commented in Sect. 4.6.

## 4.2 Scale Invariance, Key Principle of Compositions

When somebody records the composition of a product, material, shares of a market, species in an ecosystem or a kitchen recipe, he or she implicitly recognizes that the total amount is irrelevant for the description of the product, material, shares, species or recipe. This does not mean that the size or the amount is not informative, it only tells us that, whichever is the size, the elements of the total are distributed according to the specified composition. Essential are the ratios between the components of the described system. One can say that for any system that can be decomposed into parts its description has, at least, two types of information: one that is referred to as *size*, and another one that concerns the relations between the parts irrespective of the size. This latter one is called compositional information and, when the system is a geometric object, it is called *shape*. Beyond size (total amount) and composition (shape), there may be other properties of the system which can be quantified (color, sound, complexity, strength, ...) and again these additional properties may be decomposed into size and composition. Here, attention is paid to systems which are formed by parts, while their size or total amount is either analysed in another way or is irrelevant. For a discussion of a possible approach to a problem where interest lies in the relative information and in the total, see Pawlowsky-Glahn et al. (2015), Olea et al. (2016), Ferrer-Rossell et al. (2016).

Think about the map of a region; even changing the scale of the map, the same region is identified. If the distance between two mountain peaks was 12 cm, and a lake between the two was 4 cm broad, halving the scale new lengths of 6 and 2 cm will be obtained. The distance between the two peaks and the width of the lake can be identified as equal in the two maps, as the ratio is in both cases $12/4 = 6/2 = 3$. Only when the maps are to be transformed into an actual region, the size becomes relevant and it is revealed taking into account the scale of the maps. Note that in the case of the peaks and the lake, the considered parts, the distance between peaks and the width of the lake, are not disjoint, as the first includes the second. In fact, the previous comments did not imply that the parts of the system had to be non-overlapping or disjoint.

The irrelevance of the total led J. Aitchison (1986) to introduce the principle of scale invariance for compositions. A composition is assumed to be represented by an array of positive numbers which quantitatively represent the parts of the system. Let $\mathbf{x} = (x_1, x_2, \ldots, x_D)$, $x_i > 0$ for $i = 1, 2, \ldots, D$, be such a composition. Consider any positive constant $c > 0$. The scale invariance principle can be stated as: $\mathbf{x}$ and $c\mathbf{x}$ contain the same compositional information. From this point of view, *compositional equivalence* can be defined (Aitchison 1997; Barceló-Vidal et al. 2001; Barceló-Vidal and Martín-Fernández 2016; Pawlowsky-Glahn et al. 2015).

**Definition 4.2.1** (*Compositional equivalence*) Let $\mathbf{x} = (x_1, x_2, \ldots, x_D)$ and $\mathbf{y} = (y_1, y_2, \ldots, y_D)$ be two arrays of $D$ positive components. They are compositionally equivalent if there exists a positive constant $c$ such that, for $i = 1, 2, \ldots, D$, $y_i = cx_i$.

Two equivalent arrays **x**, **y** represent the same composition. Both the equivalence class generated and its representative are called compositions.

Figure 4.1 shows some artificial, arbitrary data of Ca and Mg in mg/l from a fictitious water analysis (circles). Each pair (Ca,Mg) can be considered as a two part composition. A line from the origin through each data point consists of compositionally equivalent points, thus visualising a composition, strictly speaking an equivalence class. Any point on these rays can be chosen as a representative of the composition. Particularly, they can be selected so that the sums of the two components add to 100, which correspond to the triangles on the 2-part simplex (full line). This means that compositions are equivalence classes of compositionally equivalent arrays. Equivalence classes are handled by selecting a representative of each class and operating with these representatives. The selection of representative of a class is arbitrary, but imposes a condition on any further analysis. This condition is the principle of scale invariance formulated in Aitchison (1986).

**Principle 4.2.1** (Scale invariant analysis) *Any analysis or operation with compositions must be expressed by scale invariant functions of the components. Scale invariant functions are identified with real, 0-degree homogeneous functions, that is, satisfying the condition* $f(\mathbf{x}) = f(c\mathbf{x})$ *for any positive constant c and for any composition* **x**.

Consequently, for any composition given by the array **x** it is possible to choose another compositionally equivalent array, denoted $C\mathbf{x}$, such that it is in the simplex, that is, it fulfills the CSC (4.1). To this end, the constant in CSC (4.1) $\kappa = 1$ is chosen, thus yielding

$$C\mathbf{x} = \left( \frac{x_1}{\sum_{i=1}^{D} x_i}, \frac{x_2}{\sum_{i=1}^{D} x_i}, \ldots, \frac{x_D}{\sum_{i=1}^{D} x_i} \right).$$

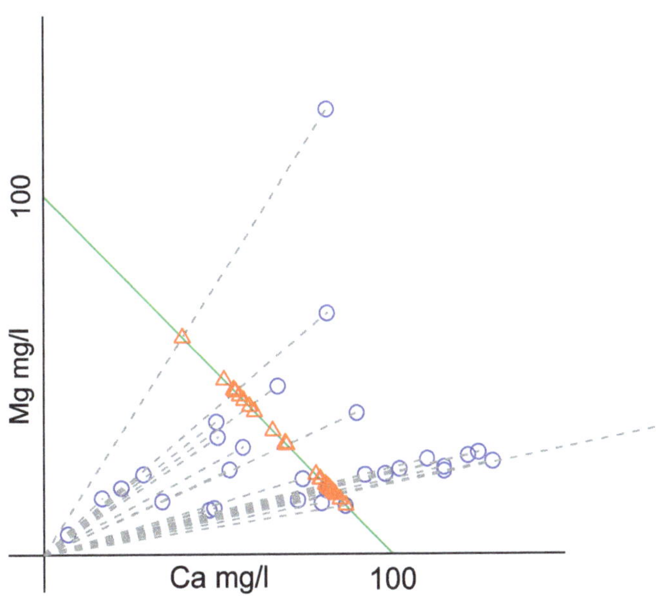


**Fig. 4.1** Some two-component data points with positive components (circles), are compositionally equivalent to all points on the dashed lines from the origin through the data points. Triangles are the representatives of each equivalence class on the 2-part simplex in which components add to 100

The symbol $C$ is called closure operator. It assigns a representative in the simplex (closed form of $\mathbf{x}$, satisfying the CSC) to the equivalence class where $\mathbf{x}$ is included. Due to the scale invariance analysis principle, any analysis on the elements in the simplex (closed) must lead to identical results as that performed using the non-closed representatives.

The scale invariance principle is familiar to any scientist. For instance, an array of probabilities as $(0.1, 0.3, 0.2)$, originally expressed as values between 0 and 1, can be expressed in percentages as $(10, 30, 20)$ without any confusion; a set of concentrations given in percentages of mass can be translated into ppm (parts per million of mass) just multiplying by $10,000$ and the geologist does not get confused provided that he/she is informed about which units are in use.

Despite the intuitive character of the scale invariance principle, in practice it is frequently violated. For instance, when performing a cluster analysis of geochemical samples given in ppm using the Euclidean distance between the samples. In fact, assume that we have two samples $\mathbf{x}$ and $\mathbf{y}$, and the square distance between them is taken as the square-Euclidean distance $\mathrm{d}^2(\mathbf{x}, \mathbf{y}) = \sum_{i=1}^{D}(x_i - y_i)^2$. Imagine that $\mathbf{y}$ is now expressed in ppb (parts per billion). This is a valid operation as $\mathbf{y}$ in ppm and in ppb are compositionally equivalent, but $\mathrm{d}^2(\mathbf{x}, \mathbf{y})$ changes dramatically as the square-differences $(x_i - y_i)^2$ become $(x_i - 1000 \cdot y_i)^2$ which constitutes a violation of the scale invariance principle.

Similarly, given a set of geochemical samples in ppm, $\mathbf{x}_1, \mathbf{x}_2, \ldots, \mathbf{x}_n$, the Pearson correlation coefficient between two components also violates the principle of scale invariance. This coefficient between $x_{\cdot 1}$ and $x_{\cdot 2}$ is

$$r_{12} = \frac{\sum_{j=1}^{n}(x_{j1} - \bar{x}_1)(x_{j2} - \bar{x}_2)}{\sqrt{\sum_{j=1}^{n}(x_{j1} - \bar{x}_1)^2 \sum_{j=1}^{n}(x_{j2} - \bar{x}_2)^2}}, \tag{4.2}$$

where $\bar{x}_k$ is the average of the $k$-th component along the sample. Now suppose that the first sample $\mathbf{x}_1$ is expressed in ppb. This should not change the analysis as preconized by the scale invariance principle. However, everything changes: the average values $\bar{x}_k = (1/n)\sum_{j=1}^{n} x_{jk}$ are now dominated by the first term $1000 \cdot x_{1k}$ which replaced the initial term $x_{1k}$. The global effect is evident after a simple inspection of Eq. (4.2). When the change of closure affects all the samples, the effect is the *spurious correlation* studied by Chayes (1960), although without any successful solution. Nowadays, after J. Aitchison's work, spurious correlation just corresponds to a violation of the scale invariance principle. Or, in other words, if a data set is assumed scale invariant, covariance or Pearson correlation are meaningless and spurious, and should not be used.

## 4.3 The Simplex as Sample Space of Compositions

In any data analysis, the first modeling step is to establish an appropriate sample space. In general, this step conditions all subsequent steps, and may affect dramatically the conclusions. Dealing with compositional data is not an exception. However, the choice and structure of the sample space is usually not explicit, and its consequences remain hidden in practice. Even the analyst is frequently not aware of the choice he or she has made when taking a decision on which methodology to apply.

The sample space of an observation (variable, vector, function or, in general, object) is a set where all the possible outcomes can be represented. However, the sample space may contain elements which do not correspond to any possible observation. When the considered object is a random one, the sample space must contain subsets, called events, which can be assigned a probability. Technically, if $\mathcal{S}$ is the sample space, a $\sigma$-field in $\mathcal{S}$ (e.g. Ash 1972; Feller 1968) needs to be defined. This is the minimum structure of a sample space for a random object. There are many qualitatively different random objects in practice. Multivariate real random $d$-vectors may be thought of as taking values in real space $\mathbb{R}^d$; a discrete time, real valued stochastic process, can be represented in the space $\ell^\infty$ of all real, bilaterally bounded sequences; if the observation is a random set on a plane, like paint stains on the floor, the sample space can be the set of compact sets in the plane; there are many more examples. It should be noted that the sample space is a choice of the analyst and it must be selected according to the stated questions from the beginning of the analysis. Commonly, beyond probability statements, the data analysis requires performing operations (sums, differences, averages, scaling), metric computations (distances or divergences, projections, approximations), or computing functionals (averages of components, extraction of extremes). All these procedures must be defined on the sample space. Consequently, the structure of the sample space is richer than that provided by the $\sigma$-field of events.

When dealing with $D$-part compositional data, the simplex $\mathbb{S}^D$ as the sample space is a valid choice, given that any composition can be assigned a representative in it. However, there are many alternatives. Figure 4.1 suggests that any curve intersecting once, and only once, all rays from the origin in the positive orthant might be taken as sample space. For instance, for two dimensional data points like those shown in Fig. 4.1, a possible choice is a quarter of a circumference, or two segments completing a square with the axes, as shown in Fig. 4.2. In the case of compositional data, the analyst is mainly interested in proportions and ratios, thus suggesting the choice of the simplex as an appropriate and intuitive representation. However, a key point for the choice of an adequate sample space is the decision on which is a translation or shift relevant for the analysis.

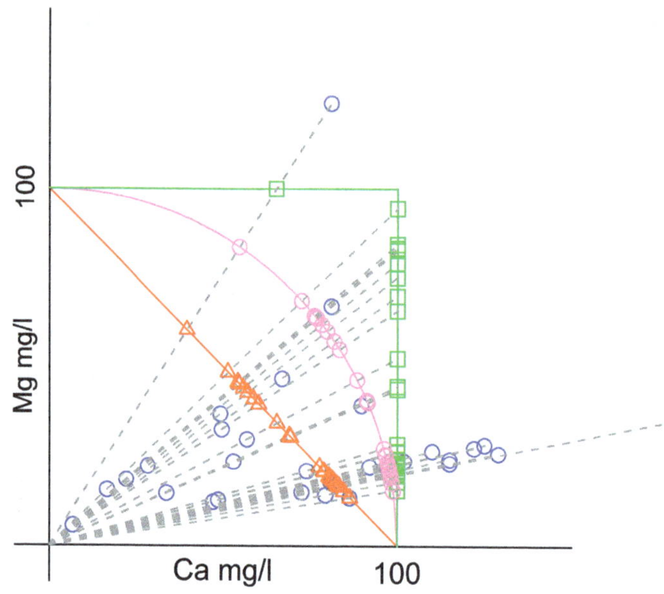


**Fig. 4.2** Some two-component data points with positive components (blue circles), are compositionally equivalent to all points on the dashed lines from the origin through the data points. Red triangles are the representatives of each equivalence class on the 2-part simplex in which components add to 100. Violet circles are representatives of these data points on a quarter of circumference. Green squares are representatives on the 100-square

## 4.4 Perturbation, a Natural Shift Operation on Compositions

Perturbation, as operation in the simplex, was introduced by Aitchison (1986) on an intuitive basis. It can be stated as follows.

**Definition 4.4.1** (*perturbation*) Let $\mathbf{x}$, $\mathbf{y}$ be two elements in the $D$-part simplex $\mathbb{S}^D$, $\mathbf{x} = (x_1, x_2, \ldots, x_D)$, $\mathbf{y} = (y_1, y_2, \ldots, y_D)$. The perturbation between them is

$$\mathbf{x} \oplus \mathbf{y} = \mathcal{C}(x_1 y_1, x_2 y_2, \ldots, x_D y_D)\,. \tag{4.3}$$

Some properties of perturbation are quite immediate. They can be summarized as that perturbation is a commutative group operation in $\mathbb{S}^D$ (Aitchison 1997). The neutral element is the composition with equal components $\mathbf{n} = \mathcal{C}(1, 1, \ldots, 1)$. The opposite to $\mathbf{x}$ is

$$\ominus\mathbf{x} = \mathcal{C}((1/x_1), (1/x_2), \ldots, (1/x_D)),$$

where each component is inverted.

Repeated perturbation, like $\mathbf{x} \oplus \mathbf{x} \oplus \mathbf{x}$, suggests the definition of a multiplication by a real scalar, so that $\mathbf{x} \oplus \mathbf{x} \oplus \mathbf{x} = 3 \odot \mathbf{x}$. Following this idea, multiplication by real scalars, called powering, is defined as follows.

**Definition 4.4.2** (*powering*) Let $\mathbf{x} = (x_1, x_2, \ldots, x_D)$ be an element in the $D$-part simplex $\mathbb{S}^D$ and let $\alpha$ be a real scalar. The powering of $\mathbf{x}$ by $\alpha$ is

$$\alpha \odot \mathbf{x} = \mathcal{C}(x_1^{\alpha}, x_2^{\alpha}, \ldots, x_D^{\alpha})\,. \tag{4.4}$$

These definitions present perturbation and powering as operations on elements of the simplex. However, as the simplex can be taken as the sample space of compositions and its elements are representatives of compositions, perturbation and powering are also operations on compositions. The simplex, endowed with perturbation and powering is a $(D-1)$-dimensional vector space. Perturbation plays the role of the sum in real space, and powering is multiplication by a real scalar. Perturbing a composition $\mathbf{x}$ by another composition $\mathbf{y}$ is thus a shift of $\mathbf{x}$ in the direction of $\mathbf{y}$.

Despite the mathematical aspect of Definition 4.4.1, perturbation is a common place in real life and scientific activity. To begin with, imagine a water filtering device which is fed with an inflow with disolved matter characterised by the concentrations (mg/l) of the major ions specified in Table 4.1, first row. Suppose that the filtering device has been designed to filter out sulphur, $SO_4$, iron, Fe, and phosphorus, P; $SO_4$ is ideally reduced by 75%, Fe by 10%, and P by 5%, meanwhile other ions remain unaltered. In order to compute the outflow concentrations, the filter factor or transfer function (4th row) is computed as $1-(10/100)=0.9$ in the case of Fe. Then, the filter factor multiplies the inflow concentrations to obtain the outflow concentrations in mg/l. Notably, when the inflow concentrations are represented in closed form, as percentages (second row), then, once multiplied by the filter factor, the same outflow concentrations in percent are obtained. In fact, the outflow concentrations in mg/l, when closed to 100, are those in the last row of the table. The closed form of the filter factor, labelled filter perturbation, can be used to obtain the same outflow concentrations. That is the filter factor is a composition. Although elementary, this example shows that inflow and outflow concentrations and the filter factor can be represented by different, but compositionally equivalent, arrays; and that the traditional form of expressing change of concentrations by percentages is nothing else than a way of expressing a perturbation. Also, one may be confronted with the estimation of the filter factor (perturbation) from the inflow and outflow concentrations. From the example, it is clear that a ratio of outflow over inflow concentrations gives a factor compositionally equivalent to the filter perturbation. This suggests the

**Table 4.1** Inflow concentrations of some ions disolved in water are filtered reducing Fe, $SO_4$ and P by a given percentage. Outflow concentrations are obtained by multiplication of inflow concentration by the filter factor (closed or not). Inflow, outflow concentrations and filter factor are presented also in closed form as they are treated as compositions

| | Ca | Fe | K | Mg | Na | P | $SO_4$ |
|---|---|---|---|---|---|---|---|
| Inflow (mg/l) | 0.760 | 0.225 | 5.30 | 1.54 | 2.00 | 0.079 | 2.40 |
| Inflow (closed to 100) | 6.177 | 1.829 | 43.08 | 12.52 | 16.25 | 0.642 | 19.51 |
| Filter effect (%) | 0 | −10 | 0 | 0 | 0 | −5 | −75 |
| Filter factor | 1 | 0.9 | 1 | 1 | 1 | 0.95 | 0.25 |
| Filter perturbation | 0.164 | 0.148 | 0.164 | 0.164 | 0.164 | 0.156 | 0.041 |
| Outflow (mg/l) | 0.760 | 0.203 | 5.30 | 1.54 | 2.00 | 0.075 | 0.60 |
| Outflow (closed to 100) | 7.254 | 1.933 | 50.58 | 14.70 | 19.09 | 0.716 | 5.73 |

definition of the difference-perturbation, the opposite operation to perturbation, as

$$\mathbf{y} \ominus \mathbf{x} = \mathcal{C}\left(\frac{y_1}{x_1}, \frac{y_2}{x_2}, \ldots, \frac{y_D}{x_D}\right),$$

which is the natural difference for perturbation as a group operation.

In the context of probability theory, arrays of probabilities can be considered as compositions. Consider a family of non overlapping events $A_i$, $i = 1, 2, \ldots, D$, which are assigned probabilities $p_i = \mathrm{P}[A_i]$. Observing the result $R$ of an experiment, the conditional probabilities $q_i = \mathrm{P}[R|A_i]$ allow to update the probabilities $p_i$ —according to the information obtained from the observation $R$— using Bayes' formula

$$\mathrm{P}[A_i|R] = \frac{\mathrm{P}[A_i] \cdot \mathrm{P}[R|A_i]}{\sum_{j=1}^{D} \mathrm{P}[A_j] \cdot \mathrm{P}[R|A_j]} = \mathcal{C}(\mathbf{p} \oplus \mathbf{q}),$$

where $\mathbf{p} = (p_1, p_2, \ldots, p_D)$ and $\mathbf{q} = (q_1, q_2, \ldots, q_D)$. Bayes' formula states that the final probabilities, conditioned to the result $R$, are the perturbation of the initial or prior probabilities $\mathbf{p}$ and the probabilities of the result given the events $A_i$, denoted $q_i$, also known as the likelihood of $R$. In this way perturbation becomes a very natural way of operating vectors of probabilities and likelihood, as it is the paradigm of incorporating information from observations. This interpretation of perturbation was proposed in Aitchison (1986, 1997) and developed in other contexts (Egozcue and Pawlowsky-Glahn 2011b; Egozcue et al. 2013).

Perturbation also appears as a natural operation on compositions when changing units. For instance, consider a grain size distribution for different sieve diameters. It may be expressed as proportions of volume corresponding to each sieve or as proportions of mass assigned to the same sieves. Both distributions can be considered as compositions. Transforming volume to mass consists of multiplication by the density of the material in each sieve, possibly different from one sieve to the other. This componentwise multiplication is a perturbation (Parent et al. 2012). Also, changing the concentrations of chemical elements from mg/kg to molar concentration consists of dividing each component by its molar mass, thus performing a perturbation. In all these examples, the secondary role of the closure and the CSC is remarkable: closure might only be necessary to facilitate interpretation.

Exponential decay of mass is frequent in nature. The typical example is the decay of mass of radioactive isotopes in time. These type of processes describe straight lines in the simplex (Egozcue et al. 2003; Pawlowsky-Glahn et al. 2015; Tolosana-Delgado 2012). This supports that perturbation is a natural operation in the simplex and between compositions. To sketch the argument, consider the masses of $D = 3$ fictitious radioactive isotopes $\mathbf{x}(t) = (x_1(t), x_2(t), x_3(t))$, which decay rates in time are $\lambda_1 = 3$, $\lambda_2 = 0.5$, $\lambda_3 = 0.1$, respectively. Initially, at $t = 0$, there are masses $\mathbf{x}(0) = (0.9, 0.04, 0.01)$ which disintegrate into other non considered isotopes. The total mass decreases in time, and the mass of each isotope changes as

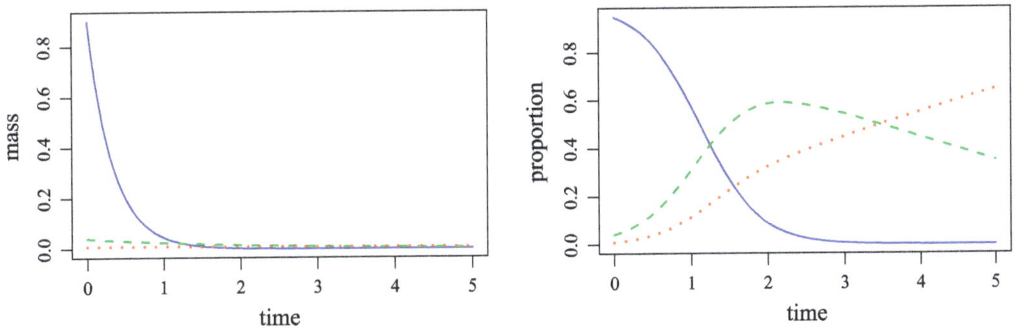


**Fig. 4.3** Evolution of masses (left panel) and proportions (right panel) of three isotopes which disintegrate at rates $3, 0.5, 0.1$ in time, respectively. Initial masses are $0.9, 0.04, 0.01$

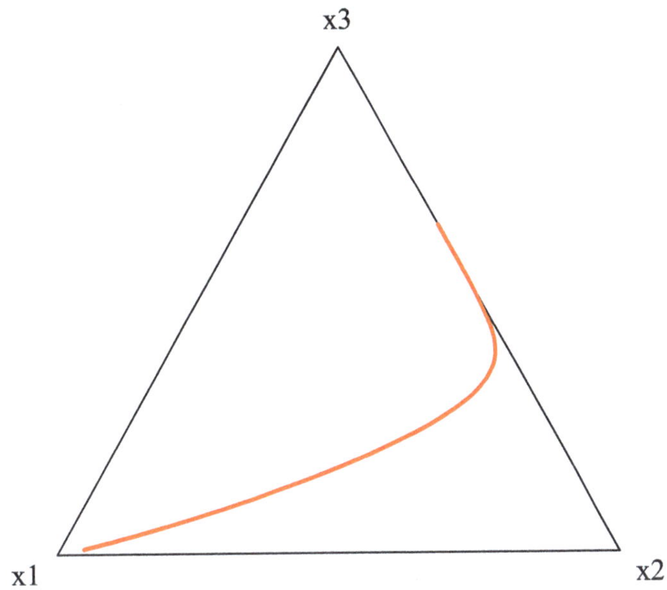


**Fig. 4.4** Evolution of proportions in time of three isotopes which disintegrate at rates $3, 0.5, 0.1$, respectively, represented in a ternary diagram. The initial masses are $0.9, 0.04, 0.01$, and they change as a function of time

$$x_i(t) = x_i(0) \cdot \exp[-\lambda_i\, t]\,, \quad i = 1, 2, 3\,. \tag{4.5}$$

This evolution of mass is shown in Fig. 4.3, left panel, where the decreasing mass is clearly observed. Figure 4.3, right panel, shows the evolution of proportions of the isotopes after the closure, which corresponds to

$$C\mathbf{x}(t) = C\left(\mathbf{x}(0) \oplus (-t \odot \exp[\lambda])\right)\,, \tag{4.6}$$

where $\exp[\lambda] = (\exp(\lambda_1), \exp(\lambda_2), \exp(\lambda_3))$. Figure 4.4 shows the evolution of the isotopes in a ternary diagram. The main fact on this exponential decay of isotopes is that it is naturally expressed using perturbation and powering, as in Eq. (4.6). In the simplex, this compositional evolution is a linear one. If proportions are thought as real variables, as they are shown in Fig. 4.3 (right panel), or in Fig. 4.4, then they are taken as non-linear thus ignoring their simplicity as compositional evolution.

The fact that perturbation is easily interpreted on vectors of proportions supports the idea that the simplex is a suitable sample space for compositions. Think, for instance, how perturbation could be interpreted when taking representatives of compositions as projections on the positive orthant of a hypersphere, or on the surface of a unit hypercube. It is not intuitive at all. Obviously, if the operation that is considered relevant for the stated problem is a rotation, the representation on the hypersphere may be a sensible choice of sample space.

## 4.5 Conditions on Metrics for Compositions

In many applications a distance between data points is a central issue. Cluster analysis is a typical example of this. Other metric concepts are crucial, like the size of a vector, the norm, or the possibility of performing orthogonal projections. Note that all these metric concepts are used in the omnipresent regression analysis. Compositional data analysis has the same need of introducing metrics, distances, norms and orthogonality. From the early developments by J. Aitchison (1983), a distance between compositions was introduced and developed (Aitchison 1992; Aitchison et al. 2000). Nowadays, that distance between compositions is called Aitchison distance, and the corresponding Euclidean geometry is named Aitchison geometry (Pawlowsky-Glahn and Egozcue 2001).

The need of a distance between compositions can be motivated from the most basic statistics. For instance, concepts as elementary as mean and variance are based on a choice of a distance in the sample space. Following Fréchet (1948) (see also Pawlowsky-Glahn et al. 2015, Chap. 6), mean and variance of a sample can be introduced in a metric space (sample space endowed with a distance). Consider a compositional sample $\mathbf{x}_i$, $i = 1, 2, \ldots, n$, represented in the $D$-part simplex $\mathbb{S}^D$. The data matrix $\mathbf{X}$ has the compositions $\mathbf{x}_i$ as rows. Suppose that a distance in $\mathbb{S}^D$ is $\mathrm{d}_a(\cdot,\cdot)$ (this notation corresponds to the Aitchison distance, although here it is used in a generic sense). A first step is to define variability of the sample with respect to a given composition $\mathbf{z}$ as

$$\mathrm{Var}[\mathbf{X}, \mathbf{z}] = \frac{1}{n} \sum_{i=1}^{n} \mathrm{d}_a^2(\mathbf{x}_i, \mathbf{z}), \quad \mathbf{z} \in \mathbb{S}^D. \tag{4.7}$$

The sample mean, called center for compositions, and the total variance are then defined as

$$\mathrm{Cen}[\mathbf{X}] = \operatorname*{argmin}_{\mathbf{z} \in \mathbb{S}^D} \{ \mathrm{Var}[\mathbf{X}, \mathbf{z}] \}, \tag{4.8}$$

$$\mathrm{totVar}[\mathbf{X}] = \min_{\mathbf{z} \in \mathbb{S}^D} \{ \mathrm{Var}[\mathbf{X}, \mathbf{z}] \} = \mathrm{Var}[\mathbf{X}, \mathrm{Cen}[\mathbf{X}]]. \tag{4.9}$$

Equations (4.7), (4.8) and (4.9) show that elementary statistics like mean and variance depend critically on the distance used in the sample space.

The Aitchison distance can be defined in different ways (see Pawlowsky-Glahn et al. 2015). One of them is

$$\mathrm{d}_a^2(\mathbf{x},\mathbf{y}) = \frac{1}{2D}\sum_{j=1}^{D}\sum_{k=1}^{D}\left(\ln\frac{x_j}{x_k} - \ln\frac{y_j}{y_k}\right)^2 , \tag{4.10}$$

where it is worth to realize that $\ln(x_k/x_k) = 0$. The distance has been subscripted as $\mathrm{d}_a$ to emphasize that it is the Aitchison distance. The first observation on the Aitchison distance is that it is scale invariant, as required by Principle 4.2.1. In fact, any multiplicative constant in $\mathbf{x}$ or $\mathbf{y}$ cancels out in the log-ratios in Eq. (4.10). After accepting the Aitchison distance as a proper one for compositions, a simple but tedious computation drives us to the expression of the sample center

$$\mathrm{Cen}[\mathbf{X}] = \frac{1}{n}\odot\bigoplus_{i=1}^{n}\mathbf{x}_i ,$$

where $\bigoplus$ stands for repeated perturbation, similar to a summation for real addition. At a first glance, just dropping the circles in the signs $\oplus$ and $\odot$, this expression is an average where the traditional sum has been changed to perturbation. Thus, the computation of Cen[**X**] consists of computing the geometric mean of the columns of **X** and closing the resulting vector if a representation on the simplex is desired.

An interesting question is which are desirable and intuitive properties of a metric (distance, norm, inner product) for compositions. Our geometric intuition comes from our experience in the Euclidean space $\mathbb{R}^3$ and we try to translate these observations to a geometry of the simplex. In this way, if we have a rigid object on the table and we move this to another position, for instance on the floor, we expect that distances between points of the object are equal to those observed previous to the movement. Also, we observe that projecting a segment on the floor ($\mathbb{R}^2$), perhaps the edge of a roof, produces a segment with length shorter than the original one. If the points delimiting the segment are expressed in Cartesian coordinates, $x$ and $y$, on the floor, and $z$ vertical or orthogonal to the floor, the projection of the points consists in suppressing the $z$-coordinate. That is, our experience tells us that suppressing coordinates makes the resulting projected distances shorter than or equal to the original ones. Being a little bit more subtle, we realize that suppressing the $z$-coordinate is a special projection (orthogonal projection), but there are other kinds of projections. For instance, the shadow projected by the edge of the roof on the floor may be larger than the length of the edge depending on the position of the sun. This is because the shadow is not an orthogonal projection unless the floor is tilted orthogonal to the sun rays. These daily experiences with Euclidean geometry may inspire the following properties of the geometry in the simplex that we take as requirements.

A. **Equidistance on shift**: The distance between two compositions $\mathbf{x}_1$ and $\mathbf{x}_2$ in $\mathbb{S}^D$ is equal to their distance after a shift $\mathbf{z}$, that is

$$\mathrm{d}_a(\mathbf{x}_1 \oplus \mathbf{z}, \mathbf{x}_2 \oplus \mathbf{z}) = \mathrm{d}_a(\mathbf{x}_1, \mathbf{x}_2) \; ; \tag{4.11}$$

B. **Dominance on subcompositions**: From the composition $\mathbf{x}_k = (x_{k1}, x_{k2}, \ldots, x_{kD})$, a subcomposition $\mathbf{x}_k^{sub}$ is extracted by suppressing some components, for instance, $\mathbf{x}_k^{sub} = (x_{k1}, x_{k2}, \ldots, x_{kd})$, with $D > d > 1$. Then, for $k = 1, 2$, compositional distance should satisfy $\mathrm{d}_a(\mathbf{x}_1, \mathbf{x}_2) \geq \mathrm{d}_a(\mathbf{x}_1^{sub}, \mathbf{x}_2^{sub})$;
C. **Subcomposition as orthogonal projection**: The geometry on the simplex is an Euclidean geometry, that is, there is an inner product from which the norm and distance derive. Particularly, geometry on subcompositions in $\mathbb{S}^d$, $D > d > 1$, is equivalent to that of the orthogonal projection of $\mathbb{S}^D$ onto $\mathbb{S}^d$.

Point A is essential for defining sensible elementary statistics as shown in Eqs. (4.8) and (4.9). To show the importance of this property a subset of water analyses in Bangladesh has been selected. It comes from a survey conducted in the 1990s as a joint effort by the British Geological Survey and the Department of Public Health Engineering of Bangladesh (British Geological Survey 2001a, b). The subset, called hereafter Northern Bangladesh data, includes 13 disolved ions in Northern Bangladesh (latitude greater than 26 °N) and has been selected with the only purpose to serve as illustration. This data set was also used in several studies (see Pawlowsky-Glahn et al. 2015 and references therein). Concentrations of As, Fe and P (mg/l) are shown in a ternary diagram (Fig. 4.5). In the left panel they appear close to the border Fe-P due to the small concentrations of As relative to Fe and P. Right panel of Fig. 4.5 shows the same data set after centering it, that is $\mathbf{X} \ominus \mathrm{Cen}[\mathbf{X}]$. Now details are made visible; for instance, the rounding of As to 1 $\mu$g/l is now visible in form of straight bands extending from the Fe vertex. Although the aspect of the data points is more disperse in the left panel than the right one, the total variance is equal in the two representations, as perturbation does not change the total variance; that is, $\mathrm{totVar}[\mathbf{X}] = \mathrm{totVar}[(\mathbf{X} \ominus \mathrm{Cen}[\mathbf{X}])]$. This points out the inconvenience of using the visual distance (Euclidean distance) in the ternary diagram.

Requirement B is a consequence of point C, and is to be discussed at the end of this section. Requirement C is a bit technical but is again inspired by the real multivariate geometry. Suppose that a sample of $d$ real variables has been observed and the corresponding data set is arranged in an $(n, d)$ matrix. One may be interested in a multiple scatter-plot of each couple of variables, similar to that shown in Fig. 4.6. The fact that the axes of such plots are perpendicular does not surprise anybody. The assumption is that adding a real variable to a previous set is naturally represented by adding a new coordinate on an axis orthogonal to the previous ones.

Requirement C is implicitly claiming for an orthogonality relation, usually given by an inner product between compositions, namely $\langle \mathbf{x}, \mathbf{y} \rangle_a$, where $\mathbf{x}$ and $\mathbf{y}$ are compositions represented in the same simplex, say $\mathbb{S}^D$. From this inner product two

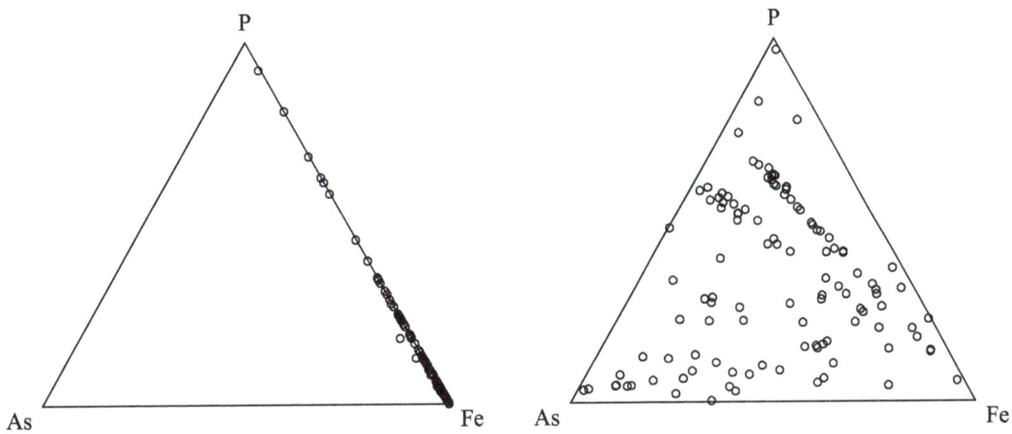


**Fig. 4.5** Disolved As, Fe, P data set. Left panel, data expressed in mg/l. Right panel, same data after centering

**Fig. 4.6** Disolved As, Fe, P data set represented in orthonormal coordinates. Triangles: original data; Circles: centered data. The arrow indicates the centering perturbation and it is anchored in the sample mean of coordinates

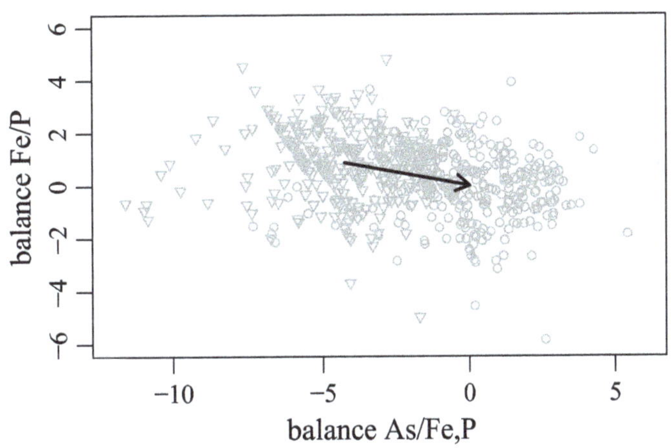


compositions are orthogonal if they satisfy $\langle \mathbf{x}, \mathbf{y} \rangle_a = 0$. All metric elements can be derived from the inner product. The square-norm (square size) is $\|\mathbf{x}\|_a^2 = \langle \mathbf{x}, \mathbf{x} \rangle_a$; and square-distance is $\mathrm{d}_a^2(\mathbf{x}, \mathbf{y}) = \|\mathbf{x} \ominus \mathbf{y}\|_a^2$. A general property of Euclidean spaces (Queysanne 1973) is that there exists an orthonormal basis constituted by $D-1$ compositions $\mathbf{e}_1, \mathbf{e}_2, \ldots, \mathbf{e}_{D-1}$. Orthonormal coordinates are then computed as

$$\phi_k(x_1, x_2, \ldots, x_D) = \langle \mathbf{x}, \mathbf{e}_k \rangle_a \,, \quad k = 1, 2, \ldots, D-1 \,,$$

and, consequently,

$$\|\mathbf{x}\|_a^2 = \sum_{k=1}^{D-1} \phi_k^2(x_1, x_2, \ldots, x_D) \,.$$

The question is which form can the coordinates $\phi_k$ take, so that they satisfy requirements A, B, C, and so that they are compatible with perturbation and powering. These latter conditions lead to the following additional requirement.

D. The coordinates in $\mathbb{S}^D$, $\phi_k$, $k = 1, 2, \ldots, D-1$ satisfy

$$\phi_k(\mathbf{x} \oplus (\alpha \odot \mathbf{y})) = \phi_k(\mathbf{x}) + \alpha \cdot \phi_k(\mathbf{y}), \tag{4.12}$$

for any compositions $\mathbf{x}, \mathbf{y}$, and any real constant $\alpha$.

From requirements A and D, the $\phi_k$ can be deduced. Consider first a two part subcomposition of $\mathbf{x}$, denoted $\mathbf{x}^{(2)}$. These subcompositions constitute a Euclidean space of dimension 1, and two part compositions can be represented by a single coordinate $\phi_1 = \phi_1(x_1^{(2)}, x_2^{(2)})$. This function must be scale invariant and such that it can take all real values. A simple log-ratio, $\phi_1 = a_1 \ln(x_1^{(2)}/x_2^{(2)})$, where $a_1$ is a real constant to be determined, is a possible choice. The ratio argument within the logarithm guarantees scale invariance, and the logarithm allows $\phi_1$ to range over all real numbers. The superscripts denoting the number of parts of the subcomposition are superfluous due to the scale invariance property and, from now on, it is assumed that $x_i^{(k)} = x_i$, being the latter the value of the $i$-th component in the large composition $\mathbf{x}$.

Consider now a 3-part subcomposition $\mathbf{x}^{(3)} = (x_1, x_2, x_3)$ in a 2-dimensional subspace which includes subcompositions $\mathbf{x}^{(2)}$, that is $(x_1^{(3)}, x_2^{(3)}) = (x_1, x_2)$. The additional dimension corresponds to a new coordinate $\phi_2$ in an orthogonal direction to that $\phi_1$ as proposed by requirement C. Again this coordinate needs to be scale invariant and taking any real value. A simple choice can be $\phi_2 = a_2 \ln(x_3/\mathrm{g_m}(\mathbf{x}^{(2)}))$ where $\mathrm{g_m}$ denotes geometric mean of the arguments. Iterating the reasoning for increasing number of parts of the subcomposition the $k$-th coordinate takes the form

$$\phi_k = a_k \ln \frac{x_{k+1}}{\mathrm{g_m}(\mathbf{x}^{(k)})}, \quad k = 1, 2, \ldots, D-1 .$$

These expressions for the coordinates fulfill conditions A–D.

The inner product in a Euclidean space can be expressed using Cartesian coordinates as

$$\langle \mathbf{x}, \mathbf{y} \rangle_a = \sum_{k=1}^{D-1} \phi_k \psi_k , \tag{4.13}$$

where $\phi_k$ and $\psi_k$ are the coordinates of the $D$-part compositions $\mathbf{x}$, $\mathbf{y}$ respectively. A tedious exercise consists of substituting the expression of the coordinates in Eq. (4.13) and carrying out the sum for values of $a_k$ such that all components of $\mathbf{x}$, $\mathbf{y}$ appear in a symmetric way. Up to a multiplicative constant, the result is

$$\langle \mathbf{x}, \mathbf{y} \rangle_a = \sum_{j=1}^{D} \ln \frac{x_j}{\mathrm{g_m}(\mathbf{x})} \ln \frac{y_j}{\mathrm{g_m}(\mathbf{y})}, \quad a_j = \sqrt{\frac{j}{j+1}},$$

where the $a_j$s appear as normalizing constants homogenizing the scale of the different axes. The inner product $\langle \mathbf{x}, \mathbf{y} \rangle_a$ is the ordinary inner product of the $\mathbb{R}^D$ vectors $\mathrm{clr}(\mathbf{x})$ and $\mathrm{clr}(\mathbf{y})$, which are

$$\mathrm{clr}(\mathbf{x}) = \left( \ln \frac{x_1}{\mathrm{g_m}(\mathbf{x})}, \ln \frac{x_2}{\mathrm{g_m}(\mathbf{x})}, \ldots, \ln \frac{x_D}{\mathrm{g_m}(\mathbf{x})} \right),$$

and analogously for $\mathrm{clr}(\mathbf{y})$.

The square Aitchison distance expressed in coordinates is the ordinary Euclidean distance in $\mathbb{R}^{D-1}$, which can be compared to the expression using the clr coefficients in $\mathbb{R}^D$:

$$\mathrm{d}_a^2(\mathbf{x}, \mathbf{y}) = \sum_{k=1}^{D-1} (\phi_k - \psi_k)^2 = \sum_{j=1}^{D} (\mathrm{clr}_j(\mathbf{x}) - \mathrm{clr}_j(\mathbf{y}))^2 . \tag{4.14}$$

Requirement B on dominance of distance of a subcomposition is now evident. From the expression of the distance in coordinates (Eq. 4.14, central term), computing distances within a subcomposition consists of removing some positive terms from the sum.

Apparently, there are many possible choices for the form of coordinates $\phi_k$, but most of them are discarded by requirements A and D on compatibility with perturbation (Eqs. 4.11, 4.14). For instance, $\phi_k = \ln(x_{k+1}/(x_1 + x_2 + \cdots + x_k))$, implicitly proposed in Aitchison (1986), Sect. 10.3, does not lead to a distance and coordinate expressions satisfying A and D. The critical point is that amalgamation or sum of compositional parts is not a linear operation for compositions.

Figure 4.6 shows the sample of disolved As, Fe, P previously represented in Fig. 4.5 in ilr-coordinates. These coordinates are the balances

$$\phi_1 = \sqrt{\frac{2}{3}} \ln \frac{\mathrm{As}}{(\mathrm{Fe} \cdot \mathrm{P})^{(1/2)}}, \quad \phi_2 = \sqrt{\frac{1}{2}} \ln \frac{\mathrm{Fe}}{\mathrm{P}} .$$

The visual distances between the data points are now the Aitchison distances. The triangles correspond to the original data set. Its center, expressed in coordinates, is the point where the arrow is anchored. A shift (perturbation) is applied in order to center the data set (circles), so that the new center is the origin of coordinates (end of the arrow). Importantly, the distances between data points after shifting (requirement A) are equal to the previous ones. The fact that the axes are drawn orthogonally, exactly corresponds to the fact that these coordinates are orthogonal in the Aitchison geometry for compositional data.

The historical way of defining the centered log-ratio transformation of $\mathbf{x}$ and the whole structure was the reverse of the one here presented. The definitions of perturbation, powering and clr can be found in Aitchison (1986), although the Aitchison distance was already introduced in Aitchison (1983) and discussed in Aitchison et al. (2000). The inner product as such, and the corresponding Euclidean space structure (Aitchison geometry), was introduced independently in Pawlowsky-Glahn and

Egozcue (2001), and in Billheimer et al. (2001), although there is a previous definition of orthogonal log-contrasts in Aitchison (1986). Orthogonal coordinates were introduced in Egozcue et al. (2003), and in Egozcue and Pawlowsky-Glahn (2005).

## 4.6 Consequences of the Aitchison Geometry in the Sample Space of Compositional Data

The consequences of the Euclidean character of the Aitchison geometry for compositional data are multiple and relevant. Once the principles and requirements on the sample space are assumed, they appear as a guidance in most, if not all, statistical models. The main idea is that compositions are advantageously represented as vectors in coordinates, better than as proportions. Standard operations, sum and multiplication, on appropriate coordinates are equivalent to perturbation and powering on compositions in the simplex. The fact that Aitchison distances, norms and orthogonal projections are transformed into the ordinary Euclidean distances, norms and orthogonal projections opens the door to use on ilr coordinates all mathematical and statistical methods designed for real variables. The recommendation of working on coordinates has been formulated as *the principle of working on coordinates* (Mateu-Figueras et al. 2011). The specific exploratory tools for compositional data are examples of the usefulness of ilr coordinates.

Principal component analysis for compositional data (CoDa-PCA) and its graphical representation, the CoDa-biplot, were studied before ilr-coordinates were available (Aitchison 1983; Aitchison and Greenacre 2002), but they are a wonderful example of their usefulness. A $D$-part compositional data set, $\mathbf{X}$ in a $(n, D)$-matrix, is clr-transformed and centered; then, the singular value decomposition is carried out. This can be summarized as

$$\mathrm{clr}(\mathbf{X}_c) = \mathrm{clr}(\mathbf{X} \ominus \mathbf{1}_n \mathrm{Cen}[\mathbf{X}]) = \mathbf{U}\boldsymbol{\Lambda}\mathbf{V}^\top , \tag{4.15}$$

where clr is applied to each composition (row) of the centered matrix, and $\mathbf{1}_n$ is a column vector of $n$ ones. The diagonal matrix $\boldsymbol{\Lambda}$ contains $D-1$ singular values ordered from the largest one to the smallest. The $D$-th singular value is always null, since the rows of $\mathrm{clr}(\mathbf{X}_c)$ add to zero, and can be removed. The $(D, D-1)$-matrix $\mathbf{V}$ (loadings matrix), once the last column corresponding to the null singular value is removed, is orthogonal and satisfies $\mathbf{V}^\top\mathbf{V} = \mathbf{I}_{D-1}$, $\mathbf{V}\mathbf{V}^\top = \mathbf{I}_D - (1/D)\mathbf{1}_D\mathbf{1}_D{}^\top$. Therefore, it is a contrast matrix like that used to compute ilr-coordinates of a composition $\mathbf{x}$ (column vector) (Egozcue et al. 2011)

$$\mathbf{z} = \mathrm{ilr}(\mathbf{x}) = \mathbf{V}^\top \mathrm{clr}(\mathbf{x}) , \quad \mathbf{x} = c \cdot \exp[\mathbf{V}\mathbf{z}] .$$

This means that the rows of the $(n, D-1)$-matrix $\mathbf{U}\boldsymbol{\Lambda}$ are ilr-coordinates of the centered compositional data set. A form biplot represents simultaneously the rows of

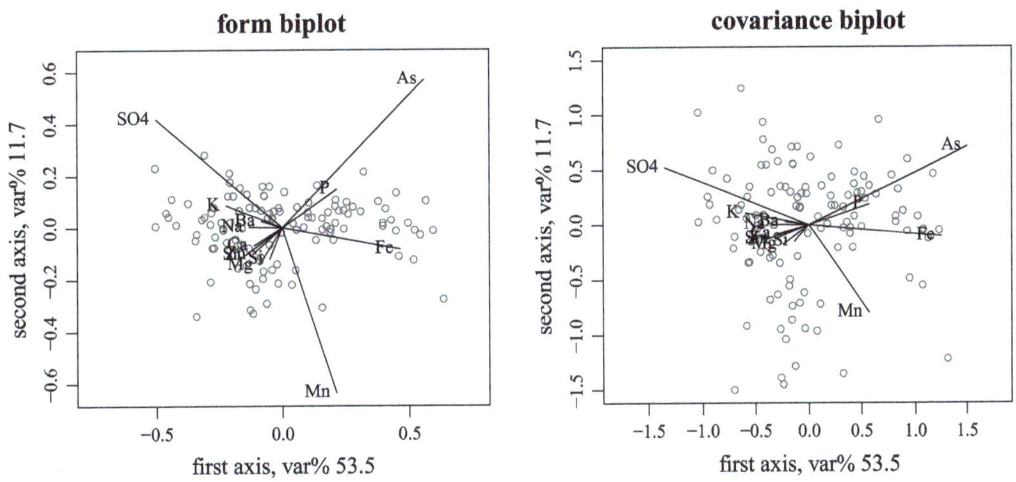


**Fig. 4.7** Biplots of Northern Bangladesh data set, representing 13 disolved ions. Left: form biplot showing that the projection is mainly dominated by the clr coefficients of As, Mn, and $SO_4$; up to the projection (65.2% of total variance), Aitchison distances between data points are approximately those visualized. Right: covariance biplot adequate for interpretation. Up to the projection, length of links between vertices of rays are proportional to the standard deviation of the corresponding logratio. The length of the rays are approximately proportional to the standard deviation of the corresponding clr-coefficients. Variability is largely dominated by the log ratios of $SO_4$ over As, Fe and Mn

$\mathbf{U}\boldsymbol{\Lambda}$ (coordinates of the compositions) and the columns of $\mathbf{V}$ (clrunitary vectors of the ilr-basis) in an optimal bi-dimensional projection for visualization.

Figure 4.7 shows the form biplot of the Northern Bangladesh data set. Form biplots (Fig. 4.7, left) and scatter-plots of coordinates (Fig. 4.6) can replace plots on ternary diagrams, as distances between compositions are not distorted in an uncontroled manner. They are only affected by the orthogonal projections.

The ilr coordinates are real variables and their exploratory analysis relies on standard exploratory analysis tools (mean, standard deviation, quantiles, correlations). However, interpretable coordinates are desirable. They can be designed by the analyst to get insight in some aspects of the data he/she may be interested in. Other times a data driven technique may be used to design suitable coordinates (Pawlowsky-Glahn et al. 2011; Martín-Fernández et al. 2017). In these cases, the CoDa-dendrogram (Pawlowsky-Glahn and Egozcue 2011) can be useful to summarize properties of the coordinate sample jointly with an interpretable description of the coordinates used. The definition of the coordinates is based on a sequential binary partition (SBP) of the parts of the composition (Egozcue and Pawlowsky-Glahn 2005, 2006). Each coordinate is associated with a partition of a group of parts into two new groups. For instance, Table 4.2 shows this kind of partitions for the Northern Bangladesh data set. The second row of Table 4.2, indicates the separation of As (+1) from the group constituted by Fe, Mn and P (−1). This separation is associated with the second ilr coordinate

**Table 4.2** Sign code for a SBP of the 13 disolved ions, obtained by clustering variables of the Northern Bangladesh data set

| As | Ba | Ca | Fe | K | Mg | Mn | Na | P | Si | SO4 | Sr | Zn |
|---|---|---|---|---|---|---|---|---|---|---|---|---|
| +1 | −1 | −1 | +1 | −1 | −1 | +1 | −1 | +1 | −1 | −1 | −1 | −1 |
| +1 | 0 | 0 | −1 | 0 | 0 | −1 | 0 | −1 | 0 | 0 | 0 | 0 |
| 0 | 0 | 0 | −1 | 0 | 0 | +1 | 0 | −1 | 0 | 0 | 0 | 0 |
| 0 | 0 | 0 | +1 | 0 | 0 | 0 | 0 | −1 | 0 | 0 | 0 | 0 |
| 0 | −1 | −1 | 0 | −1 | −1 | 0 | −1 | 0 | −1 | +1 | −1 | −1 |
| 0 | 0 | 0 | 0 | 0 | 0 | 0 | +1 | 0 | −1 | 0 | +1 | −1 |
| 0 | 0 | 0 | 0 | 0 | 0 | 0 | +1 | 0 | 0 | 0 | −1 | 0 |
| 0 | 0 | 0 | 0 | 0 | 0 | 0 | 0 | 0 | +1 | 0 | 0 | −1 |
| 0 | −1 | −1 | 0 | −1 | −1 | 0 | +1 | 0 | +1 | 0 | +1 | +1 |
| 0 | −1 | −1 | 0 | +1 | −1 | 0 | 0 | 0 | 0 | 0 | 0 | 0 |
| 0 | +1 | −1 | 0 | 0 | −1 | 0 | 0 | 0 | 0 | 0 | 0 | 0 |
| 0 | 0 | +1 | 0 | 0 | −1 | 0 | 0 | 0 | 0 | 0 | 0 | 0 |

$$z_2 = \sqrt{\frac{3}{4}} \ln \frac{\text{As}}{(\text{Fe} \cdot \text{Mn} \cdot \text{P})^{1/3}} .$$

These kinds of coordinates are called balances between two groups of parts (Egozcue and Pawlowsky-Glahn 2005) as they are logratios of the geometric mean of the elements in each group; the coefficient in front of the logarithm is a normalization coefficient which takes into account the number of elements in each group of parts. Figure 4.8 shows the CoDa-dendrogram for the Northern Bangladesh data set. The tree-dendrogram itself follows the partition in Table 4.2. The length of the lines perpendicular to the labels, say vertical lines, are proportional to the variance of the balance separating the groups of elements at left and right hand sides. These vertical lines are anchored to horizontal segments joining the two groups of parts. All these segments are scaled in such a way that the zero value is placed in the center of the segment, and the length represents the same length in all cases. The fulcrum of the vertical line is placed at the average value of the balance; it can be compared to the median indicated in the box-plot under the horizontal line. In this way, the CoDa-dendrogram combines the interpretation of the balance-coordinates given by the SBP and their mean, variance and quantiles (box-plots).

In Fig. 4.8, the variances within the subcomposition (Zn, Si, Sr, Na, $SO_4$) are small compared to other variances, thus pointing out a possible compositional association between these elements; it suggests that these elements change proportionally along the considered sample. At the same time, most of the total variance is driven by As, Fe, Mn and P, as indicated by longer vertical lines.

The explanatory power of the CoDa-biplot and the CoDa-dendrogram relies on the fact that they are based on Cartesian coordinates for plotting data-points and that the represented variables are orthonormal in a geometric sense. The key in interpret-

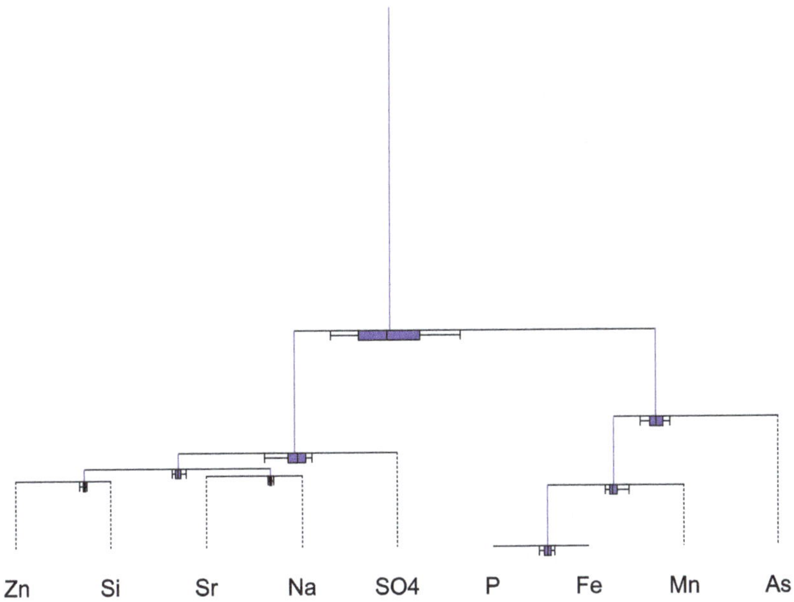


**Fig. 4.8** CoDa-dendrogram following the sign code in Table 4.2 obtained by clustering variables of the Northern Bangladesh data set. Vertical bars describe the decomposition of the total variance given in Eq. (4.16). Anchoring points of vertical bars indicate the mean value of the corresponding coordinate

ing the results is the decomposition of the total variance of the data set into variances of the ilr-coordinates (Egozcue and Pawlowsky-Glahn 2011a)

$$\operatorname{totVar}[\mathbf{X}] = \sum_{k=1}^{D-1} \operatorname{Var}[\phi_k] \,. \tag{4.16}$$

## 4.7 Conclusions

The first step in any data modelling is to establish a sample space able to give answers to the questions stated by the analyst. If these questions involve probabilistic statements, the sample space needs a sigma field of events for which probabilities can be defined. However, most analysts search for statements implying operations, distances, projections between data points or variables. All these concepts need to be defined in the sample space for useful computations and interpretations. These definitions are not intrinsic, but are adapted to the questions stated by the analyst in a subjective way. Therefore, the choice of a sample space has always a subjective character, which is only validated by the ability in giving useful answers to sound questions.

Compositional data require defining a sample space with a rich structure. The log-ratio approach to the analysis of compositional data is based on a set of principles and conditions. The approach here presented is a modification of the standard principles introduced by J. Aitchison in the eighties and reformulated afterwards. Scale invariance and compositional equivalence are maintained exactly as they were introduced, but additional conditions are to be discussed in relation to perturbation, which is assumed to be the main operation between compositions. The Euclidean structure of compositional data represented in the simplex, called Aitchison geometry, is here motivated using the idea that reduction to a subcomposition should be an orthogonal projection.

The Aitchison geometry is thought as a powerful mathematical tool which consistently completes the previous Aitchisonian ideas on the log-ratio approach. The main points are the conception of compositions as equivalence classes (Barceló-Vidal and Martín-Fernández 2016) thus overcoming the early definitions based on the constant sum constraint; and the introduction of coordinates in the Aitchison geometry (Pawlowsky-Glahn and Egozcue 2001; Egozcue et al. 2003; Egozcue and Pawlowsky-Glahn 2005) thus overcoming the idea that taking log-ratios is just a transformation which circumvents the constant sum constraint.

**Acknowledgements** The research on compositional data analysis has been continuously supported by the *International Association for Mathematical Geosciences* (formerly *for Mathematical Geology*), *IAMG*. The authors appreciate this support, in many cases unconditional, and recognize that the development of the compositional data analysis would have been a lot harder without it. This research has been funded by the Spanish Ministry of Economy and Competitiveness under the project CODA-RETOS/TRANS-CODA (Ref: MTM2015-65016-C2-1/2-R).

## References

Aitchison J (1982) The statistical analysis of compositional data (with discussion). J R Stat Soc Ser B (Stat Methodol) 44(2):139–177

Aitchison J (1983) Principal component analysis of compositional data. Biometrika 70(1):57–65

Aitchison J (1986) The statistical analysis of compositional data. Monographs on statistics and applied probability, p 416. Chapman & Hall Ltd, London (Reprinted in 2003 with additional material by The Blackburn Press)

Aitchison J (1992) On criteria for measures of compositional difference. Math Geol 24(4):365–379

Aitchison J (1997) The one-hour course in compositional data analysis or compositional data analysis is simple. In Pawlowsky-Glahn V (ed) Proceedings of IAMG'97, pp 3–35 Barcelona (E). CIMNE, Barcelona, Spain ISBN 978-84-87867-76-7

Aitchison J, Barceló-Vidal C, Martín-Fernández JA, Pawlowsky-Glahn V (2000) Logratio analysis and compositional distance. Math Geol 32(3):271–275

Aitchison J, Egozcue JJ (2005) Compositional data analysis: where are we and where should we be heading? Math Geol 37(7):829–850

Aitchison J, Greenacre M (2002) Biplots for compositional data. J R Stat Soc, Ser C 51(4):375–392

Aitchison J, Shen S (1980) Logistic-normal distributions. Some properties and uses. Biometrika 67(2):261–272

Ash RB (1972) Real analysis and probability. Academic Press Inc, New York, NY (USA), p 476

Barceló-Vidal C, Martín-Fernández JA (2016) The mathematics of compositional analysis. Aust J Stat 45:57–71

Barceló-Vidal C, Martín-Fernández JA, Pawlowsky-Glahn V (2001) Mathematical foundations of compositional data analysis. In: Ross G (ed) Proceedings of IAMG'01 – The VII annual conference of the international association for mathematical geology, p 20. Cancun (Mex)

Billheimer D, Guttorp P, Fagan W (2001) Statistical interpretation of species composition. J Am Stat Assoc 96(456):1205–1214

British Geological Survey (2001a) Arsenic contamination of groundwater in Bangladesh. Technical Report (WC/00/019), Dep. of Public Health Engineering (Bangladesh). p 630

British Geological Survey (2001b). Arsenic contamination of groundwater in bangladesh: data. Technical Report (WC/00/019), Dep. of Public Health Engineering (Bangladesh)

Chayes F (1960) On correlation between variables of constant sum. J Geophys Res 65(12):4185–4193

Chayes F (1962) Numerical correlation and petrographic variation. J Geol 70(4):440–452

Connor RJ, Mosimann JE (1969) Concepts of independence for proportions with a generalization of the Dirichlet distribution. J Am Stat Assoc 64(325):194–206

Egozcue JJ (2009) Reply to "On the Harker variation diagrams;..." by J. A. Cortés. Math Geosci 41(7):829–834

Egozcue JJ, Barceló-Vidal C, Martín-Fernández JA, Jarauta-Bragulat E, Díaz-Barrero JL, Mateu-Figueras G (2011). Elements of simplicial linear algebra and geometry. See Pawlowsky-Glahn Buccianti, pp 141–157, p 378

Egozcue JJ, Pawlowsky-Glahn V (2005) Groups of parts and their balances in compositional data analysis. Math Geol 37(7):795–828

Egozcue JJ, Pawlowsky-Glahn V (2006) Simplicial geometry for compositional data. Compositional data analysis in the geosciences: from theory to practice, vol 264, pp 145–159, Special Publication, Geological Society, London

Egozcue JJ, Pawlowsky-Glahn V (2011a) Basic concepts and procedures. See Pawlowsky-Glahn and Buccianti, pp 12–28

Egozcue JJ, Pawlowsky-Glahn V (2011b) Evidence information in bayesian updating. In: Egozcue JJ, Tolosana–Delgado R, Ortego MI (eds.)Proceedings of CoDaWork-2011, Sant Feliu de Guixols, Girona, Spain

Egozcue JJ, Pawlowsky-Glahn V, Mateu-Figueras G, Barceló-Vidal C (2003) Isometric logratio transformations for compositional data analysis. Math Geol 35(3):279–300

Egozcue JJ, Pawlowsky-Glahn V, Tolosana-Delgado R, Ortego MI, van den Boogaart KG (2013) Bayes spaces: use of improper distributions and exponential families. Revista de la Real Academia de Ciencias Exactas, Físicas y Naturales, Serie A. Matemáticas (RACSAM) 107:475–486. https://doi.org/10.1007/s13398-012-0082-6

Feller W (1968) An introduction to probability theory and its applications, p 501 1950 (1st edn.), 1968 (3rd edn.), Vol I. Wiley, New York, NY (USA)

Ferrer-Rossell B, Coenders G, Mateu-Figueras G, Pawlowsky-Glahn V (2016) Understanding low-cost airline users' expenditure patterns and volume. Tour Econ 22(2):269–291

Fréchet M (1948) Les éléments Aléatoires de Nature Quelconque dans une Espace Distancié. Annales de l'Institut Henri Poincaré 10(4):215–308

Martín-Fernández JA, Barceló-Vidal C, Pawlowsky-Glahn V (2003) Dealing with zeros and missing values in compositional data sets using nonparametric imputation. Math Geol 35(3):253–278

Martín-Fernández J-A, Pawlowsky-Glahn V, Egozcue JJ, Tolosana-Delgado R (2017) Principal balances for compositional data. Math Geosci under review

Mateu-Figueras G, Pawlowsky-Glahn V, Egozcue JJ (2011) The principle of working on coordinates. See Pawlowsky-Glahn and Buccianti, pp 31–42

Olea RA, Luppens JA, Egozcue JJ, Pawlowsky-Glahn V (2016) Calorific value and compositional ultimate analysis with a case study of a Texas lignite. J Coal Geol 162:27–33

Parent LE, de Almeida CX, Hernandes A, Egozcue JJ, Gülser C, Bolinder MA, Kätterer T, Andrén O, Parent SE, Anctil F, Centurion JF, Natale W (2012) Compositional analysis for an unbiased measure of soil aggregation. Geoderma 179–180:123–131

Pawlowsky-Glahn V, Buccianti A (eds) (2011) Compositional data analysis: theory and applications, p 378. Wiley

Pawlowsky-Glahn V, Egozcue JJ (2001) Geometric approach to statistical analysis on the simplex. Stoch Environ Res Risk Assess (SERRA) 15(5):384–398

Pawlowsky-Glahn V, Egozcue JJ (2011) Exploring compositional data with the coda-dendrogram. Aust J Stat 40(1 & 2):103–113

Pawlowsky-Glahn V, Egozcue JJ, Olea RA, Pardo-Igúzquiza E (2015) Cokriging of compositional balances including a dimension reduction and retrieval of original units. J South Afr Inst Min Metal, SAIMM 115:59–72

Pawlowsky-Glahn V, Egozcue JJ, Tolosana-Delgado R (2011) Principal balances to analyse the geochemistry of sediments. In Marschallinger R, Zobel F (eds) Proceedings of IAMG 2011–The XVth annual conference of the international association for mathematical geology, p 10. University of Salzburg, Austria

Pawlowsky-Glahn V, Egozcue JJ, Tolosana-Delgado R (2015) Modeling and analysis of compositional data. Statistics in practice. Wiley, Chichester UK, p 272

Pearson K (1897) Mathematical contributions to the theory of evolution. On a form of spurious correlation which may arise when indices are used in the measurement of organs. In: Proceedings of the Royal Society of London LX, pp 489–502

Queysanne M (1973) Álgebra Básica. Editorial Vicens Vives, Barcelona (E), p 669

Scealy JL, Welsh AH (2014) Colors and cocktails: compositional data analysis. Aust New Zealand J Stat 56(2):145–169

Tolosana-Delgado R (2012) Uses and misuses of compositional data in sedimentology. Sediment Geol 280:60–79

# Chapter 5
# Properties of Sums of Geological Random Variables

G. M. Kaufman

"All models are wrong. Some are useful" George E. P. Box.

**Abstract** In the absence of empirical data that allows resolution of the vexing problem of how to address probabilistic dependencies among and between elements of large sets of geologic random variables data we need methods that refocus and streamline expert geological judgment inputs along with analytical methods for modeling dependencies that go beyond pairwise correlation and its cousins. Some possibilities are reviewed.

## 5.1 Introduction

Suppose that you are given the marginal distribution of each of a set of n random variables but no other information. What can be said about the behavior of their sum? This is an old problem, extensively studied by probability theorists and statisticians (Hoeffding 1940; Frèchet 1951). There is a rich probabilistic finance and actuarial risk analysis literature devoted to calculation of bounds on sums of random variables. This question motivates our review of state of the art methods designed to reduce geologists' cognitive load when asked to assign judgmental probabilities to uncertain geologic variables.

In a wide range of settings geologists are asked to provide personal probability judgments about a collection of uncertain quantities and, in particular, about sums of them. Probabilistic assessments of oil and gas in unexplored petroleum plays and basins are recurring examples. In the absence of hard data they deal rather well with the cognitive task of providing personal judgments about marginal distributions of geologic attributes; i.e. their assessments are, in the large, reasonably well calibrated. Geologists' personal judgments about dependencies among uncertain geologic quantities are more problematic.

G. M. Kaufman (✉)
Management Emeritus, E62-437, Sloan School of Management MIT,
50 Memorial Drive, Cambridge, MA 92142, USA
e-mail: gkaufman@mit.edu



B. S. Daya Sagar et al. (eds.), *Handbook of Mathematical Geosciences*,
https://doi.org/10.1007/978-3-319-78999-6_5

It is worthwhile to distinguish micro-assessments—assessment of dependencies among individual reservoir attributes for example—from macro-assessments—assessment of dependencies among assessment units, each of which may be a collection of anomalies, reservoirs and fields. Measurable data bearing directly on probabilistic dependencies at the micro-assessment level is often available but precise measurable data bearing on dependencies among elements in a macro-assessment is seldom available. Chen et al. (2012) point out that

> Although efforts have been made to address variable dependence in both methodology and tool development, the greatest emphasis and attention have been given to resource aggregation. Until now, the impact of interdependencies among variables in volumetric resource calculations has been mostly ignored, and the implementation of variable dependency remains a challenge to petroleum resource appraisal. In practice, inadequate data commonly exist to either specify a standard multivariate distribution with an appropriate correlation structure or to quantify the resource aggregation correlation matrices. However, variable correlations are so common among geologic variables that ignoring their interdependence may lead to serious bias, affecting both the resulting resource potential estimation.

Most geologists with some training and experience in probability assessment can provide reasonable responses to questions about marginal distributions of individual attributes of a target entity. Few if any are well equipped to provide sharp coherent judgments about possible dependencies among them. Some progress has been made in understanding how to elicit sensible, coherent judgements about second order co-variability of petroleum assessment units—the recent USGS study of $CO_2$ sequestration in depleted oil and gas reservoirs is an example. However, specification of marginal distributions along with second order moments is not sufficient for identification of a joint distribution of a set of uncertain quantities. This matters when interest centers on the right tail of a sum of magnitudes of petroleum in assessment units. Excepting special cases—joint lognormality for example—the right tail of a sum of jointly dependent uncertain quantities can, both in principle and in practice differ meaningfully from the right tail of an approximation based on marginal distributions and second moment properties alone. Lillestøl and Sinding-Larsen's (2017) study of giant field probabilities based on 182 North Sea discoveries highlights the importance of accurate modeling of tail probabilities. For economists, bureaucrats and politicians right tail probabilities are often the most interesting feature of a probabilistic oil and gas assessment. What, for example, is the probability of finding at least one more giant field in a given mature petroleum province? Objectives here are first, to outline how methods currently used by geologists to impute probabilistic dependencies among uncertain geologic quantities fit (or don't fit) into a conceptual framework developed by probabilists to answer the question posed at the outset and second, to review how the probability distribution of a sum of such quantities can be bounded given knowledge of marginal distributions alone assuming they are governed by a type of functional dependency called co-monotonicity. Co-monotonicity and cupolas are conceptual twins.

Section 5.2 lays out necessary theory and definitions and calls attention to co-monotonic upper bounds on sums of random variables and lower bounds expressed in terms of conditional expectations. Section 5.3 addresses geologic case studies in two of which geologists compute a probability distribution of a sum of random geologic magnitudes in three steps: first, specify marginal distributions of each magnitude, second, elicit judgmental appraisals of pairwise correlations among magnitudes and third, combine the two using Monte Carlo simulation to arrive at a distribution of the sum. This approach might be labelled "incomplete specification" (not to be confused with the econometric definitions of just-, over- and under-specification.). Iman and Conover's (1982) ingenious method for imputing dependencies among a set of random variables requiring only pairwise correlations among elements of that set and marginal distributions is deployed in the $CO_2$ sequestration study cited above (Sect. 5.3.2). Chen et al. (2012) use of cupolas to capture probabilistic dependencies in geologic micro-assessments is reviewed in Sect. 5.3.3. Brief concluding remarks appear in Sect. 5.4. Blondes et al. (2013a, b) offer a sensible rationale for careful attention to dependencies:

> In the Circum-Arctic aggregation of the 48 AUs, the 90-percent uncertainty interval for recoverable gas is 1,471, 2,009, or 3,515 tcf for assumptions of independence, assessor specified dependency (correlation), or total dependence respectively. Clearly, decision makers who rely on assessment results need accurate interval projections. Too broad an interval provides little information; too narrow an interval gives a false sense of precision.

Spatial modeling provides important insights into the structure of probabilistic dependencies among petroleum play attributes and deserves careful attention in parallel with methods and models discussed here. It is a topic for another day.

## 5.2 Preliminaries

Define $F_X$ to be the distribution function of a random vector $\mathbf{X} = (X_1, \ldots, X_n)^t$ with domain $\mathbf{R}^n$ and marginal distributions $F_i, i = 1, \ldots, n$. Set $F_X(\mathbf{x}) = Prob\{X_1 \le x_1, \ldots, X_n \le x_n\}$. Assume that each $F_i$ is continuous and possesses a one to one inverse. Define the $p$th fractile of $X_i$ as the value in the domain of $X_i$ such that $Prob\{X_i \le x_p\} = p$ and its inverse as $F_i^{-1}(p) = x_i(p)$. In turn the $p$th fractile of the sum $S_n = X_1 + \cdots + X_n$ is $s_p$ such that $Prob\{S_n \le s_p\} = p$ or $F_{S_n}^{-1}(p) = s_p$.

What conditions guarantee that fractiles are strictly additive? That is that for all $p \in (0, 1) s_p = x_1(p) + \cdots + x_n(p)$? Imposition of functional dependencies among $X_1, \ldots, X_n$ is one route to sufficient conditions for this to be true. To divide difficulties suppose that $X_1, \ldots, X_n$ share a common domain $D_X$ and consider $n$ continuous invertible functions $h_i$, each with domain $D_X$. Suppose that $x_i = h_i(x_1)$ for all $x_i \in D_X$, $i = 2, .., n$. Then $Prob\{S_n < s\} = Prob\{X_1 + h_2(X_1) + \cdots + h_n(X_1) < s\}$. The omnibus function $g(x_1) = x_1 + h_2(x_1) + \cdots + h_n(x_1), x_1 \in D_X$ is continuous and invertible so $Prob\{g(X_1) < s\} = Prob\{X_1 < g^{-1}(s)\}$. The $p$th fractile of $S_n$ is $s_p$ such that $Prob\{g(X_1) < s_p\} = p$ or $Prob\{X_1 < g^{-1}(s_p)\} = p$ leading to $x_1(p) = g^{-1}(s_p)$.

Equivalently $g(x_1(p)) = s_p$. Functional dependencies of this type are too strong to survive the rigors of modeling most real world data. In the absence of complete knowledge of a joint distribution co-monotonicity is a more flexible approach to modeling joint behavior of dependent random variables.

**Definition** The random vector $\mathbf{X} = (X_1, \ldots, X_n)^t$ is co-monotonic if and only if $(X_1, \ldots, X_n) =_d (F_1^{-1}(U), \ldots, F_n^{-1}(U))$, $U$ a uniform random variable with domain (0, 1).

Here $=_d$ means agreement in distribution. Intuitively each element of a co-monotonic random vector is a functional of a single random variable $U$ so all elements of $\mathbf{X}$ exhibit strong positive dependency. McNeil et al. (2005) provide a more general definition: $\mathbf{X}$ is co-monotonic if and only if it agrees in distribution with a random vector, each of whose components is a non-decreasing function of a single random variable. If elements of $\mathbf{X}$ are co-monotonic increasing one element of $\mathbf{X}$ increases all others. Goovaerts et al. (2000) provide a clear readable account of properties of sums of co-monotonic random variables in an actuarial context. Deelstra et al. (2009) offer a literature review of co-monotonicity in financial economics.

Foreshadowing a possible critique by geologists that in their setting, some elements of $\mathbf{X}$ may be independent or possibly negatively dependent (rather rare), co-monotonicity and its consequences provide upper and lower bounds on a sum of random variables with specified marginal distributions that embrace a wide range of dependence structures. When these bounds are judged to be tight enough, reasonable projections of probability distributions of aggregates can be made using marginal distributions along with specification of certain conditional expectations. (See 5.1, 5.5). They provide useful information about projections made based on information elicited from geologists about dependencies and police reasonableness of geologic probabilistic projections of uncertain geologic resources made using other methods.

### 5.2.1 Bounds

A random variable $X$ precedes a random variable $Y$ in convex order, denoted by $X \geq_{cx} Y$ if and only if $E(g(X)) \geq E(g(Y))$ for all real convex functions $g$ for which expectations are finite. Kaas et al. (2009) use convex order to show that fractiles of co-monotonic random variables can be added in the following sense: for any random vector $\mathbf{X} = (X_1, \ldots, X_n)$ possessing marginal cumulative distribution functions $F_1, \ldots, F_n$ and $U$ a uniform (0, 1) random variable

$$(X_1 + \cdots + X_n) \leq_{cx} S_u \equiv F_1^{-1}(U) + \cdots + F_n^{-1}(U). \tag{5.1}$$

If $S_u =_d F_1^{-1}(U) + \cdots + F_n^{-1}(U)$ it follows immediately that the $p$th fractile of $S_u$ is $F_{S_u}^{-1}(p) = F_1^{-1}(p) + \cdots + F_n^{-1}(p),\ \mathit{for\,all}\, p \in (0,1)$. They point out that (5.1) is a supremum in terms of convex order and is a best bound for marginal distributions in a Fréchet space. It is well known that if a random vector $\mathbf{X}$ with marginal distributions $F_1, \ldots, F_n$ belong to a Fréchet space $\mathsf{F}_n$ the joint cumulative distribution function $Prob\{X_1 \leq x_1, \ldots, X_n \leq x_n\}$ of $\mathbf{X}$ is bounded from above by $M_n \equiv \min\{F_1(x_1), \ldots, F_n(x_n)\}$. Goovarts et al. note that $M_n$ is reachable in $\mathsf{F}_n$.

For sums of elements of $\mathbf{X}$ introduction of a random variable $Z$ such that distribution functions of each $X_i$ given $Z$ are known with certainty leads to refined upper and lower bounds. In a geologic context $Z$ is interpretable as a latent (background) variable describing gross geologic characteristics of, for example, a petroleum assessment unit. The conditioning variable $Z$ might be regression dependent on geologic attributes of an assessment unit and need not be scalar. These authors define $F_{X_i|Z}^{-1}(U)$ to be a random variable $f_i(U,Z)$ that for $(U,Z) = (u,z)$ assumes value $F_{X_i|z}^{-1}(u)$ and prove that for $U$ uniform $(0,1)$ and $Z$ independent of $U$

$$(X_1 + \cdots + X_n) \leq_{cx} S_u^* \equiv F_{X_1|Z}^{-1}(U) + \cdots + F_{X_n|Z}^{-1}(U). \tag{5.2}$$

Jensen's inequality leads to a lower bound

$$E(X_1|Z) + \cdots + E(X_n|Z) \leq_{cx} (X_1 + \cdots + X_n). \tag{5.3}$$

Kaas et al. (2009) point out that (a) the random vector $E(X_1|Z) + \cdots + E(X_n|Z)$ will not in general have marginal distributions $F_1, \ldots, F_n$ (b) If $E(X_1|Z), \ldots, E(X_n|Z)$ are either jointly non-increasing or non-decreasing functions of $Z$ the LHS in (5.3) is a sum of co-monotonous random variables and (c) $Var(E(X_i|Z)) < Var(X_i)$ unless $Var(E(X_i|Z)) = 0$. In order to create a path to direct computation of the cdf of the LHS of (5.4) suppose that (b) obtains and that each of the random variables $E(X_1|Z), \ldots, E(X_n|Z)$ are non-decreasing functions of increasing $Z = z$. Write the lower bound as $E(X_1|Z) + \cdots + E(X_n|Z) = E(S|Z)$ and define $F_{E(X_i|Z)}(x) = Prob\{E(X_i|Z) \leq x\}$. They show that, provided that the cdf of $E(X_i|Z)$ is continuous and increasing

$$F_{E(X_1|Z)}^{-1}(F_{E(S|Z)}(x)) + \cdots + F_{E(X_n|Z)}^{-1}(F_{E(S|Z)}(x)) = x, \tag{5.4}$$

a prescription for calculating a lower bound. The quality of the lower bound (5.3) depends of course on the choice of a model for Z. Kaas et al. (2002) and Goovarts et al. (2000) demonstrate that upper and lower bounds (5.1) and (5.3) provide reasonable bounds on the cumulative distribution function of certain sums of discounted cash flows as well as for the cumulative distribution function of sums of dependent lognormal random variables. Lux and Papantoleon (2017) show that upper and lower Fréchet–Hoeffding bounds such as those described above can be tightened. They demonstrate that other types of information, knowledge of functionals of lower dimensional marginals of an n-dimensional cupola for example, also lead to improvements. The tradeoff is that the improved bounds are quasi-cupolas but not cupolas.

Comparison of predictive distributions of undiscovered mineral resources derived by conventional methods currently in use with co-monotonic bounds on them is a promising avenue of research.

## 5.3 Thumbnail Case Studies

Thumbnail sketches of three case studies serve as a template for discussion of probabilistic dependence issues discussed above: examples of the USGS approach to probabilistic dependencies among oil and gas assessment units, the USGS probabilistic assessment of $CO_2$ sequestration in mature oil and gas reservoirs in the United States and a Canadian Geological Survey study of use of cupolas to capture probabilistic dependencies among accumulations in individual oil and gas plays.

### *5.3.1 USGS Oil and Gas Resource Projections*

The USGS developed an assessment system in the 1980s with the acronym FASP (fast appraisal system for petroleum resources). FASP incorporated perfect positive correlation between micro-level reservoir attributes but allowed specification of any positive correlation in the course of aggregating play resources. However, the USGS 2000 World Petroleum Assessment aggregates undiscovered resource volumes from assessment unit level to regional level using perfect correlation as the argument for adding assessment unit fractiles to arrive at regional level aggregates. Recognizing that at the global level dependencies among large regional aggregates of resources are unlikely to be perfectly correlated they adopt pairwise correlation of 0.5 between pairs of eight regions (Klett et al. 2000). No sensitivity analysis of how aggregate projections vary with these particular choices is provided.

Many USGS assessment studies present tables of fractiles of individual assessment units and then add them to arrive at a fractile assessment of total resources. Addition is qualified by the statement that "Fractiles are additive under assumption of perfect positive correlation" allowing avoidance of direct assessment of dependencies among units. Table 2 in "Assessment of Undiscovered Continuous Oil and Gas Resources in the Monterey Formation, San Joaquin Basin Province, California" USGS Fact Sheet 2015-3058 September 2015 and Table 2 in USGS Fact Sheet 2014–3082 "Assessment of Potential Shale-Oil and Shale-Gas Resources in Silurian shales of Jordan" September 2014 are examples. Chen et al. (2012) cite additional examples (Klett et al. 2000, 2005; Klett 2004). It is easy to show that "perfect correlation" is not robust to variations in specification of the functional form of marginal distributions elicited from geologists. Worse, addition of fractiles without careful attention to properties of the joint distribution of a set of uncertain quantities can lead to incoherence. On the other hand mutual independence allows specification of arbitrary marginal probability distributions without doing violence to coherence but often leads to an unacceptably narrow probability projection of sums of oil and gas magnitudes.

A salient feature of Pearson's correlation coefficient is that random variables $X$ and $Y$ possess correlation 1.0 or $-1.0$ only if $X$ and $Y$ are linearly dependent. As Denuit and Dehaene (2003) point out, a limiting case is a bivariate normal pair of random variables for which the variance of one member of the pair is zero. If $X$ and $Y$ are jointly lognormal and $\log X$ is a linear function of $\log Y$ the Pearson correlation of $\log X$ and $\log Y$ is either 1.0 or $-1.0$. However, the Pearson correlation of $X$ and $Y$ is then less than 1.0. Denuit and Dehaene provide a more nuanced treatment. Suppose $F_1$ and $F_2$ are marginal cumulative distribution functions of $X$ and $Y$ respectively, each concentrated on $(0, \infty)$ and $U$ is a uniform random variable independent of $X$ and $Y$. Using super-modularity these authors prove that if $F_1$ and $F_2$ lie in a Fréchet space the Pearson correlation coefficient $r(X, Y)$ of $X$ and $Y$ is bounded by

$$\frac{Cov(F_1^{-1}(U), F_2^{-1}(1-U))}{\sqrt{Var(X)}\sqrt{Var(Y)}} \leq r(X, Y) \leq \frac{Cov(F_1^{-1}(U), F_2^{-1}(U))}{\sqrt{Var(X)}\sqrt{Var(Y)}}. \tag{5.5}$$

In this setting perfect correlation is not achievable. They also prove that it is possible for a pair of co-monotonic lognormal random variables to have pairwise correlation close to zero, contradicting the intuitive notion that small correlation implies weak dependence. Denuit and Dehane call attention to Shih and Huang (1992) and Schechtman and Yitzhaki's (1999) observation that, for any two random variables, the achievable range of Pearson's correlation coefficient is (−1, 1) only if the functional form of the two marginal distributions differ solely in values of location and/or scale parameters. If not, the range of Pearson's r is narrower than (−1, 1) and depends on the shape of the two marginal distributions.

These authors document several important features of Kendall's $\tau$ and Spearman's $\rho$. (Spearman's $\rho$ is at the center of the Iman and Conover method deployed in the USGS (2013) study of $CO_2$ sequestration to compute predictive probability distributions of aggregates). First, both are invariant with respect to strictly monotone transformations. Second, when one variable is a non-decreasing (non-increasing) transformation of the other they equal 1 (or $-1$) at the Fréchet upper (resp. lower) bound. They note that at a value of 1.0 or $-1.0$ Kendall's $\tau$ and Spearman's $\rho$ achieve Fréchet bounds. According to them Kendall's $\tau$ and Spearman's $\rho$ are more desirable measures of association for non-normal multivariate distributions than Pearson's $r$ because the latter does not share Kendall and Spearman's correlation invariance properties. These invariance properties come into play in Iman and Conover's method discussed below. Denuit and Dehane prove the non-obvious fact that if positively or negatively quadrant dependent random couples are jointly uncorrelated they are mutually independent.

All of this emphasizes that "perfect correlation" as an omnibus argument for adding fractiles has many pitfalls. Co-monotonic bounds on random sums are a conceptually satisfactory alternative that deserves much future study.

### 5.3.2 USGS Probabilistic Assessment of $CO_2$ Storage Capacity

A recent USGS probabilistic assessment of $CO_2$ sequestration in mature petroleum reservoirs (Blondes et al. 2013a, b) is based on both micro- and macro-assessments by geologists. Their macro-assessment aggregates storage assessment units (SAUs) at basin, regional and national levels. An objective was to provide probabilistic assessments that take into account dependencies among assessment units arising from "overlap of geologic analogs, assessment methods and assessors" using individual SAU marginal probability distributions and "…a correlation matrix obtained by expert elicitation describing interdependencies between pairs of SAUs". The correlation matrix dimension is $192 \times 192$. Because a menagerie of marginal distributions—Beta-PERT, lognormal, truncated lognormal—were deployed at the micro-level use of standard multivariate distribution theory is not appropriate. Dependencies among storage capacity magnitudes are induced using an innovative distribution free method developed by Iman and Conover (1982) that allows marginal distribution shapes to be estimated from data sets distinct from data sets used to estimate dependency structure. Their method is designed to provide rank correlations that match assessed correlations and to translate the match into a predictive probability distributions for individual assessment units and larger aggregates. (See Blondes et al. 2013a for informative examples).

How to aggregate from basin, to region and then to a national scale is an issue. Should this be done in a single stage using the correlation matrix for all SAUs in the study or successively aggregate subsets of SAUs in multiple stages? Blondes et al. (2013b) conclude that

> Although the single-stage approach requires determination of significantly more correlation coefficients, it captures geologic dependencies among similar units in different basins and it is less sensitive to fluctuations in low correlation coefficients than the multiple stage approach. Thus, subsets of one single-stage correlation matrix are used to aggregate to basin, regional, and national scales.

Successive aggregation in multiple stages drastically reduces the number of pairwise correlations that must be elicited from geologists at the expense of requiring each assessor to appraise pairwise correlations of sums of assessment unit magnitudes. Although there are no studies comparing how well geologists' assessments calibrate when asked to appraise dependencies among sums of SAU magnitudes relative to appraisal of dependencies among individual SAUs it is reasonable to conjecture that individual SAU appraisals are much more likely to be well calibrated. Properties of single and multi-stage appraisal methods are studied in Kaufman et al. (2018).

### *5.3.3 Cupolas and Oil and Gas Resource Assessment*

Chen et al. (2012) emphasize that at an assessment micro-level, reservoir attributes such as porosity, permeability, pressure and temperature are often decisively dependent and that empirical data suggest dependencies are present among more aggregate assessment units in mature provinces—among fields in a mature play or basin for example. Their argument is that a basin's tectonic framework exerts "strong geographic control" over many geological features and leads to geographic and spatial dependencies and that because plays in a given basin share "...petroleum system elements, such as source rocks, regional top seal, migration fairways, timing, regional tectonics for trap formation, and accumulation preservation factors" a probabilistic model of pools or fields in a play in a given basin should incorporate probabilistic dependencies among these attributes as well as between plays. They are the first to use copulas in this setting.

Sklar (1959) proved that, subject to mild restrictions a multivariate cumulative distribution can be mapped into a joint cumulative distribution of uniform random variables called a cupola. As with Iman and Conover's method, adoption of a cupola model allows marginal distribution shapes to be estimated from data sets distinct from those used to estimate dependency structure.

Suppose as in Sect. 5.2 above that $F_X$ is the distribution function of a random vector $\mathbf{X}=(X_1,\ldots,X_n)^t$ with domain $\mathbf{R}^n$ and marginal cumulative distributions $F_i, i=1,\ldots,n$. Let $\mathbf{U}_n=(U_1,\ldots,U_n)$ be a vector of independent uniform $(0,1)$ random variables and $\mathbf{u}_n=(u_1,\ldots,u_n)$ be a realization of $\mathbf{U}_n$. Then with $u_i=F_i(x_i), i=1,\ldots n$ $Prob\{X_1\le x_1,\ldots,X_n\le x_n\}=Prob\{U_1\le u_1,\ldots,U_n\le u_n\}$.

**Definition** $C(u_1,\ldots,u_n)=Prob\{U_1\le u_1,\ldots,U_n\le u_n)\}$ is the cupola of $F_X$.

Set $dF_i=f_i$, $i=1,\ldots,n$ and $dC(u_1,\ldots,u_n)=c(u_1,\ldots,u_n)du_1\ldots du_n$. The joint density of $\mathbf{X}$ can be written as $c(u_1,\ldots,u_n)\times f_1(x_1)\times\ldots\times f_n(x_n)$. The term $c$ in the joint density captures the dependency structure of elements of $\mathbf{X}$. Because $Prob\{X_1\le x_1,\ldots,X_n\le x_n\}=Prob\{U_1\le u_1,\ldots,U_n\le u_n\}$ a procedure for generating samples from $C$ produces samples of $\mathbf{X}$ by inversion of $u_i=F_i(x_i), i=1,\ldots n$.

Computation requires choice of a cupola functional form. Among a variety of choices Chen et al. chose the bivariate normal cupola, a popular choice closely tied to standard multivariate normal distribution theory.

Their regional resource assessment of the Canadian Arctic's Beaufort-McKenzie Basin is based on analysis of 48 "significant" oil and gas discoveries containing 53 distinct accumulations. Empirical data is sufficiently detailed to allow study and estimation of pairwise correlations among reservoir attributes—area, porosity, oil saturation, net pay—for plays in the three major petroleum systems. The authors treat geologic risk factors as probabilistically independent because the data is not sufficient to allow empirical estimation of them and restrict their study of dependencies to reservoir volume attributes within each play and through them to the impact of probabilistic dependencies on the distribution of total resource volumes.

Four plays, Ivik, Taglu, Kugmallit (East) and Kugmallit (West) are used to illustrate how to incorporate dependencies among individual play resources. Although no systematic method for eliciting geologists' judgments about between play dependencies are discussed the authors motivate their choice of a rather large correlations between plays (0.6) and perfect correlation (1.0) by noting that all four plays share the same source rock and petroleum system: "The resource richness of each play is basically a function of both the oil charge and the preservation of accumulations that are mostly controlled by common petroleum system elements… we infer that the resources in the four plays are highly correlated, although the pool size distributions among the four plays vary considerably." Pairwise correlations between area, net pay, porosity and oil saturation vary from a low of 0.20 to a high of 0.86. The authors call attention to the substantial difference between total ultimate oil resource medians under the assumption of independence and under the assumption of within and between play correlations: the latter is 1.6 times the former.

Principal messages are that to be realistic, probabilistic appraisal of oil and gas resources in unexplored and partially explored regions must account for multiple sources of dependencies and that cupolas are useful for doing so.

## 5.4 Concluding Remarks

In the absence of empirical data that allows resolution of the vexing problem of how to address probabilistic dependencies among and between elements of large sets of geologic random variables we need methods that refocus and streamline expert geological judgment inputs as well as analytical methods for modeling dependencies that go beyond pairwise correlation and its cousins. One promising avenue is the theory of vines proposed by Bradford and (2002). Their theory broadens the range of allowable dependency structures beyond Bayesian belief networks and exploits properties of rank correlations in a fashion that leads to efficient computation.

## References

Blondes MS, Schuenemeyer JH, Olea RA, Drew LJ (2013a) Aggregation of carbon dioxide sequestration storage assessment units. Stoch Environ Res Risk Assess 27:1839–1859

Blondes MS, Scheunemeyer JH, Drew LJ, Warwick PD (2013b) Probabilistic aggregation of individual assessment units in the U.S. Geological Survey national $CO_2$ sequestration Assessment. Energy Procedia 37:5110–5117

Chen Z, Osadetz KG, Dixon J, Deitrich J (2012) Using copulas to implement variable dependencies in petroleum resource assessment: example from Beaufort-Mackenzie Basin. Can AAPG Bull 96(30):439–457

Deelstra G, Jan Dhaene J, Vanmaelez M (2009) An overview of comonotonicity and its applications in finance and insurance, working paper

Denuit and Dehaene (2003) Simple characterizations of co-monotonicity and counter-monotonicity by extremal correlations, Univ of Louvain working paper

Fréchet M (1951) Sur les tableaux de correlation dont les marges sont donnés. Ann Univ Lyon Sect A, Ser 3, 14:53–77

Goovaerts et al (2000) Research report 9914 U. of Leuven

Hoeffding W (1940) Masstabinvariante Korrelationstheorie. In: Schriften des mathematischen Instituts und des Instituts fur angewandte Mathematik des l.hii-versitat Berlin 5, pp 179–233

Iman RL, Conover WJ (1982) A distribution-free approach to inducing rank correlation among input variables. Commun Stat—Simul Comput 11(3):311–334

Kaufman GM, Olea RA, Blondes MS, Faith R (2018) Probabilistic aggregation of geologic resources. Forthcoming in Math Geosci

Klett TR (2004) Future petroleum supply: exploration or development? 2nd U.S. Geological Survey Conference on Reserve Growth Denver Federal Center, Lakewood, Colorado, May 6, 2004

Klett TR, Charpentier RR, Schmoker JW (2000) Assessment operational procedures: U.S. Geological Survey Digital Data Series 60, http://energy.cr.usgs.gov/WEcont/chaps/OP.pdf. Accessed 1 Aug 2011

Klett TR, Gautier DL, Ahlbrandt TS (2005) An evaluation of the U.S. Geological survey world petroleum assessment 2000. AAPG Bulletin 89(8):1033–1042

Lillestøl J, Sinding-Larsen R (2017) Creaming and the likelihood of discovering additional giant petroleum fields. Math Geosci 49(1):67–83

Lux T, Papapantoleon A (2017) Improved Fréchet-Hoeffding Bounds on dcopulas and applications in model-free finance. Ann Appl Probab 27(6):3633–3671

McNeil AJ, Frey R, Embrechts P (2005) Quantitative risk management. concepts, techniques and tools. Princeton series in finance. Princeton University Press, Princeton, NJ. ISBN 0-691-12255-5, MR 2175089, Zbl 1089.91037

Schechtman E, Yitzhaki S (1999) On the proper bounds of the Gini correlation. Econ Lett 63 (2):133–138

Sklar M (1959) Fonctions de répartition à n dimensions et leur marges. Publ Inst Statist Univ Paris 8:229–231

Shih WJ, Huang WM (1992) Evaluating correlation with proper bounds. Biometrics.1207–1213

# Chapter 6
# A Statistical Analysis of the Jacobian in Retrievals of Satellite Data

**Noel Cressie**

**Abstract** Remote sensing has become an essential component of the geosciences (the study of Earth and its system components). Remote sensing measurements are almost always energies measured in selected parts of the electro-magnetic spectrum. That is, the geophysical variable of interest is only observed indirectly; a forward model relates the energies to the variable(s) of interest and other elements of the state. The first derivative of that forward model with respect to the state is known as the Jacobian. In this chapter, we review the importance of the Jacobian to inferring the state, and we use it to diagnose which state elements may be difficult to estimate. We develop the Statistical Significance Filter and flag those state elements that consistently fail to get through the filter.

## 6.1 Introduction

Remote sensing of the environment is a fundamentally important part of humans' quest to understand the Earth system and how the different components interact (e.g., climate, water, carbon). In the future, this knowledge may be critical to our survival. Satellite and aircraft campaigns allow a "bird's-eye view" of large parts of Earth, but not all campaigns are alike. For example, polar-orbiting satellites allow global coverage, passive instruments rely on the sun's reflected light and do not take measurements when there are clouds or when it is night, and programs such as NASA's ASCENDS will measure day or night, anywhere on the orbit track.

In this chapter, a passive instrument on a polar-orbiting satellite, namely Japan's Greenhouse Gases Observing Satellite (GOSAT), will be used as a leading example. However, the idea behind what I shall present is general and could apply to many remote sensing inversion problems involving a non-linear forward model. In such

N. Cressie (✉)
Distinguished Professor, National Institute for Applied Statistics
Research Australia (NIASRA), School of Mathematics and Applied Statistics,
University of Wollongong, Wollongong, Australia
e-mail: ncressie@uow.edu.au



B. S. Daya Sagar et al. (eds.), *Handbook of Mathematical Geosciences*,
https://doi.org/10.1007/978-3-319-78999-6_6

problems, the goal is to infer a hidden state from energies detected by an instrument sensitive to certain known bands of the electro-magnetic spectrum.

Section 6.2 of this chapter gives a statistical framework behind the problem of uncertainty quantification of retrieved states. Section 6.3 calls out the Jacobian matrix as an important component of the retrieval algorithm and defines a unit-free Jacobian for subsequent statistical analysis. That analysis is described in Sect. 6.4, where a Statistical Significance Filter is defined. In Sect. 6.5, this methodology is applied to a number of retrievals taken over Australia, where certain state elements are flagged as being potentially difficult to estimate. The last section, Sect. 6.6, finishes with a discussion of the results obtained.

## 6.2 A Statistical Framework for Satellite Retrievals

The biases, variances, and mean squared prediction errors of retrievals need to be calculated in the general setting of a nonlinear forward model. The book by Rodgers (2000) has a section on error analysis, but it approaches the problem mostly from a numerical-sensitivity viewpoint. The strongly statistical viewpoint given here calculates the first two moments of a retrieval and the distribution of elements of the associated Jacobian matrix (defined below as $\mathbf{K}$). In the case where relationships are non-linear, the well known "delta method" (based on Taylor-series expansions; e.g., Meyer 1975, Chap. 10) gives *approximate* (to leading orders) biases and mean squared prediction errors of the estimators (Cressie and Wang 2013).

The $n_\varepsilon$-dimensional radiances $\mathbf{Y}$ are related to the $n_\alpha$-dimensional state $\mathbf{X}$ through a non-linear forward model,

$$\mathbf{Y} = \mathbf{F}(\mathbf{X}) + \boldsymbol{\varepsilon}, \tag{6.1}$$

where the state vector $\mathbf{X}$ includes volume mixing ratios of $CO_2$ at prespecified geopotential heights, the error vector $\boldsymbol{\varepsilon} \sim \text{Gau}(\mathbf{0}, \mathbf{S}_\varepsilon)$, and $\mathbf{X}$ and $\boldsymbol{\varepsilon}$ are statistically independent. Further, there is an *a priori* assumption that

$$\mathbf{X} = \mathbf{X}_\alpha + \boldsymbol{\alpha}, \tag{6.2}$$

where $\boldsymbol{\alpha} \sim \text{Gau}(\mathbf{0}, \mathbf{S}_\alpha)$. Notice that if there is consistent bias present in the retrieval, this can be accounted for by adding it to $\mathbf{X}_\alpha$, leaving the assumption, $\boldsymbol{\alpha} \sim \text{Gau}(\mathbf{0}, \mathbf{S}_\alpha)$, intact. Define the matrices,

$$\mathbf{K}(\mathbf{x}) \equiv \frac{\partial \mathbf{F}(\mathbf{x})}{\partial \mathbf{x}} \equiv \left( \frac{\partial F_i(\mathbf{x})}{\partial x_j} : i = 1, \dots, n_\varepsilon; j = 1, \dots, n_\alpha \right) \tag{6.3}$$

$$\mathbf{G}(\mathbf{x}) \equiv \{\mathbf{S}_\alpha^{-1} + \mathbf{K}(\mathbf{x})'\mathbf{S}_\varepsilon^{-1}\mathbf{K}(\mathbf{x})\}^{-1}\mathbf{K}(\mathbf{x})'\mathbf{S}_\varepsilon^{-1} \tag{6.4}$$

$$\mathbf{A}(\mathbf{x}) \equiv \mathbf{G}(\mathbf{x})\mathbf{K}(\mathbf{x}), \tag{6.5}$$

where $\mathbf{x}$ is any atmospheric state. (Recall that the true state is denoted as $\mathbf{X}$.)

The $n_\varepsilon \times n_\alpha$ matrix $\mathbf{K}(\cdot)$ is called the *Jacobian*. Partial derivatives of $\mathbf{K}(\cdot)$ represent the degree of non-linearity in the forward model. In the case of a *linear* forward model, $\mathbf{K}$ is constant, and any partial derivatives of it are zero.

An estimate of $\mathbf{X}$, sometimes called a *retrieval*, is often obtained by choosing an $\hat{\mathbf{X}}$ that allows $\mathbf{F}(\hat{\mathbf{X}})$ to be "close to" $\mathbf{Y}$, subject to smoothness conditions on $\hat{\mathbf{X}}$. This regularisation is usually defined as follows: Minimise

$$(\mathbf{Y} - \mathbf{F}(\mathbf{X}))'\mathbf{S}_\varepsilon^{-1}(\mathbf{Y} - \mathbf{F}(\mathbf{X})) + (\mathbf{X} - \mathbf{X}_\alpha)'\mathbf{S}_\alpha^{-1}(\mathbf{X} - \mathbf{X}_\alpha) \tag{6.6}$$

with respect to $\mathbf{X}$, which results in the retrieval $\hat{\mathbf{X}}$.

The $n_\alpha \times n_\varepsilon$ matrix $\mathbf{G}(\cdot)$ represents a type of *"gain" matrix* in the relationship between retrieval $\hat{\mathbf{X}}$ and data $\mathbf{Y}$; that is,

$$\hat{\mathbf{X}} = \mathbf{X}_\alpha + \mathbf{G}(\hat{\mathbf{X}})(\mathbf{Y} - \mathbf{F}(\mathbf{X}_\alpha) - \mathbf{K}(\hat{\mathbf{X}})\mathbf{X}_\alpha) + \text{"remainder"}.$$

In the linear case, $\mathbf{G}$ is constant and the "remainder" term is zero.

The $n_\alpha \times n_\alpha$ matrix $\mathbf{A}(\cdot)$ yields the *averaging kernel matrix* in the relation between retrieval and true state; that is,

$$\hat{\mathbf{X}} = \mathbf{X}_\alpha + \mathbf{A}(\hat{\mathbf{X}})(\mathbf{X} - \mathbf{X}_\alpha) + \text{"remainder"}.$$

In the linear case, $\mathbf{A}$ is constant, the "remainder" term is $\mathbf{G}\boldsymbol{\varepsilon}$, and recall that $\boldsymbol{\varepsilon}$ is independent of $\mathbf{X}$.

In this section, I discuss the bias vector and the mean-squared-prediction-error (MSPE) matrix of the retrieval, $\hat{\mathbf{X}}$. The bias vector is defined as:

$$E(\hat{\mathbf{X}} - \mathbf{X}) = E(\hat{\mathbf{X}}) - E(\mathbf{X}) = E(\hat{\mathbf{X}}) - \mathbf{X}_\alpha,$$

where recall that $\mathbf{X}_\alpha$ is the prior mean of the state vector $\mathbf{X}$.

The MSPE matrix is defined as:

$$E((\hat{\mathbf{X}} - \mathbf{X})(\hat{\mathbf{X}} - \mathbf{X})') = \text{var}(\hat{\mathbf{X}} - \mathbf{X}) + (E(\hat{\mathbf{X}}) - \mathbf{X}_\alpha)(E(\hat{\mathbf{X}}) - \mathbf{X}_\alpha)',$$

where $\text{var}(\hat{\mathbf{X}} - \mathbf{X})$ is the covariance matrix of the retrieval error, $\hat{\mathbf{X}} - \mathbf{X}$. The MSPE matrix can be a more appropriate statistical measure of uncertainty than the covariance matrix of retrieval error when there is bias present. When the bias is zero, the two measures of uncertainty are the same.

When the forward model is *linear*, it is easily seen (e.g., Rodgers 2000) that the bias vector,

$$E(\hat{\mathbf{X}} - \mathbf{X}) = \mathbf{0}. \tag{6.7}$$

That is, in the linear case, $\hat{\mathbf{X}}$ is *unbiased*. Further, in the linear case, the MSPE matrix can be derived *exactly* and written in a number of equivalent ways. From Connor et al. (2008), Cressie and Wang (2013),

$$E((\hat{\mathbf{X}} - \mathbf{X})(\hat{\mathbf{X}} - \mathbf{X})') = E(\mathrm{var}(\mathbf{X}|\mathbf{Y})) \equiv \hat{\mathbf{S}}\,, \tag{6.8}$$

where the MSPE matrix is given by

$$\hat{\mathbf{S}} = \{\mathbf{S}_\alpha^{-1} + \mathbf{K}'\mathbf{S}_\varepsilon^{-1}\mathbf{K}\}^{-1} = (\mathbf{A} - \mathbf{I})\mathbf{S}_\alpha(\mathbf{A} - \mathbf{I})' + \mathbf{G}\mathbf{S}_\varepsilon\mathbf{G}'\,. \tag{6.9}$$

When the forward model is *nonlinear*, the bias of $\hat{\mathbf{X}}$ is *nonzero*, and the equalities in (6.9) are no longer true. However, from the "delta method," Cressie et al. (2016) show that (6.7) and (6.9) hold, *to leading order*. In what follows, a leading-order analysis is carried out. This amounts to assuming the forward model to be locally linear, which is a weaker assumption than assuming global linearity, namely $\mathbf{Y} = \mathbf{c} + \mathbf{K}\mathbf{X} + \varepsilon$, across the whole state space defined by all possible values of $\mathbf{X}$.

The locally linear forward model is derived using a Taylor-series expansion:

$$\begin{aligned}\mathbf{Y} &= \mathbf{F}(\mathbf{X}) + \varepsilon \\ &= \mathbf{F}(\mathbf{X}_0) + \left.\frac{\partial \mathbf{F}(\mathbf{x})}{\partial \mathbf{x}}\right|_{\mathbf{x}=\mathbf{X}_0} \times (\mathbf{X} - \mathbf{X}_0) + \lambda \\ &\equiv \mathbf{c}(\mathbf{X}_0) + \mathbf{K}(\mathbf{X}_0)\mathbf{X} + \lambda\,,\end{aligned}$$

where $\lambda$ models the lack of fit of the local linear model (about the linearisation point $\mathbf{x} = \mathbf{X}_0$) to $\mathbf{F}(\mathbf{X})$. The linearisation point $\mathbf{X}_0$ is often chosen to be the prior mean $\mathbf{X}_\alpha$, but I want to emphasise here that it need not be.

## 6.3 The Jacobian Matrix and its Unit-Free Version

The Jacobian matrix is the first derivative of the $n_\varepsilon$-dimensional forward function vector, $\mathbf{F}(\mathbf{x})$, with respect to the $n_\alpha$-dimensional state $\mathbf{x}$. From the definition given in (6.3), it is an $n_\varepsilon \times n_\alpha$ matrix. Write the matrix as $(K_{ij})$, and note that the units of $K_{ij}$ are radiance (energy) per unit of state-space element $j$.

Define the vectors,

$$\begin{aligned}(\sigma^2_{\varepsilon,1}, \ldots, \sigma^2_{\varepsilon,n_\varepsilon})' &\equiv \mathrm{diag}(\mathbf{S}_\varepsilon) \\ (\sigma^2_{\alpha,1}, \ldots, \sigma^2_{\alpha,n_\alpha})' &\equiv \mathrm{diag}(\mathbf{S}_\alpha)\,,\end{aligned}$$

where diag(·) is a matrix operator that extracts a vector made up of the matrix's diagonal elements. Then the *unit-free Jacobian* is defined as follows:

$$\phi_{ij} \equiv K_{ij}\sigma_{\alpha,j}/\sigma_{\varepsilon,i}\ ;\ i = 1, \ldots, n_\varepsilon, j = 1, \ldots, n_\alpha\ . \tag{6.10}$$

During the retrieval, the most difficult and time-consuming part is to minimise (6.6); for example, using a Levenberg-Marquardt algorithm requires evaluation of the Jacobian matrix at each iteration of the minimisation. Let $\hat{K}_{ij}$ be a generic Jacobian element used during the retrieval. Then define the corresponding unit-free version as,

$$\hat{\phi}_{ij} \equiv \hat{K}_{ij}\sigma_{\alpha,j}/\sigma_{\varepsilon,i}\ , \tag{6.11}$$

and denote $\hat{\mathbf{\Phi}} \equiv (\hat{\phi}_{ij})$ as the $n_\varepsilon \times n_\alpha$ *unit-free Jacobian matrix*.

For satellite retrievals, the data vector $\mathbf{Y}$ can often be partitioned as

$$\mathbf{Y} = (\mathbf{Y}_1', \ldots, \mathbf{Y}_K')'\ ,$$

where

$$\mathbf{Y}_k \equiv (Y_i\ :\ i \in \mathrm{band}_k)'\ , \tag{6.12}$$

and $\mathrm{band}_1, \ldots, \mathrm{band}_K$ are mutually exclusive index sets that represent a grouping of radiances according to which bands of the electro-magnetic spectrum they belong. For example, Japan's GOSAT and NASA's Orbiting Carbon Observatory-2 (OCO-2) instruments have $K = 3$ bands, corresponding to the oxygen A band (OA), the weak carbon dioxide band (WC), and the strong carbon dioxide band (SC); our analysis in Sect. 6.5 uses data from GOSAT's three bands. Another example is from NASA's Atmospheric Infrared Sounder (AIRS) instrument flying on the Aqua satellite, which has $K = 4$ bands, corresponding to four geophysical variables, namely temperature, water vapour, ozone, and carbon dioxide.

In what follows, we abbreviate "$\mathrm{band}_k$" to "$b_k$." Because the unit-free Jacobian has elements that are potentially comparable, we can partition it and analyse it in comparable ways. Recall that the index $j$ corresponds to a given element of the state vector, for example, a water-vapour scale factor or a near-surface carbon-dioxide volume mixing ratio. Then fix the state element $j$, and consider the behaviour of the $j$th column as row $i$ varies within individual bands. That is, for a fixed $j$, consider

$$\{\hat{\phi}_{ij}\ :\ i \in b_k\} \tag{6.13}$$

to be a random sample from a distribution indexed by $k$, for bands $k = 1, \ldots, K$.

Consequently, instead of thinking about $n_\varepsilon \cdot n_\alpha$ entries in the Jacobian, attention turns to $n_\alpha \cdot K$ *distributions*. For example, for the retrievals from GOSAT data that are being considered here, $n_\varepsilon = 2240$, $n_\alpha = 112$, and $K = 3$. Hence, the pair $(j, k)$ indexes one of 336 possible distributions, whose mean, $\mu_{jk}$, is of primary interest. For $j$ a fixed element of the state vector, if $\mu_{j1} = \mu_{j2} = \cdots = \mu_{jK} = 0$, then that element is poorly determined by the data alone; see Sect. 6.4. This is a flag that says the (prior) mean and precision of the $j$th state element need to be specified very carefully in the second term of (6.6) in order to obtain an acceptably precise retrieval $\hat{X}_j$.

## 6.4 Statistical Significance Filter

To leading order, the forward model (6.1) can be written as,

$$\mathbf{Y} = \mathbf{c} + \mathbf{K}_1 X_1 + \cdots + \mathbf{K}_{n_\alpha} X_{n_\alpha} + \boldsymbol{\varepsilon}, \tag{6.14}$$

which is a multiple-regression model with known, typically different, intercepts given by the elements of $\mathbf{c}$; known covariates $\mathbf{K}_1, \ldots, \mathbf{K}_{n_\alpha}$ (the $n_\alpha$ columns of $\mathbf{K}$); and unknown regression coefficients $X_1, \ldots, X_{n_\alpha}$. Clearly, if $\mathbf{K}_j$ is zero, then $X_j$ will not be estimable. Further, if for a given $j$, $\{|K_{ij}| : i = 1, \ldots, n_\varepsilon\}$ are uniformly "small," then the uncertainty associated with the estimate of $X_j$ will be large.

In the previous section, we noted that for remote sensing retrievals, the $n_\varepsilon$ elements in $\mathbf{Y}$ can be partitioned into $K$ bands, $\mathbf{Y}_1, \ldots, \mathbf{Y}_K$. Then write (6.14) equivalently as $K$ equations. In obvious notation that respects the partitioning,

$$\mathbf{Y}_k = \mathbf{c}_k + \mathbf{K}_{1k} X_1 + \cdots + \mathbf{K}_{n_\alpha k} X_{n_\alpha} + \boldsymbol{\varepsilon}_k \,;\; k = 1, \ldots, K, \tag{6.15}$$

where $\{\mathbf{K}_{jk} : j = 1, \ldots, n_\alpha\}$ are the $n_\alpha$ vectors corresponding to the $k$th band.

Clearly, if $\mathbf{K}_{jk} = \mathbf{0}$, then its unit-free version, $\boldsymbol{\Phi}_{jk}$, is also $\mathbf{0}$. Hence, the problem of whether $X_j$ is poorly determined in the forward model (6.1) can be addressed in a statistical manner by considering the retrieval's unit-free Jacobian entries $\{\hat{\phi}_{ij} : i = 1, \ldots, n_\varepsilon\}$ as $K$ arrays of random variables, $\{\hat{\phi}_{ij} : i \in b_k\}$, for $k = 1, \ldots, K$. If, for a fixed $j$, the means $\mu_{j1}, \ldots, \mu_{jK}$ of these $K$ arrays are all zero, then $X_j$ will be difficult to estimate.

### 6.4.1 Hypothesis Tests

Consider (6.13) and make the following assumption: For a given retrieval, a given state element $j$, and a given band $k$,

$$\{\hat{\phi}_{ij} : i \in b_k\} \overset{iid}{\sim} \mathrm{Dist}(\mu_{jk}),$$

where "iid" denotes "independent and identically distributed," and "Dist($\mu$)" denotes a probability distribution with mean $\mu$. For this retrieval, the idea is to flag those state elements and bands for which the null hypothesis, $H_{0,jk} : \mu_{jk} = 0$, is not rejected. In particular, failure to reject the composite hypothesis,

$$H_{0,j} : \mu_{j1} = \mu_{j2} = \cdots = \mu_{jK} = 0\,, \tag{6.16}$$

implies that the $j$th state element will be difficult to estimate in the given retrieval.

Since the elements of $\{\hat{\phi}_{ij} : i \in b_k\}$ are considered to be a sample from a distribution with mean $\mu_{jk}$, I shall construct a test statistic from these unit-free Jacobian values. A considerable amount of exploratory data analysis showed the common distributional assumption within the partitioned arrays to be largely correct, with occasional gross outliers that would challenge many statistical testing procedures. Those were controlled by transforming each $\hat{\phi}_{ij}$ to $|\hat{\phi}_{ij}|^{1/2}$, and the robust test statistic,

$$\tilde{\phi}_{jk} \equiv \text{med}\{|\hat{\phi}_{ij}|^{1/2} : i \in b_k\}\,, \tag{6.17}$$

was used to test $H_{0,jk} : \mu_{jk} = 0$. The composite hypothesis test $\{H_{0,j} : j = 1, \ldots, n_\alpha\}$, where $H_{0,j}$ is given by (6.16), is then carried out using a Bonferroni adjustment (Sect. 6.4.3).

### 6.4.2 *Distribution Theory for the Robust Test Statistic*

Consider generic iid random variables $W_1, \ldots, W_m$ distributed according to a Gaussian distribution with mean $\mu_W$ and variance $\sigma_W^2$, which is written as $\text{Gau}(\mu_W, \sigma_W^2)$. To test

$$H_0 : \mu_W = 0 \text{ versus } H_1 : \mu_W \neq 0\,, \tag{6.18}$$

consider the robust test statistic,

$$\tilde{X} \equiv \text{med}\{|W_i|^{1/2} : i = 1, \ldots, m\}. \tag{6.19}$$

I now obtain distribution theory for $\tilde{X}$ under the null hypothesis in order to carry out a significance test.

If $Y \sim \text{Gau}(0, 1)$, then $E(|Y|^{1/2}) = 0.82216$ and $\text{var}(|Y|^{1/2}) = 0.12192$, which was derived by Cressie and Hawkins (1980). Then under $H_0 : \mu_W = 0$, $|W_i|^{1/2} \dot{\sim} \text{Gau}(0.82216 \cdot \sigma_W^{1/2}, 0.12192 \cdot \sigma_W)$, where "$\dot{\sim}$" denotes "is approximately distributed as," and the approximation is established by Cressie and Hawkins (1980). Now the distribution of the median $\tilde{X}$ from a random sample $X_1, \ldots, X_m$ of Gaussian random variables can be approximated as Gaussian with mean $E(\tilde{X}) = E(X_1)$, and variance $\text{var}(\tilde{X}) = \pi \text{var}(X_1)/2m$. If all these results are combined, then under the null hypothesis $H_0$ in (6.18),

$$\tilde{X} \dot{\sim} \text{Gau}(0.82216 \cdot \sigma_W^{1/2}, 0.12192 \cdot \pi\sigma_W/2m)\,.$$

Clearly, the alternative hypothesis $H_1$ in (6.18) is accepted if the test statistic $\tilde{X}$ is large. At significance level $\alpha$, $H_1$ is accepted if

$$\tilde{X} > 0.82216 \cdot \sigma_W^{1/2} + \Phi^{-1}(1-\alpha)(0.12192 \cdot \pi\sigma_W/2m)^{1/2}, \tag{6.20}$$

where $\Phi^{-1}(\cdot)$ is the inverse cumulative distribution function of a Gau(0, 1) random variable. In practice, an estimate of $\sigma_W$ will be needed.

Continuing with the same approach as above, an asymptotically unbiased, robust estimator of $\sigma_W$ is used. Now, $\sigma_W = \text{var}(|W_i|^{1/2})/0.12192$, and hence $\text{var}(|W_i|^{1/2})$ can be estimated using the median absolute deviation (MAD):

$$\text{MAD} \equiv \text{med}\{||W_i|^{1/2} - \tilde{X}| : i = 1, \ldots, m\}.$$

Then an asymptotically unbiased estimator of $\text{var}(|W_i|^{1/2})$ is

$$\widehat{\text{var}}(|W_i|^{1/2}) = (1.4826 \cdot \text{MAD})^2,$$

from which the estimator

$$\tilde{\sigma}_W \equiv (1.4826 \cdot \text{MAD})^2/0.12192 \tag{6.21}$$

is obtained and substituted into (6.20).

My approach to constructing this robust statistic to test whether a mean is zero, using data that may contain large, unpredictable outliers, is somewhat unusual, but it is statistically advantageous. First, the data $\{W_1, \ldots, W_m\}$ are made resistant by transforming to the square-root scale where variability is dampened. Then the transformed data $\{|W_1|^{1/2}, \ldots, |W_m|^{1/2}\}$ are used to define a robust test statistic, given here by the median; see (6.19). Finally, the null distribution is derived, resulting in a critical region given by (6.20) with the robust estimator (6.21) substituted in. In the next subsection, the distribution theory derived in this subsection is used in the context of multiple hypothesis testing, resulting in the *Statistical Significance Filter*.

### 6.4.3 *Multiple Hypothesis Tests Define the Statistical Significance Filter*

The elements of the unit-free Jacobian are considered as replicates within bands, which results in $n_\alpha$ (number of state elements) times $K$ (number of bands) hypothesis tests of $\{H_{0jk} : \mu_{jk} = 0, \text{ for } j = 1, \ldots, n_\alpha \text{ and } k = 1, \ldots, K\}$. To test $H_{0j}$ given by (6.16), jointly for $j = 1, \ldots, n_\alpha$, I use a family-wise error rate of 1% and conservative Bonferroni adjustments to obtain a level of significance, $\alpha = .01/(n_\alpha \cdot K)$, that is used in each individual hypothesis test of the null hypotheses, $\{H_{0jk}\}$.

The *Statistical Significance Filter* only allows estimates $\{\tilde{\phi}_{jk}\}$ to get through the filter if $\{H_{0jk}\}$ are rejected, respectively. A given state element, $j$ say, is flagged as problematic in a given retrieval if, simultaneously, the hypotheses $H_{0j1}, \ldots, H_{0jK}$ are not rejected. If it consistently happens that under similar (or different) geophysical conditions, the $j$th element's bands fail to get through the Statistical Significance Filter, that element $X_j$ is flagged as being weakly sensitive to the radiance measurements $\mathbf{Y}$. Hence, estimation of $X_j$ would be difficult if a very disperse prior distribution in (6.2) were chosen for it.

In the next section, I apply the Statistical Significance Filter to 30 retrievals from Japan's GOSAT instrument that measures atmospheric carbon dioxide, here over central Australia.

## 6.5 ACOS Retrievals of the Atmospheric State from Japan's GOSAT Satellite

Shown in Fig. 6.1 are 30 locations of retrievals from Japan's GOSAT satellite, where the ACOS (Atmospheric CO2 Observations from Space) retrieval algorithm was used. Specifically, ACOS Version B2.8 was used here, for which $n_\alpha = 112$ state elements were retrieved from $n_\varepsilon = 2240$ radiances spread roughly equally between the $K = 3$ bands, namely the OA band, the WC band, and the SC band; see Sect. 6.3. The soundings are over an arid part of Australia with uniformly high albedo, during the period from 5 June 2009–26 July 2009 (Source: CIRA, Colorado State University). The methodology and inference is illustrated on the retrieval at one of those locations, hereafter referred to as Location 1. Results from the other 29 retrievals are summarised at the end of this section.

A number of the state elements in B2.8 are functions of geopotential height, here labelled as 1 (top of atmosphere) down to 20 (surface of Earth). Figure 6.2 shows unit-free ice-cloud Jacobian values in a column of the atmosphere for Location 1; only those values that got through the Statistical Significance Filter are shown. It can be seen that for the ice-cloud variable, Jacobian values in the OA band are not statistically significant at higher altitudes in the atmospheric column, and hence they are potentially difficult to estimate. Figure 6.3 shows that the Statistical Significance Filter applied to water vapour ($H_2O$) in the column results in a similar set of plots. Contrast these to Fig. 6.4, which is for the all-important carbon-dioxide ($CO_2$) variable; only values in the SC band get through the Statistical Significance Filter.

The analysis of the retrieval for Location 1 yields non-significant Jacobian entries (i.e., forward-model derivatives near zero) *in all three bands* for the following state elements:

| | |
|---|---|
| 1, 2, 3 | $CO_2$ values near the top of the atmosphere |
| 21 | $H_2O$ scale factor |
| 23 | Temperature offset |
| 105, 107, 109 | Albedo slope for the three bands |
| 110, 111, 112 | Spectral dispersion offset for the three bands |

This behavior is visualised in Fig. 6.5; there, a light (green) stripe in a given band for a given state element indicates that the corresponding mean is not significantly different from zero. A light stripe in every band for the given state element indicates that extra care will be needed when specifying a prior for that element. Each of the 11 elements listed above have a light stripe in every band.

The analysis was carried out on all 30 retrievals, and eight elements of the 112-dimensional state vector emerged as always having non-significant Jacobian values in all three bands for all 30 retrievals. They were:

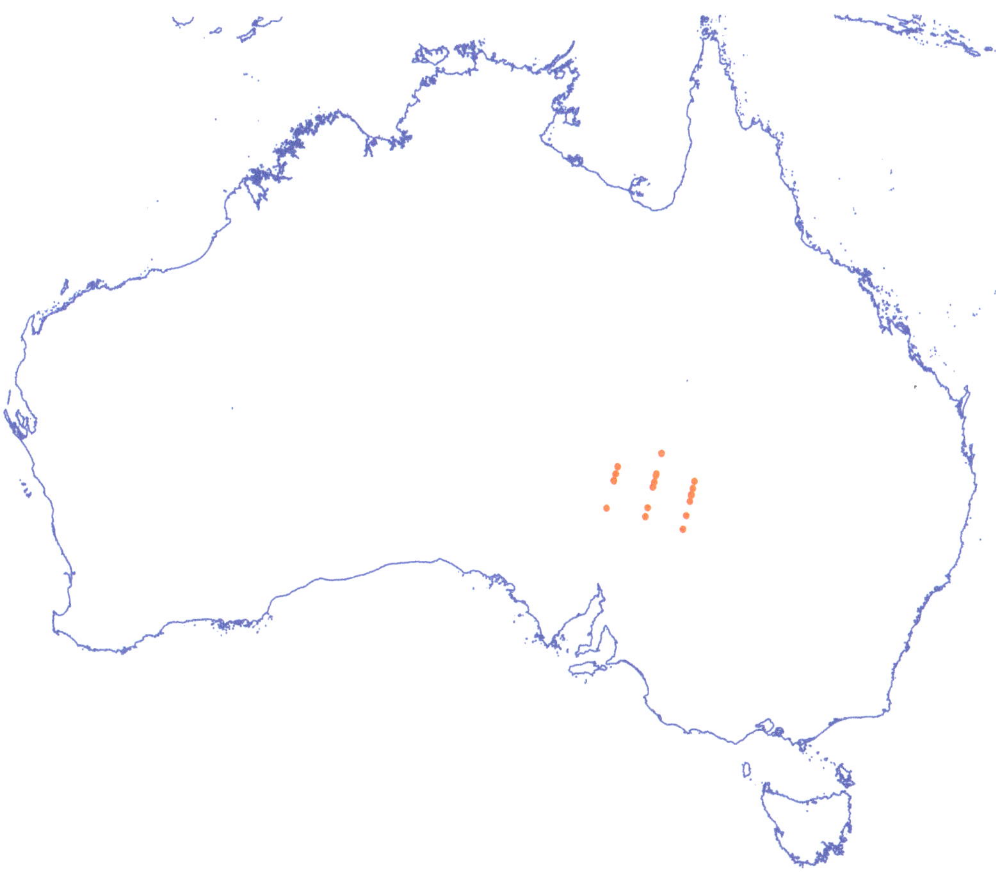

**Fig. 6.1** Locations of 30 retrievals from GOSAT using the ACOS Version B2.8 retrieval: 5 June 2009–26 July 2009

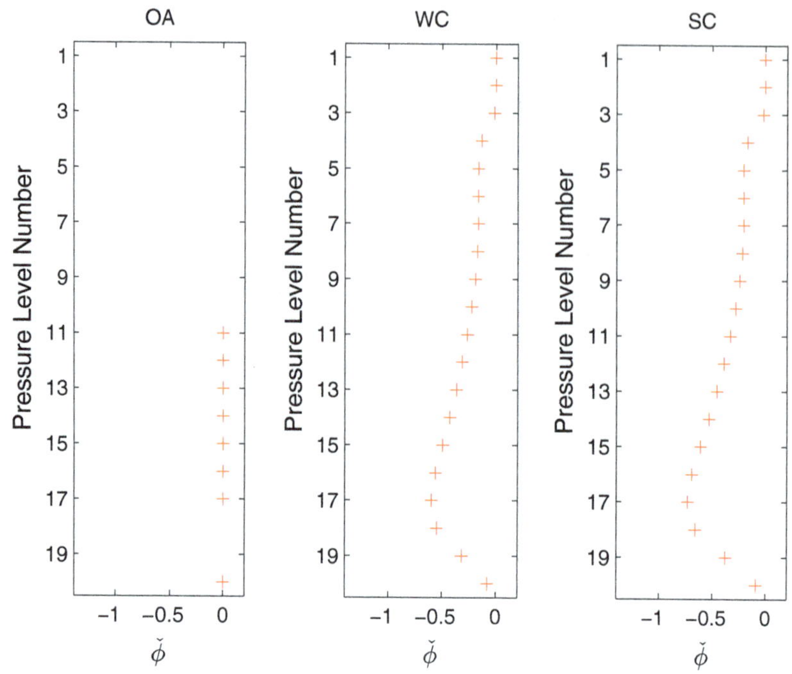


**Fig. 6.2** Unit-free Jacobian ice-cloud values that pass through the statistical significance filter in the OA, WC, and SC bands. Values that did not pass through the filter are not plotted. Location 1 (out of 30 locations)

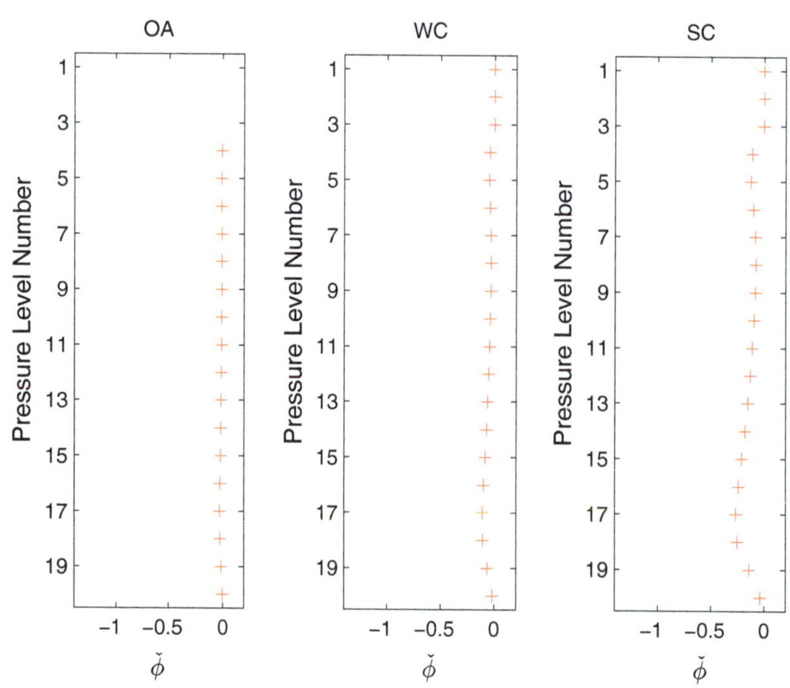


**Fig. 6.3** Unit-free Jacobian $H_2O$ values that pass through the statistical significance filter in the OA, WC, and SC bands. Values that did not pass through the filter are not plotted. Location 1 (out of 30 locations)

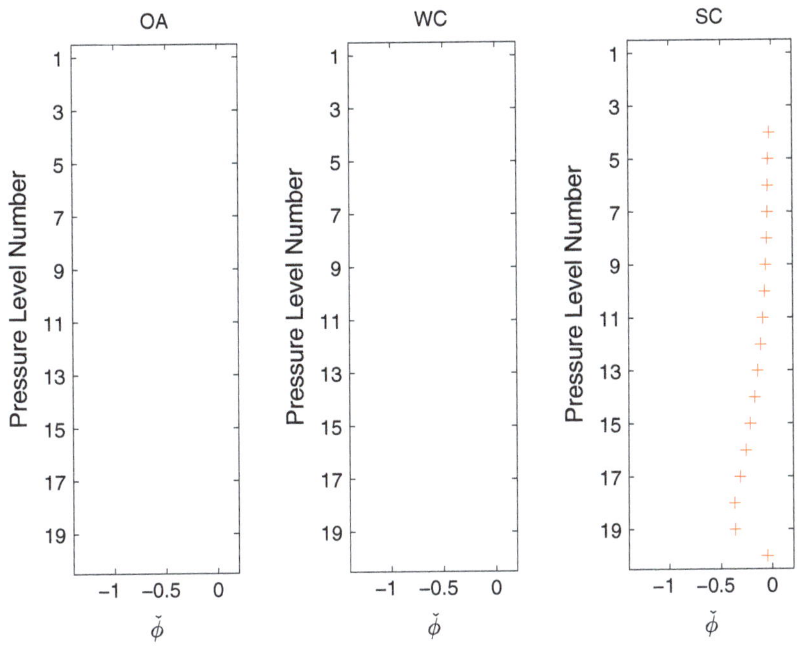


**Fig. 6.4** Unit-free Jacobian $CO_2$ values that pass through the statistical significance filter in the OA, WC, and SC bands. Values that did not through pass the filter are not plotted. Location 1 (out of 30 locations)

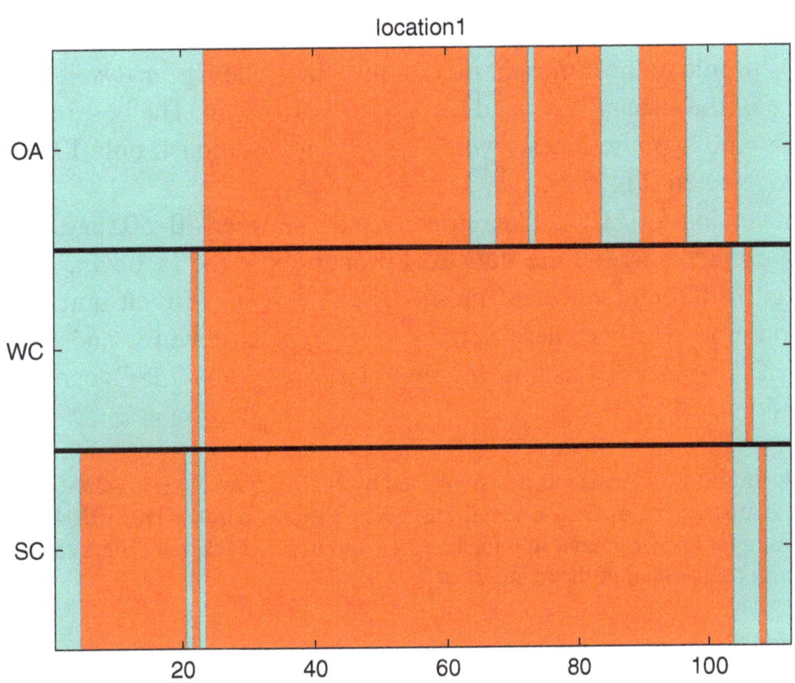


**Fig. 6.5** A graphic showing which of the 112 elements of the state vector (horizontal axis) pass through the statistical significance filter (dark, red colour) and which do not (light, green colour), for "band" = OA, WC, and SC. Location 1 (out of 30 locations)

| | |
|---|---|
| 21 | $H_2O$ scale factor |
| 23 | Temperature offset |
| 105, 107, 109 | Albedo slope for the three bands |
| 110, 111, 112 | Spectral dispersion offset for the three bands |

The results indicate a lack of sensitivity of these eight elements in the forward equation **F** given in (6.1), for the dry, bright, flat-terrain conditions found over central Australia. Different land surfaces and atmospheric states would almost certainly result in different elements being identified.

## 6.6 Discussion

The Jacobian matrix **K** is the first derivative of a vector-valued function **F**(**x**) of a state vector **x**. Consistently small elements in the *j*th column of **K** indicate that the *j*th element will be difficult to estimate (predict) based on data, **Y**, alone.

If prior information, as well as the data, is used to predict the state vector, this research indicates that acceptable precision for estimating this *j*th element may require the prior variance to be tightly constrained. For example, the element that is the $H_2O$ scale factor is tightly constrained physically in the prior. Thus, a retrieval of that element may cause no problem, even though its column in **K** fails to get through the Statistical Significance Filter. Regarding the 20 $CO_2$ elements that make up the $CO_2$ profile in the atmospheric column, the retrievals analysed here show the importance of the strong $CO_2$ band (SC) to its estimation. The best result would be if all $20 \cdot 3 = 60$ hypothesis tests were rejected; at Location 1, only 17, all in the SC band, were rejected (Fig. 6.4).

Current versions of ACOS-like retrievals have between 40–50 state elements. The research presented here, on the statistical properties of the Jacobian, would allow a comparison of different versions through the behaviour of their unit-free Jacobian values. Common to all of these versions is 20 $CO_2$ elements, and the respective estimates of the means in each of the three bands (OA, WC, SC) can be compared across versions.

**Acknowledgements** This research was supported by NASA grant NNH11-ZDA001N-OCO2 and a 2015–2017 Australian Research Council Discovery Project, number DP150104576. My thanks go to Rui Wang for his early input into the research and to Ben Maloney for his careful and timely assistance with preparation of the manuscript.

## References

Connor BJ, Boesch H, Toon G, Sen B, Miller C, Crisp D (2008) Orbiting Carbon Observatory: inverse method and prospective error analysis. J Geophys Res Atmos 113:D055305

Cressie N, Hawkins DM (1980) Robust estimation of the variogram: I. J Int Assoc Math Geol 12:115–125

Cressie N, Wang R (2013) Statistical properties of the state obtained by solving a nonlinear multivariate inverse problem. Appl Stoch Models Bus Ind 29:424–438

Cressie N, Wang R, Smyth M, Miller CE (2016) Statistical bias and variance for the regularized inverse problem: Application to space-based atmospheric $CO_2$ retrievals. J Geophys Res Atmos 121:5526–5537. https://doi.org/10.1002/2015JDO24353

Meyer SL (1975) Data analysis for scientists and engineers. Wiley, New York

Rodgers CD (2000) Inverse methods for atmospheric sounding. World Scientific Publishing, Singapore

# Chapter 7
# All Realizations All the Time

**Clayton V. Deutsch**

**Abstract** Geostatistical simulation of mineral deposits is becoming commonplace. The methodology and software are well established and professionals have access to the training and checking steps required for reliable application. Managing multiple realizations, however, remains daunting and unclear for many: (1) the non-uniqueness of multiple realizations is disturbing; (2) many calculations including mine planning algorithms are aimed at a single block model; and (3) there are concerns of excessive computational requirements. The correct approach to managing multiple realizations is reviewed: consider all realizations all the time and base decisions on the appropriate expected value. The principles of simulation and decision making are reviewed for resource management.

## 7.1 Introduction

In the context of modern geostatistics, Monte Carlo Simulation (MCS) or simply simulation can be summarized by (1) the formulation of a problem with input variables, a transfer function and response variables, (2) the simulation of realizations of the input variables, (3) the application of the transfer function to compute the response variables of interest, and (4) the assembly of the simulated response variables into a probability distribution. The distribution of response variables can be used to understand uncertainty and, perhaps, for decision making.

The input variables could be the rock type and grade on a suitable grid, the transfer function could be the calculation of resources and the response variables could be the resources or reserves expressed as tonnages, grade and quantity of metal. A comprehensive simulation study could expand the input variables to include modeling parameters, price, costs and other economic and engineering parameters. The transfer function could be a model of the entire mine planning and economic forecasting process. The response variables could be key performance

C. V. Deutsch (✉)
University of Alberta, Edmonton, Canada
e-mail: cdeutsch@ualberta.ca



B. S. Daya Sagar et al. (eds.), *Handbook of Mathematical Geosciences*,
https://doi.org/10.1007/978-3-319-78999-6_7

indicators such as net present value. The probability distributions of the input variables must be established prior to simulation; typically by a mathematical model such as the multivariate Gaussian model. The transfer function must be known to process realizations of the input variables to response variables of interest.

The key operation in simulation is the drawing of realizations from a specified probability distribution. This is done in a fair manner for unbiased results. Pseudorandom number generators generate numbers that have properties very close to random numbers, but are indexed to a seed. These numbers are uniform between 0 and 1, yet our input distributions are rarely uniform so the corresponding quantile is drawn from the distribution we are simulating from, $z = F^{-1}(r)$ where $F(\bullet)$ is the cumulative distribution, $z$ is the simulated value and $r$ is the random number.

Consider a simple example of three dice. The input variables are the three numbers showing on the faces of three fair cubic dice. The cumulative distribution of each input variable has six equal steps. The transfer function is the summation operator. The response is the sum. As illustrated on Fig. 7.1, one realization is generated by three random numbers, e.g., 0.69, 0.062 and 0.78 leading to a realization of 5, 1 and 5. Simulation is repeated for multiple realizations. The response distribution shown is the result for 100 realizations. There are many points that could be reinforced from this small example. The distribution of the input variables must be known prior to simulation; simulation is primarily transferring input variable uncertainty through a transfer function to response variable uncertainty. The space of uncertainty in this example is only $6^3 = 216$, which is a very small number, but the space of uncertainty is practically infinite in geological modeling where there are many variables at many locations. Categorical variables require an arbitrary ordering. Finally, it would be wrong to focus on one realization; in this example we should not conclude that the first and third dice are likely high numbers and the second die is a low number. We only understand the result of simulation by considering an ensemble of realizations. This is a critical point.

Although many theoreticians and practitioners understand this point, it is not emphasized enough. Most software is aimed at processing one block model at a time. Resources are often presented as a single value instead of a distribution.

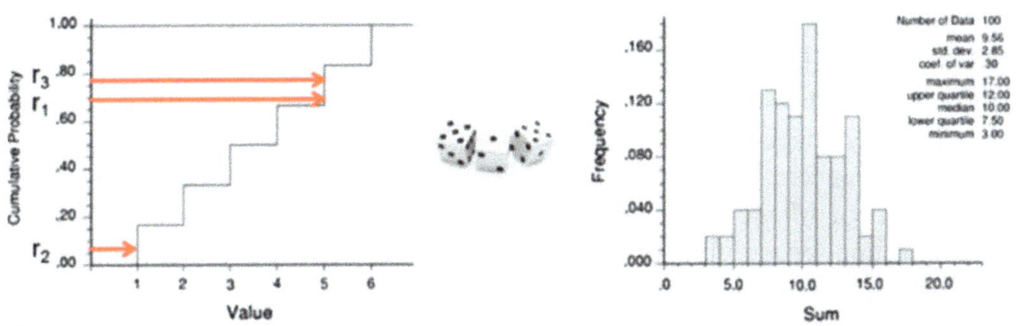


**Fig. 7.1** Simulation of the outcome on three dice (left). Histogram of the sum of the outcomes on three dice (right). One hundred realizations are shown

There are many examples of experimental mathematics in history. The scientists on the Manhattan Project are credited with the formulation and popularization of Monte Carlo simulation (MCS) or simply simulation. There are interesting historical references and internet resources. The framework of transferring input uncertainty to response variable uncertainty is often referred to as simulation. Adjectives such as Monte Carlo, stochastic or conditional are sometimes added. The outcomes of simulating are called simulations or realizations.

The pioneers of simulation suspected where we would take the method. The closing paragraph in Hammersley and Handscomb (1964) is telling: *Usually there are many nodes and possible paths, so many that a complete enumeration of the situation is impossible. This suggests a fruitful field for sampling and search procedures, but as yet little Monte Carlo work has been done here. There are challenging problems here for research into Monte Carlo techniques on multi-variable problems.* They knew we had to sample a reasonable set of realizations from the practically infinite space of uncertainty. They knew we would be challenged by multiple dependent variables. They did not know that 50 years later many practitioners would still struggle managing an ensemble of realizations.

This chapter is organized into five main sections supporting a case to use *all realizations all the time*. First, some principles of simulation are presented to set the context. Second, principles of decision making in presence of uncertainty are discussed to establish that earth scientists are not alone. Thirdly, some details of geostatistical simulation are presented to highlight important differences from simulation of independent variables. Fourthly, some details of resource decision making are presented to highlight important differences from the general principles including the information effect. Finally, some possible alternatives to using all realizations all the time are reviewed. A case is made to consider the correct approach, that is, consider all realizations all the time and base decisions on the appropriate expected value when required.

## 7.2 Simulation

In the early days of simulation there was a particular concern related to the pseudorandom numbers applied in the simulation. A large part of early texts on simulation is devoted to the generation of pseudorandom numbers. This concern has largely been addressed and there is little practical concern with the pseudorandom number generators used in most software.

Another concern is in replacing the reality with a numerical model. Many early applications of Monte Carlo simulation were directed at solving integration and other equations where the transfer function is a very close representation of the physical situation. Examples of well represented physical systems are the study of radiation shielding and reactor criticality. The simulation tracks simulated particles through collisions where the particles are absorbed, scattered or split according to physical principles. There were few concerns about this simulation due to the close correspondence between the numerical setup and the physical reality. Increasingly, complex non-linear systems are modeled with empirical statistical models causing more concern.

It is impossible to model the details of the natural geological processes that led to the deposit under study. Empirical statistical models are required. Geostatistical models do not represent the original depositional and diagenetic processes. Although all models are wrong (Box and Draper 1987) they can be useful if assembled carefully with established workflows and appropriate checking.

The premise of simulation is to construct many realizations that are equally likely to be drawn. Realizations and responses more probable than others will be drawn more often. A fundamental principle of simulation is to consider many realizations. One hundred realizations may not be enough. The average of the one hundred realizations on Fig. 7.1 was 9.6, yet the true expected response for that particular process is 10.5. This suggests that the number of realizations should be quite large. Indeed, early practitioners of simulation considered that thousands of realizations were required unless some form of stratified or directed sampling could be implemented. Of course, the problems considered early on were small compared to the complexity of modern geological modeling where 10s of variables at 10s of millions of locations are considered. In many cases, the professional and computational effort of generating more than 100s of realizations would be better spent improving the model. This claim is supported by two observations: (1) the variability at multiple locations partially cancels out, and (2) there is too much uncertainty in the model to expend resources on thousands of realizations.

Another fundamental principle of simulation is that all realizations are considered in downstream calculations. One application is to pass all realizations through the transfer function to construct a distribution of responses, for example, resource estimates. The realizations could be passed through a decision tree structure to help support a decision. Finally, the realizations could all be used in the optimization of decision variables. Incorrect or suboptimal decisions could be taken if too few realizations are considered.

The concept or ranking and choosing a few realizations is motivated by the large computational cost running realizations through a complex full physics transfer function. The processing the realizations through a simplified transfer function could rank the realizations and permit choosing a smaller number for the complex full physics transfer function. Decision making and optimization applied with one or a few realizations leads to over fitting to those realizations.

In some cases, the transfer function and decision variables are known. For example, calculating the recoverable reserves above a specified economic cutoff. In other cases, aspects of the decision must be optimized. For example, deciding the ultimate pit limits, choosing drill hole locations or deciding on the destination of mined material. If the transfer function and decision variables are known, then a probability distribution of each critical response variable is assembled from the realizations where the result of each realization is equally weighted. This distribution provides a direct understanding of uncertainty. There are many ways of summarizing the uncertainty. Considering the 0.1, 0.5 and 0.9 quantiles is common in petroleum applications, but considering the probability to be within 15% of expected is a reasonable measure of uncertainty.

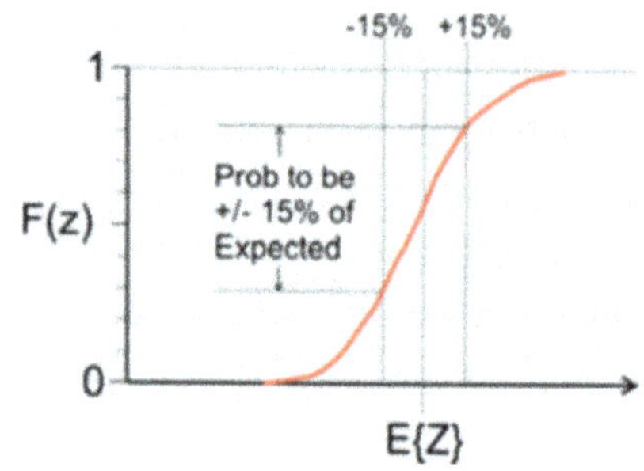


If aspects of the decision are not finalized, then decision making and optimization must be considered before calculating a distribution of the critical response variables.

## 7.3 Decision Making

Decision making in presence of uncertainty has long been studied (Bernoulli 1954; Kochenderfer 2015). The general framework of decision making could be summarized by (1) define clearly stated objectives within a value system, that is, a measure of utility (often profit), (2) enumerate the alternative decisions that could be taken—perhaps in a decision tree, (3) compute the expected utility for all alternatives, and (4) choose the alternative that maximizes expected utility. This framework becomes confounded with large one-time decisions or significant unknown unknowns that defy straightforward quantification. Grade control and mine planning decisions are made repeatedly within a clear economic framework.

Consider a recently loaded truck. The expected profit of the material if the truck goes to the mill would be computed by the average over all realizations, say $6.75 per tonne. The expected profit if the material goes to the waste dump is the average of a similar calculation over all realizations, say −$2.00 per tonne. With no other information, the truck should be sent to the mill. There are complicating factors including sequencing, stockpiling, limited milling capacity, but the principle stands. Decisions should be based on expected values as late as possible.

Decisions are based on the average over all realizations and not on one particular realization. The realizations are simply a means to represent uncertainty. One realization should not be chosen for decision making because that would mean ignoring other equally likely possibilities; the expectation is the only way to resolve the ambiguity of multiple realizations. The decision is also made as late as possible. Calling a block of material in a long term resource model ore may be convenient as an interim decision for planning purposes, but this decision would certainly be revisited with production sampling at the time of grade control.

There is another aspect to taking the expected value as late as possible. The expected value is calculated with the last numbers considered: utility or profit. The expected value should not be taken earlier. The correct decision would not always be found if the grades were averaged and the decision based on the utility computed from the expected grade. Many calculations are non-linear and the utility computed on the average of realizations is not the average (expected) utility computed on the realizations.

The distributions of payoff/utility for each possible decision are evaluated to determine the best decision. Some decisions may be completely dominated by others, that is, the best possible payoff of a dominated decision is less than the worst payoff of an alternative. All dominated decisions should be rejected. Some decisions are stochastically dominated by others (Levy 2016). That is, each quantile on the payoff distribution is less than the same quantile on an alternative. Decision makers should also reject all stochastically dominated decisions. The expected utility would be considered when multiple decisions remain to establish the optimal one.

A challenge in many geological resource application problems is that the decision involves many different options. The precise sequence of extraction or the position of all production wells is combinatorial and all options cannot be considered. Optimization algorithms are implemented where the objective function is the appropriate expected value of profit or utility over all realizations. The distribution of uncertainty in utility is only known once optimization is complete.

The utility function quantifies our position on risk; however, it is not simple to establish the utility function in practice. One approach based on the idea of the efficient frontier could be considered (Francis, and Dongcheol 2013; Hanoch and Levy 1969). Decisions are optimized based on maximum expected profit and minimum risk. The ones that are not dominated are retained as the efficient frontier. Judgement could be used to evaluate the differences between these decisions and to choose a path forward.

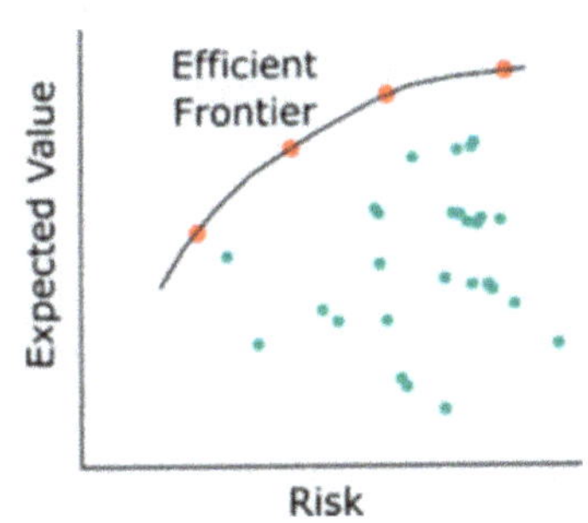


## 7.4 Geostatistical Simulation

The simulation of mineral deposits has evolved significantly over the last twenty years. The simulation is often hierarchical and multivariate with unequally sampled data and parameter uncertainty. A variety of techniques are used to create realizations that reproduce all available data and represent the variability that may influence the planning and decision making process (Caers 2011; Chilès and Delfiner 2012).

The scope of this chapter is not to present details of geostatistical simulation (Deutsch and Journel 1998; Goovaerts 1997). The main steps in managing the results will be reviewed. The transfer functions of greatest interest are resources and reserves within reasonably large volumes, uncertainty versus data spacing, uncertainty and variability in mine planning and sometimes optimization of blending and other engineering designs. Parameter uncertainty is important for the resources within large volumes. Data uncertainty is important with unequally sampled variables (common with geometallurgical and geomechanical variables). The steps in geostatistical simulation could be divided into five unit operations.

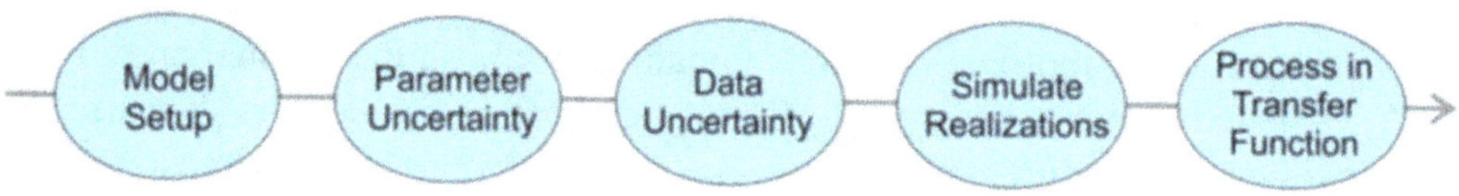


- **Model Setup** involves the formulation of the modeling workflow. A hierarchical modeling of the deposit limits, rock types and multiple grades is specified. The grid node spacing relative to production volumes of interest must be chosen. The model setup defines the software algorithms and input parameters for each step. The number of realizations is chosen at the start to ensure that there is one realization of parameters for each realization of data for each realization of the deposit. 100 or 200 realizations is common.
- **Parameter Uncertainty** amounts to simulating realizations of all of the modeling parameters identified in the Model Setup including those for gross volume uncertainty, rock type proportion uncertainty, histogram uncertainty, variogram uncertainty and multivariate relationship uncertainty. The multivariate spatial bootstrap is widely used. The uncertainty in some global parameters may be specified by experience.
- **Data Uncertainty** involves two main aspects. The first is sampling realizations of the available data if the uncertainty in the data is considered important. For example, there may be a 10% relative error in the data based on the data collection and processing. A spatial bootstrap could be considered to get the uncertainty in the mean error, then realizations of the data would be assembled. The second aspect, if required, is to fill missing data (data imputation) and downscale data with a larger support than the rest of the data (Barnett and Deutsch 2015).
- **Simulate Realizations** is the operation where deposit models of all variables are assembled. These would follow the steps identified in the Model Setup. There would be one realization for every data and parameter realization. These are constructed hierarchically and with the correct dependencies by rock-type and between all variables. These realizations have to be checked to the greatest

extent possible. The process of conditioning the realizations will update the prior uncertainty quantified in the second step. A schematic illustration of the realizations is shown below.

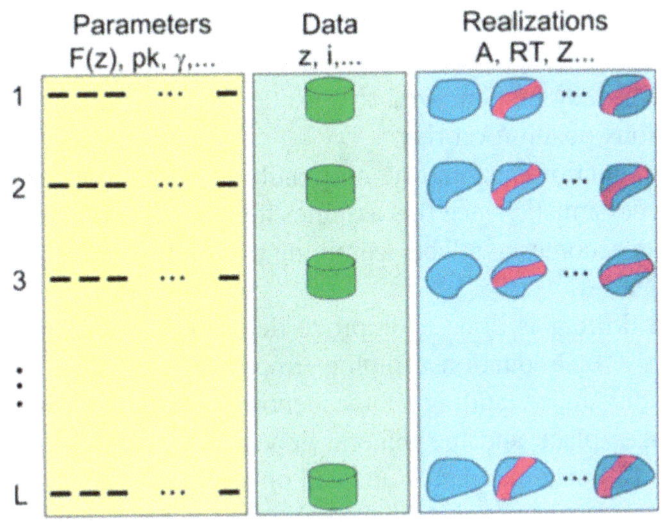


- **Process in Transfer Function** involves evaluating every realization for all calculations of interest. Local uncertainty can be computed for any block size. Resources can be computed for the entire deposit, within a mine plan or for different elevations. An ultimate pit could be computed for every realization. The economic performance of each realization could be evaluated. The uncertainty in each response variable is known non-parametrically through the distribution of responses. The expected response can be computed as an average of the responses.

The uncertainty is directly observed. It is common to assess sensitivity by indexing each realization by summary input parameters, for example, the gross rock volume, proportions of rock types, average grades, variogram ranges, and correlation coefficients. Then, the relationship between the input parameters and the response variables can be fit by a response surface and the sensitivity evaluated and presented by tornado charts. Further post processing is discussed below.

## 7.5 Resource Decision Making

All realizations should be used all the time. Anything that can be computed on one block model can be computed on one hundred, then the distribution of the response variable of interest can be assembled and summarized by expected value and other statistics. If a decision must be made, then the decision variable (economic value for ore, leach, dump…) can be computed on all realizations (Da Cruz 2000; Tversky and Kahneman 1992). The expected response determines the optimal decision.

When a mine plan is specified, then it is straightforward to evaluate all realizations through the plan and observe the uncertainty in key response variables due to the present state of incomplete knowledge. Sometimes the plan is not fixed and the realizations are to be used for planning and optimization. In principle this is not difficult. The objective function is the expected performance over all realizations. Some realizations may perform poorly with a particular plan and some better, but it is the expected value of the performance over all realizations that is the function to optimize (Pyrcz and Deutsch 2014). Considering the concept of the efficient frontier, the risk may be penalized to consider decisions that more reasonably suit the organizations position on risk.

Fixing a production plan and running multiple realizations through the plan can be somewhat pessimistic since this assumes the plan cannot change in the future. In fact, more data becomes available as mining proceeds and the plan can adapt to the new knowledge.

Additional drilling is done to improve delineation ahead of production (Damsleth et al. 1992). Production sampling improves short-term mine planning and leads to a better understanding of the deposit. Uncertainty will resolve itself as production takes place and the mineral deposit is exposed for our greater understanding. The life-of-mine plan is updated on a regular basis (often yearly). A base case long term plan can be established with the current uncertainty and different options explored. The value of future information could be determined by simulating the additional data; this was the idea of the Simulated Learning Model (Cuba et al. 2014). There is flexibility for the plan to adapt to the future, but not change the past.

Flexibility is reduced as mining takes place. There is value in future flexibility (Stirling 2012). A slightly poorer decision, based on currently expected performance, with greater future flexibility may be better than a slightly better decision with less flexibility. The simultaneous optimization over multiple realizations should consider this flexibility.

Optimizing over all realizations simultaneously and considering all realizations through all engineering designs is correct, but difficult for some practitioners to accept (Bratvold et al. 2003; Guyaguler and Horne 2001; Wang et al. 2012). The computational challenges are exaggerated. The computers now are more than 100 times faster than they were about 10 years ago. Also, the ability to use multiple cores and GPUs means that we do not need to compromise much on the complexity of our calculations to consider all realizations all the time. The attraction of a single numerical geological model is undeniable. Most software does not permit easy visualization of multiple realizations. Although the ensemble of realizations should be managed together, the non-uniqueness of multiple realizations is disturbing. The simplest alternative is to use a kriged model for planning and all reporting purposes; the simulated realizations are reserved for uncertainty statements and an understanding of variability.

## 7.6 Alternatives to All Realizations

Some simple summary models are useful. The probability to meet an economic threshold is useful; high probability is good. The local probability to exceed, say, the global 0.75 quantile is also useful to identify the areas that are surely high: if this probability is high (say over 0.9), then the area is surely high. The local probability to be below, say, the global 0.25 quantile is useful to identify areas that are surely low: if this probability is high (say over 0.9), then the area is surely low. The local variance or the probability to be within 15% of expected are also useful summary measures.

Another approach is to collapse uncertainty into a few summary measures and base planning on them. For example, multiple realizations could be summarized by proportions of ore and waste over multiple realizations within reasonable planning volumes. One could even consider that each block has a proportion of ore and a proportion of waste. The block will be found to be all ore or all waste in the future; the proportions are simply used to collapse uncertainty.

Summarizing multiple realizations is useful. The summaries make use of the multiple realizations. Plans optimized on a summary are never as good as plans optimized over all realizations simultaneously (primarily due to the complexity and non-linearity of most planning operations); however, it may be the only practical approach offered by the available software.

The realizations are equally probable; there is no right one and there is no P50 one and we have no idea if one is closer to the truth than the others. A dangerous practice emerged in the early days of simulation: run the realizations through a quick to calculate transfer function, rank the realizations by the quick-to-calculate response, then consider only selected realizations (say, the P10, P50 and P90) in the "real" more complicated transfer function.

In general, individual realizations should never be singled out for calculations. There is much about a single realization that depends on the random number generator and that is not real. Any one realization could be misleading. There are some specific calculations that could be done with one realization because the variability at specific locations (that we do not trust) averages out over multiple realizations. Blending studies and drilling spacing studies are two examples. It may be enough to run one or a few realizations through a simulation of the homogenization steps to understand the probability of plant upsets and undesirable circumstances. The variability at multiple locations reflects the overall variability and the specific location/time is not critical.

In almost all cases, the simplest and most robust approach is to consider all realizations and take expected values at the end to report a single result.

## 7.7 Concluding Remarks

Monte Carlo Simulation is a well-established experimental mathematical approach to transfer uncertainty in input geological and engineering variables through to response variables. The primary aim of this chapter was to point out the danger of using one realization instead of an ensemble of realizations. One realization may fall near the middle based on a quick-to-calculate response variable and yet it could be unusually high in some places and low in others. Planning on one realization could be misleading. The nonlinearity and complexity of many real response variables requires the ensemble of realizations to be considered for proper planning and uncertainty assessment. All realizations all the time – anything less will not give correct results.

**Acknowledgements** The author thanks the sponsors of the Centre for Computational Geostatistics (CCG) for supporting this work.

## References

Barnett RM, Deutsch CV (2015) Multivariate imputation of unequally sampled geological variables. Math Geosci. https://doi.org/10.1007/s11004-014-9580-8

Bernoulli D (1954) Exposition of a new theory on the measurement of risk. Econometrica 22 (1):23–36 (originally published in 1738)

Box GEP, Draper NR (1987) Empirical model building and response surfaces. Wiley

Bratvold RB, Begg SH, Campbell JM (2003) Even optimists should optimize. Soc Pet Eng. https://doi.org/10.2118/84329-ms

Caers J (2011) Modeling uncertainty in the earth sciences. Wiley

Chilès J, Delfiner P (2012) Geostatistics: modeling spatial uncertainty, 2nd edn. Wiley

Cuba MA, Boisvert JB, Deutsch CV (2014) Simulated learning model for mineable reserves evaluation in surface mining projects. SME Trans

Da Cruz PS (2000) Reservoir management decision-making in the presence of geological uncertainty. PhD thesis, Stanford University

Damsleth E, Hage A, Volden R (1992) Maximum information at minimum cost: a North Sea field development study with an experimental design. J Pet Technol

Deutsch CV, Journel AG (1998) GSLIB: geostatistical library and users guide, 2nd edn. Oxford University Press

Francis JC, Dongcheol K (2013) Modern portfolio theory. Wiley

Goovaerts P (1997) Geostatistics for natural resources evaluation. Oxford University Press

Guyaguler B, Horne RN (2001) Uncertainty assessment of well placement optimization. Soc Pet Eng. https://doi.org/10.2118/71625-ms

Hammersley JM, Handscomb DC (1964) Monte Carlo methods, Methuen's monographs on applied probability and statistics, Methuen

Hanoch G, Levy H (1969) The efficiency analysis of choices involving risk. Rev Econ Stud 36:335–346

Kochenderfer MJ (2015) Decision making under uncertainty: theory and application. MIT Press

Levy H (2016) Stochastic dominance: investment decision making under uncertainty, 3rd edn. Springer

Pyrcz M, Deutsch CV (2014) Geostatistical reservoir modeling, 2nd edn. Oxford University Press

Stirling WC (2012) Theory of conditional games. Cambridge University Press

Tversky A, Kahneman D (1992) Advances in prospect theory: cumulative representation of uncertainty. J Risk Uncertain

Wang H, Echeverría-Ciaurri D, Durlofsky L, Cominelli A (2012) Optimal well placement under uncertainty using a retrospective optimization framework. Soc Pet Eng. https://doi.org/10.2118/141950-pa

# Chapter 8
# Binary Coefficients Redux

**Michael E. Hohn**

**Abstract** Paleoecologists and paleogeographers still make use of binary coefficients in multivariate analysis decades after being introduced to the geosciences. Among the main groups, similarity, matching and association, selecting a particular coefficient remains a confusing and sometimes empirical process. Coefficients within groups tend to correlate highly when applied to datasets. With increasing interest in a probabilistic approach to grouping taxa or faunal lists, the Raup-Crick measure of association is closely related in purpose and empirically to coefficients of association and works well in cluster analysis and ordination. A reasonable strategy is to compare dendrograms and ordinations calculated with several coefficients, care being taken to select coefficients with different performance characteristics. Above all, the practitioner should understand the purpose of each coefficient.

## 8.1 Introduction

Founding of the International Association for Mathematical Geology resulted in part from the increased use of quantitative methods in the geosciences and simultaneously with developments in computer hardware and availability. This is no less true for paleontology and paleoecology, fields of endeavor characterized by observing, describing, and synthesizing. With the 1960s and 70s came the development of large databases of fossil occurrences from which researchers could formally infer periods of rapid evolution and episodes of major extinction. Patterns of extinction through time could be simulated with random number generators. Paleoecologists studied whether fossil communities persisted through time and the structure of these communities.

This was a period of synthesis. The *Treatise on Invertebrate Paleontology* (Moore et al. 1953–2015) provided a need for stable taxonomies, a confidence that

M. E. Hohn (✉)
West Virginia Geological and Economic Survey, Morgantown, USA
e-mail: hohn@geosrv.wvnet.edu; mehohn@frontier.com



B. S. Daya Sagar et al. (eds.), *Handbook of Mathematical Geosciences*,
https://doi.org/10.1007/978-3-319-78999-6_8

such a classification could be created, and the motivation for explaining trends in evolution.

Multivariate statistical methods developed in other fields became tools for reducing large datasets to manageable size while providing some degree of objectivity in the analysis. Cluster analysis, multidimensional scaling, factor analysis, and related eigenvector methods became familiar tools to the quantitatively-inclined geologist.

All methods required a measure of similarity, correlation, distance, dissimilarity, or association expressed as a coefficient. Eigenvector-based methods such as factor analysis and principal components analysis (PCA) by definition utilize covariance or correlation among variables (R-mode) and implicitly Euclidean distance in displays of sample coordinates (Q-mode). In contrast, cluster analysis, multidimensional scaling, and principal coordinates analysis (Gower 1971) allow use of a wide range of coefficients, but also require the user to decide which coefficient to use.

Multivariate statistical methods introduced to paleontologists in the 1960s and 70s continue to be used for studying the distribution of fossils in space and through time. With the existence of large databases of fossil occurrences, binary coefficients remain important for comparing collections made over many decades by many individuals.

It still remains to the practitioner to select one coefficient out of the many proposed over the course of more than a century now. Given no clear criterion, some elect to use several coefficients to see whether they affect results.

There is certainly a rich and extensive literature related to the purpose and performance of coefficients, both within the paleoecology literature and in the scientific and engineering literature at large as new applications are found for these coefficients. Surveys of existing measures range in approach, from considerations of the conceptual basis for each, to how well they satisfy the purpose, how they behave relative to each other, to how well they behave relative to a goal set by the author, whether or not they achieve a clear criterion such as satisfying metric properties (Gower and Legendre 1986), which seem to give similar results with each other or are correlated; and above all, whether coefficients include mutual absences. That last criterion might appear to be a small detail compared to the other comparisons but it introduces a fundamental question about the role that chance plays in the distribution of fossils in the collection under study.

This chapter reviews the criteria and arguments used in the past four decades in comparing binary coefficients. In this chapter, I will first group coefficients into three families based on shared formulations and behavior; discuss how such factors as abundance of taxa or poor sampling can affect coefficients; consider metric properties of coefficients; look at probability-based coefficients; apply several coefficients to paleoecological data; and sum up where we are today compared to four decades ago.

I will introduce coefficients as I go along, using what has become standard notation for binary coefficients. Assume we have sampled taxa from $N$ locations. Then for a given pair of taxa:

$a$ = number of co-occurrences
$b$ = number of locations where taxon 1 occurs and taxon 2 does not
$c$ = number of locations where taxon 2 occurs and taxon 1 does not
$d$ = number of locations where neither taxon is observed
$N = a + b + c + d$.

## 8.2 Empirical Comparisons and a Taxonomy

As the use of cluster analysis has expanded beyond the biological and geological sciences, papers have appeared in the literature that try to get a handle on the multitude of coefficients by comparing the way they behave relative to each other or to a criterion based on an application. Although outside the field of paleoecology, these publications often cast a wide net in gathering coefficients and present surveys purely empirical in nature. In the general area of pattern recognition, Choi et al. (2009) compute correlation coefficients among 76 binary coefficients for several types of random or structured datasets, observing that pairwise correlations between coefficients can be very high, depending in part on the pattern and number of presences.

In a companion paper, Choi et al. (2010) created random binary datasets, computed values for each coefficient, averaged the trials to create a dendrogram of the 76 coefficients. They identify eleven clusters, some with only a single coefficient, several with two to six members, and two large clusters with over twenty members. The second largest includes such frequently-used coefficients as the Jaccard, Otsuka, Dice, and the Bray and Curtis, where:

$$C_{\text{Jaccard}} = a/(a + b + c)$$
$$C_{\text{Otsuka}} = a/\sqrt{[(a + b)(a + c)]}$$
$$C_{\text{Dice}} = 2a/(2a + b + c)$$
$$C_{\text{Bray\&Curtis}} = (b + c)/(2a + b + c)$$

That these coefficients are correlated highly in an absolute sense should come as no surprise given the algebraic relationships between several. For example:

$$C_{\text{Dice}} = 1 - C_{\text{Bray and Curtis}}$$

converting a dissimilarity coefficient (Bray and Curtis) into a coefficient expressing similarity. The difference between the Dice and Jaccard coefficients is in weighting the mutual occurrences. Remember that many coefficients were defined as measures of similarity, dissimilarity, or association rather than as input to clustering and ordination routines. Their creators had specific reasons for selecting and weighting the terms—*a*, *b*, *c*, or *d*—in the context of a study and according to some research goal. In many cases they might have been fully aware that their coefficient was similar to one in the literature, but their coefficient measured what they wanted to measure.

The largest group of coefficients includes a subset with among others the Simple Matching (called the Sokol and Michener in their paper), Rogers and Tanimoto, and Hamann coefficients, where:

$$C_{\text{SimpleMatching}} = (a+d)/(a+b+c+d)$$
$$C_{\text{Rogers and Tanimoto}} = (a+d)/[a+2(b+c)+d]$$
$$C_{\text{Hamann}} = [(a+d)-(b+c)]/(a+b+c+d)$$

Notice that these coefficient include the term $d$ for mutual absence. The Rogers and Tanimoto coefficient is the same as the Simple Matching but for increased weighting for mismatches in the denominator and the Hamann can be expressed in terms of the Simple Matching by substituting $N - (a + d)$ for $(b + c)$.

A third, small group includes three similar coefficients, two derived from the familiar $\chi^2$ statistic, including the Phi coefficient:

$$C_{\text{Phi}} = (ad-bc)/\sqrt{[(a+b)(a+c)(b+d)(c+d)]} = \sqrt{(\chi^2/N)}$$

These coefficients express correlation; in fact $C_{\text{Phi}}$ is the correlation coefficient for binary data and can be calculated in the same way as a correlation coefficient for non-binary data.

Related to these coefficients is a large cluster characterized by a numerator containing the term $(ad - bc)$ or ad or $(a + d)$. Examples are the Yule's Q (or simply Yule), Ochiai 2, and Gower:

$$C_{\text{Yule}} = (ad-bc)/(ad+bc)$$
$$C_{\text{Ochiai2}} = ad/\sqrt{[(a+b)(a+c)(b+d)(c+d)]}$$
$$C_{\text{Gower}} = (a+d)/\sqrt{[(a+b)(a+c)(b+d)(c+d)]}$$

Similar to the matching coefficients, these and the Phi express agreement between two entities based on mutual presence and absence, but adjusted for relative abundance of the entities, analogous to the centering and scaling in calculating the correlation coefficient and Phi.

These four groups account for most of the binary coefficients one is likely to encounter in the geosciences, including ones discussed below. If we lump the last two clusters, a simple taxonomy of coefficients has as groups:

1. **Similarity coefficients**, computed by the number of mutual occurrences, scaled by the total number of features occurring in one or the other entities. In paleoecology, entities can be taxa and features can be locations. Some coefficients can express similarity by calculating $b + c$ rather than $a$, but the coefficient can be converted to similarity by subtracting from 1.

2. **Matching coefficients**, generally expressing agreement by including both mutual occurrences and mutual absences d. Generally these are scaled to yield values between 0 and 1, but not always. For instance, the City Block or Hamming coefficient, $b + c$ is not scaled. These are sometimes called distances (e.g. Hohn 1976) but as they are not Euclidean, this is not strictly correct. Although expressing disagreement, coefficients such as the City Block can be converted to matching coefficients.
3. **Coefficients of association**, expressing how entities tend to vary together, adjusted for abundance or rarity.

This taxonomy agrees with that in Hohn (1976) except I am using a more rigorous definition of a distance coefficient by not including the City Block metric in that group. However, $\sqrt{(b + c)}$ is a distance.

Even when two coefficients are not mathematically equivalent, they can be related monotonically (Gower and Legendre 1986) and give virtually the same results when used in cluster analysis or nonmetric multidimensional scaling. In lieu of selecting a single best coefficient, many researchers perform multiple cluster analyses or ordinations to observe whether results change with choice of coefficient. In such an exercise, one wants to make sure to select coefficients with different properties or behaviors.

## 8.3 Effects of Rare and Endemic Taxa

In an empirical study of eight similarity coefficients, Jackson et al. (1989) used a dataset comprised of 25 species of fish observed in 52 lakes in south-central Ontario, Canada. One feature that distinguishes this dataset is that species range from very common to rare, from as many as 47 lakes to as few as 2. The eight coefficients are the Jaccard, Dice, Simple Matching, Rogers and Tanimoto, Otsuka ("Ochiai" in their paper), Phi, Yule, and the Russell and Rao:

$$C_{\text{Russell and Rao}} = a/(a+b+c+d)$$

Unsurprisingly, the Jaccard and Dice gave nearly identical results in a cluster analysis. The same held for the Simple Matching and Rogers and Tanimoto coefficients. Results for the Otsuka were close to the Jaccard and Dice. The dendrogram for Russell and Rao coefficient shows almost no clusters although the general ordering of the species was very similar to the Jaccard, Dice, Simple Matching, Rogers and Tanimoto, and Otsuka.

They also performed principal coordinates analysis for each of the eight coefficients. They observed that the order of species on the first axis correlated highly with the number of lakes in which each occurred for all but the Otsuka, Phi, and Yule coefficients. Some of the correlations are very high, over 0.99 for the Simple Matching and Rogers and Tanimoto. In other words, the first axis corresponded to the frequency of each species, a general "size" factor in their words. Species abundance correlated poorly with the two major principal coordinates axes for the two coefficients of association, the Phi and Yule. The Otsuka showed some effect of species frequency. Nonmetric Multidimensional Scaling gave similar results. The order of species in dendrograms from cluster analysis also showed this frequency effect for the similarity coefficients; not so for the two coefficients of association.

They concluded that similarity coefficients—what they term co-occurrence coefficients—are heavily influenced by frequency, whereas the implicit centering that takes place in calculating the Phi and Yule mitigate this effect. They also conclude that the Otsuka formulation does a centering that partially eliminates the frequency effect.

## 8.4 Adjusting for Poor Sampling

In the context of Q-mode analysis—that is, the comparison of samples rather than the R-mode comparison of taxa—Alroy (2015a, b) looks at the effect of uneven sampling and consequent uneven sample size on four binary coefficients: the Forbes, a modified Forbes coefficient, Simpson's coefficient, and the Dice, where the Forbes coefficient is:

$$C_{\text{Forbes}} = aN/[(a+b)(a+c)]$$

and the Simpson:

$$C_{\text{Simpson}} = a/[\min(a+b),(a+c)]$$

Alroy modifies the Forbes coefficient in two ways. First, he argues against including mutual absences and therefore substitutes $n$ for $N$ where $n = a + b + c$. Secondly, he adds constants to correct for an upward bias in the coefficient:

$$C_{\text{ForbesMod}} = a(n+\sqrt{n})/[(a+b)(a+c)\ +a\sqrt{n}+1/2\,bc)]$$

Although there is no theoretical basis for these constants, the resulting coefficient does accomplish what he sets out to do. In several analyses of real and simulated datasets, he shows that both versions of the Forbes coefficient and the Simpson far outperform the Dice. This is consistent with results obtained by Jackson et al. (1989) in which coefficients such as the Dice are influenced very much by species frequency in R-mode analysis.

Alroy clearly favors the modified Forbes over Simpson's coefficient. However results for both in cluster analysis and principal components analysis are very similar and would probably lead to the same conclusions based on the relative positions of samples on dendrograms and principal coordinates axes. This is no surprise given that the Simpson was formulated to account for uneven sample size.

Although Alroy dismisses probabilistic coefficients and coefficients of association in part for including mutual absences, it would be interesting to compare them with the two Forbes and the Simpson coefficients with his datasets.

These papers address the problem of working with datasets of mixed, perhaps unknown sampling regimen. The difference between otherwise identical faunal lists might be the time or skill in observation. This is perhaps less of an issue when a dataset comes from a single sampling campaign, but in these days of large databases compiled from many studies this is a problem to be taken seriously. Alroy's results argue for careful selection of a coefficient and suggest that analysis with multiple coefficients might be beneficial if sampling issues are suspected.

Alroy points out that the Forbes coefficient has fallen out of use over time. However, since the publication of his papers, Halliday et al. (2017) used his modified form of the Forbes coefficient in cluster analysis of Late Cretaceous vertebrates across India. Although the papers by Alroy and by Halliday et al. describe ordination and cluster analysis of localities, the same problem of uneven sampling exists in analysis of taxa and their arguments and findings should have application in R-mode analysis as well.

## 8.5 Metric? Euclidean?

Some attention has been paid in the past with the question whether a dissimilarity coefficient is metric, Euclidean, or neither. A coefficient is metric if for every triplet $(i, j, k)$ the following inequality holds:

$$D_{ij} + D_{ik} \geq D_{jk}$$

On the face of it, methods such as principal coordinates analysis require a dissimilarity that is Euclidean. In actuality, Gower and Legendre (1986) and others have observed that departures from strict Euclidean geometry for many coefficients are generally small. Adding a constant to a distance can sometimes take care of this problem. It sometimes works to use the square root of the distance. They include a table showing that many familiar similarity coefficients, $C$, are metric but not Euclidean if converted to a dissimilarity coefficient $1 - C$ and even more are metric and many Euclidean if $\sqrt{(1 - C)}$ is calculated. They consider most of the binary coefficients listed above with the notable exception of Yule's coefficient.

Zhang and Srihari (2003) discuss the properties and behavior of similarity, matching, and coefficients of association, including metric properties, equivalent measures of similarity and dissimilarity, discriminatory capability of the coefficients, and the effect of weighting mutual absences. Like many authors they prefer metric coefficients. A large proportion of papers in the geosciences utilize non-metric multidimensional scaling or cluster analysis with no requirement for the coefficient to be Euclidean or even metric. Reasons for selecting a method for multivariate analysis no doubt vary among authors, ranging from convenience or familiarity, available methods in a statistical package, to wanting to avoid the stronger requirements of eigenvector-based methods. However as Gower and Legendre (1986) point out, proportionally small deviations from geometric assumptions of an eigenvector method affects the results very little.

## 8.6 From Expected Values to Null Association

We can look at the diversity of coefficients along a spectrum from similarity coefficients at one end to coefficients of association at the other. In comparing faunal lists, for instance, similarity coefficients count the number of species in common between two locations normalized by the number of species found in one or the other. In other words, they can be said to measure overlap in faunal lists in a Q-mode analysis or geographic overlap of two taxa in an R-mode analysis.

Midway along the spectrum are coefficients that compare an observed value with the expected value. As described by Alroy (2015a), the chance of a species appearing in the faunal list at one site is (a + b)/N, the chance at a second site $(a + c)/N$, and the chance of being found in both is $[(a + b)(a + c)]/N^2$. Therefore, the number of species expected to be found in both is $[(a + b)(a + c)]/N$ and the ratio of the observed number $a$ to the expected number is $aN/[(a + b)(a + c)]$.

Hohn (1976), Raup and Crick (1979) and others have argued that cluster analysis or ordination should consider whether observed overlaps in faunal lists in paleogeographic studies or occurrence of taxa in paleoecological studies represent anything more than a random distribution of taxa through space. Of course there is no denying that species respond to environmental and geographic variables, but the question is how to separate similar distributions that arose by chance from those that represent nonrandom processes.

Within a biological context, Hubálek (1982) surveyed forty-three coefficients, eliminated about half based on algebraic equivalence, mere difference in scale, or failure to meet several criteria, and compared the rest through product-moment correlation and cluster analyses. Although one of these criteria is monotonicity with $\sqrt{(\chi^2)}$, Hubálek stops short of recommending a coefficient such as Phi that is related directly to a test of significance in association.

In contrast, I proposed (Hohn 1976) that we should pay more attention to the Phi coefficient. Raup and Crick (1979) derive the formula for exact probabilities equal to Fisher's Exact Test for independence in 2 by 2 tables, an alternative to the usual $\chi^2$ test. They modified what is essentially a Phi coefficient in comparing faunal lists by using a Monte Carlo method to weight taxa according by abundance. The result is a coefficient that like Phi and similar coefficients includes mutual absences, but represents a further refinement by taking relative abundance of taxa into account.

Winrow and Sutton (2014) calculated five coefficients—Raup-Crick, Simpson, Jaccard, Dice, and Otsuka (Ochiai)—in a paleogeographic study of lingulate brachiopods during the Early Paleozoic. Unable to determine a single best coefficient, they opted to calculate and compare several. Unsurprisingly, the Jaccard, Dice, and Otsuka gave very similar results. Raup-Crick and Simpson coefficients showed different patterns among pairs of faunal lists representing different paleocontinents. They do not explain why coefficients would give different results other than attributing several anomalously-high values of the Simpson to small sample sizes.

Zhang and Srihari (2003) survey binary dissimilarity coefficients in the context of character recognition; some of their results are instructive. In their look at nine familiar coefficients they define relative discriminatory power in terms of entropy, itself proportional to the variance of dissimilarities in multivariate space. They consider coefficients with a wide range of values to have potentially greater discriminatory power, finding that the Russell and Rao coefficient had the poorest discriminatory power and the Jaccard and related coefficients moderate power. Highest discriminatory power was shared by the correlation coefficient, Yule and Rogers and Tanimoto. In the study by Winrow and Sutton, the similarity coefficients had a narrow range of values compared with the Raup-Crick and Simpson.

## 8.7 Illustrative Example

Both R-mode and Q-mode analysis were performed on presence-absence data collected from five outcrops of the Middle Devonian Hamilton Group in New York State, although only ordinations of taxa will be shown here for reasons of space. Lithology of the interval sampled included thin limestones, mudstones, silty mudstones, and calcareous siltstones. The data matrix comprises 43 samples and 32 taxa identified to species when possible (Hohn 1975).

Cluster analysis, principal components analysis, and principal coordinates analysis were carried out; results of principal coordinates analysis best illustrate similarities and differences among the coefficients used. The statistical package PAST (Hammer et al. 2001) offers a wide range of multivariate methods and coefficients including similarity, matching, and association. I looked at results for the Phi (Correlation Coefficient in PAST) and Raup-Crick coefficients to observe their near-equivalence; the Jaccard as representative of similarity coefficients; Simpson's coefficient as an unusual asymmetric coefficient used with some frequency; and to represent matching coefficients, the Hamming normalized to lie between 0 and 1:

$$C_{\text{Hamming}} = (b + c)/N$$

In signal processing and information theory, Richard Hamming is known for the Hamming distance and Hamming window in addition to other contributions. Note the simple relationship between the normalized Hamming and Simple Matching coefficients:

$$C_{\text{SimpleMatching}} = 1 - C_{\text{Hamming}}$$

Looking at plots of the first two principal coordinate axes (Figs. 8.1, 8.2, 8.3, 8.4 and 8.5), one might be struck by the how similar they appear. However, most of us would probably consider the results from the Hamming (Simple Matching) coefficient in Fig. 8.1 difficult to interpret. The Jaccard is a great improvement (Fig. 8.2) as indeed is Simpson's coefficient (Fig. 8.3). The Phi coefficients of association and Raup-Crick probabilistic measure give almost identical results with each other (Figs. 8.4 and 8.5).

The biggest differences among the five plots are positions of the most abundant taxa such as the brachiopod *Tropidoleptus* and bivalve *Paleoneilo*. They occur in a large proportion of samples (Table 8.1) and provide little discriminatory power among assemblages. Relatively abundant taxa score highly in an absolute sense on the second principal coordinate axis (vertical axis) for the Hamming and Jaccard coefficients, less so for the Raup-Crick and Phi. There is a clear correlation between principal coordinate scores on this axis with taxon count for the Hamming and Jaccard coefficients (Fig. 8.6). This observation agrees with the findings of Jackson et al. (1989).

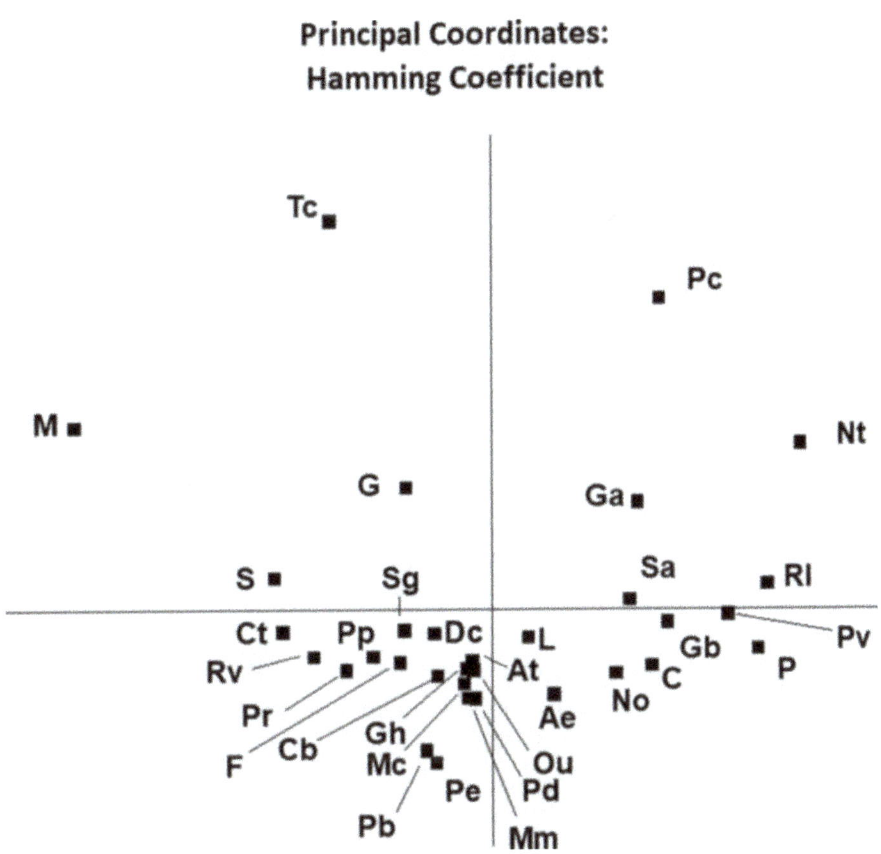


Fig. 8.1 Principal coordinates analysis with Hamming coefficient of dissimilarity

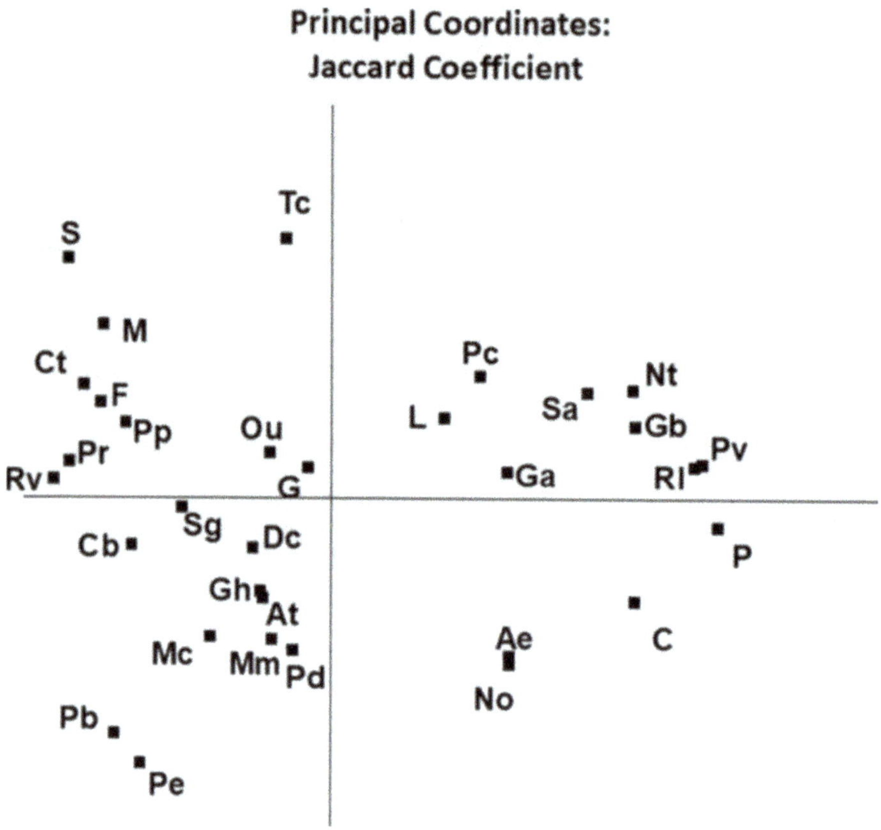


**Fig. 8.2** Principal coordinates analysis with Jaccard coefficient

Based on percent of variance explained by the first three principal coordinate axes (Table 8.2), the Hamming coefficient would appear to perform best. Similar results were obtained from nonmetric multidimensional scaling of each coefficient matrix (Table 8.3). But we already know that a portion of the variance correlates with taxon abundance. This observation suggests that selecting a coefficient based by variance explained has limited value if the coefficient measures the wrong thing.

Q-mode analyses showed similar correlation of abundance with principal coordinate scores calculated from Hemming and Jaccard coefficients. The relationship is not as strong because no sample contained more than 26% of the taxa, whereas *Tropidoleptus* in the R-mode analysis occurred in 84% of samples.

Note that the Raup-Crick procedure does not yield a binary coefficient in the sense of all the others, but rather accomplishes through Monte Carlo sampling, a similar measure as the correlation coefficient. Practitioners use the Raup-Crick measure in the same way as any of the other binary coefficients for cluster analysis and ordination. However there is no guarantee that it has strictly metric properties, and indeed, principal coordinates analysis with the Raup-Crick statistic yielded a large proportion of negative eigenvalues.

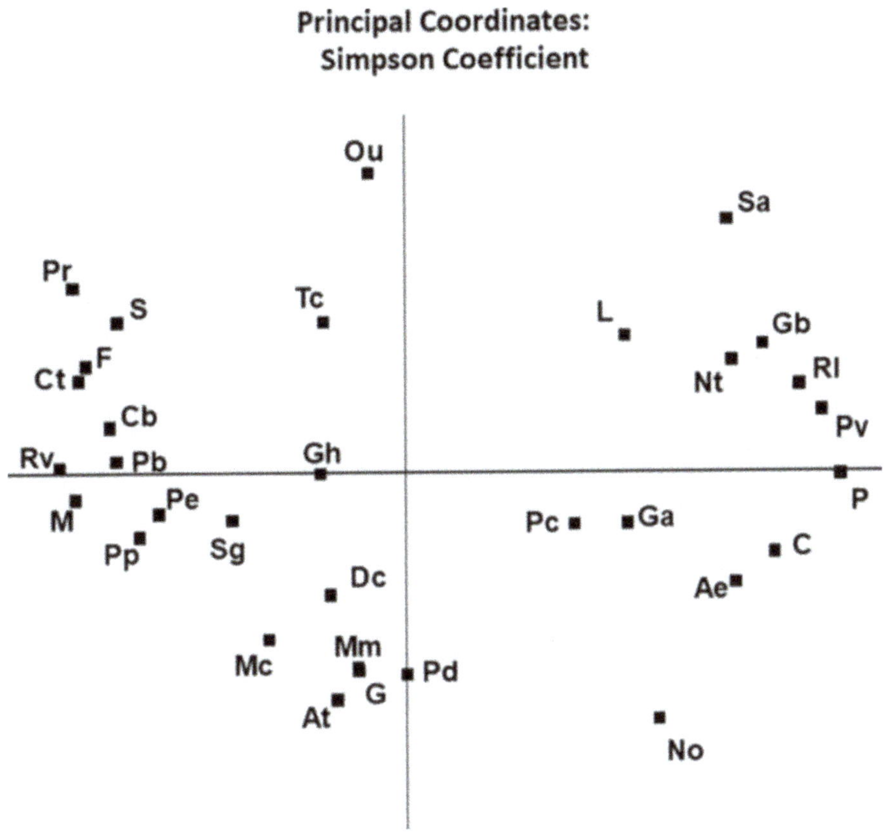


Fig. 8.3 Principal coordinates analysis with Simpson's coefficient

## 8.8 Discussion and Conclusions

Studies published over the past decade give a taste of the application of binary coefficients of all types.

Brayard et al. (2007) used distances $1 - S_{\text{Dice}}$ in Q-mode cluster analysis and ordination of Early Triassic ammonoid faunas, citing the double weight given to mutual presences, thus downweighting the influence of unique species occurrences and not giving any weight to mutual absences. They used the square root of the dissimilarity matrix so that the resulting distances would be metric and Euclidean (Gower and Legendre 1986).

In studies of faunal lists of bivalves from around the globe, Schmachtenberg (2008) compared four coefficients: Jaccard, Simpson, Raup-Crick, and a measure of endemism. He did not do any cluster analyses or ordinations, but rather regressed value of each coefficient on geographic distance. The Simpson, Raup-Crick, and natural log of the Jaccard coefficient performed almost equally well.

Huang et al. (2012) considered the performance of five coefficients—Jaccard, Dice, Cosine, Yule's Y, and Raup-Crick—in cluster analysis and nonmetric multidimensional scaling of Silurian brachiopod assemblages representing time after the Late Ordovician extinction events. They preferred the Raup-Crick coefficient for

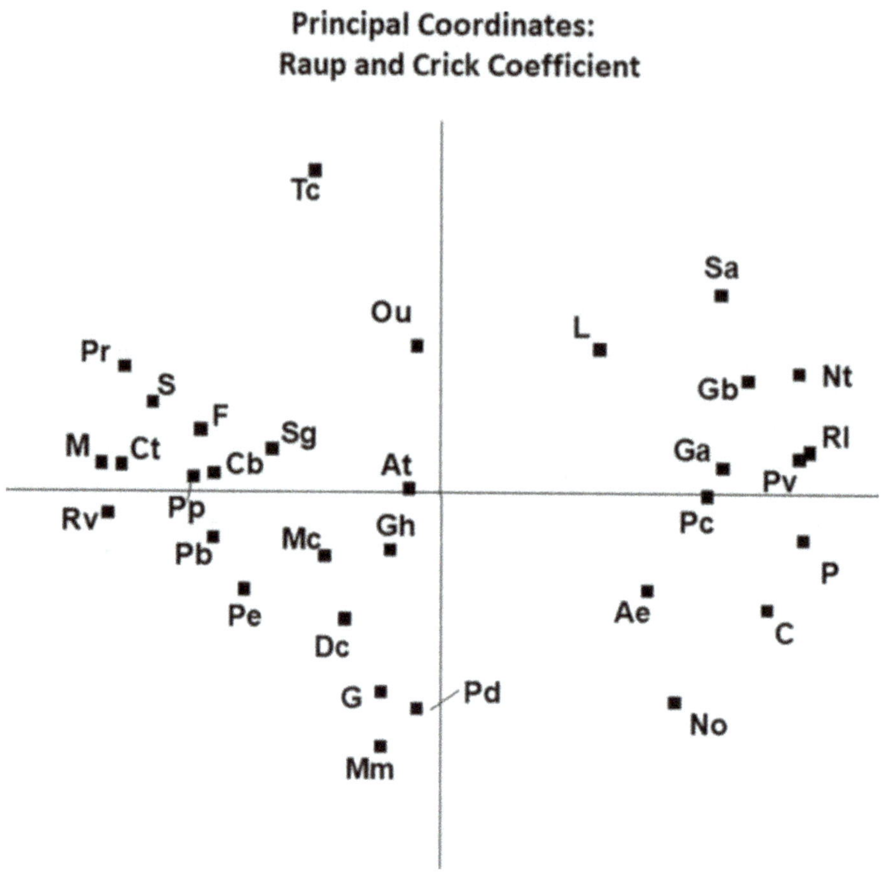


**Fig. 8.4** Principal coordinates analysis with Raup-Crick Coefficient

ordination because it yielded the lowest stress value. On the other hand, they primarily used Yule's Y in their cluster analyses, where:

$$C_{\text{YuleY}} = (\sqrt{ad} - \sqrt{bc})/(\sqrt{ad} + \sqrt{bc})$$

In a paleoecological and paleogeographical analysis of Late Ordovician cephalopods, Kröger and Ebbestad (2013) used the Raup-Crick and Bray and Curtis coefficients in cluster analysis of assemblages and concluded that the Raup-Crick dissimilarity index gave better-resolved groups.

Balseiro (2016) studied changes in composition and diversity of brachiopods and bivalves in western Argentina during the main Carboniferous glacial event. The author observed few differences among results from several types of ordination and choice of coefficients, including the modified Forbes coefficient of Alroy (2015b) and Bray and Curtis dissimilarity.

Many reviewers of binary coefficients note the controversy that surrounds the question whether mutual absences should be included in a coefficient. Some authors categorically reject coefficients that include $d$ (e.g. Shi 1993). Reasons cited include: mutual absences do not contain information; we can never know the total number of taxa $N$ in a paleogeographic study; we can inflate differences through

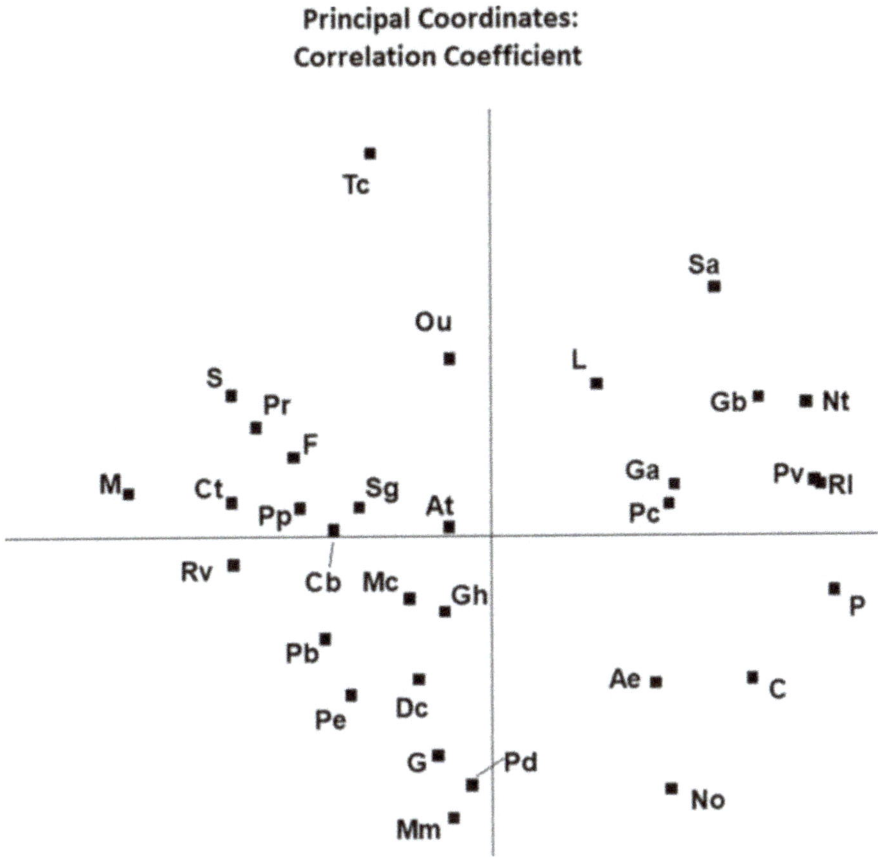


**Fig. 8.5** Principal coordinates analysis with correlation (Phi) coefficient

inappropriate inclusion of taxa or samples; or sampling effort or success is uneven and therefore the appropriate $N$ is unknown. There are counterarguments for each one of these objections and the user is left to decide for his or herself. For instance, knowledge of mutual absences is necessary to evaluate the probability of an observed pattern of occurrences, and therefore it conveys information. While we cannot know $N$ exactly, we have ways to access completeness of sampling, and after all, any statistic is based on samples and $N$ is no exception.

In contrast to the other objections to the use of mutual absences, uneven sampling among locations appears to be a real problem and the effect on even probabilistic measures of association is not well understood. Simpson's coefficient and modified Forbes coefficient of Alroy (2015a, b) attempt to correct for this problem. Neither coefficient conveys any probabilistic information. This is the price one pays when sampling is less than optimal. To draw strong conclusions sampling methods are all-important.

**Table 8.1** Number of sample occurrences by taxon

| Taxon | Count |
|---|---|
| *Tropidoleptus carinatus* | 27 |
| *Paleoneilo constricta* | 24 |
| *Mucruspirifer mucronatus* | 22 |
| *Greenops boothi* | 20 |
| *Nuculites triqueter* | 19 |
| *Grammysioidea cf. arcuata* | 15 |
| *Retispira leda* | 13 |
| *Pholadella radiata* | 11 |
| *Cypricardella tenuistriata* | 11 |
| *Goniophora cf. hamiltonensis* | 11 |
| *Pterinopecten vertumnus* | 10 |
| *"Stictopora"* | 10 |
| *Rhipidomella vanuxemi* | 10 |
| *"Chonetes"* sp. | 9 |
| *Grammysia bisulcata* | 8 |
| *Schizodus apressus* | 8 |
| *Nuculites oblongatus* | 8 |
| *Modiomorpha mytiloides* | 8 |
| *Spinocyrtia granulosa* | 8 |
| *Phacops cf. rana* | 8 |
| *Devonochonetes coronatus* | 7 |
| *Pseudaviculopecten princeps* | 6 |
| *Lingula* sp. | 5 |
| *Paleaeozygopleura delphicola* | 5 |
| *Ptychopteria boydi* | 5 |
| *Platyceras erectum* | 5 |
| *Modiomorpha concentrica* | 5 |
| *Actinodesma erectum* | 4 |
| *"Fenestella"* | 4 |
| *Cypricardella bellistriata* | 4 |
| *Orthonota undulata* | 2 |
| *Alanella tullius* | 2 |

## 8.9 Summary

1. A review of the literature in recent years shows that binary coefficients are still very much used, even given the advantages of abundance information.
2. Choice of coefficient remains a confusing and sometimes empirical process, often leading practitioners to examine results from several coefficients.
3. If a large contrast exists in abundance of taxa or length of faunal lists, one should use care in using similarity coefficients. Comparing dendrograms or ordinations

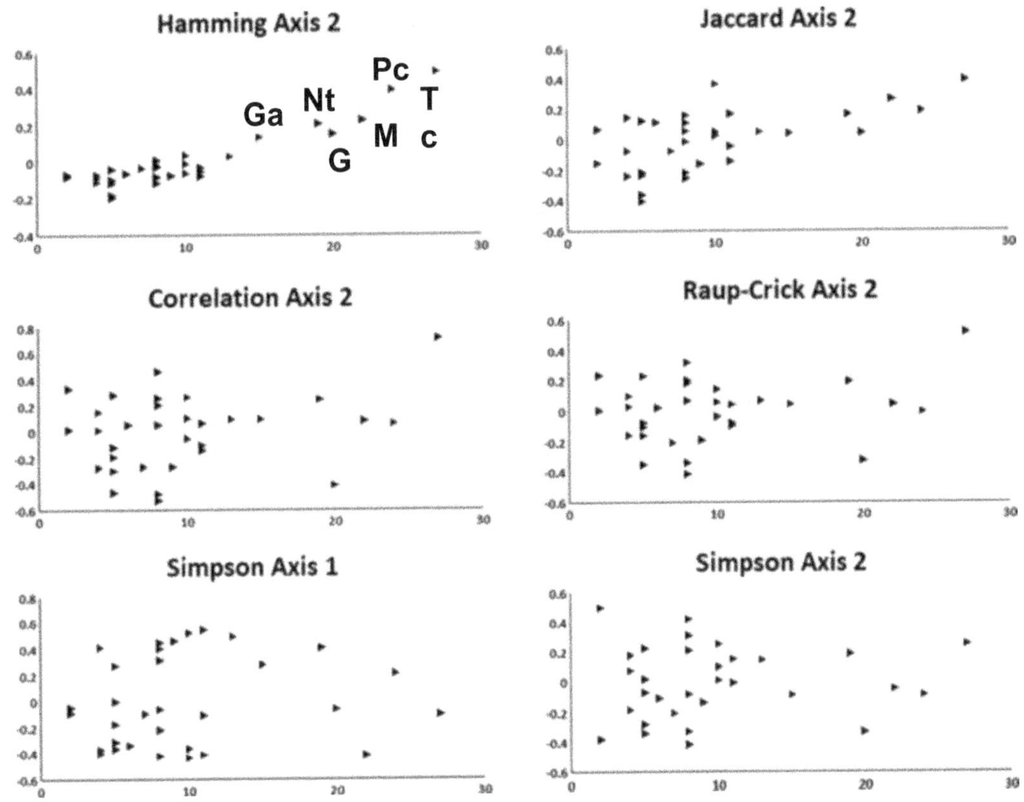


**Fig. 8.6** Bivariate plots of number of samples in which each of the 32 taxa occurred (horizontal axes) and scores on principal coordinate axes (vertical axes). Only scores on the second axis are shown for Hamming, Jaccard, Correlation coefficient (Phi), and Raup-Crick coefficients

**Table 8.2** Percent of total variance explained by each axis in principal coordinates analyses by coefficient

| Coefficient | Percent of variance | | | Total First 3 Axes |
|---|---|---|---|---|
| | Axis 1 | Axis 2 | Axis 3 | |
| Hamming | 26.4 | 21.1 | 10.3 | 57.8 |
| Jaccard | 19.7 | 9.5 | 8.8 | 38.0 |
| Raup-Crick | 15.6 | 5.7 | 3.8 | 25.1 |
| Correlation | 24.8 | 10.9 | 9.2 | 44.9 |
| Simpson | 20.6 | 9.9 | 9.0 | 39.5 |

**Table 8.3** Variance along each axis and stress for nonmetric multidimensional scaling

| Coefficient | $R^2$ | | | |
|---|---|---|---|---|
| | Axis 1 | Axis 2 | Axis 3 | Stress |
| Hamming | 0.38 | 0.12 | 0.53 | 0.1315 |
| Jaccard | 0.32 | 0.12 | 0.14 | 0.2729 |
| Raup-Crick | 0.47 | 0.04 | 0.05 | 0.2613 |
| Correlation | 0.45 | 0.05 | 0.06 | 0.2679 |
| Simpson | 0.30 | 0.07 | 0.12 | 0.3072 |

obtained using more than one coefficient could help the practitioner partition out the least informative occurrences.

4. The problem of uneven sampling has not been fully addressed in the literature.
5. There continues to be interest—perhaps growing—in evaluating occurrence data in a probabilistic context.
6. In addition to theoretical considerations, some authors have found empirically that coefficients of association and the related Raup-Crick coefficient work well in clustering and ordination.

In conclusion, it remains a reasonable strategy to compare dendrograms and ordinations calculated with several coefficients. Care should be taken to select coefficients with different performance characteristics. Finally, the practitioner should understand the purpose of each coefficient.

**Acknowledgements** Gordon Baird helped extensively in the early stages of this project by sharing his knowledge of outcrop locations and updated correlations of the Hamilton Group. Thomas Kammer made valuable suggestions to an early draft of this paper.

## References

Alroy J (2015a) A new twist on a very old binary similarity coefficient. Ecology 96:575–586

Alroy J (2015b) A simple way to improve multivariate analyses of paleoecological data sets. Paleobiology 41:377–386

Balseiro D (2016) Compositional turnover and ecological changes related to the waxing and waning of glaciers during the late Paleozoic ice age in ice-proximal regions (Pennsylvanian, western Argentina). Paleobiology 42:335–357

Brayard A, Escarguel G, Bucher H (2007) The biogeography of early Triassic ammonoid faunas: clusters, gradients, and networks. Geobios 40:749–765

Choi S-S, Cha S-H, Tappert, CC (2009) Correlation analysis of binary similarity and distance measures on different binary database types. In: Proceedings of the international conference on artificial intelligence and pattern recognition, Orlando, Florida, USA, July 2009

Choi S-S, Cha S-H, Tappert CC (2010) A survey of binary similarity and distance measures. J Syst Cybern Inf 8:43–48

Gower JC (1971) A general coefficient of similarity and some of its properties. Biometrics 27:857–871

Gower JC, Legendre P (1986) Metric and Euclidean properties of dissimilarity coefficients. J Classif 3:5–48

Halliday TJD, Prasad GVR, Goswani A (2017) Faunal similarity in Madagascan and south Indian Late Cretaceous vertebrate faunas. Palaeogeogr Palaeoclimatol Palaeoecol 468:70–75

Hammer Ø, Harper DAT, Ryan PD (2001) PAST: paleontological statistics software package for education and data analysis. Palaeontol Electron 4:9 pp

Hohn ME (1975) Palecology and biostratigraphy of the Portland Point and Kashong Members, Hamilton Group (Middle Devonian) of New York State. Unpublished Master's Thesis, Indiana University, Bloomington, Indiana, USA, 69 pp

Hohn ME (1976) Binary coefficient: a theoretical and empirical study. Math Geol 8:137–150

Huang B, Rong J, Cocks LRM (2012) Global palaeobiogeographical patterns in brachiopods from survival to recovery after the end-Ordovician mass extinction. Palaeogeogr Palaeoclimatol Palaeoecol 317–318:196–205

Hubálek Z (1982) Coefficients of association and similarity, based on binary (presence-absence) data: an evaluation. Biol Rev 57:669–689

Jackson DA, Somers KM, Harvey HH (1989) Similarity coefficients: measures of co-occurrence and association or simply measures of occurrence? Am Nat 133:486–453

Kröger B, Ebbestad JOR (2013) Palaeoecology and paleogeography of Late Ordovician (Katian-Hirnantian) cephalopods of the Boda Limestone, Siljan district, Sweden. Lethaia 47:15–30

Moore RC et al (1953–2015) Treatise on invertebrate paleontology, parts A through W. Boulder, Colorado: Geological Society of America; and Lawrence, Kansas: University of Kansas Press

Raup DM, Crick RE (1979) Measurement of faunal similarity in paleontology. J Paleontol 53:1213–1227

Schmachtenberg WF (2008) Resolution and limitations of faunal similarity indices of biogeographic data for testing predicted paleogeographic reconstructions and estimating intercontinental distances: a test case of modern and Cretaceous bivalves. Palaeogeogr Palaeoclimatol Palaeoecol 265:255–261

Shi GR (1993) Multivariate data analysis in palaeoecology and palaeobiogeography—a review. Palaeogeogr Palaeoclimatol Palaeoecol 105:199–234

Winrow P, Sutton MD (2014) Lingulate brachiopods and the early Paleozoic history of the Iapetus Ocean. Lethaia 47:456–468

Zhang B, Srihari SN (2003) Properties of binary vector dissimilarity measures. In: Proceedings of the JCIS international conference on computer vision, graphics and image processing, Cary, North Carolina, USA, 26–30 Sept 2003

# Chapter 9
# Tracking Plurigaussian Simulations

**M. Armstrong, A. Mondaini and S. Camargo**

**Abstract** The mathematical method called Plurigaussian Simulations was invented in France in the 1990s for simulating the internal architecture of oil reservoirs. It rapidly proved useful in other domains in the earth sciences: mining, hydrology and history matching. In this chapter we use complex dynamic networks first developed in statistical mechanics to track the diffusion of the method within academia, using citation data from Google Scholar. Since governments and funding agencies want to know whether ideas developed in research projects have a positive effect on the economy, we also studied how plurigaussian simulations diffused from academia to industry. The literature on innovation usually focusses on patents but as there were few on plurigaussian simulations, we needed criteria for deciding whether an innovation had been adopted by industry. Three criteria were identified:

- Repeat co-authorship. Many published papers were co-authored by mining or oil companies, or by consulting firms. While this demonstrates interest from industry, in some cases it seemed to be "window-shopping" but companies that continued to publish on this topic (i.e. "repeat co-authors") had clearly adopted the method.
- Specialized training. Companies that wanted to build-up inhouse competency, sent their personnel for postgraduate training or to specialized short courses.
- Bringing in consultants. Rather than investing the time and effort in building up competency in-house, other companies got studies carried out by consulting firms.

M. Armstrong (✉) · S. Camargo
Escola de Matemática Aplicada, Fundação Getulio Vargas, Rio de Janeiro, Brazil
e-mail: margaret.armstrong@mines-paristech.fr; margaret.armstrong@fgv.br

M. Armstrong
MINES Paristech, PSL Research University, CERNA – Centre for Industrial Economy, i3, CNRS UMR 9217, Paris, France

A. Mondaini
Department of Physics, UERJ, Rio de Janeiro, Brazil


B. S. Daya Sagar et al. (eds.), *Handbook of Mathematical Geosciences*,
https://doi.org/10.1007/978-3-319-78999-6_9

The second criterion revealed how important master's level courses are in training geoscientists in the latest techniques. Their role in transferring knowledge to industry is undervalued in current procedures for evaluating university departments.

**Keywords** Complex dynamic networks • University-industry interaction
Technology diffusion • Google Scholar citations

## 9.1 Introduction

In August 2015 Fundacao Getulio Vargas (FGV)[1] organized a 3 day seminar on applied research and invited Jane Tinkler from the London School of Economics to give a plenary lecture on how to assess the impact of research in the social sciences on policy decisions. She stressed the fact that it often takes 15–20 years to see the effects of academic research in the real world. Her talk inspired the lead author to ask how research in the geosciences diffuses within academia and from there, into industry. Why is it important to understand how ideas are adopted by industry? Because in the future, in addition to publishing in top journals, academics will probably need to demonstrate that their research is generating innovations to fuel national economies. For example, the Australian government has been funding a national survey since 2001 to collect data on the commercialization of the results of publicly funded research, especially their impact on intellectual property.

Since the pioneering work of Schumpeter in the 1940s, economists have agreed that a large component of modern economic growth has been driven by "innovation", that is, the arrival of new ideas. Nowadays, most papers on the relationship between scientific research and innovation use citation data to measure the production of new ideas in science and patent data to measure the creation of new potentially successful commercial ideas. Patents have become particularly important in this context for three reasons (Agrawal and Henderson 2002):

- The patenting process requires that inventors' names, dates, assignee institutions, locations and detailed descriptions of the invention's claims be recorded. Innovation-related details are rarely recorded systematically outside of patent records.
- Innovations that are patented are expected to be commercially useful.
- Patenting data has recently become available in machine-readable form.

This approach has proved very fruitful in fields where the technology is evolving rapidly and where patents protect their inventors, for example, pharmacy and biotechnology, nanotechnologies, and wind and solar power generation. But it is

[1]FGV is a private university and think tank located in Rio de Janeiro, that has internationally recognized research groups in economics, law, public administration, and management, and more recently an energy group and an applied maths department.

not pertinent in sectors where patents are less common and where the transfer of new ideas from academia to industry follows different channels (Zellner 2003; Martin and Tang 2007; Moser 2012; Maietta 2015). Geosciences is one such domain.

In order to discover how ideas diffuse within academia and from there into industry, we chose to focus on a specific new method (plurigaussian simulations) which was invented in France in the 1990s for simulating the internal architecture of oil reservoirs (Galli et al. 1994; Armstrong et al. 2011). It rapidly proved useful in other domains in the earth sciences: mining, hydrology and history matching. In the first part of this chapter, after collecting citation data from Google Scholar, we use complex dynamic networks first developed in statistical mechanics to track the diffusion of the method in the academic world. In the second half of the chapter we study how this method moved into industry.

The chapter is divided into five sections. The next one (Sect. 9.2) is a literature review on complex dynamic networks, especially citation networks. In Sect. 9.3 this technique is applied to our citation network for plurigaussian simulations. As only 9 out of the 550 citations were patents, these were not the vector in transferring the method into industry. In Sect. 9.4 we identify three key indicators showing how this innovation was incorporated in industry. Our conclusions follow in Sect. 9.5.

## 9.2 Review of Complex Networks

Over the past 30 years the methods developed by physicists for studying networks in statistical mechanics have been adapted to analyzing other types of networks including the world-wide web (Broder et al. 2000; Albert 1999, 2000), power grids (Watts and Strogatz 1998), telephone call grids (Abello et al. 1998) and airline timetables (Amaral et al. 2000). Newman (2001) and Barabasi et al. (2002) both studied citation networks in which the authors were the nodes in the network and a link was formed between two authors when they co-authored a paper. Newman (2001) studied four such collaboration networks:

1. Los Alamos e-print Archive: a database of unrefereed preprints in physics submitted by the authors from 1992 to 2000;
2. Medline: a database of articles on biomedical research published in refereed journals from 1961 to 2000. The entries are submitted by maintainers, rather than the papers authors, giving it a greater coverage;
3. Stanford Public Information Retrieval System (SPIRES): a database of preprints and published papers in high-energy physics;
4. Networked Computer Science Technical Reference Library (NCSTRL): a database of preprints in computer science, submitted by participating institutions and stretching back about 10 years from 2000.

Although the databases went back earlier Newman limited his study to the window from 1995 to 1999 in order to obtain a good static photo of the conditions at that time. In contrast Barabasi et al. (2002) studied the evolution over time of patterns of collaboration in two specific fields: mathematics and neuro-science, over the period from 1991 to 1998, using databases consisting of 70,975 different authors and 70,901 papers for mathematics and 209,293 authors 210,750 papers for neuroscience.

By 2000, theoretical and empirical studies had uncovered three important results: firstly, most networks have the so-called small-world property which means that the average separation between nodes is rather small; secondly, real networks display a higher degree of clustering than expected for purely random networks and finally, the degree distribution follows a scale-free power-law form (Barabasi et al. 2002). Initially it had been expected that the Web would be a random network like those characterized by Erdos and Renyi (1959). In that case the probability of any two nodes being connected is constant, and most nodes have a degree (number of connections) that is close to the average and the degree distribution is exponential. Albert et al. (1999) showed that the distribution for the Web is a power-law, which means that a few nodes are highly connected while the vast majority have a smaller degree than average.

By computing the statistics of the number of authors per paper, the number of papers per author and the number of collaborators per author in various fields, Newman (2001) confirmed that their distributions follow a power-law form. All the networks contain a giant component of scientists, any two of whom can be connected by a shortest path of intermediate collaborators.

## 9.3 Network Analysis of Google Citations of Plurigaussian Simulations

The first step in our study consisted of collecting all the publications up to December 2015, found by Google Scholar for the term "Plurigaussian simulations". A total of 555 references were obtained. Google Scholar had ordered them from the most relevant to the least (as determined by its algorithm). They include journal articles, working papers, doctoral and master's theses, final year projects, patents and the two books on Plurigaussian Simulations together with chapters from the books which are sold separately by the publishers. These citations can be split into four groups:

(1) Pertinent documents which develop the theory, or report case studies;
(2) Papers which mention that plurigaussian simulations could be used to model the internal architecture of reservoirs or orebodies but which prefer to use another method (usually multipoint geostatistics);
(3) Papers which mention plurigaussian simulations briefly. For example, Laigle et al. (2013) commented in their concluding paragraph that "*Another use would*

**Table 9.1** Information noted for each of the 473 documents

| Year of publication |
|---|
| Type of application (oil, mining, water resources, history matching, theory) |
| Number of authors |
| Whether the document is a patent |
| Whether any of the authors works for a company and if so the company's name |
| Whether any of the authors works for a consulting group or a software vendor, and if so, the consultant's name |
| Country of the lead author |
| Whether the authors came from more than 1 country |
| Whether the paper is directly relevant; that is, belongs to category (1) |

*be to constrain geostatistical simulations by the model results, e.g., training maps for multipoint or plurigaussian methods*".

(4) Papers which do not mention plurigaussian simulations at all.

Of the original 555 references, 307 fell into the first category, 166 into either the second or third while 82 fell into the fourth category. The last group were eliminated from further study. For the 473 references in the first 3 categories, we noted the information listed in Table 9.1. Table 9.2 summarises the statistics of applications in the four main domains.

- Most papers were written by teams of authors (more than 3 per paper on average, up to 10 for some oil papers). Papers by single authors were usually dissertations. This confirms the finding by Wuchty et al. (2007) that papers are now being produced by teams of authors; solo papers are getting rarer.
- International cooperation was a common feature: 28% of papers on oil, 21% for mining and 17% for water and history matching.
- Many papers had authors from companies or consulting firms (57.8% for oil; 35.2% for mining; 23.8% for history matching) but far fewer for water (only 9.2%), probably because water is a public good whereas mining and oil companies are designed to make a profit.

**Table 9.2** Results for the four main applied fields

| | Oil | Mining | Water resources | History matching |
|---|---|---|---|---|
| Total N° | 116 | 71 | 65 | 101 |
| Patents | 5 | 0 | 0 | 4 |
| Company | 41 | 11 | 2 | 16 |
| Consultant | 26 | 14 | 4 | 8 |
| International | 26 | 20 | 11 | 17 |
| Irrelevant | 24 | 25 | 31 | 35 |

- Countries with strong mining and petroleum industries were well-represented amongst the papers.
- Migration by scientists was a factor that accounted for the excellence of some countries.
- Surprisingly only 9 of the documents were patents and these were all in the petroleum sector (either oil or history matching).

### 9.3.1 Building a Citation Network

In contrast to Newman (2001) and Barabasi (2002) who built their citation network by considering authors as nodes and linking those who had joint papers, we constructed the plurigaussian network by considering each publication as a node with an edge between two of them when one publication cites the other one, producing a directed network. Our network (Fig. 9.1) is displayed with different colours for the different fields of application: black for oil, mauve for mining, blue for water, red for history matching, green for agriculture, mustard for soil science and white for others. As expected, publications in the same field tend to be clustered together in the network.

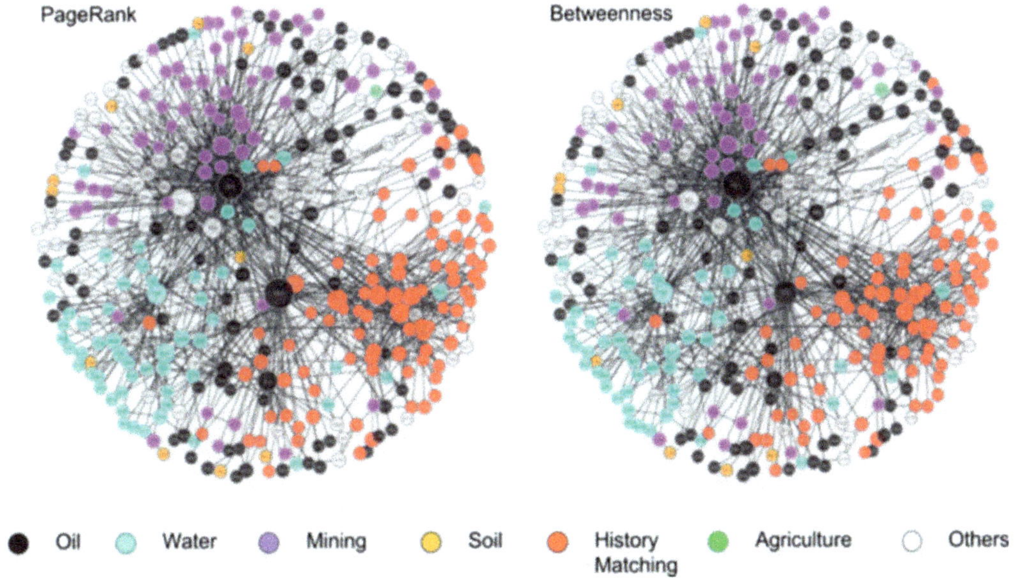


**Fig. 9.1** The citation network for plurigaussian simulations, with different colours indicating the different fields of application: black for oil, mauve for mining, blue for water, red for history matching, green for agriculture, mustard for soil science and white for others. The size of the nodes are proportional to their rank according to PageRank and Betweenness centrality

**Table 9.3** Rank of publications according centrality measures, namely Pagerank and Betweenness

| Rank | PageRank | Betweenness |
|---|---|---|
| 1 | Stochastic Modeling and Geostatistics: Principles, Methods, and Case Studies, Vol. II, AAPG Computer Applications in Geology 5 | Multivariate geostatistics |
| 2 | Multivariate geostatistics | Basic linear geostatistics |
| 3 | Truncated plurigaussian method: theoretical and practical points of view | Dealing with spatial heterogeneity |
| 4 | Geostatistics Wollongong &96. 1 (1997) | Advance supply of emergency contraception: effect on use and usual contraception—a randomized trial |
| 5 | Gradual deformation and iterative calibration of Gaussian-related stochastic models | Critical evaluation of the ensemble Kalman filter on history matching of geologic facies |
| 6 | Error propagation in environmental modelling with GIS | Gradual deformation and iterative calibration of Gaussian-related stochastic models |
| 7 | Basic linear geostatistics | Error propagation in environmental modelling with GIS |
| 8 | Advance supply of emergency contraception: effect on use and usual contraception—a randomized trial | Geostatistics for seismic data integration in Earth models: 2003 Distinguished Instructor Short Course |
| 9 | The FFT moving average (FFT-MA) generator: An efficient numerical method for generating and conditioning Gaussian simulations | Achievements and challenges in petroleum geostatistics |
| 10 | Gaussian Markov random fields: theory and applications | Real-time reservoir model updating using ensemble Kalman filter |

As the network is composed of about 500 publications, it is interesting to know which nodes are the most important, and centrality measures are a good way to provide such answers based on the topology of the network. Here we used two measures: PageRank and Betweenness centrality. PageRank (Page et al. 1999) evaluates the importance of a node based on how many edges point to it, Betweenness centrality (Freeman 1977) estimates whether a node is likely to be placed between other pairs of vertices. Figure 9.1 shows the network of plurigaussian simulations when the node size is proportional to PageRank centrality (left panel) and Betweenness centrality (right panel). At first glance the figures look very similar but there are differences in the importance of some of the nodes as can be seen in Table 9.3 which lists the ten most important publications according to these two centrality measures.

## 9.4 Diffusion of the New Method into Industry

In our analysis of the citation network we had been surprised to find so few patents (only 9 out of 550). Moreover these only started in 2006 (i.e. 10 years after the invention of the method). This was because software could not be patented software before then (See Appendix 9.1 for more detail on this). As patent data could not be used to determine when the method actually reached industry, we need some other criteria. Based on Tijssen et al. (2009), we used the following:

- One of the authors comes from a mining or an oil company, or
- One of the authors comes from a software vendor or a consulting group

It is important to distinguish between the two. Resource companies like Shell or Chevron, or Rio Tinto or Anglo-American are "end-users" whereas consultants and software vendors transfer the idea to end-users, so their business plans are quite different.

The citations came from four main applied fields[2] (oil, mining, water resources and history matching). Looking back at Table 9.2, very few papers in water resources had an author from a company or a consulting firm (only 9.2%) compared to 57.8% for oil, 35.2% for mining and 23.8% for history matching. This is probably because water is a public good that generates relatively small profits compared to the oil industry or mining.

### 9.4.1 *Co-authors and Repeat Co-authors from Industry*

Although having a co-author from a company or a consulting group shows that the company is interested in the new technique, it does not tell us whether they have effectively adopted it. In some cases, co-authoring a paper with an academic is rather like "window-shopping". It allows the company to test a new method on a case-study but adopting it as a standard procedure requires more time and effort (Martin and Tang 2007). Table 9.4 lists the companies and consultants which had co-authored more than 1 paper together with the number of papers, for each type of application. In applications to oil, seven companies and consulting groups had co-authored two or more papers, compared to 11 which had contributed to only 1; similarly five mining companies had co-authored two or more papers, compared to 8 which contributed to only 1 paper. It would be interesting to know what happened to the 11 oil companies that only participated in 1 paper, and likewise for the 8 mining companies. Did they lose interest in the method after an initial test study? Or did they decide to train their personnel or to outsource studies to consultants?

[2]Among the other papers, some were theoretical; a few were applications to precision agriculture or soil science. Plurigaussian simulations were even used to map the soil layers in archeological sites in ancient Rome (Folle 2009; Raspa 2000).

**Table 9.4** Companies and consultants which had co-authored more than 1 paper together with the number of papers, for each type of applications

| *Oil* | | *Mining* | |
|---|---|---|---|
| Beicip-Franlab | 5 | Areva | 2 |
| Geovariances | 3 | De Beers | 2 |
| Halliburton | 7 | Geovariances | 4 |
| Petrobras | 2 | QG (Aust) | 2 |
| Statoil | 4 | Rio Tinto | 2 |
| Sonangol | 2 | | |
| Total Oil Co | 7 | | |
| Only 1 Paper | 11 | Only 1 Paper | 8 |
| *Water* | | *History matching* | |
| Colenco | 3 | Shell | 2 |
| Geovariances | 2 | Statoil | 2 |
| | | Total Oil Co | 4 |
| Only 1 Paper | 1 | Only 1 Paper | 12 |

### *9.4.2 Surveys of Academics and Consultants*

The last part of the study consisted of a survey to find out (a) which companies had started training their personnel by sending them to short courses or to postgraduate and masters courses, and (b) which were outsourcing studies. While there are clear limitations to what can be obtained from voluntary declarations because people tend to bias their answers and while our survey was far from exhaustive, the results give us some ideas about what has happened.

Three groups (the IFP at Rueil-Malmaison, the CG at Fontainebleau and Jeffrey Yarus and Rich Chambers, in the USA) ran extensive programs of short courses. Table 9.5 lists the short courses on truncated gaussian and plurigaussian simulations given by Christian Ravenne[3] and Brigitte Doligez, both of the IFP. The Centre de Géostatistique was also active in giving short courses, often as pre-conference courses or in-house for oil companies, and the consulting and software company, Geovariances, regularly gives a 5 day course on conditional simulations applied to mining and has a 3 day course on advanced geostatistics for reservoir characterization. Both have modules on plurigaussian simulations. From 2000 to 2006, Jeffrey Yarus and Rich Chambers gave 4–5 courses per year through the Nautilus Training Organization and two more per year in Abu Dhabi for Schlumberger. After joining Landmark, they continued giving courses in Houston and London each year.

Most postgraduate geostatistics courses have modules on simulation. Some students choose this topic for their project/thesis. The Ecole des Mines de Paris has

[3]The list is available in his HDR thesis (Ravenne 2001). At the time he was Directeur Associé de Recherche at the IFP. He subsequently retired in 2008.

**Table 9.5** Short Courses on the truncated gaussian method and on plurigaussian simulations by Christian Ravenne who was a geologist at the IFPEN before his retirement, and more recently by Brigitte Doligez, who is also a geologist at the IFPEN

| Year | Teacher | Institute | Time | Company | Place |
|---|---|---|---|---|---|
| 1989 | Ravenne | IFP | 12 h | ARCO, Exxon, Conoco | USA |
| 1990 | Ravenne | IFP | 4 h | SAGA | Norway |
| | Ravenne | IFP | 3 h | Schlumberger | Norway |
| 1991 | Ravenne | IFP | 4 h | Petrofina | Belgium |
| | Ravenne | IFP | 6 h | JNOC | Japan |
| | Ravenne | IFP | 8 h | Statoil | Norway |
| | Ravenne | IFP | 6 h | Agip | Norway |
| 1992 | Ravenne | IFP | 3 h | Maersk | Denmark |
| 1993 | Ravenne | IFP | 5 h | Petrofina | Belgium |
| | Ravenne | IFP | 8 h + 4 days | Aramco | Saudi-Arabia |
| | Ravenne | IFP | 8 h | Petrobras | Brazil |
| | Ravenne | IFP | 2 h | Intervep | Venezuela |
| 1994 | Ravenne | IFP | 3 days | | China |
| 1995 | Ravenne | IFP | 2 h | Mobil | USA |
| | Ravenne | IFP | 2 h | Amoco | USA |
| 1996 | Ravenne | IFP | 4 h | Anadarko | USA |
| | Ravenne | IFP | 3 days | Chevron (Nigeria) | CG, France |
| 1997 | Ravenne | IFP | 2 h | PDVSA | Venezuela |
| 1998 | Ravenne | IFP | 3 h | Banoco | Bahrein |
| 2011 | Ravenne | IFP | | PDVSA | Venezuela |
| 2012 | Ravenne | IFP | | PDVSA | Venezuela |
| 2013 | Ravenne | IFP | | PDVSA | Venezuela |
| 2000 | Doligez | IFP | 20 h | PDVSA | Venezuela |
| | Doligez | IFP | 12 h | Chevron | Nigeria |
| 2006 | Doligez | IFP | 35 h | PDVSA | Venezuela |
| 2011 | Doligez | IFP | 30 h | PDVSA | Venezuela |
| | Doligez | IFP | 20 h × 2 | PUT | Iran |

been running a 9 month postgraduate geostatistics course called the CFSG[4] since 1980. The last 3 months are devoted to a personal project on a real case-study, usually provided by the company sponsoring the student. Similarly, final year undergraduates and masters students have carried out studies on plurigaussian simulations at the University of Chile, at Edith Cowan University (Western Australia), at the University of Adelaide (South Australia), at the federal university

[4]CFSG = Cycle de Formation Spécialisée en Géostatistique.

**Table 9.6** List of the titles of confidential reports on plurigaussian simulations by students at various universities

| Date | Student's Name | Country | Title of Project |
|---|---|---|---|
| *CFSG, Mines-Paristech, Fontainebleau, France* | | | |
| 2002 | A.S. Wain | Canada | Plurigaussian lithofacies modelling of Sparky Member, lower cretaceous Mannville Group, Saskatchewan |
| 2011 | E.M. Muller | Brazil | Simulation of orebody domains using the truncated plurigaussian method in a copper deposit |
| 2012 | C. Goncalves Monteiro Filho | Brazil | Modelling complex lithology indicators in the presence of border effects (Conceicao iron mine) |
| 2015 | S. Petiteau | France | Review of the procedure for pluri-gaussian simulations for roll-front uranium deposits, and updating an existing study |
| 2015 | Alan Rojas Kari | | Stochastic Geological Modeling and Multivariate Recoverable Resources Evaluation in a Lateritic Nickel Deposit |
| | C.C. Bohorquez Urdaneta | Venezuela | Plurigaussian study of an oil reservoir |
| *University of Chile, Santiago, Chile* | | | |
| 2006 | Karina Gonzalez | Chile | Modelamiento probabilístico de unidades geológicas y su aplicación a la evaluación de recursos minerals (Master's Mining) Data from Codelco-radomiro Tomic |
| 2008 | Alvaro de la Quintana | Chile | Simulación de unidades litológicas en el yacimiento Mansa Mina. (Undergraduate, Mining)Data from Codelco-Ministro Hales |
| 2008 | Daniel Silva | Chile | Control geológico en la simulación geoestadística de leyes. (Masters Mining) Data from Codelco-Andina |
| 2010 | Alejandro Caceres | Chile | Simulación conjunta de unidades geológicas y leyes de cobre en el sector Sur-Sur del depósito Río Blanco – Los Bronces; (Masters Geology) Codelco-Andina |
| 2014 | Ignacio Moscoso | Chile | Simulación Gaussiana truncada con incertidumbre en proporciones; (Undergraduate, Mining) Data from Codelco-Andina |
| 2014 | Giovanni Pernigotti | Chile | Simulación plurigaussiana usando proporciones locales; (Undegraduate, Mining) Data from Codelco-Andina |
| 2015 | Pia Leyton | Chile | Simulación Gaussiana truncada utilizando información de proporciones locales; (Undergraduate, Mining) Codelco-Ministro Hales |
| 2015 | Nadia Mery | Chile | Modelamiento y cosimulación de leyes en un yacimiento ferrífero; (Undergraduate Mining) Data from Vale |
| *Edith Cowan University, Perth, Australia* | | | |
| 2011 | Robin Dunn | Australia | Plurigaussian Simulation of Rock Types Using Data from a Gold Mine in Western Australia, (Masters Mathematics) Big Bell Mine |
| *University of Tehran, Iran* | | | |
| 2013 | Hassan Talebi | Iran | Separation of Rock Units and Alteration zones in Sungun poryphyry copper deposit using Plurigaussian Simulations |

**Table 9.7** Consulting studies involving plurigaussian simulations carried out by the consulting arm of the IFPEN, Beicip-Franlab, from 2000

| From | To | Company | Country |
|---|---|---|---|
| 2000 | 2015 | PDVSA | Venezuela |
| 2002 | 2007 | Sonatrach | Algeria |
| 2004 | 2015 | Petrobras | Brazil |
| 2005 | 2005 | Pemex | Mexico |
| 2005 | 2005 | Agoco | Libya |
| 2008 | 2008 | Foxtrot | Ivory Coast |
| 2012 | 2014 | KOC | Kuwait |
| 2014 | 2014 | ADCO | Abu Dhabi |

UFRGS (Rio Grande do Sul, Brazil), to mention just a few. As most of these are confidential, Google Scholar cannot find these. Table 9.6 lists the titles of projects that involved plurigaussian simulations and were carried out at various universities. One interesting feature is the number of studies that used data from the South American mining companies, Codelco and Vale, which were absent from the list of "repeat co-authors".

Lastly, the consulting arm of the IFPEN, Beicip-Franlab, kindly provided us with a list of the consulting projects involving plurigaussian simulations that they have carried out for clients (Table 9.7). The range of companies involved is striking. Almost all of them are national oil companies, many located in the Middle East.

Looking through these three tables, it is clear that the publications found by Google Scholar are really only the tip of the iceberg. Underneath, there are many unpublished dissertations and project reports carried out by final year and masters level students which remain confidential—in contrast to PhD theses which are usually available on the internet. Most of these final year and masters dissertations were carried out on company data by a student who had been given time off work to study. We believe that these studies are a key step in getting new methods into to regular use in industry. This suggests that university assessments should take account of final year projects and master's level dissertations, which is not the case at present in most countries, because this is one of the key channels for transferring new innovations into industry—at least as far as the earth sciences are concerned.

## 9.5 Conclusions and Perspectives for Future Work

Plurigaussian simulations were developed in France in the mid-1990s for simulating the internal architecture of oil reservoir in order to better predict oil and gas production. Although they were originally designed for the petroleum industry, they rapidly found applications in mining and hydrology and then for history matching in the oil industry. From France the technique diffused to other European countries, then to countries like the USA, Brazil and Chile.

This chapter uses complex dynamic networks to describe how the method diffused within the academic community. Citations found using Google scholar corresponding to the term "plurigaussian simulations" were used to track its diffusion within academia. In contrast to most citation networks where the nodes are the authors of papers and the link corresponds to co-authoring, in our network the papers themselves are the nodes which are linked when one paper cites another.

Papers were split according to the domain of the application: oil, mining, water or history matching. As expected, we found that

- Most papers were written by teams of authors (more than 3 per paper on average). Papers by single authors were usually dissertations.
- International cooperation was a common feature: 28% of papers on oil, 21% for mining and 17% for water and history matching.
- Many papers had authors from companies or consulting firms (57.8% for oil; 35.2% for mining; 23.8% for history matching) but far fewer for water (only 9.2%), probably because water is a public good whereas mining and oil companies are designed to make a profit.
- Countries with strong mining and petroleum industries were well-represented amongst the papers.
- Migration by scientists was a factor that accounted for the excellence of some countries.

To our surprise there were few patents (only 9 out of 550) and these only started in 2006 (i.e. 10 years after the initial discovery). It turned out that software could not be patented software before then. Studies on innovation consider that the presence of an author from industry demonstrates that company's interest in the innovation under study. In the earth sciences, companies often co-author papers in order to test new methods on their own data.

One of the main contributions of our chapter is to identify this "*window-shopping effect*". We consider that co-authoring a single paper does not necessarily mean that the company has really adopted the method. More effort is required to absorb new methods. Instead, we postulate that co-authoring a second paper indicates a more serious interest: we call this "*repeat co-authoring*". We found that seven oil companies and consulting groups had co-authored two or more papers compared to 11 which had contributed to only 1; similarly five mining companies had co-authored two or more papers compared to 8 which contributed to only 1 paper. It was surprising not to see South American mining companies such Codelco and Vale among the mining companies. We were also curious to find out whether the 11 oil companies and 8 mining that only co-authored 1 paper had lost interest in the method or had trained staff to carry out studies for them or had commissioned consultants to do them.

To find out what happened we carried out a survey of academics, end-users in companies and consultants. Clearly there are limitations to what can be obtained from voluntary declarations; people may bias their answers but the survey gave us some ideas about what had happened. The key results were:

- Companies like Codelco and Vale had been active in providing data for final year and master's level projects, but had not shown up as "repeat co-authors".
- A wide range of oil companies that had not published papers had chosen to provide in-house courses for personnel or had commissioned studies from Beicip-Franlab, the Consulting division of the IFP.

### *9.5.1 What Lessons Can Be Learned from the Study for Policy-Makers*

Firstly, while studies on patents can be very effective for assessing the industrial impact of new discoveries in some fields, they would have completely missed the target in this field, for two reasons: it was not possible to patent software developments until after 2005, and secondly even after that date, the new developments in mining software for these simulations were not patented.

Citation networks proved to be more effective than patents in this field. They allowed us to track the development of plurigaussian simulations within four different but inter-related academic domains and to industrial partners who publish in journals with academics. But even citations do not really allow us to get past the superficial "window-shopping" aspect of publications. Studying "repeat co-authoring" provides more in-depth insights; surveys of users give a clearer picture of whether companies are actually implementing new methods.

As Martin and Tang (2007) noted, firms and other users need to expend considerable effort to exploit scientific knowledge. In order to develop the in-house capability to carry out plurigaussian simulations, they need to acquire software and to train personnel. This study highlights the importance postgraduate training and masters' theses in transferring know-how and implicit knowledge to industry. The role of these courses in technology transfer to industry is undervalued in the current procedures for evaluating university departments.

**Acknowledgements** Many people have kindly answered our questions and provided feedback at seminars. The main ones are listed here in alphabetical order: Denis Allard, Hélène Beucher, Romain Bizet, Rich Chambers, Jean-Marc Chautru, Joao Felipe Costa, Sebastien Delamarre, Brigitte Doligez, Peter Dowd, Xavier Emery, Daiane Folle, Gaëlle Le Loc'h, Ute Mueller, Dean Oliver, Guiseppe Raspa, Christian Ravenne, Philippe Renard, Denis Schiozer, Olinto de Souza Gomes and Jeffrey Yarus.

## Appendix 9.1

When analyzing the citations we had been surprised that only 9 of the documents were patents (Table 9.8). Moreover these were all in the petroleum sector (either oil or history matching) and were lodged more than a decade after initial discovery of the method (1996). Why so few patents and why so late? One possible reason for this is that after the method had been published, the tacit knowledge was partly encapsulated in software and partly in knowing how to use the software. Firms of consultants who had acquired this knowledge, made a living carrying out case-studies for oil companies. Research universities which are also repositories of this knowledge, transmit it to students via postgraduate diploma courses, or masters or doctoral programs.

But the main reason for the lack of patents before 2006 (10 years after the initial discovery) is that oil companies and service providers only started patenting programs then. Until the late 1960s, computer programs were not considered patentable (Bender 1968); they could only be protected by copyright law. By the 1990s, it had become critical in the information economy to be able to protect IP on computer programs (Thurlow 1997). Ten years later the problem had been resolved. Merges (2007) commented: *the legal system is integrating software into the fabric of patent law, and software firms are integrating patents into the competitive fabric of the industry*. So this explains why patents only started to appear so late.

**Table 9.8** Patents

| Date | Field | Inventors | Assigned to |
|---|---|---|---|
| 2007 | Oil | Gunning, Glinsky and White | BHP Billiton |
| 2007 | Oil | Nivlet and Lucet | Nivlet and Lucet |
| 2009 | Oil | Le Ravalec-Dupin, Hu and Roggero | IFP |
| 2010 | History | Tillier, Enchery, Gervais-Couplet and Le Ravalec | IFP |
| 2012 | Oil | Maucec and Cullick | Landmark |
| 2014 | Oil | Biver, Henrion, D'or and Allard | Total SA |
| 2014 | History | Da Veiga and Le Ravalec-Dupin | IFPEN |
| 2014 | History | Tillier, Enchery and Gervais-Couplet | IFP |
| 2015 | Oil | Da Veiga and Le Ravalec-Dupin | IFPEN |
| 2015 | History | Heidari, Gervais-Couplet, Le Ravalec-Dupin and Wackernagel | IFPEN |

## References

Abello J, Buchsbaum A, Westbrook J (1998) A functional approach to external graph algorithms. In: Bilardi G (ed) Proceedings of the sixth European symposium on algorithms. Springer, Berlin

Agrawal A, Henderson R (2002) Putting patents in context: exploring knowledge transfer from MIT. Manag Sci 48(1):44–60. https://doi.org/10.1287/mnsc.48.1.44.14279

Albert R, Jeong H, Barabasi A-L (1999) Diameter of the world wide web. Nature (London) 401:130

Albert R, Jeong H, Barabasi A-L (2000) Error and attack tolerance of complex networks. Nature (London) 406:378–382

Amaral LAN, Scala A, Barthélémy M, Stanley HE (2000) Classes of small worlds. Proc Natl Acad Sci USA 97:11149–11152. https://doi.org/10.1073/pnas.200327197

Armstrong M, Beucher H, Doligez B, Galli A, Geffroy F, Le Loc'h G, Renard D, Eschard R (2011) Plurigaussian simulations in geosciences, 2nd edn. Springer, Berlin 200p. ISBN 3-540-42390-7

Barabasi AL, Jeong H, Neda Z, Ravasz E, Schubert A, Vicsek T (2002) Evolution of social networks of scientific colloaborations. Phys A 311(3–4):590–614

Bender D (1968) Computer programs: should they be patentable? Columbia Law Rev 68(N °2):241–259

Broder A, Kumar R, Magoul F, Raghavan P, Rajagoplan S, Stata R, Tomkins A, Wiener J (2000) Graph structure in the web. Comput Netw 33:309–320

Erdos P, Renyi A (1959) On random graphs I. Publicationes Mathematicae (Debrecen) 6:290–297 (1959)

Folle D (2009) Analise e aplicacoes da geostatistica no context geologico-geotechnico urbano, Doctoral Thesis, Civil Engineering, University of Rio Grande do Sul. http://www.lume.ufrgs.br/bitstream/handle/10183/15707/000689401.pdf?sequence=1

Freeman LC (1977) A set of measures of centrality based on betweenness. Sociometry 35–41

Galli A, Beucher H, Le Loc'h G, Doligez B, Heresim Group (1994) The pros and cons of the truncated Gaussian method. In: Armstrong et al (eds) Geostatistical simulations. Kluwer, Dordrecht, pp 217–233

Laigle L, Joseph P, de Marsily G, Violotte S (2013) 3-D process modelling of ancient storm-dominated deposits by an event-based approach: application to Pleistocene-to-modern Gulf of Lions deposits. Mar Geol 335:177–199

Maietta OW (2015) Determinants of university-firm R&D collaboration and its impact on innovation: a perspective from a low-tech industry. Res Policy 44(2015):1341–1359

Martin BR, Tang P (2007) The Benefits from publicly funded research, SPRU electronic working paper series N° 161, University of Sussex, June 2007. https://core.ac.uk/download/files/153/7372156.pdf

Merges RP (2007) Software and patent scope: report from the middle innings, Texas law review, vol 85, N° 1627, UC Berkeley public law research paper N° 994372. http://papers.ssrn.com/sol3/papers.cfm?abstract_id=994372

Moser P (2012) Innovation without patents: evidence from world fairs. J. Law Econ. 55(1):43–74 (2012)

Newman MEJ (2001) Scientific collaboration networks: network construction and fundamental results. Phys Rev E 64:016131

Page L, Brin S, Motwani R, Winograd T (1999) The PageRank citation ranking: bringing order to the web. Stanford InfoLab

Raspa G (2000) Il ruolo della geostatistica nella modellistica ambientale, Atti del Convegno Geostatistica per lo studio e la gestione della variabilità, Milano. http://users.unimi.it/agroecol/workshop/geostat2000.php. Accessed 16 Feb 2000

Ravenne C (2001) Thèse d'Habilitation à Diriger des Recherches, Université Pierre & Marie Curie, Paris VI, 21 Nov 2001

Thurlow LC (1997) Needed: a new system of intellectual property rights. Harv Bus Rev 94–103

Tijssen RJW, Van Leeuwen TN, van Wijk E (2009) Benchmarking university-industry research cooperation worldwide: performance measurements and indicators based on co-authorship data for the world's largest universities. Res Eval 18(1):13–24

Watts DJ, Strogatz SH (1998) Collective dynamics of small-world networks. Nature (London) 393:440–442

Wuchty S, Jones BF, Uzzi B (2007) The increasing dominance of teams in production of knowledge. Science 316:1036–1039

Zellner C (2003) The economic effects of basic research: evidence for embodied knowledge transfer via scientists' migration. Res Policy 32(2003):1881–1895

# Chapter 10
# Mathematical Geosciences: Local Singularity Analysis of Nonlinear Earth Processes and Extreme Geo-Events

**Qiuming Cheng**

**Abstract** In the first part of the chapter, the status of the discipline of mathematical geosciences (MG) is reviewed and a new definition of MG as an interdisciplinary field of science is suggested. Similar to other disciplines such as geochemistry and geophysics, mathematical geosciences or geomathematics is the science of studying mathematical properties and processes of the Earth (and other planets) with prediction of its resources and changing environments. In the second part of the chapter, original research results are presented. The new concepts of fractal density and local singularity are introduced. In the context of fractal density and singularity a new power-law model is proposed to associate differential stress with depth increments at the phase transition zone in the Earth's lithosphere. A case study is utilized to demonstrate the application of local singularity analysis for modeling the clustering frequency—depth distribution of earthquakes from the Pacific subduction zones. Datasets of earthquakes with magnitudes of at least 3 were selected from the Ring of Fire, subduction zones of Pacific plates. The results show that datasets from the Pacific subduction zones except from northeastern zones depict a profound frequency—depth cluster around the Moho. Further it is demonstrated that the clusters of frequency—depth distributions of earthquakes in the colder and older southwestern boundaries of the Pacific plates generally depict stronger singularity than those obtained from the earthquakes in their hotter and younger eastern boundaries.

## 10.1 Introduction

When this handbook is published, the International Association for Mathematical Geosciences (IAMG) is celebrating its 50th anniversary. Mathematical geosciences as a scientific discipline has become mature after half a century of development since the IAMG was established in 1968 at the 23rd International Geological

Q. Cheng (✉)
State Key Lab of Geological Processes and Mineral Resources,
China University of Geosciences, Beijing 100083, China
e-mail: Qiuming.cheng@iugs.org



B. S. Daya Sagar et al. (eds.), *Handbook of Mathematical Geosciences*,
https://doi.org/10.1007/978-3-319-78999-6_10

Congress (IGC) in Prague. It had grown from mathematical geology to mathematical geosciences by the time its name was changed at the 32th IGC held in Oslo in 2008. Not only has the subject been accepted widely within the geoscience community but the association has also been recognized for its reputation and significant influence on the earth sciences in general. IAMG has affiliations with several major geoscience organizations including the International Union of Geological Sciences (IUGS), International Statistical Institute (ISI), and the International Union of Geodesy and Geophysics (IUGG). Diverse earth science topics have been published in IAMG conference proceedings and the IAMG journals (*Mathematical Geosciences*, *Computers & Geosciences* and *Natural Resources Research*). However, we have to realize that as a relatively young discipline, MG still has not been very widely accepted and is often ignored by main stream geoscientists. While several definitions and terminologies were proposed to describe mathematical geology, there have been few attempts to define mathematical geosciences. For example, mathematical geosciences have often simply been referred as applications of mathematical and statistical methods for the analysis of geological (earth science) data and the development of quantitative predictive models (Howarth 2017). The mission of the IAMG as shown on the IAMG website was defined as promoting the development and application of mathematics, statistics and informatics in the geosciences. Whether MG should be defined as a formal discipline of science or simply as applications of mathematics in the geosciences is a fundamental question with critical impact on the development of the subject. In this chapter, I will review the status of the discipline and suggests a new definition for MG followed by examples to demonstrate what contributions of MG have been made to earth science and what the current developments in the field are. For the first part I will elaborate on MG on the basis of literature review and for the second on my own research in nonlinear MG as an example of a new field of MG.

## 10.2 What Is Mathematical Geosciences or Geomathematics?

One of the original definitions of mathematical geology was given by Vistelius (1962) and used in the name of the association: International Association for Mathematical Geology (IAMG) when it was first established in 1968. Geostatistics is one of the successful fields of IAMG, which originally was developed by MG scientists within the IAMG community. It has been used not only in the geosciences but later in many other fields of science as well. Geostatistics focuses on application of statistical methods in the earth sciences (e.g. Merriam 1970; McCammon 1975a, b) and still appears to be used by many in that sense. The term geomathematics was also used by several authors including Agterberg (1974) who used the term as the title of his two books (Agterberg 1974, 2014). After the name of the association was changed from mathematical geology to mathematical geosciences in 2008, the term mathematical geosciences more often appears in the literature of the IAMG and also in the titles of conferences, as well as in the name of its journal *Mathematical Geosciences*. When the author of the current chapter served as president of IAMG (2012–2016), dedication to IAMG was given by promoting the discipline of mathematical geosciences. Several notes on this were published in the President's Forums in IAMG newsletters (Issues 76–79th). The distinction between mathematical geology and mathematical geosciences is not simply in terminology but also in the scope of the discipline. While mathematical geology refers to a branch of geology, mathematical geosciences must be a subdiscipline of the geosciences which includes geology as one of its subfields. Other relevant subjects covered in the geosciences include but are not limited to geochemistry, geophysics, geobiology, and hydrology. Mathematical geosciences should be a discipline parallel to other subdisciplines in the geosciences such as geochemistry, geophysics and geobiology rather than a branch of geology. In the author's personal view this distinction is critical for the development of the discipline. Under the concept of mathematical geology, the subject is limited to the application of mathematics in geology but as mathematical geosciences just like geochemistry and geophysics, it serves the entire earth science. So, what should be the definition of mathematical geosciences or geomathematics and what are the roles mathematical geosciences should play in the family of geosciences? Here I will briefly elaborate on these questions and introduce several major contributions of MG to earth science. In order to provide a proper definition of mathematical geosciences, we should look at the definitions of other relevant disciplines such as geochemistry, geophysics and geobiology:

- *Geophysics as a science of "the study of the earth's physical properties and of the physical processes acting upon, above, and within the earth."* (Collins English Dictionary)
- *Geochemistry as a science that deals with the chemical composition of and chemical changes in the solid matter of the earth or a celestial body (Unabridged dictionary).*
- *Biogeosciences as an interdisciplinary field of study integrating geoscience and biological science: the study of the interaction of biological and geological processes (Unabridged dictionary).*

The definitions of the preceding relevant disciplines share the common concept of an interdisciplinary geoscience field. A similar definition was proposed by the author in 2014 with consultation of the IAMG Executive Committee Members and published in the President's Forum of IAMG newsletter (Issue No. 79).

- *As an interdisciplinary field merging mathematics, computer science and geosciences, MG is the science of studying mathematical properties and processes of the Earth (and other planets) with prediction and assessment of its resources and environments*

The ultimate question arising from this definition is what are the mathematical properties and processes of the Earth, with prediction and assessment of its resources and environments which have to be dealt with by mathematical geoscience for integration with other geoscience subdisciplines. Similar to other interdisciplinary fields including geochemistry and geophysics, mathematical subjects such as geometry, calculus, functional analysis, morphology, probability and mathematical statistics provide essential theory and methods for quantitative study of the Earth ranging from geometry and dynamics of the Earth, uncertainties of measurements, and observations for the prediction of Earth events.

## 10.3 What Contributions Has MG Made to the Geosciences?

There are many examples demonstrating that MG has made indispensable contributions to the geosciences. For example, the mathematical model of the Earth's shape (e.g. Clark ellipsoid, and Hayford ellipsoid) which serves as the foundation of geodesy, navigation systems (e.g. GPS), remote sensing technology (RS) and geographical information systems (GIS), and the fast growing field of geomatics; the mathematical model of mantle convection and models for plate motions (McKenzie and Parker 1967) serve as foundation of plate tectonics, the most notable development of earth science in the last century; mathematical symmetry and symmetry operations as principles of crystallography and optical mineralogy (e.g. in 1830, Hessel proved the existence of the 32 groups of crystal symmetry) which constitute a foundation of solid earth science; the mathematical topological model as foundation of geographical information systems (e.g. as basis of spatial data modeling in ArcGIS), one of the most useful technologies in geoscience; mathematical and statistical theories providing foundations for describing the spatial distribution and correlation of elements, uncertainty and error bars in geochemistry including isotope geochemistry and geochronology as are also used for the geological time scale; mathematical modeling and uncertainty of prediction of climate change, a pressing issue of the geosciences; probability theory and stochastic models for prediction of energy and mineral resources, highly demanded by many nations for economic and societal development; geo-complexity theory such as fractals, multifractals, chaos and self-organized criticality for modeling and predicting singular events and extreme phenomenal issues; and information extraction (big data mining, machine learning, geo-intelligence) in the geosciences, just to name a few.

As the International Association for Mathematical Geosciences, IAMG has earned its reputation by promoting and fostering its members to make contributions to science. Original and significant studies have been published in IAMG journals, books and conference proceedings. However, a large amount of work is documented elsewhere in publications which cover almost every mathematical subject and aspects of geosciences ranging from statistical data analysis, geometrical modeling, dynamics and processes simulation, to prediction and assessment of Earth system. MG theories and methods have been applied not only in tackling conventional solid earth issues such as assessment of mineral and energy resources, but also in other fields including hydrology, climate change, water resources, alternative energy resources and environmental issues. While the importance of MG in the geosciences has been increasingly demonstrated, the discipline of MG has not yet been fully recognized and, to some extent, buried in oblivion. There is hardly any hiring of highly qualified personal (HQP) in academic institutions or industry with as job title Mathematical Geoscientist or Geomathematician. As a matter of fact, most of our IAMG members are employed with job titles such as geologists, geophysicists, geochemists, geodesists, computer scientists, mathematicians and geoinformatical specialists instead of MG or GM. University students who are talented in mathematics and geosciences wanting to pursue mathematical geoscience have to enroll in geophysics or other fields simply because MG does not exist as such in university programs, at least in most of the programs in developed nations. There are very few interdisciplinary university programs except actuarial science, mathematical physics and mathematics for business, which have mathematics as integral part of their subject. A common misconception is that learning mathematics either can only result in kinds of two jobs: pure mathematician or mathematics teacher, or as a prerequisite for other careers in engineering, science or business. This might be one of the reasons there are not so many students wanting to pursue mathematics related subjects in their choice of career. Thus, MG faces significant challenges when promoting MG as a discipline and for facilitating training and education of future generations. This presents the bottleneck for the IAMG to grow further and to become a more successful and influential association.

The International Year of Mathematics of Planet Earth (MPE) celebrated in 2013 generated a much needed publicity of mathematics in geoscience. Mathematical courses are offered in all schools from primary to high school to university. Earth science is also a common choice of topic in essays by students. Integration of math and earth subjects must provide proper and interesting topics for students' math or science projects. The mathematical and geoinformatical techniques learned by students early on are already powerful tools for exploring the Earth. An excellent example is the work headlined in the media with publication by a high school student Alice R. Zhai who has analyzed 73 tropical cyclones that made landfall in US and used multivariate regression to examine the dependence of hurricane economic loss on maximum wind speed and storm size. This study (Zhai and Jiang 2014) not only proposes a new model by which hurricane damage might be predicted but also provides new evidence showing the area-density power law property of extreme events which, as is to be introduced in the remainder of this chapter, has deep origins in nonlinear dynamics.

The development of modern information technology enables everyone to easily retrieve big data to support their studies via internet and web services in a cloud environment. To access and process huge amounts of data is no longer only for paid professionals. More and more specialized software packages and multi-media teaching and training materials or online courses available in the public domain with Twitter, Facebook and You Tube, provide new ways for self-learning. Online communication, discussion and consultation through the internet in and out of the classroom have become common for students. It should encourage middle school, high school and university students to develop their curiosity in, passion for, and dedication to mathematical geosciences.

## 10.4 Frontiers of Earth Science and Opportunities of MG

IAMG has been rapidly expanding its scope from traditional geostatistics or statistical geology to more comprehensive interdisciplinary sciences for mathematically studying properties and processes of the Earth with prediction and assessment of its resources and environments. What are the current trends of MG and how are they associated with the Earth Science frontiers? It is impossible to create an accurate list of frontiers for MG. Of course, there exist several previous publications by IAMG members that have discussed past, current and future trends for the IAMG (Agterberg 2003). Here I will just share some thoughts based on my personal observations of several recent events and activities. Several international organizations have developed and published white papers illustrating prospective review on trends of scientific research within their organizations and strategic plans for the next 5–10 years; for example, the International Council for Science Union (ICSU) published its strategic research agenda for Future Earth 2025 Vision (http://www.futureearth.org/sites/default/files/future-earth_10-year-vision_web.pdf); the International Union of Geological Sciences (IUGS) is jointly with UNESCO offering the International Geological Correlation Program (IGCP) in addition to various other big science programs and new initiatives such as the Resourcing Future Generations (RFG), an international collaborative program (http://iugs.org/uploads/RFG.pdf); the US National Science Foundation (NSF) has published a strategic plan for 2014–2018 (https://www.nsf.gov/publications/pub_summ.jsp?ods_key=nsf14043); the American Geophysics Union (AGU) produced a scientific trends report (https://about.agu.org/trends-earth-space-science/); the American Natural Science Foundation published its strategic plan for tectonics (https://docgo.net/national-science-foundation-nsf-strategic-plan-fy-2006-2011-nsf-06-48); a white paper resulting from NSF sponsored workshops on "mathematics in geosciences" was published by a group of geoscientists in 2012 (https://cpb-us-e1.wpmucdn.com/sites.northwestern.edu/dist/8/1676/files/2017/10/agenda-xwphux.pdf), just to name a few. Relevant publications resulting from international conferences such as the International Geological Congress (IGCs), AGU, EGU, GSA as well as special articles in several journals such as Nature and Science have also been concerned with these issues. The following summary of key topics can be extracted from the preceding sources of information to reflect current trends and frontiers of the earth sciences. These key topics include but are not limited to data science, data analysis, big data and geo-intelligence, computation, inter-/multi-/cross-/transdisciplinary science, integrated models, uncertainty relative to observations and predictions, properties and dynamics of the planet, climate change, disruptive processes such as earthquakes and storms, and special studies of the Arctic, Antarctic and Tibet Plateau. The fundamental issues are for understanding Earth and environmental systems and their interactions with human activities, and for developing reliable monitoring systems, models, and information technologies for predictions and early warnings of large-scale and rapid change. The current challenges facing earth scientists are understanding and modeling the geo-complexity of the Earth and environmental systems with their interactions, chaotic nature and predictability of geo-processes, Earth singularity and human mitigation and adaptation to extreme events, plus observation and monitoring multiple-scale mixing nonlinear processes. Although most organizations neither recognize nor explicitly mention this, the majority of these frontiers are fundamentally related to MG. A long period of incremental advances of new mathematical theories and models in conjunction with modern technologies for solving these earth science problems may lead to creative leaps of innovation. MG has huge challenges and responsibilities facing the earth science frontiers. MG scientists are indeed at the frontier of earth science tackling fundamental problems of the Earth as can be evidenced by the recent advancements reflected in the topics of plenary presentations at IAMG conferences and in the best papers published in IAMG journals; for example, on multi-point geostatistics—a new field of spatial-temporal modeling (Mariethoz and Caers 2014); compositional data analysis—a new way to explore the composites of the Earth (Pawlowsky-Glahn et al. 2015); singularity analysis and singularity physics—new theory and methods of studying geodynamics and geo-complexity (Cheng 2007, 2017a; Agterberg 2017); big data visual analytics for exploratory data analysis; semantic web technology for geoinformation; uncertainty in ecosystem mapping by remote sensing; integrating structural geological data into inverse modeling frameworks; stationary and isotropic vector random fields on spheres; and mathematical morphology modeling, just to name a few.

## 10.5 Fractal Density and Singularity Analysis of Nonlinear Geo-Processes and Extreme Geo-Events

For the past several decades nonlinear theory and geocomplexity marked an era of new geoscience that deals with nonlinear processes and extreme phenomena which occurred in the evolution of earth systems. Irregular geometry was not popularized in the past until the term “fractal” was coined by Mandelbrot in the 1970s. Fractal geometry rapidly became a new field of mathematics dealing with roughness and irregularity of geometries. For example, fractals have been used for modeling complex and self-similar patterns generated by nonlinear processes (Mandelbrot 1972; Feder 1988). The concept of fractals and fractal dimension was further extended to multifractals involving self-similar measures defined on support which can be fractal itself (Mandelbrot 1972; Meakin 1987; Schertzer and Lovejoy 1987). Multifractal measures have been further extended to fractal density in local singularity analysis (Cheng 1999a, 2001). In the following sections the concept of fractal density will be introduced and followed by discussion and application of new methods for fractal differential operation and fractal integration (Cheng 2017a).

### 10.5.1 Fractal Density

Since the principle of density was discovered by the Greek scientist Archimedes approximately 2000 years ago, the well-known physical concept of density has become a fundamental property of mass or energy with a variety of applications. The density, or volumetric mass density, of a substance is its mass per unit volume. Density thus is a scale-independent property of material or energy treated as representing a fundamental physical parameter and variable in many physical models with applications in nearly all fields of study, ranging from physics to engineering, economics and the social sciences. Density often is characterized as unit of mass over volume (e.g., $g/cm^3$, $kg/m^3$) or energy over volume ($J/cm^3$, $w/L^3$). For example, the density of pure gold is 19.32 $g/cm^3$, which is approximately 19 times as much as for an equal volume of water. The density of quartz is 2.65 $g/cm^3$, which is much less than the density of gold. Therefore, the density of gold-mineralized quartz veins in hydrothermal mineral deposits is variable depending upon the concentration and distribution of gold in the quartz veins. Similarly, continental crust, which consists mostly of granitic rock, has a density of about 2.7 $g/cm^3$ and the Earth's mantle of ultramafic rock has a density of about 3.3 $g/cm^3$. The density of seawater varies with temperature and salinity of the water. Although the density of seawater varies at different points in the ocean, a good estimate of its density at the ocean's surface is 1025 $kg/m^3$ or 1.025 $g/cm^3$. Density of air is a temperature and pressure dependent parameter. For given temperature and pressure the density of air is independent of the volume of air. For a pure substance the density is independent of the volume of substance. However, for a heterogeneous substance density usually assumes different values depending upon purity and packaging. For example, rocks consisting of minerals with different densities have variable densities depending upon the proportions of the minerals. For a quartz vein with pure $SiO_2$ the density of the vein should be equal to the density of quartz, 2.65 $g/cm^3$. However, if the quartz vein involves gold mineralization, then the density of the quartz will be different from that of pure quartz relating to how the gold is distributed in the vein. At a location of higher concentration where a cluster of gold occurs in the quartz vein, the density of the vein is higher than that of pure quartz. From a fractal point of view, the structure of these types of gold distribution can be very irregular and then has to be described by using a non-integer or fractal dimension. Accordingly, the value of "volume" of the substance is lost. Instead the size of fractal is measurable only if it is measured in fractal dimensional space or as Hausdorff measure (Cheng 2017a). This means the ratio of mass over volume does not converge; and the density does not exist according to the ordinary density definition. In the following section it will be demonstrated that the concept of ordinary density of substance is only valid for substances with regular or ordinary structure. For substances packaged in a fractal manner, a new form of density is needed and the concept of ordinary density has to be generalized to a new form capable for quantifying the density of complex objects. It will also be demonstrated that the end products for many types of singular processes possess fractal mass density or energy density. The concepts of fractal density and local singularity analysis have been utilized in several dynamic models involving extreme processes (Cheng 2012, 2016, 2017b; Cheng and Agterberg 2009; Cheng and Sun 2017).

### 10.5.2 Density-Scale Power-Law Model and Singularity

According to the concept of ordinary density, the mass density of an object (ρ) can be calculated by the following equation:

$$\rho = \frac{m(v)}{v}, \tag{10.1}$$

where $m$(v) represents the mass contained in a volume (v) and $\rho$ is the average density of an object. If the object is homogenous then the density calculated in Eq. (10.1) becomes independent of volume. The unit of the density is determined by the ratio of the mass and volume; for example, g/cm$^3$. However, if the object has heterogeneous properties, the density may vary from place to place and the average density in Eq. (10.1) varies with different size of v, then a localized density must be calculated using the derivative of the mass over volume:

$$\rho = \frac{dm(v)}{dv} = \lim_{v \to 0} \frac{m(v)}{v}. \tag{10.2}$$

The density in Eq. (10.2) exists only if the limit converges when the volume becomes infinitesimal. If the limit does not converge, then the density doesn't exist. As a generalization of Eq. (10.2), the following new Eq. (10.3) was introduced (Cheng 1999b, 2001) in which there exists a parameter α (with positive value) so that the limit converges:

$$\rho_\alpha = \lim_{v \to 0} \frac{m(v)}{v^{\frac{\alpha}{3}}}. \tag{10.3}$$

The value of $\rho_\alpha$ can be considered as a generalized density because the ordinary density defined in Eq. (10.2) becomes a special case of Eq. (10.3) when α = 3, the normal dimension of volume. This new density was named fractal density since it is defined as mass or energy per unit of "fractal set" (Cheng 1999b, 2001). The fractal density defined in Eq. (10.3) has as unit the ratio of mass to a fractal set of α dimensions; for example, g/cm$^{\alpha}$ or kg/m$^{\alpha}$. Similarly, the units of fractal energy density can be J/cm$^{\alpha}$ or w/L$^{\alpha}$. Combining Eqs. (10.2) and (10.3) yields the following relationship between ordinary density and fractal density:

$$\rho(v) = \rho_{\alpha} v^{-[1-\alpha/3]}. \tag{10.4}$$

The notation of fractal density used in Eqs. (10.3) and (10.4) can be replaced by the following general model associating the fractal density with the ratio of mass and scale (ε—linear size of an E-dimensional set):

$$\rho(\varepsilon) = \rho_{\alpha} \varepsilon^{-[\mathrm{E}-\alpha]}. \tag{10.5}$$

This power-law relation between the ordinary density and scale is determined by two parameters: the fractal density $\rho_{\alpha}$ which is independent of scale and the exponent–singularity index α (fractal dimension), or $\Delta\alpha = \mathrm{E} - \alpha$; the latter is also known as the co-dimension of fractal density. The singularity index (Δα) measures the deviation of the fractal dimension from the dimension of normal density. These two parameters ($\rho_{\alpha}$ and Δα) can be estimated from observed data by measuring the intercept and slope of a straight line on the log-log plot of m against ε (Cheng 1999b, 2007).

### *10.5.3 Multifractal Density*

If fractals refer to geometry with irregular shapes and self-similar geometrical properties, multifractals refer to self-similar measures defined on support which can be fractal (Mandelbrot 1983). Multifractals are defined as spatially intertwined fractals with variable fractal dimensions (e.g., Mandelbrot 1972; Cheng 1997). According to the distribution of measures (similar to the mother functions of sets) the support can be grouped into subsets which can be fractal with specific fractal dimension. Accordingly, there are two types of multifractal measures: continuous and discrete multifractals, the former refers to multifractals corresponding to an infinite number of intertwined fractals with continuous fractal dimension spectrum, whereas the latter refers to the limit number of intertwined fractals with discrete fractal dimensions (Cheng 1997). Multifractal measures are self-similar measures with multiple scale singularities which can be characterized by the Hőlder exponent (Mandelbrot 1989). In the multifractal paradigm the measure defined on a support can be expressed as

$$\langle \mu(\varepsilon) \rangle \propto \varepsilon^{\alpha}, \tag{10.6}$$

where μ(ε) represents the measure defined on a set of linear scale ε, ∝ stands for ''proportional to'' when cell size ε approaches to zero, and α is the singularity index also known the Hőlder exponent (Mandelbrot 1989). This power law exists usually in a statistical sense and is represented as expectation <>. According to the distribution of α values, the entire support can be classified into subsets or fractals each with different singularity and accordingly different fractal dimensions. This is why it has been termed "multifractal". The distribution of singularity α in the mapped area can be described by the fractal dimension spectrum function *f*(α). The values of singularity and multifractal spectra can be estimated by several methods including box-counting and gliding-box based moment methods, and the wavelet method (Cheng 1999b). Singularity property has been commonly observed in geochemical and geophysical quantities (Cheng et al. 1994; Cheng 1999b, 2007). Since the common moment-based multifractal models are implemented according to partition functions of measures with additive property, most literature about multifractals focuses on the power law relations of multifractal measures and self-similarity of multifractal measures and few have neither emphasized the physical meaning nor the property of density of the multifractal measure. A density—area fractal model was proposed (Cheng et al. 1994) to associate the concentration with area of multifractal measure as

$$A(\geq C) \propto C^{-\beta}, \tag{10.7}$$

where the area (A) is a function of element concentration above the threshold C. The model has also been applied to characterize other types of "concentration" such as density of faults per area (Agterberg et al. 1996), density of mineral deposits per area (Cheng and Agterberg 1996), stream density per drainage area (Cheng et al. 2001), and digital number of remote sensing images (Cheng and Li 2002), just to name a few. Further utilizing the idea of C-A model locally, the following power law relation was introduced to associate the density of multifractal measures with scale (Cheng 1999b)

$$\rho(\varepsilon, x) = c(x)\varepsilon^{-[E-\alpha(x)]}, \tag{10.8}$$

where E is the Euclidean dimension of the support (e.g., E = 1 for line, 2 for area and 3 for volume), x indicates the location, and c(x) and α(x) are constants with respect to scale ε but varying with location. The values of α(x) and c(x) can be estimated from the values $\rho(\varepsilon, x)$ calculated for different sizes ε around the location x by means of least squares using log-log paper. Both values can be mapped for visualization and interpretation. For convenience without loss of generality, in the rest of the paper the notation of x will be dropped from the formulation and the equation is assumed to hold locally. The singularity index α and constant c have the following properties (Cheng 1999b): if α = E, then $\rho(\varepsilon)$ = constant, independent of vicinity (scale) size ε; if α > E then $\rho(\varepsilon)$ is a decreasing function of ε which implies the convex property of $\rho(\varepsilon)$; and α < E then $\rho(\varepsilon)$ is an increasing function of ε which implies the concave property of $\rho(\varepsilon)$. Thus, the ordinary density obeys a power-law relationship with scale which has the following properties (Cheng 1999b, 2007):

$$\lim_{\varepsilon \to 0} \rho = \begin{cases} 0, & if\ \alpha > \mathrm{E}, \\ \infty, & if\ \alpha < \mathrm{E}, \\ c, & if\ \alpha = \mathrm{E}. \end{cases} \tag{10.9}$$

In accordance with these properties, ordinary density becomes volume dependent when $\alpha \neq \mathrm{E}$ and it tends to either zero or infinity when the scale ε becomes infinitesimal. The constant c in Eq. (10.8) can be expressed in the following form:

$$c = \lim_{\varepsilon \to 0} \rho(\varepsilon)\varepsilon^{E-\alpha} = \lim_{\varepsilon \to 0} \frac{\mu(\varepsilon)}{\varepsilon^{\alpha}}, \tag{10.10}$$

The constant c indeed is a convergent value of the ratio of measure (μ) over scale (ε) with fractal dimension. This quantity is usually termed scaling factor but it can be termed as a fractal density or Hausdorff density in analogy to the mass density which corresponds to ratio of measure over ordinary geometry with integer dimension (Cheng 2015). Therefore, while a unit of ordinary density is $\mathrm{g/m^E}$, the unit of fractal density becomes $\mathrm{g/m^{\alpha}}$.

### 10.5.4 Fractal Density Structure and Clustering Distribution

The terminology of fractal density has been explained in several papers with different emphases, but the meanings of the concepts used are variable. For example, the term "fractal density" has been used to refer the number of fractals per area (Hou and Wu 1989) which does not mean the same as the concept introduced in the current paper. Tatekawa and Maeda (2001) analyzed time evolution of fractal density perturbations in the Einstein-de Sitter universe, in which the emphasis is on how the perturbation evolves and what kind of nonlinear structure will come out. Similarly, Federrath et al. (2009) has used fractal density structure in supersonic isothermal turbulence when referring to density structure. Gromov et al. (2001) used fractal density to describe fractal galaxy distribution. Carpinteri et al. (2009) used the term to describe the mean fractal density of microcrack barycenters. Pope and Mackenzie (1988) introduced the concept of fractal density for describing the morphology of fractal growth model in the evolution of gels from solution. They define the fractal density ρ which follows the relation

$$F = \frac{\rho}{\rho_0} = \left(\frac{r_0}{r}\right)^{3-D}, \tag{10.11}$$

where D is the fractal dimension of fractal growth, the F is the relative fractal density at radius r ($r \geq r_0$), with $r_0$ and $\rho_0$ being the core radius and core density, respectively. The core acts mathematically as a reference point for calculating the decrease in density as the fractal increases in size. A similar clustering fractal growth density function was used to describe tumor growth in fractal space-time with temporal density (Paramanathan and Uthayakumar 2011).

From the preceding publications we can see that in earlier studies by other authors the term of fractal density was introduced mainly for description of morphology and patterns of fractals and fractal growth modeling. The current research introduces the fractal density as a generalization of ordinary density of substance or energy to represent a fundamental new parameter or variable involved in dynamic systems.

## 10.6 Fractal Integral and Fractal Differential Operations of Nonlinear Functions

As mentioned in Eq. (10.2) for heterogenetic matter or substances, the derivative of mass over scale can be used for defining localized density of substance. Accordingly, the mass or volume of a heterogenetic substance can be calculated using integration. Obviously, integration and differentiation are two fundamental operations in calculus and used for many mathematical and physical subjects. The traditional integral and differential operations are defined on the basis of additive property of Lebesgue measure. When the measure no longer possesses additive property, then the classical integral and differential may not exist. Therefore, the ordinary integral and differential operations are not applicable to fractal density with singularity. The author has proposed the following fractal integral and differential (Cheng 2017a)

$$f_{\alpha}^{'}(x_0) = \frac{df(x)}{dx^{\alpha}} = \lim_{\Delta x \to 0} \frac{\Delta f(x)}{(\Delta x)^{\alpha}} = \lim_{x \to x_0} \frac{f(x) - f(x_0)}{(x - x_0)^{\alpha}}, \tag{10.12}$$

where $\Delta f(x)$ and $\Delta x$ represent the increments of a function $f(x)$ for an increment of x. The convergence of the limit in Eq. (10.12) can be defined as the α-fractal derivative of the function $f(x)$. Similarly, we can define the fractal integral of the function $f(x)$ as follows

$$\int f(x) dx^{\alpha} = \lim_{\Delta x \to 0} \sum f(x_i)(\Delta x)^{\alpha}, \tag{10.13}$$

where $f(x_i)$ is the magnitude of the function $f(x)$ over the small range $[x_i, x_i + \Delta x]$. If the limit of Eq. (10.13) converges, then it can be named the α-fractal integral of the function $f(x)$. It must be kept in mind that the fractal derivative defined in this paper is different from the fractional derivative (fractional order) known in the literature as $f^{(v)}(x)$, where v can be a non-integer order. The fractional derivative assumes that the normal integer order derivative $f^{(n)}(x)$ does exist. The fractal derivative is based on fractal dimension of the measure whereas the fractional derivative is based on fractional order of derivative defined on normal measure. As an example, let us take a power-law function to demonstrate the fractal derivative. Assume a power law function, $f(x) = c(x - x_0)^b$, with ordinary derivative of the function $f'(x) = cb(x - x_0)^{b-1}$, which does not exist at $x = x_0$ if $0 < b < 1$. The integral of the function then is $\int f(x)\,dx = c/(b+1)(x-x_0)^{b+1}$, which does not converge if $b < -1$ at $x = x_0$. According to Eq. (10.12), the fractal derivative at $x = x_0$ exists and $f_\alpha'(x) = c$, if $\alpha = b$, or $f_\alpha'(x) = 0$, if $\alpha < b$ and $f_\alpha'(x) = \infty$ if $\alpha > b$.

A new concept of Hausdorff derivative underlying the Hausdorff dimension of metric space/time was proposed by Chen (2006) who introduced the systematic mathematical operation of Hausdorff derivative with applications to derive a linear anomalous transport–diffusion equation underlying an anomalous diffusion process. The Hausdorff derivative operation proposed by Chen (2006) is expressed as follows

$$\frac{\partial f(x)}{\partial x^\alpha} = \lim_{x \to x_0} \frac{f(x) - f(x_0)}{x^\alpha - x_0^\alpha} = \frac{\partial f(\hat{x})}{\partial \hat{x}}, \tag{10.14}$$

This formalism was termed the Hausdorff derivative of a function $f(x)$ with respect to fractal measure $x^\alpha$.

It has to be pointed out that the fractal derivation defined in Eq. (10.12) is different from that defined in Eq. (10.14) considering that, in general, if $x_0 \neq 0$, then

$$(\Delta x^\alpha) = (x - x_0)^\alpha \neq \Delta x^\alpha = x^\alpha - x_0^\alpha. \tag{10.15}$$

The two sides in Eq. (10.15) become equal only if $x_0 = 0$. Otherwise, according to Taylor expansion, we can obtain $\Delta x^\alpha = x^\alpha - x_0^\alpha = \alpha x_0^{\alpha-1}\Delta x + o(\Delta x)$, so substitution into Eq. (10.14) gives

$$\frac{\partial f(\hat{x})}{\partial \hat{x}} = \lim_{x \to x_0} \frac{f(x) - f(x_0)}{x^\alpha - x_0^\alpha} = \frac{1}{\alpha x_0^{\alpha-1}} \frac{\partial f(x)}{\partial x}, \tag{10.16}$$

which implies that the derivative of $f(x)$ defined in Eq. (10.14) is indeed corresponding to the ordinary derivative except for the factor $\frac{1}{\alpha x_0^{\alpha-1}}$. Reconsidering the example used previously with $f(x) = c(x - x_0)^b$, the derivative of Eq. (10.14) at $x = x_0$ does not exist if $b < 1$.

## 10.7 Earth Dynamic Processes and Extreme Events

In the remainder of this chapter I demonstrate that fractal density ($\Delta\alpha \neq 0$) characterizes anomalous mass accumulation or energy release caused by extreme geo-processes, which occurred in the Earth's lithosphere originated from cascade earth dynamics (plumes, mantle convection and plate tectonics) and self-organized criticality involved in phase transitions (avalanches of slab breakoffs, faults, and volcanic eruptions).

Mantle convection at high Rayleigh number generates thermal plumes episodically which upon arrival in the crust could cause major flood basalt events, igneous provinces as well as spreading of continents and mid ocean ridges (Richards et al. 1989; White and McKenzie 1989). On a larger scale, Wilson cycles (Wilson 1966) corresponding to the periodic fragmentation and reformation of supercontinents could be linked to temporal variability in plate tectonics. Numerous studies have revealed that mantle convections can induce exchange of mass between upper and lower mantle across the endothermic phase transition zone at about 660 km. The cold downwelling material penetrates into the lower layer and, simultaneously, the hot upwelling fluid is pushed into the upper layer. The exchange of mass between the upper and lower mantle layers can occur in short bursts (often described with superlatives such as "catastrophic", overturn, "avalanche" subduction, or "super-plumes") (Zhong and Gurnis 1994). The quick injection of lower mantle hot fluid into the upper mantle can cause not only mantle heterogeneity but also anomalous thermal distribution near the surface (Le Bars and Davaille 2004). This has been considered to be the first order cause of vigorous magmatism. Deep subductions of continental crust into the deep earth interior and rebounded back to the surface of the Earth have been ascertained by the discoveries of regional metamorphic coesite (Chopin 1984; Smith 1984), and subsequently by unusual ultrahigh pressure (UHP) terranes (Hacker and Gerya 2013).

Within the lithosphere there are various types of "catastrophic" events occurring during plate subduction. Formation of magmatic arc can be caused by subduction in which the subducting or subducted oceanic crust material releases volatiles (e.g. $H_2O$ and $CO_2$) which cause partial melting of the mantle and form magma at depth under the overriding plate. Earthquakes occur at certain depths at the edges of three types of plate boundaries: convergent (subductions and collisions), divergent, and transformative.

### *10.7.1 Phase Transition*

From mathematical and physical points of view, the mechanisms that have been proved to exist correspond to the generation of power-law distributions including but not limited to phase transition (PT), self-organized criticality (SOC) and multiplicative cascade processes (MCP) (Newman 2005; Lovejoy et al. 2009). I will elaborate on each of these mechanisms in relation to mantle convections, plumes and lithosphere rheology induced tectonic events. The phase of a thermodynamic system and the state of matter in a normal system have uniform physical properties. Common phases include liquid phase, solid phase and vapor phase of chemical components which exist under certain pressure and temperature (P-T) conditions. Materials in different phases have their distinct properties such as liquid usually having higher density and smaller specific volume in comparison with gas. However, in phase transition conditions, multiple phases coexist within the same system such as liquid and vapor in magma and hydrothermal systems under proper P-T conditions. At a critical condition (critical point on phase diagram) liquid and vapor become indistinguishable and beyond this point the fluid and gas become so-called supercritical fluid, representing a special phase of matter which can effuse through solids like a gas, and dissolve materials like a liquid (McMillan and Stanley 2010). The critical point for water occurs at temperature (374 °C) and pressure (22 MPa). It has been found that the critical point is so peculiar that close to it, small changes in pressure or temperature result in large changes in density and other density related properties such as viscosity, relative permittivity, heat capacity and solubility. The special critical point phenomena can be expressed by the following empirical power law functions (Sengers and Levelt Sengers 1968, 1986):

$$\Delta\rho = c(\Delta P)^{1/3}, \quad \Delta\rho = c(\Delta \mathrm{T})^{1/2}, \tag{10.17}$$

where Δρ, ΔP, and ΔT represent the changes of density, pressure and temperature, respectively along the coexistence curve. These power-law relations hold for small changes of temperature or pressure from the condition at the critical point of the system. Although the two functions of Eq. (10.17) show continuity at zero increment with $\Delta\rho = 0$, $\Delta P = 0$, and $\Delta T = 0$, the first order derivatives of density versus either temperature or pressure (change rate of density difference) do not exist or show singularity at $\Delta P = 0$ and $\Delta T = 0$ as shown in the following forms

$$\frac{\Delta\rho}{\Delta \mathrm{T}} = c\Delta T^{-1/2}, \frac{\Delta\rho}{\Delta \mathrm{P}} = c\Delta P^{-2/3} \tag{10.18}$$

These properties describe the phenomena of property change such as fractal density (density jump) at the phase transition zone. In addition, the ratio of increments of temperature and pressure depict power-law relations $\frac{\Delta \mathrm{P}}{\Delta \mathrm{T}} = c\Delta P^{-1/3}$. Such power-law relation implies that the Clapeyron slope could become infinity or a singularity when approaching the coexistence curve. Clapeyron slope and density jump are critical parameters in numerical simulation of mantle convection; for example, Korenaga (2004) developed a numerical model to simulate mantle mixing and continental breakup magmatism by assigning a Clapeyron slope of −2 MPa/K and a density jump of 10% for the endothermic phase transition at 660 km depth. The episodicity of convection induced by the endothermic phase changes strongly depends on plate length, rheology, and Clapeyron slope (Zhong and Gurnis 1994).

Ogawa and Yanagisawa (2014) have developed models with small Clapeyron slope −0.2 to −1 MPa/K for simulating convections from punctuated layered convection to whole-mantle convection in modeling mantle evolution on Venus due to magmatism and phase transitions. Their models indicate that the earlier stage layered mantle convection is punctuated by repeated bursts of hot material from the deep mantle to the surface. Other phenomena of phase transition may occur at the boundary of deeply subducted slabs. Due to subduction of oceanic lithosphere underneath the continental lithosphere, solid phase lithosphere can be partially melted to facilitate formation of magma. During the progress of subduction, $H_2O$ and other volatile components contained in the rocks are progressively released from the slab at different depths. Fluids or melts released at greater depths will be in supercritical fluid phase which hydrates the mantle and causes partial mantle melting. This eventually leads to deeply rooted magma which provides the source for magmatic and volcanic arcs located above the subduction zones. Partial melting in lower crust and mantle also causes strain rate change of the lithosphere which facilitates formation of intermediate and deep earthquakes (Dimanov et al. 2000). The processes of fluid release and migration are complex and, to a large extent, their details still remain unknown. Due to the great depth of subduction the fluid released may be in supercritical condition with, as mentioned earlier, fractal density with strong solvent strength facilitating the hydration and metasomatism of mantle rocks. When the pressure and temperature are reduced to around the critical point, the system goes through a great reduction of gradient of density, accordingly increasing the specific volume which further enlarges porous space and fractures rocks thus in turn facilitating the formation of magma and earthquakes through positive feedback processes.

### 10.7.2 *Self-organized Criticality*

The phenomena associated with continuous phase transitions are called critical phenomena, and these are often related to so-called self-organized criticality (SOC). SOC is commonly illustrated conceptually with avalanches resulting from piles of sand which generate a power-law number-size distribution of avalanche magnitudes (Bak et al. 1987). At the criticality point in a SOC phenomenon a small continuous input to the system can cause sudden and discontinuous outputs or avalanches. For example, a fault occurs in broken brittle rock strata when an extra stress is added to change the system at the criticality point. The size and number of faults generated may follow a power law distribution with a small number of large faults and a large number of small faults. SOC is similar to critical point phase transition since both processes involve anomalous state change caused by a minor continuous input pulse at the critical condition point. Numerous studies have also pointed out the effect of the 660-km endothermic phase transition on convection. This could actually generate the periodic occurrence of abrupt changes in convective mode (660-km layered/whole mantle), consecutive with the sudden flushing of oceanic plates previously accumulated above the transition zone (e.g., Le Bars and Davaille 2004).

Many numerical simulations have demonstrated multiple scale and sizeable whole mantle convection, and sublithospheric convection can bring up dense fertile mantle materials from the lower mantle to the upper mantle (Korenaga 2004). Cold downwellings are temporarily stopped by the 660 km endothermic phase change but sink rapidly into the lower mantle (Tackley et al. 1993). The intermittence of layering reflects accumulation and release of negative buoyancy above the endothermic phase boundary (Machetel and Weber 1991; Tackley et al. 1993). The exchange of mass between upper and lower layers can occur in short bursts (Zhong and Gurnis 1994). Although these types of avalanching behaviors are not as easy to test as those of sand piles, one might reasonably assume that these types of processes with SOC nature can generate end products with power law distributions. As a matter of fact, SOC phenomena have been commonly considered to describe extreme geo-events in plate tectonics. Such examples may include but are not limited to earthquakes (Gutenberg and Richter 1944; Turcotte 1997), volcanic eruption durations (Cannavò and Nunnari 2016), plate sizes (Sornette and Pisarenko 2003), slab breakoff (Condie 1998), areal size of magmatism (Pelletier 1999), mineral deposits (Agterberg 1995; Cheng 1999b; Maier and Groves 2011), heat flow over mid-ocean ridges (Cheng 2016), episodic evolution of supercontinents and crustal growth (Cheng 2017b), and energy—probability of earthquakes (Cheng and Sun 2017). Other examples can be found in the book authored by Sornette (2004). The processes involved in response to the preceding extreme events create end products which can be described by frequency—size or frequency—time power law relations. Based on the above reasoning, we may expect lithospheric root detachments and slab breakoffs that occurred during subduction are of difference sizes which follow power-law distributions. Some of these small-sized events may not be noticeable on the surface due to small impact on the global system, but the large detachments and slab breakoffs can cause significant impact on syn- to post-collisional magmatism and metamorphism. The size—frequency distribution of these types of events can be modelled by the following general power-law relation

$$\mathrm{N}(>\mathrm{A}) = c\mathrm{A}^{-\mathrm{b}}, \tag{10.19}$$

where A represents the size of event and N(>A) the cumulative number of events with size greater than the threshold A. This power-law function involves two constant values: c and b. For example, the well-known Gutenberg-Richter power-law distribution relates the number of large earthquakes to their sizes (Gutenberg and Richter 1944; Turcotte 1997). The exponent, b-value, has been commonly used for predictive purposes. The exponential b-value was found to be internally related to singularity in terms of fractal probability density (Cheng and Sun 2017) with

$$E(<P) = E_0 P^{-\frac{1}{\beta}}, \tag{10.20}$$

where E(<P) represents the minimum energy released by large earthquakes, with occurrence probability less than P. This equation indicates that the minimum energy released by large earthquakes follows a power-law relation $(\beta = \frac{2}{3}b)$ for probability of earthquake occurrence with energy greater than E. This model implies that the smaller the probability (P) of a large earthquake, the larger its energy release (E).

### 10.7.3 Multiplicative Cascade Processes

Multiplicative cascade processes (MCP) are iterative multiplicative processes across multiple scales, which involve positive or negative feedback to generate extreme values that follow multifractal power-law distributions (power-law distributions with multiple exponents) with self-similarities and singularities (Meakin 1987; Scherzter and Lovejoy 1987; Agterberg 2007; Cheng 2014). Examples of MCP are common in the study of geocomplexity such as formation of clouds, severe weather and storms (Scherzter and Lovejoy 1987; Malamud et al. 1996; Turcotte 1997; Veneziano and Furcolo 2002), to just name a few. In terms of mantle convection, the convection processes can be viewed as multiplicative cascade processes that create heterogeneity of the mantle by recycling the materials from upper crust to mantle. On a large scale, Wilson cycle cascade evolution involves the opening and closing of an individual oceanic basin, plate drift, plate subduction and plate collision, involving the recycling of lithosphere material and causing extreme events at the interface of phase transition zones or zones around plate boundaries. Depending on the properties of subduction and other factors, plate subduction may cause slab deformation, erosion and breakoff, deep subduction, and collision of continents. These events are responsible for formation of extreme events such as magmatism and earthquakes. During such processes changes of pressure and temperature as well as water content often provides a positive feedback effect on causes of melting or partial melting of lithosphere and the generation of magma reservoirs and seismicity. In the context of multiplicative cascade processes, the mass and energy distribution resulting from these processes often are proved to have self-similarity and singularity which can be modelled by multifractal distributions (Meakin 1987; Schertzer and Lovejoy 1987; Cheng and Agterberg 2009).

The aforementioned mechanisms (PT, SOC and MCP) can coexist in the evolution of earth dynamics systems which cause cascade effects for anomalous diffusion and strain rate originating earthquakes or magmatism creating flare up formation of magmatic activity or cluster frequency-depth distribution of earthquakes. Based on possible mechanisms (PT, SOC and MCP) corresponding to power-law distributions, the fractal density (power-law density) and the singularity analysis method can be used to characterize the causational relations between extreme events such as magmatic activities and earthquakes and the aforementioned nonlinear mechanisms. In the following section a case study of earthquakes will be used to demonstrate the effect of phase transition on formation and distribution of earthquakes that occur along Pacific plate subduction zones.

## 10.8 Fractal Density of Lithosphere Rheology in Phase Transition Zones and Association with Earthquakes

### *10.8.1 Rheology Constitutive Equation*

In the study of earth tectonics, rheology is an important concept describing rock properties with respect to flow behavior which can be characterized through the following empirical constitutive equation associating stress and strain rate (e.g., Dimanov et al. 1998).

$$\dot{\varepsilon} = A\sigma^n d^{-m} f_{H_2O}^r e^{-\frac{Q+PV}{RT}}, \tag{10.21}$$

where $\dot{\varepsilon}$ represents the strain rate, σ—the stress, n—the stress exponent; d represents the grain size, m is the grain-size exponent, $f_{H_2O}$—the water fugacity, and r—the fugacity exponent, Q—the activation energy, P—the pressure, V—the activation volume, T—the absolute temperature, while R is the molar gas constant, and A—a material constant. The constitutive Eq. (10.21) is often utilized in the literature for describing rheology of ductile crust and since it is so well-known it often is provided without citation and reference. Several authors have investigated this equation by various methods such as by physical experiments (Pharr and Ashby 1983; Dimanov et al. 1998). The parameters involved in the equation can be estimated using a log-linear model except for the last combined term

$$\log(\dot{\varepsilon}) = logA + nlog(\sigma) - mlog(d) + rlog(f_{H_2O}) - \frac{Q+PV}{RT}. \tag{10.22}$$

Effects of some of the parameters have been summarized by several authors (e.g., Bürgmann and Dresen 2008). For example, diffusion-controlled deformation is linear in stress with n = 1. Different inverse dependencies on grain size have been predicted for lattice diffusion– and grain boundary diffusion–controlled creep with m = 2 and m = 3, respectively. Creep of fine-grained materials involves grain boundary sliding, which may be controlled by grain boundary diffusion (n = 1) or by dislocation motion (n = 2). For climb-controlled dislocation creep, deformation is commonly assumed to be grainsize insensitive (m = 0) with a stress exponent of n = 3–6 (e.g., Bürgmann and Dresen 2008). Materials for which strain rate is proportional to stress raised to a power n > 1 are referred to as having a power-law rheology, whose effective viscosity ($\mu = \sigma/\dot{\varepsilon} \propto \sigma^{1-n}$) decreases when stress increases. The significant effects of melt distribution on the rheology of rocks have been reported by many authors (e.g., Dimanov et al. 1998, 2000). In general, the strain rate is proportional to the water fugacity. The general bivariate relations between the strain rate and other factors considered in the equation are valid and can be applied to characterize the general associations of factors considered in the system (Wang 2016; Dimanov et al. 2000). However, the equation is valid for normal media that generally do not possess singularity for non-zero values of the factors. It is neither possible to use this equation to describe the singular behaviors of constitutive equation in phase transition nor to directly use it to delineate zones of phase transition. Variable depth-frequency distribution of crustal earthquakes and lithological compositions are often integrated to characterize crust deformation in relation to variations of tectonic styles (Mouthereau and Petit 2003). In the following section my attempt is to derive a proper equation to characterize the rheology in phase transition zones.

### 10.8.2 Rheology and Phase Transition

In order to explain the phase transition zones in the lithosphere associating the effect of phase transition with origin of seismicity and magmatism, one needs to link the rheology to depth of lithosphere. It has been generally accepted that in the brittle crust, frictional strength increases linearly with depth. Phase transitions separate regions into groups of rocks dominated by quartz, feldspar and olivine, respectively; and these regions are characterized by brittle or plastic properties of lithosphere (e.g., Jackson 2002; Bürgmann and Dresen 2008). It was suggested by Sibson (1974) that brittle strength in the crust can be approximated by the Sibson's formulation in which the coefficients of friction and cohesion for pre-fractured rocks are equal to internal friction and cohesion for intact samples:

$$\sigma = \sigma_1 - \sigma_3 = \beta \rho g z (1 - \lambda), \tag{10.23}$$

where $\sigma = \sigma_1 - \sigma_3$ represents differential stress, z is depth, ρ is average density of the overburden, g is acceleration of gravity, β is a coefficient which depends on the type of faulting, and λ represents the pore fluid ratio. Under hydrostatic pressure, λ is 0.36, and it is 0 and 0.7 for dry and wet conditions, respectively (Mouthereau and Petit 2003). In order to discuss the behavior of rheology around phase transition, let us define depth at the center of the phase transition zone as $z_0$, which will serve as reference of coordinate for further comparison. Let us also denote a small distance increment (in depth) around the phase transition zone as $\Delta z = \text{abs}(z - z_0)$, and the corresponding increment of differential stress around the phase transition zone as $\Delta\sigma = \text{abs}\{(\sigma_1 - \sigma_3)(z) - (\sigma_1 - \sigma_3)(z_0)\}$. When Δz is very small around the phase transition center $z_0$, then we can derive the following approximation assuming changes of depth z, β and λ are neglectable:

$$\Delta\sigma \propto \Delta\rho, \tag{10.24}$$

According to the phase transition property of density and temperature or pressure similar to Eq. (10.17) we can assume the mass density of lithosphere around the phase transition center to be

$$\Delta\rho \propto (\Delta T)^b. \tag{10.25}$$

Further assuming that the temperature and depth increments are linearly associated when the depth increment is very small, we obtain

$$\Delta\rho \propto (\Delta T)^b \propto (\Delta z)^b, \tag{10.26}$$

Therefore, the derivative of Eq. (10.26) satisfies

$$\frac{\Delta\rho}{\Delta z} \propto (\Delta z)^{b-1}, \tag{10.27}$$

This result implies that change rate ($\frac{\Delta\rho}{\Delta z}$) of density with depth follows a power-law relation with the increment of depth (Δz). If the exponent b is less than 1, the change rate approaches infinity when $\Delta z \to 0$, which implies that the change rate of differential stress, according to Eqs. (10.27) and (10.24), can become infinitely large. Assuming the other factors to be negligibly small in Eq. (10.21) when Δz is very small, we obtain

$$\frac{\Delta\dot{\varepsilon}}{\Delta z} \propto (\Delta z)^{b-1}, \tag{10.28}$$

If the exponent b is less than 1, then the change of strain rate per increment of depth approaches infinity when $\Delta z \to 0$. It must be reminded that the derivation of the new Eqs. (10.24–10.28) is based on several assumptions involving first order approximations of factors which may need further theoretical justification (detailed discussion will be published elsewhere). Nevertheless, the results obtained here might be the first power-law model providing possible quantitative description of the singularities of differential stress at the phase transition as indicated in the schematic diagram (Fig. 10.1).

### *10.8.3 Frequency—Depth Fractal Density Distribution and Singularity Analysis of Earthquakes*

In order to demonstrate the effect of differential stress caused by phase transition on formation and distribution of earthquakes, several datasets of earthquakes with magnitudes three or above were selected for several small regions along the Ring of Fire, the Pacific plate boundaries. Data were downloaded from the USGS website under the section of USGS Earthquake Hazards Program (https://earthquake.usgs.gov/earthquakes/map/). The locations of the 30 small areas selected from Aleutian Islands, Kuril Islands, Mariana, Tonga Trench, Mexico, northern Chile and southern Chile are shown in Fig. 10.2. Several hundreds to thousands of earthquakes are

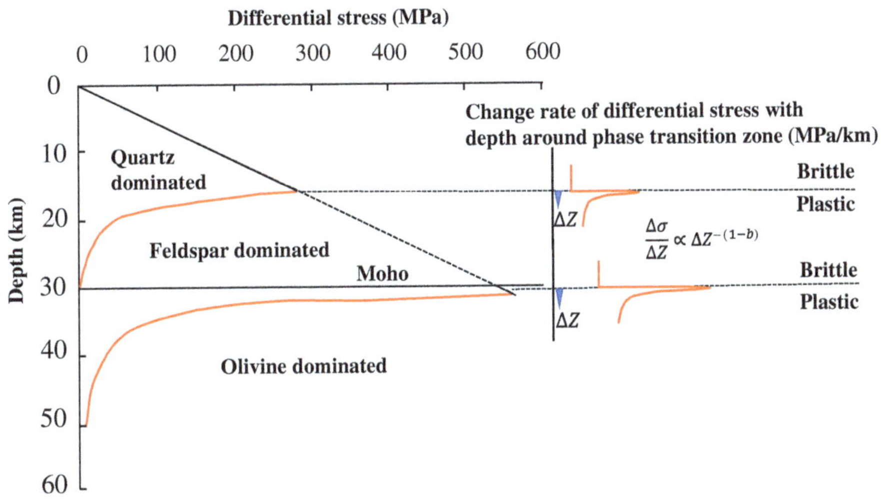


**Fig. 10.1** Strength envelopes of differential stress versus depth for a general lithospheric condition to illustrate the potential effects of phase transition. The equations are about increment rate of differential stress around the depth of phase transition zone. Notations and discussions about the equations are given in the text

selected in each area. These areas were chosen within a short range from the plate boundaries to ensure they contain enough large earthquakes which occurred along subduction zones with similar properties.

The main purpose of the case study here is to validate whether earthquakes that occurred in the subduction zones possess clustering with fractal density; therefore, we choose earthquakes in the depth around the Moho ranging from 30 to 100 km. Considering the issue of depth of shallow earthquakes being set a "normal" depth of 33 km or default depths of 5 or 10 km when depths are poorly constrained by available seismic data, we only analyze the earthquakes with occurring depth ranging 34 to 100 km. The numbers of earthquakes in each dataset were grouped on the basis of 10-km depth frequency bins. A profound peak of frequency distributions can be observed around 33 km in all datasets except for western California. To reduce the effect of the "default peak" at depth 33 km, further analysis of the frequency data will be based on earthquakes with depth from 34 km downward. As an example, the frequency—depth distribution of 1263 earthquakes with magnitude greater or equal to 3 and depths between 34 to 100 km from the Tonga region are shown in Fig. 10.3a with the data grouped in a bin of 10 km (frequency—depth distributions for other datasets are not shown here). This graph shows a profound frequency peak at 34–44 km. By eye examination one can see the frequency around the peak within 60 km (from 34 to 94 km) decaying rapidly from the location of the peak at 34 km downward. To validate the fractal density of frequency clustering distribution, the following local number-depth density of earthquakes around the peak $z_0$ was constructed

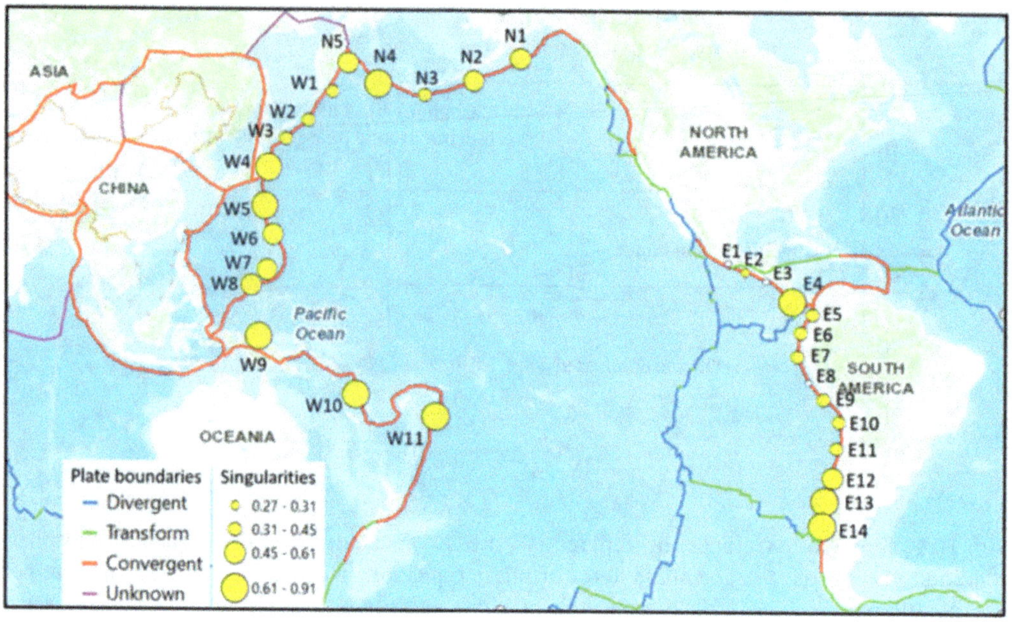


**Fig. 10.2** Study areas located along the Pacific plate boundaries. Data containing earthquakes with magnitudes M $\geq$ 3, and their depths were downloaded from the USGS website. The yellow dots represent the location of study area and the size of the dot represent level of singularity calculated using the model introduced in the current paper

$$\rho(\Delta z) = \frac{total\, number\, of\, earthquakes\, in\, depth\, range\, z_0 + \Delta z}{\Delta z} = c\Delta z^{-b}, \quad (10.29)$$

where $\Delta z$ is the window size from $z_0$, c and b are two parameters to be estimated using the local singularity analysis method (LSA) with windows of multiple sizes: $\Delta z$ = 10, 20, …, 60 km. The results are calculated for all 30 datasets. Several selected examples are shown in Fig. 10.3b–h. There is no significant peak at 33 km in the datasets from the areas of western California. The decay curves in Fig. 10.3 are least squares fittings to the data with power-law functions. The results estimated from the six datasets give b = 0.90 (E13), 0.44 (E7), 0.27 (E2), 0.49 (N2), 0.55 (N5), 0.69 (W4) and 0.74 (W11) respectively. Coefficients of determination for the least squares fittings to all six datasets are high with $R^2 > 0.98$ (student t-value > 14), indicating statistically significant power-law models fitted to the data.

The results obtained by local singularity analysis of all 30 datasets (except E1, E3, E8) demonstrate that the frequency—depth distributions for large earthquakes (M $\geq$ 3) are not uniformly distributed but show clustering which can be modelled by using the local fractal density model of Eq. (10.29). The datasets E1, E3 and E8 show linear decay instead of power-law decay. Moreover, the results (shown as yellow dots in Fig. 10.2) demonstrate that the frequency—depth density distributions of earthquakes from the southwestern boundaries of the Pacific plates depict stronger singularities than those of earthquakes from the southeastern boundaries of

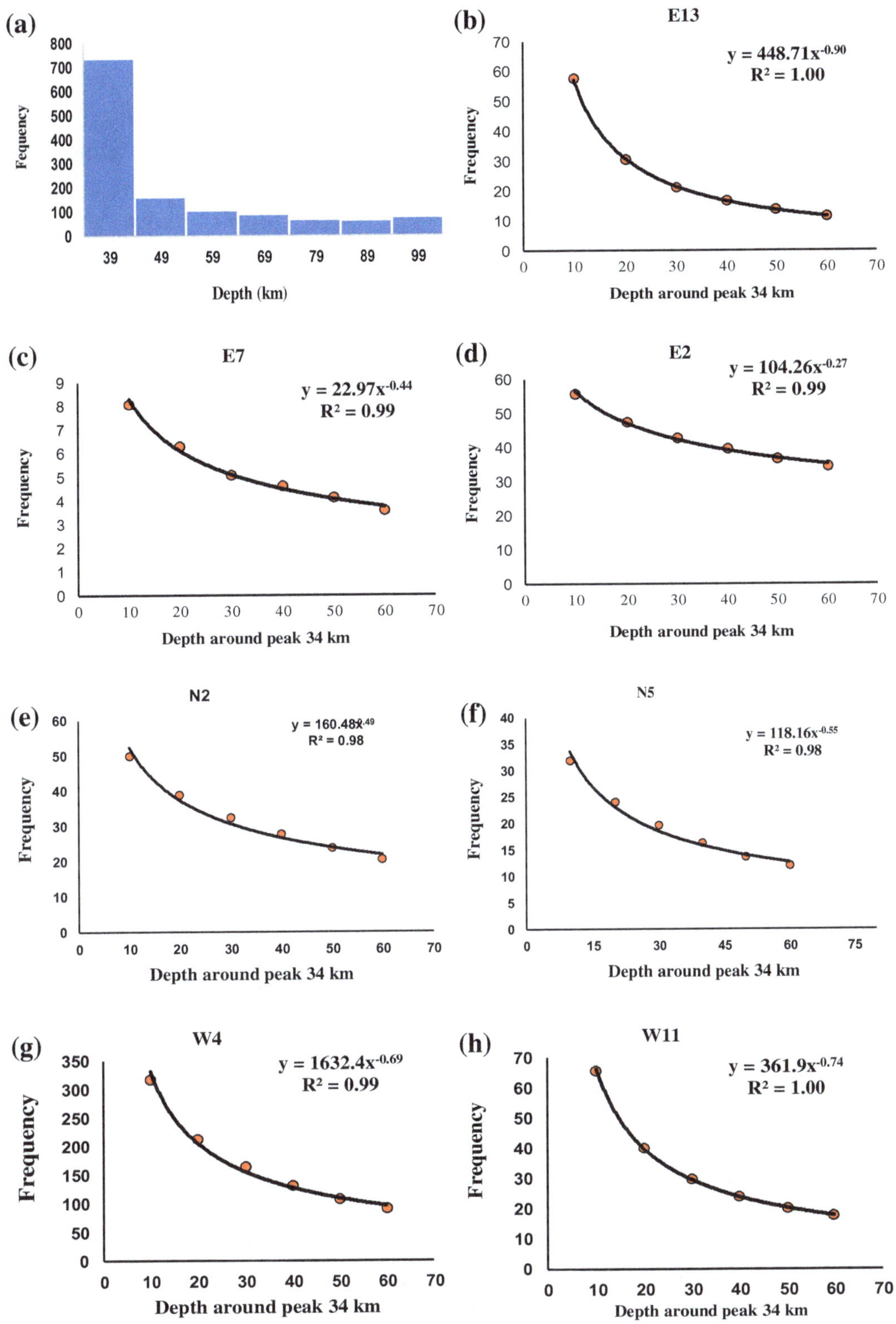


**Fig. 10.3** Distribution of frequency density of earthquakes with magnitudes equal to or greater than 3 from around Moho at 34 km downward. **a** Frequency—depth distribution of earthquakes from Tonga region; **b–h** Distribution of decay of frequency density of earthquakes (#/km) with depths from around peak at 34 km downward; Power-law functions were fitted to the observed data by least squares

the Pacific Plates except the earthquakes in conjugation regions of three plate boundaries (e.g., N4, N5, W4, W5, W9-W11, E4, E13, E14) that depict stronger singularity. This finding might be significant for understanding the different mechanisms causing earthquakes between the eastern and western Pacific plate boundaries. As reported in the literature, the western boundaries of the Pacific plates are generally colder and older in comparison with the eastern boundaries (Kong et al. 2016; Okazakl and Hirth 2016). Low slab temperatures resulting from faster subduction cause deeper earthquakes (Wei et al. 2017). Omori et al. (2004) have studied association of the distribution of dehydration events with earthquakes and found non-linear correlation between maximum depth of earthquake and temperature of the slab, with lack of deep earthquakes in young subduction-zones. Their work showed that deeper earthquakes (> 300 km) are mostly located in the selected areas along the western subduction zones of Pacific plates whereas fewer deep earthquakes occurred at the eastern boundaries of Pacific plates. The results of the current research may provide supplementary information about singularity of frequency-depth distribution of shallow earthquakes around Moho in the subductions zones of the Pacific plates. The local singularity analysis may provide a new tool for characterization and distinguishing between earthquakes from a fractal and self-similarity point of view. Further work will extend the analysis to cover more areas and other depths of earthquakes. Other sizes of earthquakes will also be considered.

## 10.9 Discussion and Conclusions

In the first part of the chapter, the purpose of including suggestions about mathematical geosciences or geomathematics as a discipline and introduction to examples of significant contributions of mathematical geoscience scientists to science was to appeal to the public and geoscientists to appreciate the indispensable role that MG can play in the family of geosciences. In the second part of the chapter, the fractal density model was introduced and used for characterizing the power-law rheology of phase transition, and singularity analysis of earthquakes from subduction zones of Pacific plates was demonstrated to be a new and promising nonlinear MG method for modeling extreme and "avalanche" geo-events. Examples of application of singularity analysis not only include earthquakes as introduced in the current chapter but also other types of extreme events such as magmatic flare ups (Cheng 2017a), mid ocean ridge anomalous heat flow (Cheng 2016), flooding caused by tropic storms (Cheng 2008), and mineral deposits as well as ore-caused anomalies in surface media (Cheng 2007). Further comprehensive analysis of earthquakes from other regions and clustering depths will be published in separate papers.

**Acknowledgements** Thanks are due to Professor Frits Agterberg for critical review of the paper and constructive comments. I thank professor B. S. Daya Sagar and Petra van Steenbergen for allowing the extra time to complete the manuscript. Mr. Shubing Zhou is thanked for assistance on preparing the datasets. The research has been jointly supported by the National Key Technology R&D Program of China (No. 2016YFC0600501) and the State Key Program of National Natural Science of China (41430320).

## References

Agterberg FP (2003) Past and future of mathematical geology. J China Univ Geosc 14(3):191–198

Agterberg FP (1974) Geomathematics: mathematical background and geo-science applications. Elsevier Science Ltd., p 596

Agterberg FP (1995) Multifractal modeling of the sizes and grades of giant and supergiant deposits. Int Geol Rev 37:1–8. https://doi.org/10.1080/206819509465388

Agterberg FP (2007) Mixtures of multiplicative cascade models in geochemistry. Nonlinear Process Geophys 14:201–209

Agterberg FP (2014) Geomathematics: theoretical foundations, applications and future developments. Springer, p 553

Agterberg FP (2017) Pareto–lognormal modeling of known and unknown metal resources. Nat Resour Res 26:3–20

Agterberg FP, Cheng Q, Brown A, Good D (1996) Multifractal modeling of fractures in the Lac du Bonnet batholith, Manitoba. Comput Geosci 22:497–507

Bak P, Tang C, Wiesenfeld K (1987) Self-organized criticality: an explanation of 1/f noise. Phys Rev Lett 59:381–384

Bürgmann R, Dresen G (2008) Rheology of the lower crust and upper mantle: evidence from rock mechanics, geodesy, and field observations. Annu Rev Earth Planet Sci 36:531–567

Cannavò F, Nunnari G (2016) On a possible unified scaling law for volcanic eruption durations. Sci Rep 6(22289). https://doi.org/10.1038/srep22289

Carpinteri A, Lacidogna G, Niccolini G, Puzzi S (2009) Morphological fractal dimension versus power-law exponent in the scaling of damaged media. Int J Damage Mech 18:259–282

Chen W (2006) Time–space fabric underlying anomalous diffusion. Chaos, Solitons Fractals 28:923–929

Cheng Q (1997) Discrete multifractals. Math Geol 29:245–266

Cheng Q (1999a) Multifractal interpolation. In: Lippard SJ, Naess A, Sinding-Larsen R (eds) Proceedings of the fifth annual conference of the international association for mathematical geology, Trondheim, Norway, vol 1, pp 245–250

Cheng Q (1999b) Multifractality and spatial statistics. Comput Geosci 25:949–961

Cheng Q (2001) Singularity analysis for image processing and anomaly enhancement. In: Proceedings IAMG, 2001. http://mmc.geofisica.unam.mx

Cheng Q (2007) Mapping singularities with stream sediment geochemical data for prediction of undiscovered mineral deposits in Gejiu, Yunnan Province, China. Ore Geol Rev 32:314–324

Cheng Q (2008) A combined power-law and exponential model for streamflow recessions. J Hydrol 352:157–167

Cheng Q (2012) Singularity theory and methods for mapping geochemical anomalies caused by buried sources and for predicting undiscovered mineral deposits in covered areas. J Geochem Explor 122:55–70

Cheng Q (2014) Generalized binomial multiplicative cascade processes and asymmetrical multifractal distributions. Nonlinear Process Geophys 21:477–487

Cheng Q (2016) Fractal density and singularity analysis of heat flow over ocean ridges. Sci Rep 6:19167. https://doi.org/10.1038/srep19167

Cheng Q (2017a) Fractal density and singularity analysis of extreme Geo-processes. First Complex Systems Digital Campus World E-Conference 2015, pp 395–405

Cheng Q (2017b) Singularity analysis of global zircon U-Pb age series and implication of continental crust evolution. Gondwana Res 51:51–63

Cheng Q, Agterberg FP (1996) Multifractal modeling and spatial statistics. Math Geol 28:1–16

Cheng Q, Li Q (2002) A fractal concentration–area method for assigning a color palette for image representation. Comput Geosci 28:567–575

Cheng Q, Agterberg F (2009) Singularity analysis of ore-mineral and toxic trace elements in stream sediments. Comput Geosci 35:234–244

Cheng Q, Agterberg FP, Ballantyne SB (1994) The separation of geochemical anomalies from background by fractal methods. J Geochem Explor 51:109–130

Cheng Q, Russell H, Sharpe D, Kenny F, Qin P (2001) GIS-based statistical and fractal/multifractal analysis of surface stream patterns in the Oak Ridges Moraine. Comput Geosci 27:513–526

Cheng Q, Sun H (2017) Variation of singularity of earthquake-size distribution with respect to tectonic regime. Geosci Front. https://doi.org/10.1016/j.gsf.2017.04.006

Chopin C (1984) Coesite and pure pyrope in high-grade blueschists of the western Alps: a first record and some consequences. Contrib Miner Petrol 86:107–118

Condie KC (1998) Episodic continental growth and supercontinents: a mantle avalanche connection? Earth Planet Sci Lett 163:97–108

Dimanov A, Dresen G, Wirth R (1998) High-temperature creep of partially molten plagioclase aggregates. J Geophys Res 103(B5):9651–9664

Dimanov A, Wirth R, Dresen G (2000) The effect of melt distribution on the rheology of plagioclase rocks. Tectonophysics 328:307–327

Feder J (1988) Fractals. Plenum Press, New York, London

Federrath C, Klessen RS, Schmidt W (2009) The fractal density structure in supersonic isothermal turbulence: solenoidal versus compressive energy injection. Astrophys J 692:364–374

Gromov A, Baryshev Y, Suson D, Teerikorpi P (2001) Tolman-Bondi model, fractal density and Hubble law. Gravit Cosmol 7:140–148

Gutenberg B, Richter CF (1944) Frequency of earthquakes in California. Bull Seismol Soc Am 4:185–188

Hacker BR, Gerya TV (2013) Paradigms, new and old, for ultrahigh-pressure tectonism. Tectonophysics 603:79–88

Hou J, Wu Z (1989) Temperature dependence of fractal formation in ion-implanted a-Ge/Au bilayer thin films. Phys Rev B 40:2008–2012

Howarth RJ (2017) Dictionary of mathematical geosciences: with historical notes. Springer, p 893

Jackson J (2002) Strength of the continental lithosphere; time to abandon the jelly sandwich? GSA Today 12:4–10

Kong XC, Li SZ, Suo YH, Guo LL, Li XY, Liu X, Somerville ID, Dai LM, Zhao SJ (2016) Hot and cold subduction systems in the Western Pacific Ocean: insights from heat flows. Geol J 51:593–608

Korenaga J (2004) Mantle mixing and continental breakup magmatism. Earth Planet Sci Lett 218:463–473

Le Bars M, Davaille A (2004) Whole layer convection in a heterogeneous planetary mantle. J Geophys Res 109:B03403. https://doi.org/10.1029/2003JB002617

Lovejoy S et al (2009) Nonlinear geophysics: why we need it? EOS 90:455–456

Machetel P, Weber P (1991) Intermittent layered convection in a model mantle with an endothermic phase change at 670 km. Nature 350:55–57. https://doi.org/10.1038/350055a0

Malamud BD, Turcotte DL, Barton CC (1996) The 1993 Mississippi flood: a one hundred or one thousand year event. Environ Eng Geosci II:479–486

Mandelbrot BB (1972) Possible refinement of the lognormal hypothesis concerning the distribution of energy dissipation in intermittent turbulence. In: Rosenblatt M, Van Atta C (eds) Statistical models and turbulence, pp 331–351, La Jolla, California. Lecture notes in physics, vol 12. Springer, New York

Mandelbrot BB (1983) The fractal geometry of nature. WH Freeman and Co., New York, p 495
Mandelbrot BB (1989) Multifractal measures, especially for the geophysicist. Pure Appl Geophys 131:5–42
Maier W, Groves D (2011) Temporal and spatial controls on the formation of magmatic PGE and Ni-Cu deposits. Miner Deposita 46:841–857
Mariethoz G, Caers J (2014) Multiple-point geostatistics: stochastic modeling with training images. Wiley, p 376
McCammon RB (ed) (1975a) Concepts in geostatistics. Springer, New York, NY
McCammon RB (1975b) On the efficiency of systematic point-sampling in mapping facies. J Sediment Petrol 45:217–229
McKenzie D, Parker RL (1967) The north pacific: an example of tectonics on a sphere. Nature 216:1276–1280
McMillan PF, Stanley HE (2010) Going supercritical. Nat Phys 6:479–480. https://doi.org/10.1038/nphys1711
Meakin P (1987) Diffusion-limited aggregation on multifractal lattices. A model for fluid-fluid displacement in porous media. Phys Rev A 36:2833–2837. https://doi.org/10.1103/physreva.36.2833. PMID 9899187
Merriam DF (ed) (1970) Geostatistics: a colloquium. In: Proceedings of a colloquium on geostatistics held on campus at the University of Kansas, Lawrence on 7–9. Plenum Press, New York, NY
Mouthereau F, Petit C (2003) Rheology and strength of the Eurasian continental lithosphere in the foreland of the Taiwan collision belt: Constraints from seismicity, flexure, and structural styles. J Geophys Res 108(B11):2512. https://doi.org/10.1029/2002JB002098
Newman MEJ (2005) Power laws, Pareto distributions and Zipf's law. Contemp Phys 46:323–351
Ogawa M, Yanagisawa T (2014) Mantle evolution in Venus due to magmatism and phase transitions: from punctuated layered convection to whole-mantle convection. J Geophys Res Planets 119:867–883. https://doi.org/10.1002/2013JE004593
Okazakl K, Hirth G (2016) Dehydration of lawsonite could directly trigger earthquakes in subducting oceanic crust. Nature 530:81–84
Omori S, Komabayashi T, Maruyama S (2004) Dehydration and earthquakes in the subducting slab: empirical link in intermediate and deep seismic zones. Phys Earth Planet Inter 146:297–311
Paramanathan P, Uthayakumar R (2011) Tumor growth in the fractal space-time with temporal density. In: Balasubramaniam P (eds) Control, computation and information systems. Communications in computer and information science, vol 140. Springer, Berlin, Heidelberg, pp 166–173
Pawlowsky-Glahn V, Egozcue JJ, Tolosana-Delgado R (2015) Introduction. In: Modelling and analysis of compositional data. Wiley, Chichester, UK. https://doi.org/10.1002/9781119003144.ch
Pelletier JD (1999) Statistical self-similarity of magmatism and volcanism. J Geophys Res 104:15,425–15,438
Pharr GM, Ashby MF (1983) On creep enhanced by a liquid phase. Acta Metall 31:129–138
Pope EJA, Mackenzie JD (1988) Theoretical modelling of the structural evolution of gels. J Non-Cryst Solids 101:198–212
Richards MA, Duncan RA, Courtillot VE (1989) Flood basalts and hotspot tracks: plume heads and tails. Science 246:103–107
Schertzer D, Lovejoy S (1987) Physical modeling and analysis of rain and clouds by anisotropic scaling of multiplicative processes. J Geophys Res 92:9693–9714
Sengers JV, Levelt Sengers A (1968) The critical region. Chem Eng News 46:104–121. https://doi.org/10.1021/cen-v046n025.p104
Sengers JV, Levelt Sengers A (1986) Thermodynamic behavior of fluids near the critical point. Annu Rev Phys Chem 37:189–222
Sibson RH (1974) Frictional constraints on thrust, wrench and normal faults. Nature 249:542–544
Smith DC (1984) Coesite in clinopyroxene in the Caledonides and its implications for geodynamics. Nature 310:641–644

Sornette D (2004) Critical Phenomena in Natural Sciences: Chaos, Fractals, Self-organization and Disorder, 2nd edn. Springer, New York, p 528

Sornette D, Pisarenko V (2003) Fractal plate tectonics. Geophys Res Lett 30:1105. https://doi.org/10.1029/2002gl015043

Tackley PJ, Stevenson DJ, Glatzmeir GA, Schubert G (1993) Effects of an endothermic phase transition at 670 km depth on spherical mantle convection. Nature 361:137–160

Tatekawa T, Maeda K (2001) Primordial fractal density perturbations and structure formation in the Universe: 1-Dimensional collisionless sheet model. Astrophys J 547:531–544

Turcotte DL (1997) Fractals and chaos in geology and geophysics, 2nd edn. Cambridge University Press, 398 pp

Veneziano D, Furcolo P (2002) Multifractality of rainfall and scaling of intensity duration frequency curves. Water Resour Res 38:42-1–42-12

Vistelius AB (1962) Problemy matematičeskoj geologii. Vklad v istoriju voprosa [Problems of mathematical geology. A contribution to the history of the problem]. Geologiya i Geofizika 12:3–9

Wang Q (2016) Homologous temperature of olivine: Implications for creep of the upper mantle and fabric transitions in olivine. Sci China Earth Sci 59:1138–1156

Wei SS, Wiens DA, van Keken PE, Cai C (2017) Slab temperature controls on the Tonga double seismic zone and slab mantle dehydration. Sci Adv 3:e1601755

White R, McKenzie D (1989) Magmatism at rift zones: The generation of volcanic continental margins and flood basalts. J Geophys Res 94:7685–7729

Wilson JT (1966) Did the Atlantic close and then re-open? Nature 211:676–681

Zhai AR, Jiang JH (2014) Dependence of US hurricane economic loss on maximum wind speed and storm size. Environ Res Lett 9:064019

Zhong S, Gurnis M (1994) Role of plates and temperature-dependent viscosity in phase change dynamics. J Geophys Res 99:15903–15917

# Part II
# General Applications

# Chapter 11
# Electrofacies in Reservoir Characterization

**John C. Davis**

**Abstract** Electrofacies are numerical combinations of petrophysical log responses that reflect specific physical and compositional characteristics of a rock interval; they are determined by multivariate procedures that include principal components analysis, cluster analysis, and discriminant analysis. As a demonstration, electrofacies were used to characterize the Amal Formation, the clastic reservoir interval in a giant oil field in Sirte Basin, Libya. Five electrofacies distinguish categories of Amal reservoir rocks, reflecting differences in grain size and intergranular cement. Electrofacies analysis guided the distribution of properties throughout the reservoir model, in spite of the difficulty of characterizing stratigraphic relationships by conventional means.

## 11.1 Introduction

The primary responsibility for reservoir modeling is in the hands of petroleum engineers, but the most successful reservoir modeling projects have included quantitative input from geologists and geophysicists. However, geologists with the necessary mathematical and computer skills are scarce, so there has been a tendency to rely instead on commercial software that runs factory-set defaults to perform geological and petrophysical modeling, even though statistical software can readily be adapted to perform many of the operations that are useful for geological reservoir modeling. These include statistical analyses of properties derived from well logs, cores and downhole measurements and investigations to determine the best geostatistical parameters for static modeling, evaluating relative effectiveness of seismic attributes, and estimating reservoir fluid properties such as hydraulic flow units. As an example, we will consider the calculation and use of electrofacies in the characterization of a giant clastic reservoir, the Amal field of Libya.

---

J. C. Davis (✉)
Heinemann Oil GmbH, 918 Jersey, Box 353, Baldwin City, KS 66006, USA
e-mail: jdavis@h-oil.com; jcdbaldwin@gmail.com


B. S. Daya Sagar et al. (eds.), *Handbook of Mathematical Geosciences*,
https://doi.org/10.1007/978-3-319-78999-6_11

## 11.2 The Amal Field of Libya

The first commercial discoveries of oil in the Sirte Basin of Libya were made in 1958, and in 1959 the first giant field in Libya was found in the Sirte Basin. Five more giant fields were discovered in the same year, including the Amal field discussed here. By the end of the 1960s, the Sirte Basin was established as one of the premier oil provinces of the world (Hallett 2002).

Most reservoirs in the major fields of the Sirte Basin have been in production for 50 years or more and are now nearing depletion. In an effort to extend the lives of fields, the Libyan National Oil Company (NOC) has authorized numerous reservoir studies in the hope that they will disclose previously untapped reserves or suggest improved production strategies. Fortunately, seismic, well, and production information is available for many fields, which permits detailed modeling of reservoirs and the investigation of production alternatives.

The Amal field is located on a wedge-shaped tilted fault block called the Rakb High, one of a series of elongated, subparallel horsts and grabens in the eastern part of the Sirte Basin. The primary reservoir interval is the Amal Formation, a typical transgressive clastic sediment composed of weathered material derived from the underlying basement. Most of the formation is a "tight, hard, quartzose, irregularly feldspathic sandstone" (Roberts 1970). Radiometric studies date the Amal Formation as Cambro-Ordovician to Permian, although a few Triassic fossils have been recovered from lacustrine shales within the formation. Elsewhere in Libya similar transgressive basal sandstones overlying the Hercynian unconformity are called the "Nubian Sandstone" and assigned a Lower Cretaceous age (El-Hawat et al. 1996). The Amal clastics were deposited in continental environments, with some small irregular intervals of possibly lacustrine and shallow marine origin. Thin volcanic sills and flows of Permian age also occur sporadically in the formation, as do local unconformities. The Amal is present everywhere on the Rakb High except at the south end of the uplift where it has been removed by erosion.

## 11.3 Electrofacies Analysis

"Electrofacies" are unique combinations of petrophysical log responses that reflect specific physical and compositional characteristics of a rock interval cut by a borehole. The term "electrofacies" was coined by Serra and Abbot (1980), who considered electrofacies to be proxies for lithofacies. An important advantage of electrofacies over alternative types of facies classifications of rocks in the subsurface is that electrofacies can be defined solely on the basis of well log responses, without reliance on cores, cuttings or outcrops. Although electrofacies are empirical, they are also objective; no subjective interpretations of sediment genesis or inferences about depositional environments are required.

There is no specific procedure for defining electrofacies. The general requirements are that they be determined from a consistent set of petrophysical log measurements; that the similarities between down-hole intervals are expressed quantitatively from the log responses; that the intervals are consistently divided into subsets that have similar responses; and that the distinctions between subsets are expressed as mathematical functions. Because of the enormous amount of data contained in the log suites from a collection of wells, it is necessary that electrofacies be determined by computer (Kiaei et al. 2015). This introduces the practical requirement that electrofacies be defined by a programmable algorithm.

Many procedures for determining electrofacies have been proposed in the literature (Berteig et al. 1985; Busch et al. 1987; Delfiner et al. 1987; Tetzlaff et al. 1989; Anxionnaz et al. 1990; Hernandez–Martinez et al. 2013; Euzen and Power 2014) and most commercial software packages for subsurface modeling have electrofacies functions. Unfortunately, details about how these functions perform are seldom revealed, and the procedures operate as "black boxes." (Exceptions are the description of Schlumberger's FACIOLOG procedure given by Wolff and Pelissier-Combescure 1982, and the software provided by Lee et al. 2002). Almost all commercial implementations consist of a combination of principal components analysis, cluster analysis, and discriminant analysis. These underlying methodologies can be duplicated using a multivariate statistical package, which has the advantages of flexibility and transparency, although perhaps less convenient for routine electrofacies calculations. Dubois et al. (2007) provide a comparison of alternative statistical methodologies for electrofacies analyses. Perez et al. (2005) have demonstrated that electrofacies are superior to other types of reservoir characterizations such as lithofacies or hydraulic flow units (HFU).

The general definition of "facies" is "the aspect, appearance, and characteristics of a rock unit, usually reflecting the conditions of its origin; especially as differentiating the unit from adjacent or associated units" (Neuendorf et al. 2005). The definition continues to more specialized varieties of facies, noting that "sedimentary facies" consist of a restricted part of a lithostratigraphic body with a unique lithology or fossil content, or a certain environment or mode of origin such as "red-bed facies." A "petrographic facies" is a body of rock of a distinctive lithology, while a "biofacies" contains a unique assemblage of fossil organisms. "Environmental facies" consist of a body of rock formed in a specific environmental setting, such as a "fluvial facies" or a "near-shore facies." The term "facies" may also refer to rocks defined on a paleogeographic or paleotectonic basis, such as a "geosynclinal facies" or a "continental margin facies."

Note that all of these definitions require either information that can only be obtained from direct observation of the rocks themselves (lithologies, fossils), or subjective interpretations about the origins or depositional environments in which the rocks were formed. In contrast, electrofacies are based solely on the "…aspect, appearance, and characteristics…" of petrophysical logs, and not of the rocks which the logs represent. The basic assumption in electrofacies interpretation is that a unique combination of log properties represents a rock that exhibits a unique combination of physical properties—in other words, the rock is unique in terms of its composition and fluid content.

### *11.3.1 Choice of Log Traces for Electrofacies Calculation*

Ideally, there will be a large suite of logs available for calculating electrofacies and the tool responses to be used can be chosen based on resolution and response to properties of primary interest. In practice, especially in areas where drilling and logging has taken place over many years, finding a common set of logs that is available in all (or most) wells severely limits the choice. In the electrofacies study discussed here, only the DT, GR and ILD logs were common to all wells in the field. However, by removing a small number of wells from consideration, the suite of logs could be expanded to include the SN and SP logs.

### *11.3.2 Standardization of Log Traces*

It is essential that the log measurements used in electrofacies calculations be consistent throughout the stratigraphic section in the well being analyzed, and from one well to another. This can be done in a variety of ways. Some commercial programs such as Schlumberger's *Petrel* do this by converting the data into principal component scores and then computing electrofacies from scores rather than from the log data itself. Although principal components were calculated here for display purposes, we prefer to compute electrofacies directly from the original log variables after appropriate transformations.

Log standardization consists of subtracting the mean log response over an interval of interest from every log reading in the interval and dividing the remainder by the standard deviation of the response in the interval. This converts the reading into dimensionless units of standard deviation, most of which will range in value from –3 to +3 (Davis 2002). Each log trace is standardized independently of all other log traces in a well, and the traces in each well are standardized independently of all other wells. This (1) removes any effects caused by differences in measurement units (ohm-meters, millivolts, microseconds/ft, etc.). It also insures (2) that all logs used in the analysis equally influence the classification of the electrofacies because all the logs have the same average value (their means are all 0.0) and their spreads in values are approximately the same (their standard deviations are all equal to 1.0). Furthermore, (3) any differences between wells caused by different hole conditions or different logging parameters are removed. In petrophysical terms, standardization of the log tracks for individual wells can be regarded as an ultimate form of well log normalization.

We can regard the transformed well log data as consisting of a matrix or flat file whose columns contain the standardized well log traces and whose rows are measured depths or elevations in specific wells. Further computations are done treating the row vectors as individual multivariate "objects" to be classified.

### 11.3.3 Estimating the Number of Distinct Electrofacies

Because electrofacies are defined empirically, the number of different electrofacies is somewhat arbitrary. The number of useful electrofacies is partly dependent on the number of log properties used in their calculation and the joint nature of the statistical distributions of the log measurements. It also reflects the purpose of electrofacies classification and the manner in which the final classification will be evaluated and used. A simple distinction between reservoir and non-reservoir rock may be made with an electrofacies classification of only two classes, while a study for environmental interpretation may require a dozen or more classes.

Because there is a limited number of well logs that measure different physical properties in the example used here, we anticipate that an effective electrofacies interpretation will not involve many facies classes. Determining the appropriate number requires trial-and-error, starting with many classes and reducing the number to eliminate trivial categories that include only a few rare observations, or to combine ill-defined classes that have very similar properties. The same trial-and-error process can be used to evaluate alternative procedures such as different clustering algorithms.

Figure 11.1 is a cross-plot of the first and second principal components of log responses from the Amal Formation. The scatter diagram represents 12,535 well log observations classified into seven electrofacies; each electrofacies category is indicated by a color (1 = red; 2 = green; 3 = blue; 4 = orange; 5 = light blue; 6 = purple; 7 = yellow). Categories 3 and 4 are relatively small and consist of scattered observations located on the periphery of the main cloud of observations; a classification with fewer categories might be better. The classification procedure was repeated with six categories, then with five, and finally with only four. Five electrofacies seemed to be an optimal compromise in which the facies are general enough to include significant thicknesses of intervals, but not so detailed that they defy interpretation (Fig. 11.2). The distribution of observations among the five classes is shown in a principal component scatter plot in Fig. 11.3.

### 11.3.4 Assigning Well Log Intervals to Electrofacies

There are two basic approaches to the assignment of log intervals to electrofacies, referred to generally as *supervised* and *unsupervised classification*. The first requires prior definition of the facies categories, which is usually done by

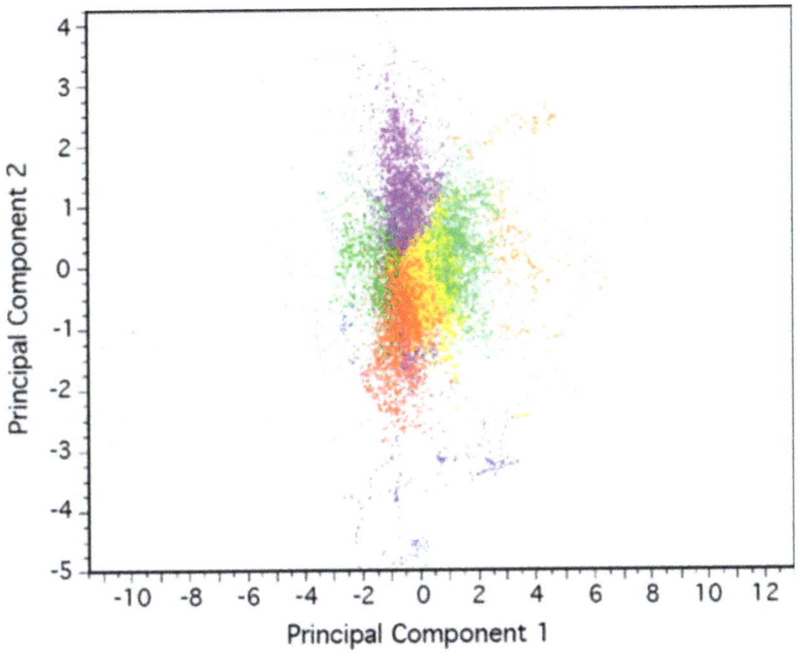


**Fig. 11.1** Cross plot of first two principal component scores of GR, DT, ILD, SN and SP log responses from Amal Formation in 15 wells of the Amal field, Libya. Points are color coded to represent seven electrofacies calculated by k-means cluster analysis

identifying unique lithologies in cores. The log traces for the corresponding intervals are then used as a training set for discriminant analysis or another classification procedure that yields equations used to discriminate between the facies in uncored intervals. Although this approach has the advantage that interpreting the "meaning" of the electrofacies categories is obvious, it has a severe disadvantage in that cores or other training materials are required. An example of a supervised electrofacies classification is given by Barthelmy (2000), who classified 360,000 feet of log from the Smackover Formation in 364 North American wells, using 47,000 feet of core as training material. In the Amal field, very few cores have been taken and not all the rock types in the Amal Formation have been sampled in a representative manner.

If adequate training materials are not available, the analyst must resort to unsupervised classification. This involves subdividing the set of log measurements into subsets that are as unique as possible in their log characteristics, and as distinct as possible from other subsets. There are many procedures that attempt to achieve this objective—their effectiveness depends on the statistical distributions of the petrophysical logs that are used.

The classification procedure used in this study is *k-means clustering*, which assigns each observation (a row vector in the data set) to the "nearest" cluster based on the multidimensional distance between the observation and the cluster centroid. The multivariate Euclidian distance, $d_{ij}$, between an observation and a cluster centroid is

$$d_{ij} = \sqrt{\frac{\sum_{p=1}^{q} \left(z_{ip} - \bar{Z}_{jp}\right)^2}{q}}$$

where $z_{ip}$ is the standardized response of log track $p$ at a well depth $i$ and $\bar{Z}_{jp}$ is the average response of log $p$ in cluster $j$. There are $q$ different standardized log traces per observation.

The $k$-means method first selects a set of $k$ points called *cluster seeds* as a first guess at the means of the clusters. Each observation is assigned to the nearest seed to form a set of temporary clusters. The seeds are then replaced by the cluster means, the points are reassigned, and the process continues until no further changes occur in the clusters. The $k$-means approach is a special case of a general approach called the *EM algorithm* (Dempster et al. 1977), where $E$ stands for *Expectation* (the cluster means in this implementation) and the $M$ stands for *maximization*, which is the assignment of observations to the closest clusters in this implementation. The algorithm will produce maximum likelihood estimates of the probability that a log reading belongs to a specific electrofacies. The procedure is widely used in computer vision and portfolio management, in addition to electrofacies

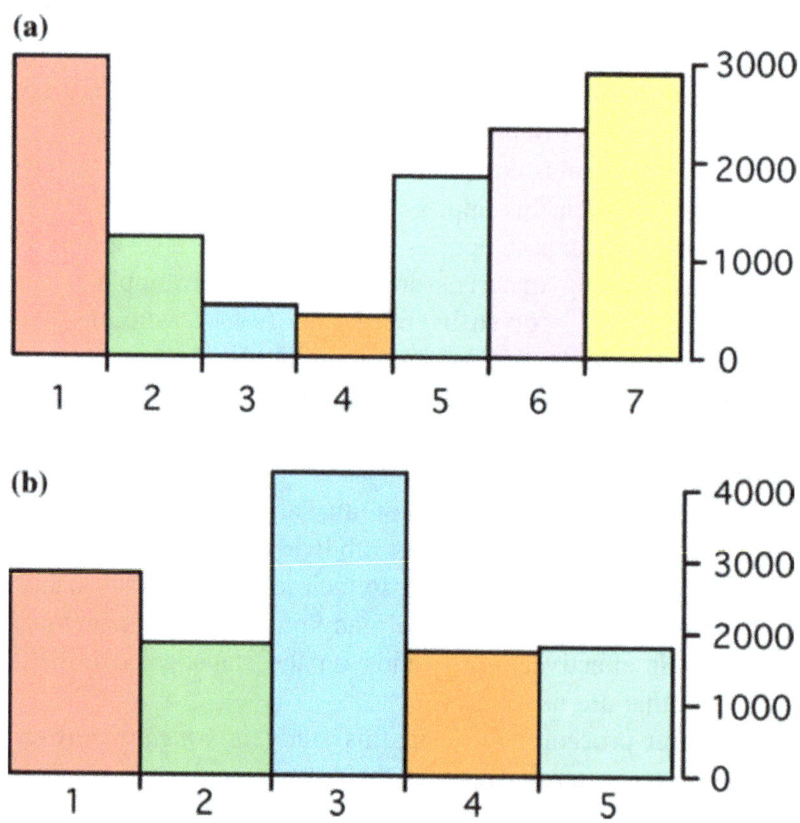


**Fig. 11.2** Histograms of the number of log readings in each electrofacies class in 15 wells of the Amal field, Libya. **a** Categorized into seven electrofacies classes. **b** Categorized into five electrofacies classes

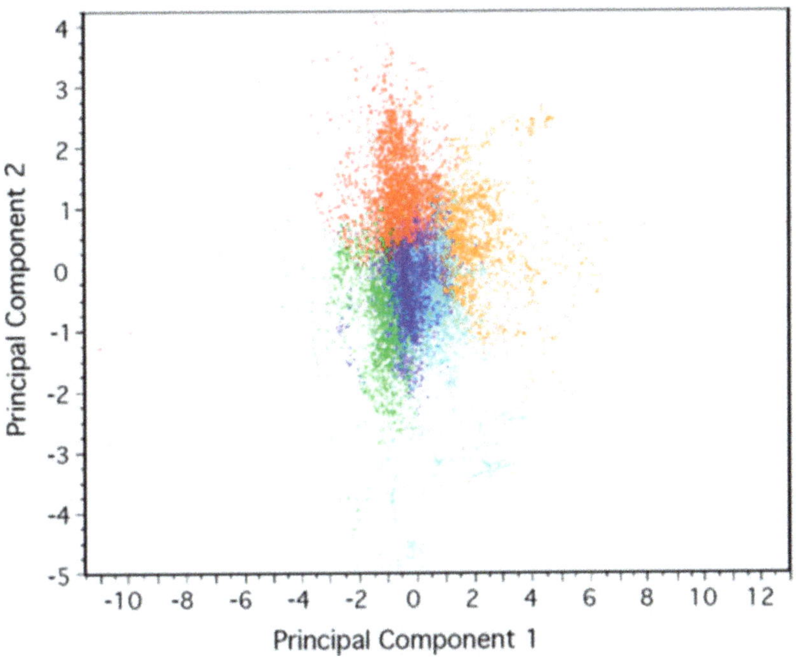


**Fig. 11.3** Cross plot of first two principal component scores of log responses from Amal Formation in 15 wells of the Amal field, Libya. Points are color coded to represent five electrofacies calculated by k-means cluster analysis

classification. Fifty-one iterations were required by the *k*-means algorithm to converge on a stable five-cluster configuration of the 12,535 log responses used here.

### *11.3.5 Converting the Electrofacies Classification into a Prediction Function*

Although the *k*-means clustering algorithm can successfully classify a collection of log responses into an arbitrary number of electrofacies, it does not produce a posterior classifier. That is, it does not create a classification rule or mathematical function that can be used to assign additional log readings to the electrofacies categories it has found. An additional step is necessary.

Canonical discriminant analysis can be used to find a set of linear functions that will separate all possible pairs of electrofacies clusters—in effect, dividing up multivariate space so only one electrofacies occupies each partitioned cell. The computations involve dividing the variance-covariance matrix of the five log properties into components that represent the variation of each observation around the grand mean, the variation of each observation around its electrofacies group mean, and the variation of the electrofacies means around the grand mean. Computational details are given in Davis (2002). Mulhern et al. (1986) discuss the application of discriminant functions to electrofacies determination.

In discriminant analysis, the distance from a log reading to the multivariate mean of the *i*-th electrofacies group is the Mahalanobis distance, $D^2$, and is computed as

$$D^2 = (z - \bar{Z}_i)' \mathbf{S}^{-1} (z - \bar{Z}_i) = z' \mathbf{S}^{-1} z - 2z' \mathbf{S}^{-1} \bar{Z}_i + \bar{Z}_i' \mathbf{S}^{-1} \bar{Z}_i$$

where **S** is the covariance matrix. The distance is divided into a portion, *dist*[0], that does not vary across groups and a portion that is the Mahalanobis distance of an observation from the centroid of the *i*-th electrofacies, *dist*[*i*]:

$$dist[0] = z' \mathbf{S}^{-1}$$
$$dist[i] = dist[0] - 2z' \mathbf{S}^{-1} \bar{Z}_i + \bar{Z}_i'$$

Assuming that each group follows a multivariate normal distribution, the posterior probability that a well log interval belongs to the *i*th electrofacies is

$$\Pr[i] = \frac{\exp dist[i]}{\Pr[0]}$$

where

$$\Pr[0] = \sum e^{-0.5 dist[i]}$$

The distances from every log observation to each electrofacies centroid is first calculated, then turned into probabilities. Each observation is then assigned to the electrofacies to which its probability of membership is the highest. Observations from other wells can also be assigned electrofacies by entering their standardized measurements into the distance and probability equations.

The assignment of individual well log observations to electrofacies by canonical discriminant analysis is not perfect, primarily because of overlapping of the original clusters. This can be evaluated by comparing the original electrofacies assignments from clustering to the results of discrimination. Figure 11.4 shows the first two principal components for 12,535 log readings in the Amal Formation in 15 wells. The points have been color-coded according to the maximum probability assignment of electrofacies by the canonical discriminant function. Compare this illustration to the original electrofacies assignments in Fig. 11.3. Contingency analysis shows that the overall correct classification rate is approximately 89%. Correct classification rates for individual electrofacies groups ranges from a low of 93.1% to a high of 97.9%.

However, the primary motivation for introducing a discrimination step in electrofacies analysis is to create numerical expressions that can be used to classify intervals in wells that were not included in the original clustering. This may be necessary if it is not possible to cluster all observations (that is, all depth intervals of

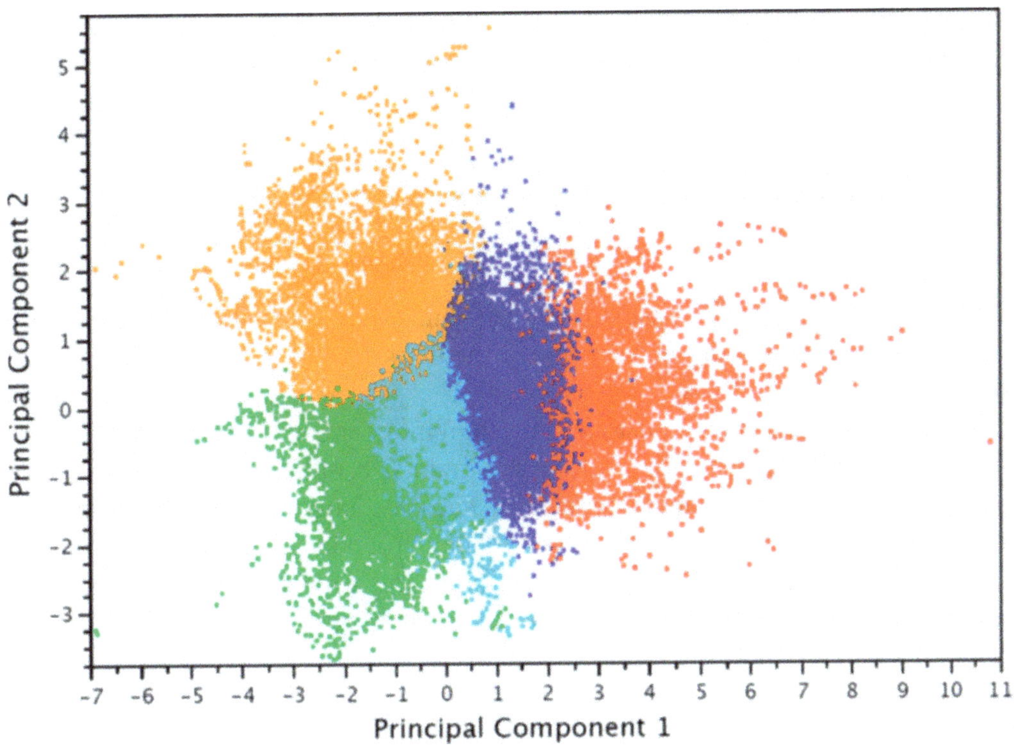


**Fig. 11.4** Cross plot of first two principal component scores of standardized log responses from Amal Formation in 15 wells of the Amal field, Libya. Points are color coded to represent maximum probability assignment into five electrofacies classes

interest in all wells) because of computer or software limitations. (A large oil field may include millions of log measurements, so such limitations may significantly constrain an electrofacies study.) Fortunately, in the Amal study it was possible to perform cluster analyses using all of the data of interest, so a discrimination step could be avoided. This not only simplifies the procedure, but also results in a slight but significant improvement in electrofacies classification.

## 11.4 What Do Amal Electrofacies Mean?

An empirical interpretation of Amal electrofacies has been made by comparing the electrofacies classifications to core descriptions for a set of wells in which extensive sets of cores were taken. The interpretations are necessarily somewhat ambiguous because of the circumstance mentioned in the preceding paragraph, and because the core descriptions were written by different geologists who may have emphasized different aspects of the rock or who used different definitions of their descriptive terms. The following lithologic descriptions represent an amalgam of the written words assigned to numerous intervals in different wells where the Amal has been given the same electrofacies classification. The lithologic distinction between Amal electrofacies is especially difficult because almost all of the formation is composed of sandstones and conglomerates of varying grain size but similar composition.

### *11.4.1 Lithologic Description of Amal Electrofacies*

Electrofacies 1 = Quartz sandstone with abundant kaolinite cement, traces of chlorite, mica and/or feldspar, very fine to medium grain size, subangular, medium to well sorted.
Electrofacies 2 = Quartz sandstone with kaolinite cement, common biotite, very thin bedded and/or crossbedded, silt to fine grain size, subangular, medium sorted.
Electrofacies 3 = Quartz conglomerate with kaolinite and/or anhydrite cement, very fine to very coarse grain size with large (>1 inch) rounded quartz pebbles, round to subround grains, unsorted. Also, quartz sandstone with silica cement, common biotite and/or hematite, silt to coarse grained, alternating sorted and unsorted layers, round to subround, no visible porosity, hard.
Electrofacies 4 = Quartz sandstone with minor kaolinite cement, traces of chlorite, mica and/or feldspar, silt to medium grain size, subangular to subround, medium sorted.
Electrofacies 5 = Igneous rock, weathered, microcrystalline to acicular, with muscovite mica and/or feldspar phenocrysts.

The lithologies corresponding to Amal electrofacies perhaps can best be understood in terms of two-way variation (Fig. 11.5). Along one axis, the electrofacies represent differences in grain size and sorting; along the other axis the electrofacies reflect the nature of the intergranular cement in the sandstone, which tends to be either kaolinite (occasionally calcite or anhydrite) or silica. Kaolinite probably has resulted from the decay of feldspar grains in what was originally an arkosic sandstone. Silica cement probably is the result of pressure solution of quartz grains and redeposition.

## 11.5 Conclusions

Electrofacies have proved to be a useful procedure for identifying and distinguishing intervals with similar petrophysical log responses and approximately equivalent lithologies within a formation that is nearly homogeneous in composition and devoid of biostratigraphic indicators or marker beds. Because the Amal Formation was mostly deposited in a terrestrial environment, facies change rapidly both laterally and vertically and conventional lithostratigraphic correlations cannot be made. Electrofacies analysis provides a framework for modeling that can guide the distribution of reservoir properties throughout the model, in spite of the

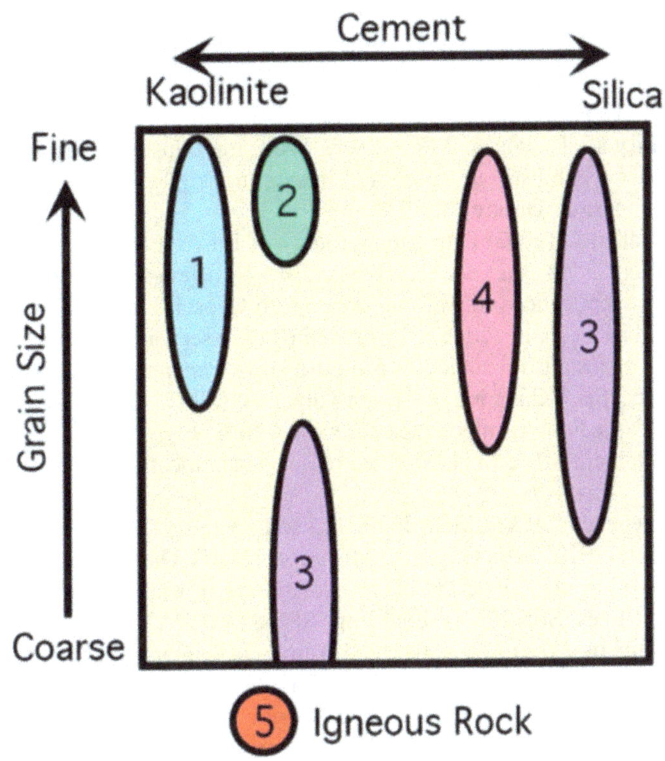


**Fig. 11.5** Amal electrofacies classes as a function of grain size from coarse to fine, and nature of matrix or cement, either clay (kaolinite) or quartz (silica)

difficulty of characterizing stratigraphic relationships by conventional means. This is one example of the type of contributions that can be made to reservoir modeling by geoscientists using a quantitative approach.

**Acknowledgements** Data from the Amal field were provided by the operator, Harouge Oil Operations of Tripoli, Libya. Electrofacies analyses were part of a much larger study of the reservoir by Heinemann Oil GmbH, Leoben, Austria. The assistance of the HOL staff, especially Stephan Egger, is gratefully acknowledged.

## References

Anxionnaz H, Delfiner P, Delhomme JP (1990) Computer-generated corelike descriptions from open-hole logs. Am Assoc Petrol Geol 74(4):375–393

Barthelmy D (2000) Multivariate analyses using geologic information from cores (MAGIC). http://www.barthelmy.com/consulting/projects/proj_magic_logs.htm

Berteig V et al (1985) Lithofacies prediction from well data. In: Transactions of SPWLA 26th logging symposium Paper TT, 25 pp

Busch JM, Fortney WG, Berry LN (1987) Determination of lithology from well logs by statistical analysis. SPE Form Eval 2(4):412–418

Davis JC (2002) Statistics and data analysis in geology, 3rd edn. Wiley, New York

Delfiner PC, Peyret O, Serra O (1987) Automatic determination of lithology from well logs. SPE Form Eval 2(3):303–310

Dempster A, Laird N, Rubin D (1977) Maximum likelihood from incomplete data via the EM algorithm. J R Stat Soc Ser B 39(1):1–38

Dubois MK, Bohling GC, Chakrabarti S (2007) Comparison of four approaches to a rock facies classification problem. Comput Geosci 33(5):599–617

El-Hawat AS et al (1996) The Nubian Sandstone in Sirt Basin and its correlatives. In: Salem MJ, El-Hawat AS, Sbeta AM (eds) Geology of the Sirt Basin, vol 2. Elsevier, Amsterdam, pp 3–30

Euzen, T, Power MR (2014) Well log cluster analysis and electrofacies classification: a probabilistic approach for integrating log with mineralogical data. Am Assoc Petrol Geol Search Discov 41277

Hallett D (2002) Petroleum geology of Libya. Elsevier, Amsterdam

Hernandez–Martinez E et al (2013) Facies recognition using multifractal Hurst analysis: applications to well-log data. Math Geosci. https://doi.org/10.1007/s1100401394456

Kiaei H et al (2015) 3D modeling of reservoir electrofacies using integration clustering and geostatistic method in central field of Persian Gulf. J Petrol Sci Eng 135:152–160

Lee SH, Khargoria A, Datta-Gupta A (2002) Electrofacies characterization and permeability predictions in complex reservoirs. SPE Reserv Eval Eng 5(3), SPE-78662-PA

Mulhern ME et al (1986) Electrofacies identification of lithology and stratigraphic trap. Geobyte 1 (4):48–56

Neuendorf KKE, Mehl Jr JP, Jackson JA (eds) (2005) Glossary of geology. Am Geol Inst

Perez HH, Datta-Gupta A, Mishra, S (2005) The role of electrofacies, lithofacies, and hydraulic flow units in permeability prediction from well logs: a comparative analysis using classification trees. SPE Reserv Eval Eng, SPE-84301-PA, 143–155

Roberts JM (1970) Amal field, Libya. In: Halbouty MT (ed) Geology of giant petroleum fields. Am Assoc Petrol Geol Mem 14:438–448

Serra O, Abbott HT (1980) The contribution of logging data to sedimentology and stratigraphy. In: SPE 9270, 55th Technical Conference, Dallas, TX, 19 pp

Tetzlaff DM, Rodriguez E, Anderson HT (1989) Estimating facies and petrophysical parameters from integrated well data. In: SPWLA log analysis software evaluation and review (LASER) symposium, Paper 8, London, 22 pp

Wolff, M, Pelissier-Combescure, J (1982) FACIOLOG-automatic electrofacies determination. In: Transactions of SPWLA 23rd logging symposium, Paper FF, Corpus Christi, TX, 22 pp

# Chapter 12
# Shoreline Extrapolations

**Jean Serra**

**Abstract** A morphological approach for studying coast lines time variations is proposed. It is based on interpolations and forecasts by means of weighted median sets, which allow to average the shorelines at different times. After a first translation invariant method, two variants are proposed. The first one enhances the space contrasts by multiplying the quench function, the other introduces homotopic constraints for preserving the topology of the shore (gulfs, islands).

**Keywords** Median sets · Binary interpolation · Hausdorff distances · Shoreline Time forecasting

## 12.1 Three Problems, One Theoretical Tool

The following study holds on lagoon inlets movements. It extends and develops an experimental study made by N.V. Thao and X. Chen about Thuan An Inlet Area (Thao and Chen 2005). The predictions proposed by these authors were obtained by averaging over the time the successive positions of a complex shoreline, including lagoon inlets, which results in a prediction of the coast line. J. Chaussard showed, in Chaussard (2006), that this prediction correctly fits with ulterior data from Google Earth (see Fig. 12.1).

In Thao and Chen (2005), the authors used a popular way to estimate accretions (Srivastava et al. 2005). Figure 12.2 depicts this semi-manual approach: the shoreline has been discretized into segments which are shifted upwards according a given accretion law (here the linear law $y = ax + b$, where $x$ stands for the time). Indeed, this is nothing but a sampled version of the dilation the shoreline by the disc of radius $ax + b$. Such a circular dilation of a shoreline turns out to be the simplest expression of its evolution under an accretion process, since it is uniform everywhere and does not take the previous stages of the shoreline into account. As a matter of fact, the

J. Serra (✉)
Ecole des Mines de Paris, Paris, France
e-mail: jean.serra@cmm.ensmp.fr



B. S. Daya Sagar et al. (eds.), *Handbook of Mathematical Geosciences*,
https://doi.org/10.1007/978-3-319-78999-6_12

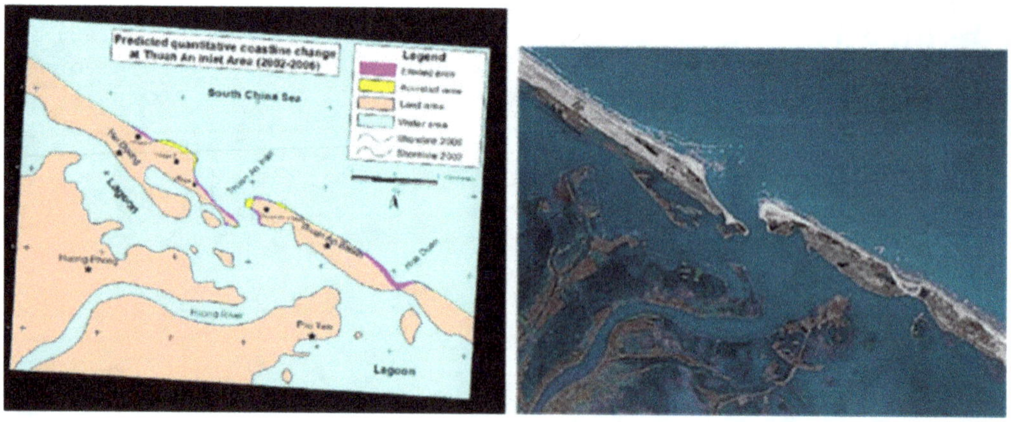


**Fig. 12.1** Left: Lagoon Inlets forecast by N.V. Thao and X. Chen; right: Current Google earth view of the same area

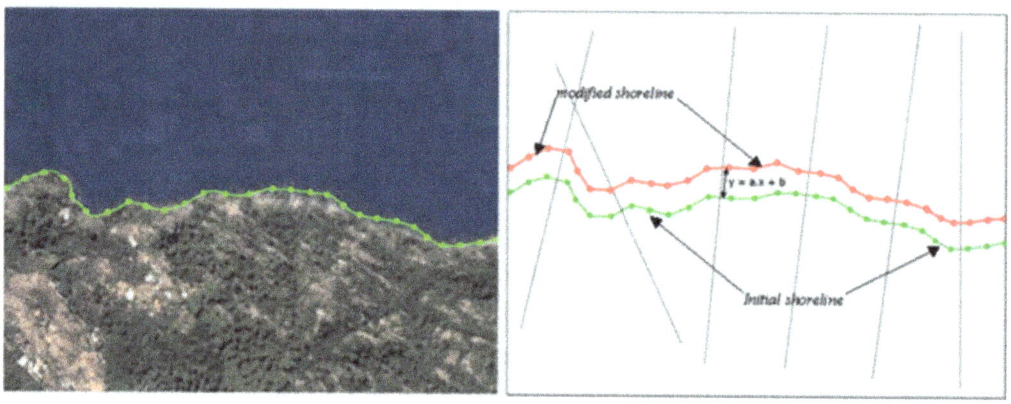


**Fig. 12.2** Classical semi-manual technique of extrapolation

notion of a set extrapolation is not straightforward, and depends considerably on the features one wishes to preserve or to emphasize.

1. If, by comparing the shorelines at years $n$ and $n - 1$, there appear zones of erosion[1] and zones of accretion, we may require a forecast of the shoreline, at year $n + 1$, to pursue erosions and accretions, but always in the same zones as previously; moreover, we must be able to express several laws for this time evolution (for example, in Thao and Chen 2005, a linear and a logarithmic laws are discussed);
2. if we know the movements of the shore during the last ten years, with one map per year, we can average these ten sets independently of their dates and base the extrapolation on this average only, or we can alternatively emphasize the more recent maps, considering that the last one, or the last two ones, carry most of the information;

---

[1]The shoreline context, the two words of "erosion" and "accretion" refer to the two types of changes depicted in Fig. 12.3. The word "erosion" also appears in the context of mathematical morphology, for naming the operation $\ominus$ involved in Eq. 12.1. It is pure coincidence.

3. if the shore exhibits small gulfs, islands and lagoon lakes, we may require from the extrapolation to preserve their homotopy, i.e. neither to create new islands (new gulfs, new lakes) nor to suppress the existing ones.

The first two questions can be treated within the framework of the median set theory, and the third one reduces to a small variant. Though median elements were thoroughly studied for interpolation problems, by M. Iwanowski in particular Iwanowski and Serra (2000) no attention was paid to their potentialities for generating averages and extrapolations. We believe nevertheless median sets turn out to be convenient tools for shorelines forecast, which in addition extend directly to numerical functions (however, we shall not treat the numerical extension here, and restrict ourself to the binary approach).

What follows is an attempt in this direction. After a presentation of the median set, that we adapt to shorelines in Sect. 12.2, we analyze in Sect. 12.3 a series of derived notions, such as weighted median set, quench function and quench stripe, and averages. The heart of the matter is treated in Sect. 12.4, where various laws are proposed for the dynamics of the coast movements. A short section on homotopy preservation precedes the conclusion. All images of coasts which are used below are *simulations*, and have the same digital size of $512 \times 320$ pixels.

## 12.2 Median Set

In literature, median set appears as an interpolation algorithm in Casas (1996) and in Meyer (1996), and was extended to partitions in Beucher (1998). Its formal definition and its basic properties were given in Serra (1998). Since, the approach has been developed by several authors (Angulo and Meyer 2009; Charpiat et al. 2006). In what follows, the geographical space is modelled by the Euclidean plane, but the approach applies as well to any metric space, including the digital ones. The model of Euclidean median sets does not concern the *lines* of the shores, but the *whole landsets*, whose the shorelines are the boundaries. These landsets, denoted below by $A_1$, $A_2$, etc., are depicted for example in Fig. 12.3 left, whereas the only shorelines boundaries, in another example, are depicted in Fig. 12.5 left. The basic results we need to start with are the Definition 1 of a median set, and the two properties 2 and 3, drawn from Serra (1998).

Hausdorff distance $\rho$ concerns the class $\mathscr{K}'$ of the noncompact sets of $R^n$ (here of $R^2$). It is the mapping $\rho : \mathscr{K}' \times \mathscr{K}' \rightarrow R_+$

$$\rho(X, Y) = \inf\{\lambda : X \subseteq Y \oplus \lambda B ; Y \subseteq X \oplus \lambda B\} \tag{12.1}$$

where $B$ designates the unit disc centered at the origin, and where $\oplus$ and $\ominus$ designate Minkowski addition (or dilation) and substraction (or erosion) respectively.

Consider now an *ordered* pair of closed sets $\{X, Y\}$, with $X \subseteq Y$, and such that the numerical value $\rho(X, Y)$, as given by Eq. (12.1), is finite. Their median element is defined as follows:

**Definition 1** The median element between the two ordered sets $X, Y \in \mathcal{K}'$, with $X \subseteq Y$, is the compact set $M(X, Y)$, comprised between $X$ and $Y$ and whose boundary points are equidistant from $X$ and $Y^c$.

In other words, the boundary $\partial M$ of $M$ is nothing but the skeleton by zone of influence, or *skiz*, between $X$ and $Y^c$.

**Proposition 1** *The median set between X and Y is obtained by taking the union*

$$M(X, Y) = \cup\{(X \oplus \lambda B) \cap (Y \ominus \lambda B)\ \lambda \geq 0\} \tag{12.2}$$

*where the $\lambda$ can be limited to the values smaller or equal to*

$$\mu = \inf\{\lambda : \lambda \geq 0,\ X \oplus \lambda B \supseteq Y \ominus \lambda B\} \tag{12.3}$$

*and where the equality is reached for at least one point of $\partial M$.*

*Proof* A point $m$ at a distance $\leq \lambda$ from $X$ and $\geq \lambda$ from $Y^c$ belongs to set $(X \oplus \lambda B) \cap (Y \ominus \lambda B)$, hence to set of Eq. (12.1). Conversely, as every point $m \in M$ belongs to at least one term of the union, there exists a $\lambda \geq 0$ with $d(m, X) \leq \lambda$ and $d(m, Y^c) \geq \lambda$, which results in Eq. (12.1). As for Eq. (12.2), we observe that for $\lambda$ large enough we have $(X \oplus \lambda B) \cup (Y^c \oplus \lambda B) = R^2$ because set $Y$ is bounded. These $\lambda$ bring no contribution to set $M(X, Y)$, since $X \oplus \lambda B \supseteq Y \ominus \lambda B$. Finally, for $\lambda = \mu$, we obtain a point of the boundary $\partial M$ because $X$ and $Y$ are closed, which achieves the proof.

Here is now an instructive property which shows how both Hausdorff distances by dilation and by erosion[2] are involved in the median $M(X, Y)$ (Serra 1998).

**Proposition 2** *Given $X, Y \in \mathcal{K}'(R^n)$, the median element $M(X, Y)$ is at Hausdorff dilation distance $\mu$ from $X$ and from the closing $X \bullet \mu B = (X \oplus \mu B) \ominus \mu B$, and at Hausdorff erosion distance $\mu$ from $Y$ and from the opening $Y \circ \mu B = (Y \ominus \mu B) \oplus \mu B$.*

---

[2]Hausdorff distance $\sigma$ for erosion, introduced in by the relation

$$\sigma(X, Y) = \inf\{\lambda : X \ominus B_\lambda \subseteq Y\ ;\ Y \ominus B_\lambda \subseteq X\}$$

concerns the subclass $\mathcal{A}$ of $\mathcal{K}'(E)$ of the regular compact sets, i.e. such that $\overline{X^o} = X$. It is indeed a distance on $\mathcal{A} \times \mathcal{A}$. If $\sigma(X, Y) = 0$, then we have

$$Y \supseteq \bigcup_{\lambda > 0} X \ominus B_\lambda = X^o \Rightarrow Y \supseteq \overline{X^o} = X \qquad X, Y \in \mathcal{A}$$

and similarly $X \supseteq Y$, hence$X = Y$ (the other two axioms are proved as for distance $\rho$) (Serra 1998).

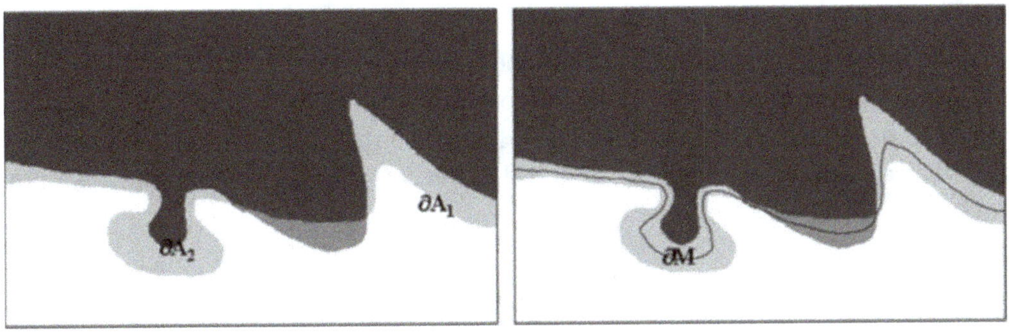


**Fig. 12.3** Left: two simulated shore images $A_1$ and $A_2$. The older is supposed to be $A_1$ (the white one).The zones of accretion from $A_1$ to $A_2$ are in light grey, those of erosion in dark grey; right: the boundary of the median set $M$ between $A_1$ and $A_2$

The Hausdorff distance applies to non empty compact sets. But clearly, the landsets under study are not empty, and the above assumption that $\rho(X, Y) < \infty$ comes back to say that all involved distances are bounded.

## 12.3 Median and Average for Non Ordered Sets

**Non ordered sets** In general, two successive shores $A_1$ and $A_2$ are not ordered, i.e. their change comprises both erosions and accretion areas. If so, the previous results do not apply to two $A_1$ and $A_2$ directly, but to their intersection $X = A_1 \cap A_2$ and their union $Y = A_1 \cup A_2$ which are ordered since $X \subseteq Y$. Equation (12.1) of the median element becomes

$$M(A_1, A_2) = \bigcup_{\lambda \geq 0} [A_1 \cap A_2) \oplus \lambda B] \cap [(A_1 \cup A_2) \ominus \lambda B] \tag{12.4}$$

Figures 12.3 depicts an example of median set $M$. One observes that $\partial M$ goes through all points where the two coastlines intersect. The property is general, since these points belong to both $A_1 \cap A_2$ and $A_1 \cup A_2$.

**Weighted median** Set $M$ is said to be *median* because each point of $\partial M$ is equidistant from $X$ and $Y^c$, which is a consequence of the same weight given to dilation and erosion in Eq. (12.2). By changing this weight, i.e. by replacing $M$ by

$$M_\alpha(X, Y) = \bigcup_{\lambda} \{(X \oplus \alpha\lambda B) \cap (Y \ominus (1 - \alpha)\lambda B)\} \tag{12.5}$$

for a $\alpha \in [0, 1]$, we generate another interpolation, and by making $\alpha$ vary, a series of progressive interpolations from $X$ to $Y$ (Huttenlocher 1995), all the closer to set $Y$ since $\alpha$ is high. One will notice that when the two shores $A_1$ and $A_2$ are *not* nested in each other, then one takes for the two operands of Eq. (12.5) $X = A_1 \cap A_2$ and

$Y = A_1 \cup A_2$. This provides interpolators such as those of Fig. 12.4. Unfortunately, these interpolators are closer to the highest or to the lowest line, no matter these lines are portions of $\partial A_1$ or of $\partial A_2$. For correcting this drawback, one must take the interpolator $M_\alpha$ in the zones where $A_1$ is larger than $A_2$ (for example), and $M_{1-\alpha}$ in the other ones. Denoting by $N(A_1, A_2)$ the correct weighted interpolator, we now have

$$\begin{aligned} N_\alpha(A_1, A_2) &= M_{1-\alpha}(A_1, A_2)\ when\ A_1 \backslash A_2 \neq \emptyset \\ &= M_\alpha(A_1, A_2) when A_2 \backslash A_1 \neq \emptyset \end{aligned} \quad (12.6)$$

Figure 12.5 depicts such corrected interpolators.

**The physical equation of the phenomenon** Physically speaking, the accretion/erosion process evolves at each instant from the stage it has reached before. It takes some $M_\alpha(X, Y)$, with $\alpha \in [0, 1]$, as starting point and moves to $M_\beta[M_\alpha(X, Y), Y]$, for some value $\beta \in [0, 1]$. The weighted medians $M_\alpha$ do model this evolution because they form a semi-group. By calculating firstly the set $M_\alpha(X, Y)$ median between $X$ and $Y$, and then the set $M_\beta[M_\alpha(X, Y), Y]$ between $M_\alpha(X, Y)$ and $Y$, we obtain indeed the same result as by calculating directly $M_\gamma(X, Y)$ for the weight $\gamma = \alpha + (1 - \alpha)\beta = \alpha + \beta - \alpha\beta$, i.e.

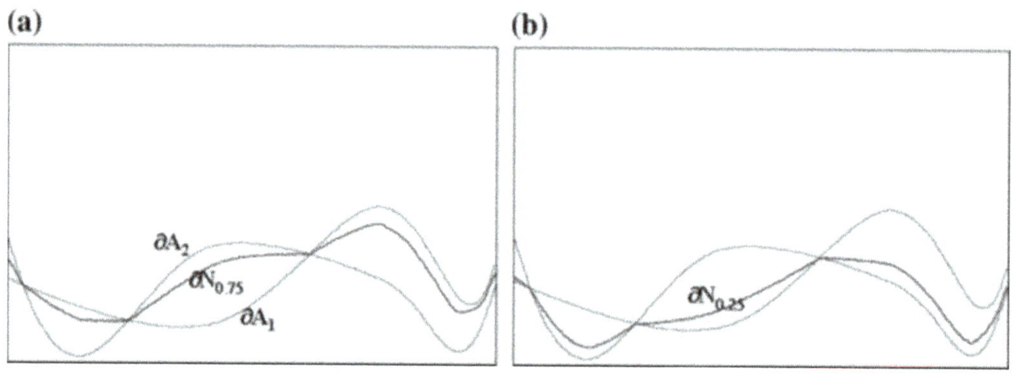


**Fig. 12.4** Raw weighted median lines

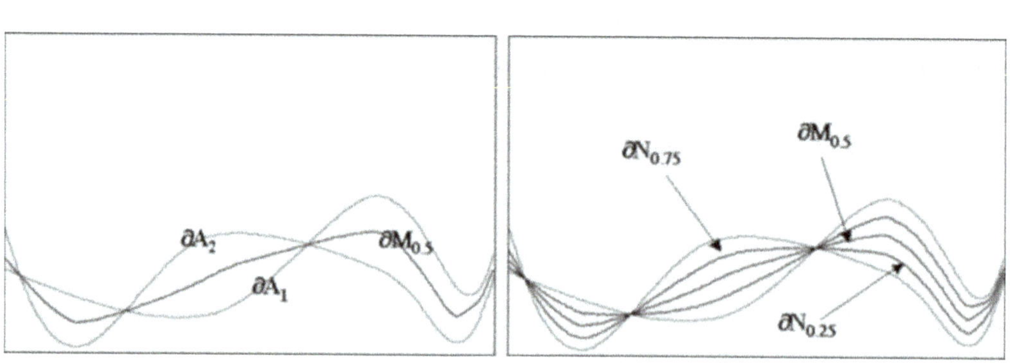


**Fig. 12.5** Left: two shores $A_1$ and $A_2$, of boundaries $\partial A_1$ and $\partial A_2$, and their median line of boundary $\partial M_{0.5}$; right: the same, plus two additional weighted median lines according to Eq. (12.5)

$$M_\beta[M_\alpha(X,\ Y),\ Y] = M_{\alpha+\beta-\alpha\beta}(X,\ Y) \tag{12.7}$$

For example, in Fig. 12.5 right, the three median sets correspond to $\alpha = 0.75, 0.5$, and 0.25, and the weighted median $M_{0.75}$ is *also* the median element between $M_{0.5}$ and $A_1 \cup A_2$.

**Proposition 3** *Given* $X,\ Y \in \mathcal{K}'(R^n)$*, the family* $\{M_\alpha(X,\ Y), 0 \leq \alpha \leq 1\}$ *of median elements form an additive semi-group for the addition* $\alpha \otimes \beta = \alpha + \beta - \alpha\beta$*.*

*Proof* Clearly, $\alpha \otimes \beta \in [0, 1]$, *thus Eq.* (12.7) *defines a commutative semi-group. The operation* $\alpha \otimes \beta$ *is also associative, since*

$$\gamma \otimes (\alpha + \beta - \alpha\beta) = \gamma + \alpha + \beta - \alpha\beta - \gamma\alpha - \gamma\beta + \gamma\alpha\beta$$

*is symmetrical in* $\alpha, \beta, \gamma$, *therefore* $\alpha \otimes \beta$ *is an algebraic addition.*

**Quench function and quench stripe** As a matter of fact, the median operator provides *two outputs*, since we have on the one hand the (weighted or not) *median set* $M$, whose contour $\partial M$ is the dark middle line in Fig. 12.5 left, or Fig. 12.6 left, and the *quench function* $q$, defined on $\partial M$ and which gives at each the radius of the minimum disc hitting the two contours $\partial A_1$ and $\partial A_2$.

$$q(z) = \inf\{r\ :\ B_z(r) \cap \partial A_1 \neq \emptyset \textit{ and } B_z(r) \cap \partial A_2 \neq \emptyset\} \tag{12.8}$$

A few of such discs, for the two inputs $A_1$ and $A_2$ of Fig. 12.3 left, are depicted in Fig. 12.6 left, and their union for the whole quench function gives the *quench stripe* $w$, i.e. the dark grey stripe $W$ around the black line $\partial M$ in Fig. 12.6 right, with

$$W = \cup\{B_z(q(z)),\ z \in M(A_1,\ A_2)\} \tag{12.9}$$

Note hat this dark grey stripe does not reach the edges of input sets $A_1$ and $A_2$, but an open version of their union, and a closed version of their intersection.

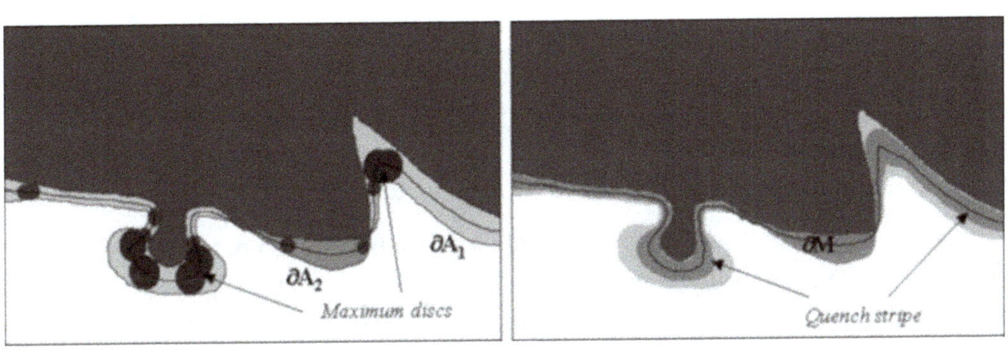


**Fig. 12.6** Left: a few maximum discs centered on the median line; right: the dark grey stripe is the union of all maximum discs, or "quench stripe"

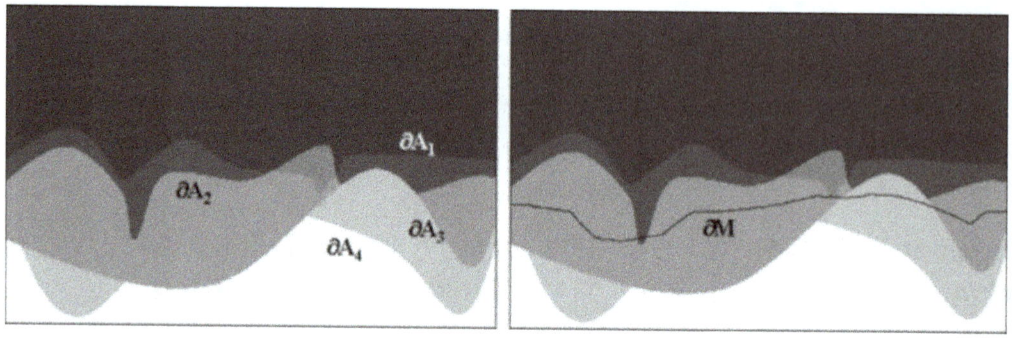


**Fig. 12.7** Left: four shores; right in dark, their median line

**Averages** The structure of Eq. (12.7) suggests a technique for extending the median element to more than two input sets. Starting for example from the triplet $\{A_1, A_2, A_3\}$, we can calculate $M_{0.5}(A_1, A_2)$ in a first stage, and then $M_{0.33}[M_{0.5}(A_1, A_2), A_3]$. The resulting median element averages the three inputs, in a median sense. Figure 12.7 depicts an example of such an average for the four inputs $\{A_1, ..A_4\}$ shown in Fig. 12.7 left (two of them are the sets involved in Fig. 12.5 left). The initial stage consists in calculating $M_{0.5}(A_1, A_2)$ and $M_{0.5}(A_3, A_4)$, and the final one in calculating $M_{0.5}[M_{0.5}(A_1, A_2), M_{0.5}(A_3, A_4)]$, a set whose contour is drawn in black in Fig. 12.7 right. This final result is independent of the choice of the sets in the initial stage, and we could start as well from $M_{0.5}(A_1, A_3)$ and $M_{0.5}(A_2, A_4)$.

The averages obtained this way blur the structural features of the shores. Imagine for example that $A_2$, $A_n$ are shifted versions of $A_1$ in the horizontal direction. As $n$ increases, the median average contour tends towards an horizontal line: all features, gulfs, capes, etc. are lost. We meet here the same trouble as in interpolating moving objects, with translation and rotation. In case of shore movements, the translations are probably less intense, but the problem still remains. Remark also that this drawback is the counterpart of the advantage of preserving accretion and erosion zones.

## 12.4 Extrapolations via the Quench Function

In this section and the next one, we focus on the extrapolation of two shores at most, $A_1$ and $A_2$ say. If we dispose of a chronological sequence of the coast movements, $A_1$ and $A_2$ stand for the last two observations, $A_2$ being the more recent. The principle of the extrapolation consists in two possible changes:

1. that of the quench function according to a given law, which models the dynamics of the movement, and which results in a new quench stripe $W$;
2. that of the respective importances of $A_1$ and $A_2$. If we take the median $M_{0.5}(A_1, A_2)$, then both shores are given the same weight, but if we consider that $A_2$, more recent, is two times more significant than $A_1$, then we can take $N_{0.66}(A_1, A_2)$.

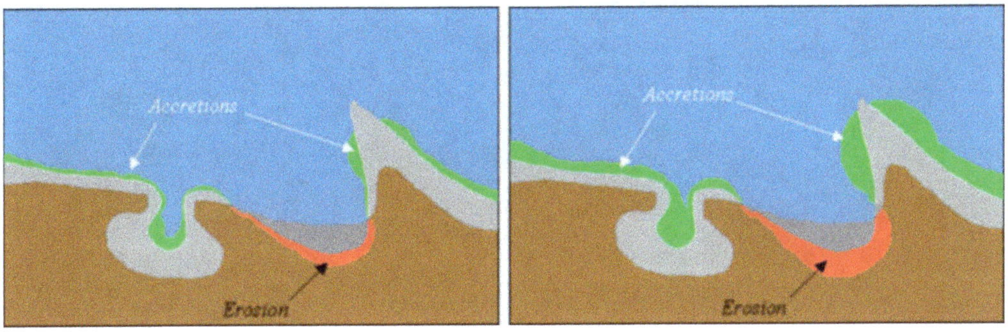


**Fig. 12.8** Two extrapolations of the shoreline of Fig. 12.3; both are centered on $\partial M_{0.5}(A_1, A_2)$; the quench function is multiplied by 2 in the left image and by 3 in the right one

Fig. 12.8 depicts two extrapolations where the median element equals $M_{0.5}(A_1, A_2)$, hence where the two input shores are given the same importance, but where the quench stripe $W$ of Eq. (12.9) is replaced by

$$W = \cup\{B_z(kq(z)),\ z \in M_{0.5}(A_1, A_2)\}$$

The radius of the disc centered at each point of $M_{0.5}(A_1, A_2)$ is quench value multiplied by factor $k$, with $k = 2$ for Fig. 12.8 left and $k = 3$ for Fig. 12.8 right. We see that, as $k$ increases, both accretion and erosion zones are developed. We can also notice that the shape of the cape provokes a bizarre inflation in Fig. 12.8 right.

This swelling may be due to the great distance from the median line to extremity of the cape, as shown in Fig. 12.6 right, so that we can try to avoid it by making the median line closer to contour $\partial A_2$ which delineates the cape. Replace then the median set $M_{0.5}(A_1, A_2)$ by $N_\alpha(A_1, A_2)$, in the sense of Eq. (12.6), with $\alpha = 0.75$, so that the quench stripe becomes

$$W = \cup\{B_z(kq(z)),\ z \in N_{0.75}(A_1, A_2)\}.$$

The resulting changes are depicted in Fig. 12.9, left for $k = 3$, and right for $k = 4$. By comparing Figs. 12.8 right and 12.9 left where the quench function is multiplied by the same value $k = 3$, we see that the cape inflates less, but in compensation the erosion zone vanished. The erosion can reappear by taking $k = 4$ (Fig. 12.9 right), but again the cape inflates as strongly as in the previous extrapolation of Fig. 12.8 right.

In fact, transforming a quench function according to pure magnification is probably too poor. One can easily imagine more sophisticated laws such as the two following ones:

1. the median line is slightly moved toward the second contour, by taking $N_{0.66}(A_1, A_2)$, and the quench stripe $W$ is obtained by dilating each point $z$ of the median line by the disc of radius $2q(z)$ and by the segment $L_\alpha(2q(z))$ of length $2q(z)$ in the main direction $\alpha$ of the cape, which gives

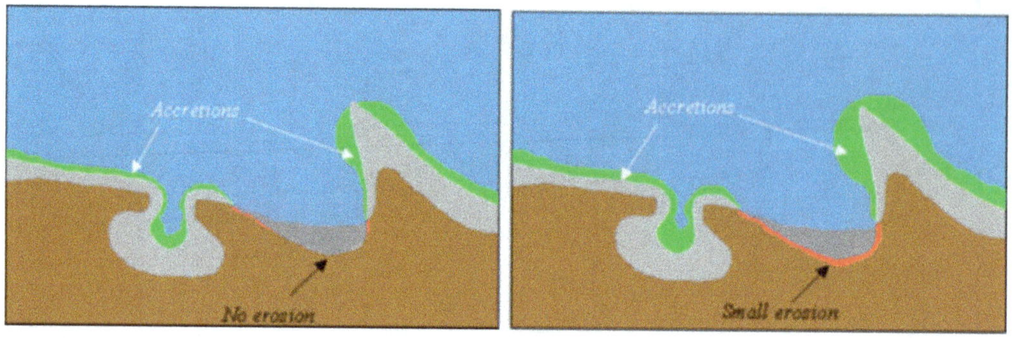


**Fig. 12.9** Two other extrapolations of the shoreline of Fig. 12.3; both are centered on $\partial N_{0.75}(A_1, A_2)$; the quench function is multiplied by 3 in the left image and by 4 in the right one

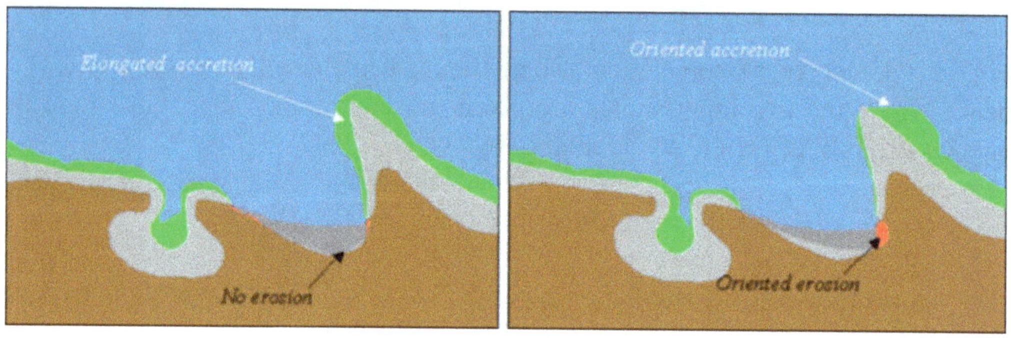


**Fig. 12.10** Two extrapolations of the shoreline of Fig. 12.3, by emphazising the new capes in the left image, and by introducing an east-west trend in the right one

$$W = \cup\{[B_z(2q(z) \oplus L_\alpha(2q(z))], \ z \in N_{0.66}(A_1, A_2)\}$$

and which is depicted in Fig. 12.10 left. The accretion around the cape turns out to be now more realistic, but the erosion zone has disappeared.

2. The median set $N_{0.66}(A_1, A_2)$ is left unchanged, and a supplementary trend in the horizontal direction is introduced by a dilating points $z$ by the horizontal segment $L_0(3q(z))$. For avoiding too fast changes, the parameters of the two other dilations are divided by 2. The shifting effect of the trend operation appears clearly in Fig. 12.10 right, where the accretion forms a deposit at the east of the cape. Similarly, the directional effect of the erosion holds for west oriented regions.

Unlike the previous models, which all are invariant under rotation of the map, these last two laws, which model marine currents, depend on the North direction (see Fig. 12.10).

## 12.5 Accretion and Homotopy

It may happen that, for some reasons, one wishes to preserve the homotopy of the shore, which excludes the creation, or the suppression, of lakes and islands. Now, by dilating enough the shore of Fig. 12.10, we risk to close the gulf on the left and to generate in internal island. An easy way to protect the gulf as such consists in replacing the dilation w.r.t. the unit disc by a cycle of elementary homotopic thickenings in the eight directions of the square grid, or the six ones in the hexagonal case (Serra 1982). The circular dilation of size $n$ becomes the series of $n$ thickening cycles. One can see in Figure ll, left and right, the results of two thickenings of sizes 25 and 33 respectively (for a $512 \times 320$ digital image). The gulf is preserved by a narrow channel, which could be enlarged by modifying the homotopy preservation algorithm. This conceptually simple method is not the only possible one. In Vidal et al. (2005) the authors propose a median set based interpolation that preserves particles by marking them by a homotopic thinning, and translating them during the interpolation process.

## 12.6 Conclusion

Our purpose was to demonstrate the physical sense of the median set approach and its flexibility. In the first section, we indicated three features to be respected by interpolations. According to the first one, an accretion (resp. erosion) zone must continue to evolve by accretion (resp. erosion). This basic modality is fulfilled by all models of Sect. 12.4. The laws proposed in this section are far from being the only possible ones. In particular, each of the six examples of the section is given a same law for accretion and erosion, which is not at all an obligation. The second feature holds for the role of the past. In the approach of Sect. 12.4, this past reduces to the last two stages: they suffice to determine the starting shoreline, the "gradient", and the location of accretion/erosion (Fig. 12.11).

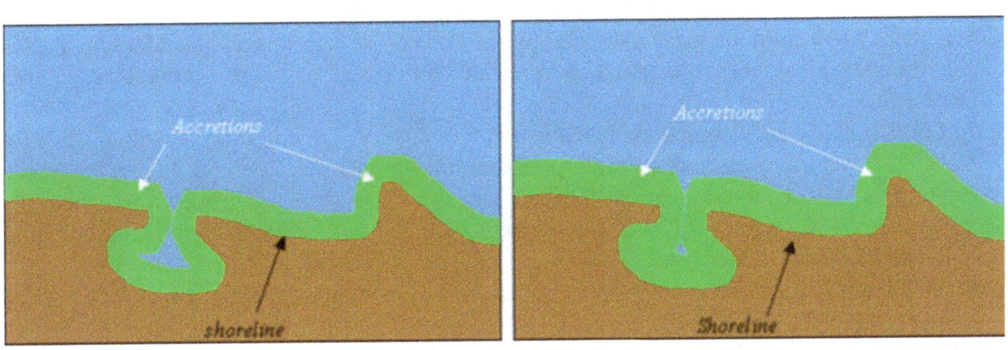


**Fig. 12.11** Two extrapolations of the shoreline of Fig. 12.3 by homotopic thickenings of sizes 25 (left) and 33 (right)

The third feature was the subject of Sect. 12.5, where a thickening is substituted for the dilation in the extrapolator, in order to preserve homotopy. Indeed, all extrapolation equations, from sections two to four, can be rewritten by replacing the unit disc erosion and dilation by unit cycles of thinnings and thickenings, and the linear dilations by unidirectional thickenings. It would result in a series of algorithms where increasingness is lost (non direct extension to numerical functions) but where topological features are preserved.

Finally, as the weighted median of Eq. (12.4) is an increasing function of its two operands, it extends to numerical functions by means of their subgraphs, and allows to process colour images (Daya Sagar 2007).

**Acknowledgements** I am extremely grateful to Dr B.R. Kiran for his precious help in preparing this chapter.

## References


Angulo J, Meyer F (2009) Morphological exploration of shape spaces. In: Wilkinson MHF, Roerdink JBTM (eds) Mathematical morphology and its applications to signal and image processing. Springer, Berlin, pp 226–237

Beucher S (1998) Interpolation of sets, of partitions and of functions. In: Heijmans H, Roerdink J (eds) Mathematical morphology and its applications to image and signal processing. Kluwer, Dordrecht

Casas JR (1996) Image compression based on perceptual coding techniques, PhD thesis, UPC Barcelona, March 1996

Charpiat G, Faugeras O, Keriven R, Maurel P (2006) Distance-based shape statistics. IEEE ICASP 5:925–928

Chaussard J (2006) Surveillance cotire l'aide de la morphologie, a bibliographical study, A2SI, ESIEE France

Daya Sagar BS (2007) Universal scaling laws in surface water bodies and their zones of influence. Water Resour Res 43(2):W02416

Huttenlocher DP, Klunderman GA, Rucklidge WJ (1995) Comparing images using the Hausdorff distance. IEEE PAMI. 15(9)

Iwanowski M, Serra J (2000) The morphological-affine object deformation. In: Goutsias J, Vincent L, Bloomberg DS (eds) Mathematical morphology and its applications to image and signal processing. Kluwer Academic Publisher, Dordrecht, pp 81–90

Meyer F (1996) A morphological interpolation method for mosaic images. In: Maragos P et al. (eds) Mathematical morphology and its applications to image and signal processing. Kluwer, Dordrecht

Serra J (1982) Image analysis and mathematical morphology. Academic Press, London

Serra J (1998) Hausdorff distance and Interpolations. In: Heijmans H, Roerdink J (eds) Mathematical morphology and its applications to image and signal processing. Kluwer, Dordrecht

Srivastava A, Niu X, Di K, Li R (2005) Shoreline modeling and erosion prediction. In: proceedings of the ASPRS annual conference

Thao NV, Chen X (2005) Temporal GIS for monitoring lagoon inlets movements. In: Chen X (ed) DMAI' 2005. AIT Bangkok

Vidal J, Crespo J, Maojo V (2005) Recursive interpolation technique for binary images based on morphological median sets. In: Ronse Ch, Najman L, Decencire E (eds) Mathematical morphology: 40 years on. Springer, Berlin, pp 53–65

# Chapter 13
# An Introduction to the Spatio-Temporal Analysis of Satellite Remote Sensing Data for Geostatisticians

**A. F. Militino, M. D. Ugarte and U. Pérez-Goya**

**Abstract** Satellite remote sensing data have become available in meteorology, agriculture, forestry, geology, regional planning, hydrology or natural environment sciences since several decades ago, because satellites provide routinely high quality images with different temporal and spatial resolutions. Joining, combining or smoothing these images for a better quality of information is a challenge not always properly solved. In this regard, geostatistics, as the spatio-temporal stochastic techniques of geo-referenced data, is a very helpful and powerful tool not enough explored in this area yet. Here, we analyze the current use of some of the geostatistical tools in satellite image analysis, and provide an introduction to this subject for potential researchers.

## 13.1 Introduction

The spatio-temporal analysis of satellite remote sensing data using geostatistical tools is still scarce when comparing with other kinds of analyses. In this chapter we provide an introduction to this field for geostatisticians, empathising the importance of using the spatio-temporal stochastic methods in satellite imagery and providing

A. F. Militino (✉) · M. D. Ugarte · U. Pérez-Goya
Department of Statistics and O.R., Public University of Navarra (Spain), Pamplona, Spain
e-mail: militino@unavarra.es

U. Pérez-Goya
e-mail: unai.perez@unavarra.es

A. F. Militino · M. D. Ugarte
InaMat (Institute for Advanced Materials), Pamplona, Spain
e-mail: lola@unavarra.es



B. S. Daya Sagar et al. (eds.), *Handbook of Mathematical Geosciences*,
https://doi.org/10.1007/978-3-319-78999-6_13

a review of some applications (Sagar and Serra 2010). We explain how to proceed for accessing remote sensing data, and which are the common tools for downloading, pre-processing, analysing, interpolating, smoothing and modeling these data. The chapter encloses six additional sections where a short explanation of the state of the art in the analysis of remote sensing data using free statistical software is given. Particular attention is devoted to the use of geostatistical tools in this subject. Section 13.2 explains the profile and the main features of the most popular satellites. It also encompasses Sect. 13.2.1 for describing some R packages for importing, analysing, and managing satellite images. Section 13.3 explains how to retrieve two derived variables, the normalized difference vegetation index (NDVI) and the land surface temperature (LST). In Sect. 13.4 some common methods of pre-processing data after downloading satellite images are reviewed. Section 13.5 explains the importance of the spatial interpolation in remote sensing data and reviews the most popular interpolation methods. The actual scenario of the spatio-temporal geostatistics is reviewed in Sect. 13.6, where an additional subsection describes some *R* packages for using spatial and spatio-temporal geostatistics techniques with satellite images. The paper ends up with some conclusions in Sect. 13.7.

## 13.2 Satellite Images

Satellite images are available since more than four decades ago, and since then there has been a notable improvement in quality, quantity, and accessibility of these images, making it easier to extract huge amounts of data from all over the Earth. We can retrieve data from the land or the ocean, from the coast or the mountains, and also from the atmosphere where advanced sensors give the opportunity of monitoring meteorological variables that are crucial for the study of the climatic change, the phenology trend, the changes in vegetation or many other environmental processes.

Remote sensing refers to the process of acquiring information from the Earth or the atmosphere using sensors or space shuttles platforms. Therefore, remote sensing is born as a crucial necessity when using satellite images for analyzing and converting them into different frames of data that can be managed with specific software. Nowadays, Landsat, Modis, Sentinel or Noaa are some of the most popular satellite missions among researchers and practitioners of remote sensing data because of the free accessibility. Next, we summarize the main characteristics of these missions:

1. LANDSAT, meaning Land+Satellite, represents the world's longest continuously acquired collection of space-based moderate-resolution land remote sensing data. See GLCF (2017) for details. It is available since 1972 from six satellites in the Landsat series. These satellites have been a major component of NASA's Earth observation program, with three primary sensors evolving over thirty years: MSS (Multi-spectral Scanner), TM (Thematic Mapper), and ETM+

(Enhanced Thematic Mapper Plus). Landsat supplies high resolution visible and infrared imagery, with thermal imagery, and a panchromatic image also available from the ETM+ sensor. Landsat also provides land cover facility to complement overall project goals of distributing a global, multi-temporal, multi-spectral and multi-resolution range of imagery appropriate for land cover analysis.

2. The SENTINEL satellites were launched from 2013 onwards and include radar, spectrometers, sounders, and super-spectral imaging instruments for land, ocean and atmospheric applications (Aschbacher and Milagro-Pérez 2012). In particular, the multispectral instrument on-board Sentinel-2 aims at measuring the Earth reflected radiance through the atmosphere in 13 spectral bands spanning from the Visible and Near Infra-Red to the Short Wave Infra-Red. The main goal of this satellite is the monitoring of rapid changes such as vegetation characteristics during growing seasons with improved change detection techniques.
3. NOAA is the acronym of National Oceanic and Atmospheric Administration. The satellite observations of the atmosphere on a global scale began more than 40 years ago. In the URL (NOAA 2017), it is said that over 150 data variables from satellites, weather models, climate models, and analyses are available to map, interact with, and download using NOAA View's Global Data Explorer. NOAA generates more than 20 terabytes of daily data from satellites, buoys, radars, models, and many other sources. All of that data are archived and distributed by the National Centers for Environmental Information.
4. Moderate Resolution Imaging Spectroradiometer (MODIS) is a key instrument aboard the TERRA and AQUA satellites. See the URL (MODIS 2017) for details. TERRA's orbit around the Earth is timed so that it passes from north to south across the equator in the morning, while AQUA passes south to north over the equator in the afternoon, providing a high temporal resolution of images all over the world. TERRA MODIS and AQUA MODIS are viewing the entire Earth's surface every 1 to 2 days, acquiring data in 36 spectral bands, or groups of wavelengths. These data facilitate the global dynamics and processes occurring on the land, in the oceans, and in the lower atmosphere.

Remote sensing data of some of these missions can be accessed via the free statistical software R, publicly accessible in R Core Team (2017).

### 13.2.1 Access and Analysis of Satellite Images with R

This subsection provides a summary of some R packages that can be used for downloading, importing, accessing, processing, and smoothing remote sensing data from satellite images.

1. dtwSat (Maus et al. 2016) implements the Time-Weighted Dynamic Time Warping (TWDTW) method for land use and land cover mapping using satellite image time series. TWDTW is based on the Dynamic Time Warping technique and it

has achieved high accuracy for land use and land cover classification using satellite data.
2. gapfill (Gerber et al. 2016) fills missing values in satellite data and develops new gap-fill algorithms. The methods are tailored to images observed at equally-spaced points in time.
3. gdalUtils (R Core Team 2017) gives wrappers for the geospatial data abstraction library (GDAL) utilities.
4. gimms (R Core Team 2017) provides a set of functions to retrieve information about GIMMS NDVI3g files
5. landsat (Goslee 2011) includes relative normalization, image-based radiometric correction, and topographic correction options.
6. landsat8 (Survey 2015) provides functions for converted Landsat 8 multispectral satellite imagery rescaled to the top of atmosphere (TOA) reflectance, radiance and/or at satellite brightness temperature using radiometric rescaling coefficients provided in the metadata file (MTL file).
7. mod09nrt (R Core Team 2017) processes and downloads MODIS Surface reflectance Product HDF files. Specifically, MOD09 surface reflectance product files, and the associated MOD03 geo-location files (for MODIS-TERRA).
8. MODIS (R Core Team 2017) allows for downloading and processing functionalities for the Moderate Resolution Imaging Spectroradiometer (MODIS)
9. modiscloud (Nicholas J. Matzke 2013) is designed for processing downloaded MODIS cloud product HDF files and derived files
10. raster (R Core Team 2017) is a very powerful library for the geographic data analysis and modeling
11. rgdal (R Core Team 2017) provides bindings for the geospatial data abstraction library.
12. satellite (Nauss et al. 2015) provides a variety of functions which are useful for handling, manipulating, and visualizing remote sensing data.

## 13.3 Derived Variables from Remote Sensing Data

When a satellite image is accessed, an assorted number of bands are provided. The combination of these bands can facilitate different types of remote sensing data. For example, extracting the Normalized Difference Vegetation Index (NDVI) can be done by a simple combination of bands. NDVI is an important index that reflects vegetation growth and it is closely related to the amount of photosynthetically absorbed active radiation as indicated by Slayback et al. (2003) and Tucker et al. (2005). It is calculated using the radiometric information obtained for the red (R) and near-infrared (NIR) wavelengths of the electromagnetic spectrum in the following way: $NDVI = ((NIR) - R)/((NIR) + R)$ (Rouse Jr et al. 1974). As mentioned in Sobrino and Julien (2011), this parameter is sensitive to the blueness of the observed area,

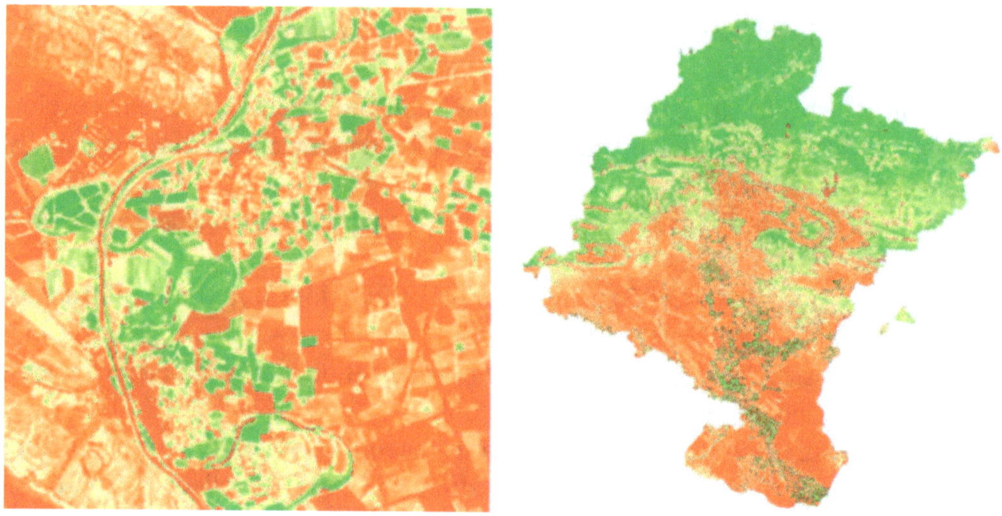

**Fig. 13.1** (Left) NDVI Sentinel image of Funes village in Navarra, and (Right) NDVI for the whole Navarra (Spain)

which is closely related to the presence of vegetation. Although numerical limits of NDVI can vary for the vegetation classification, it is widely accepted that negative NDVI values correspond to water or snow. NDVI values close to zero could correspond to bare soils, yet these soils can show a high variability. Values between 0.2 and 0.5 (approximately) to sparse vegetation, and values between 0.6 and 1.0 conform to dense vegetation such as that found in temperate and tropical forests or crops at their peak growth stage. Therefore, NDVI provides a very valuable instrument for monitoring crops, vegetation, and forestry, and it is directly calculated in specific images by the aforementioned satellites missions. On the left of Fig. 13.1 a Sentinel NDVI satellite image of Funes, a village of Navarra (Spain) is shown, and on the right of the same Figure, the NDVI for the whole region of Navarra.

Another important variable derived with satellite images is the land surface temperature (LST), that can be retrieved with different algorithmic procedures. As an example Sobrino et al. (2004) compare three methods to retrieve the LST from thermal infrared data supplied by band 6 of the Thematic Mapper (TM) sensor onboard the Landsat 5 satellite. The first is based on the radiative transfer equation using in situ radiosounding data. The others are the mono-window algorithm developed by Qin et al. (2001) and the single-channel algorithm developed by Jiménez-Muñoz and Sobrino (2003). Many satellites platforms provide specific images of LST all over the Earth, because it is also a very outstanding variable for many environmental process. Figure 13.2 shows the daily land surface temperature in Navarra (Spain) the 13th of July 2015 from TERRA satellite.

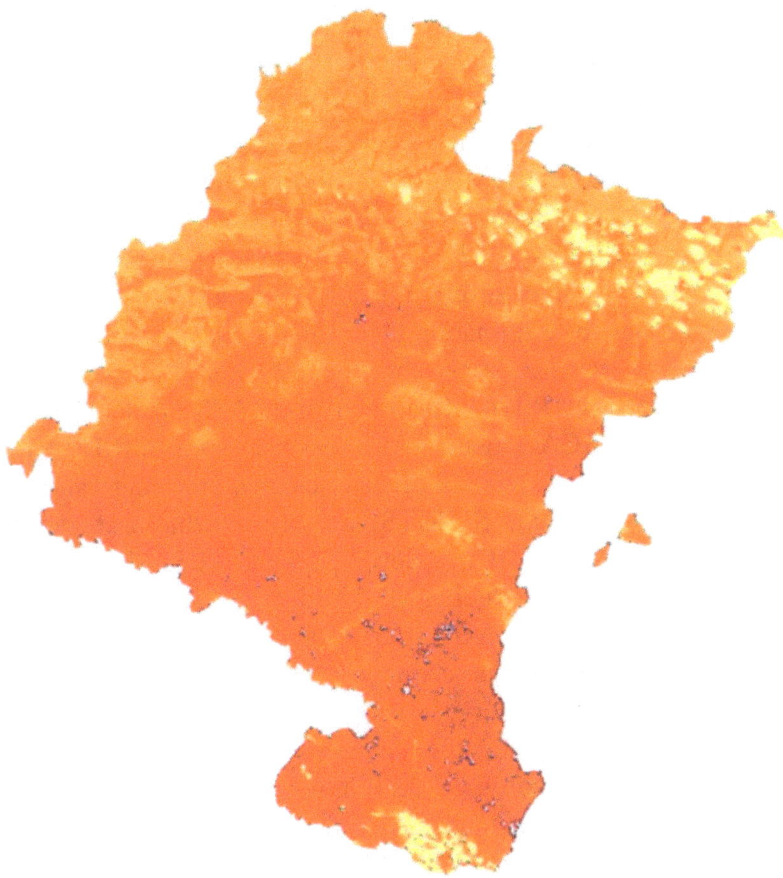

**Fig. 13.2** Land Surface Temperature of Navarra the 13th of July 2015

## 13.4 Pre-processing

The atmosphere is between the satellite and the Earth, and its effects over the electromagnetic radiation caused by the satellite can distort, blur or degrade the images. These effects must be corrected before the image processing. The correction consists of composing several images into a new single one. Different algorithms have been developed in the literature according to the derived variable. The most common method with NDVI is the maximum value composite (MVC) procedure (Holben 1986) that assigns the maximum value of the time-series of pixels across the composite period. Alternative techniques include using a bidirectional reflectance distribution function (BRDF-C) to select observations and the constraint view angle maximum value composite (CV-MVC) (MODIS 2017). For LST day/night it is common to average the cloud-free pixels over the compositing period (Vancutsem et al. 2010). Nowadays, many composite images can be directly downloaded with different spatial and temporal resolutions. For example, raw daily images can be downloaded from AQUA or TERRA satellites all over the world, but usually composite images are at least of weekly or bi-weekly temporal resolution.

Spatial and temporal resolutions are also different from the same or different satellites. High temporal resolution can be useful when tracking seasonal changes in vegetation on continental and global scales, but when downscaling to small regions, a higher spatial resolution is needed, and frequently with lower temporal resolution. At this step, numerical, physical or mechanical analyses solve the image pre-processing. Later, removing the effect of clouds or other atmospheric effects is also required, otherwise remote sensing data can be inaccurate. Sometimes, the highest presence of clouds determine the dropout of several images, but if they are only partially clouded, different approaches for eliminating these effects can be used. Noise reduction in image time series is neither simple nor straightforward. Many alternatives have been provided. For example R.HANTS macro of GRASS, SPIRITS, BISE, TIMESAT, GAPFILL or the CACAO methods are very well spread. R.HANTS performs an harmonic analysis of time series in order to estimate missing values and identify outliers (Roerink et al. 2000). SPIRITS is a software that processes time series of images (Eerens et al. 2014). It was developed by PROBA-V data provider and gives four smoothing options, including MEAN (Interpolate missing values & apply Running Mean Filter RMF) and BISE (Best Index Slope Extraction), (Viovy et al. 1992). TIMESAT uses numerical procedures based on Fourier analysis, Gauss, double logistic or SavitzkyGolay filters (Jönsson and Eklundh 2004). GAPFILL uses quantile regression to produce smoothed images where the effect of the clouds have been reduced. Usually, every software has different requirements with regard to the number of images necessary for smoothing (Atkinson et al. 2012). Finally, CACAO software (Verger et al. 2013) provides smoothing, gap filling, and characterizing seasonal anomalies in satellite time series.

All these procedures give composite images that are smoothed versions of the raw images, but very often they are not completely free of noise. Many of the attributes that can be extracted from the combination of satellite image bands are still vulnerable to many atmospheric or electronic accidents. For example, highly reflective surfaces, including snow and clouds, and sun-glint over water bodies may saturate the reflective wavelength bands, with saturation varying spectrally and with the illumination geometry (Roy et al. 2016). Land surface temperature or normalized vegetation index are examples of attributes where these type of errors can be present. Therefore, after pre-processing is done, interpolation and smoothing methods can be very useful for drawing or detecting trend changes, clustering or many other processes on remote sensing data.

## 13.5 Spatial Interpolation

Likely, interpolation and classification are among the most used tools with remote sensing data. Classification of satellite images in supervised or unsupervised versions are important research areas not only with satellite images but also with big data and data mining where there are a great number of algorithmic procedures (Benz et al. 2004). Here, we are more interested in interpolation as it is more closely related to geostatistics.

Interpolation has been widely used in environmental sciences. Li and Heap (2011) revise more than 50 different spatial interpolation methods that can be summarized in three categories: non-geostatistical methods, geostatistical methods, and combined methods. All of them can be represented as weighted averages of sampled data. A-mong the non-geostatistical methods the authors find: nearest neighbours, inverse distance weighting, regression models, trend surface analysis, splines and local trend surfaces, thin plate splines, classification, and regression trees. The different versions of simple, ordinary, disjunctive or model-based kriging are among the geostatistical methods. The combined methods include: trend surface analysis combined with k-riging, linear mixed models, regression trees combined with kriging or regression kriging.

Recently, Jin and Heap (2014) present an excellent review of spatial interpolation methods in environmental sciences introducing 10 methods from the machine learning field. These methods include support vector machines (SVM), random forest-s (RF), neural networks, neuro-fuzzy networks, boosted decision trees (BDT), the combination of SVM with inverse distance weighting (IDW) or ordinary kriging (OK), the combination of RF with IDW or OK (RFIDW, RFOK), general regression neural network (GRNN), the combination of GRNN with IDW or OK, and the combination of BDT with IDW or OK. Although all these methods were not developed specifically for remote sensing data, nowadays the majority of them have been implemented in different packages of the free statistical software R, and can be used with satellite images. Many of these methods are ready to use and interpret, but the family of kriging methods as the core of geostatistics, are preferred and widely used.

## 13.6 Spatio-Temporal Interpolation

Since the publication of the seminal book *Spatial Autocorrelation* (Cliff and Ord 1973), and at latter date *Spatial Statistics* (Ripley 1981), *Statistics for Spatial Data* (Cressie and Wikle 2015), and *Multivarate Geostatistics* (Wackernagel 1995) books, there has been a rapid growth of spatial geostatistical methods, as they are essential tools for interpolating meteorological, physical, agricultural or environmental variables in locations where these variables are not observed.

The use of spatial geostatistics with remote sensing data is also very well widespread, and its procedures are present in many specific softwares of satellite image analysis (Stein et al. 1999). Geostatistics techniques can help to explore and describe the spatial variability, to design optimum sampling schemes, and to increase the accuracy estimation of the variables of interest. These models can be enriched with auxiliary information coming from classified land cover or historical information (Curran and Atkinson 1998). Kriging is the most popular geostatistical method with several versions such as block kriging, universal kriging, ordinary kriging, regression kriging or indicator kriging. It provides the spatial interpolation of different spatial variables through the use of spatial stochastic models, and it is the best linear unbiased predictor under normality assumptions when using spatially dependent data.

However, the extension to the spatio-temporal geostatistics methods is more complicated. Time series models typically assume a regularly sampling over time, but the temporal lag operator cannot be easily generalized to the spatial domain, where data are likely irregularly sampled (Phaedon and André 1999). Scales of time and space are different, therefore defining joint spatio-temporal covariance functions is not a trivial task (De Iaco et al. 2002). Recently, Cressie and Wikle (2015) show the state of the art in this area and explain the difficulties of inverting covariance matrices in spatio-temporal kriging, because it becomes problematic without some form of separable models or dimension reduction. Modelling the spatio-temporal dependence is frequently case-specific. Therefore, yet the presence of the spatio-temporal keyword is abundant in many satellite imagery papers, the use of spatio-temporal stochastic models is scarce. Very often, spate-time refers only to descriptive analyses of time series of satellite images where every image is analyzed as a set of separate pixels, i.e., when estimating trends, or trend changes, statistical methods of univariate time series are used for every pixel. For example, when completing, reconstructing or predicting the spatial and temporal dynamics of the future NDVI distribution many papers use a time series of images (Forkel et al. 2013; Tüshaus et al. 2014; Klisch and Atzberger 2016; Wang et al. 2016; Liu et al. 2015; Maselli et al. 2014). These studies include temporal correlation of individual pixels at different resolutions but ignoring spatial dependence among them.

Spatio-temporal stochastic models use the spatial or temporal dependence to estimate optimally local values from sampled data. In satellite images, sampled data can be a huge amount of spatially and temporally dependent pixels, if a sequence of images is involved. We briefly review in what follows some stochastic spatio-temporal models that can be used when analysing remote sensing data.

1. Spatio-temporal kriging (Gasch et al. 2015). This paper uses spatio-temporal R packages for fitting some of the following spatio-temporal covariance functions: separable, product-sum, metric and sum-metric classesin a spatio-temporal kriging model, and a random forest algorithm for modeling dynamic soil properties in 3-dimensions.
2. State-space models (Cameletti et al. 2011). The authors apply a family of state-space models with different hierarchical structure and different spatio-temporal covariance function for modelling particular matter in Piemonte (Italy).
3. Hierarchical spatio-temporal model (Cameletti et al. 2013). The paper introduces a hierarchical spatio-temporal model for particulate matter (PM) concentration in the North-Italian region Piemonte. The authors use stat-space models involving a Gaussian Field (GF), affected by a measurement error, and a state process characterized by a first order autoregressive dynamic model and spatially correlated innovations. The estimation is based on Bayesian methods and

consists of representing a GF with Matérn covariance function as a Gaussian Markov Random Field (GMRF) through the Stochastic Partial Differential E-quations (SPDE) approach. Then, the Integrated Nested Laplace Approximation (INLA) algorithm is proposed as an alternative to MCMC methods, giving rise to additional computational advantages (Rue et al. 2009).

4. Spatio-temporal data-fusion (STDF) methodology (Nguyen et al 2014). This method is based on reduced-dimensional Kalman smoothing. The STDF is able to combine the complementary GOSAT and AIRS datasets to optimally estimate lower-atmospheric CO2 mole fraction over the whole globe.
5. Hierarchical statistical model (Kang et al. 2010). This model includes a spatio-temporal random effects (STRE) model as a dynamical component, and a temporally independent spatial component for the fine-scale variation. This article demonstrates that spatio-temporal statistical models can be made operational and provide a way to estimate level-3 values over the whole grid and attach to each value a measure of its uncertainty. Specifically, a hierarchical statistical model is presented, including a spatio-temporal random effects (STRE) model as a dynamical component and a temporally independent spatial component for the fine-scale variation. Optimal spatio-temporal predictions and their mean squared prediction errors are derived in terms of a fixed-dimensional Kalman filter.
6. Three-stage spatio-temporal hierarchical model (Fassò and Cameletti 2009). This work gives a three-stage spatio-temporal hierarchical model including spatio-temporal covariates. It is estimated through an EM algorithm and bootstrap techniques. This approach has been used by (Militino et al. 2015) for interpolating daily rainfall data, and for estimating spatio-temporal trend changes in NDVI with satellite images of Spain from 2011-2013 (Militino et al. 2017).
7. Space-varying regression model (Bolin et al. 2009). In this space-varying regression model the regression coefficients for the spatial locations are dependent. A second order intrinsic Gaussian Markov Random Field prior is used to specify the spatial covariance structure. Model parameters are estimated using the Expectation Maximisation (EM) algorithm, which allows for feasible computation times for relatively large data sets. Results are illustrated with simulated data sets and real vegetation data from the Sahel area in northern Africa.

### *13.6.1 Geostatistical R Packages*

In this section we briefly describe some of the most useful R packages for geostatistical analysis, including spatial and spatio-temporal interpolation in satellite imagery.

1. FRK (Cressie and Johannesson 2008) means fixed rank kriging and it is a tool for spatial/spatio-temporal modelling and prediction with large datasets.

2. geoR (Ribeiro Jr et al. 2001) offers classical geostatistics techniques for analysing spatial data. The extension to generalized linear models was made in geoRglm package (Christensen and Ribeiro 2002).
3. georob (R Core Team 2017) fits linear models with spatially correlated errors to geostatistical data that are possibly contaminated by outliers.
4. geospt (Melo et al. 2012) estimates the variogram through trimmed mean and does summary statistics from cross-validation, pocket plot, and design of optimal sampling networks through sequential and simultaneous points methods.
5. geostatsp (Brown 2015) provides geostatistical modelling facilities using raster. Non-Gaussian models are fitted using INLA, and Gaussian geostatistical models use maximum likelihood estimation.
6. gstat (Pebesma 2004) does spatio-temporal kriging, sequential Gaussian or indicator (co)simulation, variogram and variogram map plotting utility functions.
7. RandomFields (Schlather et al. 2015) provides methods for the inference on and the simulation of Gaussian fields.
8. spacetime (Pebesma et al. 2012) gives methods for representations of spatio-temporal sensor data, and results from predicting (spatial and/or temporal interpolation or smoothing), aggregating, or sub-setting them, and to represent trajectories.
9. spatial (Venables and Ripley 2002) provides functions for kriging and point pattern analysis.
10. spatialEco (Evans 2016) does spatial smoothing, multivariate separability, point process model for creating pseudo- absences and sub-sampling, polygon and point-distance landscape metrics, auto-logistic model, sampling models, cluster optimization and statistical exploratory tools. It works with raster data.
11. SpatialTools (R Core Team 2017) contains tools for spatial data analysis with emphasis on kriging. It provides functions for prediction and simulation.
12. spBayes (Finley et al. 2007) fits univariate and multivariate spatio-temporal random effects models for point-referenced data using Markov chain Monte Carlo (MCMC).

## 13.7 Conclusions

The multitemporal Earth observation satellites have been very well developed since the seventies, and along with the free availability of millions of satellite images, the number of publications of remote sensing data with geostatistical techniques has been rapidly increased. But unfortunately, not all published papers deriving, analysing or monitoring spatio-temporal evolutions, spatio-temporal trends or spatio-temporal changes are necessarily geostatistical papers, because they do not really use spatio-temporal stochastic models. These models are still scarce in remote sensing data because many of these models are computationally very intensive, or because they are not so broadly applicable as the spatial models are. The solutions found in the literature are very well fitted to specific problems, but we cannot always plug-in to other applications. The use of time series analysis in remote sensing opens a great window of opportunities for monitoring, smoothing, and detecting changes in large series of satellite images, but there are still many remote sensing papers ignoring the spatial dependence when analysing time series of images (Ban 2016). Instead, a huge discretization of the problem is presented where time-series of pixels are treated as spatially independent.

Nowadays, the upcoming opportunities for geostatisticians in remote sensing data are not based on the use of spatial models and time series separately, but on the use of spatial, temporal, or spatio-temporal stochastic models embedding both types of dependencies when necessary. Moreover, a single free statistical software like R is a powerful tool for downloading, importing, accessing, exploring, analysing and running advanced statistical modelling with remote sensing data in a row.

**Acknowledgements** This research was supported by the Spanish Ministry of Economy, Industry and Competitiveness (Project MTM2017-82553-R), the Government of Navarra (Project PI015, 2016 and Project PI043 2017), and by the Fundación Caja Navarra-UNED Pamplona (2016 and 2017).

## References

Aschbacher J, Milagro-Pérez MP (2012) The European earth monitoring (GMES) programme: s-tatus and perspectives. Remote Sens Environ 120:3–8

Atkinson PM, Jeganathan C, Dash J, Atzberger C (2012) Inter-comparison of four models for s-moothing satellite sensor time-series data to estimate vegetation phenology. Remote Sens Environ 123:400–417

Ban Y (2016) Multitemporal remote sensing. Methods and applications, vol 1. Remote sensing and digital image processing. Springer, Berlin

Benz UC, Hofmann P, Willhauck G, Lingenfelder I, Heynen M (2004) Multi-resolution, object-oriented fuzzy analysis of remote sensing data for GIS-ready information. ISPRS J Photogramm Remote Sens 58(3):239–258

Bolin D, Lindström J, Eklundh L, Lindgren F (2009) Fast estimation of spatially dependent temporal vegetation trends using Gaussian Markov random fields. Comput Stat Data Anal 53(8):2885–2896

Brown PE (2015) Model-based geostatistics the easy way. J Stat Softw 63(12):1–24. http://www.jstatsoft.org/v63/i12/

Cameletti M, Ignaccolo R, Bande S (2011) Comparing spatio-temporal models for particulate matter in Piemonte. Environmetrics 22(8):985–996

Cameletti M, Lindgren F, Simpson D, Rue H (2013) Spatio-temporal modeling of particulate matter concentration through the SPDE approach. AStA Adv Stat Anal 97(2):109–131

Christensen O, Ribeiro PJ (2002) geoRglm - a package for generalised linear spatial models. R-news 2(2):26–28. http://cran.R-project.org/doc/Rnews. ISSN 1609-3631

Cliff AD, Ord JK (1973) Spatial autocorrelation, vol 5. Pion, London

Cressie N, Johannesson G (2008) Fixed rank kriging for very large spatial data sets. J R Stat Soci: Ser B (Stat Methodol) 70(1):209–226

Cressie N, Wikle CK (2015) Statistics for spatio-temporal data. Wiley, New York

Curran PJ, Atkinson PM (1998) Geostatistics and remote sensing. Prog Phys Geogr 22(1):61–78

De Iaco S, Myers DE, Posa D (2002) Nonseparable space-time covariance models: some parametric families. Math Geol 34(1):23–42

Eerens H, Haesen D, Rembold F, Urbano F, Tote C, Bydekerke L (2014) Image time series processing for agriculture monitoring. Environ Model Softw 53:154–162

Evans JS (2016) spatialEco. http://CRAN.R-project.org/package=spatialEco, R package version 0.0.1-4

Fassò A, Cameletti M (2009) The EM algorithm in a distributed computing environment for modelling environmental space-time data. Environ Model Softw 24(9):1027–1035

Finley AO, Banerjee S, Carlin BP (2007) spBayes: an R package for univariate and multivariate hierarchical point-referenced spatial models. J Stat Softw 19(4):1–24. http://www.jstatsoft.org/v19/i04/

Forkel M, Carvalhais N, Verbesselt J, Mahecha MD, Neigh CS, Reichstein M (2013) Trend change detection in NDVI time series: effects of inter-annual variability and methodology. Remote Sens 5(5):2113–2144

Gasch CK, Hengl T, Gräler B, Meyer H, Magney TS, Brown DJ (2015) Spatio-temporal interpolation of soil water, temperature, and electrical conductivity in 3D + T: the cook agronomy farm data set. Spat Stat 14:70–90

Gerber F, Furrer R, Schaepman-Strub G, de Jong R, Schaepman ME (2016) Predicting missing values in spatio-temporal satellite data. arXiv:1605.01038

GLCF (2017) Global land cover facility. http://glcf.umd.edu/data/landsat/

Goslee SC (2011) Analyzing remote sensing data in R: the landsat package. J Stat Softw 43(4):1–25. http://www.jstatsoft.org/v43/i04/

Holben BN (1986) Characteristics of maximum-value composite images from temporal AVHRR data. Int J Remote Sens 7(11):1417–1434

Jiménez-Muñoz JC, Sobrino JA (2003) A generalized single-channel method for retrieving land surface temperature from remote sensing data. J Geophys Res: Atmos 108 (D22)

Jin L, Heap AD (2014) Spatial interpolation methods applied in the environmental sciences: a review. Environ Model Softw 53:173–189

Jönsson P, Eklundh L (2004) Timesat a program for analyzing time-series of satellite sensor data. Comput Geosci 30(8):833–845

Kang EL, Cressie N, Shi T (2010) Using temporal variability to improve spatial mapping with application to satellite data. Can J Stat 38(2):271–289

Klisch A, Atzberger C (2016) Operational drought monitoring in Kenya using modis NDVI time series. Remote Sens 8(4):267

Li J, Heap AD (2011) A review of comparative studies of spatial interpolation methods in environmental sciences: performance and impact factors. Ecol Inform 6(3):228–241

Liu Y, Li Y, Li S, Motesharrei S (2015) Spatial and temporal patterns of global NDVI trends: correlations with climate and human factors. Remote Sens 7(10):13,233–13,250

Maselli F, Papale D, Chiesi M, Matteucci G, Angeli L, Raschi A, Seufert G (2014) Operational monitoring of daily evapotranspiration by the combination of MODIS NDVI and ground meteorological data: application and evaluation in Central Italy. Remote Sens Environ 152:279–290

Matzke NJ (2013) modiscloud: an R Package for processing MODIS Level 2 Cloud Mask products. University of California, Berkeley, Berkeley, CA, http://cran.r-project.org/web/packages/modiscloud/index.html, this code was developed for the following paper: Goldsmith, Gregory; Matzke, Nicholas J.; Dawson, Todd (2013). The incidence and implications of clouds for cloud forest plant water relations. Ecol Lett, 16(3), 307–314. https://doi.org/10.1111/ele.12039

Maus V, Camara G, Cartaxo R, Sanchez A, Ramos FM, de Queiroz GR (2016) A time-weighted dynamic time warping method for land-use and land-cover mapping. IEEE J Sel Top Appl Earth Obs Remote Sens 9(8):3729–3739. https://doi.org/10.1109/JSTARS.2016.2517118

Melo C, Santacruz A, Melo O (2012) geospt: an R package for spatial statistics. http://geospt.r-forge.r-project.org/, R package version 1.0-0

Militino A, Ugarte M, Goicoa T, Genton M (2015) Interpolation of daily rainfall using spatiotemporal models and clustering. Int J Climatol 35(7):1453–1464

Militino AF, Ugarte MD, Pérez-Goya U (2017) Stochastic spatio-temporal models for analysing NDVI distribution of GIMMS NDVI3g images. Remote Sens 9(1):76

MODIS (2017) https://modis.gsfc.nasa.gov/about/

Nauss T, Meyer H, Detsch F, Appelhans T (2015) Manipulating satellite data with satellite. www.environmentalinformatics-marburg.de

Nguyen H, Katzfuss M, Cressie N, Braverman A (2014) Spatio-temporal data fusion for very large remote sensing datasets. Technometrics 56(2):174–185

NOAA (2017) National Oceanic and Atmospheric Administration. https://www.nesdis.noaa.gov/

Pebesma E et al (2012) Spacetime: spatio-temporal data in R. J Stat Softw 51(7):1–30

Pebesma EJ (2004) Multivariable geostatistics in S: the gstat package. Comput Geosci 30(7):683–691

Phaedon CK, André GJ (1999) Geostatistical spacetime models: a review. Math Geol 31(6):651–684

Qin Z, Karnieli A, Berliner P (2001) A mono-window algorithm for retrieving land surface temperature from landsat TM data and its application to the Israel-Egypt border region. Int J Remote Sens 22(18):3719–3746

R Core Team (2017) R: a language and environment for statistical computing. R Foundation for Statistical Computing, Vienna, Austria, https://www.R-project.org/

Ribeiro PJ Jr, Diggle PJ et al (2001) geoR: a package for geostatistical analysis. R news 1(2):14–18

Ripley BD (1981) Spatial statistics, vol 575. Wiley, New York

Roerink G, Menenti M, Verhoef W (2000) Reconstructing cloudfree NDVI composites using Fourier analysis of time series. Int J Remote Sens 21(9):1911–1917

Rouse J Jr, Haas R, Schell J, Deering D (1974) Monitoring vegetation systems in the Great Plains with ERTS. NASA special publication 351:309

Roy D, Kovalskyy V, Zhang H, Vermote E, Yan L, Kumar S, Egorov A (2016) Characterization of landsat-7 to landsat-8 reflective wavelength and normalized difference vegetation index continuity. Remote Sens Environ 185:57–70

Rue H, Martino S, Chopin N (2009) Approximate Bayesian inference for latent Gaussian models by using integrated nested laplace approximations. J R Stat Soc: Ser B (Stat Methodol) 71(2):319–392

Sagar DB, Je Serra (2010) Spacial issue on spatial information retrieval, analysis, reasoning and modelling. Int J Remote Sens 31(22):5747–6032

Schlather M, Malinowski A, Menck PJ, Oesting M, Strokorb K (2015) Analysis, simulation and prediction of multivariate random fields with package random fields. J Stat Softw 63(8):1–25. http://www.jstatsoft.org/v63/i08/

Slayback DA, Pinzon JE, Los SO, Tucker CJ (2003) Northern hemisphere photosynthetic trends 1982–99. Glob Chang Biol 9(1):1–15

Sobrino J, Julien Y (2011) Global trends in NDVI-derived parameters obtained from gimms data. Int J Remote Sens 32(15):4267–4279

Sobrino JA, Jiménez-Muñoz JC, Paolini L (2004) Land surface temperature retrieval from Landsat TM 5. Remote Sens Environ 90(4):434–440

Stein A, van der Meer FD, Gorte B (1999) Spatial statistics for remote sensing, vol 1. Springer Science & Business Media

Survey UG (2015) Landsat 8 (l8) data users handbook. US geological survey, Version 10(97p):1–97

Tucker CJ, Pinzon JE, Brown ME, Slayback DA, Pak EW, Mahoney R, Vermote EF, El Saleous N (2005) An extended avhrr 8-km NDVI dataset compatible with MODIS and spot vegetation NDVI data. Int J Remote Sens 26(20):4485–4498

Tüshaus J, Dubovyk O, Khamzina A, Menz G (2014) Comparison of medium spatial resolution ENVISAT-MERIS and terra-MODIS time series for vegetation decline analysis: a case study in Central Asia. Remote Sens 6(6):5238–5256

Vancutsem C, Ceccato P, Dinku T, Connor SJ (2010) Evaluation of MODIS land surface temperature data to estimate air temperature in different ecosystems over Africa. Remote Sens Environ 114(2):449–465

Venables WN, Ripley BD (2002) Modern applied statistics with S, 4th edn. Springer, New York. http://www.stats.ox.ac.uk/pub/MASS4. ISBN 0-387-95457-0

Verger A, Baret F, Weiss M, Kandasamy S, Vermote E (2013) The CACAO method for smoothing, gap filling, and characterizing seasonal anomalies in satellite time series. IEEE Trans Geosci Remote Sens 51(4):1963–1972

Viovy N, Arino O, Belward A (1992) The best index slope extraction (bise): a method for reducing noise in NDVI time-series. Int J Remote Sens 13(8):1585–1590

Wackernagel H (1995) Multivariate geostatistics: an introduction with applications. Springer Science & Business Media

Wang R, Cherkauer K, Bowling L (2016) Corn response to climate stress detected with satellite-based NDVI time series. Remote Sens 8(4):269

# Chapter 14
# Flint Drinking Water Crisis: A First Attempt to Model Geostatistically the Space-Time Distribution of Water Lead Levels

**Pierre Goovaerts**

**Abstract** The drinking water contamination crisis in Flint, Michigan has attracted national attention since extreme levels of lead were recorded following a switch in water supply that resulted in water with high chloride and no corrosion inhibitor flowing through the aging Flint water distribution system. Since Flint returned to its original source of drinking water on October 16, 2015, the State has conducted eleven bi-weekly sampling rounds, resulting in the collection of 4,120 water samples at 819 "sentinel" sites. This chapter describes the first geostatistical analysis of these data and illustrates the multiple challenges associated with modeling the space-time distribution of water lead levels across the city. Issues include sampling bias and the large nugget effect and short range of spatial autocorrelation displayed by the semivariogram. Temporal trends were modeled using linear regression with service line material, house age, poverty level, and their interaction with census tracts as independent variables. Residuals were then interpolated using kriging with three types of non-separable space-time covariance models. Cross-validation demonstrated the limited benefit of accounting for secondary information in trend models and the poor quality of predictions at unsampled sites caused by substantial fluctuations over a few hundred meters. The main benefit is to fill gaps in sampled time series for which the generalized product-sum and sum-metric models outperformed the metric model that ignores the greater variation across space relative to time (zonal anisotropy). Future research should incorporate the large database assembled through voluntary sampling as close to 20,000 data, albeit collected under non-uniform conditions, are available at a much greater sampling density.

---

P. Goovaerts (✉)
BioMedware, Inc, 11487 Highland Hills Drive, Jerome, MI 49249, USA
e-mail: goovaerts@biomedware.com



B. S. Daya Sagar et al. (eds.), *Handbook of Mathematical Geosciences*,
https://doi.org/10.1007/978-3-319-78999-6_14

## 14.1 Introduction

The drinking water contamination crisis in Flint, Michigan has attracted national attention since extreme levels of lead were recorded in local water supplies and the percentage of children with elevated blood lead levels (BLL) increased in neighborhoods with the highest water lead levels (WLL). Problems started when the City of Flint, Michigan adopted the cost-saving decision of drawing and treating water from the Flint River instead of relying on the Detroit Water and Sewerage Department's system (DWSD) for its public water supply. A few months later, in December 2014, water samples showed elevated levels of trihaloethanes (THMs) a disinfection byproduct of chlorine, as well as high levels of lead and copper. A public health emergency was declared and residents were told to avoid drinking the water until it was tested or approved water filters were installed. In July 2015, public concerns were raised that lead and copper were being leached from corrosion (chlorine-induced) in the underground lead service lines and home plumbing fixtures as a result of not using corrosion control treatment (CCT). In August and September 2015, 16.6% of the 271 water samples collected by a Virginia Tech's team were found to exceed the EPA action level of 15 μg/L (ATSDR 2010). In September and October 2015, elevated childhood blood lead levels were confirmed and an emergency response was initiated (Hanna-Attisha et al. 2016), leading the city to switch back to the DWSD water supply on October 16, 2015.

Starting in February 2016, samples were collected bi-weekly at more than 600 sentinel sites chosen by the EPA and MDEQ (Michigan Department of Environmental Quality) across the city to determine the general health of the distribution system and to track changes in lead concentrations over time (Flint Safe Drinking Water Task Force 2016). After five rounds of sentinel sampling, a new sentinel program called "Extended Sentinel Site Program" started in June 2016, targeting specifically sites with high WLL during previous rounds or located in the highest-risk areas. Six additional sampling rounds were conducted for this smaller network including fewer than 200 sites. Overall these 11 sampling rounds resulted in the collection of 4,120 data at 819 different sites over a 40-week time period. This State-controlled monitoring program was supplemented by a voluntary or homeowner-driven sampling whereby concerned citizens received a testing kit and conducted sampling on their own (Goovaerts 2017a, b). Despite the larger size of this database (18,760 samples collected over 53 weeks at 10,341 sites), its heterogeneity and lack of systematic sampling across time prohibited its use in the present space-time analysis.

Except for a few graphs and location maps, the database assembled by the City of Flint and made available online has not undergone any rigorous statistical treatment by State employees and only a few studies have been published so far. Using a data-driven approach Abernethy et al. (2016) developed an ensemble of predictive models (e.g., random forest, logistic regression, linear discriminant analysis) to assess the risk of lead contamination in individual homes and neighborhoods in Flint. They trained these models using a wide range of data sources, including residential water tests, historical records, and city infrastructure data. Their analysis however ignored the spatial correlation among data and did not include a temporal component. A time trend analysis was conducted by Goovaerts (2017a) who used joinpoint regression to model time series of lead levels collected by the state-controlled and voluntary sampling programs. This analysis carried out at the city and ward levels still ignored the spatial correlation among data and did not provide any tax parcel-based prediction. A space-time analysis of these data should however provide important information to identify residences where high levels of lead are expected. It would also support any assessment of past and current lead exposures among the population at risk, particularly pregnant women and children.

Geostatistical techniques have been routinely used to analyze and map the spatial variability of soil and sediment lead concentrations (Goovaerts et al. 1997; Cattle et al. 2002; Solt et al. 2015), yet their application to lead in drinking water is far less common and mainly concerns groundwater quality (Siddique et al. 2012). A recent study (Wang et al. 2014) applied geographic information systems (GIS) and a hydraulic model of distribution systems to test the influences of pipe material, pipe age, water age, and other water quality parameters on lead/copper leaching in Raleigh (NC). In Symanski et al. (2004), mixed effect models were used to assess spatial fluctuations, temporal variability, and errors due to sampling and analysis for levels of disinfection by-products in water samples collected in households within the same distribution system. To the author's knowledge, the present study is however the first application of geostatistics to lead in drinking water within a distribution system.

This chapter describes a new methodology to predict lead level in tap water, accounting for WLL measurements collected in neighboring houses, housing characteristics (e.g., age of the house or presence of lead pipes), and temporal trends (e.g., decline since return to pre-crisis source of drinking water). Linear regression was used to model temporal trends at sentinel sites, accounting for the composition of service line (SL), construction year, poverty level, and census tracts as covariates. Cross-validation analysis allowed one to assess the benefit of this approach and compare the results obtained using three different types of space-time covariance models. Both the cases of predicting unsampled times at monitored locations (i.e., filling gaps in time series) and making predictions at unsampled locations were investigated.

## 14.2 Materials and Methods

### *14.2.1 Datasets*

4,150 WLL measurements recorded over the period 2/20/2016-11/20/2016 were downloaded from http://www.michigan.gov/flintwater (residential testing results).

**Table 14.1** Datasets available for the space-time analysis: 4,120 water lead levels measured over 11 sampling rounds. Statistics include the number of data available, the sampling period, the percentage of WLL above 15 μg/L, the mean of logtransformed concentrations, and the composition of service line that was recorded for each sentinel site (three main categories besides plastic, unknown, and other)

| Sampling round | Data (n) | Sampling period | %WLL > 15 μg/L | Mean $Log_{10}$ (μg/L) | Composition of SL | | |
|---|---|---|---|---|---|---|---|
| | | | | | Lead | Galvanized | Copper |
| Round S1 | 610 | 2/16/2016–2/29/2016 | 9.51 | 0.487 | 5.90 | 20.66 | 68.20 |
| Round S2 | 606 | 2/24/2016–3/13/2016 | 8.42 | 0.465 | 8.91 | 19.97 | 67.00 |
| Round S3 | 654 | 3/15/2016–3/24/2016 | 8.26 | 0.480 | 11.62 | 19.57 | 63.91 |
| Round S4 | 644 | 3/29/2016–4/5/2016 | 7.14 | 0.457 | 13.66 | 17.39 | 64.29 |
| Round S5 | 622 | 4/13/2016–4/15/2016 | 6.43 | 0.427 | 14.31 | 15.27 | 65.43 |
| Round X1 | 170 | 5/23/2016–6/7/2016 | 7.06 | 0.604 | 45.88 | 9.41 | 44.71 |
| Round X2 | 178 | 6/14/2016–6/30/2016 | 8.99 | 0.638 | 49.44 | 7.87 | 42.70 |
| Round X3 | 167 | 7/19/2016–7/22/2016 | 6.59 | 0.557 | 46.11 | 8.38 | 45.51 |
| Round X4 | 162 | 8/18/2016–8/22/2016 | 9.88 | 0.579 | 45.06 | 9.26 | 45.68 |
| Round X5 | 158 | 9/19/2016–9/27/2016 | 6.33 | 0.522 | 45.57 | 9.49 | 44.94 |
| Round X6 | 149 | 11/17/2016–11/23/2016 | 6.71 | 0.532 | 45.64 | 9.40 | 44.97 |

Data were then allocated to an individual tax parcel unit on the basis of their postal address. Data with incomplete address (two samples) or duplicates (e.g., samples taken from two different faucets on the same day in the same house) were discarded, leading to a total of 4,120 samples collected at 819 different sites; see Table 14.1. Because of their strongly positively skewed distribution (concentrations range from 0 to 5,986 μg/L) and large proportion of zero values (34.6%), data were transformed using the following formula $Log_{10}(z+1)$.

Sentinel sites were initially selected from a pool of 1,951 volunteer sites identified during door-to-door water distribution; in particular it included all 156 sites with lead or lead combination service lines according to City records. Other sites were added according to several criteria: (i) spatial distribution to ensure coverage of all nine City wards, (ii) measurements of high blood levels (Hanna-Attisha et al. 2016), and (iii) environmental justice considerations (e.g. presence of houses with lead-based paint, minority population, and lower socio-economic households). This

**Table 14.2** Statistics computed for time series of different lengths: number of sentinel sites, percentage of WLL above 15 µg/L, the mean of logtransformed concentrations, and the composition of service line

| Length | Sites | %WLL > 15 µg/L | Mean $Log_{10}$ (µg/L) | Composition of SL | | |
|---|---|---|---|---|---|---|
| | | | | Lead | Galvanized | Copper |
| 1 | 80 | 7.50 | 0.413 | 6.25 | 22.50 | 66.25 |
| 2 | 33 | 6.06 | 0.475 | 12.12 | 18.18 | 63.64 |
| 3 | 36 | 6.48 | 0.433 | 22.22 | 25.00 | 46.30 |
| 4 | 95 | 4.74 | 0.411 | 5.26 | 18.95 | 70.53 |
| 5 | 409 | 3.52 | 0.358 | 2.93 | 17.85 | 73.59 |
| 6 | 41 | 8.54 | 0.530 | 21.95 | 4.88 | 73.17 |
| 7 | 19 | 9.77 | 0.651 | 89.47 | 5.26 | 5.26 |
| 8 | 10 | 11.25 | 0.705 | 68.75 | 7.50 | 23.75 |
| 9 | 23 | 11.59 | 0.693 | 84.54 | 4.35 | 11.11 |
| 10 | 32 | 18.75 | 0.750 | 38.75 | 15.63 | 45.63 |
| 11 | 41 | 19.82 | 0.793 | 33.92 | 20.26 | 45.81 |

initial set evolved between sampling rounds as some residents stopped participating, while others asked to be included in the network (Goovaerts 2017b), which explains the fluctuation in the number of sampled sites during the first five rounds S1-S5: 607–621 (Table 14.1). Fewer sites (149–178) were then part of the "Extended Sentinel Site Program". Table 14.2 indicates that only 41 sites were sampled in all 11 rounds, while 80% of time series included five observations or less.

Each house selected to be part of the sentinel network was visited by a licensed plumber who classified the material of the service line coming into the home (i.e., customer-side service line) into six categories: lead, galvanized, copper, plastic, other, and unknown. Galvanized refers to iron pipe with a protective "galvanized" surface coating composed of zinc, lead, and cadmium, and therefore can be a long-term source of lead (Clark et al. 2015). The term "unknown" was used whenever the SL material could not be confirmed because, for example, the line was behind a wall or way back in a crawl space.

City records were the only source of service line data available for the majority of 56,039 tax parcels which were not part of the sentinel sampling program. These records are however inaccurate and lead to the over-identification of lead SLs, likely because old records were not updated as these lines were being replaced (Goovaerts 2017c). The same author found that construction year was a good predictor of service line material: galvanized lines were mostly found in pre-1934 houses, while the frequency of lead service lines (LSLs) peaked for houses built around World War II. This information was combined with field inspection data and city records to predict by indicator kriging the likelihood that a home has lead or galvanized SL (Goovaerts 2017c).

Besides service lines, lead in drinking water mainly comes from lead-based solder and lead-containing plumbing fixtures (Lee et al. 1989; Cartier et al. 2011). Plumbing material is usually related to the installation year of a plumbing system, which can be approximated by the year of construction. For example, most faucets purchased prior to 1997 were made of brass or chrome-plated brass containing up to 8 percent lead (Rabin 2008). Construction year was retrieved from the 2016 Parcels GIS layer. The attribute "Year_built" was missing for 20,372 parcels and was estimated by ordinary kriging (Goovaerts 1997) with a mean absolute error of prediction of 6.43 years. Based on its relationship to water lead levels (Goovaerts 2017a), construction year was discretized into three classes: pre-1940, 1940–1959, and post-1959.

Poor workmanship as well as lack of regular maintenance can also lead to more corrosion and leaching, and the presence of lead particulates, such as disintegrating brass or detaching pieces of old solder (Wang et al. 2014). Socio-economic status was here assessed using 2015 ACS (American Community Survey) 5-year estimates of the percentage of the block group population living in households where the income is less than or equal to twice the federal "poverty level".

There are many other variables known to influence lead in drinking water. For example, longer water age (i.e., water travel time between the treatment plant and home plumbing system) can decrease the effectiveness of corrosion control; increasing leaching and water lead levels (US EPA 2002; Wang et al. 2014). This information was however unavailable for this study.

### 14.2.2 Space-Time Kriging and Covariance Models

Let $z(\mathbf{u}_\alpha;t)$ denote the water lead level recorded on time $t$ at sentinel site $\alpha$ georeferenced by the geographical coordinates $\mathbf{u}_\alpha = (x_\alpha, y_\alpha)$ of the corresponding tax parcel centroid. Prediction of z-value at unsampled time $t_0$ and location $u_0$ was conducted using the following kriging estimator:

$$Z^*(u_0;t_0) = \sum_{t=t_0-\Delta t}^{t_0+\Delta t} \sum_{\alpha=1}^{n(t)} \lambda_{\alpha t} \times z(u_\alpha;t) \quad (14.1)$$

$n(t)$ is the number of observations recorded at time $t$, within the time window $2\Delta t$, that were retained for estimation. The weights $\lambda_{\alpha t}$ are solution of the following space-time (ST) kriging system:

$$\begin{aligned}&\sum_{t=t_0-\Delta t}^{t_0+\Delta t}\sum_{\alpha=1}^{n(t)}\lambda_{\alpha t}C\left(u_\alpha-u_\beta;t-t'\right)+\mu=C\left(u_0-u_\beta;t_0-t'\right) \quad \beta=1,\cdots,n\left(t'\right)\\&\sum_{t=t_0-\Delta t}^{t_0+\Delta t}\sum_{\alpha=1}^{n(t)}\lambda_{\alpha t}=1 \qquad t'=t_0-\Delta t,\cdots,t_0+\Delta t\end{aligned} \tag{14.2}$$

The parameter $\mu$ is a Lagrange multiplier accounting for the constraint on the weights. The term $C\left(u_\alpha-u_\beta;t-t'\right)$ is the ST covariance between any two observations recorded at locations $u_\alpha$ and $u_\beta$ at times $t$ and $t'$, respectively. Euclidian distances were used here since most lead in drinking water comes from premise plumbing materials and service lines instead of being transported through water mains (Del Toral et al. 2013; EET Inc. 2015).

One challenge associated with the application of ST kriging is the choice of a ST covariance model within the ever growing class of models (Montero et al. 2015). The following three non-separable ST covariance models were compared in the present study:

- The generalized product-sum model (De Iaco et al. 2002):

$$C(h,\tau)=k_1C_s(h)+k_2C_t(\tau)+k_3C_s(h)C_t(\tau) \tag{14.3}$$

  where $k_1, k_2,$ and $k_3$ are non-negative (strictly positive for $k_3$) coefficients estimated from the sills of the spatial, temporal, and spatio-temporal semivariograms (De Cesare et al. 2002).
- The metric model (Dimitrakopoulos and Luo 1994):

$$C(h,\tau)=C_{st}\left(\sqrt{\left(\frac{h}{a_s}\right)^2+\left(\frac{\tau}{a_t}\right)^2}\right) \tag{14.4}$$

  where a normalized space-time distance measure is created by rescaling the spatial and temporal lags, $h$ and $\tau$, by the ranges of the spatial and temporal semivariograms, $a_s$ and $a_t$ (case of geometric anisotropy).
- The sum-metric model (Heuvelink and Griffith 2010):

$$C(h,\tau)=C_s(h)+C_t(\tau)+C_{st}\left(\sqrt{\left(\frac{h}{a_s}\right)^2+\left(\frac{\tau}{a_t}\right)^2}\right) \tag{14.5}$$

  This model combines characteristics of the two previous models: (i) sum of spatial and temporal covariances allowing for the presence of zonal anisotropies (i.e., semivariogram sills are not the same in all directions), and (ii) a metric ST model for the residual variability (geometric anisotropy).

Two other classes of non-separable ST covariance models, Cressie-Huang model (Cressie and Huang 1999) and Gneiting models (Gneiting 2002), were not considered because: (1) the fitting of these models needs a complex iterative parameter optimization technique (De Iaco 2010), whereas the three selected models can be fitted using straightforward techniques similar to those already used for spatial-only and temporal-only semivariograms, and (2) recent studies (Guo et al. 2015) indicate that these two more complex models provide similar fits to experimental ST semivariograms and comparable prediction accuracy as the product-sum model, confirming previous findings (De Iaco 2010).

The main difficulty in the practical implementation of the product-sum and sum-metric models is the inference of the sill of the ST semivariogram model, $C_{st}(0)$, which is most often estimated visually from the 3D plot of the experimental ST semivariogram $\hat{\gamma}_{st}(h,\tau)$ (e.g., De Cesare et al. 2002; Heuvelink and Griffith 2010). In order to make the fitting procedure more user-friendly, the space-time sill $C_{st}(0)$ was here computed as the following weighted average of experimental space-time semivariogram values:

$$C_{st}(0) = \frac{1}{\sum_h \sum_\tau w_{h,\tau}} \sum_h \sum_\tau w_{h,\tau}\hat{\gamma}_{st}(h,\tau) \quad \text{if } \hat{\gamma}_{st}(h,\tau) \geq g_c \tag{14.6}$$

where the weight $w_{h,\tau}$ is the number of data pairs falling into the class of spatial and temporal lags $(h,\tau)$. Only the classes where the ST semivariogram values exceed a critical sill $g_c$, defined as the maximum of the spatial and temporal sills, were used.

### 14.2.3 Accounting for Secondary Information

Lead service lines are widely considered the main source of lead in drinking water (Lee et al. 1989; Clark et al. 2015). Another culprit is lead fixtures and pipes present within old houses (premises plumbing), and poverty can compound the problems through the lack of maintenance. Goovaerts (2017a) also found that temporal trends can vary greatly across the city. This secondary information was here incorporated in the definition of a stochastic trend model $M(u;t)$, leading to the following decomposition of the space-time random function (RF) (Kyriakidis and Journel 1999):

$$Z(u;t) = M(u;t) + R(u;t) \tag{14.7}$$

where $M(u;t)$ is a nonstationary spatiotemporal RF modeling the space-time distribution of the mean process, with $E[M(u;t)] = m(u;t)$ and $R(u;t)$ is a zero mean stationary spatiotemporal RF modeling space-time fluctuations around $M(u;t)$.

The trend component at each sentinel site $\mathbf{u}_\alpha$ was fitted using a linear model including six fixed factors: presence/absence of LSL, presence/absence of galvanized service line (GSL), time since first sample was collected (TIME), poverty level (POV), house age (AGE), and census tract (CT). The model takes the following form:

$$
\begin{aligned}
M(u;t) = {} & LSL(u) \times TIME + CT(u) \times TIME + LSL(u) \times CT(u) \\
& + GSL(u) \times CT(u) + AGE(u) \times CT(u) \\
& + POV(u) \times CT(u)
\end{aligned}
\tag{14.8}
$$

This model naturally handles uneven spacing of repeated measurements within each time series, as well as their correlation which was modeled using a spherical variance-covariance structure. Once the trend model was fitted, regression residuals were interpolated using space-time simple kriging and the ST covariance models introduced in Sect. 14.2.2.

### 14.2.4 Cross-Validation

The accuracy of the predictive models created by the different approaches (e.g., three types of ST covariance models, univariate vs incorporation of secondary information) was assessed by cross-validation whereby each observation or time series (i.e., all data collected at the same site) was removed at a time and re-estimated using data collected at neighboring sentinel sites. The following performance criteria were then computed from $n$ kriging estimates:

- the mean error (ME) of prediction as:

$$ME = \frac{1}{n} \sum_{t=1}^{T} \sum_{\alpha=1}^{n(t)} \left(z^*(u_a;t) - z(u_a;t)\right) \tag{14.9}$$

- the mean absolute error (MAE) of prediction as:

$$MAE = \frac{1}{n} \sum_{t=1}^{T} \sum_{\alpha=1}^{n(t)} \left|z^*(u_a;t) - z(u_a;t)\right| \tag{14.10}$$

- the mean square standardized residual (MSSR) as:

$$MSSR = \frac{1}{n} \sum_{t=1}^{T} \sum_{\alpha=1}^{n(t)} \frac{\left(z^*(u_a;t) - z(u_a;t)\right)^2}{\sigma_K^2(u_a;t)} \tag{14.11}$$

where $\sigma_K^2(u_a;t)$ is the kriging variance.

A mean error close to zero indicates a lack of bias, while the mean absolute error should be as small as possible. If the actual estimation error is equal, on average, to the error predicted by the model, the MSSR statistic should be about one (Wackernagel 1998, p. 91).

One application of the predictive models is to prioritize any further sampling or intervention by ranking tax parcels from highly hazardous to less hazardous on the basis of kriging estimates. The ability of this ranking to identify successfully sites where WLL is greater or equal to the EPA action level of 15 μg/L was assessed using Receiver Operating Characteristics (ROC) curves which plot the probability of false positive versus the probability of detection (Swets 1988; Fawcett 2006; Goovaerts et al. 2016). The accuracy of the classification was quantified using the relative area under the ROC curve (AUC statistic), which ranges from 0 (worst case) to 1 (best case). The AUC is equivalent to the probability that the classifier will rank a randomly chosen positive instance (e.g., $z_c \geq 15$ μg/L) higher than a randomly chosen negative instance (e.g., $z_c < 15$ μg/L).

## 14.3 Results and Discussion

### *14.3.1 Spatial Distribution*

Figure 14.1a shows the location of all 819 sentinel sites within the nine wards in the city of Flint. Site-specific statistics such as number of observations and average log concentrations recorded for each time series, as well as composition of service line (GSL vs. LSL), were aggregated at the census tract level for better visualization. Geographical clusters of sentinel sites can be distinguished in several census tracts (e.g. border of wards 2 and 6, wards 7 and 9) which tend to be the tracts with the largest WLLs (Fig. 14.1c) and percentages of sampled LSLs (Fig. 14.1d). There is also a clear spatial trend with fewer lead service lines (e.g., none in Ward 1) and shorter time series (Fig. 14.1b) sampled in the Northern part of the city. Ward 5 includes the oldest neighborhood where GSLs are prevalent (Fig. 14.1e), while LSLs appear as small clusters, in particular in wards 6, 7 and 9 (Goovaerts 2017c).

### *14.3.2 Temporal Trend Modeling*

Temporal trends for the three major types of service line were visualized by aggregating observations within non-overlapping 14-day windows, which corresponds to the average time interval between sampling rounds during the first phase (Round S) of the sentinel monitoring program (Table 14.1). Except for LSLs water lead levels do not appear to have declined over the 40-week sampling period; actually they seem to have slightly increased for GSLs (Fig. 14.2a). These results are however a direct artifact of the sampling strategy whereby 80% of sentinel sites

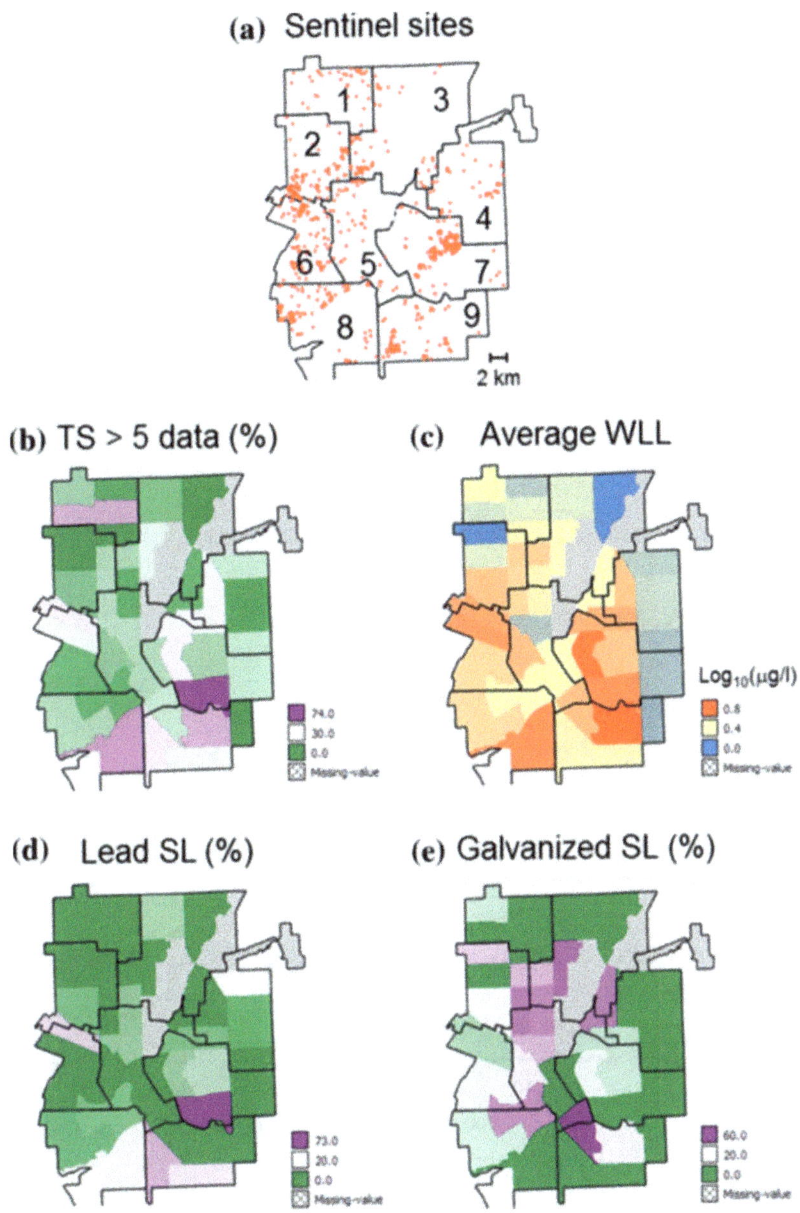


**Fig. 14.1** **a** Location of sentinel sites in each of the nine wards, and several census tract-level statistics: **b** percentages of time series (TS) including more than five observations, **c** average water lead levels, **d** percentage of sites with lead service lines, **e** percentage of sites with galvanized service lines. Shaded polygons indicate census tracts that do not include any sentinel site (missing values)

were not sampled beyond week 16, while sampling continued at sites where the risk of exceeding the EPA action level of 15 μg/L was the greatest (Table 14.2).

After elimination of all sites where fewer than six observations were collected, the averaged time series display the expected decline (Fig. 14.2b). The impact is minimal for LSLs since most of these sites are considered at risk and were sampled during both the initial and extended sentinel sampling programs (Rounds S and X).

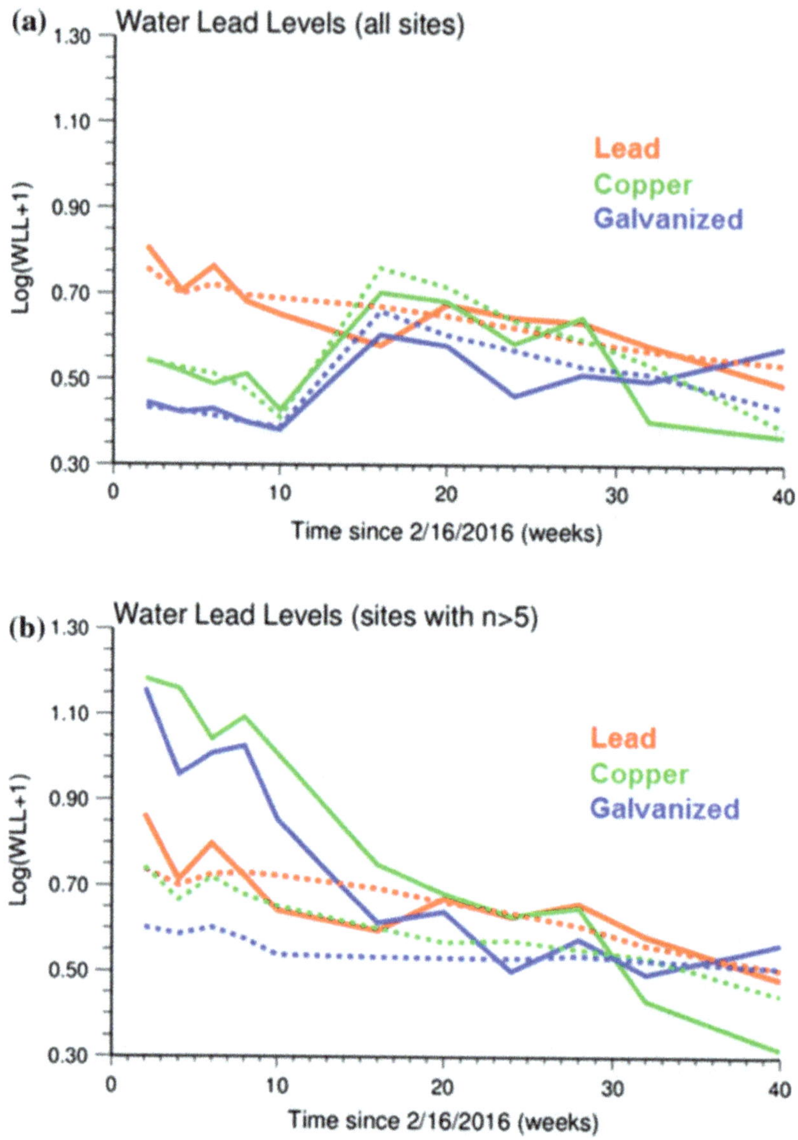


**Fig. 14.2** Time series of observed (solid line) and predicted by regression (dashed line) water lead levels computed on average for the three major types of service line: lead, galvanized, and copper. Results (log transformed concentrations) are calculated from: **a** all sites, and **b** subset of sites where at least six observations were recorded

The selection bias is stronger for copper and galvanized lines, which explains the larger water lead levels recorded during the first 16 weeks relative to LSLs.

This sampling bias complicated greatly the modeling of temporal trends by regression. Indeed using all the data would underestimate the weekly rate of decline of water lead levels, whereas subsetting the dataset (e.g., using only time series including more than five data points as in Fig. 14.2b) will result in overestimating the concentrations at a majority of sites. In addition, the time series length cannot be used as covariate in the model to allow its application at unmonitored locations. Two modeling strategies were considered in this chapter. First, because of its relationship with time series length (Fig. 14.1) census tract was used as covariate in the regression model (Eq. 14.8). The second more complicated approach was to allow the intercept to fluctuate among sentinel sites, even when located within the same tract; i.e., use a mixed model where the intercept is modeled as a random effect. The trade-off cost for this added flexibility was the need to estimate the intercept at unmonitored locations, which was accomplished using ordinary kriging. Despite providing a better fit than the first alternative, the mixed model did not lead to more accurate kriging estimates, hence only the first option is discussed hereafter.

All six interaction terms in the trend model (Eq. 14.8) were highly significant ($\alpha = 0.01$). The correlation between predicted and observed WLL is however rather weak ($r = 0.47$), which illustrates the challenge of predicting spatial and temporal variations in lead for drinking water (Bailey and Russell 1981; Del Toral et al. 2013). While the output of the regression model provides a reasonable fit to the SL-specific time series computed using all the data (Fig. 14.2a), it underestimates water lead levels for LSL and GSL when using only time series including more than five data points (Fig. 14.2b).

### *14.3.3 Variography*

Semivariograms helped quantifying the scale and magnitude of the space-time variability displayed by the maps and time series of Figs. 14.1 and 14.2. The spatial semivariogram (Fig. 14.3a) shows three nested scales of spatial variability: (1) a long range (2.35 km) caused by the neighborhood effect since houses in the same neighborhood tend to be built at the same time (i.e., similar plumbing system) and have similar water age, (2) a short range (200 m) corresponding to variability between adjacent houses, and (3) a nugget effect or discontinuity at the origin which represents the variability among samples taken within the same tax parcel (i.e. different apartments and/or measurement error for samples taken within the same residence). The substantial short-range variability (71% of total sill) likely reflects the heterogeneity in housing conditions (e.g., renovated houses) as well as the lack of uniformity of sampling conducted by homeowners since even with simple instructions it is difficult to ensure strict adherence to any sampling protocol (Del Toral et al. 2013). This interpretation is confirmed by the similar short-range variability displayed by the semivariogram of regression residuals (Fig. 14.3a, lower blue curve) since the regression model (Eq. 14.8) does not account for sampling characteristics. It is noteworthy that the longer range of 2.35 km is still fairly small relative to the size of the city (see legend of Fig. 14.1a), while the average separation distance between each sentinel site and the closest neighbor (293 m) exceeds the shortest range (200 m) that encapsulates 71% of the total spatial variability.

The temporal semivariogram (Fig. 14.3b) also displays three nested scales of variability although the longer range structure (110 days) represents here 53% of the total variability. Another difference with the spatial case is the overlap of

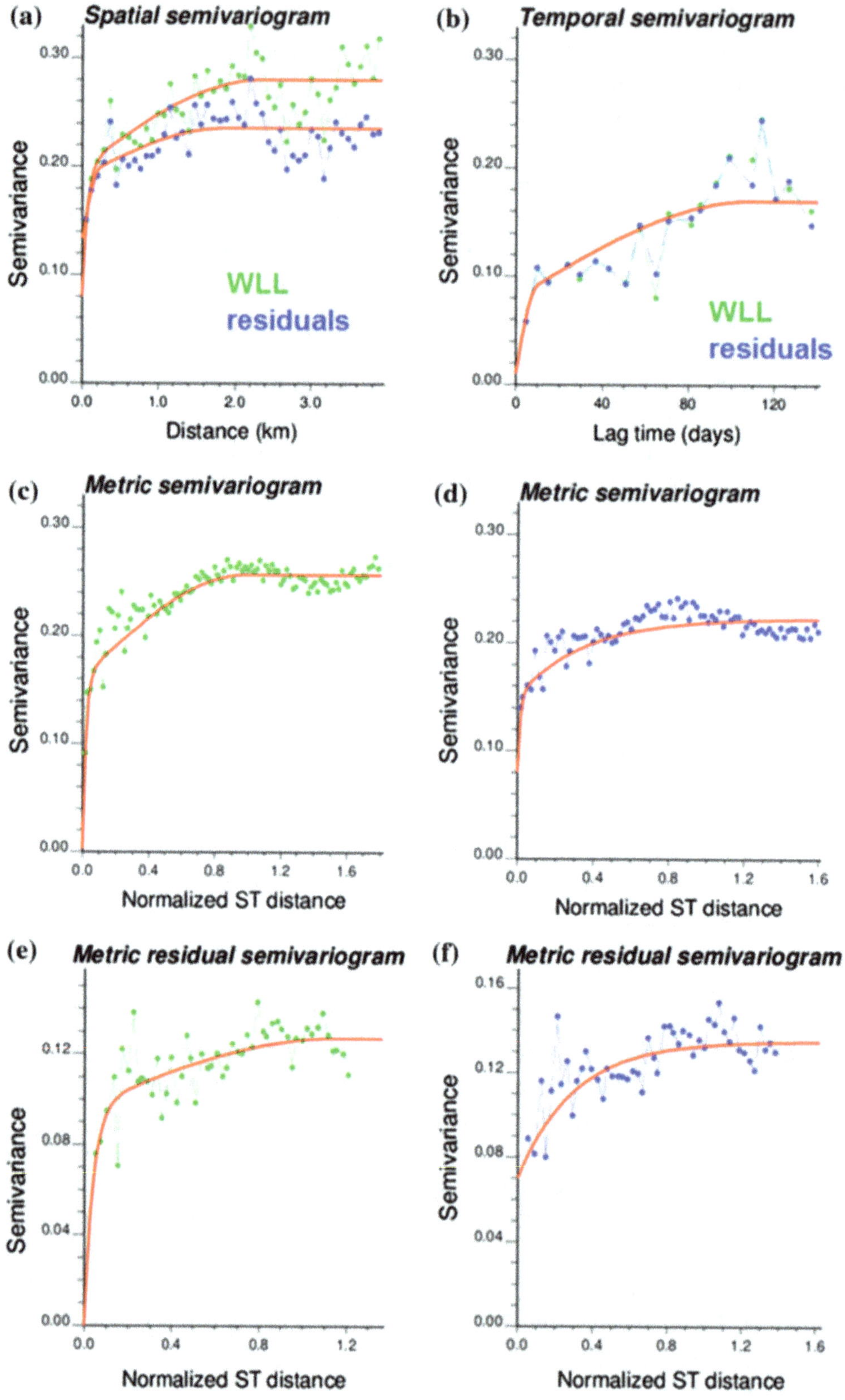
(a)
Spatial semivariogram
Semivariance
Distance (km)
WLL
residuals
(b)
Temporal semivariogram
Semivariance
Lag time (days)
WLL
residuals
(c)
Metric semivariogram
Semivariance
Normalized ST distance
(d)
Metric semivariogram
Semivariance
Normalized ST distance
(e)
Metric residual semivariogram
Semivariance
Normalized ST distance
(f)
Metric residual semivariogram
Semivariance
Normalized ST distance

**Fig. 14.3** Experimental semivariograms with the model fitted that were used to form the three types of ST covariance models (Eqs. 14.3–14.5) **a** spatial semivariogram (lower curve is for residuals), **b** temporal semivariogram, **c** metric semivariogram for WLLs, **d** metric semivariogram for regression residuals, **e** metric residual semivariogram (sum-metric model) for WLLs, **f** metric residual semivariogram for regression residuals

temporal semivariograms for WLLs and regression residuals, illustrating the inability of the trend model (Eq. 14.8) to capture purely temporal changes. This result is in agreement with the small magnitude of changes displayed by the time series of predicted values in Fig. 14.2 (dashed line). Comparison of the total sills of spatial and temporal semivariograms (Fig. 14.3a–b) indicates that the variability observed across space is greater than the temporal variability. Such zonal anisotropy is in conflict with the assumption underlying the metric ST covariance model (Eq. 14.4).

Figure 14.3c–d show the semivariograms computed using a normalized space-time distance (metric model). Because the spatial and temporal lags were rescaled using different constants for the WLL and residual semivariograms, these two curves are plotted separately. The vertical axis is however comparable and illustrates the smaller variability of residuals (i.e., lower sill for the semivariogram of Fig. 14.3d). Once again, both semivariograms display substantial short-range variability. The last two semivariograms (Fig. 14.3e–f) represent the metric space-time model that captures the residual variability in the sum-metric model (Eq. 14.5).

### *14.3.4 Cross-Validation Analysis*

The semivariogram models of Fig. 14.3 were used to conduct a cross-validation analysis whereby one observation (LOO approach) or one time series (LTO approach) was removed at a time and re-estimated using data collected at neighboring sentinel sites. Based on a sensitivity analysis using ST ordinary kriging and MAE criterion, 48 observations with a maximum of three data points per site were retained for the estimation by univariate and residual ST kriging. Results obtained for predictions by the time trend model were also included as reference in Table 14.3.

The first three rows in Table 14.3 indicate that all algorithms give unbiased predictions (ME close to zero). As expected, the best prediction scores (i.e., lower MAE and higher AUC) are obtained when using data from the same time series (LOO approach) instead of relying solely on non-colocated data (LTO approach). Except for MSSR the product-sum model performs best, with the sum-metric model being a close second. The metric model underperforms the other two models because the combination of both spatial and temporal dimensions through a normalized space-time distance leads one to underestimate the correlation among observations of the same time series. In other words, the assumption underlying the

**Table 14.3** Results of cross-validation analysis conducted by leaving one observation out (LOO) or one time series out (LTO) at a time. The four performance criteria described in Sect. 14.2.4 were computed for three types of space-time covariance models (generalized product-sum, metric, and sum-metric) and three space-time interpolation algorithms (ST ordinary kriging, trend model fitted by linear regression with and without interpolation by ST residual kriging)

| Algorithm | Performance criteria | | | | | |
|---|---|---|---|---|---|---|
| | Product-sum model | | Metric model | | Sum-metric model | |
| | LOO | LTO | LOO | LTO | LOO | LTO |
| Mean error of prediction (ME) | | | | | | |
| ST ordinary kriging | −0.001 | 0.009 | 0.003 | 0.007 | −0.001 | 0.008 |
| ST residual kriging | 0.0 | 0.008 | 0.003 | 0.005 | 0.001 | 0.008 |
| Trend model[a] | 0.0 | | | | | |
| Mean absolute error of prediction (MAE) | | | | | | |
| ST ordinary kriging | 0.257 | 0.375 | 0.336 | 0.384 | 0.263 | 0.378 |
| ST residual kriging | 0.251 | 0.337 | 0.318 | 0.346 | 0.254 | 0.343 |
| Trend model[a] | 0.331 | | | | | |
| Mean square standardized residual (MSSR) | | | | | | |
| ST ordinary kriging | 1.326 | 0.954 | 1.026 | 1.208 | 1.190 | 1.111 |
| ST residual kriging | 1.119 | 0.912 | 0.957 | 1.086 | 1.015 | 1.086 |
| Trend model[a] | 74.9 | | | | | |
| Area under the ROC curve for 15 μg/L (AUC) | | | | | | |
| ST ordinary kriging | 0.832 | 0.615 | 0.743 | 0.598 | 0.829 | 0.613 |
| ST residual kriging | 0.839 | 0.707 | 0.768 | 0.692 | 0.836 | 0.697 |
| Trend model[a] | 0.713 | | | | | |

[a]value for trend model is the same for all six combinations

metric model is incompatible with the zonal anisotropy detected on Fig. 14.3. Accounting for secondary information through residual kriging slightly improves the prediction relative to ST ordinary kriging; both kriging algorithms outperformed the trend model.

These results however apply only to the narrow situation where exposure to lead in drinking water is reconstructed at the sole sentinel sites. For prediction at sites where no data was collected, LTO results indicate that differences between ST covariance models are much smaller as purely temporal correlations are not used in the kriging system. Nevertheless, the product-sum model still performs best. The LTO approach also emphasizes the benefit of using trend models that account for secondary information (i.e., larger differences between residual kriging and ordinary kriging). Yet, prediction performances actually deteriorate when kriged residuals are added to the trend model: the sole trend model gives better prediction than residual kriging. It is however noteworthy that the trend model was not cross-validated, hence the observation being predicted was used to create the model.

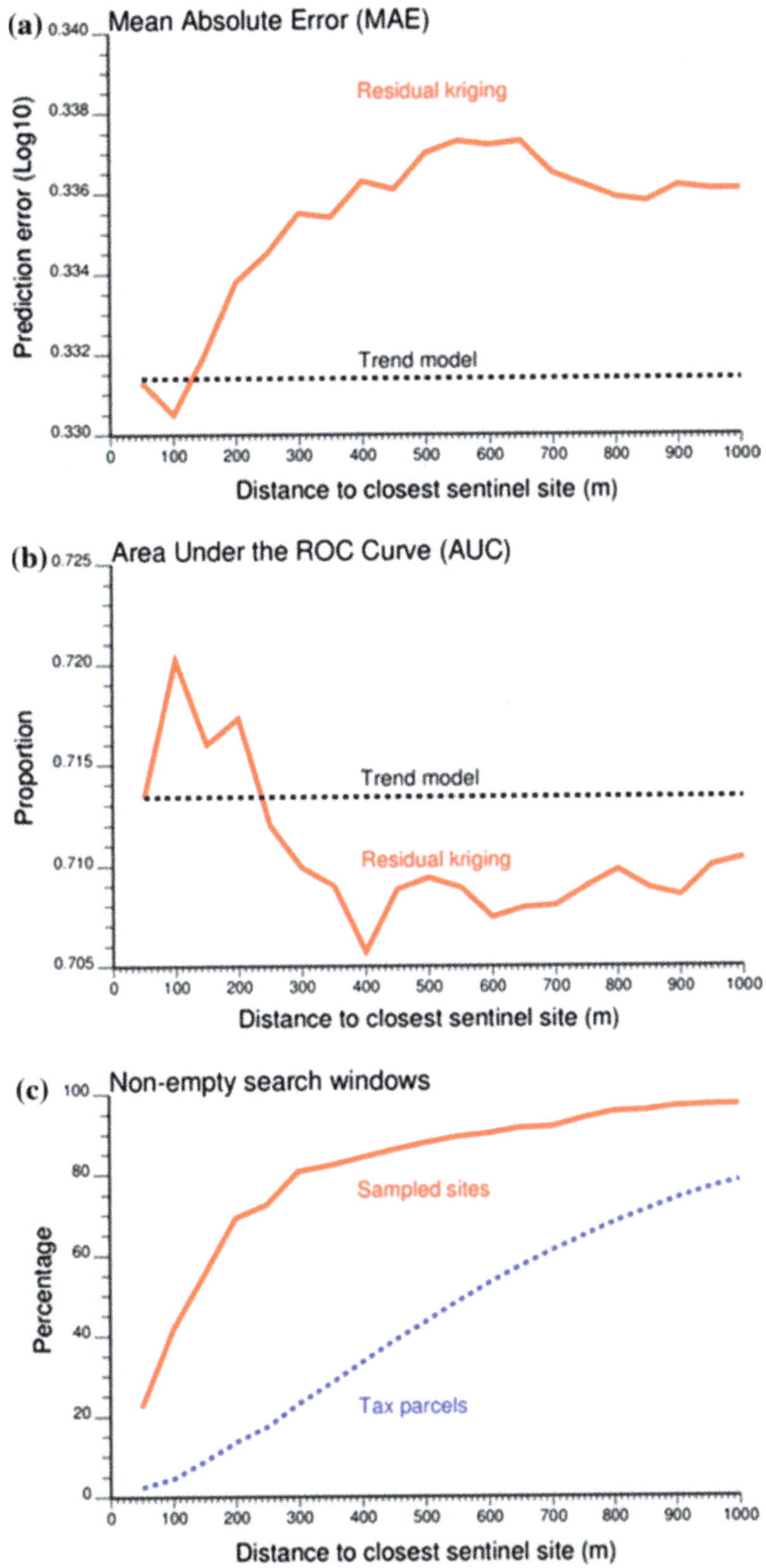


**Fig. 14.4** Impact of the size of kriging search window on several statistics computed by the leave one time series out (LTO) approach: **a** mean absolute error of prediction, and **b** area under the ROC curve. Horizontal dashed lines represent the values obtained for the time trend model created by linear regression. **c** percentages of search windows that include at least one observation when centered on sampled sentinel sites or tax parcels

Because of the substantial short-scale spatial variability retaining increasingly distant data is expected to add more and more noise to the kriging estimate. This was investigated by changing the search strategy and selecting only sentinel sites located within a given distance of the site being predicted. If no data was located within the search radius, the kriged residual was zero and the residual kriging estimate was simply the value of the trend model. Figure 14.4 shows results of this sensitivity analysis conducted for the product-sum model over distances ranging from 50 m to 1 km. For the mean error of prediction the little benefit of residual kriging vanishes as soon as data beyond 100 m are used in the estimation (Fig. 14.4a), while this distance is 200 m for the area under the ROC curve (Fig. 14.4b). Figure 14.4c indicates that 42% of sentinel sites have another sentinel site within 100 m, while this percentage is only 4.6% for tax parcels (Fig. 14.4c). In other words, there is little benefit in applying geostatistics to model the space-time distribution of WLL over the 56,039 tax parcels in Flint using the data collected at sentinel sites.

## 14.4 Conclusions

This chapter presented the first application of space-time geostatistics to lead levels recorded in drinking water of a public distribution system. The methodology was illustrated using 4,120 water samples that were collected at 819 "sentinel" sites over a 40-week period in the city of Flint. Despite a sizable database assembled by the State of Michigan, the geostatistical analysis was hampered by a temporal sampling bias and the existence of substantial variability over a few hundred meters. Unlike other countries such as Canada or France, sampling is not conducted by a trained technician in the US. Instead, homeowners are expected to collect water samples after a minimum of 6 h. of stagnation (e.g., overnight stagnation) following specific instructions (US EPA 2016), which can cause substantial variability among households. Other sources of fluctuation include heterogeneity in the plumbing system (e.g., renovation, installation of a new meter), location of sampled faucets (e.g., bathroom vs. kitchen), or water temperature (e.g., lead solubility increases with water temperature), to name a few.

In the present case-study, space-time kriging proved beneficial only in the situation where observations had been collected at the site being predicted; i.e., to fill the gaps in time series. The generalized product-sum and sum-metric space-time covariance models then outperformed the metric model that ignores the greater variation across space relative to time (zonal anisotropy). Sentinel sites represent however only 1.5% of tax parcels in the city of Flint. At unsampled sites the kriging prediction was no better than the temporal trend estimated by linear regression and it turned out to become less accurate if no data was collected within 100 meters. Although the regression model included site-specific characteristics, such as construction year and composition of service lines, it was unable to explain the short-range variability, leaving 78% of the total variance unaccounted for ($R^2 = 22\%$).

In the future, several approaches will be investigated to tackle the impact of short-range variability on prediction. First, the data analyzed in this chapter represent less than 20% of the water samples available for the city of Flint. The majority of samples were collected by voluntary sampling whereby concerned citizens received a testing kit and conducted sampling on their own (Goovaerts 2017a, b). Despite the lack of periodic sampling in time and existence of temporal bias (e.g., houses with low lead levels were less likely to be tested again) the greater spatial coverage (i.e., more than 18% of tax parcels sampled) will reduce substantially the average distance between a tax parcel and the closest observation. However, spatial heterogeneity will likely still be present over short distances, leading one to question our ability to make prediction at the tax parcel level. More appropriate spatial supports for prediction could be census block groups which are statistical divisions of census tracts and are generally defined to contain between 600 and 3,000 people. The city of Flint includes 132 block groups and 40 census tracts. Such spatial aggregation or upscaling would be a way to filter between-household fluctuations which appears to be mainly noise. As more US cities are facing similar drinking water crisis, reliable techniques for sampling and modeling spatial and temporal changes in water lead levels will be sorely needed.

**Acknowledgements** This research was funded by grant R44 ES022113-02 from the National Institute of Environmental Health Sciences. The views stated in this publication are those of the author and do not necessarily represent the official views of the NIEHS.

## References

Abernethy J, Anderson C, Dai C et al (2016) Flint water crisis: data-driven risk assessment via residential water testing. arXiv:1610.00580. https://arxiv.org/abs/1610.00580. Accessed 26 May 2017

Agency for Toxic Substances and Disease Registry (2010) Case studies in environmental medicine (CSEM) lead toxicity. Course: WB 1105 Original Date August 15. US Department of Health and Human Services, Public Health Service: Atlanta

Bailey RJ, Russell PF (1981) Predicting drinking water lead levels. Environ Technol Lett 2:57–66. https://doi.org/10.1080/09593338109384023

Cartier C, Laroche L, Deshommes E et al (2011) Investigating dissolved lead at the tap using various sampling protocols. J Am Water Works Assoc 103:55–67

Cattle JA, McBratney AB, Minasny B (2002) Kriging method evaluation for assessing the spatial distribution of urban soil lead contamination. J Environ Qual 31:1576–1588

Clark BN, Masters SV, Edwards MA (2015) Lead release to drinking water from galvanized steel pipe coatings. Environ Eng Sci 32:713–721. https://doi.org/10.1089/ees.2015.0073

Cressie N, Huang H-C (1999) Classes of nonseparable, spatio-temporal stationary covariance functions. J Am Stat Assoc 94:1330–1340

De Cesare L, Myers DE, Posa D (2002) FORTRAN programs for space–time modeling. Comput Geosc 28:205–212

De Iaco S (2010) Space–time correlation analysis: a comparative study. J Appl Stat 37:1027–1041

De Iaco S, Myers DE, Posa D (2002) Nonseparable space-time covariance models: some parametric families. Math Geol 34:23–42

Del Toral MA, Porter A, Schock MR (2013) Detection and evaluation of elevated lead release from service lines: a field study. Environ Sci Technol 47:9300–9307

Dimitrakopoulos R, Luo X (1994) Spatiotemporal modeling: covariances and ordinary kriging systems. In: Dimitrakopoulos R (ed) Geostatistics for the next century. Kluwer, Dordrecht, pp 88–93

EET Inc (2015) Evaluation of lead sampling strategies. WRF, Final grant report 4569. http://www.waterrf.org/PublicReportLibrary/4569.pdf. Accessed 26 May 2017

Fawcett T (2006) An introduction to ROC analysis. Pattern Recogn Lett 27:861–874

Flint Safe Drinking Water Task Force Recommendations on MDEQ's Draft Sentinel Site Selection (2016). https://www.epa.gov/sites/production/files/2016-02/documents/task_force_recommendations_on_sentinel_site_selection_2-16.pdf. Accessed 26 May 2017

Goovaerts P (1997) Geostatistics for natural resources evaluation. Oxford University Press, New-York

Goovaerts P (2017a) The drinking water contamination crisis in Flint: modeling temporal trends of lead level since returning to Detroit water system. Sci Total Environ 581–582:66–79

Goovaerts P (2017b) Monitoring the aftermath of Flint drinking water contamination crisis: another case of sampling bias? Sci Total Environ 590–591:139–153

Goovaerts P (2017c) How geostatistics can help you find lead and galvanized water service lines: the case of Flint, MI. Sci Total Environ 599–600:1552–1563

Goovaerts P, Webster R, Dubois J-P (1997) Assessing the risk of soil contamination in the Swiss Jura using indicator geostatistics. Environ Ecol Stat 4:31–48

Goovaerts P, Wobus C, Jones R et al (2016) Geospatial estimation of the impact of deepwater horizon oil spill on plant oiling along the Louisiana shorelines. J Environ Manag 180:264–271

Gneiting T (2002) Nonseparable, stationary covariance functions for space-time data. J Am Stat Assoc 97:590–600

Guo L, Lei L, Zeng Z et al (2015) Evaluation of spatio-temporal variogram models for mapping Xco2 using satellite observations: a case study in China. IEEE J Sel Topics Appl Earth Observ Remote Sens 8:376–385

Hanna-Attisha M, LaChance J, Sadler RC et al (2016) Elevated blood lead levels in children associated with the Flint drinking water crisis: a spatial analysis of risk and public health response. Am J Public Health 106:283–290

Heuvelink GBM, Griffith DA (2010) Space–time geostatistics for geography: a case study of radiation monitoring across parts of Germany. Geogr Anal 42:161–179

Kyriakidis PC, Journel AG (1999) Geostatistical space-time models: a review. Math Geol 31:651–684

Lee RG, William CB, David WC (1989) Lead at the tap: sources and control. J Am Water Works Ass 81:52–62

Montero JM, Fernandez-Aviles G, Mateu J (2015) Spatial and spatio-temporal geostatistical modeling and kriging. Wiley, New York

Rabin R (2008) The lead industry and lead water pipes "A Modest Campaign". Am J Public Health 98:1584–1592

Siddique A, Zaigham NA, Mohiuddin S et al (2012) Risk zone mapping of lead pollution in urban groundwater. J Basic Appl Sci 8:91–96

Solt MJ, Deocampo DM, Norris M (2015) Spatial distribution of lead in Sacramento, California, USA. Int J Environ Res Public Health 12:3174–3187

Symanski E, Savitz DA, Singer PC (2004) Assessing spatial fluctuations, temporal variability, and measurement error in estimated levels of disinfection by-products in tap water: implications for exposure assessment. Occup Environ Med 61:65–72

Swets JA (1988) Measuring the accuracy of diagnostic systems. Science 240:1285–1293

US Environmental Protection Agency, Office of Ground Water & Drinking Water (2002) Effect of water age on distribution system water quality. https://www.epa.gov/sites/production/

files/2015-09/documents/2007_05_18_disinfection_tcr_whitepaper_tcr_waterdistribution.pdf. Accessed 26 May 2017

US Environmental Protection Agency, Office of Ground Water & Drinking Water (2016) Memorandum: clarification of recommended tap sampling procedures for purposes of the lead and copper rule. https://www.epa.gov/sites/production/files/2016-02/documents/epa_lcr_sampling_memorandum_dated_february_29_2016_508.pdf. Accessed 26 May 2017

Wackernagel H (1998) Multivariate geostatistics, 2nd completely revised edition. Springer, Berlin

Wang Z, Devine H, Zhang W et al (2014) Using a GIS and GIS-assisted water quality model to analyze the deterministic factors for lead and copper corrosion in drinking water distribution systems. J Environ Eng 140:A4014004

# Chapter 15
# Statistical Parametric Mapping for Geoscience Applications

**Sean A. McKenna**

**Abstract** Spatial fields represent a common representation of continuous geoscience and environmental variables. Examples include permeability, porosity, mineral content, contaminant levels, seismic impedance, elevation, and reflectance/absorption in satellite imagery. Identifying differences between spatial fields is often of interest as those differences may represent key indicators of change. Defining a significant difference is often problem specific, but generally includes some measure of both the magnitude and the spatial extent of the difference. This chapter demonstrates a set of techniques available for the detection of anomalies in difference maps represented as multivariate spatial fields. The multiGaussian model is used as a model of spatially distributed error and several techniques based on the Euler characteristic are employed to define the significance of the number and size of excursion sets in the truncated multiGaussian field. This review draws heavily on developments made in the field of functional magnetic resonance imaging (fMRI) and applies them to several examples motivated by environmental and geoscience problems.

## 15.1 Introduction

A general problem in geological and environmental investigations is rapid and accurate identification of anomalous measurements from one, two or three-dimensional data. Example applications include cluster identification in spatial point processes (e.g., Byers and Raftery 1998; Cressie and Collins 2001) detection of anomalies in remotely sensed imagery (e.g., Stein et al. 2002) and identification of anomalous clusters in lattice data (e.g. Goovaerts 2009).

S. A. McKenna (✉)
IBM Research, Dublin, Ireland
e-mail: seanmcke@ie.ibm.com



B. S. Daya Sagar et al. (eds.), *Handbook of Mathematical Geosciences*,
https://doi.org/10.1007/978-3-319-78999-6_15

The problem of anomaly detection is complicated when the data set is composed of more than a handful of variables (multi-variate) and becomes even more complex when the multiple variables comprise a random field exhibiting spatial correlation.

The temporal and/or spatial correlation of the data rules out the application of standard statistical tests for change detection and has also limited the development of hypothesis testing techniques for correlated data (Gilbert 1987). For applications with correlated data, simulation techniques can often be used to develop the null distribution, but development of closed form hypothesis tests for analysis of the spatial random fields associated with geostatistics has remained sparse.

One approach to detection of anomalies in spatially correlated data are Local Indicators of Spatial Association (LISA) statistics (Anselin 1995; Goovaerts et al. 2005; Goovaerts 2009). These tests focus on the local relationships between adjacent cells and explore combinations of cells defined with an adjacency matrix and or a moving window visiting all cells in a lattice. A very different approach is to model the difference between images as a continuous random field and use properties of an underlying random field model to identify anomalies.

Change detection in spatial-temporal data sets has received considerable attention over the past 15–20 years within the medical imaging research community (Brett et al. 2003; Friston et al. 1994, 1995; Worsley et al. 1992, 1996) and a significant development of this research has been Statistical Parametric Mapping (SPM).

The practice of statistical parametric mapping has been developed in the field of medical imaging, particularly in brain imaging, and in the practice of functional magnetic resonance imaging (fMRI) of the brain while the subject is performing various tasks (functions). Friston et al. (1995, p. 190) provide a concise definition of SPM: "*one proceeds by analyzing each voxel using any (univariate) statistical parametric test. The resulting statistics are assembled into an image, that is then interpreted as a spatially extended statistical process*". In other words, at each pixel (voxel) in an image, a univariate statistical test (e.g., *t*-test) is applied and the resulting values of the test statistic at each pixel are then displayed as a map. The underlying spatial correlation of the map is used in creating a multivariate statistical model that describes that map and this model can be used for inference. Typically, the resulting map is analyzed using theory that underlies stationary Gaussian fields and techniques developed for excursion sets of these fields. Properties of truncated Gaussian fields (e.g., Adler and Hasofer 1976; Adler 1981; Adler and Taylor 2007; Adler et al. 2009) serve as the basis of the SPM techniques.

To date, the SPM approach has not been applied outside of medical imaging, but it appears to be a technique that could be successfully applied in a number of areas of interest in the earth and environmental sciences. The goals of this work are to both describe the basis of SPM and then apply SPM to example problems.

## 15.2 Anomaly Detection with Statistical Parametric Mapping

Anomaly detection is defined here as the identification of a region in time and/or space that is anomalous in its shape, size (duration) and/or values within the region (intensity). Two modes to anomaly detection in spatial-temporal data sets can be defined: (1) Anomaly detection in an online mode where prior data are used to predict future values of the measured variable and anomalies occur in areas and/or times where the predictions are inconsistent with the corresponding measurements; (2) Anomaly detection as the difference between two classes of data where differences in some treatment or external forcing condition is suspected to cause a difference in the measured variable. The anomalies in this case are significant differences in measured variables observed with and without activation of the external condition. This latter case is the focus of the work in this chapter.

Specifically, an ensemble of geologic models can be created in 1, 2 or 3 dimensions where each member of the ensemble is associated with a specific "treatment" or "result" that can be used to group ensemble members into separate classes. As examples:

- An ensemble of 3D geostatistical realizations of porosity can be created conditioned to a single set of observations where two different variogram models, both of which fit the available data equally well, are used. The different variogram models constitute a "treatment" and the question arises as to whether or not the treatments create significant differences (anomalies) in the resulting realizations and where, spatially, those differences occur.
- Petrophysical logs from different wells intersecting the same reservoir constitute an ensemble of 1-D measurements. When split into groups based on the result of which wells produced a threshold amount of petroleum and which did not, the question arises as to whether or not the petrophysical log profiles are significantly different between the groups? If they are, what portions of the log create this significant difference?

Two measures of anomaly detection can be employed: omnibus and localized (Worsley et al. 1992). Omnibus detection uses a set of calculations to determine if the current curve, map or volume, taken as a whole, is anomalous. Localized detection determines the specific location(s) within the study domain where the anomaly occurs and are the focus of this work.

Anomaly detection is not done directly on the observed generated or observed ensemble members, but on a difference between groups of members as defined through the treatment or result. Here, the differences are calculated as the differences of two average values. The averages are calculated at each point, pixel or voxel within the domain using standard univariate statistical tests (e.g., $t$-test). Each pixel-wise average is calculated over a set of ensemble members created under a specific condition (treatment) or generating a specific result. For example, in studies of the human brain, images are often collected under "resting" and "stimulated" conditions and the average image from each condition is then used to create a difference map.

SPM was developed to directly address the problem of spatial correlation in statistical testing. Direct application of most statistical tests requires independence of the observations, but for many problems, including those studied here, correlation between adjacent observations is the norm. Therefore, the results of the statistical tests for adjacent, or even nearby, pixels cannot be effectively evaluated using standard techniques. SPM considers a single map comprised of the results of all local (pixel-wise) statistical tests and provides several measures for comparison of the values in the map to critical threshold levels.

### *15.2.1 MultiGaussian Fields*

The basis of the SPM approach is the analysis of the number, size and degree of excursions from a multiGaussian (mG) random field. For a concise, statistical description of mG fields, see Adler et al. (2009, p. 27). Stationary multiGaussian fields are fully defined by a mean and covariance matrix. In a practical sense, values at each pixel are defined with a Gaussian distribution. The correlation between those multiple distributions is defined by the covariance. Spatial correlation can be added to an uncorrelated field through the convolution of a smoothing kernel with an uncorrelated (white noise) field. As an example, the 2D Gaussian kernel is defined as

$$\mathrm{G}(\mathrm{x},\mathrm{y}) = \frac{1}{2\pi|\Sigma|^{1/2}} exp\left(-\frac{1}{2}d\Sigma^{-1}\mathrm{d}^{\mathrm{T}}\right)$$

where $d$ is the distance vector containing distances $d_x$ and $d_y$ from any location $(x, y)$ to the origin of the Gaussian function $x_O$, $y_O$ (here (0, 0) for the standard normal distribution). In this work, the covariance matrix, $\Sigma = \sigma^2 I$, ($I$ = identity matrix) is diagonal for the specific case of the kernel being aligned with the grid axes.

An often-used measure of the spatial bandwidth of a smoothing kernel in the image processing literature is the "full width at half maximum" (FWHM). For the Gaussian kernel above, the FWHM is:

$$FWHM = \sigma\sqrt{8\ln(2)}$$

If the mG field is not created, but is obtained from some type of imagery or other analyses, then there is no known underlying kernel and it is necessary to estimate the FWHM directly from the image. Estimation can be done using the covariance matrix of the partial derivatives of the image values, $T$, with respect to the discretization of the image. In 2D, the covariance matrix is:

$$\Lambda = \begin{bmatrix} Var\left(\frac{\partial T}{\partial x}\right) & Cov\left(\frac{\partial T}{\partial x}, \frac{\partial T}{\partial y}\right) \\ Cov\left(\frac{\partial T}{\partial x}, \frac{\partial T}{\partial y}\right) & Var\left(\frac{\partial T}{\partial y}\right) \end{bmatrix}$$

This covariance matrix can be interpreted as a measure of the roughness/smoothness of the image.

Estimation of $\Lambda$ can be achieved through several approaches and here the simple relationship defined by Worsley et al. (1992) between the FWHM values in each of the principal directions and $\Lambda$ is utilized. The derivatives in the covariance matrix of an image can be approximated numerically in each spatial dimension with differences between adjacent pixels are calculated as:

$$Z_{xi}(x, y) = \{T_i(x + \delta x, y) - T_i(x, y)\}/\delta_x$$
$$Z_{yi}(x, y) = \{T_i(x, y + \delta y) - T_i(x, y)\}/\delta_y$$

where $\delta_x$ and $\delta_y$ are the dimensions of the image pixels in the x and y directions. The variances and covariances of the differences are then used to approximate the variances and covariances of the derivatives:

$$V_{xx} = \sum_{i,x,y,z} Z_{xi}(x, y, z)^2/N(n-1)$$
$$V_{yy} = \sum_{i,x,y,z} Z_{yi}(x, y, z)^2/N(n-1)$$
$$V_{xy} = \sum_{i,x,y,z} \{Z_{xi}(x, y, z) + Z_{xi}(x, y + \delta_y, z)\}\{Z_{xi}(x, y, z) + Z_{xi}(x + \delta_x, y, z)\}/4N(n-1)$$

These variance and covariance estimates are used to estimate $\Lambda$:

$$\Lambda = \begin{bmatrix} V_{xx} & V_{xy} \\ V_{xy} & V_{yy} \end{bmatrix}$$

Finally, the FWHM in the X and Y directions are calculated as:

$$FWHM_x = \sqrt{\frac{4\ln(2)}{V_{xx}}}$$
$$FWHM_y = \sqrt{\frac{4\ln(2)}{V_{yy}}}$$

## *15.2.2 Calculating the SPM*

The Statistical Parametric Map is the difference image between individual pairs of images or average images, which is typically transformed from a map of t-statistics to a map of Gaussian *Z*-score values. The different methods used in this study for calculating the SPM are described in this section.

### 15.2.2.1 Conditional Differences

The *t*-test and *t*-statistic are used exclusively in this chapter for the conditional differences between two ensembles and a review of the *t*-statistic is provided in the Appendix. It is noted that other statistical tests and their resulting test-statistics, e.g., $\chi$, *Z*, *f*, as well as measures of correlation can also be used as the basis of an SPM. For the *t*-tests employed here, a location (pixel)-specific calculation of the standard deviation is used. Another approach is to calculate the pooled standard deviation across the image (image-based) and arguments for using the image-based standard deviation are given by Worsley et al. (1992). In typical applications, the number of observations under each condition is small, near a dozen, and therefore the effective degrees of freedom for $T(x, y)$ is generally small and needs to be used in the transformation of the *t*-field to a standard normal Gaussian *Z*-field.

The cumulative probability of a *t*-statistic is found from the *t*-distribution function with the appropriate degrees of freedom. This probability is then used with the inverse of the Gaussian distribution function to get the z-score value:

$$P(Y \leq y) = T(y; v)$$
$$z = G^{-1}(P(Y \leq y))$$

The resulting fields are now multiGaussian SPM's and the anomaly detection algorithms developed for SPM analysis can be applied.

### 15.2.2.2 Isolated Regions of Activation

Anomaly detection here is focused on the number, size and location of regions within an SPM that is a curve/image/volume that exceed a given threshold level, *u*. These regions are known as "regions of activation", "regions of exceedance" or "excursions". The numbers, sizes and locations of these excursions are then compared against a reference model of the expected expression of such regions. Truncation of a Gaussian field at a threshold *u* defines the u-level excursion set:

$$X_u = \{x \in R^D : Y(x) \geq u\}$$

A large body of literature on the properties of excursion sets (regions of exceedence) in Gaussian random fields is available (e.g., Adler et al. 2009; Friston et al. 1994; Lantuejoul 2002). Friston et al. (1994) characterize three related properties of excursion sets in truncated Gaussian random fields:

$N$ the number of pixels above the truncation threshold, $u$,
$m$ the number of distinct regions (inclusions) above the threshold, and
$n$ the number of pixels in each region,

with expectation relationship $E[N] = E[m]E[n]$. For threshold value, $u$, the number of cells above that threshold, $N$, is provided by the Gaussian cdf and the size of the domain, $S$:

$$\mathrm{E[N]} = S\int_u^\infty (2\pi)^{-1/2} e^{-z^2/2} dz$$

A measure of the number of isolated regions above the threshold can be obtained from the Euler Characteristic, EC. In two dimensions, the EC represents the number of connected excursion sets in the domain minus the total number of holes within those sets. Therefore, EC goes to 0.0 at $u = 0$ and EC becomes negative when $u < 0.0$ as the truncated field represents a single domain-spanning set containing a large number of holes. In 2D, and at relatively high truncation thresholds, EC is equivalent to the number of regions above the threshold, E[m].

$$\mathrm{E}[m] = EC = \left|(2\pi)^{-((D-1))/2} W^{-D} S u^{(D-1)} e^{u^2/2}\right|$$

where $D$ is the dimension of the domain and $W$ is an alternative measure of the spatial correlation of the mG field defined as a fraction of the FWHM:

$$\mathrm{W = FWHM}/\sqrt{4\ln(2)}$$

For a given threshold, $u$. the average area of the individual regions is found from the expectation relationship:

$$\mathrm{E[n] = E[N]/E[m] = E[N]/[EC]}$$

Figure 15.1 compares a direct calculation of EC on a multiGaussian field using the Matlab Image Processing toolbox (Matlab 2009) with estimates made using the Euler characteristic equation above across a range of $u$ values increasing from left to right. Deviations between the calculated and estimated number of excursions indicate deviations from the definition of a multiGaussian field. The corresponding binary fields (500 × 500 cells) are also shown for several representative threshold

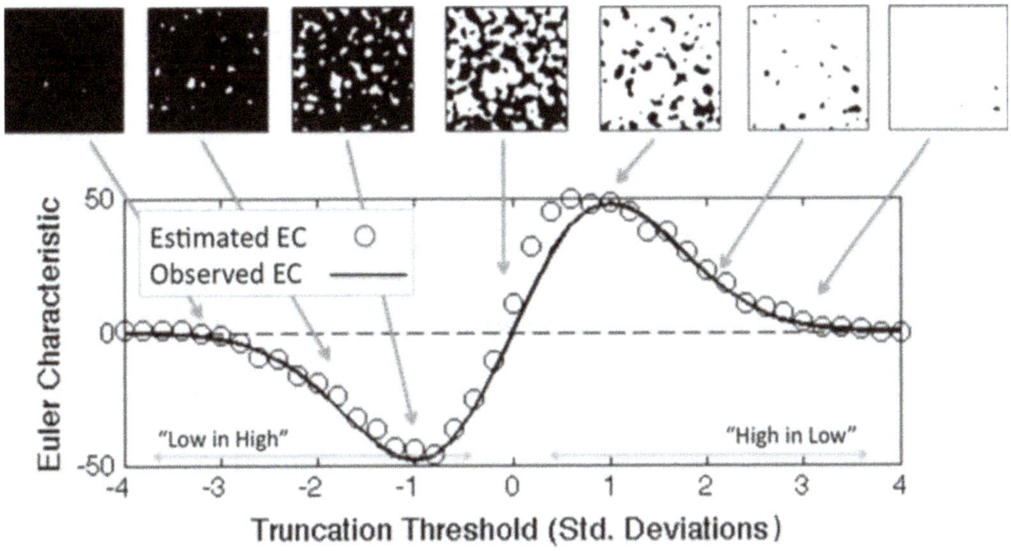


**Fig. 15.1** Observed (calculated) and estimated Euler characteristic for a mG field as a function of the truncation threshold, u. The excursion sets for u > 0 are black regions in the binary fields at the top of the image (after McKenna et al. 2011)

values. Note, that typically the extreme ends of the graph corresponding to $u$ values (truncation thresholds) with absolute values of 2.5 or greater are of interest.

### *15.2.3 Localized Anomaly Detection*

Further analysis of the excursion sets is focused on the size and location of the detected anomalies. The excursion set maps themselves can be examined to determine the location of where the excursions are occurring. An extremely localized, yet very strong anomaly will be of interest. An anomaly with a much lower amplitude but greater spatial extent may also be of interest. The definition of spatial extent (size) of any anomaly is defined relative to the spatial correlation length of the field in which it is detected. The size of the anomaly is expressed through truncation of the field at a threshold value and defining the size of the excursion regions above that threshold.

In general, the significance of any anomaly in a spatial field is a function of its amplitude (intensity or strength) and its spatial extent (size). The observed SPM is compared against a specified multivariate spatial random field with a defined correlation length that serves as the model of the null hypothesis for the differences between two ensembles of spatial fields. Truncation of the observed SPM at a given threshold level creates regions of excursions above that threshold and the significance of the number and size of these excursions relative to the model of the null hypothesis is calculated. As in classical statistical hypothesis testing, the $p$-value defines the chance that the observed anomaly would occur under the null hypothesis. Here, the focus is on identifying the largest region of excursion for a specified threshold and calculating the chances of that anomaly occurring under the null hypothesis.

The pre-processing steps and the approach used for application of statistical parametric mapping to detection of significant excursion sets is outlined here and these steps are then applied to an example problem. The focus is on the approach used for calculation of the probability that one or more regions of activation of a certain area, or larger, could have occurred by chance under the constructed mG model. The full development of this approach for medical imaging is provided by Friston (1994) and Worsley et al. (1996). Additionally, Adler (2000) and Taylor and Adler (2003) provide further development of level crossing in random fields and the relationship to the Euler characteristic.

Steps:

(1) Create an SPM through pixel by pixel application of 1-D (pixelwise) univariate statistical tests. The test statistic values resulting from this test at every point may be distributed as $\chi^2$, $t$, $F$, or other and can be transformed into a Gaussian Z-score to create a Gaussian SPM.
(2) Smooth the resulting SPM using a Gaussian kernel. The resulting SPM created in step 1 may be coarse and noisy. A small amount of smoothing using a Gaussian filter is enough to create a smoothed SPM.
(3) Reinflate the variance. The smoothing process in Step 2 decreases the variance of the SPM and a reinflation process is used here to transform the SPM to a unit variance (1.0) for easier interpretation of results. Here, the empirical probability distribution function of the SPM after Step 2 is fit with a Gaussian Mixture Model (GMM) having three components. A quantile-preserving transform is used to transform the empirical cumulative distribution as modeled with the GMM to a cumulative Gaussian distribution. Use of the GMM better preserves the original shape of the distribution relative to simpler transforms such as the normal-score. No translation, or recentering, of the resulting Gaussian distribution is done.
(4) Calculate the characteristics of the SPM, and choose an exceedance threshold to identify regions of exceedance.

    a. Calculate the FWHM of the smoothed and transformed SPM created in Steps 1–3. The FWHM is derived from the variances and covariances of the spatial derivatives of the SPM. The resulting FWHM values are typically 5–15 times the size of the smoothing kernel used in Step 2.
    b. Identify pixels that are above/below the $\pm$ threshold value.
    c. Employ a flood-fill algorithm to determine the sizes of the separate regions of connected pixels, or regions of exceedance and label each region for both positive and negative excursions.

(5) Apply a hypothesis test to determine the probability of a particular result having occurred under the null hypothesis of the mG model. Here, a test of the chance of obtaining the size of the of largest region of exceedance (excursion) under

the null hypothesis of a Gaussian SPM with calculated FWHM is calculated. The significance of the maximum excursion size is calculated using the methods of Friston et al (1994):

a. The three main features of an SPM are: (1) the number of pixels, $N$, exceeding a threshold, $u$. Or the number of pixels in the excursion set; (2) the number, $m$, of regions above the threshold—the number of connected subsets of the excursion set; (3) the number, $n$, of pixels in each of the $m$ subsets. The expectations of these three features are related as: $E\{N\} = E\{n\} \bullet E\{m\}$.
b. Eq. 14 of Friston et al. (1994) gives the probability of at least one excursion region having a size $\geq$k pixels:

$$\begin{aligned} P(n_{\max} \geq k) &= \sum_{i=1}^{\infty} P(m=i) \cdot \left[1 - P(n<k)^{i}\right] \\ &= 1 - e^{-E[m] \cdot P(n \geq k)} \\ &= 1 - \exp(-E[m] \cdot e^{-\beta k^{2/D}} \end{aligned}$$

where $\beta = [\Gamma(D/2 + 1) \bullet E[m]/E[N]]^{2/D}$ and $D$ is the dimension of the domain.

Calculations of $P(n_{max} \geq k)$ within the $(k, u)$ parameter space for spatial fields with two different correlation lengths (FWHM) are shown in Fig. 15.2. The role of the correlation length of the null hypothesis model is clear from Fig. 15.2 where the probability of an excursion region of 60 pixels or more is approximately 0.001 for a field with a FWHM of 9.0, but is essentially zero ($\sim 10 \times 10^{-12}$) for a field with a FWHM of 3.0.

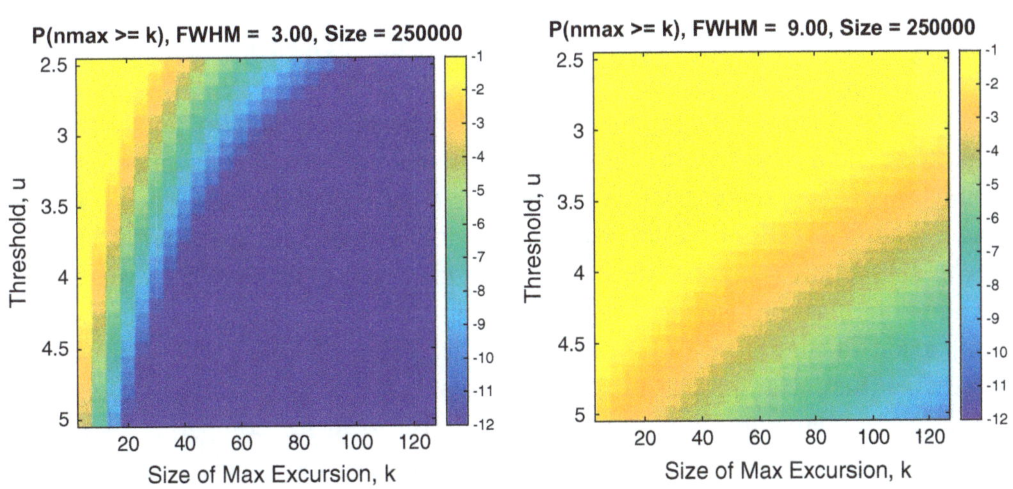


**Fig. 15.2** $P(n_{max} \geq k)$ as a function of size of the excursion region, k, and the truncation threshold, u, for fields of size 500 × 500 with an isotropic FWHM of 3.0 pixels (left) and 9.0 pixels (right). The color scale is $\log10(P(n_{max} \geq k))$

## 15.3 Example Problems

Two example problems are used here to demonstrate the calculations and application of SPM to detecting anomalies in spatial random fields. Both example problems are two-dimensional, but the same approaches are applicable to anomaly detection in 1-D and 3-D domains.

### 15.3.1 Anomaly Detection in Images

A simple simulation study designed to mimic the detection of anomalous regions in either remote sensing or geophysical imagery is used here to test a few of the SPM calculations. The focus is on identifying the largest anomaly above a specified threshold and the significance of that anomaly.

A multiGaussian field is created through geostatistical simulation. The field is comprised of square, 5 × 5 m pixels, and has an isotropic Gaussian variogram with a range of 150 pixels. The field is created in standard normal space, $N(0, 1)$ and the simulated values serve as the observed image. Measurement noise is added to the image by considering the simulated realization value, $z(x)$ to be the mean value of a local Gaussian distribution at every pixel. The standard deviation of the Gaussian at every pixel, $\sigma_Z(x)$, is set to 2.0 and a Gaussian random deviate is drawn and added to $z(x)$ to create the final image. This measurement noise is added independently at every pixel (i.i.d.) and then smoothed prior to adding to the observed image. The amount of spatial smoothing of the noise term is varied and the impact on anomaly detection is examined.

Anomalies are added to the observed image within a circular region having a radius of 90 pixels and centered at the center of the image. Background values within the anomaly region are multiplied by 1.5 creating stronger negative and positive values within the region depending on the sign of the original observed values. The area of the anomaly region is 5027 pixels.

Figure 15.3 shows background images (left column) at two levels of noise smoothing and the background images with the anomalies added (right column). As would occur in any image capture process, the noise values added to each image are drawn randomly and independently from any other image prior to smoothing. This creates subtle differences between the images in each row of Fig. 15.3 even without the addition of anomalies. Detection of the presence of the anomalies through visual comparison of the left and right images in each row of Fig. 15.3 is not obvious, even when the location of the anomaly is known.

The SPM's are calculated through a pixelwise $t$-test for comparing two means (Appendix) between the image with and without the anomalies. These $t$-statistic maps are transformed to Gaussian $Z$ maps that are the SPM (Fig. 15.4). The large anomalies in the center of the image are readily seen along with the dramatic changes in the results due to the increased spatial correlation of the noise

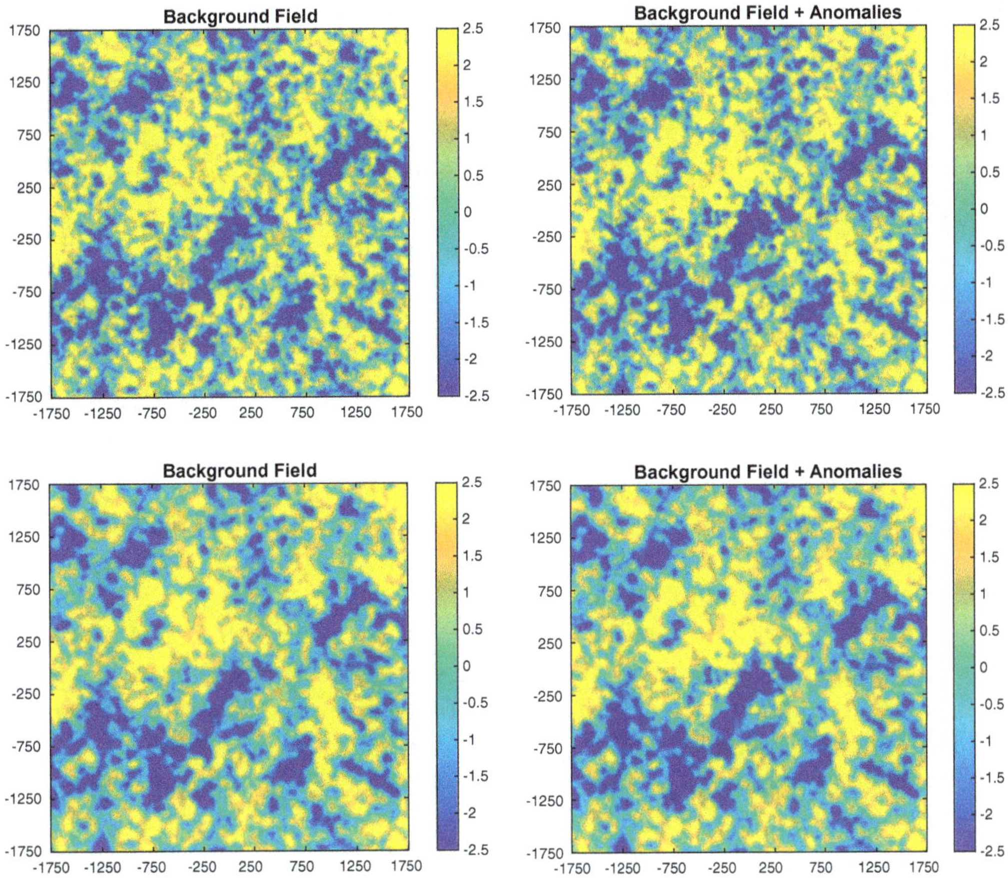


**Fig. 15.3** Background fields without (left column) and with (right column) added anomalies with a smoothing kernel size $\sigma = 1.5$ pixels (top row) and $\sigma = 7.5$ pixels (bottom row). Color scale units are arbitrary in this example

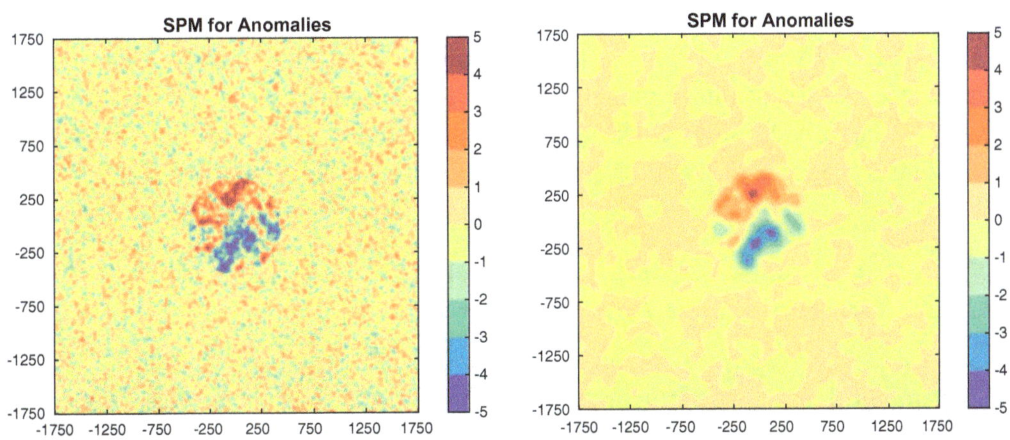


**Fig. 15.4** SPM's for the case of smoothing with a filter bandwidth of $\sigma = 1.5$ (left) and $\sigma = 7.5$ pixels (right). The color scale is in standard deviations away from the mean of zero

Table 15.1 Results of SPM analysis for four levels of noise smoothing

| Bandwidth of smoothing filter (pixels) | 1.5 | 3.5 | 5.5 | 7.5 |
|---|---|---|---|---|
| FWHM | 5.0 | 23.5 | 45.0 | 71.6 |
| Size Max Positive (pixels) | 2061 | 2886 | 1878 | 2075 |
| $P(n_{max} > k)$ Positive | $<1.0 \times 10^{-16}$ | $1.3 \times 10^{-06}$ | 0.237 | 0.405 |
| Size Max Negative (pixels) | 3640 | 3664 | 3772 | 3417 |
| $P(n_{max} > k)$ Negative | $<1.0 \times 10^{-16}$ | $1.5 \times 10^{-08}$ | 0.015 | 0.200 |
| # regions >+ threshold | 238 | 5 | 5 | 1 |
| # regions <− threshold | 212 | 5 | 2 | 1 |
| Max SPM value (standard deviations) | 8.28 | 6.45 | 5.25 | 4.55 |
| Min SPM value (standard deviations) | −7.83 | −6.23 | −5.43 | −4.58 |

component with increased smoothing. Additional SPM's are created at intermediate levels of smoothing but are not shown here. Results from all levels of noise smoothing are shown in Table 15.1.

A threshold of ±2.5 standard deviations is applied to the SPM's and the excursion regions for the two extreme levels of noise smoothing are shown in Fig. 15.5. There are over 200 positive and 200 negative excursions for the smallest amount of noise smoothing and only 1 positive and 1 negative excursion at the largest amount of smoothing. The size of the excursions that are due to the added anomalies clearly stands out in the left image of Fig. 15.5. Table 15.1 also shows how the maximum and minimum images in the SPM decrease with increased levels of noise smoothing.

With increased smoothing of the noise, the FWHM of the image increases from 5.0 to ~ 72 pixels (Table 15.1). While the size of the largest positive and negative excursions remains approximately constant near 2000 and 3600 pixels, respectively, the *p*-value for excursions of that size occurring in the image changes dramatically. At the lowest level of smoothing, the chances of getting excursions of size 2061 or 3640 pixels under the Gaussian random field model with a FWHM of 5.0 are essentially zero ($< 1.0 \times 10^{-16}$). However, getting excursion regions of a similar size occurring under greater smoothing of the noise and a FWHM of 71.6 pixels is relatively common at 40 and 20%, respectively. These results demonstrate the strong dependence of $P(n_{max} >= k)$ on the spatial correlation of the field.

### 15.3.2 Ground Water Pumping

A general problem in a number of geoscience disciplines is the case where an ensemble of inputs is used in a calculation to provide a probabilistic result to a particular question. The calculation can be relatively simple or complex, but acts as a transfer function to transfer uncertainty in spatially distributed physical properties to uncertainty in an outcome of interest. Examples include groundwater models

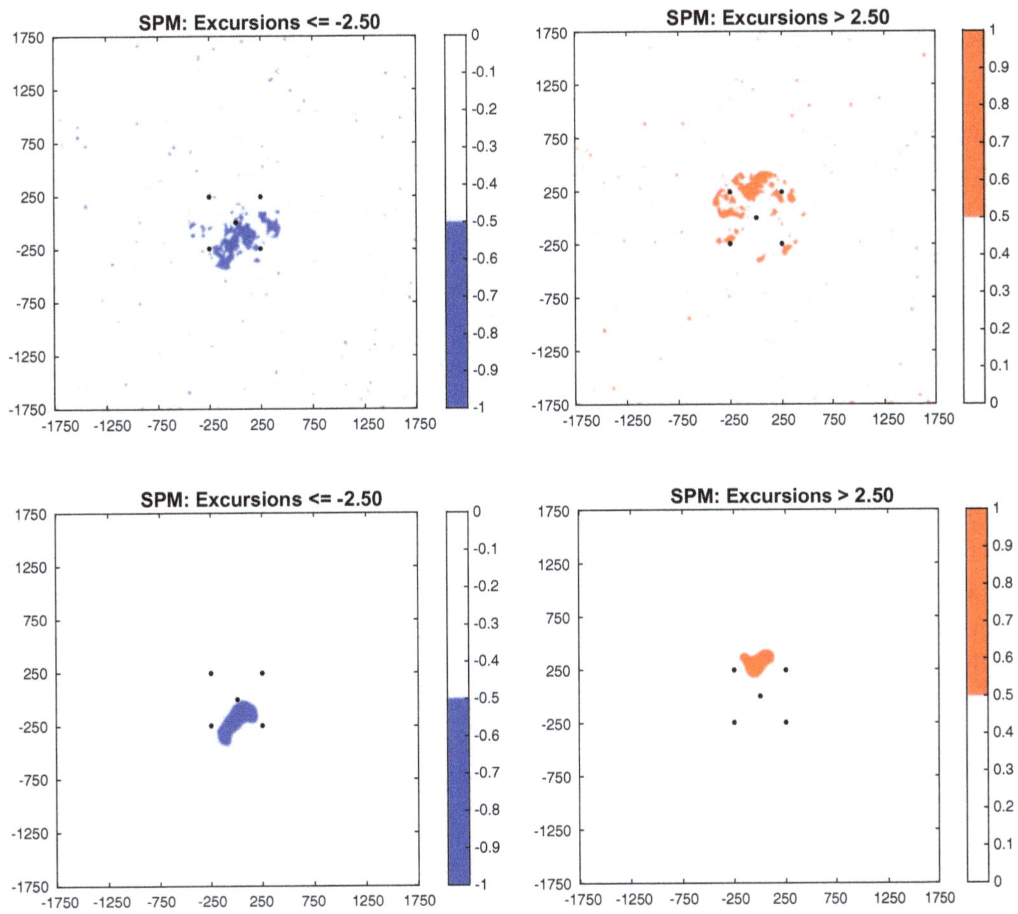


**Fig. 15.5** Regions of excursion below a threshold of −2.5 (left column) and above 2.5 (right column) for images with noise smoothed using a filter of σ = 1.5 (top row) and σ = 7.5 (bottom row)

transferring uncertainty in hydraulic conductivity and recharge into radionuclide transport times; reservoir simulators transferring uncertainty in permeability and porosity into estimated recoverable oil; and simple spatial integration to transfer uncertainty in soil nutrient levels into estimating total crop yield for an agricultural field.

Here, a ground water example problem is used with the SPM approach to detect significant differences between two groups of an ensemble of spatial random fields of transmissivity. The ensemble is split into groups that create high results and all others. The SPM approach is used here to identify statistically significant features within the ensemble of input fields responsible for the specific results. This approach can be considered identification of the significant features in the random fields responsible for a specific result of a process that integrates across the entire field.

#### 15.3.2.1 Problem Setup

The ground water problem is motivated by the regulatory issue of impacts on a nearby wetland due to pumping from a planned water supply well. Well test criteria dictate that the pressure drop (drawdown) at a location 353 m to the northwest of the pumping well must be <2.00 m after pumping at a rate of 250 $m^3/h$ for 48 h. To simulate the aquifer test, a 12 × 12 km square domain, with zero-flux boundaries on the north and south and constant-head boundaries on the east and west is defined. Prior to pumping, the fixed head boundaries create steady state flow across the domain. A constant transmissivity, T, of 10.0 $m^2/h$ is assumed across the majority of the domain. This constant value is replaced by a heterogeneous T field within the center of the domain. The heterogeneous field is 3500 × 3500 m with 5 × 5 m cells. A large pumping well is set in the center of the domain.

The aquifer is confined in this area and the mean and spatial co-variance of the transmissivity can be estimated from other studies in aquifers of similar age and depositional history. The log10 values of transmissivity within the heterogeneous domain are simulated as a multiGaussian field with an isotropic Gaussian variogram with range 250 m and nugget of 5% of the sill. Transmissivity at the well location is considered known and provides the only conditioning point within the domain. A total of 200 realizations are created, and the 2D, confined, transient ground water flow equation is solved using finite differences on each realization:

$$\frac{\partial h(x,y)}{\partial t} = \frac{1}{S(x,y)} \cdot \left(\frac{\partial}{\partial x} T(x,y) \frac{\partial h}{\partial x}\right) + \left(\frac{\partial}{\partial y} T(x,y) \frac{\partial h}{\partial y}\right) \pm Q(x,y)$$

where (x, y) indicates the spatial location, $h$ (L) is the head (pressure), $t$ is time and $Q$ ($L^3/T$) are sources or sinks—here the pumping rate at the well. Transmissivity, $T$ ($L^2/T$), is spatially heterogeneous within the central domain and for the calculations here, storativity, $S$ (–) is set to a single value of $1.0 \times 10^{-05}$ across all locations in the aquifer. The initial conditions for the transient simulation are taken from a steady state head solution using the same input $T$ field. Three example transmissivity realizations and maps of the resulting drawdowns after 48 h of pumping are shown in Fig. 15.6. Figure 15.6 demonstrates that the heterogeneous T field strongly impacts the resulting pressure response in a non-linear manner.

#### 15.3.2.2 Results

For each ground water simulation, the drawdown at the test location (353 m NW of the pumping well) at 48 h is recorded and compared to the regulatory limit, $R$, of 2.00 m. The $T$ realization is placed into one of two classes: those that meet the pressure drop limit, drawdown <= $R$, and those that exceed the limit. After 200 ground water simulations, the pixelwise mean and standard deviation within each

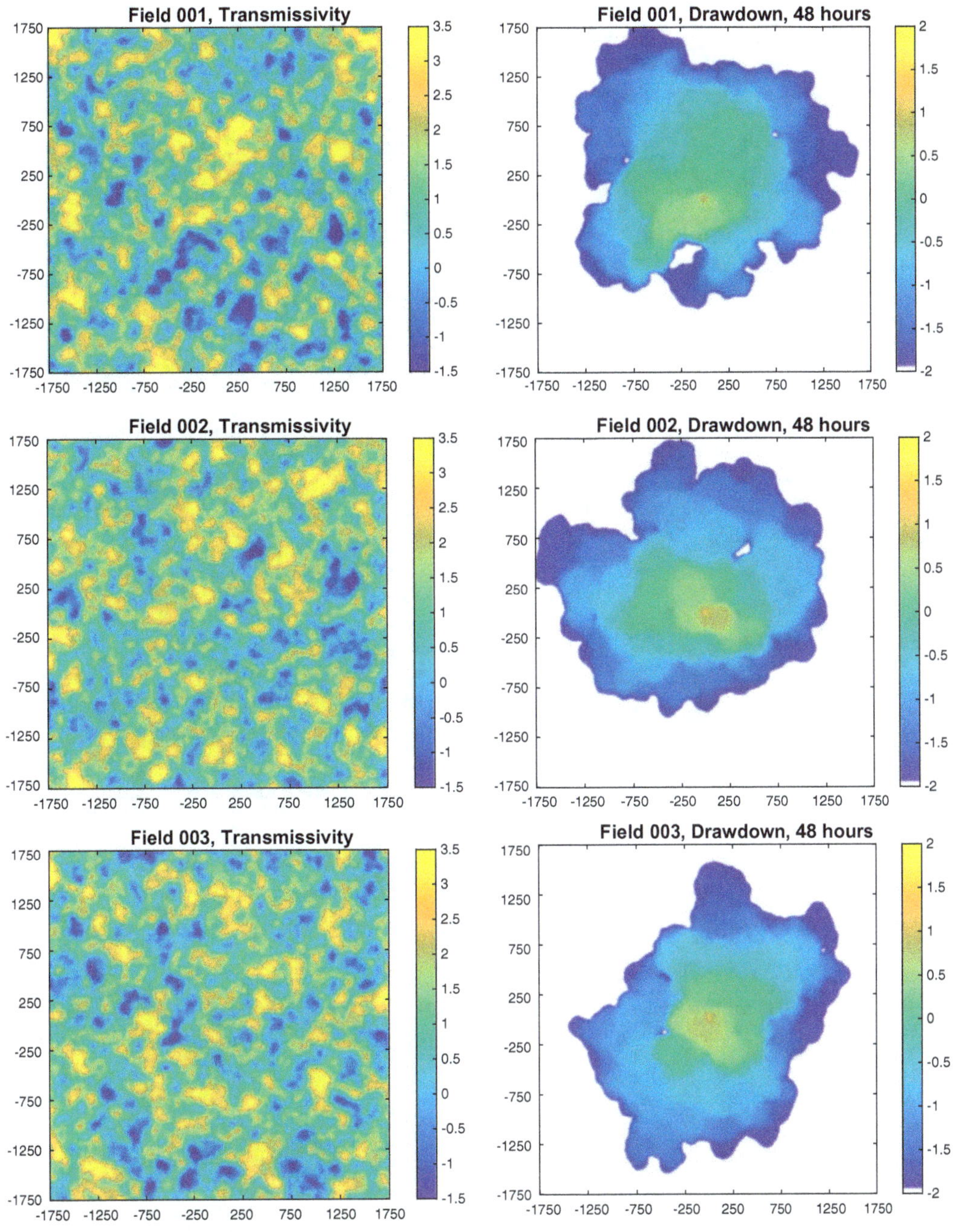


**Fig. 15.6** Three example transmissivity fields (left column) and the corresponding ground water drawdown levels after 48 h of pumping (right column). The color scales define log10 T in $m^2/h$ and log10 drawdown in meters

class are calculated (Fig. 15.7). These four maps provide the input to a two-sample $t$-test to determine the difference between two means. The resulting map of $t$-statistics is the SPM. Here the $t$-statistics are smoothed with Gaussian kernel and transformed to $Z$-statistics and the $Z$-score SPM is shown in Fig. 15.8.

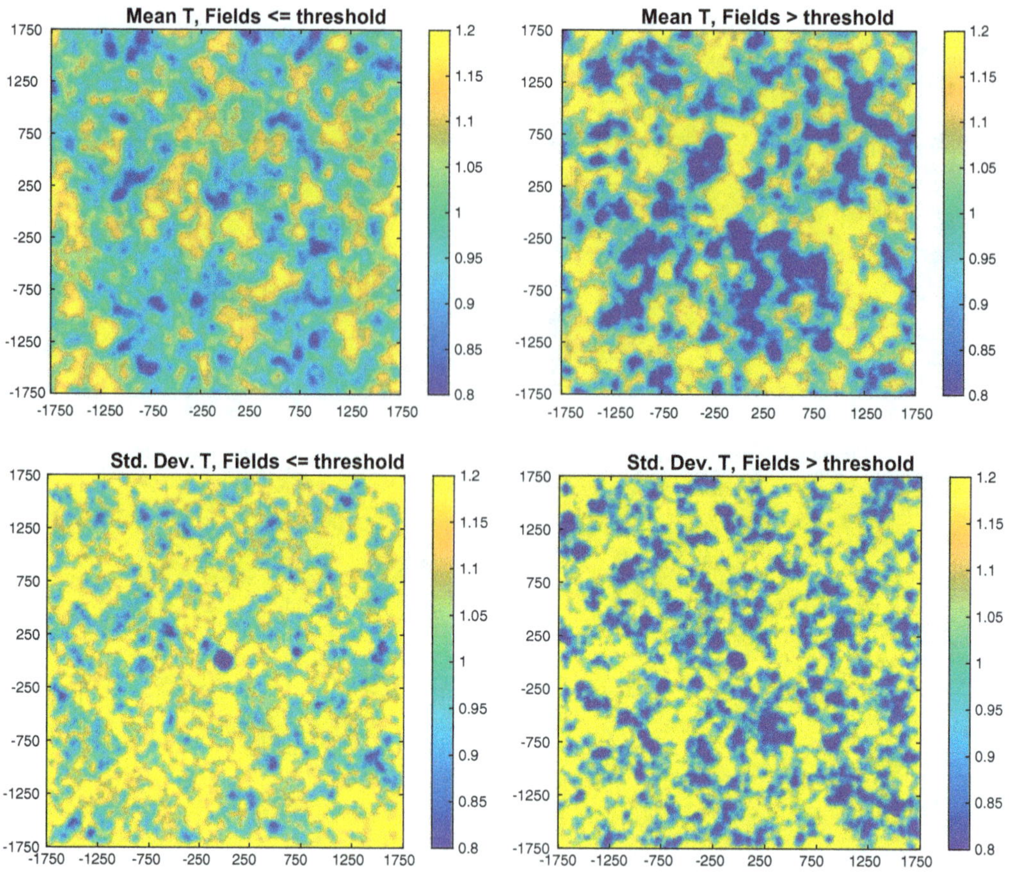


**Fig. 15.7** The mean (top row) and standard deviation (bottom row) fields for the transmissivity realizations that create drawdown <=2.00 m (left column) and drawdown >2.00 m (right column). The color scales show log10(T) in $m^2/h$

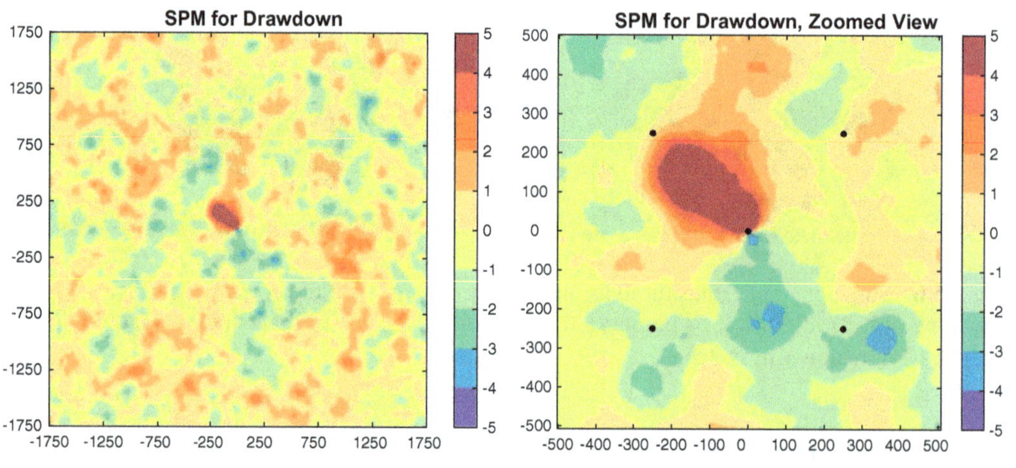


**Fig. 15.8** SPM for the difference between realizations. Full field is shown on the left and a zoomed in view of the central field on the right. Color scale is in standard deviations away from the mean of zero

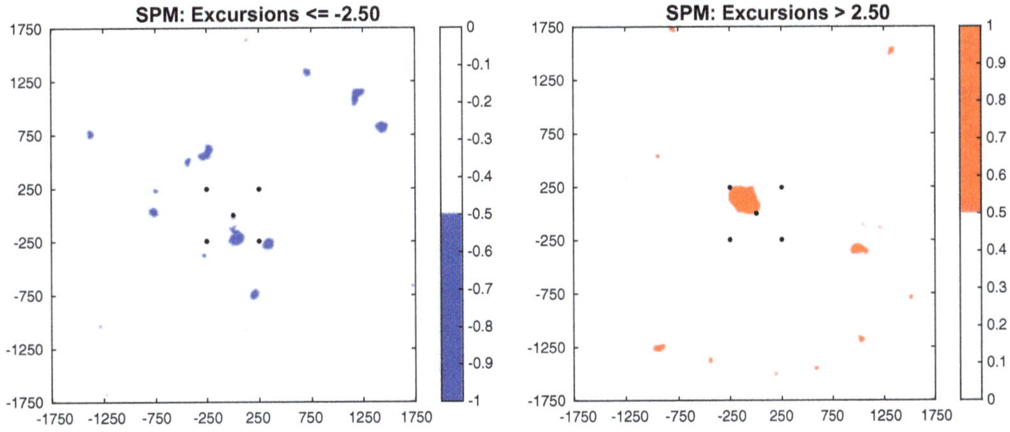


**Fig. 15.9** Regions of excursion below a threshold of −2.5 (left) and above 2.5 (right)

The SPM is calculated at every pixel as the mean *T* value of the fields that created drawdowns exceeding the regulatory threshold, *R*, minus those that resulted in drawdowns less than or equal to the threshold: $T_{>R} - T_{\leq R}$. This convention creates a positive value in the SPM in an area where higher *T* values are associated with realizations that created exceedance of *R* and negative values where higher *T* values created drawdowns $\leq R$. Figure 15.8 shows regions of positive and negative values, but the dominant anomaly is a high SPM value between the pumping well and the observation point to the northwest. For this example, 155 realizations (77.5%) created drawdowns $\leq R$ and 45 (22.5%) created drawdowns that exceeded *R*.

The SPM is truncated at a threshold of $\pm 2.5\sigma$ and the excursion regions are defined (Fig. 15.9). The size of the largest excursions and the probability of them occurring under the mG model are shown in Table 15.2. The SPM has a FWHM of 111.5 m (22.3 pixels). The large positive excursion between the pumping well and the monitoring point is significant with a *p*-value near $1.0 \times 10^{-04}$ while the largest negative excursion is not.

Here the SPM approach also serves as a means of determining the regions of increased sensitivity of drawdown to the *T* values. As expected, when viewed from the perspective of influencing extreme drawdown values, the *T* values in the area between the pumping well and the monitoring point are significantly more important than other values in the *T* field. The remaining regions of excursion do not have any readily discernible connection to the ground water flow dynamics and are consistent with expected excursions in a mG field with this amount of

**Table 15.2** Largest positive and negative excursions in ground water example SPM

| FWHM (m) | 111.5 | |
|---|---|---|
| Excursions | Positive | Negative |
| Largest excursion (pixels) | 2369 | 713 |
| $P(n_{max} > k)$ | ~1.0E$^{-04}$ | 0.198 |

correlation. In practice, the large positive excursion region in the SPM can be used to focus resources for additional data collection, e.g., geophysical survey and/or additional wells.

## 15.4 Summary

There is a large amount of work reported in the functional MRI literature on the detection of anomalies in spatially correlated fields using SPM. Apart from some work in astrophysics, this SPM work has generally been restricted to medical imaging. The body of knowledge around SPM and the statistical approaches developed for fMRI can be readily applied to problems in the earth and environmental sciences. This chapter reviews some of the major developments from the fMRI literature and demonstrates their application with an image anomaly detection problem and a ground water modelling problem. A strong advantage of SPM is that it directly addresses the challenge of enabling hypothesis testing, including calculation of the significance of the results, in spatially correlated fields.

The example problems chosen here emphasized defining the significance of the largest, positive and negative, anomaly in each SPM. The SPM framework also supports hypothesis testing on non-localized, "omnibus", features such as the maximum/minimum value of the SPM, the number pixels exceeding the threshold and the number of excursion regions within the SPM. Additionally, hypothesis testing of localized, "focal", features is also supported including hypothesis testing of the occurrence of any size excursion.

The example problems used here relied on the underlying images being realizations of mG fields, but that is not a requirement. It is the map of the test statistic values defining the differences between fields that is modelled as a mG field, and that flexibility makes SPM applicable to a very general set of problems as the mG model is a standard for differences between images. For example the same approach could be used to compare geologic models with discrete features. Future work will consider the application of other statistical tests within the SPM framework.

## Appendix: Conditional Differences

The $t$-test is a traditional measure of the difference between two means (e.g., Walpole and Myers 1989). Quite simply, the $t$-statistic is the difference between two values, at least one of which is a population or sample mean, normalized by the standard error of the mean:

$$t = \frac{\overline{X} - \mu}{s_e} = \frac{\overline{X} - \mu}{s\sqrt{1/n}}$$

where $\overline{X}$ is a sample mean, $\mu$ is a population mean, $s_e$ is the standard error of the mean which is the standard deviation of the observations, $s$, that make up the data vector $X$ multiplied by the square root of 1 over the number of samples within $X$. The cumulative probabilities for any value of $t$ are available from the Student's $t$ distribution and require knowledge of the degrees of freedom, $\nu$, within the test. For the analyses done here, $\nu$ is generally $n - 1$.

In the case of comparing two sample means to each other at each location, i.e., $A(x, y)$ and $B(x, y)$, instead of comparing a sample mean to a theoretical population mean, the value of $s_e$ must be calculated from both sample sets as:

$$S_e = S_p\sqrt{\frac{1}{n_1} + \frac{1}{n_2}}$$

where $n_1$ and $n_2$ are the number of images that were used in calculating the average maps $A$ and $B$ and $s_p$ is the average pooled standard deviation:

$$S_p(x, y) = \sqrt{\frac{(n_1 - 1)s_1^2(x, y) + (n_2 - 1)s_2^2(x, y)}{n_1 + n_2 - 2}}$$

Here we are assuming that $n_1$ and $n_2$ are constant for all locations and therefore not a function of $(x, y)$. The $t$-statistic image (map), based on the pooled standard deviation, is:

$$t(x, y) = \frac{\Delta(x, y)}{S_p(x, y)\sqrt{\frac{1}{n_1} + \frac{1}{n_2}}}$$

## References

Adler RJ, Hasofer AM (1976) Level crossings for random fields. Ann Probab 4:1–12

Adler RJ (1981) The geometry of random fields. Wiley, New York

Adler RJ (2000) On excursion sets, tube formulas and maxima of random fields. Ann Probab 10(1):1–74

Adler RJ, Taylor JE (2007) Random fields and geometry. Springer

Adler RJ, Taylor JE, Worsley KJ (2009) Applications of random fields and geometry: foundations and case studies. http://webee.technion.ac.il/people/adler/publications.html. Accessed Jan 2009

Anselin L (1995) Local indicators of spatial association—LISA. Geogr Anal 27(2):94–115

Brett M, Penny W, Kiebel S (2003) An introduction to random field theory, chapter 14. In: Frackowiak RSJ, Friston KJ, Frith C, Dolan R, Friston KJ, Price CJ, Zeki S, Ashburner J, Penny WD (eds) Human brain function, 2nd edn. Academic Press

Byers S, Raftery AE (1998) Nearest-neighbor clutter removal for estimating features in spatial point processes. J Am Stat Assoc 93:577–584

Cressie N, Collins LB (2001) Patterns in spatial point locations: local indicators of spatial association in a minefield with clutter. Nav Res Logist. https://doi.org/10.1002/nav.1022

Friston KJ, Worsley KJ, Frackowiak RSJ, Mazziotta JC, Evans AC (1994) Assessing the significance of focal activations using their spatial extent. Hum Brain Mapp 1:210–220

Friston KJ, Holmes AP, Worsley KJ, Poline J-P, Firth CD, Frackowiak RSJ (1995) Statistical parametric maps in functional imaging: a general linear approach. Hum Brain Mapp 2:189–210

Gilbert RO (1987) Statistical methods for environmental pollution monitoring. Van Nostrand Reinhold, New York, p 320

Goovaerts P, Jacquez GM, Marcus A (2005) Geostatistical and local cluster analysis of high resolution hyperspectral imagery for detection of anomalies. Remote Sens Environ 95(3): 351–367

Goovaerts P (2009) Medical geography: a promising field of application for geostatistics. Math Geosci 41:243

Lantuejoul C (2002) Geostatistical simulation: models and algorithms. Springer, Berlin, p 256

Matlab (2009) Image processing toolbox. Mathworks Inc, Natick, MA, USA

McKenna SA, Ray J, van Bloemen Waanders B, Marzouk Y (2011) Truncated multigaussian fields and effective conductance of binary media. Adv Water Resour 34:617–626

Stein DWJ, Beaven SG, Hoff LE, Winter EM, Schaum AP, Stocker AD (2002) Anomaly detection from hyperspectral imagery. IEEE Signal Process Mag 19(1):58–69

Taylor JE, Adler RJ (2003) Euler characteristics for Gaussian fields on manifolds. Ann Probab 31(2):533–563

Walpole RE, Myers RH (1989) Probability and statistics for engineers and scientists, 4th edn. MacMillan Publishing Co., New York, p 765

Worsley KJ, Evans AC, Marrett S, Neelin P (1992) A three-dimensional statistical analysis for rCBF activation studies in human brain. J Cereb Blood Flow Metab 12:900–918

Worsley KJ, Marrett S, Neelin P, Vandal AC, Friston KJ, Evans AC (1996) A unified statistical approach for determining significant signals in images of cerebral activation. Hum Brain Mapp 4:58–73

# Chapter 16
# Water Chemistry: Are New Challenges Possible from CoDA (Compositional Data Analysis) Point of View?

Check for updates

**Antonella Buccianti**

**Abstract** John Aitchison died in December 2016 leaving behind an important inheritance: to continue to explore the fascinating world of compositional data. However, notwithstanding the progress that we have made in this field of investigation and the diffusion of the CoDA theory in different researches, a lot of work has still to be done, particularly in geochemistry. In fact most of the papers published in international journals that manage compositional data ignore their nature and their consequent peculiar statistical properties. On the other hand, when CoDA principles are applied, several efforts are often made to continue to consider the log-ratio transformed variables, for example the centered log-ratio ones, as the original ones, demonstrating a sort of resistance to thinking in relative terms. This appears to be a very strange behavior since geochemists are used to ratios and their analysis is the base of the experimental calibration when standards are evolved to set the instruments. In this chapter some challenges are presented by exploring water chemistry data with the aim to invite people to capture the essence of thinking in a relative and multivariate way since this is the path to obtain a description of natural processes as complete as possible.

## 16.1 Water Chemistry Data as Compositional Data

When geochemical data are analysed by using statistical methods, several units can be used to express concentrations and a first discussion of their compositional nature is reported in Buccianti and Pawlowsky-Glahn (2005). The usual units of measurement include milligrams per liter (mg/L), parts per million by weight (ppm), parts per billion by weight (ppb), millimole per liter (mmol/L), and

A. Buccianti (✉)
Department of Earth Sciences, University of Florence, Via G. La Pira 4, 50121 Florence, Italy
e-mail: antonella.buccianti@unifi.it

A. Buccianti
CNR-IGG, Unit of Florence, Florence, Italy



B. S. Daya Sagar et al. (eds.), *Handbook of Mathematical Geosciences*,
https://doi.org/10.1007/978-3-319-78999-6_16

milliequivalent per liter (meq/L). The ppm and mg/L units are numerically equal if the density of the water sample is 1 g/cm$^3$, as in pure water. Samples can be converted from mg/L to ppm by multiplying each component by the density of water. The term mmol/L indicates the number of ions or molecules in the water when multiplied by Avogadro's number (the number of molecules in a mole of material, $6.023 \times 10^{23}$). The measure mg/L is converted to mmol/L by dividing by the atomic or molecular weight. To express concentration by meq/L (electrical charges are considered), mmol/L is multiplied by the charge of the ions. In each case the base of the calculus is given by the content of some chemical species referred to a given weight or volume then multiplied by a constant (atomic or molecular weight, electrical charges).

These types of data describe parts of some whole and even if proportions are expressed as real numbers, they cannot be interpreted, or even analysed, as real data. It is well known that this practice can lead to paradoxes and/or misinterpretations (e.g. intervals covering negative proportions, spurious correlations) already discussed a century ago (Pearson 1897), but mostly forgotten and neglected over the years (Chayes 1960).

No other ways are possible to compare different samples from dissimilar sites and times, as is usually required. Thus the compositional nature of the experimental data is an intrinsic property related to their origin (e.g. instrument calibration) and to the necessity of making comparisons to investigate the genesis of environmental variability. As directional (circular) observations (Fisher 1995) compositional data move in a constrained sample space called *simplex* (Aitchison 1986):

$$S^D = \{\mathbf{x} = [x_1, x_2, \ldots, x_D] | x_i\}, > 0,\ i = 1, 2, \ldots, D;\ \sum_{i=1}^{D} x_i = \kappa \qquad (16.1)$$

where the $D$ components of the vector $S^D$ are called parts (variables) of the composition. The value of κ depends of the units of the measurement or rescaling procedure, and usual values are 1 (proportions), 100 (%), $10^6$ (ppm) or similar. Note that it is not necessary to have $\sum_{i=1}^{D} x_i = \kappa$ (closed data) to obtain compositional observations. In fact, a (row) vector $\mathbf{x} = [x_1, x_2, \ldots, x_D]$ is a $D$-part composition when all its components are strictly positive real numbers and carry only *relative information*. This means that the message about what is occurring is mainly contained in the ratios between the parts since the numerical value of each variable by itself is not relevant. A recent thorough analysis of the "compositional problem" can be found in Pawlowsky-Glahn and Buccianti (2011) and Pawlowsky-Glahn et al. (2015). On the other hand interesting applications on water chemistry can be found in literature (e.g. Engle and Rowan 2013, 2014; Engle and Blondes 2014; Buccianti and Zuo 2016; Owen et al. 2016; Buccianti et al. 2018; Shelton et al. 2018) where the different potentialities of the family of the log-ratio transformations are differently exploited posing at the central point of the analysis the relativity of the values and the multivariate vision. The cited papers are not exhaustive but have been chosen since they successfully focus on the use of the isometric log-ratio transformation as a way to describe the dynamics of geochemical processes.

## 16.2 Isometric-Log Ratio Transformation: Is This the Key to Decipher the Dynamics of Geochemical Systems?

### *16.2.1 Coordinates as Balances*

Water present below the land surface and running above it tells the history of the environment with which it has been in contact. Rainfall and snowmelt interact with the rock of the Earth surface and percolate through the soil zone where chemical reactions with gases, minerals and organic compounds take place. Chemical reactions occur because the composition of the water is not in equilibrium with the solid phases or the gaseous component (Kleidon 2010). Thus disequilibrium drives the reactions and solutes in the water are derived from the dissolution or leaching of the solid phases and from the dissolution of gases from the air or from the oxidation of organic matter. Most of the natural systems are open and according with Nicolis and Prigogine (1989) they are characterized by dissipative structures and presence of irreversible processes. Dissipative structures contain subsystems, which permanently fluctuate until the fluctuation becomes so strong that it breaks the original system to generate a new condition, more complex and characterized by a higher level of order. The dynamics of systems being far from equilibrium requires a continuous self-organization and to maintain this condition the energy flux from the environment is higher than required for the initial state and irreversible processes can be a source of order rather than chaos. Most of the geological systems are open and dynamic, characterized by a great number of components and develop in a nonlinear way far from equilibrium (Shvartsev 2009). Particularly interesting from this point of view is the water-rock system where also synergetic properties can be found, with respect to the thermodynamical equilibrium where elements (molecules) behave independently of one another (Shvartsev 2013).

The use of the isometric log-ratio coordinates (Egozcue et al. 2003) not only allows us to manage compositional data with classical statistical tools, but also could offer a powerful tool to probe the level of self-organization of a geochemical system as a whole. When coordinates are obtained by using the sequential binary partition method (Egozcue and Pawlowsky-Glahn 2005), guided by a geochemical criterion, the analysis of their frequency distribution may represent an interesting way to understand the laws governing randomness and variability. By taking into account this consideration, an improvement of the *balance dendrogram* (Pawlowsky-Glahn and Egozcue 2001) is here presented with the aim to investigate the behavior of aqueous systems.

The sample space of *D*-part compositional data, the simplex, being a subset of the real space $R^D$, has a real Euclidean vector space structure (Billheimer et al. 2001; Pawlowsky-Glahn and Egozcue 2001; Buccianti and Magli 2011). This situation allows the representation of data in coordinates with respect to an orthonormal basis, for example following the Gram-Schmidt orthonormalization process or a Singular Value Decomposition (Egozcue et al. 2003). Since these methods often reveal coordinates not easy to interpret, *balances*, a specific type of orthonormal coordinates associated with groups of parts, have been proposed (Egozcue and Pawlowsky-Glahn 2005). This method is based on a sequential binary partition of a $D$-part composition into non-overlapping groups and when the procedure is geochemically guided it leads to coordinates easy to interpret. Moreover, it allows understanding of how the total variance is decomposed into marginal variances, thus pointing out the relationship between intra-group and inter-group compositional parts variability. For the *i-th* order of partition, the balance is

$$b_i = \sqrt{\frac{r_i \cdot s_i}{r_i + s_i}} log \frac{\left(\prod_{x_j \in G_{i1}} x_j\right)^{1/r_i}}{\left(\prod_{x_l \in G_{i2}} x_l\right)^{1/s_i}} \quad (16.2)$$

where $r_i$ and $s_i$ are the number of parts in the groups of numerator ($G_{i1}$) and denominator ($G_{i2}$), respectively. As we can see, the balance is defined as the natural logarithm of the ratio of geometric means of the parts in each group, normalized by the coefficient needed to obtain unit length of the vectors of the basis.

### *16.2.2 Behavior of Self-organizing Systems and CoDA Phylosophy*

A general characteristic of self-organizing systems is robustness and resilience (Dakos et al. 2014; Dai et al. 2015). This means that they are relatively insensitive to perturbations or errors, and can show a strong capacity to restore themselves after changes (Scheffer et al. 2009, 2012). One reason for this fault-tolerance is the redundant, distributed organization so that the non-damaged regions can usually make up for the damaged ones. Within certain limits, another reason for the intrinsic robustness is that self-organization is facilitated by randomness, fluctuations or "noise" while the stabilizing effect of feedback loops guarantee resilience. The presence of feedback mechanisms generates systems that can be responsible for their own maintenance, and thus largely independent from the environment. Although in general there will still be exchange of matter and energy between systems and surroundings, the organization is determined purely internally. Thus the system is thermodynamically open, but organizationally closed. Organizational closure turns a collection of interacting elements into an individual, coherent whole. This whole has properties that arise out of its organization that can be described by the probability laws that govern the relative behaviour of its elements (van Rooij 2013). From this point of view CoDA theory appears to capture the philosophy of this condition and the analysis of the shape of the frequency distribution of isometric coordinates should be the adequate tool (Allegre and Lewin 1995; Seely et al. 2012; Holden and Rajaraman 2012; Buccianti and Zuo 2016).

As reported in Scheffer et al. (2012) the probability density distribution of some variables describing the state of a system can be used to estimate how the potential landscape is reflecting its stability properties. The shape of the probability density function indicates where the data are more aggregated and which laws are governing the variability, giving us fundamental information about the genesis of randomness (Agterberg 2014). In our case it will be the shape of the frequency distribution of isometric log-ratio coordinates representing some geochemical process that will inform us about dynamic properties of the system. In Fig. 16.1 some examples of a non-equilibrium dynamics are reported (Scheffer et al. 2009). Conditions represented in (a) are far from a bifurcation point. The pothole in the potential line corresponds to an area where data tend to aggregate in the density probability distribution function. Here resilience is large since the basin of attraction is wide and the rate of recovery from perturbations is relatively high. If the system is stochastically forced, the resulting dynamics will be characterised by low correlation between states at subsequent time intervals. In (b) the system is closer to the transition point and resilience decreases due to the shrinking of the attraction basin and the low rate of recovery from small perturbations. Here the slight depression could be related to presence of bimodality indicating presence of alternative states. In this case the system in a stochastic environment will have a long memory for perturbations and its dynamics will be governed by high variance and stronger correlations between subsequent states.

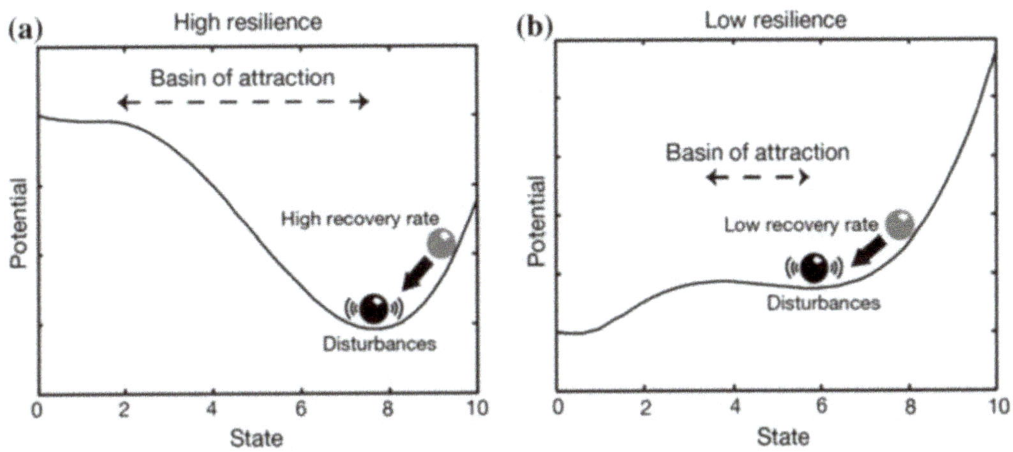


**Fig. 16.1** Example of non-equilibrium dynamics (from Scheffer et al. 2009, modified). The pothole in the potential line of diagram **a** corresponds to an area where data tend to aggregate in the density probability distribution function. The slight depression in **b** could be related to presence of bimodality indicating presence of alternative states (Scheffer et al. 2012)

## 16.3 Improving CoDA-Dendrogram: Checking for Variability, Resilience and Stability

The chemical composition of groundwaters from the Arezzo basin aquifer (Tuscany, central Italy) was analysed, as an application example, to obtain information about the dynamics of the aqueous geochemical system. The Arezzo Basin (Fig. 16.2), formed since Upper Pliocene, is a structural depression bordered to the North and to the East by the Pratomagno and Chianti belts, respectively, and to the South and to the East by two tectonic lineaments (Val d'Arbia-Val Marecchia transversal and Chitignano normal faults). Along these tectonic discontinuities $CO_2$-rich manifestations either seep out or are exploited by private companies down to the depth of 1000 m. Three main aquifers are recognized: (i) a relatively deep aquifer hosted in Tertiary sandstone formations; (ii) an intermediate aquifer hosted in Quaternary fluvio-lacustrine sediments and (iii) a shallow aquifer in recent alluvial sediments. The available geochemical data-base consists of about 500 samples that were collected in different dry and rainy seasons in recent years from 80 wells diffused in all the basin area. Depth of the sampling is, unfortunately, not always known and few differences can be related to seasonal changes. Physical parameters (temperature and electrical conductivity), major, minor and trace dissolved species (pH, Ca, Mg, Na, K, $NH_4$, $HCO_3$, $SO_4$, $NO_3$, $NO^2$, Cl, Br, F and heavy metals), oxygen and hydrogen isotopes in the water molecules and dissolved gases (including $^{13}C$-$CO_2$) were analyzed. On the basis of Total Dissolved Solids (TDS) the waters from Arezzo aquifer can be considered mainly oligomineral and medium-mineral, whereas mineral waters are almost exclusively associated with

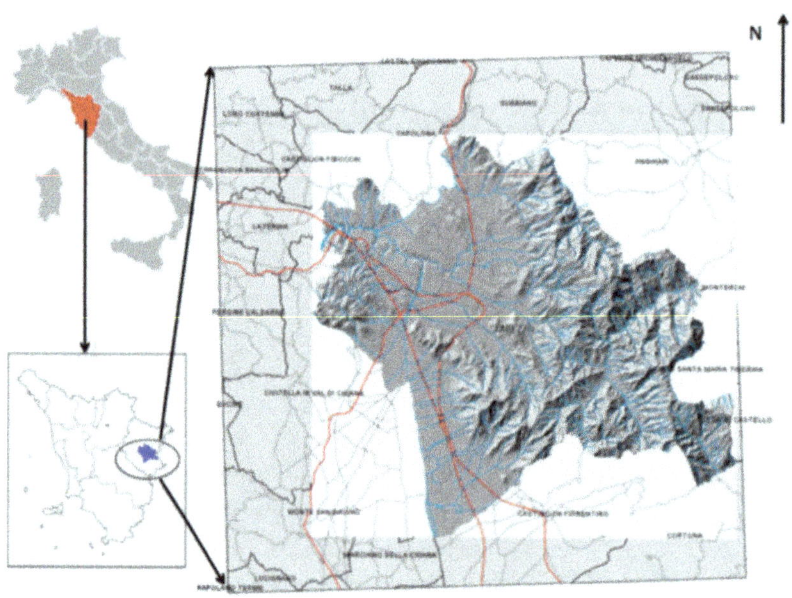


**Fig. 16.2** The hydrographic system of the Arezzo basin (Tuscany, central Italy) (http://sit.comune.arezzo.it/normativa/index.php?normativa=_ps&mappa=ps_b11a)

$CO_2$-rich wells. From a classification point of view, Ca(Mg)-$HCO_3$ is by far the most representative geochemical facies, followed by Na(K)-$HCO_3$, Ca(Mg)-$SO_4$ and Na(K)-Cl types. It is noteworthy to point out here that the Na(K)-$HCO_3$ waters, whose origin is related to the presence of $CO_2$-rich waters that favor cation exchange processes with clay minerals contained in the sedimentary formations, are aligned along the Val d'Arbia-Val Marecchia transversal tectonic system.

In Table 16.1 the sequential binary partition process to construct the isometric log-ratio coordinates is reported. The first coordinate could represent the balance between the most important chemical reactions involving carbonatic and silicatic rocks ($Ca^{2+}$, $Mg^{2+}$, $Na^+$, $K^+$, $HCO_3^-$ and $H^+$) versus elements and chemical species whose sources could be different, including pollution ($Cl^-$, $SO_4^-$, $NO_3^-$). The second coordinate is an analysis inside the carbonatic and silicatic cycle, balancing cations and anions. The third compares the behaviour of the involved bivalent versus monovalent elements while the fourth and the fifth compare their relative behaviour. The sixth coordinate analyses the anions giving us information about the pH water conditions. Finally, the remaining coordinates investigate the behaviour of variables whose source may be related to pollution. Considering $Cl^-$ in absence of atmospheric cyclic salts and evaporates about 30% of its amount is related to pollution, 54% in case of $SO_4^{2-}$, while for nitrate the most important anthropogenic sources are septic tanks, application of nitrogen-rich fertilizers to turf grass, and intensive agricultural processes (Berner and Berner 1996; Liu et al. 2011; Menció et al. 2016).

As we can see variance is higher for the first balance comparing natural and anthropic processes, and the last one, comparing $SO_4^{2-}$ and $NO_3^-$ whose ratio variability is a further witness of the presence of numerous sources/fluctuations. A first result here reveals that when elements are more related to natural weathering processes their balance variability appears to be reduced, probably indicating that the same processes have been working through time in a similar way. By taking into account the previous discussion about the dynamics of geochemical systems more information should be obtained by the analysis of the frequency distribution of the balances.

To achieve this aim in Fig. 16.3 an improved version of the *balance dendrogram* is reported where the original boxplots (Pawlowsky-Glahn and Egozcue 2011) are associated with the frequency distribution of the coordinates. Histograms have the same horizontal and vertical scale so they are comparable. Red line is related to the Gaussian distribution, black treated line to the Kernel density estimation.

Application of several normality tests indicates that under no circumstances the Normal distribution can be considered as model for the log-ratio coordinates; the consequence is that the log-normal model cannot be used to describe ratios between parts or group of parts. In most of the cases it appears to be due to some bimodality or to the presence of a heavy tail in the right-hand part of the distribution. The presence of power laws is associated with complex systems composed of processes that interact to self-organize their behavior across multiple temporal and/or spatial scales. Both fractals and multifractals are commonly associated with local self-similarity or scale-independence, generally leading to power-law relations

**Table 16.1** Sequential binary partition process for the groundwater chemistry of the Arezzo basin (Tuscany, central Italy). Units of the chemical concentrations are mol/L while the variance expresses the contribution of each balance in explaining the total variability (% in parenthesis)

| Balance | mol_Ca | mol_Mg | mol_Na | mol_K | $mol\_HCO_3$ | mol_Cl | $mol\_SO_4$ | $mol\_NO_3$ | $mol\_H^+$ | Variance |
|---|---|---|---|---|---|---|---|---|---|---|
| $ilr_1$ | 1 | 1 | 1 | 1 | 1 | −1 | −1 | −1 | 1 | 1.23 (20.6) |
| $ilr_2$ | 1 | 1 | 1 | 1 | −1 | 0 | 0 | 0 | −1 | 0.29 (4.86) |
| $ilr_3$ | 1 | 1 | −1 | −1 | 0 | 0 | 0 | 0 | 0 | 0.54 (9) |
| $ilr_4$ | −1 | 1 | 0 | 0 | 0 | 0 | 0 | 0 | 0 | 0.18 (3) |
| $ilr_5$ | 0 | 0 | 1 | −1 | 0 | 0 | 0 | 0 | 0 | 0.94 (15.7) |
| $ilr_6$ | 0 | 0 | 0 | 0 | 1 | 0 | 0 | 0 | −1 | 0.23 (3.85) |
| $ilr_7$ | 0 | 0 | 0 | 0 | 0 | 1 | −1 | −1 | 0 | 0.99 (16.6) |
| $ilr_8$ | 0 | 0 | 0 | 0 | 0 | 0 | 1 | −1 | 0 | 1.57 (26.3) |

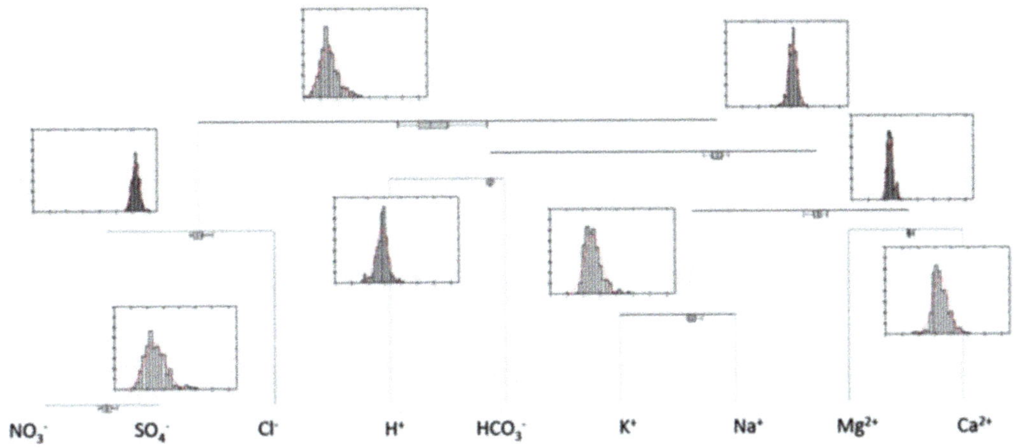


**Fig. 16.3** Balance dendrogram (Thió-Henestrosa et al. 2008) with associated histograms. Red line corresponds to the Gaussian model, black treated line to the Kernel density estimation. The length of the vertical bar represents the proportion of the sample total variance

(Agterberg 2014). On the other hand the lognormal shape represents a special condition in which the interdependencies among processes are minimized or absent and repeated fragmentation (or dilution) dominates. As we can see in Fig. 16.3 the presence of heavy tails characterizes coordinates that mainly balance weathering of silicate and carbonates ($K^+$, $Na^+$, $Mg^{2+}$, $Ca^{2+}$, $H^+$, $HCO_3^-$) versus other environmental processes ($NO_3^-$, $SO_4^-$, $Cl^-$). Moreover, considering the internal partition of the previous balances, $K^+/Na^+$, $Mg^{2+}/Ca^{2+}$ and, in particular, $NO_3^-/SO_4^-$ ratios repeat this type of behavior.

The use of the complementary distribution function reveals the presence of power laws more clearly. In this plot, reported in Fig. 16.4, if $X$ has a power law distribution the behavior of the $Prob[X \geq x]$ will be a straight line (Mitzenmacher 2004). As we can see, linear models can well describe several portions of curves for all the coordinates. This condition asks for multifractality perhaps associated to the space-time heterogeneity of the aquifer structure. Here a sudden change in the number of data with given concentration values is expected, particularly for pollution processes (Agterberg 2014). The fractal dimension of the phenomena, related to the slope of the straight lines, indicates how much more often there are low differences between the data rather then high differences.

On the whole the aquifer system appears to be governed by an interaction-dominant dynamics but it does not present a clear multimodality (or bimodality) that could be associated to different states. By considering Fig. 16.1 and the information deduced by the shape of the frequency distribution (Figs. 16.3 and 16.4) the aquifer could be associated with a sufficient resilience and recovery state (Scheffer et al. 2009, 2012). Of notice here is that the most important contribution to variability appears to be related to chemicals such as $NO_3^-$ and $SO_4^-$ suggesting the weight and the intermittency of the anthropic pressure. The multifractality revealed in Fig. 16.4 could indicate that in the dynamical system the energy

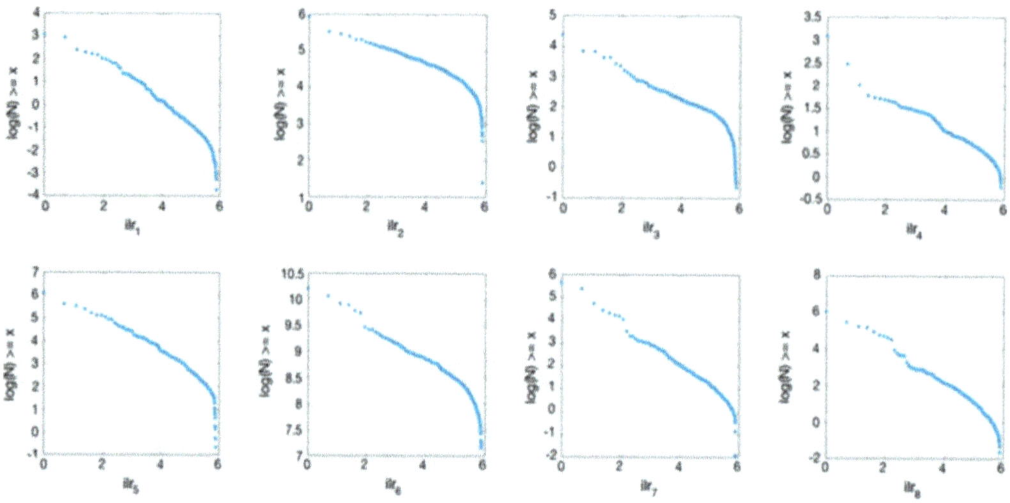

**Fig. 16.4** Complementary distribution function to reveals the presence of power laws. If $X$ has a power law distribution the behavior of the $Prob[X \geq x]$ will be a straight line (Mitzenmacher 2004)

dissipation cannot be neglected and that extended areas (intervals) of low fluctuations intermittent with small areas of extremely large fluctuations are to be expected. Moreover, the system as a whole is undergoing a non-linear dissipation with the energy interchange on different scales.

## 16.4 Conclusions

Starting from Garrels and Christ (1965) equilibrium in the water-rock system is usually analysed through the application of thermodynamic methods. In this context the statistical analysis of water concentrations, opportunely transformed into isometric logratio coordinates, could be an effective approach to understand where the randomness in nature comes from (Agterberg 2014) and if equilibrium conditions are really encountered.

The frequency distribution of the ratio of the compositional parts of Arezzo aquifer chemistry exhibits an overlapping between log-normal and power-law probability distributions when silicate and carbonate weathering ($K^+$, $Na^+$, $Mg^{2+}$, $Ca^{2+}$, $H^+$, $HCO_3^-$) is balanced versus other environmental processes ($NO_3^-$, $SO_4^-$, $Cl^-$). Similar results are obtained when the partition to generate new balances is applied to the previous group of parts ($NO_3^-$ versus $SO_4^-$, $K^+$ versus $Na^+$ or $Mg^{2+}$ versus $Ca^{2+}$). The result indicates a system subjected to nonlinear compositional changes due to presence of feedback effects attributable in a porous medium to change in porosity causing a remarkable change in permeability, in the pore-fluid flow and in the chemical-species concentration (Zhao 2014). Since thermodynamic equilibrium represents a homogeneous distribution of the parts, the obtained results indicate that the system is able to create and maintain a given amount of gradient, generating heterogeneity. However no clear multimodality is present and for the span of time here analysed different steady states (basins of attraction for concentration values) have not yet clearly emerged. Thus, from a compositional point of view, the system could be characterised by sufficient resilience and recovery rate from disturbances since the dissipative behaviour appears to be able to adsorb fluctuations. New progress would be made in this direction by exploiting the capacity of CoDA to capture the interdependence of concentration values, thus describing the water system and the surrounding as a whole, as in reality.

**Acknowledgements** This research was supported by the University of Florence (2016 funds) and by the GEOBASI project financed in 2015 by Tuscany Region through CNR-IGG. Frits Agterberg and B.S. Daya Sagar are warmly thanked for their support as well as IAMG for the 2003 Felix Chayes Prize and the constant sustain to my research activity. Lisa Merli helped me for the English language.

## References

Aitchison J (1986) The statistical analysis of compositional data. In: Monographs on statistics and applied probability. Chapman and Hall, Ltd, London, UK, 416 pp (reprinted 2003 with additional material by The Balckburn Press)

Agterberg F (2014) Geomathematics: theoretical foundations, applications and future developments. Quantitative geology and geostatistics series, vol 18. Springer, 553 pp

Allegre CJ, Lewin E (1995) Scaling laws and geochemical distributions. Earth Planetary Sci Lett 132:1–13

Berner EK, Berner RA (1996) Global environment. In: Water, air, and geochemical cycles. Prentice Hall, Upper Saddle River, New Jersey, 07458, 376 pp

Billheimer D, Guttorp P, Fagan W (2001) Statistical interpretation of species composition. J Am Stat Assoc 96(456):1205–1214

Buccianti A, Magli R (2011) Metric concepts and implications in describing compositional changes for world river's chemistry. Comput Geosci 37(5):670–676

Buccianti A, Pawlowsky-Glahn B (2005) New perspectives on water chemistry and compositional data analysis. Math Geol 37(7):703–727

Buccianti A, Zuo R (2016) Weathering reactions and isometric log-ratio coordinates: do they speak each other? Appl Geochem 75:189–199

Buccianti A, Nisi B, Raco B (2016) Towards the concept of background/baseline compositions: a practicable path? In: Springer Proceedings in Mathematics and Statistics, vol 187, pp 31–43

Buccianti A, Lima A, Albanese S, DeVivo B (2018) Measuring the change under compositional data analysis (CoDA): insight on the dynamics of geochemical systems. J Geochem Explor 189:100–108

Cardenas MB (2008) Surface water-groundwater interface geomorphology leads to scaling of residence times. Geophys Res Lett 35(L08402):1–5

Cardenas MB, Jiang XW (2010) Groundwater flow, transport, and residence times through topography-driven basins with exponentially decreasing permeability and porosity. Water Resour Res 46(W11538):1–9

Chayes F (1960) On correlation between variables of constant sum. J Geophys Res 65(12): 4185–4193

Dai L, Korolev KS, Gore J (2015) Relation between stability and resilience determines the performance of early warning signals under different environmental drivers. PNAS 112(32): 10056–10061

Dakos V, Carpenter SR, van Nes EH, Scheffer M (2014) Resilience indicators: prospects and limitations for early warnings of regime shifts. Philos Trans B 370(20130263):1–10

Egozcue JJ, Pawlowsky-Glahn V, Mateu-Figueras V (2003) Isometric logratio transformations for compositional data analysis. Math Geol 35(3):279–300

Egozcue JJ, Pawlowsky-Glahn V (2005) Groups of parts and their balances in compositional data analysis. Math Geol 37(7):795–828

Engle MA, Rowan EL (2013) Interpretation of Na-Cl-Br systematics in sedimentary basin brines: comparison of concentration, element ratio, and isometric log-ratio approaches. Math Geosci 45(1):87–101

Engle MA, Blondes MS (2014) Linking compositional data analysis with thermodynamic geochemical modeling: oilfield brines from the Permian Basin, USA. J Geochem Explor 141:61–70

Engle MA, Rowan EL (2014) Geochemical evolution of produced waters from hydraulic fracturing of the Marcellus Shale, northern Appalachian Basin: a multivariate compositional data analysis approach

Fisher NI (1995) Statistical analysis of circular data. Cambridge University Press, UK, p 277

Garrels RM, Christ CL (1965) Solutions, minerals and equilibrium. Harper and Row, NY

Holden JG, Rajaraman S (2012) The self-organization of a spoken word. Front Psychol 3(209): 1–24

Kleidon A (2010) Life, hierarchy, and the thermodynamic machinery of planet Earth. Phys Life Rev 7:424–460

Liu Z, Dreybrodt W, Liu H (2011) Atmosperic $CO_2$ sink: silicate weathering or carbonate weathering? Appl Geochem 26:S292–S294

Menció A, Mas-Pla J, Otero N, Regàs O, Boy-Roura M, Puig R, Bach J, Domènech C, Zamorano M, Brusi D, Folch A (2016) Nitrate pollution of graoudwater; all righ…, but nothing else? Sci Total Environ 539:241–251

Mitzenmacher M (2004) A brief history of generative models for power law and lognormaldistributions. Internet Math 1(2):226–251

Nicolis G, Prigogine I (1989) Exploring complexity: an introduction. Freeman and Company, New York, p 343

Owen DDR, Pawlwosky-Glahn V, Egozcue JJ, Buccianti A, Bradd JM (2016) Compositional data analysis as a robust tool to delineate hydrochemical facies within and between gas-bearing aquifers. Water Resour Res 52(8):5771–5793

Pawlowsky-Glahn V, Buccianti A (2011) Compositional data analysis: theory and applications. Wiley Ltd, 378 p

Pawlowsky-Glahn V, Egozcue JJ (2001) Geometric approach to statistical analysis on the simplex. Stoch Environ Res Risk Assess (SERRA) 15(5):384–398

Pawlowsky-Glahn V, Egozcue JJ (2011) Exploring compositional data with the CoDa dendrogram. Austrian J Stat 40(1/2):103–113

Pawlowsky-Glahn V, Egozcue JJ, Tolosana-Delgado R (2015) Modeling and analysis of compositional data, statistics in practice series. Wiley Ltd, 247 pp

Pearson K (1897) Mathematical contributions to the theory of evolution. On a form of spurious correlation which may arise when indices are used in the measurement of organs. Proc R Soc Lond, LX, 489–502

Seely AJ, Macklem P (2012) Fractal variability: an emergent property of complex dissipative systems. Chaos 22:0131081–013108-7

Scheffer M, Bascompte J, Brock WA, Brovkin V, Carpenter SR, Dakos V, Held H, van Nes EH, Rietkerk M, Sugihara G (2009) Early-warning signlas for critical transitions. Nature 461:53–59

Scheffer M, Carpenter SR, Lenton TM, Bascompte J, Brock W, Dakos V, van de Koppel J, van de Leemput IA, Levin SA, van Nes EH, Pascual M, Vandermeer J (2012) Anticipating critical transitions. Science 338:344–348

Shelton JL, Engle MA, Buccianti A, Blondes M (2018) The isometric log-ratio (ilr)-ion plot: a proposed alternative to the Piper diagram. J Geochem Explor 190:130–141

Shvartsev SL (2009) Self-organizing abiogenic dissipative structures in the geologic history of the earth. Earth Sci Front 16(6):257–275

Shvartsev SL (2013) Water-rock interaction: implications for the origin and program of global evolution. Int J Sci 2(10):26–30

Thió-Henestrosa S, Egozcue JJ, Pawlowsky-Glahn V, Kovacs LO, Kovacs G (2008) Balance dendogram, a new routine of codapack. Comput Geosci 34(12):1682–1696

van Rooij MMJW, Nash BA, Rajaraman S, Holde JG (2013) A fractal approach to dynamic inference and distribution analysis. Front Physiol 4:1–16

Zhao C (2014) Physical and chemical dissolution front instability in porous media. Lecture notes in earth system sciences. Springer International Publishing Switzerland, 354 pp

# Chapter 17
# Analysis of the United States Portion of the North American Soil Geochemical Landscapes Project—A Compositional Framework Approach

**E. C. Grunsky, L. J. Drew and D. B. Smith**

**Abstract** A multi-element soil geochemical survey was conducted over the conterminous United States from 2007–2010 in which 4,857 sites were sampled representing a density of 1 site per approximately 1,600 km$^2$. Following adjustments for censoring and dropping highly censored elements, a total of 41 elements were retained. A logcentred transform was applied to the data followed by the application of a principal component analysis. Using the 10 most dominant principal components for each layer (surface soil, A-horizon, C-horizon) the application of random forest classification analysis reveals continental-scale spatial features that reflect bedrock source variability. Classification accuracies range from near zero to greater than 74% for 17 surface lithologies that have been mapped across the conterminous United States. The differences of classification accuracy between the Surface Layer, A- and C-Horizons do not vary significantly. This approach confirms that the soil geochemistry across the conterminous United States retains the characteristics of the underlying geology regardless of the position in the soil profile.

---

**Electronic supplementary material** The online version of this chapter (https://doi.org/10.1007/978-3-319-78999-6_17) contains supplementary material, which is available to authorized users.

---

E. C. Grunsky (✉)
Department of Earth and Environmental Sciences, University of Waterloo, Waterloo, ON, Canada
e-mail: egrunsky@gmail.com

E. C. Grunsky
China University of Geosciences, Beijing, China

L. J. Drew
United States Geological Survey, Reston, VA, USA

D. B. Smith
United States Geological Survey, Denver, CO, USA



B. S. Daya Sagar et al. (eds.), *Handbook of Mathematical Geosciences*,
https://doi.org/10.1007/978-3-319-78999-6_17

## 17.1 Introduction

A continental-scale soil geochemical survey was conducted over the conterminous United States from 2007 to 2010 by the U.S. Geological Survey (Smith et al. 2011, 2012, 2013, 2014). The survey collected samples at 4857 sites (Fig. 17.1), representing a density of 1 site per approximately 1600 km$^2$. The sampling protocol included, at each site, a sample from a depth of 0–5 cm (referred to as the surface soil for the remainder of this paper), a composite of the soil A horizon (the uppermost mineral soil), and a sample from the soil C horizon (generally the partially weathered parent material). If the top of the C horizon was at a depth greater than 1 m, a sample over a 20 cm interval was collected at a depth of approximately 1 m.

Studies on the geochemistry of two transects (east-west and north-south) across the United States and Canada, conducted as pilot studies in preparation for the continental-scale survey (Smith 2009; Smith et al. 2009) showed variability of soil geochemistry and mineralogy along both directions (Garrett 2009; Eberl and Smith 2009; Woodruff et al. 2009). As well, Drew et al. (2010) studied the two transects and demonstrated that the geochemical variability of soil is also closely associated with ecoregions (CEC 1997), which reflect continental scale features such as soil, landform, major vegetation types and climate. These studies indicate that the soil geochemistry is useful for mapping both geological and ecological domains.

Soil geochemistry, from a geological context, reflects a range of mineralogy, as a function of weathering of different parent materials, along with organic content due to biological activity. Ideally, soil geochemistry will represent underlying parent material and processes associated with the modification of those parent materials through comminution, weathering, ground water activity and biogenic processes. Grunsky et al. (2012, 2014) smf Mueller and Grunsky (2016) demonstrated that the

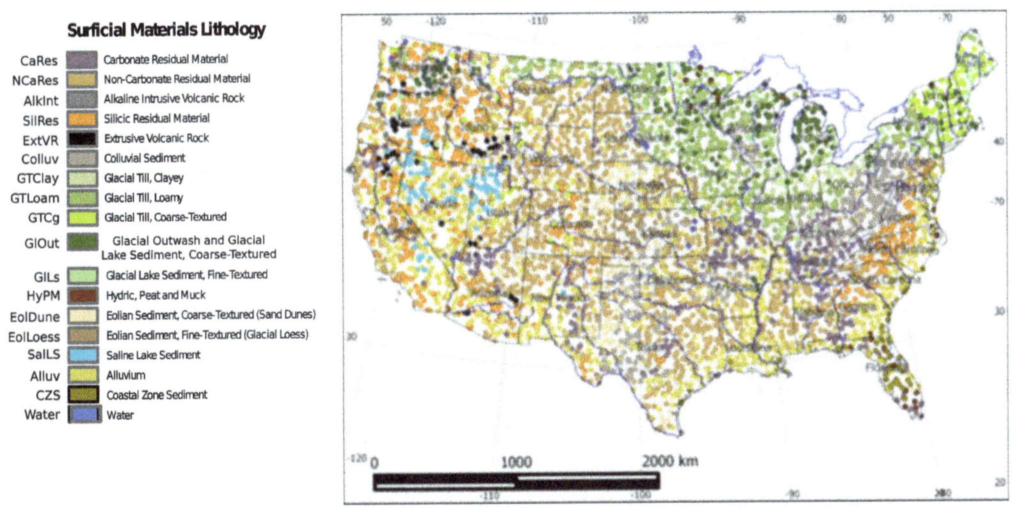


**Fig. 17.1** Soil sample sites over the conterminous United States. Samples were taken at the (0–5) cm layer, the A- and C-horizons

geochemistry of lake sediment and glacial till in northern Canada can be used to predict the underlying lithologies. As part of the North American Soil Geochemistry Landscape Project (Smith et al. 2009), Grunsky et al. (2013) used soil geochemistry collected over the Maritime Provinces of Canada and the northeast United States to demonstrate that A-, B- and C-horizon soils geochemistry is useful for mapping the underlying lithologies. More recently, Grunsky et al. (2017) have shown that geochemistry of surficial soils can identify and classify underlying crustal blocks across the Australian continent, even after extended periods of weathering, transport and reworking.

The approach is based on the use of training sets of representative lithologies. Unfortunately, there are no continental-scale lithologic maps or representative training sets which can be used for predictive bedrock lithologic mapping in Canada or the United States. Sayre et al. (2009) classified the land surface of the conterminous United States according to surficial materials lithology, terrestrial ecosystems and isobioclimate. Isobioclimatic zones were subdivided into thermotypes, (temperature) and ombrotypes (moisture). It follows that soil geochemistry is a proxy for processes controlled by climatic factors. A key question that arises from this is can any of these processes be identified uniquely in the soil geochemistry and, if so, how can these processes be identified in terms of spatial continuity and distinctive chemistry? Drew et al. (2010) studied two transects across the US and demonstrated that the soil geochemistry is closely tied to zones that define the terrestrial ecosystems intersected by these transects. The objective of the current study is to address this question through the use of multivariate statistical analysis and Bayesian-based classification in conjunction with geostatistical methods that accurately describe processes in terms of distinctive geochemistry and spatial continuity.

## 17.2 Methods

### *17.2.1 Sampling and Analysis*

The soil samples were analysed for geochemistry and mineralogy as described by Smith et al. (2011, 2012, 2013, 2014). The samples were air-dried and sieved to <2 mm after which the material was crushed in a ceramic mill prior to chemical analysis. Concentrations of Ag, Al, Ba, Be, Bi, Ca, Cd, Ce, Co, Cr, Cs, Cu, Fe, Ga, In, K, La, Li, Mg, Mn, Mo, Na, Nb, Ni, P, Pb, Rb, S, Sb, Sc, Sn, Sr, Te, Th, Ti, Tl, U, V, W, Y, Zn in all the soil samples (14,434) were determined using a near-total digestion using $HCl$-$HNO_3$-$HClO_4$-$HF$ followed by inductively coupled plasma-mass spectrometry and inductively coupled plasma-atomic emission spectrometry. Mercury values were obtained using cold-vapor atomic absorption spectrometry following dissolution in a mixture of HCl and $HNO_3$ and Se was determined by hydride-generation atomic absorption spectrometry (HGAAS) following dissolution in a mixture of $HNO_3$, HF, and $HClO_4$. Arsenic was also determined by HGAAS following fusion in a mixture of sodium peroxide and sodium hydroxide at 750 °C. Total carbon was determined by combustion. Smith et al. (2013) provides details on the analytical methods and quality control protocols. Silicon was not determined.

All A-horizon and C-horizon samples (9575) were analysed by X-ray diffraction, and the percentages of major mineral phases were calculated using a Rietveld refinement method. Splits of the <2 mm fraction were used for analysis. Complete details of the technique and quality control protocols are provided in Smith et al. (2013).

### 17.2.2 Data Screening and the Compositional Nature of Geochemical Data

Geochemical analyses require screening and adjustment prior to any application of statistical methods and interpretation. A generalized sequence of data screening and adjustment strategies is documented in Grunsky (2010). The data were evaluated and analysed using the R programming and statistical environment (R Core Team 2013).

Major element concentrations, reported as percentages, were converted to ppm, by multiplying the values by a factor 10,000. Summary statistics for the data are given in Smith et al. (2013). The data were screened to determine the number of values that were reported at less than the lower limit of detection. Data that are reported at less than the lower limit of detection are termed as “censored”. Censored data, when used in the application of statistical procedures, can influence estimates of mean and variance and therefore a replacement value that accurately reflects an estimate of the true mean is preferred. Furthermore, geochemical data are, by definition, compositions and as such the issue of closure becomes important (Aitchison 1986). Egozcue et al. (2003) describe various transformations that assist in evaluating data that are constrained by the effect of closure. For censored geochemical data, replacement values can be determined using the several methods based on maximum likelihood estimates of replacements values (Palarea-Albaladejo et al. 2014). Elements in which >80% of the values were censored were dropped from further evaluation, which included Ag, Cs and Te.

The data were also screened for sample sites where a large number of elements were reported at less than the lower limit of detection (<LLD). In the surface soil, 8 sites were found to have more than 25 elements reported at <LLD (3 from Florida). For the A horizon, 2 sites, all from Florida, were found to have more than 25 elements reported at <LLD. For the C horizon, 3 sample sites, in Florida, were found to more that have more than 25 elements reported at <LLD. These sites were dropped from further evaluation.

Summary statistics for the elements are provided by Smith et al. (2013, 2014). The remaining 43 elements: Al, As, Ba, Be, Bi, total C, Ca, Cd, Ce, Co, Cr, Cs, Cu, Fe, Ga, Hg, In, K, La, Li, Mg, Mn, Mo, Na, Nb, Ni, P, Pb, Rb, S, Sb, Sc, Se, Sn, Sr, Th, Ti, Tl, U, V, W, Y, Zn were then evaluated for the estimate of replacement values for those results that were reported at less than the lower limit of detection. The method of nearest neighbour replacement estimates (R package: zCompositions, function **lrEM**) was used on the censored data (Palarea-Albaladejo et al. 2014). The adjusted data were then used for subsequent multivariate statistical analysis.

### *17.2.3 Integration of Land Surface Parameters with Soil Geochemistry*

Land surface maps of the conterminous United States (Sayre et al. 2009) were used to test the effectiveness of the soil geochemistry for revealing information on surficial materials lithology, terrestrial ecosystems and isobioclimate. Isobioclimatic zones were subdivided into thermotypes, (temperature) and ombrotypes (moisture). In this study, only the surface lithologies were studied in further detail. The results of the evaluation of the soil geochemistry in the context of terrestrial ecosystems, thermotypes and ombrotypes will be provided at a later time.

The maps were obtained as raster images with a pixel resolution of 1 km and a geodetic projection of decimal degrees using the North American Datum of 1983 (NAD83). These images were re-projected to the Lambert Conformal Conic projection using the following parameters (Spheroid—GRS 1980; Central Meridian: 96° West; Standard Parallels of 32° and 44°; Latitude of Origin: 38°; False Eastings and Northings of 0 m). This projection was used throughout the study.

The Quantum Geographic Information Systems (QGIS) (QGIS Development Team 2016) was used for the integration of various data sources and the geospatial rendering of the results. Within QGIS, two procedures were used from the Geospatial Data Abstraction Library (GDAL) procedure, "**warp (reprojection)**" and "**point sampling tool**". The map images were initially re-projected to the Lambert Conformal Conic (**lcc**) projection listed above using the "**warp**" procedure. The point dataset of the geochemical sampling sites were also reprojected from latitude/longitude coordinates to the **lcc** projection. The **lcc** image of the surface lithology was then sampled at the geochemical site coordinates using the "**point sampling tool**" and the surface lithology value was integrated into the geochemical database. This methodology was carried out for the other land surface maps (terrestrial ecosystems, surface lithologies, thermotypes and ombrotypes). The values of these features were integrated into the soil geochemistry dataset for further evaluation. It should be noted that the maps produced by Sayre et al. (2009) are generalizations and expressed at a resolution of 30 m (landforms, topographic moisture), 1 km (biogeographic regions) and 15 km for the surface lithology. It is possible that the class defined at any given point on the maps produced by Sayre does not correspond with the surface lithology, biogeographic, landform or topographic classes that were encountered during the soil survey sampling program.

For geospatial rendering purposes (interpolation), the Level 1 Ecology map of the conterminous United States was used to create a grid with a cell size of 40 km × 40 km.

Interpolation of principal component scores, posterior probabilities and measures of typicality were carried out using a geostatistical framework. The gstat package (Pebesma 2004) was used to generate and model semi-variograms with sufficient parameters to generate interpolated images through kriging. The cell size used for image interpolation was chosen as 40 km, the approximate spacing of the site sampling locations.

### *17.2.4 Process Discovery—Empirical Investigation of Soil Geochemistry*

After screening the data for detection limit issues and missing values, the geochemical data were then subjected to an empirical investigation in which the assumptions about the data are minimal. To deal with the effect of closure, the data for 41 elements (Al As Ba Be Bi Ca Cd Ce Co Cr Cu Fe Ga Hg In K La Li Mg Mn Mo Na Nb Ni P Pb Rb S Sb Sc Se Sn Sr Th Ti Tl U V W Y Zn) were log-centred transformed after which a principal component analysis (PCA) was carried out using the methodology of Zhou et al. (1983) and Grunsky (2001). PCA was carried out on the entire set of multi-element data for the surface soil, the A and C horizons combined. PCA was also carried out on the multi-element data individually for the surface soil, A and C horizons. The rationale for this is based on enhancement of the multi-element signature for each layer rather than a principal component signature derived from the combined layers. The principal component biplots and corresponding maps of the component scores were subsequently generated for the surface soil, the A- and C-horizons independently. The biplots and interpolated maps provide insight into the orthogonal linear relationships that can reflect dominant geochemical processes that are influenced by mineral stoichiometry. The three soil layers were evaluated together in order to show any possible relationships between the two soil horizons (A and C) and the surface soil layer. To assist with insight into processes that influence the relationship of the elements and patterns of the scores of the observations, the loadings of the elements were coloured according to the classification of Goldschmidt (1937) into lithophile. siderophile or chalcophile affinity Elements associated with the atmophile affinity were not considered in this study.

### 17.2.5 Process Validation—Modelled Investigation of Soil Geochemistry

Using the classified information derived from the land surface maps of Sayre et al. (2009), the geochemical data were used to establish the ability to predict these classifications using a cross-validation approach in which the data are repeatedly sub-sampled as part of the classification process.

Previous studies (Grunsky et al. 2012, 2014) demonstrated that the use of multivariate statistical methods was able to classify bedrock lithologies based on lake sediment and glacial till geochemical data using discriminant analysis. The methodology employed the results of principal component analysis (described above), followed by an analysis of variance and the application of linear discriminant analysis (Venables and Ripley 2002) to determine which principal components were best at classifying and predicting the bedrock lithologies. This approach relies on having a sufficient number of degrees of freedom and homogeneity of covariance between the classes of the training sets. An alternative to linear discriminant analysis is quadratic discriminant analysis (Venables and Ripley 2002), which compensates for the classes where the condition of homogeneity of covariance cannot be met. The results of applying these methods includes measures of posterior probability in which each site is assigned a measure of probability of belonging to each of the classes and the class with the highest posterior probability is assigned to that site. Posterior probabilities are also compositions, as the sum of the probabilities for all of the classes for each site must sum to 1.0 and are, therefore, compositional in nature.

Both methods were tested for discriminating between the surface lithologies in this study. However, a comparison of results between linear discriminant and quadratic discriminant analysis showed little difference in the results and some classes had to be omitted because of an insufficient number of training sites.

To overcome some of the problems of applying classification methods in previous studies, we employed the statistical method, Random Forests (Breiman 2001) as employed by Harris and Grunsky (2015) and used as part of a remote predictive mapping strategy (Harris et al. 2008). The Random Forest method is based on the construction of classification trees (Venables and Ripley 2002, Chap. 9) in which nodes (splits in classes) are based on continuous variables from which a series of branches in the tree will correctly classify (categorical variables) all of the data. The Random Forest method "grows" many trees and each tree provides a classification. Each classification is termed a vote and a classification is assigned to the forest with the most votes. A useful description of the methodology is provided in Breiman and Cutler (2016). The function "**randomForest**", herein referred to as "RF", from the package **randomForest** (Breiman and Cutler 2016) was used for the analysis.

For each tree that is created, a training set of approximately one-third of the data is drawn, with replacement and are left out of the sample population. This is known as the out-of-bag (oob) data and is used to get a running unbiased estimate of the classification error, as trees are added to the forest. Variable importance is also determined from the out-of-bag data. For each tree, all of the data are applied to the tree and "proximities" are determined for each pair of cases. If two cases occur at the same node, then the proximity of that pair is increased by one. When all of the trees have been estimated, the proximities are normalized by dividing by the number of trees. These proximities can be used for replacing missing data, identifying outliers and creating lower dimensional views of the data. Each tree is constructed from bootstrapping the original sample population and about one third of the data are left out from each bootstrap sample and not used in tree construction but are then classified from the tree created from the other two thirds of the sample population. An unbiased estimate of the classification error is determined from each case that is oob and did not classify correctly. Variable importance is determined by comparing oob classification results and the non-oob classification results after random permutations of each of the variables. Another measure of variable importance is determined by the Gini measure that is determined by the number of splits that are made for a given variable over all of the trees in the forest. Variables do not need to be pre-selected using techniques such as analysis of variance as the RF procedure determines which variables are the best classifiers.

Maps of the normalized votes, which are equivalent to posterior probabilities, can be created using geostatistical methods such as kriging. However, since the posterior probabilities are compositions and sum to 1.0, these values must be logratio transformed, followed by subsequent co-kriging, and then back transformed for subsequent geographic rendering (Pawlowsky-Glahn and Egozcue 2015; Mueller and Grunsky 2016). Instead, maps of the posterior probabilities for each of the classes were created by posting the sample sites with points and colours. An alternative to this would be to consider the un-normalized (raw) votes as independent and carry out kriging on these estimations. The results of these interpolations are provided in the Supplementary Annex.

## 17.3 Results

### *17.3.1 Process Discovery—Principal Component Analysis*

A logcentred transform was applied to the adjusted data after which a principal component analysis was carried out. An examination of an ordered plot of eigenvalues in the form of a screeplot (Jolliffe 2002) are shown in Fig. 17.2a–d for (a) all of the data, (b) Surface Soil, (c) A horizon only and (d) C horizon only. Figure 17.2a–d display two important inflection points; at PCs 3 and 9. The first three eigenvalues define the dominant structure in the data and the next 5 display lesser but significant structure also. This is also expressed numerically in Table 17.1 where the first 10 eigenvalues are listed along with the associated cumulative contribution to the structure in the data. As shown in the screeplots of Fig. 17.2, a comparison of the first four successive eigenvalues between the C-horizon,

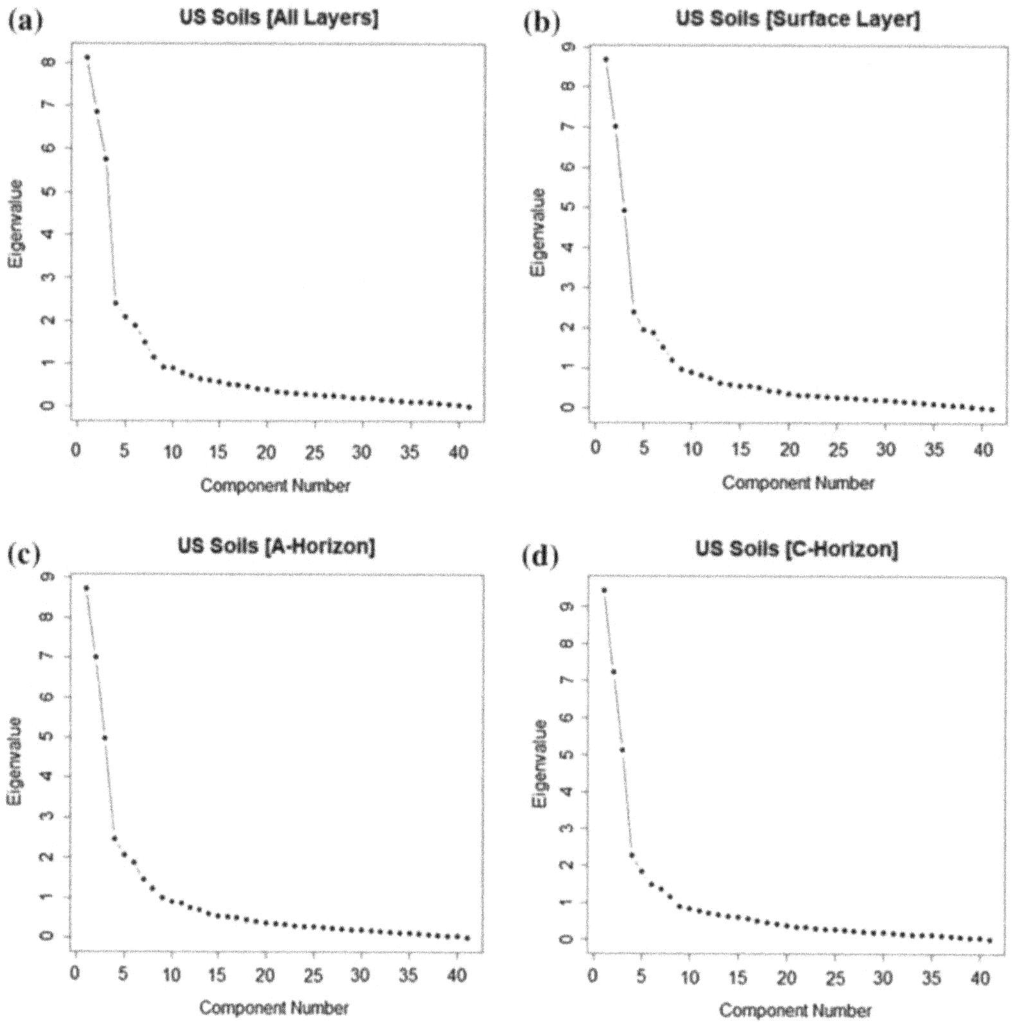


**Fig. 17.2** **a**—Screeplot of eigenvalues of the soil geochemistry for the combined Surface Soil (0–5) cm layer, the A- and C- horizons, from the application of a principal component analysis to logcentred transformed data. **b**—Screeplot of eigenvalues of the soil geochemistry for the Surface Soil (0–5) cm layer from the application of a principal component analysis to logcentred transformed data for the top layer only. **c**—Screeplot of eigenvalues of the soil geochemistry for the A-horizon from the application of a principal component analysis to logcentred transformed data for the A-horizon only. **d**—Screeplot of eigenvalues of the soil geochemistry for the C-horizon from the application of a principal component analysis to logcentred transformed data for the C-horizon only

A-horizon and Surface Soil is slightly greater for the C-horizon. This implies that the linear combinations of the elements are stronger for the C-horizon than for the other two. Eigenvalues with values less than 1 and are interpreted to represent under-sampled processes or random effects (noise).

The largest eigenvalues signify that the linear combinations of the elements for these components are significant and defines "structure" in the data. This structure can be interpreted as the influence of stoichiometric control of mineralogy.

**Table 17.1** Principal Component Analysis results for logcentred transformed soil geochemistry

| RQPCA [clr] All layers | | | | | | | | | | |
|---|---|---|---|---|---|---|---|---|---|---|
| | PC1 | PC2 | PC3 | PC4 | PC5 | PC6 | PC7 | PC8 | PC9 | PC10 |
| λ | 8.13 | 6.87 | 5.76 | 2.39 | 2.08 | 1.88 | 1.50 | 1.15 | 0.92 | 0.89 |
| λ% | 19.83 | 16.76 | 14.05 | 5.83 | 5.07 | 4.59 | 3.66 | 2.80 | 2.24 | 2.17 |
| Σλ% | 19.83 | 36.59 | 50.63 | 56.46 | 61.54 | 66.12 | 69.78 | 72.59 | 74.83 | 77.00 |
| RQPCA [clr] surface soil | | | | | | | | | | |
| | PC1 | PC2 | PC3 | PC4 | PC5 | PC6 | PC7 | PC8 | PC9 | PC10 |
| λ | 8.70 | 7.01 | 4.93 | 2.41 | 1.96 | 1.89 | 1.53 | 1.21 | 0.98 | 0.90 |
| λ% | 21.19 | 17.08 | 12.01 | 5.87 | 4.77 | 4.60 | 3.73 | 2.95 | 2.39 | 2.19 |
| Σλ% | 21.19 | 38.27 | 50.28 | 56.15 | 60.93 | 65.53 | 69.26 | 72.20 | 74.59 | 76.78 |
| RQPCA [clr] A horizon | | | | | | | | | | |
| | PC1 | PC2 | PC3 | PC4 | PC5 | PC6 | PC7 | PC8 | PC9 | PC10 |
| λ | 8.73 | 7.00 | 4.97 | 2.47 | 2.07 | 1.88 | 1.45 | 1.22 | 1.00 | 0.90 |
| λ% | 21.29 | 17.07 | 12.12 | 6.02 | 5.05 | 4.59 | 3.54 | 2.98 | 2.44 | 2.20 |
| Σλ% | 21.29 | 38.37 | 50.49 | 56.51 | 61.56 | 66.15 | 69.68 | 72.66 | 75.10 | 77.29 |
| RQPCA [clr] C horizon | | | | | | | | | | |
| | PC1 | PC2 | PC3 | PC4 | PC5 | PC6 | PC7 | PC8 | PC9 | PC10 |
| λ | 9.45 | 7.22 | 5.12 | 2.29 | 1.84 | 1.50 | 1.36 | 1.17 | 0.89 | 0.82 |
| λ% | 23.02 | 17.59 | 12.47 | 5.58 | 4.48 | 3.65 | 3.31 | 2.85 | 2.17 | 2.00 |
| Σλ% | 23.02 | 40.61 | 53.08 | 58.66 | 63.14 | 66.80 | 70.11 | 72.96 | 75.13 | 77.13 |

### 17.3.2 *PCA of the Combined Surface Soil, A-Horizon, C-Horizon*

Figures 17.3a, and 17.4a shows biplots (PC1-PC2 and PC2-PC3) for the principal component scores and loadings for the combined data from the surface soil, A- and C-horizons Table 17.1 shows that the first three principal components for the combined data (All Layers) account for 50.6% of the overall variation in the data.

Figure 17.3a shows the mass of data points defined by two vertices: (1) Cr-V-Ni-Co-Fe-Sc-Mn-P-Zn; (2) Hg-In-Ti-Se-Mo-As-Sb-Sn-Bi (chalcophile) and a trend of element associations: Mg-Ca-Na-Sr-Ba-K-Be-Rb-Tl that are inversely associated with the vertex defined by (2) above. The chalcophile elements are grouped along the +PC1 axis. Siderophile elements are associated with the +PC2 axis and the lithophile elements are distributed around the ±PC1/−PC2 axes and the −PC1/+PC2 axes.

Figure 17.4a shows the three sets of data (Surface Layer, A- and C-horizon) combined onto a biplot of PC2–PC3. The PC scores along the PC2 axis define a contrast between mafic (+ scores) and felsic (−scores) source material. Siderophile (Fe, Co, Ni), lithophile (Cr, V, Sc, Ti) and chalcophile elements (Cu, In) are associated along the +PC2 axis and lithophile elements (Rb, K, Tl, Ba, Th, La, Be, Ce) are concentrated along the −PC2 axis.

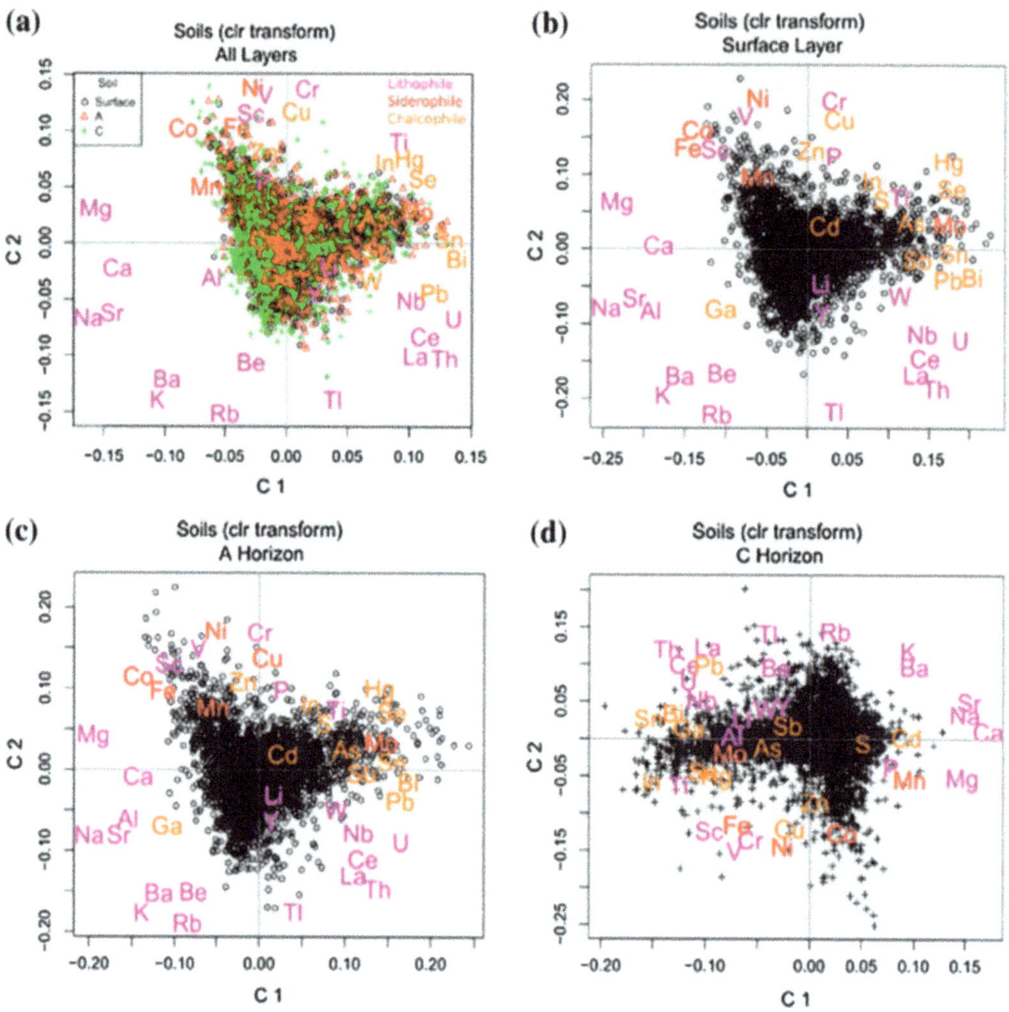


**Fig. 17.3** **a**—Biplot of principal components 1 and 2 for the soil geochemistry for the combined Surface Layer, A, and C horizon soil geochemical data based on a log centred transform. The colours and symbols represent the surface soil and the soil A and C horizons. **b**—Biplot of principal components 1 and 2 for the Surface Soil geochemistry data based on a log centred transform. **c**—Biplot of principal components 1 and 2 for the A-horizon soil geochemistry data based on a log centred transform. **d**—Biplot of principal components 1 and 2 for the C-horizon soil geochemistry data based on a log centred transform

An association of chalcophile elements (Cd, S, Sb, As, Hg, Pb) occurs along the +PC3 axis with a corresponding concentration of sample sites associated with the surface layer and A-horizon, most likely representing complexing with organic rich soils. PC scores for the C-horizon are concentrated along the ±PC2 axis, which may represent a range of source material from mineral soils that are low in organic material (−PC3) to soils that are rich in organic material or derived from shales/ weathered materials (+PC3).

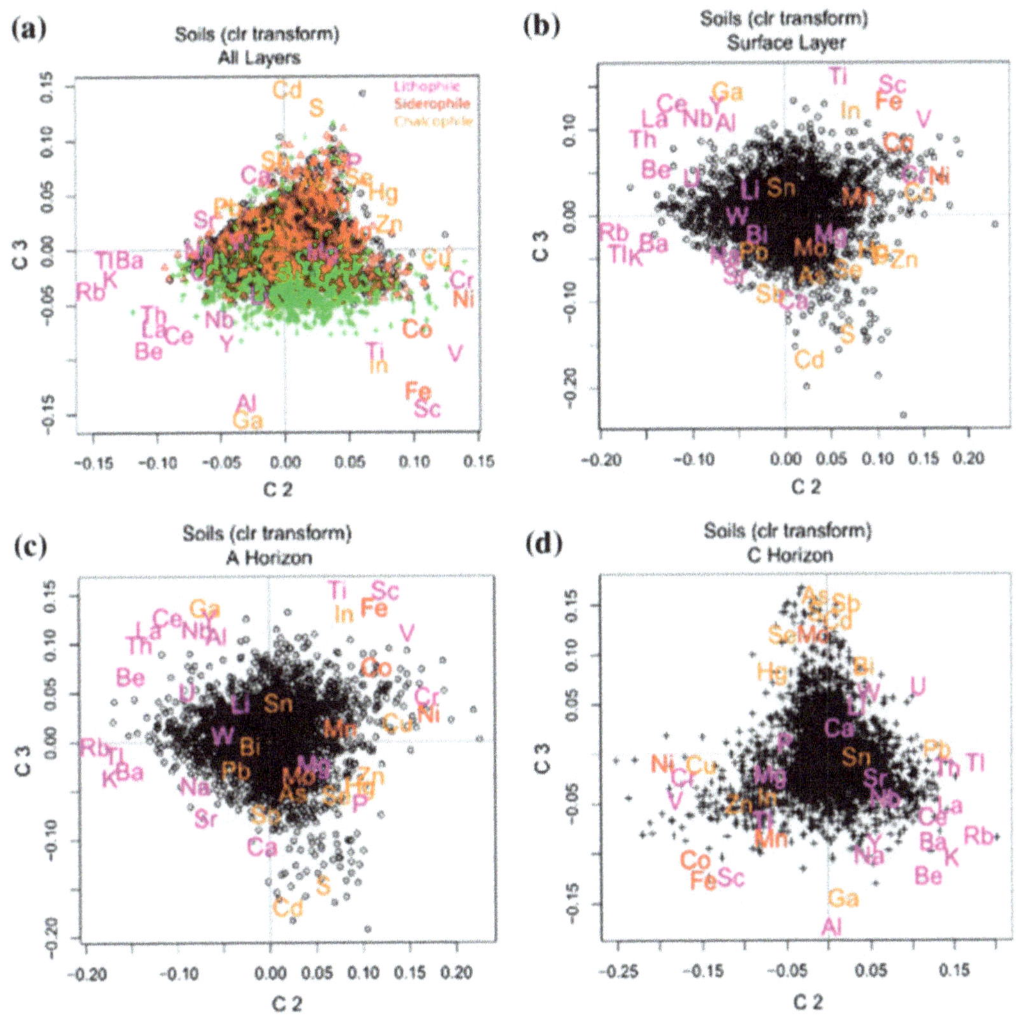


**Fig. 17.4** **a**—Biplot of principal components 2 and 3 for the soil geochemistry for the combined Surface Soil, A, and C horizon soil geochemical data based on a log centred transform. The colours and symbols represent the surface soil and the soil A and C horizons as shown in Fig. 17.3a. **b**—Biplot of principal components 2 and 3 for the top layer soil geochemistry data based on a log centred transform. **c**—Biplot of principal components 2 and 3 for the A-horizon soil geochemistry data based on a log centred transform. **d**—Biplot of principal components 2 and 3 for the C-horizon soil geochemistry data based on a log centred transform

### *17.3.3 PCA of the Surface Soil, A-Horizon, C-Horizon*

The biplots of Fig. 17.3a–c for all of the data, the surface soil data and the A-horizon data, show similar patterns in terms of the relationships of the elements with each other and the shape of the data cloud for the projection of the principal component scores onto the PC1 and PC2 axes. The biplots exhibit a range of lithophile loadings that define materials derived from mafic, feldspathic, carbonate and REE-enriched sources within the quadrants described previously. Similarly, the chalcophile element association is concentrated along the +PC1 axis for both

Fig. 17.3b, c, likely representing weathered and organic-rich material, which adsorb chalcophile elements.

The biplot of Fig. 17.3d (C-horizon) displays a different pattern in comparison with Fig. 17.3a–c. The +PC1 axis shows an association of lithophile elements (Ca-Mg-Na-Sr-P) and chalcophile elements (S-Cd), possibly representing a mix of feldspathic and/or carbonate source material. Along the PC1 axis and on the +PC2 domain, there is a contrast between (Ca-Na-Mg-S-Ba-K) and (Th-Ce-U-La-Nb-Al-Li) that may reflect a feldspathic/carbonate source environment from an environment with relative enrichment in heavy minerals.

Figure 17.4a shows a pattern and association of elements that displays a contrast of the C-horizon data with the surface soil and A-horizon data. Figure 17.4a shows a siderophile and mafic lithophile pattern of Cr-Ni-Cu-V-Co-Fe-Sc along the +PC2 axis. Along the −PC2 axis of Fig. 17.4a there is a lithophile association of Rb-K-Ti_Ba-Ce-La-Tl. The +PC3 axis in Fig. 17.4a shows a chalcophile/lithophile association of Cd-S-Sb-Ca-P-Se-Hg-As-Mo-Pb-Sr-Zn. This region of the plot is dominated by surface soil and A-horizon data although some C-horizon data are also present. A similar pattern is observed in Figs. 17.4b, c although the groups of the elements are at opposite ends of PC3 (a sign switch). In Fig. 17.4b, c, transitional between the siderophile/lithophile elements (Fe-Sc-Co-Cr-Ni) and the lithophile elements (Rb-Tl-K-Ba) is the grouping of Al-Ga-Nb-Y-Ce-La-Th-U that represents feldspars, clays and heavy minerals. As in Figs. 17.3d and 17.4d, representing the C-horizon data, shows the chalcophile enrichment trend along the +PC3 axis and a siderophile/lithophile trend along the PC2 axis. Transitional between the trend along the PC2 axis is an association of Al-Ga, likely representing feldspars and clays.

### *17.3.4 Mapping the Components*

The first three principal components for the surface soil, the A- and the C-horizons were interpolated using the geostatistical package, gstat (Pebesma 2004). Experimental semi-variograms were generated followed by variogram model fitting with subsequent kriging. The images for the three principal components are shown in Figs. 17.5a–c, 17.6a–c and 17.7a–c.

**Principal Component 1**

Geospatially these patterns are observed in Figs. 17.5, 17.6 and 17.7. Figure 17.5a–c show interpolated images based on kriging of the first principal component for the surface soil, A- and C-horizons respectively. The patterns observed in Fig. 17.5a and b are consistent with the patterns observed in Fig. 17.3b and c. The +PC1 axis in Fig. 17.3b and c show relative enrichment of the previously identified chalcophile elements and relative enrichment of the mafic lithophile and siderophile elements along the −PC1 axis. In Fig. 17.5a and b, the positive scores of PC1 appear to correspond with the region in the southeast US and the negative scores of PC1

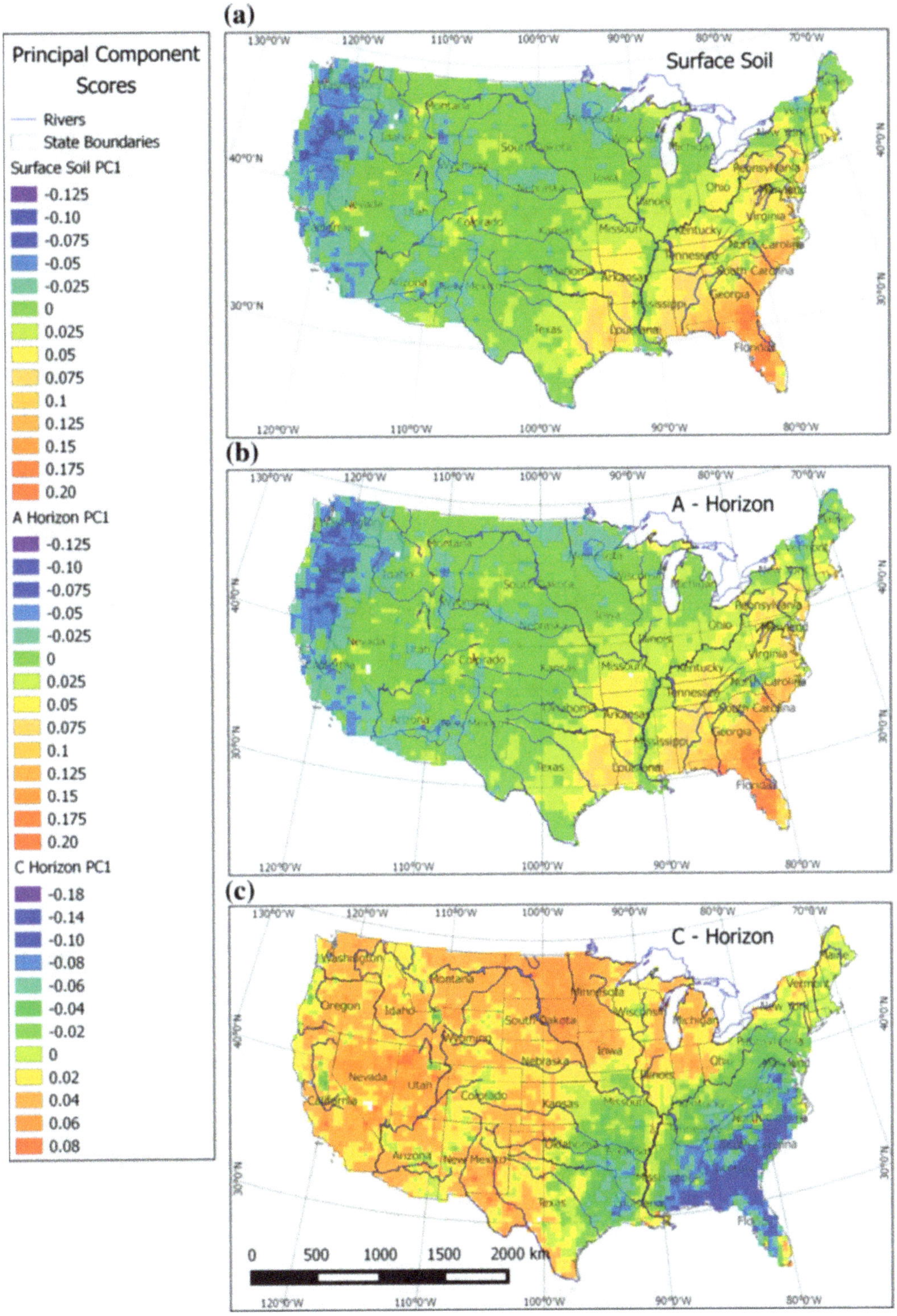


**Fig. 17.5** **a–c** Map of kriged principal component 1 for the Surface Soil, A- and C-horizon data. Figures 17.4b–d provide the context for relative element enrichment/depletion associated with each of the layers

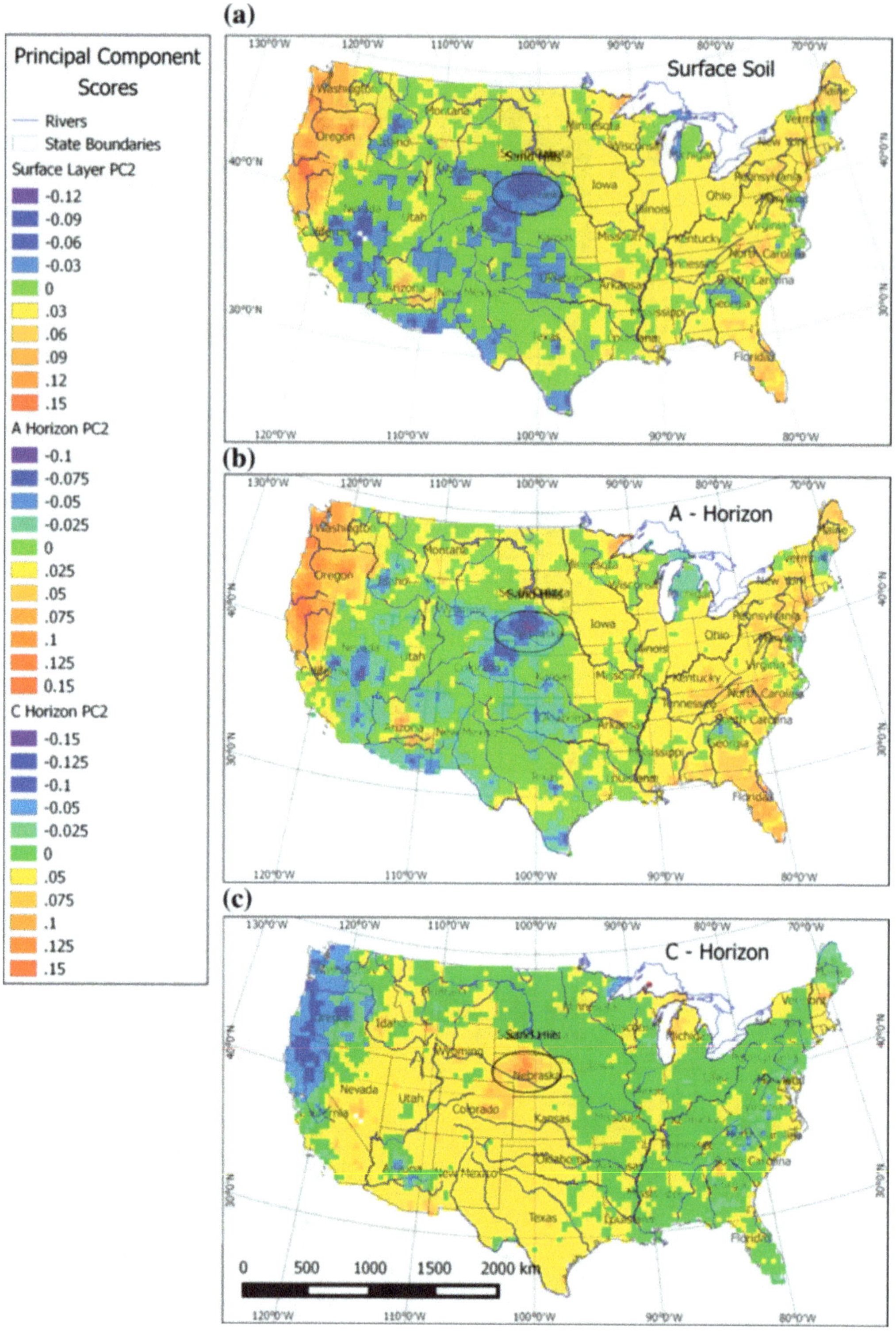


**Fig. 17.6 a–c** Map of kriged principal component 2 for the Surface Soil, A- and C-horizon data. Figures 17.4b–d provide the context for relative element enrichment/depletion associated with each of the layers

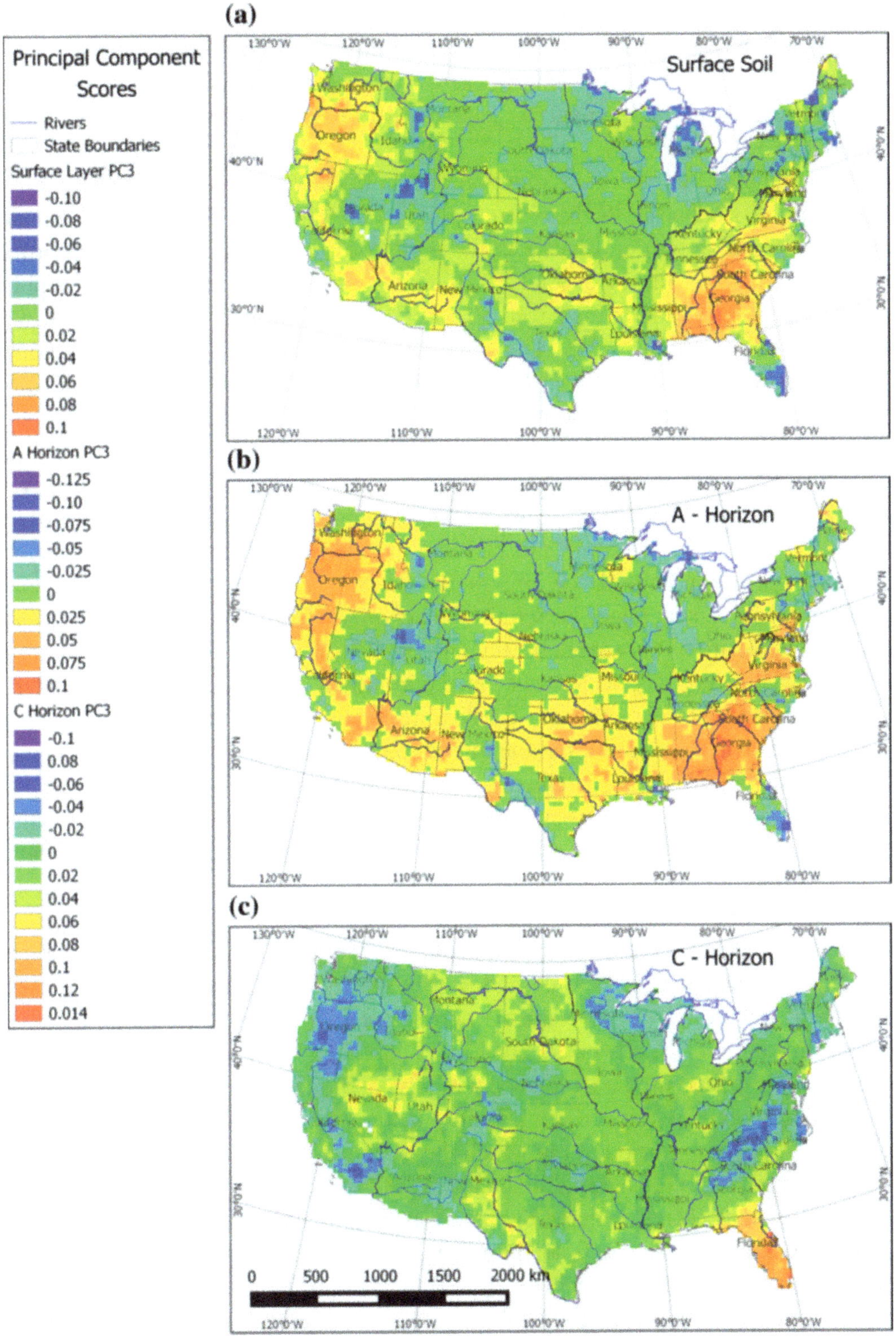


**Fig. 17.7** **a–c** Map of kriged principal component 3 for the Surface Soil, A- and C-horizon data. Figures 17.4b–d provide the context for relative element enrichment/depletion associated with each of the layers

appear to occur in the northwest US and west of Lake Superior. All three figures show a pattern that coincides with the banks of the Mississippi River. Negative PC1 scores for the surface layer and A-horizon correspond to relative enrichment in Na-Sr-Al-Ca-Mg-K-Ba element associated with feldspars and/or carbonate source material.

The image of PC1 for the C-horizon data (Fig. 17.5c) shows a strong negative region in the southeast US that corresponds to the chalcophile group of elements along the negative portion of PC1 in the biplot of Fig. 17.3d. The positive portion of PC1 in Fig. 17.3d corresponds to the dominantly lithophile and siderophile groups of elements and is displayed as a large region throughout the US, with the exception of the southeast US. The same "corridor" pattern along the Mississippi River is observed in Fig. 17.5c, for the C-horizon results and represent the same relative concentration of lithophile elements observed in the surface layer and A-horizon.

Figure 17.5c shows the kriged image for the first principal component derived from the C-horizon data. In this case, the negative scores are restricted to the eastern US and reflect the chalcophile and rare earth elements indicative of detrital heavy minerals corresponding to the region of quartz enrichment accompanied with weathered and detrital materials within the erosional and weathering domain of the eastern US. Positive PC1 scores reflect a lithophile association of Ca-Na-Sr-Cd-Mg-Ba-K-Mn (Fig. 17.3d) and suggest an environment that is likely dominated by Ca-Na-K-Ba-Sr feldspars and Mg-Ca bearing ferromagnesian minerals.

An important consideration in the interpretation of the biplots is the significance of the associations of the elements. An initial interpretation of the biplots of Fig. 17.3a–d was that the associations of the chalcophile groups indicated relative enrichment of these elements (Hg-Se-As-Sb-Sn-Bi-Pb-S-In) that represent weathered materials along with the accumulation of detrital minerals within the erosional and weathering domain of the southeastern US. In fact, these elements do not reflect relative enrichment but rather relative depletion with respect to the other groups of elements, notably the siderophile and lithophile elements. Geospatially, the chalcophile association of these elements corresponds to the region of a high quartz content in the soil (Smith et al. 2014) and has been termed the "quartz dilution effect". This effect in the soil geochemistry and the subsequent multi-element associations would likely be significantly different had Si been included in the analysis. A test was carried out in which the Si content of the data was simulated as the difference from the potential total (1,000,000 ppm) from the summed content of the compositions. This simulated Si value was then included in the composition and a PCA was carried out. The first component identified the relative Si enrichment as occurring in the southeast US. The simulated value of Si was not included in this study because other elements should also be considered in a total composition, including oxygen and nitrogen.

**Principal Component 2**

As shown in Fig. 17.3b, c, the multi-element signature of tpc2 is nearly the same for the surface soil and A-horizon. The patterns in both figures show two trends, one with relative enrichment in Cr-Ni-Co-Cu-V-Fe-Sc (siderophile/lithophile + Cu-Zn) and the other with relative enrichment in Hg-Se-As-Sn-Sb-Pb-Bi-In-S. (chalcophile) These two multi-element associations reflect the chemistry of mafic minerals and elements that are associated with weathering and organic complexing. This is reflected in the maps of Fig. 17.6a, b in which high PC2 values are noted in the eastern and south eastern US and the western US. The negative PC scores for the surface soil and A-horizon show relative enrichment in Rb-K-Tl-Ba-Be-Na-Sr-Al-Ga and, as shown in Fig. 17.6a, b are geospatially concentrated in the central US corresponding to the location of the Sand Hills of Nebraska, (~105° W/ 42° N), which is comprised of sand-sized particles of quartz and feldspar (Smith et al. 2014). There are also areas of negative PC2 scores, most likely representing feldspars associated with granitoid rocks in southern Nevada, California, Arizona, Texas, New Hampshire and Maine (Smith et al. 2014).

The map of PC2 (Fig. 17.6c) for the C-horizon data shows positive scores associated with the mafic volcanic rocks of the northwest US and corresponds to the relative enrichment of siderophile (Fe-Ni-Co), lithophile (Cr-V-Sc), chalcophile (Cu-Zn) elements as shown in Figs. 17.3d and 17.4d. The negative scores for PC2 show a similar pattern to those of the surface soil and A-horizon; relative enrichment in alkali lithophile elements (Rb-K-Ba-Be-Na-Sr) with Al-Ga representing feldspars and REE lithophile elements (U-Th-La-Ce-Ng-Tl) that represents heavy minerals and quartz (as explained previously). The geochemical expression of these minerals in PC2, which are resistant to weathering, are reflected in both horizons and the surface soil.

**Principal Component 3**

The positive scores for the PC3 show relative enrichment of siderophile, mafic lithophile, and light REE elements for both the surface soil and A-horizon; whereas this pattern is represented by negative scores for the C-horizon. As shown in Fig. 17.4b–d, for all three layers, there is a continual transition from relative enrichment in alkali lithophile and REE elements, including Al and Ga, representing feldspars and minerals associated with felsic domains to relative enrichment in Cr-Ni-V-Cu-C-Fe-Sc-Ti-In-Zn that represents minerals associated with mafic domains. Figures 17.7a–c show the kriged images for the third principal component. The negative scores show relative enrichment of Cd-S-Ca-Sr-Sb-P-As, which may reflect the processes of organic complexing and sulphates. Negative scores noted in Utah, Nevada, west Texas, the Mississippi delta and south Florida may have a greater component of S. Negative scores that occur in Minnesota, Michigan, Indiana and the coast of New England may reflect the presence of shales, clays and organic accumulations. The negative PC3 scores of Fig. 17.4b exhibit a bimodal pattern of relative enrichment of Fe-Sc-In-Ti and Ga-Al-Y-Nb-Ce-La. The Fe-rich pattern is associated with the mafic volcanic rocks in the northwest and southwest US and the Ga-rich pattern occurs in the eastern US and reflects the presence of feldspars in the weathering of granitoid rocks in the southern Appalachians.

As seen in Fig. 17.4c, and nearly identical to that the of surface soil, the positive scores of PC3 exhibit a bimodal pattern for the A-horizon and indicate relative enrichment of Ti-Sc-Fe-In-V and Ga-Al-Th-La-Nb-Ce. These two groups reflect both a mafic and feldspathic/heavy mineral rich environment. Figure 17.7b shows the mafic association (Ti-Sc-Fe-In-V) in the northwest US. The positive scores in the eastern, southern, and in particular, the southeast US reflect elements associated with feldspars and heavy minerals, which reflects the concentration of minerals through the weathering process, which may be due more to gravitational effects than chemical breakdown. As in Fig. 17.7a, the negative scores of PC3 in the A-horizon demonstrate the same patterns and processes.

The C-horizon map shows two distinct geospatial patterns. The positive scores of Fig. 17.4d show relative enrichment in the chalcophile group, Sb-As-S-Mo-Se-B-Cd-Hg-U-Li-W and occur primarily in the southeast US. This pattern likely reflects both the quartz dilution effect and the presence of chalcophile elements relative to other areas throughout the US. The negative scores, which show relative enrichment of the lithophile elements Al-Ga-Na-Y-K-Be-Ba-Mn-Ti-Fe-Sc-Co, reflect a combination of mafic minerals and feldspars. These patterns are observed in the western US, Minnesota-Wisconsin, central Appalachia and the northeast US. Patterns associated with the elements that reflect mafic domains are the northwest US and Wisconsin-Minnesota. Patterns that reflect the feldspathic domains are Nebraska-Colorado, central Appalachia and the northeast US.

Evaluation of the soil geochemistry for the surface soil, the soil A horizon and the soil C horizon using a principal component approach reveals that there are continental-scale geochemical patterns that appear to be associated with the composition of the underlying soil parent material, climate, and weathering. At the scale of evaluation, details on specific lithologies are difficult to resolve, but the patterns are consistent with those mineralogical patterns delineated by Smith et al. (2014).

***Process Validation Predictive Mapping of Surface Lithologies***

The lithology of surficial materials by Sayre et al. (2009) is represented by 18 classes plus unknowns and listed in Table 17.2. A total of 17 classes were selected for further study. The classes "unknown" and "water" were not used as they were not considered suitable for classification.

Figure 17.8 shows a map of the sampling sites with the surface materials lithology from Sayre et al. (2009). The patterns of surface materials on the map show some similarities with the patterns observed from the first three principal components for the surface soil, A- and C-horizons. Figure 17.9 shows a biplot of the first two principal components that are coded according to the surface lithologies. The pattern of the mafic lithophile elements (Cr-Ni-Cu-V-Co-Fe-Sc) in Fig. 17.9a, b are dominated by silica-rich residual soils (SilRes), whereas the chalcophile enrichment pattern (Hg-Se-Mo-Sn-Bi-Pb-Sb-As-Ti-S-In) appears to be associated mostly with alluvium (Alluv) and coastal zone sediments (CZS). The lithophile element grouping in the negative portion of the PC2 shows a mix of several lithologies. The results of the PCA suggest that the linear combinations of elements from the PCA are related to the patterns observed in Surface Materials Lithologies of Fig. 17.8.

**Table 17.2** List of surface lithologies across the conterminous United States

| Mnemonic | Class description | Surface layer | A-horizon | C-horizon | Total |
|---|---|---|---|---|---|
| AlkInt | Alkaline intrusive/volcanic rocks | 6 | 7 | 6 | 19 |
| Alluv | Alluvium and fine-textured coastal Zone sediment | 994 | 989 | 984 | 2967 |
| CaRes | Carbonate residual material | 265 | 263 | 260 | 788 |
| Colluv | Colluvial sediment | 379 | 379 | 366 | 1124 |
| CZS | Coastal zone sediment, coarse-textured | 44 | 45 | 43 | 132 |
| EolDune | Eolian sediment, coarse-textured (Sand Dunes) | 152 | 151 | 151 | 454 |
| EolLoess | Eolian sediment, fine-textured (Glacial Loess) | 156 | 155 | 155 | 466 |
| ExtVR | Extrusive volcanic Rock | 50 | 51 | 51 | 152 |
| GlLs | Glacial lake sediment, fine-textured | 89 | 85 | 86 | 260 |
| GlOut | Glacial outwash and Glacial lake sediment, coarse-textured | 221 | 220 | 221 | 662 |
| GTCg | Glacial till, coarse-textured | 114 | 111 | 111 | 336 |
| GTClay | Glacial till, Clayey | 61 | 61 | 61 | 183 |
| GTLoam | Glacial till, Loamy | 529 | 528 | 526 | 1583 |
| HyPM | Hydrick peat muck | 25 | 25 | 26 | 76 |
| NCaRes | Non-carbonate residual material | 1188 | 1174 | 1170 | 3532 |
| SalLS | Saline lake sediment | 78 | 82 | 79 | 239 |
| SilRes | Silicic residual material | 457 | 456 | 452 | 1365 |
| Water[a] | | 22 | 21 | 21 | 64 |
| Unknown[a] | | 6 | 6 | 6 | 18 |
| Total | | 4836 | 4809 | 4775 | 14420 |

[a]Not Used

From the application of the random forest classification, the Gini Index (significance of the variables) for the surface soil, A- and C-horizons are listed in Table 17.3 and shown graphically in Fig. 17.10. The significance uses the Gini Index, which is a measure of purity based on the success of a variable in distinguishing between classes. Table 17.3 shows that generally, PC's 4, 5, 1, 2, 3 and 6 are the best variables for classification of the surface lithologies for the surface soil, A- and C-horizons. Maps of the normalized votes in point form and interpolated (kriged) maps of the raw votes are shown in the Supplementary Annex (Supplementary Figs. 1–15).

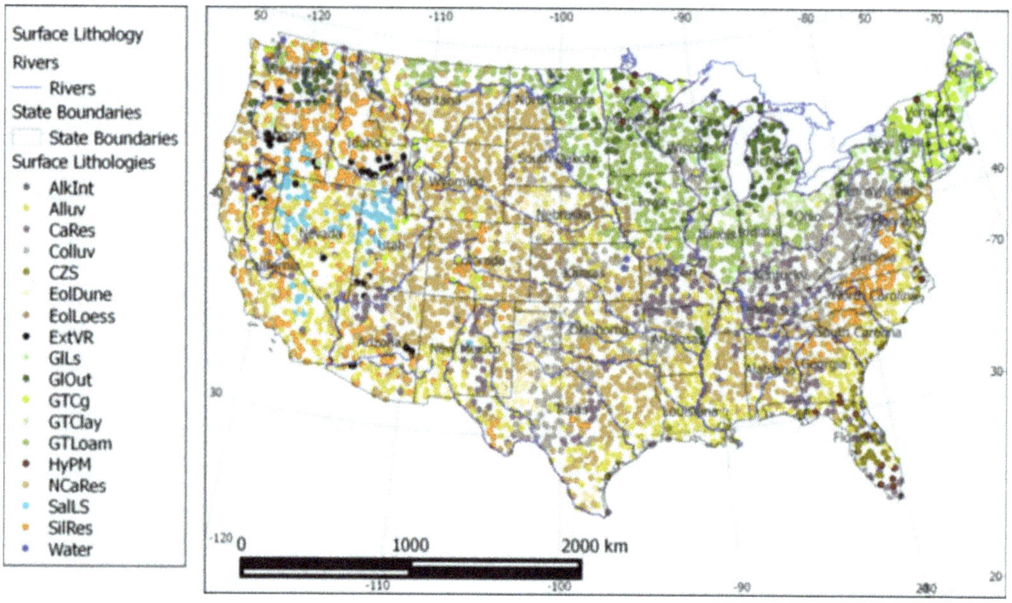


Fig. 17.8 Map of soil sample sites coded by the Surface Lithology classification. This map represents the actual classification based on the maps of Sayre et al. (2009). Colours used in this figure are the same colours used in Sayre's maps. See text for details on how the sites were selected

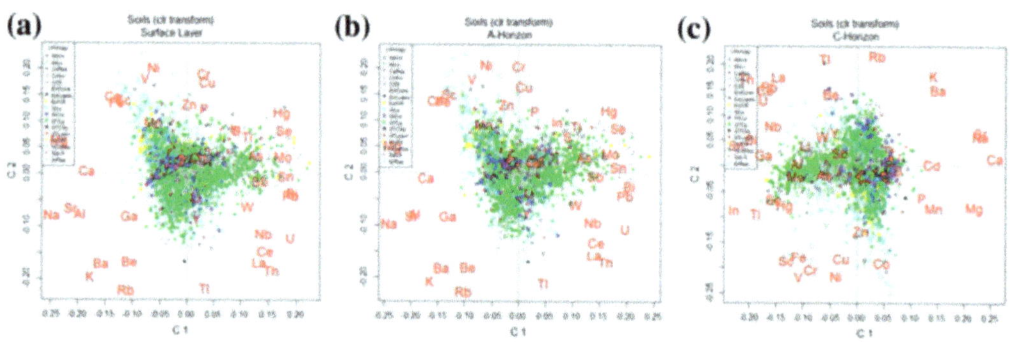


Fig. 17.9 a–c Principal component biplot of the surface layer (a), A-horizon (b) and C-horizon (c) scores that are coded and coloured according to the surface lithologies

Table 17.4 shows the accuracy of prediction for each of the surface lithologies based on the Random Forest out-of-bag classification methodology for each of the surface soil, A- and C-Horizons. The table has been ordered from the highest to the lowest prediction accuracies based on the surface soil. It is worth noting that the depth of soil has only a minor influence in the prediction accuracies, suggesting that the geochemical signature of the underlying material persists throughout the soil column. Non-carbonate residual soils (NCaRes) (~74%), loam associated with glacial till (GTLoam) (66–72%), siliceous residual soils (SilRes) (48–56%), alluvium (Alluv) (~50%) and coastal zone sediments (CZS) (45–48%) have the highest prediction accuracies, whereas the lowest accuracies are associated with hydric peat and muck (HyPM) (0%), alkalic intrusions (AlkInt) (0%), glacial lake sediments (GlLs)

**Table 17.3** List of variable importance for the surface layer, A- and C-horizons as determined from Random Forest classification of the principal component results applied to the clr-transformed data. Colours reflect the most significant PCs (red) to least significant PCs (blue)

| Surface Layer PC | Importance | A Horizon PC | Importance | C Horizon PC | Importance |
|---|---|---|---|---|---|
| 4 | 198.35 | 4 | 185.34 | 2 | 165.83 |
| 5 | 180.88 | 5 | 172.04 | 4 | 156.36 |
| 1 | 163.70 | 1 | 170.09 | 1 | 154.76 |
| 2 | 155.81 | 3 | 150.05 | 3 | 152.11 |
| 3 | 152.73 | 2 | 148.14 | 6 | 131.94 |
| 9 | 129.17 | 9 | 127.18 | 16 | 128.21 |
| 12 | 109.74 | 6 | 126.50 | 5 | 115.55 |
| 6 | 108.61 | 20 | 119.91 | 11 | 113.04 |
| 32 | 106.28 | 29 | 110.74 | 8 | 109.54 |
| 23 | 102.15 | 13 | 102.50 | 7 | 109.05 |
| 30 | 100.84 | 12 | 100.25 | 10 | 106.46 |
| 8 | 98.87 | 11 | 98.07 | 14 | 101.91 |
| 20 | 98.77 | 8 | 96.86 | 13 | 96.87 |
| 40 | 98.19 | 23 | 94.89 | 12 | 96.57 |
| 10 | 96.90 | 10 | 94.48 | 28 | 95.32 |
| 21 | 95.83 | 7 | 93.99 | 9 | 94.97 |
| 11 | 93.68 | 16 | 92.35 | 18 | 93.88 |
| 15 | 91.76 | 18 | 92.21 | 17 | 92.57 |
| 24 | 91.73 | 19 | 91.46 | 31 | 92.53 |
| 13 | 91.49 | 21 | 89.23 | 34 | 92.34 |

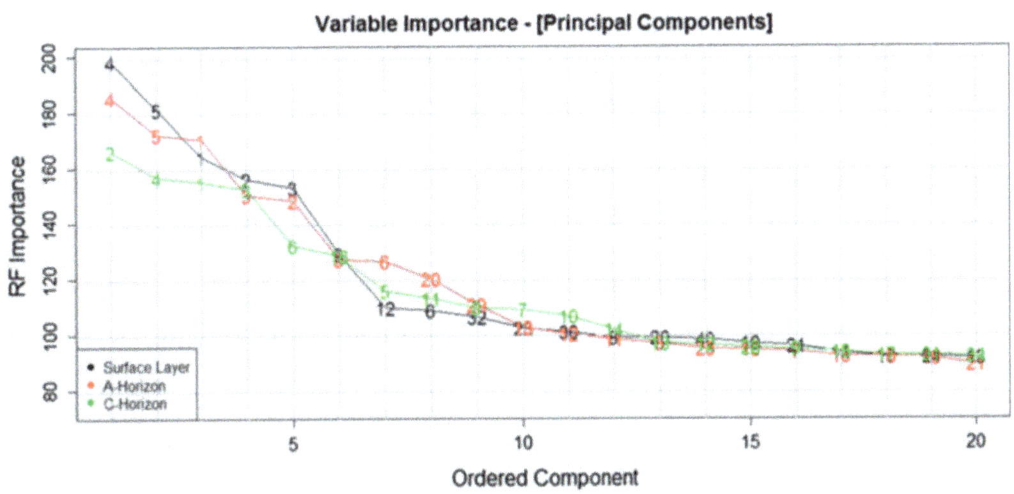


**Fig. 17.10** Plot of the significance of the principal components used in the random forest classification based on the Gini Index for the Surface Layer, A- and C-horizons. See the text for a detailed explanation

**Table 17.4** Measures of ordered predictive accuracy for the surface lithologies for the surface layer, the A- and C-horizons based on a Random Forest classification of the principal component results applied to the clr-transformed data

| | Surface Layer | A-Horizon | C-Horizon |
|---|---|---|---|
| NCaRes | 74.82 | 73.66 | 74.51 |
| GTLoam | 71.61 | 68.90 | 65.74 |
| SilRes | 52.02 | 47.97 | 55.92 |
| Alluv | 50.38 | 50.63 | 49.77 |
| CZS | 44.90 | 48.34 | 48.26 |
| Colluv | 37.14 | 38.20 | 32.73 |
| GlOut | 28.41 | 32.63 | 29.32 |
| GTClay | 27.54 | 37.32 | 42.23 |
| GTCg | 22.65 | 21.47 | 20.57 |
| EolDune | 22.25 | 22.40 | 16.47 |
| EolLoess | 21.05 | 26.97 | 30.19 |
| CaRes | 19.19 | 15.16 | 9.58 |
| SalLS | 15.22 | 15.69 | 1.25 |
| ExtVR | 1.96 | 5.78 | 0.00 |
| GlLs | 1.11 | 0.00 | 0.00 |
| AlkInt | 0.00 | 0.00 | 0.00 |
| HyPM | 0.00 | 0.00 | 0.00 |
| Overall Accuarcy | 49.92 | 49.37 | 48.61 |

(0–1%) and extrusive volcanics (ExtVR) (0–6%). The prediction accuracy is sensitive to the initial representation of each class in the dataset. This sensitivity is partly due to the masking and swamping effect that a large population of sites for one type of surface lithology over another (i.e. Alluvium vs. Hydric Peat and Muck).

Supplementary Tables 2, 3 and 4 provide a complete summary of the prediction accuracies for the surface soil, A- and C-horizons, respectively. The diagonal of each upper table (Tables 2a, 3a, 4a) indicates how many sample sites were classified correctly. Each row of the off-diagonal elements indicates the misclassification of the sites for each of the classes. The lower tables in Tables 2b, 3b, 4b show the classification accuracies as expressed in percentages. The overall classification accuracy is shown at the bottom of each table. Scanning the columns of Tables 2a, 3a, and 4a reveals that many classes are confused with alluvium (Alluv), siliceous residual material (SilRes), loam derived from glacial till (GTLoam) and non-carbonate residual material (NCaRes). Alluvium and non-carbonate residual material appear to overlap with almost all of the classes. The overall prediction accuracies for the surface soil, A- and C-Horizons are 50%, 49% and 49%, respectively.

The R package "**randomForests**" produces raw and normalized votes for each of the classes. Votes are a record of the number of times a site is correctly classified. As described above, normalized votes are the equivalent of a posterior probability and are therefore compositions. Classes such as AlkInt, HyPM and other classes that have low abundance in the data create problems in the creation of co-regionalization that is required for co-kriging. Examples of the spatial distribution of the normalized and raw votes are shown below. The Supplementary Annex provides predictive maps for all of the surface lithologies, based on the normalized votes, for the surface soil, A- and C-horizons. Predictive maps for AlkInt and HyPM are not shown because the normalized votes for these two surface lithologies were very low and do not show any geospatial patterns. The prediction accuracies for the three media from Table 17.4 are: 49.9%, 49.4% and 48.6% respectively. Supplementary Tables 2, 3 and 4 provide details on the overlap of predictions for each surface lithology. In most cases, overlap is associated with non-carbonate residual soils, glacial till derived loam and alluvium. These three classes have the broadest range of compositional variation and occupy a significant amount of area across the conterminous US.

Figure 17.11 shows a map of normalized votes of Non-carbonate residual soils (NCaRes) derived from the random forest classification. Normalized votes >0.3 occur throughout the Midwest states from the Canadian border in the north to the Gulf of Mexico in the south. From Table 17.4, the overall classification accuracy is approximately 75% for the surface soil and the two soil horizons. Supplementary Tables 2, 3 and 4 show that compositional overlap occurs primarily with alluvium, which is also shown in the maps of Fig. 17.11 where a large number of sample sites show low normalized votes (~0.2–0.3). Supplementary Fig. 13a, b show the normalized and raw vote maps of the NCaRes prediction.

Figure 17.12 shows a map of normalized votes for loam derived from Glacial Till (GTLoam). The overall classification accuracy ranges from 65.7 to 71.6% over the three soil layers. Supplementary Tables 2, 3 and 4 show the overlap of the GTLoam composition is associated with non-carbonate residual material (NCaRes) and alluvium (Alluv) for the surface soil, A- and C-horizons (Supplementary Tables 2, 3, 4). The pattern of elevated normalized votes coincides with the region described by Sayre et al. (2009) that is located in the north central US and south of the Great Lakes. The pattern of elevated GTLoam follows the course of the Mississippi River, which highlights the erosional path of this material. Supplementary Figs. 12a, b show the normalized and raw vote maps of the GTLoam prediction.

Normalized votes for the prediction of alluvium (Alluv) are shown in Fig. 17.13 (Supplementary Fig. 1). The overall prediction accuracy is ~50% (Table 17.4) and compositional overlap is observed with the surface lithology non-carbonate residual soil (NCaRes) (Supplementary Tables 2, 3, 4). High predictions of alluvium are located in Nevada, western Texas and the southeast US states. The dispersed prediction of 0.2–0.3 represents the regions of compositional overlap with NCaRes, which can be seen on the map of Fig. 8. Supplementary Figs. 1a, b show the normalized and raw vote maps of the Alluv prediction and supplementary Figs. 13a, b show the normalized and raw votes of the NCaRes prediction.

Figure 17.14 shows prediction based on the normalized votes for the Eolian Dunes (EolDune) of Nebraska, southward into Texas. The patterns are the same for the surface soil, A- and C-horizon maps. The highest values of normalized votes

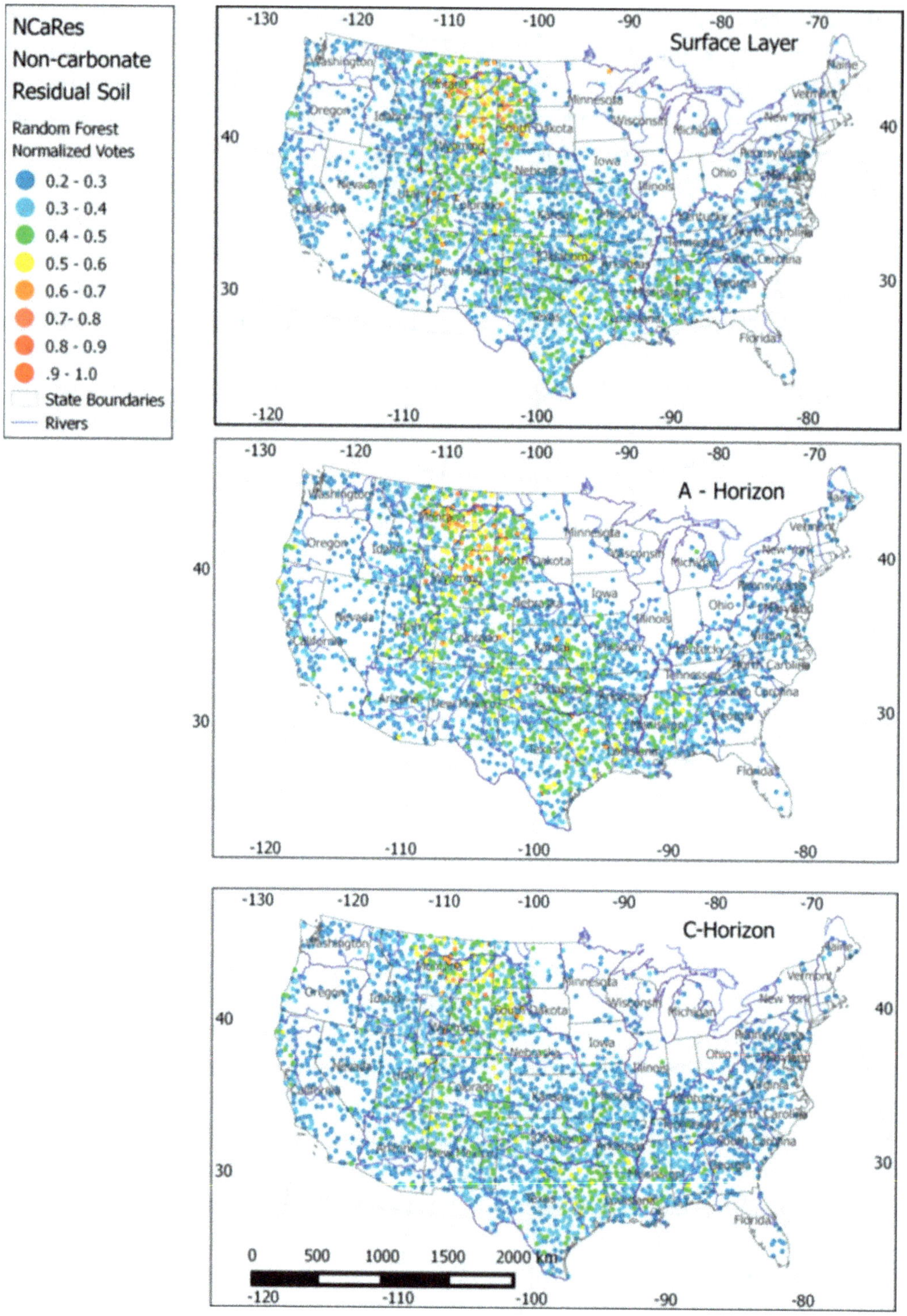


Fig. 17.11 Map of normalized votes for the surface lithology class, non-calcium residual soil (NCaRes). Sites with a normalized vote of less than 0.2 are omitted

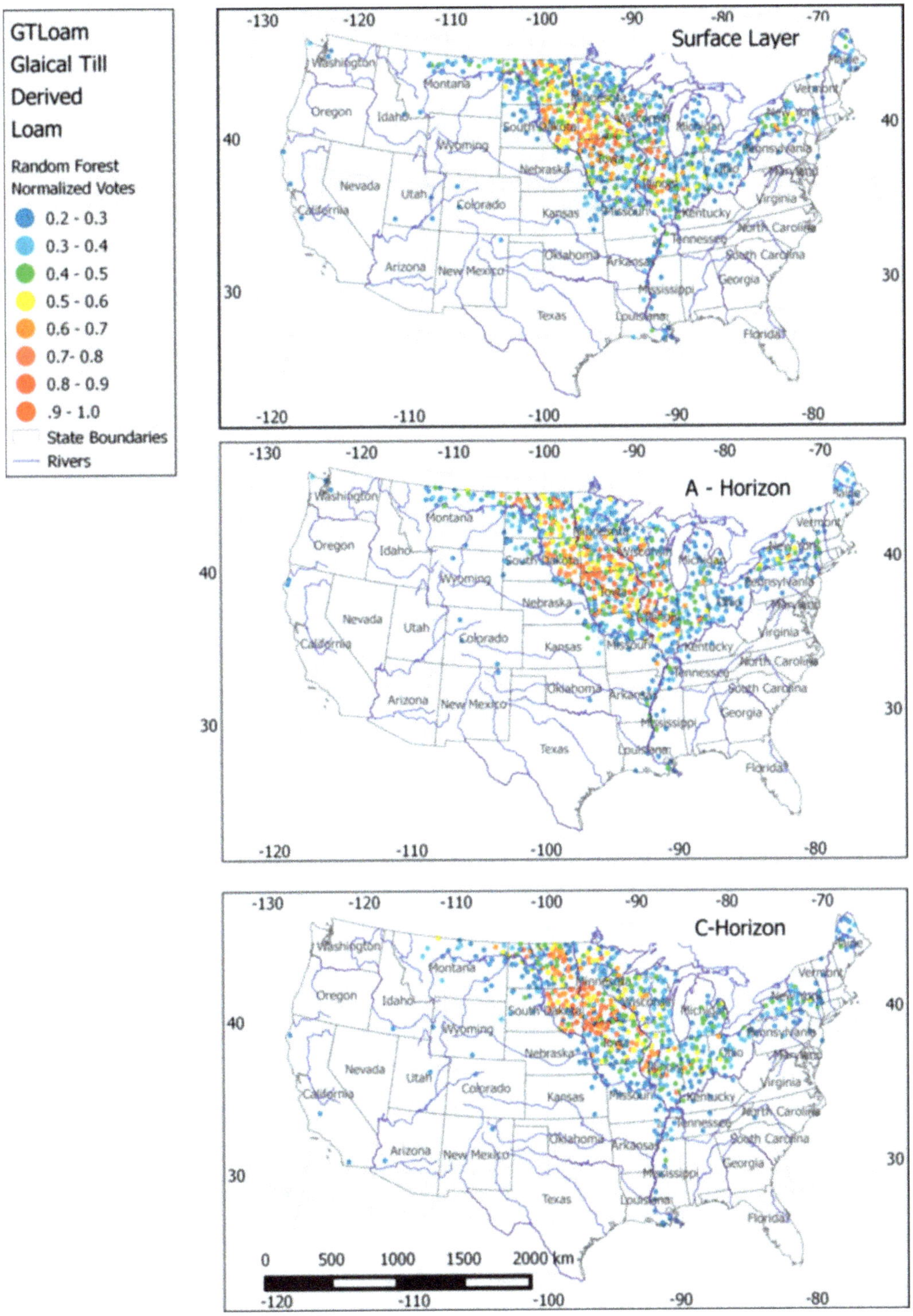


**Fig. 17.12** Map of normalized votes for the surface lithology class, loam derived from glacial till (GTLoam). Sites with a normalized vote of less than 0.2 are omitted

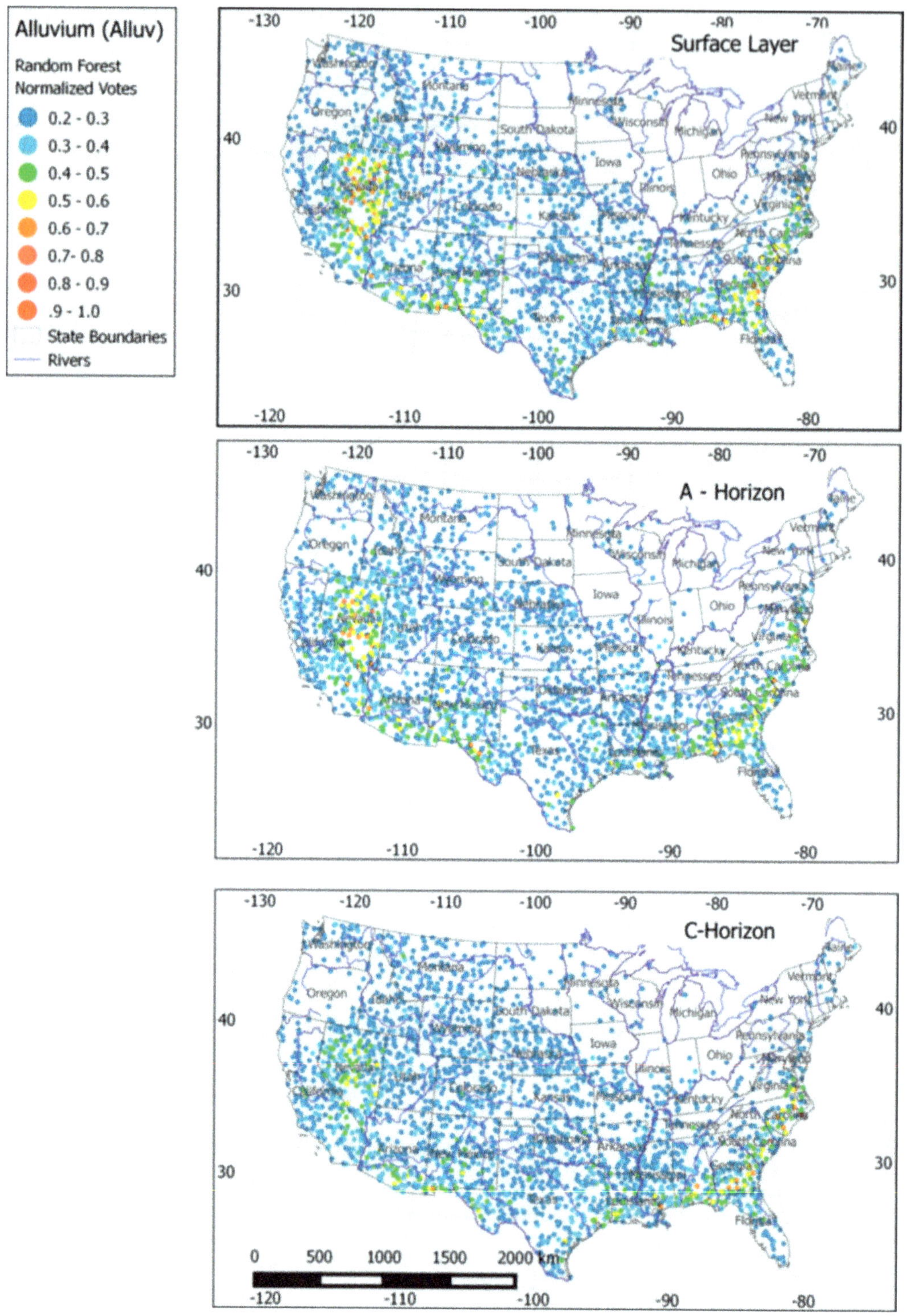


**Fig. 17.13** Map of normalized votes for the surface lithology class, alluvium (Alluv). Sites with a normalized vote of less than 0.2 are omitted

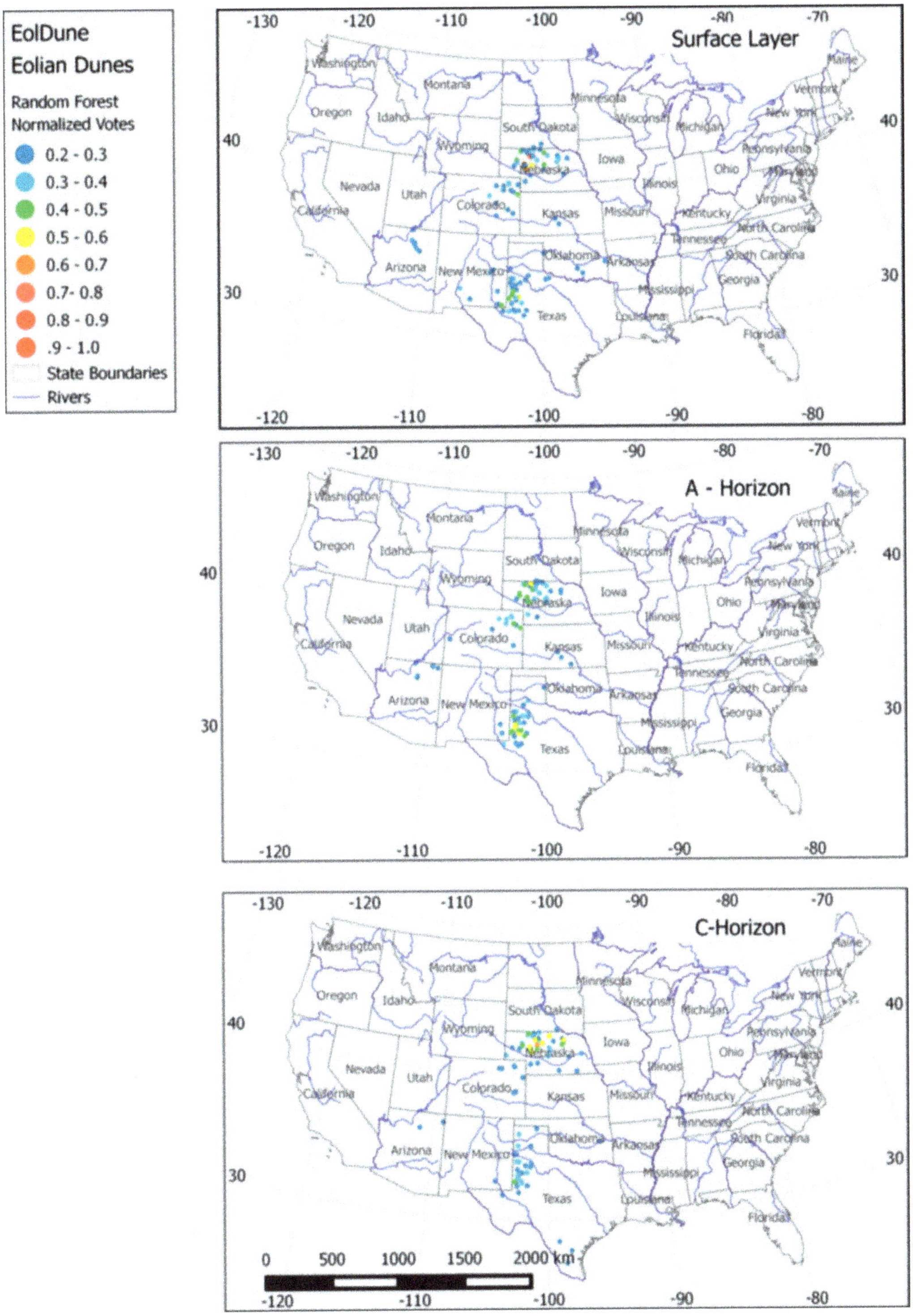


Fig. 17.14 Map of normalized votes for the surface lithology class, eolian dunes (EolDune). Sites with a normalized vote of less than 0.2 are omitted

occur in Nebraska and west-central Texas. The map of Sayre et al. (2009) shows EolDune in northern Texas and the Oklahoma Panhandle, although these two regions are not predicted in the surface soil, A- or C-Horizon results. Table 17.4 shows predictive accuracies of 22.3, 22.4 and 16.5% for the surface soil, A- and C-horizons, respectively. Supplementary Tables 2, 3 and 4 show that compositional overlap occurs with alluvium (Alluv) and non-carbonate residual soil (NCaRes). Supplementary Figs. 5a, b show the normalized and raw vote maps of the EolDune prediction.

The effects of erosion and subsequent re-deposition along the banks of the Mississippi River is observed for several of the surficial lithologies. NCaRes, CaRes and Colluv exhibit an erosional pattern along the Mississippi River, while EolLoess, GlLS, GlOut and GTLoam exhibit depositional patterns. This suggests that the recent deposition of the sediments along the banks of the Mississippi River has modified the composition of the upper layers of the soil. These classes (EolLoess, GlLS, GlOut, GTLoam—Supplementary Figs. 6a, b, 8a, b, 9a, b, 12a, b) show a distinct compositional presence down the length of Mississippi River starting from the northern Midwest states and reflecting continued transport of these materials at a continental scale.

A brief description of the maps for the surface soil, A and C-horizon data that are displayed in the Supplementary Annex are discussed in the section, Supplementary Material.

## 17.4 Discussion

Examination of the principal component biplots (Figs. 17.3 and 17.4) show that the multi-element patterns are very similar for the surface soil and A-horizon data. The C-horizon biplots show similar multi-element groupings, but the shape of the point patterns (Figs. 17.3d and 17.4d) are different from those of the surface soil and A-horizon (Figs. 17.3b, c and 17.4 b, c). As described previously, the element groupings for the three sampling layers are:

(1) Group 1: Tl-Rb-Be-Ba-K-Ga-Al-Sr-Na-Ca-Mg [felsic and mafic lithophile elements (silicates)]
(2) Group 2: Ni-Cr-V-Fe-Sc-Co-Cu-Zn-Mn [Ferromagnesian silicates and clays]
(3) Group3: Hg-Se-Mo-Sn-Bi-Pb-Sb-As-Ti-S-In. [Shales and organic material with adsorbed elements]

These associations are slight variants on Goldschmidt's classification of elements; lithophile (Group 1), siderophile (Group 2) and chalcophile (Group 3).

The principal component biplots, along with the maps of the dominant principal components (Figs. 17.5, 17.6 and 17.7), indicate that there is strong stoichiometric and geospatial control on the patterns that are observed. These patterns, both in the biplots and the kriged map images, provide the justification to use the soil geochemical data to predictively map (validate) the surface lithology classification of Sayre et al. (2009). It should be noted that Sayre's map of surface lithologies does not distinguish lithologies with different mineralogies, and, hence there is considerable overlap between some of the classes defined by Sayre.

The results of the random forest classification show that for most of the surface lithology classes, the accuracy of prediction and spatial coherence of the predicted sites is variable, as shown in Table 17.4 and Figs. 17.11, 17.12, 17.13 and 17.14 and the Supplementary Tables and Figures. The surface lithologies with the lowest predictions are: Hydric Peat and Muck (HyPM), Alkalic Intrusives (AlkInt), Glacial Lake Sediments (GlLS), Extrusive Volcanic Rocks (ExtVR) and Saline Lake Sediments (SalLS). Two factors influence the classification accuracy. The first is the areal extent that a given class occupies. The compositional range of a class of small spatial extent may be swamped or masked by the compositional range of a class that is geographically adjacent to it and has a much larger areal extent. Surface lithologies such as AlkInt, HyPM ExtVr, SalLS and GlLS have limited geospatial extent and the compositions of these lithologies are similar to several other lithologies, including Alluv GTLoam and NCaRes. The second factor that influences the prediction accuracy is the common compositions of several of the surface lithology classes namely, alluvium (Alluv), non-carbonate residual soil (NCaRes), and silica-rich residual soil (SilRes). These surface lithologies are comprised of similar mineralogies and are, therefore, compositionally similar and result in compositional overlap in the statistically based prediction process.

Silicate mineralogy, including quartz, is under-represented in the data used for this study. As discussed previously, the quartz dilution effect has an influence on how the various relationships of the elements are observed, particularly in the methods that are part of the "Process Discovery" component of this study. The absence of silicon in the geochemical analysis in terms of the classifications may have some effect on the ability to distinguish between the different surface lithologies, but the exact effect is unknown at this time and further studies where Si is included and subsequently excluded in process discovery studies are warranted.

The validation of surface lithologies using soil geochemistry highlights some of the limitations on predicting distinct surface lithologies that have similar geochemical compositions but represent different processes. Despite this confusion of compositions between surface lithology classes, the predictive maps render a close representation of the maps of Sayre et al. (2009).

## 17.5 Concluding Remarks

The multi-element soil geochemistry over the conterminous United States contains a rich set of information that reflects the original source material and subsequent modification through weathering, mass transport, climate and biological activities.

As a result, continental-scale geochemistry may represent many processes. In this study, we have focused on the evaluation and interpretation of the multi-element soil geochemistry from the surface soil, A- and C-horizons in the context of predicting the surface lithologies.

Process discovery makes use of multivariate methods such as principal component analysis, which creates orthogonal linear combinations of the elements that often reflect processes controlled by mineral stoichiometry that comprise the parent material. This parent material may be bedrock (igneous, metamorphic, sedimentary), glacial deposits, loess or fluvial deposits. Ideally, soil geochemistry can be used to predict the composition of the underlying soil parent material. As demonstrated in this study, multivariate methods such as principal component analysis cannot decouple all of these processes. Processes such as igneous and metamorphic mineral reactions share similar mineral stoichiometry, making them indistinguishable from a geochemical perspective. Many distinct sedimentary assemblages are comprised of similar lithologies with similar mineralogy, and are thus difficult to distinguish solely on a geochemical basis.

With the exception of the surface lithology map of Sayre et al. (2009), a continental-scale map of lithology does not exist, which creates difficulty in an attempt to predictively map at large scales. However, the availability of the maps by Sayre et al. (2009) that include terrestrial ecosystems, thermoclimate, soil moisture and surface lithologies provides an opportunity to test the capacity of soil geochemistry to uniquely define these features. Although not presented here, the soil geochemistry has the ability to uniquely define terrestrial ecosystems and regional climate indicators. We intend to publish the results of using soil geochemistry to uniquely identify the terrestrial ecosystems, thermoclimatic zones and soil moisture (ombrotype) as defined by Sayre et al. (2009).

With few exceptions, there are only minor differences between the geochemical compositions of the surface soil and the A-horizon. The geochemistry of the C-horizon displays a distinct geochemical difference between the surface soil and A-horizon as it has not undergone the degree of weathering as the near-surface soils and contains less organic material.

The overall predictive accuracies for the predicting the surface lithologies for the surface soil, A- and C-horizons are 49.9%, 49.4% and 48.6%, respectively. As described above, the reasons for these low accuracies are due to the overlap of many of the lithologies with Alluvium, Non-carbonate residual soils, Siliceous soils, Eolian Dunes, Eolian Loess and materials deposited from glaciation. However, the spatial continuity of the posterior probabilities confirm the distinctiveness of these lithologies and demonstrate the effectiveness of soil geochemistry in recognizing the differences between the classes.

The geochemistry of soils represents modification of the initial parent material through weathering in response to varying precipitation and temperature, groundwater effects, meteoric water effects, biologic activity and geologic complexity.

Thus, geochemistry is a rich source of information that can be used in many ways to describe, monitor and predict processes derived from natural and anthropogenic events (Grunsky et al. 2013).

The results from the statistical evaluation of the geochemical data in the context of predicting surface lithologies across the conterminous US indicates that soil geochemistry reflects a number of physical processes. Further studies of the soil geochemistry across the US will evaluate the ability to predict terrestrial ecosystems and indicators of climate.

**Acknowledgements** The authors thank Karl Ellefson of the United States Geological Survey for his thoughtful and helpful review of the manuscript.

## References

Aitchison J (1986) The statistical analysis of compositional data. Chapman and Hall, New York, p 416

Breiman L (2001) Randomforests. Machine. Learning 45:5–32

Breiman L, Cutler A (2016) Random forests. https://www.stat.berkeley.edu/~breiman/RandomForests/cc_home.htm#intro

Commission for Environmental Cooperation (CEC) (1997) Ecological regions of North America—toward a common perspective. Montreal: commission for Environmental Cooperation, p 71

Drew LD, Grunsky EC, Sutphin DM, Woodruff LG (2010) Multivariate analysis of the geochemistry and mineralogy of soils along two continental-scale transects in North America. Sci Total Environ 409:218–227. https://doi.org/10.1016/j.scitotenv.2010.08.004

Eberl DD, Smith DB (2009) Mineralogy of soils from two continental-scale transects across the United States and Canada and its relation to soil geochemistry and climate. Appl Geochem 24(8):1394–1404

Egozcue JJ, Pawlowsky-Glahn V, Mateu-Figueras G, Barcelo-Vidal C (2003) Isometric logratio transformations for compositional data analysis. Math Geol 35(3):279–300

Garrett RG (2009) Relative spatial soil geochemical variability along two transects across the United States and Canada. Appl Geochem 24(8):1405–1415

Goldschmidt VM (1937) The principal of distribution of chemical elements in minerals and rocks. The seventh Hugo Muller lecture, delivered before the Chemical Society on March 17th, 1937. J Chem Soc 1937, 665–673. https://doi.org/10.1039/jr9370000655

Grunsky EC (2001) A program for computing rq-mode principal components analysis for S-Plus and R. Comput Geosci 27:229–235

Grunsky EC (2010) The interpretation of geochemical survey data. Geochem Explor Environ, Anal 10(1):27–74

Grunsky EC, de Caritat P, Mueller UA (2017) Using surface regolith geochemistry to map the major crustal blocks of the Australian continent. Gondwana Res 46:227–239. https://doi.org/10.1016/j.gr.2017.02.011

Grunsky EC, McCurdy MW, Perhsson SJ, Peterson TD, Bonham-Carter GF (2012) Predictive geologic mapping and assessing the mineral potential in NTS 65A/B/C, Nunavut, with new regional lake sediment geochemical data. Geological Survey of Canada, Open File 7175, 1 sheet. https://doi.org/10.4095/291920

Grunsky EC, Mueller UA, Corrigan D (2014) A study of the lake sediment geochemistry of the Melville Peninsula using multivariate methods: applications for predictive geological mapping. J Geochem Explor. https://doi.org/10.1016/j.gexplo.2013.07.013

Grunsky EC, Drew LJ, Woodruff LG, Friske PWB, Sutphin DM (2013) Statistical variability of the geochemistry and mineralogy of soils in the maritime provinces of Canada and part of the northeast United States. Geochem Explor Environ Anal 13(2013):249–266. https://doi.org/10.1144/geochem2012-138

Harris JR, Schetselaar EM, Lynds T, deKemp EA (2008) Remote predictive mapping: a strategy for geological mapping of Canada's North, chapter 2. In: Harris JR (ed) Remote predictive mapping: an aid for Northern mapping. Geological Survey of Canada, Ottawa, Ontario, Canada, pp 5–27 (OpenFile5643)

Harris JR, Grunsky EC (2015) Predictive lithological mapping of Canada's North using random forest classification applied to geophysical and geochemical data. Comput Geosci 80:9–25

Jolliffe IT (2002) Principal component analysis, 2nd edn. Springer, New York, p 487

Mueller UA, Grunsky EC (2016) Multivariate spatial analysis of lake sediment geochemical data. Melville Peninsula, Nunavut, Canada, Applied Geochemistry, https://doi.org/10.1016/j.apgeochem.2016.02.007

Palarea-Albaladejo J, Martín-Fernández JA, Buccianti A (2014) Compositional methods for estimating elemental concentrations below the limit of detection in practice using R. J Geochem Explor. ISSN 0375–6742. https://doi.org/10.1016/j.gexplo.2013.09.003. Accessed 28 Sept 2013

Pawlowsky-Glahn V, Egozcue J-J (2015) Spatial analysis of compositional data: a historical review. J Geochem Explor 164:28–32. https://doi.org/10.1016/j.gexplo.2015.12.010

Pebesma EJ (2004) Multivarilabe geostatistics in S: the gstat package. Comput Geosci 30:683–691

QGIS Development Team (2016) QGIS geographic information system. Open Source Geospatial Foundation Project. http://www.qgis.org/

R Core Team (2013) R: a language and environment for statistical computing. R foundation for statistical computing, Vienna, Austria. http://www.r-project.org

Sayre R, Comer P, Warner H, Cress J (2009) A new map of standardized terrestrial ecosystems of the conterminous United States: U.S. Geological Survey Professional Paper 1768, 17 p. http://pubs.usgs.gov/pp/1768

Smith DB (2009) Geochemical studies of North American soils: results from the pilot study phase of the North American Soil Geochemical Landscapes Project. Appl Geochem 24(8):1355–1356. https://doi.org/10.1016/j.apgeochem.2009.04.006

Smith DB, Woodruff LG, O'Leary RM, Cannon WF, Garrett RG, Kilburn JE, Goldhaber MB (2009) Pilot studies for the North American Soil Geochemical Landscapes Project—site selection, sampling protocols, analytical methods, and quality control protocols. Appl Geochem 24(8):1357–1368

Smith DB, Cannon WF, Woodruff LG (2011) A National–scale geochemical and mineralogical survey of soils of the conterminous United States. Appl Geochem 26:S250–S255

Smith DB, Cannon WF, Woodruff LG, Rivera FM, Rencz AN, Garrett RG (2012) History and progress of the North American Soil Geochemical Landscapes Project, 2001–2010. Earth Sci Front 19(3):19–32

Smith DB, Cannon WF, Woodruff LG, Solano F, Kilburn JE, Fey DL (2013) Geochemical and mineralogical data for soils of the conterminous United States. U.S. Geological Survey Data Series 801, 19 p. http://pubs.usgs.gov/ds/801/

Smith DB, Cannon WF, Woodruff LG, Solano F, Ellefsen KJ (2014) Geochemical and mineralogical maps for soils of the conterminous United States. U.S. Geological Survey Open-File Report 2014–1082, 386 p. https://pubs.usgs.gov/of/2014/1082/

Venables WN, Ripley BD (2002) Modern applied statistics with S, 4th edn. Springer, Berlin

Woodruff LG, Cannon WF, Eberl DD, Smith DB, Kilburn JE, Horton JD, Garrett RG, Klassen RA (2009) Continental-scale patterns in soil geochemistry and mineralogy: results from two transects across the United States and Canada. Appl Geochem 24(8):1369–1381

Zhou D, Chang T, Davis JC (1983) Dual extraction of R-mode and Q-mode factor solutions. Math Geol 15(5):581–606

# Part III
# Exploration and Resource Estimation

# Chapter 18
# Quantifying the Impacts of Uncertainty

**Peter Dowd**

**Abstract** This chapter reviews the general concepts of uncertainty and probabilistic risk analysis with a focus on the sources of epistemic and aleatory uncertainty in natural resource and environmental applications together with examples of quantifying both types of uncertainty. The initial uncertainty in these applications arises from the in-situ spatial variability of variables and the relatively sparse data available to model this variability. Subsequent uncertainty arises from processes applied either to extract the in-situ variables or to subject them to some form of flow and/or transport. Various approaches to quantifying the impacts of these uncertainties are reviewed and several practical mining and environmental examples are given.

## 18.1 Introduction

This chapter provides an overview of the quantification of uncertainty with a focus on mineral and energy resources and environmental applications drawing on the work of the author and his co-authors over the past 30 years. Rarely in mining applications do initial estimates reconcile with production—there is almost always some reverse calibration or model revision to achieve an operationally acceptable agreement. This feedback approach can be a useful means of model calibration but the production 'reality' is an outcome conditional on the model and data used to make the production decision and may be biased. The resort to post hoc empirical calibration is due partly to insufficient data and partly to inadequate accounting for all sources of uncertainty. This situation will worsen as, increasingly, mineral resources will be extracted from deeper and/or lower grade deposits, which will require new technologies and new types of indirect sampling. In applications such as hydrocarbon extraction, the feedback reconciliation approach is essential because

P. Dowd FREng, FTSE (✉)
The University of Adelaide, Adelaide, Australia
e-mail: peter.dowd@adelaide.edu.au



B. S. Daya Sagar et al. (eds.), *Handbook of Mathematical Geosciences*,
https://doi.org/10.1007/978-3-319-78999-6_18

the in-situ variables can never be directly observed; Caers (2011) gives a comprehensive account of uncertainty quantification for these types of application.

The focus here is on geological applications in which the purpose is to extract material, store material or monitor the flow of fluids or contaminants. In these applications, uncertainty arises from two sources of variability: the in-situ variability of the geology and associated quantitative variables and the variability that is generated by applying processes to the in-situ resource. The basic approach is to combine data with a model to make predictions. Such predictions are meaningless unless accompanied by quantitative measures of the uncertainty of the prediction.

The general focus, particularly in mining applications, has been on the uncertainty arising from sparse data and not on uncertainty arising from the model, even though the model is inferred, and its parameters are estimated, from the sparse data. Variability arising from processes applied to the in-situ resource is either quantified in an overly simplistic manner or is ignored. The additional aspect in these and most spatial applications is that variability (and, therefore, uncertainty) is scale-dependent and may be relevant on multiple scales depending on the application.

## 18.2 Sources of In-Situ Uncertainty

In the field of uncertainty and probabilistic risk analysis two types of uncertainty are identified: aleatory and epistemic uncertainty (or irreducible and reducible uncertainty). In the generally accepted definitions (e.g., Bedford and Cooke 2001), aleatory uncertainty arises from the inherent variability of a phenomenon and cannot be reduced; epistemic uncertainty arises from incomplete knowledge of the phenomenon and can be reduced by more data, analysis or research. As both types of uncertainty are expressed in terms of probabilities, some authors question the necessity to distinguish between them. Others (e.g. Hora 1996; Winkler 1996) prefer sources of uncertainty rather than types, "the distinction between uncertainties is a matter of choice of scale and is, therefore, mutable." In the geostatistics context, Matheron (1975, 1976, 1978), notes that the empirical basis of uncertainty is the same in both cases and there is no objective criterion to distinguish them. Journel (1994) gives guidelines for modelling uncertainty on which Srivastava (1994) provides critical comment. However, as Winkler (1996) noted "uncertainty is uncertainty but the distinctions are related to very important practical aspects of modelling and obtaining information". This is especially so in the applications given here.

A fundamental difference between geological applications and many others is that each occurrence (orebody, karst system) is unique and, apart from measurement error, once a physical sample is taken at a location and the required variable is measured directly from the sample, there is no longer any uncertainty about the value of the variable at that location. The general geostatistical model includes stationarity, which allows for repeated sampling of the same random variable at different locations. In principle (but not in practice), all locations in an orebody could be sampled and aleatory uncertainty would be eliminated. Thus, in these applications aleatory uncertainty is entirely a function of the amount and quality of data. Epistemic uncertainty arises from the assumed or inferred geological model (e.g., type, or style, of mineralisation). In mining applications, at least in terms of a general model, there may be significant epistemic uncertainty during early stages of proving a deposit when geological models are inferred from sparse data. Model uncertainty may persist in later stages in terms of the specific characteristics or parameters of the model.

In some natural resource applications, the variables that define the resource can never be directly observed. For example, in hot dry rock (HDR) enhanced geothermal systems, the variable of interest is the combination of natural and stimulated fractures that form connected networks to extract heat. These fractures, at depths of up to 4.5 km, can never be directly observed or measured; their locations, extents and characteristics can only be inferred from micro-seismic events generated by fracture movement, stimulation and propagation (e.g., Xu and Dowd 2014). In these applications, the detailed model can never be known irrespective of the amount of data available. As mineral resources are extracted from increasingly deeper deposits there will be a move from physical samples, from which variables are directly measured, to sensed proxy variables and a move from traditional mining methods to in-situ recovery. For indirectly sensed variables, the aleatory uncertainty of the required variable (e.g., porosity) is largely due to the quality of the relationship with the directly sensed proxy variable (e.g., acoustic impedance), which could be classified as measurement, or interpretation, error.

Thus, although both sources of in-situ uncertainty in these applications are functions of the amount of data, it is useful to distinguish between them in quantifying uncertainty. Hereafter, epistemic uncertainty is used to mean conceptual or descriptive geological models as well as quantitative parametric models that describe spatial variability and in which parameter values are calculated or inferred from data.

Although epistemic uncertainty is recognised, it is largely ignored in practice. Once a model is assumed or inferred and/or its parameters are inferred or estimated from the available data, all measures of uncertainty are based on the data; in most applications, the model of spatial variability is implicitly assumed to be known with certainty. In other fields, there has been a longstanding recognition of the importance of identifying and quantifying both sources of uncertainty and of propagating them into a complete systems model (e.g., Bedford and Cooke 2001; Helton et al. 2004; Oberkampf et al. 2002, 2004). In natural resource applications, particularly mining, the emphasis has largely been on aleatory uncertainty with implicit acceptance that epistemic uncertainty is negligible. Geostatistical simulation is widely used to quantify the effects of limited data on resource modelling and estimation (aleatory uncertainty) but the model (e.g., variogram, spatial pattern) is generally assumed to be perfectly known (no, or negligible, epistemic uncertainty).

## 18.3 Transfer Uncertainty

A further complication in mineral and energy resources is that there are additional significant sources of uncertainty in extraction and processing to produce a final product. To borrow a petroleum industry term these might be called transfer, or process, functions and the associated uncertainties, transfer or process uncertainty. A general approach to integrating this source of uncertainty is to quantify all sources of in-situ uncertainties and propagate them into simulated transfer processes (e.g., blasting, selective loading, transport, mineral processing).

In resource extraction applications, it is useful to distinguish two broad types of process (or transfer) uncertainty:

(1) The uncertainty associated with in-situ variables that is propagated into processes applied to them. This might be termed passive in the sense that it does not change spatial variability. An example is the impact of grade uncertainty on mine design, which could be assessed by applying the same design process (e.g., optimal open-pit) to a range of simulated realisations of grades.
(2) The uncertainty transferred, or propagated, to in-situ variables by applying processes to them. This might be termed active as the process changes spatial variability. Changes in spatial variability can be predicted by modelling the process. An example is blasting a block of ground from which ore is selected.

## 18.4 Consequences of In-Situ Uncertainty

There are broadly two aspects of a geological model used in mineral resource applications: the generic type (e.g., stratiform silver/lead/zinc orebody) and the unique aspects that distinguish a specific orebody within the type (e.g., faulting, folding, degree of spatial continuity and of regularity of orebody boundaries). In general, for mineral deposits the first of these is known with near certainty at a relatively early stage but the distinguishing aspects and the relevant scales on which these aspects occur may not be known until much later. In these applications, the two types of in-situ uncertainty are not independent. The sampling scale (e.g., drilling grid) is determined, or at least significantly informed by, the geological model; the sampling scale determines the data, the spatial variability of which is the aleatory uncertainty; the parameters of the model are estimated by the data.

The Stekenjokk mine in Sweden provides a striking example of the consequences of epistemic uncertainty. Boliden Mineral AB mined this massive copper-zinc-silver orebody from 1976 to 1988 and processed a total of 8 M tonnes of ore. Prior to mine development the drilling grid was 20 m × 20 m and, in places, 20 m × 10 m. Figure 18.1 is an idealised, but typical, vertical cross-section through the orebody showing the drill-hole intersections with the ore. Drilling data were combined with the assumed geological model to generate the estimated orebody boundaries.

Figure 18.1 shows the complex, multi-directional folding of ore zones encountered in mining. The practical consequences of these predictions were significant (Hoppe 1978):

- Inappropriate choice of mining methods and mining equipment.
- Increased ore dilution, mining costs, development and processing provisions.
- Complications of highly mechanised equipment purchased for a simpler mine.

In principle, the problem could have been resolved by more appropriate sampling but the "appropriateness" of sampling was determined by the assumed geological model. In addition, sampling is constrained by cost (relative to the value of the mined product) and the cost of a drilling grid capable of capturing the folding may well have been prohibitive.

Geological models are only as good as the quality and interpretation of the data and the appropriateness of the scale on which the data are collected. Stekenjokk is an extreme (but not unique) example of epistemic uncertainty that could only be

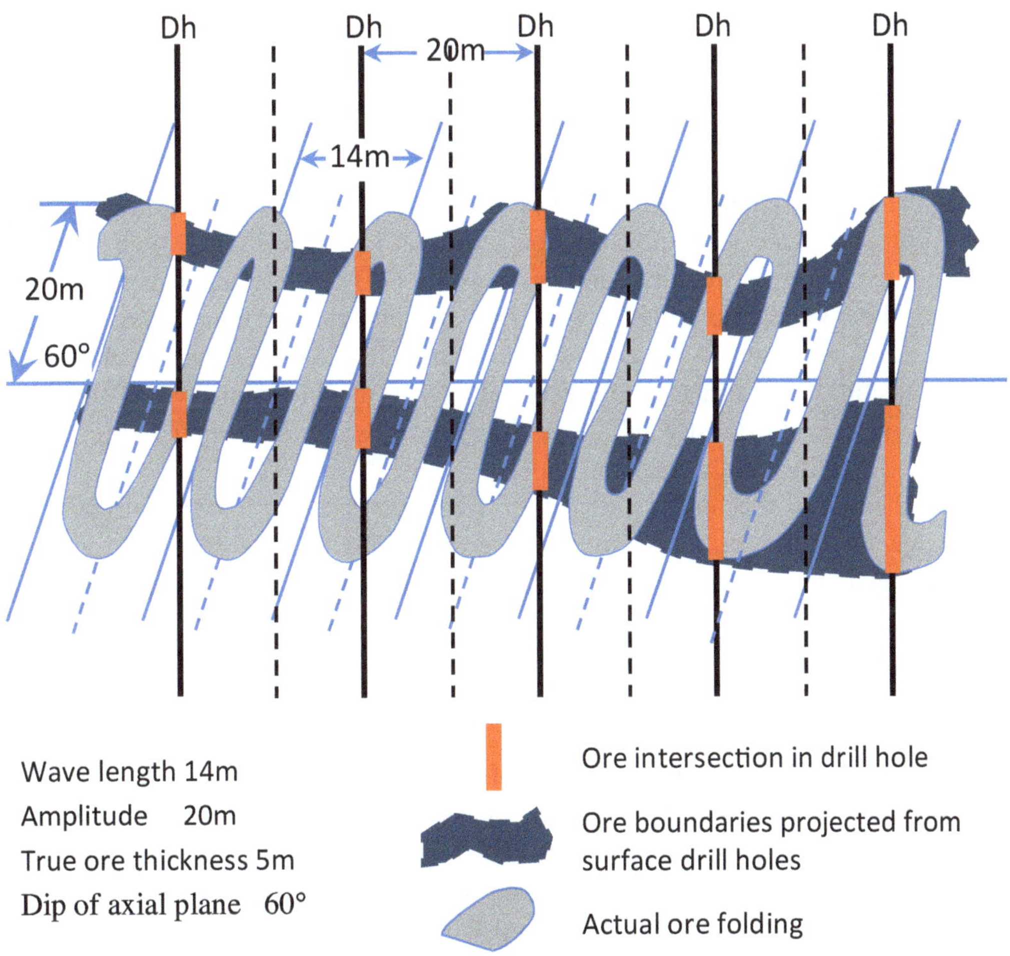


**Fig. 18.1** Interpolation of ore continuity from surface drilling data prior to mine development; adapted from Hoppe (1978)

reduced to an acceptable level by more data. However, this observation is somewhat circular: the geological model depends on the amount of data/information available but the data type and collection are informed by the assumed model.

### *18.4.1 Scale and Variability Example: Hilton Orebodies Australia*

This example is from a study of a complex group of three silver/lead/zinc orebodies at what, at the time, was known as the Hilton mine in north-western Queensland, Australia. The full study is given in Dowd and Scott (1984) with a later study in Dowd et al. (1989).

The Hilton orebodies are 22 km north of Mt Isa, one of the world's largest stratiform base metal deposits. The Hilton orebodies have a similar diagenesis to the Mt Isa orebodies with mineralisation occurring in the same dolomitic shale. The study was undertaken at the pre-feasibility stage and all original drilling, sampling and interpretation were influenced by 50 year's mining experience at Mt Isa. Although the Mt Isa and Hilton styles of mineralisation are similar, the Hilton orebodies are structurally more complex and less continuous.

Two test areas were extensively drilled to provide detailed information for a geostatistical study to determine optimal drilling densities for mine planning purposes. The holes were drilled from access drives as fans on cross-sections spaced 10 and 20 m apart. One such cross-section is shown in Fig. 18.2 in which the holes intersect the main 2 orebody footwall lens (2 O/B FW) at approximately 5 m centres. The dark blue outlines in Fig. 18.2 are the orebody boundaries estimated from the drill-hole data on the cross-section and on the cross-sections on either side. In the feasibility stage cost would prohibit such a drilling density over the entire orebody. Given the density of the drilling these estimated boundaries could be regarded as reality on all practical scales.

The effects of other drilling densities were assessed by removing drill data to create new datasets; e.g., removing every second drill-hole on a cross-section yields a 10 m spacing. Datasets for 5, 10, 20 and 40 m drill spacing were used in the study. Orebody boundaries were estimated for each drilling density and the results were given to mining engineers to design stopes. As an example, the estimated orebody boundaries for 20 m drill spacing is shown in Fig. 18.3. As expected, these boundaries are much smoother (less variable, more continuous) than the "reality" represented by the boundaries estimated from the 5 m spacing dataset. The variability of the boundaries is critical in the choice of mining method: the variability of boundaries and their exact delineation are less critical if a bulk mining method is adopted than if more selective methods are used. The original mining method was cut and fill followed later by sub-level open stoping and bench mining.

Figure 18.4 shows the 5 m interpolation overlaid on the 20 m interpolation. Taking the 5 m interpolated boundaries as reality, all visible light blue areas

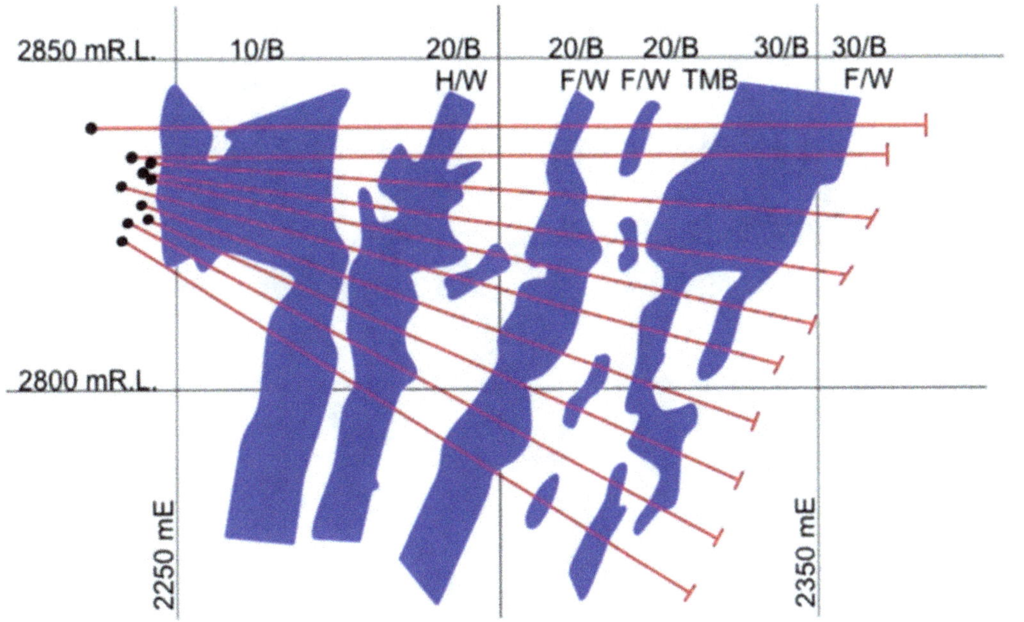


**Fig. 18.2** Cross-sectional interpretation based on 5 m drill spacing

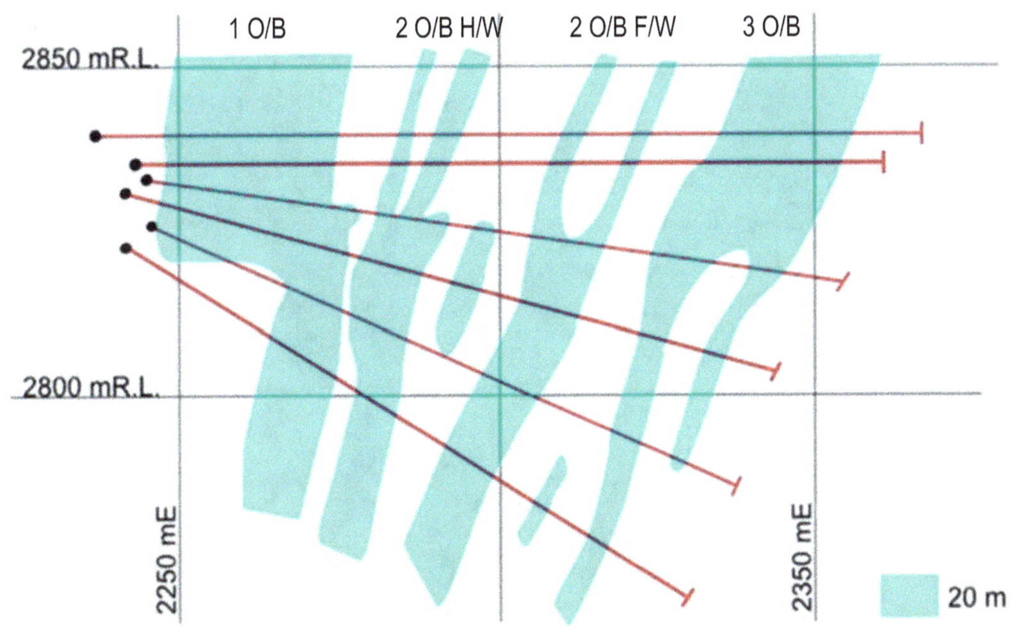


**Fig. 18.3** Cross-sectional interpretation based on 20 m drill spacing

represent ore dilution arising from planning and extraction based on the 20 m interpolated boundaries.

Figure 18.5 shows the 20 m interpolation overlaid on the 5 m interpolation. Again, taking the 5 m boundaries as reality, all visible dark blue areas represent the ore loss arising from planning and extraction based on the 20 m interpolated

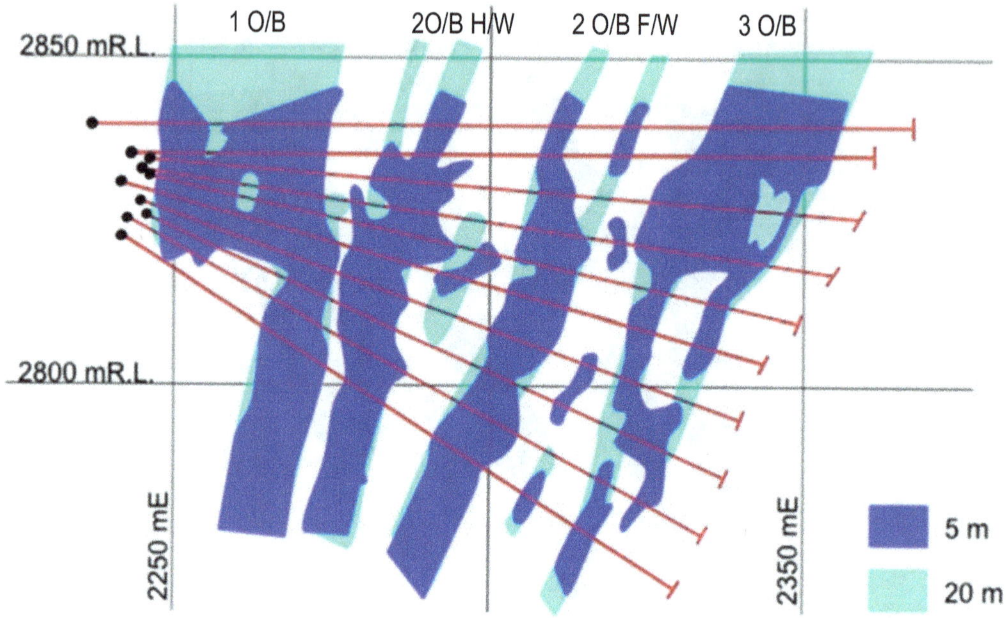


**Fig. 18.4** Overlay of 5 m interpolation on 20 m interpolation

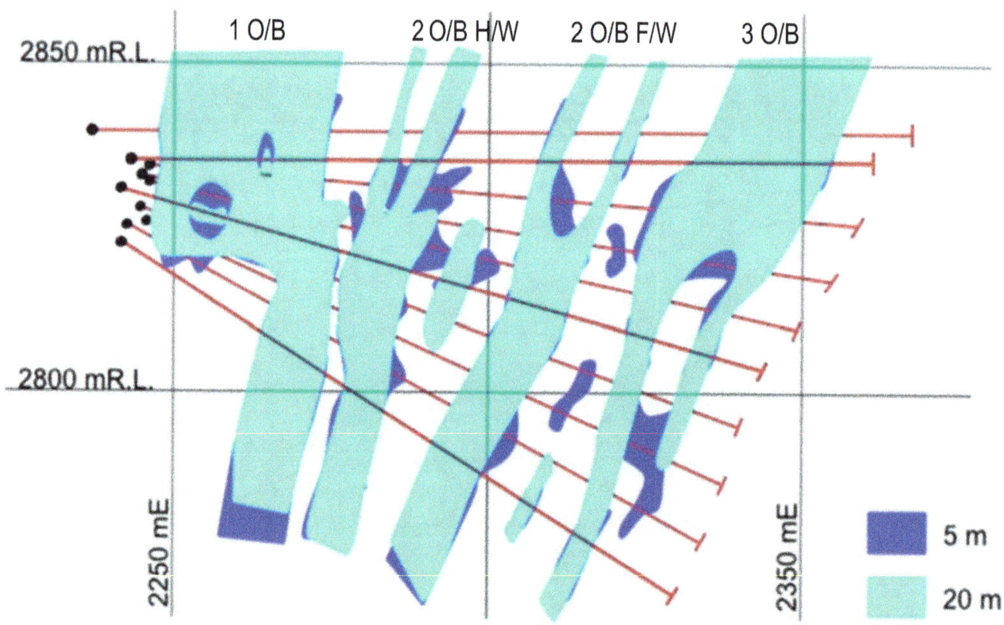


**Fig. 18.5** Overlay of 20 m interpolation on 5 m interpolation. Based on 20 m model, all visible dark blue areas represent ore loss

boundaries. Of course, the perfect selection and the adherence to estimated boundaries during production implied by this exercise are not entirely realistic. However, the impact on the choice of mining method, on the predicted grades and tonnages, and on economic outcomes is real.

The outputs from the stope design exercise are summarised in Fig. 18.6 for 5, 10 and 20 m drill spacing. Orebodies 1 and 2 H/W (hanging wall) are mined in a single stope and orebodies 2F/W and 3 are mined in separate stopes. Grades were estimated by kriging and are in metal equivalents of lead (weighted sum of lead, zinc and silver grades); intervals are $\pm 2\sigma_K$ where $\sigma_K$ is the square root of the kriging variance and is used as an index of uncertainty rather than a confidence interval. Taking the 5 m designs as actual boundaries, the stope designs based on 10 and 20 m drilling show the effects of decreasing amounts of data on planned tonnage and average grade.

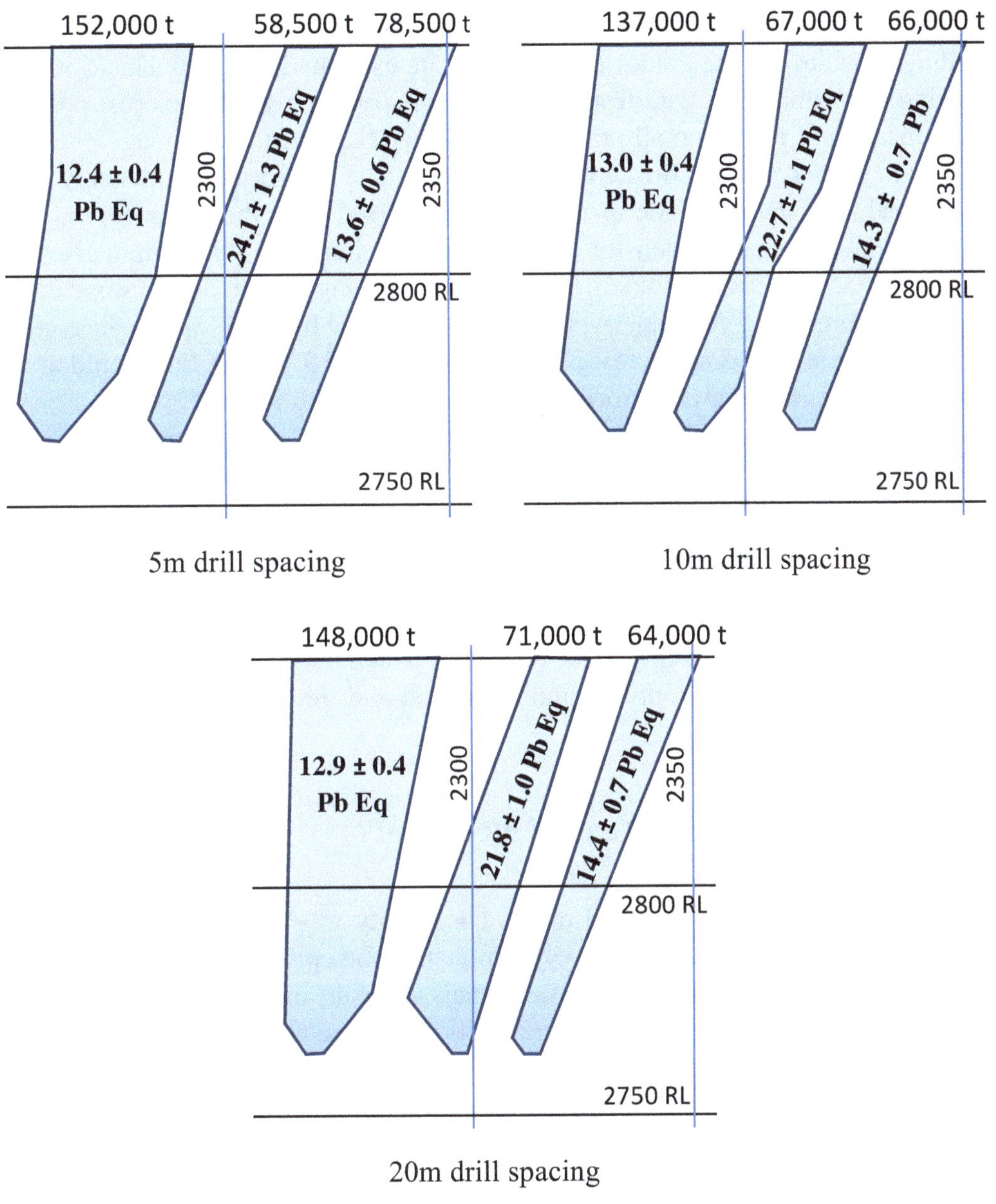


**Fig. 18.6** Stope designs with contained tonnages and grades for 5, 10 and 20 m drill spacing for orebodies 1 and 2 HW (left); 2 FW (centre) and 3 (right)

**Table 18.1** Differences in tonnes and grades of stopes compared with 5 m designs

| Stope | Drill spacing (m) | Change in ore tonnes (%) | Change in grade (%) | Change in metal tonnes (%) |
|---|---|---|---|---|
| 1 | 10 | −10.2 | +4.2 | −6.4 |
| | 20 | −3.0 | +3.6 | −0.5 |
| 2 | 10 | +14.5 | −5.9 | +7.8 |
| | 20 | +21.4 | −9.6 | +9.7 |
| 3 | 10 | −15.9 | +5.0 | −11.8 |
| | 20 | −18.5 | +5.5 | −14.0 |

The stope designs are based on the data and interpretations from the respective drilling densities but the grades and tonnages are estimated using all data (5 m drill spacing). Assuming the data from the 5 m drill spacing gives the closest possible quantification of reality on all practical scales then the grade and tonnage of the 10 and 20 m stope designs estimated from all data can be regarded as sufficiently close to the real tonnage and grade that could be recovered from the designs.

The effects of data density on grades and tonnages are summarised in Table 18.1. As an example, using the 20 m drill spacing data to design stope 2 (the high-grade orebody 2 footwall) would increase tonnage by 21.4% and reduce grade by 9.6%. There would an increase in metal tonnage of 9.7% but this would at the cost of mining, hauling and processing the additional ore tonnage.

Whilst the effects of data on a specific type of mining are of interest, the more important issue is the effect of the assumed geological model on the choice of mining method. The initial geological model was influenced by the knowledge accumulated over a long period of mining in the neighbouring Mt Isa orebodies. The detailed analysis described here enabled the effects of the greater complexity and less continuity of the Hilton orebodies to be systematically quantified, thereby significantly reducing the impact of epistemic uncertainty and contributing to the selection of the most appropriate mining method and mine design.

## 18.5 Quantifying Epistemic Uncertainty

In the Hilton example, geological model uncertainty was addressed at the significant cost of more samples—effectively eliminating the epistemic uncertainty on the operational scale through more data and analysis. With the hindsight of the additional data and analysis, and on the assumption that the test volume is sufficiently representative of the remainder of the orebodies, the epistemic uncertainty associated with various drilling grids could be quantified. This would allow assessment of the value of additional information against the cost of collecting it and/or the operational cost of not collecting it. Stekenjokk is an example of the practical consequences of proceeding with an unacceptable level of epistemic uncertainty.

There is an extensive literature on using Bayesian probability to quantify epistemic uncertainty particularly to combine sources of uncertainty (e.g., Winkler 1981; Sankararaman and Mahadevan 2011) and to incorporate expert knowledge and informed guesses in the form of subjective probabilities. It can be argued that subjective probabilities are used implicitly throughout geostatistical analysis, modelling, estimation and simulation irrespective of the amount of data. Expert knowledge/judgment guides variogram calculation and interpretation, choice of training images, domaining, sample differentiation, choice of estimation or simulation method and validity of outputs. There is, however, a distinction between the explicit subjective probability of informed guesses and possible geological models and the implicit subjectivity in inferring model parameters from quantitative data.

In the remainder of this chapter, a distinction is made between model uncertainty and uncertainty of the parameters of a specific model. Many authors do this although in some cases the former may be a case of the latter e.g., it might be argued (with some difficulty) that Stekenjokk was a matter of incorrect structural parameters (degree of folding). A more convincing argument could be made for the Hilton case—the initial assumed model was a Mt Isa type stratiform orebody and the final agreed version was a more complex and less continuous version of the latter.

In addition to Bayesian approaches, others include evidence theory: Shafer (1976) and Dempster (1968); fuzzy sets: (Zadeh 1965); and possibility theory: Zadeh (1978) and Dubois and Prade (2001). These and other approaches are extensively used to quantify uncertainty in risk analysis and a good coverage of probabilistic risk analysis is given in Bedford and Cooke (2001).

Over the past 30 years, all these approaches have been used to incorporate model uncertainty in geostatistical estimation and simulation and the following list is intended as representative rather than exhaustive. Omre (1987) used Bayesian kriging to include qualified guesses when few data are available; the weight assigned to the guess increases as the amount of data decreases.

Fuzzy kriging has been proposed as a means of including aleatory uncertainty (in the sense of inaccurate or imprecise measurements) and epistemic uncertainty (imprecise variogram parameters) in estimation. Uncertain data will, of course, lead to an uncertain variogram but certain (accurate, error-free) data will not necessarily lead to a certain variogram. Diamond (1989) proposed fuzzy kriging to deal with uncertain or imprecise data. Bardossy et al. (1988, 1990a, b) proposed fuzzy kriging for dealing with both sources of uncertainty but the computational cost hindered its use. More recently, Loquin and Dubois (2010a, b) have developed these approaches in computationally feasible forms. Bandemar and Gebhardt (2000) combine fuzzy kriging with Bayesian incorporation of prior knowledge. Bardossy and Fodor (2004) provide a comprehensive coverage of the use fuzzy set theory to quantify geological uncertainty and consequent risk.

Srivastava (2005) used probabilistic modelling of ore lenses to account for uncertainty in the boundaries of geological domains that constrain grade occurrence. Dowd (1986, 1994) and Dowd et al. (1989) used deterministic and probabilistic methods for the same purpose in estimating and simulating grades.

Verly et al. (2008) quantified geological model uncertainty in a porphyry copper deposit by simulating the four principal characteristics of porphyry models: faults defining fault blocks; faulted rock types within fault blocks; un-faulted intrusive and breccia bodies and alteration and copper grade shells.

Maximum likelihood estimation of spatial model parameters has been widely reported in geostatistical applications: Mardia and Marshall (1984), Kitanidis and Lane (1985), Zimmerman (1989), Dietrich and Osborne (1991) among others. Pardo-Igúzquiza and Dowd (1997a, b, c, 2003, 2013), Dowd and Pardo-Igúzquiza (2002) and Pardo-Igúzquiza et al. (2013) used maximum likelihood estimates of variogram parameters and associated uncertainties to incorporate the effects of model uncertainty in simulation and estimation.

For categorical variables, such as geological shapes and surfaces, multiple point statistics simulation provides a means of specifying possible geological scenarios in the form of alternative training images. Caers (2011) uses different training images to introduce geological model uncertainty into the simulation of oil reservoirs. Park et al. (2013) use history matching to quantify the uncertainty of facies models in the form of alternative training images. Hermans et al. (2014) choose among several geological scenarios in the form of possible training images using geophysical data and Bayes rule to compute the conditional probabilities of the alternative training images given the geophysical data.

With a few notable exceptions, in most mining applications the geological (model) uncertainty from the feasibility stage onwards can be limited to uncertainty in model parameters rather than uncertainty about the general model (e.g., strati-form, vein, disseminated). However, for cases where fundamental (and a priori, unverifiable) assumptions are/must be made about the general model, as in oil and gas applications or applications in which physical processes give rise to the vari-ables (e.g., HDR fracture occurrence and propagation), it is essential to test the sensitivity of these assumptions by reconciling the consistency of outputs (e.g., heat production from a geothermal reservoir) with predicted responses to inputs (e.g., fluid flow through fracture networks). The fundamental difference between these cases and mining applications is that ultimately the latter can be directly observed.

On the assumption that the most important characteristics of the underlying model can be captured in several parameters of a broad model, the uncertainty in the parameter estimates can be quantified by generating a set of parameter values using an appropriate set of rules; simulating the spatial random variable(s) using these parameter values; and repeating this process a sufficiently large number of times. Methods for sampling parameter values include Maximum Likelihood, Bootstrap methods (Olea et al. 2015), Bayesian analysis (Kitanidis 1986) and, in multiple point statistics simulations, Bayesian selection of alternative templates or training images (Park et al. 2013; Hermans et al. 2014) and clustering combined with system responses (Caers 2011).

The following two examples illustrate the use of maximum likelihood in model selection and parameter inference and the propagation of the associated uncer-tainties into geostatistical simulation for environmental and mining applications.

### 18.5.1 Example: Transmissivity Uncertainty

This example is taken from Dowd and Pardo-Igúzquiza (2002). The data are from Gotway (1994) and comprise 41 transmissivity measurements in the Culebra Dolomite formation in New Mexico. The original application was for nuclear waste site assessment, where uncertainty in the groundwater travel time of a particle is assessed through its probability density function, which is estimated by running groundwater flow and transport programs with different transmissivity field inputs. These inputs are generated by conditional simulations of transmissivity.

The data are the logarithms of transmissivity in $m^2 s^{-1}$ and the data locations are shown in Fig. 18.7 together with a histogram of the log-transmissivity data.

Maximum Likelihood was used to estimate the parameters of an exponential covariance model of the residuals for drift orders 0, 1 and 2. Although drift is a deterministic component of the universal model, in practice the coefficients are estimated from the available data and are thus random variables with the means and standard errors given in Table 18.2 for the optimal (determined by the Akaike information criterion) drift model of order 1: drift $(x, y) = \beta_0 + \beta_1 x + \beta_2 y$. The estimated covariance parameters for $k = 1$ are given in Table 18.3 and the variogram is shown in Fig. 18.8.

In this case, as there is no nugget variance, the range and sill are estimated independently. The correlation between range and sill is thus zero and any combination of values of the two parameters inside their respective intervals is inside the 95% confidence region as shown in Fig. 18.9a. The drift coefficients are also independent of the sill and the range. As the estimated drift coefficients are correlated, not every combination of the three parameter values is equally reliable, i.e. values inside the 95% confidence interval of the parameters taken together may not be inside the 95% confidence interval for each individual parameter. The confidence interval is an ellipsoid. Figure 18.9b shows the 95% confidence region for $(\beta_1, \beta_2)$ when the third coefficient the model is set to the estimated value given in Table 18.3.

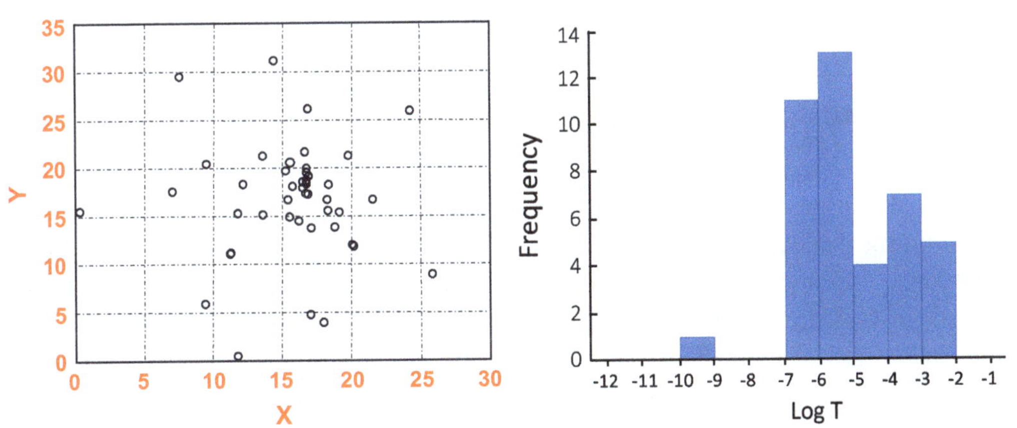


**Fig. 18.7** Data locations (distances in km) and histogram of log transmissivity data

**Table 18.2** Maximum likelihood estimates of drift coefficients

| Parameter | Estimate | Stand. error |
|---|---|---|
| $\beta_0$ | −1.6062 | 0.8653 |
| $\beta_1$ | −0.2245 | 0.0426 |
| $\beta_2$ | −0.0141 | 0.0323 |

**Table 18.3** ML estimates of range and sill: exponential covariance

| Sill | Stand. error | Range | Stand. error |
|---|---|---|---|
| 1.28 | 0.284 | 1.99 | 0.667 |

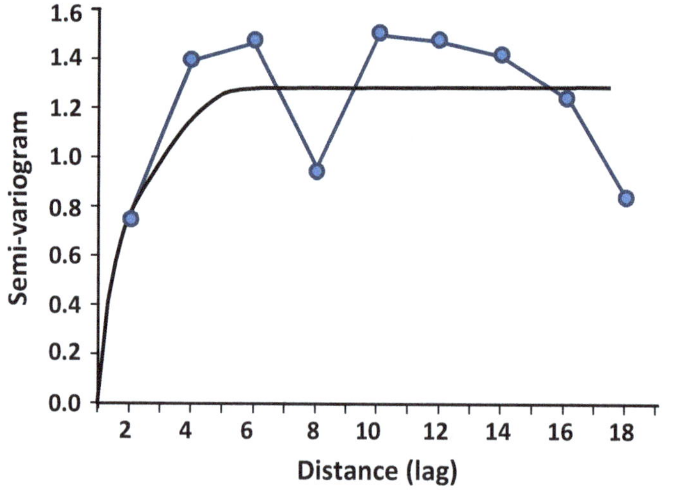


**Fig. 18.8** Semi-variogram of the residuals for $k = 1$ and maximum likelihood model fitted: sill 1.28, range 1.99 km (effective range ~6 km)

The effects of model uncertainty on simulation outputs are illustrated by generating six simulations for each pair of values A, B, C, D and E in Fig. 18.9; each set of simulations was started with the same random number seed. The simulations are shown in Fig. 18.10. The differences between corresponding simulations (e.g., first simulation in each of A, B, C, D and E) for the five sets of parameters reflect the model uncertainty, which could be quantified further by simulating flow and transport through the simulated transmissivity realisations.

### *18.5.2 Example: Coal Resource Risk Assessment*

One of the most significant contributors to the total risk in the evaluation of coal-mining projects is the uncertainty of the resource tonnage and quality characteristics, often called the resource risk. This example is from the As Pontes deposit in Galicia, Spain (Pardo-Igúzquiza et al. 2013). The most significant variable in the assessment of resource uncertainty is the thickness of the coal seam. Figure 18.11 shows the data locations at which seam thickness is measured together with the estimated variogram values and the manually fitted (isotropic) variogram model.

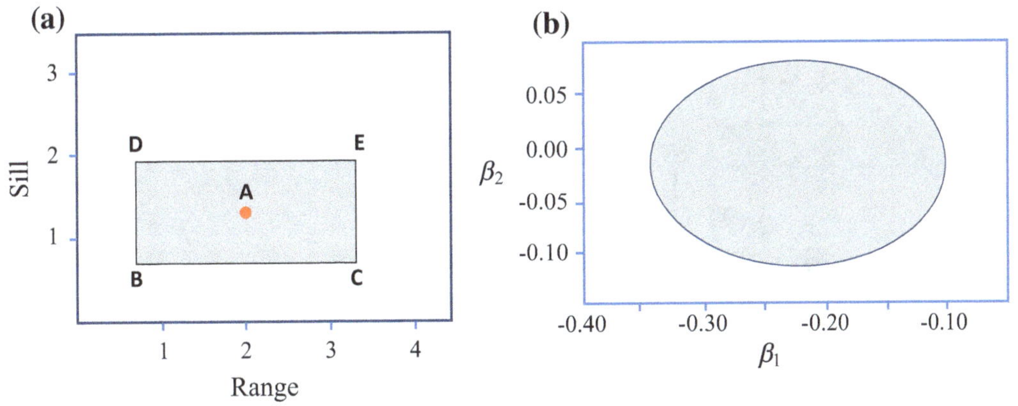


**Fig. 18.9** **a** (left) 95% confidence region for sill and range; **b** (right) confidence region for drift parameters $\beta_1$ and $\beta_2$ with $\beta_0 = -1.6062$

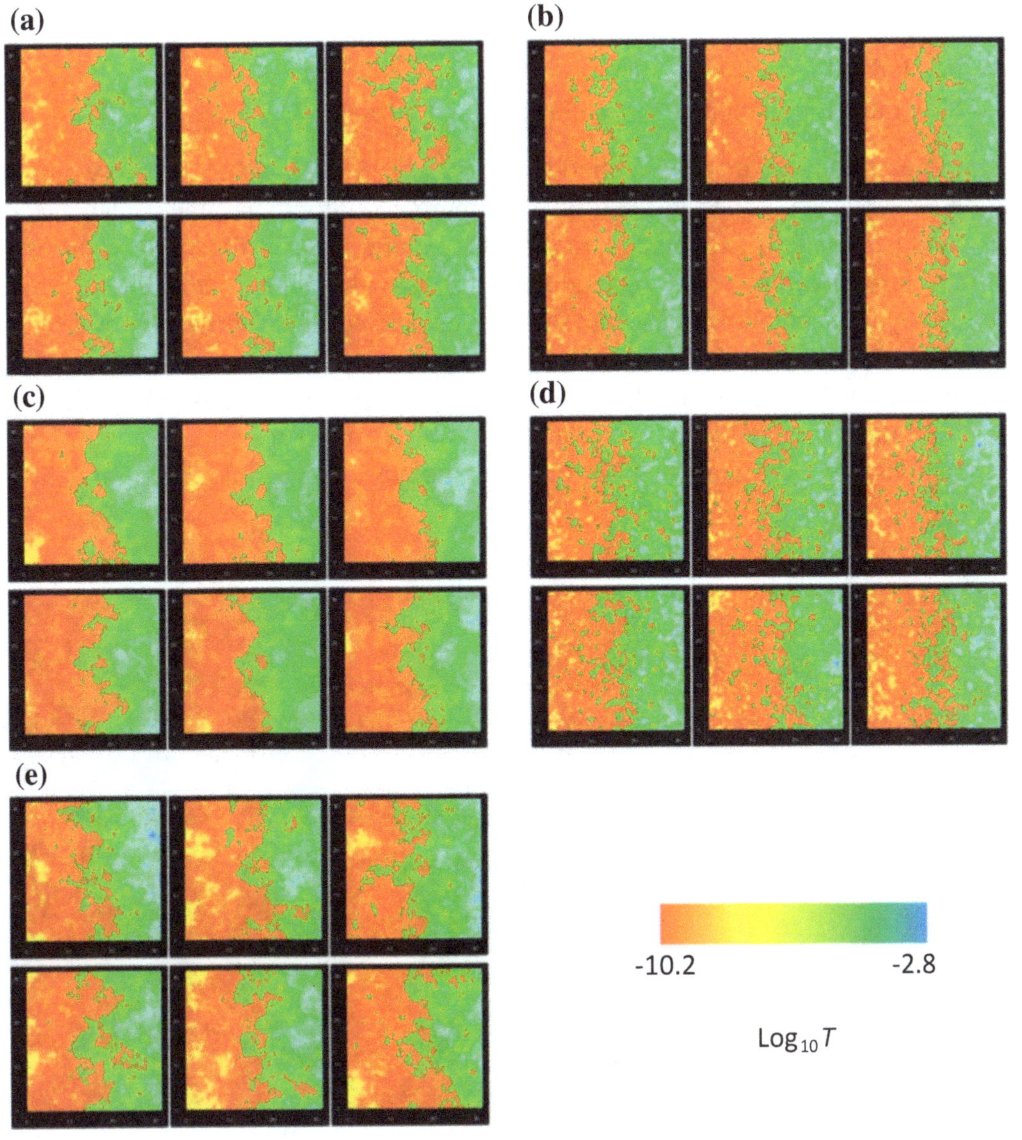


**Fig. 18.10** Outputs from six simulations using the variance and range parameters denoted by the mean values A and the extreme values B, C, D and E in Fig. 18.9

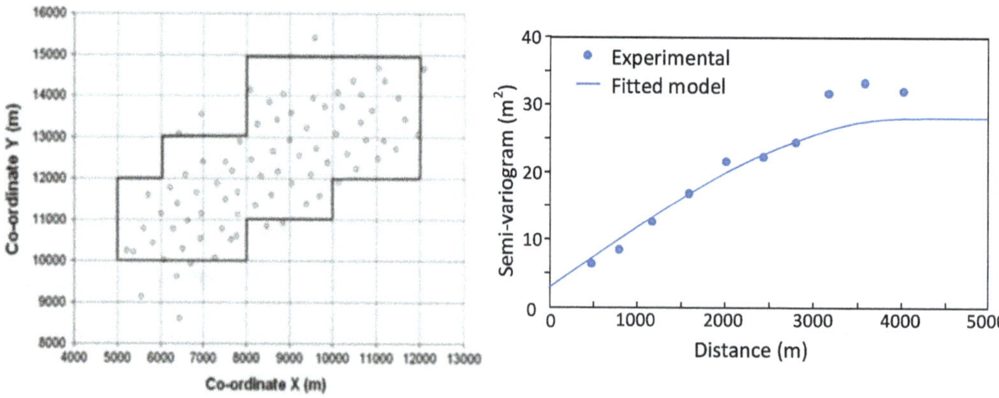


**Fig. 18.11** (Left) drill-hole locations and boundary of the study area. (Right) Variogram and manually fitted model for seam thickness

Spherical model variograms for seam thickness:

- Manual fitting: $a = 4128$ m, $C_0 = 3$ m$^2$ and $C = 25$ m$^2$.
- Maximum Likelihood: $a = 4460$ m, $C_0 = 4$ m$^2$ and $C = 23$ m$^2$.

Although the maximum likelihood estimates of the parameters are very similar to those estimated by visual fitting, maximum likelihood has the advantage of providing estimates of the uncertainty of the parameters. For illustrative purposes, resources were computed as tonnage from panels with thickness above a threshold defined by the 25th percentile of the sample data and equal to a thickness of 8.65 m. The kriged resource volume is $1.97 \times 10^8$ m$^3$.

Sequential Gaussian simulation was used to generate realisations of the thickness of the seam. To quantify the uncertainty in the estimated resource, a total of 870 simulations were generated using the 'certain' variogram (maximum likelihood

**Fig. 18.12** Conditionally simulated realisation of coal seam thickness

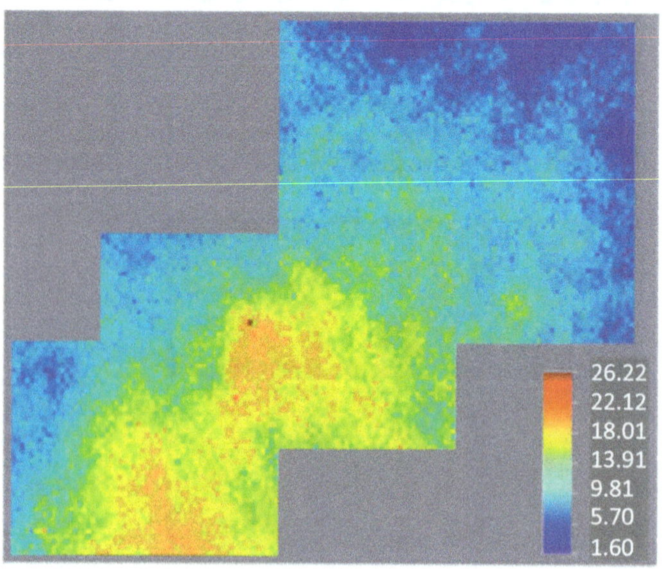

parameters) and the total resource was calculated for each simulation. The histogram of the 870 simulated resources quantifies the uncertainty of the estimated resources. An example simulation is shown in Fig. 18.12.

The parameter space $\{r_0, a, \sigma^2\}$ comprising respectively the nugget/variance ratio, range and variance, is used to quantify the uncertainty in the model. The parameter values were divided into discrete steps of 0.05 for $r_0$ in the interval [0, 1]; 700 m for $a$ in the interval [1,000, 15,000] and 0.1 for $\sigma^2$ in the interval [0.6, 2.6]. There are 268 models of triplets $\{r_0, a, \sigma^2\}$ that lie inside the 75% confidence region. As these models are not equally probable, the probabilities are normalised so that they sum to 1.0 and each model is included as many times as indicated by its normalised probability (i.e., probability sampling in which, for example, a model with a normalised probability of 0.35 comprises 35% of the total simulated triplets). A total of 870 simulations were used.

Histograms of the total resources for the 870 simulations, with and without the uncertainty of the variogram model parameters, are given in Fig. 18.13. There is no significant difference in mean resource values for the certain and uncertain values.

The 95% confidence interval for the total resource assuming the variogram is known with certainty is $[1.88 \times 10^8, 2.19 \times 10^8]$ m$^3$ and $[1.90 \times 10^8, 2.23 \times 10^8]$ m$^3$, when the uncertainty of the variogram model is included. The latter is slightly higher than the same interval calculated under the assumption that the variogram is known with certainty. However, the probability that the total resource will be greater than $2.0 \times 10^8$ m$^3$, is 0.59 when the uncertainty of the variogram parameters is ignored and 0.75 when the uncertainty of the variogram parameters is propagated into the simulated realisations. In other words, whilst there is no significant difference in the mean resource for the two sets of simulations, the difference in the two distributions (because of different variances) is sufficient to generate significantly different resource estimates above selected cut-offs.

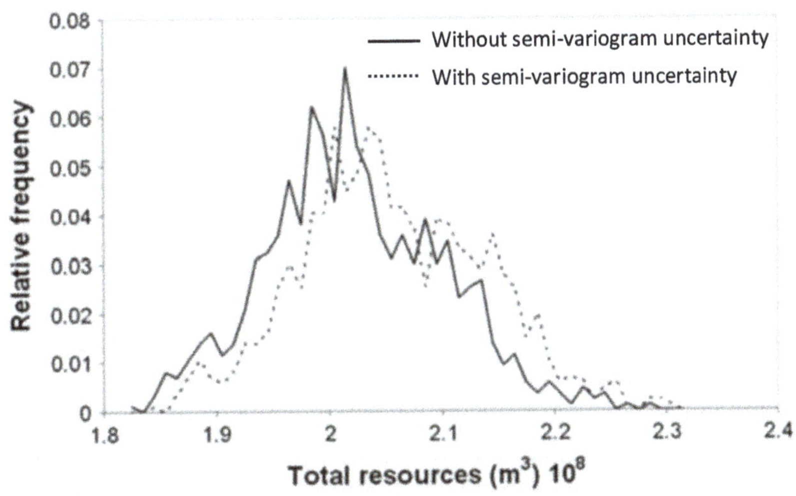


**Fig. 18.13** Histograms of total resources calculated by geostatistical simulation assuming the variogram model parameters are known with certainty (solid line) and including the uncertainty of the semi-variogram model parameters (dashed line)

In this case, the differences in the total volume of resources, with and without quantification of semi-variogram uncertainty, are small but the consequence of selecting from the distribution of possible resources is significant. This illustrates a general principle: the estimated total resource and the mean simulated resource, with and without semi-variogram uncertainty, may not differ significantly but the distributions of the two simulations will differ because of the different variances. Similarly, selecting panel values above a threshold from the set of estimated panel thicknesses or from a set of simulated panel thicknesses will yield different results.

In general, the outcome from the simulations with and without semi-variogram uncertainty depends on the deposit and the amount of data available. Evaluation of model uncertainty is critical in resource risk assessment even if it is ultimately found that there is no practical difference between resource estimates obtained by ignoring or including semi-variogram uncertainty. This example also has important implications for compliance with resource and reserve reporting codes, most of which use terms such as, or equivalent to, *the amount of error* [associated with an estimate], *the level of accuracy* [of an estimate], the *level of confidence* [in a reserve statement], and *levels of geological confidence* (words in italics are quoted from JORC 2012). Whilst all reporting codes currently use these terms qualitatively they all have specific quantitative meanings in statistics, probability and risk assessment and are increasingly being referred to explicitly in reporting codes.

## 18.6 Quantifying the Effects of Transfer Uncertainty

An example of passive transfer uncertainty is the variation in open-pit size and shape as a function of grade uncertainty as shown in Fig. 18.14 taken from a study of a small gold orebody (Dowd 1995, 1997). The impacts of these types of uncertainty can be quantified by standard applications of geostatistical simulation. Dimitrakopoulos and co-workers have made significant contributions to the integration of in-situ grade and geological uncertainty into optimization algorithms (e.g., Dimitrakopoulos et al. 2002; Goodfellow and Dimitrakopoulos 2013).

More challenging is the impact of propagating in-situ uncertainty through the mining (extraction) process. The critical component of most metalliferous open-pit mining operations is ore selection, i.e. the minimisation of ore loss and ore dilution during extraction. In general, extraction comprises drilling, blasting and loading, all of which are planned and designed on uncertain models of local geology and grade. The conversion of the in-situ block model resource to a realistically recoverable reserve may, in many instances, be the most significant source of uncertainty in reserve estimation. The usual assessment of recoverable reserves, for example, is limited to a simple volumetric exercise in which ore recovery is assessed as a function of applying a range of selection volumes to a simulated orebody or an even simpler volume-based adjustment of the variance of estimated block values. These simplistic approaches ignore the practicalities of the mining, selection and loading processes—blast design, behaviour and performance; equipment type, size and

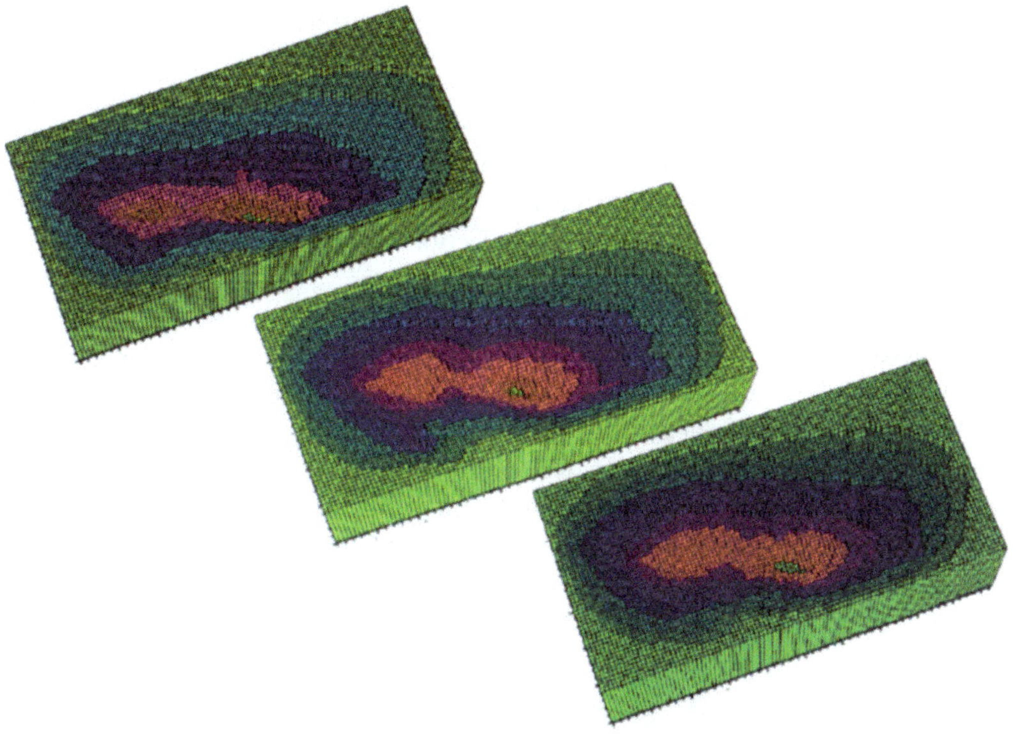

**Fig. 18.14** Optimal open pits generated from 100 simulations of a small gold orebody. Top: maximum volume; centre: median volume; bottom: minimum volume

operation; ore displacement during blasting and loading; and ability to identify ore zones within a blast muck pile. In many applications, the uncertainties introduced by these technical processes are at least as significant as those that derive from the in-situ spatial characteristics of grades and geology.

An approach to quantifying transfer process uncertainty for blasting and loading comprises:

- generation of an in-situ model of the orebody comprising the grade, geology, geomechanical properties and grade control variables within small volumes determined by the smallest selectable volume within a blast muck-pile;
- definition of a blast volume comprising a large number of in-situ model volumes, and subjecting it to a blast simulator, which effectively moves each component model volume to its final resting place in the blast muck-pile; and
- application of simulated selective loading processes to the simulated blast muck-pile to determine the selectivity that can be achieved by various sizes of loader and types of loading and to quantify ore dilution and ore loss.

The in-situ model, representing perfect knowledge at all relevant scales, is obtained by geostatistical simulation. An in-situ model that represents the reality of knowing only the data and information that are available from specific grade control drilling and sampling grids can be obtained by sampling the geostatistically simulated model on a specified grid. The volumes comprising the in-situ model are then populated by estimates based only on the data corresponding to the specified grade-control drilling and sampling grids. Different drilling and sampling grids can be used to generate different models, each reflecting the levels of data and information available. Selectivity can then be assessed as a function of the drilling and sampling grids as well as the size and type of loader. Performance is assessed against the ideal selectivity that can be achieved on the perfect knowledge model, comprising the simulated values of each component volume. Applying costs, prices and financial criteria enables an optimal selection of the grade control drilling grid, size of loader, type of loading and even blast design.

The following case study (Dowd and Dare-Bryan 2004) is based on the Minas de Rio Tinto SAL open-pit copper mine at Rio Tinto, southern Spain, which is typical of a low-grade operation in the later stages of its life. Ore/waste delineation for selective mining is difficult because the head grades are near the economic cut-off grade and there are no clear geological controls on the mineralisation.

Sequential Gaussian simulation, with the blast-hole grades as conditioning data, was used to generate realisations of each mining bench on a block grid of 0.5 m × 0.5 m × 0.5 m, the grid determined based on blast and selection criteria.

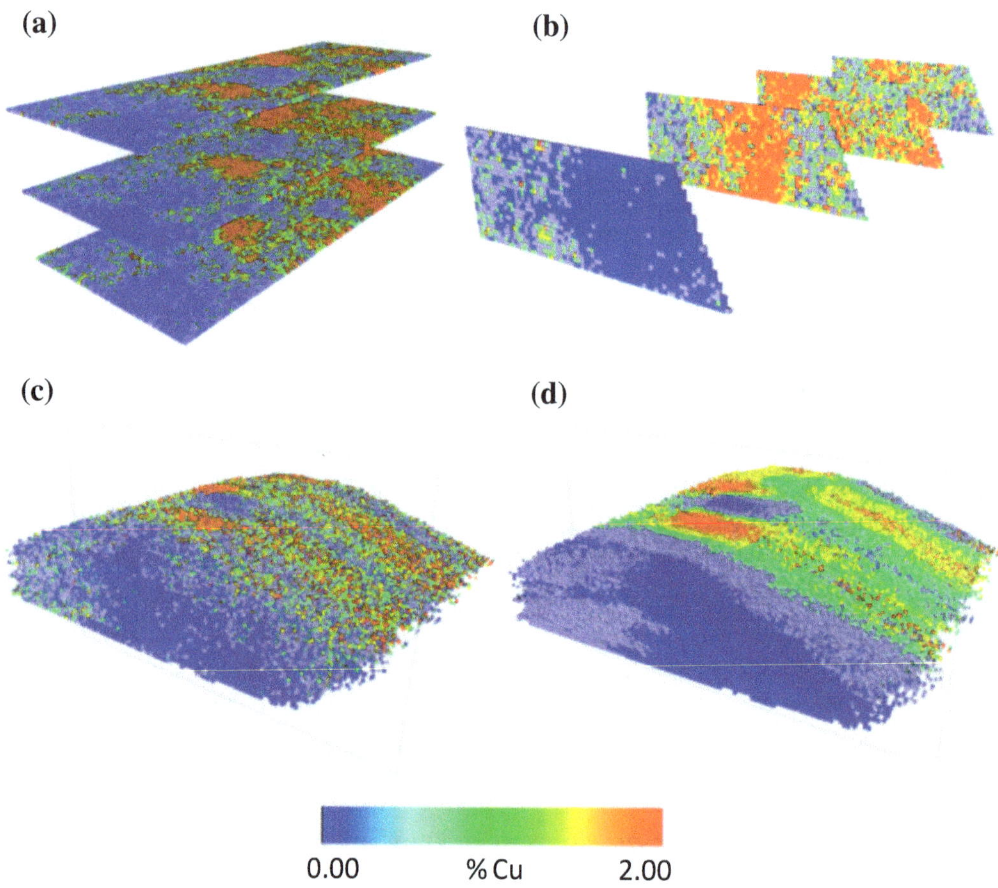


**Fig. 18.15** **a** simulated copper grades in a bench: three horizontal sections; **b** four vertical sections; **c** blast profile resulting from simulated blast applied to simulated grades; **d** predicted composition of blast profile from simulated blast applied to in-situ grades estimated from samples taken from blast-holes on 8 m spacing

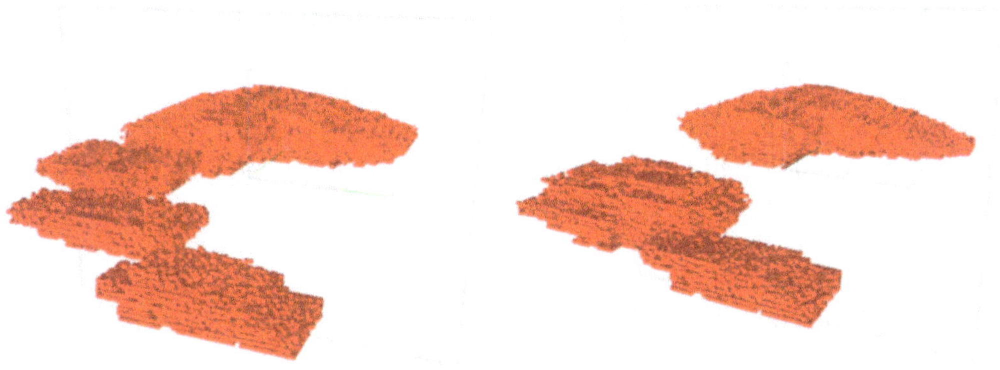

**Fig. 18.16** (Left) selected ore volumes based on estimates (Right) actual ore volumes

The first aspect of predicting recovery is the in-situ heterogeneity of the ore and the extent to which it forms contiguous 'parcels' of a size relative to the selection size (capacity and size of loading equipment). The second aspect is the heterogeneity of the ore after it has been subjected to blasting (i.e., the in-situ geological spatial variability and the post-transfer in-situ blast-pile spatial variability).

Figure 18.15 shows horizontal and vertical cross-sections through a simulated bench of dimensions 80 m × 40 m × 12 m (height) simulated copper grades on horizontal planes at the top and bottom of a 12 m bench height and a 6 m mid-plane. The vertical cross-sections of the bench are extremities (0 and 80 m) and intermediate planes at 28 m intervals.

Figure 18.16 shows the assumed contiguous parcels of ore in the blast pile based on estimated in-situ grade values together with the actual (simulated) parcels of ore. A comparison of the two sets of ore volumes in Fig. 18.16 would quantify ore loss and ore dilution. Blast movement sensors, inserted in drill holes and detected in the blast-pile, are widely used to identify post-blast ore parcels. In such cases, this process would quantify the uncertainty associated with the initial placement of sensors based on estimated in-situ ore locations and a grade continuity model.

Among other examples, Goodfellow and Dimitrakopoulos (2017) describe an approach that integrates sources of uncertainties arising from the combined production of several mines. The in-situ orebody uncertainties are integrated with process uncertainties from extraction to processing to marketing as the basis of modelling and stochastically optimising the value chain of a mining complex.

## 18.7 Conclusion

There is a growing requirement for integrated frameworks for uncertainty quantification in all geologically based applications. Quantified uncertainty and geostatistical methods are increasingly being referenced explicitly in mineral resource and reserve codes. This does not require rewriting the reporting codes but it does mean that there is a need to establish a general accepted framework for the quantification of all sources of uncertainty.

Quantified risk assessments for environmental applications are now required in many jurisdictions for applications such as waste burial and the treatment, storage and disposal of radioactive material. These assessments are required to cover time periods that range from around 200 years for household wastes to thousands of years for the underground storage or disposal of radioactive wastes.

The management of groundwater resources, especially karst systems in environmentally vulnerable coastal areas, requires the integration of flow, extraction, seawater intrusion, contamination from agriculture and other activities.

In these and all such applications the identification and quantification of all sources of uncertainty is critical to ensuring reliable estimation, planning, design and, for resource extraction, production and to managing associated risks. As summarised here, many methods and approaches have been developed by many authors but most are limited to aleatory uncertainty.

The work summarised here provides examples of methods that have been successfully applied to identify and quantify all sources of uncertainty in mineral resource and environmental applications. They provide a contribution to the need, and the increasing requirement, to develop integrated frameworks for uncertainty quantification in all geologically based applications.

**Acknowledgements** I am grateful to my co-authors of our cited joint publications and particularly to Eulogio Pardo-Igúzquiza with whom I have collaborated for over 20 years.

## References

Bandemer H, Gebhart A (2000) Bayesian fuzzy kriging. Fuzzy Sets Syst 112:405–418

Bardossy A, Bogardi I, Kelly WE (1988) Imprecise (fuzzy) information in geostatistics. Math Geol 20:287–311

Bardossy A, Bogardi I, Kelly WE (1990a) Kriging with imprecise (fuzzy) variograms. I: theory. Math Geol 22:63–79

Bardossy A, Bogardi I, Kelly WE (1990b) Kriging with imprecise (fuzzy) variograms. II: application. Math Geol 22:81–94

Bardossy G, Fodor J (2004) Evaluation of uncertainties and risks in geology: new mathematical approaches for their handling. Springer. ISBN: 978-3-642-05833-2

Bedford T, Cooke R (2001) Probabilistic risk analysis: foundations and methods. Cambridge University Press. ISBN: 978-052-1773-20-1

Caers J (2011) Modelling uncertainty in the earth sciences. Wiley-Blackwell. ISBN: 978-111-9992-63-9

Dempster AP (1968) A generalisation of Bayesian inference. J Roy Stat Soc B 30:205–247

Diamond P (1989) Fuzzy kriging. Fuzzy Sets Syst 33:315–332

Dietrich CR, Osborne MR (1991) Estimation of covariance parameters in kriging via restricted maximum likelihood. Math Geol 23(7):655–672

Dimitrakopoulos R, Farrelly CT, Godoy M (2002) Moving forward from traditional optimization: grade uncertainty and risk effects in open-pit design. Trans Inst Min Metall Sect A Min Technol 111:82–88

Dowd PA (1986) Geometrical and geological controls in geostatistical estimation and orebody modelling. In: Ramani RV (ed) Proceedings of 19th APCOM conference, Jostens Publications, pp 81–94. ISSN: 0741-0603; ISBN: 0-87335-058-8

Dowd PA (1994) Geological controls in the geostatistical simulation of hydrocarbon reservoirs. Arab J Sci Eng 19(2B):237–247

Dowd PA (1995) Björkdal gold-mining project, northern Sweden. Trans Inst Min Metall Sect A Min Ind 104:149–163

Dowd PA (1997) Risk in minerals industry projects: analysis, perception and management. Trans Inst Min Metall Sect A Min Ind 106:9–18

Dowd PA, Dare-Bryan PC (2004) Planning, designing and optimising production using geostatistical simulation. In: Proceedings of the international symposium on orebody modelling and strategic mine planning, AusIMM (Melbourne). ISBN: 1-920806-22-9; 321-338

Dowd PA, Scott IR (1984) The application of geostatistics to mine planning in a structurally complex silver/lead/zinc orebody. In: Jones MJ (ed) Proceedings of the 18th APCOM conference; pub. institution of mining and metallurgy, London. ISBN: 0-900488-73-5. 255-264

Dowd PA, Johnstone SAW, Bower J (1989) The application of structurally controlled geostatistics to the Hilton orebodies, Mt. Isa, Australia. In: Weiss A (ed) Proceedings of 21st APCOM conference, society of mining engineers, Colorado, USA. pp 275-285. ISBN 0-87335-079-0

Dowd PA, Pardo-Igúzquiza E (2002) Incorporation of model uncertainty in geostatistical simulation. Geogr Environ Model 6(2):149–171

Dubois D, Prade H (2001) Possibility theory, probability theory and multiple-valued logics: a clarification. Ann Math Artif Intell 32:35–66

Goodfellow R, Dimitrakopoulos R (2013) Algorithmic integration of geological uncertainty in pushback designs for complex multi-process open pit mines. Trans Inst Min Metall Sect A Min Technol 122(2):67–77

Goodfellow R, Dimitrakopoulos R (2017) Simultaneous stochastic optimization of mining complexes and mineral value chains. Math Geosci 49:341–360

Gotway CA (1994) The use of conditional simulation in nuclear waste site performance assessment. Technometrics 36(2):129–141

Helton JC, Johnson JD, Oberkampf WL (2004) An exploration of alternative approaches to the representation of uncertainty in model predictions. Reliab Eng Syst Saf 85:39–71

Hermans T, Caers J, Nguyen F (2014) Assessing the probability of training image-based geological scenarios using geophysical data. In: Pardo-Igúzquiza E, Guardiola-Albert C, Heredia J, Moreno L, Durán J, Vargas-Guzmán J (eds) Mathematics of planet earth. Lecture notes in earth system sciences. Springer, pp 679–682

Hoppe RW (1978) Stekenjokk: a mixed bag of tough geology and good mining and milling practices. Engineering and mining journal operating handbook of underground mining, pp 270–274. ISBN: 0-0709-9928-7

Hora SC (1996) Aleatory and epistemic uncertainty in probability elicitation an example from hazardous waste management. Reliab Eng Syst Saf 54:217–223

JORC Code (2012) Australasian code for reporting of exploration results, mineral resources and ore reserves. http://www.jorc.org

Journel AG (1994) Modelling uncertainty; some conceptual thoughts. In: Dimitrakopoulos R (ed) Geostatistics for the next century; Kluwer quantitative geology and geostatistics series, vol 6, pp 30–43. ISBN: 0-7923-2650-4

Kitanidis PK (1986) Parameter uncertainty in estimation of spatial functions: Bayesian analysis. Water Resour Res 22(4):499–507

Kitanidis PK, Lane RW (1985) Maximum likelihood parameter estimation of hydrologic spatial processes by the Gauss-Newton method. J Hydrol 79(1–2):53–71

Loquin K, Dubois D (2010a) Kriging and epistemic uncertainty. In: Jeansoulin R, Papini O, Prade H, Shockaert S (eds) Methods for handling imperfect spatial information. Studies in fuzziness and soft computing, vol 256. Springer, pp 269-305. ISSN: 1434-9922; ISBN: 978-3-642-14754-8

Loquin K, Dubois D (2010b) Kriging with ill-known variogram and data. In: Deshpande A and Hunter A (eds) Scalable uncertainty management. Lecture notes in computer science, vol 6379. Springer, Berlin, pp 219-235. ISSN: 0302-9743; ISBN: 978-3-642-15950-3

Mardia KV, Marshall RJ (1984) Maximum likelihood estimation of models for residual covariance in spatial regression. Biometrika 71(1):135–146

Matheron G (1975) Hasard, échelle et structure. Ann des Min

Matheron G (1976) Le choix des modèles en géostatistique. In: Guarascio M, David M, Huijbregts C (eds) Advanced geostatistics in the mining industry, NATO A.S.I. Series C: Mathematical and physical sciences, vol 24. D. Reidel Pub. Co, pp 11–27. Print ISBN: 978-940-1014-72-4. Online: 978-940-1014-70-0

Matheron G (1978) Estimer et choisir. Centre de Géostatistique et de Morphologie Mathématique, Fontainebleau. English translation: Hasofer AM (1989) Estimating and Choosing: an essay on probability in practice. Springer. ISBN: 978-3-540-50087-2. Republished 2013 by Presses des Mines, France; ISBN: 978-2-35671-056-7

Oberkampf WL, DeLand SM, Rutherford BM, Diegert KV, Alvin KF (2002) Error and uncertainty in modelling and simulation. Reliab Eng Syst Saf 75:333–357

Oberkampf WL, Helton JC, Joslyn CA, Wojtkiewicz SF, Ferson S (2004) Challenge problems: uncertainty in system response given uncertain parameters. Reliab Eng Syst Saf 85:11–19

Olea RA, Pardo-Igúzquiza E, Dowd PA (2015) Robust and resistant semi-variogram modelling using a generalized bootstrap. J South Afr Inst Min Metall 115:37–44

Omre H (1987) Bayesian kriging—merging observations and qualified guesses in kriging. Math Geol 19(1):25–39

Pardo-Igúzquiza E, Dowd PA (1997a) Statistical inference of covariance parameters by approximate maximum likelihood estimation. Comput Geosci 23(7):793–805

Pardo-Igúzquiza E, Dowd PA (1997b) A case study of model selection and parameter inference by maximum likelihood with application to uncertainty analysis. Non-renew Resour 7(1):63–73

Pardo-Igúzquiza E, Dowd PA (1997c) Maximum likelihood inference of spatial covariance parameters of soil properties. Soil Sci 163(3):212–219

Pardo-Igúzquiza E, Dowd PA (2003) Assessing the uncertainty of spatial covariance parameters of soil properties estimated by maximum likelihood. Soil Sci 168(11):769–782

Pardo-Igúzquiza E, Dowd PA (2013) Comparison of inference methods for estimating semi-variogram model parameters and their uncertainty: the case of small data sets. Comput Geosci 50:154–164

Pardo-Igúzquiza E, Dowd PA, Baltuille JM, Chica-Olmo M (2013) Geostatistical modelling of a coal seam for resource risk assessment. Intern J Coal Geol 112:134–140

Park H, Scheidt C, Fenwick D, Boucher A, Caers J (2013) History matching and uncertainty quantification of facies models with multiple geological interpretations. Comput Geosci 17 (4):609–621

Sankararaman S, Mahadevan S (2011) Model validation under epistemic uncertainty. Reliab Eng Syst Saf 96:1232–1241

Shafer G (1976) A mathematical theory of evidence. Princeton University Press. ISBN: 978-069-1100-42-5

Srivastava RM (1994) Comments on modelling uncertainty: some conceptual thoughts. In: Dimitrakopoulos (ed) Geostatistics for the next century; Kluwer quantitative geology and geostatistics series, vol 6. ISBN: 0-7923-2650-4. 44-45

Srivastava RM (2005) Probabilistic modelling of ore lens geometry: an alternative to deterministic wireframes. Math Geol 37(5):513–544

Verly G, Brisebois K, Hart W (2008) Simulation of geological uncertainty, resolution porphyry copper deposit. In: Proceedings of the eighth geostatistics congress, vol 1, Santiago, Chile. pub. Gecamin Ltd, pp 31–40. ISBN: 978-956-8504-18-2

Winkler RL (1981) Combining probability distributions from dependent information sources. Manag Sci 27(4):479–488

Winkler RL (1996) Uncertainty in probabilistic risk assessment. Reliab Eng Syst Saf 54:127–132

Xu C, Dowd PA (2014) Stochastic fracture propagation modelling for enhanced geothermal systems. Math Geosci 46(6):665–690

Zadeh LA (1965) Fuzzy sets. Inf Control 8:338–353

Zadeh LA (1978) Fuzzy sets as a basis for a theory of possibility. Fuzzy Sets Syst 1:3–28

Zimmerman DL (1989) Computationally efficient restricted maximum likelihood estimation of generalised covariance functions. Math Geol 21(7):655–672

# Chapter 19
# Advances in Sensitivity Analysis of Uncertainty to Changes in Sampling Density When Modeling Spatially Correlated Attributes

**Ricardo A. Olea**

**Abstract** A comparative analysis of distance methods, kriging and stochastic simulation is conducted for evaluating their capabilities for predicting fluctuations in uncertainty due to changes in spatially correlated samples. It is concluded that distance methods lack the most basic capabilities to assess reliability despite their wide acceptance. In contrast, kriging and stochastic simulation offer significant improvements by considering probabilistic formulations that provide a basis on which uncertainty can be estimated in a way consistent with practices widely accepted in risk analysis. Additionally, using real thickness data of a coal bed, it is confirmed once more that stochastic simulation outperforms kriging.

## 19.1 Introduction

In any form of sampling, there is always significant interest in establishing the reliability that may be placed on any conclusions extracted from a sample of certain size. In the earth sciences and engineering, such conclusions can be the extension of a contamination plume or the in situ resources of a mineral commodity. Increases in sample size result in monotonic improvements with diminishing returns: up to measuring the entire population, the benefits increase with the number of observations. In the classical statistics of independent random variables, the number of observations is all that counts. In spatial statistics, however, the locations of the data are also important.

Early on in spatial sampling, it was recognized that sampling distance was a factor in determining the reliability of estimations. However, insurmountable difficulties of incorporating other factors led to the reliability of spatial samplings

R. A. Olea (✉)
U.S. Geological Survey, 12201 Sunrise Valley Drive, Mail Stop 956,
Reston, VA 20192, USA
e-mail: rolea@usgs.gov





B. S. Daya Sagar et al. (eds.), *Handbook of Mathematical Geosciences*,
https://doi.org/10.1007/978-3-319-78999-6_19

being determined solely by geographical distance, particularly for the public disclosure of mineral resources (e.g., USBM and USGS 1976).

Significant advances in the determination of spatial uncertainty did not take place until the advent of digital computers and the formulation of geostatistics (e.g., Matheron 1965). Geostatistics introduced the concept of kriging variance, which was a significant improvement over the relatively simplistic distance criteria for determining reliability. The third generation of methods to determine reliability of spatial sampling came with the development of spatial stochastic simulation shortly after the formulation of kriging (Journel 1974).

Although there are several reports in the literature about applications of distance methods (e.g., USGS 1980; Wood et al. 1983; Rendu 2006) and kriging (e.g., Olea 1984; Bhat et al. 2015), the mere fact that distance methods are still being used indicates that the merits of the geostatistical methods remain unappreciated. This chapter is an application of the three families of methods for conducting sensitivity analyses on the reliability of the assessment of geologic resources due to variations in sample spacing. The simulation formulation given here is novel as it is an illustrative example used for comparing all three approaches.

## 19.2 Data

The data in Fig. 19.1 and Table 19.1 of the Appendix will be used to anchor the presentation. They are thickness measurements for the Anderson coal bed in a central part of the Gillette coal field of Wyoming taken from a more extensive study (Olea and Luppens 2014). A conversion factor could have been used to transform

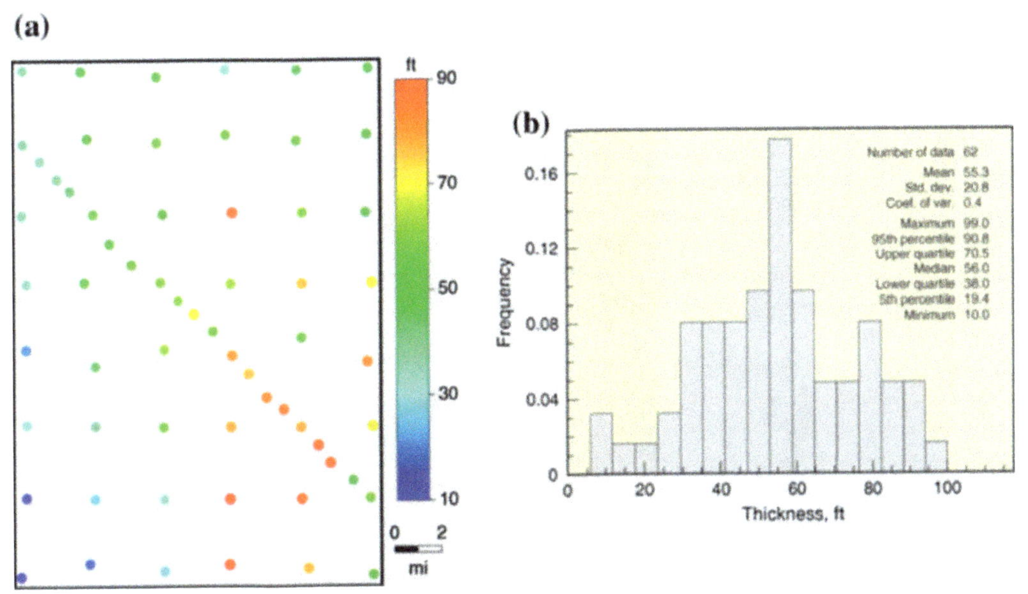


**Fig. 19.1** Measurements of thickness for the Anderson coal bed in a central part of the Gillette coal field, Wyoming, USA: **a** posting of values; **b** histogram

all the thickness values to tonnage, but it was decided to perform the analysis in terms of the attribute actually measured. The reader may want to know, however, that a density of 1,770 short tons per acre-foot for subbituminous coal is a good average value to estimate tonnage values and that the cell size used here is 400 ft by 400 ft.

With resources of more than 200 billion short tons of coal in place, the Gillette coal field is one of the largest coal deposits in the United States (Luppens et al. 2008). There are eleven beds of importance in the field. The Anderson coal bed, in the Paleocene Tongue River Member of the Fort Union Formation, is the thickest and most laterally continuous of the six most economically significant beds. This low sulfur, subbituminous coal has a field average thickness of 45 ft. Hence, it is the main mining target.

## 19.3 Traditional Uncertainty Assessment

For a long time, the prevailing practice has been the determination of uncertainty in mining assessments based on distance between drill holes. Figure 19.2 shows an example following U.S. Geological Survey Circular 891 (Wood et al. 1983), hereafter referred to as Circular 891. This example uses the drill holes in Fig. 19.1a after eliminating the holes along the diagonal. Circular 891 classifies resources into four categories according to the distance from the estimation location to the closest drill hole:

- 0 to ¼ mi: measured
- ¼ to ¾ mi: indicated
- ¾ to 3 mi: inferred
- More than 3 mi: hypothetical

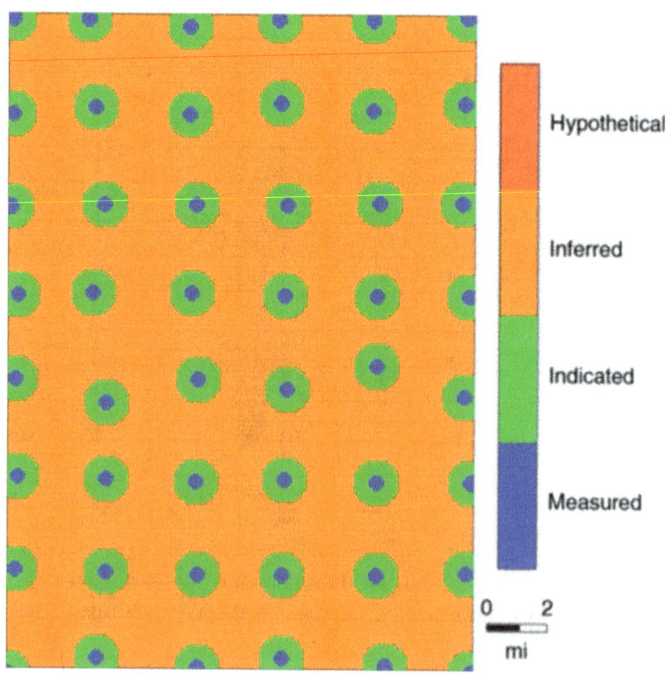


**Fig. 19.2** Classification of in situ resources according to Circular 891 for the data in Fig. 19.1a after eliminating the drill holes along the diagonal

Classification schemes like this are fairly simple and gained popularity prior to the advent of computers. Evaluating the degree of uncertainty of a magnitude or an event is the domain of statistics (e.g., Caers 2011). The standard approach for analyzing uncertainty consists of listing all possible values or events and then assigning a relative frequency of occurrence. A simple example is the tossing of a coin, where the outcomes are head and tail. For a fair coin, these two events occur with the same frequency, which is called probability when normalized to vary from 0 to 1. The same concept can be applied to any event or attribute, including coal bed thickness. For example, the outcome at a site not yet drilled could be modeled as the following random variable:

- 5–10 ft, probability 0.3
- 10–15 ft, probability 0.4
- 15–21 ft, probability 0.2
- 21–28 ft, probability 0.1

Note that the sum of the probabilities of all possible outcomes is 1.0. Random variables rigorously allow answering multiple questions about unknown magnitudes, in this case, the likely thickness to penetrate. A sample of just three assertions would be: (a) coal will certainly be intersected because the value zero is not listed among the possibilities; (b) it is more likely that the intersected thickness will be less than 15 ft than greater than 15 ft; and (c) odds are 6 to 4 that the thickness will be between 10 and 21 ft, or to put it differently, the 11 ft interval between 10 and 21 ft has a probability of 0.6 of containing the true thickness. These are the standard concepts and tools used universally in statistics to characterize uncertainty.

The classification system established by Circular 891 does not use probabilities and lacks the predictive power of a random variable approach. In particular,

- The classification uses an ordinal scale (e.g., Urdan 2017), supposedly ranked, but the classification does not indicate how much more uncertain one category is relative to another. In practice, it has been found that errors may not be significantly different among categories (Olea et al. 2011).
- The results of a distance classification are difficult to validate. The tonnage in a class denotes an accumulated magnitude over an extensive volume of the deposit. The entire portion of the deposit comprising a class would have to be mined in order to determine the exact margin of error in the classification for such a class. In practical terms, the classification is not falsifiable, thus it is unscientific (Popper 2000). Moreover, there is little value in determining the reliability of a prediction post mining.
- The classification fails to consider the effect of geologic complexity. Coal deposits ordinarily contain several geologically different beds that may be penetrated drilling a single hole. When all beds are penetrated by the same vertical drill holes, the drilling pattern is the same for all beds. Using the Circular 891 classification method, the areal extension of each category is the same for the resources of each coal bed separately and for the accumulated

resources considering all coal beds, while logic indicates that the extension of true reliability classes should be all different.

- For similar reasons, in a multi-seam deposit, increasing the drilling density results in the same reduction in uncertainty for all coalbeds, which is also unrealistic.
- The number of methods for estimating resources is continuously growing, hopefully for the better. Considering that not all methods are equally powerful, independently of the data, different methods offer varying degrees of reliability. The uncertainty denoted by the Circular 891 classification is insensitive to the methods used in the calculation of the tonnage. For example, inferred resources remain as inferred resources independently of the nature and quality of the methods used in the assessment.

Despite these drawbacks and the formulation of the superior alternatives below, Circular 891 and similar approaches remain the prevailing methods worldwide for the public disclosure of uncertainty in the assessment of mineral resources and reserves (JORC 2012; CRIRSCO 2013).

## 19.4 Kriging

Kriging is a family of spatial statistics methods formulated for the improvement in the reporting of uncertainty and in the estimation of the attributes of interest themselves. Although it is possible to establish links between kriging and other older estimation methods in various disciplines, mining was the driving force behind the initial developments of kriging and other related methods collectively known today as geostatistics (Cressie 1990).

Kriging is basically a generalization of minimum mean square error estimation taking into account spatial correlation. Kriging provides two numbers per location $(\mathbf{s}_o)$ conditioned to some sample of the attribute $(z(\mathbf{s}_i), i = 1, 2, \ldots, N)$: an estimate of the unknown value $(z^*(\mathbf{s}_o))$ and a standard error $(\sigma(\mathbf{s}_o))$. The exact expression for these results depends on the form of kriging. For ordinary kriging, the most commonly applied form and the one used here, the equations are:

$$z^*(\mathbf{s}_o) = \sum_{i=1}^{n} \lambda_i \cdot z(\mathbf{s}_i) \quad (19.1)$$

$$\sigma^2(\mathbf{s}_o) = \left( \sum_{i=1}^{n} \lambda_i \cdot \gamma(\mathbf{s}_o, \mathbf{s}_i) \right) - \mu \quad (19.2)$$

where:

$n \leq N$ is a subset of the sample consisting of the observations closest to $\mathbf{s}_o$;
$\gamma(\mathbf{d})$ is the semivariogram, a function of the distance $\mathbf{d}$ between two locations;

$\lambda_i$ is a weight determined by solving a system of linear equations comprising semivariogram terms; and

$\mu$ is a Lagrange multiplier, also determined by solving the same system of equations.

The method presumes knowledge of the function characterizing the spatial correlation between any two points, which is never the case. A structural analysis must be conducted before running kriging to estimate this function: a covariance or semivariogram. The semivariogram can be regarded as a scaled distance function. The weights and the Lagrange multiplier depend on the semivariogram for multiple drill-hole to drill-hole distances and estimation location to drill-hole distances. For details, see for example Olea (1999).

The two terms, $z^*(\mathbf{s}_o)$ and $\sigma^2(\mathbf{s}_o)$, are the mean and the variance of the random variable modeling the uncertainty of the true value of the attribute $z(\mathbf{s}_o)$, terms that are compatible with all that is known about the attribute through the sample of size $N$. Variance is a measure of dispersion, in this case, dispersion of possible values around the estimate, which is the most likely value. Hence, changing the sample, a sensitivity analysis of kriging variance is a sensitivity analysis of variations in uncertainty due to changes in the sampling scheme. From Eq. 19.2, the kriging variance does not depend directly on the observations. The dependence is only indirect through the semivariogram, which is based on the data. Considering that there is one true semivariogram per attribute, changes in adequate sampling should not result in significant changes in the estimated semivariogram, which is kept constant. This independence between data and standard error facilitates the application of kriging to the sensitivity analysis in the reliability of an assessment due to changes in sampling strategy because mathematically actual measurements are not necessary to calculate standard errors; the modeler only has to specify the semivariogram and the sampling locations.

Figure 19.3 shows the set of estimated semivariogram values obtained using the sample in Fig. 19.1 plus a model fitting the points for the purpose of having valid semivariogram values for any distance. In this case, the fitted curve is called a spherical model with a nugget of 20 sq ft, sill of 595 sq ft and a spatial correlation range of 88,920 ft. Geologically, the nugget is related to the variance of short scale fluctuations; the sill is of the same order of magnitude as the sample variance, and the correlation range is equal to half the average geographical size of the anomalies. For details on structural analysis, see for example Olea (2006).

Figure 19.4 shows the results of applying ordinary kriging to the sample in Fig. 19.1a and Table 19.1 in the Appendix. As expected, the standard error is zero at the drill holes because there is no uncertainty where measurements have been taken.

Although kriging can analyze any configuration, Fig. 19.5 only relates to additions or eliminations to the basic sample in Fig. 19.1a. Values along the diagonal were used only for modeling the semivariogram and producing Fig. 19.4. Figure 19.5a also has every other row and column eliminated. Estimates could be produced for the first two configurations because thickness is known at each drill hole. The other maps were produced by interpolating locations in the sample with

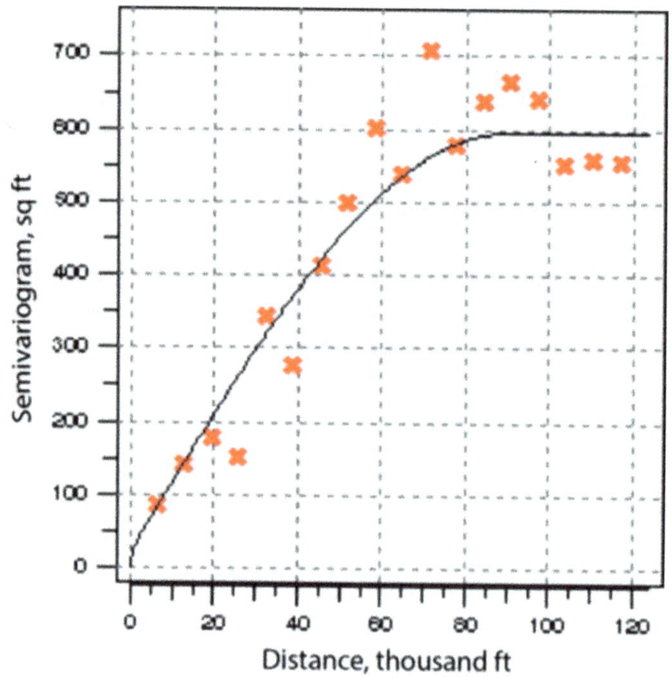


**Fig. 19.3** Semivariogram for the Anderson coal bed thickness. The crosses denote estimated values and the curve is a model fitting the values

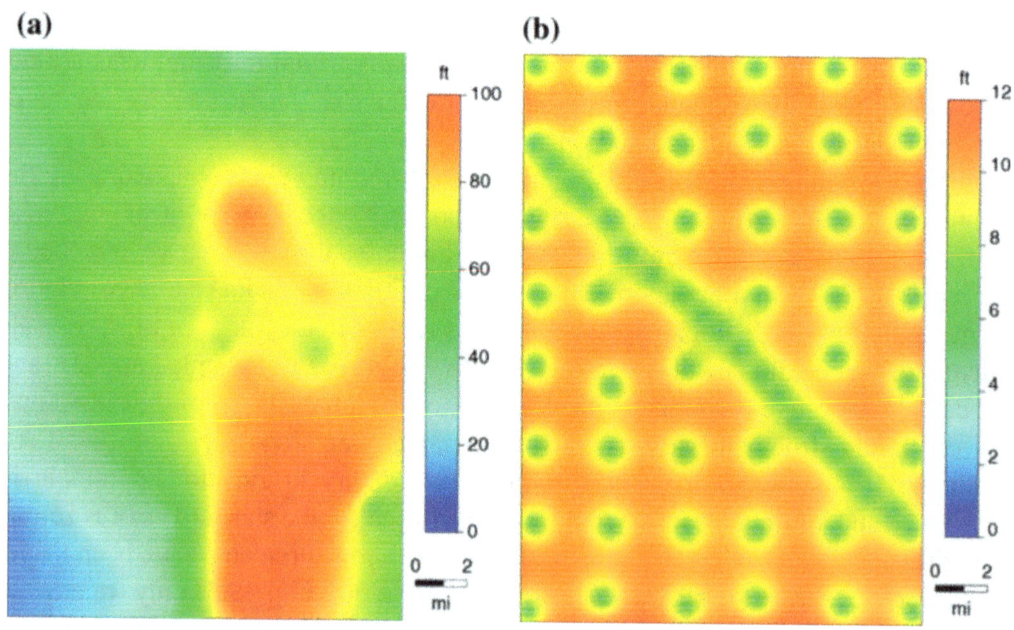


**Fig. 19.4** Ordinary kriging maps for the Anderson coal bed in a central part of the Gillette coal field (Wyoming) using the sample in Fig. 19.1: **a** thickness; **b** standard error

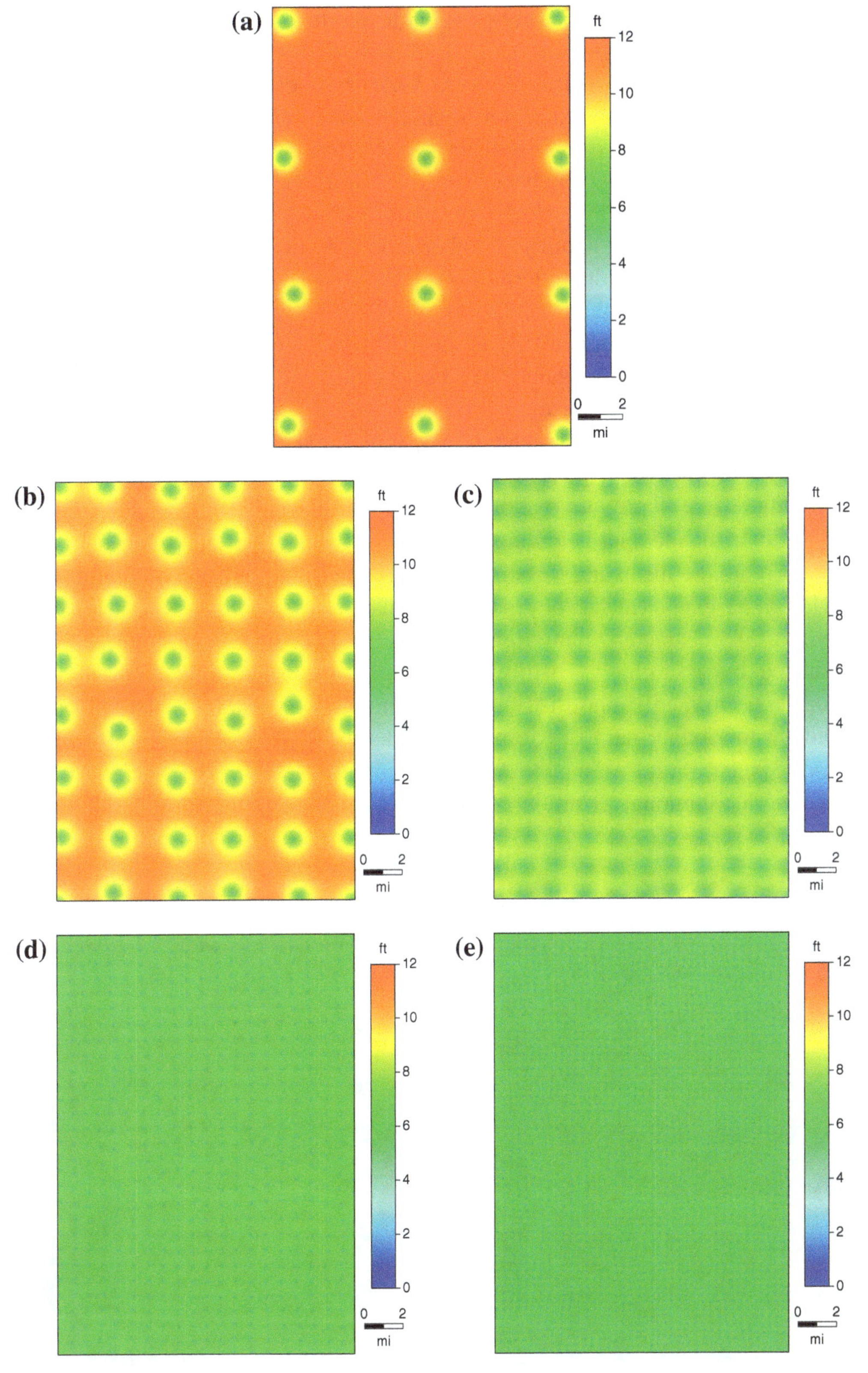


**Fig. 19.5** Ordinary kriging standard error for the same configuration in Fig. 19.2 for several average spacings: **a** 6 mi; **b** 3 mi; **c** 1.5 mi; **d** 3/4 mi; **e** 3/8 mi

the next largest spacing; it is only possible to produce the standard error map for Fig. 19.5c–e.

The similarity between Figs. 19.2 and 19.5b may lead to incorrect conclusions. Although the location and extension of similar colors are approximately the same, what is important is the meaning of the colors. Figure 19.2 does not provide any numerical information that can be associated with the accuracy and the precision of the estimated values. In Fig. 19.5b the numbers are standard errors, a direct measurement of estimation reliability. In other more irregular configurations, there will not be similarity in color patterns no matter how the colors are selected. For example, by expanding the boundary of the study area, Fig. 19.6 shows how the Circular 891 classification is totally insensitive to the fact that, along the periphery, there is an increase in uncertainty because the data are now to one side, not surrounding the estimation locations. Instead, kriging accounts for the fact that extrapolation is always a more uncertain operation than interpolation, an important capability when accounting for boundary effects.

Kriging is able to provide random variables for the statistical characterization of uncertainty if the modeler is willing to introduce a distributional assumption. $z^*(\mathbf{s}_o)$ and $\sigma^2(\mathbf{s}_o)$ are the mean and the variance of the distribution of the random variable providing the likely values for $z(\mathbf{s}_o)$. These parameters are necessary but not sufficient to fully characterize any distribution. However, this indetermination can be eliminated by assuming a distribution that is fully determined with these two parameters. Ordinarily, the distribution of choice is the normal distribution, followed by the lognormal. The form of the distribution does not change by subtracting $z(\mathbf{s}_o)$ from all estimates. As the difference $z^*(\mathbf{s}_o) - z(\mathbf{s}_o)$ is the estimation error, the distributional assumption also allows characterizing the distribution for the error at $\mathbf{s}_o$.

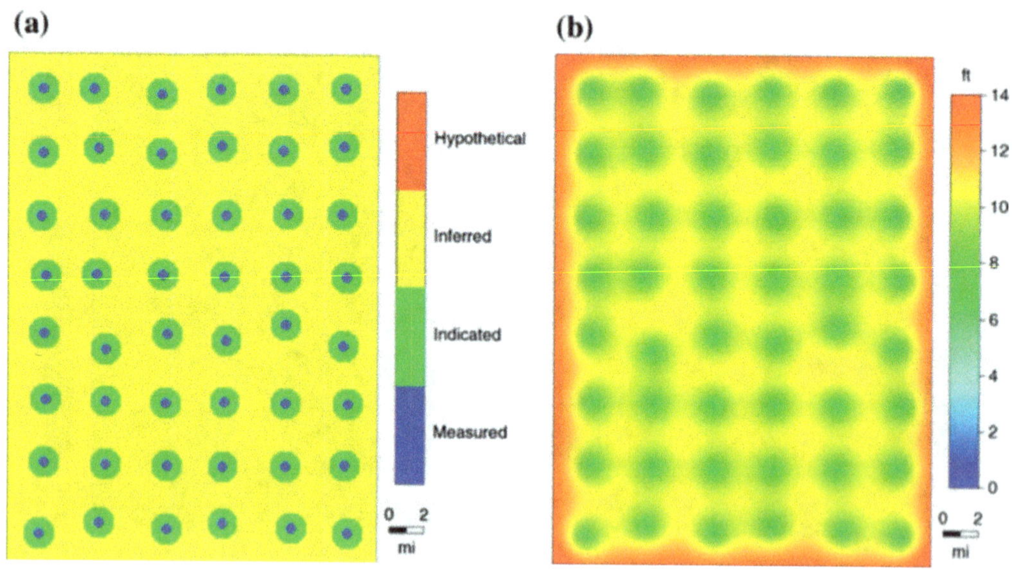


**Fig. 19.6** Comparison of results when expanding the boundaries of the study area: **a** Circular 891 classification; **b** ordinary kriging standard error

Kriging with a distribution for the errors overcomes all the disadvantages of the distance methods listed in the previous section:

- It is possible to calculate the probability that the true value of the attribute lies in any number of intervals. Probabilities are a form of a ratio variable, for which zero denotes an impossible event and, say, a 0.2 probability denotes twice the likelihood of occurrence of an event than 0.1.
- Validation is modular. An adequate theory assures that, on average, $z^*(\mathbf{s}_o)$ and $\sigma^2(\mathbf{s}_o)$ are good estimates of reality. Yet, as illustrated by an example in the last Section, if going ahead with validation of the uncertainty modeling primarily to check the adequacy of the normality assumption, it is not necessary to validate all possible locations throughout the entire deposit to evaluate the quality of the modeling.
- The effect of complexity in the geology is taken into account by the semivariogram.
- In general, the thickness of every coal bed or the accumulated values of thickness for several coal beds has a different semivariogram. Thus, even if the sampling configuration is the same, the standard error maps will be different.
- The characterization of uncertainty is specific to the estimation method because the results are valid only for estimated values using the same form of kriging used to generate the standard errors.

Figure 19.7 summarizes the results of the maps in Fig. 19.5. Display of the 95th percentile is based on the assumption that all random variables follow normal distributions. The curves clearly outline the consequences of varying the spacing in

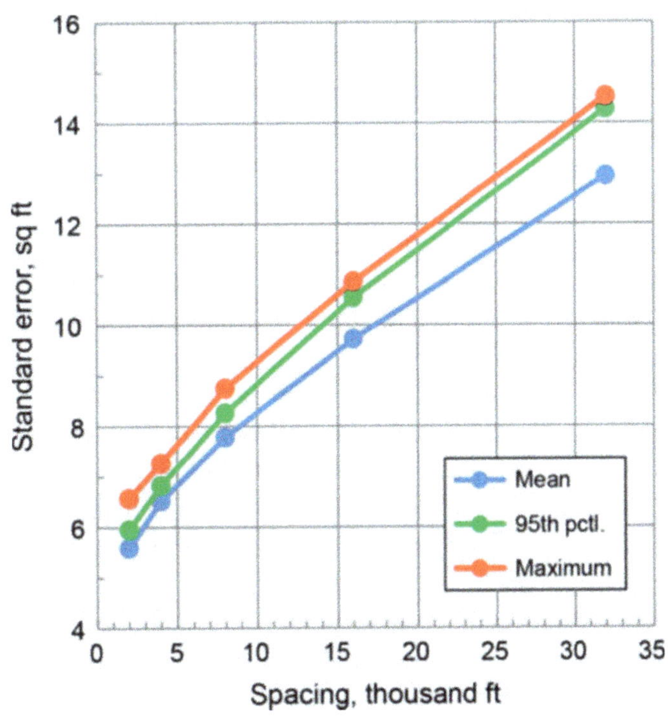


**Fig. 19.7** Sensitivity of ordinary kriging to spacing of mean standard error, its 95th percentile, and the maximum standard error based on the Fig. 19.5 configurations

a square sampling pattern from 2,000 to 32,000 ft. So, for example, if it is required that all estimates in the study area must have a standard error less than 10 ft, then the maximum spacing must be at most 12,500 ft. The validity of the results, however, is specific to the attribute and sampling pattern: thickness of the Anderson coal bed investigated with a square grid. Any change in these specifications requires preparation of another set of curves.

## 19.5 Stochastic Simulation

Despite limited acceptance, the kriging variance has been in use for a while in the sensitivity analysis of uncertainty to changes in sampling distances and configurations (e.g., Olea 1984; Cressie et al. 1990). Kriging, like any mathematical method, has been open to improvements. One result has been the formulation of another family of methods: stochastic simulation.

Relative to the topic of this chapter, stochastic simulation offers two improvements: (a) it is no longer necessary to assume the form for the distribution providing all possible values for the true value of the attribute $z(\mathbf{s}_o)$; and (b) the standard error is sensitive to the data.

As seen in Fig. 19.4, for every attribute and sample, kriging produces two maps, a map of the estimate and a map of the standard error. The idea of stochastic simulation is to characterize uncertainty by producing instead multiple attribute maps, all compatible with the data at hand and each representing one possible outcome of reality—realization, for short. From among the many available methods of geostatistical simulation, sequential Gaussian simulation has been chosen for this study because of its simplicity, versatility and efficiency (Pyrcz and Deutsch 2014). Figure 19.8 shows four simulated realizations, each of which is a possible reality in the sense that the values have the same statistics and spatial statistics (semivariogram) and the simulation reproduces the known sample values (i.e., the sample used to prepare Fig. 19.5b).

Generation of significant results needs preparation of more realizations than the four in Fig. 19.8. An estimation of uncertainty requires summarizing the fluctuations from realization to realization, either at local or global scales. Figure 19.9 is an example of local fluctuation summarizing all values of thickness at the same location for 100 realizations. This histogram is the numerical characterization of uncertainty through a random variable. There is one random variable for each of the 57,528 pixels (cells) comprising each realization. As clearly implied by the selected values in the tabulation, this collection of 100 maps provides multiple predictions of the true thickness value that should be expected at this location. For example, the most likely value (mean) is 65.75 ft; the standard error is 13.47 ft; and there is a 0.95 probability that the coal bed will be less than 87.8 ft thick.

Maps can be generated for various statistics across the study area to display fluctuations in their values. Figure 19.10 shows a map of the mean and a map of the standard error. Note that the map for the mean is quite similar to the ordinary kriging map in Fig. 19.4a. More importantly, the maps for the standard errors in

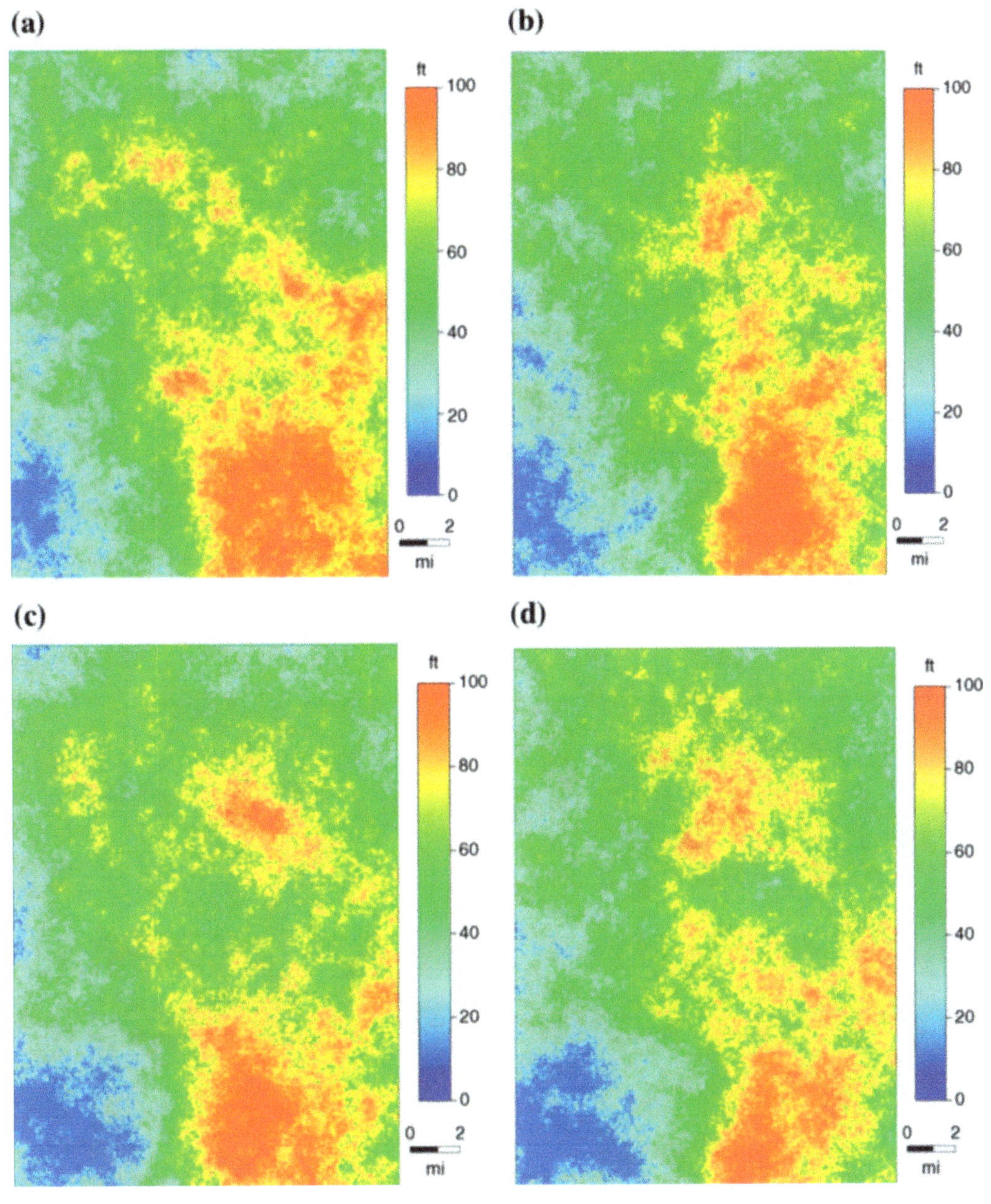


**Fig. 19.8** A sample of four sequential Gaussian realizations using the same data used in the preparation of Fig. 19.5b

Figs. 19.5b and 19.10b are significantly different. The differences in the standard errors are primarily the result of the dependency of the standard error not only on the semivariogram and the drill hole locations, but also on the values of thickness as well. For example, comparing Figs. 19.1a and 19.10b, despite the regularity in the drilling, there is less uncertainty in the southwest corner where all values are low as well as in the south central part where all values are consistently high.

Production of a display of the standard error equivalent to that in Fig. 19.5 is more challenging now that the standard deviation must be extracted from multiple realizations and the preparation of each realization requires a value at each drill hole

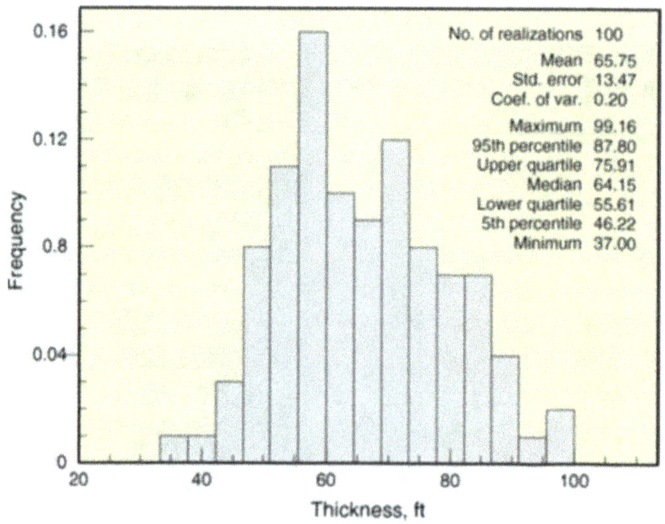


**Fig. 19.9** Example of the numerical approximation to the random variable modeling uncertainty in the value of thickness at a site not yet drilled

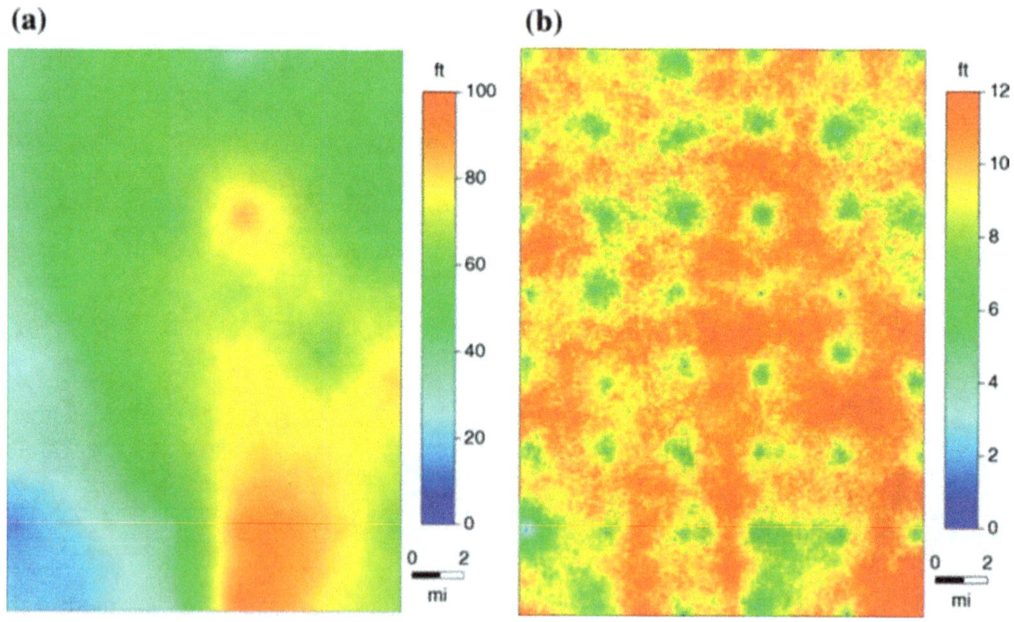


**Fig. 19.10** Anderson coal bed thickness according to 100 sequential Gaussian simulations: **a** expected value of thickness; **b** standard error

in the configuration of interest to complete the analysis. Figure 19.11 shows the equivalent results to Fig. 19.5 for the same drill holes, but now produced after applying sequential Gaussian simulation. The additional data necessary to prepare the maps in Fig. 19.11c–e where obtained by randomly selecting 10 of the 100 realizations used to prepare the maps in Figs. 19.8 and 19.10. The data for the hypothetical drill holes were taken from the values at the collocated nodes in these selected 10 realizations, thus obtaining 10 datasets consisting partly of the 48 actual data in Fig. 19.11b plus the artificial data obtained by "drilling" the realizations.

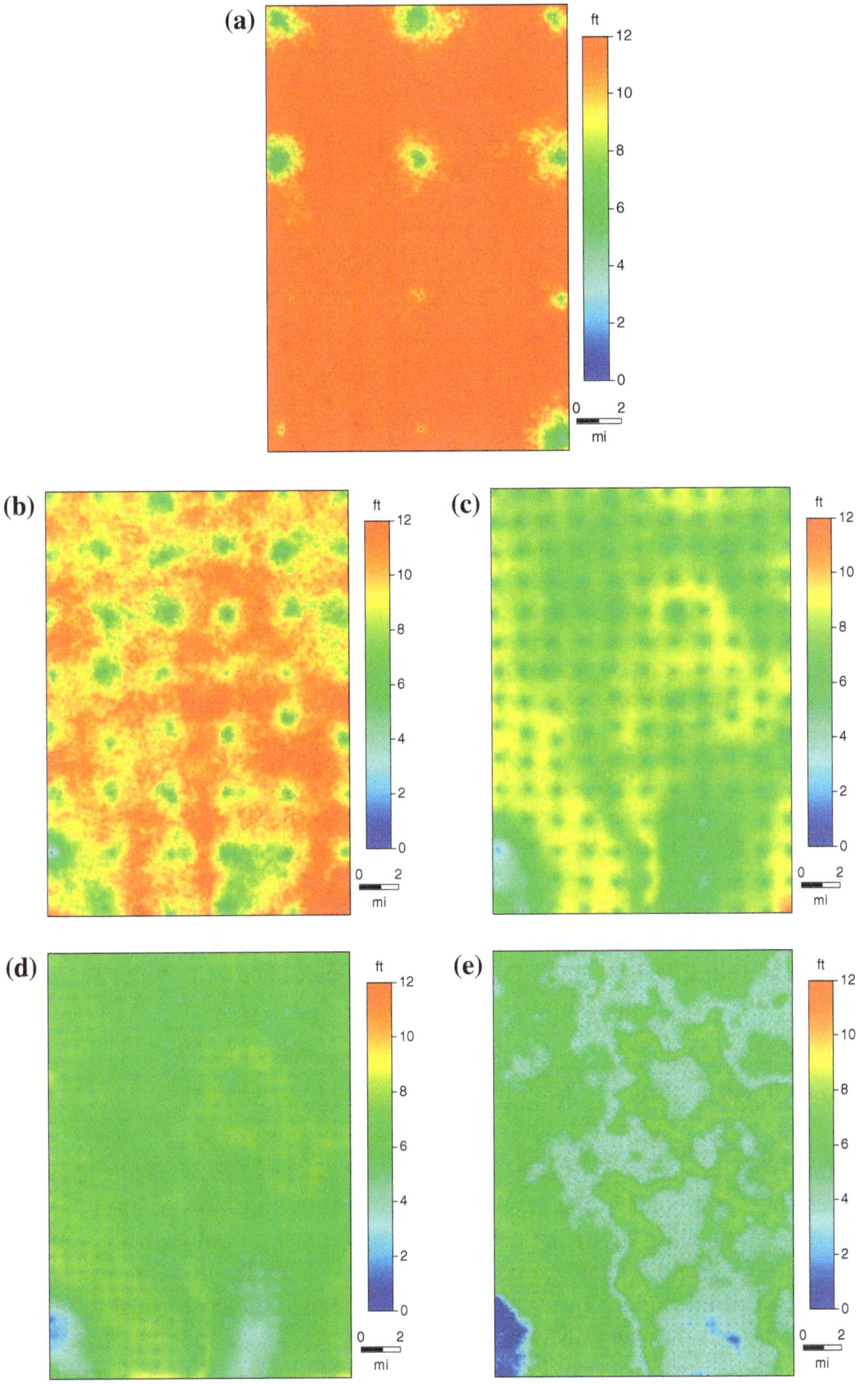


**Fig. 19.11** Sequential Gaussian simulation standard error for the same configuration on Fig. 19.2 for several average spacings: **a** 6 mi; **b** 3 mi; **c** 1.5 mi; **d** 3/4 mi; **e** 3/8 mi

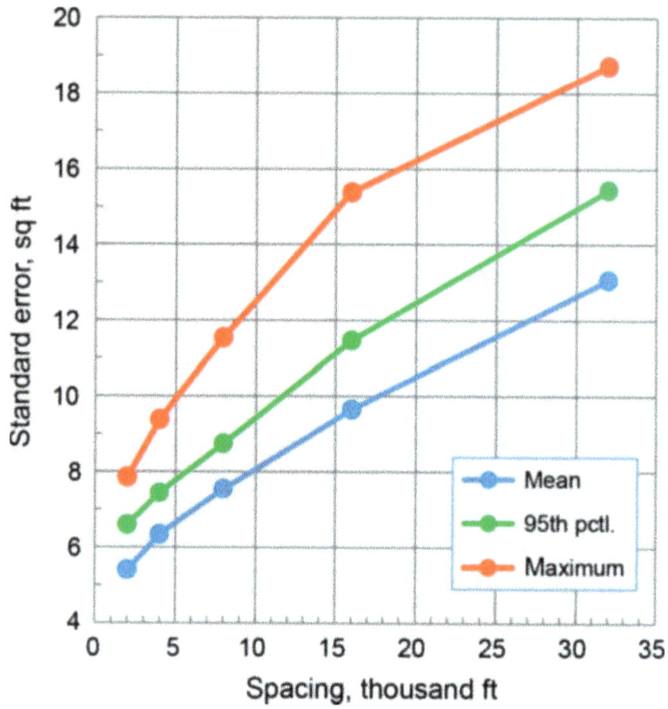


**Fig. 19.12** Sequential Gaussian simulation sensitivity to spacing of mean standard error, its 95th percentile, and the maximum standard error based on the configurations in Fig. 19.11

Finally, each dataset was used to generate 100 realizations, for a total of 1,000 realizations per configuration. As mentioned for Fig. 19.10b, despite the regularity of the drill hole pattern, the fluctuations in standard error are no longer completely determined by the drilling pattern.

Figure 19.12 is the summary equivalent to that in Fig. 19.7. Considering the completely different methodologies behind both sets of curves, the results are quite similar, particularly the curves for the mean standard error, which are almost identical. The more extreme standard errors of the sequential Gaussian simulation are larger than those for ordinary kriging in the case of the 95th percentile and the maximum value. The remaining question is: Which approach produces the most realistic forecasts of uncertainty?

## 19.6 Validation

Figure 19.13 provides an answer to the question above in terms of percentiles. A percentile is a number that separates a set of values into two groups, one below and the other one above the percentile. The percentage of values below gives the name to the percentile. For example, in Fig. 19.9, the value 46.22 ft separates the 100 values of thickness into two classes, those below and those equal to or above 46.22 ft. It turns out that only 5 of the 100 values are below 46.22 ft. Hence, 46.22 ft is the 5th percentile of that dataset. Accepting only integer values of percentages, there are 99 percentiles in any dataset. The quality of a model of uncertainty can be validated by checking the proportion of true values that are actually below the percentiles of the prediction random variables collocated with data not used in the

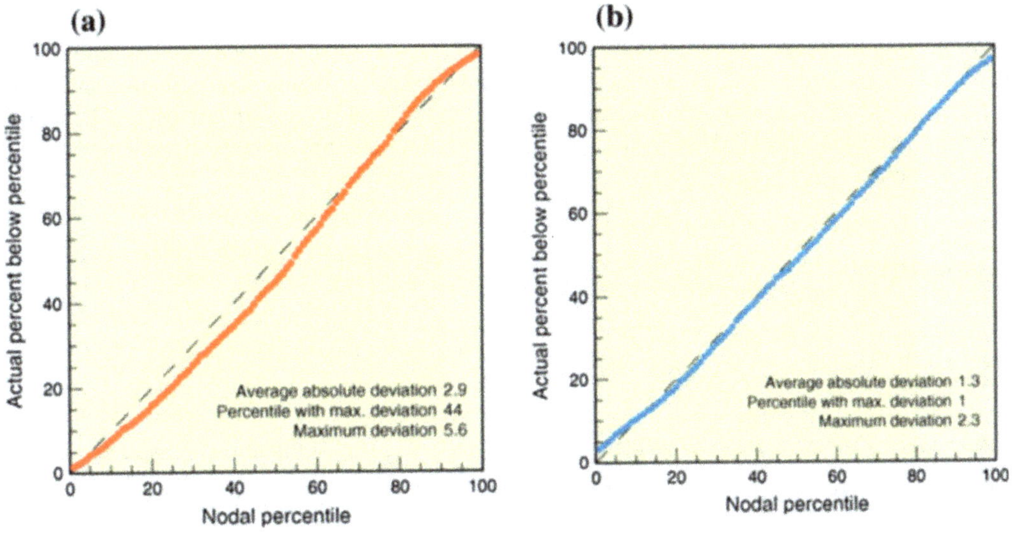


**Fig. 19.13** Validation of the uncertainty predictions made for the 3 mi spacing samples: **a** ordinary kriging; **b** sequential Gaussian simulation

modeling. One of the reasons for selecting the Anderson coal thickness for the study is that there are much more data than the 48 values used to generate the realizations, a generous set of 2,136 additional values to be precise. This larger number of values has been used for checking the accuracy of the percentiles, not only the 5th percentile, but all 99 percentiles. In the graphs, the actual percentage shows, on average, the proportion of times the true value was below the percentile of a random variable at the location of a censored measurement. For example, in Fig. 19.13a, 641 times out of 2,136 (i.e., 30%) the true value was indeed below the 35th percentile. Ideally, all dots should lie along the main diagonal. The clear winner is sequential Gaussian simulation.

## 19.7 Conclusions

Distance methods, kriging and stochastic simulation rank, in that order, in terms of increasing detail and precision of the information that they are able to provide concerning the uncertainty associated to any spatial resource assessment.

The resource classification provided by distance methods is completely independent of the geology of the deposit and the method applied to calculate the mineral resources. The magnitude of the resource per class has no associated quantitative measure of the deviation that could be expected between the calculated resource and the actual amount in place.

The geostatistical methods of kriging and stochastic simulation base the modeling on the concept of random variable used in statistics, which allows the same type of probabilistic forecasting used in other forms of risk assessments. Censored data were used for validating the accuracy of the probabilistic predictions that can be made using the geostatistical methods. The results were entirely satisfactory, particularly in the case of stochastic simulation.

**Acknowledgements** This contribution completed a required review and approval process by the U.S. Geological Survey (USGS) described in Fundamental Science Practices (http://pubs.usgs.gov/circ/1367/) before final inclusion in this volume. I wish to thank Brian Shaffer and James Luppens (USGS), Peter Dowd (University of Adelaide) and Josep Antoni Martín-Fernández (Visiting Fulbright Scholar, USGS) for suggestions leading to improvements to earlier versions of the manuscript.

# Appendix

See Table 19.1.

**Table 19.1** Thickness data. ID = identification number; Thick. = thickness; ft = feet

| ID | Easting (ft) | Northing (ft) | Thick. (ft) | ID | Easting (ft) | Northing (ft) | Thick. (ft) |
|---|---|---|---|---|---|---|---|
| 2 | 431,326 | 1,316,298 | 49.0 | 39 | 399,741 | 1,236,607 | 77.0 |
| 3 | 398,753 | 1,316,124 | 32.0 | 40 | 432,107 | 1,236,582 | 70.5 |
| 4 | 352,156 | 1,316,015 | 37.0 | 41 | 384,280 | 1,236,527 | 58.0 |
| 5 | 365,531 | 1.315,818 | 49.0 | 42 | 415,737 | 1,236,459 | 78.0 |
| 7 | 382,816 | 1,314,601 | 55.0 | 44 | 352,743 | 1,221,026 | 10.0 |
| 9 | 430,850 | 1,301,568 | 48.0 | 45 | 368,483 | 1,220,742 | 26.0 |
| 10 | 398,805 | 1,301,506 | 57.0 | 46 | 431,473 | 1,220,645 | 59.0 |
| 11 | 352,234 | 1,299,533 | 37.0 | 47 | 399,596 | 1,220,598 | 92.0 |
| 12 | 366,769 | 1,300,871 | 50.0 | 48 | 415,871 | 1,220,477 | 86.0 |
| 13 | 414,876 | 1,300,240 | 56.0 | 49 | 384,411 | 1,220,477 | 32.0 |
| 14 | 382,892 | 1,299,775 | 58.0 | 51 | 367,180 | 1,206,180 | 17.0 |
| 16 | 416,097 | 1,284,247 | 60.0 | 52 | 399,353 | 1,205,960 | 99.0 |
| 17 | 430,593 | 1,284,243 | 47.0 | 53 | 417,304 | 1,204,922 | 76.0 |
| 18 | 400,291 | 1,284,132 | 87.0 | 54 | 384,456 | 1,204,470 | 28.0 |
| 19 | 384,138 | 1,283,859 | 53.0 | 55 | 432,027 | 1,203,507 | 52.0 |
| 20 | 368,123 | 1,283,849 | 56.0 | 56 | 351,466 | 1,203,245 | 11.0 |
| 21 | 351,956 | 1,283,728 | 36.0 | 123 | 356,115 | 1,295,788 | 35.0 |
| 23 | 366,138 | 1,268,773 | 55.0 | 145 | 360,095 | 1,291,759 | 38.0 |
| 24 | 383,559 | 1,268,661 | 60.0 | 166 | 362,980 | 1,289,047 | 42.0 |
| 25 | 431,915 | 1,268,363 | 70.0 | 216 | 371,863 | 1,277,272 | 50.0 |
| 26 | 415,962 | 1,268,347 | 75.0 | 234 | 377,019 | 1,272,660 | 57.0 |
| 27 | 399,884 | 1,268,270 | 63.0 | 282 | 387,755 | 1,264,534 | 60.0 |
| 28 | 352,933 | 1,268,254 | 34.0 | 299 | 391,477 | 1,261,727 | 70.0 |
| 30 | 352,738 | 1,253,951 | 21.0 | 318 | 395,814 | 1,257,798 | 58.0 |
| 31 | 384,499 | 1,253,969 | 62.0 | 380 | 403,832 | 1,248,290 | 75.0 |
| 32 | 400,076 | 1,252,554 | 79.0 | 406 | 407,848 | 1,243,143 | 81.0 |
| 33 | 415,868 | 1,256,420 | 57.0 | 427 | 411,790 | 1,240,470 | 84.0 |
| 34 | 368,579 | 1,250,159 | 44.0 | 470 | 419,690 | 1,232,449 | 92.0 |
| 35 | 430,979 | 1,251,072 | 81.5 | 497 | 422,447 | 1,228,465 | 90.0 |
| 37 | 352,979 | 1,237 155 | 30.0 | 512 | 427,604 | 1,224,598 | 46.0 |
| 38 | 368,493 | 1,236,862 | 37.0 | 1001 | 415,000 | 1,316,000 | 45.0 |

## References


Bhat S, Motz LH, Pathak C, Kuebler L (2015) Geostatistics-based groundwater-level monitoring network design and its application to the Upper Floridan aquifer, USA. Environ Monit Assess 187(1):4183, 15

Caers J (2011) Modeling uncertainty in the earth sciences. Wiley-Blackwell, Chichester, UK, p 229

Cressie N (1990) The origins of kriging. Math Geol 22(3):239–252

Cressie N, Gotway CA, Grondona MO (1990) Spatial prediction from networks. Chemometr Intell Lab Syst 7(3):251–271

CRIRSCO (Combined Reserves International Reporting Standards Committee) (2013) International reporting template for the public reporting of exploration results, mineral resources and mineral reserves, pp 41. http://www.crirsco.com/templates/international_reporting_template_november_2013.pdf

JORC (Joint Ore Reserves Committee) (2012) Australasian code for reporting of exploration results, mineral resources and ore reserves. http://www.jorc.org/docs/jorc_code2012.pdf

Journel AG (1974) Geostatistics for conditional simulation of ore bodies. Econ Geol 69(5):673–687

Luppens JA, Scott DC, Haacke JE, Osmonson LM, Rohrbacher TJ, Ellis MS (2008) Assessment of coal geology, resources, and reserves in the Gillette coalfield, Powder River Basin, Wyoming. U.S. Geological Survey, Open-File Report 2008-1202, pp 127. http://pubs.usgs.gov/of/2008/1202/

Matheron G (1965) Les variables régionalisées et leur estimation; Une application de la théorie des fonctions aléatories aux Sciences de la Nature. Masson et Cie, Paris, pp 305

Olea RA (1984) Sampling design optimization for spatial functions. J Int Assoc Math Geol 16 (4):369–392

Olea RA (1999) Geostatistics for engineers and earth scientists. Kluwer Academic Publishers, Norwell, MA, p 303

Olea RA (2006) A six-step practical approach to semivariogram modeling. Stoch Env Res Risk Assess 39(5):453–467

Olea RA, Luppens JA (2014) Modeling uncertainty in coal resource assessments, with an application to a central area of the Gillette coal field, Wyoming. U.S. Geological Survey Scientific Investigations Report 2014-5196, pp 46. http://pubs.usgs.gov/sir/2014/5196/

Olea RA, Luppens JA, Tewalt SJ (2011) Methodology for quantifying uncertainty in coal assessments with an application to a Texas lignite deposit. Int J Coal Geol 85(1):78–90

Popper KR (2000) The logic of scientific discovery. Routledge, London, p 480

Pyrcz MJ, Deutsch CV (2014) Geostatistical reservoir modeling, 2nd edn. Oxford University Press, New York, p 433

Rendu JM (2006) Reporting mineral resources and mineral reserves in the United States of America—Technical and regulatory issues. In: Proceedings, Sixth International Mining Geology Conference: Australasian Institute of Mining and Metallurgy, pp 11–20

Urdan TC (2017) Statistics in plain English. Routledge, New York, p 265

USBM and USGS (U.S. Bureau of Mines and U.S. Geological Survey) (1976) Coal resource classification system of the U.S. Bureau of Mines and U.S. Geological Survey. Geological Survey Bulletin 1450-B, pp 7. https://pubs.usgs.gov/bul/1450b/report.pdf

USGS (U.S. Geological Survey) (1980) Principles of resource/reserve classification for minerals. U.S. Geological Survey Circular 831, pp 5. http://pubs.usgs.gov/circ/1980/0831/report.pdf
Wood GH Jr, Kehn TM, Carter MD, Culbertson WC (1983) Coal resources classification system of the U.S. Geological Survey. U.S. Geological Survey Circular 891, pp 65. http://pubs.usgs.gov/circ/1983/0891/report.pdf

# Chapter 20
# Predicting Molybdenum Deposit Growth

**John H. Schuenemeyer, Lawrence J. Drew and James D. Bliss**

**Abstract** In the study of molybdenum deposits and most other minerals deposits, including copper, lead and zinc, there is speculation that most undiscovered ore results from an increase (or "growth") in the estimated size of a known deposit due to factors such as exploitation and advances in mining and exploration technology, rather than in discovering wholly new deposits. The purpose of this study is to construct a nonlinear model to estimate deposit "growth" for known deposits as a function of cutoff grade. The model selected for this data set was a truncated normal cumulative distribution function. Because the cutoff grade is commonly unknown, a model to estimate cutoff grade conditioned upon the deposit grade was constructed using data from 34 deposits with reported data on molybdenum grade, cutoff grade, and tonnage. Finally, an example is presented.

**Keywords** Porphyry molybdenum · Deposit growth · Cutoff grade
Truncated cumulative distribution model fitting and estimation · Confidence and prediction intervals for nonlinear estimation

## 20.1 Introduction

Initial estimates of a mineral deposit size based on limited data usually underestimate the ultimate size of a mineral deposit, often by a significant amount. The initial size estimate may be of only marginal interest but the size estimate after some exploration and development can be of significant interest. The steps in this process are the subject of this chapter. "Mineral resources" are defined as concentrations or occurrences of material of economic interest in or on the Earth's crust in such form, quality, and quantity that there are reasonable prospects for eventual economic extraction (Zientek and Hammarstrom 2014), and the term "mineral reserves" is restricted to the economically mineable part of a mineral resource.

J. H. Schuenemeyer (✉) · L. J. Drew · J. D. Bliss
Southwest Statistical Consulting, LLC, Cortez, CO, USA
e-mail: jackswsc@q.com





B. S. Daya Sagar et al. (eds.), *Handbook of Mathematical Geosciences*,
https://doi.org/10.1007/978-3-319-78999-6_20

The reported size of known mineral or oil and gas deposit reserves recorded in the mining literature typically increases through time as subsequent development drilling and mining enlarge the deposit's footprint. This phenomenon is referred to as "deposit growth". In a sense, a deposit is never finished "growing" until it is completely mined out. Research on the growth of a deposit's reserves has been a topic of investigation for many years within the United States Geological Survey. Drew (1997) illustrated the growth of oil and gas fields over time in the United States and determined that a large percentage of the ultimate production of a region could come from deposit growth, if the forecast was made early enough in the discovery process. Long (2008) defined reserve growth as the ratio of current reserves plus past production to original reserves. He examined reserve growth in porphyry copper deposits and found that about 20% of porphyry copper mines in the Western Hemisphere had experienced reserve growth of a factor of 10 or better over initial reserves. Reserve growth at these mines added reserves comparable in size to reserves added through discovery of new deposits during the same time period.

Three variables are required to estimate the ultimate size of a deposit: (1) the grade of the deposit, (2) cutoff grade of the deposit, and (3) associated tonnage of ore at successive points in the development of the deposit (Long 2008). The grade of a deposit is defined as the relative quantity of ore mineral within the orebody, typically expressed as a percentage (or g/t). The grade may vary across an orebody, but commonly an average grade may be applied to the orebody as a whole. A cutoff grade is the lowest grade of mineralized material that qualifies as economically mineable and available in a given deposit (Committee for Mineral Reserves International Reporting Standards 2006). Mined material with a grade below the cutoff grade is not processed into metal but is set aside. As deposit development and mining progress, over time the cutoff grade usually declines in an orderly manner. Tonnage is typically reported in metric tons (mt) and includes the mass of total production, reserves and resources of pre-mined material.

The purpose of this study was to construct a nonlinear model to estimate the incremental deposit "growth" for known mineralized areas as a function of cutoff grade, using porphyry molybdenum deposits as an example. Porphyry molybdenum deposits are related to granitic plutons, mostly of Tertiary age, and are formed by hydrothermal fluids associated with the emplacement of granites. They typically occur as large tonnage, low-grade deposits that are commonly mined using open-pit methods.

Two issues must be addressed to predict porphyry molybdenum deposit growth. The first is that, in many instances, the cutoff grade is not available for a given deposit and thus must be estimated. Thus, the first part of this study uses the known molybdenum grade of a deposit to predict probable cutoff grade. The second part of this study in turn uses this predicted cutoff grade to estimate deposit growth as a function of cutoff grade. Two data sets were used in this study. Nearly all porphyry molybdenum deposits used in this study are for unworked deposits; that is, deposits that have been delineated by drilling but are yet unmined. The first data set (Appendix 1) consists of 34 porphyry molybdenum deposits used to model molybdenum cutoff grade in percent (COG) as a function of molybdenum deposit grade, also expressed in percent. The second data set (Appendix 2) is used to model the deposit growth as a function of cutoff grade. The references to Appendices 1 and 2 are Barnes et al. (2009), Baudry (2009), Becker et al. (2009), British Columbia Ministry of Energy and Mines (2012, 2014a, b), Chen and Wang (2011), Ewert et al. (2008), General Moly (2012), Geological Survey of Finland (2011), Geoscience Australia (2012), Kramer (2006), Lowe et al. (2001), Ludington and Plumlee (2009), Mercator Minerals (2011), Mindat.org (1992, 2011), Nanika Resources Inc (2012), Northern Miner (2010), Raw Minerals Group (2011), RX Exploration Inc (2010), Singer et al. (2008), Smith (2009), Taylor et al. (2012), Thompson Creek Metals Company Inc (2011), TTM Resources Inc (2009), US Geological Survey (2011), Wu et al (2011), Yukon Geological Survey (2005). The authors know of no subset of publications that cite the deposits presented in Appendices 1 and 2.

## 20.2 Cutoff Grade as a Function of Deposit Grade

The first and most straightforward of the two models to analyze is the relationship between molybdenum cutoff grade (Mo COG, %) as a function of molybdenum deposit grade (Mo Grade, %) for the 34 deposits shown in Appendix 1. A scatter plot between these two variables plus a fitted linear regression line, 95% confidence intervals, and 95% prediction intervals are shown in Fig. 20.1.

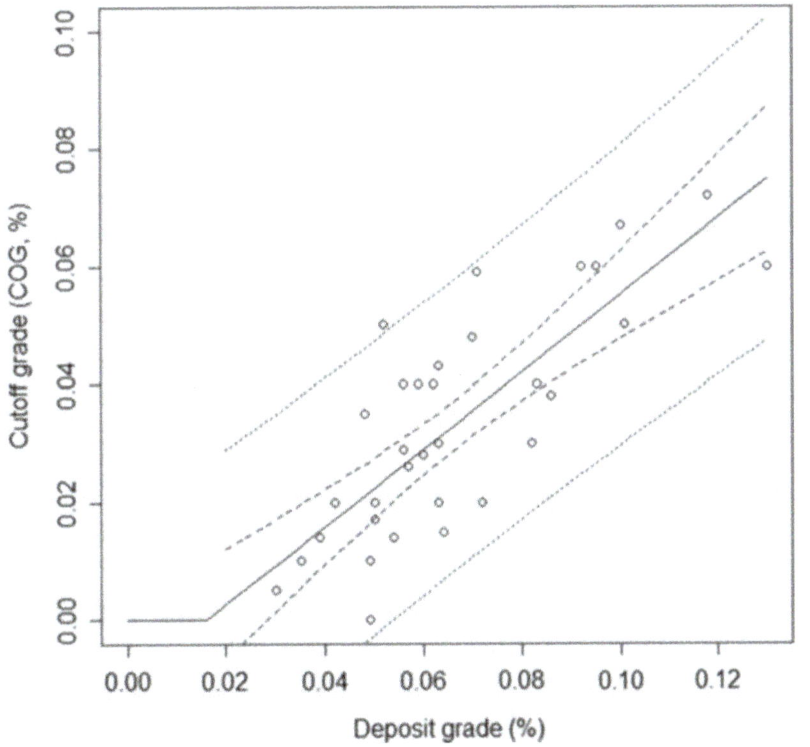


**Fig. 20.1** Cutoff grade (COG, %) versus deposit grade (%) plus a fitted linear model and the 95% confidence intervals (dashed lines) and corresponding prediction intervals (dotted lines) for the 34 deposits (Appendix 1)

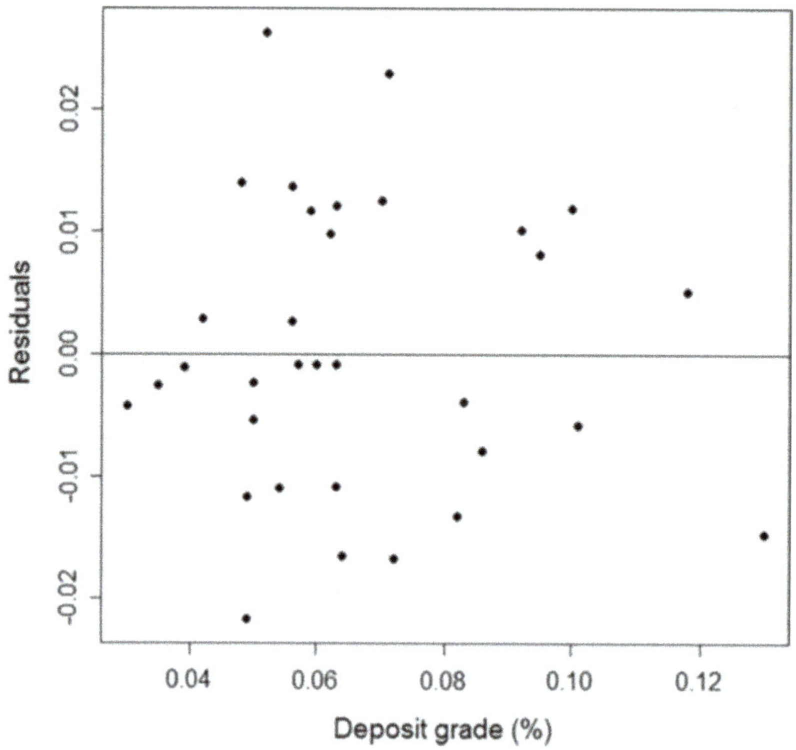


**Fig. 20.2** Residuals versus deposit grade for the linear model fit (Fig. 20.1)

The model to fit cutoff grade $U$ as a function of deposit grade $D$ is

$$U = \begin{cases} 0 & 0 \le D < c \\ \beta_0 + \beta_1 D + \varepsilon & D \ge c \end{cases}$$

where $\varepsilon$ is the random error, assumed to be normal $N(0, \sigma^2)$. The constant $c$ is determined from the linear regression fit since the COG $\ge 0$.

The fitted model is:

$$\hat{U} = \begin{cases} 0 & 0 \le D < c = 0.0159 \\ \beta_0 + \beta_1 D = -0.01042 + 0.6553D & D \ge 0.0159 \end{cases}$$

where $\hat{U}$ is the estimated cutoff grade in percent and $D$ is the deposit grade in percent. The residual standard error is 0.012 on 32 degrees of freedom and the adjusted $R^2 = 0.61$. The model is statistically significant and reasonable for the given data set. The residual plot is shown in Fig. 20.2.

There is no evidence to suggest that the residuals are non-normal. Thus, within the domain of the deposit grade, namely from 0.03 to 0.13, the linear model shown above appears to be appropriate. Predictions outside of this interval will depend on the same linear relationship holding.

## 20.3 Deposit Growth as a Function of Cutoff Grade

The second model is the fraction of growth as a function of estimated cutoff grade. In this example the growth data (Fig. 20.3) consists of 58 observations from eight deposits (Appendix 2). The inverse S shaped form of the data corresponds to an inverse cumulative distribution function. Therefore, this relationship is modeled as an inverse cumulative distribution function, since the fraction growth is a number between 0 and 1, inclusive. Several models including the gamma, lognormal, normal and their left truncated forms were candidates to fit this data. Of these, the left truncated normal was the best fit by visual inspection and by a nonlinear least squares fit. The form of the left truncated normal probability distribution function is:

$$f_L(x|\Theta) = \frac{f(x|\Theta)}{1 - F(\lambda|\Theta)} \qquad x > \lambda$$

where $\Theta' = (\mu, \sigma^2)$ and the left truncation point $\lambda$ is assumed known. The probability density function for the normal distribution with mean $\mu$ and standard deviation $\sigma$ is:

$$f(x|\mu, \sigma^2) = \frac{e^{-(x-\mu)^2/2}}{\sqrt{2\pi\sigma}}$$

The corresponding left truncated cumulative distribution function, cdf, is:

$$F_L(x|\Theta) = \frac{F(x|\Theta) - F(\lambda|\Theta)}{1 - F(\lambda|\Theta)}, \qquad x > \lambda$$

The truncated distributions' models used for model fitting are from the package truncdist (r-project.org) by Novomestky and Nadarajah (2012) based upon work by Nadarajah and Kotz (2006).

As Fig. 20.1 shows, there is uncertainty in the COG when estimated from the deposit grade. However, when estimating the left truncated normal cumulative distribution function (cdf), the estimates are conditioned upon the COG being known. A possible alternative is an errors-in- variables approach (Schennach 2004) where both the fraction growth and cutoff grade are considered to be random variables.

The chosen optimization criterion to estimate the fraction growth (Fig. 20.3) is

$$\min\left(\sum_{i=1}^{n} (F(x_i|\Theta) - \hat{F}(x_i)\right)^2,$$

where $x_i$ is the $i$th COG and $F$ is the cumulative distribution function. $\Theta$ contains the estimated parameters. If $F$ is a normal distribution the parameters would be $\hat{\mu}$ and $\hat{\sigma}$.

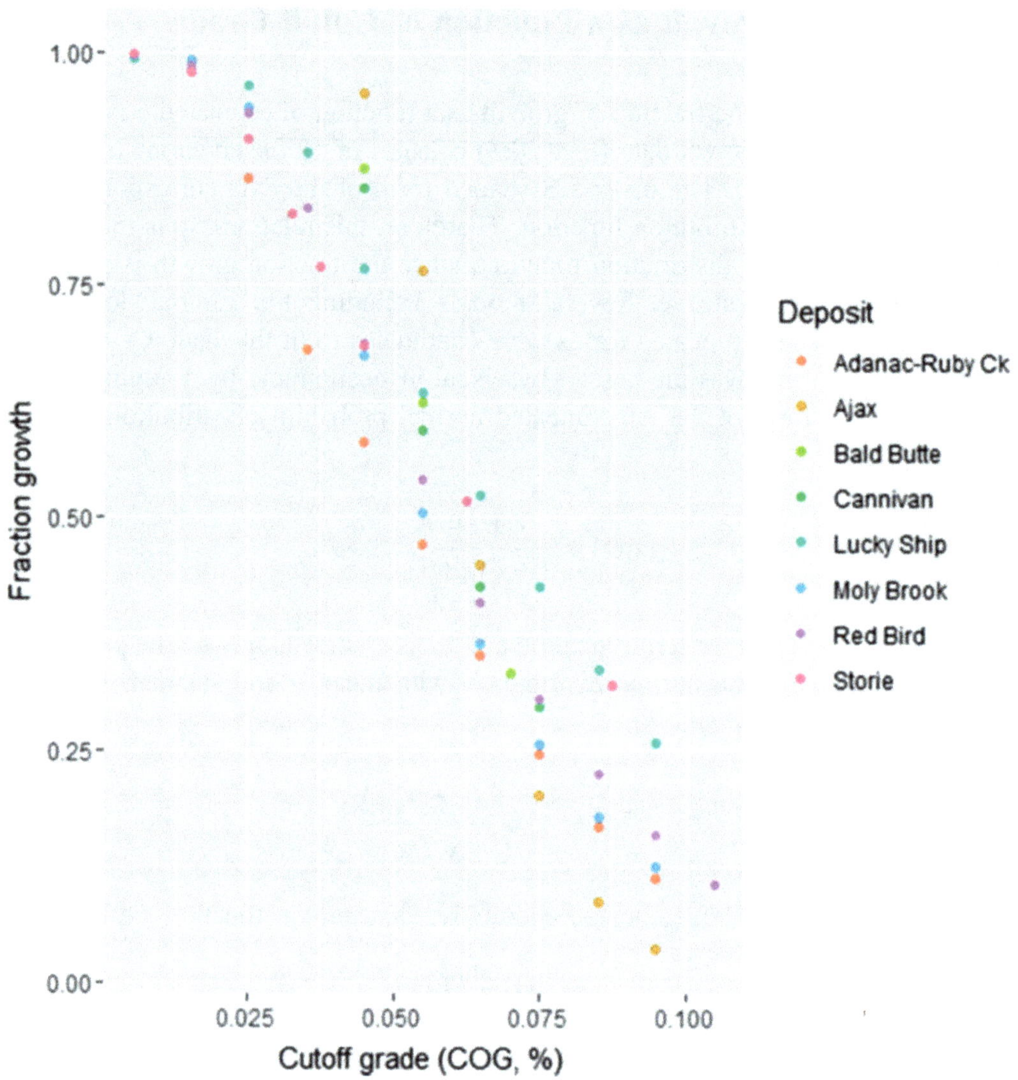


**Fig. 20.3** Deposit fraction growth plotted against cutoff grade (COG) in percent for the 8 deposits used in this study

The *i*th COG is represented by $x_i$ and $\hat{F}(x_i)$. Note that $\hat{F}(x_i) = 1 - \hat{G}(x_i)$ where $\hat{G}(x_i)$ is the fraction growth. The nonlinear least squares package used to estimate the left truncated normal model parameters is nls2 (r-project.org). See Grothendieck (2013). The left truncation point is $\lambda = 0$.

Deposit growth as a function of cutoff grade was modeled for each of the eight deposits (not shown). These results indicate that the data could have been generated from the same population Thus, the observations were pooled and a single model was fit. The reason to fit a cumulative distribution function was twofold. One was that eight deposits were used so the data was not in the form of a stepwise function. The second was that the data were not randomly or systematically spaced across the domain of the empirical distribution. The data, expressed as an empirical distribution function, together with the cumulative left truncated normal distribution fit and confidence intervals, are shown in Fig. 20.4. The results of the least square fit were $\hat{\mu} = 0.0609$ and $\hat{\sigma} = 0.0282$. The residual sum of squares, RSS = 0.3631.

The 95% confidence and prediction intervals for nonlinear estimation are approximate. The confidence interval shown in Fig. 20.4 (dashed lines) is from package propagate, r-project library predictNLS programmed by Spiess (2014) based upon work by Bates and Watts (2007), and others. It uses a second-order Taylor series expansion and Monte Carlo simulation. The second order approximation captures the nonlinearities around $f(x)$. A corresponding algorithm for the prediction interval has not been developed. The prediction interval shown in Fig. 20.4 (dotted lines) is based upon a linear model of the form $H = \alpha_0 + \alpha_1 U + \varepsilon$ where $U$ was the COG. $H$ is a linear estimate of growth. The next step was to estimate the upper and lower prediction intervals for the linear model with $U$ = 0, 0.001, 0.002, …, 0.150. These are vectors **LPIu** and **LPIl** respectively. The upper and lower 95% nonlinear confidence interval vectors estimated above are **CIu** and **CIl** respectively. The differences between the linear prediction intervals and the nonlinear confidence intervals are computed as follows. Let **Lud** = **LPIu** − **CIu** and **Lld** = **CIl** − **LPIl**. The estimated upper and lower predictions intervals, **UP** and **LP**, for the nonlinear fit (Fig. 20.4) are **UP** = **CIu** + **Lud** and **LP** = **CIi** − **Lld.** These estimates appear reasonable in the given domain, namely for COG between 0.04 and 0.10.

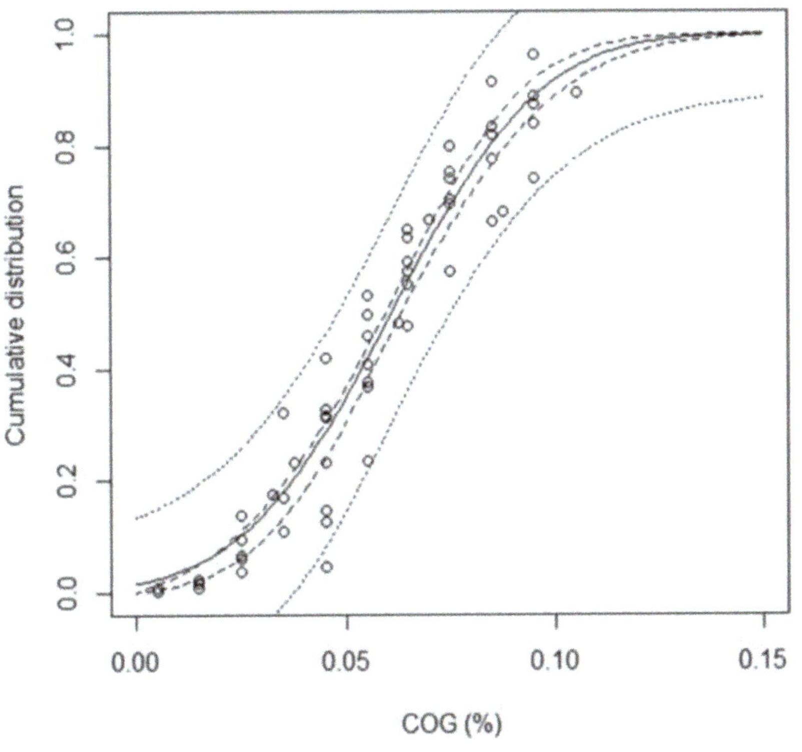


**Fig. 20.4** Data fit to a left truncated (at 0) normal distribution is the solid line. The approximate 95% confidence interval is the dashed line. The approximate 95% prediction interval is the dotted line

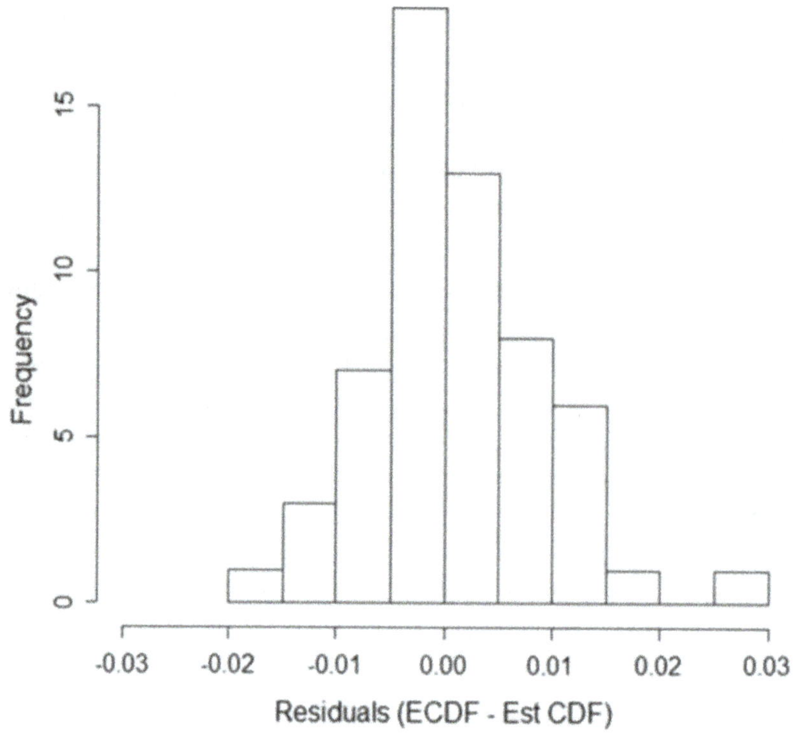


**Fig. 20.5** Histogram of residuals for fit to a left truncated normal distribution

A histogram of the residuals, which appear normal, is shown in Fig. 20.5. The truncated normal probability density function corresponding to the cumulative distribution function (Fig. 20.4) and COG data are shown in Fig. 20.6.

Figure 20.7 is like Fig. 20.4 except that the variable plotted on the vertical axis is the fraction growth as opposed to the cumulative distribution. There is no suggestion that the model illustrated in Fig. 20.7 is universal, even for molybdenum deposits. Clearly different deposits may require different models.

## 20.4 An Example

Suppose the problem is to estimate the fraction growth corresponding to a COG (%) = 0.06 using the model shown in Fig. 20.7. Then, given that the assumed distribution is a truncated normal at zero with estimated model parameters, $\hat{\mu} = 0.0609$ and $\hat{\sigma} = 0.0282$, the results are shown in Table 20.1. The point estimate of fraction growth, namely 0.479, is straightforward to compute. Namely it is:

$$\hat{F}_L(x|\hat{\Theta}) = \frac{F(x|\hat{\Theta}) - F(\lambda|\hat{\Theta})}{1 - \hat{F}(\lambda|\hat{\Theta})}, \quad x > 0, \hat{\Theta}' = (\hat{\mu}, \hat{\sigma}^2)$$

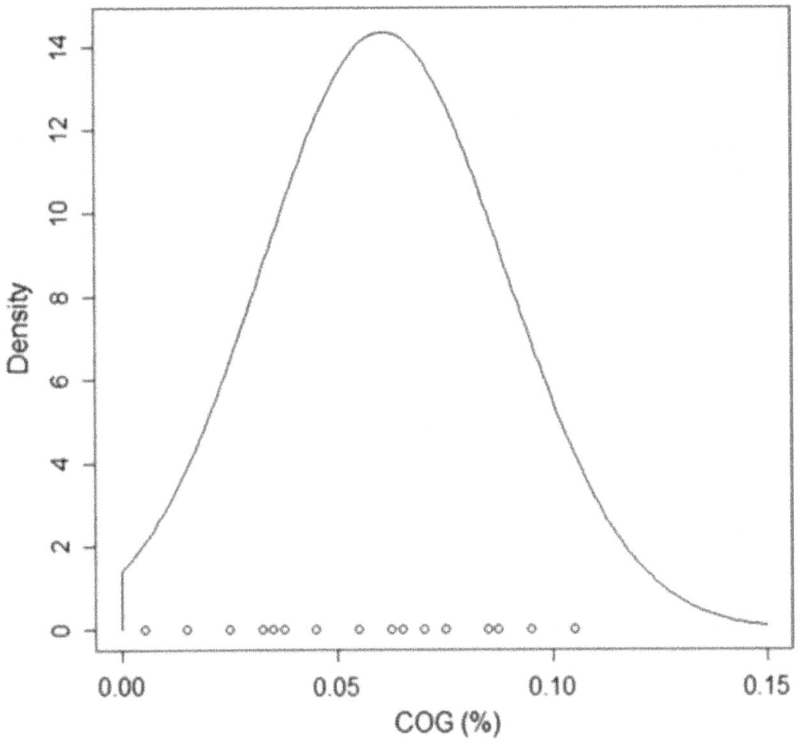


**Fig. 20.6** The fitted truncated normal probability density function and COG data (the circles)

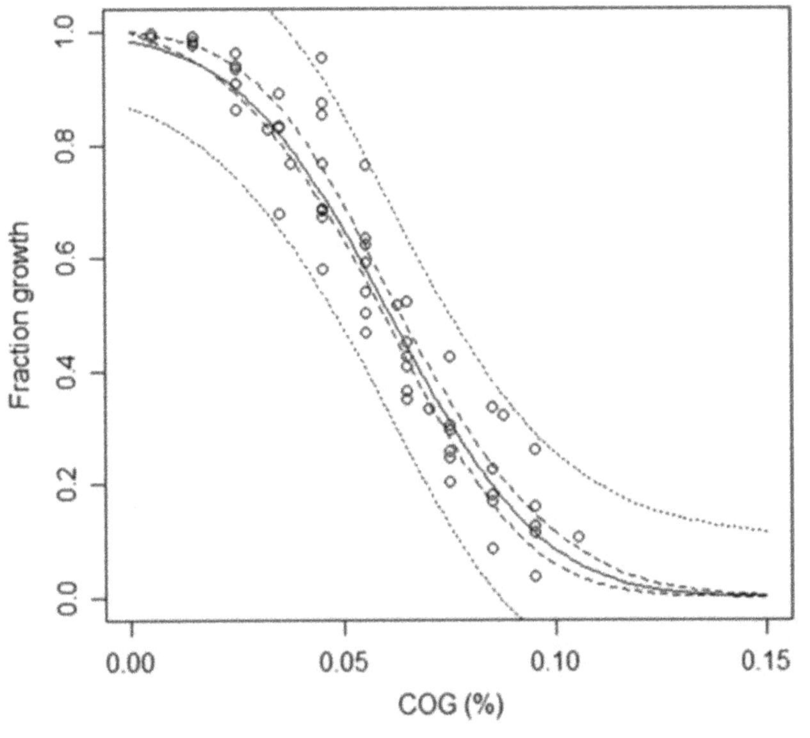


**Fig. 20.7** Fraction growth as a function of COG (%) and corresponding fitted values (solid line), 95% confidence interval (dashed line) and 95% prediction interval (dotted line)

**Table 20.1** Estimated fraction growth, 95% confidence and prediction intervals for COG (%) = 0.06

| | | Confidence interval | | Prediction interval | |
|---|---|---|---|---|---|
| COG (%) | Fraction growth | 2.50% | 97.50% | 2.50% | 97.50% |
| 0.06 | 0.479 | 0.450 | 0.507 | 0.291 | 0.666 |

The confidence and prediction intervals are more difficult to compute; however, the R code is available on request from John Schuenemeyer.

## 20.5 Conclusions

Mineral deposit growth commonly constitutes most unknown resources. The growth considered in this study is due to a progressively lower cutoff grade, which may be unknown. In this study, a statistical model was constructed to model cutoff grade as a function of deposit grade, followed by construction of a model to estimate the fraction growth as a function of cutoff grade. This latter model involves estimation of a truncated normal distribution and second order Taylor series estimates to characterize uncertainty.

**Acknowledgements** Data used in this chapter represent part of an extensive and ongoing data compilation effort on porphyry molybdenum deposit types. This study evolved over several years through discussions with current and former U.S. Geological Survey employees including Arthur A. Bookstrom, Mark D. Cocker, Robert J. Kamilli, Keith R. Long, Steve Ludington, Barry C. Moring, Greta J. Orris, Ryan D. Taylor, Jay A. Sampson, and Gregory T. Spanski. Eric Seedorf, University of Arizona, Tucson provided his bibliography on porphyry molybdenum deposits, which was of considerable use in this study.

## Appendix 1

Porphyry molybdenum data for 34 selected deposits used to model molybdenum cutoff grade as a function of deposit grade.

[Country and state codes: AUQL = Australia, Queensland; CHHN = China; CHNA = China; CNBC = Canada, British Columbia; CNNF, Canada, Newfoundland and Labrador; CNON = Canada, Ontario; CNYT = Canada, Yukon Territory; GRLD = Greenland; MCDA = Macedonia; MNGA = Mongolia; MXCO = Mexico; RUSA = Russia; USAK = USA, Alaska; USID = USA, Idaho; USMT = USA, Montana; USNV = USA, Nevada; USWA = USA, Washington]

| Name | ID | Country-State | Mo grade (%) | Deposit size (Mt) | Mo COG (%) |
|---|---|---|---|---|---|
| Ada nac-Ruby Creek | 101 | CNBC | 0.042 | 791 | 0.020 |
| Adjax-Le Roy | 102 | CNBC | 0.062 | 552 | 0.040 |
| Anduramba | 103 | AUQL | 0.054 | 32 | 0.014 |
| Bald Butte | 106 | USMT | 0.059 | 176 | 0.040 |
| Big Ben | 108 | USMT | 0.092 | 245 | 0.060 |
| Buckingham | 110 | USNV | 0.063 | 1800 | 0.043 |
| Cannivan Gulch-White Cloud | 111 | USMT | 0.056 | 327 | 0.040 |
| Carmi | 113 | CNBC | 0.057 | 40 | 0.026 |
| Cave Creek | 114 | USTX | 0.130 | 28 | 0.060 |
| Chu | 115 | CNBC | 0.050 | 673 | 0.017 |
| Creston | 118 | MXCO | 0.071 | 215 | 0.059 |
| Endako | 124 | CNBC | 0.050 | 1232 | 0.020 |
| Jiguanshan (Jiganshuan) | 130 | CHNA | 0.095 | 100 | 0.060 |
| Joem-Haskin Mountain | 131 | CNBC | 0.101 | 11 | 0.050 |
| Kitsault (Updated 11/2015) | 132 | CNBC | 0.070 | 688 | 0.048 |
| Lobash | 140 | RUSA | 0.063 | 365 | 0.030 |
| Lone Pine | 143 | CNBC | 0.072 | 179 | 0.020 |
| Lucky Ship | 144 | CNBC | 0.064 | 85 | 0.015 |
| Mac | 145 | CNBC | 0.048 | 248 | 0.035 |
| Malmbjerg | 148 | GRLD | 0.118 | 229 | 0.072 |
| Moly Brook | 151 | CNNF | 0.049 | 199 | 0.010 |
| Mount Hope | 152 | USNV | 0.039 | 1148 | 0.014 |
| Mount Tolman | 156 | USWA | 0.056 | 2200 | 0.029 |
| Pidgeon-Lateral Lake | 163 | CNON | 0.083 | 59 | 0.040 |
| Pine Nut | 165 | USNV | 0.060 | 181 | 0.028 |
| Qua rtz Hill | 167 | USAK | 0.082 | 1310 | 0.030 |
| Red Bird-Haven Lake | 169 | CNBC | 0.049 | 201 | 0.010 |
| Red Mountain | 170 | CNYT | 0.100 | 187 | 0.067 |
| Sphinx | 178 | CNBC | 0.035 | 62 | 0.010 |
| Storie Molie | 180 | CNBC | 0.049 | 105 | 0.000 |
| Sudulica-Mackatica-Kucisnjak-Groznatova Dolina | 146 | MCDA | 0.030 | 383 | 0.005 |
| Tangjiaping | 183 | CHHN | 0.063 | 373 | 0.020 |
| Thompson Creek | 184 | USID | 0.086 | 575 | 0.038 |
| Zuun Mod | 195 | MNGA | 0.052 | 408 | 0.050 |

# Appendix 2

Molybdenum data for estimating fraction deposit from cutoff grade; $n = 58$

| Deposit name | COG (%) | Fraction growth | Deposit name | COG (%) | Fraction growth |
|---|---|---|---|---|---|
| Adanac-Ruby Creek | 0.095 | 0.113 | Moly Brook | 0.095 | 0.126 |
| Adanac-Ruby Creek | 0.085 | 0.168 | Moly Brook | 0.085 | 0.180 |
| Adanac-Ruby Creek | 0.075 | 0.247 | Moly Brook | 0.075 | 0.258 |
| Adanac-Ruby Creek | 0.065 | 0.351 | Moly Brook | 0.065 | 0.365 |
| Adanac-Ruby Creek | 0.055 | 0.470 | Moly Brook | 0.055 | 0.504 |
| Adanac-Ruby Creek | 0.045 | 0.581 | Moly Brook | 0.045 | 0.673 |
| Adanac-Ruby Creek | 0.035 | 0.679 | Moly Brook | 0.035 | 0.831 |
| Adanac-Ruby Creek | 0.025 | 0.864 | Moly Brook | 0.025 | 0.941 |
| Ajax | 0.095 | 0.037 | Moly Brook | 0.015 | 0.991 |
| Ajax | 0.085 | 0.087 | Red Bird | 0.105 | 0.107 |
| Ajax | 0.075 | 0.202 | Red Bird | 0.095 | 0.160 |
| Ajax | 0.065 | 0.450 | Red Bird | 0.085 | 0.226 |
| Ajax | 0.055 | 0.765 | Red Bird | 0.075 | 0.305 |
| Ajax | 0.045 | 0.956 | Red Bird | 0.065 | 0.409 |
| Bald Butte | 0.070 | 0.333 | Red Bird | 0.055 | 0.540 |
| Bald Butte | 0.055 | 0.623 | Red Bird | 0.045 | 0.687 |
| Bald Butte | 0.045 | 0.875 | Red Bird | 0.035 | 0.833 |
| Cannivan | 0.075 | 0.296 | Red Bird | 0.025 | 0.935 |
| Cannivan | 0.065 | 0.426 | Red Bird | 0.015 | 0.984 |
| Cannivan | 0.055 | 0.593 | Storie | 0.088 | 0.320 |
| Cannivan | 0.045 | 0.854 | Storie | 0.063 | 0.518 |
| Lucky Ship | 0.095 | 0.260 | Storie | 0.045 | 0.685 |
| Lucky Ship | 0.085 | 0.337 | Storie | 0.038 | 0.768 |
| Lucky Ship | 0.075 | 0.426 | Storie | 0.033 | 0.827 |
| Lucky Ship | 0.065 | 0.523 | Storie | 0.025 | 0.907 |
| Lucky Ship | 0.055 | 0.634 | Storie | 0.015 | 0.977 |
| Lucky Ship | 0.045 | 0.767 | Storie | 0.005 | 0.998 |

(continued)

(continued)

| Deposit name | COG (%) | Fraction growth | Deposit name | COG (%) | Fraction growth |
|---|---|---|---|---|---|
| Lucky Ship | 0.035 | 0.891 | | | |
| Lucky Ship | 0.025 | 0.963 | | | |
| Lucky Ship | 0.015 | 0.985 | | | |
| Lucky Ship | 0.005 | 0.993 | | | |

## References


Barnes A, Thomas D, Bowell RJ et al (2009) The assessment of the ARD potential for a 'Climax' type porphyry molybdenum deposit in a high Arctic environment: Skellefteå, Sweden, In: 8th international conference on acid rock drainage securing the future (ICARD), p 10

Baudry P (2009) Zuun Mod porphyry molybdenum-copper project, South-Western Mongolia: China: Minarch-Mineconsult Independent technical report dated June 2009 prepared for Erdene Resource Development Corporation (Project No. 3421 M), p 147

Becker LA, Gustin MM, Drielick PE et al (2009) NI 43-101 Technical Report, Creston Project, Pre-feasibility study, Sonora, Mexico: Tucson, Ariz., M3 Engineering & Technology Corporation and Golder Associates, and Reno, Nev., Mine Development Associates, 23 March 2008, to Creston Moly Corp., Vancouver, British Columbia, p 273

British Columbia Ministry of Energy and Mines (2012) Ajax, Le Roy: British Columbia Ministry of Energy and Mines MINFILE No. 103P 223

British Columbia Ministry of Energy and Mines (2014) Huber, Mineral Hill, Butte, Granby, Lone Pine, Independent: British Columbia Ministry of Energy and Mines MINFILE No. 093L 027

British Columbia Ministry of Energy and Mines (2014) Kitsualt, Clary Creek, B.C. Molybdenum, Alice, Lime Creek Lynx, Cariboo, MINFILE Record No. 103P 120

Bates D, Watts D (2007) Nonlinear regression analysis and its applications. Wiley-Interscience

Chen Y, Wang Y (2011) Fluid inclusion study of the Tangjiaping Mo deposit, Dabie Shan, Henan Province: implications for the nature of the porphyry systems of post-collisional tectonic settings. Int Geol Rev 53(5–6):635–655

Committee for Mineral Reserves International Reporting Standards (2006) International reporting template for the reporting of exploration results, mineral resources, and mineral reserves

Drew LJ (1997) Undiscovered petroleum and mineral resources. Assessment and controversy. Plenum Press, New York, pp xiii + 210

Ewert WD, Puritch EJ, Armstrong TJ et al (2008) Technical report and resource estimate on the Carmi molybdenum deposit kettle river property, Greenwood Mining Division, British Columbia: Brampton, Ontario, P&E Mining Consultants Inc., Effective date: 4 August 2008, Signing data: 25 September, 2008; for Hi Ho Silver Resources Inc., Mississauga, Ontario, p 97

General Moly (2012) Mt. Hope. web pages @ http://www.generalmoly.com/properties_mt_hope.php. Accessed 8 Nov 2010

Geological Survey of Finland (2011) Large unexploited deposits in Fennoscandia. web pages @ http://en.gtk.fi/ExplorationFinland/fodd/largeunexpl_060508.htm. Accessed 15 March 2011

Geoscience Australia (2012) Australian Mines Atlas of minerals resources, mines and processing centres, Geoscience Australia. web pages @ http://www.australianminesatlas.gov.au/?site=atlas&tool=search. Accessed 12 June 2012

Grothendieck G (2013) Non-linear regression with brute force. R package nls2. contact: ggrothendieck@gmail.com

Kramer B (2006) Mountain of controversy, tribe's history, future clash in face of mine proposal: SpokesmanReview.com, Web pages @ http://www.spokesmanreview.com/tools/story_pf.asp?ID=116309. Accessed 16 Dec 2010

Long K (2008) Economic life-cycle of porphyry copper mining, In: Spencer JE, Titley SR (eds) Ores and orogenesis; Circum-Pacific tectonics, geologic evolution, and ore deposits: Arizona Geological Society Digest 22, pp 101–110. Ores and orogenesis; Circum-Pacific tectonics, geologic evolution, and ore deposits: Arizona Geological Society Digest 22, pp 101–110

Lowe C, Enkin RJ, Struik LC (2001) Tertiary extension in the central British Columbia intermontane belt: magnetic and paleomagnetic evidence from the Endako region. Can J Earth Sci, vol 38, pp 657–678

Ludington S, Plumlee GS (2009) Climax-type porphyry molybdenum deposits. U.S. Geological Survey Open-File Report 2009–1215, p 16

Mercator Minerals (2011) Molybrook: mercator minerals. http://www.mercatorminerals.com/s/OtherProjects.asp. Accessed 8 May 2013

Mindat.org (1992) Buckingham mine (Hardy mine; Bentley mine; O'Leary mine), Battle Mountain District, Lander Co., Nevada, USA: Mindat.org. web page @ http://www.mindat.org/loc-60116.html. Accessed 29 Oct 2010

Mindat.org (2011) Mačkatica ore field, Čemernik Mts., Serbia; Mindat.org. web page @ http://www.mindat.org/loc-40789.html. Accessed 21 Jan 2011

Nadarajah S, Kotz S (2006) R programs for computing truncated distributions. J Stat Softw http://www.jstatsoft.org/v16/c0. Accessed 16 Aug 2006

Nanika Resources Inc (2012) Management discussion and analysis for the six months ended 31 March 2012. Form 51-102F1, Nanika Resources Inc., filed 30 May 2012, p 12

Northern Miner (2010) Canadian & American MINESCAN Folio, 2009–2010, CD ROM

Novomestky F, Nadarajah S (2012) R-project, package truncdist, updated 20 Feb 2015. contact: fnovomes@poly.edu

Raw Materials Group (2011) Sphinx molybdenum deposit, Canada. Raw Materials Group. Web page @ http://www.rmg.se/RMDEntities/S2/Sphinx_Molybdenum_Deposit_SPHIMO.html. Accessed 26 April 2011

RX Exploration Inc (2010) Drumlummon gold mine, Marysville, Montana, USA. RX Exploration Inc. Web pages @ http://www.pdac.ca/pdac/conv/2009/pdf/core-shack/cs-drumlummon-gold.pdf. Accessed 3 March 2010

Schennach SM (2004) Estimation of nonlinear models with measurement error. Econometrica 72 (1):33–75

Singer DA, Berger VI, Moring BC (2008) Porphyry copper deposits of the world: database and grade and tonnage models, 2008. U.S. Geological Survey Open-File Report 2008-1155, p 45

Smith JL (2009) A study of the Adanac Porphyry Molybdenum Deposit and surrounding placer gold mineralization in Northwest British Columbia with a comparison to porphyry molybdenum deposits in the North American Cordillera and Igneous geochemistry of the Western United States, University of Nevada, unpublished Master's Thesis, p 198

Spiess AN (2014) Package propagate. contact: a.spiess@uke.uni-hamburg.de

Taylor RD, Hammarstrom J, Piatak NM, Seal RR II (2012) Arc-related porphyry molybdenum deposit model, chap. D of Mineral deposit models for resource assessment. U.S. Geological Survey Scientific Investigations Report 2010–5070–D, p 64

Thompson Creek Metals Company Inc (2011) Annual report pursuant to section 13 or 15(d) of the Securities Exchange Act of 1934, for the fiscal year ended 31 December 2011: United States Securities and Exchange Commission Form 10-K, p 140. Web Pages @ http://www.thompsoncreekmetals.com/s/Annual_Report.asp?DateRange=2011/01/01...2011/12/31. Accessed 4 June 2012

TTM Resources Inc (2009) TTM Chu Molybdenum Project accepted by BC and Canadian Environmental Assessment Agencies & SGS Lakefield preliminary metallurgical results: TTM Resources Inc. Press Release dated 4 May 2009. Web pages @ http://ttmresources.ca/english/wp-content/documents/09-05-04.pdf. Accessed 26 March 2010

U.S. Geological Survey (2011) Buckingham molybdenum deposit: mineral resource data system (MRDS) deposit ID 10155557. Web pages @ http://tin.er.usgs.gov/mrds/show-mrds.php?dep_id=10310305. Accessed 17 Dec 2010

Wu H, Zhang L, Wan B et al (2011) Re-Os and 40Ar/39Ar ages of the Jiguanshan porphyry Mo deposit Xilamulun metallogenic belt, NE China, and constraints on mineralization events. Miner Deposita 46:171–185

Yukon Geological Survey (2005) Red Mountain: Yukon MINFILE No. 105C 009. Web pages @ http://data.geology.gov.yk.ca/Occurrence/12735. Accessed 1 April 2011

Zientek ML, Hammarstrom JM (2014) Mineral resource assessment methods and procedures used in the global mineral resource assessment. In: Zientek ML, Bliss J, Broughton DW et al (eds) Sediment-hosted stratabound copper assessment of the Neoproterozoic Roan Group, Central African Copperbelt, Katanga Basin, Democratic Republic of the Congo and Zambia: U.S. Geological Survey Scientific Investigations Report 2010–5090–T, Appendix A, pp 54–64

# Chapter 21
# General Framework of Quantitative Target Selections

**Guocheng Pan**

**Abstract** Mineral target selection has been an important research subject for geoscientists around the world in the past three decades. Significant progress has been made in development of mathematical techniques and estimation methodologies for mineral mapping and resource assessment. Integration of multiple data sets, either by experts or statistical methods, has become a common practice in estimation of mineral potentials. However, real effect of these methodologies is at best very limited in terms of uses for government macro policy making, resource management, and mineral exploration in commercial sectors. Several major problems in data integration remain to be solved in order to achieve significant improvement in the effect of resource estimation. Geoscience map patterns are used for decision-making for mineral target selections. The optimal data integration methods proposed so far can be effectively applied by using GIS technologies. The output of these methods is a prognostic map that indicates where hidden ore bodies may occur. Issues related to randomness of mineral endowment, intrinsic statistical relations, exceptionalness of ore, intrinsic geological units, and economic translation and truncation, are addressed in this chapter. Moreover, a number of specific important technical issues in information synthesis are also identified, including information enhancement, spatial continuity, data integration and target delineation. Finally, a new concept of dynamic control areas is proposed for future development of quantification of mineral resources.

## 21.1 Introduction

Instead of elaboration of new techniques, this chapter focuses on fundamental aspects in mineral resources assessment (Pan et al. 1992). Some of the critical issues are reconsidered here with respect to new understanding of basic geo-relations

G. Pan (✉)
China Hanking Holdings, 227 Qingnian Avenue, Hanking Tower 22nd Floor, Shenyang 110016, Liaoning, People's Republic of China
e-mail: gpan100@yahoo.com; pangc@hanking.com



B. S. Daya Sagar et al. (eds.), *Handbook of Mathematical Geosciences*,
https://doi.org/10.1007/978-3-319-78999-6_21

between resource descriptors and geological processes. Various multivariate models and techniques have been used over the past two decades to relate geological variables to some aspects of mineral occurrence or deposits. Conventional objective methods for mineral resource assessment have estimated either mineral endowment or discoverable mineral resources of a particular type of deposit in a region. The mineral endowment of a region usually refers to that quantity of mineral in accumulations meeting specified physical characteristics, such as grade, size, and depth. A multivariate endowment model is essentially characterized by a particular information extraction strategy for the so-called optimum combination of those geological features most related to spatial variations of endowment (Pan and Harris 1991). Most of these models estimate mineral resources based upon the principle of analogy, i.e., the resources in a study region are estimated by a model that is established on a control area by assuming different regions with similar geological environments have similar endowment (Pan and Harris 1991; Harris 1984; Harris and Pan 1991; Pan and Harris 2000; Agterberg 1981, 2014).

Most of these models have employed as information reference a grid of regularly spaced cells (inter-grid areas) and have dealt in one way or another with either mineral favorability, probability, mineral wealth or density of mineral occurrence (deposit). Of special interest have been those models that describe uncertainty about these estimates, such as the probability for occurrence of mineral deposits within a cell. These studies seem to have been a necessary step in the evolution of the science of mineral resources prediction, because geologists in general have been slow to adapt quantitative methods, and even reluctant to substitute objective and quantitative analysis for all or part of subjective analysis. Thus, there was a need to demonstrate quantitative methods that could be used to estimate undiscovered mineral resources. However, to some extent, this reluctance represented the dissatisfaction by geologists for the at-best low, and sometimes trivial, level of geoscience information captured by the quantitative variables and related to mineral occurrence by the multivariate models. Simply stated, mineral resource estimates by quantitative and objective methods will not improve significantly until more geoscience information is related in more appropriate ways to the various descriptors of mineral resources.

Supplying worldwide demand of metallic raw materials throughout the rest of this century may require multiple times the amount of metals contained in known ore deposits (Patiño Douce 2016a, b). Sustainability of resource supply is a key task for scientific mineral assessments. The concept of mineral resource is many faceted, including physical and chemical properties of mineral deposits, as they occur naturally in the earth's crust and economic properties created by man's socio–technical production system and the demands for mineral materials derived there from. The discussion presented here focuses upon several aspects of mineral resources that are fundamental considerations in the effective information synthesis for mineral resource estimation: randomness of mineral endowment, basic statistical relations, scarceness, geological foundations, economic truncation and translation, and spatial continuity. Some major issues in quantitative mineral resource estimation are addressed, including information enhancement, information synthesis, as well as target identification. Information synthesis is a central task in both mineral exploration and resource estimation.

## 21.2 Randomness of Mineral Endowment

Most of the past and current studies on mineral resource estimation have been constructed and applied on the basis of a common assumption that mineral endowment descriptors and at least some of the related geologic processes behave more or less according to certain stochastic rules. The assumption is seldom challenged, although controversies have continued over four decades, for example, the types of the stochastic laws that govern the true distributions of geochemical element concentrations (Harris 1984; Vistelius 1960; Brinck 1972). This seems to indicate that the assumption that some geological processes are to some extent stochastic and follow certain stochastic laws has been widely accepted, although it is premature to assert that all of the geoscience features are stochastic. It is useful to examine this notion before investigating specific stochastic laws for particular geologic events, the use of statistical models to estimate mineral resources, and probabilistic descriptions of resource descriptors.

In his famous 'Ideal Granite Model', Vistelius (1972) showed that the crystallization of minerals, such as potassium feldspar, quartz, as well as plagioclase contained in the 'ideal granite' can be modeled by some stochastic functions that vary in space and time. It has been proved mathematically that there is a three-dimensional 'packing of particles' such that the three mutually perpendicular directions can be described according to the Markov property in each direction with identical transition probability matrices in the three directions (Vistelius and Harbaugh 1980). Another example due to Vistelius is his gravitational stratification package model (Visteluus 1981). In the study of red beds of the Cheleken Peninsula, under certain assumptions, Vistelius showed that the sequence of red beds with two distinct states, S (arenaceous beds) and A (argillaceous beds), can be treated as a homogenous reversible Markov chain of second order, with the partial transition through A being first order Markov and the partial transitions through S being second-order Markov.

Sedimentary sequences have been regarded generally as some types of cyclic processes which are associated with certain Markov properties (Schwarzacher 1969; Hattori 1976; Pan 1987; Kantsel 1967; Pan and Porterfield 1995). Pan (1987) demonstrated that many sedimentary sections can be treated as homogeneous stochastic processes if no significant depositional discontinuities or structural unconformities occur in the sequences and that homogeneous sedimentary processes can be decomposed uniquely into the sum of independent reversible and unidirectional stochastic flows.

The process of ore deposition was closely examined by Kantsel (1967) based upon the function of metal distribution in ores. The process of hydrothermal mineralization during a single stage can be treated as a continuous stationary process of the Markov type. The resulting concentration of metal can be represented by a distribution function, the most important characteristic reflecting speed of the mineralization process. Stochastic modeling methods and uncertainty quantification are important tools for gaining insight into the geological variability of subsurface structures and formation of mineral deposits (Wang et al. 2017). Modeling of 3D geological processes helps reveal hidden information on the variability of controlling factors, which defines likelihood of occurrence of mineralization processes.

These contributions are informative about some fundamental and crucial controversial issues regarding the application of stochastic models to mineral exploration, although some concerns cannot be satisfactorily resolved without more research. A partial conclusion drawn from these preliminary works should be that *at least under certain conditions some of the geologic or earth processes can be modeled by stochastic laws*. However, it would be incorrect to associate the earth processes with the stochastic laws through one to one relations, since the random properties of geologic events generally are space and time dependent.

## 21.3 Fundamental Geo-process Relations

Observations on geologic features in certain spatial and temporal settings are the outcomes of a sequence of geologic processes superimposed during crustal evolution and initiated by inner energies of the earth, biosphere, hydrosphere, atmosphere, as well as other universal forces. Conceptually, there should be two levels of cause–effect relations among the geologic events, crustal evolution and initial forces, that created the earth. The earth commonly represents the entity of earth processes, e.g., crustal movement, magmatic intrusion, migration of ore-bearing fluids, erosions, etc., while geologic entities, such as lithologic phases, hydrothermal alterations, geologic structures, ore deposits, etc., are outcomes of the processes. Let $\mathrm{o}_1$, $\mathrm{o}_2$, …, $\mathrm{o}_\mathrm{k}$ denote the $k$ initial forces, $\mathrm{f}_1$, $\mathrm{f}_2$, …, $\mathrm{f}_\mathrm{p}$ the $p$ earth processes, and $\mathrm{z}_1$, $\mathrm{z}_2$, …, $\mathrm{z}_\mathrm{m}$ the $m$ geological features, including resource descriptors. Then, the cause–effect relations may be conceptualized as follows:

$$f_j = g_j(o_1, o_2, \ldots, o_k), \quad j = 1, 2, \ldots, p, \tag{21.1a}$$

$$z_i = h_i(f_1, f_2, \ldots, f_p), \quad i = 1, 2, \ldots, m. \tag{21.1b}$$

The conceptual model (21.1a, 21.1b) implies that the original forces are direct causes of the crustal evolution represented by a series of geologic processes which in turn are the direct causes of the geologic features (outcomes). Since some of these geologic features are resource descriptors, such as number of deposits, quantity of endowment, etc., relation (21.1a, 21.1b) states that a mineral deposit is the result of a sequence of superimposed geologic processes. The functions $g_j$'s and

$h_i$'s may be assumed to be random, provided that the original causes or geologic processes are considered to be stochastic.

A relevant question in statistical estimation of resources concerns basic statistical models useful for describing inherent relations between the geodata and resource descriptors given that geoscience information is stochastic. One should keep in mind the basic cause-effect relations (21.1a, 21.1b) and that these cause-effect relations do not imply any cause-effect between the resource descriptors and other geological features, although syngenetic or parallel relations do exist because both of these are outcomes of some common earth processes. For example, both argillic alteration and copper mineralization result from the same process of magmatic intrusion. Since the current knowledge on the original causes is very limited, it is not realistic to discover relations $g_j$'s in (21.1a, 21.1b). Assuming that the random portions of the earth's processes can be isolated from the deterministic part, the following two sets of auxiliary relations should be essential:

$$r_l = \varphi_l(f_1, f_2, \ldots, f_p) + v_l, \quad l = 1, 2, \ldots, d, \tag{21.2a}$$

$$z_i = \psi_i(f_1, f_2, \ldots, f_p) + e_i, \quad i = 1, 2, \ldots, m, \tag{21.2b}$$

where $r_l$'s are the resource descriptors, $z_i$'s are other geologic features and $v_l$'s and $e_i$'s are the random errors. However, a further difficulty arises because our knowledge of earth processes is also limited. What one can observe in practice are only the geological features $z_j$'s and maybe part of the resource descriptors. Although there is no direct causal relation between the mineral resource descriptors and other geologic features, their syngenetic and concurrent relations will assure some indirect information from the geologic features about the resources. Hence, the geological processes, and thus the mineral resource descriptors, can be mathematically reconstructed through a reverse functional estimation:

$$f_j = \Psi_j(z_1, z_2, \ldots, z_m) + \omega_j, \quad j = 1, 2, \ldots, p, \tag{21.3a}$$

$$r_l = \Phi_l(f_1, f_2, \ldots, f_p) + \varepsilon_l, \quad l = 1, 2, \ldots, d, \tag{21.3b}$$

where $\omega_j$ and $\varepsilon_l$ are the random error terms for the geological process and resource descriptor estimates.

Accordingly, if m is much greater than $d$, a feasible solution for mineral resource estimate may be completed in two steps:

(a) Factor out the $f_1$, $f_2$, etc. from relations (21.3a) based upon the known information on the geological features $z_i$'s;
(b) Substitute these estimates of the factors into relations (21.3b) and derive the estimates for the multivariate resource descriptors.

The first step of the manipulation is exactly analogous to factor-type analysis, constructing significant geologic factors (causes) from observable geological features, whereas the second step is regression-type analysis, predicting the resource descriptors (effects) from the geological factors. Consequently, *factor-type and regression-type models should be fundamental multivariate statistical models for quantitative mineral resource estimation, and other relevant statistical methods may be considered as variations and combinations of the two types of method.* That's why the mineral resource descriptors ($r$) can be statistically estimated through the geological features by the following function:

$$r_l = \Theta_l(z_1, z_2, \ldots, z_m) + \vartheta_l, \quad l = 1, 2, \ldots, d, \tag{21.4}$$

where $\theta_l$ the random error. The geological processes are directly created by the initial forces of earth movement, while accumulation of mineral resources is directly resulted from complex interactions of the geological processes. Since the geological processes cannot be directly measured, they must be reconstructed by observable geological features, which can be, in turn, indirectly used to estimate mineral resource descriptors through relation (21.4).

## 21.4 Scarceness, Rareness, and Exceptionalness

The activities of mineral exploration have been motivated chiefly by economic and social pursuits (Pan et al. 1992). Constantly growing economic and social demands require greater amounts of raw material, including nonrenewable mineral commodities. The conduct of mineral resource exploration is predicated upon the economic return expected from the discovery of new deposits. An increase in the price of a mineral product, which is equivalent to the sum of the marginal rent and marginal extraction cost, indicates that the mineral resource has become scarce. A basic perspective of both geologists and economists is that mineral resources are scarce materials in the crust as they occupy only an insignificant portion of crustal material.

Any major ore deposit may be regarded in principle as an anomalous or rare phenomenon commonly characterized by one or more geological, geochemical, and geophysical features. Consequently, signatures of significant endogenic mineralization are anomalous and exceptional geologic settings (Gorelov 1982). In particular, the formation of a giant deposit is an extremely rare event created by an exceptional combination of earth processes. Rareness of the giant deposits is reflected in both spatial and temporal dimensions. Significant concentrations of a metal usually have a strong affinity or correlation with particular geologic formations and epochs, as well as metallogenic environments. The genesis of giant deposits may be controlled by particular regularities that differ from those controlling the formation of medium and small-size deposits of the same composition. It is also thought that the formation of huge deposits appears to be controlled by a so-called 'ore-controlling structure' (Tomson and Polyakova 1984).

Giant deposits often dominate reserves and production. It is not uncommon for a few supergiant and giant deposits to constitute over 50% of the total metal recoverable under current economic and technological conditions; accordingly, the metal quantity in small size deposits is almost negligible (Laznicka 1983). Conversely, giant deposits typically constitute an insignificant part of the total number of ore deposits.

Thus, the scarcity of a mineral resource is essentially determined by the fact that few giant deposits exist in the crust, but the few that do exist strongly dominate reserves and production. Accordingly, the economic viability of mineral exploration is strongly predicated upon its capability of locating the giant or large mineral deposits through delineating the associated geologically anomalous regions of the crust. Unfortunately, conventional quantitative techniques employed have failed to deal with these important particulars satisfactorily, mainly owing to inability to capture the nature of these exceptional constraints, since these unique deposits rarely exhibit common statistical properties.

The discovery process for some deposit types, e.g., those for which structural, geochemical, alteration, or geophysical signatures are correlated to deposit size or those for which discovery is primarily by drilling and for which size is strongly related to areal extent, is size biased, meaning that large, high-grade deposits tend to be discovered in early stages of the exploration of regions (Chung et al. 1992; Pan and Harris 1991). For such deposit types, the prognostication of exploration outcomes or the estimation of additional resources in undiscovered deposits should take into account the implication of this bias to the tonnages and grades of the undiscovered deposits. However, representing the discovery process of other deposit types, such as vein deposits with great vertical extent or those for which size is only weakly related to exploration anomalies, as size bias sampling may not be appropriate (Stanley 1992). Improvement in locating deposits or in estimating probabilities for their occurrence requires consideration of the exploration effect and the conjunction of improved genetic, tectonic, and other unifying geoscience theories with improved synthesis methods for the effective extraction of information from diverse geodata and improved quantitative models for inference or estimation.

Considering the low concentration of many elements, e.g., 65 ppm for copper, in common crust rock, the presence of a large accumulation (1 to 10 million tons for copper) of metal at concentrations that are mined today requires enrichments by 100 or 1000 s times crustal concentrations and the accumulation of metal from a large amount of common crustal materials into a relatively small volume. Typically, this concentration or accumulation is seen as requiring the successive operations of several enrichment-depletion stages. Since these sub-processes rarely take place at the scale and strength required to form an ore deposit, their joint (sequential) occurrence could be an extremely rare event in both space and time. If each of these processes is assumed to be stochastic, the mineralization process is also stochastic, and thus the formation of ore deposits is deemed to be a rare, random event. To the extent that this assumption is acceptable, the concept of rareness of ore deposits is equivalent to the smallness of the probability for the formation of an economic deposit.

The concept of rareness can be compared to that of exceptionalness described by Gorelov (1982) and the conditional exceptionalness proposed by Pan (1989). Some other terms found in literature carrying similar meanings include atypicality, uniqueness, anomaly, etc. The concept of exceptionalness is important and useful in quantitative mineral exploration. The most general feature of major commercial ore deposits is that the geological structures of their ore fields are exceptional and anomalous compared with those of neighboring areas.

It is noted that scarceness is a term relevant to economic aspects of resources, rareness is more closely associated with statistical (probabilistic) characteristics of mineral occurrences; and exceptionalness should be used in a geological context. More specifically, one would say that ore deposits are probabilistically rare and geologically exceptional, even though the metal derived from them may not be scarce in the economic sense described by Barnett and Morse (Barnett and Morse 1963). These terms are often used to describe the status of mineralization events in a relative sense, but they can be statistically quantified in a rigorous framework.

## 21.5 Intrinsic Geological Unit

Most traditional resource estimations have been made on the basis of regular inter-grids or cells as the sampling scheme and estimation unit. The "cell" approach is associated with a number of drawbacks. The most significant problem is that geological processes can be reconstructed through observable geoscience features, which are measurable in geological units, not artificial cells. The cell-based measurements tend to distort the intrinsic relations between geological features and mineral resource descriptors. Secondly, quantification of the geological features, spatially correlated and even connected, is difficult to capture essential genetic factors that played key roles of metal enrichment. Finally, the cell-approach easily ignores exceptional conditions for formation of large deposits, which cannot be readily quantified through grids.

### *21.5.1 IGU Definition*

In contrast with a population of cells having multiple attributes, consider a population in which each member consists of a set of genetically related objects, e.g., igneous intrusives and associated altered host rock, and each member is described by fields of the related geologic objects. Here, mineral resource descriptors and geoscience measures are attributes of a group of geoscience fields which in turn are attributes of a set of genetically related geologic bodies. Such a scheme employs a sampling reference for quantification and integration of geoscience information that is *intrinsic* to the deposit type being sought. That is why the Intrinsic Geological Units (IGU) was proposed by Pan (1989) and Harris and Pan (1990).

The concept of intrinsic geological units, formally documented in Pan and Harris (1993), has evolved from the notion of *intrinsic samples* (IS), or *consistent geological area*. The basic ideas behind both notions are identical and a minor difference lies in the procedure for delineation. This concept has some common characteristics with the notion of "geological anomalies" proposed by Zhao (2007) (also see Zhao and Chi 1991), although the procedure of unit delineations differs significantly.

An appropriately delineated IGU is at once a great improvement over the traditional inter-grid area or cell because it represents the joint occurrence of geologic bodies that are genetically related to the mineral resources of interest. Thus, even before geological attributes of the IGU are quantified, the very presence of an IGU implies highly significant geoscience information about geology and mineral resources. In contrast, the cell is simply a geometric reference. Therefore, it is inevitably true that geological attributes of an IGU carry far more geoscience information than do the geological attributes of a cell.

IGUs may be formally defined as *members of a population consisting of sets of genetically related geologic objects that are usually defined by their geofields* (Pan 1989). Each member (IGU) of the population of IGUs constitutes an independent set of geologic objects that are genetically related to each other and to mineral deposits, although generally only some of these members contain ore deposits and mineral resources. Moreover, although a particular member of a population of IGUs contains mineral deposits, it may not be uniformly mineralized everywhere within its volume. In other words, a mineral resource unit generally is a subset of an intrinsic geologic unit.

### *21.5.2 Critical Genetic Factor*

Any mineral deposit or mineralization can be considered as an anomalous concentration of one or more elements or their chemical compounds when compared to crustal materials. This anomalous region originated from anomalous genetic processes or their superposition during certain geological epochs. Usually, a genetic model consists of a hierarchy of earth processes—from preconditions to post mineralization preservation—which acted during one or more previous time spans, and as such, these processes are not observable. Instead, the geologist must infer their previous existence and operation using observable indirect evidence, e.g., geologic features, geochemical suites, hydrothermal alteration, aeromagnetic and gravity anomalies, etc.

Since particular genetic processes were initiated and developed under certain specialized circumstances, existence of mineralization, as a significant outcome of the processes, must also be conditional upon these relevant circumstances. In other words, whether an anomalous concentration of a metal exists in a region depends solely upon the existence of certain necessary conditions during crustal evolution. Although there might exist a number of such necessary conditions for a particular genetic process or mineralization, one, or at most a few of them, is referred to as critical. For convenience, this (these) critical or necessary condition(s) is called the *Critical Genetic Factor*(s)

(CGF). The idea of CGF does not rest solely upon one factor being more important or critical than another in the formation of a mineral deposit, because unless all genetic factors are present, there is no mineral deposit or mineral endowment. Criticality, as used here, rests more upon the idea that the CGF arises from few, preferably only one, earth process and that those features formed by that process can be detected reasonably well by conventional sensing technologies, e.g., magnetics, gravity, geochemistry, and geology mapping. If this CGF is not present, the intrinsic geological unit is considered to be absent. For example, for a mineral deposit related to magmatic fluids, the heat source that drives intrusion may be treated as the CGF for identification of the IGUs associated with the deposits of this type. Practically, only a single CGF is necessary for identifying spatial units that are intrinsic for mineral deposits of a single genetic type, but more than one CGF may be necessary when there is more than one genetic type of interest.

An IGU can be further understood to be a member of a population consisting of sets of geologic objects genetically associated with the CGF, each set being a member of the IGU population. Individuals from the population are called *known* IGUs if the related CGF is directly observed, while others are unknown or predicted when the CGF cannot be observed directly, but is inferred to exist because of the presence of geologic fields related to the CGF and to recognition criteria.

### 21.5.3 Critical Recognition Criteria

The CGF often may be identified as a process, based upon geoscience; conceptually, it may be an abstraction, instead of an observable feature. In order to make the CGF concept workable in practice, a set of special geologic features which give firm evidence of the previous existence and operation of the CGF are established. Such a feature is here termed a *Critical Recognition Criterion* (CRC). Each of these CRCs constitutes a sufficient condition for existence of the CGF. Any spatial location at which one or more CRCs occur is by definition a location within an intrinsic unit.

Although the concepts of CRC make it possible for identification of CGF, the occurrence of CRCs known at the time of application may not represent the entire picture of a CGF. In other words, estimation of the presence of a CGF based upon only CRCs could be biased due to imperfect knowledge on the spatial distribution of CRCs. For example, a CRC might exist underneath the sedimentary cover, even though it is not found by surface geological mapping. This fact dictates that the identification of CRCs beyond surface observation is an important step in the appropriate prediction of the distribution of the CGF. This can be done by establishing statistical relations of each CRC to a set of selected geological, geochemical, and geophysical fields, which provide indirect evidence for the presence of the CGF.

Although the existence of a recognition criterion at a spatial location almost surely indicates that the location is within an IGU, the boundary of the IGU still is unknown. Consider, for example, the outcrop of a Tertiary intrusive assumed to be a CRC. Then, the outcrop area is surely within an IGU, but probably, some of the area around the outcrop also is within the same IGU because of the likelihood that at depth the intrusive extends laterally underneath the surface rocks. Consequently, the boundary of an IGU is usually uncertain. One way of representing such uncertainty is to assign each spatial location a probability for presence of one or more recognition criteria based upon a collection of geological observations at that location.

### 21.5.4 IGU Delineation

At a known location (with at least one observed CRC), the probability for the CGF should be one or very close to one. This implies that the point is almost surely within an IGU. At an unknown location (with no observed CRCs), all of the CRC probabilities estimated from geoscience fields will provide a measure of the likelihood of the presence of the CGF.

Several methods have been proposed and employed for delineating IGUs. One such example is that which consists of three steps developed by Pan and Harris (1993). The method delineates IGUs by estimating and combining probabilities of CRCs. Another example is given by Pan (1989) and Harris and Pan (Harris and Pan 1991) based on the union of marginal field anomalies. As discussed, the presence of a CRC gives evidence for the existence of an IGU; delineation of the boundary of the IGU is made by resolution of the geoscience fields associated with the CRCs. In this approach, the key step is to establish a procedure to identify the anomalies in terms of CRCs for each geosciences field. These anomalies (called marginal anomalies) are then combined into one anomaly through spatial union. This is similar to the concept of using the maximum CRC probability to represent the probability for CGF.

As we know, genetic theories are most useful for grass-roots exploration or reconnaissance programs, where deposit information is not abundant. Without the guidance of genetic models, it is unsafe to select an area for a massive investment. Hence, the concept of IGU is most useful for regional mineral exploration, because it provides a quantitative framework for delineation of those areas having the conditions necessary for the presence of deposit. In large-scale exploration, such as deposit or district scale, the methodology of IGU is still useful if detailed aspects of deposit genetic models can be specified. With abundant occurrence information, it is possible to extract genetic factors as necessary conditions for the localization of deposit. However, in most cases, this detailed information is not available or not in a usable form. In general, a mining district is already a known IGU defined by broad genetic models. Unless refined genetic models are available, IGU will not provide additional power to identify areas for the potentials of deposit or district scale.

### *21.5.5 Relations Between IGU and Mineral Target*

As discussed, CGF serves as the necessary condition for presence of an IGU, but it is not a sufficient condition for the boundary definition of the IGU. The purpose of IGU proposal is to improve methodology of target identification and delineation, which, in turn, improves the effect of mineral resource assessment. The IGU theory creates a new platform on which new approach to mineral target identification can be constructed. A critical question to ask would be what is the relation between IGU and mineral targets?

Theoretically, an IGU is a necessary condition for presence of mineralization of interest. The concept of IGU provides a precursor to the identification of mineralization or deposits. However, presence of an IGU does not necessarily serve as sufficient conditions to the presence of mineralization or deposit. Presence of an IGU is a necessary condition of presence of mineral target. In general, an IGU is much broader in areal or volumetric extents than a mineral target. Mineral targets are defined in the IGU areas where additional necessary and even sufficient conditions are observable or inferable from maps or data collected from various sensing or engineering technologies. Instead of using an inter-grid sampling scheme, the framework of IGU provides a more practical and useful approach for extraction of sufficient conditions for identification of mineralization events through reconstruction of geological processes that resulted in the occurrence of mineralization.

For mineral resources appraisal, the concept of IGU establishes a theoretical base for definitions of necessary and sufficient conditions of mineralization or deposit. It has radically changed the conventional methodology for estimation of mineral potentials. The relationships of IGU, target, occurrence, and deposit are depicted as follows:

$$Deposit \subseteq Mineral\,Target \subseteq IGU \subseteq Working\,Area$$

Clearly, an IGU is not a mineral target, but a mineral target must be enclosed in an existing IGU. Similarly, a mineral target is not a deposit, but a deposit must be localized inside an existing mineral target. Therefore, identification and delineation of IGUs is a necessary step for definition of mineral targets. This new approach will play a revolutionary role in improvement of mineral resources assessment.

## 21.6 Economic Truncation and Translation

Mineral deposit is not a purely geological concept when it is linked to resources and reserves. The effects of economic truncation and translation on mineral deposits have been recognized several decades ago, and a thorough discussion of these has been given by Harris (1984). These phenomena reflect an important fact that mineral resources generally are a dynamic function of relevant economic and technologic constraints, including price of product and costs associated with various production phases, such as mining, milling, smelting, as well as refining. Available data on mineral deposits generally are truncated by a cost surface which is defined in terms of physical features of the deposits and technological states. In other words, the collection of mineral deposits reported reflects only the truncated fraction of the entire population of mineral deposits. Thus, use of these data directly and unavoidably results in biased estimates of mineral resources, as the characteristics of the resource distribution derived from the partial data set only are a distorted representation of deposits as they occur in nature.

Translation refers to the fact that commonly reported deposit grades and tonnages are for ore reserves and that these tonnages and grades generally differ from those for the total mineralized material for the deposit as a geologic phenomenon. For deposit types having great lateral or vertical gradation in mineralization, economic rents may lead to the selection of a cutoff grade that leaves part of the deposit in the ground. When this is the case, reported ore tonnage is smaller than deposit tonnage and average grade is higher than deposit average grade.

The importance of translation as a distortion varies with the mineral commodity and the maturity of the exploration activity. In general, the greater variation of the grade within a deposit (intra deposit grade variance), the stronger the translation effect, and vice versa. For those deposit types having sharp boundaries or a uniform grade distribution, the translation effect may be negligible. For some deposit types, it is also true that the longer the deposit has been mined, the greater the reserve additions and the more representative the revised ore tonnage and grade data are of the geologic deposit.

The truncation and translation effects are related to some degree when production costs are strongly influenced by ore tonnage and ore average grade, provided that intra deposit grade variation and the spatial distribution of grades permit the effective use of cutoff-average grade relations to maximize the net present value of economic rents. However, translation occurs mainly in mine development and subsequent mining, while truncation reflects both exploration and mining. Conversion of resources to reserves involves using cutoffs for grades that define boundaries of ore economic portions in the deposits. This procedure involves both translation and truncation.

In order to resolve these difficulties, Harris (1984) suggested a possible remedy: treating the truncation effect requires first identifying the truncation relationship, and second the explicit consideration of this relationship in the estimation of parameters, one of which is the correlation of deposit tonnage with grade. Although several attempts have been made to mitigate the difficulty in practical studies by employing more sophisticated mathematical methods in mineral endowment estimation, the problem remains to be explored further, as estimation of the cost relation is still based on the truncated data. Thus, the cost relation must be reconstructed from a truncated surface before estimation is carried out.

The importance of truncation and translation effects on a quantitative estimate of mineral resources depends to some degree upon the means of estimation and upon the objective of the estimation. For example, when estimation is to be done using analogue or control regions and the objective is to estimate the magnitude of resources for price, cost, and technology similar to those of the analogue regions, the effect of truncation and translation on the estimate may be minor. But, when the objective is to estimate the magnitude of resources for improved exploration and production technology, the effect of truncation and translation upon the estimate may be very significant.

## 21.7 Information Synthesis

The geologist's view of an ore deposit may differ from that of the economist. Economists tend to consider an ore deposit as being a continuous geologic phenomenon that is discretized by applying a set of economic regularities, while geologists tend to perceive a deposit to be a discrete geologic phenomenon with anomalous concentration of one or more valuable elements (Agterberg 1981). Physical mechanisms of ore genesis suggest that the continuity of ore concentration is meaningful mainly in a relative sense. A high magnitude of element concentration in host rocks often contrasts sharply with concentrations in surrounding wall rocks. This perspective may be partially illustrated by the DeWijs' scheme of element enrichment in a deposit, which was extended by Brinck (1972) to describe element concentrations within the crust. Another well-known hypothesis is Skinner's bimodal proposition of element distribution which asserts that a gap exists between the grades of mineralized rock and the grades of common crustal material (Skinner 1976).

### 21.7.1 *Spatial Continuity*

Although the continuity of the statistical distribution of grades seems to differ conceptually from that of spatial and temporal distributions, they are in fact closely related. For example, if the proposition is accepted that the grades of an element are continuously distributed in space and time, the continuity of the statistical distribution of these grades can be automatically invoked in certain environments, and vice versa. This assertion may be explained by the requirement that samples must be taken in a uniform and regular manner from the population of interest.

Metallogenic and tectonic studies depict elements to be concentrated in geologic terrains of different scales, such as ore shoot, ore body, ore district, ore belt, ore province, etc. (Laznicka 1983). This hierarchical structure of ore formation seems to indicate that continuity exists within each of these scales, while discreteness of ore concentrations can be seen between these different scales. For instance, an ore district may be viewed as a continuously anomalous region within an ore belt, but the individual deposits included in that same district are discrete geological phenomena. This perspective carries strong implications as to sampling procedures and the organization of data for the estimation of mineral potentials.

Thus, a specific mineral exploration project focused upon the ore deposits of certain valuable elements formed and confined in a particular dimensional scale requires an appropriate sampling scheme of that same scale. For example, a new ore body developed within a deposit may be considered as mineral potential at the deposit scale, while a new ore deposit discovered in a district is regarded as mineral potential at a district scale. When estimation is aimed at predicting the mineral potentials at the district scale, the sampling scheme must accommodate the geological and mineral continuity at the corresponding hierarchical level. The match in scale is a prerequisite in mineral resource estimations.

### 21.7.2 Information Enhancement

Although in one sense considerable progress is apparent in the use of quantitative techniques for mineral exploration and resource estimation since the early work in the 1950s and 1960s (Allais 1957; Harris 1965), much less success has been made in creating estimates that are or have been used in mineral exploration and mineral policy decisions. Even though quantitative estimation of local/drilling targets may require the detailed quantitative characterization of favorable geological, geochemical, and geophysical information, many explorationists still favor subjective and qualitative methods for the integration of geodata. Concurrent with these applications, mathematical methods were designed and demonstrated, but few were adopted. Perhaps, this is a natural evolution of the science of quantitative mineral exploration in terms of data integration, because geologists in general have been slow to adopt quantitative techniques. However, this reluctance is at least partly related to ineffective integration of geodata and insufficient extraction of geoscience information by quantitative models. Mineral resources cannot be satisfactorily estimated until more geoscience information is related by improved methods to mineral occurrence. Major difficulties that have hindered further development have been far from fully attacked, and some of them are even completely ignored.

A common practice in quantitative mineral exploration is to collect all relevant geoscience data available in the study region, including numerical observations, digitized maps, and remotely sensed images. These data are then compiled, digitized, resorted, and formatted in a readily manageable data base. Each record is usually stored as a row, while each geologic attribute occupies a column. In standard statistical terms, each record in a data base is called a sample and each attribute is referred to as a variable. A sample in mineral exploration can be a spatial point or a one-, two-, or three-dimensional block. Most data in regional mineral exploration are interpreted in two dimensional areas.

Sampling schemes are considered to be an important factor in data interpretation and target identification. A viable sampling scheme should be able to cope with the hierarchical structures of mineralization or ore concentration. Mineralized geological bodies in different hierarchical scales correspond to different domains in space and time, which are generally defined by particular tectonic settings and geological formations. Statistically, samples should be randomly taken in the population of mineralized and non-mineralized geological blocks of the same scale. Furthermore, spatial characterization of geological features is another criterion for reasonable representation of the resource variability. A reliable sampling scheme should also result in a sample distribution which portrays closely the 'true' population distribution of geological and mineralized bodies. Our experience has shown that quantities measured on the basis of equal area cells might lead to distorted probability distributions.

The original data may include geological, geochemical, geophysical, as well as remote sensing information in diverse modes. For example, geological data can be hydrothermal alteration, faults, and lithology, which are typically considered as non-numerical attributes. Geochemical data can be collected from a rock outcrop, stream sample survey, or a soil grid survey. Magnetics data can be obtained from an airborne geophysical survey. It is readily seen that all these types of geodata are diverse not only in terms of sampling methods, but also the presentation of quantities. Different sampling schemes create different data densities, inconsistent spatial locations, disconnectivity, as well as uneven precisions. Different quantity presentations may give rise to even more serious problems in data integration. The most difficult problem is dealing with the correlation of different variables, which is the most critical step in geological information synthesis, especially when some data are non-numerical. The first step in overcoming these difficulties is the quantification and unification of different data sets.

The quantification of non-numerical attributes refers to assignment of a numerical value to each sample location; of course, the numerical value must convey explicit geological information. For example, a binary assignment gives 1 or 0 to the attributes to represent presence or absence. When each data set is 'quantitative', the next step is to enhance geological information of each individual data set before they are compared, correlated, and integrated. As a matter of fact, enhancement of information from original and individual data is the most critical step towards a successful information synthesis for mineral target selection. Unfortunately, geologists traditionally tend to place too much emphasis on the original data and denigrate the importance and necessity of data filtering, cleaning, and enhancing. Conversely, some geomathematicians devote too much attention to processing of data and give too little regard to fundamental characteristics of the original data and the useful information of the data. Original data carry the most genuine information, but they may be 'contaminated' or masked by noise and even distorted due to inadequate sampling or analytical methods.

Filtering and enhancing of useful information is important to remove noise and reveal signals, such as separation of soil geochemical anomalies from background values. Furthermore, one data set may carry information on several geological aspects. Some of these signals are not the major interests and their presence sometimes masks or distracts from the information useful in identifying mineral targets. These signal components are unwanted, even though they are not noise, and should be filtered out, or at least suppressed. However, many filtering, enhancing, and other data processing techniques can easily introduce artifacts or false signatures. For instance, a magnetic anomaly map generated from a short-wavelength filter can exhibit many high-amplitude, single-grid-point anomalies, which are known as the aliasing effect in the geophysical literature. Another example is interpolation which has been commonly used in data interpretation and quantitative mapping. All interpolation algorithms, e.g., minimum curvature and kriging, which can be considered as low pass filters, are notorious in that they tend to produce overly smoothed surfaces and quite often cause a loss of important detailed features. It is our opinion that some applications of quantitative analysis in mineral exploration have either failed to extract the important geoscience information or have created too many artifacts relative to signals; these effects are believed to be among the major reasons underlying the reluctance of geologists to replace qualitative judgment by quantitative analysis.

The above discussion suggests that filtering and enhancing is necessary for geological data interpretation and integration, but care is warranted in the use of enhancing techniques. Also, enhancement of a geological attribute includes identification and description of spatial structural characteristics, which constitute useful information about spatial auto-correlation of the attribute. More specifically, the objective of information enhancement is to maximize the signal relative to noise. By analogy, the best picture of an object taken by a camera requires a correct focus on the object; either too short or too long of a focus will blur the picture. Moreover, one should keep in mind that any enhancement technique cannot create information that is not present; instead, it is only able to reveal important features of the information carried by the attribute. But, without enhancement, some important features may not be identified nor employed in subsequent analyses. Since the amount of information in each attribute is limited, enhancement also is limited. A minimum level is necessary, for an insufficient removal of noise fails to reveal the signals to be extracted and used in subsequent analyses. Generally, the tendency of analysts is to ignore or inadequately remove noise and to over-enhance the signals. Of course, intense enhancement of data that contain noise leads to enhancement of noise as well as the signal and to false patterns and inter-relations with other information.

### 21.7.3 Data Integration

Synthesis of geoscience information includes the quantification of geological observations, maps, and other geological images; extraction of quantitative variables; statistical preprocessing; filtering and enhancement; estimation of statistical relations among variables; and the combination of different data sets (layers). Clearly, most of the components require some amount of computation which can be performed more efficiently by using a computer. There is an obvious advantage of using a computer when many variations of the same type of analysis are required

(Green 1991) or when important information includes the computer interaction of several large sets of geodata. This additional information helps to reduce uncertainties and ambiguities in geological interpretation and mineral potential estimation. Furthermore, some effective and sophisticated statistical techniques which generally prohibit manual calculations can be readily implemented on a computer.

Mineral exploration generally deals with diverse geological data in various chemical and physical forms. Appropriate information synthesis should reflect the types of information contained in each data set and their geological implications. For example, geochemical information is generally different than geophysical data. Even the same type of data, e.g., geochemical, may require different interpretation when it is obtained through different sampling techniques. For instance, soil geochemical samples are processed in different ways from stream samples. Geophysical data are rich in depth information and are capable of locating blind targets, but the extraction of such information requires appropriate processing and analysis. It is important to note that any data set has its limitations in the diagnosis of geologic favorability for mineralization, and interpretation and information synthesis must recognize these limits. Because of vast differences in geoscience content, precisions of measurement, and scales of reference among diverse geologic data, integration of these data directly cannot constitute their optimum use in mineral exploration unless the data are appropriately preprocessed and unified. Unfortunately, these problems are far less than adequately treated in traditional exploration applications.

Geoscience attributes are usually processed, correlated, and integrated to produce some estimates which characterize the favorability or probability of mineral occurrence. A more comprehensive approach treats each of the various kinds of geoscience information as a field of a particular type, e.g., geochemical fields, magnetic fields, etc. (Harris and Pan 1990, 1991). Mineralization may also be viewed as an ore field. The notion of field enriches useful information about three dimensional characteristics of geological bodies. Such a field is generally more expressive of meaningful geoscience information relevant to mineral resources than are ‘man-made’ variables, e.g., measurements quantified with regard to an artificial reference, such as a grid.

A major objective of information synthesis is to maximize the extraction of relevant geoscience information in terms of mineral potentials. Geological measurements in mineral exploration are commonly multivariate in terms of either several variables (fields) measured at same sample locations, or different variables measured in different sample locations but in the same study region. In the latter case, synthesis may require an appropriate interpolation of the data before they can be jointly analyzed. When strong correlations exist among the variables, multivariate techniques are necessary to capture the joint information from multiple associations as well as the marginal contributions from individual attributes. A multivariate exploration system sometimes can be decomposed into several less significantly correlated sub systems with smaller dimensions. This partitioning may reduce the complexity of modeling and possibly permit more robust estimates at the expense of decreasing the degrees of freedom in the system.

Optimum combination of different geological data sets (layers) has been a central task in data integration and information synthesis. Agterberg (1989) gives a comprehensive review on some major integration methods developed in recent years. Two major types of models notable in literature include favorability analyses and probability methods. Pan and Harris (1992) propose a weighted canonical correlation method for the estimation of a favorability function. These methods are most suitable for combining continuous geological attributes. Agterberg (1992) provides probabilistic techniques for combining indicator patterns in weights of evidence modeling. Both types of models, however, are deficient in some regards. Favorability methods often carry ambiguities in predicting mineral potentials, whereas evidence combination techniques are subject to strong constraints on the independency of different attributes. Moreover, as an information synthesis method, weight of evidence is simplistic. Another useful combination approach is color (RGB) image composition (Sabins 1987). This type of technique also bears some serious limitations, since most current image processing software systems are only capable of combining a very limited number of 'layers'. Therefore, there is a need for development of more effective combination methods.

Geologic information about mineral occurrence may be roughly grouped into two categories: marginal information contributed from individual variables or fields and joint information contributed from the cross correlations between different variables or fields. The first category of information has been extensively quantified and interpreted in most of the traditional studies on mineral exploration. The second category, however, has been inadequately treated due to complexities and ambiguities. Information from the inter-dependencies of variables can be an important factor in improving the definition of exploration targets, if single exploration variables are ambiguous, noisy, and/or uncertain as to mineral occurrence. Thus, an effective synthesis technique must be able to efficiently quantify and extract the cross-correlation information.

Intuitively, there should exist a combination of variables in multivariate mineral exploration that is sufficient to capture the majority of useful information and at the same time to minimize the effort of manipulation. It is probably incorrect to think that more variables are always preferred. On the contrary, a large set of data almost always contains redundant information which, if not appropriately eliminated, can result in unstable solutions and create noisy estimates. Therefore, another important problem in information synthesis is to select and refine variable sets such that redundant and trivial variables are excluded from consideration.

### 21.7.4 Target Delineation

Mineralization is considered as an anomalous geologic event, because the element is either present in anomalous grades, rare minerals, or in anomalous quantities. The purpose of mineral exploration is to locate economic mineral deposits in such anomalous regions based on direct and most often indirect information (chemical, physical, structural, etc.) and ore genetic theories. Since the direct information, e.g., the concentration of the metal of interest, is usually meager in the early stages of exploration, indirect information (e.g., geological, geophysical, geochemical, remote sensing, etc.) is commonly employed to identify mineral exploration targets. However, the mineralized anomalies, which are distinctive from the surrounding areas in terms of the accumulated metal(s), are typically fuzzy or ambiguous in terms of indirect information. Therefore, ambiguities of information raise an intricate question, i.e., how to 'best' define targets in terms of the maximum inclusion of mineralized rock and exclusion of non-mineralized rock.

Information synthesis produces either a set of processed (enhanced, quantified, integrated) geological, geochemical, geophysical fields, or a single synthesized index characterizing the favorability/probability of mineral occurrence. Based upon the derived grids, maps, or images, all of which are commonly referred to as 'layers', mineral exploration targets can be delineated by overlaying or combining the different layers. Since the synthesized results, however, are generally continuous, some threshold values are necessary to define the boundaries of targets. The traditional approaches to determine the boundaries are generally subjective and tend to introduce too many uncertainties. Obviously, a precise definition of a target is an important exploration problem to be solved.

Delineation of potential mineral targets has been a central task especially in the earlier phases of a mineral exploration program. Target areas have been identified by either subjective or objective analysis. Subjective methods provide opportunity for the maximum use of genetic theories of ore deposits and connect genetic knowledge and geological observations either intuitively by expert geologists or formally by a computer system (Harris and Carrigan 1981; Finch and McCammon 1987; McCammon 1990; Koch and Papacharalampos 1988). Subjective methods have been generally formulated as follows: (i) formulate genetic models, (ii) relate geological observations to genetic processes, and (iii) estimate subjective probabilities of mineral occurrence. Objective (mathematical) methods attempt to maximally use various existing mineral occurrence data and quantified geological variables (Botbol et al. 1978; Chung and Agterberg 1980; Agterberg 1988; McCammon et al. 1983; Singer and Kouda 1988). An objective approach generally consists of three major steps: (i) quantification of geological variables, (ii) estimation of mathematical models, and (iii) extrapolation of the estimated models to identify target areas.

Ore genesis models are crucial in mineral exploration and resource evaluation. Since genetic models of ore deposits are usually constructed on the basis of man's past experience, imagination, and logical inference, they have a natural connection to subjective probability analyses and expert systems, giving such an approach great potential for prediction. However, in practice this approach also is subject to some limitations. First, expert systems are costly to build and to validate; second, the full potential of such systems requires the construction and incorporation of extensive data bases. Without such data bases, estimates may be associated with large uncertainties. Furthermore, genetic models change as knowledge is acquired and geologists often disagree on at least some points of a genetic model; this creates uncertainty about the identification of mineral targets. An obvious advantage of objective methods is the production of relatively robust estimates of mineral potentials by extensively using geological, geochemical, and geophysical data. However, these methods also are deficient in some regards. Without using genetic theories, geoscience information content of the variables may be low and may have poor predicting power, i.e., the estimates often 'at best' reproduce what an expert geologist had recognized.

A useful procedure as a link between the two types of model is outlined as follows. First, based upon genetic theories, identify one or more critical genetic factors which are considered as necessary conditions for ore formation. A mineral deposit is believed to be absent if these genetic factors do not exist. Second, identify a set of recognition criteria that offer 'almost sure' existential evidence for critical genetic factors. Third, estimate the favorabilities or probabilities of occurrence of these recognition criteria based upon multiple geodata sets. Fourth, generate a synthesized favorability or probability measure for the occurrence of critical genetic factor(s) based upon the probabilities estimated in the third step. Finally, potential exploration targets are delineated from the synthesized favorability or probability measure through optimum discretization (Pan and Harris 1990). These targets have been referred to as *intrinsic geological units* with respect to the chosen critical genetic factor(s) (Pan and Harris 1993). These targets are so-called chiefly because they are not delineated directly in terms of mineral deposits, but in terms of the critical genetic factor that is a necessary condition for formation of the mineral deposits.

Upon the completion of target delineation, a decision needs to be made as to which targets should receive high priority to be drilled, as different targets vary in the degrees of favorability of mineral occurrence. This need requires the ranking of the targets in the sequence of drilling plans. Rank estimates may be derived directly from the synthesized fields or index. When a reasonable amount of known information on the metal(s) of interest is available in the study region, the rank estimation can be substantially improved by using a functional relation between the synthesized index and the quantity of metal. Of course, estimation of metal quantities is a difficult task, if not impossible. Such a function for estimation of metal quantities is valid only in a sense of pseudo terms, meaning that the results are meaningful only in a statistical sense. Verification for the results is necessary in later stages of exploration and estimation.

## 21.8 Prediction with Dynamic Control Samples

Most conventional resource analyses are constructed on the basis of extrapolation of some mathematical relations established in control areas into unknown areas (Pan and Harris 2000). Control areas are commonly employed in geodata integration and for the estimation of mineral resources of a relatively unexplored region. As such estimation is predicated upon the principle of *analogy*, the properties of the estimates are heavily reflective of (1) how good of a geological analogue the control area is of the unexplored region and (2) the economic reference for the estimated resources. When analogue and desired resource estimate is for economic and technologic conditions similar to those that induced the exploration and resource development of the control area, resource estimates produced by a mathematical model estimated on a control area may be unbiased. However, when economic or technologic references for the estimates differ or when the control area is not a good geologic analogue, resource estimates are biased and even totally wrong.

Two different approaches to improvement of estimation by mathematical models estimated on control areas are: (1) use only control areas that are exhaustively explored and (2) extend the mathematical model to include exploration variables (such as those defined in Pan and Harris (1991). Both of these solutions present difficulties however: (1) except for very small regions, there are few regions large enough to make good control areas that are exhaustively explored and (2) information on exploration activities generally is not available for regions large enough to make good control areas. When exploration variables are not explicitly included in the model, identification of an appropriate control area presents a difficult problem, for it must represent an unbiased sample of deposit occurrence and nonoccurrence for the relevant geologic environment. As noted by Chung et al. (1992), to compute unbiased estimates of the probability for deposit occurrence conditional upon a set of geologic attributes, it is necessary to know not only the distribution of various attributes in and near mineral deposits, but also the distribution of the same attributes away from mineral deposits (Cox 1990; Agterberg 2015).

Given the issues presented above, it is necessary to solve the dilemma in the selection of control areas and even method of extrapolations of these control areas into unknown regions. The nature of control areas so far is static, meaning that the control areas are fixed when a mathematical model established from these control areas is extended into unexplored regions. Clearly, this static model is hardly adequate for prediction of a large region with complex variability of geological conditions and mineralization characteristics. In other words, the mathematical model built on a basis of samples collected from a control area is only appropriate when the extrapolated areas have geological conditions identical to those in the control areas. It is deemed invalid when the geological conditions in the estimated areas differ from those in the control areas. Hence, a new concept is proposed here: dynamic control areas, which are characterized as self-improvement of the mathematical models through information gains of extrapolated areas away from the initial control areas. The methodology of dynamic control areas and extrapolation of mathematical models are implemented in three steps as follows:

(1) Select the best explored areas in the working region as the initial control area, from which control samples are collected. On the basis of this sample data, a mathematical model is established through data enhancement, combination of different datasets, and techniques of information synthesis. This mathematical

model is then used as the initial model for extrapolation and prediction of unknown areas in the working region.

(2) Update the mathematical model when the model is used for prediction of an unknown unit based on an expanded control sample through addition of new information of exploration variables and target variables (if any) in the predicted unit. The new mathematical model will be more appropriate to the estimation of unknown units. The decision of model update is predicated upon availability of new known target variable information and variability of geological and mineralization conditions from the initial control areas.

(3) Tests are performed with the updated model with respect to its effect in prediction of known units in the initial control areas and the unknown unit. The updated model would be accepted if the test results are satisfied; otherwise, the models will be reconstructed. Quantification of variability of geological and mineralization conditions in the unknown units plays a key role in the predicting power of the updated mathematical models.

The model update above is in nature an iterative process, which improves predictability of the model in the unknown units. The initial control sample is only used for establishment of the initial mathematical model, which is then updated and optimized as it is extended into the predicted areas through incorporation of new information on the variability of geological environments.

**Acknowledgements** The author wishes to thank for the guidance of Dr. D. P. Harris for the subject and the useful comments provided by Dr. Frits Agterberg of Geological Survey of Canada and Dr. B. S. Daya Sagar of Indian Statistical Institute-Bangalore Centre.

## References

Agterberg FP (1981) Application of image analysis and multivariate analysis to mineral resource appraisal. Econ Geol 76:1016–1031

Agterberg FP (1988) Application of recent developments of regression analysis in regional mineral resource evaluation. In: Chung CF, Fabbri AG, Sinding–Larsen R (eds) Quantitative analysis of mineral and energy resources. D. Reidel Publishing Company, p 1–28

Agterberg FP (1989) Computer programs for mineral exploration. Science 245:76–81

Agterberg FP (1992) Combining indicator patterns in weights of evidence modeling for resource evaluation. Nonrenewable Resour 1:1–16

Agterberg FP (2014) Geomathematics: theoretical foundations, applications and future developments. Quantitative geology and geostatistics, vol 18. Springer, Heidelberg

Agterberg FP (2015) Self-similarity and multiplicative cascade models. J South Afr Inst Min Metall 115:1–11

Allais M (1957) Method of appraising economic prospects of mining exploration over large territories: Algerian Sahara case study. Manag Sci 3:285–347

Barnett HJ, Morse C (1963) Scarcity and growth: in the economics of natural resource availability. Johns Hopkins Press, Baltimore

Brinck JW (1972) Prediction of mineral resources and long–term price trends in the nonferrous metal mining industry: in section 4-mineral deposits. In: Twenty-fourth session international geological congress, Montreal, Ottawa, Canada, p 3–15

Botbol JM, Sinding-Larsen R, McCammon RB, Gott GB (1978) A regionalized multivariate approach to target selection in geochemical exploration. Econ Geol 73:534–546

Chung CF, Agterberg FP (1980) Regression models for estimating mineral resources from geological map data. Math Geol 12:473–488

Chung CF, Jefferson CW, Singer DA (1992) A quantitative link among mineral deposit modeling, geoscience mapping, and exploration-resource assessment. Econ Geol 87:194–197

Cox DP (1990) Development and use of deposit models in the U.S. Geological Survey. In: 8th IAGOD symposium program with abstracts, Ottawa, 12–18 Aug 1990, p A99

Finch WI, McCammon RB (1987) Uranium resource assessment by the Geological Survey. Methodology and plan to update the national resource base. U.S. Geological Survey Circular 994, p 22

Gorelov DA (1982) Quantitative characteristics of geologic anomalies in assessing ore capacity. Intern Geol Rev 24:457–466

Green WR (1991) Exploration with a computer: geoscience data analysis and applications. Pergamon Press, Oxford, p 225

Harris DP (1965) An application of multivariate statistical analysis to mineral exploration. PhD dissertation, The Pennsylvania State Univ., University Park, Pennsylvania, p 261

Harris DP (1984) Mineral resources appraisal–mineral endowment, resources, and potential supply. Concept, methods, and cases. Oxford University Press, New York, p 455

Harris DP, Carrigan FJ (1981) Estimation of uranium endowment by subjective geological analysis—a comparison of methods and estimates for the San Juan Basin, New Mexico. Econ Geol 76:1032–1055

Harris DP, Pan GC (1990) Subdividing consistent geological areas by relative exceptionalness of additional information—methods and case study. Econ Geol 85:1072–1083

Harris DP, Pan GC (1991) Consistent geological areas for epithermal gold-silver deposits in the Walker Lake quadrangle of Nevada and California, delineated by quantitative methods. Econ Geol 86:142–165

Hattori I (1976) Entropy in Markov chains and discrimination of cyclic patterns in lithologic successions. Math Geol 8:477–497

Kantsel AV (1967) Function of metal distribution in ores, as genetic characteristics of mineralization process. Int Geol Rev 9:669–676

Koch GS Jr, Papacharalampos D (1988) GEOVALUATOR, an expert system for resource appraisal: a demonstration prototype for Kalin in Georgia, USA. In: Chung CF (ed) Quantitative analysis of mineral and energy resources. D. Reidel Publication Company, p 513–527

Laznicka P (1983) Giant ore deposits—a quantitative approach. Glob Tectonics Metallogeny 2:41–63

McCammon RB (1990) Prospector III: in statistical applications in the earth sciences. In: Agterberg FP, Bonham-Carter GF (eds) Geological Survey of Canada, Paper 89-9, p 395–404

McCammon RB, Botbol JM, Sinding-Larsen R, Bowen RW (1983) Characteristic analysis—1981—final program and a possible discovery. Math Geol 15:59–83

Pan GC (1987) A stochastic approach to optimum decomposition of cyclic patterns in sedimentary processes. Math Geol 19:503–521

Pan GC (1989) Concepts and methods of multivariate information synthesis for mineral resources estimation. PhD dissertation, University of Arizona, Tucson, p 302

Pan GC, Harris DP (1990) Three nonparametric techniques for optimum discretization of quantitative geological measurements. Math Geol 22:699–722

Pan GC, Harris DP (1991) Geology-exploration endowment models for simultaneous estimation of discoverable mineral resources and endowment. Math Geol 23:507–540

Pan GC, Harris DP (1992) Estimating a favorability equation for the integration of geodata and selection of mineral exploration target. Math Geol 24:177–202

Pan GC, Harris DP (1993) Delineation of intrinsic geological units. Math Geol 25:9–39

Pan GC, Harris DP (2000) Information synthesis for mineral exploration. Oxford University Press, p 450

Pan GC, Harris DP, Heiner T (1992) Fundamental issues in quantitative estimation of mineral resources. Nonrenewable Resour 1:281–292

Pan GC, Porterfield B (1995) Large-scale mineral potential estimation for blind precious metal ore bodies. Nonrenewable Resour 4:187–207

Patino Douce AE (2016a) Metallic mineral resources in the twenty first century. I. Constraints on future supply. Nat Resour Res 25:71–90

Patino Douce AE (2016b) Metallic mineral resources in the twenty first century. II. Constraints on future supply. Nat Resour Res 25:97–124

Sabins FF Jr (1987) Remote sensing. Principles and interpretation, 2nd edn. W. H. Freeman and Company, New York, p 449

Schwarzacher W (1969) The use of markov chains in the study of sedimentary cycles. Math Geol 1:17–39

Singer DA, Kouda R (1988) Integrating spatial and frequency information in the search for Kuroko deposits of the Hokuroku district, Japan. Econ Geol 83:18–29

Skinner BJ (1976) A second iron age ahead? Am Sci 64:258–269

Stanley M (1992) Statistical trends and discoverability modeling of gold deposits in the Arbitibi greenstone belt. In: the 23rd international symposium of APCOM, Ontario, p 17–28

Tomson IN, Polyakova OP (1984) Mineralogical and geochemical indicators of large ore deposits. Glob Tectonics Metallogeny 2:183–186

Vistelius AB (1960) The skew frequency distributions and the fundamental law of geochemistry. J Geol 68:1–22

Vistelius AB (1972) Ideal granite and its properties. I. The stochastic model. Math Geol 4:89–102

Vistelius AB, Harbaugh JW (1980) Granitic rocks of Yosemite Valley and ideal granite model. Math Geol 12:1–24

Vistelius AB (1981) Gravitational stratification. In: Graid RG, Labovitz ML (eds) Future trends in geomathematics, p 134–158

Wang H, Wellmann J, Li Z, Wang X, Liang R (2017) A segmentation approach for stochastic geological modeling using hidden markov random fields. Math Geosci 49:145–177

Zhao PD (2007) Quantitative mineral prediction and deep mineral exploration. Earth Sci Front 14:1–10

Zhao PD, Chi S (1991) Discussion of geo-anomaly. Earth Sci 16:241–248

# Chapter 22
# Solving the Wrong Resource Assessment Problems Precisely

**Donald A. Singer**

**Abstract** Samples are often taken to test whether they came from a specific population. These tests are performed at some level of significance (α). Even when the hypothesis is correct, we risk rejecting it in α percent of the cases—a Type I error. We also risk accepting it when it is not correct—a Type II error at β probability. In resource assessments much of the work is balancing these two kinds of errors. Remarkable advances in the last 40 years in mathematics, statistics, and computer sciences provide extremely powerful tools to solve many mineral resource problems. It is seldom recognized that perhaps the largest error—a third type—is solving the wrong problem. Most such errors are a result of the mismatch between information provided and information needed. Grade and tonnage or contained models can contain doubly counted deposits reported at different map scales with different names resulting in seriously flawed analyses because the studied population does not represent the target population of mineral resources. Among examples from mineral resource assessments are providing point estimates of quantities of recoverable materials that exist in Earth's crust. What decision is possible with that information? Without conditioning such estimates with grades, mineralogy, remoteness, and their associated uncertainties, costs cannot be considered, and possible availability of the resources to society cannot be evaluated. Examples include confusing mineral occurrences with rare economically desirable deposits. Another example is researching how to find the exposed deposits in an area that is already well explored whereas any undiscovered deposits are likely to be covered. Some ways to avoid some of these type III errors are presented. Errors of solving the wrong mineral resource problem can make a study's value negative.

**Keywords** Quantitative resource assessment • Decision analysis
Uncertainty • Lognormal

---

D. A. Singer (✉)
10191 N. Blaney Ave., Cupertino, CA 95014, USA
e-mail: singer.finder@comcast.net



B. S. Daya Sagar et al. (eds.), *Handbook of Mathematical Geosciences*,
https://doi.org/10.1007/978-3-319-78999-6_22

## 22.1 Introduction

Howard Raiffa (1968, p. 264) noted that statistics students learn the importance of constantly balancing making an error of the first kind (that is, rejecting the null hypothesis when it is true) and an error of the second kind, that is, accepting the null hypothesis when it is false (Fig. 22.1). Raiffa thought it was John Tukey who suggested that practitioners all too often make errors of a third kind: of solving the wrong problem. Raiffa nominated a candidate for the error of the fourth kind: solving the right problem too late. John Tukey believed that it was better to find an approximate answer to the right question, than the exact answer to the wrong question, which can always be made precise. More recently, Mitroff and Silvers (2009) focused mostly on social questions where type III errors occurred and provided many examples of developing good answers to the wrong questions (type III error). Unfortunately concerns of Raiffa, Tukey, Mitroff, Silvers, and others are appropriate for mineral resource assessments. And the concerns should not be limited to classical statistics.

Supply of minerals to society is dependent not only on the total amount of mineral material but also on quality or concentrations, spatial distributions or how scattered the material is, whether it has been found, whether it is remote from infrastructure, and a whole host of other issues such as government policies, production technologies, and market structures. Decision-makers, whether concerned about development of a technology, development of a region, exploration, or land management, are faced with the dilemma of obtaining new information, or allowing or encouraging others to obtain it, and the possible benefits and costs of development if mineral deposits of value are discovered. Decisions about exploration for these resources and their possible development require awareness of various kinds and the import of errors that can be made by analysts in their studies.

A type I error is the rejection of the null hypothesis when it is true. In some fields a type I error is called a false positive. The risk of this error is α, the level of significance. A type II error is the acceptance of the null hypothesis when it is false, also known as a false negative error. The probability of making a Type II error, β,

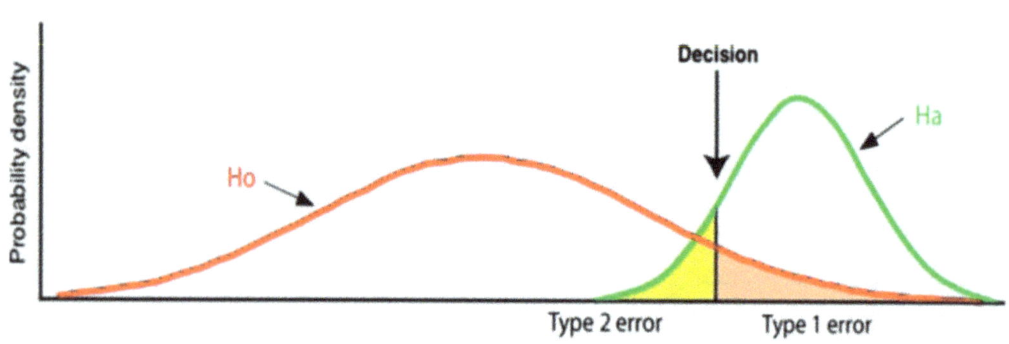


**Fig. 22.1** Type I error is the rejection of the null hypothesis (Ho) when it is true. The risk of this is α, the level of significance. Type II error is the acceptance of the null hypothesis when it is false

depends on the alternative value and its distribution. The most important question of the analyst and decision-maker should be: Are we solving the right problem? It is the need to consider this source of error in mineral resource studies that is the focus of this chapter. Common to many of the errors of solving the wrong problem is a mismatch of the studied population and the population that is central to the decisions—this topic is presented first. Next, effects of mismatches of populations to some mineral resource assessments are discussed. Possible ways to avoid some of these type III errors are finally presented.

## 22.2 Target Population

Type III errors are fundamental and should be considered before errors of types I and II. Type III errors stem from improper definition of the problem and therefore are not strictly a statistical issue, but one of critical thinking. It does no good to minimize the expected costs of type I and type II errors if the wrong problem is being solved. In mineral resource assessments, careless problem definition is the primary source of type III errors. For almost all resource assessment problems, the fundamental sample is the mineral deposit.

The idea of a mineral resource involves both geologic and economic aspects and because knowledge about the earth and future economic conditions is limited, should recognize uncertainty. Mineral deposits are the geologic entities containing resources. Mineral deposits and their contents are the fundamental target populations that are estimated. So what is a mineral deposit? Mineral deposits are defined as mineral occurrences of sufficient size and grade that they might, under favorable circumstances, be economic.

A map of some volcanogenic massive sulfide deposits from Northern Japan is used to clarify our understanding of what is a deposit (Fig. 22.2). From this plot one can see that some of the deposits are just a few meters apart from each other. Grade and tonnages are available for 23 of these named deposits from the western part of the Hokuroku district, Japan (Ohmoto and Takahashi 1983). It is important that if a different map scale were used, this part of the district might have three or four named deposits with grades and tonnages. This well-studied district has more detailed maps than many other volcanogenic massive sulfide districts around the world. If one gathered all available data on the names and grades and tonnages of volcanogenic massive sulfide deposits and built grade and tonnage or contained metal models, the models would contain metals double counted from deposits reported at different map scales and from the same deposits with different names due to grouping. To have a consistent sampling unit that can be applied in statistical analysis and in assessments of undiscovered deposits it is necessary to have spatial rules to help define a deposit. In addition, mine names and deposit names do not always match, mine names sometimes change over time, and district and deposits can be reported with different names and numbers. For example, careless data gathering might contain the grades and tonnage of the total Sudbury Ni-Cu District

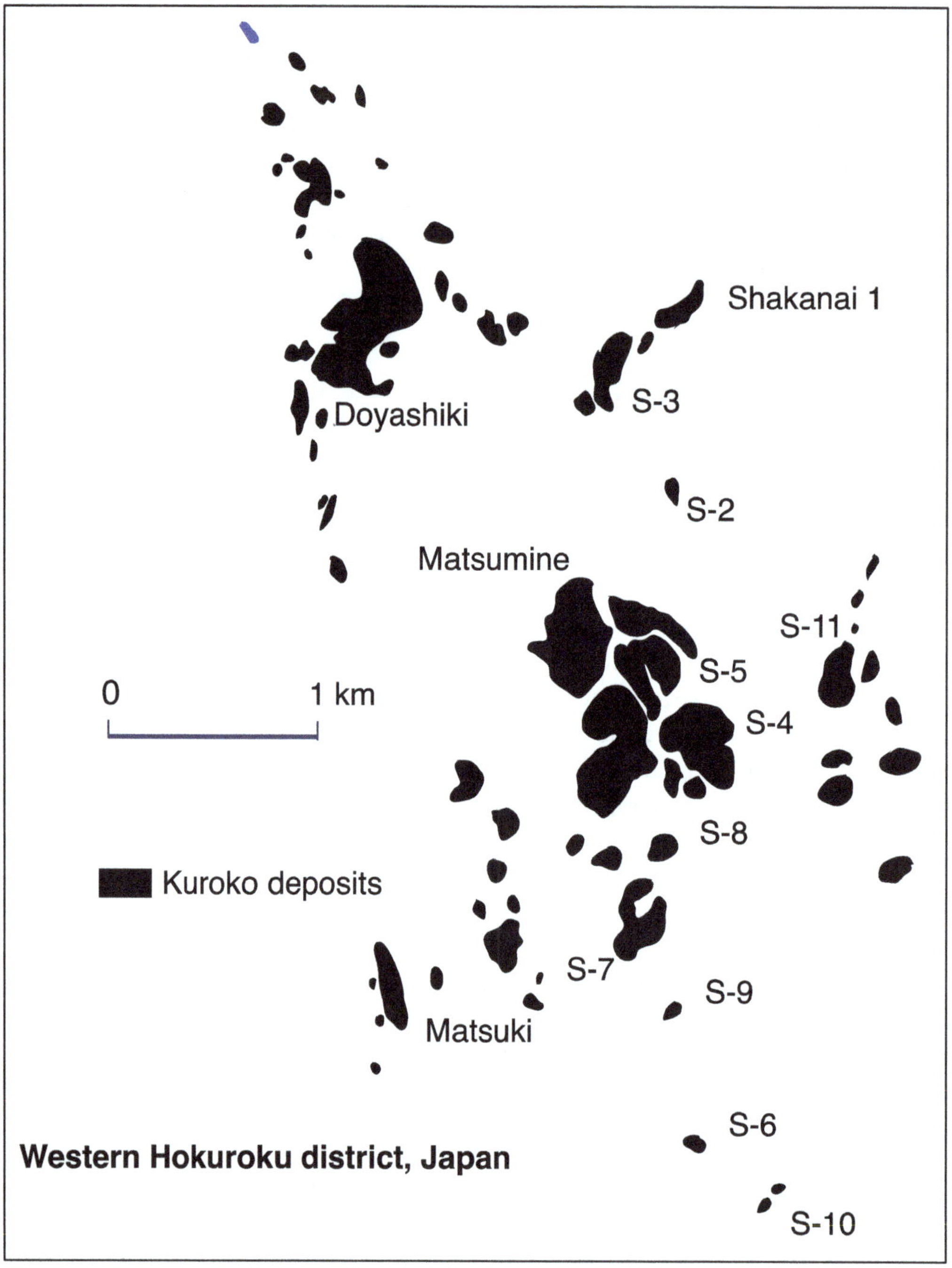


**Fig. 22.2** Kuroko volcanogenic massive sulfide deposits of the western part of the Hokuroku district in Northern Japan (after Ohmoto and Takahashi 1983)

in Canada and also contain grades and tonnages of the many mines thus double counting and generating biased metal statistics and frequency distributions of questionable value. There are databases in which spatial rules for combining adjacent deposits have been consistently applied and multiple names have been eliminated (e.g., Mosier et al. 2009). Compilations that use the above sources combined with other sources of data on, for example, volcanic-hosted massive sulfide deposits very likely contain deposits and prospects counted twice (e.g., Patiño-Douce 2016), resulting in statistical analyses that are seriously flawed because the studied population does not represent the target population of mineral resources. Operational rules defining deposits need to account for these map scale effects and for the fact that some deposits have multiple names, mines and separate reported tonnages (Singer 2017).

Mineral occurrences or prospects which are the focus of prospectivity analysis do not qualify as economic mineral deposits because they are typically quite small and incompletely explored. Because number of undiscovered deposits estimates must be defined in a way that is linked to the grade-tonnage or contained metal models, estimates of number of deposits made using models based on such flawed grade-tonnage models must also be a mismatch with the target population.

## 22.3 Examples of Mismatches in Assessments

Solving the wrong problem due to mismatches of the target population with the studied or estimated population abound in mineral resource assessments. Examples of mismatches include issues of not understanding where the undiscovered resources might exist and estimating something other than mineral deposits that might be economic to mine (De Young and Singer 1981).

In one example, five or more epithermal gold vein deposits were estimated at the 90% level but no grade-and-tonnage model was provided, so the estimated deposits could be any size (Singer and Menzie 2010). To provide critical information to decision-makers, a grade-and-tonnage or contained metal model is key, and the estimated number of deposits that might exist must be from the linked grade-and-tonnage frequency distributions. Estimates of number of undiscovered deposits are completely arbitrary unless tied to a grade-and-tonnage or contained metal model that has been defined in a consistent operational manner.

In an unpublished study, four geoscientists made subjective probabilistic estimates of the number of undiscovered hot-spring mercury deposits in a 1:250,000 scale quadrangle in Alaska. They made independent estimates at the 90th, 50th, and 10th percentiles (Table 22.1). The 10th percentile, for example, is the number of deposits for which there is at least a 10% chance of that number of deposits or more exist.

It was pointed out to participant D that because the number of deposit estimates must be consistent with the grade and tonnage model, his estimates imply that there is more undiscovered mercury in this quadrangle than has been found in the world

**Table 22.1** Independent estimates by four scientists of the number of undiscovered hot-spring Hg deposits in a quadrangle in Alaska

| Participant | A | B | C | D | |
|---|---|---|---|---|---|
| 90% chance of at least | 1 | 1 | 2 | 9,000 | Deposits |
| 50% chance of at least | 3 | 2 | 4 | 10,000 | Deposits |
| 10% chance of at least | 6 | 6 | 7 | 11,000 | Deposits |

in this deposit type. He responded that he was estimating wisps of cinnabar, not deposits consistent with the grade and tonnage model. In this case, the population considered by participant D did not match the target population. Using a variety of different guidelines such as deposit densities (Singer 2008) for estimates of the number of undiscovered deposits provides a useful crosscheck of assumptions that may have been relied upon and discourages mismatches between target and estimated populations. In these examples of errors in estimating the number of undiscovered deposits, the key is the difference between the understanding of what was being estimated and the population of interest.

In Harris's landmark study (1965), multiple discriminate analysis was used to predict value of mineral production—among the best predictors was geologic cover with a negative value. In a study by Singer (1971), multiple regression was used to predict mineral production and again, cover with a negative value was an important variable. Unlike in petroleum exploration, minerals exploration under cover is a developing technology. Most commonly, mineral exploration under cover results from trying to extend known deposits, that is, additions to reserves. More difficult discovery and higher costs relative to exposed deposits, tend to reduce interest in covered areas. Covered areas tend to be poorly explored and, consequently, deposits under cover tend to be underreported.

In situations where resource assessments are made based on local information, the possibility of solving the wrong problem is high. For example, if the mapped geology were used to predict where and how many undiscovered orogenic gold deposits might in the Bendigo Zone of Victoria Australia, one would conclude that deposits are clustered in space and gold deposits are related to older rocks and covered areas would be worst place to look (Fig. 22.3). Even if we use some modern tools like weights of evidence or neural networks, we would predict no undiscovered deposits under cover. Yet, because geology permissive for the gold deposits is known under cover, and exposed permissive geology is thoroughly explored, most experts would recommend exploration under cover (Lisitin et al. 2007).

Each of these examples demonstrates mismatches of the target population and the studied population. Type III errors in these cases could produce useless or, even worse, misleading assessments.

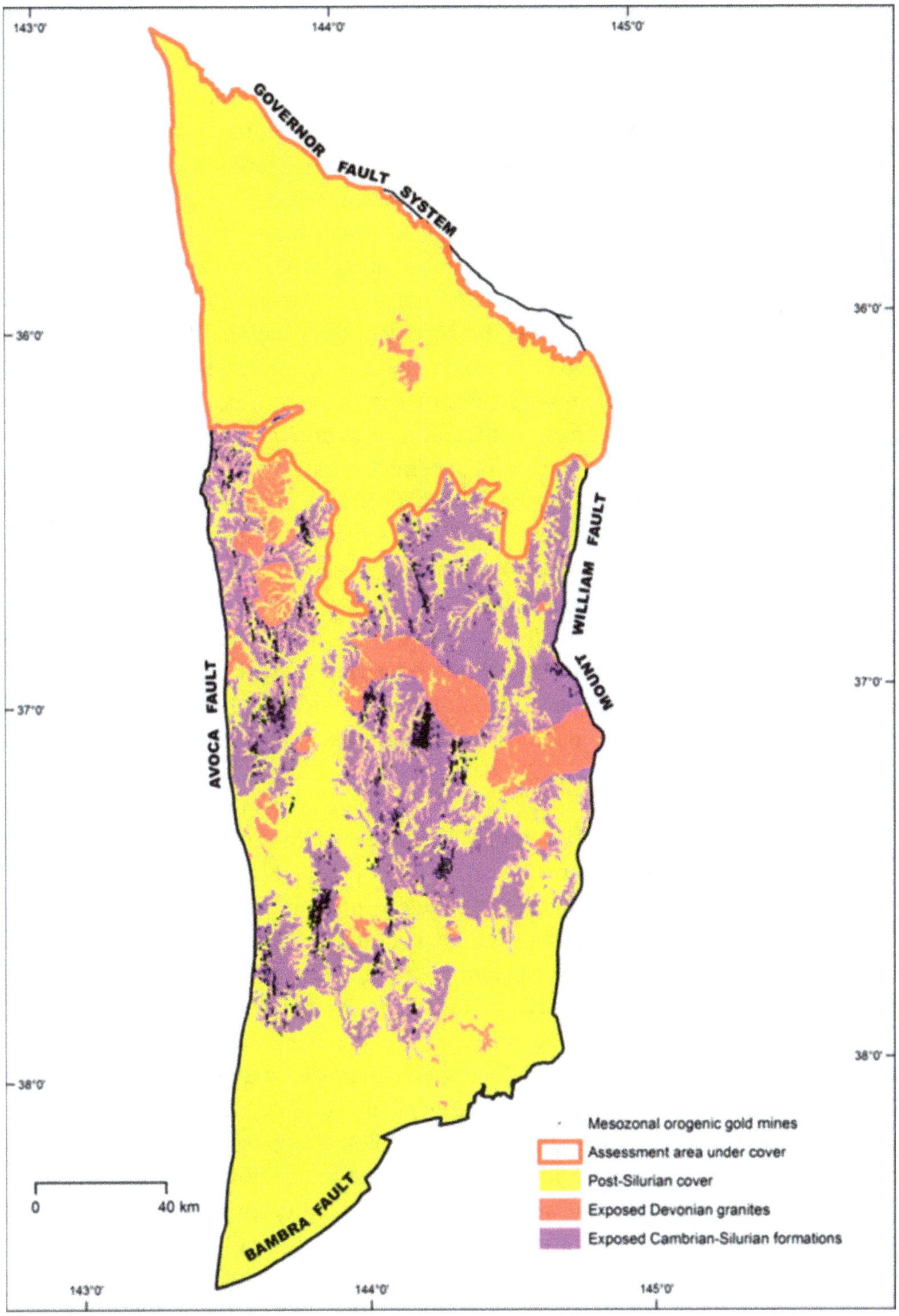


**Fig. 22.3** Geology and known orogenic gold deposits (black) in the Bendigo Zone of Victoria, Australia (modified after Lisitsin et al. 2007)

## 22.4 How to Correct Type III Errors

The problems of mineral resource assessment can only be solved if they are formulated in a way consistent with the decision-maker's language and understanding of the problem. The questions need to be asked: Why perform an assessment? Who is the study being done for and what are the problems they are trying to resolve?

We start with the question of what kinds of issues decision makers are trying to resolve and what types and forms of information would aid in resolving these issues. Unfortunately, the decision-maker may not be available for the needed insight or may not be able to clearly state the information needs. Because the primary purpose of the kinds of assessments recommended here is to help decision-makers determine consequences of economic and policy decisions about tracts of land, regions, countries, or the earth, it is critical that the assessments be unbiased. For example, if the question concerns the long-term supply of a metal, the data used should not contain biased information such as grades and tonnages on multiple versions of the same deposits. These situations require care in compiling data and using sources that report locations, other names of deposits and names of deposits that have been combined with the primary deposit to meet spatial combination rules. A reliable source (e.g., Mosier et al. 2009) has specific information about locations, rules used to combine deposits and specific names that were combined for each deposit. These kinds of data provide a reliable basis for testing statistical distributions of metals in mineral deposits such as the lognormal distribution (Singer 2013).

It is important to recognize that success of assessments depends on the assessments following an integrated approach. This means that no part of the models and methods of estimation have any meaning in isolation. For instance, estimates of number of undiscovered deposits are completely arbitrary unless tied to a grade and tonnage or contained metal model. The goal should be to make explicit the factors that can affect a mineral-related decision so that the decision-maker can clearly see what are the possible consequences of decisions (Singer and Menzie 2010).

To avoid situations where occurrences are the basis of information used to discriminate barren areas from the economic deposits sought, it is necessary to construct models based on the economic deposits sought. Mineral deposit models can be based on data gathered from well-explored deposits of each type from around the world. This would allow the determination of how commonly different attributes and combinations of attributes occur. Quantifying mineral deposit attributes is the necessary and sufficient next step in statistically classifying known deposits by type. Quantified deposit attributes also can provide a firm foundation to identify which observations on geologic and other maps should be effective in delineation of tracts and perhaps identifying sites for detailed exploration. The kind of digital models advocated here would require the recording of both absolute time units and the relative time units of spatially related mineral deposits, rocks, geochemistry, geophysics, and tectonics. The scale of the observations is critical to proper application of such models. This is required to properly apply the models in new geologic settings. Information in these models about the attributes associated with known deposits is necessary but not sufficient to discriminate barren from mineralized environments; quantifying the attributes of barren environments also is necessary for this task. Such digital models could be the foundation for identifying the discriminating functions that could remove many type III errors in assessments.

The exploration department of a major zinc producer found it essential to document a robust decision-making process to maintain internal and investor support (Penney et al. 2004). Zinc deposits from around the world were classed by type, grade, and tonnage models developed for each, cost filters were applied to each, and tracts around the world were delineated where the types could occur (Penney et al. 2004). This study was designed to aid the exploration decision-makers plan the search for economic deposits. Their process was the same as that recommended in three-part assessments (Singer and Menzie 2010), with the exception that they ranked or scored tracts rather than estimating the number of undiscovered deposits.

## 22.5 Conclusions

Errors of solving the wrong mineral resource problem can make a study's value negative. Type III errors, solving the wrong problem, can be avoided by using care in matching the information needed to solve the decision-maker's problem with information provided in the study. In some cases, we know how to solve the wrong problem but not the real one. It is not uncommon to get rewarded for publishing an answer—not THE answer. With some care and critical thinking in the planning stages, it is possible to provide information useful to decision-makers and to be rewarded for a publication.

## References

DeYoung JH Jr, Singer DA (1981) Physical factors that could restrict mineral supply. In: Skinner BJ (ed) 75th anniversary volume. Economic geology, pp 939–954

Harris DP (1965) An application of multivariate statistical analysis to mineral exploration. PhD dissertation, The Pennsylvania State University, 265 pp

Lisitsin V, Olshina A, Moore DH, Willman CE (2007) Assessment of undiscovered mesozonal orogenic gold endowment under cover in the northern part of the Bendigo Zone. GeoScience Victoria Gold Undercover Report 2, Department of Primary Industries, 98 pp. www.dpi.vic.gov.au/minpet/store

Mitroff II, Silvers A (2009) Dirty rotten strategies—how we trick ourselves and others into solving the wrong problems precisely. Stanford University Press, Stanford, 232 pp

Mosier DL, Berger VI, Singer DA (2009) Volcanogenic massive sulfide deposits of the world—database and grade and tonnage models. U.S. Geological Survey Open-file Report 2009-1034. http://pubs.usgs.gov/of/2009/1034/

Ohmoto H, Takahashi T (1983) Submarine calderas and Kuroko genesis. In: Omoto H, Skinner BJ (eds) The Kuroko and related volcanogenic massive sulfide deposits. Econ Geol Monogr 5:38–54

Patiño-Douce AE (2016) Metallic mineral resources in the twenty first century. II. Constraints on future supply. Nat Resour Res 25(1):97–124

Penney SR, Allen RM, Harrisson S, Lees TC, Murphy FC, Norman AR, Roberts PA (2004) A global-scale exploration risk analysis technique to determine the best mineral belts for exploration. Trans Inst Min Metall Sect B. Appl Earth Sci 113:B183–B196

Raiffa H (1968) Decision analysis—introductory lectures on choices under uncertainty. Random House, New York, 309 pp

Singer DA (1971) Multivariate statistical analysis of the unit regional value of mineral resources. Unpublished PhD thesis, The Pennsylvania State University, 210 pp

Singer DA (2017) Future copper resources. Ore Depos Rev 86:271–278. http://dx.doi.org/10.1016/j.oregeorev.2017.02.022

Singer DA (2008) Mineral deposit densities for estimating mineral resources. Math Geosci 40(1):33–46

Singer DA (2013) The lognormal distribution of metal resources in mineral deposits. Ore Geol Rev 55:80–86. http://dx.doi.org/10.1016/j.oregeorev.2013.04.009

Singer DA, Menzie WD (2010) Quantitative mineral resource assessments—an integrated approach. Oxford University Press, New York, 219 pp

# Chapter 23
# Two Ideas for Analysis of Multivariate Geochemical Survey Data: Proximity Regression and Principal Component Residuals

**G. F. Bonham-Carter and E. C. Grunsky**

**Abstract** Proximity regression is an exploratory method to predict multielement haloes (and multielement 'vectors') around a geological feature, such as a mineral deposit. It uses multiple regression directly to predict proximity to a geological feature (the response variable) from selected geochemical elements (explanatory variables). Lithogeochemical data from the Ben Nevis map area (Ontario, Canada) is used as an example application. The regression model was trained with geochemical samples occurring within 3 km of the Canagau Mines deposit. The resulting multielement model predicts the proximity to another prospective area, the Croxall property, where similar mineralization occurs, and model coefficients may help in understanding what constitutes a good multielement vector to mineralization. The approach can also be applied in 3-D situations to borehole data to predict presence of multielement geochemical haloes around an orebody. Residual principal components analysis is another exploratory multivariate method. After applying a conventional principal components analysis, a subset of PCs is used as explanatory variables to predict a selected (single) element, separating the element into predicted and residual parts to facilitate interpretation. The method is illustrated using lake sediment data from Nunavut Territory, Canada to separate uranium associated with two different granites, the Nueltin granite and the Hudson granite. This approach has the potential to facilitate the interpretation of multielement data that has been affected by multiple geological processes, often the situation with surficial geochemical surveys.

---

G. F. Bonham-Carter (✉)
110 Aaron Merrick Drive, Merrickville, ON K0G 1N0, Canada
e-mail: graeme.bc1@gmail.com

E. C. Grunsky
Department of Earth and Environmental Sciences, University of Waterloo, Waterloo, ON N2L 3G1, Canada



B. S. Daya Sagar et al. (eds.), *Handbook of Mathematical Geosciences*,
https://doi.org/10.1007/978-3-319-78999-6_23

**Keywords** Ben Nevis area • Nunavut • Multivariate geochemistry
Lake sediment surveys • Regression • Principal components
Spatial modelling • Proximity • Residuals

## 23.1 Introduction

Proximity to selected spatial features on geological maps has been used in the analysis of multivariate data in several ways, but usually as a weighting function not as a variable to be directly predicted. For example, Cheng et al (2011) describe "spatially weighted principal component analysis" to emphasize proximity to selected intrusions in the analysis of geochemical patterns. This involves using spatial weights (in range 0–1) to calculate weighted correlation coefficients, before the usual eigenvector determinations of principal components analysis. The resulting weighted principal component scores were mapped to predict element associations related to intrusions. Brunsdon et al. (1998 and other papers) have used "geographically weighted regression" to analyze long-term illness data from a UK census. This approach recognizes that a regression may often not be spatially stationary, but will show changes geographically. Again, the regression equations use spatial variables as weights. In both these examples, proximity to some feature is introduced as a spatial weight, not as a response variable for direct prediction.

In the first part of this chapter we suggest that proximity to a geological feature can be more directly studied by using proximity itself as a response variable in a regression using a collection of geochemical elements as explanatory variables. In regional geochemical surveys, one may be interested in understanding which variables are good predictors of proximity to a mineral deposit, or to some other selected feature with known location. This is frequently referred to in mineral exploration as finding good 'vectors' to mineralization, but as far as we are aware direct prediction of proximity from multielement data has not been published, although plots of single elements, or element ratios, on profiles showing distance to known mineralization are often used. If a good predictive suite of elements can be determined (either from understanding a genetic model or from empirical tests) and based on a training set of samples relatively close to the geological feature of interest, the resulting predictive equation can be used to look for similar associations outside the training area. If the feature of interest is a mineral deposit, this approach may be useful in finding new deposits. This may be used both for 2-D regional geochemical surveys, and in 3-D geochemical data from borehole data.

The second part of the chapter is about using residual principal components analysis (PCA) of multielement geochemical data. PCA has been widely used by exploration geochemists and others to understand multielement geochemical processes, particularly in surficial geochemical surveys, but also in lithogeochemical data collected at surface or in boreholes. This literature is large, and here we refer as an example to a study of soil geochemistry as measured along two continental scale transects of North America. PCA of logratio-transformed variables revealed the effects of soil-forming processes, including soil parent material, weathering, and soil age as interpreted from PCs (Drew et al. 2010). There are many examples of successful geological interpretations by PC analysis. Individual PCs can often be interpreted both from variable loadings, from biplots and from spatial patterns seen by mapping PC scores (e.g. Grunsky 2010).

Sometimes, however, one may be interested in the spatial distribution of a single geochemical element, and it is desirable to remove the effect of some particular geological process or processes that are reflected in one or more PCs. For example, in the analysis of till geochemical surveys, the first PC is often interpreted as due to the effect of till transport. Thus it may be desirable to look at the element distribution after removing PC1. Usually this is carried out by progressively examining element loadings and the spatial patterns seen by mapping PC scores. However, there may be situations where it is helpful to examine spatial patterns of a single element after removing PC1 (or several PCs). This can be achieved what we are terming here as "principal component regression". This is a straightforward regression using the selected element as the response variable, and PC1 (or PC combination) as the explanatory variable(s). The residuals (the observed response variable minus the predicted response variable) provide the desired element distribution after removing the effect of PC1 (or PC combination). If PC1 is interpreted as due to till transport, then the residuals represent the element values after removing the effect of till transport.

This approach represents a process that is somewhat analogous to a geochemical selective leach separating a mineral phase or perhaps several mineral phases. A 'total' analysis is designed to dissolve all mineral phases, whereas a partial leach targets a selected mineral phase. The element under study can thereby be partitioned into phases by selective leaching. Residual PCA also separates the element under study into parts, although the partitions are not the same as those targeted in selective leaches. The partitions in residual PCA are related to proportions of an element quantity that can be 'explained' by different multivariable associations as determined by PCA. Residual PCA was first used by Bonham-Carter and Hall (2010) in a study of uranium in soils in the Athabasca Basin. Residual U, after removing the effect of till transport (as determined by PCA), was a better predictor of buried mineralization than raw U values in A-horizon soils.

In this chapter, we use a lithogeochemical dataset from the Ben Nevis area of Ontario to illustrate proximity regression, and a lake-sediment dataset from southern Nunavut to illustrate residual principal components analysis.

## 23.2 Method 1: Direct Prediction of Spatial Proximity

Suppose we have an array of geochemical data, with rows being samples, and elements as columns. In addition, we have distance measurements for each sample reflecting the shortest distance from the sample to some geological feature (mineral deposit, an intrusion, a fault, etc.). Before multivariate analysis, it will be important to transform the element variables by a centred logratio, to overcome the effects of closure (Aitchison 1986; Buccianti et al. 2006; and many other papers).

Although distances may be used directly, we have found that transforming distance to proximity gives somewhat better predictions. If for example the goal is to model the dispersion 'halo' around a deposit, the decay of the halo effect with distance from the contact may be exponential, or may follow a power law. Thus, instead of using distance as a response variable, we often get better results by transforming distances inversely to proximities. Here we have used a simple exponential decay of proximity with distance, that assumes that the rate of decay of proximity with distance is constant, similar to the familiar model of decay of a radioactive element with time. Let distance be denoted as $Z$ (metres from feature) and proximity by $Y$ (in range 1, 0 where 1 is at zero distance decreasing to zero at infinitely large distances), then the rate of decay of proximity with distance is assumed to be a constant

$$\frac{\mathrm{d}Y}{\mathrm{d}Z} = -\alpha. \tag{23.1}$$

Integrating (23.1) from distance 0 to $Z$ leads to:

$$Y(Z) = Y(0)\mathrm{e}^{-\alpha z}. \tag{23.2}$$

The value of proximity at zero distance $Y(0) = 1$, so this term drops out. It is also convenient to define the 'half distance' $Z_{0.5}$ where proximity $Y$ equals 0.5, then by rearranging Eq. 23.2 we can express α in terms of the half-distance:

$$\alpha = \frac{-\ln 0.5}{Z_{0.5}}. \tag{23.3}$$

Substituting for α in (23.1), distance can then be transformed to proximity from

$$Y(Z) = \exp\left(\frac{\ln 0.5}{z_{0.5}} \cdot Z\right) \tag{23.4}$$

We note that an alternative approach was used by Cheng et al. (2011) in the spatially weighted principal components to determine spatial weights W (equivalent to proximities) using a power relation:

$$W = \left(\frac{1 - Z}{Z_{max}}\right)^{\gamma} \tag{23.5}$$

where $\gamma$ is a power parameter, and $Z_{\max}$ is a selected maximum distance for modelling. For $\gamma = 0$, all weights = 1, with $\gamma = 1$, weights are a linear inverse of distance, but positive values of gamma such as 2, 8, 16 define a power-law decrease of proximity with increasing distance.

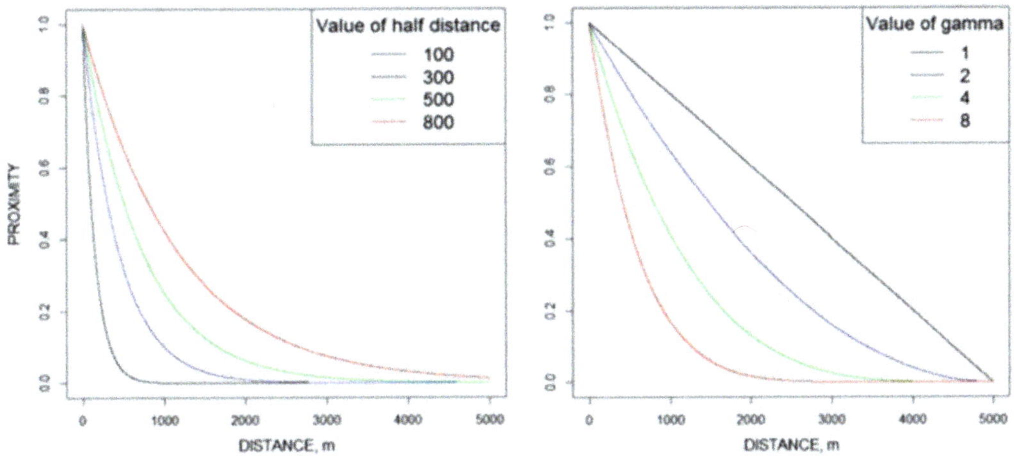


**Fig. 23.1** Left. Example of relationship between proximity and distance using exponential decay with a 'half-distance' parameter. Proximity = 1 at distance = 0, proximity = 0.5 at distance = half-distance. Right. Similar to left diagram, but using power law model with gamma parameter

Typical exponential curves and power law curves using Eqs. 23.4 and 23.5 are shown in Fig. 23.1.

We now model proximity with a training set of samples (chosen within some arbitrary but reasonable distance from the selected feature) using selected geochemical variables.

Then let *X* be the matrix of CLR-transformed element values, with rows as samples, columns as elements. The geochemical elements are the explanatory variables, and the column vector Y contains the proximity values, the response variable. The geochemistry is used to 'explain' the response. Here we used multiple linear regression to model this relationship, although other approaches could be taken.

$$Y = X\beta + \in \tag{23.6}$$

where $\beta$ is a column vector of coefficients to be determined by least squares, and $\epsilon$ is the vector of errors. The coefficients are solved from the normal equations

$$\beta = (X'X)^{-1}(X'Y) \tag{23.7}$$

where X' is the transpose of X and $(X'X)^{-1}$ is the inverse of $X'X$.

If inspection of the coefficients and goodness of fit are satisfactory, the predicted values of proximity, $\hat{Y}$, are calculated from

$$\hat{Y} = X\beta. \tag{23.8}$$

### 23.2.1 Application of Proximity Regression with Ben Nevis Lithogeochemical Data

#### 23.2.1.1 Background Geology

The Ben Nevis Township area is part of the Blake River Group (Fig. 23.2) a calc-alkaline volcanic sequence. The same sequence extends eastward to the Noranda area of Quebec where major Cu-Zn-Ag deposits are located. Extensive alteration and mineralization was recognized in the Ben Nevis area (Jensen 1975; Wolfe 1977), which led to a later geochemical study by Wolfe (1977) with emphasis on the metal distribution of stratiform volcanogenic sulphide deposits in Archean volcanic rocks. Lithogeochemical sampling was undertaken across the area by Jensen (1975) and Wolfe (1977) followed by additional sampling by Grunsky (1986a, b). Grunsky and Agterberg (1988) and Grunsky (1986a, b) carried out a detailed a multivariate geostatistical investigation of these data. A regional multi-element geochemical study over the Abitibi Greenstone Belt was later undertaken by Grunsky (2013) in which multivariate statistical methods were applied to recognize lithological variation, areas of alteration and potential base-metal mineralization.

The principal lithologies of the study area are basaltic pillowed flows, pillow breccias and breccias of calc-alkaline affinity (Grunsky 1986a). Two felsic volcanic units comprised of tuff, tuff breccia and flows of rhyolitic and dacitic composition occur within the basaltic sequence. The volcanic sequence has been intruded by tholeiitic gabbroic and diorite bodies throughout (Fig. 23.3). More recent studies of

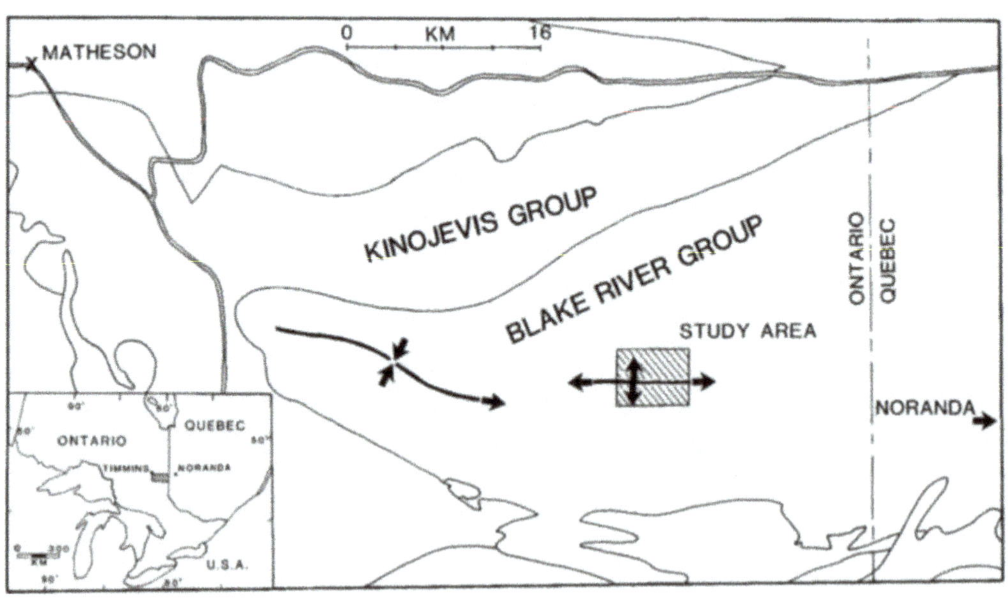


**Fig. 23.2** Location map of Ben Nevis study area adapted from Grunsky (1986a)

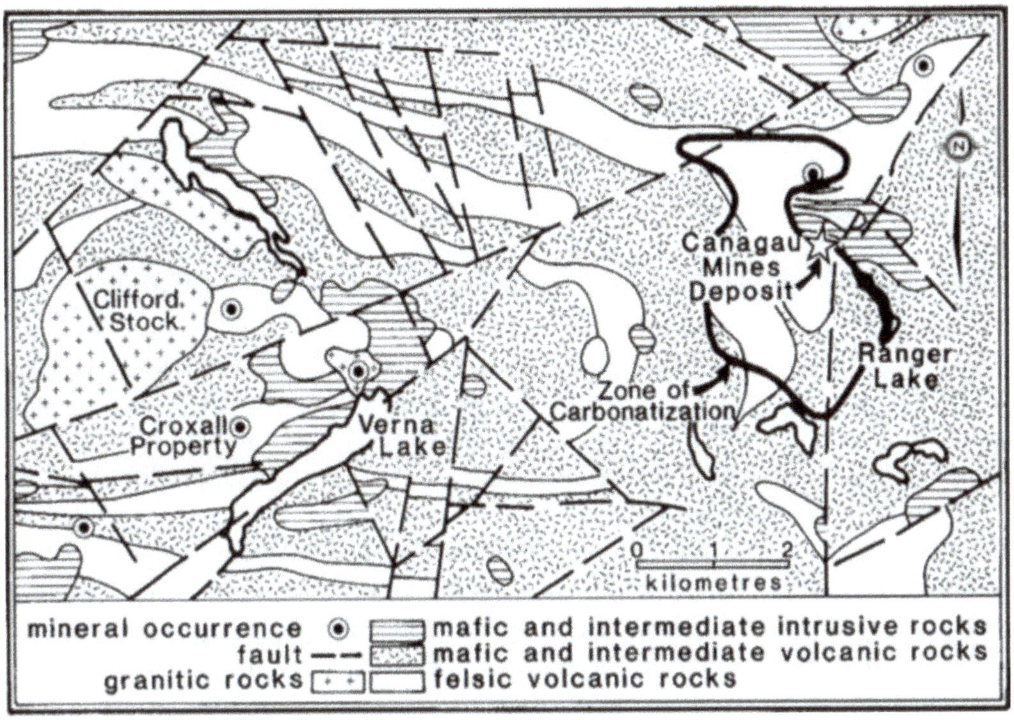


**Fig. 23.3** Geology of Ben Nevis area, adapted from Grunsky (1986a). Note locations of Canagau Mines deposit and Croxall property. Figure from Grunsky (1986b)

the volcanic assemblage in the context of the Abitibi Greenstone Belt are described by Pelogquin et al. (2008).

Within the area, the two most significant mineral occurrences are the Canagau Mines deposit and the Croxall property. The Canagau Mines deposit is dominated by strongly carbonatized, sericitized, and silicified mafic and felsic volcanic rocks. Mineralization consists of sphalerite, gold, silver, galena, chalcopyrite, and pyrite within east-trending fractures and shear zones that dip 40–60° south. Tonnages are unknown, and the grade is as high as 11 ppm gold and 22 ppm silver. The area was extensively explored by Wallbridge Mining in 2004 (Wallbridge 2004) and a report on exploration activities by Meyer et al. (2004). The deposit is currently considered to be uneconomic. The Au-Ag-Cu-Pb-Zn style of mineralization is typical of an epithermal system.

The Croxall property consists of a zone of brecciated and sheared rhyolite with interstitial pyrite, chalcopyrite, chlorite, calcite and quartz. Gold assays have been reported up to 1 ppm.

Grunsky (1986a, b) showed that multivariable data analysis techniques distinguish the altered from unaltered volcanic rocks.

#### 23.2.1.2 Application

The purpose of this application is to determine whether a multielement signature can be identified related to proximity to the Canagau Mines deposit, then use this signature to look for other places with similar patterns.

The distances between each sample and the Canagau Mines deposit was calculated using the eastings and northings associated with each sample, plus the known location of the deposit. Distances were converted to proximities using Eq. (23.4). Different proximity vectors were calculated for half-distances of 100, 300, 500, 800, 1000 and 1500 m so that an optimal half distance parameter could be determined. Figure 23.4 shows the sample points with proximity (half distance equal to 800 m) classified by colour and dot size. The training set comprises all points lying within 3 km of the deposit (equivalent to points with proximity greater than exp(ln(0.5) * 3000/800) = 0.074).

There are 26 geochemical variables in the dataset—a mixture of trace elements and major oxides. After converting all elements to a common unit of measurement (ppm), all chemical variables were transformed by centred logratios (CLR) to avoid the problem of closure. Using the training samples, correlation coefficients were calculated between each element (CLR-transforms) and proximity. These correlations were sorted by magnitude and used to reduce the number of elements selected to predict proximity by multiple regression analysis. Elements were selected for Model 1 if the absolute value of correlation (Pearson's r) with proximity was greater

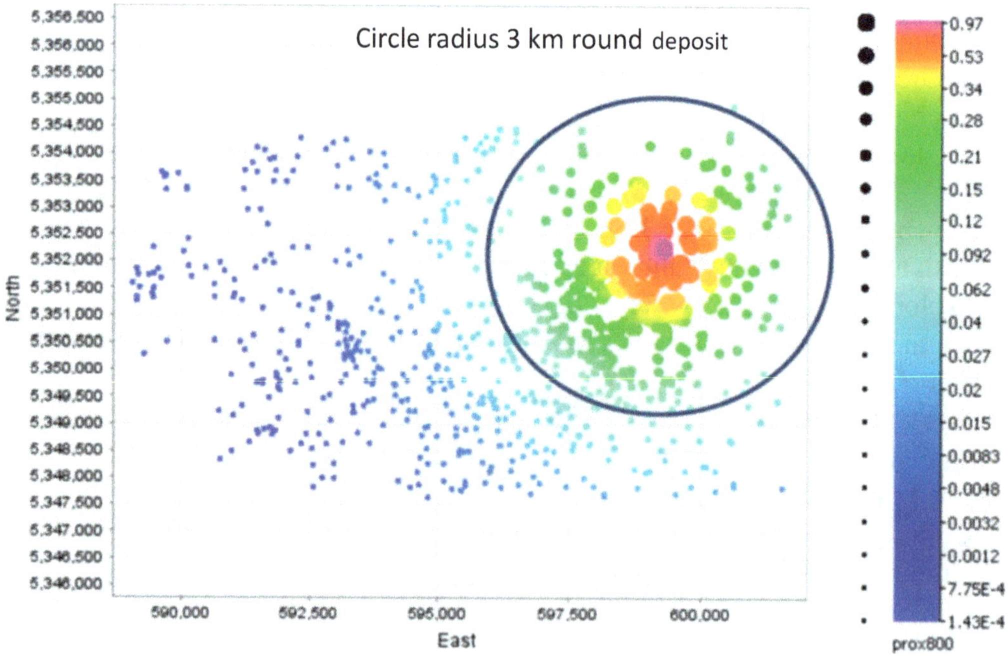


**Fig. 23.4** Map showing locations of lithogeochemical samples, with size and colour of dots related to proximity to Canagau Mines deposit (Fig. 23.3). Training set for regression model includes only those samples within 3 km of deposit (within circle)

**Table 23.1** Result of multiple linear regression. Variables selected for regression against proximity (Model 1) by selecting those with abs (correlation coefficient) > 0.2. The explanatory variables are CLR-transformed geochemical element values, the response variable is proximity to the Canagau Mines deposit, using n = 278 samples that lie within 3 km of the deposit for training. Variables selected for Model 2 based on p-values < 0.03 from Model 1

| | Correlation coefficient, r | Model 1 | | Model 2 | |
|---|---|---|---|---|---|
| Element | Proximity | Coefficient | p-value | Coefficient | p-value |
| Co-CLR | −0.34 | 0.0489 | 0.3242 | | |
| Li-CLR | 0.35 | 0.0285 | 0.2841 | | |
| Ni-CLR | −0.26 | −0.0164 | 0.5226 | | |
| Pb-CLR | 0.38 | 0.0081 | 0.5816 | | |
| Sr-CLR | −0.23 | −0.0227 | 0.2633 | | |
| V-CLR | −0.35 | −0.0681 | 0.0101 | −0.0646 | 0.0001 |
| CaO-CLR | −0.48 | −0.1100 | 0.0000 | −0.1206 | 0.0000 |
| Na2O-CLR | −0.30 | −0.0728 | 0.0000 | −0.0801 | 0.0000 |
| K2O-CLR | 0.26 | −0.0285 | 0.0205 | −0.0287 | 0.0046 |
| TiO2-CLR | −0.43 | −0.0721 | 0.2082 | | |
| CO2-CLR | 0.26 | 0.0330 | 0.0012 | 0.0417 | 0.0000 |
| S-CLR | 0.24 | −0.0123 | 0.2411 | | |
| Constant | | 1.1344 | 0.0000 | 0.8083 | 0.0000 |
| Adjusted $R^2$ | | 0.3991 | | | 0.3942 |

than 0.2 (Table 23.1). This reduced the number of elements to be used as explanatory variables from 26 to 11.

CLR variables were not further transformed, and the coefficients and associated probabilities obtained by using Eq. (23.7) are shown in Table 23.1. Note that Co, Li, Pb and $CO_2$ have positive coefficients, whereas Ni, Sr, V, CaO, $Na_2O$, $K_2O$, $TiO_2$ and S have negative coefficients. This model has a goodness-of-fit of about 40% (adjusted $R^2 = 0.399$). A second model was then run to remove those variables in Model 1 with p-values greater than 0.03. In Model 2, $CO_2$ is the only variable with a positive coefficient, and V, CaO, $Na_2O$ and $K_2O$ have negative coefficients. The goodness-of-fit of Model 2 is almost the same as Model 1, with adjusted $R^2 = 0.394$. Although not shown here, a plot of predicted values from Model 1 and Model 2 are highly correlated, and maps of each are virtually indistinguishable.

The predicted values of proximity are shown in Fig. 23.5 for both the training and non-training samples. As expected, the Canagau Mines deposit shows up as a 'bullseye' at the centre of the training sample area. Notice that the Croxall property shows as another less prominent bullseye to the west, in the non-training sample area. Other high values of predicted proximity to the south of the Canagau Mines deposit and northeast of the Croxall property are associated with known sulphide occurrences as shown in Fig. 23.3. Thus, we can conclude that proximity regression led to the selection of a suite of useful explanatory variables that, after training on the Canagau Mines deposit, was able to 'discover' the Croxall property.

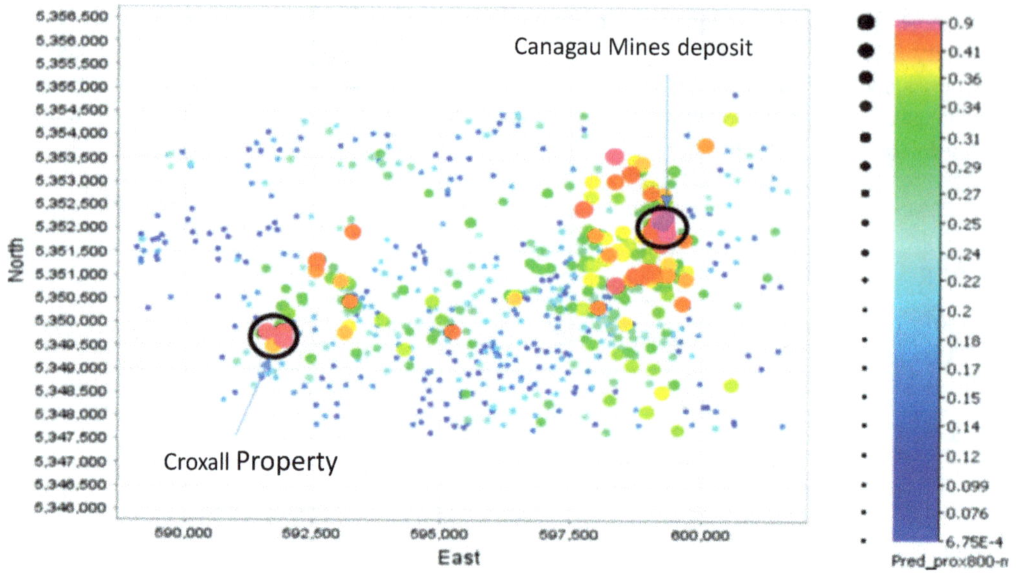


**Fig. 23.5** Map showing predicted proximity to Canagau Mines deposit. Plot includes both points used in training (those within 3 km of deposit) and other sample points. Croxall property is identified with large proximity values by this model

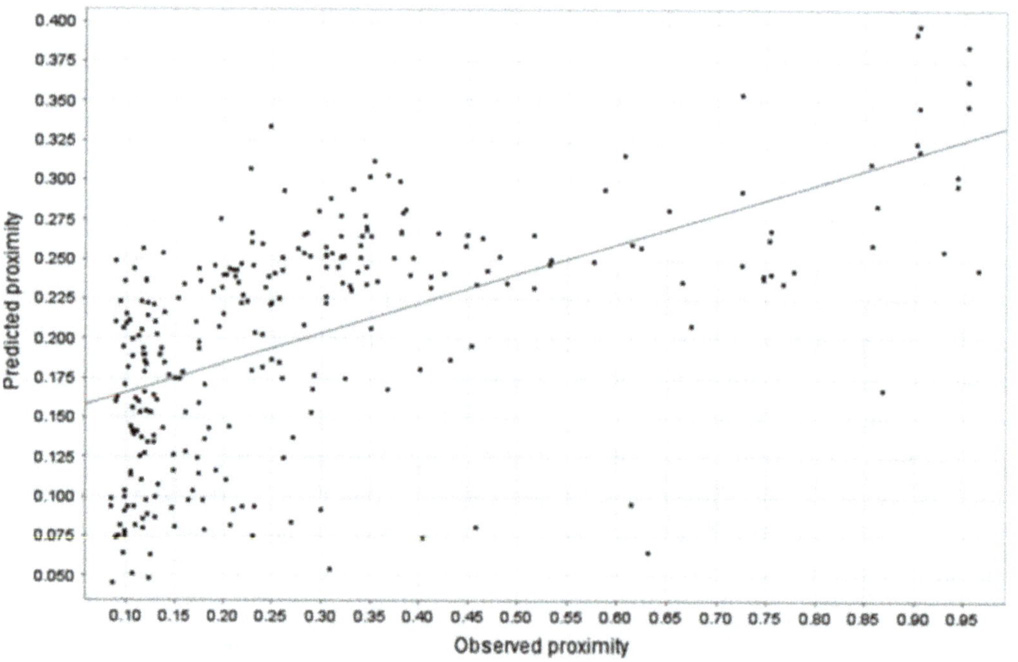


**Fig. 23.6** Plot of observed proximity versus predicted proximity, with best fit line, training points only. In general, fit is noisier at lower values of proximity. Points with proximity >0.5 (i.e. within the 'half-distance' of 800 m of the Canagau Mines deposit) show a stronger relationship

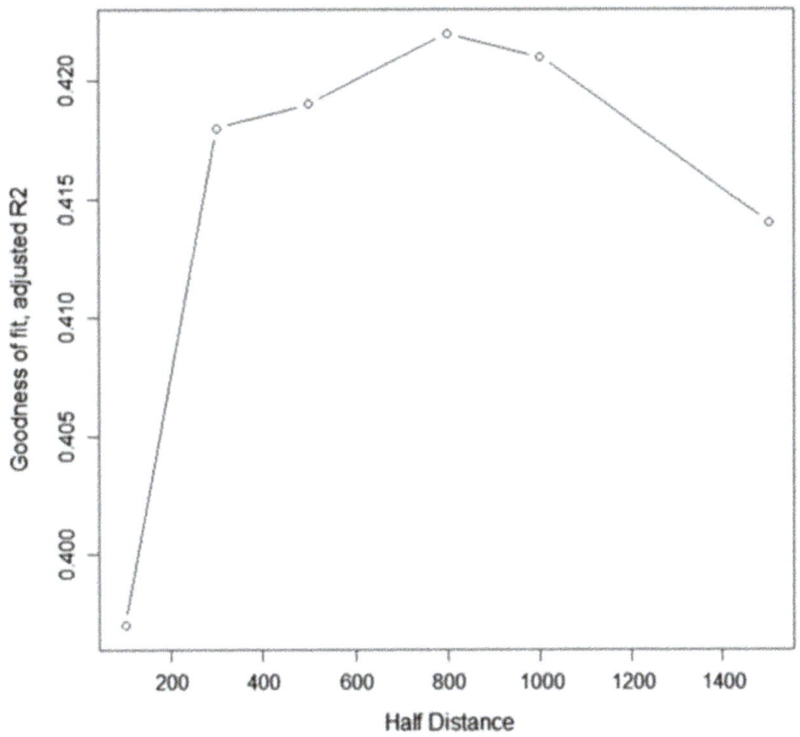


**Fig. 23.7** Variation in goodness of fit (adjusted $R^2$) with changes in 'half distance', the parameter used to control rate of exponential decay of proximity with increasing distance (23.4). Note that curve shows that relationship is strongest using half-distance parameter = 800 m

A bivariate plot of observed versus predicted proximity, training points only, (Fig. 23.6) shows that the relationship is noisier far away from the deposit than closer to it, consistent with the proximity response weakening at increasing distance.

Experimental results show that an optimum half distance for modelling proximity as an inverse function of distance is 800 m, although the results are not very sensitive to changes in the 300–1000 m range (Fig. 23.7). It is not clear how useful this parameter might be in describing the geometry of the 'halo' effect around the deposit.

## 23.3 Method 2: Principal Component Residuals

Many geochemical survey data are difficult to interpret, because multiple overlapping processes affect element levels in space and time. In some situations, a principal component will show a composition (based on element loadings) and a spatial pattern reflecting an interpretable geological process, but usually interpretation is complex because of interacting processes.

Residual principal components analysis is an exploratory approach that can sometimes be helpful in sorting out complex multielement interactions. The method is a straightforward extension of applying principal components, followed by a series of multiple linear regressions. As with the proximity regression method, it is important first to carry out a centred log ratio transform of all the elements, otherwise distortions may occur in principal component (and subsequent multiple regression) results due to constant sum 'closure' effects.

Regular PCA is carried out in the usual way on the correlation matrix calculated from CLR-transformed element variables (e.g. Davis 2002, ch. 6).

Inspection of the eigenvectors for each PC, inspecting biplots, and mapping PC scores for the at least the first few PCs can then lead to an interpretation of PCs in terms of geological processes (Grunsky 2010). Here the objective is to focus on a selected element to separate out ('partition') this element compositionally and spatially using the principal component results.

For the element of interest, the next step is to inspect the corresponding row of the eigenvector matrix (the 'loadings') to understand better in which components the element occurs. It may be decided to predict the element from PC1 only, or from PC1 and PC2, or PC1, PC2 and PC3, and so on. For each of these selections, a multiple regression is carried out with the selected PCs as explanatory variables, and the chosen element as the response variable. For example, if the response variable is $V$ and the explanatory variables are PCs 1 to PC3, then

$$V = \beta_0 + \beta_1 PC_1 + \beta_2 PC_2 + \beta 3 PC_3 + \in \qquad (23.9)$$

can be solved as before for the coefficients β by least squares. If the predicted values of $V$ are $V^*$, then the residuals $V_R$ are simply

$$V_R = V - V^* \qquad (23.10)$$

computed over all sample locations.

The choice of PCs in Eq. (23.9) may be as simple or as complex as needed. We have had good results by successively adding PCs, inspecting the goodness of fit at each stage and mapping the predicted and residual values at each step. Inspection of residual patterns may reveal, spatially, where concentrations of that particular element are distributed, facilitating interpretation.

In this method, there is no training set, calculations are carried out on all samples.

## *23.3.1 Application to Nunavut Lake Sediment Data*

### 23.3.1.1 Geological Background

The lake sediment survey was carried out over three 1:250,000 scale map areas (NTS 65A, 65B, 65C) in southern Nunavut Territory, Canada (McCurdy et al. 2012). The geology of two of the NTS sheets (65A, 65B) were mapped by Eade (1973) and is shown in Fig. 23.8. Of particular interest to this study, we notice that there are two important granitic intrusion types: the Hudson granite (1.83 Ga) and the Nueltin granite (1.75 Ga) suites as identified and characterized by Peterson et al. (2015).

This area lies within the southern Hearne Province, a poorly understood terrane. The domain is dominantly comprised of Archean tonalitic and charnokitic gneisses, approximately 2.8 Ga in age. However, strong evidence for fragments of much older crust, up to 3.3 Ga, has been found in the form of inherited Archean zircons and Sm–Nd model ages obtained from Proterozoic post-orogenic plutons of the Hudson granite, intruded at about 1.83 Ga. Nueltin rapakivi granite (ca. 1.75 Ga) is also present in the area.

A comprehensive multielement study of the lake sediment data was carried out by Grunsky et al. (2012a, b), and by Grunsky and Kjarsgard (2016). One of the results of those studies was to show that the multivariate geochemistry could be used to map the various rock types using a variety of methods including PCA.

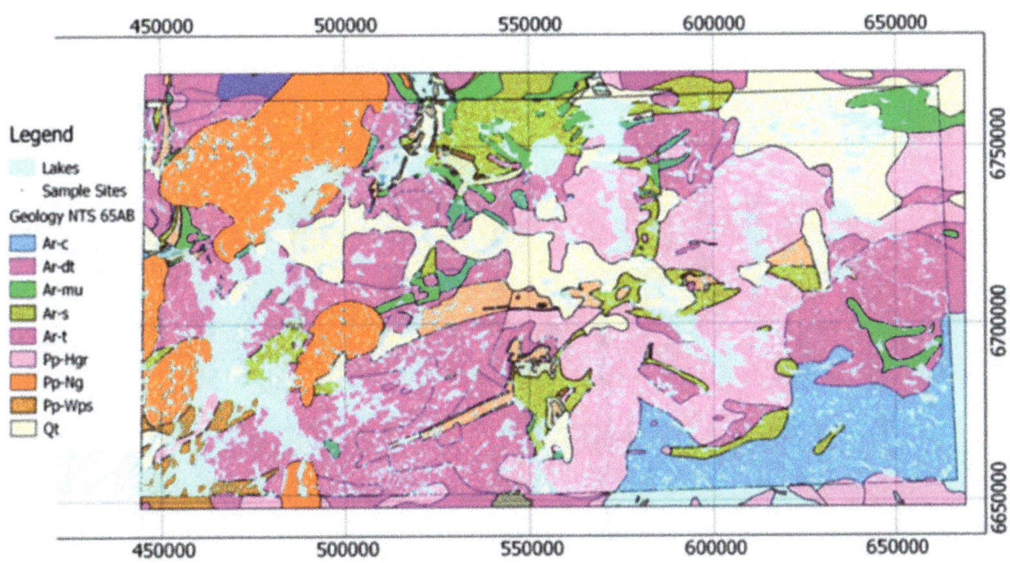


**Fig. 23.8** Geological map of NTS sheets 65A, 65B and 65C, with coordinates shown for UTM Zone 14, Nunavut Territory, adapted from Grunsky et al. (2012a, b). Two units noted in text are Nueltin granite (Pp-Ng shown in orange) occurring in west and Hudson granite (Pp-Hgr shown in light pink) occurring in east

#### 23.3.1.2 Application

The data consists of 1611 samples and 48 geochemical elements—both major and traces. Prior to CLR transformation, all variables were converted to ppm. PCA was carried out on all 48 elements. The objective was to understand better how uranium is partitioned between the two granites: the Nueltin and the Hudson.

PCA analysis was calculated on all 48 CLR transformed variables. A scree plot (Fig. 23.9a) shows that the first 15 PCs (out of the full 48) account for almost 85% of the total variation in the data, and the first 5 PCs account for over 60%. Inspection of the uranium loadings (Fig. 23.9b) shows that PCs 2 and 3 both have high positive loadings, whereas PC 5 has a strong negative loading. Multiple regressions were carried out (using U-CLR, **not** untransformed U) starting with

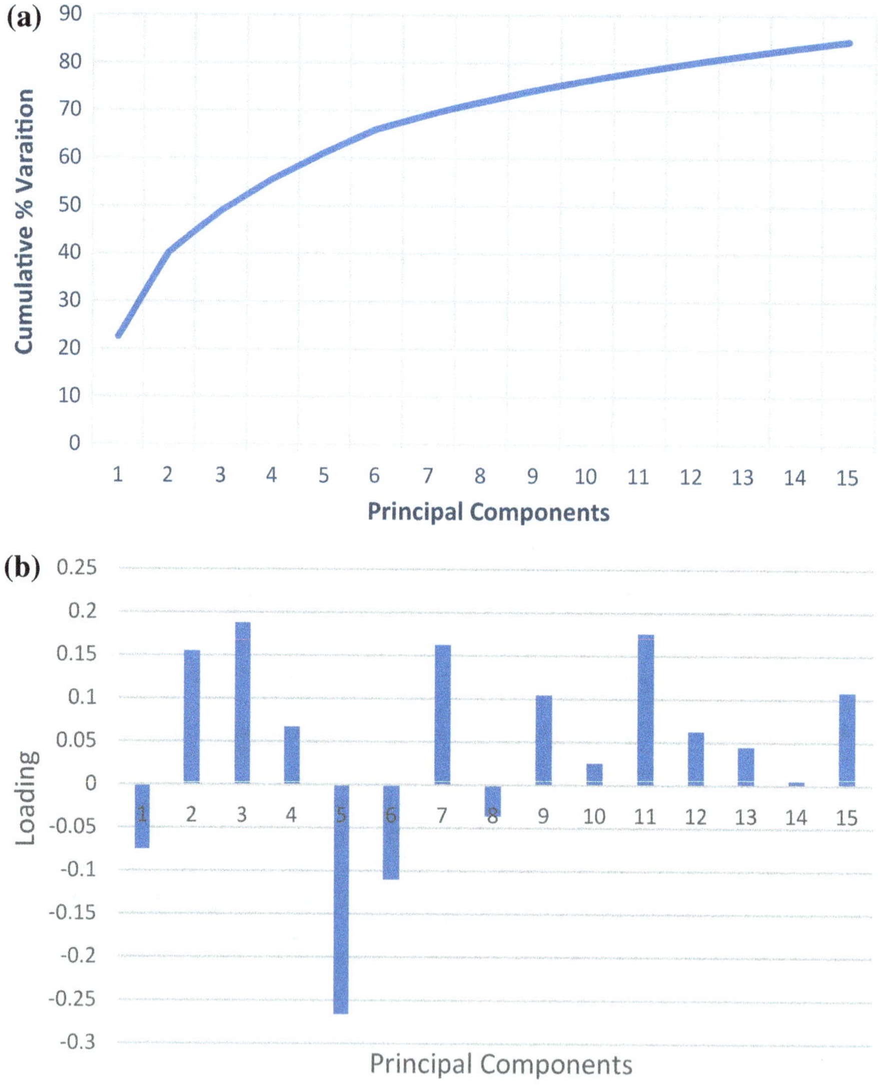


**Fig. 23.9** **a** Scree plot showing cumulative variation explained by first 15 PCs. **b** Values of loadings for U-CLR on first 15 PCs

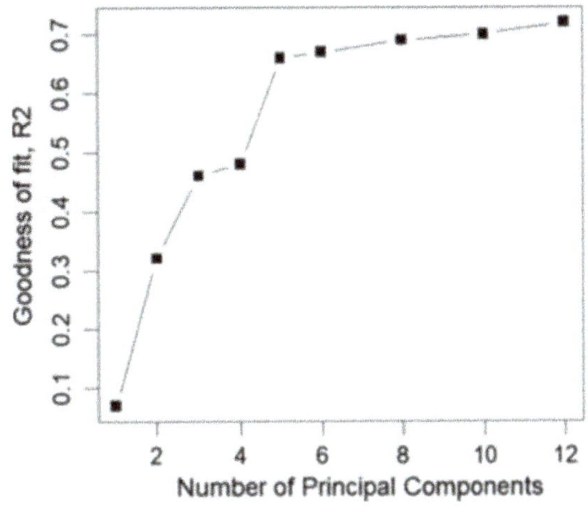


**Fig. 23.10** Goodness of fit ($R^2$) for successive multiple regressions with U-CLR as response variable and an increasing number of PCs as explanatory variables. Note that after adding PC 5, there is little change in $R^2$ values

PC1, then successively adding PCs up to 12. For each regression, predicted U and residual U were calculated and mapped (not shown here), and a record made of the goodness of fit (Fig. 23.10). This graph shows that PC1 does not account for much U variation, but PCs 2 and 3 show marked increases in goodness of fit. PC 4 shows a minor increase, and PC5 shows a major increase. After PC5, improvements in goodness of fit are minor.

Figure 23.11 shows maps of U-CLR predicted from PCs 1-5, and U-CLR residuals. Not shown is the unmodified U-CLR map (which sums these two parts). Notable here is that the predicted map shows a pattern strongly correlated with the Nueltin granite, whereas the residual map is strongly correlated with the Hudson granite. PCs 1-5 'explain' the uranium in the Nueltin granite, whereas the residual uranium is that which occurs in the Hudson granite. The residual PC analysis has partitioned uranium into two parts that have a distinct geological interpretation.

This is confirmed in Fig. 23.12 which shows for the successive regressions results of t-tests on the mean U-residual in the Nueltin and Hudson granites. The value of t increases up to PC5, then decreases. This confirms that, for partitioning uranium between the two granites, regression against PC1-5 gives the best result.

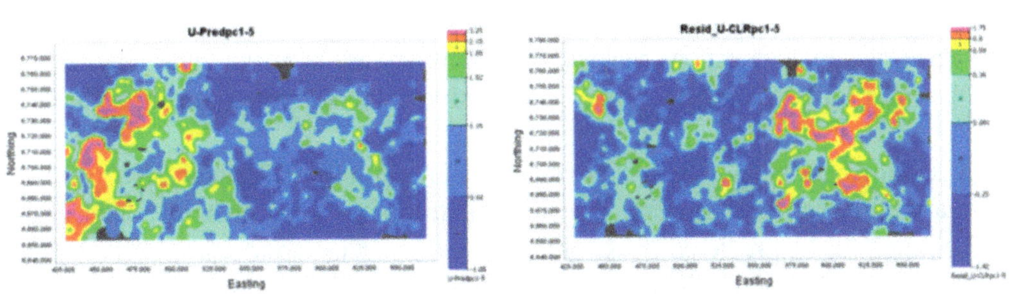

**Fig. 23.11** Left. Map of U-CLR predicted from PCs 1-5 using lake sediment data. Right. Map of residual U-CLR unexplained by PCs 1-5. Predicted uranium is strongly related to presence of Nueltin granite, whereas residual uranium is strongly related to presence of Hudson granite. Map of total U-CLR does not distinguish between these two granites

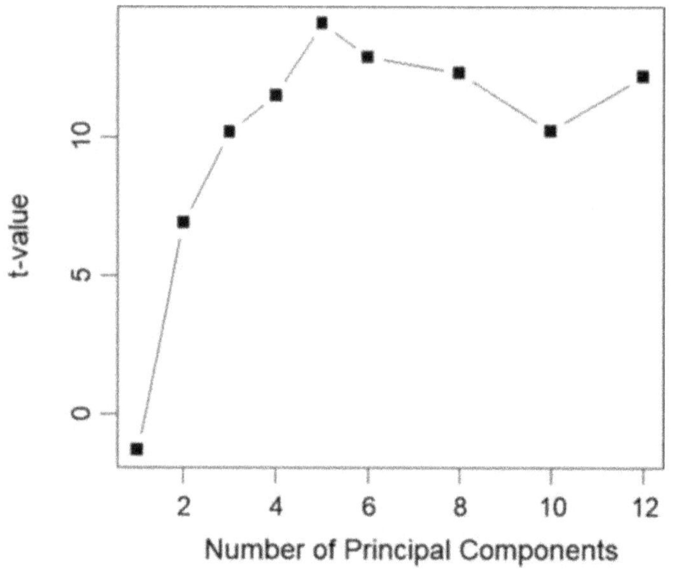


**Fig. 23.12** t-test values for U-CLR residuals obtained from successive regressions with U-CLR as response variable and increasing number of PCs as explanatory variables. These are '2-sample' t-tests comparing means between two groups: Nueltin granite samples and Hudson granite samples. *Note* that regression to predict U-CLR using PCs 1-5 gives best separation of two granites by U residuals. Adding more PCs reduces t-value and significance of difference between U residual means

## *23.3.2 Discussion*

These two methods add to the already large basket of multivariate methods useful for interpreting regional geochemical surveys.

With the wide use of GIS, spatial information is now easily determined for many features of map data. Distance calculations from points to points, points to lines and points to polygons are now routine, allowing the spatial characterization of proximity of geochemical samples to mineral deposits (points—depending on map scale), to faults of specified contacts (lines), or to rock units (polygons). In 3-D, proximity of geochemical samples to an orebody using borehole data is also straightforward. There are therefore many potential applications of proximity regression for a variety of situations involving multivariate geochemical data.

One particular idea that may be worthy of investigation is the application of this approach to prospectivity mapping. Instead of treating known mineral occurrences as binary points to be predicted from a series of evidential layers (weights of evidence, logistic regression, neural networks, etc.), a response variable could be constructed showing distance (or proximity) to the nearest mineral occurrence. The explanatory variables can be various evidential layers, as usual. The result would not be the probability of occurrence of a mineral occurrence, but rather the predicted proximity to the nearest mineral occurrence.

It should also be noted that proximity regression as described here has used ordinary multiple linear regression, so although the observed proximity measure in in the range (0, 1), predicted proximities are unconstrained and may be greater than 1 or negative. There might be some advantage to using logistic regression, that would automatically constrain the expected proximity to the range (0, 1), and would also allow the use of non-numeric explanatory variables (e.g. presence/absence of geological units, etc.). Alternatively, there are several neural network approaches that could also be tried for predicting proximity.

It should be noted that when doing a residual PCA on geochemical data, that logratio transforms are essential, because the effect of closure for introducing artefacts in PCA results is well known. Experience has also shown that residual analysis requires that the geochemical element used as a response variable must also be CLR transformed, as regression results are poor if untransformed response variables are used in the analysis.

In the separation of uranium between the Nueltin and Hudson granites, it would be most interesting to determine whether this partition was also related to isotopic differences. But this would require isotopic analyses of the lake sediment samples, an expensive proposition.

## 23.4 Conclusions

Proximity analysis allows for the use of multielement geochemical data for direct prediction of proximity to geological features, such as mineralization, faults and intrusions.

Application of proximity analysis to lithogeochemical data from the Ben Nevis area showed that a suite of elements provided a good prediction of proximity to the Canagau Mines deposit, and that this model also predicted the Croxall property and other nearby sulphide occurrences.

Residual principal components analysis is a useful way to partition particular geochemical elements that can facilitate geological interpretation.

For example, uranium in a lake sediment survey could be partitioned into two groups based on PCs. Uranium associated with PCs 1-5 is strongly correlated with the Nueltin granite, whereas, residual uranium, after removing the effects of PC 1-5, is strongly correlated with the Hudson granite.

## References

Aitchison J (1986) The statistical analysis of compositional data. Chapman and Hall, London, New York, 416 pp

Bonham-Carter GF, Hall GEM (2010) Multi-media techniques for direct detection of covered unconformity uranium deposits in the Athabasca Basin: results of soil geochemistry using selective leaches. Final report of Phase III of CAMIRO Project 08E01. www.appliedgeochemists.org/images/stories/Publications_and_Reports/Athabasca%20Final%20Report-Nov%202013.pdf

Buccianti A, Mateu-Figueras G, Pawlowsky-Glahn V (2006) Compositional data analysis in the geosciences: from theory to practice. Geological Society Special Publications, no 264, 212 pp

Brunsdon C, Fotheringham S, Charlton M (1998) Geographically weighted regression: modelling spatial non-stationarity. Statistician 47(3):431–443

Cheng Q, Bonham-Carter G, Wang W, Zhang S, Li W, Qinglin X (2011) A spatially weighted principal component analysis for multi-element geochemical data for mapping. Comput Geosci 37(5):662–669

Davis JC (2002) Statistics and data analysis in geology, 3rd edn. Wiley, New York, NY, 638 pp

Drew LJ, Grunsky EC, Sutphin DM, Woodruff LG (2010) Multivariate analysis of the geochemistry and mineralogy of soils along two continental-scale transects in North America. Sci Total Environ 409(1):218–227

Eade KC (1973) Geology of Nueltin Lake and Eddelson Lake (west half), District of Keewatin. Geol Surv Can Pap 69–51:27

Grunsky EC (1986a) Recognition of alteration in volcanic rocks using statistical analysis of lithogeochemical data. In: Nichols CE (ed) Exploration for ore deposits of the North American Cordillera. J Geochem Explor 25:157–185

Grunsky EC (1986b) Recognition of alteration and compositional variation pattern in volcanic rocks using statistical analysis of lithogeochemical data, Ben Nevis Township Area, District of Cochrane, Ontario. Ontario Geological Survey Open File Report 5628, 187 pp

Grunsky EC (2010) The interpretation of geochemical survey data. Geochem Explor Environ Anal 10(1):27–74

Grunsky EC (2013) Predicting Archean volcanogenic massive sulfide deposit potential from lithogeochemistry: application to the Abitibi Greenstone Belt. Geochem Explor Environ Anal 13:317–336

Grunsky EC, Agterberg FP (1988) Spatial and multivariate analysis of geochemical data from metavolcanic rocks in the Ben Nevis area, Ontario. Math Geol 20:825–861

Grunsky EC, Kjarsgaard BA (2016) Recognizing and validating structural processes in geochemical data: examples from a diamondiferous kimberlite and a regional lake sediment geochemical survey. In: Martin-Fernandez JA, Thio-Henestrosa S (eds) Compositional data analysis, vol 187. Springer Proceedings in Mathematics and Statistics, pp 85–116

Grunsky EC, McCurdy MW, Perhsson S, Peterson TA, Bonham-Carter GF (2012a) Predictive geologic mapping and assessing the mineral potential in NTS 65A/B/C with new regional lake sediment geochemical data. Society of Economic Geologists Student Chapter Minerals Colloquium, Prospectors and Developers Association Convention, Toronto, 5 March 2012

Grunsky EC, McCurdy MW, Pehrsson S, Peterson TA, Bonham-Carter GF (2012b) Predictive geologic mapping and assessing the mineral potential in NTS 65A/B/C with new regional lake sediment geochemical data. Geological Survey of Canada, Open File 7175, 1 sheet

Jensen LS (1975) Geology of Clifford and Ben Nevis Townships, District of Cochrane. Ontario Division of Mines, GR 132, 55 pp, accompanied by Map 2283, scale 1 inch to 1/2 mile

McCurdy MW, McNeil RJ, Day SJA, Pehrsson SJ (2012) Regional lake sediment and water geochemical data, Nueltin Lake area, Nunavut (NTS 65A, 65B and 65C). Geological Survey of Canada, Open File 6986, 2012; 13 pp, 1 CD-ROM

Meyer G, Grabowski GPB, Guindon DL, and Chaloux EC (2004) Report of activities 2003, Resident geologist program, Kirkland Lake Regional Resident Geologist Report: Kirkland Lake District; Ontario Geological Survey, Open File Report 6131, 52 pp

Peloquin AS, Piercey SJ, Hamilton MA (2008) The Ben Nevis volcanic complex, Ontario, Canada: Part of the Late Volcanic Phase of the Blake River Group, Abitibi Subprovince. Econ Geol 103(6):1219–1241

Peterson TD, Scott JMJ, LeCheminant AN, Jefferson CW, Pehrsson SJ (2015) The Kivalliq igneous suite: anorogenic bimodal magmatism at 1.75 Ga in the western Churchill Province, Canada. Precambrian Res 262:101–119

Wallbridge Mining Company Limited (2004) Wallbridge Resumes Exploration on the Ben Nevis Property, Near Kirkland Lake, Press Release, June 04, 2004. http://www.wallbridgemining.com/s/press-releases.asp?ReportID=680563&_Type=Press-Releases&_Title=Wallbridge-Resumes-Exploration-on-the-Ben-Nevis-Property-Near-Kirkland-Lake. Accessed 27 June 2017

Wolfe WJ (1977) Geochemical exploration of early Precambrian sulphide mineralization in Ben Nevis Township, District of Cochrane. Ontario Geological Survey Study 19, 39 pp

# Chapter 24
# Mathematical Minerals: A History of Petrophysical Petrography

**John H. Doveton**

**Abstract** The quantitative estimation of mineralogy from wireline petrophysical logs began as an analytical stepchild. The calculation of porosity in reservoir lithologies is affected by mineral variability, and methods were developed to eliminate these components. Simple inversion methods were applied in pioneer applications by mainframe computers to a limited suite of digital log data. Over time, the value of lithological characterization of reservoirs and resource plays has been recognized. At the same time, the introduction of newer petrophysical measurements, particularly geochemical logs, in conjunction with increasingly sophisticated algorithms, has increased confidence in mineral profiles from logs as a routine evaluation tool.

## 24.1 Pioneering Computer Methods

The volumetric determination of mineral composition from petrophysical logs originated in efforts to estimate reliable porosity estimates that were confounded by variations in rock mineralogy. When Archie (1950) introduced the term 'petrophysics' he framed it in terms of "the physics of particular rock types" and then elaborated on the petrophysics of reservoir rocks. The petrophysical properties that he considered were restricted entirely to those "related to the pore and fluid distribution". The reason was obvious in that almost all boreholes were drilled for the location of either hydrocarbons or useable water in commercial quantities. The mineralogy of the pore framework complemented the fluid content of the pore network, but estimations would be focused on the evaluation of pore volume, permeability, and fluid content. In monomineralic rocks, pore volumes could be estimated very simply by interpolating between two endpoints of mineral and fluid.

---

J. H. Doveton (✉)
Kansas Geological Survey, Lawrence, KS 66047, USA
e-mail: doveton@kgs.ku.edu



B. S. Daya Sagar et al. (eds.), *Handbook of Mathematical Geosciences*,
https://doi.org/10.1007/978-3-319-78999-6_24

In multimineralic rocks, porosity estimates became more difficult and significant errors were introduced if the mineral properties were radically different from one another.

Probably the earliest application of a mathematical solution to the resolution of porosity in a multimineralic rock was directed to Permian carbonate reservoirs in West Texas. Petrophysicists were frustrated by complex mineralogy in their attempts to obtain reliable porosity estimates from logs as described by Savre (1963). Porosities had been commonly estimated from neutron logs, but values were excessively high in zones that contained gypsum, caused by the hydrogen within the water of crystallization. If the density log was used, then porosity estimation was compromised by the occurrence of either anhydrite or gypsum. Collectively, the mix of dolomite, anhydrite, gypsum, and porosity meant that pore volumes could not be resolved by graphical methods such as crossplots and nomograms that were the standard procedures of that time.

It was recognized that lithologies composed of several minerals would require several porosity logs to be run in combination in order to estimate volumetric porosity. In the most simple solution model, the proportions of multiple components together with porosity could be estimated from a set of simultaneous equations for the measured log responses. These equations can be written in matrix algebra form as:

$$CV = L$$

where $C$ is a matrix of the component petrophysical properties, $V$ is a vector of the component unknown proportions, and $L$ is a vector of the log responses of the evaluated zone. The equation set describes a linear model that links the log measurements with the component mineral properties. Although porosity represents the proportion of voids within the rock, the pore space is filled with fluid whose physical properties make it a "mineral" component. The set of equations is then solved as an "inverse problem", in which rock composition is deduced from the logging measurements. As a closed system of dolomite, anhydrite, gypsum, and porosity, a deterministic solution is possible from three log inputs, which were chosen as neutron, density, and acoustic velocity log measurements. The solution for the unknown vector, $V$ is:

$$V = C^{-1}L$$

where $C^{-1}$ is the inverse of the $C$ matrix.

Savre (1963) described how this procedure was coded in a computer program, as a pioneer application of computers to petrophysics. An example of the graphical output drafted from one of the earliest computer runs is shown in Fig. 24.1 (Alger et al. 1963), where profiles of porosity, dolomite, anhydrite, and gypsum are shown from a Permian San Andres Formation section in West Texas. At the time that this early application was made, computing power was typically provided by a single

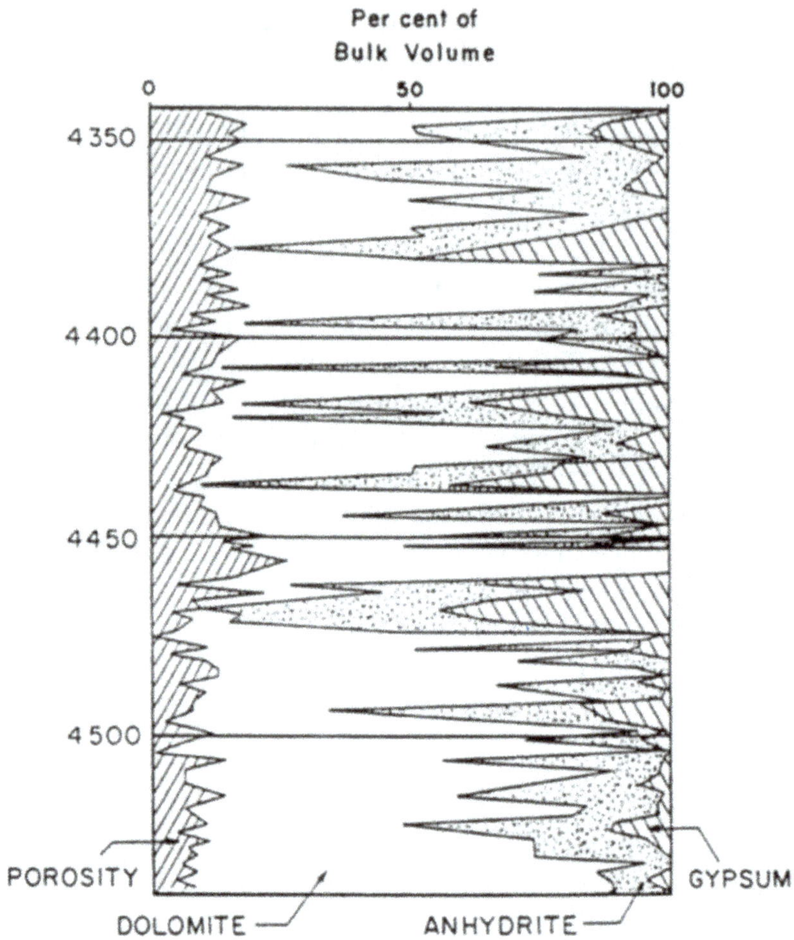


**Fig. 24.1** Graphical output profiles of porosity, dolomite, anhydrite, and gypsum from one of the earliest computer runs that processed neutron, sonic, and density logs of a Permian San Andres Formation section in West Texas (from Savre 1963)

mainframe computer in the company or university which had extended computing times and limited memory, while programming code was a specialized and time-consuming task. The same application is very easy to implement today as a spreadsheet procedure, using standard matrix functions and graphical outputs.

The inverse solution is a simple and powerful procedure for compositional analysis, but its simplicity carries certain assumptions that must be considered carefully. In particular, the basic model contains no intrinsic constraint to preclude negative estimates of compositional proportions. A unity equation dictates the closure of the system so that the proportions collectively sum to unity. However, individual proportions can have a negative value or one that exceeds unity. Rather than representing mathematical error, apparently anomalous zones are located outside the composition space defined by the mineral endmembers as vertices. Consequently, the generation of negative proportions is a perfectly natural consequence of the model and can contain useful feedback information. If the negative values are small, then this is usually called by the stochastic nature of the input nuclear logs coupled with borehole rugosity perturbations. If large, the possibility of washouts and gas effects should be examined before evaluating the possibility of another mineral that is not included in the composition model.

If these explanations are not sufficient, then negative proportions of components have a role as a basic check on the validity of the model used for compositional analysis. As such, they are diagnostic errors with an information content to be used to guide the analysis to a better solution. The distinction between errors that are acceptable as minor, random measurement noise and systematic deviations is best made by a comparison between the original logs and the logs predicted by the model solution. The predictions are given by:

$$\hat{L} = CV$$

If the inverse procedure has generated zone solutions with proportions that are negative or exceed unity, then the adjustment to rational proportions will result in log predictions that will deviate from the original logs. The deviations between measurements and predictions can then be examined to differentiate minor measurement error from systematic perturbations that require intervention and correction. In the more sophisticated models to be reviewed, tool response errors are actively incorporated within the solution algorithm, together with constraints that preclude irrational compositional proportions.

However, if the solution results in compositional proportions that are all positive, then there will be an exact match between the logs and model predictions. This equivalence does not imply that the result is geologically correct; it simply means that the solution is rational and consistent with the choice of components and their properties. There may be other satisfactory solutions based on alternative mineral suites.

## 24.2 Mineralogy of Underdetermined Systems

The basic compositional inversion procedure requires a precise match between the number of knowns and unknowns. This situation is a "determined system". The alternative possibilities are that the number of logs is insufficient to provide a unique resolution of the proportions of the components (an underdetermined system) or that the number of logs exceeds the number of components (an overdetermined system). In reality, it is likely that most formations present underdetermined compositional problems, if all the constituents are counted and matched against the number of logs run in a typical borehole. As counterpoint, many of the minerals will be found in small quantities and the overall composition dominated by a few components.

McCammon (1970) and Harris and McCammon (1971) considered alternative model procedures to the estimation of mineral compositions from logs in underdetermined cases. Although their algorithms have been superseded by optimization procedures, their approach is instructive concerning the role of information in log compositional analysis and the potentially competing criteria of mathematical optimality and geological reality. McCammon (1970) considered the underdetermined system. In terms of classical information theory, which proposes that the least biased solution is the one that maximizes the entropy function:

$$E = \sum p_i \log p_i$$

where *pi* is the proportion of the *i*th component. This equation for entropy is closely approximated by that for proportional variance:

$$\begin{aligned} P &= \left(\frac{n}{n-1}\right) \sum v_i(1 - v_i) \\ &= \left(\frac{n}{n-1}\right) \left(1 - \sum v_i^2\right) \end{aligned}$$

The maximum of the variance function, *P*, is close to the condition of maximum entropy, and the resulting optimal solution is easier to compute using the matrix algebra equation:

$$V = C^t (CC^t)^{-1} L$$

where *V* is the vector of unknown proportions, *C* is the matrix of component log properties, *t* signifies a matrix transpose, and *L* is the vector of zone log responses (Doveton and Cable 1979).

The compositional solution from the proportional variance algorithm is optimal from a classical statistical viewpoint: the average squared errors between estimates and real compositions should be the minimum possible.

This is a conservative philosophy that aims to be least wrong or risk-averse with a minimum error as penalty. However, mineral proportions are frequently distributed in a highly unequal manner. Therefore the real rock composition will often be one of several extreme possibilities, rather than the less likely seemingly homogeneous composition that can result from a minimum variance solution. The correct interpretation of a bland compositional solution is that it represents the average of a range of possibilities. As such, it is a good estimate of the average, but may be a very poor prediction of the particular: the composition of the zone in question. Such a result is a useful diagnostic that suggests that several extreme alternatives should be reviewed and that extra information is required. The information can take a variety of forms, such as explicit geological knowledge of the range of actual compositions, or the use of additional constraints that preclude impossible solutions.

## 24.3 Mineralogy of Overdetermined Systems

Many rocks are dominated by a relatively small number of components, so that the number of logging tool measurements may exceed the number of significant lithological components. The situation becomes overdetermined when the number of log response equations is greater than the number of components. The appropriate solution is then one that most accurately reproduces the original logs when logs are calculated as predictions from the compositional solutions. Using conventional statistical theory, this solution is the one that minimizes the sums of squares of the deviations between the original logs and their predictions. The least-squares solution is given readily by the matrix algebra equation:

$$V = (C^t C)^{-1} C^t L$$

where the terms are the same as those in both the determined and underdetermined matrix algorithms written earlier. The matrix formulation requires some additional weighting function to allow for the fact that the logging measurements are recorded in radically different units. Without any weighting, the error minimization is predicated on equal units and results in a solution which preferentially honors logs with the highest data ranges. The modified least-squares algorithm is then:

$$V = (C^t W C)^{-1} C^t W L$$

where $W$ is a diagonal matrix that contains the elements of a weight vector (Harvey et al. 1990). The weights may be assigned based on physical first principles or by a standardization scheme, such as transformation from the original measurement to a scale anchored to the mean and counted in standard deviation units.

For any given zone, the sum of squares error is given by:

$$e = \left(L - \hat{L}\right)^t \left(\left(L - \hat{L}\right)\right)$$

where $\hat{L}$ is the vector of log responses associated with the least-squares solution. The error term can be plotted as a monitor log to highlight zones where there are striking inconsistencies between the model and the log responses. The overall performance of an algorithm may be judged from the standard error, computed from the summed zone errors as:

$$s_e = \sqrt{\frac{\Sigma e}{(n - m - 1)}}$$

where $n$ is the number of observations and $m$ is the number of logs.

## 24.4 Optimization Methods

Current compositional analysis procedures has moved beyond simple inversion algorithms described, so that constraints and tool error functions have been incorporated as part of the solution process. The methodology was first developed by Mayer and Sibbit (1980) who applied modified steepest-descent strategies to hunt for an optimal solution that minimized the "incoherence" between the logs and their predicted values. For any given log, the incoherence function is given by:

$$I_A = \frac{(a - \hat{a})^2}{(\sigma_A^2 + \tau_A^2)}$$

where $I_A$ is the incoherence for log $A$, a is the log response for the zone and $\hat{a}$ is its prediction, $\sigma_A^2$ and $\tau_A^2$ are the uncertainties associated with the log measurement and the response equation, respectively.

The uncertainty term for each log measurement is compounded from the sources of sensor error, data acquisition, and the dispersions associated with environmental corrections. Response equation dispersion represents the uncertainties introduced by linear approximations, erroneous choices of component log responses, and hidden factors such as the influence of textural parameters. It seems reasonable to suppose that these two types of uncertainty are independent, so that they can be summed as one total error term for each tool:

$$u_A^2 = \sigma_A^2 + \tau_A^2$$

The total log incoherence for any particular depth zone is the sum of the separate log incoherences:

$$I_t = I_A + I_B + I_C + \cdots$$

The form of the equations shows that the solution will tend to be most strongly influenced by the logs to which the most confidence can be attributed. Logs with large errors will have greater incoherences and will contribute more to the total incoherence term.

Constraints are also included and take the general form of:

$$g_i(v_i) \geq 0$$

where *gi* is some function that constrains the value of the unknown proportion of the *i*th component. Rigid, mathematical constraints are those that preclude the occurrence of proportions that are negative or those that exceed unity. Geological and local constraints incorporate relations that conform to general geological principles or prior knowledge of local geology. These geological constraints are more generalized, so that appropriate uncertainties are assigned to them. The constraint dispersions generate additional incoherence terms to be considered. A combined incoherence function is then the sum of the log and constraint incoherences:

$$I_t = \sum \frac{(a_i - \hat{a}_i)^2}{\sigma_i^2 + \tau_i^2} + \sum \frac{g_i(v_j)^2}{\tau_j^2}$$

Notice that if the system is fully determined, then the total incoherence will be zero, provided that no constraints are violated. This special situation is the limiting case of applications which are otherwise presumed to be overdetermined. In a routine application of the optimization algorithm, the number of logs would be expected to exceed the number of components. In part, this is feasible because the bulk of rock compositions tend to be dominated by relatively few components. In addition, the range of wireline measurements used today typically extends beyond the traditional porosity logs to resistivity, spectral gamma ray and geochemical logs.

The optimization method of Mayer and Sibbit (1980) is an iterative search procedure. The system model of input logs and output components are first defined. The incoherence values associated with each log type are entered, together with the constraints to be met. For each zone, an initial composition is estimated by an approximate method and used as the starting point for a sequence of intermediate solutions. At each step, the incoherence is calculated between the input log responses and those predicted from the solution. A gradient is also computed as the means to generate the next solution, using a steepest descent technique. The process terminates when it is determined that convergence has been satisfied, at which time there is no appreciable difference between successive solutions. The final solution will be approximate, but the total incoherence between the logs and the compositional estimate will be the minimum possible. The combined display of real and theoretical logs is invaluable as a quality control mechanism to alert the user to problem zones which may be optimal, but are flatly wrong. The generality of the approach allows alternative and remedial attempts to be made without major difficulty.

In further refinements, Gysen et al. (1987) described an extension of the method to the simultaneous optimization of component proportions and response parameters. Moss and Harrison (1985) also reported a technique to solve for the uncertainty multipliers which contain the total error associated with each tool. Although the errors cannot be solved for every depth zone, they can at least be estimated for selected intervals and assumed to be effectively constant between zones.

Phyllosilicate minerals pose a difficult problem because their composition is so variable. However, the clay mineral properties listed provide a useful reference standard in the estimation of hypothetical composition volumes in the absence of explicit information keyed to the formation that is analyzed. The estimates can be considered as normative, as contrasted with modal predictions of clay mineral proportions based on X-ray diffraction analyses from core.

Optimal, minimum error solutions are worthless if the component model is incorrectly specified. Meaningful results are best obtained by patient geological evaluation of a sequence of solutions where the results of each are used to an improvement of the successive solution. Modern compositional analysis software utilizes the power of the error minimization method, but allows user interaction so that alternative geological models can be compared.

Quirein et al. (1986) described the use of quadratic programming techniques and linearized response equations, as an improvement on the penalty constraint approach used by earlier methods. In addition, they incorporated a program to solve for poorly known log responses of a component subset, as an optimization procedure applied to specific depths that could be used for calibration. These calibration intervals are those where both logs and compositions are known and are most typically those that have been cored. In addition, knowledge of composition could be utilized from other sources. Not all component log responses need to be estimated since their properties are restricted to a limited range. However, a subset of mineral components have ambiguous and locally variable properties. The most notorious example of such components are clay minerals, and these will be discussed more fully in the following section.

In common with earlier optimization methodologies, the system is assumed to be either determined or overdetermined. The use of multiple alternative models then allows a more realistic treatment of this assumption, in which common associations can be modeled in parallel and a final selection made between them at any depth. Wherever possible, each separate model is designed to be close to fully determined in an attempt to find a good match and to sidestep problems associated with the estimates of log and equation dispersions (Marett and Kimminau 1990). The appropriate logs for each model are clearly those that discriminate well between the separate components. If a poor choice of logs is made, then the model is ill-conditioned. The model structure can be checked through the computation of the condition number of:

$$C^{t}DC$$

where $C$ is the matrix of component log responses and $D$ is a matrix of uncertainty values. The condition number is higher for ill-conditioned models and gives a measure of the sensitivity of proportion estimates to small changes in component log responses (Quirein et al. 1986). The choice between alternative models for any zone can be made by the user based on an assessment of the relative incoherence of the solutions and their feasibility as reasonable geological descriptions. Alternatively, the decision can be made on the basis of probability established either from comparison of alternative solutions or the use of a Bayesian prior probability.

While generally still applied to an overdetermined system, the multiple models are not far removed from determined matches of components and logs. Where a model becomes determined, the solution is that of a simple and fast matrix inversion with zero incoherence, provided that the non-negative constraint is not violated. The analysis of the relative conditioning of the model system is a valuable mathematical contribution to the determination of which logs provide the maximum discrimination of model components that will lead to the most stable estimates of volumetric proportions.

## 24.5 Clay Component Estimation

Shales are composed typically of a mixture of clay minerals, quartz, carbonates, and iron minerals, as well as other accessory components. Clay minerals are markedly different from other rock-forming minerals in terms both of their complexity and variability. Shales present special problems for log interpretation and while many algorithms have been designed for their volumetric estimation, the meaning and limitations of their results should be understood.

In more detailed work, the older and broader methods of shale evaluation have been expanded to the quantitative assessment of clay mineral species. Clay minerals show differing degrees of variability, but are generally subdivided between four major types: illite, smectite, kaolinite, and chlorite. Clay mineral typing is based on several log criteria which must be considered carefully and collectively. Ellis (1987, pp. 460–461) noted that the four principal clay mineral types could be combined into two types, based on their hydroxyl content. Kaolinite and chlorite have eight hydroxyls, as contrasted with four for smectite and illite. The neutron log is sensitive to this difference, which can be used as one diagnostic guide, through comparison of the neutron and density porosities when they are both scaled with respect to a quartz matrix. The photoelectric factor is also a useful clay discriminator because of its control by the aggregate atomic number. Ellis (1987, pp. 451–454) pointed out that iron-free aluminosilicate clays would have photoelectric absorption characteristics that are virtually the same as for quartz. Therefore, variations in the photoelectric factor within shales are primarily a reflection of iron content. Overall, there is a tendency for a progressive increase in iron from low values in kaolinite, through smectite and illite, to high values for iron-bearing chlorite. Distinctions between clay minerals can also be made on the basis of spectral gamma-ray logs, particularly in the differentiation of relatively potassium–rich illites from low-potassium kaolinite and chlorite.

The quantitative estimation of clay mineral abundances from the neutron, density, photoelectric factor, and spectral gamma ray measurements is fraught with difficulties. Wide compositional changes within clay mineral groups pose special problems. Useful quantitative models are not easy to define and are frequently ambiguous in their interpretation. The most realistic approach would be to coordinate log measurements with laboratory analyses of core samples. The core values may be idealized as a calibration standard in the development of a statistical prediction model for clay minerals from logs. Even this strategy must be considered thoughtfully and honestly. The most widely used laboratory method to estimate quantities of clay minerals is that of X-ray diffraction. Even with careful sample preparation procedures, the error of clay mineral estimates from X-ray diffraction can be routinely expected to be 50% or more of the reported value (Eslinger and Pevear 1988, p. A-24). Nevertheless, an important result is that at least the appropriate mineral subset can be identified with some confidence. This ensures that the correct components will be selected for compositional analysis from logs. Reconciliation of the log estimates with X-ray diffraction analyses should then be made within a model that attributes appropriate error magnitudes to both data sources.

## 24.6 Normative Estimation by Geochemical Logs

Geochemical logging tools measure induced gamma-ray spectra that are created when a formation is bombarded by high energy neutrons from an electronic pulsed source. A matrix inversion spectral fit algorithm then separates the spectrum into individual elemental sources. The major rock composition elements of silicon, calcium, magnesium, iron, sulfur, titanium and carbon are estimated together with the rare earth, gadolinium. In addition, potassium, thorium, uranium can be estimated from the natural gamma rays emitted by formations and measured by the spectral gamma-ray log. As a consequence of the direct relationship between elemental data and mineral compositions more realistic mineral transforms have been developed that are a major improvement on models based on mineral properties. However, a distinction must be made between normative minerals that are computed from transforms of elemental data and modal minerals that are observed visually or by petrographic laboratory methods such as X-ray diffraction or infra-red spectroscopy. Clearly, the fundamental goal of an effective transform is to provide a close match between normative mineral solutions and modal mineral suites.

"Normative" minerals calculated from oxide analyses have been a standard procedure in igneous petrology since the CIPW (Cross-Iddings-Pirsson-Washington) norm was introduced by Cross et al. (1902). These normative minerals are contrasted with modal compositions that are commonly measured by point-counting of minerals in thin-sections of rock. The normative concept has also been extended to sedimentary rocks in attempts to compute realistic mineral assemblages. Krumbein and Pettijohn (1938) pp. 490–492 explained the molecular ratio method to calculate the probable mineral composition of a rock, based on chemical analyses of oxide percentages. As a first step, the minerals to be resolved are first identified from thin-section observation or other sources of information. The molecular ratios are then assigned in a stepwise fashion to the minerals. The process consists of a logical order of steps that first accommodates unique associations between oxides and certain minerals, and then allocates the remainder to other components. Imbrie and Poldervaart (1959) described a commonly used method of sedimentary normative analysis and then compared the results with modal estimates of mineralogy. From a detailed study of the Permian Florena Shale, they concluded that estimates of the chert, calcite, dolomite, and clay had errors of less than 5%. However, there was little agreement between computed clay mineral proportions and those produced from X-ray diffraction analysis. Imbrie and Poldervaart (1959) were not surprised by this discrepancy, but attributed it to the known high variability of clay mineral compositions through isomorphous substitution.

Essentially the same problems are tackled in the computation of sedimentary normative minerals, when based on elements measured by geochemical logs (Herron 1986). However, many of the older normative methods predated computers. The classical norm calculation is subtractive, deterministic and rigidly leveraged. As discussed by Harvey et al. (1990), the method can be useful when certain elements can be assigned totally to single individual minerals. These assignations can then be made in an ordered protocol of analysis partition between mineral species. Otherwise, the use of simultaneous equations to link mineral compositions with elemental measures is a much more general and powerful method. The speed of modern software also allows real-time interaction between petrophysicist and machine, so that alternative models can be evaluated quickly and decisions made that blend mathematical optimality with geological credibility. Any analysis should be preceded by some notion of what constitutes a fit-for-purpose estimation. Less accuracy is needed if the intent is for a generalized semi-quantitative description of variation rather than more rigorous estimates for use in quantitative basin modeling or physical property predictions (Harvey et al. 1998).

The model that links minerals with elements can be set up as a fully determined system and solved by standard matrix inversion using methods described earlier. Whenever the components are computed as positive proportions, then the compositional solution is rational and honors the analysis perfectly. However, in common with the normative model, any apparent precision read into the result is illusory because the determined system makes no allowance for analytical error. It is usually practical to model a rock with a set of minerals that are fewer in number than the elements available from geochemical logging. The system is then overdetermined and can be resolved by one or other of a variety of optimization techniques. The additional complexity in computation is offset by several distinct advantages. The overdetermination allows constraints and error functions to be incorporated, both for optimal solution control and diagnostic evaluation of sources of analytical error. The choice of an overdetermined system also provides better assurance of a stable solution in situations where the mineral response matrix becomes sparse or there are potential compositional colinearities that link some of the mineral subsets (Harvey et al. 1990).

Strictly speaking, there will almost always be more minerals than elements to solve for them, so that the problem is always underdetermined. However, as Herron (1988) noted, the overwhelming majority of sedimentary rocks are composed of only ten minerals: quartz, four clays, three feldspars, and two carbonates. In practice, reasonable compositional solutions can be generated using relatively small mineral sub-sets, provided that they have been identified correctly and that the compositions used are both fairly accurate and constant. Alternatively, the inversion procedure can be run as an unconstrained procedure and components with negative proportions eliminated from the model. Harvey et al. (1998) found this approach to be successful, but cautioned that negative components should be eliminated one at a time, starting with the largest negative component, because of interactions between the components.

Mineral solutions may be calculated by two alternative strategies. In the first, the average chemical compositions of minerals drawn from a large data-base are used as endmember responses and resolved by standard matrix inversion procedures. This result is normative and generic in the sense that it is based on a sample drawn from a universal mineral reference set and applied to a specific sequence where local mineral compositions may deviate from the global average. The result is hypothetical, but has the particular advantage that comparisons can be made between a variety of locations and do not require expensive ancillary core measurements. New methods of classification may also be necessary as discussed by Herron (1988) in his study of terrigenous sands and shales in terms both of core and geochemical log data.

In a second approach, the solution is calibrated to core data, where laboratory determinations of mineralogy and elemental geochemistry are analyzed by multiple regression techniques to determine local mineral compositions. This result is linked to petrography and so is philosophically closer to an estimated modal solution, rather than the more hypothetical normative model. As mentioned earlier, realistic statistical calibration models should incorporate error terms from all sources of measurement. When geochemical logging was first introduced, several detailed studies were made to assess the strengths and limitations of borehole geochemistry through exhaustive comparisons with core elemental and mineralogical analyses. These included comparisons in the Conoco Research well, Ponca City, Oklahoma by Hertzog et al. (1987); the discussion of the results from an Exxon research well which penetrated Upper Cretaceous siliciclastic rocks in Utah by Wendlandt and Bhuyan (1990); and an assessment of data from three Shell wells in the Netherlands, Oman, and the U.S. by van den Oord (1990).

There are several ways to assess modal mineralogy, so which constitutes the most accurate method to use as a standard for the real mineral composition? Harvey et al. (1998) addressed this problem when they compared core data from the spectral measurements of quantitative X-ray diffraction and infrared spectroscopy, as well as micrometric analysis from thin section point counts. Overlapping peaks and poor resolution at low resolution pose special problems for the spectral methods, while appropriate sample sizes must be observed for robust statistics in micrometric analysis. Also, the distinction between volume percentage and weight percentage must be observed when interrelating modal and normative compositions. Harvey et al. (1998) concluded that the results of their study did not favor one method over another, but pointed out that their comprehensive analysis demonstrated the difficulty of obtaining accurate modal estimates and even the notion of what constitutes the "real" mineral composition. This is certainly worth bearing in mind when making a judgement about the "accuracy" of a normative mineral solution from inversion of log responses. So, for example, mismatches in clay mineral estimates by log inversion represents a failure to reproduce the results of quantitative X-ray diffraction which are themselves only estimates of the true composition.

A major obstacle to the production of unique mineral transformations from element concentrations has been the problem of compositional colinearity. If precisely colinear, then an infinite range of solutions is possible, causing a matrix singularity and a breakdown of an inversion procedure. If average mineral compositions are used, a solution becomes possible, but may be unstable (Harvey et al. 1998). Wendlandt and Bhuyan (1990) found that the use of silicon, potassium and aluminum tended to result in overestimates of kaolinite; the use of iron to predict illite content caused underestimates of kaolinite. However, effective discrimination between illite and kaolinite contents became possible when dry density was applied as an extra constraint.

## 24.7 Conclusion

The estimation of mineral composition from petrophysical logs is now a standard feature on any log analysis software package. However, the degree to which these estimates match reality is highly variable and requires a knowledgeable and experienced user to work with powerful procedures. The identification of the major mineral suite that actually occurs in the rock is an important first step. As the old Chinese proverb says, "The beginning of wisdom is calling a thing by its right name." In the end, the solution of "mathematical minerals" will often come down to a choice between an acceptable estimate of an unreachable modal mineralogy or the realization of a useful, but hypothetical, normative assemblage.

## References

Alger RP, Raymer LL, Hoyle WR, Tixier MP (1963) Formation density log applications in liquid-filled holes. SPE AIME Trans 228:321–332

Archie GE (1950) Introduction to petrophysics of reservoir rocks. AAPG Bull 34(5):943–961

Cross W, Iddings JP, Pirsson LV, Washington HS (1902) A quantitative chemico-mineralogical classification and nomenclature of igneous rocks. J Geol 10:555–690

Doveton JH, Cable HW (1979) Fast matrix methods for the lithological interpretation of geophysical logs. In: Gill D, Merriam DF (eds) Geomathematical and petrophysical studies in sedimentology. Pergamon Press, Oxford, Computers in Geology Series, pp 101–116

Ellis DV (1987) Well logging for earth scientists. Elsevier Science Publishing Company, New York, 532 pp

Eslinger E, Pevear D (1988) Clay minerals for petroleum geologists and engineers. SEPM Short Course (2), 410 pp

Gysen M, Mayer C, Hashmy KH (1987) A new approach to log analysis involving simultaneous optimization of unknowns and zoned parameters, paper B. In: 11th formation evaluation symposium transactions. Canadian Well Logging Society, 20 pp

Harris MH, McCammon RB (1971) A computer-oriented generalized porosity-lithology interpretation of neutron, density and sonic logs. J Pet Technol 23:239–248

Harvey PK, Bristow JF, Lovell MA (1990) Mineral transforms and downhole geochemical measurements. Sci Drill 1(4):163–176

Harvey PK, Brewer TS, Lovell MA (1998) The estimation of modal mineralogy: a problem of accuracy in core-log calibration. In: Harvey PK, Lovell MA (eds) Core-log integration, vol 136. Geological Society, London Special Publication, pp 25–38

Herron MM (1986) Mineralogy from geochemical well logging. Clays Clay Miner 34(2):204–213

Herron MM (1988) Geochemical classification of terrigenous sands and shales from core or log data. J Sediment Petrol 58(5):820–829

Hertzog R, Colson L, Seeman B, O'Brien M, Scott H, McKeon D, Grau J, Ellis D, Schweitzer J, Herron M (1987) Geochemical logging with spectrometry tools. SPE-16792. In: SPE annual technical conference and exhibition, proceedings, volume omega. Formation evaluation and reservoir geology: Society of Petroleum Engineers, pp 447–460. Later published in 1989, SPE Formation Evaluation 4(2):153–162

Imbrie J, Poldervaart A (1959) Mineral compositions calculated from chemical analyses of sedimentary rocks. J Sediment Petrol 29(4):588–595

Krumbein WC, Pettijohn FJ (1938) Manual of sedimentary petrography. Appleton-Century-Crofts, New York, 549 pp

Marett G, Kimminau S (1990) Logs, charts, and computers–the history of log interpretation modeling. Log Anal 31(6):335–354

Mayer C, Sibbit A (1980) GLOBAL: a new approach to computer-processed log interpretation. SPE Paper 9341, 12 pp

McCammon RB (1970) Component estimation under uncertainty. In: Merriam DF (ed) Geostatistics, A Colloquium. Plenum, New York, pp 45–61

Moss B, Harrison R (1985) Statistically valid log analysis method improves reservoir description. SPE-13981: Society of Petroleum Engineers, Offshore Europe conference, Aberdeen, preprint, 32 pp

Quirein J, Kimminau S, Lavigne J, Singer J, Wendel F (1986) A coherent framework for developing and applying multiple formation evaluation models, paper DD. In: 27th annual logging symposium transactions. Society of Professional Well Log Analysts, 16 pp

Savre WC (1963) Determination of a more accurate porosity and mineral composition in complex lithologies with the use of the sonic, neutron, and density surveys. J Pet Technol 15:945–959

van den Oord RJ (1990) Experience with geochemical logging, paper T. In: 31st annual logging symposium transactions. Society of Professional Well Log Analysts, 25 pp

Wendlandt RF, Bhuyan K (1990) Estimation of mineralogy and lithology from geochemical log measurements. AAPG Bull 74(6):837–856

# Chapter 25
# Geostatistics for Seismic Characterization of Oil Reservoirs

**Amílcar Soares and Leonardo Azevedo**

**Abstract** In the oil industry, exploratory targets tend to be increasingly complex and located deeper and deeper offshore. The usual absence of well data and the increase in the quality of the geophysical data, verified in the last decades, make these data unavoidable for the practice of oil reservoir modeling and characterization. In fact the integration of geophysical data in the characterization of the subsurface petrophysical variables has been a priority target for geoscientists. Geostatistics has been a key discipline to provide a theoretical framework and corresponding practical tools to incorporate as much as possible different types of data for reservoir modeling and characterization, in particular the integration of well-log and seismic reflection data. Geostatistical seismic inversion techniques have been shown to be quite important and efficient tools to integrate simultaneously seismic reflection and well-log data for predicting and characterizing the subsurface lithofacies, and its petro-elastic properties, in hydrocarbon reservoirs. The first part of this chapter presents the state of the art and the most recent advances of geostatistical seismic inversion methods, to evaluate the reservoir properties through the acoustic, elastic and AVA seismic inversion methods with real case applications examples. In the second part we present a methodology based on seismic inversion to assess uncertainty and risk at early stages of exploration, characterized by the absence of well data for the entire region of interest. The concept of analog data is used to generate scenarios about the morphology of the geological units, distribution of acoustic properties and their spatial continuity. A real case study illustrates the this approach.

A. Soares (✉) · L. Azevedo
CERENA, Instituto Superior Técnico, Universidade de Lisboa,
Av. Rovisco Pais, 1049-001 Lisbon, Portugal
e-mail: asoares@tecnico.ulisboa.pt

L. Azevedo
e-mail: leonardo.azevedo@tecnico.ulisboa.pt



B. S. Daya Sagar et al. (eds.), *Handbook of Mathematical Geosciences*,
https://doi.org/10.1007/978-3-319-78999-6_25

## 25.1 Integration of Geophysical Data for Reservoir Modeling and Characterization

One of the main challenges regarding hydrocarbon reservoir characterization has been the integration of different types of data—geological conceptual models, well-log data, geophysical data, production data—for modelling the subsurface properties of interest while assessing the corresponding uncertainty and risk. Although well data provides certain 'hard' measures of the subsurface properties, given the usual lack of such data and, consequently, its limited spatial representativeness, the corresponding models normally provide little understanding of the complex and heterogeneous subsurface geology of the entire reservoir area. Since the eighties, Geostatistics has been a key discipline to provide a theoretical framework and corresponding practical tools to incorporate as much as possible different types of data for reservoir modeling and characterization, in particular the seismic reflection data (Dubrule 2003). One of the most important contributions of geostatistical methods for seismic data integration in reservoir modelling, has been the development of stochastic seismic inversion techniques.

Seismic reflection data, since it has high spatial representativeness, by covering the full spatial extent of the reservoir volume, is a different and privileged window for targeting the subsurface petro-elastic properties of interest. However, seismic reflection data represents an indirect measurement of these properties and has a poor spatial resolution along the vertical direction (temporal domain). This is translated in a much greater support compared with the well-log data and much greater uncertainty derived both from measurement errors and the nonlinear relationship between the recorded seismic signal and the subsurface properties one wishes to describe (Tarantola 2005). This has been the most serious limitation of direct use of seismic data as secondary information either in methods using it as local trends or in joint simulation methods (Dubrule 2003), or even accounting for the different support of both data (Liu and Journel 2009).

To overcome such limitations, an alternative approach has been widely used. Seismic inversion methods are based on the following rational: subsurface petrophysical properties (such as facies, porosity and saturation), can have a relationship to other seismic attributes, such as acoustic and/or elastic impedances; hence, one wishes to know the model parameters $\mathbf{r}$ (reflectivity coefficients derived from the subsurface elastic properties), which convolved with a known wavelet $\mathbf{w}$ give rise to the known solution $\mathbf{A}$ (i.e. the recorded seismic amplitudes):

$$\mathbf{A} = \mathbf{r} * \mathbf{w}. \tag{25.1}$$

The theoretical solutions for seismic inversion are stated in Tarantola (2005). The seismic inversion problem began to be tackled with deterministic methodologies (Lindseth 1979; Lancaster and Whitcombe 2000; Russell 1988; Coléou et al. 2005). Later, this framework was extended into a statistical domain. Among the many statistical inverse approaches, two different stochastic approaches for solving the seismic inversion are worth mentioning. The first group of stochastic methodologies approach the seismic inversion as an optimization problem in an iterative and convergent process. This includes what are traditionally designated by iterative geostatistical seismic inversion methods, from the seminal work by Bortolli et al. (1993), until the most recent geostatistical inversion methods (Soares et al. 2007; Nunes et al. 2012; Azevedo et al. 2015; Azevedo and Soares 2017). The second group of stochastic seismic inversion algorithms is known by linearized Bayesian inverse methodologies. These are based on a particular solution of the inverse problem using the Bayesian framework and assuming the model parameters and observations as multi-Gaussian distributed as well as the data error, which allows the forward model to be linearized (Buland and Omre 2003). Several authors have recently contributed towards overcoming some of the limitations of this method, particularly the multi-Gaussian assumption, by using Gaussian Mixture Models (Grana and Della Rossa 2010).

This chapter summarizes some iterative geostatistical modeling techniques dealing with the integration of seismic reflection and well-log data, through seismic inversion procedures, for characterizing hydrocarbon reservoirs with high spatial resolution models of main properties of interest, such as lithologies, facies and fluid saturations.

Uncertainty and risk assessment at different stages of exploration are also important targets of the proposed methodologies approached in this chapter. Hence, this chapter finishes with the introduction of recent advances of geostatistical seismic inversion methods for the uncertainty and risk assessment at early stages of exploration.

## 25.2 Iterative Geostatistical Seismic Inversion Methodologies

The aim of seismic inversion is the inference of the subsurface elastic or acoustic properties from recorded seismic reflection data. The retrieved inverse models can be acoustic and/or elastic impedance for post-stack seismic data, or density, P-wave and S-wave models if the inversion algorithm is used to invert pre-stack seismic reflection data (Francis 2006).

Seismic inversion might be described as an ill-posed and nonlinear problem with multiple solutions that can be summarized by (Tarantola 2005):

$$\mathbf{d_{obs}} = \mathbf{F}(\mathbf{m}) + \mathbf{e}. \tag{25.2}$$

The goal is to estimate a subsurface Earth model, $\mathbf{m}$, that after being forward modelled, $\mathbf{F}$, produces synthetic seismic data showing a good correlation with the recorded seismic data, the observed data, $\mathbf{d_{obs}}$, which are normally contaminated by measurement errors $\mathbf{e}$. The match between observed and synthetic seismic is achieved by the maximization (or minimization) of an objective function measuring the mismatch between inverted and real seismic. For example, the objective function can be as simple as the Pearson's correlation coefficient:

$$\rho_{X,Y} = \frac{cov(X,Y)}{\sigma_X \sigma_Y}, \tag{25.3}$$

where *cov* is the centered covariance between variables *X* and *Y*, which are the synthetic and real seismic volumes, respectively, and σ the individual standard deviations of each variable. More complex objective functions integrate Pearson's correlation coefficient with least-square errors calculated between the synthetic and the recorded seismic reflection data in terms of amplitudes.

A geostatistical seismic inversion framework consists on an iterative procedure in which a set of realizations of parameters, **m**, are generated by using stochastic sequential simulation methods (Deutsch and Journel 1996) and optimized until the match of the objective function reaches a given user-defined value, or a certain number of fixed iterations. Geostatistical inversion techniques are based on the use of stochastic sequential simulation as the model perturbation technique, ensuring in this way the reproduction of the main spatial continuity patterns and the joint distribution functions of the acoustic and/or elastic properties of interest as retrieved from the existing well-log data in all the models generated during the iterative procedure, while simultaneously allowing access to the uncertainty attached to the retrieved inverse models.

Within this framework there are two traditional approaches for integrating seismic reflection and well-log data for hydrocarbon reservoir modeling.

## 25.3 Trace-by-Trace Geostatistical Seismic Inversion

Geostatistical seismic inversion was introduced by the seminal papers of Bortoli et al. (1993) and Haas and Dubrule (1994). These authors proposed a sequential trace-by-trace approach in which each seismic trace, or location within the inversion grid, is visited individually following a pre-defined random path within the seismic volume. At each step along the random path a set of *Ns* realizations of one acoustic impedance trace is simulated using sequential Gaussian simulation (Gómez-Hernández and Journel 1993; Deutsch and Journel 1996), taking the well-log data and previously visited/simulated nodes into account. Then, for each individual simulated impedance trace, the corresponding reflection coefficient is derived and convolved by a wavelet, resulting in a set of *Ns* synthetic seismic traces. Each of the *Ns* synthetic traces is compared in terms of a mismatch function with the recorded/real seismic trace. The acoustic impedance realization that produces the best match between the real and the synthetic seismic traces is retained in the reservoir grid as conditioning data for the simulation of the next acoustic impedance trace at the new location following the pre-defined random path. One of the main drawbacks of trace-by-trace stochastic seismic inversion methodologies concerns those areas of the record seismic reflection data with low signal-to-noise ratio. In areas of poor seismic signal, the sequential trace-by-trace approaches impose inverted models fitting the observed noisy seismic reflection data. As the simulated trace is assumed to be 'real' data for subsequent steps, this can lead to the spread of unreliable impedance values that are related with noisy seismic samples. Noisy areas should be interpreted as high uncertainty areas with very low influence throughout the inversion process. More recent versions of trace-by-trace models try to overcome this drawback by avoiding noisy areas in the early stages of the inversion procedure (Grijalba-Cuenca and Torres-Verdín 2000).

## 25.4 Global Geostatistical Seismic Inversion Methodologies

To overcome these limitations, Soares et al. (2007) introduced the global stochastic inversion methodology that, contrary to trace-by-trace approaches, uses a global approach during the stochastic sequential simulation stage of the inversion procedure: at each iteration a set of *Ns* impedance models is generated at once for the entire inversion grid. The general outline of this family of geostatistical inversion algorithms is depicted in Fig. 25.1. Briefly, this group of iterative inverse approaches uses the principle of cross-over genetic algorithms as the global optimization technique driving the convergence of the procedure from iteration to iteration, while the model perturbation is performed using direct sequential simulation and co-simulation (Soares 2001). The global optimizer uses the trace-by-trace correlation coefficients between the different simulated synthetic seismic data and the real model as the affinity criterion to create the next generation of models for the next iteration, by using stochastic sequential co-simulation. The iterative procedure continues until a stopping criterion is reached: frequently the global correlation coefficient between real and inverted seismic reflection data.

In global iterative geostatistical seismic inversion procedures, areas of low signal-to-noise ratio remain poorly matched throughout the entire iterative inversion

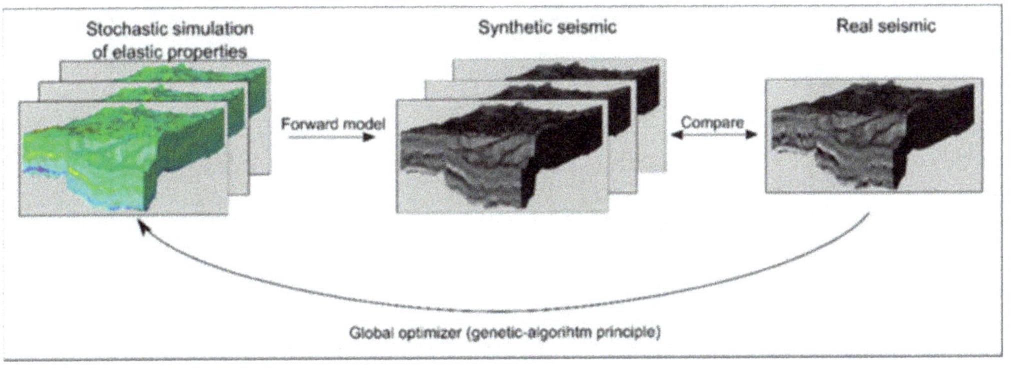


**Fig. 25.1** General outline for global iterative geostatistical seismic inversion

procedure: an ensemble of best-fit inverted models will always present high variability, or high uncertainty, for those noisy areas where the signal-to-noise ratio is low.

This framework was generalized for the inversion of seismic reflection data for acoustic and elastic impedance, direct inversion of petrophysical properties and seismic AVA inversion. These methods are introduced with more detail in the following sections.

### 25.4.1 Global Geostatistical Acoustic Inversion

The global stochastic inversion (GSI; Soares et al. 2007; Caetano 2009) is one of the existing methods to invert fullstack seismic reflection data for acoustic impedance (Ip) models. The general outline of this iterative geostatistical methodology can be described in the following sequence of steps, summarized in Fig. 25.2:

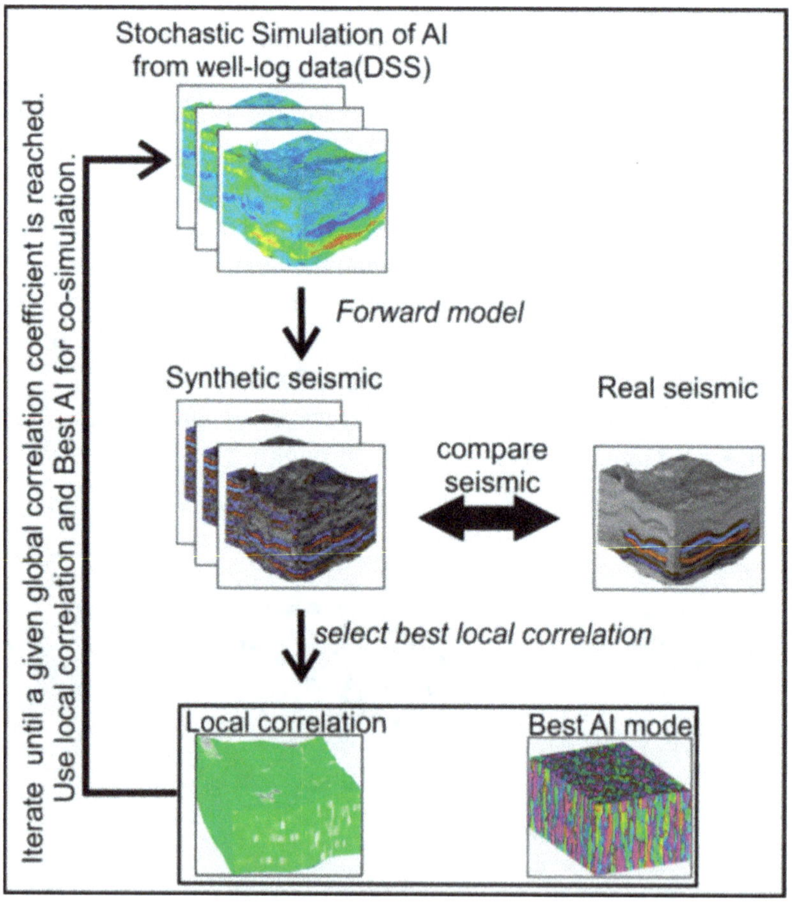


**Fig. 25.2** Outline of geostatistical acoustic inversion (adapted from Azevedo and Soares 2017)

(1) Simulate with direct sequential simulation (Soares 2001) for the entire seismic grid a set of *Ns* acoustic impedance models, conditioned to the available acoustic impedance well-log data and assuming a spatial continuity pattern as revealed by a variogram model;
(2) From the impedance models simulated in the previous step, derive a set *Ns* synthetic seismic volumes by computing the corresponding normal incidence reflection coefficients (RC) (Eq. 25.4):

$$RC = \frac{Ip_2 - Ip_1}{Ip_2 + Ip_1}, \tag{25.4}$$

where the indexes 1 and 2 correspond to the layer above and below a given reflection interface.

(3) The resulting RC are convolved by an estimated wavelet for that particular seismic dataset in order to compute synthetic seismic volumes (Eq. 25.1).
(4) Each seismic trace from the *Ns* synthetic seismic volumes is compared in terms of correlation coefficient against the real seismic trace from the same location. From the ensemble of simulated Ip models, the acoustic impedance traces that produce synthetic seismic with the highest correlation coefficient are stored in an auxiliary volume along with the value of the correlation coefficient.
(5) These auxiliary volumes, the one with the best acoustic impedance traces and the other with the corresponding local correlation coefficients, are used as secondary variables and local regionalized models for the generation of the new set of acoustic impedance models for the next iteration. The new set of *Ns* acoustic impedance models is built using direct sequential co-simulation (Soares 2001) conditioned to the available acoustic impedance well-log data, and using the best Ip volumes as secondary variable and local correlation coefficients to condition the co-simulation.
(6) The iterative procedure stops when the global correlation coefficient between the full synthetic and real stacked seismic volumes is above a certain threshold.

Synthetic and real case applications of geostatistical acoustic inversion can be found in several studies; for example, Soares et al. (2007) and Caetano (2009). A summary of a real application example, using a fullstack seismic volume acquired offshore Brazil, illustrates herein the method (a detailed description of the dataset is available in Azevedo et al. 2015). The best-fit Ip model (Fig. 25.3) was retrieved after 6 iterations where on each iteration an ensemble of 32 realizations of Ip were generated. The use of stochastic seismic inversion allows retrieving high resolution (with high variability) acoustic impedance models. The synthetic full-stack seismic data computed from this model (Fig. 25.4) do match the observed seismic reflection data in both the spatial extent of the main seismic reflection and its amplitude content. This is of great importance for this case study since the reservoir areas are related with those spatially constrained amplitude anomalies

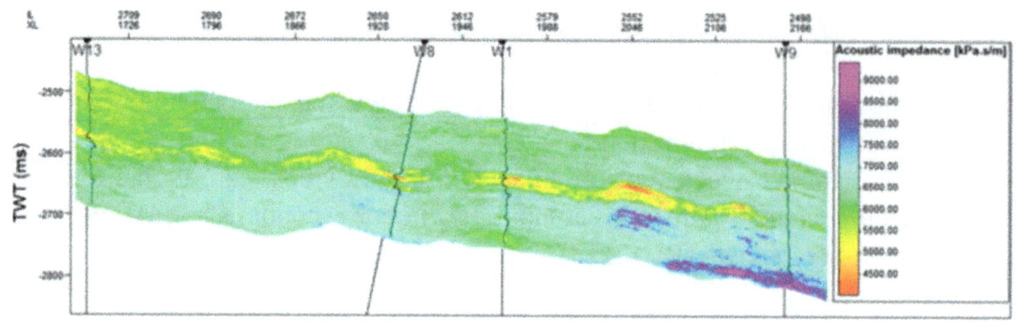


**Fig. 25.3** Vertical well-section extracted from the best-fit P-impedance volume retrieved from the global stochastic inversion after six iteration with thirty-two realizations generated at each iteration

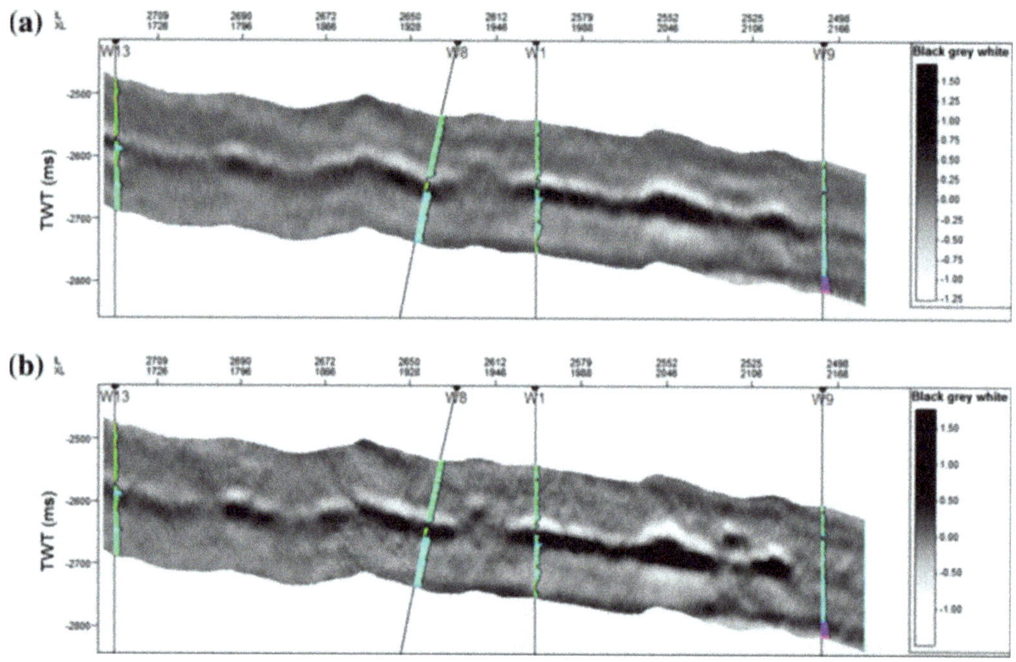


**Fig. 25.4** Comparison between vertical well sections extracted from: **a** synthetic seismic reflection data computed from the best-fit inverse Ip model shown in Fig. 25.3 and **b** real seismic volume. The log curve plotted on top of the seismic data represents Ip (same color scale as shown in Fig. 25.3)

observed in the real seismic volume. The global correlation between the inverted and the real seismic volumes is 87%.

### *25.4.2 Global Geostatistical Elastic Inversion*

The acoustic inversion algorithm was extended for the inversion of partial angle stacks directly, and simultaneously, for acoustic and elastic impedance (Is) models

(Nunes et al. 2012; Azevedo et al. 2013b). The main purpose of this development was the integration of more information, related with the elastic domain (Is), to enrich the final elastic reservoir models allowing better lithofacies prediction. Two main differences compared with acoustic inversion summarize this elastic inversion method (Azevedo and Soares 2017):

(i) Acoustic and elastic impedances, Ip and Is, are jointly simulated (step 1) of previous outline and co-simulated (step 5) by using the direct sequential simulation with joint distributions of probability (Horta and Soares 2010). This simulation method succeeds in reproducing the bivariate distribution function (Ip, Is) as it was estimated from the experimental log data.

(ii) The reflectivity coefficients (step 9) are obtained with the *Ns* pairs of Ip and Is, simulated at each iteration, using the approximation outlined in Fatti et al. (1994) (Eq. 25.5) for the calculation of the corresponding angle-dependent reflection coefficient volumes:

$$
\begin{aligned}
R_{pp}(\theta) &\approx (1+tan^2\theta)\frac{\Delta I_p}{\Delta I_s} - 4\left(\frac{I_s}{I_p}\right)^2 sin^2\theta\frac{\Delta I_s}{2I_s},\\
\Delta I_p &= I_{p2} - I_{p1},\\
I_p &= \frac{I_{p2}+I_{p1}}{2},\\
\Delta I_s &= I_{s2} - I_{s1},\\
I_s &= \frac{I_{s2}+I_{s1}}{2}.
\end{aligned}
\tag{25.5}
$$

The index 1 refers to the vertical location in which the calculation of the reflection coefficient is carried out, the layer above the reflection interface; and 2 refers to the sample immediately below, the layer below the reflection interface.

Detailed application examples of this method can be found in the following studies: Nunes et al. (2012), Azevedo et al. (2013b), Azevedo and Soares (2017). For illustrative purpose, here we show the application of this methodology to the same case study shown in the previous section. The best-fit Ip and Is models that jointly produce the highest value of correlation coefficient between synthetic and real seismic reflection data are shown in Fig. 25.5. Comparing the Ip models derived from the acoustic and elastic inversion it is clear that the introduction of more information using different angles of incidence brings more detail for the retrieved inverse model. The comparison between real and synthetic seismic reflection data derived from the best-fit elastic models is shown in Fig. 25.6.

Due to the use of direct sequential simulation with joint probability distributions (Horta and Soares 2010) the relationship between Ip and Is as observed in the well-logs is reproduced for all pairs of models generated during the inversion procedure (Fig. 25.7). Besides the richness of the inverted models, this is a key step of the proposed inversion technique since it allows, for example, more reliable facies classification, and consequently a better reservoir description, over the inverted elastic models.

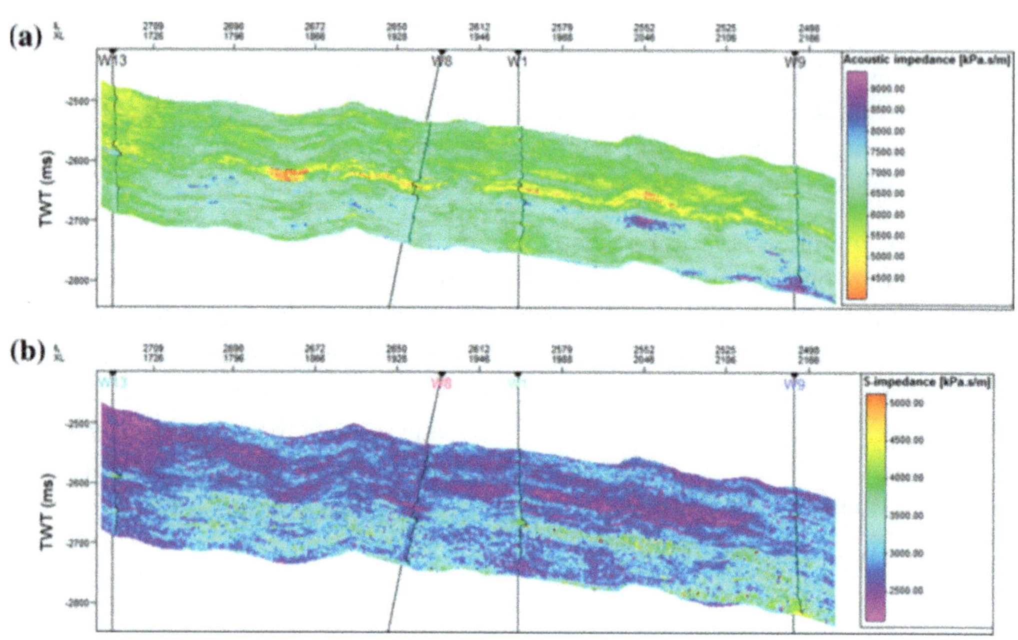


**Fig. 25.5** Comparison between vertical well sections extracted from: **a** best-fit Ip model and **b** best-fit Is model

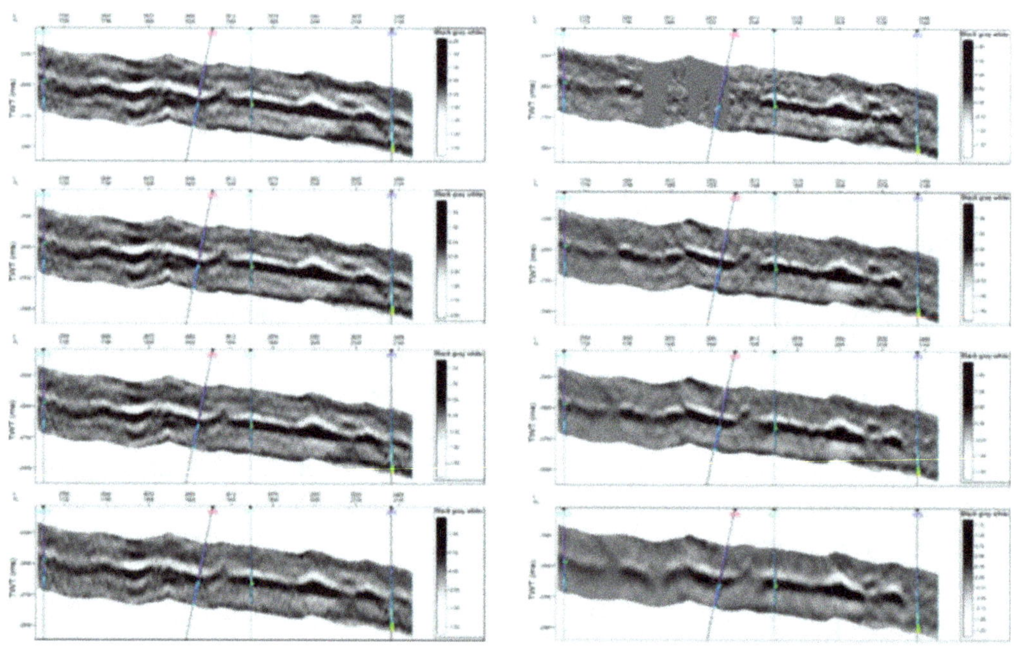

**Fig. 25.6** Comparison between vertical well sections extracted from: (left) synthetic seismic reflection data computed from the best-fit inverse Ip and Is models and (right) real seismic volume. From top to bottom: nearstack, near-mid stack, far-mid stack and farstack. The log curve plotted on top of the seismic data represents Is (same color scale as shown in Fig. 25.5)

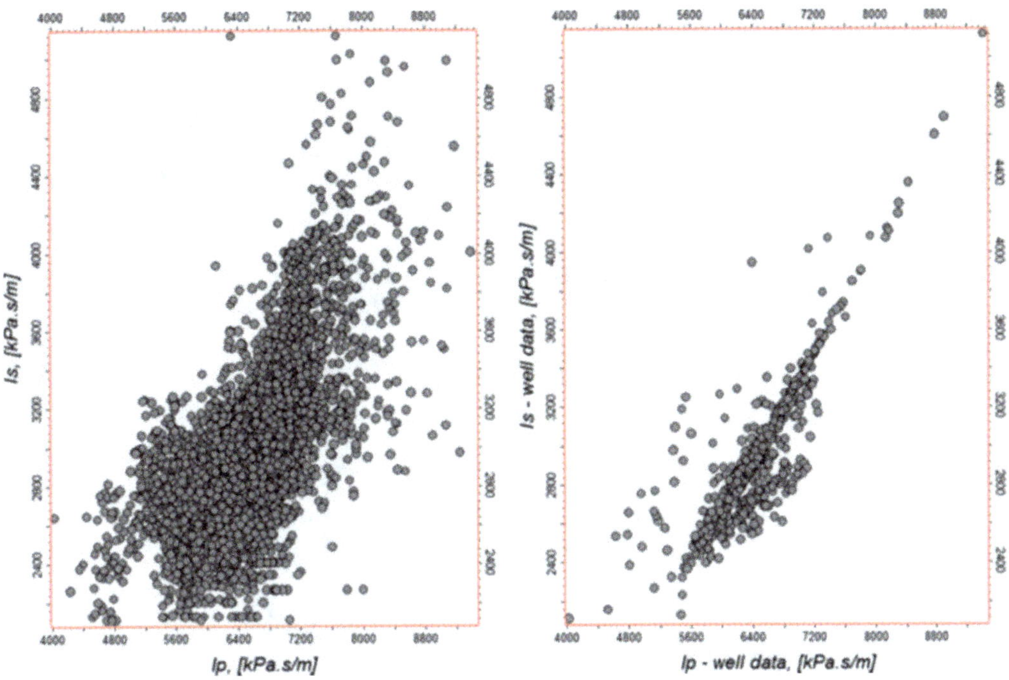


**Fig. 25.7** Comparison between the joint distribution of Ip and Is as retrieved from the best-fit inverse pair of Ip and Is and from the well-logs

### *25.4.3 Geostatistical Seismic AVA Inversion (Pre-stack Inversion)*

During the last decades, the quality of seismic reflection data has increased tremendously, together with the decreasing of its acquisition costs. Pre-stack seismic data with high signal-to-noise ratio and high fold number is nowadays a reality, increasing this data's use in seismic reservoir characterization even within early exploratory stages. The better subsurface characterization using pre-stack seismic data is achieved by interpreting the changes of amplitude versus the offset (AVO), or with the angle of incidence (AVA; Castagna and Backus 1993; Avseth et al. 2005). The use of pre-stack seismic reflection data allows the inference of density, P-wave and S-wave velocity models, instead of the traditional impedance models. The availability of the three properties individually is a clear enhancement in what reservoir modelling and characterization are concern with.

Stochastic seismic inversion methodologies for pre-stack seismic data, commonly called seismic AVA inversion, are being proposed based on different assumptions and frameworks (Mallick 1995; Ma 2002; Buland and Omre 2003; Contreras et al. 2005). Here we refer to geostatistical seismic AVA inversion (Azevedo et al. 2013a), which relies on the same general framework of global iterative geostatistical seismic inversion methodologies but with the following main characteristics of pre-stack inversion (see outline of Fig. 25.8; Azevedo and Soares 2017):

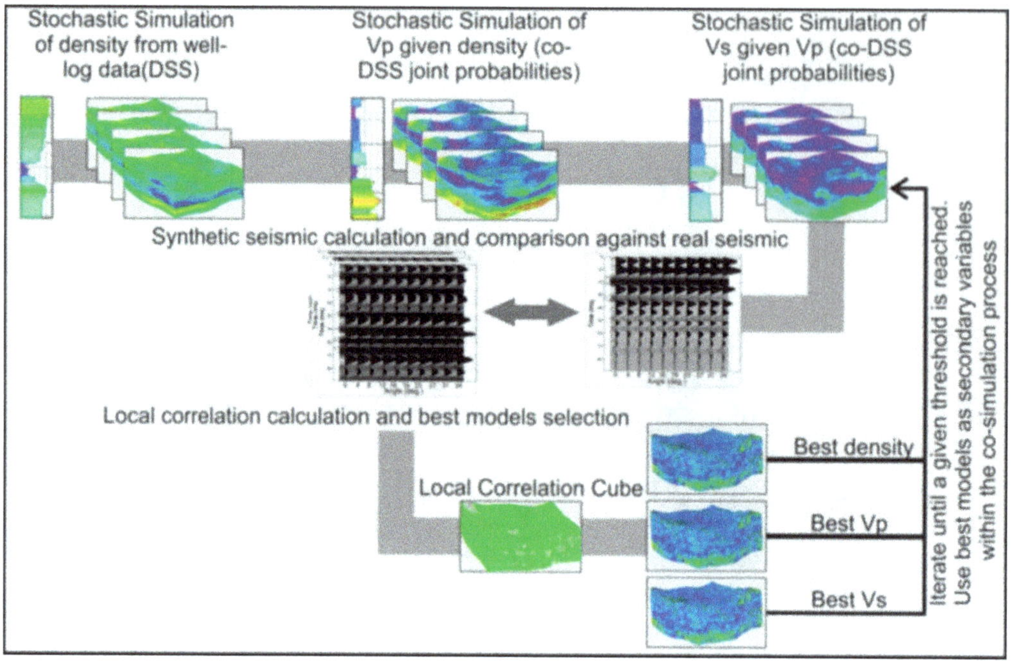


**Fig. 25.8** Schematic representation of the global iterative geostatistical seismic AVO inversion methodology (adapted from Azevedo and Soares 2017)

(i) the perturbation of the model parameters for density, P-wave and S-wave velocities is performed sequentially using stochastic sequential co-simulation with joint distributions (Horta and Soares 2010);
(ii) forward modeling is computed using an angle-dependent approximation when computing the reflection coefficients that can be modified according to the complexity of the subsurface geology;
(iii) the mismatch evaluation between the observed and the inverted seismic data and selection of the conditioning data for the generation of the next set of elastic models during the next iteration by multi-variable optimization.

In this approach, each elastic property is generated sequentially. Density is first simulated because it is the property associated with a higher degree of uncertainty since its contribution to the recorded seismic reflection data is small, i.e. the component of the seismic reflection data related with density is low and mostly related to the signal received at the far angles (Avseth et al. 2005). Also, density is the most spatially homogeneous variable and consequently most convenient to be used as secondary variable for the co-simulation with joint probability distributions of Vp. The resulting Vp models are then used as auxiliary variable for the co-simulation with joint probability distributions of Vs. At the end of the iterative inversion procedure, the reproduction of the joint distribution densities, Vp and Vs, allows a distinction to be made between any litho-fluid facies previously identified from the original well-log data within the inverted set of elastic models. As well as the spatial interpretation of these litho-fluid facies, the stochastic approach allows the assessment of the spatial uncertainty related with each facies of interest.

After the sequential simulation of *Ns* elastic models, density, Vp and Vs, an ensemble of synthetic pre-stack seismic volumes are calculated. The angle-dependent RC ($R_{pp}(\theta)$) may be calculated, for example, following Shuey's (1985) three-term approximation:

$$R_{pp}(\theta) \approx R(0) + G sin^2\theta + F\left(tan^2\theta - sin^2\theta\right), \tag{25.6}$$

with the normal incidence, R(0), reflection as defined by:

$$R(0) = \frac{1}{2}\left(\frac{\Delta Vp}{Vp} + \frac{\Delta\rho}{\rho}\right),$$

and the variation of the reflectivity versus the angle, the AVO gradient, G:

$$G = R(0) - \frac{\Delta V\rho}{V\rho}\left(\frac{1}{2} + \frac{2\Delta Vs^2}{Vs^2}\right) - \frac{4\Delta Vs^2}{Vp^2}\frac{\Delta Vs}{Vs},$$

and *F*, the reflectivity at the far angles (reflection angles higher than 30°), defined as:

$$F = \frac{1}{2}\frac{\Delta Vp}{Vp}.$$

Each elastic property is defined on each side of the interface where the reflection is happening as follows:

$$\begin{aligned}
\Delta V_p &= V_{p2} - V_{p1}, \\
V_p &= \frac{V_{p2} + V_{p1}}{2}, \\
\Delta V_s &= V_{s2} - V_{s1}, \\
V_s &= \frac{V_{s2} + V_{s1}}{2}, \\
\Delta V_\rho &= V_{\rho 2} - V_{\rho 1}, \\
V_\rho &= \frac{V_{\rho 2} + V_{\rho 1}}{2}.
\end{aligned}$$

Indexes 1 and 2 have the same meaning as in Eq. 25.4.

Each angle gather is composed by *n* seismic traces, equal to the number of reflection angles considered. The *Ns* angle-dependent reflection coefficient traces are convolved by estimated angle-dependent wavelets for each particular incident angle θ (Fig. 25.9) to obtain *Ns* synthetic angle gathers. The best elastic models, created at the end of each iteration, are composed by the portions of the elastic traces from the ensemble of density, P-wave and S-wave velocity models simulated at the current iteration, that jointly produce synthetic seismic reflection data with the highest correlation coefficient compared with the real seismic volume. Hence, the

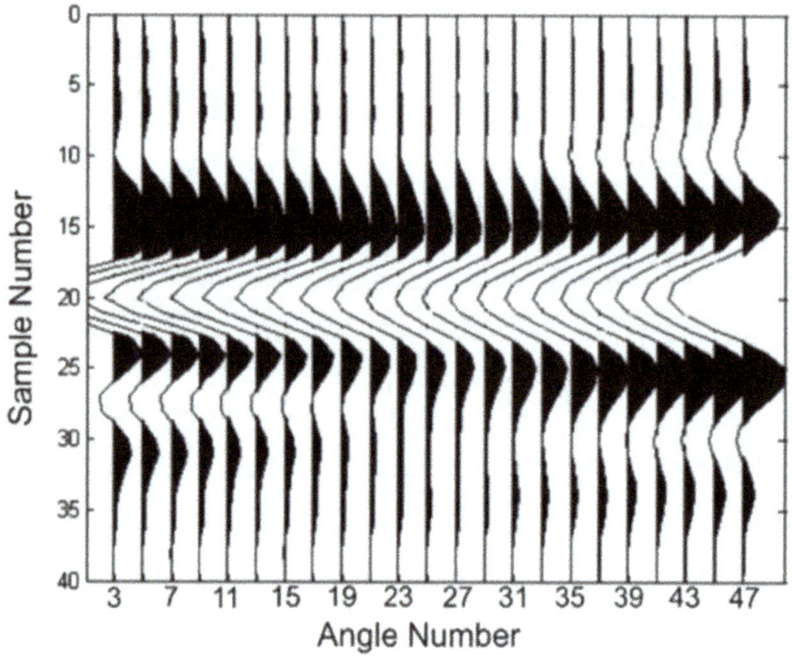


**Fig. 25.9** Example of an angle-dependent wavelet, for 23 angles, used for the convolution of the angle-dependent reflection coefficients ($R_{pp}(\theta)$) to generate pre-stack seismic reflection data

best models are selected by using a multivariate (traces for each angle) objective function (Azevedo and Soares 2017 illustrate an example of multivariate objective function).

As an application example, Fig. 25.10 shows vertical well sections extracted from the triplet of elastic models that produced synthetic pre-stack seismic reflection data with the maximum correlation coefficient during the iterative procedure. The inverted density, Vp and Vs models show high variability and agree with the expected spatial extent of the anomalies of interest as inferred from previous studies (Azevedo et al. 2015).

By comparing the inverse elastic inversion, shown in the previous sections for the different geostatistical seismic inversion techniques (Figs. 25.3, 25.5 and 25.10) it is clear that introducing more information within the inversion procedure, i.e. moving from the fullstack into the pre-stack domain, allows retrieving more detailed and variable inverse models. Usually, such models allow for a better understanding of the reservoir and identify and assess the main uncertainties related with its subsurface properties.

### 25.4.4 Recent Developments of Iterative Geostatistical Seismic Inversion

The global iterative geostatistical inversion techniques presented in the previous sections have been extended to allow inferring the subsurface petrophysical

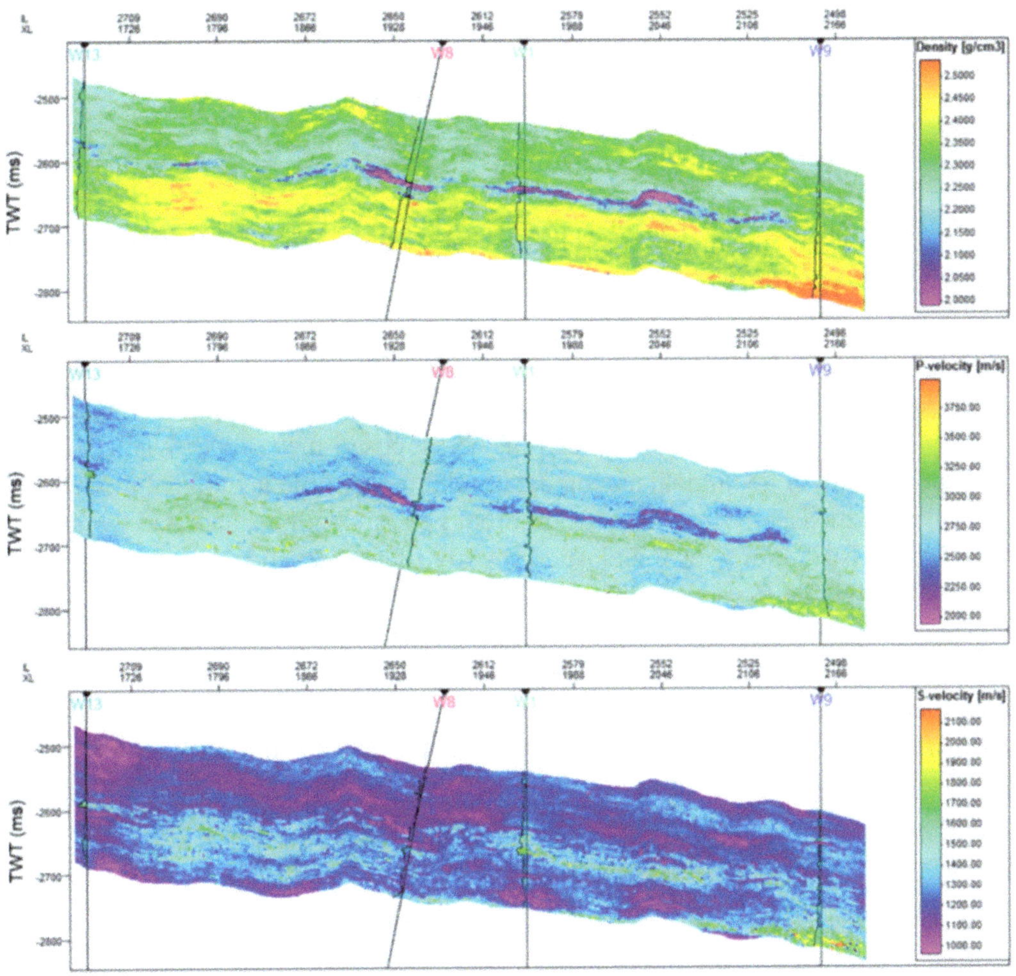


**Fig. 25.10** Vertical well section extracted from the best-fit models of: (from top to bottom) density, Vp and Vs

properties of interest, directly from the existing seismic reflection data: direct geostatistical seismic inversion to porosity (Azevedo and Soares 2017); and integration of rock physics into geostatistical seismic AVA inversion for simultaneous characterization of facies (Azevedo et al. 2015). In addition, the potentiality of these methodologies is enormous in what concerns the very different data integration like for example the electromagnetic data (CSEM). Application example of the joint inversion of seismic and electromagnetic data is illustrated in the study of Azevedo and Soares (2014).

The integration of dynamic production data with seismic data is another important and very promising field of application of these methodologies. In fact the integration of dynamic production data in reservoir modelling (commonly designated as history matching) is an even more complex inverse problem (e.g. Oliver and Chen 2011; Oliver et al. 2008; Mata-Lima 2008; Demyanov et al. 2011; Caeiro et al. 2015). If this is approached by a geostatistical iterative outline, the integration of both inverse methods can lead to a very rich model able to characterize geological complex structures and, simultaneously, reproduce the geological conceptual model, the seismic data and the dynamic data at the production wells (Marques et al. 2015; Azevedo and Soares 2017).

## 25.5 Uncertainty and Risk Assessment at Early Stages of Exploration

This section introduces a recent development of using seismic inversion for uncertainty and risk assessment at early stages of reservoir exploration characterized by the lack of well data. The idea of the proposed methodology is to account with the concept of geological analog data to define possible geological models of a given target, such as the geometry of different geological units, and also the a priori probability distributions for the elastic property of interest. An a priori uncertainty space is first built from plausible geological scenarios, generated from different sources of knowledge about the area of interest. For each scenario the corresponding elastic properties are computed and existing seismic reflection data is integrated, through a geostatistical seismic inversion, giving rise to an uncertainty space of petro-elastic properties. The first steps towards this direction correspond to the case study presented below.

### *25.5.1 Characterization of Different Scenarios with Analogue Data*

Due to the lack of data, several authors use analog data to constrain and integrate regional geological knowledge into reservoir models (e.g. Martinius et al. 2014; Grammer et al. 2004). The use of analog fields, and/or sedimentary basins, can help understand and predict the behavior of a reservoir since they are natural systems that may have similarity with the unknown study area. For example, one of the most valuable information that analogs can give to reservoir modelling, normally obtained from outcrop studies (Howell et al. 2014), is related to the geometry and the relation between the different geological units and their elastic properties.

This section proposes the extension of a traditional geostatistical seismic inversion methodology to integrate data from analogs (Pereira et al. 2017). In this application example the analog information is provided by well-logs located very far from the exploration area but somehow geologically related with the area of study. This iterative geostatistical seismic inversion methodology integrates a priori knowledge from the regional geology and the information from analogs, such as existing well-logs far from the region of interest (illustrated in Fig. 25.11).

One of the mandatories steps of this procedure, consists in dividing the area of interest in regional geological units based on conventional seismic interpretation

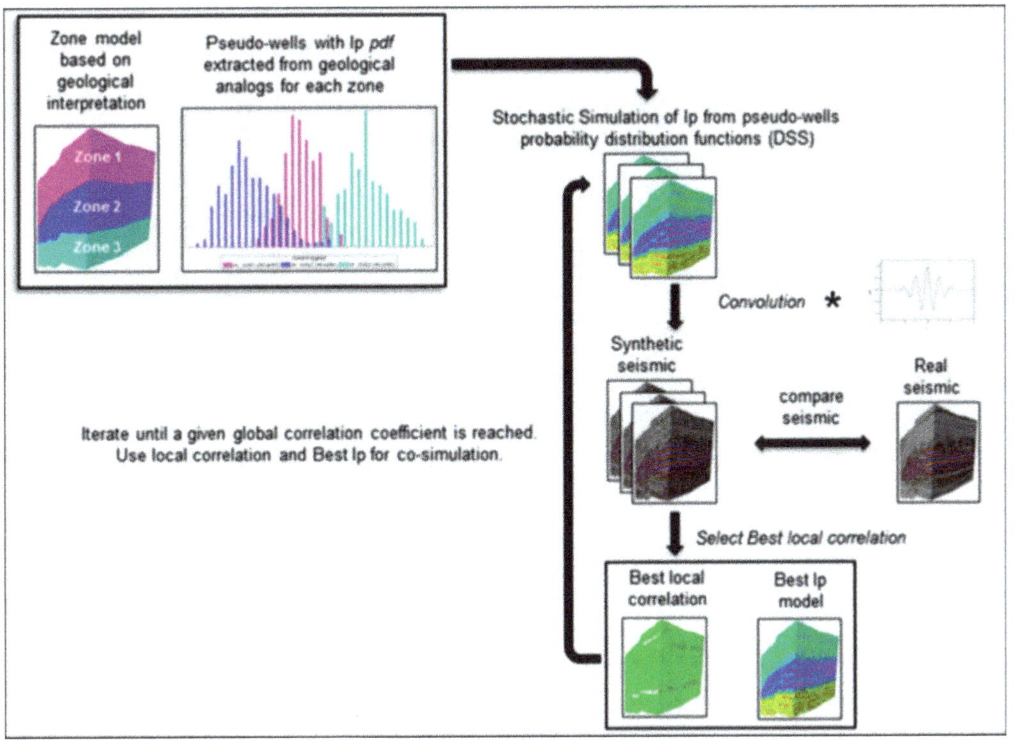


**Fig. 25.11** Schematic representation of the workflow to integrate geological analogue data into geostatistical seismic inversion, for each scenario

and the current knowledge of the prospect under study. The interpretation of the available seismic reflection data should be such that the interpreted seismic units are consistent with the stratigraphy of the region. The geological regionalization model of the area of study should be based not only on available seismic reflection data but include information from outcrop analogs or based on the geological knowledge of the sedimentary basin.

After the definition of the geological regionalization model, one needs to establish different scenarios, for each geological unit, about its elastic responses. These can be inferred from for example analogue data. This critical step should be done by integrating expertizes from different fields. The correlation between the elastic and rock properties should result in probability distribution functions of the elastic property of interest per region. The resulting distributions should be representative of the elastic properties of the geological region, and also of the relationship between the different geological regions. Meaning that if there is a progressive transition between geological regions (i.e. geological transition in terms of facies), this relationship should be expressed in the distributions of each region.

This approach is illustrated here with a real case study located in an offshore unexplored basin. The available data of this basin comprises a 3D seismic reflection and three appraisal wells drilled outside the main region of interest. The existing appraisal wells show evidences that suggest hydrocarbon generation, migration and possibly accumulation. Within this unexplored basin a promising prospect was

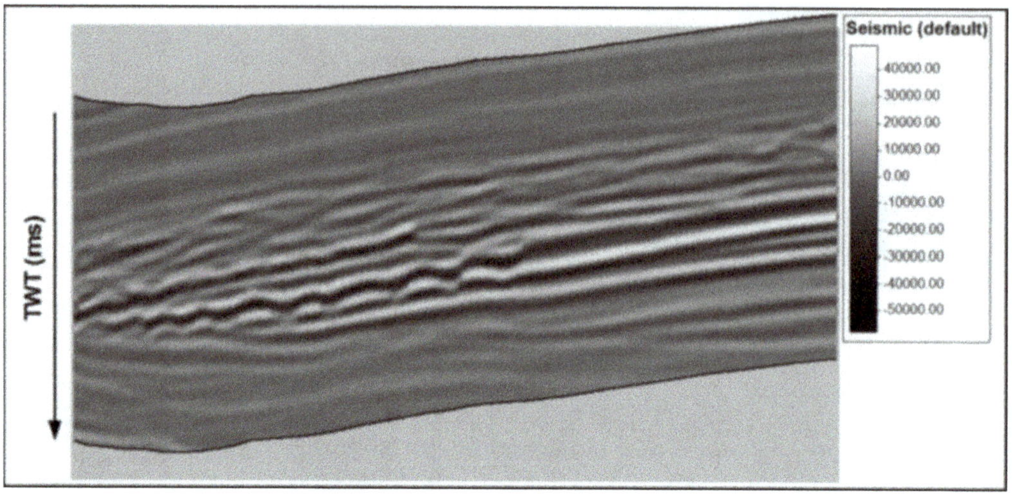


**Fig. 25.12** Real Seismic data for the area of interest showing the seismic signature of the prospect of interest. Lighter values indicate positive polarity and darker values indicate negative polarity

identified associated with a turbidite system, corresponding to a classic clastic sedimentary unit. This can be recognized and interpreted from the available seismic reflection data (Fig. 25.12). A detailed description about the geology of this basin can be found in Pereira et al. (2017).

The interpretation of the existing seismic reflection data resulted in three main geological units. For each region, probability distribution functions of Ip were assumed, taking into account the geological knowledge of the region of interest and from the Ip-logs available at the three neighbor wells. A representative wavelet of the time interval of interest was extracted exclusively from the available seismic reflection data using conventional wavelet extraction techniques based on statistical procedures (i.e. Weiner-Levinson filters). One of the main difficult steps of this methodology is the validation of the wavelet scale. A possible approach to tackle this issue can be selecting the distribution function of Ip for each region, making them plausible, by comparing the amplitude values of the synthetic seismic against the observed one.

### *25.5.2 Geostatistical Seismic Inversion of Each Scenario*

The previous step of this approach results in a set of geological models that represent the uncertainty about the prospect to be modelled. In order to reduce this space, the purpose of this step is based in the following rationale:

(i) for each one of the a priori chosen scenarios, in terms of geological regionalization model, one intends to access the models of acoustic and/or petrophysical properties, that match the known seismic, by running a conventional iterative geostatistical seismic inversion;

(ii) The match of each scenario synthetic seismogram with the real seismic can be used to validate or falsify them and build an uncertainty space of those properties.

Here, we show an example for one of the scenarios considered. The iterative geostatistical seismic inversion ran with six iterations, where on each sets of thirty-two realizations of Ip were generated conditioned simultaneously by the regionalization model (i.e. the three main seismic units resulting from seismic interpretation (Fig. 25.12)) and the individual Ip distributions as inferred from the nearby analog wells and published data.

The seismic inversion converged after six iterations when a global correlation coefficient between real seismic and synthetic seismic reflection data reached 85%. For region 1, the overburden region the correlation coefficient was 80%; for region 2, the potential reservoir region the correlation coefficient was 89% and for region 3, the underburden region the correlation coefficient was 70%. The synthetic seismic data was able to reproduce the real observed seismic reflection data in terms of the location and spatial distribution of the main geological features of interest.

The best-fit inverse Ip model (Fig. 25.13) allows the interpretation of the turbidite feature of interest in both vertical and horizontal slices. It also shows a reasonable spatial continuity pattern where it is possible to identify both large and subtle features of potential interest when appraising an unexplored sedimentary basin. Moreover it is clear that each region of the inversion grid is constrained individually by a given distribution function of Ip values. In this way we are constraining the spatial distribution of the simulated values. Since the regionalization of the area of interest is done using a geological criterion, the resulting best-fit inverse models are therefore geological consistent with the geological knowledge.

Uncertainty and risk of this unexplored area could be accessed by doing identical exercise but for different scenarios regarding the geometry of different geological units (regions) and, as well as, the Ip distributions for each one of them.

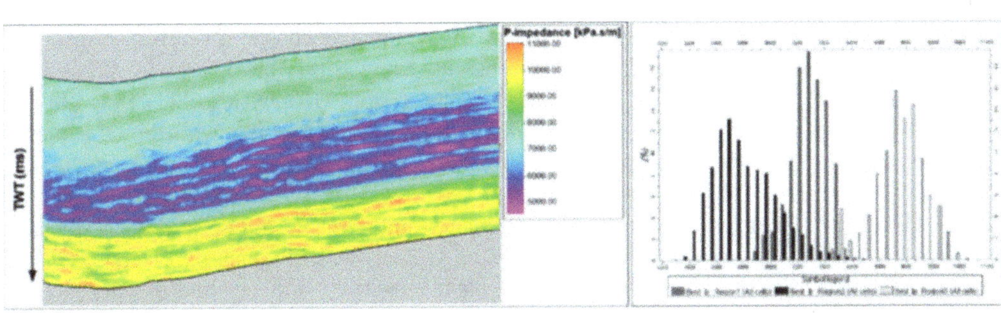


**Fig. 25.13** Best-fit inverse model of Ip retrieved after 6 iterations (left). It is possible to identify the turbidite system of interest corresponding to lower acoustic impedance values (purple). At right is the distribution function of the Best-fit inverse model of Ip, which reproduces the initial distribution function of Ip

## 25.6 Final Remarks

This chapter presents the state of the art and the most recent advances in geostatistical seismic inversion. The promising results of presented and also referenced case studies clearly show an evident maturity of these methods as privileged instruments for the integration of different types of data, particularly seismic reflection data, for the characterization and modeling of hydrocarbon reservoirs.

Very recent studies, regarding the integration of electromagnetic data and production data, show the inversion methodologies as important new paths on geostatistical tools for modelling complex geological structures.

The methodology introduced for the characterization of uncertainty and risk in early stages of exploration integrates two important components: (i) the use of analog data to generate scenarios of uncertainty regarding the morphology of geological units and the distribution of acoustic and petrophysical properties; (ii) the stochastic inversion methodologies evaluate the most probable images within each scenario and also validate (or falsify) these scenarios regarding the known seismic reality.

## References

Avseth P, Mukerji T, Mavko G (2005) Quantitative seismic interpretation. Cambridge University Press

Azevedo L, Soares A (2014) Geostatistical joint inversion of seismic and electromagnetic data. Geosciencias Aplicadas LatinoAmerica 1:45–52

Azevedo L, Soares A (2017) Geostatistical methods for reservoir geophysics. Springer

Azevedo L, Nunes R, Soares A, Neto GS (2013a) Stochastic seismic AVO inversion. In: 75th EAGE conference & exhibition, June 2013, pp 10–13

Azevedo L, Nunes R, Correia P, Soares A, Guerreiro L, Neto GS (2013b) Multidimensional scaling for the evaluation of a geostatistical seismic elastic inversion methodology. Geophysics 79(1):M1–M10

Azevedo L, Nunes R, Soares A, Mundin EC, Neto GS (2015) Integration of well data into geostatistical seismic amplitude variation with angle inversion for facies estimation. Geophysics 80(6):M113–M128

Bortoli LJ, Alabert F, Haas A, Journel AG (1993) Constraining stochastic images to seismic data. In A. Soares (ed) Geostatistics Troia'92. Dordrecht, Kluwer, pp 325–337

Buland A, Omre H (2003) Bayesian linearized AVO inversion. Geophysics 68(1):185–198

Caeiro MH, Demyanov V, Soares A (2015) Optimized history matching with direct sequential image transforming for non-stationary reservoirs. Math Geosci. https://doi.org/10.1007/s11004-015-9591-0

Caetano H (2009) Integration of seismic information in reservoir models: global stochastic inversion. PhD thesis, Instituto Superior Técnico, Universidade Técnica de Lisboa, Portugal

Castagna JP, Backus M (eds) (1993) Offset-dependent reflectivity—theory and practice of AVO analysis. Investigations in geophysics, no 8. Society of Exploration Geophysicists, Tulsa

Coléou T, Allo F, Bornard R, Hamman J, Caldwell D (2005) Petrophysical seismic inversion. SEG annual meeting 2005, Houston, US

Contreras A, Torres-Verdin C, Kvien K, Fasnacht T, Chesters W (2005) AVA stochastic inversion of pre-stack seismic data and well logs for 3D reservoir modelling. In: 67th EAGE conference & exhibition—Madrid, 13–16 June

Demyanov V, Foresti L, Christie M, Kanevski M (2011) Reservoir modelling with feature selection: a kernel learning approach. In: Proceedings of SPE reservoir simulation symposium. https://doi.org/10.2118/141510-ms

Deutsch CV, Journel AG (1996) GSLIB: geostatistical software library and user's guide, 2nd edn. Oxford University Press, Oxford

Dubrule O (2003) Geostatistics for seismic data integration in earth models. Tulsa, OK: SEG/ EAGE Distinguished Instructor Short Course Number 6

Fatti JL, Smith GC, Vail PJ, Strauss PJ, Levitt PR (1994) Detection of gas in Sandstone Reservoirs using AVO analysis: a 3-D seismic case history using the geostack technique. Geophysics 59 (9):1362–1376. https://doi.org/10.1190/1.1443695

Francis AM (2006) Understanding stochastic inversion: Part 1. First Break 24:79–84

Gómez-Hernández JJ, Journel AG (1993) Joint sequential simulation of multigaussian field, Geostatistics Troia'92, Kluwer, pp 85–94

Grammer GM, Harris PM, Eberli GP (2004) Integration of outcrop and modern analogs in reservoir modeling. AAPG Memoir 80:1–22

Grana D, Della Rossa E (2010) Probabilistic petrophysical-properties estimation integrating statistical rock physics with seismic inversion. Geophysics 75(3):O21–O37. https://doi.org/10.1190/1.3386676

Grijalba-Cuenca A, Torres-Verdin C (2000) Geostatistical inversion of 3D seismic data to extrapolate wireline petrophysical variables laterally away from the well. SPE 63283

Haas A, Dubrule O (1994) Geostatistical inversion—a sequential method of stochastic reservoir modeling constrained by seismic data. First Break 12:561–569

Horta A, Soares A (2010) Direct sequential co-simulation with joint probability distributions. Math Geosci 42(3):269–292. https://doi.org/10.1007/s11004-010-9265-x

Howell JA, Martinius AW, Good TR (2014) The application of outcrop analogues in geological modelling: a review, present status and future outlook. In: Martinius AW, Howell JA, Good TR (eds) Sediment-body geometry and heterogeneity: analogue studies for modelling the subsurface. Geological Society Special Publications, London, p 387

Lancaster S, Whitcombe D (2000) Fast-track 'coloured' inversion000. SEG Expanded Abstracts: 3–6

Lindseth RO (1979) Synthetic sonic logs—a process for stratigraphic interpretation. Geophysics 44:3–26

Liu Y, Journel AG (2009) A package for geostatistical integration of coarse and fine scale data. Comput Geosci 35(3):527–547. https://doi.org/10.1016/j.cageo.2007.12.015

Ma X-Q (2002) Simultaneous inversion of prestack seismic data for rock properties using simulated annealing. Geophysics 67(6):1877–1885. https://doi.org/10.1190/1.1527087

Mallick S (1995) Model based inversion of amplitude variations with offset data using a genetic algorithm. Geophysics 60(4):939–954. https://doi.org/10.1190/1.1443860

Marques C, Azevedo L, Demyanov V, Soares A, Christie M (2015) Multiscale geostatistical history matching using block direct sequential simulation. In: Petroleum geostatistics 2015, Biarritz, France

Martinius AW, Howell JA, Good TR (2014) Sediment-body geometry and heterogeneity: analogue studies for modelling the subsurface. Geological Society Special Publications, London, p 387

Mata-Lima H (2008) Reservoir characterization with iterative direct sequential co-simulation: integrating fluid dynamic data into stochastic model. J Pet Sci Eng 62(3–4):59–72. https://doi.org/10.1016/j.petrol.2008.07.003

Nunes R, Soares A, Neto GS, Dillon L, Guerreiro L, Caetano H, Maciel C, Leon F (2012) Geostatistical inversion of prestack seismic data. In Ninth international geostatistics congress, Oslo, Norway, pp 1–8

Oliver DS, Chen Y (2011) Recent progress on reservoir history matching: a review (February 2010), 185–221. https://doi.org/10.1007/s10596-010-9194-2

Oliver DS, Reynolds A, Liu N (2008) Inverse theory for petroleum reservoir characterization and history matching. Cambridge University Press

Pereira A, Nunes R, Azevedo L, Guerreiro L, Soares A (2017) Geostatistical seismic inversion for frontier exploration. Interpretation

Russell BH (1988) Introduction to seismic inversion methods. SEG

Shuey RT (1985) A simplification of the Zoeppritz Equations. Geophysics 50(4):609–614. https://doi.org/10.1190/1.1441936

Soares A (2001) Direct sequential simulation and cosimulation. Math Geol 33(8):911–926

Soares A, Diet JD, Guerreiro L (2007) Stochastic inversion with a global perturbation method. In: Petroleum geostatistics. EAGE, Cascais, Portugal, pp 10–14, Sept 2007

Tarantola A (2005) Inverse problem theory. Methods for data fitting and model parameter estimation. Elsevier

# Chapter 26
# Statistical Modeling of Regional and Worldwide Size-Frequency Distributions of Metal Deposits

**Frits Agterberg**

**Abstract** Publicly available large metal deposit size data bases allow new kinds of statistical modeling of regional and worldwide metal resources. The two models most frequently used are lognormal size-grade and Pareto upper tail modeling. These two approaches can be combined with one another in applications of the Pareto-lognormal size-frequency distribution model. The six metals considered in this chapter are copper, zinc, lead, nickel, molybdenum and silver. The worldwide metal size-frequency distributions for these metals are similar indicating that a central, basic lognormal distribution is flanked by two Pareto distributions from which it is separated by upper and lower tail bridge functions. The lower tail Pareto distribution shows an excess of small deposits which are not economically important. Number frequencies of the upper tail Pareto are mostly less than those of the basic lognormal. Parameters of regional metal size-frequency distributions are probably less than those of the worldwide distributions. Uranium differs from other metals in that its worldwide size-frequency distribution is approximately lognormal. This may indicate that the lognormal model remains valid as a standard model of size-frequency distribution not only for uranium but also for the metals considered in this chapter, which are predominantly mined from hydrothermal and porphyry-type orebodies. A new version of the model of de Wijs may provide a framework for explaining differences between regional and worldwide distributions. The Pareto tails may reflect history of mining methods with bulk mining taking over from earlier methods in the 20th century. A new method of estimating the Pareto coefficients of the economically important upper tails of the metal size-frequency distributions is presented. A non-parametric method for long-term projection of future metal resource on the basis of past discovery trend is illustrated for copper.

**Keywords** Pareto-lognormal distribution · Size-frequency distributions
Worldwide metal resources · Future metal supply · Model of de Wijs

---

F. Agterberg (✉)
Geological Survey of Canada, 601 Booth Street, Ottawa, ON K1A 0E8, Canada
e-mail: frits.agterberg@canada.ca; frits@rogers.com



B. S. Daya Sagar et al. (eds.), *Handbook of Mathematical Geosciences*,
https://doi.org/10.1007/978-3-319-78999-6_26

## 26.1 Introduction

Most models for regional or worldwide mineral or hydrocarbon resource appraisal assume either a lognormal or a Pareto model for the size-frequency distribution of the deposits considered. It can also be assumed that both models apply with the lognormal distribution providing a good fit to all sizes except for the smallest and largest deposits that satisfy fractal/multifractal Pareto distributions. The largest deposits obviously are rare and may be too few in number for adequate modeling in regional studies. However, recently, very large data bases have become available for metal deposits (Patiño Douce 2016a, b, c, 2017). In a newly proposed Pareto-lognormal model for worldwide metal deposit size-frequency distributions (Agterberg 2017a, b, in press), a basic lognormal distribution is flanked by two Pareto distributions. In this chapter this model is applied to copper, zinc, lead, nickel, molybdenum and silver. The upper and lower tail Pareto's are separated from the central lognormal by bridge functions to ensure continuity. An improved version of the Pareto-lognormal model will be applied to the upper tails of the size-frequency distributions for the six metals considered.

Previously, this approach was also applied to gold and uranium (Agterberg 2017b). For gold, the Pareto-lognormal model is not fully satisfied in that there is a shortage of gold deposits with sizes in the vicinity of the median of the worldwide gold size-frequency distribution. For uranium (size measured in tons of $U_3O_8$), a lognormal size-frequency distribution without Pareto tails provides a good fit. In the earlier publications (Agterberg 2017a, b, in press) comparisons were made between regional and worldwide size-frequency distributions for copper and gold. Logarithmic variances of worldwide size-frequency distributions exceed those of regional distributions and worldwide separate mineral deposit-type distributions. This observation also applies to the upper tail Pareto size-frequency distributions. A new variant of the model of de Wijs, to be discussed in more detail later in this chapter, can provide a partial explanation of the fact that the worldwide basic lognormal can be regarded as a mixture of regional lognormal distributions with parameters less than those of the worldwide basic lognormals and Pareto's. For example, within the Abitibi volcanic belt on the Precambrian Canadian Shield, the largest deposits for copper and gold satisfy Pareto size-frequency distributions with Pareto parameters ($\alpha_{Cu} = 0.45$; $\alpha_{Au} = 0.88$) that are less than those of their worldwide distributions ($\alpha_{Cu} = 1.21$; $\alpha_{Au} = 1.16$) illustrating that upper tail size parameter estimates for individual metal deposits are not stochastically independent data but subject to spatial correlation.

It should be pointed out that worldwide size-frequency distributions for some metals including copper (2541 deposits) are sufficiently large so that original data (without use of parametric statistical models) can be employed for long-term projections into the future at specific cut-off metal sizes (Agterberg 2017b; also see later in this chapter). Main emphasis in this chapter will be on size-frequency distribution modeling of the upper tail Pareto distribution and its transition into the basic lognormal. This is because total amount of metal in the lower tail of each Pareto-lognormal distribution is negligibly small. For example, 1340 copper deposits with greater than median size contain 99.7% of all copper in the complete data set of 2541 deposits so that information provided by the approximately 50% smaller deposits can be neglected (cf. Patiño Douce 2016c).

Patiño Douce (2016a, b, c, 2017) has published four important papers that are helpful in planning future metal supply; showing, for example, that for copper there would be a deficit of about 2.39 $\times$ $10^9$ t (tonnes) by the end of this century if recent discovery rates are maintained. For comparison, according to the USGS Mineral Commodity Summaries (2015), proven copper reserves currently are 0.68 $\times$ $10^9$ t. According to Patiño Douce (2017), current copper resources including the estimated reserves are 2.32 $\times$ $10^9$ t whereas new demand by 2100 will be 4.70 $\times$ $10^9$ t. Consequently, estimated future copper deficit is approximately equal to currently known copper resources. Using a non-parametric statistical method, this forecast was confirmed by Agterberg (2017b) who estimated copper resources to be discovered by the end of this century at 2.77 $\times$ $10^9$ t with 95% confidence interval of $\pm$0.994 $\times$ $10^9$ that contains Patiño Douce's estimate (also see Sect. 26.5).

Patiño Douce (2016b) is accompanied by a supplementary database with sizes and grades for 20 metals. For example, his data on 2541 copper deposits were compiled from as many as 49 different sources. Patiño Douce (2016b) initially fitted lognormal distributions to the metal deposit size-frequency distributions in this data base pointing out that the logarithmic (base $e$) standard deviation ranges from about 2 to 3 for different metals, although average metal deposit sizes are greatly different. Both Patiño Douce (2016c) and Agterberg (2017a) showed that the largest deposits for different metals can be described by means of Pareto distributions. In the Pareto-lognormal metal size-frequency distribution model of Agterberg (2017a, b) the lognormal has a Pareto upper tail separated from the central lognormal by a bridge zone. This model recognizes both (1) lognormality of metal content of ore deposits from within smaller regions and those belonging to different mineral deposit types (see, e.g. Singer 2013), and (2) Pareto size-frequency distribution of the largest deposits but also for the economically unimportant smallest metal deposits that exhibit Pareto size-frequency distributions as well.

The Pareto-lognormal model for metal deposits provides an alternative to other size-frequency distribution models, which until about 30 years ago almost exclusively were based on the lognormal model. Mandelbrot (1983, p. 263) stated that oil and other natural resources have Pareto distributions and "this finding disagrees with the dominant opinion, that the quantities in question are lognormally distributed. This difference is extremely significant, the reserves being much higher under the hyperbolic than under the lognormal law." It will be seen in this chapter that size frequencies in the upper Pareto tails of the worldwide metal deposits taken for example are less than those of the basic lognormals when these are projected to the largest sizes. In this sense, the metal size frequency distributions are not "heavy-tailed". It can, however, be assumed that the Pareto represents a stable limiting form for the largest as well as the smallest metal deposits. Pareto size-frequency distribution modeling of the largest deposits has during the past 35 years been used by many authors including Drew et al. (1982) and Crovelli (1995) for oil and gas fields, and Cargill (1981), Cargill et al. (1980, 1981) and

Turcotte (1997) for metal deposits. The latter author has developed a modification of the model of de Wijs (1951) that results in a Pareto instead of a lognormal distribution. Turcotte (1997) based this model on original publications by Cargill et al. (1980, 1981) and Cargill (1981) who had assumed power-law instead of lognormal models for U.S. mercury, lode gold and copper production. Like the lognormal, the Pareto-lognormal distribution is not universally applicable to all elements, which show bimodal or multimodal size-frequency distributions when all the many different rock bodies within the Earth's crust would be considered.

The fact that uranium has lognormal distribution without Pareto tails suggests that a multiplicative form of the central limit theorem is applicable for this metal and possibly for other metals in different kinds of mineral deposits as well. A new variant of the model of de Wijs described in the next section provides a partial explanation of the fact that the basic lognormal probably can be regarded as a mixture of regional lognormals with parameters that are less than those of the worldwide basic lognormal.

## 26.2 Modified Version of the Model of de Wijs Applied to Worldwide Metal Deposits

In the original model of de Wijs (1951) for metal concentration values in blocks of rock, any block with metal concentration model $\zeta$ is repeatedly divided into halves with concentration values $(1 + d) \cdot \zeta$ and $(1 - d) \cdot \zeta$ where $d$ is the coefficient of dispersion which us assumed to be independent of block size. The frequency distribution for metal concentration values in increasingly smaller blocks then satisfies the so-called logbinomial distribution that rapidly approaches lognormal form. If there are $p$ subdivisions, the logbinomial distribution of the $\binom{p}{K}$ concentration values of the resulting $n = 2^p$ blocks is

$$X(p, K) = \zeta \cdot (1 + d)^{p-K}(1 - d)^K$$

where $K$ satisfies the binomial distribution with $\mu(K) = p/2$ and variance $\sigma^2(K) = p/4$ (cf. Agterberg 1974, p. 322). This logbinomial has $\mu(X) = \zeta$ and variance $\sigma^2(X)$ approaching to:

$$\sigma^2(X) = \frac{p}{4} \cdot \left[\ln \frac{1-d}{1+d}\right]^2$$

Various modifications of the original model of de Wijs (1951) were developed by Matheron (1962), Turcotte (1997) and Agterberg (2007). These modifications were primarily concerned with randomizing the model of the Wijs (e.g. in the random-cut model), spatial realizations to account for spatial autocorrelation, maximizing $p$ (three-parameter model of de Wijs) and producing a Pareto tail (or other types of tail) on the logbinomial (e.g., as in the accelerated dispersion model, Agterberg (2007)). As discussed by Mandelbrot (1983), the model of de Wijs was the earliest example of a multifractal cascade. Lovejoy and Schertzer (2007) have pointed out that this original cascade is micro-canonical in that average metal concentration value is preserved locally at every cut. In universal multifractal theory these authors have generalized the cascade-type approach by preserving regionalized instead of strictly local averages. Their approach can result in a cascade that is largely lognormal but generates tails which are exactly Pareto-type. Here, another modified version of the original model of de Wijs (1951) is introduced as follows.

Suppose that the sizes of all deposits are combined with one another into a single very large block which is assigned to an arbitrary point in the upper part of the Earth's crust that contains metal deposits that have been or can be discovered. Suppose further that this block is divided into halves and the two smaller blocks are assigned to two points randomly located within halves of the upper part of the Earth's crust. This process can be repeated $2^p$ times. At each step, the two resulting half-blocks of metal are further divided into halves that, after every cut, are randomly assigned to successively smaller segments of the upper Earth's crust. If there are $n$ known deposits the cascade process is repeated until $n \leq 2^p$. For example, in relatively well-known parts of the Earth's crust there occur 2541 copper deposits. Suppose that $p = 12$ so that total number of subdivisions would be 4096. The 2541 copper deposits then can be regarded as a random subset of this larger population, so that the overall mean copper content value $\zeta$ and the coefficient of dispersion $d$ can be estimated. From the parameters of the straight line representing the basic copper lognormal distribution (Fig. 26.2a, see later) it follows that the logarithmic (base $e$) mean and standard deviation are $\mu = 10.445$ and $\sigma = 3.1062$. Consequently, $\zeta = \exp(\mu + \sigma^2/2) = 4.277 \times 10^9$. It then follows that $d = 0.7276$.

"Observed" frequencies satisfying the log-binomial model are shown in Fig. 26.1. The best-fitting straight line ($y = 0.755x - 3.8123$) in this diagram has coefficients corresponding to mean $\mu' = 11.627$ and standard deviation $\sigma' = 3.050$ which are relatively poor estimates in comparison to the values to derived later for the basic lognormal for copper in Fig. 26.2a. Main reason for this minor discrepancy is relatively strong influence on the best-fitting regression line of logbinomial frequencies represented by first and last points which are for single blocks only. Positions of these two points illustrate that the logbinomial produces slightly weaker upper and lower tails in comparison with the lognormal. On the whole, the logbinomial very closely approximates the lognormal in this application.

The preceding model would allow for spatial autocorrelation of metal deposit size observations, which is known to exist. For example, the largest copper deposits are porphyry type and largely clustered in the Andes mountain chain of South America. On the other hand, the largest copper deposits in the Abitibi volcanic belt on the Canadian Shield are volcanogenic massive sulphide deposits which are smaller than the South American porphyry coppers. Because of the close resemblance of the

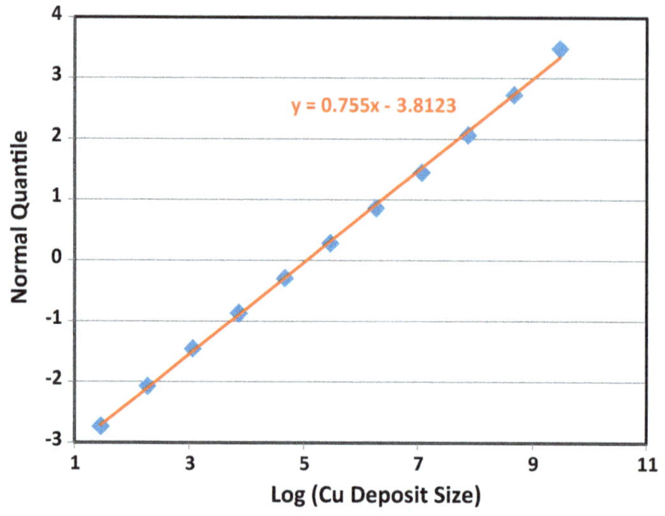


**Fig. 26.1** Model of de Wijs applied to worldwide copper deposit size-frequency distribution. Overall mean set equal to $\zeta = 4.277$ Mt copper; dispersion index $d = 0.7276$; number of subdivisions $p = 12$. "Observed" frequencies satisfy log-binomial model. Best-fitting straight line represents lognormal distribution. Logbinomial frequencies represented by first and last point are for single blocks only (*Source* Agterberg, in press)

logbinomial to the lognormal, preceding results also can be represented as follows. The characteristic function of a random variable $X$ is:

$$g(u) = E(e^{iux}) = \int_{-\infty}^{\infty} e^{-iux} f(x)dx$$

where $f(x)$ is the probability density function of $X$. Characteristic functions are discussed in statistical textbooks including Billingsley (1986) and Bickel and Doksum (2001). For a normal distribution:

$$g(u) = e^{i\mu u - \sigma^2 u^2/2}$$

If $Z$, with mean $\mu_z$ and variance $\sigma_z^2$, represents the sum of two random variables $X$ and $Y$, then the respective three characteristic functions satisfy:

$$g_z(u) = g_x(u) \cdot g_y(u)$$

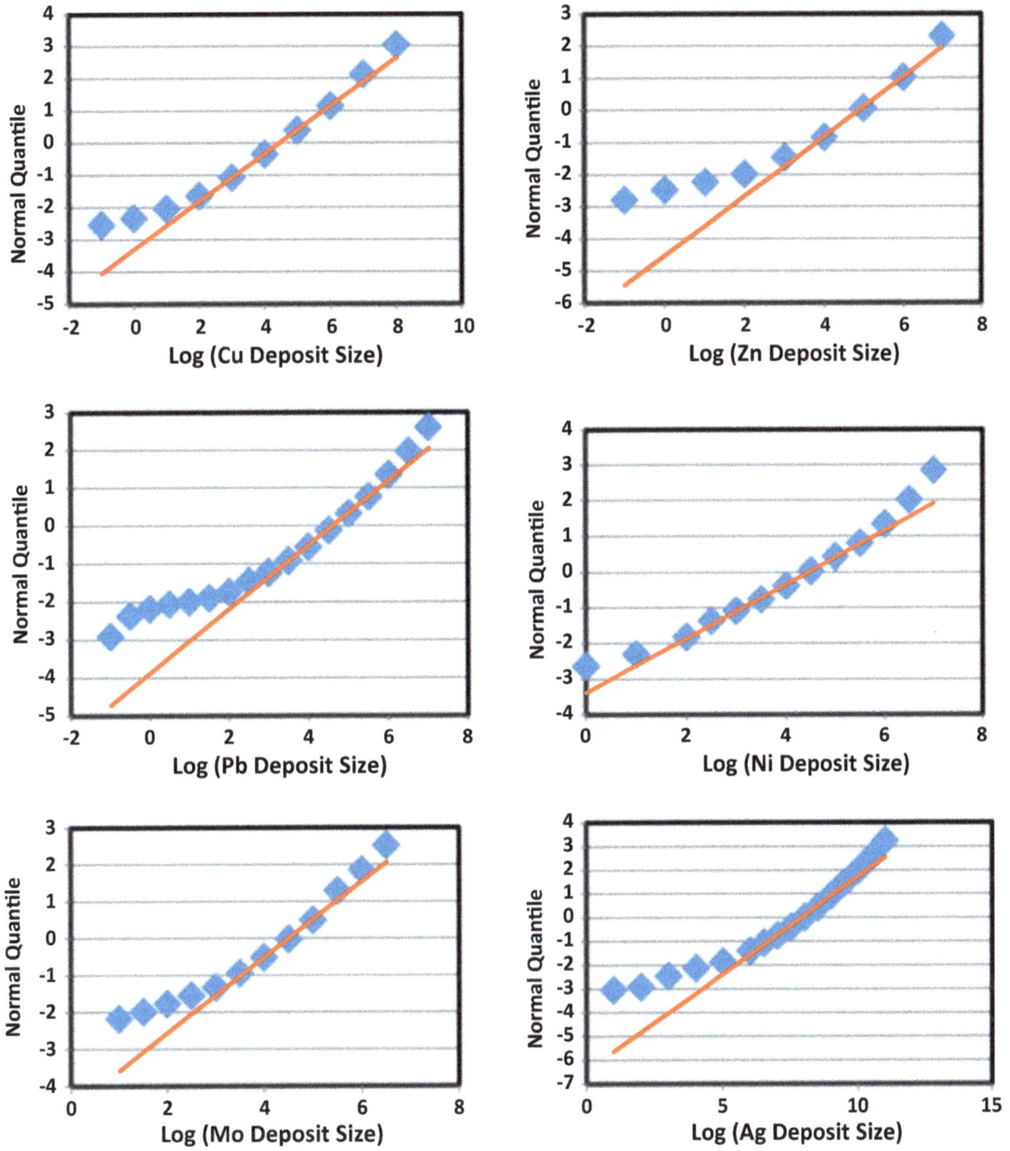


**Fig. 26.2** Lognormal *Q-Q* plots for six metals (Cu, Zn, Pb, Ni, Mo and Ag). Coefficients of straight lines representing truncated lognormal distributions are shown in Table 26.1. Sample sizes are shown in Table 26.2. In each case, frequencies for the largest and smallest deposits deviate from the straight-line pattern indicating lower and higher number frequencies than expected on the basis of the lognormal size frequency distribution models represented by the straight lines

If $X$ is normal with zero mean and variance $\sigma_x^2$, and $Y$ is normal as well with mean $\mu_y$ and variance $\sigma_y^2$, then $Z$ is normal with:

$$g_z(u) = e^{\left[i(\mu_x + \mu_y) \cdot u - (\sigma_x^2 + \sigma_y^2) \cdot u^2/2\right]}$$

Consequently, the probability density function of $Z$ is:

$$f(z) = \frac{1}{\sqrt{\sigma_x^2 + \sigma_y^2} \cdot \sqrt{2\pi}} e^{-\{z-(\mu_x+\mu_y)\}^2 \cdot \{2 \cdot (\sigma_x^2+\sigma_y^2)\}^{-1}}$$

Interpretation of this result in the context of worldwide metal deposits can be as follows. Suppose that log $Z$ represents the basic lognormal metal deposit size-frequency distribution with logarithmic mean $\mu_z = \mu_x + \mu_y$ and logarithmic variance $\sigma_z^2 = \sigma_x^2 + \sigma_y^2$. Then log $Z$ can be regarded as a composite of many regional lognormal distributions with different means and lesser logarithmic variances, much like as in the previous version of the model of de Wijs the overall logbinomial would consist of regional logbinomials with different parameters.

## 26.3 Theory and Applications of the Pareto-Lognormal Model

The cumulative frequency distribution for the Pareto-lognormal distribution $F(x) = F(\log x)$ can be written as

$$F(\log x) \approx \Phi\left(\frac{\log x - \mu}{\sigma}\right) + H(\log x - \mu) \cdot B_1(\log x) \cdot (\log x - \mu)^{-\alpha}$$
$$+ H(\mu - \log x) \cdot B_2(\log x) \cdot (\mu - \log x)^{-\kappa}$$

where $\Phi\left(\frac{\log x - \mu}{\mu}\right)$ represents the basic lognormal (logs base 10). $H$ (…) is the Heaviside function that applies to two filtered Pareto distributions, for positive and negative values of (log $x$ - $\mu$), respectively; it signifies that values at the other side of $\mu$ are set equal to zero when the equation is applied to either the upper tail or the lower tail of the Pareto-lognormal distribution.. The bridge functions $B_1$(log $x$) and $B_2$(log $x$) span relatively short intervals between the basic lognormal and the Pareto distributions for the largest and smallest values, respectively. They satisfy $\lim_{x\to\infty} B_1(\log x) = \lim_{x\to 0} B_2(\log x) = 1$ and $\lim_{x\to 0} B_1(\log x) = \lim_{x\to\infty} B_2(\log x) = 0$.

The Pareto-lognormal probability density function $f$(log $x$) corresponding to $F$(log $x$) can be written as

$$f(\log x) \approx \varphi\left(\frac{\log x - \mu}{\sigma}\right) + H(\log x - \mu) \cdot B'_1(\log x) \cdot (\log x - \mu)^{-\alpha-1}$$
$$+ H(\mu - \log x) \cdot B'_2(\log x) \cdot (\mu - \log x)^{-\kappa-1}$$

It may be useful for prediction of resources to be discovered in the future. The exponents in $(\log x - \mu)^{-\alpha-1}$ and $(\mu - \log x)^{-\kappa-1}$ reflect the fact that the Pareto probability density function remains linear on a plot with logarithmic scales for both frequency and deposit size, but has a steeper dip.

The lognormal *QQ*-plot (logarithmic probability paper) provides a useful first step in fitting the Pareto-lognormal distribution. Figure 26.2 contains results for the six metals. Original data were taken from Patiño Douce (2016b). Each graph shows a straight-line pattern with departures from lognormality in the upper and lower frequency distribution tail. Relatively, there are too many smallest deposits and too few largest deposits. In the Pareto-lognormal model both the upper and lower tail distributions have transitions to the central lognormal that are gradual and described by the two bridge functions. For projections into the future (or for global downward projections into the Earth's crust) only the upper tails of the size-frequency distributions are of economic interest. In the next section, a new, relatively simple method will be described for fitting the upper tail Pareto distributions. The upper tail bridge function will be fitted empirically by connecting this Pareto to the central lognormal distribution. Copper can be used for illustrating details of the methods used. The straight line $y = bx + a$ in Fig. 26.2a for copper represents the basic lognormal with coefficients $a = -3.314$ and $b = 0.741$ derived from the logarithmic mean $\mu = -a/b = 4.469$ and standard deviation $\sigma = 1/b = 1.349$ of a truncated lognormal for which 10% (or 254 values) in both upper and lower tail were excluded from the sample of 2541 original copper deposit size values. The mean $\mu$ of this truncated distribution is only slightly different from 4.403 representing the logarithmic mean of all values. The basic lognormal standard deviation $\sigma = 1.349$ is slightly less than 1.423 representing the standard deviation based on all values because there are relatively many copper deposit size values in the lower tail. It was obtained by dividing 0.893 representing the standard deviation of the truncated copper data set by 0.662, representing a value taken from Johnson and Kotz (1970, Table 10, p. 84). Other published truncation correction factors were used for metals with wider upper or lower tails. Coefficients for all six straight lines shown in Fig. 26.2 are given in Table 26.1. The basic statistics estimated for all six metals shown in Table 26.2 were taken from Agterberg (2017a, b and in press) except for the upper tail Pareto coefficients with slightly different values newly derived by the method to be described in the next section.

**Table 26.1** Constants $a$ and $b$ in equations $y = a + bx$ for straight lines shown in Fig. 26.2 representing truncated lognormals for six metals

| Metal | $a$ | $b$ |
|---|---|---|
| Cu | −3.314 | 0.741 |
| Zn | −4.538 | 0.924 |
| Pb | −3.894 | 0.847 |
| Ni | −3.400 | 0.758 |
| Mo | −4.616 | 1.030 |
| Ag | −6.501 | 0.819 |

**Table 26.2** Comparison of basic statistics for eight metals including the six metals represented in Table 26.1 and Figs. 26.2, 26.3 and 26.4. N—number of deposits; Mt—million tons, t metric tons; LM, LS—logarithmic mean and standard deviation; $\mu$, $\sigma$—ditto for truncated lognormal; $\alpha$, $\kappa$—upper and lower tail Pareto coefficients

| Metal | N | Total metal | Mean | LM | LS | $\mu$ | $\sigma$ | $\alpha$ | $\kappa$ |
|---|---|---|---|---|---|---|---|---|---|
| Cu | 2541 | 2319.11 Mt | 0.9127 Mt | 4.403 | 1.423 | 4.469 | 1.349 | 1.206 | 0.332 |
| Zn | 1476 | 1111.51 Mt | 0.7531 Mt | 4.821 | 1.215 | 4.910 | 1.082 | 1.162 | 0.318 |
| Pb | 1102 | 481.43 Mt | 0.4367 Mt | 4.479 | 1.337 | 4.596 | 1.180 | 1.654 | 0.340 |
| Ni | 464 | 171.05 Mt | 0.3686 Mt | 4.384 | 1.261 | 4.484 | 1.319 | 1.352 | 0.515 |
| Mo | 343 | 59.76 Mt | 0.1742 Mt | 4.371 | 1.131 | 4.480 | 0.971 | 1.093 | 0.358 |
| Ag | 1644 | 1899.43 t | 1.1554 t | 7.832 | 1.342 | 7.936 | 1.221 | 1.382 | 0.361 |
| Au | 2106 | 284.33 t | 0.1350 t | 6.551 | 1.168 | 6.629 | 0.994 | 1.164 | 0.297 |
| U | 172 | 59.76 Mt | 0.3474 Mt | 2.979 | 1.177 | 2.979 | 1.177 | 0.000 | 0.000 |

## 26.4 Upper Tail Pareto Distribution and Its Connection to the Basic Lognormal Distribution

The cumulative Pareto distribution function satisfies

$$F(x) = 1 - \left(\frac{k}{x}\right)^{\alpha}$$

where $\alpha > 0$ and $k > 0$ are its two parameters. The following maximum likelihood estimator of the Pareto coefficient $\alpha$ has been used in several publications (Clauset et al. 2009; Patiño Douce 2016c; Agterberg 2017b) in various ways:

$$\alpha = \frac{n}{\sum_{i=1}^{n} \ln \frac{x_i}{k}}$$

where $n$ represents number of metal deposits selected in an ordered sequence of values $x_i$ ($i = 1, 2, \ldots, n$), and $k$ is the critical size parameter representing the truncation point at which the maximum value of the Pareto probability—density drops to zero. In the original algorithm of Clauset et al. (2009), which was used by Patiño Douce (2016c), all possible values of $k$ are tested for sizes $x_1 < x_2 < x_3 \cdots < x_n$. Minimum size ($x_1$) was set at median size and $x_n$ at maximum size. Each sample of $n$ sizes provides a different estimate of $k$ and $\alpha$. The Kolmogorov-Smirnov test was used to find the Pareto distribution that provides the best fit.

In Agterberg (2017b)'s application, $x_1 > x_2 > x_3 \cdots > x_n$, was used instead. This reversal of order was based on the following three premises: (1) worldwide metal deposit size sample sizes are very large ensuring that cumulative frequencies become increasingly precise when $n$ is increased, regardless of whether or not the Pareto distribution model is satisfied; (2) starting with the largest deposits and increasing sample size by including progressively more deposits improves results if the Pareto distribution model would indeed be satisfied; and (3) for increasingly large values of $n$, observed frequencies become increasingly less than expected Pareto distribution model frequencies because the upper tail Pareto gradually passes into the lower frequency density basic lognormal via the upper tail bridge function. Theoretically, if $\alpha$ is known, $k$ could be derived from $\alpha$ by using the preceding equation for the maximum likelihood estimator. In Agterberg (2017a, b, in press), $\alpha$ was pre-determined by visual inspection for 7 metals that all show approximately linear patterns in log rank—log size plots for their largest deposits.

Initially, for small values of $n$, the resulting patterns for copper and other metals show large random fluctuations. For larger values of $n$ the plots develop multi-peak patterns for $\alpha$ that are superimposed on a gradational decrease. In Agterberg (2017b) a straight line was fitted by least squares for copper and gold avoiding the large small-sample fluctuations at the largest size values end capture the downward bend of log rank values toward lower log size values. This procedure produced estimates of $k_{Cu} = 6.98$ and $k_{Au} = 8.98$. Both estimates were confirmed by more detailed analysis of cumulative frequencies for largest deposits yielding $k_{Cu} \approx 7.0$ and $k_{Au} \approx 9.2$.

However, the preceding method does not work very well for some metals with fewer data than copper and gold. The following relatively simple method gave good results for six metals as shown in Figs. 26.3 and 26.4. The value of $n$ was set equal to 20 in each application for a window that was slid along the series of ordered metal deposit size values from the largest deposit downward. Initial random fluctuations connected to the largest values were avoided and so were windows on the upper bridge function transition zone toward the basic lognormal size-frequency distribution. For copper this procedure gives $\alpha = -1.2059$ for $k = 6.996$. The straight line with slope $\alpha$ passing through the point with average log size and average log rank for the 20 pairs of copper deposit size values used is shown in Fig. 26.3a. Similar results for the other five metals are shown in Figs. 26.3 and 26.4. According to the Pareto-lognormal model, a decrease of estimated values of $\alpha$ at the point where the upper tail Pareto ceases to be applicable is indeed expected. However, it is not clear why there is an equally strong decrease of estimated values of $\alpha$ in the patterns of Fig. 26.3 from the peak outward toward increasing values of log (deposit size). Very large random fluctuations are known to exist for the largest deposits. However, the upper tail downward trends in Fig. 26.3 could mean larger sizes than expected for the largest deposits although there are no indications of this in Fig. 26.4. Neither are there obvious deviations from linearity in log rank—log size plots that include the largest deposits for various metals (Agterberg 2017a, b). Residuals from the straight lines representing the Pareto distributions show relatively strong autocorrelation. Because of this uncertainty, it remains important to look for alternative upper tail models like the lognormals proposed by Patiño Douce (2016c, 2017) and shown for copper and gold in Agterberg (2017b). These alternate lognormals differ from the basic lognormals primarily in that they have much large mean deposit size values.

In order to fully represent the upper tail cumulative size-frequency distribution, the Pareto's have to be connected to the basic lognormals. Taking copper for

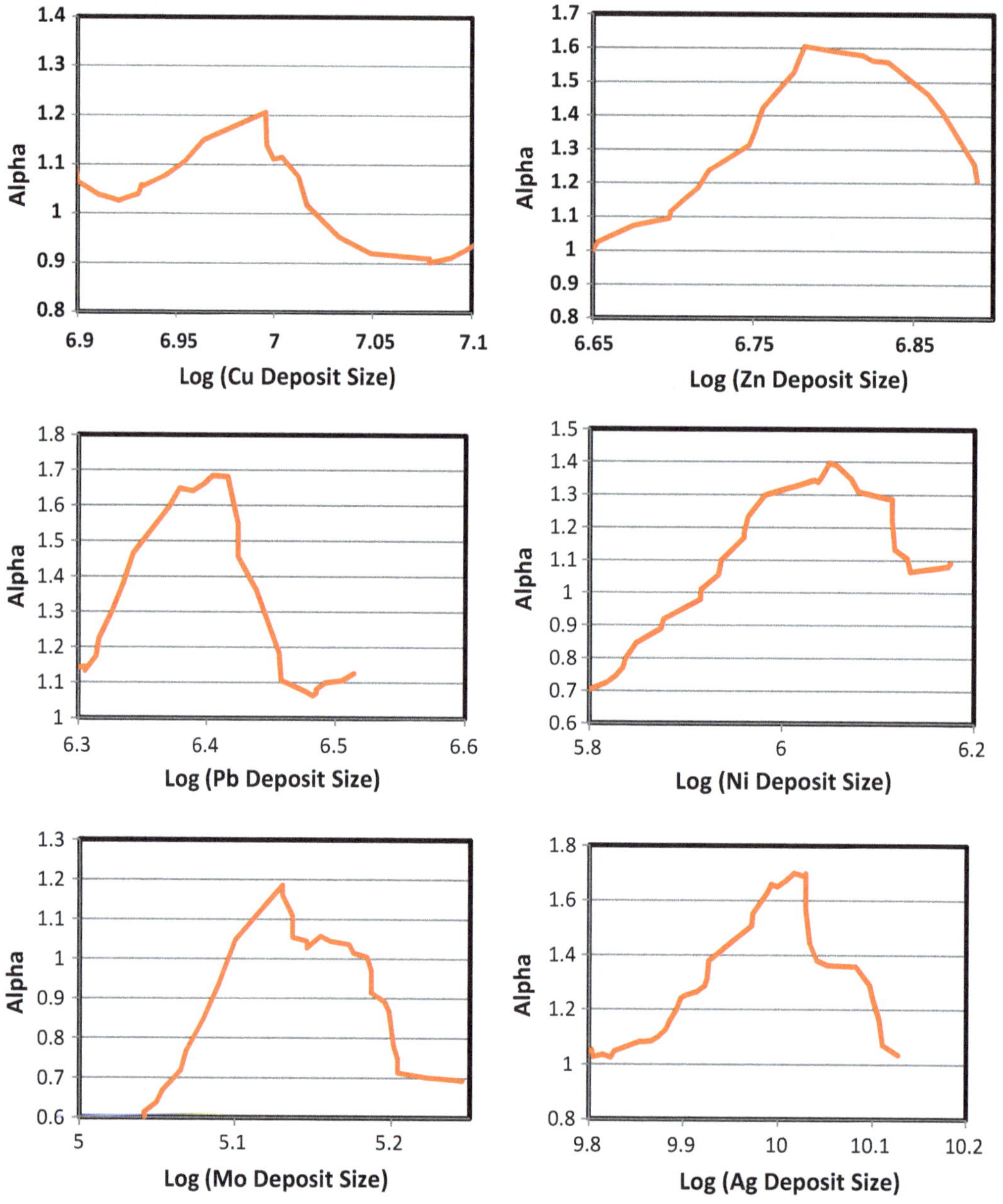


**Fig. 26.3** Pareto coefficient ($\alpha$) for log of metal deposit size as obtained in the text, setting $n$ equal to 20 for overlapping data sets moving from larger to smaller log (deposit size) values. Maximum $\alpha$ will be taken as optimum value with data sets, on which it is based, for the six metals shown in Fig. 26.4

example again, it can be seen in Fig. 26.2a that observed frequencies deviate from the best-filling straight line for log Cu deposit size values greater than 6. In total 42 deposits have log Cu deposit size values greater than 7 and their observed cumulative frequency of 42/2524 can be used as an anchor point to connect the upper tail Pareto to the upper bridge function which represents the transition zone between the basic lognormal (for values $< 6$) and the Pareto (for values $\geq 7$). Table 26.3 shows anchor points used for all six metals considered. Figure 26.5 shows best-fitting

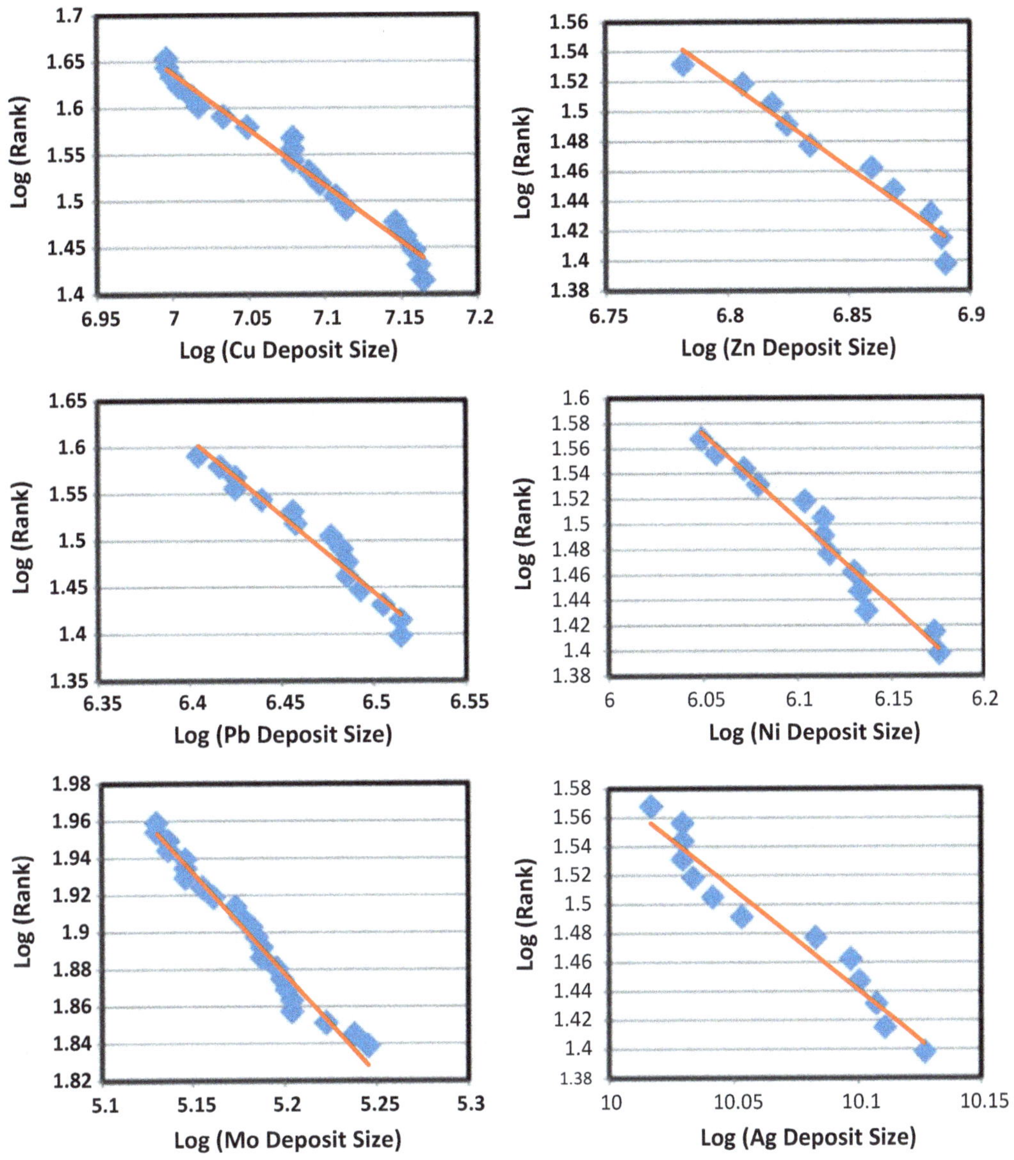


**Fig. 26.4** Sets of twenty log (metal deposit size) values corresponding to maximum value of $\alpha$ in Fig. 26.3 for the six metals. Corresponding Pareto distribution functions are shown as straight lines on these log rank—log size plots

**Table 26.3** Pareto coefficients $\alpha$ and $k$ corresponding to peaks for six metals in Fig. 26.3. Anchor point is log metal deposit size on upper tail Pareto distribution with relative frequency (Rel Freq) and observed $y$-value—$\log_{10}$ (1—cumulative frequency)

| Metal | α | $k$ | Anchor pnt | Rel Freq | Obs $y$-value |
|---|---|---|---|---|---|
| Cu | 1.2059 | 6.996 | 7 | 42 | −1.7818 |
| Zn | 1.1615 | 6.782 | 7 | 15 | −1.9930 |
| Pb | 1.6535 | 6.405 | 6.5 | 27 | −1.6108 |
| Ni | 1.3523 | 6.049 | 6 | 42 | −1.0433 |
| Mo | 1.0926 | 5.130 | 5 | 106 | −0.5286 |
| Ag | 1.3820 | 10.017 | 10 | 38 | −1.6361 |

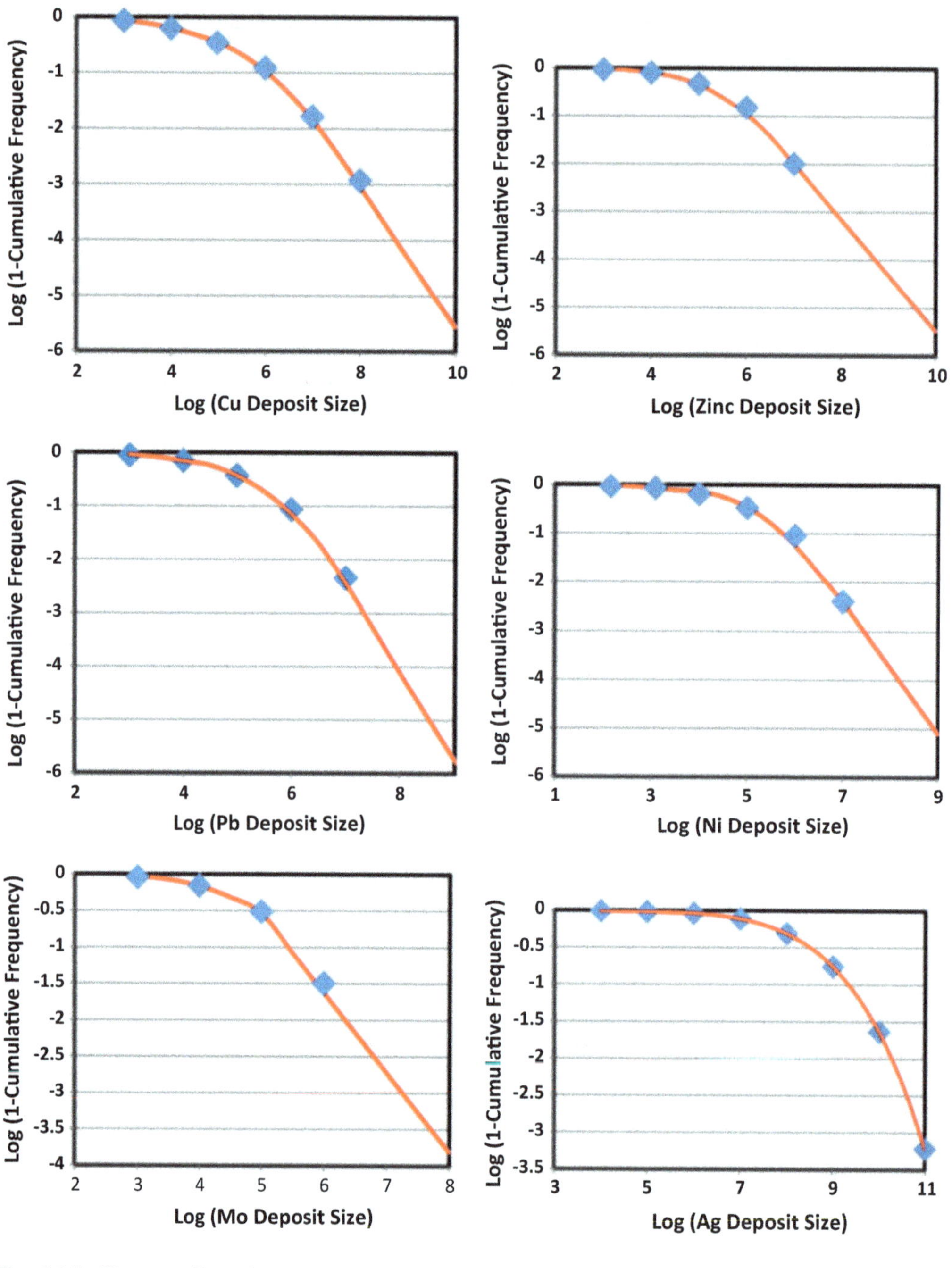


**Fig. 26.5** Upper tails of Pareto-lognormal size-frequency distributions for the six metals constructed by using the method explained in Table 26.4. Upper tails bridge functions are smooth curves that satisfy quartic polynomials fitted by least squares to log size values satisfying basic lognormal on the left and upper tail Pareto on the right side. For copper the result does not differ significantly from sextic polynomial previously shown in Agterberg (2017b, Fig. 14). For molybdenum no bridge function was fitted. Points with log (Mo deposit size) $\leq 5$ satisfy basic lognormal shown as straight line on Fig. 26.2e; points with log (Mo deposit size) $\geq 5$ belong to the upper tail Pareto distribution

frequency distribution curves that are Pareto-type for log deposit size values exceeding the anchor points. Some anchor points slightly exceed the estimated values of the truncation parameters $k$ without significantly changing the results. Quartic polynomials were used to approximate the smooth shapes of each frequency distribution within the upper tail bridge function that connects the Pareto with the basic lognormal. Table 26.4 shows results of this interpolation procedure for copper. The curve in Fig. 26.5a resembles the curve previously shown in Agterberg (2017a) where it was a best-fitting sextic polynomial. Contrary to the fitting of sextic polynomials to other metals, the method using a quartic explained in Table 26.4 gave good results for the other metals considered with the exception of molybdenum that does not seem to need a bridge function to pass from the Pareto into the basic lognormal. It is the only metal for which the upper tail Pareto and the central lognormal almost continuously pass into one another. Molybdenum, therefore, almost exactly satisfies the model proposed by Patiño Douce (2016b, Appendix 1) in which the probability density function of the lognormal as well as its first derivative pass continuously into the density function of the Pareto. The value at log (Mo deposit size) = 5 predicted by the basic lognormal is equal to the value of the Pareto at this point. Figure 26.5e, however, shows that there is a slight change of dip of the curve for log (1—cumulative frequency) at this point. All frequency distribution curves in Fig. 26.5 are close to their observed cumulative frequencies also shown in these diagrams.

## 26.5 Prediction of Future Copper Resources

As previously pointed out in Agterberg (2017b; in press), one of the purposes of developing statistical models of the size-frequency distributions of worldwide metal deposits is to use these models for prediction purposes either spatially (e.g., from relatively well-explored regions to unexplored regions, or deeper down from the Earth's surface), or in time. For multifractal modeling of the spatial distribution of mineral deposits, see Cheng (1994) or Cheng and Agterberg (1995). Use of parametric models is discussed by many authors including Agterberg (1974), Patiño Douce (2017) and Agterberg (2017b). The following non-parametric approach was first presented in the latter paper.

Suppose that $X$ is a continuous random variable denoting mineral deposit size and that $K$ is a discrete random variable for number of deposits per unit of area, volume or time; then the continuous random variable $Y$ representing the sum of the sizes of the $K$ deposits satisfies:

$$Y = X_1 + X_2 + \cdots + X_K$$

**Table 26.4** Curve connecting smoothed $y$-values for $\log_{10}$ (1—cumulative frequency) in Fig. 26.5a in comparison with observed $y$-values for copper. Commencing as lognormal, the curve passes gradually into the straight line for its upper tail Pareto. Smoothed $y$-values for the intermediate bridge function satisfy a quartic polynomial equation fitted by least squares to lognormal values for $x \leq 5$ and $x \geq 7$. Smoothed values include quartic polynomial for $x = 5$ and $x = 7$

| $x$ | Lognormal | Pareto | Quartic | Smoothed | Observed |
|---|---|---|---|---|---|
| 4 | −0.196 | | −0.199 | −0.196 | −0.198 |
| 4.5 | −0.309 | | −0.298 | −0.309 | |
| 5 | −0.460 | | −0.449 | −0.449 | −0.460 |
| 5.5 | −0.653 | | −0.671 | −0.671 | |
| 6 | −0.892 | | −0.974 | −0.974 | −0.908 |
| 6.5 | −1.180 | | −1.364 | −1.364 | |
| 7 | −1.518 | −1.782 | −1.838 | −1.838 | −1.782 |
| 7.5 | −1.909 | −2.411 | −2.388 | −2.411 | |
| 8 | −2.354 | −3.041 | −2.999 | −3.041 | −2.928 |
| 8.5 | −2.853 | −3.670 | −3.650 | −3.670 | |
| 9 | −3.408 | −4.299 | −4.313 | −4.299 | |

The mean $E(Y)$ and variance $\sigma^2(Y)$ satisfy:

$$E(Y) = E(K) \cdot E(X); \quad \sigma^2(Y) = E(K) \cdot \sigma^2(X) + \sigma^2(K) \cdot E^2(X)$$

These equations were previously used in Agterberg (1974, Eq. 7.72) who had adopted them from Feller (1968, Chap. 12) where they are derived for $K$ and $X$ both representing integral-valued random variables. The approach also is applicable when $X$ is a continuous random variable. The variance equation can be found in an online article on compound distributions (Lin 2014, Eq. (4)) with many additional references. Specific distribution models can be assumed to hold true for $K$ and $X$. However, as shown earlier in this chapter, significant uncertainties remain in modeling the upper tail of worldwide metal size-frequency distributions that contain most metal. Fortunately, samples now available for statistical modeling are so large that the following non-parametric approach can be used.

Patiño Douce (2017) contains tables with statistics based on number of 1950–2007 copper deposit discoveries originally derived from a plot by Schodde (2010) for copper deposits with size > 3 × 10$^5$ t Cu. Mean and variance of yearly number of discoveries are 8.621 and 14.304, respectively. Extrapolation of these two parameters over 85 years, toward the end of this century, would yield an expected number of 732.8 discoveries with variance of 12.158 × 10$^3$. Patiño Douce (2016b)'s original data base contains 591 copper deposits with sizes > 3 × 10$^5$ t Cu resulting in estimated values of $E(X) = 3.784 \times 10^6$ t and $\sigma^2(X) = 1.135 \times 10^{14}$. Because of the large sample size, the 95% confidence limits on the estimated mean value are $3.784 \times 10^6 \pm 0.859 \times 10^6$ t with the large sample ensuring approximate normality of the frequency distribution of this mean. Consequently, this estimate is rather precise. Using the preceding equations for mean $E(Y)$ and variance $\sigma^2(Y)$, it follows that estimated total tonnage copper value amounts to $732.8 \times 3.784 \times 10^6 = 2.773 \times 10^9$ t. The corresponding variance amounts to $25.726 \times 10^{16}$, from which it follows that the 95% confidence limits on the estimated mean value are $2.773 \times 10^9 \pm 0.994 \times 10^9$ t. This mean value is approximately normally distributed as well. Although the method for deriving this result differs significantly from the computer simulation method used by Patiño Douce (2017), the end result is only $0.654 \times 10^9$ t greater and the difference between the two estimates is not statistically significant. These results confirm Patiño Douce (2017)'s conclusion that there would be a significant shortage of copper if current rates of discovery will be maintained. The problem would become even worse if future rates would decrease.

## 26.6 Concluding Remarks

In this chapter it was argued that publicly available large metal deposit size data bases (especially Patiño Douce 2016b) allow new kinds of statistical modeling of regional and worldwide metal resources. The two models most frequently used in the past are lognormal size-grade and Pareto upper tail modeling. Both approaches are probably valid for several metals including copper, zinc, lead, nickel, molybdenum and silver taken for example because the upper tails of their mostly lognormal size frequency distributions satisfy the Pareto distribution model. The worldwide metal size-frequency distributions for these metals are similar indicating that a central, basic lognormal distribution is flanked by two Pareto distributions from which it is separated by upper and lower tail bridge functions. The lower tail Pareto distribution shows an excess of small deposits which are not economically important. Number frequencies of the upper Pareto are mostly less than those of the basic lognormal. A new method for fitting the upper tail Pareto was introduced and produces good results for the six metals taken for example. Parameters of regional metal size-frequency distributions as well as those of mineral deposit type distributions are less than those of the worldwide distributions. Uranium differs from other metals in that its worldwide size-frequency distribution is approximately lognormal. This may indicate that the lognormal model remains a standard model of size-frequency distributions of metals predominantly mined from hydrothermal and porphyry-type orebodies. A new version of the model of de Wijs may provide a framework for explaining the differences between regional and worldwide distributions. Further research on this topic remains to be carried out. The Pareto tails may reflect historical mining methods with bulk mining becoming prevalent in the 20th century. A new method of estimating the Pareto coefficients of the economically important upper tails of the size-frequency distributions was presented, and a non-parametric method for long-term projection of future metal resource on the basis of past discovery trend was illustrated for copper.

## References

Agterberg FP (1974) Geomathematics. Elsevier, Amsterdam

Agterberg FP (2007) Mixtures of multiplicative cascade models in geochemistry. Nonlinear Process Geophys 14:201–209

Agterberg FP (2017a) Pareto-lognormal modeling of known and unknown metal resources. Nat Resour Res 26:3–20 (with erratum on p 21)

Agterberg FP (2017b) Pareto-lognormal modeling of known and unknown metal resources. II. Method refinement and further applications. Nat Resour 26:265–283

Bickel PJ, Doksum KA (2001) Mathematical statistics, 2nd edn. Prentice-Hall, Upper Saddle River, New Jersey

Billingsley P (1986) Probability and measure, 2nd edn. Wiley, New York

Cargill SM (1981) United States gold resource profile. Econ Geol 76:937–943

Cargill SM, Root DH, Bailey EH (1980) Resources estimation from historical data: mercury, a test case. Math Geol 12:489–522

Cargill SM, Root DH, Bailey EH (1981) Estimating unstable resources from historical industrial data. Econ Geol 76:1081–1095

Cheng Q (1994) Multifractal modeling and spatial analysis with GIS: gold mineral potential estimation in the Mitchell-Sulphurets area, northwestern British Columbia: unpublished doctoral dissertation. University of Ottawa, Canada

Cheng Q, Agterberg FP (1995) Multifractal modeling and spatial statistics. Math Geol 28(1):1–16

Clauset A, Shalizi CR, Newman MEJ (2009) Power-law distributions in empirical data. SIAM Rev 51:661–703

Crovelli RA (1995) The generalized 20/80 law using probabilistic fractals applied to petroleum field size. Nonrenewable Resour 4(3):223–241

De Wijs HJ (1951) Statistics of ore distribution, I. Geol Mijnbouw 30:365–375

Drew LJ, Schuenemeyer JH, Bawlee WJ (1982) Estimation of the future rates of oil and gas discoveries in the western Gulf of Mexico, US Geological Survey Professional Paper 1252. GPO, Washington, DC, US

Feller W (1968) An introduction to probability theory and its applications, vol. I, 3rd edn. Wiley, New York

Johnson NL, Kotz S (1970) Continuous univariate distributions-1. Wiley, New York

Lin XS (2014) Compound distributions. Wiley StatsRef: Statistics Reference Online http://dx.doi.org/10.1002/9791118445112.stat04411

Lovejoy S, Schertzer D (2007) Scaling and multifractal fields in the solid Earth and topography. Nonlinear Process Geophys 14:465–502

Mandelbrot BB (1983) The fractal geometry of nature. Freeman, San Francisco

Matheron G (1962) Traité de Géologique Appliquée. Mémoire BRGM 14. Paris, Éditions Technip

Patiño Douce AE (2016a) Metallic mineral resources in the twenty first century. I. Historical extraction trends and expected demand. Nat Resour Res 25:71–90

Patiño Douce AE (2016b) Metallic mineral resources in the twenty first century. II. Constraints on future supply. Nat Resour Res 25:97–124

Patiño Douce AE (2016c) Statistical distribution laws for metallic mineral deposit Sizes. Nat Resour Res 25:365–387

Patiño Douce AE (2017) Loss distribution model for metal discovery probabilities. Nat Resour Res 26:241–263

Singer DA (2013) The lognormal distribution of metal resources in mineral deposits. Ore Geol Rev 55:80–86

Singer DA, Menzie DW (2004) Quantitative mineral resource assessments. Oxford University Press

Schodde R (2010) The declining discovery rate—What is the real story? https://www.minexconsulting.com/publications/mar2010a.html

Turcotte DL (1997) Fractals and chaos in geology and geophysics, 2nd edn. Cambridge University Press

USGS Mineral Commodity Summaries (2015). http://minerals.usgs.gov/minerals/pubs/mcs2015.pdf

# Part IV
# Reviews

# Chapter 27
# Bayesianism in the Geosciences

**Jef Caers**

**Abstract** Bayesianism is currently one of the leading ways of scientific thinking. Due to its novelty, the paradigm still has many interpretations, in particular with regard to the notion of "prior distribution". In this chapter, Bayesianism is introduced within the historical context of the evolving notions of scientific reasoning such as inductionism, deductions, falsificationism and paradigms. From these notions, the current use of Bayesianism in the geosciences is elaborated from the viewpoint of uncertainty quantification, which has considerable relevance to practical applications of geosciences such as in oil/gas, groundwater, geothermal energy or contamination. The chapter concludes with some future perspectives on building realistic prior distributions for such applications.

## 27.1 Introduction

Much of the topic of research within the IAMG community involves developing tools for prediction: what is the grade? The volume of Oil in Place? The spatio-temporal changes of a contaminant plume? Making realistic predictions, meaning providing realistic uncertainty quantification, is key to making informed decisions. Decisions and their consequences are what matters in the end, not the kriging map of gold, or simulated permeability, or hydraulic conductivity. These are only intermediate steps to decision-making. In this chapter, I focus on a fundamental discussion on how we make predictions in the Geosciences and about the current leading paradigm: Bayesianism. This chapter is a revised version of the book "Quantifying Uncertainty in Subsurface Systems", Scheidt et al. Wiley Blackwell, 2018. The term UQ is therefore used for "Uncertainty Quantification"

Most of our applications involve three major components: data, a model and a decision. For example, in contaminant hydrology, we need to decide on a

J. Caers (✉)
Stanford University, Stanford, USA
e-mail: jcaers@stanford.edu



B. S. Daya Sagar et al. (eds.), *Handbook of Mathematical Geosciences*,
https://doi.org/10.1007/978-3-319-78999-6_27

remediation strategy or simply a decision to clean or not. We collect data: geochemical samples, geological studies, possibly even some geophysical surveys. We build models: a reactive transport model, a geostatistical model of spatial properties, a geochemical model. How does this all come together? Bayesian modeling is usually invoked as a way of integrating all these components. But what really constitutes "Bayesian" modeling? Thomas Bayes did not write Bayes' rule in the form we often see it in textbooks. However, after a long period of being mostly ignored in history, his idea of using a "prior" distribution heralded a new way of scientific reasoning which can be broadly classified as: Bayesianism. The aim of this chapter is to frame Bayesianism within the historical context of other forms of scientific reasoning such as induction, deduction, falsification, intuitionism and others. The application of Bayesianism is then discussed in the context of uncertainty quantification and specific to the Geosciences. This makes sense since quantifying uncertainty is about quantifying a lack of understanding or lack of knowledge. Science is all about creating knowledge. But then, what do we understand and what exactly is knowledge (the field of epistemology)? How can this ever be quantified with a consistent set of axioms and definitions, that is, if a mathematical approach is taken? Is such quantification unique? Is it rational at all to quantify uncertainty? Are we in agreement as to what Bayesianism really is?

These questions are not just practical questions towards engineering solutions, but to a deeper discussion around uncertainty. This discussion is philosophical, a discussion at the intersection of philosophy, science and mathematics. The science of studying knowledge and as a result, uncertainty. In many papers published journals that address uncertainty in subsurface systems, or in any system for that matter, philosophical views are rarely touched upon. Many such publications would start with the "we take the Bayesian approach…" or, "we take a fuzzy logic approach to…." But what entails making this decision? Papers quickly become about algebra and calculus. Bayes or any other way of inferential reasoning is simply seen as a set of methodologies, technical tools and computer programs. The emphasis lies on the beauty of the calculus, solving the puzzle, improving "accuracy" not on any desire of deeper understanding to what exactly one is quantifying. A pragmatic realist may state that in the end, the answer is provided by the computer codes, based on the developed calculus. Ultimately, everything is about bits and bytes and transistors amplifying or switching electronic signals; inputs and outputs. The debate is then which method is better, but such debate is only within the choices of the particular way of reasoning about uncertainty. That particular choice is rarely discussed. The paradigm is blindly accepted.

Bayes is like old medicine, we know how it works, what the side effects are and has been debated, tweaked, improved, discussed since Reverend Bayes' account was published by Price (1763). Our discussion will start with a general overview of the scientific method and the philosophy of science. This discussion will be useful in the sense that it will help introduce Bayesianism, as a way of inductive reasoning, compared to very different ways of reasoning. Bayes is popular, but not accepted by all (Earman 1992; Wang 2004; Gelman 2008; Klir 1994).

## 27.2 A Historical Perspective

In the philosophy of sciences, fundamental questions are posed such as: what is a "law of nature"? How much evidence and what kind of evidence should we use to confirm a hypothesis? Can we ever confirm hypotheses as truths? What is truth? Why do we appear to rely on inaccurate theories (e.g. Newtonian physics) in the light of clear evidence that they are false and should be falsified? How does science and the scientific method work? What is science and what is not (the demarcation problem)? Associated with the philosophy of science are concepts such as epistemology (study of knowledge), empiricism (the importance of evidence), induction and deduction, parsimony, falsification, paradigm…. all of which will be discussed in this chapter.

Aristotle (384-322 BC) is often considered to be the founder of both science and the philosophy of science. His work covers many areas such as physics, astronomy, psychology, biology, and chemistry, mathematics, and epistemology. Attempting to not solely be Euro-centric, one should also mention the scientist and philosopher Ibn al-Haytham (Alhazen), who could easily be called the inventor of the peer-review system, on which this chapter too is created. In the modern era, Galileo Galilei and Francis Bacon take over from the Greek philosophy of thought (rationality) over evidence (empiricism). Rationalism was continued by Rene Descartes. David Hume introduced the problem of induction. A synthesis of rationalism and empiricism was provided by Emanuel Kant. Logical positivism (Wittgenstein, Bertrand Russel, Carl Hempel) ruled much of the early twentieth century. For example, Bertrand Russel attempted to reduce all of mathematics to logic (logicism). Any scientific theory then requires a method of verification using a logic calculus in conjunction with the evidence, to prove such theory true of false. Karl Popper appeared on the scene as a reaction to this type of reasoning, replacing verifiability with falsifiability, meaning that for a method to be called scientific, it should be possible to construct an experiment or acquire evidence that can falsify it. More recently Thomas Kuhn (and later Imre Lakatos) rejected the idea that one method dominates science. They see the evolution of science through structures, programs and paradigms. Some philosophers such as Feyerabend go even further ("Against method", Feyerabend 1993) stating that no methodological rules really exist (or should exist).

The evolution of the philosophy of science has relevance to UQ. Simply replace the concept of "theory" with "model", and observations/evidence with data. There is much to learn from how people's viewpoints towards scientific discovery differs; how they have changed and how such change has affected our ways of quantifying uncertainty. One of the aims of this chapter therefore is to show that there is not really a single objective approach to uncertainty quantification based on some laws or rules provided by a passive, single entity (the truth-bearing clairvoyant God!). Uncertainty quantification just like science is dynamic, relies on interaction between data, models and predictions and evolving views on how these components interact. It is with high certainty that few methods covered in this chapter will not be used in 100 years; just consider the history of science as evidence.

## 27.3 Science as Knowledge Derived from Facts, Data or Experience

Science has gained considerable credibility, including in everyday life, because it is sold as "being derived from facts". It provides an air of authority, of truth to what are mainly uncertainties in daily life. This was basically the view with the birth of modern science in the seventeenth century. The philosophies that exalt this view are empiricism and positivism. Empiricism states that knowledge can only come from sensory experience. The common view was that (1) sensory experience produces facts to objective observers, (2) facts are prior to theories (3) facts are the only reliable basis for knowledge.

Empiricism is still very much alive in the daily practice of data collection, model building and uncertainty quantification. In fact, many scientists find UQ inherently "too subjective" and of lesser standing than "data", physical theories or numerical modeling. Many claim that decisions should be based merely on observations, not models.

*Seeing is believing*. "Data is objective, models are subjective". If facts are to be derived from sensory experience, mostly what we see, then consider Fig. 27.1. Most readers see a panel of squares, perhaps from a nice armoire. Others (very few) see circles and perhaps will interpret this as an abstract piece of art with interesting geometric patterns. Those who don't see circles at first, need to simply look longer, with different focusing of their retinas. Hence, there seems to be more than meets the eyeball (Hanson 1958). Consider another example in Fig. 27.2. What do you see? Most will recognize this as a section of a geophysical image (whether seismic, radar etc.…). A well-trained geophysicist will potentially observe a "bright spot" which may indicate the presence of a gas (methane, carbon dioxide) in the subsurface formations. A sedimentologist may observe deltaic formations consisting of channel stacks. Hence, the experience in viewing an object is highly dependent on the interpretation of the viewer and not the pure sensory light perceptions hitting one's retina. In fact, Fig. 27.2 is a modern abstract work of art by Mark Bardford (1963) on display in the San Francisco Museum of Modern Art (September 2016).

**Fig. 27.1** How many circles do you see?

**Fig. 27.2** What do you see?

Anyone can be trained to make interpretations, and this is usually how education proceeds. Even pigeons can be trained to spot cancers as well as humans, Levenson et al., PLOS ONE (18 November 2015) http://www.sciencemag.org/news/2015/11/pigeons-spot-cancer-well-human-experts. But this idea may also backfire. First off, the experts may not do better than random (Financial times, March 31, 2013: "Monkey beats man on stock market picks", based on a study by the Cass Business School in London), or worse produce cognitive biases, as pointed out by a study of interpretation seismic images (Bond et al. 2007).

*First facts, then theory*. Translated to our UQ realm as "first data, then models". Let's consider another example in Fig. 27.3, now with actual geophysical data and not a painting. A statement of fact would then be "this is a bright spot". Then, in the empiricist view, deduction, conclusions can be derived from it ("It contains gas").

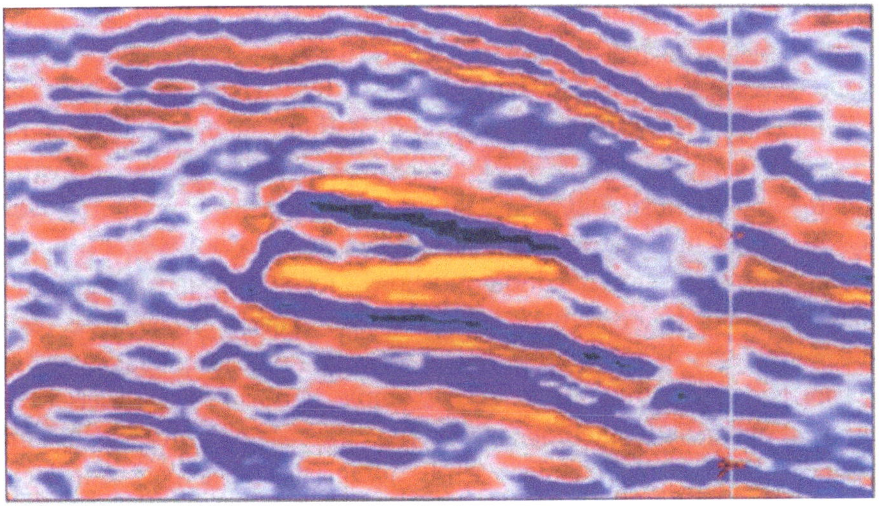

**Fig. 27.3** Not art, just a geophysical image

However, what is relevant here is the person making this statement. A lay person will state as fact “There are squiggly lines”. This shows that any observable fact is influenced by knowledge (“the theory”) of the object of study. Statements of fact are therefore not simply recordings of visual perceptions. Additionally, quite an amount of knowledge is needed to consider taking the geophysical survey in the first place, hence facts do not proceed theory. This is the case for the example here but a reality for many scientific discoveries (we need to know where to look). A more nuanced view therefore is that data and models interact with each other.

*Facts as the basis for knowledge*. “Data precedes the model”. If facts depend on observers resulting in statements that depend on such observers, and if such statements are inherently subjective, then can we trust data as a prerequisite to models (data precede models)? It is now clear that data does not come without a model itself, and hence if the wrong “data model” is used, then the data will be used to build incorrect models. “If I jump in the air and observe that I land on the same spot, then ‘obviously’ the Earth is not moving under my feet”. Clearly the “data model” used here is lacking the concept (theory) of inertia. This again reinforces the idea that in modeling, and in particular UQ, data does not and should precede the model, or that one is subjective and the other somehow is not.

## 27.4 The Role of Experiments—Data

Progress in science is usually achieved by experimentation, the acquisition of information in a laboratory or field setting. Since “data” is central to uncertainty quantification, we spend some time on what “data” is, what “experiments” aim to achieve and what the pitfalls are in doing so.

First, the experiment is not without the “experimenter”. Perceptual judgements may be unreliable, and hence such reliance needs to be minimized as much as possible. For example, in Fig. 27.4, the uninformed observer may notice that the moon is larger when on the horizon, compared to higher up in the sky, which is merely an optical illusion (on which there still is no consensus as to why). Observations are therefore said to be both objective as well as fallible. Objective in the sense that they are shared (in public, presentations, papers, online) and subject to further tests (such measuring of the actual moon size by means of instruments, revealing the optical illusion). Often such progress happens when more advances in the ways of testing or gathering data occur.

Believing that a certain acquisition of data will resolve all uncertainty and lead to determinism on which “objective” decisions is an illusion because the real world involves many kinds of physical/chemical/biological processes that cannot be captured by one way of experimentation. For example, performing a conservative tracer test, to reveal better hydraulic conductivity, may in fact be influenced by the reactions in the subsurface taking place while doing such an experiment. Hence the

**Fig. 27.4** The harvest moon appearing gigantic as compared to the moon in the high sky (https://commons.wikimedia.org/wiki/File:Harvest_Moon_over_looking_vineyards.jpg)

hydraulic conductivity measured and interpreted through some modeling without geochemical reactions may provide a false sense of certainty about the information deduced from such an experiment. In general, it is very difficult to isolate a specific target of investigation in the context of one type of experiment or data acquisition. A good example is in the interpretation of 4D geophysics (repeated geophysics). The idea of the repetition is to remove the influence of those properties that do not change in time, and therefore reveal only those that do change, for example, a change in pressure, a change in saturation, etc. … However, many processes may be at work at the same time, a change in pressure, in saturation, rock compressibility, even porosity and permeability, geomechanical effects, etc. … Hence someone interested in the movement of fluids (change in saturation) is left with a great deal of difficulty in unscrambling the time signature of geophysical sensing data. Furthermore, the inversion of data into a target of interest often ignores all these interacting effects. Therefore, it does not make sense to state that a pump test or a well test reveals permeability, it only reveals a pressure change under the conditions of the test and of the site in question, and many of these conditions may remain unknown or uncertain.

An issue that arises in experimentation is the possibility of a form of circular reasoning that may exist between an experimental set-up and a computer model aiming to reproduce the experimental set-up. If experiments are to be conducted to reveal something important about the subsurface (e.g. flow experiments in a lab), then often the results of such experiments are "validated" by a computer model. Is the physical/chemical/biological model implemented in the computer code derived from the experimental result, or, are the computer models used to judge the adequacy of the result? Do theories vindicate experiments and do experiments vindicate the stated theory? To study these issues better, we introduce the notion of induction and deduction.

## 27.5 Induction Versus Deduction

Bayesianism is based on inductive logic (Howson 1991; Howson et al. 1993; Chalmers 1999; Jaynes 2003; Gelman et al. 2004), although some argue that it is based both on induction and deduction (Gelman and Shalizi 2013). Given the above consideration (and limitations) of experiments (in a scientific context) and data (in a UQ context), the question now arises on how to derive theories from these observations. Scientific experimentation, modeling, studies often rely on a logic to make certain claims. Induction and deductions are such kinds of logic. What such logic offers, is a connection between premises and conclusions:

1. All deltaic systems contain clastic sands.
2. The subsurface system under study is deltaic.
3. The subsurface system contains clastic sands.

This logical deduction is obvious, but such logic only establishes a connection between premises 1 and 2 and the conclusion 3, it does not establish the truth of any of these statements. If that would be the case, then also:

1. All deltaic systems contain steel;
2. The subsurface system under study is deltaic;
3. The subsurface system contains steel.

is equally "logic". The broader question therefore is if scientific theories can be derived from observations. The same question occurs in the context of UQ: can models be derived from data. Consider an experiment in a lab doing a set of n experiments.

Premises:

1. The reservoir rock is water-wet in sample 1.
2. The reservoir rock is water-wet in sample 2.
3. The reservoir rock is water-wet in sample 3.

…

20. The reservoir rock is water-wet in sample 20.

Conclusion: the reservoir is water-wet (and hence not oil-wet).

This simple idea is mimicked from Bertrand Russel's Turkey argument (in his case it was a chicken). "I (the turkey) am fed at 9 am" day after day, hence "I am always fed at 9 am", until the day before Thanksgiving (Chalmers 1999). Another form of induction occurred in 1907: "But in all my experience, I have never been in any accident … of any sort worth speaking about. I have seen but one vessel in distress in all my years at sea. I never saw a wreck and never have been wrecked nor was I ever in any predicament that threatened to end in disaster of any sort. (E. J. Smith 1907, Captain, RMS Titanic)".

Any model or theory derived from observations can never be proven in the sense as being derived from it (David Hume).

This does not mean that induction (deriving models from observations) is completely useless. Some inductions are more warranted than others. Specifically, in the case when the observations set is "large", performed and under a "wide variety of conditions", although these qualitative statements depend clearly on the specific case. "When I swim with hungry sharks, I get bitten", needs really be asserted only once.

The second qualification (variety of conditions) requires some elaboration because we will return to it when discussing Bayesianism. Which conditions are being tested is important (the age of the driller for example is not), hence in doing so we rely on some prior knowledge of the particular model or theory being derived. Such prior knowledge will determine which factors will be studied, which are influencing the theory/model and which not. Hence the question is to how this "prior knowledge" itself is asserted by observations. One runs into the never-ending chain of what prior knowledge is used to derive prior knowledge. This point was made clear by David Hume, an eighteenth-century Scottish philosopher (Hume 2000, originally 1739). Often the principle of induction is argued because it has "worked" from experience. The reader needs simply to replace the example of the water-wet rocks with "Induction has worked in case *j*" etc.... to understand that induction is, in this way, "proven" by means of induction. The way out of this "mess" is to not make true/false statements, but to use induction in a probabilistic sense (probably true), a point to which we will return when addressing Bayesianism.

## 27.6 Falsificationism

### A Reaction to Induction

Falsificationism, as championed by Karl Popper (1959) starting in the 1920s was born partly as a reaction to inductionism (and logical positivism). Popper claimed that science should not involve any induction (theories derived from observations). Instead, theories are seen as speculative or tentative, as created by the human intellect, usually to overcome limitations of previous theories. Once stated, such theories need to be tested rigorously with observations. Theories that are inconsistent with such observation should be rejected (falsified). The theories that survive are the best theories, currently. Hence, falsificationism has a time component and aims to describe progress in science, where new theories are born out of old ones by a process of falsification.

In terms of UQ, one can then see models not as true representations of actual reality but as hypotheses. One has as many hypotheses as models. Such a hypothesis can be constrained by previous knowledge, but real field data should be used not to confirm a model (it confirms this with data) but to falsify a model (reject, the model does not confirm with data). A simple example illustrates the difference:

*Induction*:
Premise: All rock samples are sandstones.
Conclusion: The subsurface system contains only sandstone.
*Falsification*:
Premise: A sample has been observed that is shale.
Conclusion: The subsurface system does not consist just of sandstone.

The latter is clearly a logically valid deduction (true). Falsification therefore can only proceed with hypotheses that are falsifiable (this does not mean that one has to falsify the observations, but that such observation could exist). Some hypotheses are not falsifiable; for example, "the subsurface system consists of rock that are sandstone or not sandstone". This then raises the question of the degree of falsifiability of a hypothesis and the strength (precision) of the observation in falsifying. Not all hypotheses are equally falsifiable and not all observations should be treated on the same footing. A strong hypothesis is one that makes strong claims, there is a difference between:

1. Significant accumulation in the Mississippi delta requires the existence of a river system; and
2. Significant accumulation in all deltas require the existence of a river system.

Clearly 2 has more consequences than 1. Falsification therefore invites stating bold conjectures rather than safe conjectures. Science advances through a large number of bold conjectures that would be easily falsifiable. As a result, a hypothesis *B* that is offered after hypothesis *A* should also be more falsifiable.

The latter has considerable implications in UQ and model building. Inductionists tend to bet on one model, the best possible, best explaining most observations, within a static context, without the idea that the model they are building will evolve. Inductionists do evolve models, but that is not the outset of their viewpoint, there is always the hope that the best possible will remain the best possible. The problem with this inductionist attitude is that new observations that cannot be fitted into the current model are used to "fix" the model with ad hoc modifications. A great example of this can be found in the largest oil reservoir in the world, namely the Ghawar field (see *Twilight in the Desert: The Coming Saudi Oil Shock and the World Economy*, Matt Simmons). Before 2000, most modelers (geologists, geophysicist, engineers) did not consider fractures as being a driving heterogeneity for oil production. However, flow meter observations in wells indicated significant permeability. To account for this data, the existing models with already large permeabilities (1000–10.000mD) where modified to 200D, see Fig. 27.5. While this dramatic increase in permeability in certain zones did lead to explaining the flow meter data, the ad hoc modification cannot be properly tested with the current observations. It is just a fix to the model (the current "theory" of no fractures). Instead, a new test would be needed, such as new drilling to confirm or not the presence of a gigantic cave that can explain such ridiculous permeability values. Today, all models built of the Ghawar field contain fractures.

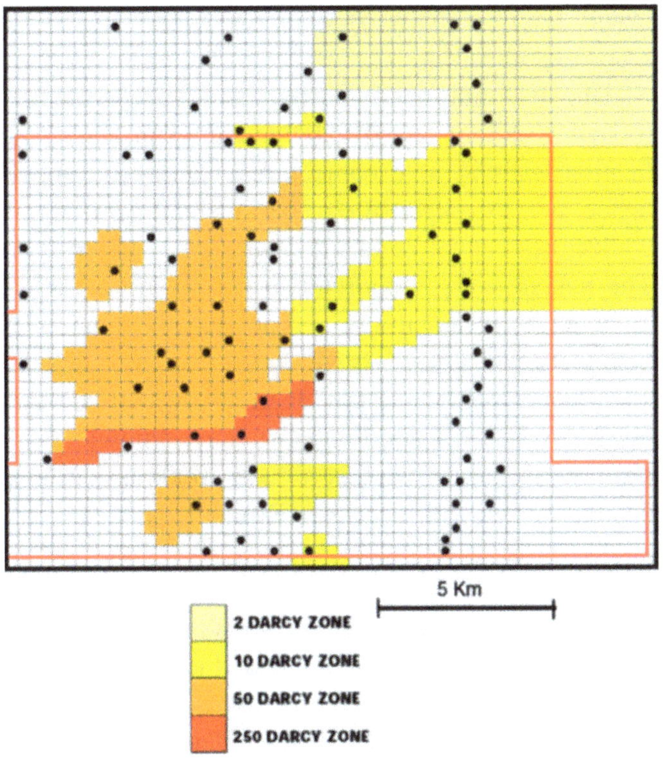


**Fig. 27.5** A reservoir model developed to reflect super permeability channels; note the legend with permeability values (Valle et al. 1993)

Falsificationism does not use ad hoc modification, because the ad hoc modification cannot be falsified. In the Ghawar case, the very notion of fluid flow by means of large matrix permeability tells the falsificationist that bold alternative modifications to the theory are needed and not simple ad hoc fixes, in the same sense that science does not progress by means of fixes. An alternative therefore to the inductionist approach in Ghawar could be as follows: most fluid flow is caused by large permeability, except in some area where it is hypothesized that fractures are present despite the fact that we have not directly observed then. The falsificationist will now proceed by finding the most rigorous (new) test to test this hypothesis. This could consist of acquiring geomechanical studies of the system (something different than flow) or by means of geophysical data that aims to detect fractures (AVOZ data). New hypotheses also need to lead to new tests that can falsify them. This is how progress occurs. The problem is often "time"; a falsificationist takes the path of high risk, high gain, but time may run out on doing experiments that falsify certain hypothesis. "Failures" are often seen as that and not as lessons learned. In the modeling world one often shies away from bold hypothesis (certainly if one wants to obtain government research funding!) and that modelers, as a group tends to gravitate towards some consensus under the banner of being good at "team-work". It is the view of the authors that such practice is however the death of any realistic UQ. UQ needs to include bold hypothesis, model conjectures that are not the norm, or based on any majority vote, or by playing it safe, being conservative. Uncertainty cannot be reduced by just great team-work, it will require equally rigorous observations (data) that can falsify any (preferably bold) hypothesis.

This does not mean that inductionist type of modeling and falsification type of modeling cannot co-exist. If inductionism leads to cautious conjectures and falsification leads to bold conjectures. Cautious conjectures may carry little risk, and hence, if they are falsified, then insignificant advance is made. Similarly, if bold conjectures cannot be falsified with new observations, significant advance is made. The matter that is important in all this however is the nature of the background knowledge (recall, the prior knowledge), what is currently known about what is being studied. Any "bold" hypothesis is measured against such background knowledge. Likewise, the degree to which observations can falsify hypothesis needs to be measured against such knowledge. This background knowledge changes over time (what is bold in 2000 may no longer be bold in 2020), and such change, as we will discuss is explicitly modeled in Bayesianism.

### Falsificationism in Statistics

Schools of statistical inference are sometimes linked to the falsificationist views of science, in particular the work of Fischer, Neyman and Pearson; all well-known scientists in the field of (frequentist) statistics (Fisher and Fisher 1915; Fisher 1925; Rao 1992; Pearson et al. 1994; Berger 2003; Fallis 2013 for overviews and original papers). Significance tests, confidence intervals $p$-values are associated with a hypothetico-deductive way of reasoning. Since these methods are pervasive in all areas of science, particularly in UQ, we present some discussion on its rationality as well as the opposing views of inductionism within this context.

Historically, Fisher can be seen as the founder of classical statistics. His work has a falsificationist foundation, steeped in statistical "objectivity" (lack of necessary subjective assumption, which is the norm in Bayesian methods). The now well-known procedure starts by stating a null-hypothesis (a coin is fair), then defines an experiment (flipping), a stopping rule (e.g. number of flips) and a test-statistic (e.g. number of heads). Next, the sampling distribution (each possible value of the test-statistic), assuming the null-hypothesis is true, is calculated. Then, we calculate a probability $p$ that our experiment falls in an extreme group (e.g. 4 heads or less which hypothesis has only a probability of 1.2% for 20 flips). Then a convention is taken to reject (falsify) the hypothesis when the experiment falls in the extreme group, say $p \leq 0.05$.

Fisher's test works only on isolated hypotheses, which is not how science progresses; often many competing hypotheses are proposed that require testing under some evidence. Neyman and Pearson developed statistical methods that involve rival hypotheses, but again reasoning from an "objective" perspective, without relying on priors or posteriors of Bayesian inductive reasoning. For example, in the case of two competing hypotheses $H_1$ and $H_2$, Neyman-Pearson reasoned that either of the hypotheses are accepted or rejected, leading to two kinds of errors (stating that one is false, while the other is false and vice versa), better known as type I and II errors. Neyman and Pearson improved on Fischer in better defining "low probability". In the coin example, a priori, any combination of 20 tosses has a probability of $2^{-20}$, even under a fair coin, most tosses have small probability. Neyman-Pearson provide some more definition of this critical region (where hypotheses are rejected). If $X$ is the random variable describing the outcome (e.g. a combination of tosses), then the outcome space is defined by the following inequality:

$$L(X) = \frac{P(X|H_1)}{P(X|H_2)} \leq \delta \quad P(L(X) \leq \delta | H_1) = \alpha \tag{27.1}$$

with $\delta$ depending on the significance level $\alpha$ and the nature of the hypothesis. This theorem known as the Fundamental Lemma (Neyman and Pearson 1933) defines the most powerful test to reject $H_1$ in favor of $H_2$ at significance level $\alpha$ for a threshold $\delta$. The interpretation of likelihood ratio was provided by Bayesianists as the Bayes' factor (the evidential force of evidence). This was however not the interpretation of Neyman-Pearson, who rejected subjective models.

What then does a significance test tell us about the truth (or not) of a hypothesis? Since the reasoning here is in terms of falsification (and not induction), the Neyman-Pearson interpretation is that if a hypothesis is rejected, then "one's actions should be guided by the assumption that it is false" (Lindgren 1976). Neyman-Pearson gladly admit that significance tests tell nothing about whether a hypothesis is true or not. However, they do attach the notion of "in the long run", interpreting the significance level as, for example, the number of times in 1000 times that the same test is being done. The problem here is that no testing can be done and will be done in exactly the same fashion, under the exact same circumstances. This idea would also invoke the notion that under a significance level of 0.05, a *true* hypothesis would be rejected with a probability of 0.05. The latter violates the very reason on which significance tests were formed: events with probability $p$ can never be proven to occur (that requires subjectivity!), let alone with the exact frequency of $p$.

The point here is to show that classical statistics should not be seen as purely falsificationist, a logical hypothetic-deductive way of reasoning. Reasoning in classical statistics comes with its own subjective notions of personal judgements (choosing which hypothesis, what significance level, stopping rules, critical regions, iid assumptions, Gaussian assumptions etc. …). This was in fact later acknowledged by Pearson himself (Neyman and Pearson 1967, p. 277).

#### Limitations of Falsificationism

Falsificationism comes with its own limitations. Just as induction cannot be induced, falsificationism cannot be falsified, as a theory. This becomes clearer when considering real-world development of models or theories. The first problem is similar to the one discussed in using inductive and deductive logic. Logic only works if the premises are true, hence falsification, as a deductive logic cannot distinguish between a faulty observation and a faulty hypothesis. The hypothesis does not have to be false when inconsistent with observations, since observations can be false. This is an important problem in UQ that we will revisit later.

The real world involves considerably more complication than "the subsurface system is deltaic". Let's return to our example of monitoring heat storage using geophysics. A problem that is important in this context is to monitor whether the heat plume remains near the well and is compact, so that it does not start to disperse, since then recovery of that heat becomes less efficient. A hypothesis could then be "the heat plume is compact", geophysical data can be used to falsify this by, for example, observing that the heat plume is indeed influenced by heterogeneity. Unfortunately, such data does not directly observe "temperature", instead it measures resistivity, which is related to temperature and other factors. Additionally, because monitoring is done at a distance from the plume (at the surface), the issue of limited resolution occurs (any "remote sensing" suffers from this limited resolution). This is then manifested in the inversions of the ERT data into temperature, since many inversion techniques result in smooth versions of actual reality (due to this limited resolution issue), from which the modeler may deduce that homogeneity of the plume is not falsified. How do we find where the error lies? In the instrumentation? In the instrumentation set-up? In the initial and boundary conditions that are required to model the geophysics? In the assumptions about geological variability? In the smoothness of the inversion? Falsification does not provide a direct answer to this. In science, this problem is better known as the Duhem–Quine thesis after Pierre Duhem and Willard Quine (Ariew 1984). This thesis states that it is impossible to falsify a scientific hypothesis in isolation, because the observations required for such falsification themselves rely on additional assumptions (hypothesis) than cannot be falsified separately from the target hypothesis (or vice versa). Any particular statistical method that claims to do so, ignores the physical reality of the problem.

A practical way to deal with this situation is not consider just falsification, but sensitivity to falsification. What impacts the falsification process? Sensitivity, even with limited or approximate physical models provide more information that can lead to (1) changing the way data is acquired (the "value of information") changing the way the physics of the problem (e.g. the observations) is modeled by focusing on what matters most towards testing the hypothesis.

More broadly, falsification does not really follow the history of the scientific method. Most science has not been developed by means of bold hypothesis that are then falsified. Instead, theories that are falsified are carried through history; most notably, because observations that appear to falsify the theory can be explained by means of causes other than the theory that was the aim of falsification. This is quite common in modeling too: observations are used as claims that a specific physical model does not apply, only to discover at a later time that the physical model was correct but that the data could be explained by some other factor (e.g. a biological reason, instead of a physical reason). Popper himself acknowledged this dogmatism (hanging onto models that have "falsified" to "some degree"). As we will see later, one of the problems in the application of probability (and Bayesianism) is that zero probability models are deemed "certain" not to occur. This may not reflect the actual reality that models falsified under such Popper-Bayes philosophy become "unfalsified" later by new discoveries and new data. Probability and "Bayesianism" are not at fault here, but the all too common underestimation of uncertainties in many applications.

## 27.7 Paradigms

### Thomas Kuhn

From the previous presentation, one may argue that both induction and falsification provide too much of a fragmented view of the development of scientific theory or methods that often do not agree with reality. Thomas Kuhn, in his chapter "The Structure of Scientific Revolution" (Kuhn 1996) emphasizes the revolutionary character of scientific methods. During such revolution one abandons one "theoretical" concept for another, which is incompatible with the previous one. In addition, the role of scientific communities is more clearly analyzed. Kuhn describes the following evolution of science:

paradigm → crises → revolution → new paradigm → new crisis.

Such a single paradigm consists of certain (theoretical) assumptions, laws, methodologies and applications adapted by members of a scientific community. Probabilistic methods, or Bayesian methods, can be seen as such paradigms: they rely on axioms of probability and the definition of a conditional probability, the use of prior information, subjective beliefs, maximum entropy, principle of indifference, algorithms of McMC, etc. ... Researchers within this paradigm do not question the fundamentals of such paradigm, the fundamental laws or axioms. Activities within the paradigm are then puzzle-solving activities (e.g. studying convergence of a Markov chain) governed by the rules of the paradigm. Researchers within the paradigm do not criticize the paradigm. It is also typical that many researchers within that paradigm are unaware of the criticism on the paradigm or ignorant as to the exact nature of the paradigm, simply because it is a given: who is really critical of the axioms of probability when developing Markov chain samplers? Or, who questions the notion of conditional probability when performing stochastic inversions? Puzzles that cannot be solved are deemed to be anomalies, often attributed to the lack of understanding of the community about how to solve the puzzle within the paradigm, rather than a question about the paradigm itself. Kuhn considers such unsolved issues as anomalies rather than what Popper would see as potential falsifications of the paradigm. The need for greater awareness and articulation of the assumptions of a single paradigm becomes necessary when the paradigm requires defending against offered alternatives.

Within the context of UQ, a few such paradigms have emerged reflecting the concept of revolution as Kuhn describes. The most "traditional" of paradigms for quantifying uncertainty is by means of probability theory and its extension of Bayesian probability theory (the addition of a definition of conditioning).

We provide here a summary account of the evolution of this paradigm, the criticism leveled, the counter-arguments and the alternatives proposed, in particular possibility theory.

## Is Probability Theory the Only Paradigm for Uncertainty Quantification?

### The Axioms of Probability: Kolmogorov—Cox

The concept of numerical probability emerged in the mid-seventeenth century. A proper formalization was developed by (Kolmogoroff 1950) based on classical measure theory. A comprehensive study of its foundations is offered in Fine (1973). The treatment is vast and comprises many works of particular note (Gnedenko et al. 1962; Fine 1973; de Finetti 1974, 1995; de Finetti et al. 1975; Jaynes 2003; Feller 2008). Also of note is the work of (Shannon 1948) on uncertainty-based information in probability. In other words, the concept of probability has been around for three centuries. What is probability? It is now generally agreed (the fundamentals of the paradigm) that the axioms of Kolmogorov form the basis, as well as the Bayesian interpretation by Cox (1946). Since most readers are unfamiliar with the Cox theorem and the consequences for interpreting probability, we provide some high-level insight.

Cox works from a set of postulates for example (we focus on just two of three postulates)

- "A proposition $p$ and its negation $\neg p$ is certain" or $plaus(p \cap \neg p) = 1$ which is also termed the logical principle of the excluded middle. *plaus* stands for plausibility.
- Consider now two propositions $p$ and $q$ and the conjunction between them $p \cap q$. This postulate states that the plausibility of the conjunction is the only function of the plausibility of $p$ and the plausibility of $q$ given that $p$ is true. In other words

$$plaus(p \vee q) = f(plaus(p), plaus(q|p)).$$

The traditional laws are recovered when setting *plaus* to be a probability measure or $P$ or stating as per the Cox theorem "any measure of belief is isomorphic to a probability measure". This seems to suggest that probability is sufficient in dealing with uncertainty, nothing else is needed (due to this isomorphism). The consequence is that one can now perform calculations (a calculus) with "degrees of belief" (subjective probabilities) and even mix probabilities based on subjective belief with probabilities based on frequencies. The question is therefore whether these subjective probabilities are the only legitimate way of calculating uncertainty? For one, probability requires that either the fact is there, or it is not there, nothing is left in the "middle". This then necessarily means that probability is ill-suited in cases where the excluded middle principle of logic does not apply. What are those cases?

## Intuitionism

Probability theory is truth driven. An event occurs or does not occur. The truth will be revealed. From a hard scientific, perhaps engineering approach this seems perfectly fine, but it is not. A key figure in this criticism is the Dutch mathematician and philosopher Jan Brouwer. Brouwer founded the mathematical philosophy of intuitionism countering the then-prevailing formalism, in particular of David Hilbert as well as of Bertrand Russell, claiming that mathematics can be reduced to logic; the epistemological value of mathematical constructs lies in the fundamental nature of this logic.

In simplistic terms perhaps, intuitionists do not accept the law of excluded middle in logic. Intuitionism reasons from the point that science (in particular mathematics) is the result of the mental construction performed by humans rather than principles founded in the actual objective reality. Mathematics is not "truth", rather it constitutes applications of internally consistent methods used to realize more complex mental constructs, regardless of their possible independent existence in an objective reality. Intuition should be seen in the context of logic as the ability to acquire knowledge without proof or without understanding how the knowledge was acquired.

Classic logic states that existence can be proven by refuting non-existence (the excluded middle principle). For the intuitionist, this is not valid; negation does not entail falseness (lack of existence), it entails that the statement is refuted (a counter example has been found). For an intuitionist a proposition $p$ is stronger than a statement of not (not $p$). Existence is a mental construction, not proof of non-existence. One specific form and application of this kind of reasoning is fuzzy logic.

## Fuzzy Logic

It is often argued that epistemic uncertainty (or knowledge) does not cover all uncertainty (or knowledge) relevant to science. One such particular form of uncertainty is "vagueness" which is borne out of the vagueness contained in language (note that other language dependent uncertainties exists such as "context-driven"). This may seem rather trivial to someone in the hard sciences, but it should be acknowledged that most language constructs ("this is air", meaning 78% nitrogen, 21% oxygen, and less than 1% of argon, carbon dioxide, and other gases) are a purely theoretical construct, of which we still may not have incomplete understanding. The air that is outside is whatever that substance is, it does not need human constructs, unless humans use if for calculations, which are themselves constructs. Unfortunately (possibly flawed) human constructs is all that we can rely on.

The binary statements "this is air" and "this is not air" are again theoretical human constructs. Setting that aside, most of the concepts of vagueness are used in cases with unclear borders. Science typically works with classification systems ("this is a deltaic deposit", "this is a fluvial deposit"), but such concepts are again man-made constructs. Nature does not decide to "be fluvial", it expresses itself through laws of physics, which are still not fully understood.

A neat example presents itself in the September 2016 edition of EOS: "What is magma?" Most would think this is a problem which has already been solved, but it isn't, mostly due to vagueness in language and the ensuing ambiguity and difference in interpretation by even experts. A new definition is offered by the authors: "*Magma*: naturally occurring, fully or partially molten rock material generated within a planetary body, consisting of melt with or without crystals and gas bubbles and containing a high enough proportion of melt to be capable of intrusion and extrusion."

Vague statements ("this may be a deltaic deposit") are difficult to capture with probabilities (it is not impossible, but quite tedious and construed). A problem occurs in setting demarcations. For example, in air pollution, one measures air quality using various indicators such as PM2.5, meaning particles which pass through a size-selective inlet with a 50% efficiency cut-off at 2.5 μm aerodynamic diameter. Then standards are set, using a cut-off to determine what is "healthy" (a green color) and what is "not so healthy" (orange color) and "unhealthy" (a red color) (the humorous reader may also think of terrorist alert levels). Hence, if the particular matter changes by one single particle, the air goes suddenly from "healthy" to "not so healthy"?

In several questions of UQ, both epistemic and vagueness-based uncertainty may occur. Often vagueness uncertainty exists at a higher-level description of the system, while epistemic uncertainty may then deal with questions of estimation because of limited data within the system. For example, policy makers in the environmental sciences may set goals that are vague, such as "should not exceed critical levels". Such a vague statement then needs to be passed down to the scientist who is required to quantify risk of attaining such levels by means of data and numerical models, where epistemic uncertainty comes into play. In that sense there is no need to be rigorously accurate, for example according to a very specific threshold, given the above argument about such thresholds and classification systems.

Does probability easily apply to vagueness statements? Consider a proposition "the air is borderline unhealthy". The rule of the excluded middle no longer applies because we cannot say that the air is either not unhealthy or unhealthy. Probabilities no longer sum to one. It has therefore been argued that the propositional logic of probability theory needs to be replaced with another logic: fuzzy logic (although other logics have been proposed such as intuitionistic, trivalent logic, we will limit the discussion to this one alternative).

Fuzzy logic relies on fuzzy set theory (Zadeh 1965, 1975, 2004). An example of fuzzy set $A$ such as "deltaic" is said to be characterized by a membership function $\mu_{deltaic}(u)$ representing the degree of membership given some information $u$ on the deposit under study, for example $\mu_{deltaic}(deposit) = 0.8$ for a deposit with info $u$ under study. Probabilists often claim that such membership function is nothing more than a conditional probability $P(A|u)$ in disguise (Loginov 1966). The link is made using the following mental construction. Imagine 1000 geologists looking at the same limited info $u$ and then voting whether the deposit is "deltaic" or "fluvial". Let's assume these are the two options available. $\mu_{deltaic}(deposit) = 0.832$ means that 832 geologists picked "deltaic" and hence a vote picked at random has 83.2% chance of being deltaic. However, the conditional probability comes with its limitations as it attempts to cast a very precise answer into what is still a very vague concept. What really is "deltaic"? Deltaic is simply a classification made by humans to describe a certain type of depositional system subject to certain geological processes acting on it. The result is a subsurface configuration, termed architecture of clastic sediments. In modeling subsurface systems, geologists do not observe the processes (the deltaic system) but only the record of it. However, there is still no full agreement as to what is "deltaic" or when "deltaic" ends and "fluvial" starts as we go more upstream? (Recall our discussion on "magma") What are the processes which are actually happening and how all this gets turned into a subsurface system? Additionally, geologist may not have a consensus on what "deltaic" is, where "fluvial" starts, or, may classify based on personal experiences, different education (schools of thought about "deltaic"), and different education levels. What then does 0.832 really mean? What is the meaning of the difference between 0.832 and 0.831? Is this due to education? Misunderstanding or disagreement on the classification? Lack of data provided? It clearly should be a mix of all this, but probability does not allow an easy discrimination. We find ourselves again with a Duhem–Quine problem.

Fuzzy logic does not take the binary route of voting up or down, but allows a grading in the vote of each member, meaning that it allows for more gradual transition between the two classes for each vote. Each person takes the evidence at his/her value and makes a judgement based on their confidence and education level: I don't really know, hence 50/50; I am pretty certain, hence 90/10. (More advanced readers in probability theory may now see a mixture of the models of probability stated based on the evidence of what the $u$ is. However, because of the overlapping nature of how evidence is regarded by each voter, these prior probabilities are no longer uniform).

### The Dogma of Precision

Clearly probability theory (randomness) does not work well when the event itself is not clearly defined, subject to discussion. Probability theory does not support the concept of a fuzzy event, hence such information (however vague and incomplete) becomes difficult and non-intuitive to account for. Probability theory does not provide a system for computing with fuzzy probabilities expressed as likely, unlikely and not very likely. Subjective probability theory relies on the elicitation rather than the estimation of a fuzzy system. It cannot address questions of the nature "What is the probability that the depositional system *may* be deltaic". One should question, under all this vagueness and ambiguity what is really the meaning of the digit "2" or "3" is in $P(A|u) = 0.832$. The typical reply of probabilists to possibilists is to "just be more precise" and the problem is solved. But this would ignore a particular form of lack of understanding, which goes to the very nature of UQ. Precision is required that does not agree with the realism of vagueness on concepts, which are as yet imprecise (such as in subsurface systems).

The advantage and the disadvantage of the application of probability to UQ are that, dogmatically, it requires, precision. It is an advantage in the sense that it attempts to render subjectivity into quantification, that the rules are very well understood, the methods deeply practiced, because of the nature of the rigor of the theory, the community (of 300 years of practice) is vast. But, this rigor does not always jive with reality. Reality is more complex than "Navier Stokes" or "Deltaic", so we apply rigor to concepts (or even models) that probably deviate considerably from the actual processes occurring in nature. Probabilists often call this "structural" error (yet another classification and often ambiguous concept, because it has many different interpretations) but provide no means of determining what exactly this is and how it should be precisely estimated, as is required by their theories. It is left as a "research question", but can this question be truly answered within probability theory itself? For the same reasons, probabilistic method (in particular Bayesian, see the following sections are computationally very demanding, exactly because of this dogmatic quest for precision.

**Possibility Theory: Alternative or Compliment?**

Possibility theory has been popularized by Zadeh (1978), also by Dubois and Prade (1990). The original notion goes back further to the economist (Shackle 1962) studying uncertainty based on degrees of potential surprise of events. Shackle also introduces the notion of conditional possibility (as opposed to conditional probability). Just as probability theory, possibility theory has axioms. Consider $\Omega$ to be a finite set, with subsets $A$ and $B$ that are not necessarily disjoint:

axiom 1: $pos(\varnothing) = 0$ ($\Omega$ is exhaustive)

axiom 2: $pos(\Omega) = 1$ (no contradiction)

axiom 3: $pos(A \cup B) = \max(pos(A), pos(B))$ ("additivity")

Noticeable difference with probability theory is that addition is replaced with "max" and the subsets for axiom 3 need not be disjoint. Additionally, probability theory uses a single measure, the probability, whereas possibility theory uses two concepts, the possibility and the necessity of the event. This necessity, another measure is defined as:

$$nec(A) = 1 - pos(\bar{A}) \tag{27.2}$$

If the complement of an event is impossible, then the event is necessary. $nec(A) = 0$ means that $A$ is unnecessary. One should not be "surprised" if $A$ does not occur, it says nothing about $pos(A).nec(A) = 1$ means that $A$ is certainly true, which implies $pos(A) = 1$. Hence $nec$ carries a degree of surprise: $nec(A) = 0.1$ a little bit surprised, $nec(A) = 0.9$ very surprised if $A$ is not true. Possibility also allows for indeterminacy (which is not allowed in epistemic uncertainty), this is captured by $nec(A) = 0, pos(A) = 1$.

Logically then

$$nec(A \cap B) = \min(nec(A), nec(B)) \tag{27.3}$$

Possibility does not follow the rule of the excluded middle because

$$pos(A) + pos(\bar{A}) \geq 1 \tag{27.4}$$

Take the following example. Consider a reservoir. It either contains oil $(A)$ or contains no oil $(\bar{A})$ (something we like to know!). $pos(A) = 0.5$ means that I am willing to bet that the reservoir contains oil so long as the odds are even *or better*. I would not bet that it contains oil. Hence this describes a degree of belief very different from subjective probabilities.

Possibilities are sometime called "imprecise probabilities" (Hand and Walley 1993) or are interpreted that way. "Imprecise" need not be negative, as discussed above, it has its own advantages, in particular in terms of computation. In probability theory, information is used to update degrees of belief. This is based on Bayes' rule whose philosophy will be studied more closely in the next section. A counterpart to Bayes' rule exists in possibility theory, but because of the imprecision of possibilities over probabilities, no unique way exists to update possibilities into a new possibility, given new (vague) information. Recall that Bayes' rule relies on the product (corresponding to a conjunction in classical logical)

$$P(A|B) = \frac{P(B|A)}{P(B)} P(A) \tag{27.5}$$

Consider first the counterpart of the probability density function $f_X(x)$ in possibility theory: namely the possibility distribution $\pi_X(x)$. Unlike probability densities which could be inferred from data, possibility distributions are always specified by users, and hence take simple form (constant, triangular) functions. Densities express likelihoods, a ratio of the densities assessed in two outcomes denotes how much more (or less) likely one outcome is over the other. A possibility distribution simply states how possible an outcome $x$ is. Hence a possibility distribution is always equal or less than unity (not the case for a density). Also, note that $P(X = x) = 0$, always if $X$ is a continuous variable, while $pos(X = x)$ is not zero everywhere. Similarly, in the case of a joint probability distribution, we can define a joint possibility distribution as $\pi_{X,Y}(x, y)$ and conditional possibility distributions as $\pi_{X|Y}(x|y)$. The objective now is to infer $\pi_{X|Y}(x|y)$ from $\pi_{Y|X}(y|x)$ and $\pi_X(x)$.

As mentioned above, probability theory relies on a logical conjunction, see Fig. 27.6. This conjunction has the following properties:

$$a \cap b = b \cap a \quad \text{(commutativity)}$$

$$\textit{if } a \leq a' \textit{ and } b \leq b' \textit{ then } a \cap b \leq a' \cap b' \quad \text{(monotonicity)}$$

$$(a \cap b) \cap c = a \cap (b \cap c) \quad \text{(associativity)}$$

$$a \cap 1 = a \quad \text{(neutrality)}$$

Possibility theory, as it is based on fuzzy sets, rather than random sets, relies on an extension of the conjunction operation. This new conjunction is termed a

| a | b | $a \cap b$ |
|---|---|---|
| 0 | 0 | 0 |
| 1 | 0 | 0 |
| 0 | 1 | 0 |
| 1 | 1 | 1 |

| a | b | $T_1(a,b)$ | $T_2(a,b)$ | $T_3(a,b)$ |
|---|---|---|---|---|
| 0.1 | 0.1 | 0.1 | 0.01 | 0.053 |
| 0.9 | 0.1 | 0.1 | 0.09 | 0.098 |
| 0.1 | 0.1 | 0.1 | 0.09 | 0.098 |
| 0.9 | 0.9 | 0.9 | 0.81 | 0.82 |

$T_1(a,b) = min(a,b);\ T_2(a,b) = a.b;\ \ T_2(a,b) = \frac{ab}{a+b-ab}$

**Fig. 27.6** Example of t-norms for conjunction operations

triangular norm (T-norm) (Jenei and Fodor 1998; Höhle 2003; Klement et al. 2004) because it follows the following four properties:

$$\begin{aligned}
&T(a,b) = T(b,a) \quad \text{(commutativity)}\\
&if\ a \leq a'\ and\ b \leq b'\ then\ T(a,b) = T(a',b')\,\text{(monotonicity)}\\
&T(a, T(b,c)) = T(T(a,b),c) \quad \text{(associativity)}\\
&T(a,1) = a \quad \text{(neutrality)}
\end{aligned}$$

Recall that Cox relied on the postulate that $plaus(p \cap q) = f(plaus(p), plaus(q|p))$. Similarly, possibility theory relies on:

$$\pi_{Y|X}(y|x) = T\left(\pi_X(x), \pi_{Y|X}(y|x)\right) = T\left(\pi_Y(x), \pi_{X|Y}(x|y)\right) \tag{27.6}$$

For example, for the minimum triangular norms we get

$$\pi_{X|Y}(x|y) = \begin{cases} 1 & if\ \pi(x) = \min\{\pi(x), \pi_{Y|X}(x|y)\} \\ \min\{\pi(x), \pi_{Y|X}(x|y)\} & if\ \pi(x) > min\{\pi(x), \pi_{Y|X}(x|y)\} \end{cases} \tag{27.7}$$

and for the product triangular norm, we get something that looks Bayesian

$$\pi_{X|Y}(x|y) = \frac{\pi_{Y|X}(x|y)\pi(x)}{\pi(y)} \tag{27.8}$$

## 27.8 Bayesianism

### Thomas Bayes

Uncertainty quantification, today often has a Bayesian flavor. What does this mean? Most researchers simply invoke Bayes' rule, as a theorem within probability theory.

They work within the paradigm. But what is really the paradigm of Bayesianism? It can be seen as a simple set of methodologies, but it can also be regarded as a philosophical approach to doing science, in the same sense as empiricism, positivism, falsificationism or inductionism. The reverend Bayes' would perhaps be somewhat surprised by the scientific revolution and main stream acceptance of the philosophy based on his rule.

Thomas Bayes was a statistician, philosopher and Reverend. Bayes presented a solution to the problem of inverse probability in "An Essay towards Solving a Problem in the Doctrine of Chances". This essay was read after his death, by Richard Price for the Royal Society of London, a year after his death. Bayes' theorem remained in the background until reprinted in 1958, and even then it took a few more decades before an entirely new approach to scientific reasoning, Bayesianism was created (Howson et al. 1993; Earman 1992).

Prior to Bayes' most works on chance were focused on direct inference, such as the number of replications needed to calculate a desired level of probability (how many flips of the coin are needed to assure 50/50 chance?). Bayes' treated the problem of inverse probability: "given the number of times an unknown event has happened and failed: required the chance that the probability of its happening in a single chance lies between any two degrees of probability that can be named" (see the Biometrika publication of Bayes' essay). Bayes' essay has essentially four parts. Part 1 consists of a definition of probability and some basic calculation which are now known as the axioms of probability. The second part uses these calculations in a chance event related to a perfectly leveled billiard table, see Fig. 27.7. Part 3 consists of using the equations obtained from the analysis of the billiard problem to his problem of inverse probability. Part 4 consists of more numerical studies and applications.

Bayes, in his essay, was not concerned with induction and the role of probability in it. Price, however, in the preface to the essay did express a wish that the work would in fact lead to a more rational approach to induction than was then currently available. What is perhaps less known is that "Bayes' theorem" in the form that we now know it, was never written by Bayes'. However, it does occur in the solution to his particular problem. As mentioned above, Bayes' was interested in a chance event with unknown probability (such as in the billiard table problem), given a

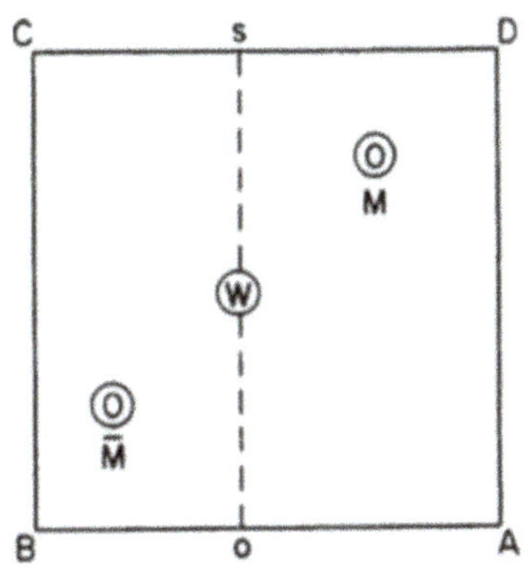


**Fig. 27.7** Bayes' billiard table: "to be so made and leveled that if either of the ball O and W thrown upon it, there shall be the same probability that it rests upon any one equal part of the plane as another" (Bayes and Price 1763)

number of trials. If $M$ counts the number of times that an event occurs in $n$ trials, then the solution is given through the binomial distribution

$$P(p_1 \leq p \leq p_2 | M = m) = \frac{\int_{p_1}^{p_2} \binom{n}{m} p^m (1-p)^{n-m} P(dp)}{\int_0^1 \binom{n}{m} p^m (1-p)^{n-m} P(dp)} \tag{27.9}$$

where $P(dp)$ is the prior distribution over p. Bayes' insight here is to "suppose the chance is the same that it $(p)$ should lie between any two equi-different degrees". $P(dp) = dp$, in other words the prior is uniform, leading to

$$P(p_1 \leq p \leq p_2 | M = m) = \frac{(n+1)!}{m!(n-m)!} \int_{p_1}^{p_2} \binom{n}{m} p^m (1-p)^{n-m} dp \tag{27.10}$$

Why uniform? Bayes' does not reason from the current principle of indifference (which can be debated, see later), but rather from an operation characterization of an event concerning the probability which we know absolutely nothing about prior to the trials. The use of prior distributions however was one of the key insights of Bayes' that very much lives on.

**Rationality for Bayesianism**

Bayesians can be regarded more as relativists than absolutists (such as Popper). They believe in prediction based on imperfect theories. For example, they will take an umbrella on their weekend, if their ensemble Kalman filter prediction of the weather at their trip location puts a high (posterior) probability of rain in 3 days. Even if the laws involved are imperfect and probably can be falsified (many weather predictions are completely wrong!), they rely on continued learning from future information and adjustments. Instead of relying on Popper's zero probability (rejected or not), they rely more on an inductive inference yielding non-zero probabilities.

If we now take the general scientific perspective (and not the limited topic of UQ), then Bayesians see science progress by hypothesis, theories and evidence offered towards these hypotheses as all quantified using probabilities. In this general scientific context, we may therefore state hypothesis $H$, gather evidence $E$, with $P(H|E)$ the probability of the hypothesis in the light of the evidence, $P(E|H)$ the probability that the evidence occurs when the hypothesis is true, $P(H)$ the probability of the hypothesis without any evidence and $P(E)$ the probability of the evidence, without stating any hypothesis being true.

$$P(H|E) = \frac{P(E|H)}{P(E)} P(H) \tag{27.11}$$

$P(H)$ is also termed the prior probability and $P(H|E)$ the posterior probability. We provided some discussion on a logical way of explaining this theorem (Cox 1946) and the subsequent studies that showed this was not quite as logical as it seems (Halpern 1995, 2011). Few people today know that Bayesian probability has 6 axioms (Dupré and Tiplery 2009). Despite these perhaps rather technical difficulties, a simple logic underlies this rule. Bayes' theorem states that the extent to which some evidence supports a hypothesis is proportional to the degree to which the evidence is predicted by the hypothesis. If the evidence is very likely ("Sandstone has lower acoustic impedance than shale) then the hypothesis ("Acoustic impedance depends on mineral composition") is not supported significantly when indeed we measure that "Sandstone has lower acoustic impedance than shale". If, however, the evidence is deemed very unlikely, (e.g. "Shale has higher acoustic impedance than sandstone"), then the hypothesis of another theorem ("acoustic impedance depends not only on mineralization, but also fluid content") will be highly confirmed (have high posterior probability).

Another interesting concept is how Bayes deals with multiple evidences of the same impact on the hypothesis. Clearly, more evidence leads to an increase in the probability of a hypothesis supported by that evidence. But evidences of the same impact will have a diminishing effect. Consider that a hypothesis has as equal probability as some alternative hypothesis:

$$P(H) = 0.5$$

Now consider multiple evidence sources such that

$$P(H|E_1) = 0.8; P(H|E_2) = 0.8; P(H|E_3) = 0.8;$$

Then according to a model of conditional independence and Bayes' theorem (Bordley 1982; Journel 2002; Clemen and Winkler 2007):

$$P(H|E_2, E_1) = 0.94; P(H|E_3, E_2, E_1) = 0.98;$$

Compounding evidence leads to increasing probability of the hypothesis.

### Objective Versus Subjective Probabilities

In the early days of the development of Bayesian approaches, several general principles were stated under which researchers "should" operate, resulting in an "objective" approach to the problem of inference, in the sense that everyone is following that same logic. One such principle is the principle of maximum entropy (Jaynes 1957), of which the principle of indifference (Laplace) is a special case. Subjectivists do not see probabilities as objective (leading to prescribing zero probabilities to well-confirmed ideas). Rather, subjectivists (Howson et al. 1993) see Bayes' theorem as an objective theory of inference. Objective is the sense that *given* prior probabilities and evidence, posterior probabilities are calculated. In that sense, subjective Bayesian make no claim on the nature of the propositions on which inference is being made (in that sense, they are also deductive).

One interesting application of reasoning in this way results when disagreement occurs on the same model. Consider modeler A (the conformist) who assigns a high probability to some relatively well-accepted modeling hypothesis and low probability to some rare (unexpected) evidence. Consider modeler B (the skeptic) who assigns low probability to the norm and hence high probability to any unexpected evidence. Consequently, when the unexpected evidence occurs and hence is confirmed $P(E|H)=1$, then the posterior of each is proportional to $1/P(E)$. Modeler A is forced to increase their prior more than the Modeler B. Some Bayesians therefore state that the prior is not that important as continued new evidence is offered. The prior will be "washed out" by cumulating new evidence. This is only true for certain highly idealized situations. It is more likely that two modelers will offer two hypotheses, hence evidence needs to be evaluated against each other. However, there is always a risk that neither model can be confirmed, regardless how much evidence is offered, hence the prior model space is incomplete, which is the exact problem of the objectivist Bayes. Neither objective nor subjective Bayes' addresses this problem.

**Bayes with Ad Hoc Modifications**

Returning now to the example of Fig. 27.5. Bayesian theory, if properly applied allow for assessing these ad hoc model modifications. Consider that a certain modeling assumption $H$ is prevailing in multi-phase flow: "oil flow occurs in rock with permeability of 10-10000 md" $(H)$, now this modeling assumption is modified ad hoc to "oil flow occurs in rock with permeability of 10-10000md and 100-200D $(H \cap AdHoc)$. However, this ad hoc modification, under $H$, has very low probability, $P(AdHoc) \simeq 0$ and hence $P(H \cap AdHoc) \simeq 0$. The problem, in reality is that those making the ad hoc modification often do not use Bayesianism, hence never assess or use the prior $P(AdHoc)$.

**Criticism of Bayesianism**

What is critical to Bayesianism is the concept of "background knowledge". Probabilities are calculated based on some commonly assumed background knowledge. Recall that theories cannot be isolated and independently tested. This "background" consists of all the available assumptions tangent to the hypothesis at hand. The problem often resulting with using Eq. (27.11) is that such "background knowledge" $BK$ is taken implicit:

$$P_{BK_0}(H|E) \simeq P_{BK_0}(E|H)P_{BK_0}(H) \rightarrow P_{BK_1}(H) \tag{27.12}$$

where 0 indicated at time $t=0$. The posterior then includes the "new knowledge" which is included in the new background knowledge at the next stage $t=1$. A problem occurs when applying this to the real world: what is this "background knowledge"? In reality, the prior and likelihood are not determined by the same person. For example, in our application, the prior may be given by a geologist, the likelihood by a data scientist. It is unlikely that they have the same "background knowledge" (or even agree on it). A more "honest" way of conveying this issue is to make the background knowledge explicit. Suppose that $BK^{(1)}$ is the background knowledge of person 1, who deals with evidence (the data scientist) then

$$P\left(H|E\cap BK^{(1)}\right)\simeq P\left(E\cap BK^{(1)}|H\right)P\left(H|BK^{(1)}\right) \tag{27.13}$$

Suppose $BK^{(2)}$ is person 2 (geologist) who provides the "prior", meaning provides background knowledge on his/her own, without evidence. Then, the new posterior can be written as

$$P\left(H|E\cap BK^{(1)}\cap BK^{(2)}\right)\simeq P\left(E\cap BK^{(2)}|H\right)P\left(H|BK^{(2)}\right)P\left(H|BK^{(1)}\right) \tag{27.14}$$

assuming however, there is no overlap between background knowledge. In practice, the issue that different components of the "system" (model) are done by different modelers with different background knowledge is ignored. Even if one would be aware of this issue, it would be difficult to implement in practice. The ideal Bayesian approach rarely occurs. No single person understands all the detailed aspects of the scientific modeling study at hand. A problem then occurs with dogmatism. The study in Fig. 27.5 illustrates this. Hypotheses that are given very high probability (no fractures) will remain high, particularly in the absence of strong evidence (low to medium $P(E)$). Bayes' rule will keep assigning very high probabilities to such hypotheses, particularly due to the dogmatic belief of the modeler or the prevailing leading idea of what is going on. This is not the problem of Bayes', but its common (faulty) application. Bayes' itself cannot address this.

More common is to select a prior hypothesis based on general principles or mathematical convenience, for example using a maximum entropy principle. Under such a principle, complete ignorance results in choosing for uniform distribution. In all other cases, one should pick the distribution that makes the least claims, from whatever information is currently available, on the hypothesis being studied. The problem here is not so much the ascribing of uniform probabilities but providing a statement of what all the possibilities are (on which then uniform probabilities are assigned). Who chooses these theories/models/hypotheses? Are those the only ones?

The limitation therefore of Bayesianism is that no judgment is leveled to the stated prior probabilities. Hence, any Bayesian analysis is as strong as the analysis of the prior. In subsurface modeling this prior is dominated by the geological understanding of the system. Such geological understanding and its background knowledge is vast, but qualitative. Later we will provide some ideas on how to make quantitative "geological priors".

### Deductive Testing of Inductive Bayesianism

The leading paradigm of Bayesianism is to subscribe to an induction from of reasoning: learning from data. Increasing evidence will lead to increasing probabilities of certain theories, models or hypothesis. As discussed in the previous section, one of the main issues lies in the statement of a prior distribution, the initial universe of possibilities. Bayesianism assume that a truth exists, that such truth is generated by a probability model, and also than any data/evidence are generated from this model. The main issue occurs when the truth is not even with the support (the range/span) generated by this (prior) probability model. The truth is not part of this initial universe. What happens then? The same goes when the error distribution on the data is chosen at too optimistic a level, in which case the truth may be rejected. Can we verify this? Diagnose this? Figure out whether the problem lies with the data or the model? Given the complexity of models, priors, data in the real world this issue may in fact go undiagnosed if one stops the analysis with the generation of the posterior distribution. Gelman and Shalizi (2013) discuss how mis-specified prior models (the truth is not in the prior) may result in either no solution, multi-model solutions to problems that are unimodal or complete non-sense.

Recent work (Mayo 1996) started to look at these issues. They attempt to frame these tests within classical hypothesis testing. Recall that classical statistics rely on a deductive form of hypothesis testing, which is very similar in flavor to Popper's falsification. In a similar vein, some form of model testing can be performed posterior to the generation of the posterior. Note that Bayesian model averaging (Rings et al. 2012; Henriksen et al. 2012; Refsgaard et al. 2012; Tsai and Elshall 2013) or model selection are not tests of the posterior, rather, they are consequences of the posterior distribution, yet untested! Classical checks are whether posterior models match data, but these are checks based on likelihood (misfit) only.

Consider a more elaborate testing framework. These formal test rely on generating replicates of the data given some model hypothesis and parameters are the truth. Take a simple example of a model hypothesis with two faults ($H=$ two faults) and the parameters $\boldsymbol{\theta}$ representing those faults (e.g. dip, azimuth, length etc.). The bootstrap allows for a determination of achieved significance level ($ASL$) as

$$ASL(\boldsymbol{\theta}) = P\left(S\left(\mathbf{d}_{rep}\right) \geq S(\mathbf{d}_{obs})|H, \boldsymbol{\theta}\right) \tag{27.15}$$

here, we consider calculating some summary statistic of the data as represented by the function $S$. This summary statistic could be based on some dimension reduction method; for example, a first or second principal component score. The uncertainty on $\boldsymbol{\theta}$ is provided by its posterior distribution, hence we can sample various $\boldsymbol{\theta}$ from the posterior. Therefore we first sample $\mathbf{d}_{rep}$ from the following distribution (averaging out over posterior in $\boldsymbol{\theta}$)

$$P\left(\mathbf{d}_{rep}|H, \mathbf{d}_{obs}\right) = \int P\left(\mathbf{d}_{rep}|H, \boldsymbol{\theta}\right)P(\boldsymbol{\theta}|H, \mathbf{d}_{obs})d\boldsymbol{\theta} \tag{27.16}$$

and calculate average $ASL$ over the posterior distribution. Analytically this equals to

$$ASL = \int ASL(\boldsymbol{\theta})P(\boldsymbol{\theta}|H, \mathbf{d}_{obs})d\boldsymbol{\theta} \tag{27.17}$$

or for given limited sample $\boldsymbol{\theta}^{(\ell)}, \ell = 1, \ldots, L \sim P(\boldsymbol{\theta}|H, \mathbf{d}_{obs})$

$$ASL = \frac{1}{L}\sum_{\ell=1}^{L} ASL\left(\boldsymbol{\theta}^{(\ell)}\right) \tag{27.18}$$

These tests are not used to determine whether a model is true, or even should be falsified but whether discrepancies exist between model and data. The nature of the functions $S$ defines the "severity" of the tests (Mayo 1996). Numerous complex functions will allow for a more severe testing of the prior modeling hypothesis. We can learn how the model fails by generating several of these summary statistics, each representing different elements of the data (a low, a middle and some extreme case etc.…).

Within this framework of deductive tests, the prior is no longer treated as "absolute truth", rather the prior becomes a modeling assumption that is "testable" given the data. Some may however disagree on this point: why should the data be any better than the prior? In the next section, we will try to get out of this trap, by basing priors on physical processes, with the hope that such priors are more realistic representations of the universe of variability, rather than simply relying on statistical methods that are devoid of physics.

## 27.9 Bayesianism for Subsurface Systems

### What is the Nature of Geological Priors?

### Constructing Priors from Geological Field Work

In a typical subsurface system, the model variables are parameterized in a certain way, for example with a grid, or a set of objects with certain lengths, widths dips, azimuths etc. What is the prior distribution of these model variables? Since we are dealing with a geological system, e.g. a delta, a fluvial or turbidite systems, a common approach is to do geological field work. This entails measuring and interpreting the observed geological structures, on outcrops, and creating a history of their genesis, with an emphasis on generating (an often qualitative) understanding of the processes that generated the system. The geological literature contains a vast amount of such studies.

To gather all this information and render it relevant for modeling UQ, geological databases based on classification systems have been compiled (mostly by the Oil industry). Analog databases, for example, on proportions, paleo-direction, morphologies and architecture of geological bodies or geological rules of association (Eschard and Doligez 2000; Gibling 2006) for various geological environments (FAKT: Colombera et al. 2012; CarbDB: Jung and Aigner 2012; WODAD: Kenter and Harris 2006; Paleoreefs: Kiessling and Flügel 2002; Pyrcz et al. 2008) have been constructed. Such relational databases employ a classification system based on geological reasoning. For example, the FAKTS database classifies existing studies, whether literature-derived or field-derived from modern or ancient river systems, according to controlling factors, such as climate, and context-descriptive characteristics, such as river patterns. The database can therefore be queried on both architectural features and boundary conditions to provide the analogs for modeling subsurface systems. The nature of the classification is often hierarchical. The uncertain style or classification, often termed "geological scenario" (Martinius and Naess 2005) and variations within that style.

While such approach appears to gather information, it leaves the question of whether the collection of such information and the extraction of parameters values to state prior distribution produce realistic priors (enough variance, limited bias) for what is actually in the subsurface. Why?

- Objects and dimensions in the field are only apparent. An outcrop is only a 2D section of a 3D systems. This invokes stereological problems in the sense structural characteristics (e.g. shape, size, texture) of 2D outcrops are only apparent properties of the three-dimensional subsurface. These apparent properties can drastically change depending on the position/orientation of the survey (e.g. Beres et al. 1995). Furthermore, interpreted two-dimensional outcrops of the subsurface may be biased because large structures are more frequently observed than small structures (Lantuéjoul 2013). The same issue occurs for those doing 2D geophysical surveys to interpret 3D geometries (Sambrook Smith et al. 2006). For example, quantitative characterization of two-dimensional ground penetrating radar (GPR) imaging (e.g. Bristow and Jol 2003) ignore uncertainty on the three-dimensional subsurface characteristics resulting from the stereological issue.
- The database is purely geometric in nature. It records the end-result of deposition not the process of deposition. In that sense it does not include any physics underlying the processes that took place and therefore may not capture the complexity of geological processes fully to provide a "complete" prior. For that reason, the database may aggregate information that should not be aggregated, simply because each case represents different geological processes, accidently creating similar geometry. For modeling, this may appear irrelevant (who cares about the process), yet it is highly relevant. Geologists reason based on geological processes, not just the final geometries, hence this "knowledge" should be part of a prior model construction. Clearly prior should not ignore important background knowledge, such as process understanding.

The main limitation is that this pure parameterization-based view (the geometries, dimensions) lacks physical reasoning, hence ignore important prior information. The next section provides some insight into this problem as well as suggests a solution.

### Constructing Priors from Laboratory Experiments

Depositional systems are subject to large variability whose very nature is not fully understood. For example, channelized transport systems (fan, rivers, delta, etc.) reconfigure themselves more or less continually in time, and in a manner often difficult to predict. The configurations of natural deposits in the subsurface are thus uncertain. The quest for quantifying prior uncertainty necessitates understanding the sedimentary systems by means of physical principles, not just information principles (such as the principle of indifference). Quantifying prior uncertainty thus requires stating all configurations of architectures of the system deemed *physically* possible and at what frequency (a probability density) they occur. This probability density need not be Gaussian or uniform. Hence, the question arises: what is this probability density for geological systems, and how does one represent it in a form that can be used for actual predictions using Bayesianism?

The problem in reality is that we observe geological processes over a very short time span (50 years of satellite data and ground observations), while the deposition of the relevant geological systems we work with in this chapter may span 100.000 years or more. For that reason, the only way to study such system is either by computer models or by laboratory experiment. These computer models solve a set of partial differential equations that describe sediment transport, compaction, diagenesis, erosion, dissolution, etc. (Koltermann and Gorelick 1992; Gabrovsek and Dreybrodt 2010; Nicholas et al. 2013). The main issue here is that PDEs are a limited representation of the actual physical process and require calibration with actual geological observations (such as erosion rules), require boundary conditions and source terms. Often their long computing times limit their usefulness for constructing complete priors.

For that reason, laboratory experiments are increasingly used to study geological deposition, simply because physics occurs naturally, and not as constructed with an artificial computer code. Next, we provide some insight into how laboratory experiments work and how they can be used to create realistic analogs of depositional systems.

## Experimenting the Prior

We consider a delta constructed in an experimental sedimentary basin subject to constant external boundary conditions (i.e. sediment flux, water discharge, subsidence rates), see Fig. 27.8. The data set used is a subset of the data collected during

**Fig. 27.8** Flume experiment of a delta with low Froude number performed by John Martin, Ben Sheets, Chris Paola and Michael Kelberer. Image *source* https://www.esci.umn.edu/orgs/seds/Sedi_Research.htm

an experiment in the Tulane Delta Basin, conducted in 2010 (Wang et al. 2011). Basin dimensions were 4.2 m long, 2.8 m wide and 0.65 m deep. The sediment consisted of a mix of 70% quartz sand and 30% anthracite coal sand. These experiments are used for a variety of reasons. One of them is to study the relationship between the surface processes and the subsurface deposition. An intriguing aspect of these experiments is that much of the natural variability is not due to forcing (e.g. uplift, changing sediment source), but due to the internal dynamics of the system itself, i.e. it is autogenic. In fact, it is not known if the autogenic behavior of natural channels is chaotic (Lanzoni and Seminara 2006), meaning one cannot predict with certainty the detailed configuration of even a single meandering channel very far into the future. This then has an immediate impact on uncertainty in the subsurface in the sense that configuration of deposits in the subsurface cannot be predicted with certainty away from wells. The experiment therefore investigates uncertainty related to the dynamics of the system, our lack of physical understanding (and not some parameter uncertainty or observational error). All this is a bit unnerving, since this very fundamental uncertainty is *never* included in any subsurface UQ. At best, one employs a Gaussian prior, or some geometric prior extracted from observation databases, as discussed above. The fundamental questions are:

1. Can we use these experiments to construct a realistic prior, capturing uncertainty related to the physical processes of the system?
2. Can a statistical prior model represent (mimic) such variability?

To address these questions and provide some insight (not an answer quite yet!), we run the experiment under constant forcing for long enough to provide many different realizations of the autogenic variability—a situation that would be practically impossible to find in the field. The autogenic variability in these systems is due to t temporal and spatial variability in the feedback between flow and sediment transport, weaving the internal fabric of the final subsurface system.

Under fixed boundary conditions, the observed variability in deposition is therefore the result of only the autogenic (intrinsic) variability in the transport system. The data-set we use here is based on a set of 136 time-lapse overhead photographs that capture the dynamics of flow over the delta approximately every minute. Figure 27.9 shows representative images from this database. This set of images represents a little more than 2 h of experimental run time. Figure 27.9b shows the binary (wet-dry) images for the same set, which will be used in the investigation.

The availability of a large reference set of images of the sedimentary system enables testing any statistical prior by allowing a comparison of the variability of the resulting realizations, since all possible configurations of the system are known. In addition, the physics are naturally contained in the experiment (photographs are the result of the physical depositional processes). A final benefit is that a physical analysis of the prior model can be performed, which aids in understanding what depositional patterns should be in the prior for more sophisticated cases.

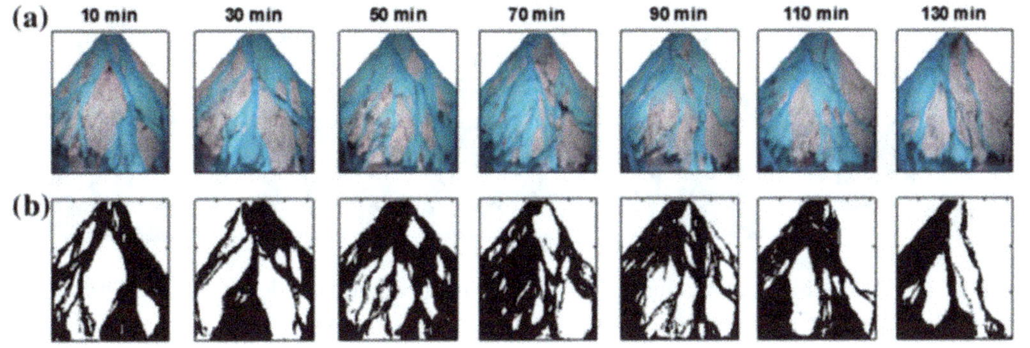


**Fig. 27.9** Examples of a few photographic images of the flume experiment for different time periods. Flow is from top to bottom. **a** Photographs of the experiments. The blue pixels indicate locations where flow moves over the surface. The black sediment is coal which is the mobile fraction of the sediment mixture, and the tan sediment is sand. **b** Binary representation of the photographs. Black represents wet (flow) pixels, white represents dry (no flow) pixels

### Reproducing Physical Variability with Statistical Models

In this study we employ a geostatistical method termed multiple-point geostatistics. MPS methods have grown popular in the last decade due to their ability to introduce geological realism in modeling via the training image (Mariethoz and Caers 2014). Similar to any geostatistics procedure, MPS allows for the construction of a set of stochastic realizations of the subsurface. Training images, along with trends (usually modeled using probability maps or auxiliary variables) constitute the prior model as defined in the traditional Bayesian framework. The choice of the initial set of training images has a large influence on the stated uncertainty, and hence a careful selection must be done to avoid artificially reducing uncertainty from the start.

It is unlikely that all possible naturally-occurring patterns can be contained in one single training image within the MPS framework (although this is still the norm; similarly, it is the norm to choose for a multi-Gaussian model by default). To represent realistic uncertainty realizations should be generated from multiple TIs. The set of all these realizations then constitutes a wide prior uncertainty model. The choice of the TIs brings a new set of questions: how many training images should one use, and which ones should be selected? Ideally, the TIs should be generated in such a way that natural variability of the system under study is represented (fluvial, deltaic, turbidite, etc.), hence all natural patterns are covered in the possibly infinite set of geostatistical realizations. Scheidt et al. (2016) use methods of computer vision to select a set of representative TIs. One such computer vision method evaluates a rate of change between images in time, and the training images are selected in periods of relative temporal pattern stability (see Fig. 27.10).

The training image set shown in Fig. 27.10 displays patterns consistent with previous physical interpretations of the fundamental modes of this type of delta system: a highly channelized, incisional mode; a poorly channelized, depositional mode; and an intermediate mode. This suggests that some clues to the selection of

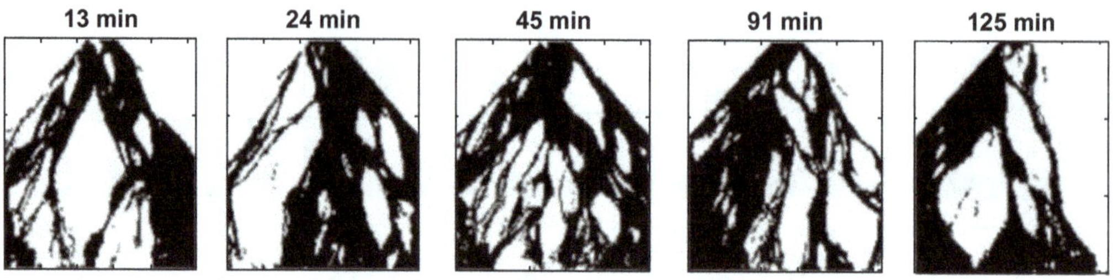


**Fig. 27.10** Selected images by clustering based on the modified Hausdorff distance. The value at the top of the image represents the time in minutes of the experiment

appropriate training images lie in the physical properties of the images from the experiment.

With a set of training images available, multiple geostatistical realization per each training image can be generated (basically a hierarchical model of realizations). These realizations can now be compared with the natural variability generated in the laboratory experiments, to verify whether such set of realizations can in any way reproduce natural variability. Scheidt et al. (2016) calculate the Modified Hausdorff Distance (MHD, a distance used in image analysis), between any two geostatistical realization and also between any two overhead shots A QQ-plot of the distribution of the MHD between all the binary snapshots of the experiment and the MPS models is shown in Fig. 27.11a, showing similarity in distribution.

The result is encouraging but also emphasizes a mostly ignored question of what a complete geological prior entails, that the default choices (one training image, one Boolean model, one multi-Gaussian distribution) make very little sense when dealing with realistic subsurface heterogeneity. The broader question remains as to how such a prior should be constructed from physical principles and how statistical models, such as geostatistics should be employed in Bayesianism when applied to

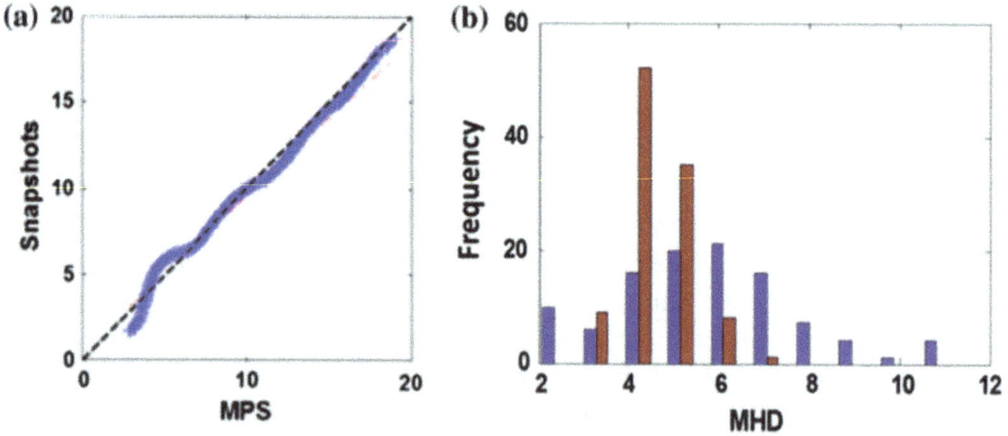


**Fig. 27.11** **a** QQ-plot of the MHD distances between the 136 images from the experiment and 136 images generated using DS. **b** Comparison of the variability, as defined by MHD, between generated realizations per each training image (red) and the images from the experiment (blue) closest (in MHD) to the selected TI

geological systems. This fundamental question remains unresolved and certainly under-researched.

**Field Application**

The above flume experiments have helped in the understanding of the nature of a geological prior, at least for deltaic type deposits. Knowledge accumulated from these experiments will create scientific understanding on the fundamental processes involved in the genesis of these deposits and thereby understand better the range of variability of the generated stratigraphic sequences.

It is unlikely, however, that laboratory experiments will be of direct use in actual applications, since they take considerable time and effort to set them up. In addition, there is a question of how these scale to the real world. It is more likely in the near future that computer models, built from such understanding, will be used in actual practice. Various such computer models exist for depositional systems (process-based, process-mimicking, etc.).

We consider here one such computer model, FLUMY (Cojan et al. 2005), which is used to model meandering channels, see Fig. 27.12. FLUMY uses a combination of physical and stochastic process models to create realistic geometries. It is not an object-based model, which would focus on the end result, but it actually creates the depositional system. The input parameters are therefore a combination of physical parameters as well as geometrical parameters describing the evolution of the deposition.

Consider a simple application to an actual reservoir system (Courtesy of ENI). Based on geological understanding generated from well data and seismic, modelers are asked to input the following FLUMY parameters: channel width, depth and sinuosity (geometric), and two aggradation parameters: (1) decrease of the alluvium thickness away from the channel, and, (2) maximum thickness deposited on levees during an overbank flood. More parameters exist but these are kept fixed for this simple application.

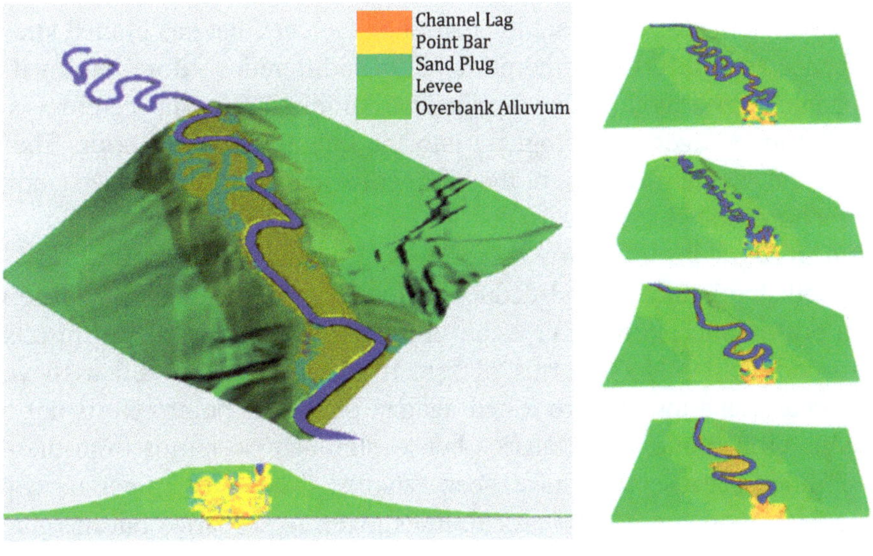


**Fig. 27.12** Example of a FLUMY model with several realizations of the prior generated from FLUMY with uncertain input parameters

The prior belief now consists of (1) assuming the FLUMY model as a hypothesis that describes variability in the depositional system and (2) prior distributions of the five parameters. After generating 1000 s of FLUMY models (see Fig. 27.12), we can run the same analysis as done for the flume experiment to extract modes in the system that can be used as training images for further geostatistical modeling.

## 27.10 Summary

Eventually philosophical principles will need to be translated into workable practices, ultimately into data acquisition, computer codes, and actual decisions. A summary of some important observations and perhaps also personal opinion based on this chapter are:

- *Data acquisition, modeling and predictions "collaborate";* going from data to models to prediction ignores the important interactions that take place between these components. Models can be used, prior to actual data acquisition to understand what role they will play in modeling and ultimately into the decision-making process. The often classical route of first gathering data, then creating models, may be completely inefficient if the data has no or little impact on any decision. This should be studied beforehand and hence requires building models of the data, not just of the subsurface.
- *Prior model generation is critical to Bayesian approaches* in the subsurface and statistical principles of indifference are very crude approximations of realistic geological priors. Uniform and multi-Gaussian distributions have been clearly falsified with many case studies (Gómez-Hernández and Wen 1998; Feyen and Caers 2006; Zinn and Harvey 2003). They may lead to completely erroneous predictions when used in subsurface applications. One can draw an analogy here with Newtonian physics: it has been falsified but it is still around, meaning it can be useful to make many predictions. The same goes with multi-Gaussian type assumptions. Such choices are logical for an "agent" that has limited knowledge and hence (rightfully) uses the principal of indifference. More informed agents will however use more realistic prior distribution. The point therefore is to use more informed agents (geologists) into the quantification of prior. The use of such agents would make use of the vast geological (physical) understanding that has been generated over many decades.
- *Falsification or prior.* It now seems logical to propose workflows of UQ that have both the induction and deduction flavors. Falsification should be part of any a priori application of Bayesianism, and also on the posterior results. Such approaches will rely on forms of sensitivity analysis as well as developing geological scenarios that are tested against data. The point here is not to state rigorous probabilities on scenarios but to eliminate scenarios from the pool of possibilities because they have been falsified. The most important aspect of geological priors are not the probabilities given to scenarios but the generation

of a suitable set of representative scenarios to represent the geological process taking place. This was illustrated in the flume experiment study.

- *Falsification of the posterior*. The posterior is the result of the prior model choice, the likelihood model choice and all of the auxiliary assumptions and choices made (dimension reduction method, sampler choices, convergence assessment etc. …). Acceptance of the posterior "as is" would follow the pure inductionist approach. Just as the prior, it would be good practice to attempt to falsify the posterior. This can be done in several ways, usual using hypothetico-deductive analysis, such as the significance tests introduced in this chapter.

## References

Ariew R (1984) The duhem thesis. Br J Philos Sci 35(4):313–325

Bayes M. & Price, M., 1763. An essay towards solving a problem in the Doctrine of chances. By the Late Rev. Mr. Bayes, F. R. S. communicated by Mr. Price, in a Letter to John Canton, A. M. F. R. S. Philos Trans R Soc Lond 53(0):370–418

Berger JO (2003) Could Fisher, Jeffreys and Neyman have agreed on testing? Stat Sci 18(1):1–32

Beres M, Green A, Huggenberger P, Horstmeyer H (1995) Mapping the architecture of glaciofluvial sediments with three-dimensional georadar. Geology 23(12):1087–1090

Bond CE et al (2007) What do you think this is? "Conceptual uncertainty" in geoscience interpretation. GSA Today 17(11):4–10

Bordley RF (1982) A multiplicative formula for aggregating probability assessments. Manage Sci 28(10):1137–1148

Bristow CS, Jol HM (2003) An introduction to ground penetrating radar (GPR) in sediments. Geol Soc 211(1):1–7. London, Special Publications

Chalmers AF (1999) What is this thing called science?, 3rd edn. Metascience, London

Clemen RT, Winkler RL (2007) Aggregating probability distributions. Adv Decis Anal 120(919): 154–176

Cojan I et al (2005) Process-based reservoir modelling in the example of meandering channel. In: Geostatistics Banff 2004. Springer, Netherlands, pp 611–619

Colombera L et al (2012) A database approach for constraining stochastic simulations of the sedimentary heterogeneity of fluvial reservoirs. AAPG Bull 96(11):2143–2166

Cox RT (1946) Probability, frequency and reasonable expectation. Am J Phys 14(1):1

Dubois D, Prade H (1990) The logical view of conditioning and its application to possibility and evidence theories. Int J Approx Reason 4(1):23–46

Dupré MJ, Tiplery FJ (2009) New axioms for rigorous Bayesian probability. Bayesian Anal 4(3): 599–606

Earman J (1992) Bayes or bust: a critical examination of Bayesian confirmation theory

Eschard R, Doligez B (2000) Using quantitative outcrop databases as a guide for geological reservoir modelling. In: Geostatistics Rio 2000. Springer, pp 7–17

Fallis A (2013) Fisher, Neyman and the creation of classical statistics

Feller W (2008) An introduction to probability theory and its applications. 2nd edn, vol 2, p xxiv-669

Feyen L, Caers J (2006) Quantifying geological uncertainty for flow and transport modeling in multi-modal heterogeneous formations. Adv Water Resour 29(6):912–929

Feyerabend, P., 1993. Against method

Fine A (1973) Probability and the interpretation of quantum mechanics. Br J Philos Sci 24(1):1–37

de Finetti B (1974) The value of studying subjective evaluations of probability. In: The concept of probability in psychological experiments, pp 1–14
de Finetti B, Machí A, Smith A (1975) Theory of probability: a critical introductory treatment
de Finetti B (1995) The logic of probability. Philos Stud 77(1):181–190
Fisher R (1925) Statistical methods for research workers. http://psychclassics.yorku.ca/Fisher/Methods
Fisher R, Fisher R (1915) Frequency distribution of the values of the correlation coefficient in samples from an indefinitely large population. Biometrika 10(4):507–521
Gabrovsek F, Dreybrodt W (2010) Karstification in unconfined limestone aquifers by mixing of phreatic water with surface water from a local input: a model. J Hydrol 386(1–4):130–141
Gelman A et al (2004) Bayesian data analysis
Gelman A (2008) Objections to Bayesian statistics Rejoinder. Bayesian Anal 3(3):467–477
Gelman A, Shalizi CR (2013) Philosophy and the practice of Bayesian statistics. Br J Math Stat Psychol 66(1996):8–38
Gibling MR (2006) Width and thickness of fluvial channel bodies and valley fills in the geological record: a literature compilation and classification. J Sediment Res 76(5):731–770
Gnedenko BV, Aleksandr I, Khinchin A (1962) An elementary introduction to the theory of probability
Gómez-Hernández JJ, Wen X-H (1998) To be or not to be multi-Gaussian? A reflection on stochastic hydrogeology. Adv Water Resour 21(1):47–61
Halpern J (2011) A counter example to theorems of Cox and Fine, vol 10, pp 67–85. arXiv:1105.5450
Halpern JY (1995) A logical approach to reasoning about uncertainty: a tutorial. In: Discourse, interaction, and communication, pp 141–155
Hand DJ, Walley P (1993) Statistical reasoning with imprecise probabilities. Appl Stat 42(1):237
Hanson (1958) Patterns of discovery. Philos Rev 69(2):247–252
Henriksen HJ et al (2012) Use of Bayesian belief networks for dealing with ambiguity in integrated groundwater management. Integr Environ Assess Manag 8(3):430–444
Höhle U (2003) Metamathematics of fuzzy logic
Howson C (1991) The “old evidence” problem. Br J Philos Sci 42(4):547–555
Howson C, Urbach P, Gower B (1993) Scientific reasoning: the Bayesian approach
Hume D (2000) A treatise of human nature. Treatise Hum Nat 26(1739):626
Jaynes ET (1957) Information theory and statistical mechanics. Phys Rev 106(4):181–218
Jaynes ET (2003) Probability theory: the logic of science. Math Intell 27(2):83
Jenei S, Fodor JC (1998) On continuous triangular norms. Fuzzy Sets Syst 100(1–3):273–282
Journel AG (2002) Combining knowledge from diverse sources: an alternative to traditional data independence hypotheses. Math Geol 34(5):573–596
Jung A, Aigner T (2012) Carbonate geobodies: hierarchical classification and database–a new workflow for 3D reservoir modelling. J Pet Geol 35:49–65
Kenter JAM, HarrisPM (2006) Web-based Outcrop Digital Analog Database(WODAD): archiving carbonate platform margins. In: AAPG international conference (November 2006). p 5–8
Kiessling W, Flügel E (2002) Paleoreefs—a database on Phanerozoic reefs
Klement EP, Mesiar R, Pap E (2004) Triangular norms. Position paper I: basic analytical and algebraic properties. In: Fuzzy sets and systems, pp 5–26
Klir GJ (1994) On the alleged superiority of probabilistic representation of uncertainty. IEEE Trans Fuzzy Syst 2(1):27–31
Kolmogoroff A (1950) Foundations of the theory of probability
Koltermann C, Gorelick S (1992) Paleoclimatic signature in terrestrial flood deposits. Science 256 (5065):1775–1782
Kuhn TS (1996) The structure of scientific revolution

Lantuéjoul C (2013) Geostatistical simulation: models and algorithms. Springer Science & Business Media
Lanzoni S, Seminara G (2006) On the nature of meander instability. J Geophys Res Earth Surf 111(4)
Lindgren B (1976) Statistical theory. MacMillan, New York
Loginov VJ (1966) Probability treatment of Zadeh membership function and their use in pattern recognition. Eng Cybern, 68–69
Mariethoz G, Caers J (2014) Multiple-point geostatistics: stochastic modeling with training images. Wiley Blackwell, Hoboken
Martinius AW, Naess A (2005) Uncertainty analysis of fluvial outcrop data for stochastic reservoir modelling. Pet Geosci 11(3):203–214
Mayo D (1996) Error and the growth of experimental knowledge. Chicago University Press, Chicago
Neyman J, Pearson ES (1967) Joint statistical papers
Neyman J, Pearson ES (1933) On the problem of the most efficient tests of statistical hypotheses. Philos Trans R Soc A Math Phys Eng Sci 231(694–706):289–337
Nicholas AP et al (2013) Numerical simulation of bar and island morphodynamics in anabranching megarivers. J Geophys Res Earth Surf 118(4):2019–2044
Pearson K, Fisher RA, Inman HF (1994) Karl Pearson and R. A. Fisher on statistical tests: a 1935 exchange from nature. Am Stat 48(1), 2–11
Popper ER (1959) The logic of scientific discovery. Hutchinson, London
Pyrcz MJ, Boisvert JB, Deutsch CV (2008) A library of training images for fluvial and deepwater reservoirs and associated code. Comput Geosci 34:542–60
Rao CR (1992) R. A. Fisher: the founder of modern statistics. Stat Sci 7(1):34–48
Refsgaard JC et al (2012) Review of strategies for handling geological uncertainty in groundwater flow and transport modeling. Adv Water Resour 36:36–50
Rings J et al (2012) Bayesian model averaging using particle filtering and Gaussian mixture modeling: theory, concepts, and simulation experiments. Water Resour Res 48(5)
Sambrook Smith GH, Ashworth PJ, Best JL, Woodward J, Simpson CJ (2006) The sedimentology and alluvial architecture of the sandy braided South Saskatchewan River, Canada. Sedimentology 53(2):413–434
Scheidt C et al (2016) Quantifying natural delta variability using a multiple-point geostatistics prior uncertainty model. J Geophys Res Earth Surf
Shackle GLS (1962) The stages of economic growth. Polit Stud 10(1):65–67
Shannon CE (1948) A mathematical theory of communication. Bell Syst Tech J 27(July 1928), pp 379–423
Tsai FTC, Elshall AS (2013) Hierarchical Bayesian model averaging for hydrostratigraphic modeling: uncertainty segregation and comparative evaluation. Water Resour Res 49(9):5520–5536
Valle A, Pham A, Hsueh PT, Faulhaber J (1993) Development and use of a finely gridded window model for a reservoir containing super permeable channels. In: Middle east oil show. Society of Petroleum Engineers
Wang P (2004) The limitation of Bayesianism. Artif Intell 158(1):97–106
Wang Y, Straub KM, Hajek EA (2011) Scale-dependent compensational stacking: an estimate of autogenic time scales in channelized sedimentary deposits. Geology 39(9):811–814
Zadeh L (1965) Fuzzy sets. Inf Control 8:338–353
Zadeh LA (1975) Fuzzy logic and approximate reasoning. D Reidel Publ Co 30(i):407–428
Zadeh LA (1978) Fuzzy sets as a basis for a theory of possibility. Fuzzy Sets Syst 1(1):3–28

Zadeh LA (2004) Fuzzy logic systems: origin, concepts, and trends. Science, pp 16–18
Zinn B, Harvey CF (2003) When good statistical models of aquifer heterogeneity go bad: a comparison of flow, dispersion, and mass transfer in connected and multivariate Gaussian hydraulic conductivity fields. Water Resour Res 39(3):1–19

# Chapter 28
# Geological Objects and Physical Parameter Fields in the Subsurface: A Review

**Guillaume Caumon**

**Abstract** Geologists and geophysicists often approach the study of the Earth using different and complementary perspectives. To simplify, geologists like to define and study objects and make hypotheses about their origin, whereas geophysicists often see the earth as a large, mostly unknown multivariate parameter field controlling complex physical processes. This chapter discusses some strategies to combine both approaches. In particular, I review some practical and theoretical frameworks associating petrophysical heterogeneities to the geometry and the history of geological objects. These frameworks open interesting perspectives to define prior parameter space in geophysical inverse problems, which can be consequential in under-constrained cases.

## 28.1 Introduction

The earth is three-dimensional, heterogeneous and, for its major part, inaccessible to direct observations. A consequence is that the static and dynamic parameters governing physical processes below the earth surface are generally poorly known. A recurrent challenge for geoscientists and engineers is, therefore, to predict the likely nature or behavior of the subsurface from limited data. In all fields of geophysics sensu lato, these forecasts may use physically and mathematically-based data processing (such as upward continuation of potential fields, seismic processing, classical processing of ground penetrating radar (Nobakht et al. 2013), reservoir production decline curves (Davis and Annan 1989; Fetkovich 1980, Fig. 28.1a), or the resolution of an inverse problem that explicitly uses physical models computing observations from some earth parameters and physical parameters (Fig. 28.1b–d, f–h). In geology, forecasts (e.g., about the location and volume of a specific formation or resource) and geological scenarios involve direct

G. Caumon (✉)
GeoRessources-ENSG, Université de Lorraine – CNRS–CREGU,
2 rue du Doyen Marcel Roubault, 54500 Vandoeuvre-lès-Nancy, France
e-mail: Guillaume.caumon@univ-lorraine.fr


B. S. Daya Sagar et al. (eds.), *Handbook of Mathematical Geosciences*,
https://doi.org/10.1007/978-3-319-78999-6_28

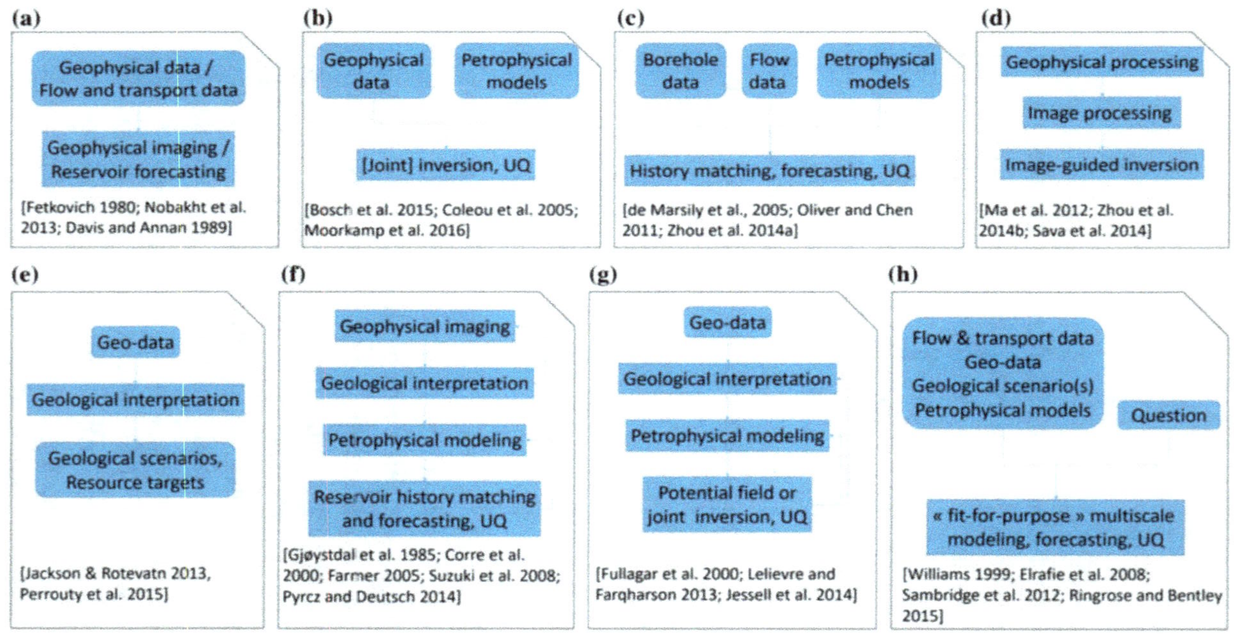


**Fig. 28.1** Examples of approaches and workflows using geological and geophysical data to make forecasts about the earth or the associated processes and some illustrative references. **a-d** workflows that use no or minimal geological prior information. **e** classical use of geophysical data in fundamental and applied geology. **f-h** workflows that explicitly incorporate geological parameters in the process

observations and geophysical images (Jackson and Rotevatn 2013; Perrouty et al. 2014; Fig. 28.1e). In this process, the loop may not always close: in the end, the interpretations are not guaranteed to be compatible with the initial geophysical observations. This may or may not be a problem, depending on the purpose of this interpretation. For example, a qualitative match between reflection seismic data and structural interpretations is probably sufficient to discuss fault growth models (Jackson and Rotevatn 2013), whereas such mismatch can be problematic in other tasks such as natural resource assessment (Caumon 2010; Jessell et al. 2014). Another practical problem is the interpretation and fusion of several independent data sets corresponding to different physical or geological observations (Corbel and Wellmann 2015; Paasche 2016). Geostatistics (Chiles and Delfiner 2012; Goovaerts 1997) was historically developed with these problems in mind, and is an attractive theoretical framework to recombine point and volume data coming from geophysical images consistently with spatial statistics. However geological reasoning and statistical reasoning are of different nature (Frodeman 1995), so honoring some spatial statistics is very useful but not always sufficient to represent geological concepts. Therefore, several methodologies have been introduced to explicitly incorporate geological knowledge in subsurface interpretation, all of them explicitly considering geological objects (Fig. 28.1f–h).

The main focus of this chapter is to review the main frameworks by which geological concepts can be represented in earth models and inverse methods addressing several types of physics. Thus, it aims at complementing the existing reviews and discussions of Linde et al. (2015) and Jessell et al. (2014), who address this problem with similar objectives but different perspectives. As the topic is very vast, the reader is also referred to previous review papers related to this topic (Farmer 2005; Lelièvre and Farquharson 2016; Linde et al. 2015; de Marsily et al. 2005; Mosegaard and Hansen 2016; Oliver and Chen 2011; Pyrcz et al. 2015; Zhou et al. 2014a). Several books also present complementary perspectives and more complete descriptions and details (Agterberg 2014; Caers 2011; Mallet 2002, 2014; Perrin and Rainaud 2013; Pyrcz and Deutsch 2014). Section 28.2 provides further motivations for considering geology in geophysical models, and tries to define what "geology" means in that sense. Then, Sect. 28.3 briefly describes the type of parameterizations classically used in computational physics. We discuss some links between these physical parameterizations and the frameworks used to represent geological domains in Sect. 28.4.

## 28.2 Motivations for Explicit Geological Parameterizations

A wealth of perspectives is essential and complementary to make progresses in the understanding of our planet and its resources. This is exemplified by the various disciplines involved in natural resource characterization, see for instance Ringrose and Bentley (2015). Feedbacks and interactions between the various approaches generate many types of possible workflows to integrate geological data and produce forecasts, as illustrated in Fig. 28.1. For example, geophysical processing and inverse methods that use minimal geological prior information (Fig. 28.1a–d) are typically considered as data for geological interpretations (Fig. 28.1e). Whereas these "minimal prior" approaches are not this chapter's focus, they are very useful and are always used to some extent in practical studies, because they provide at least a useful first-order view of the geological domain. This is illustrated in particular in deterministic workflows of Fig. 28.1h that strive for fit-for purpose, simplest as possible, subsurface models (Elrafie et al. 2008; Ringrose and Bentley 2015; Williams et al. 2004). They are also conceptually satisfying in the sense that they produce images or forecasts that mainly depend on the physics, hence can be claimed to be parsimonious and objective. As a consequence of this parsimony and of the non-linear nature of most involved physical processes, these models make it difficult to evaluate uncertainty (Watson et al. 2013). The term "objective" is also relative, as some choices are always made in these methods. In data processing methods these subjective choices relate to the underlying model assumptions (e.g., sub-horizontal layers). In inverse methods, choices must also be made about the parameterization, and a statistical model (e.g., the multi-Gaussian model) or a particular regularization (e.g., Mosegaard 2011).

Among the approaches that try to get the most out of the physics with minimal assumptions, recent and most promising developments use several types of data and petrophysical models to constrain local anisotropy, (see for instance Clapp et al. 2004; Ma et al. 2012; Sava et al. 2014; Zhou et al. 2014b) and recent reviews in geophysical imaging (Meju and Gallardo 2016), reservoir seismology (Bosch et al. 2010), hydrogeophysics (Linde and Doetsch 2016), mineral exploration (Lelièvre and Farquharson 2016), petroleum exploration (Moorkamp et al. 2016). Two main ideas underlie these approaches. First, some local structural orientations are inferred from borehole data or other geophysical data to constrain the covariance function used during inversion. Second, a petrophysical model is used to exploit the existing correlation between the physical parameters. As these correlations generally depend on the rock type, the model often includes discrete variables that estimate or sample the rock type at a given location. This notion of rock type is close to the notion of lithofacies, so it is a way to integrate geological reasoning into inverse methods.

In the field of reservoir engineering and hydrogeology, methods incorporating prior geological knowledge in flow and transport models have also been developed very early on, as discussed in several review papers (Farmer 2005; Linde et al. 2015; de Marsily et al. 2005; Oliver and Chen 2011; Zhou et al. 2014a). One fundamental reason is that flow and transport processes can be highly non-linear while pressure and concentration measurements are generally quite sparse as compared to the number of potential factors influencing fluid transfers in porous and fractured media. The same observation holds in potential field inversion, where geological prior information can significantly help addressing the ill-posedness of the inverse problem (Lelièvre and Farquharson 2016). But what does "geological prior" exactly mean?

As noted in particular by Frodeman (1995), geology is an interpretive science which includes a significant component of historical thinking. One aim of geology is to describe the earth in historical terms by identifying the main geological processes and their impact. In terms of scientific philosophy, it is interesting to highlight that geology generally produces refutable scenarios, whereas mathematics are concerned with formal and irrefutable proofs (given some hypotheses). The encounter of these two scientific methods is deeply written in the DNA of Mathematical Geosciences. Advanced methods in physically-based modeling have been developed to quantitatively model geological processes. Some very interesting inverse methods that use such models have been developed recently to quantitatively integrate spatial observations (Charvin et al. 2009; Cross and Lessenger 1999; Gallagher et al. 2009). These methods are ideal in the sense that they could in principle unify geology and geophysics rigorously. However, the interplay of multiple coupled physical and chemical processes at geological time scales remains extremely challenging to model on a computer. The use of such models in an inverse framework is also very challenging, as the number of unknown or poorly known parameters makes the inverse problem highly ill-posed and computationally intractable. This empty space problem is very general and applies to most inverse problems in geosciences, but it is critical when an explicit time dimension is considered because the density of information in time-space is very small (e.g., only a few points typically constrain pressure and temperature in basin studies). This explains why most of the methods in Fig. 28.1e–h do not explicitly consider geological time and instead use an object-based approach, a statistics-based approach or a combination of both to represent the geological prior information and make forecasts in the 3D physical space.

Classically, the object-based strategy is essential to the geological approach. For example, geological mapping typically decomposes a complex reality into discrete and interconnected tectonic, igneous, metamorphic, diagenetic, stratigraphic and sedimentological objects. These object definitions do integrate historical and process-based considerations. For instance, time is explicitly considered in the definition of the remarkable surfaces that sequence stratigraphers use to interpret geoscience data. The characterization of these objects in mathematical and computational terms has been a significant focus of the IAMG for that last 50 years. The statistics-based approach, another clear focus of the IAMG, is clearly complementary to the object-based approach. Indeed, objects are heterogeneous, boundaries between objects may be difficult to define and objects can be difficult to map from available observations. Statistical reasoning is key to address these problems. In this chapter, we will try to explain a few manners by which the object-based and statistics-based methods interact in the frame of geo-data and physical modeling integration. For this, we will start from the perspective of what physical modeling needs.

## 28.3 Parameterizations for Physical Models

Sambridge et al. (2012), among others, give a very crisp and generic summary of the parameterizations used in most numerical physical modelling methods. In this view, a model $m(\mathbf{x})$ is defined at any point $\mathbf{x}$ of the physical space by a set of basis functions:

$$m(\mathbf{x}) = \sum_{k=1}^{K} m_k \varphi_k(\mathbf{x}). \tag{28.1}$$

For example, in the finite element method with linear triangular elements, a basis function $\varphi_k$ is defined for each mesh vertex $\mathbf{x}_k$: $\varphi_k(\mathbf{x}_k)$ is equal to 1, $\varphi_k(\mathbf{x}_{j\neq k})$ is equal to 0 and $\varphi_k$ linearly decreases in the mesh elements adjacent to $\mathbf{x}_k$. The values $m_k$ are the parameter values (e.g., thermal conductivity) associated to the mesh vertices.

The general formulation (28.1) allow to compute or approximate differential operators to solve partial differential equations describing physical processes. Many recent advances in computational physics consist of particular choices of basis functions. For instance, in the extended finite element method, the use of Heaviside basis functions to represent internal discontinuities in a mesh was a step change in the computation of fracture growth (Moës et al. 2002). Another very active research field concerns the combination of basis functions at several scales (e.g., Efendiev et al. 2013). These methods have been applied for instance in finite volume modeling of flow in porous media to solve the flow and transport equations at two distinct and interacting scales (Jenny et al. 2003; Møyner and Lie 2014).

Equation (28.1) is also compatible with the theory of spatial random fields. At point scale, the values $m_k$ are seldom known below the Earth surface. Geostatistics offers many ways to estimate or simulate such values (Chiles and Delfiner 2012; Goovaerts 1997) using statistical parameters inferred from subsurface data. One of these parameters is the variogram, which models the statistical correlation between two variables as a function of the distance. In dual kriging, Eq. (28.1) is also used, as the unknown value is estimated as a linear combination of covariance functions centered on the data points. The use of point-based parameterizations is also much studied in computational physics under the term "meshless methods", see for instance Liu and Gu (2005). In the practice of geostatistical methods, the values $m_k$ are generally modeled on a Cartesian grid, but recent papers also discuss about the application of geostatistics on unstructured grids (Gross and Boucher 2015; Manchuk et al. 2005; Zaytsev et al. 2016), or directly on points (Zagayevskiy and Deutsch 2016). A major interest of these methods is to estimate or simulate values directly on the physical modeling support, and also to use adaptive resolution depending on the local information density and on the sensitivity between the model parameters and the physical process.

Last, but not least, Eq. (28.1) is compatible with a new breed of inverse methods in which the number of parameters $K$ is variable, see Sambridge et al. (2012) and references therein. These transdimensional inverse methods show much promise to address some of the challenges highlighted in this chapter.

The beauty of Eq. (28.1) lies in its potential to unify object-based geological descriptions and mathematical descriptions. In a sense, the goal of the various workflows described in Fig. 28.1 and in the associated references can be seen as a quest to find "geological basis functions" to model the earth. The purpose of Sect. 28.4 is to try to establish a more explicit correspondence between geological concepts and existing mathematical and computational models for representing geological domains in three-dimensional space. In doing so, we keep in mind that these 3D models will eventually need to be expressed by Eq. (28.1) in physical models.

## 28.4 Geological Parameterizations

As discussed in Sect. 28.2, geologists apply the divide and conquer principle to analyze the earth. Hundreds of years of geological reasoning have essentially led to identify multiple geological features at various scales, depending on their origin:

- Tectonic objects: Faults, joints, folds, cleavages.
- Sedimentary objects: stratigraphic units, horizons and unconformities, sedimentary bodies, facies, bedding structures.
- Intrusive and effusive objects: salt diapirs, salt sheets, shale diapirs, shale mounds, sills or dykes, lava flows, etc.
- Epigenetic objects (originating from chemical and mineralogical processes after rock formation): Metamorphic units, hydrothermalized facies, dissolved rocks (karsts).

These features typically exist at kilometric to micrometric scales (from plates to minerals and fluid inclusions). It is not useful (and not possible) for a model to explicitly represent all objects across these scales. Rather, most modeling approaches hierarchically subdivide the domain to represent a few nested scales (Pyrcz and Deutsch 2014; Ringrose and Bentley 2015).

Two main complementary mathematical and numerical frameworks exist to represent these geological features: spatial random fields and object-based methods. The choice of which framework is most appropriate (or whether and how these frameworks should be combined) depends on the size of the features with regard to the density of observations and on the likely impact of the features for the question at hand. Whereas the object size can be objectively discussed and characterized, the impact of features is often based on rules of thumb derived from experience (Ringrose and Bentley 2015). This may be a source of biases in forecasts. In practical studies, choices may also be constrained by very practical reasons, as some methods are implemented only in commercial software or in distinct software which are not interoperable. These problems and the need for better and abstract knowledge integration are also discussed by Perrin and Rainaud (2013).

### 28.4.1 Spatial Random Fields

As geological processes are not random and result from many physical processes, the resulting spatial fields are generally correlated in space. The characterization of the correlation structure by statistical inference is an essential aspect of geostatistics (Chiles and Delfiner 2012; Goovaerts 1997). Indeed, trust can be gained when data are numerous enough to provide robust statistics—even though the modeling assumptions themselves may remain questionable (Journel 2005). In inverse modeling of flow and transport in porous media, this has led to many approaches that perturb parameters on a grid while preserving variogram or spatial covariance models (de Marsily et al. 2005; Oliver and Chen 2011; Zhou et al. 2014a).

In geostatistics, a result of the divide-and-conquer strategy used in geology is the definition of many types of discrete categories to describe the physical world. These categories can be localized in space in the form of a geological map (or, in three dimensions, a 3D geological model). From a geostatistical standpoint, categories can be modeled with indicator variables. This has led to significant advances, in particular in the field of multiple-point geostatistics (MPS), to represent discrete facies from sparse data and analog training images. Since the seminal work of Guardiano and Srivastava (1993), a vast community of mathematical geoscientists has embraced this field and made essential advances, see Hu and Chugunova (2008), Mariethoz and Caers (2014). In particular, MPS have opened concrete and effective ways to using complex (and deliberately subjective) geological priors models in inversion (Linde et al. 2015; Melnikova et al. 2015). MPS have shown, in a number of instances, the impact of applying analog reasoning and scenarios to find sensible sets of solutions to inverse problems and to assess uncertainties. They also make up an interesting formalism to analyze complex geological systems (Scheidt et al. 2016).

However, even though progresses can still be made (see for instance Renard et al. 2011), a recurrent challenge with the indicator geostatistical approaches is to ensure that some categories are always connected or adjacent to other categories. This is why, to echo a friendly discussion we had with Andre Journel in 2005, I persist considering that there is more to geological realism than MPS (in its spatial understanding). The Truncated Gaussian method and the Pluri-Gaussian methods (Armstrong et al. 2011), even though they rely on multi-Gaussian assumptions, enforce continuity conditions that approach geological reasoning in a very interesting way. For instance, they can produce consecutive successions of facies from shallow marine to offshore environments. This type of method is appropriate when the discrete geological categories originate from an underlying continuous variable (in the previous example, this variable can be assimilated to bathymetry, all facies being defined between consecutive threshold values). In the Pluri-Gaussian approach, the application of Boolean operations on simulated random fields is also a way to emulate the succession of geological events (e.g. simulation of late diagenetic facies overprinting the depositional facies).

In general, spatial random field methods are implemented on grids of fixed resolution. As a result, the discontinuities that may exist in the medium are sampled at that particular resolution. However, some important features such as fractures or shale lenses may be much smaller than the grid resolution, hence cannot be explicitly represented in the grid. Under some hypotheses, this can be addressed by directly modeling a field of equivalent properties assumed representative of the block scale (e.g., equivalent dual porosity and dual permeability fields in fractured media). However, this can be a source of bias in a number of cases (Jackson et al. 2014). The explicit consideration of these objects generally relies on fewer assumptions and provides a way to deal with more complex geometries and with spatial observations, as will be discussed in Sect. 28.4.2. Note that these two approaches are not mutually exclusive and a combination of both equivalent and explicit approaches are, in general, relevant (Bourbiaux et al. 2002; Maier et al. 2016).

Another important aspect of geological reality is that the orientation and the magnitude of spatial correlation can vary in space. This can be modelled with random fields using locally varying anisotropy (Boisvert et al. 2009; Stroet and Snepvangers 2005; Xu 1996). In geophysics, the use of local anisotropy is illustrated for instance by Clapp et al. (2004) and by the image-guided inversion methods mentioned in Sect. 28.2 and Fig. 28.1d. In the absence of exhaustive data to constrain these orientations, one should estimate or simulate the orientations away from local observations (Gumiaux et al. 2003; Stroet and Snepvangers 2005; Xu 1996). A practical challenge in the presence of locally varying anisotropy is the inference of geostatistical parameters, as the domain is non-stationary. Object approaches offer another way of dealing with locally varying anisotropy, as will be discussed in the next section.

## 28.4.2 *Object Models*

In a general sense, object models directly represent the tectonic, sedimentological, intrusive and epigenetic features listed at the beginning of Sect. 28.4. As geological objects originate from distinct geological processes at different periods of time, they often correspond to contrasts or discontinuities of the physical parameters of interest. This explains why, beyond pure cartographic goals, so much effort is dedicated to object modeling in geosciences.

### Geometry and Topology

As discussed by Mallet (2002) and Perrin and Rainaud (2013), geological objects can be represented in geometrical and topological terms. Topology refers to essential characteristics: the dimension of objects (line, surface or volume), whether objects have inclusions or holes, and if they are connected to other objects. Depending on the type of geological objects, some topological configurations are impossible (Caumon et al. 2004). For instance, a chronostratigraphic horizon must be an open surface and may include internal holes due to faults or intrusions. More generally, the continuity of objects can have a relation to the genesis of the object, hence is a way to constrain geological models. Knowing what is topologically possible and what is not gives precious insights to design modeling methods and to test the validity of geological models (Pellerin et al. 2017; Wellmann et al. 2014). Topological analysis also provides interesting metrics to characterize and understand geological objects such as karsts (Collon et al. 2017), fracture networks (Sanderson and Nixon 2015) and structural models (Lindsay et al. 2013; Pellerin et al. 2015; Thiele et al. 2016a, b). Last, but not least, topology is very important for flow modeling, as it directly relates to the connectivity of permeability conduits and barriers. The links between connectivity and effective flow properties has been much studied at multiple scales in the frame of percolation theory (Berkowitz and Balberg 1993; King et al. 2001). In the cases where geological considerations are not sufficient to fully characterize the topology of the medium, specific methods have been proposed to find possible object geometry honoring some prescribed connectivity (Borghi et al. 2012; Collon-Drouaillet et al. 2012; Henrion et al. 2010).

Geometry concerns the embedding of the topological objects in 3D space, and is typically described either analytically (e.g., ellipses for fractures) or numerically (using a mesh). Meshes provide much flexibility to discretize the geometry of rock volumes (geological bodies), surfaces (geological boundaries) and lines (contacts between boundaries). All these geometric components are linked by topological relationships (Pellerin et al. 2017). More fundamentally, meshes are a way to define basis functions approximating the geometry of the true object. For example, one can define mathematically a triangulated surface as a set of a "hat" basis functions centered on each surface node (taking the value 1 at each node and linearly decreasing it to zero at the node's neighbors), as in Eq. (28.1). This description is very powerful to devise advanced geometry processing algorithms and reduce the dimensionality of complex geometrical shapes (Vallet and Lévy 2008). In the frame of inverse modeling, several inverse methods use the meshed model geometry as an unknown parameter (Fullagar et al. 2000; Gjøystdal et al. 1985; Mondal et al. 2010).

Over the past decade, computational advances have also made it possible to consider implicit surfaces to represent geological boundaries. In these approaches, the surfaces are considered as level sets of some three-dimensional scalar field (Calcagno et al. 2008; Cowan et al. 2003; Frank et al. 2007; Henrion et al. 2010). These methods share the same principles as the Truncated Gaussian and Pluri-Gaussian methods (Mannseth 2014), but the underlying random function model is not necessarily Gaussian, and their focus is set on the geometry of object boundaries. These level set methods are very powerful to automate geometric modeling tasks such as interpolation and extrapolation. In particular, they have shown much interest in stratigraphic modeling as one single scalar field can represent a conformable stratigraphic series at once, which opens new possibilities in structural data interpolation (Calcagno et al. 2008; Caumon et al. 2013; Hillier et al. 2014; Laurent et al. 2016). Implicit surfaces also offer very nice ways to consider geometric model perturbations needed to address inverse problems in geosciences (Cardiff and Kitanidis 2009; Caumon et al. 2007; Noetinger 2013; Zheglova et al. 2013). A major distinction between explicit and implicit surface models is about topological control: the surface topology has to be chosen before interpolation in explicit methods, whereas it emerges from the interpolation in implicit models, see also Collon et al. (2016) for more discussions.

As in Pluri-Gaussian simulation, it is possible to indirectly account for geological time in object models using the truncation between implicit or explicit objects (Calcagno et al. 2008; Caumon et al. 2009; Gjøystdal et al. 1985). Boolean operations also provide ways to obtain sharp features in object geometry using constructive solid geometry principles (Rongier et al. 2014; Ruiu et al. 2016). In terms of Eq. (28.1), Boolean operations between implicit objects can be described as indicator (or Heaviside) basis functions (Mannseth 2014; Moës et al. 2002): these functions are equal to zero on one side of the interface and equal to 1 on the other side. The representation of faults is a major challenge which is specific to geosciences. Indeed faults are not just discontinuities or sharp geometric features: they result from sliding of rocks that were previously connected. Several authors have proposed mathematical or numerical solutions to address this problem by considering directly or indirectly the displacement between either sides of a fault (Calcagno et al. 2008; Georgsen et al. 2012; Hale 2013; Holden et al. 2003; Jessell and Valenta 1996; Laurent et al. 2013; Mallet 2002, 2014).

**From Objects to Physical Parameters**

Generally, geological object geometry cannot be described analytically and determining the associated physical parameter fields is not straightforward. In most cases, objects are first discretized in space with a mesh that will support the numerical resolution of the physical equations (Kolditz et al. 2012; Pellerin et al. 2017). This mesh is a numerical translation of Eq. (28.1) discretizing the space in elementary volumes deemed representative of some effective physical properties (the values $m_k$ in Eq. (28.1)).

A possible working assumption is to consider a constant (or analytically defined) parameter value associated to each type of geological object. This principle is used for simplicity in a number of numerical models (Gjøystdal et al. 1985; Jackson et al. 2015). However, as discussed above, heterogeneity exists at many different scales and can have an impact on the physical process below the scale of the objects that are explicitly represented in a numerical model. For example, it is well known in stochastic hydrogeology and reservoir engineering that petrophysical heterogeneity exists within layers or sedimentary facies and impacts flow and transport (see for instance de Marsily et al. 2005 for a review). In many cases, the orientation of heterogeneities within a geological object depends on the object geometry (e.g., crystal orientations in a dyke may be preferentially aligned along the dyke boundaries; sedimentary heterogeneities tend to be more continuous along layers than orthogonally to layers). This can be addressed in modeling by explicitly using locally variable directions of anisotropy (Boisvert et al. 2009) or by considering a geometric transform between two spaces (Mallet 2014; Shtuka et al. 1996). This last option is very promising as it provides a way to simplify geostatistical modeling, and as it allows to define some useful geological variables such as the apparent sedimentation rate (Kedzierski et al. 2007; Mallet 2014; Massonnat 1999). Such use of indirect geological parameters is an essential and powerful way to introduce geological principles in earth models.

Nonetheless, one should not neglect that object geometry affects model predictions at the two main stages of geostatistical models: (1) geostatistical inference (distributions of continuous variables within each subdomain, multivariate relationships between different variables, trends, spatial variability) and (2) geostatistical modeling (interpolation or simulation). The separation of integrated modeling into an object-modeling phase and a petrophysical modeling phase are, therefore, relatively easy in the classical case where objects are known, when a clear separation of scales exist between representative elementary volume (REV) properties and object geometry, and when objects do not affect geostatistical parameters. However, uncertainty about object geometry and topology can have a significant impact on statistical parameters (Lallier et al. 2016), which can be a significant source of complexity in practical studies. More generally, finding at what scale explicit objects properties and REV effective properties can be separated is a fundamental problem in modeling. Therefore, more research is clearly needed to capture the interactions between object geometric (and topological) parameters and random field parameters.

### Object Uncertainty

Geometric uncertainty can be sampled by adding geometric perturbations to an existing reference model (Caumon et al. 2007; Corre et al. 2000; Lecour et al. 2001) or creating several models after perturbing data (Lindsay et al. 2013; Wellmann et al. 2010). As the very existence of some objects is also uncertain in many cases, it is also useful to consider object-based stochastic simulation. In random set theory, geometric objects are placed randomly and independently in the domain by combining the simulation of points (Poisson Point Process) and the simulation of objects shapes around these points (see Chiles and Delfiner 2012 and references therein; Lantuéjoul 2002). Classically, objects are geometric primitives defined analytically, whose shape, orientation and size parameters are simulated from some input distribution. Random set theory places a lot of emphasis on the statistical aspects of this process and on conditioning to spatial data, see in particular Lantuéjoul (2002) and Allard et al. (2006). These models, in particular the Boolean Model, have been used to simulate many types of geological objects such as fractures (Chiles 1988), shale lenses (Haldorsen and Lake 1984) or sedimentary channels (Deutsch and Wang 1996; Holden et al. 1998). Extensions of the Boolean Model have also been proposed to introduce interactions between objects such as attraction or repulsion between fractures to reproduce their mechanical interactions (Aydin and Caers 2017; Bonneau et al. 2016; Chiles 1988; Hollund et al. 2002).

From a random set perspective, a deterministic object model is a particular realization of some underlying random set process. In this case, the relatively large data density allows one to consider mainly the data conditioning problem rather than focusing on the number of objects and on their spatial density. Another focus of deterministic object modeling approaches relates to the expert-guided definition of interactions between objects using interactive editing tools to ensure that the connectivity between objects is compatible with the geological history of the domain (e.g., how faults branch one onto another and how faults displace horizons).

Yet, more and more complex geometric object parameterizations have recently been introduced in object-based simulation methods. For instance, several authors propose to anchor sedimentary channels on discrete polygonal curves (Mariethoz et al. 2014; Pyrcz et al. 2009; Rongier et al. 2017; Ruiu et al. 2016; Viseur 2004). Other variants consider the bounding surfaces of stratigraphic deposits together with some rules to mimic depositional processes (Graham et al. 2015; Labourdette 2008; Michael et al. 2010; Pyrcz et al. 2005, 2015; Rongier et al. 2017; Ruiu et al. 2016; Sech et al. 2009). As argued in the review of Pyrcz et al. (2015), these models make it possible to consider genetic principles such as erosion, progradation and aggradation of sedimentary deposits in an automatic way. Similarly, pseudo-process-based models have also been proposed in the area of fracture modeling to approximate mechanical interactions and truncations that occur during fracture growth (Bonneau et al. 2013; Davy et al. 2013; Srivastava et al. 2005). At a larger scale, a recent trend has been to simulate possible stochastic geometries where the number and the connectivity of faults is variable (Aydin and Caers 2017; Cherpeau et al. 2010, 2012; Cherpeau and Caumon 2015; Holden et al. 2003; Julio et al. 2015a). In all these approaches, the use of rules is often a means to generate realistic objects and to produce likely connectivities and spatial features without being constrained by some input grid resolution. However, conditioning to dense spatial data sets remains challenging with these approaches. A possible way forward is to consider parameter-rich object-models and to consider process-based rules backward in time (Parquer et al. 2016; Ruiu et al. 2015). In all cases, expert control of model realism is also difficult and may call for additional "geological likelihood" functions to scrutinize the realizations (Jessell et al. 2010).

Interestingly, the use of continuous functions around the Poisson points used in object-based simulation (Random Function Model (Jeulin 2002)), is a possible way to relate random sets to Eq. (28.1). However, formalizing the link between object models and basis functions used in physical models is not easy and relies on the assumption that values are analytically defined on each object, and that objects have stationary statistics (Jeulin 2012; Oda 1986). Dealing with more realistic geometries and sequential Boolean operations to reproduce the succession of geological events calls for further numerical and mathematical developments. Meanwhile, as statistical properties of random sets are not easily checked in practical cases, the numerical approach to relate objects to physics clearly remains an area of much interest (Botella et al. 2016; Cacace and Blöcher 2015; Karimi-Fard and Durlofsky 2016; Merland et al. 2014; Mustapha 2011; Pellerin et al. 2014; Zehner et al. 2015).

## 28.5 Conclusions and Challenges

Several complementary ways exist to incorporate geological information in earth models (Fig. 28.2): spatial statistics, geological variables, geometry and topology of geological objects and explicit geological process modeling. Links exist between the random field and object-based frameworks in cases where the canonical random field theory is applicable (e.g., homogeneous and stationary object densities). This forms the rationale for most modeling methods where "small" objects are treated though their (spatially correlated) equivalent properties at the representative elementary volume scale. "Large" objects are modeled explicitly using rules and parameters that incorporate geological principles and may be calibrated from data and analogs.

Although geostatistics has proven an invaluable theoretical framework to rigorously describe geological domains, it needs to be complemented by geological reasoning (*sensu* Frodeman 1995). Namely, considering discrete time steps approximating geological history and geological variables which cannot be directly measured can significantly help generating more predictive geological models, which may not always have stationary statistical properties. Geometric and topological interactions between objects have a direct connection to geological history and prove a powerful tool to characterize geological domains.

From a physical modeling perspective, geometric object models allow to represent small spatial features which can have a large impact on physical processes (Jackson et al. 2015; Julio et al. 2015b; Matthäi et al. 2007). This calls for specific developments in meshing and physical simulation, for example to better account for object features directly in the numerical code (Pichot et al. 2012). In the frame of inverse problems, sensitivity analysis is essential in practical studies. Theoretically, specific methods integrating the probability of existence of objects also need to be considered more widely, such as random vector parameterization (Cherpeau et al. 2012), reversible jump Monte-Carlo Markov Chain simulation (Green 1995; Sambridge et al. 2012) or ensemble-based methods (Scheidt and Caers 2009). Both in forward and inverse physical models, an additional and significant challenge is to better characterize the multi-sale interactions between geometrical and petrophysical parameterizations (basis functions and associated parameters values).

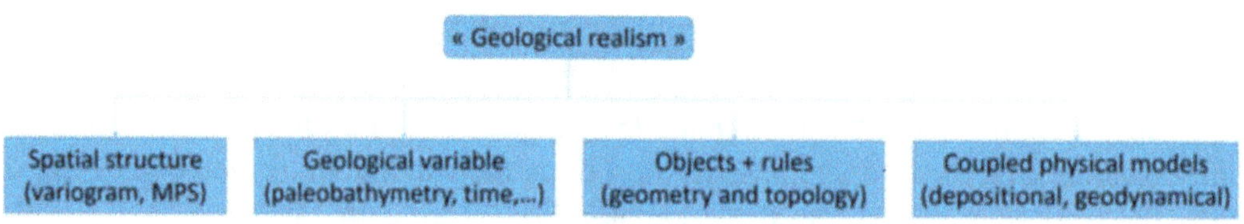


**Fig. 28.2** Summary of the various complementary ways to incorporate geological knowledge in earth models

**Acknowledgements** The ideas expressed in this chapter owe much to Bruno Lévy, Albert Tarantola, Andre Journel and Jean-Laurent Mallet. Their encouragements, trust and critical remarks have been essential influences. I am also grateful to the graduate students and colleagues of the Research for Integrative Numerical Geology (RING) Team, especially my colleagues Pauline Collon and Paul Cupillard, for their multiple contributions to such a stimulating research environment. Discussions in the frame of the HIWAI ANR project led by Yann Capdeville also fed some of the ideas about scale expressed in this chapter. Last, but not least, at a time where research funding is getting more and more complex, I express my great appreciation to the academic and industrial sponsors of RING-Gocad Consortium for their continued support and to ASGA for its effective Consortium management.

## References

Agterberg F (2014) Geomathematics: theoretical foundations, applications and future developments [Internet]. Springer International Publishing, Cham. http://link.springer.com/10.1007/978-3-319-06874-9. Accessed 17 Aug 2015

Allard D, Froidevaux R, Biver P (2006) Conditional simulation of multi-type non stationary Markov object models respecting specified proportions. Math Geol 38(8):959–986

Armstrong M, Galli A, Beucher H, Loc'h G, Renard D, Doligez B et al (2011) Plurigaussian simulations in geosciences. Springer Science & Business Media

Aydin O, Caers JK (2017) Quantifying structural uncertainty on fault networks using a marked point process within a Bayesian framework. tectonophysics [Internet], 2017 May. http://linkinghub.elsevier.com/retrieve/pii/S0040195117301610. Accessed 9 May 2017

Berkowitz B, Balberg I (1993) Percolation theory and its application to groundwater hydrology. Water Resour Res 29(4):775–794

Boisvert JB, Manchuk JG, Deutsch CV (2009) Kriging in the presence of locally varying anisotropy using non-Euclidean distances. Math Geosci 41(5):585–601

Bonneau F, Caumon G, Renard P (2016) Impact of a stochastic sequential initiation of fractures on the spatial correlations and connectivity of discrete fracture networks. J Geophys Res Solid Earth 121(8):5641–5658

Bonneau F, Henrion V, Caumon G, Renard P, Sausse J (2013) A methodology for pseudo-genetic stochastic modeling of discrete fracture networks. Comput Geosci 56:12–22

Borghi A, Renard P, Jenni S (2012) A pseudo-genetic stochastic model to generate karstic networks. J Hydrol 414–415:516–529

Bosch M, Mukerji T, Gonzalez EF (2010) Seismic inversion for reservoir properties combining statistical rock physics and geostatistics: a review. Geophysics 75(5):75A165–75A176

Botella A, Lévy B, Caumon G (2016) Indirect unstructured hex-dominant mesh generation using tetrahedra recombination. Comput Geosci 20(3):437–451

Bourbiaux B, Basquet R, Cacas MC, Daniel JM, Sarda S (2002) An integrated workflow to account for multi-scale fractures in reservoir simulation models: implementation and benefits. In: Abu Dhabi international petroleum exhibition and conference 2002

Cacace M, Blöcher G (2015) MeshIt—a software for three dimensional volumetric meshing of complex faulted reservoirs. Environ Earth Sci 74(6):5191–5209

Caers J (2011) Modeling uncertainty in the earth sciences. Wiley

Calcagno P, Chilès J-P, Courrioux G, Guillen A (2008) Geological modelling from field data and geological knowledge: Part I. Modelling method coupling 3D potential-field interpolation and geological rules. Phys Earth Planet Inter 171(1):147–157

Cardiff M, Kitanidis PK (2009) Bayesian inversion for facies detection: an extensible level set framework: level sets for facies detection. Water Resour Res 2009 Oct; 45(10):n/a–n/a

Caumon G (2010) Towards stochastic time-varying geological modeling. Math Geosci 42(5): 555–569

Caumon G, Collon-Drouaillet P, Le Carlier De Veslud C, Viseur S, Sausse J (2009) Surface-based 3D modeling of geological structures. Math Geosci 41(8):927–45

Caumon G, Gray G, Antoine C, Titeux M-O (2013) Three-dimensional implicit stratigraphic model building from remote sensing data on tetrahedral meshes: theory and application to a regional model of La Popa basin, NE Mexico. IEEE Trans Geosci Remote Sens 51(3):1613–1621

Caumon G, Lepage F, Sword CH, Mallet J-L (2004) Building and editing a sealed geological model. Math Geol 36(4):405–424

Caumon G, Tertois A-L, Zhang L (2007) Elements for stochastic structural perturbation of stratigraphic models. EAGE Pet, Geostat

Charvin K, Gallagher K, Hampson GL, Labourdette R (2009) A Bayesian approach to inverse modelling of stratigraphy, Part 1: method. Basin Res 21(1):5–25

Cherpeau N, Caumon G (2015) Stochastic structural modelling in sparse data situations. Pet Geosci 21(4):233–247

Cherpeau N, Caumon G, Caers J, Lévy B (2012) Method for stochastic inverse modeling of fault geometry and connectivity using flow data. Math Geosci 44(2):147–168

Cherpeau N, Caumon G, Lévy B (2010) Stochastic simulations of fault networks in 3D structural modeling. Comptes Rendus Geosci 342(9):687–694

Chiles JP (1988) Fractal and geostatistical methods for modeling of a fracture network. Math Geol 20(6):631–654

Chiles J-P, Delfiner P (2012) Geostatistics: modeling spatial uncertainty, 2nd edn. Wiley, Hoboken, N.J

Clapp RG, Biondi BL, Claerbout JF (2004) Incorporating geologic information into reflection tomography. Geophysics 69(2):533–546

Collon P, Bernasconi D, Vuilleumier C, Renard P (2017) Statistical metrics for the characterization of karst network geometry and topology. Geomorphology 283:122–142

Collon P, Pichat A, Kergaravat C, Botella A, Caumon G, Ringenbach J-C et al (2016) 3D modeling from outcrop data in a salt tectonic context: example from the Inceyol minibasin, Sivas Basin, Turkey. Interpretation Aug 4(3):SM17–SM31

Collon-Drouaillet P, Henrion V, Pellerin J (2012) An algorithm for 3D simulation of branchwork karst networks using Horton parameters and A*: application to a synthetic case. Geol Soc Lond Spec Publ 370(1):295–306

Corbel S, Wellmann JF (2015) Framework for multiple hypothesis testing improves the use of legacy data in structural geological modeling. GeoResJ 6:202–212

Corre B, Thore P, de Feraudy V, Vincent G (2000) Integrated uncertainty assessment for project evaluation and risk analysis. In: SPE European petroleum conference. Society of Petroleum Engineers, Paris, France

Cowan EJ, Beatson RK, Ross HJ, Fright WR, McLennan TJ, Evans TR et al (2003) Practical implicit geological modelling. In: Fifth international mining geology conference, pp 17–19

Cross TA, Lessenger MA (1999) Construction and application of a stratigraphic inverse model

Davis JL, Annan AP (1989) Ground-penetrating radar for high-resolution mapping of soil and rock stratigraphy1. Geophys Prospect 37(5):531–551

Davy P, Le Goc R, Darcel C (2013) A model of fracture nucleation, growth and arrest, and consequences for fracture density and scaling: a discrete fracture network model. J Geophys Res Solid Earth 118(4):1393–1407

Deutsch CV, Wang L (1996) Hierarchical object-based stochastic modeling of fluvial reservoirs. Math Geol 28(7):857–880

Efendiev Y, Galvis J, Hou TY (2013) Generalized multiscale finite element methods (GMsFEM). J Comput Phys 251:116–135

Elrafie EA, White JP, Awami FH (2008) The event solution–a new approach for fully integrated studies covering uncertainty analysis and risk assessment. SPE-105276-PA. Oct 1 2008

Farmer CL (2005) Geological modelling and reservoir simulation. In: Iske A, Randen T (eds) Mathematical methods and modelling in hydrocarbon exploration and production [Internet]. Springer, Berlin, Heidelberg, pp 119–212. http://dx.doi.org/10.1007/3-540-26493-0_6

Fetkovich MJ (1980) Decline curve analysis using type curves. SPE-4629-PA. June 1, 1980

Frank T, Tertois A-L, Mallet J-L (2007) 3D-reconstruction of complex geological interfaces from irregularly distributed and noisy point data. Comput Geosci 33(7):932–943

Frodeman R (1995) Geological reasoning: geology as an interpretive and historical science. Geol Soc Am Bull 107(8):960–968

Fullagar P, Hughes N, Paine J (2000) Drilling-constrained 3D gravity interpretation. Explor Geophys 31(1/2):17–23

Gallagher K, Charvin K, Nielsen S, Sambridge M, Stephenson J (2009) Markov chain Monte Carlo (MCMC) sampling methods to determine optimal models, model resolution and model choice for earth science problems. Mar Pet Geol 26(4):525–535

Georgsen F, Røe P, Syversveen AR, Lia O (2012) Fault displacement modelling using 3D vector fields. Comput Geosci 16(2):247–259

Gjøystdal H, Reinhardsen JE, Åstebøl K (1985) Computer representation of complex 3-D geological structures using a new "Solid Modeling" technique*. Geophys Prospect 33(8): 1195–1211

Goovaerts P (1997) Geostatistics for natural resources evaluation. Oxford University Press

Graham GH, Jackson MD, Hampson GJ (2015) Three-dimensional modeling of clinoforms in shallow-marine reservoirs: Part 1 concepts application. AAPG Bull. 99(06):1013–1047

Green PJ (1995) Reversible jump Markov chain Monte Carlo computation and Bayesian model determination. Biometrika 82(4):711–732

Gross H, Boucher AF (2015) Geostatistics on unstructured grid-coordinate systems, connections and volumes, petroleum geostatistics 2015 [Internet]. http://www.earthdoc.org/publication/publicationdetails/?publication=82214. Accessed 13 Jul 2016

Guardiano FB, Srivastava RM (1993) Multivariate geostatistics: beyond bivariate moments. Geostatistics Troia'92 [Internet]. Springer, pp 133–144. http://link.springer.com/chapter/10.1007/978-94-011-1739-5_12. Accessed 30 Apr 2017

Gumiaux C, Gapais D, Brun J (2003) Geostatistics applied to best-fit interpolation of orientation data. Tectonophysics 376(3–4):241–259

Haldorsen HH, Lake LW (1984) A new approach to shale management in field-scale models. SPE-10976-PA. Aug 1, 1984

Hale D (2013) Methods to compute fault images, extract fault surfaces, and estimate fault throws from 3D seismic images. Geophysics 78(2):O33–O43

Henrion V, Caumon G, Cherpeau N (2010) ODSIM: an object-distance simulation method for conditioning complex natural structures. Math Geosci 42(8):911–924

Hillier MJ, Schetselaar EM, de Kemp EA, Perron G (2014) Three-dimensional modelling of geological surfaces using generalized interpolation with radial basis functions. Math Geosci 46(8):931–953

Holden L, Hauge R, Skare Ø, Skorstad A (1998) Modeling of fluvial reservoirs with object models. Math Geol 30(5):473–496

Holden L, Mostad P, Nielsen BF, Gjerde J, Townsend C, Ottesen S (2003) Stochastic structural modeling. Math Geol 35(8):899–914

Hollund K, Mostad P, Fredrik Nielsen B, Holden L, Gjerde J, Grazia Contursi M et al (2002) Havana—a fault modeling tool. In: Norwegian petroleum society special publications, vol 11, pp 157–71

Hu LY, Chugunova T (2008) Multiple-point geostatistics for modeling subsurface heterogeneity: a comprehensive review: review of multiple POI. Water Resour Res [Internet] 44(11). http://doi.wiley.com/10.1029/2008WR006993. Accessed 30 Apr 2017

Jackson CA-L, Rotevatn A (2013) 3D seismic analysis of the structure and evolution of a salt-influenced normal fault zone: a test of competing fault growth models. J Struct Geol 2013 Sep 54:215–34

Jackson M, Percival J, Mostaghimi P, Tollit B, Pavlidis D, Pain C et al (2015) Reservoir modeling for flow simulation by use of surfaces, adaptive unstructured meshes, and an overlapping-control-volume finite-element method. SPE-163633-PA. May 1, 2015

Jackson MD, Hampson GJ, Saunders JH, El-Sheikh A, Graham GH, Massart BYG (2014) Surface-based reservoir modelling for flow simulation. Geol Soc Lond Spec Publ 387(1):271–292

Jenny P, Lee S, Tchelepi H (2003) Multi-scale finite-volume method for elliptic problems in subsurface flow simulation. J Comput Phys 187(1):47–67

Jessell M, Ailleres L, De Kemp E, Lindsay M, Wellmann JF, Hillier M et al (2014) Next generation three-dimensional geologic modeling and inversion. Econ Geol 18:261–272

Jessell MW, Ailleres L, de Kemp EA (2010) Towards an integrated inversion of geoscientific data: what price of geology? Tectonophysics 490(3–4):294–306

Jessell MW, Valenta RK (1996) Structural geophysics: integrated structural and geophysical modelling. Comput Methods Geosci 15:303–324

Jeulin D (2002) Modelling random media. Image Anal Stereol 2002 21(4):31

Jeulin D (2012) Morphology and effective properties of multi-scale random sets: a review. Comptes Rendus Mécanique 340(4–5):219–229

Journel AG (2005) Beyond covariance: the advent of multiple-point geostatistics. Geostat Banff 2004, 225–233

Julio C, Caumon G, Ford M (2015a) Sampling the uncertainty associated with segmented normal fault interpretation using a stochastic downscaling method. Tectonophysics 12(639):56–67

Julio C, Caumon G, Ford M (2015b) Impact of the en echelon fault connectivity on reservoir flow simulations. Interpretation 3(4):SAC23–SAC34

Karimi-Fard M, Durlofsky LJ (2016) A general gridding, discretization, and coarsening methodology for modeling flow in porous formations with discrete geological features. Adv Water Resour 96:354–372

Kedzierski P, Mallet JL, Caumon G (2007) Combining stratigraphic and sedimentological information for realistic facies simulations. EAGE Pet Geostat

King PR, Buldyrev SV, Dokholyan NV, Havlin S, Lee Y, Paul G et al (2001) Predicting oil recovery using percolation theory. Petroleum Geoscience 7(S):S105–107

Kolditz O, Bauer S, Bilke L, Böttcher N, Delfs JO, Fischer T et al (2012) OpenGeoSys: an open-source initiative for numerical simulation of thermo-hydro-mechanical/chemical (THM/C) processes in porous media. Environ Earth Sci 67(2):589–599

Labourdette R (2008) 'LOSCS' lateral offset stacked channel simulations: towards geometrical modelling of turbidite elementary channels: lateral offset stacked channels simulations. Basin Res 20(3):431–44

Lallier F, Caumon G, Borgomano J, Viseur S, Royer J-J, Antoine C (2016) Uncertainty assessment in the stratigraphic well correlation of a carbonate ramp: method and application to the Beausset basin, SE France. Comptes Rendus Geosci. 348(7):499–509

Lantuéjoul C (2002) Boolean model. In: Lantuéjoul C. Geostatistical simulation models and algorithms [Internet]. Springer, Berlin, Heidelberg, pp 153–66. http://dx.doi.org/10.1007/978-3-662-04808-5_13

Laurent G, Ailleres L, Grose L, Caumon G, Jessell M, Armit R (2016) Implicit modeling of folds and overprinting deformation. Earth Planet Sci Lett 15(456):26–38

Laurent G, Caumon G, Bouziat A, Jessell M (2013) A parametric method to model 3D displacements around faults with volumetric vector fields. Tectonophysics 1(590):83–93

Lecour M, Cognot R, Duvinage I, Thore P, Dulac J-C (2001) Modelling of stochastic faults and fault networks in a structural uncertainty study. Pet Geosci 7(S):S31–42

Lelièvre PG, Farquharson CG (2016) Integrated imaging for mineral exploration. In: Integrated imaging of the earth [Internet]. Wiley Inc, pp 137–66. http://dx.doi.org/10.1002/9781118929063.ch8

Linde N, Doetsch J (2016) Joint inversion in hydrogeophysics and near-surface geophysics. In: Integrated imaging of the earth [Internet]. Wiley, pp 117–135. http://dx.doi.org/10.1002/9781118929063.ch7

Linde N, Renard P, Mukerji T, Caers J (2015) Geological realism in hydrogeological and geophysical inverse modeling: a review. Adv Water Resour 86:86–101
Lindsay MD, Jessell MW, Ailleres L, Perrouty S, de Kemp E, Betts P (2013) Geodiversity: exploration of 3D geological model space. Tectonophysics 594:27–37
Liu GR, Gu YT (2005) An introduction to meshfree methods and their programming. Springer, Dordrecht; New York
Ma Y, Hale D, Gong B, Meng Z (Joe) (2012) Image-guided sparse-model full waveform inversion. Geophysics 77(4):R189–198
Maier C, Karimi-Fard M, Lapene A, Durlofsky LJ (2016) An MPFA-based dual continuum–discrete feature model for simulation of flow in fractured reservoirs. In: ECMOR XV-15th European conference on the mathematics of oil recovery [Internet]. http://www.earthdoc.org/publication/publicationdetails/?publication=86292. Accessed 14 May 2017
Mallet JL (2002) Geomodeling. Oxford University Press, USA
Mallet JL (2014) Elements of mathematical sedimentary geology: the GeoChron model. EAGE publications
Manchuk J, Leuangthong O, Deutsch CV (2005) Direct geostatistical simulation on unstructured grids. Geostat Banff 2004, 85–94 [Internet]. Springer. http://link.springer.com/chapter/10.1007/978-1-4020-3610-1_9. Accessed 30 Apr 2017
Mannseth T (2014) Relation between level set and truncated Pluri-Gaussian methodologies for facies representation. Math Geosci 46(6):711–731
Mariethoz G, Caers J (2014) Multiple-point geostatistics: stochastic modeling with training images. Wiley
Mariethoz G, Comunian A, Irarrazaval I, Renard P (2014) Analog-based meandering channel simulation: Analog-Based Meandering Channel Simulation. Water Resour Res 50(2):836–854
de Marsily G, Delay F, Gonçalvès J, Renard P, Teles V, Violette S (2005) Dealing with spatial heterogeneity. Hydrogeol J 13(1):161–183
Massonnat GJ (1999) Breaking of a paradigm: geology can provide 3D complex probability fields for stochastic facies modelling. In: SPE annual technical conference and exhibition. Society of Petroleum Engineers
Matthäi SK, Geiger S, Roberts SG, Paluszny A, Belayneh M, Burri A et al (2007) Numerical simulation of multi-phase fluid flow in structurally complex reservoirs. Geol Soc Lond Spec Publ 292(1):405–429
Meju MA, Gallardo LA (2016) Structural coupling approaches in integrated geophysical imaging. In: Integrated imaging of the earth [Internet]. Wiley Inc, pp 49–67. http://dx.doi.org/10.1002/9781118929063.ch4
Melnikova Y, Zunino A, Lange K, Cordua KS, Mosegaard K (2015) History matching through a smooth formulation of multiple-point statistics. Math Geosci 47(4):397–416
Merland R, Caumon G, Lévy B, Collon-Drouaillet P (2014) Voronoi grids conforming to 3D structural features. Comput Geosci 18(3–4):373–383
Michael HA, Li H, Boucher A, Sun T, Caers J, Gorelick SM (2010) Combining geologic-process models and geostatistics for conditional simulation of 3-D subsurface heterogeneity. Water Resour Res 46(5):W05527
Moës N, Gravouil A, Belytschko T (2002) Non-planar 3D crack growth by the extended finite element and level sets-Part I: mechanical model: non-planar 3D crack growth-Part I. Int J Num Methods Eng 53(11):2549–2568
Mondal A, Efendiev Y, Mallick B, Datta-Gupta A (2010) Bayesian uncertainty quantification for flows in heterogeneous porous media using reversible jump Markov chain Monte Carlo methods. Adv Water Resour 33(3):241–256
Moorkamp M, Heincke B, Jegen M, Hobbs RW, Roberts AW (2016) Joint inversion in hydrocarbon exploration. In: Integrated imaging of the earth [Internet]. Wiley Inc, pp 167–89. http://dx.doi.org/10.1002/9781118929063.ch9

Mosegaard K (2011) Quest for consistency, symmetry, and simplicity—the legacy of Albert Tarantola. Geophysics 76(5):W51–W61

Mosegaard K, Hansen TM (2016) Inverse methods. In: Integrated imaging of the earth [Internet]. Wiley Inc, pp 7–27. http://dx.doi.org/10.1002/9781118929063.ch2

Møyner O, Lie K-A (2014) The multiscale finite-volume method on stratigraphic grids. SPE-163649-PA. Oct 1, 2014

Mustapha H (2011) G23FM: a tool for meshing complex geological media. Comput Geosci 15 (3):385–397

Nobakht M, Clarkson CR, Kaviani D (2013) New type curves for analyzing horizontal well with multiple fractures in shale gas reservoirs. J Nat Gas Sci Eng 10:99–112

Noetinger B (2013) An explicit formula for computing the sensitivity of the effective conductivity of heterogeneous composite materials to local inclusion transport properties and geometry. Multiscale Model Simul 11(3):907–924

Oda M (1986) An equivalent continuum model for coupled stress and fluid flow analysis in jointed rock masses. Water Resour Res 22(13):1845–1856

Oliver DS, Chen Y (2011) Recent progress on reservoir history matching: a review. Comput Geosci 15(1):185–221

Paasche H (2016) Post-inversion integration of disparate tomographic models by model structure analyses. In: Integrated imaging of the earth [Internet]. Wiley Inc, pp 69–91. http://dx.doi.org/10.1002/9781118929063.ch5

Parquer M, Collon P, Caumon G (2016) Conditioning channel backward migration modeling to seismic data. In: 78th EAGE conference and exhibition 2016 proceedings, Vienna, Austria

Pellerin J, Botella A, Bonneau F, Mazuyer A, Chauvin B, Lévy B et al (2017) RINGMesh: a programming library for developing mesh-based geomodeling applications. Comput Geosci [Internet]. http://linkinghub.elsevier.com/retrieve/pii/S0098300417302637. Accessed 1 May 2017

Pellerin J, Caumon G, Julio C, Mejia-Herrera P, Botella A (2015) Elements for measuring the complexity of 3D structural models: connectivity and geometry. Comput Geosci 76:130–140

Pellerin J, Lévy B, Caumon G (2014) Toward mixed-element meshing based on restricted voronoi diagrams. Procedia Eng 82:279–290

Perrin M, Rainaud J-F (2013) Shared earth modeling: knowledge driven solutions for building and managing subsurface 3D geological models. Editions Technip

Perrouty S, Lindsay MD, Jessell MW, Aillères L, Martin R, Bourassa Y (2014) 3D modeling of the Ashanti Belt, southwest Ghana: evidence for a litho-stratigraphic control on gold occurrences within the Birimian Sefwi Group. Ore Geol Rev 63:252–264

Pichot G, Erhel J, de Dreuzy J-R (2012) A generalized mixed hybrid mortar method for solving flow in stochastic discrete fracture networks. SIAM J Sci Comput 34(1):B86–B105

Pyrcz MJ, Boisvert JB, Deutsch CV (2009) ALLUVSIM: a program for event-based stochastic modeling of fluvial depositional systems. Comput Geosci 35(8):1671–1685

Pyrcz MJ, Catuneanu O, Deutsch CV (2005) Stochastic surface-based modeling of turbidite lobes. AAPG Bull 89(2):177–191

Pyrcz MJ, Deutsch CV (2014) Geostatistical reservoir modeling. Oxford university press

Pyrcz MJ, Sech RP, Covault JA, Willis BJ, Sylvester Z, Sun T et al (2015) Stratigraphic rule-based reservoir modeling. Bull Can Pet Geol 63(4):287–303

Renard P, Straubhaar J, Caers J, Mariethoz G (2011) Conditioning facies simulations with connectivity data. Math Geosci 43(8):879–903

Ringrose P, Bentley M (2015) Reservoir model design [Internet]. Springer, Dordrecht, Netherlands. http://link.springer.com/10.1007/978-94-007-5497-3. Accessed 27 Apr 2017

Rongier G, Collon P, Renard P (2017) A geostatistical approach to the simulation of stacked channels. Mar Pet Geol [Internet]. http://linkinghub.elsevier.com/retrieve/pii/S0264817217300375. Accessed 21 Feb 2017

Rongier G, Collon-Drouaillet P, Filipponi M (2014) Simulation of 3D karst conduits with an object-distance based method integrating geological knowledge. Geomorphology 217:152–164
Ruiu J, Caumon G, Viseur S (2015) Semiautomatic interpretation of 3D sedimentological structures on geologic images: an object-based approach. Interpretation 3(3):SX63–SX74
Ruiu J, Caumon G, Viseur S (2016) Modeling channel forms and related sedimentary objects using a boundary representation based on non-uniform rational B-splines. Math Geosci 48(3): 259–284
Sambridge M, Bodin T, Gallagher K, Tkalcic H (2012) Transdimensional inference in the geosciences. Philos Trans R Soc Math Phys Eng Sci 371(1984):20110547–20110547
Sanderson DJ, Nixon CW (2015) The use of topology in fracture network characterization. J Struct Geol 72:55–66
Sava P, Revil A, Karaoulis M (2014) High definition cross-well electrical resistivity imaging using seismoelectric focusing and image-guided inversion. Geophys J Int 198(2):880–894
Scheidt C, Caers J (2009) Representing spatial uncertainty using distances and kernels. Math Geosci 41(4):397–419
Scheidt C, Fernandes AM, Paola C, Caers J (2016) Quantifying natural delta variability using a multiple-point geostatistics prior uncertainty model: delta variability and geostatistics. J Geophys Res Earth Surf 121(10):1800–1818
Sech RP, Jackson MD, Hampson GJ (2009) Three-dimensional modeling of a shoreface-shelf parasequence reservoir analog: Part 1. Surface-based modeling to capture high-resolution facies architecture. AAPG Bull 93(9):1155–1181
Shtuka A, Samson P, Mallet JL (1996) Petrophysical simulation within an object-based reservoir model. In: SPE-35480-MS. Society of Petroleum Engineers, SPE
Srivastava RM, Frykman P, Jensen M (2004) Geostatistical simulation of fracture networks. Geostat Banff 2005:295–304
te Stroet CBM, Snepvangers JJJC (2005) Mapping curvilinear structures with local anisotropy kriging. Math Geol 37(6):635–649
Thiele ST, Jessell MW, Lindsay M, Ogarko V, Wellmann JF, Pakyuz-Charrier E (2016a) The topology of geology 1: topological analysis. J Struct Geol 91:27–38
Thiele ST, Jessell MW, Lindsay M, Wellmann JF, Pakyuz-Charrier E (2016b) The topology of geology 2: topological uncertainty. J Struct Geol 91:74–87
Vallet B, Lévy B (2008) Spectral geometry processing with manifold harmonics. Comput Graph Forum, 251–60. Wiley Online Library
Viseur S (2004) Caracterisation de reservoirs turbiditiques: simulations stochastiques basees-objet de chenaux meandriformes (10 fig.). Bull Soc Geol Fr 175(1):11–20
Watson TA, Doherty JE, Christensen S (2013) Parameter and predictive outcomes of model simplification: outcomes of model simplification. Water Resour Res 49(7):3952–3977
Wellmann JF, Horowitz FG, Schill E, Regenauer-Lieb K (2010) Towards incorporating uncertainty of structural data in 3D geological inversion. Tectonophysics 490(3–4):141–151
Wellmann JF, Lindsay M, Poh J, Jessell M (2014) Validating 3-D Structural models with geological knowledge for improved uncertainty evaluations. European Geosciences Union Assembly 2014 EGU division energy resources and the environment ERE. 59(0):374–381
Williams GJJ, Mansfield M, MacDonald DG, Bush MD (2004) Top-down reservoir modelling. In: SPE-89974-MS. Society of Petroleum Engineers, SPE
Xu W (1996) Conditional curvilinear stochastic simulation using pixel-based algorithms. Math Geol 28(7):937–949
Zagayevskiy Y, Deutsch CV (2016) Multivariate grid-free geostatistical simulation with point or block scale secondary data. Stoch Environ Res Risk Assess 30(6):1613–1633
Zaytsev V, Biver P, Wackernagel H, Allard D (2016) Change-of-support models on irregular grids for geostatistical simulation. Math Geosci 48(4):353–369
Zehner B, Börner JH, Görz I, Spitzer K (2015) Workflows for generating tetrahedral meshes for finite element simulations on complex geological structures. Comput Geosci 79:105–117

Zheglova P, Farquharson CG, Hurich CA (2013) 2-D reconstruction of boundaries with level set inversion of traveltimes. Geophys J Int 192(2):688–698

Zhou H, Gómez-Hernández JJ, Li L (2014a) Inverse methods in hydrogeology: evolution and recent trends. Adv Water Resour 63:22–37

Zhou J, Revil A, Karaoulis M, Hale D, Doetsch J, Cuttler S (2014b) Image-guided inversion of electrical resistivity data. Geophys J Int 197(1):292–309

# Chapter 29
# Fifty Years of Kriging

**Jean-Paul Chilès and Nicolas Desassis**

**Abstract** Random function models and kriging constitute the core of the geostatistical methods created by Georges Matheron in the 1960s and further developed at the research center he created in 1968 at Ecole des Mines de Paris, Fontainebleau. Initially developed to avoid bias in the estimation of the average grade of mining panels delimited for their exploitation, kriging received progressively applications in all domains of natural resources evaluation and earth sciences, and more recently in completely new domains, for example, the design and analysis of computer experiments (DACE). While the basic theory of kriging is rather straightforward, its application to a large diversity of situations requires extensions of the random function models considered and sound solutions to practical problems. This chapter presents the origins of kriging as well as the development of its theory and its applications along the last fifty years. More details are given for methods presently in development to efficiently handle kriging in situations with a large number of data and a nonstationary behavior, notably the Gaussian Markov random field (GMRF) approximation and the stochastic partial differential (SPDE) approach, with a synthetic case study concerning the latter.

## 29.1 Introduction

The creation of the IAMG is a landmark of year 1968, which motivates the present book. Another important event of this year is the foundation of a research center of Ecole des Mines de Paris dedicated to geostatistics and mathematical morphology, two disciplines created by Georges Matheron. Concerning geostatistics, this research center was about to develop the applications of kriging, invented by Matheron several years earlier. The theory of kriging seems so straightforward that

J.-P. Chilès (✉) · N. Desassis
Centre of Geosciences, Mines ParisTech, Fontainebleau, France
e-mail: jean-paul.chiles@mines-paristech.fr; jean-paul@chiles.name

N. Desassis
e-mail: nicolas.desassis@mines-paristech.fr



B. S. Daya Sagar et al. (eds.), *Handbook of Mathematical Geosciences*,
https://doi.org/10.1007/978-3-319-78999-6_29

it was reasonable to imagine that, after some generalizations, kriging would become a classical tool requiring no further research. On the contrary, 50 years later it remains the subject of active research, with renewed points of view. Other paradox: originating from mining estimation problems, and very close to statistical regression from a theoretical standpoint, it was not obvious that kriging would be considered in other domains than mining and earth sciences. However applications now consider, for example, the design of aircrafts (Chung and Alonso 2002), the prediction of the mechanical properties of nanomaterials (Yan et al. 2012), the optimization of supply chain networks (Dixit et al. 2016), the construction of financial term-structures (Cousin et al. 2016), the modeling of social systems (Oliveira et al. 2013), and in all cases the quantification of the uncertainty.

It is therefore not surprising to see in Table 29.1 that the number of articles on kriging (word "kriging" or "cokriging" present in the title) published by the journals of the Scopus database doubles decade after decade. The situation is slightly different for the three journals published by the IAMG: *Mathematical Geosciences* (formerly *Journal of the International Association for Mathematical Geology*, then *Mathematical Geology*), *Computers & Geosciences*, and *Natural Resources Research*; indeed, IAMG journals played a major role in the dissemination of the geostatistical literature in English in the first decades, but have now to share this role with the journals of the new application domains. (Note incidentally that few articles were published before 1980: the literature relative to kriging was largely written in French or published in monographs and conference proceedings.)

At a closer look, the originality of kriging lies in its inclusion in the geostatistical approach, where the optimality provided by kriging rests on an analysis of the spatial variability of the phenomenon of interest. Indeed, if methods for characterizing that variability were lacking, the optimality of kriging would simply be virtual. As for the persistence of research works on kriging, it is widely bound to the evolution of the capacities of calculation and memory of computers, and to the increase of the volume of the data. At its origin kriging considered some samples in the vicinity of a target block, while it has now to take into account up to thousands or even millions of data (remote sensing, laser, seismic).

This chapter first presents the origins of kriging and its theory. It continues with further developments, roughly chronologically, up to current research. Kriging has a number of variants and generalizations. We focus here on linear kriging, moreover in a monovariate context. Cokriging and disjunctive kriging are therefore not

**Table 29.1** Articles whose title includes the word "kriging" or "cokriging": number of articles per decade for IAMG journals and for all journals of the Scopus database

| Decade | IAMG | Scopus |
|---|---|---|
| 1970–1979 | 2 | 14 |
| 1980–1989 | 63 | 136 |
| 1990–1999 | 61 | 272 |
| 2000–2009 | 53 | 512 |
| 2010–2016 | 28 | 1076 |

considered; conversely, the use of kriging to condition geostatistical simulations is acknowledged. Our aim is not a thorough presentation of kriging, which can be found in many textbooks, for example, Chilès and Delfiner (2012).

## 29.2 The Origins of Kriging

One of the tasks of the mining engineer is to select the panels to be exploited, and even to delimit them if the exploitation method lets him this freedom. Indeed, to simplify, a panel deserves to be exploited only if the cost of its extraction and processing does not exceed the value of the metal which can be extracted from it. For given technico-economic parameters, this means that the panel grade has to exceed some cutoff grade. In practice the true grade of a panel is not known before its exploitation, so that the selection is made on the basis of an estimated grade. At the beginning of the 1950s the estimate was simply the average grade of the data belonging to the panel or situated at its border. Krige (1951, 1952), studying exploitation data of several orebodies, observed that for high cutoffs the panels selected that way were on average less rich than expected.

As Fig. 29.1 shows it, this is not really surprising. Two parallel galleries in a sub-horizontal deposit present segments AB and CD with grades above the cutoff, contrarily to the neighboring parts of the galleries. Therefore the decision is made to exploit the trapezoid ABDC, and its grade is anticipated to be equal to the weighted average of the grades of segments AB and CD. In fact, segments AC and BD do not represent the real border between rich and poor ores. The true (unknown) limits look like the dotted lines. Therefore, poor ore is exploited (and rich ore abandoned), so that the grade of the exploited ore is lower than expected.

Mathematically, this expresses a conditional bias: Denoting $Z_v$ the panel grade and $\bar{Z}$ the average grade of the cores situated within the panel, the conditional expectation $E[Z_v|\bar{Z}]$ is not equal to $Z_v$.

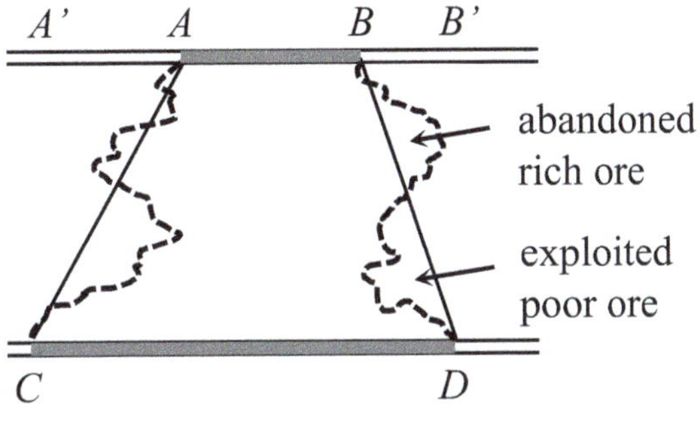


**Fig. 29.1** Illustration of the estimation bias. The panel ABDC to be exploited was delimited from the rich samples observed along AB and CD. Because the true border between rich and poor ores follows a line similar to the dotted line rather than segments AC and BD, poor ore will be exploited and rich ore abandoned. (from Matheron 1961)

To avoid this bias, Krige gives a weight $\lambda$ to the average grade of the data situated in the panel and the complementary weight $1 - \lambda$ to the average grade of the orebody, $\lambda$ being determined by linear regression (Krige in fact considered the lognormal case and worked with grade logarithm).

Also facing problems of mining estimation, Matheron studied Krige's work and generalized his approach by assigning a proper weight to each sample, these weights being determined so as to minimize the estimation variance under the condition that the weights sum to 1 (this condition simply expresses that the estimator is a weighted average of the data).

Matheron called this method "kriging" in honor to Danie Krige. To be accurate, according to Cressie (1990), the French term "krigeage" was coined by Pierre Carlier and first used at the French Commissariat à l'énergie atomique in the late 1950s, and Matheron translated it by "kriging" in Matheron (1963b) (the first appearance of "krigeage" found by the present authors in Matheron's work is Matheron 1960, where it is mentioned as an already known concept).

### 29.2.1 *Ordinary Kriging (OK)*

Geostatistics considers natural variables distributed in space, whose behavior presents a large complexity of detail. These regionalized variables cannot be adequately represented by deterministic functions and therefore methods dedicated to random functions (RF) are considered. The theory of kriging as it is usually presented appears in Matheron (1962, 1963a). It takes place in the framework of an order-2 stationary random function (SRF) model. The regionalized variable of interest (here a grade) is considered as a realization of an SRF $Z(x)$, where $x$ denotes a point in a two- or three-dimensional space. $N$ data are available, at locations $x_\alpha$, $\alpha = 1, 2, \ldots, N$, with values $Z_\alpha = Z(x_\alpha)$. The target $Z_0$ is the value $Z(x_0)$ of $Z$ at an unobserved point $x_0$, or more generally the average value $Z(v)$ of $Z$ in a given cell or block $v$. The kriging estimator of $Z_0$ is by definition of the form

$$Z^* = \sum_{\alpha=1}^{N} \lambda_\alpha Z_\alpha$$

with weights $\lambda_\alpha$ summing to 1. The weights are chosen so as to minimize the variance of the estimation error $Z^* - Z_0$ subject to the condition on their sum. This leads to a linear system of $N + 1$ equations with $N + 1$ unknowns (the $N$ weights $\lambda_\alpha$ and a Lagrange parameter $\mu$):

$$\begin{cases} \sum_\beta \lambda_\beta \sigma_{\alpha\beta} + \mu = \sigma_{\alpha 0} & \alpha = 1, \ldots, N \\ \sum_\beta \lambda_\beta \quad = 1 & \end{cases}$$

where $\sigma_{\alpha\beta}$ denotes the covariance of the observations $Z_\alpha$ and $Z_\beta$ and $\sigma_{\alpha 0}$ the covariance of $Z_\alpha$ and the target $Z_0$. This is the ordinary kriging system. The ordinary kriging variance can then be expressed as:

$$\sigma^2_{OK} = E(Z^* - Z_0)^2 = \sigma_{00} - \sum_\alpha \lambda_\alpha \sigma_{\alpha 0} - \mu$$

where $\sigma_{00}$ denotes the variance of $Z_0$.

### 29.2.2 *Simple Kriging (SK)*

Note that the kriging system and variance do not require the knowledge of the mean. If the mean $m$ were known, we would use an estimator of the form

$$Z^* = \sum_\alpha \lambda_\alpha Z_\alpha + \left(1 - \sum_\alpha \lambda_\alpha\right) m$$

without constraint on the weights, and the minimization of the estimation variance would lead to the simple kriging system

$$\sum_\beta \lambda_\beta \sigma_{\alpha\beta} = \sigma_{\alpha 0} \quad \alpha = 1, \ldots, N$$

and to the simple kriging variance

$$\sigma^2_{SK} = E(Z^* - Z_0)^2 = \sigma_{00} - \sum_\alpha \lambda_\alpha \sigma_{\alpha 0}$$

Simple kriging receives limited applications. It is, however, important, because it has nice properties that are not shared by ordinary kriging and of course universal kriging (see Chilès and Delfiner 2012, Chap. 3). From a computational point of view, the kriging matrix being positive definite, the system can be solved by the Cholesky method.

### 29.2.3 *Ordinary Kriging in the IRF Model*

Because the mean $m$ is not involved in ordinary kriging, it is possible to extend ordinary kriging to a more general random function model, the (order-2) intrinsic random function (IRF) model, characterized by

$$\begin{aligned} &\mathrm{E}[Z(x+h)-Z(x)]=0 \\ &\tfrac{1}{2}\mathrm{E}[Z(x+h)-Z(x)]^2=\gamma(h) \end{aligned}$$

The variogram $\gamma(h)$ summarizes the spatial variability of the random function. Geostatistics provides a set of consistent tools for choosing the variogram model adapted to a particular situation (e.g., Chilès and Delfiner 2012, Chap. 2). The above OK system and OK variance remain valid provided that $C(h)$ is formally replaced by $-\gamma(h)$ in the expressions of $\sigma_{\alpha\beta}$, $\sigma_{\alpha 0}$ and $\sigma_{00}$ given in the next section. This is the framework where kriging is widely used, especially in mining applications.

### 29.2.4 Discussion

Finally, kriging appears as nothing but (a straightforward generalization of) multiple linear regression on $N$ data $Z_\alpha$ that need not to be of the form $Z(x_\alpha)$. Does it deserve a special consideration?

In fact the application of this regression requires that the covariances between the observations, and between each observation and the target, are known. They can be determined experimentally when repeated measurements are available, as is the case in meteorology, but not in usual earth sciences applications, where a unique phenomenon is considered. Applying the regression formula with a priori covariances would provide an estimator that would lose any optimality, except if by chance these covariances are perfectly suited to the data.

Kriging implies a spatial context:

- The random variables $Z_\alpha$ are point values of an SRF $Z(x)$ at points $x_\alpha$.
- Structural analysis methods make it possible to determine the covariance function $C(h)$ of the SRF $Z(x)$.

The covariances $\sigma_{\alpha\beta}$ are then of the form $C(x_\beta - x_\alpha)$, and $\sigma_{\alpha 0}$ is $C(x_0 - x_\alpha)$ if the target is $Z(x_0)$ or the average value of $C(x - x_\alpha)$ when $x$ spans $v$ if the target is $Z(v)$. The variance $\sigma_{00}$ of $Z_0$ that appears in the expression of the kriging variance is $C(0)$ if the target is $Z(x_0)$ or the average value of $C(x' - x)$ when $x$ and $x'$ span $v$ independently if the target is $Z(v)$.

Several authors proposed an approach similar to simple or ordinary kriging before Matheron but not in a spatial context (see Cressie 1990). The noticeable exception is Gandin (1963), who independently developed an approach similar to Matheron's one, in meteorology. SK is called *optimal interpolation*, and OK *optimal interpolation with normalization of weighting factors*. Like Matheron, Gandin was concerned by the theory and its applications; he is, for example, the first author to define and compute a variogram cloud.

### 29.2.5 *Analytic Calculation of Average Covariances*

In the early 1960s computers were not available, at least for mining applications. It was therefore not easy to solve linear systems of equations. Even if point (or core) data could be used to determine the variogram, kriging was applied to aggregated data. In the case of Fig. 29.1, a typical situation examined by Matheron (1961), all cores along AB are represented by their average grade $Z_1$, those along CD by $Z_2$, and those belonging to A′A and BB′ by $Z_3$. The target is the average grade $Z_0$ of the trapezoid ABDC. Kriging amounts to finding the best weights $\lambda_1$ for $Z_1$, $\lambda_2$ for $Z_2$, and $\lambda_3 = 1 - \lambda_1 - \lambda_2$ for $Z_3$ minimizing the variance of $\lambda_1 Z_1 + \lambda_2 Z_2 + (1 - {}_1 - \lambda_2) Z_3 - Z_0$. Kriging amounts to solving a system of two equations, which is straightforward, but first requires to calculate the various covariances involved. For example, if the series of contiguous cores along AB is described by a three-dimensional elongated volume $s$ and the target block (the trapezoid ABDC in projection on the horizontal plane, with some thickness in the vertical direction) by $v$, $\sigma_{10}$ represents $\frac{1}{|s||v|} \int_s \int_v C(x' - x)\, dx'\, dx$, which is a sextuple integral. A special variogram model, the logarithmic or de Wijsian model, was widely used because it is very tractable for analytical calculations of average covariances with Taylor expansions (see numerous technical reports of Matheron on the internet site of Mines ParisTech, Center of Geosciences, On-line geostatistical library).

## 29.3 Development and Maturity: Trend, Neighborhood Selection

With the availability of computers in the late 1960s, it was possible to solve linear systems with about 10–20 equations. Kriging was then carried out with about ten data in and around the target block. Usually a neighborhood of one or two rings or aureolae around the target was used. If necessary, some data were grouped whose situations with respect to the target were similar. At the first international geostatistical congress in Rome in 1975, Michel David claimed that he was able to krige a mining block for a few cents, a reasonable price for real-world applications (David 1976).

In mining applications the outputs were documents with grid cells representing the blocks; the block estimates and the associated kriging standard deviations were printed in the grid cells. Very soon applications emerged in other domains than mining, with a slightly different objective: cartography, more precisely contour mapping. See, for example, Huijbregts and Matheron (1971), Chauvet and Chilès (1975) in oceanography; Delfiner (1973), Chauvet et al. (1976) in meteorology; Delfiner and Delhomme (1975), Delhomme (1978) in hydrology. Moreover, the phenomena considered in these application domains usually present a trend: the sea floor is deeper when moving away from the coast line, aquifers have a general gradient, the top of petroleum reservoirs is usually dome shaped. This called for developments in two directions: kriging theory, with universal kriging to account for trends, and kriging practice, with a careful design of kriging neighborhoods.

### 29.3.1 Universal Kriging (UK)

The assumption of a constant mean—even if unknown—became soon a limitation for the application of kriging to phenomena displaying a trend. Kriging was therefore generalized by Matheron (1969) to random functions with a polynomial drift $m(x)$ of the form

$$m(x) = \sum_{\ell=0}^{L} a_\ell f^\ell(x)$$

where the $a_\ell$ are unknown coefficients and the $f^\ell(x)$ are the $L + 1$ monomials with degree up to a given degree $k$ (in the one-dimensional case, $L = k$ and $f^\ell(x) = x^\ell$). For $\ell = 0, f^0(x) \equiv 1$. The kriging estimator remains of the form $Z^* = \sum_\alpha \lambda_\alpha Z_\alpha$ but, because the $a_\ell$ are not known, unbiasedness is ensured only under the $L + 1$ constraints

$$\sum_\alpha \lambda_\alpha f_\alpha^\ell = f_0^\ell \quad \ell = 0, \ldots, L$$

where $f_\alpha^\ell = f^\ell(x_\alpha)$ and $f_0^\ell$ is $f^\ell(x_0)$ if the target is $Z(x_0)$ or the average value of $f^\ell(x)$ when $x$ spans $v$ if the target is $Z(v)$. The minimization of the estimation variance leads to a system similar to the OK system except that there are now $L + 1$ constraints instead of a single one, and as many Lagrange parameters.

The UK kriging matrix is no more positive definite, so that the kriging system should be solved by Gaussian elimination, which is less efficient than the Cholesky method. However, UK can be expressed as simple kriging, followed by a drift correction. The second step appears as the solution of a linear system of $L + 1$ equations with $L + 1$ unknowns, whose matrix is positive definite. It is thus advantageous to exploit this result to solve the SK system and the drift correction system by the Cholesky method (an additivity property also allows the calculation of the UK variance).

The equations of UK were already presented by Goldberger (1962) but not in a spatial context and with covariances supposed to be known, whereas Matheron proposed tools for determining the underlying variogram in the presence of a drift. These tools let appear an inference problem that was adequately solved in the framework of a more general model, presented hereafter.

### 29.3.2 *Kriging in the IRF-k Model*

Like the mean for OK, the coefficients $a_\ell$ are not involved in universal kriging. This made it possible to extend it to a more general random function model, the model of intrinsic random functions of order $k$ (IRF-$k$), where a generalized covariance function $K(h)$ is substituted to $C(h)$. The RF model was first presented by Yaglom and Pinsker (1953), and the complete theory in the $n$-dimensional space by Matheron (1971, 1973). It suffices to say here that the class of GCs includes ordinary covariances and covariances of the form $-\gamma(h)$ when $k = 0$, and increases with $k$. It includes, for example the power covariances $(-1)^{p+1} |h|^{2p+1}$, $0 \leq p \leq k$, and the "spline" covariances $(-1)^{p+1} |h|^{2p} \log |h|$, $p$ integer, $1 \leq p \leq k$. The kriging system is the same as for UK, with $K$ replacing $C$.

### 29.3.3 *Kriging as an Interpolant*

In cartography, the objective of the applications of kriging was more precisely to draw maps with isolines derived from point kriging at the nodes of a regular grid. Nowadays it is possible to locally refine the grid to precisely track an isoline. In both cases, there is a requirement that kriging is not only the optimal linear estimator for a single point or block but also has nice interpolation properties.

According to theory, when kriging is considered as an interpolant, that is, as a function $z^*(x)$ of the target point $x$, the kriged map inherits from the covariance or variogram model. Indeed the universal kriging estimate can be presented in its dual form

$$z^*(x) = \sum_\alpha b_\alpha C(x - x_\alpha) + \sum_\ell c_\ell f^\ell(x)$$

with the convention that $C$ can be replaced by $-\gamma$ or by the generalized covariance $K$. The coefficients $b_\alpha$ and $c_\ell$ are linear functions of the data. They are obtained as solutions of a system of equations similar to the UK system (same kriging matrix). If the variogram is parabolic at the origin, then $z^*(x)$ is differentiable; if the variogram is linear at the origin (and thus with a cusp at the origin when considered as a function of vector $h$), $z^*(x)$ is continuous with cusps at the data points. This may not be aesthetically nice from the user's point of view, because this is not primarily the purpose of kriging. Nevertheless, a smooth map can always be obtained by applying kriging with a smooth variogram or generalized covariance model. This is the way splines were used at that time, without explicit reference to geostatistics, but Matheron (1981) showed that any spline problem is equivalent to a kriging problem in the framework of the IRF-$k$ model. For example, in 2D, interpolating with biharmonic splines is equivalent to kriging in the framework of an IRF-1 model with the generalized covariance $|h|^2 \log |h|$. Of course if the "true" covariance model does not conform to this model, kriging loses its optimality.

### 29.3.4 Neighborhood Selection

The dual kriging approach is very efficient in terms of computer time but presents two limitations: (i) it does not provide the kriging variance, and (ii) like direct kriging, its above interpolation properties are valid when working globally, that is, all data points are taken into account (global neighborhood). Due to practical limitations in memory space and calculation time, there is a limit in the number $N$ of data that can be processed (several hundreds at that time, several thousands now). Therefore, in practice kriging often continues to be used with a moving neighborhood, that is, a limited number of data points around the target point are taken into account.

Now, when kriging with a moving neighborhood, the neighborhoods of two grid nodes can differ, and this can produce spurious discontinuities, especially when an outlier data is included in the neighborhood of a grid node and not in the neighborhood of the next grid node.

The neighborhood problem is also important when building conditional simulations. The classical way at that time (and even now) for continuous variables was to work in the framework of a Gaussian RF model (if necessary after suitable transformation of the data), to generate a nonconditional simulation of the Gaussian RF, and to condition that simulation on the data with a kriging step (Journel 1974). Due to their random nature, nonconditional simulations present small-scale variations. If spurious discontinuities are added by the kriging step, it is not easy to distinguish them from natural variations, which can lead to inaccurate conclusions.

Therefore, during years, much effort was devoted by software developers to neighborhood selection (e.g., Renard and Yancey 1984). Sophisticated algorithms have been devised to reach a compromise between near and far sample points. Focusing on 2D only, neighborhoods usually include all points of the first ring and then more distant points, following a strategy that attempts to sample all directions as uniformly as possible while keeping the number of points as low as possible (octant search). Typically, 16 to 32 points are retained, from at least five octants or four noncontiguous octants. For contour mapping purposes, where continuity is important, larger neighborhoods may be considered to provide more overlap. Such an algorithm may not provide satisfactory results when data originate from profiles sampled with a short interval. The neighborhood selection then includes the requirement to have data originating from several profiles. Along years, the size of the neighborhoods increased with the improvements of computers in terms of CPU time and storage.

### 29.3.5 Maturity

In the 1980s kriging seemed to have reached maturity. It was widely used in mining projects to build block models of orebodies, even with a large number of sample data and a very large number of blocks. In civil engineering it enabled an accurate design of the Channel tunnel on the basis of a model of the geological layers obtained by kriging from about 100 000 data, with a sound evaluation of the uncertainty of the model (Blanchin and Chilès 1993; Chilès and Delfiner 2012, Sect. 3.8). There were further developments specific to nonlinear geostatistics (disjunctive kriging, indicator kriging) and to multivariate geostatistics (factorial kriging analysis) which are not considered here.

At the same period, Sacks et al. (1989) opened a completely new domain to kriging: the design and analysis of computer experiments (DACE). The coordinates of $x$ are no longer geographic but represent scalar design variables, while the variable of interest $Z$ is an objective function that depends on the design variables. A computer experiment gives the value of the objective function for chosen values of the design variables. When computer experiments are costly, kriging is used to interpolate the response surface from a limited number of data (computer experiments). Applications mainly concern engineering problems, for example, the design of aircrafts (Chung and Alonso 2002). They call for specific research works, due to the very special space considered, the sparsity of the data, the difficulty to infer the covariance. See Kleijnen (2016) for a recent review.

## 29.4 Iterative Use of Kriging to Handle Inequality Data

Up to the early 1980s, geostatistics provided direct solutions: kriging was obtained by solving a linear system of equations, (Gaussian) simulations were built by turning bands or other methods directly transforming a vector of independent standard normal random variables in a vector representing a discrete view of the random function. Iterative algorithms appeared to handle inequality data and more specifically to generate conditional simulations of truncated Gaussian RFs.

Inequality data were already considered in the 1980s, notably by Dubrule and Kostov (1986) and Kostov and Dubrule (1986), with a solution based on quadratic programming where inequality data are treated as constraints placed on the kriging estimate. At the end, the inequalities are classified either as inactive (they can be forgotten) or active, and in the latter case they are replaced by an equality to the upper or lower bound of the inequality. This classification is not trivial at all and is the value of the method, but the clamping effect produced by the replacement of some inequalities by their lower or upper bound is not really satisfactory.

An alternative approach proposed by Langlais (1990) is to regard inequalities as data and replace them by exact values. The procedure is to (i) simulate exact data satisfying the given inequalities while honoring the exact data and the spatial structure, (ii) average the results over several simulations, thus generating data that will replace the inequality data, and (iii) proceed to kriging from both actual and generated data.

At the same period, truncated Gaussian RFs were considered to represent geological facies. In its simplest form, such RF is defined by a Gaussian SRF $Y(x)$ and a threshold $s$. The truncated Gaussian RF is simply the indicator $1_{Y(x) \geq s}$. The applications account for a threshold that varies with $x$ (an ordinary function of $x$). More general models are obtained with several thresholds and possibly two or three Gaussian SRFs (plurigaussian RF). Matheron et al. (1987) proposed a method to build conditional simulations of truncated Gaussian RFs in the case of a separable exponential covariance. The method is rather simple because it fully exploits the Markov properties of that covariance model.

From that time the geostatistics community devoted a growing interest to Markov chain Monte Carlo (MCMC) methods (e.g., Tjelmeland and Holden 1993), and particularly to the Gibbs sampler (Geman and Geman 1984). Initially developed to solve optimization problems, these methods also provide useful algorithms for generating simulations of RFs at a finite number of sites (e.g., grid nodes). The Gibbs sampler gives a consistent iterative method to achieve the first step of Langlais (1990), which is the critical one: simulate exact data satisfying the inequalities. Let us consider that the inequality data are of the form $Z_\alpha \in B_\alpha$ for some values of $\alpha$, where $B_\alpha$ denotes an interval. The procedure is initialized by generating each of these $Z_\alpha$ separately, by a value $z_\alpha$ chosen in the interval $B_\alpha$. Then the following sequence is repeated:

1. Select an inequality site $\alpha$.
2. Simulate $Z_\alpha$ conditional on $Z_\alpha \in B_\alpha$ and $Z_\beta = z_\beta$ for all $\alpha \neq \beta$ ($\beta$ ranges over all sites except $\alpha$), and assign the simulated value to $z_\alpha$.

The procedure changes the simulated values at the inequality sites so that they progressively honor the spatial structure given by the covariance. This approach finds its theoretical justification in the ideal case of a Gaussian SRF with a known mean, where the conditional distribution of $Z_\alpha$ is Gaussian with mean and variance equal to the kriging estimate and the kriging variance. It is however robust and is used even in the case of an unknown mean. The same approach is used effectively to generate conditional simulations constrained by inequality data, and especially truncated Gaussian RFs (the 0 or 1 data are transformed in inequality data of the form $Y(x_\alpha) < s$ or $Y(x_\alpha) \geq s$). The algorithm should be used in global neighborhood; otherwise, care should be given to the neighborhood selection, because the algorithm may diverge.

## 29.5 Nonstationary Covariance

Up to now we have considered models with a stationary covariance. But reality does not care about our theoretical models. If a stationary covariance is often a reasonable assumption when a limited number of samples is available, large data sets usually show some lateral variations in the covariance or variogram, so that a global model with a stationary covariance would be a too crude approximation. This problem is obviously not new. A simple solution is to split the study domain into several subdomains, to determine a specific variogram in each subdomain, and to krige each subdomain with its own variogram. To avoid discontinuities at subdomains boundaries, the variogram parameters evolve progressively from one model to the next in a transition area. This ad hoc method was used, for example, for the study of the Channel tunnel where the 100 000 data clearly showed structural variations along the 60 km of the tunnel project. Machuca-Mory and Deutsch (2013) generalize and systematize this approach.

Global nonstationary covariance models are of course sounder than the previous approach from a theoretical point of view, and also from a practical one if they can adapt to actual situations. A simple global covariance model can be derived by generalization of the covariogram model, defined by autoconvolution of an integrable and square integrable function $w(u)$:

$$g(h) = \int w(u)w(u+h)du$$

If we replace $w(u)$ by a dilution or kernel function $w(x;\, u)$ also depending on $x$, integrable and square integrable in $u$ whatever $x$, and define

$$g(x, x') = \int w(x; u)w(x'; u)du$$

then $g(x,\, x')$ is a nonstationary covariance model (e.g., Higdon et al. 1999). A random function with that covariance can be obtained by the dilution method (Higdon 2002).

Let us now examine the case where $w$, considered as a function of $u$ for fixed $x$, is a Gaussian kernel with variance–covariance matrix $\Sigma_x$. The resulting correlation function can be written (e.g., Paciorek and Schervish 2006)

$$g(x, x') = |\Sigma_x|^{1/4}|\Sigma_{x'}|^{1/4}\left|\frac{\Sigma_x + \Sigma_{x'}}{2}\right|^{-1/2}\exp(-Q_{xx'})$$

with quadratic form

$$Q_{xx'} = (x' - x)^{\mathrm{T}} \left( \frac{\Sigma_x + \Sigma_{x'}}{2} \right)^{-1} (x' - x)$$

If $\Sigma_x$ is constant with respect to $x$, then $g(x, x')$ is the standard Gaussian correlation function with global anisotropy matrix $\Sigma_x$. Otherwise, if $\Sigma_x$ varies slowly, $g$ is approximately stationary in a small neighborhood of $x$. This locally stationary correlation function can be generalized by replacing $\exp(-Q_{xx'})$ by $\rho(-Q_{xx'})$ where $\rho$ is a stationary correlation function that is valid in every dimension. This class of nonstationary covariance functions can be fitted by using local variograms whose parameters are used to build local $\Sigma_x$ matrices (e.g., Fouedjio et al. 2016). Emery and Arroyo (2018) describe a spectral algorithm for simulating such models.

## 29.6 Kriging for Large Data Sets

We have seen that kriging with moving neighborhoods provides artifacts that can be limited in their amplitude by a careful design of the neighborhood selection but not eliminated. This problem is important when putting the Gibbs algorithm into practice because the procedure might diverge. The best way to avoid artifacts is to krige in global neighborhood, that is, any target point is kriged from all the data. As the capabilities of computers in terms of memory and computational performance always increase, this becomes possible for larger and larger data sets. However, the size of most data sets is also increasing with the advent of automatic measurement stations, so that the problem remains. A direct solving of the kriging system by Gaussian elimination or the Cholesky method is possible for up to several thousand equations. Several attempts were made for processing larger systems. Before presenting two truly global approaches, let us start with a method deriving from moving neighborhoods.

### *29.6.1 Continuous Moving Neighborhood*

Gribov and Krivoruchko (2004) developed an original method to ensure continuity with moving neighborhoods. The idea is to modify the kriging system so that data beyond a specified distance from the estimated point receive weights gradually approaching zero. This way, no discontinuity occurs when data points enter or exit the kriging neighborhood.

Rivoirard and Romary (2011) propose an equivalent approach from a different perspective: The idea is to introduce a penalty on the kriging weights in the objective function to be minimized. This penalty acts as a noise variance except that it varies with the target point $x_0$. It is typically equal to 0 for data points $x_\alpha$ within a distance $r$ of the estimated point $x_0$ (no penalty applied near the target point), and increases continuously to infinity as $x_\alpha$ approaches the outer boundary of the kriging neighborhood, located at a distance $R$. Data points at a distance larger than $R$ thus receive a zero weight. Because this method is solely based on the addition of a noise that increases with distance, it works for all versions of kriging algorithms: OK, UK, and even IRF-$k$. Because it is local, this method can handle lateral changes in the covariance parameters.

### *29.6.2 Covariance Tapering*

Large systems can be solved if the kriging matrix is sparse. This can be achieved by tapering the covariance function to zero beyond a certain range. Furrer et al. (2006), who proposed this approach, define the tapered covariance as the product of the true covariance $C$ by a taper covariance $K$ that has a finite range. To preserve the behavior of the true covariance $C$ near the origin, which controls the lateral continuity of the interpolant, the taper covariance $K$ should be more regular near the origin than $C$. The authors apply the method with about 6 000 data.

### *29.6.3 Fixed Rank Kriging*

In order to reduce the complexity of the kriging system when the number of data is very large, Cressie and Johannesson (2008) represent $Z(x)$ as a linear combination of $r$ given basis functions $S_k(x)$ with random coefficients $\eta_k$, plus a white noise $\varepsilon(x)$ (for simplicity, we omit the covariates considered by the authors as external drift functions):

$$Z(x) = \sum_{k=1}^{r} \eta_k S_k(x) + \varepsilon(x)$$

The basis functions need not be orthogonal. They are usually chosen so as to represent several scales of variation and, for each scale, to cover the whole study domain. A typical choice is wavelet functions.

Denoting by $\mathbf{S}(x)$ the vector of the basic functions $S_k(x)$, by $\mathbf{K}$ the variance–covariance matrix of the $\eta_k$, and assuming that the white-noise variance is constant and equal to $\sigma^2$, the covariance of $Z(x)$ and $Z(x')$ is

$$C(x, x') = \mathbf{S}(x)^{\mathrm{T}}\, \mathbf{K}\, \mathbf{S}(x') + \sigma^2\, \delta(x' - x)$$

where $\delta$ is the Kronecker function.

Given a vector **Z** of $N$ data $Z(x_\alpha)$, the kriging matrix is

$$\mathbf{\Sigma} = \mathbf{S}\,\mathbf{K}\,\mathbf{S}^{\mathrm{T}} + \sigma^2 \mathbf{I}$$

where **S** is the $N \times r$ matrix whose $(\alpha, k)$ element is $S_k(x_\alpha)$. The authors show that the inverse of $\mathbf{\Sigma}$ (an $N \times N$ positive-definite matrix) in fact only requires the inversion of **K** and $\mathbf{K}^{-1} + \mathbf{S}^{\mathrm{T}}\,\mathbf{S}/\sigma^2$ (two $r \times r$ positive-definite matrices). They also show that the inference of the positive-definite matrix **K** and the variance $\sigma^2$ can be done with the classical geostatistical approach. Therefore, kriging becomes tractable even with a very large number of data. In an application to ozone satellite data, the authors use 396 basis functions, a huge reduction in comparison with the 173 000 data.

### *29.6.4 Gaussian Markov Random Field Approximation*

The approach of Gaussian Markov random fields may be seen as the opposite of that of covariance tapering in the sense that it seeks to make the inverse of the covariance matrix—and not the covariance matrix itself—sparse. It was first used to generate simulations (Besag 1974, 1975) but offers a new approach to kriging (Rue and Held 2005). Let us consider a Gaussian random vector $\mathbf{Z} = \{Z_i: i = 1, \ldots, N\}$ with known mean **m** and variance–covariance matrix **C**. The conditional distribution of $Z_i$ given the other components $\{Z_j: j \neq i\}$ is Gaussian with mean and variance the kriging estimate $Z^*_{-i}$ of $Z_i$ (the minus sign recalls that $Z_i$ is excluded from the data used for that kriging) and the associated kriging variance $\sigma^2_{\mathrm{K}i}$. Denoting by **B** the inverse of **C**, the kriging weights are found to be equal to $\lambda_j(i) = -B_{ij}/B_{ii}$ so that we have

$$Z^*_{-i} = m_i - \frac{1}{B_{ii}} \sum_{j \neq i} B_{ij}\,(Z_j - m_j) \qquad \sigma^2_{\mathrm{K}i} = \frac{1}{B_{ii}}$$

Since $B_{ii}$ is the inverse of the conditional variance of $Z_i$ given $\{Z_j: j \neq i\}$ (all except the $i$-th), **B** is known as the precision matrix. Its off-diagonal elements are related to the conditional correlations of $Z_i$ and $Z_j$ given $\{Z_k: k \neq i, j\}$ by

$$\mathrm{Corr}(Z_i, Z_j | \{Z_k : k \neq i, j\}) = -\frac{B_{ij}}{\sqrt{B_{ii}\,B_{jj}}}$$

**B** is a symmetric positive-definite matrix. The pattern of zeroes of **B** can be used to define an undirected graph structure in which two nodes are connected by an edge when $B_{ij} \neq 0$. Let ne($i$) denote the neighborhood of node $i$, that is, the set of nodes connected to $i$ by an edge. The vector **Z** has the Markov property that $Z_i$ is conditionally independent of $\{Z_k: k \notin \mathrm{ne}(i)\}$ given $\{Z_j: j \in \mathrm{ne}(i)\}$. The discretely indexed Gaussian **Z** is called a Gaussian Markov random field (GMRF).

If the $N$ components $Z_i$ are split in $N_1$ unknown components to be estimated and $N_2 = N - N_1$ data, it can be shown that kriging can be achieved by solving a linear system of $N_1$ variables and $N_1$ equations whose system matrix is that part of the precision matrix $\mathbf{B}$ corresponding to the $N_1$ unknown components. The GMRF approach is used when this matrix is sparse, so that the system can be solved even when $N_1$ is large.

### 29.6.5 The Stochastic Partial Differential Equation (SPDE) Approach

Although the GMRF approach seems particularly appealing to deal with large data sets, its use remained limited due to the fact that the link with the geostatistical models based on covariance functions was not clear, making it difficult to parameterize the precision matrix. Nevertheless, some empirical studies showed that the commonly used covariance functions could be approximated quite closely by GMRFs (e.g., Rue and Tjelmeland 2002; Hrafnkelsson and Cressie 2003). These results spurred some authors to model the data by using a Gaussian field characterized by its covariance and then to find a discretized GRMF for which the inverse of the associated precision matrix $\mathbf{B}$ provides a good approximation of the covariance matrix of the Gaussian field (Song et al. 2008; Cressie and Verzelen 2008). Although promising, these algorithms suffer from a lack of theoretical foundations, which makes their application difficult.

In their seminal paper, Lindgren et al. (2011) propose a formal link between Gaussian field and GRMFs. They use a result established by Whittle in the 1950s linking some Gaussian fields and the solutions of a class of SPDEs. More precisely, let us consider the Matérn covariance function

$$C(h) = \frac{\sigma^2}{2^{\nu-1}\Gamma(\nu)} \left(\frac{|h|}{a}\right)^{\nu} K_\nu\left(\frac{|h|}{a}\right)$$

where $\sigma^2$ is the sill parameter, $a > 0$ is the scale parameter, $\nu > 0$ is a regularity parameter which determines the mean-square differentiability of the Gaussian field and $K_\nu$ is the modified Bessel function of the second kind and order $\nu$. The result of Whittle (1954) states that a Gaussian field $Z$ with Matérn covariance function $C$ is a solution of the linear fractional SPDE

$$(\kappa^2 - \Delta)^{\alpha/2} Z(s) = \tau W(s) \quad s \in \mathbb{R}^d$$

where $\alpha = \nu + d/2, \kappa = 1/a, \tau^2 = \frac{\Gamma(\nu + d/2)(4\pi)^{d/2}\kappa^{2\nu}}{\Gamma(\nu)}$, $\Delta$ is the Laplacian operator, and $W$ is a Gaussian white noise with unit variance. The pseudo-differential operator $(\kappa^2 - \Delta)^{\alpha/2}$ can be defined through its Fourier transform but it is simply a linear combination of iterated Laplacians when $\alpha/2$ is an integer.

Then, by using some numerical methods to solve the PDE, for example, a finite differences method (FDM) or a finite elements method (FEM), Lindgren et al. (2011) show that the resulting discretized field at the mesh points (which can include the data locations) is a discrete GRMF. The precision matrix is directly provided by the FDM or FEM implementation. It is a sparse matrix although the number of non-zero elements increases with $\nu$. Therefore, by including the target points in the mesh generation, one can perform kriging with very large data sets by using an efficient solver for sparse matrices. Note that, when $\alpha$ is not an integer, the operator $(\kappa^2 - \Delta)^{\alpha/2}$ has to be approximated by $\left(\sum_{i=0}^{p} \lambda_i \Delta^i\right)^{1/2}$, where $p$ is the smallest integer greater than $\alpha$. This operator can also be discretized by a FDM or FEM.

Anisotropies can be handled with the operator $(\kappa^2 - \text{div}(H.\nabla))^{\alpha/2}$ where **H** is a symmetric positive-definite matrix linked to the anisotropy matrix and div is the divergence operator.

An interesting feature of the SPDE approach is that it allows to easily incorporate varying coefficients. For instance, the matrix **H** can be replaced by **H**($s$) to handle a varying anisotropy (see Fuglstad et al. 2015).

Figure 29.2 presents a synthetic vertical section that could represent a variable of interest such as porosity in a sedimentary layer. The base and top of the layers were obtained by standard geostatistical simulations. The variable of interest was built according to the model of Fuglstad et al. (2015) with $\alpha = 3/2$, the matrix **H** incorporating the anisotropy model depicted in Fig. 29.3. This anisotropy model was deduced from the model of the base and top of the layer, with a constant range along the local direction of the layer, and a shorter range, varying proportionally to layer thickness, in the orthogonal direction. Figure 29.2 shows five vertical "drill-holes" considered as the data set, and Fig. 29.4 shows the kriged section obtained with the SPDE method. The latter shows the capability of this approach to account for the anisotropy model even in areas where there are no data (provided of course that information is available concerning the anisotropy). From a computational point of view, the method is extremely efficient: in 2D a data set with about 100 000 data can be processed in about 10 s on a standard computer, with possibly a number of conditional simulations nearly in the same time.

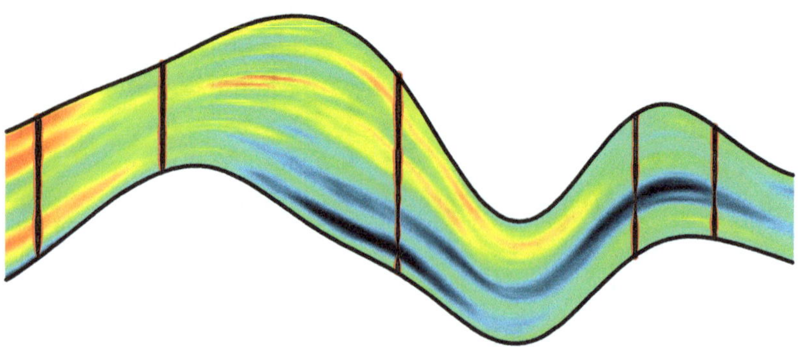

**Fig. 29.2** SPDE synthetic case study: "Reality" (in fact a simulation) and sampling of five drill-holes

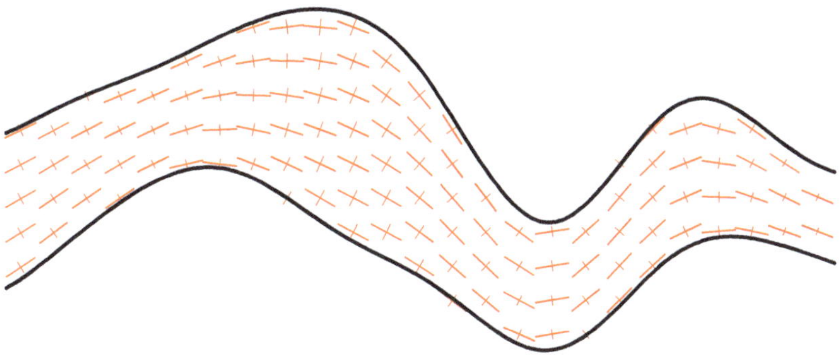

Fig. 29.3 SPDE synthetic case study: Anisotropy model

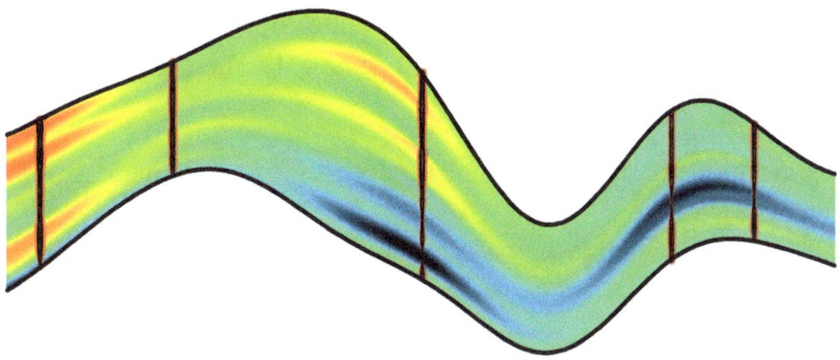

Fig. 29.4 SPDE synthetic case study: Kriging from the data of the five drill holes

## 29.7 Iterative Algorithms for Solving the Kriging System

Before to conclude, it is advisable to remind a presentation of two iterative kriging algorithms by Jean-François Royer in 1974, that is, in the early times of geostatistics. In meteorology, at that time, two main approaches were used to carry out the "objective analysis", that is, the interpolation of temperature and pressure at the nodes of a grid from the observations at time $t$, then used as input for a numerical weather forecast at time $t + 1$. One is Gandin's approach (1963), similar to simple kriging (in meteorology, the mean can be considered known thanks to a long sequence of observations). The other is an iterative approach, the method of successive corrections proposed by Cressman (1959).

Royer (1975) considers the simple monovariate situation. Rewritten with present notations, let us consider a vector $\mathbf{z}$ with $N = N_G + N_S$ components $z_i$, the first $N_G$ components corresponding to grid nodes ($i \in G = \{1; \ldots; N_G\}$) and the other $N_S$ components corresponding to observation stations ($i \in S = \{N_G + 1; \ldots; N_G + N_S\}$); $z_i$ represents the variable of interest, at location $x_i$. Because the average situation for the season or month considered is known from past observations, we can subtract it and assume that $\mathbf{z}$ has mean $\mathbf{0}$. Two iterative algorithms are proposed, depending on the set of points that drives the changes (grid nodes or observation stations). In both cases, an influence function $\rho(h)$ is used for extending a change made at location $x$ to location $x + h$ depending on separation $h$. This function satisfies $\rho(0) = 1$ and decreases to 0 when $h$ increases. When extending to $x_j$ a change made at $x_i$, the notation $\rho_{ij} = \rho(x_j - x_i)$ will be used.

*Algorithm driven by grid nodes:* As step $p = 0$, select a vector $\mathbf{z}^0$ with components $z_i^0$, for example zeroes or the values of the weather forecast for time $t$ based on the objective analysis made at time $t - 1$. Then iterate as follows:

1. increase the step number $p$ by 1
2. calculate the discrepancies of step $p - 1$ with regard to the data: $z_i - z_i^{p-1}, \; i \in S$
3. define model $p$ as $z_i^p = z_i^{p-1} + \sum_{j \in S} \rho_{ij}(z_i - z_i^{p-1}), \quad i \in G \cup S$

*Algorithm driven by the observations:* As initial state, select a vector $\mathbf{z}^{\text{new}}$ with components $z_i^{new}$, for example zeroes or the values of the weather forecast for time $t$ based on the objective analysis made at time $t - 1$. Then iterate as follows:

1. Set $z_j^{\text{current}} = z_j^{\text{new}}, \quad j \in G \cup S$
2. Select a component of $S$, say $i$, at random or by systematic scans of all the components of $S$
3. Calculate the discrepancy of the current value with regard to the observation: $z_i - z_i^{\text{current}}$
4. Define $z_i^{\text{new}} = z_i$
5. Update all other components so that $z_j^{\text{new}} - z_j^{\text{current}} = \rho_{ij}(z_i^{\text{new}} - z_i^{\text{current}}), \; j \neq i$

The convergence of both algorithms is ensured if and only if the matrix $\boldsymbol{\rho}$ defined by the $\rho_{ij}$ is positive definite, which is ensured if $\rho(h)$ is a correlogram. Moreover, in that case, the iterative process converges to the solution of dual kriging. Indeed, both approaches amount to an iterative resolution of the dual kriging system (by the Jacobi method in the first approach, by the Gauss-Seidel method in the second one), followed, after each iteration, by the propagation of the changes to the point kriging estimates.

The second algorithm is very similar to the Gibbs propagation algorithm proposed nearly 40 years later by Lantuéjoul and Desassis (2012) to simulate a Gaussian vector (this algorithm is also presented in Chilès and Delfiner 2012, Sect. 7.6.3; it constitutes a further step to an algorithm proposed by Galli and Gao 1999). It is this similarity that reminded one of the present authors the paper of Royer, not exploited by geostatisticians to our knowledge, which should deserve new consideration. These iterative algorithms have the advantage that they can be used even with a very large number of data, notably when the Cholesky method cannot be used.

## 29.8 Conclusion

We have shown the long way from Krige's regression, which took account of two average sample grades (a local one and a global one) to avoid bias in the estimation of a panel, to present applications of kriging, which can deal with few data (e.g., a limited number of computer experiments in applications to DACE) as well as several hundred thousand data (remote sensing, seismic). We have seen the large diversity of application domains of kriging, so that is it probable that many users do not know the origin of the word: this is the price of success.

We also gave a look at current research to enable a global application of kriging to large data sets, with the requirement to also benefit from nonstationary random function models. Much work remains necessary to transform them in standard methods applicable to a large variety of situations but, in view of the large community of researchers and developers in this area, no doubt that it will be done. The future will show which approaches are the most efficient ones.

## References

Besag J (1974) Spatial interaction and the statistical analysis of lattice systems (with discussion). J Royal Stat Soc Ser B 36(2):192–236

Besag J (1975) Statistical analysis of non-lattice data. J Royal Stat Soc Ser D 24(3):179–195

Blanchin R, Chilès J-P (1993) The Channel Tunnel: geostatistical prediction of the geological conditions and its validation by the reality. Math Geol 25(7):963–974

Chauvet P, Chilès J-P (1975) Kriging: a method for the cartography of the sea floor. Int Hydrogr Rev 52(1):25–41

Chauvet P, Pailleux J, Chilès J-P (1976) Analyse objective des champs météorologiques par cokrigeage. La Météorologie, Sciences et Techniques 6(4):37–54

Chilès J-P, Delfiner P (2012) Geostatistics: modeling spatial uncertainty, 2nd edn. Wiley, New York

Chung H, Alonso JJ (2002) Design of a low-boom supersonic business jet using cokriging approximation models. In: 9th AIAA/ISSMO symposium on multidisciplinary analysis and optimization, p 5598

Cousin A, Maatouk H, Rullière D (2016) Kriging of financial term-structures. Eur J Oper Res 255 (2):631–648

Cressie N (1990) The origins of kriging. Math Geol 22(3):239–252

Cressie N, Johannesson G (2008) Fixed rank kriging for very large spatial data sets. J R Stat Soc Ser B 70(1):209–226

Cressie N, Verzelen N (2008) Conditional-mean least-squares fitting of Gaussian Markov random fields to Gaussian fields. Comput Stat Data Anal 52(5):2794–2807

Cressman GP (1959) An operational objective analysis system. Mon Weather Rev 87(10):367–374

David M (1976) The practice of kriging. In: Guarascio M et al (eds) Advanced geostatistics in the mining industry. Reidel, Dordrecht, pp 31–48

Delfiner P (1973) Analyse objective du géopotentiel et du vent géostrophique par krigeage universel. New print (1975): La Météorologie 5(25):1–58

Delfiner P, Delhomme J-P (1975) Optimum interpolation by kriging. In: Davis JC, McCullagh MJ (eds) Display and analysis of spatial data. Wiley, London, pp 96–114

Delhomme J-P (1978) Kriging in the hydrosciences. Adv Water Resour 1(5):251–266

Dixit V, Seshadrinath N, Tiwari MK (2016) Performance measures based optimization of supply chain network resilience: a NSGA-II+Co-Kriging approach. Comput Ind Eng 93:205–214
Dubrule O, Kostov C (1986) An interpolation method taking into account inequality constraints: I. Methodology. Math Geol 18(1):33–51
Emery X, Arroyo D (2018) On a continuous spectral algorithm for simulating non-stationary Gaussian random fields. SERRA 32(4):905–919
Fouedjio F, Desassis N, Rivoirard J (2016) A generalized convolution model and estimation for non-stationary random functions. Spat Stat 16:35–52
Fulgstad GA, Lindgren F, Simpson D, Rue H (2015) Exploring a new class of non-stationary spatial Gaussian random fields with varying local anisotropy. Stat Sin 25:115–133
Furrer R, Genton MG, Nychka DW (2006) Covariance tapering for interpolation of large spatial datasets. J Comput Graph Stat 15(3):502–523
Galli A, Gao H (1999) Rate of convergence of the Gibbs sampler in the Gaussian case. Math Geol 33(6):653–677
Gandin LS (1963) Ob″ektivnyi analiz meteorologicheskikh polei. Gidrometeologicheskoe Izdatel'stvo, Leningrad. Translation (1965): Objective analysis of meteorological fields. Israel program for scientific translations, Jerusalem
Geman S, Geman D (1984) Stochastic relaxation, Gibbs distributions and the Bayesian restoration of images. IEEE Trans Pattern Anal Mach Intell 6(6):721–741
Goldberger AS (1962) Best linear unbiased prediction in the generalized regression model. J Am Stat Assoc 57:369–375
Gribov A, Krivoruchko K (2004) Geostatistical mapping with continuous moving neighbourhood. Math Geol 36(2):267–281
Higdon D (2002) Space and space-time modeling using process convolutions. In: Anderson CW et al (eds) Quantitative methods for current environmental issues. Springer, London, pp 37–54
Higdon D, Swall J, Kern J (1999) Non-stationary spatial modeling. In: Bernardo JM et al (eds) Bayesian statistics 6: proceedings of the sixth Valencia international meeting. Oxford University Press, New York, pp 761–768
Hrafnkelsson B, Cressie N (2003) Hierarchical modeling of count data with application to nuclear fall-out. Environ Ecol Stat 10(2):179–200
Huijbregts C, Matheron G (1971) Universal kriging (an optimal method for estimating and contouring in trend surface analysis). In: McGerrigle JI (ed) 9th international symposium on techniques for decision-making in the mineral industry. Canadian Institute of Mining and Metallurgy, pp 159–169
Journel AG (1974) Geostatistics for conditional simulation of orebodies. Economic Geol 69 (5):673–687
Kleijnen JPC (2016) Regression and Kriging metamodels with their experimental designs in simulation: A review. Eur J Oper Res 256(1):1–16
Kostov C, Dubrule O (1986) An interpolation method taking into account inequality constraints: II. Practical approach. Math Geol 18(1):53–73
Krige DG (1951) A statistical approach to some basic mine valuation problems on the Witwatersrand. J Chem Metal Min Soc S Afr, December, 119–139
Krige, DG (1952) A statistical analysis of some of the borehole values in the Orange Free State goldfield. J Chem Metal Min Soc S Afr, September, 47–64
Langlais V (1990) Estimation sous contraintes d'inégalités. Doctoral thesis, E.N.S. des Mines de Paris
Lantuéjoul C, Desassis N (2012) Simulation of a Gaussian random vector: a propagative version of the Gibbs sampler. In: The 9th international geostatistics congress, Oslo, Norway, 11–15 June 2012, pp 174–181

Lindgren F, Rue H, Lindström J (2011) An explicit link between Gaussian fields and Gaussian Markov random fields: the stochastic partial differential approach. J R Stat Soc B 73(4):423–498

Machuca-Mory DF, Deutsch CV (2013) Non-stationary geostatistical modeling based on distance weighted statistics and distributions. Math Geosci 45(1):31–48

Matheron G (1960) Krigeage d'un panneau rectangulaire par sa périphérie, Note géostatistique no 28, Centre de Géostatistique, Fontainebleau, France

Matheron G (1961) Exemple de krigeage d'un panneau riche. Note géostatistique No 31. Centre de Géostatistique, Fontainebleau, France

Matheron G (1962) Traité de géostatistique appliquée, Tome I. Mémoires du Bureau de Recherches Géologiques et Minières, no. 14. Editions Technip, Paris

Matheron G (1963a) Traité de géostatistique appliquée, Tome II: Le krigeage. Mémoires du Bureau de Recherches Géologiques et Minières, no. 24, Editions B.R.G.M., Paris

Matheron G (1963b) Principles of geostatistics. Econ Geol 58(8):1246–1266

Matheron G (1969) Le krigeage universel. Cahiers du Centre de morphologie mathématique de Fontainebleau, Fasc 1, Ecole des mines de Paris

Matheron G (1971) La théorie des fonctions aléatoires intrinsèques généralisées. Note Géostatistique no 117. Technical report N-252, Centre de Géostatistique, Fontainebleau, France

Matheron G (1973) The intrinsic random functions and their applications. Adv Appl Prob 4 (3):508–541

Matheron G (1981) Splines and kriging: their formal equivalence. In: Merriam DF (ed) Down-to-earth statistics: solutions looking for geological problems. Syracuse University Geological Contributions, pp 77–95

Matheron G, Beucher H, Fouquet C de, Fouquet C de, Galli A, Guérillot D, Ravenne C (1987) Conditional simulation of the geometry of fluvio-deltaic reservoirs. SPE Paper 16753

de Oliveira MA, Possamai O, Dalla Valentina LVO, Flesch CA (2013) Modeling the leadership—project performance relation: radial basis function, Gaussian and Kriging methods as alternatives to linear regression. Expert Syst Appl 40(1):272–280

Paciorek CJ, Schervish MJ (2006) Spatial modeling using a new class of nonstationary covariance functions. Environmetrics 17(5):483–506

Renard D, Yancey JD (1984). Smoothing discontinuities when extrapolating using moving neighbourhoods. In: Verly G et al (eds) Geostatistics for natural resources characterization, Reidel, Dordrecht, part 2, pp 679–690

Rivoirard J, Romary T (2011) Continuity for kriging with moving neighborhood. Math Geosci 43 (4):469–481

Royer J-F (1975) Comparaison des méthodes d'analyse objective par interpolation optimale et par approximations successives. Technical report no 365, Etablissement d'Etudes et de Recherches Météorologiques, Direction de la Météorologie, Paris

Rue H, Held L (2005) Gaussian Markov random fields. Chapman & Hall/CRC, Boca Raton, Florida

Rue H, Tjelmeland H (2002) Fitting Gaussian Markov random fields to Gaussian fields. Scand J Stat 29(1):31–49

Sacks J, Welch WJ, Mitchell TJ, Wynn HP (1989) Design and analysis of computer experiments. Stat Sci 4(4):409–453

Song H, Fuentes M, Gosh S (2008) A comparative study of Gaussian geostatistical models and Gaussian Markov random field models. J Multivar Anal 99(8):1681–1697

Tjelmeland H, Holden L (1993) Semi-Markov random fields. In: Soares A (ed) Geostatistics Tróia '92. Kluwer, Dordrecht, vol 1, pp 479–491

Yaglom AM, Pinsker MS (1953) Random processes with stationary increments of order *n*. Dokl Akad Nauk SSSR 90(5):731–734

Yan JW, Liew KM, He LH (2012) A mesh-free computational framework for predicting buckling behaviors of single-walled carbon nanocones under axial compression based on the moving Kriging interpolation. Comput Methods Appl Mech Eng 247–248:103–112

Whittle P (1954) On stationary processes in the plane. Biometrika 41(3–4):434–449

# Chapter 30
# Multiple Point Statistics: A Review

**Pejman Tahmasebi**

**Abstract** Geostatistical modeling is one of the most important tools for building an ensemble of probable realizations in earth science. Among them, multiple-point statistics (MPS) has recently gone under a remarkable progress in handling complex and more realistic phenomenon that can produce large amount of the expected uncertainty and variability. Such progresses are mostly due to the recent increase in more advanced computational techniques/power. In this review chapter, the recent important developments in MPS are thoroughly reviewed. Furthermore, the advantages and disadvantages of each method are discussed as well. Finally, this chapter provides a brief review on the current challenges and paths that might be considered as future research.

## 30.1 Introduction

Characterization and modeling of geological structures have been investigated for several years in geosciences. Geostatistics is one of the such methods that can be used to analyze the data effectively. Such analysis can be performed both spatially and temporally. Lack of data is one of the intrinsic issues in the earth science applications, which causes a significant uncertainty and ambiguity in these problems. Kriging, as one of the most widespread geostatistical tools, was developed for dealing with such problems. The basic mathematically equations of Kriging, after developing by Daniel Krige, was further advanced by Matheron (Journel and Huijbregts 1978; Matheron 1973). Kriging is a deterministic method, meaning that it only produces one outcome from the available sparse data, which intrinsically cannot be used to effectively quantify the uncertainty. This method requires a prior model of variability and correlation between the variables, known as the variogram

P. Tahmasebi (✉)
Department of Petroleum Engineering, University of Wyoming, Laramie, WY 82071, USA
e-mail: ptahmase@uwyo.edu



B. S. Daya Sagar et al. (eds.), *Handbook of Mathematical Geosciences*,
https://doi.org/10.1007/978-3-319-78999-6_30

(Chiles and Delfiner 2011; Cressie and Wikle 2011; Deutsch and Journel 1998; Goovaerts 1997; Kitanidis 1997).

It has been shown that Kriging produces excessively smooth results (Deutsch and Journel 1998; Journel and Zhang 2006) and it cannot represent the heterogeneity and non-smooth phenomena. One consequence is the underestimation and overestimation for low and high values, respectively. This problem becomes evident when important parameters such as water breakthrough is intended to be predicted. Thus, the results of Kriging cannot be used for these situations as they ignore the connectivity and variability.

Stochastic simulation can be used to overcome the limitations of Kriging (Goovaerts 1997; Journel and Huijbregts 1978). Several simulation methods have been proposed that can produce various equi-probable realizations. Methods such as sequential Gaussian simulation (SGSIM) and sequential indicator simulation (SISIM) have become popular among different fields of earth sciences. These methods, give a number of "realizations" or interpolation scenarios, which allow assessing the uncertainty and quantifying it more accurately. It should be noted that Kriging is still the main algorithm used in the above stochastic methods. An example for the application of Kriging and stochastic modeling is provided in Fig. 30.1.

Due to relying on variogram (i.e. covariance), kriging-based geostatistical simulations are not able to reproduce complex patterns. Clearly, considering only two points is not sufficient for reproducing complex and heterogeneous models. Thus, several attempts in the recent years in the context of multiple point geostatistics (MPS) have been made that can use more than two points simultaneously. Using the information from multiple points require a big source of data, which is not usually available in the earth science problems as they come with sparse and incomplete data. Such data, instead, can be browsed in the form a conceptual image, called training image (**TI**).

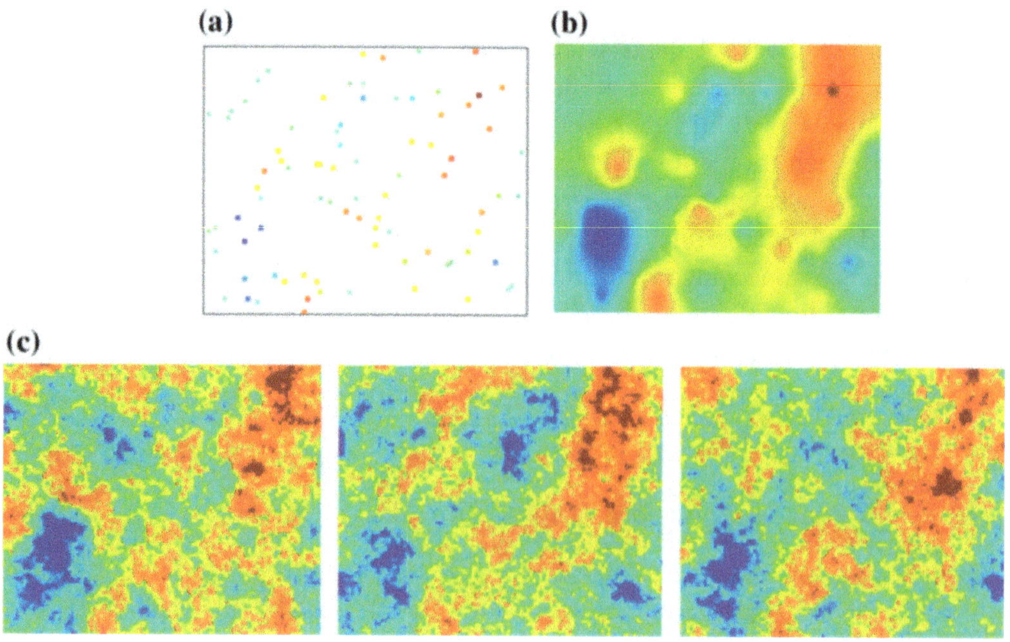


**Fig. 30.1** Comparison between the results of Kriging (**b**) and stochastic simulation (**c**) using conditioning point data in (**a**)

Technically, geostatistical methods can be divided into three main groups. Object-based (or Boolean) simulation methods are in the first group (Kleingeld et al. 1997). These methods consider the medium as a group of stochastic objects that are defined based on a specific statistical distribution (Deutsch and Wang 1996; Haldorsen and Damsleth 1990; Holden et al. 1998; Skorstad et al. 1999).

Pixel-based methods are considered in the second group. These methods are based a set of points/pixels that represent various properties of a phenomenon. Mathematically speaking, such methods vary from the LU decomposition of the covariance matrix (Davis 1987), sequential Gaussian simulation (Dimitrakopoulos and Luo 2004), frequency- domain simulation (Borgman et al. 1984; Chu and Journel 1994), simulated annealing (Hamzehpour and Sahimi 2006), and the genetic algorithm. The last two methods, namely optimization techniques, also belong to this group as they gradually change an earth model in a pixel-by-pixel manner.

Each of the above methods has some advantages and limitations. For example, geological structures can be reproduced accurately using the object- based simulations. However, conditioning in these methods to well and soft data require intensive computation.

Pixel-based methods simulation on one pixel at a time. Such techniques produce the conditioning point data exactly. One drawback of these methods is that they are based on variograms that represent two-point statistics and, thus, they cannot reproduce the complex and realistic geological structures. Consequently, the generated models using these techniques cannot represent an accurate representation of any physics-based simulations (e.g. flow, grade distribution, contaminate forecasting and etc.).

In the MPS methods, the spatial statistics are not either extracted using variogram, but a conceptual tool named training image (TI), which is an example of the spatial structure to be reproduced, is provided that can represent the necessary data. During the recent years, several MPS methods have been developed to address issues related to CPU time and improved graphical representation of the models produced. This chapter, thus, reviews the existing concepts in MPS and discusses the available methods. The main two-point based stochastic simulation methods are first reviewed. Then, the basic terminologies and concepts of MPS are demonstrated. Next, different MPS methods are explained and the advantages and disadvantages associated with each method are demonstrated. Finally, some avenues for future research are discussed.

## 30.2 Two-Point Based Stochastic Simulation

The smoothing effect of Kriging can be avoided using the sequential simulation, which helps to quantify the uncertainty accurately. Consider a set of $N$ random variables $Z(\mathbf{u}_\alpha), \alpha = 1, \ldots, N$ defined at locations $\mathbf{u}_\alpha$. The aim of sequential simulation is to produce realizations $\{z(\mathbf{u}_\alpha), \alpha = 1, \ldots, N\}$, conditioned to $n$ available data and reproducing a given multivariate distribution. For this aim, the multivariate distribution is decomposed into a set of $N$ univariate conditional cumulative distribution functions (*ccdfs*):

$$\begin{aligned} F(\mathbf{u}_1, \ldots, \mathbf{u}_N; z_1, \ldots, z_N|(n)) = {} & F(\mathbf{u}_1; z_1|(n)) \times \\ & F(\mathbf{u}_2; z_2|(n+1)) \times \cdots \times \\ & F(\mathbf{u}_{N-1}; z_N - 1|(n+N-2)) \times \\ & F(\mathbf{u}_N; z_N|(n+N-1)) \end{aligned} \tag{30.1}$$

where $F(\mathbf{u}_N; z_N|(n+N-1)) = Prob\ \{Z(\mathbf{u}_N) \leq z_N|(n+N-1)\}$ is the conditional ccdf of $Z(\mathbf{u}_N)$ conditioned to a set of $n$ original data and $(N-1)$ previously simulated values.

### *30.2.1 Sequential Gaussian Simulation (SGSIM)*

In this method, the multivariate distribution and the higher order are constructed based on the lower order statistical such as histogram and variogram. In other words, the mean and covariance matrix are used to build a Gaussian function. Therefore, along a random path, the mean and variance of the Gaussian distribution is estimated via Kriging and Kriging variance. The overall algorithm of SGSIM can be summarized as follows. First, a random path is defined over all visiting points on the simulation grid. Then, the *ccdf* at each node based on the hard data and previously simulated data are considered in Kriging. Then, a random value from the obtained Gaussian *ccdf* is drawn and added to the simulation grid. Next, based on the predefined random path, another node is chosen and simulated. Finally, another realization can be generated using a different random path.

It is worth noting that the conditioning data should be normally distributed. If it is not the case, it entails transforming them into a Gaussian distribution in order to be useable for SGSIM. Finally, the results must be back-transferred at the end of simulation. Such transformations can be accomplished using normal-score transforms or histogram anamorphous through Hermite polynomials.

### *30.2.2 Sequential Indicator Simulation (SISIM)*

Indicator simulation follows the same principle as SGSIM. This method, however, is suited for categorical data, which do not have an order relationship. Typical examples in earth science are rock type, lithology codes and some other categorical properties. The similar sequential procedure based on the estimation of the *ccdf* conditioning to neighboring data is applied here as well. This algorithm is based on two-point indicator variograms, which represent the spatial variability of each category. An indicator variable is defined for each variable, equal to 1 if at location $u$ a particular category is found, and zero otherwise. Also, $E\{I(\mathbf{u})\} = p$ is the stationary proportion of a given category. The indicator variogram can be expressed as:

$$\begin{aligned} &Prob\{I(\mathbf{u}) = 1, I(\mathbf{u}+h) = 1\} \\ &= E\{I(\mathbf{u})I(\mathbf{u}+h)\} \\ &= Prob\ \{I(\mathbf{u}) = 1 | I(\mathbf{u}+h) = 1\} \end{aligned} \tag{30.2}$$

Usually, the categorical variables expressed as a set of $K$ discrete categories that $z(\mathbf{u})\{0, \ldots, k-1\}$. Therefore, the indicator value for each of the defined classes can be expressed as:

$$I(\mathbf{u}, k) = \begin{cases} 1 & Z(\mathbf{u}) = k \\ 0 & otherwise \end{cases} \tag{30.3}$$

The aim of the indicator formulation is to estimate the probability of $Z(\mathbf{u})$ to be less than the predefined threshold for a category conditional to the data $(n)$ retained:

$$\begin{aligned} I^*(\mathbf{u}, z_k) &= E^*(I(\mathbf{u}, z_k)|(n)) \\ &= Prob^*(Z(\mathbf{u}) < z_k(n)) \end{aligned} \tag{30.4}$$

We can rewrite the above equation for categorical variables by using simple Kriging as:

$$\begin{aligned} &I^*(\mathbf{u}, z_k) - E(I(\mathbf{u}, k)) = \sum_{\alpha=1}^{n} \lambda_\alpha(\mathbf{u})(I(\mathbf{u}_\alpha, k) - E\{I(\mathbf{u}_\alpha, k)\}) \\ &\sum_{k=1}^{K} I^*(\mathbf{u}, k) = 1 \end{aligned} \tag{30.5}$$

where $E\{I(\mathrm{u}, k)\}$ is the marginal probability for category $k$.

The above formulation can be applied within the sequential scheme which known as SISIM. Indicator Kriging (IK) is used to estimate the probability of each category. This algorithm can be described as follow. Similarly, as SGSIM, a random path is defined by which all of the nodes are visited. Then, using Simple Kriging, the indicator random variable for each category is estimated for each node on the random path based on the neighboring data. Next, the conditional probability density function (*cpdf*) is obtained and a value is randomly drawn from that *cpdf* and assigned to the simulated node. This procedure is repeated sequentially for all the visiting nodes until the simulation grid is completed. By choosing another random path, one can generate another realization. More information on this method can be found in Goovaerts (1997).

## 30.3 Multiple Point Geostatistics (MPS)

One of the bottlenecks in the two-point based geostatistical simulations is their inability in dealing with complex and heterogeneous spatial structures. Such methods cannot fully reproduce the existing physics and most of their parameters usually do not have an equivalent in the reality. In particular, these methods cannot convey the connectivity and variability when the considered phenomenon contains definite patterns or structures. For example, models containing regular structures cannot be reproduced using the SGSIM method. Thus, increasing the number of points can help reproducing the connectivities and complex features. The MPS methods, indeed, intend to reproduce the physics in natural phenomena and they all are based on a set of training images. Below, some preliminary concepts are first reviewed.

### *30.3.1 Training Image*

Training image (TI) is one of the most important inputs in the MPS techniques. Thus, providing a representative TI, or a set of TIs, is the biggest challenge in the MPS applications. In general, TIs can be generated using the physics derived from process-based methods or statistical methods or by using the extracted and observed rules for each geological system. The TI can be of any type, ranging from an image to statistical properties in space and time. In fact, TIs let us to include subjectivity in the geological modeling, as they are difficult to be taken into account in the traditional statistical methods. In a broader sense, TI can be constructed based on the traditional statistical methods. These outcomes, however, do not represent the deterministic aspects of geological models, as they usually tend to signify the randomness fragment. Geologically speaking, most of the images in natural sciences represent some degree of complexity and uniqueness. Some examples of the available **TI**s are shown in Fig. 30.2.

The available methods for constructing the TIs are divided into three main groups:

- *Outcrop Data*: An example of TI is the outcrop images, which are one of the preliminary sources of information at the first step of geological modeling. They provide a unique and direct representation of geological features. They also provide a clear illustration of the geometry and spatial continuity that allow visual inspection of the existing structures in 2D sections.
- *Object-based Methods*: An alternative for constructing structured categorical models is the object-based (or Boolean) method (Deutsch and Wang 1996; Haldorsen and Damsleth 1990; Holden et al. 1998; Lantuéjoul 2002; Skorstad et al. 1999). These methods are defined based on some shape parameters (e.g. size, direction, and sinuosity). The results can be used within an iterative

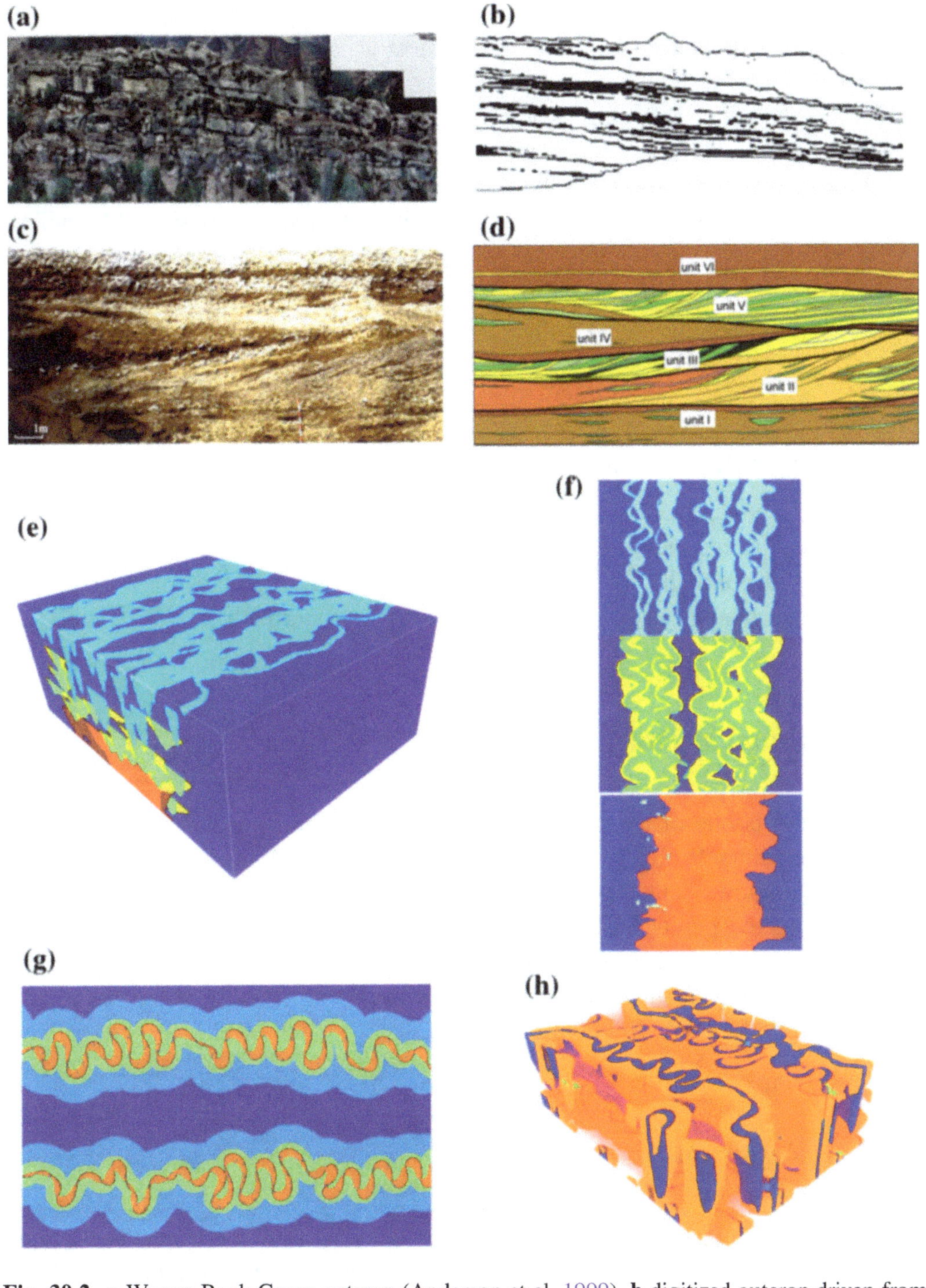


**Fig. 30.2** **a** Wagon Rock Caves outcrop (Anderson et al. 1999), **b** digitized outcrop driven from (**a**), **c** Herten gravel pit (Bayer et al. 2011), **d** litho and hydrofacies distribution extracted from (**c**), **e** a 3D object-based model (Tahmasebi and Sahimi 2016a), **f** some 2D section of the 3D model shown in (**e**), **g** a 2D model generated using the process-based techniques (Tahmasebi and Sahimi 2016a), **h** a 3D model generated by the process-based methods (Tahmasebi and Sahimi 2016a)

algorithm to provide any further alterations. The results of object-based simulation methods are one of the best and most accessible sources for TIs.

- *Process-based Methods*: Process-based methods (Biswal et al. 2007, 1999; Bryant and Blunt 1992; Gross and Small 1998; Lancaster and Bras 2002; Pyrcz et al. 2009; Seminara 2006) try to develop 3D models by mimicking the physical processes that form the porous medium. Though realistic, such methods are, however, computationally expensive and require considerable calibrations. Moreover, they are not general enough, because each of them is developed for a specific type of formation, as each type is the outcome of some specific physical processes.

## 30.4 Simulation Path

Geostatistical techniques are conducted on a simulation grid G, which is constructed on several cells. These cells are visited in diverse ways on a predefined path, either in random or in structural manner (i.e. raster path).

### *30.4.1 Random Path*

Random path is one of the most commonly used visiting path in sequential simulation algorithms. In this particular path, a series of random number equal to the number of unknown cells, based on a random seed, is generated for each realization and the unvisited points on **G** are simulated accordingly. Clearly, the number of simulated (i.e. known) points increase as the simulation proceeds. Each realization is generated using a simulation path. These paths commonly come with unbiasedness around the conditioning point data.

### *30.4.2 Raster Path*

Algorithms based on raster path are popular in the stochastic modeling. These paths are constructed based on structural 1D path, meaning that the simulation cells are visited systematically and one can predict the future visiting points. Daly (2005) presented a Monte Carlo algorithm that utilized raster path. Then, patch-based algorithm was used based on this path by El Ouassini et al. (2008). Next, Parra and Ortiz (2011) used a similar path in their study. Finally, Tahmasebi et al. (2014, 2012a, b) implemented a raster path along a fast similarity computation and achieved high-quality realizations. Such paths usually produced high quality realizations that can barely be produced using the random path algorithms.

One of the advantages in using such paths is the small number of constraints that help the algorithms to better identify the matching data (or patterns). For instance, one only deals with 1–2 overlap regions in 2D simulations, which is much more efficient when four overlaps are used in the random path algorithms. Thus, one should expect more discontinuities and artefacts when then number of overlaps are increased. Indeed, identifying a pattern from TI based on four constraints is very difficult, if not impossible. Therefore, using small number of overlaps is desirable as they result in high-quality realizations. Raster path algorithms offer such a prospect and one can achieve realizations with higher quality.

Dealing with conditioning data (e.g. point and secondary data) is one of the crucial issues in these paths. They, in fact, cannot account for the conditioning data that are ahead of them. Therefore, some biases have been observed in these algorithms, particularly around the conditioning point data. Some complementary methods such as template splitting (Tahmasebi et al. 2012a) and co-template (Parra and Ortiz 2011; Tahmasebi et al. 2014) have addressed this issue partially.

### *30.4.3 Some Other Definitions*

*Simulation Grid* (**G**): a 2D/3D computation grid on which the geostatistical modeling is performed and is composed of several cells, depending on the size of domain and simulation. It contains no information for unconditional simulation, while the hard data are distributed in their corresponding cells.

*Data-Event*: a set of points that are characterized by a distance, namely lag, which are considered around a visiting point (cell) on G.

*Template*: a set of points that are organized systematically and used for finding similar patterns in TI.

## 30.5 Current Multiple Point Geostatistical Algorithms

Generally, the MPS methods have been developed in both pixel- and pattern-based states, each of which, as discussed, have similar pros and cons. For example, the pixel-based MPS methods can perfectly match the well data, whereas, these methods, in some complex geological models, produce unrealistic structures. On the other hand, pattern-based techniques bring a more accurate representation of the subsurface model, while they usually miss the conditioning data. The pattern-based methods simulate a group of points at a time. Currently, these techniques are under different progress, due to their ability for simultaneous reproduction of conditioning data and geologically realistic structures. As mentioned, conditioning to well data is one of the critical issues in the pattern-based techniques. Thus, taking advantage of the capabilities of both pixel- and pattern-based techniques in the MPS methods through the hybrid frameworks will result in an efficient algorithm. Such a combination is reviewed thoroughly in this chapter as well.

Most of the available MPS methods can be used with non-stationary systems, the ones in which the statistical properties of a region is different from other parts (Chugunova and Hu 2008; Honarkhah and Caers 2012; Mariethoz et al. 2010; Strebelle 2012; Tahmasebi and Sahimi 2015a; Wu et al. 2008).

### 30.5.1 Pixel-Based Algorithms

i. *Extended Normal Equation Simulation (ENESIM)*

The ENESIM is the first method wherein the idea of MPS was raised (Guardiano and Srivastava 1993). This method is based on an extended concept of indicator kriging, which allows reproduction of multiple-event inferred from a TI. It first finds the data even at each visiting point and then scans the TI for identifying all occurrences. Then, a conditional distribution for all the identified occurrences is constructed. Next, a sample from the generated histogram is drawn and placed in the visiting point on **G**. One of the main drawbacks of this algorithm is scanning the TI for each visiting point, which makes it unpractical for large **G** and TI. This algorithm was later redesigned in the SNESIM algorithm by aid of search tree so one does not need to rescan the TI for each visiting point, but it can be done once before the simulation begins. Some of the results of this algorithm are presented in Fig. 30.3.

ii. *Simulated Annealing*

Simulated annealing (SA) is one of the popular methods in optimization that is used to the global minima. Suppose $E$ represent the energy:

$$E = \sum_{j=1}^{n} \left[ f(\mathbf{O}_j) - f(\mathbf{S}_j) \right]^2 \tag{30.6}$$

where $\mathbf{O}_j$ and $\mathbf{S}_j$ represent the observed (or measured) and the corresponding simulated (calculated) properties of a porous medium, respectively, with $n$ being the number of data points. If there are more than one set of data for distinct properties of the medium, the energy $E$ is generalized to

$$E = \sum_{i=1}^{m} \omega_i E_i \tag{30.7}$$

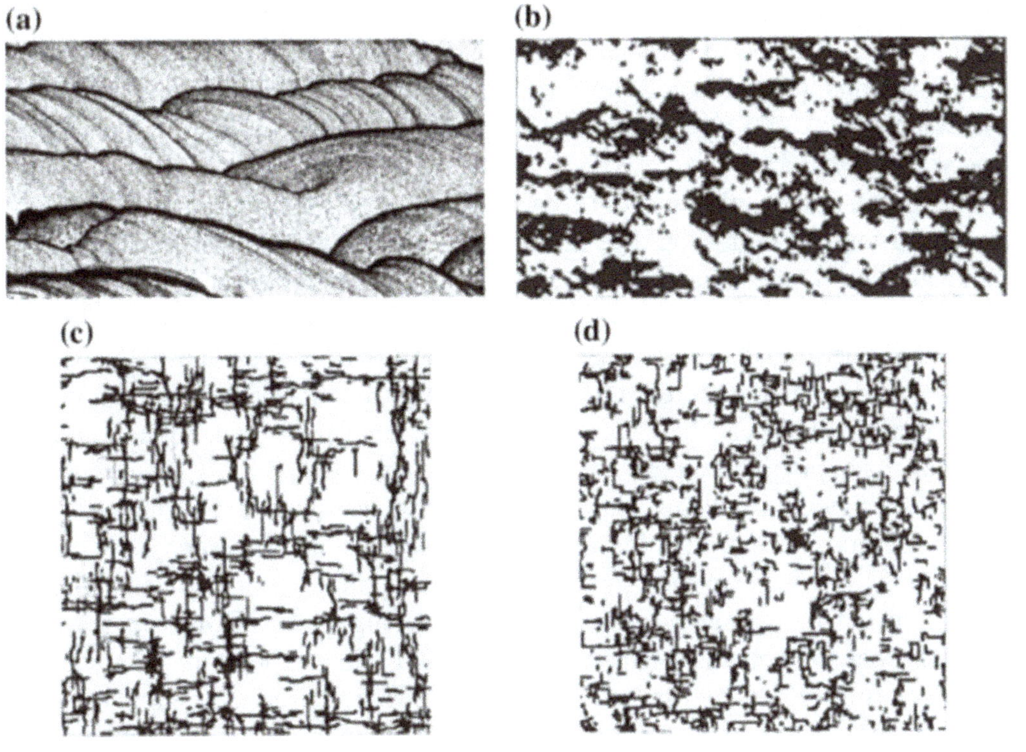

Fig. 30.3 The results of the ENESIM algorithm. **a** cross-bedded sand, which is used as TI, **b** one realization generated using ENESIM, **c** TI: fractured model generated using physical rock propagation, **d** one conditional realization based on the TI in (**c**) and 200 conditioning data points. The results are browed from Guardiano and Srivastava (1993)

where $E_i$ is the total energy for the data set $i$, and $\omega_i$ the corresponding weight, as two distinct set of data for the same porous medium do not usually have the same weight or significance.

An initial guess is usually considered as the structure of medium by which the algorithm can start. Then, a small perturbation is made on the initial model and the new energy $E^{'}$ and the difference $\Delta E = E^{'} - E$, are then computed. Based on a probability this interchange is then accepted. The interchange is then accepted with a probability $p(\Delta E)$. Then, according to the Metropolis algorithm,

$$p(\Delta E) = \begin{cases} 1, & \Delta E \leq 0, \\ \exp(-\Delta E/T) & \Delta E > 0, \end{cases} \tag{30.8}$$

where $T$ is a fictitious temperature.

Based on statistical mechanics, it is well known that the equilibrium state of ground state can be achieved when it is heated up to a high temperature $T$ and then slowly cooled down to absolute zero. The cooling is usually considered slow to allow the system to reach its true equilibrium state. It, indeed, allows the systems to not trap in a local energy minimum. At each annealing step $i$ the system is allowed to evolve long enough to "*thermalize*" at $T(i)$. Then, the temperature $T$ is decreased and continues until the true ground state of the system is reached. This process stops when the $E$ is deemed to be small enough. This method has been used widely for reconstruction of fine-scale porous media by Yeong and Torquato (1998a, b), Manwart et al. (2000) and Sheehan and Torquato (2001), as well as large-scale oil reservoirs.

Using this framework, the algorithm starts on a spatially random distribution for the simulation grid **G**. It should be noted that the hard data are placed in the **G** at the same time. Afterwards, the simulation cells are visited and the energy of realization (i.e. global energy) is calculated using the terms considered in the objective function. Then, the probability of acceptance/rejection is calculated and the new value will be used/ignored accordingly. Consequently, the objective function will be updated and next $T$ can be defined afterwards. This process continues until the predefined stopping criteria are meet.

Deutsch (1992) used this algorithm to reproduce some MPS properties. He considered an objective function that satisfies some constrains such as histogram and variogram. Furthermore, some researchers applied simulated annealing for simulation of continuous variables (Fang and Wang 1997). However, simulated annealing has drawbacks, a major one being CPU time. Therefore, one can only consider a limited number of statistics as constrains, because increasing the number of constrains has a strong effect on CPU time. In addition, this algorithm has many parameters which should be tuned and therefore need a large amount of trial and error to achieve optimal values. Peredo and Ortiz (2011) used speculative parallel computing to accelerate the simulated annealing; however the computation times are still far from what is obtained with sequential simulation methods (Deutsch and Wen 2000). A overall comparison between the SA algorithm and the traditional algorithms is presented in Fig. 30.4. In a similar fashion, the multiple point statistical methods have also used the effect of iterations on removing the artifacts (Pourfard et al. 2017; Tahmasebi and Sahimi 2016a). It should be noted that the sequential algorithms can be parallel using different strategies which are not discussed in this review chapter (Rasera et al. 2015; Tahmasebi et al. 2012b).

iii. *Markov Random Field (MRF)*

These models incorporate constraints by formulating high-order spatial statistics and enforcing them on the simulated domain using a Metropolis-Hastings algorithm. In this case, the computational problem of the previous methods remains because the Metropolis-Hastings algorithm, although always converging in theory, may not converge in a reasonable time. The model parameters are inferred from the available data, namely TI.

The Markovian properties are usually expressed as a conditional probability:

$$p(\mathbf{Z}|all\,previous\,\mathbf{Z}) = p(z_1)p(z_2|z_1)\dots\underbrace{p(z_N|z_{N-1}, z_{N-2}, \dots, z_2, z_1)}_{p\left(z_N|\mathbf{Z}_{\Phi_N}\right)} \quad (30.9)$$

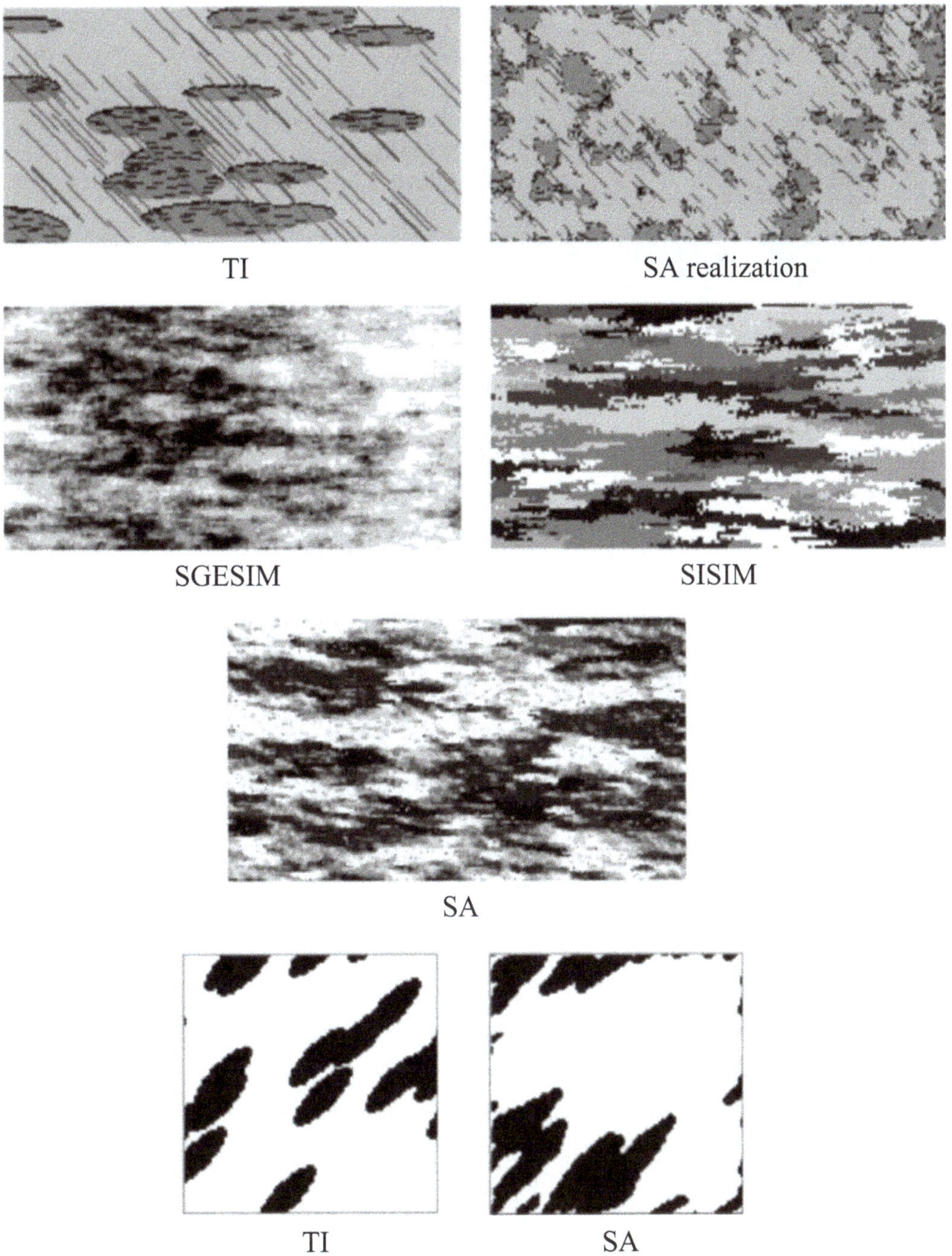


**Fig. 30.4** A comparison between the results of the SA algorithm and traditional two-point based geostatistical simulations (Deutsch 1992). It should be noted that the results of the SGESIM, SISIM and SA algorithms are generated based on the TI shown in Fig. 30.3a. The last raw is browed from Peredo and Ortiz (2011)

where $\mathbf{Z}_{\Phi_N}$ indicates the conditional probability of $z_N$ and $p(z_N) > 0 \forall z_N$.

Fully utilizing the MRF algorithm for large 3D simulation grids in earth science is not practical. Thus, researchers have focused on less computationally demanding

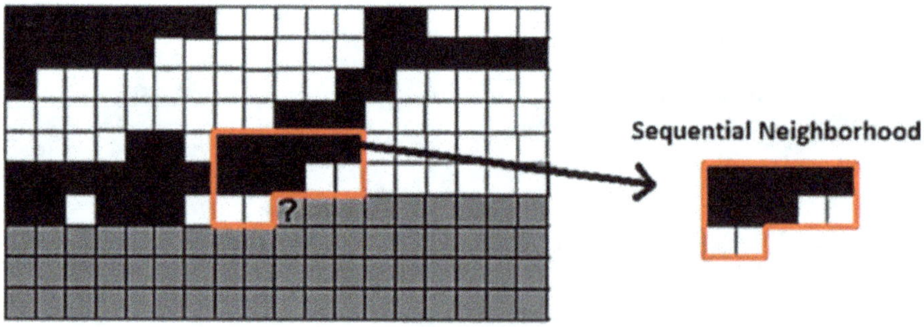


**Fig. 30.5** An illustration of the MMM method. The gray cells represent the unvisited points. The neighborhood is shown in a red polygon. This figure is taken from Stien and Kolbjørnsen (2011)

algorithms such as Markov Mesh Models (MMM) (Daly 2005; Stien and Kolbjørnsen 2011; Toftaker and Tjelmeland 2013). In this algorithm, the simulation is only restricted to a reasonable small window around the visiting point, see Fig. 30.5. Thus, Eq. (9) can be shorten as:

$$p(\mathbf{Z}) = p(z_1)p(z_2|z_1)\ldots\underbrace{p(z_n|z_{n-1}, z_{n-2}, \ldots, z_2, z_1)}_{p(z_n|\mathbf{Z}_{\Phi_n})} \tag{30.10}$$

where $n \ll N$.

Tjelmeland and Eidsvik (2005) used a sampling algorithm that incorporates an auxiliary random variable. These methods suffer from extensive CPU demand and instability in convergence. Besides, the large structures cannot be reproduced finely, a series of factors that make them difficult to use for 3D applications. Some of the results of this method are shown in Fig. 30.6.

iv. *Single Normal Equation Simulation (SNESIM)*

The single normal equation simulation (SNESIM) is an improved version of the original algorithm proposed by Guardiano and Srivastava (1993). The SNESIM algorithm scans the input TI for once and then stores the frequency/probability of all pattern occurrences in a search tree (Boucher 2009; Strebelle 2002), which reduces the computational time significantly. Then, the probabilities are retrieved from the constructed search-tree based on the existing data in the data-event. The SNESIM algorithm is a pixel-based algorithm, which can perfectly reproduce the conditioning point data.

The SNESIM algorithm is a sequential algorithm and, thus, each cell $S$ can take $k$ possible states $\{s_k, k = 1, \ldots, K\}$, which usually represents facies unit. This algorithm, like any other conditional techniques, calculates the joint probability over $n$ discrete points using:

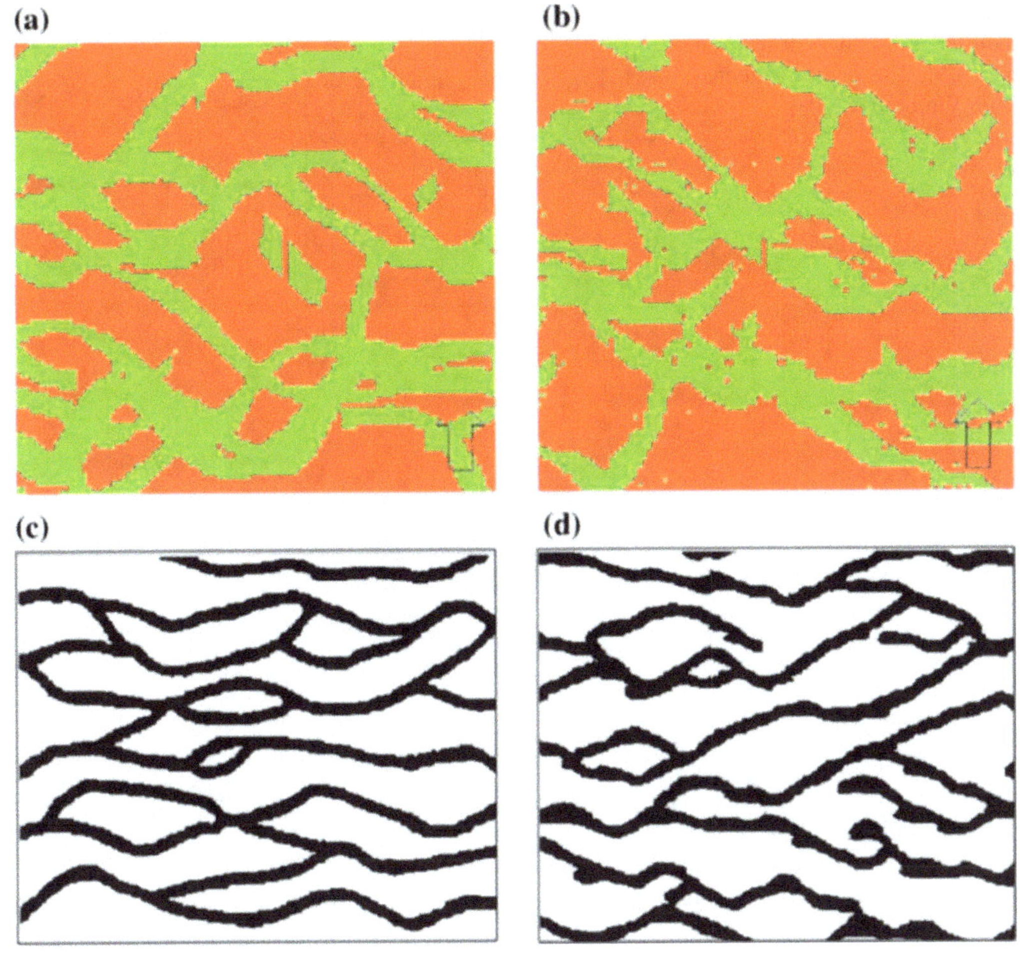


Fig. 30.6 A demonstration of the results of MRF (Daly 2005; Stien and Kolbjørnsen 2011). **a**, **c** TI and **b**, **d** realizations

$$\Phi(\mathbf{h}_1, \ldots, \mathbf{h}_n; k_1, \ldots, k_n) = E\left\{ \prod_{\alpha=1}^{n} I(\mathbf{u}+\mathbf{h}_\alpha; k_\alpha) \right\} \tag{30.11}$$

where $\boldsymbol{h}$, $k$, $\boldsymbol{u}$ and $E$ represent separation vector (lag), state value, visiting location and expected value, respectively. $I(\mathbf{u}; k)$ also denotes the indicator value at location $\boldsymbol{u}$. This equation, thus, gives the probability of having $n$ values $(k_1, \ldots, k_n)$ at the locations $s(\mathbf{u}+\mathbf{h}_1), \ldots, s(\mathbf{u}+\mathbf{h}_n)$. The above probability is replaced with the following equation in SNESIM:

$$\Phi(h_1, \ldots, h_n; k_1, \ldots, k_n) \cong \frac{c(d_n)}{N_n} \tag{30.12}$$

where $N_n$ and $c(d_n)$ denote the total number of patterns in the TD and number of replicates for the data event $d_n = \{s(\mathbf{u}+\mathbf{h}_n) = s_{k_\alpha}, \alpha = 1, \ldots, n\}$.

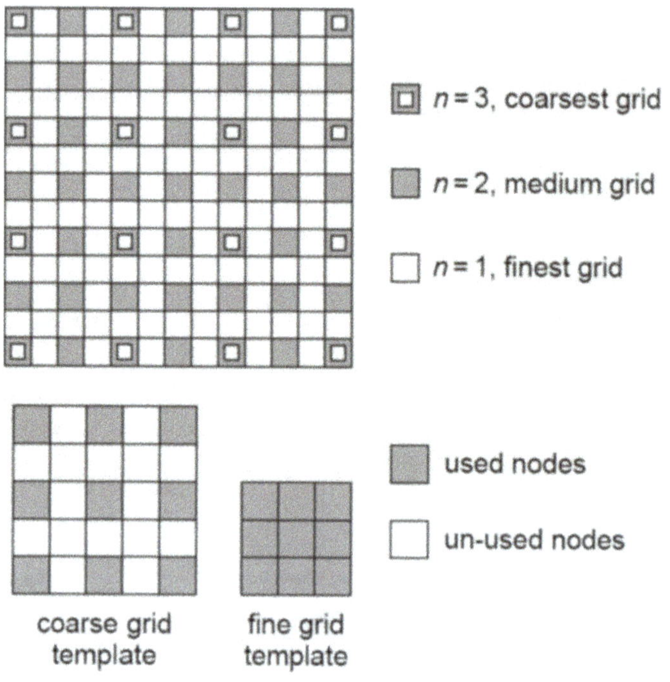


**Fig. 30.7** Demonstration of multiple-grid approach in SNESIM. The figure is taken from Wu et al. (2008)

This algorithm benefits from multiple-grid by which the large structures are first captured using a smaller number of nodes and then the details are added. This concept is illustrated in Fig. 30.7.

One of the limitations in the SNESIM algorithm is lack of producing realistic, highly connected and large-scale geological features. This algorithm, however, can be used only on categorical TIs. The SNESIM algorithm is still inefficient for the real multimillion cells applications (Tahmasebi et al. 2014). Several other methods were latter proposed to improve the efficiency and quality of the SNESIM algorithm (Cordua et al. 2015; Straubhaar et al. 2013). A new technique has recently been presented that can take the realizations and perform a structural adjustment to match the well data (Tahmasebi 2017) (Fig. 30.8).

V. *Direct Sampling*

Direct sampling method is very similar to SIMPAT algorithm (see below) in that sense it only scans a part of TI and pastes one single pixel (Mariethoz et al. 2010). Since the TI is scanned in each loop of the simulation, thus, there is no need to make any database and less RAM is required. Like the pattern-based techniques, this algorithm uses a distance function for finding the closest patterns in TI. This method can be used for both categorical and continuous variables.

The DS algorithm selects the known data at each visiting point. Then, the similarity of the data-event with the TI is calculated based on a predefined searching portion. As soon as the first occurrence of a matching data event in the TI is found (corresponding to a distance under a given threshold acceptance), the value of the central node of the data event in the TI is accepted and pasted in the simulation. It

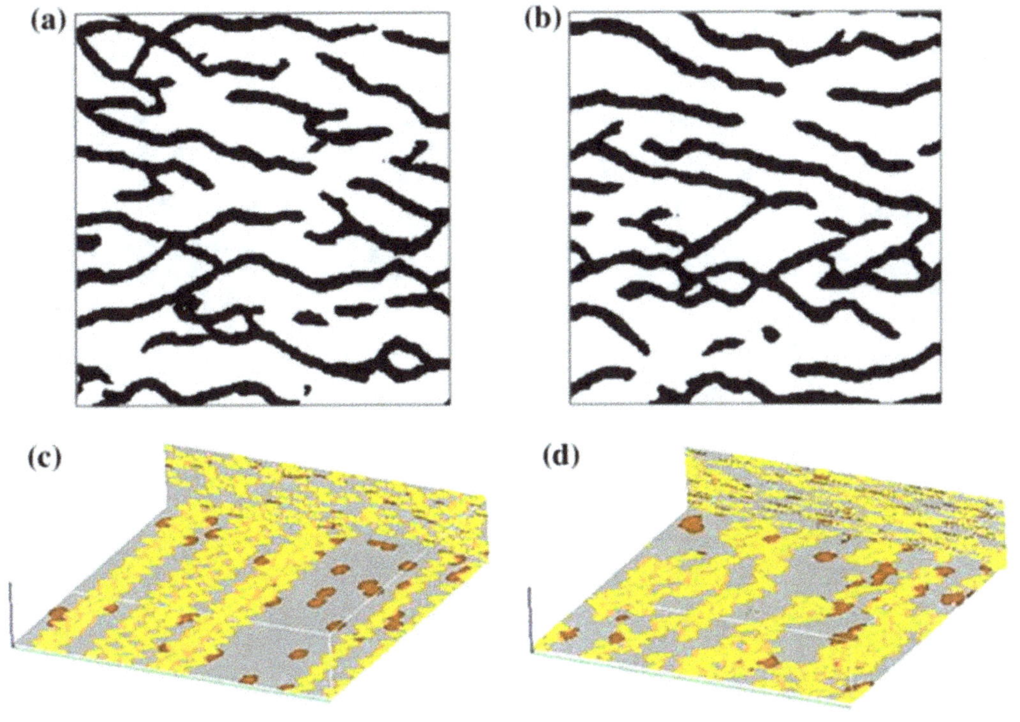

**Fig. 30.8** The results of SNESIM. The realizations shown in (**a**, **b**) are generated using the TI in Fig. 30.6c. **c** TI and **d** a realization based on the TI in (**c**)

should be noted that the searching phase stops if the algorithm finds a pattern that is similar up to a given threshold. If not, the most similar pattern found in the predefined portion of TI is selected and its central node is pasted on the simulation grid (Fig. 30.9).

vi. *Cumulants*

More information beyond two-point statistics can be inferred using the cumulants approach. This method, indeed, can extract such a higher information directly from the existing data, rather than the TI. Dimitrakopoulos et al. (2010) first used this method to simulate geological structures. The geological process, anisotropy and pattern redundancy are the important factors that should be considered in selecting the necessary cumulants (Mustapha and Dimitrakopoulos 2010). The conditional probability is first calculated based on the available data. Then, the TI is only researched if not sufficient replicates cannot be found in the data. One requires selecting appropriate spatial cumulants for each geological scenario and there is no specific strategy on this. Some of the results of this method are shown in Fig. 30.10.

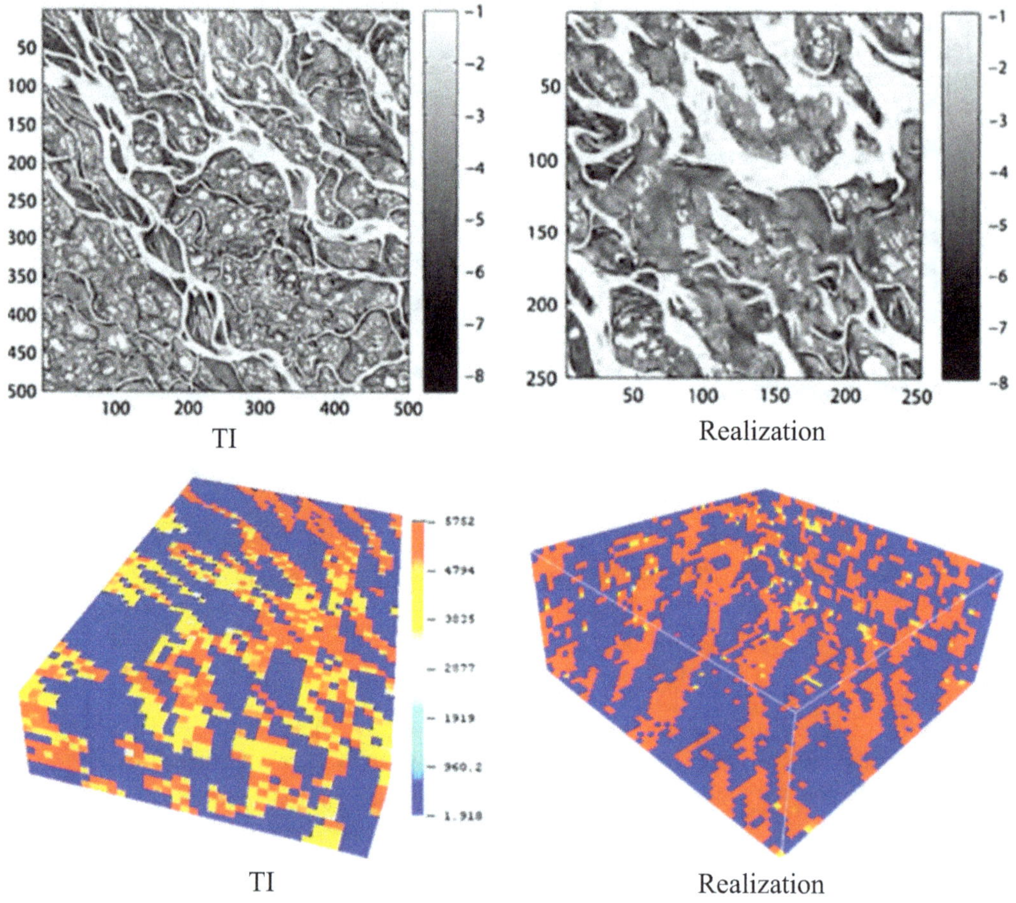


**Fig. 30.9** The results of the DS algorithm for modeling of a hydraulic conductivity field (upper row) and a continuous property (**b**). The results are taken from Mariethoz et al. (2010) and Rezaee et al. (2013)

### *30.5.2 Pattern-Based Algorithms*

Pixel-based algorithms can have problems to preserve the continuity of the geological structures. To palliate this, some pattern-based methods have been developed which briefly are introduced bellow. Their commonality is that they do not simulate one pixel at a time, but they paste an entire "patch" in the simulation. One of the main aims of using pattern based simulation methods is their ability to preserve the continuity and overall structure observed in TI.

i. *Simulation of Pattern (SIMPAT)*

The algorithm of simulation of patterns was first introduced to address some of the limitations in the SNESIM algorithm, namely the CPU time and connectivity of patterns (Arpat and Caers 2007). This method replaces the probability with a distance for finding most similar pattern. The algorithm can be summarized as follows.

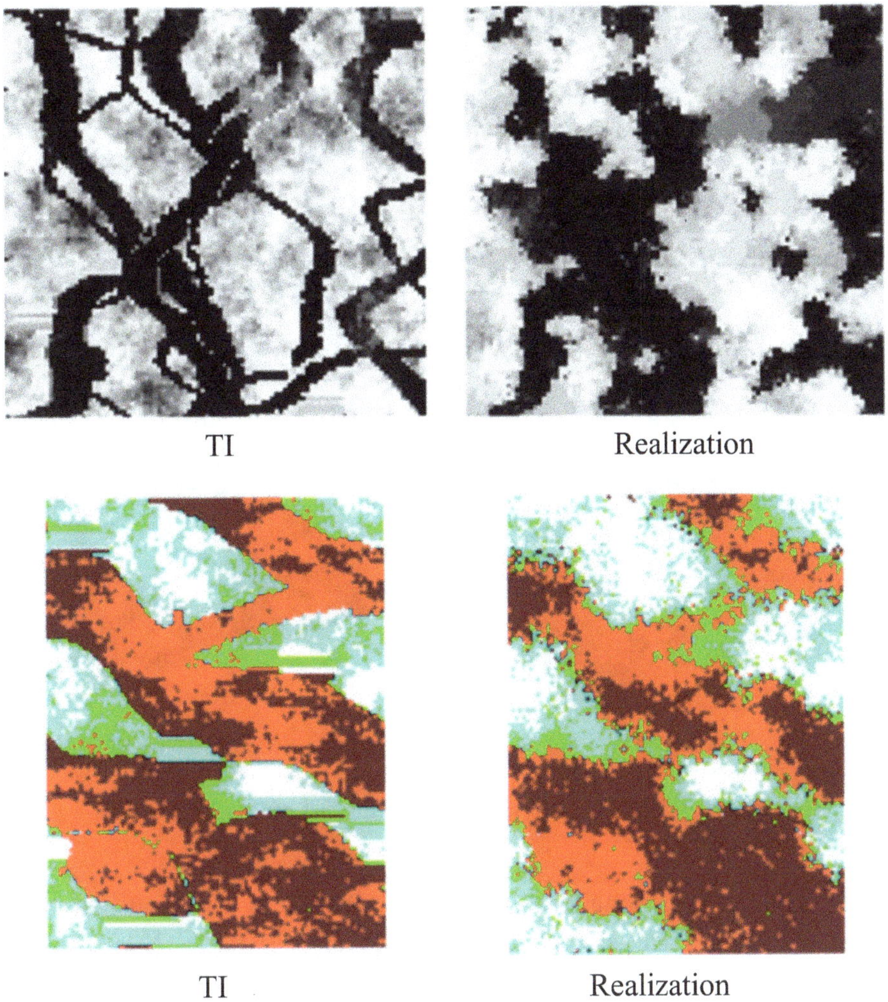


**Fig. 30.10** The results of cumulants for modeling of two complex channelized systems. The results are taken from Mustapha and Dimitrakopoulos (2010, 2011)

The TI is first scanned using a predefined template $T$ and all the extracted patterns are stored in a pattern database. Then simulation points are visited based on the given random path and the corresponding data-event is extracted accordingly. One of the patterns in pattern database is selected randomly if the data-event at the visiting point contains no data. Otherwise, the most similar pattern is selected based on the similarity between the data-event the patterns in pattern database. The above steps are repeated for all visiting points. The results of SIMPAT algorithm are realistic. However, it requires an extensive CPU time and encounters various serious issues in 3D modeling. Furthermore, the produced results manifest a considerable similarity with TI as this algorithm seeking for the best matching pattern. Thus, this method seems to underestimate the spatial uncertainty. Some of the results of SIMPAT are shown in Fig. 30.11.

In a similar fashion, the pattern-based techniques can be used within a Bayesian framework (Abdollahifard and Faez 2013). This process, however, can be very

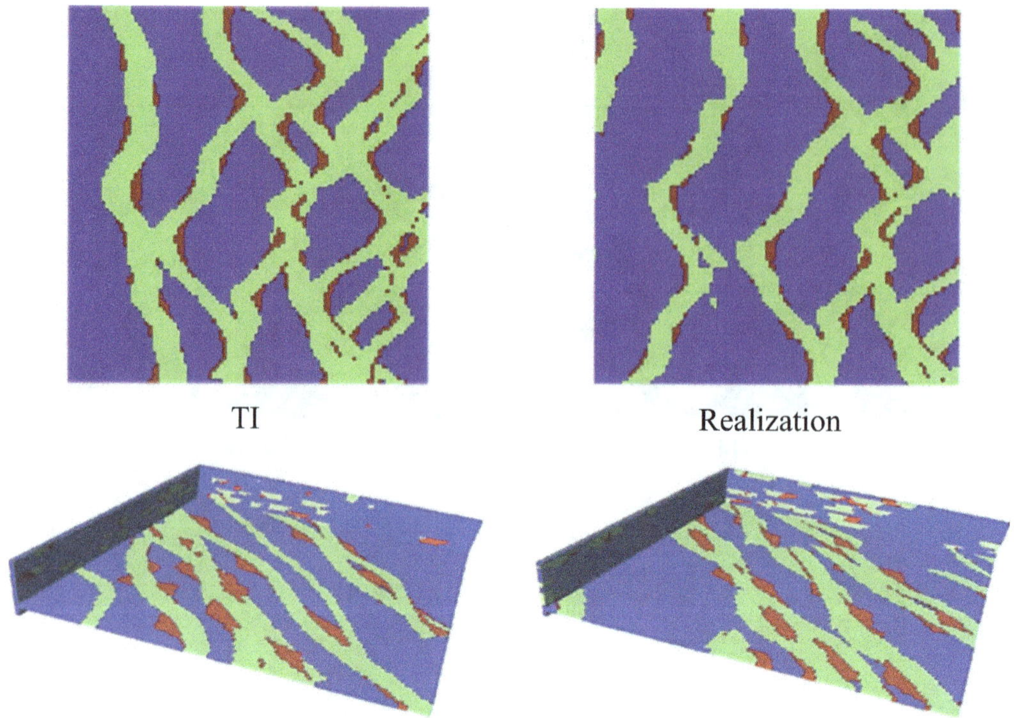


**Fig. 30.11** The results of SIMPAT. These results are taken from Arpat (2005)

CPU demanding. Other enhancements on SIMPAT was also considered later by incorporating wavelet decomposition (Gardet et al. 2016).

ii. *Filter-based Simulation (FILTERSIM)*

As pointed out, SIMPAT suffers from its computational cost, as it requires calculating the distance between the data-event the entire patterns in pattern database. One possible solution is to summarize both the TI and data-event. Zhang et al. (2006) proposed a new method, FLITERSIM, in which various filters (6 and 9 filters in 2D and 3D) have been used in order to reduce the spatial complexity and dimensions. This allows reducing the complexity and computation time. Thus, the patterns are first filtered using the pre-defined linear filters. Then, the outputs are clustered based the similarity of the filtered patterns. Next, a prototype pattern is computed for each cluster that represents the average of all the patterns in the cluster. Afterwards, similar to SIMPAT, the most similar prototype is identified using a distance function and one of its patterns is selected randomly. These steps are continued until the simulation grid is filled. Due to using a limited number of filters, this algorithm requires less computational time compared to SIMPAT. The distance function in FILTERSIM was later replace with wavelet (Gloaguen and Dimitrakopoulos 2009). The drawbacks of the wavelet-based method are that it has a lot of parameters (e.g. wavelet decomposition level) that can effect on both quality and CPU time. Such parameters require an extensive tuning in order to achieve good results in a reasonable time.

In a similar fashion, Eskandari and Srinivasan (2010) proposed Growthism to integrate the dynamic data in the simulation. This method begins with the locations of data and grows gradually and completes the simulation grid.

The most important shortcoming of FILTERSIM is that, it uses a limited set of linear filters that cannot always convey all the information and variability in the TI. Moreover, selecting the appropriate filters and several user-dependent parameters for each TI is an issue that is that common among many MPS methods. Some of the generated realizations using the FILTERSIM algorithm are shown in Fig. 30.12.

iii. *Cross-Correlation based Simulation (CCSIM)*

One of the recent algorithms of MPS is the cross correlation-based simulation (CCSIM) algorithm that utilizes a cross-correlation function (CCF) along a 1D-raster path (Tahmasebi et al. 2012a). The CCF, which represents a multi-point

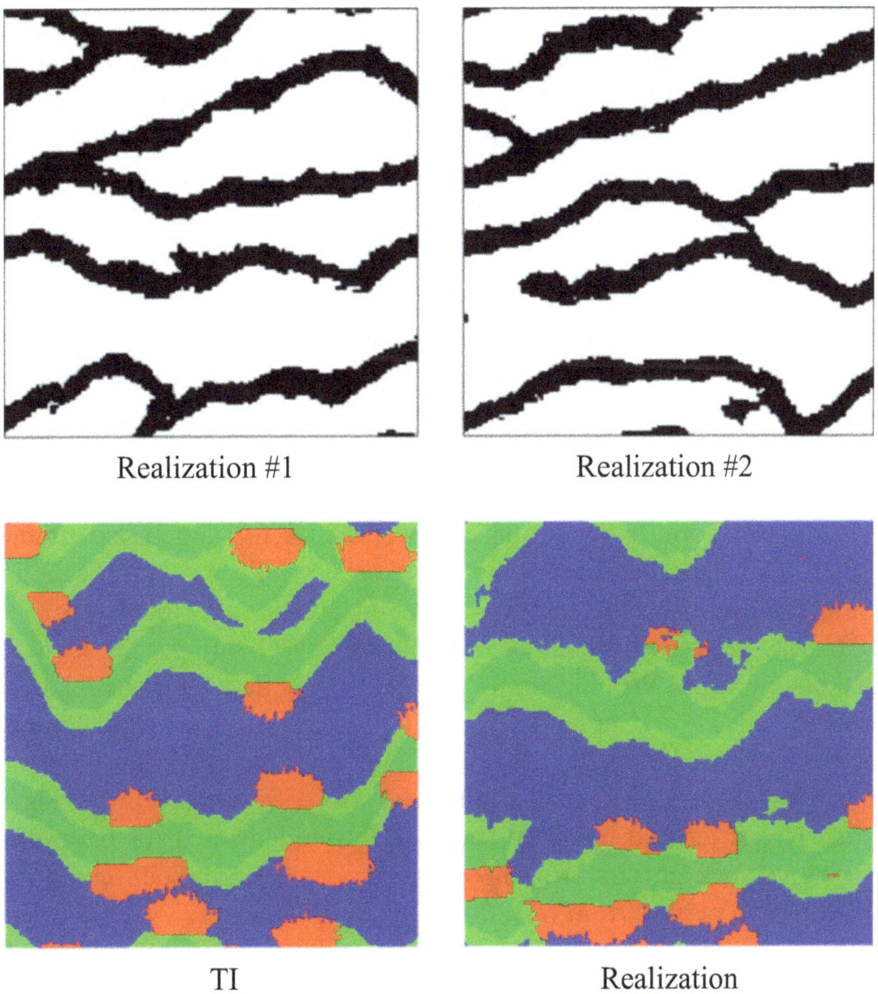


**Fig. 30.12** The results of FILTERSIM. These results are taken from Zhang et al. (2006)

characteristic function, is used to match the patterns in a realization with those in the TI. This algorithm has been adopted for different scenarios and computational grids. For example, multi-scale CCSIM (Tahmasebi et al. 2014) can be used when the simulation grid is very larger. This algorithm, similar to the SNESIM algorithm, is also based on calculating the joint probability. The SNESIM algorithm calculates the probability using a search-tree algorithm. However, calculating the above conditional probability for every single point in the simulation grid, in the presence of a large 2D/3D TD, is computationally prohibitive. Unless a very small neighborhood is used, which leads to poorly connected features in the outcome realizations.

In the CCSIM algorithm, the above limitation is addressed differently. First, the probability function is replaced with a similarity function, called cross-correlation function (CCF), which is much more efficient than drawing the probability. Secondly, based on the Markov Chain theory, the CCSIM algorithm uses a similar search template (i.e. radius). However, unlike the previous algorithms where all the data in the search template are used for simulating the visiting point, only a small data-event located in the boundaries, called overlap region OL, is considered in the calculations. Furthermore, except the point fallen in the OL region, the rest of the points are removed from the visiting points and they would not be simulated again. Thus, instead of simulating each single point, this algorithm ignores some of them and partitions the grid into several blocks. The CCF can be calculated as follow:

$$\mathrm{C}_{\mathbf{TD},\mathbf{D}_T}(i,j) = \sum_{x=0}^{D_x-1}\sum_{y=0}^{D_y-1} \mathbf{TI}(x+i,y+j)\mathbf{D}_T(x,y), \tag{30.13}$$

with $i \in [0T_x + D_x - 1)$ and $j \in [0\ T_y + D_y - 1)$ and $i, j \in Z$. The $i$ and $j$ represent the shift steps in the $x$ and $y$ directions. $\mathbf{TI}(x, y)$ represents the location at point $(x, y)$ of TD of size $L_x \times L_y$, with $x \in \{0, \ldots, D_x - 1\}$ and $y \in \{0, \ldots, D_y - 1\}$. An OL region of size $\mathrm{D}_x \times \mathrm{D}_y$ and a data event $\mathbf{D}_T$ are used to match the pattern in the TI. T represents the size of template used in CCSIM.

The CCSIM algorithm can realistically reproduce the large-scale structures in diminutive time. These techniques, however, do not fully match the well data and some artefacts are generated around the point conditioning data. Recently, this techniques has been used within an iterative framework along with boundary cutting methods by which the efficiency and conditioning data reproduction have been increased significantly (Gardet et al. 2016; Kalantari and Abdollahifard 2016; Mahmud et al. 2014; Moura et al. 2017; Scheidt et al. 2015; Tahmasebi and Sahimi 2016a, b; Yang et al. 2016). Some of the results of CCSIM are shown in Fig. 30.13. Furthermore, this method has been successfully implemented for fine-scale modeling in digital rock physics (Karimpouli et al. 2017; Karimpouli and Tahmasebi 2015; Tahmasebi et al. 2016a, b, c, 2017a, b).

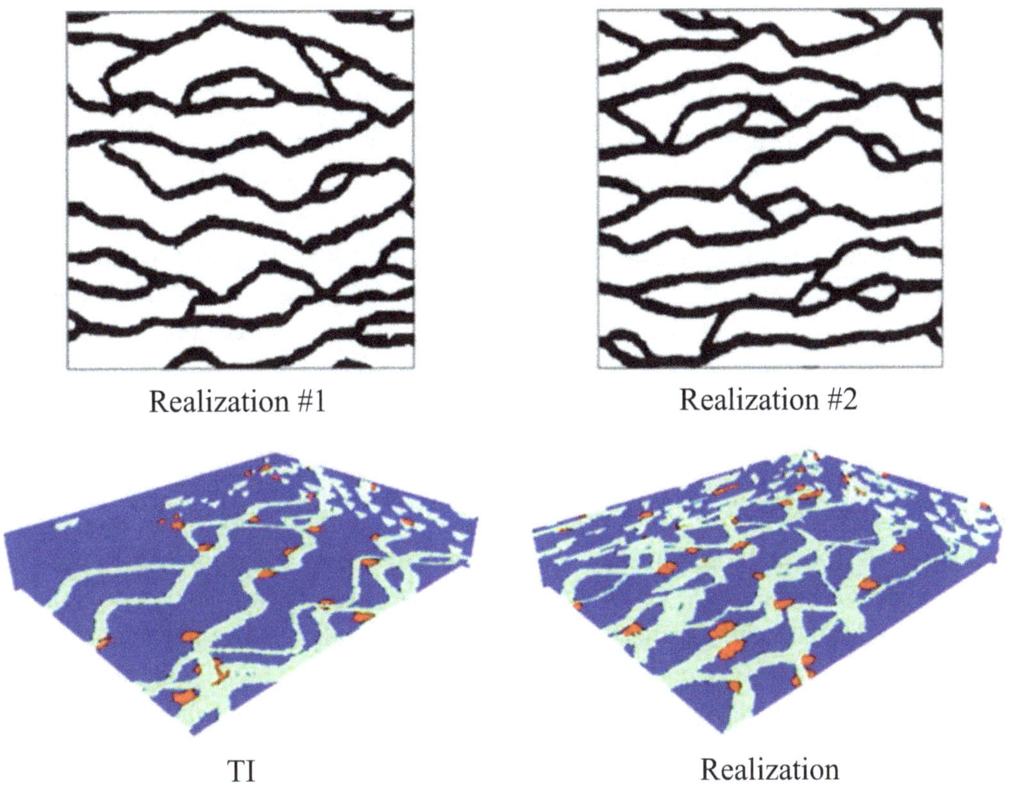


**Fig. 30.13** The results of the CCSIM algorithm

### *30.5.3 Hybrid Algorithms*

Each of the current MPS has some specific limitations. For example, the pixel-based techniques are good in conditioning the point data, while they barely can produce long-range connectivities. Similarly, the pattern-based methods can produce such structures, but they are unable to preserve the hard data. Thus, the idea of hybrid MPS method can be interesting if one uses the strength of both group effectively. Following, the available hybrid methods are reviewed and their advantages and disadvantages are discussed as well.

i. *Hybrid Sequential Gaussian/Indicator Simulation and TI*

Ortiz and Deutsch (2004), under an assumption of independence of the different data sources, integrate the indicator method with MPS. Hence, instead of using a TI, the MPS properties are obtained directly from the available hard data (variogram) and integrated with the results of indicator kriging. Finally, a value is drawn from this new distribution. These methods were further investigated by Ortiz and Emery (2005). However, in most cases, the initial results of indicator kriging highly influence final realization.

ii. *Hybrid Pixel- and Pattern-based Simulation (HYPPS)*

The strength of both the pixel- and pattern-based algorithms can be combined and make a hybrid algorithm. Tahmasebi (2017) has combined these two algorithm and proposed a new hybrid algorithm, called HYPPS. This algorithm discretizes the simulation grid into regions with/without the conditioning data. One needs to consider more attention the location containing the well data as providing them require considerable cost. Thus, the SNESIM algorithm, as a pixel-based method can be used around such locations. Regardless of the type of any geostatistical methods, reproducing of patterns and well data are the most important factors. Producing realistic models, without taking into account the conditioning data, or vise versa, is not deasirable. Any successful algorithm should be able to manintin both of the above crietra at the same time.

In the HYPPS algorithm the simulation grid is divided into two regions. Then, the geostatistical methods are applied on each of them. Following this step, the HYPPS algorithm uses the CCSIM, as a pattern-based algorithm, for the location where no HD exist and, similarly, the SNESIM algorithm is used around the well data, which can precisely reproduce the conditioning data. Thus, the hybrid state of the pixel-based and pattern-based techniques can be written as follow:

$$\Phi(\mathbf{h}_1, \ldots, \mathbf{h}_n; k_1, \ldots, k_n) = E\left\{ \prod_{\alpha=1}^{n} I(\mathbf{u} + \mathbf{h}_\alpha; k_\alpha) \right\} + \prod_{\alpha=1}^{n} \Phi(\mathbf{h}_\alpha | \mathbf{h}_{\Phi_n}) \tag{30.14}$$

which implies that the joint event over a template where both methods are working simultaneously can be expressed as the summation of the two probability distributions defined earlier (see the SNESIM algorithm). Thus, a normalization terms, namely $n_x$ and $n_p$, should be included such that $n_x + n_p = 1$. Note that $n_x$ and $n_p$ represent normalized number (or percentage) of the simulated points used in the pixel- and pattern-based methods. An equivalent form of the above probability can be expressed as:

$$\Phi(\mathbf{h}_1, \ldots, \mathbf{h}_n; k_1, \ldots, k_n) \cong \left( \begin{array}{c} n_x \left( \frac{c(d_n)}{N_n} \right) + \\ n_p \left( \sum_{x=0}^{D_x - 1} \sum_{y=0}^{D_y - 1} \mathbf{TI}(x+i, y+j) \mathbf{D}_T(x, y) \right) \end{array} \right) / n_x + n_p \tag{30.15}$$

The second term in the above equation on the right side is used for the areas where the CCSIM algorithm is utilized. While, the visiting points around the well data are evaluated jointly.

It is worth mentioning that co-template (Tahmasebi et al. 2014) can be used with the CCSIM to give the priority to the patterns that contain the conditioning data ahead of the raster path. Therefore, long-range connectivity structures are taken into

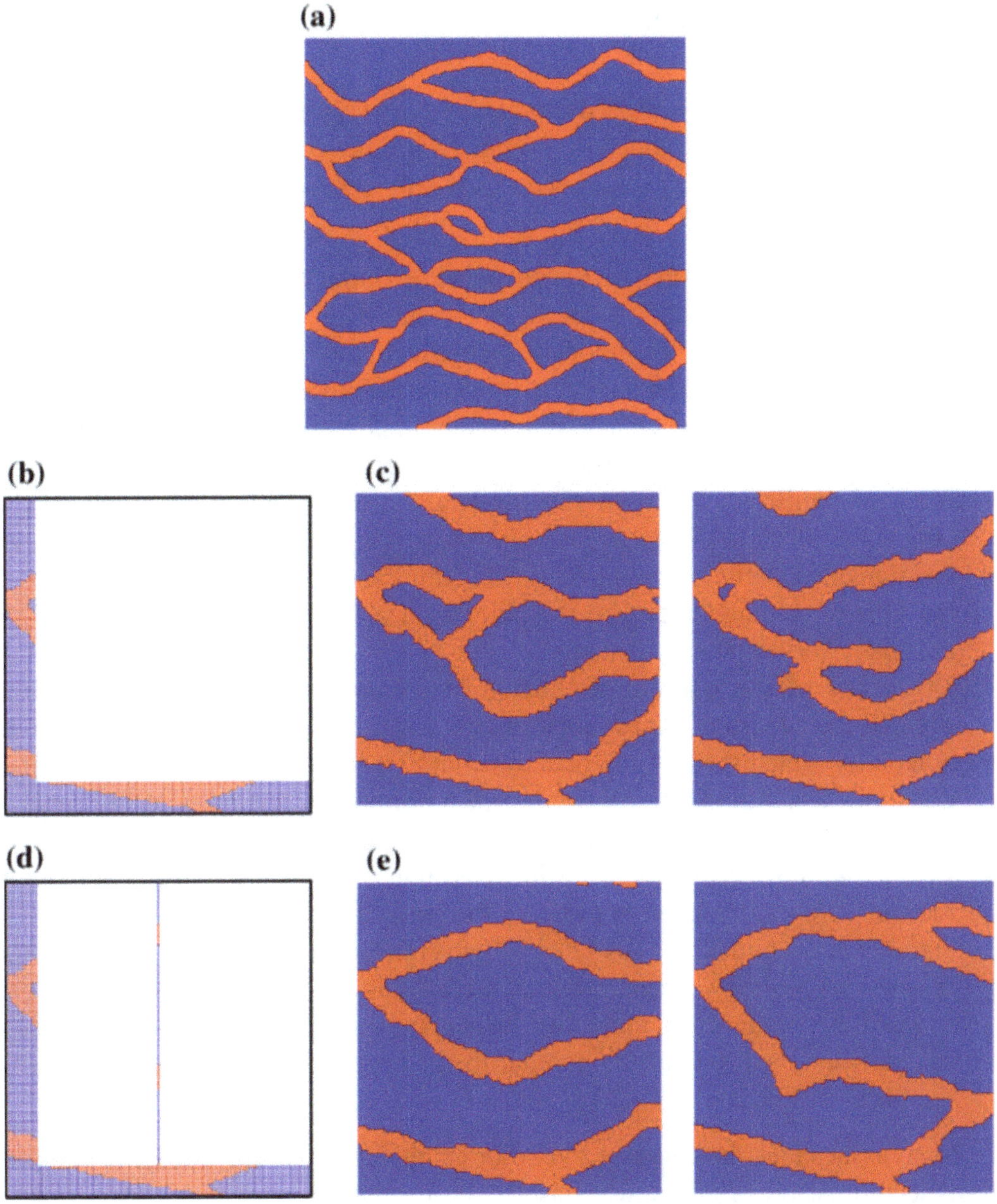


**Fig. 30.14** The application of the SNESIM algorithm for simulating a grid when the borders are conditioned (**a**): the TI, **b**: boundary data, **c** generated realizations using the boundary data, **d**: boundary data along with well data, **e** generated realizations using the boundary and well data. Note, the sizes of the TI and simulation grid in (**a**) and (**b**) are 250 × 250 and 100 × 100, respectively. The shale and sand facies are indicated by blue and red colors, individually

account even in the blocks with no conditioning data. An example is provided in Fig. 30.13e. Although the density of conditioning data, compared to the previous scenario, is increased, but the produced realizations represent the real heterogeneity represented in the TI. The HYPPS algorithm can be used to integrate data at different scales as well (Fig. 30.14).

## 30.6 Current Challenges

The MPS techniques have been developed extensively to deal with complex and more realistic problems in which the geology and other source of data such as well and secondary data are reproduced. There are still some critical challenges in MPS that require more research. Some of them are listed below:

- **Conditioning the model to dense point/secondary datasets** (e.g. well and seismic data): This issue is one of the main challenges in the pattern-based techniques. In fact, such methods require calculating distance function between the patterns of heterogeneities in the TI and the model and between the conditioning data that must be honored exactly in conditional simulation (Hashemi et al. 2014a, b; Rezaee and Marcotte 2016; Tahmasebi and Sahimi 2015b).
- **Lack of similarity between the internal patterns in the TI**: Producing a comprehensive and diverse TI is very challenging. Thus, serious discontinuity and unrealistic structures are generated using the current MPS techniques when the TI is not enough large. On the other hand, using very large or multiple TIs are costly for most of such methods.
- **Deficiency of the current similarity functions for quantitative modeling**: Most of the current distance functions are based on some simple and two-point criteria by which the optimal pattern cannot be identified. Such distance functions are very limited in conveying the information. Thus, more informative similarity functions are required.
- **Better and more realistic validation methods**: There is not so many number of methods that can be used to evaluate the performance of new developed MPS algorithms (Dimitrakopoulos et al. 2010; Tan et al. 2014). For example, the realizations can show a considerable different with each other and TI, while such a variability cannot be quantified using the current methods. Thus, visual comparison is still one of the popular method for verifying the performance of the MPS methods.

Many issues must be addressed yet. For example, the current MPS methods are designed for stationary TIs, whereas the properties of many large-scale porous media exhibit non-stationary features. Some progress has recently been made in this direction (Chugunova and Hu 2008; Honarkhah and Caers 2012; Tahmasebi and Sahimi 2015a). In addition, associated with every TI is large uncertainties. Thus, if several TIs are available, it is necessary to design methods that can determine which TI(s) to use in a given context.

## References


Abdollahifard MJ, Faez K (2013) Stochastic simulation of patterns using Bayesian pattern modeling. Comput Geosci 17:99–116. https://doi.org/10.1007/s10596-012-9319-x

Anderson KS, Hickson TA, Crider JG, Graham SA (1999) Integrating teaching with field research in the wagon rock project. J Geosci Educ 47:227–235. https://doi.org/10.5408/1089-9995-47.3.227

Arpat B (2005) Sequential simulation with patterns. Stanford University

Arpat B, Caers J (2007) Stochastic simulation with patterns. Math Geol 39:177–203

Bayer P, Huggenberger P, Renard P, Comunian A (2011) Three-dimensional high resolution fluvio-glacial aquifer analog: Part 1: field study. J Hydrol 405:1–9. https://doi.org/10.1016/j.jhydrol.2011.03.038

Biswal B, Manwart C, Hilfer R, Bakke S, Øren PE (1999) Quantitative analysis of experimental and synthetic microstructures for sedimentary rock. Phys Stat Mech Appl 273:452–475. https://doi.org/10.1016/S0378-4371(99)00248-4

Biswal B, Øren P-E, Held RJ, Bakke S, Hilfer R (2007) Stochastic multiscale model for carbonate rocks. Phys Rev E 75:61303. https://doi.org/10.1103/PhysRevE.75.061303

Borgman L, Taheri M, Hagan R (1984) Three-dimensional, frequency-domain simulations of geological variables. In: Geostatistics for natural resources characterization. Springer, Netherlands, Dordrecht, pp 517–541. https://doi.org/10.1007/978-94-009-3699-7_30

Boucher A (2009) Considering complex training images with search tree partitioning. Comput Geosci 35:1151–1158

Bryant S, Blunt M (1992) Prediction of relative permeability in simple porous media. Phys Rev A 46:2004–2011. https://doi.org/10.1103/PhysRevA.46.2004

Chiles J-P, Delfiner P (2011) Geostatistics : modeling spatial uncertainty. Wiley-Blackwell

Chu J, Journel AG (1994) Conditional fBm simulation with dual kriging. Springer Netherlands, pp. 407–421. https://doi.org/10.1007/978-94-011-0824-9_44

Chugunova TL, Hu LY (2008) Multiple-point simulations constrained by continuous auxiliary data. Math Geosci 40:133–146. https://doi.org/10.1007/s11004-007-9142-4

Cordua KS, Hansen TM, Mosegaard K (2015) Improving the pattern reproducibility of multiple-point-based prior models using frequency matching. Math Geosci 47:317–343. https://doi.org/10.1007/s11004-014-9531-4

Cressie NAC, Wikle CK (2011) Statistics for spatio-temporal data. Wiley

Daly C (2005) Higher order models using entropy, markov random fields and sequential simulation. Springer, Netherlands, pp 215–224. https://doi.org/10.1007/978-1-4020-3610-1_22

Davis MW (1987) Production of conditional simulations via the LU triangular decomposition of the covariance matrix. Math Geol 19:91–98. https://doi.org/10.1007/bf00898189

Deutsch CV (1992) Annealing techniques applied to reservoir modeling and the integration of geological and engineering (well test) data/. Stanford University

Deutsch CV, Journel AG (1998) GSLIB: geostatistical software library and user's guide, 2nd edn. Oxford University Press, New York

Deutsch CV, Wang L (1996) Hierarchical object-based stochastic modeling of fluvial reservoirs. Math Geol 28:857–880. https://doi.org/10.1007/BF02066005

Deutsch CV, Wen XH (2000) Integrating large-scale soft data by simulated annealing and probability constraints. Math Geol 32:49–67. https://doi.org/10.1023/A:1007502817679

Dimitrakopoulos R, Luo X (2004) Generalized sequential gaussian simulation on group size and screen-effect approximations for large field simulations. Math Geol 36:567–591. https://doi.org/10.1023/B:MATG.0000037737.11615.df

Dimitrakopoulos R, Mustapha H, Gloaguen E (2010) High-order statistics of spatial random fields: exploring spatial cumulants for modeling complex non-gaussian and non-linear phenomena. Math Geosci 42:65–99. https://doi.org/10.1007/s11004-009-9258-9

El Ouassini A, Saucier A, Marcotte D, Favis BD (2008) A patchwork approach to stochastic simulation: a route towards the analysis of morphology in multiphase systems. Chaos Solitons Fractals 36:418–436. https://doi.org/10.1016/j.chaos.2006.06.100

Eskandari K, Srinivasan S (2010) Reservoir modelling of complex geological systems–a multiple-point perspective. J Can Pet Technol 49:59–69. https://doi.org/10.2118/139917-PA

Fang JH, Wang PP (1997) Random field generation using simulated annealing vs. fractal-based stochastic interpolation. Math Geol 29:849–858. https://doi.org/10.1007/BF02768905

Gardet C, Le Ravalec M, Gloaguen E (2016) Pattern-based conditional simulation with a raster path: a few techniques to make it more efficient. Stoch Environ Res Risk Assess 30:429–446. https://doi.org/10.1007/s00477-015-1207-1

Gloaguen E, Dimitrakopoulos R (2009) Two-dimensional conditional simulations based on the wavelet decomposition of training images. Math Geosci 41:679–701. https://doi.org/10.1007/s11004-009-9235-3

Goovaerts P (1997) Geostatistics for natural resources evaluation. Oxford University Press

Gross LJ, Small MJ (1998) River and floodplain process simulation for subsurface characterization. Water Resour Res 34:2365–2376. https://doi.org/10.1029/98WR00777

Guardiano FB, Srivastava RM (1993) Multivariate geostatistics: beyond bivariate moments. In: Geostatistics Troia'92. Springer, pp 133–144

Haldorsen HH, Damsleth E (1990) Stochastic Modeling (includes associated papers 21255 and 21299). J Pet Technol 42:404–412. https://doi.org/10.2118/20321-PA

Hamzehpour H, Sahimi M (2006) Development of optimal models of porous media by combining static and dynamic data: the porosity distribution. Phys Rev E 74:26308. https://doi.org/10.1103/PhysRevE.74.026308

Hashemi S, Javaherian A, Ataee-pour M, Tahmasebi P, Khoshdel H (2014a) Channel characterization using multiple-point geostatistics, neural network, and modern analogy: a case study from a carbonate reservoir, southwest Iran. J Appl Geophys 111. https://doi.org/10.1016/j.jappgeo.2014.09.015

Hashemi S, Javaherian A, Ataee-pour M, Tahmasebi P, Khoshdel H (2014b) Channel characterization using multiple-point geostatistics, neural network, and modern analogy: a case study from a carbonate reservoir, southwest Iran. J Appl Geophys 111:47–58. https://doi.org/10.1016/j.jappgeo.2014.09.015

Holden L, Hauge R, Skare Ø, Skorstad A (1998) Modeling of fluvial reservoirs with object models. Math Geol 30:473–496. https://doi.org/10.1023/A:1021769526425

Honarkhah M, Caers J (2012) Direct pattern-based simulation of non-stationary geostatistical models. Math Geosci 44:651–672. https://doi.org/10.1007/s11004-012-9413-6

Journel A, Zhang T (2006) The necessity of a multiple-point prior model. Math Geol 38:591–610

Journel AG, Huijbregts CJ (1978) Mining geostatistics. Academic Press

Kalantari S, Abdollahifard MJ (2016) Optimization-based multiple-point geostatistics: a sparse way. Comput Geosci 95:85–98. https://doi.org/10.1016/j.cageo.2016.07.006

Karimpouli S, Tahmasebi P (2015) Conditional reconstruction: an alternative strategy in digital rock physics. Geophysics 81. https://doi.org/10.1190/geo2015-0260.1

Karimpouli S, Tahmasebi P, Ramandi HL, Mostaghimi P, Saadatfar M (2017) Stochastic modeling of coal fracture network by direct use of micro-computed tomography images. Int J Coal Geol 179. https://doi.org/10.1016/j.coal.2017.06.002

Kitanidis PK, Peter K (1997) Introduction to geostatistics: applications to hydrogeology. Cambridge University Press

Kleingeld WJ, Thurston ML, Prins CF, Lantuéjoul C (1997) The conditional simulation of a Cox process with application to deposits with discrete particles. Geostat Wollongong 96:683–694

Lancaster ST, Bras RL (2002) A simple model of river meandering and its comparison to natural channels. Hydrol Process 16:1–26. https://doi.org/10.1002/hyp.273

Lantuéjoul C (2002) Geostatistical simulation: models and algorithms, in: geostatistical simulation: models and algorithms. Springer, Berlin, Heidelberg, pp 1–6. https://doi.org/10.1007/978-3-662-04808-5_1

Mahmud K, Mariethoz G, Caers J, Tahmasebi P, Baker A (2014) Simulation of earth textures by conditional image quilting. Water Resour Res 50:3088–3107. https://doi.org/10.1002/2013WR015069

Manwart C, Torquato S, Hilfer R (2000) Stochastic reconstruction of sandstones. Phys Rev E 62:893–899. https://doi.org/10.1103/PhysRevE.62.893

Mariethoz G, Renard P, Straubhaar J (2010) The direct sampling method to perform multiple-point geostatistical simulations. Water Resour Res 46

Matheron G (1973) The intrinsic random functions and their applications. Adv Appl Probab 5:439–468. https://doi.org/10.1017/S0001867800039379

Moura P, Laber E, Lopes H, Mesejo D, Pavanelli L, Jardim J, Thiesen F, Pujol G (2017) LSHSIM: a locality sensitive hashing based method for multiple-point geostatistics. Comput Geosci https://doi.org/10.1016/j.cageo.2017.06.013

Mustapha H, Dimitrakopoulos R (2010) High-order stochastic simulation of complex spatially distributed natural phenomena. Math Geosci 42:457–485

Mustapha H, Dimitrakopoulos R (2011) HOSIM: a high-order stochastic simulation algorithm for generating three-dimensional complex geological patterns. Comput Geosci 37:1242–1253. https://doi.org/10.1016/j.cageo.2010.09.007

Ortiz JM, Deutsch CV (2004) Indicator simulation accounting for multiple-point statistics. Math Geol 36:545–565. https://doi.org/10.1023/B:MATG.0000037736.00489.b5

Ortiz JM, Emery X (2005) Integrating multiple-point statistics into sequential simulation algorithms. Springer, Netherlands, pp 969–978. https://doi.org/10.1007/978-1-4020-3610-1_101

Parra Á, Ortiz JM (2011) Adapting a texture synthesis algorithm for conditional multiple point geostatistical simulation. Stoch Environ Res Risk Assess 25:1101–1111. https://doi.org/10.1007/s00477-011-0489-1

Peredo O, Ortiz JM (2011) Parallel implementation of simulated annealing to reproduce multiple-point statistics. Comput Geosci 37:1110–1121. https://doi.org/10.1016/j.cageo.2010.10.015

Pourfard M, Abdollahifard MJ, Faez K, Motamedi SA, Hosseinian T (2017) PCTO-SIM: Multiple-point geostatistical modeling using parallel conditional texture optimization. Comput Geosci 102:116–138. https://doi.org/10.1016/j.cageo.2016.12.012

Pyrcz MJ, Boisvert JB, Deutsch CV (2009) ALLUVSIM: a program for event-based stochastic modeling of fluvial depositional systems. Comput Geosci 35:1671–1685. https://doi.org/10.1016/j.cageo.2008.09.012

Rasera LG, Machado PL, Costa JFCL (2015) A conflict-free, path-level parallelization approach for sequential simulation algorithms. Comput Geosci 80:49–61. https://doi.org/10.1016/j.cageo.2015.03.016

Rezaee H, Marcotte D (2016) Integration of multiple soft data sets in MPS thru multinomial logistic regression: a case study of gas hydrates. Stoch Environ Res Risk Assess 1–19. https://doi.org/10.1007/s00477-016-1277-8

Rezaee H, Mariethoz G, Koneshloo M, Asghari O (2013) Multiple-point geostatistical simulation using the bunch-pasting direct sampling method. Comput Geosci 54:293–308. https://doi.org/10.1016/j.cageo.2013.01.020

Scheidt C, Tahmasebi P, Pontiggia M, Da Pra A, Caers J (2015) Updating joint uncertainty in trend and depositional scenario for reservoir exploration and early appraisal. Comput Geosci 19. https://doi.org/10.1007/s10596-015-9491-x

Seminara G (2006) Meanders. J Fluid Mech 554:271. https://doi.org/10.1017/S0022112006008925

Sheehan N, Torquato S (2001) Generating microstructures with specified correlation functions. J Appl Phys 89:53–60. https://doi.org/10.1063/1.1327609

Skorstad A, Hauge R, Holden L (1999) Well conditioning in a fluvial reservoir model. Math Geol 31:857–872. https://doi.org/10.1023/A:1007576801266

Stien M, Kolbjørnsen O (2011) Facies modeling using a markov mesh model specification. Math Geosci 43:611–624. https://doi.org/10.1007/s11004-011-9350-9

Straubhaar J, Walgenwitz A, Renard P (2013) Parallel multiple-point statistics algorithm based on list and tree structures. Math Geosci 45:131–147. https://doi.org/10.1007/s11004-012-9437-y

Strebelle S (2012) Multiple-point geostatistics: from theory to practice. Ninth international geostatistics congress. Springer, Oslo, Norway, pp 11–15

Strebelle S (2002) Conditional simulation of complex geological structures using multiple-point statistics. Math Geol 34:1–21

Tahmasebi P (2017) Structural adjustment for accurate conditioning in large-scale subsurface systems. Adv Water Resour 101. https://doi.org/10.1016/j.advwatres.2017.01.009

Tahmasebi P, Hezarkhani A, Sahimi M (2012a) Multiple-point geostatistical modeling based on the cross-correlation functions. Comput Geosci 16:779–797. https://doi.org/10.1007/s10596-012-9287-1

Tahmasebi P, Sahimi M, Mariethoz G, Hezarkhani A (2012b) Accelerating geostatistical simulations using graphics processing units (GPU). Comput Geosci 46:51–59. https://doi.org/10.1016/j.cageo.2012.03.028

Tahmasebi P, Javadpour F, Sahimi M (2016a) Stochastic shale permeability matching: three-dimensional characterization and modeling. Int J Coal Geol 165:231–242. https://doi.org/10.1016/j.coal.2016.08.024

Tahmasebi P, Javadpour F, Sahimi M, Piri M (2016b) Multiscale study for stochastic characterization of shale samples. Adv Water Resour 89:91–103. https://doi.org/10.1016/j.advwatres.2016.01.008

Tahmasebi P, Sahimi M (2015a) Reconstruction of nonstationary disordered materials and media: watershed transform and cross-correlation function. Phys Rev E 91:32401. https://doi.org/10.1103/PhysRevE.91.032401

Tahmasebi P, Sahimi M (2015b) Geostatistical simulation and reconstruction of porous media by a cross-correlation function and integration of hard and soft data. Trans Porous Media 107:871–905. https://doi.org/10.1007/s11242-015-0471-3

Tahmasebi P, Sahimi M (2016a) Enhancing multiple-point geostatistical modeling: 2. Iterative simulation and multiple distance function. Water Resour Res 52:2099–2122. https://doi.org/10.1002/2015WR017807

Tahmasebi P, Sahimi M (2016b) Enhancing multiple-point geostatistical modeling: 1. Graph theory and pattern adjustment. Water Resour Res 52:2074–2098. https://doi.org/10.1002/2015WR017806

Tahmasebi P, Sahimi M, Andrade J (2017a) Direct modeling of granular materials. In: Poromechanics VI. American society of civil engineers, Reston, VA, pp. 1436–1442. https://doi.org/10.1061/9780784480779.178

Tahmasebi P, Sahimi M, Andrade JE (2017b) Image-based modeling of granular porous media. Geophys Res Lett https://doi.org/10.1002/2017gl073938

Tahmasebi P, Sahimi M, Caers J (2014) MS-CCSIM: accelerating pattern-based geostatistical simulation of categorical variables using a multi-scale search in fourier space. Comput Geosci 67:75–88. https://doi.org/10.1016/j.cageo.2014.03.009

Tahmasebi P, Sahimi M, Kohanpur AH, Valocchi A (2016) Pore-scale simulation of flow of $CO_2$ and brine in reconstructed and actual 3D rock cores. J Pet Sci Eng. https://doi.org/10.1016/j.petrol.2016.12.031

Tan X, Tahmasebi P, Caers J (2014) Comparing training-image based algorithms using an analysis of distance. Math Geosci 46. https://doi.org/10.1007/s11004-013-9482-1

Tjelmeland H, Eidsvik J (2005) Directional metropolis : hastings updates for posteriors with nonlinear likelihoods. Springer, Netherlands, pp. 95–104. https://doi.org/10.1007/978-1-4020-3610-1_10

Toftaker H, Tjelmeland H (2013) Construction of binary multi-grid markov random field prior models from training images. Math Geosci 45:383–409. https://doi.org/10.1007/s11004-013-9456-3

Wu J, Boucher A, Zhang T (2008) A SGeMS code for pattern simulation of continuous and categorical variables: FILTERSIM. Comput Geosci 34:1863–1876. https://doi.org/10.1016/j.cageo.2007.08.008

Yang L, Hou W, Cui C, Cui J (2016) GOSIM: a multi-scale iterative multiple-point statistics algorithm with global optimization. Comput Geosci 89:57–70. https://doi.org/10.1016/j.cageo.2015.12.020

Yeong CLY, Torquato S (1998a) Reconstructing random media. II. Three-dimensional media from two-dimensional cuts. Phys Rev E 58:224–233. https://doi.org/10.1103/PhysRevE.58.224
Yeong CLY, Torquato S (1998b) Reconstructing random media. Phys Rev E 57:495–506. https://doi.org/10.1103/PhysRevE.57.495
Zhang T, Switzer P, Journel A (2006) Filter-based classification of training image patterns for spatial simulation. Math Geol 38:63–80

# Chapter 31
# When Should We Use Multiple-Point Geostatistics?

Gregoire Mariethoz

**Abstract** Multiple-point geostatistics should be used when there is either too little or too much information available for other types of geostatistics.

## 31.1 Under-Informed Versus Over-Informed Models

For a long time, the classical geostatistical framework required moderate amounts of knowledge. Too little knowledge (few hard data, poorly distributed, absence of auxiliary information), makes it difficult to infer the parameters of a covariance model. In the other extreme, too much knowledge risks revealing characteristics of the underlying field that are too complex to be represented by a handful of covariance model parameters. These two situations can be denoted respectively under-informed and over-informed models. In-between these extremes, we have the moderately informed case where it is convenient to use the covariance-based geostatistical framework, which has been—and still is—a very solid basis for building models that incorporate spatial and temporal variability.

Extreme under-informed and over-informed cases have often presented technical challenges, for which practical workarounds are used. For under-informed cases, standard geostatistical practice consists for example in including interpretative knowledge to guide variogram fitting when too few hard data are available. This is one of the reasons for the common recommendation to fit variograms by hand (e.g. Olea 1999). The question of designing spatial models for over-informed cases (i.e., when large amounts of data are available) is relatively recent, with the development of improved sensors and high-resolution numerical models that triggered the era of "big data".

The concept of multiple-point statistics (MPS) appeared in the early 1990s, initially as a means of overcoming extreme under-informed situations. The idea, at

G. Mariethoz (✉)
Institute of Earth Surface Dynamics (IDYST), University of Lausanne,
Lausanne, Switzerland
e-mail: gregoire.mariethoz@unil.ch



B. S. Daya Sagar et al. (eds.), *Handbook of Mathematical Geosciences*,
https://doi.org/10.1007/978-3-319-78999-6_31

the time developed by Guardiano and Srivastava (1993) under the impulsion and guidance of A. Journel, was to give the modeler improved tools to include interpretative knowledge in spatial models. The fundamental novelty of the MPS framework was to encapsulate in a training image the interpretative knowledge on the spatial structure of the modeled phenomenon. Since an image is an object most people are familiar with, it allows combining different types of expertise and data, in particular from people who are not familiar with geostatistics.

This approach naturally leads to disregarding hard data as a tiny fraction of the information to include in a model, implying that data alone are not enough. Then, an important part of the modeling work resides in the design of the training image, which can be difficult as natural images are typically not sufficiently repetitive or stationary. Unsurprisingly, the first successful applications of MPS took place in fields where data are typically few, uncertain and expensive, such as reservoir modeling, soil science or mining. In those domains, MPS is often seen as an alternative to object-based methods. Later, it was found that the concept of training image could also be used to incorporate large amounts of information in a model, and therefore address over-informed and data-rich situations, where an increasing number of applications are taking place.

## 31.2 MPS Versus Covariance-Based Geostatistics

These different aspects have resulted in MPS being seen as in opposition with covariance-based geostatistics. Indeed, from a traditional statistics point of view, MPS is not rigorous in many respects: for instance there is no real model inference, the uncertainty that can be estimated based on a set of MPS realizations is poorly defined, and extreme events cannot be produced beyond those found in the training image. Emery and Lantuéjoul (2014) have shown, based on thorough numerical and theoretical investigations, that MPS only produces random fields when the size of the training image tends to infinity. With a finite training image, MPS algorithms do no longer approximate a random function. Their value then lies in their capability to automatically generate realistic model realizations, but without control of the underlying statistical model. These issues make MPS methodologically close to machine learning and computer graphics. As a result, when using MPS, one often has to make compromises with random function theory and model consistency. In return, it may be possible to explore the data in new ways and obtain, in some cases, models that are more in line with the unobserved physical reality (Journel 1993).

While the hypotheses and tools used are very different, the domains of application of MPS are essentially the same as traditional geostatistics, consisting in the simulation of either conditional or unconditional random fields, mainly for geoscience applications. As such, MPS and covariance-based geostatistics can be seen as competing, and it is not very surprising that in the last decade there have been many cases of fierce debate between the promoters of these two concurrent approaches (Journel and Zhang 2006; Li et al. 2015). My view is that in fact, the two sets of methods should not be seen as opposed, but as complementary approaches. They are complementary because they are able to solve different types of problems which can be distinguished by the nature and amount of information at hand. Seeing the covariance-based and the algorithm-based approaches as opposed can distract from the higher goal of building on the strengths of each approach. The risk has been stated by Breiman (2001) on the topic of machine learning methods: "*statisticians have ruled themselves out of some of the most interesting and challenging statistical problems that have arisen out of the rapidly increasing ability of computers to store and manipulate data*".

When the available data and knowledge on the studied phenomenon allow building a random function model, using covariance-based geostatistics is usually appropriate. There are numerous examples of successful models designed in this framework for which it would be very difficult to apply MPS (e.g. Diggle et al. 1998; Goovaerts 2005). Conversely, there are applications where the use of training images and MPS algorithms are better able to address some practical questions. In the next sections, I will show two such examples where the available information is either extremely poor or extremely rich. Applying covariance-based geostatistics to these examples would likely yield unsatisfactory results. I emphasize here that for the purpose of demonstration, I am exclusively focusing on examples that are tailored for the application of MPS. Countless examples can be found for which covariance models are perfectly applicable, but it is beyond the scope of this short chapter to show them here.

## 31.3 Examples for Which MPS Works Well

### *31.3.1 MPS Can Be Used in Extreme Under-Informed Situations*

An example of extreme under-informed model is the common problem of interpolating rainfall data over a given area based on a small number of rain gauges. Rainfall is an inherently intermittent and highly spatially variable process (Benoit and Mariethoz 2017). Moreover, in some cases rain gauge data can be of poor quality, and it is not uncommon to only have binary wet/dry information (as opposed to rainfall accumulation). An example of such poor dataset is shown in Fig. 31.1, with synthetic rain observations consisting of 30 rain gauges. While this case is synthetic, the setting is relatively standard in terms of data density and heterogeneity. It is quite clear that 30 observation points are insufficient to properly infer a spatial model, which is confirmed by the experimental variogram that shows no spatial structure (and wild fluctuations when the number of lags is varied).

In such a setting, the MPS approach starts by stating that the information contained in the hard data is insufficient. At best, the data points can be used for conditioning, but not for inferring any kind of structural model. Instead, one has to

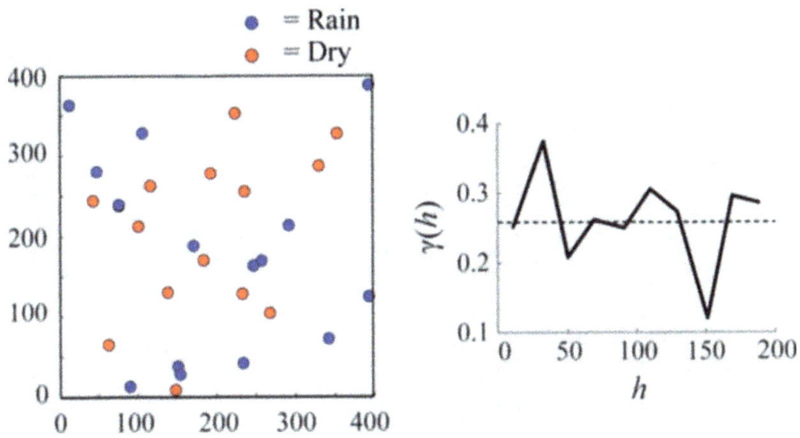


**Fig. 31.1** Under-informed setting. Left: synthetic rain gauge network made of 30 points with only wet/dry information. Right: experimental omnidirectional indicator variogram of the probability of rainfall, computed on 10 lags

supplement the insufficient data by resorting to external knowledge of the modeled process. For example, one may know the type of rainfall for that specific day. Based on this knowledge, it is possible to collect radar images of rain events of the same type. Rainfall radar images, either ground-based or satellite-based, are typically collected by national weather agencies and made available to the scientific community. Then, using these representative radar images as training images, MPS can be used to generate rain fields conditioned to the gauge data.

Figure 31.2 shows the results of using two different training images to interpolate the data shown in Fig. 31.1, by considering as training image alternatively a cyclone (left) or a tropical storm (right). It is obvious here that the choice of the training image has a strong influence on the results as it determines the types of patterns found in the simulations, as well as global statistics such as the proportion of wet areas.

This example illustrates the conceptual differences between MPS and covariance-based geostatistics. These differences extend beyond the formalism or the algorithms used. While classical geostatistics infer a model based on data, MPS generates additional data based on external knowledge, in this case through the search for and the selection of an appropriate radar image.

### *31.3.2 MPS Can Be Used in Extreme Over-Informed Situations*

The most common situation in geostatistics is to have a handful of data points, and based on these, to estimate the target variable on a large grid. Increasingly in recent years, the opposite situation occurs with a large number of data used to predict the value at a smaller set of locations. One prime example of such over-informed

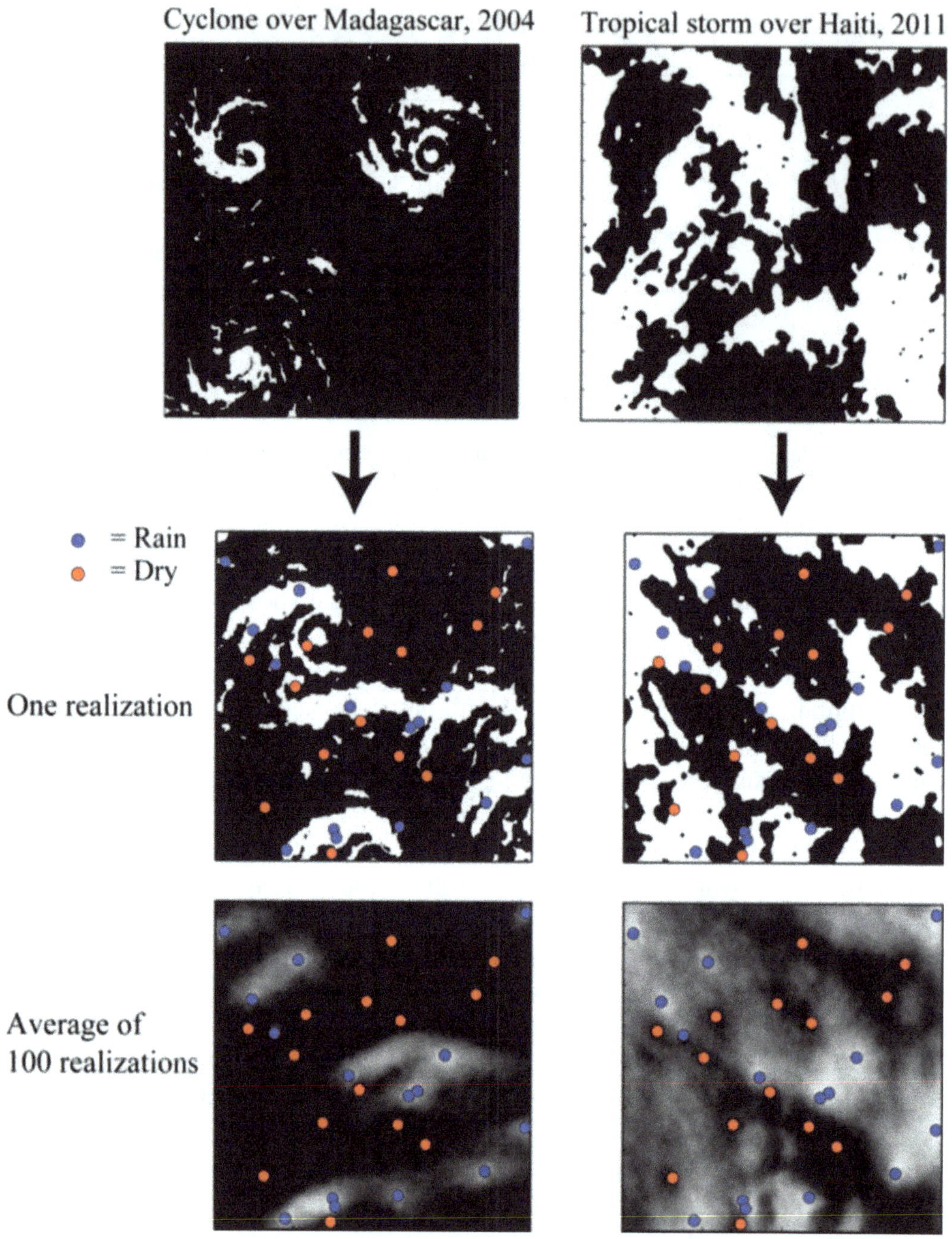


**Fig. 31.2** Application of MPS for rain occurrence simulation. Left: simulation of binary rainfall based on a training image of a cyclone. Right: same setting based on a training image of a tropical storm. Size of training images: 572 × 584 pixels. Size of simulation grid: 400 × 400 pixels. The Direct Sampling MPS algorithm was used

problems is applications to satellite imagery, which typically consist in large spatial datasets (typically the entire Earth is covered at high spatial resolution) that also present a temporal aspect since the same location is imaged at regular intervals. Here we look at the Landsat 7 ETM + sensor, which has the characteristic that it

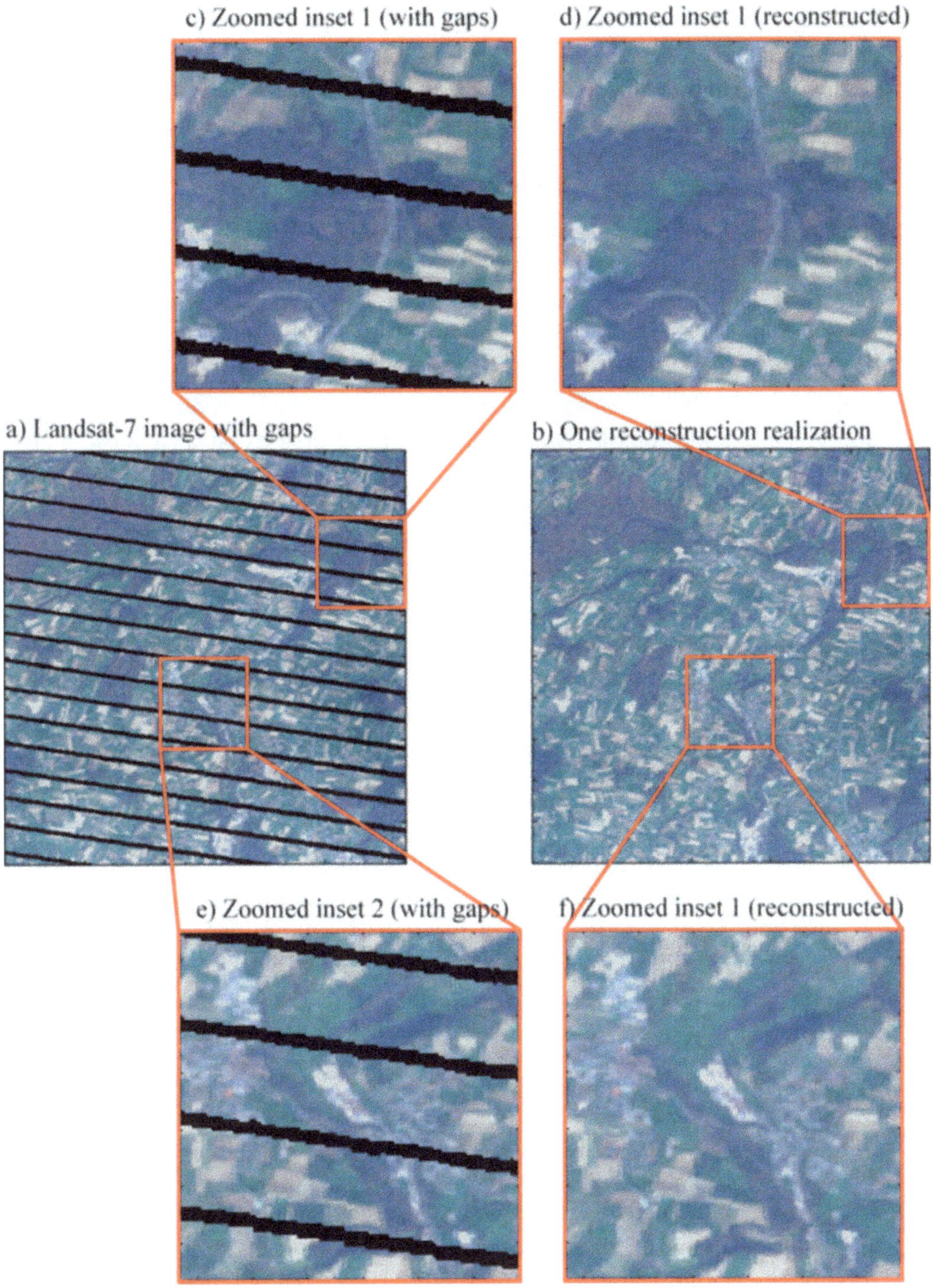


**Fig. 31.3** MPS applied to gap-filling of a 5-band Landsat 7 image. Scene acquired on March 22, 2017 in Western Switzerland. Image size: 500 × 500 pixels. The Direct Sampling MPS algorithm was used. Image shown in natural colors

partially failed in 2003, and since then the images it acquires present gaps (as shown on Fig. 31.3a). The goal here is to fill these gaps with simulated values. In such an image, the regions to reconstruct typically represent about 20% of the domain, the rest consisting of conditioning data. These data contain not only local information, but also very rich structural information such as the type of land surface features (fields, forests, cities), the connectivity of the different objects (roads, water bodies), and their spatial arrangement (see details shown in Fig. 31.3c, e).

The application of covariance-based geostatistics is in this case difficult, not because of challenges related to model inference and identification (as in Fig. 31.1), but because standard simulation techniques, such as Sequential Gaussian Simulation or turning bands, will likely result in artifacts that are clearly visible to the eye. Indeed, the complex land surface information cannot be entirely represented by covariance models which are typically represented by a small number of parameters. Furthermore, although interpolation artifacts are sometimes obvious to the eye, they are typically undetectable by standard statistical metrics because these metrics are based on covariance (or two-point statistics) and cannot identify complex patterns such as connectivity, for which the human eye is very well suited. It can of course be argued that there are applications where these complex properties do not matter; but if they do, the covariance-based framework is inappropriate (Zinn and Harvey 2003).

In contrast, applying MPS to this gap-filling problem is straightforward. The MPS approach used here for the simulation of gaps is the one presented by Yin et al. (2017a, b). Each color channel is co-simulated and no auxiliary variables are used. Contrarily to the data-poor case, there is no need here to infer, construct or hypothesize a training image. The training image is given by the 80% of the domain that is known. While the training image size is far from infinity, it is a little closer to the ideal situation outlined by Emery and Lantuéjoul (2014). The gap-filling results (Fig. 31.3b, d, f) present very few visual artifacts. In certain places, it is possible to see that some reconstructed elongated features are discontinuous (e.g. the road near the center of Fig. 31.3d). However in most cases it is difficult to distinguish the reconstructed and the original areas (e.g. in Fig. 31.3f).

## 31.4 Conclusion

Often the debate around MPS and covariance-based approaches has been centered on the dichotomy between multiGaussianity or non-multiGaussianity of the variable to simulate (Gómez-Hernández and Wen 1998). The choice of a simulation approach or algorithm should certainly be driven by the nature of the variable of interest: is it non-multiGaussian? is it non-stationary? is it channelized? do these characteristics matter for a given problem? I argue here that the question of the amount of information at hand is also a critical factor to consider when choosing which simulation framework to use, and this question has often been overlooked. It may make sense to also base this choice on the quantity of information available: do I have a conceptual model? do I have enough hard or soft data to infer a covariance? do I have so much data that I am able to detect non-multiGaussian behavior?

To summarize, one can say that different tools are available, and those should be chosen according to the problem to be solved. While no example with moderate amount of information has been shown in this chapter, it is understood that it is generally the realm of covariance-based geostatistics. Under-informed situations are always going to be difficult because there are important modelling choices to make. For over-informed cases, relatively few assumptions are needed and, with some precautions, it can be possible to rely on algorithms such as MPS.

Better defining the role of MPS in the galaxy of existing spatial modeling tools can potentially help narrowing areas where future MPS research should focus. So far, there has been a strong emphasis on the development of simulation algorithms. The different algorithms available can reproduce spatial features with various degrees of faithfulness, they may need different computing resources or may offer specific options. While developments in MPS are still needed (in particular regarding training image selection and manipulation, as well as parametrization), the simulation algorithms are becoming quite mature. Moving beyond the dichotomy between covariance-based geostatistics and MPS can enable the development of new hybrid approaches. For example, using distance-based (also known as convolution-based) MPS algorithms can be seen as bootstrapping the training image. However, the link with bootstrapping theory (e.g. Davison and Hinkley 1997) has not yet been fully explored. Similarly, the MPS framework is currently unable to simulate extreme values. Combining MPS with more standard statistical approaches may open new fields of applications, in particular in domains such as climate science, hydrology or earth surface observation where increasingly rich space-time datasets are now available.

## References

Benoit L, Mariethoz G (2017) Generating synthetic rainfall with geostatistical simulations. Wiley Interdiscip Rev Water 4(2):e1199-n/a

Breiman L (2001) Statistical modeling: the two cultures. Stat Sci 16(3):199–231

Davison A, Hinkley D (1997) Bootstrap methods and their application. Cambridge University Press

Diggle PJ, Tawn JA, Moyeed RA (1998) Model-based geostatistics. J R Stat Soc Ser C Appl Stat 47(3):299–325

Emery X, Lantuéjoul C (2014) Can a training image be a substitute for a random field model? Math Geosci 46(2):133–147

Gómez-Hernández JJ, Wen XH (1998) To be or not to be multi-Gaussian? A reflection on stochastic hydrogeology. Adv Water Resour 21(1):47–61

Goovaerts P (2005) Geostatistical analysis of disease data: estimation of cancer mortality risk from empirical frequencies using poisson kriging. Int J Health Geogr 4(31):1–33

Guardiano F, Srivastava M (1993) In: Soares A (ed) Geostatistics-Troia. Kluwier Academic, Dordrecht, pp 133–144

Journel A, Zhang T (2006) The necessity of a multiple-point prior model. Math Geol 38(5): 591–610

Journel AG (1993) Geostatistics: roadblocks and challenges. Geostatistics Troia '92, vol 1, pp 213–224

Li L, Romary T, Caers J (2015) Universal kriging with training images. Spat Stat 14:240–268

Olea RA (1999) Geostatistics for engineers and earth scientists. Springer, New York

Yin G, Mariethoz G, McCabe M (2017a) Gap-filling of landsat 7 imagery using the direct sampling method. Remote Sens 9(1):12

Yin G, Mariethoz G, Sun Y, MCabe MF (2017b) A comparison of gap-filling approaches for landsat-7 satellite data. Int J Remote Sens. https://doi.org/10.1080/01431161.01432017.01363432

Zinn B, Harvey C (2003) When good statistical models of aquifer heterogeneity go bad: a comparison of flow, dispersion, and mass transfer in connected and multivariate Gaussian hydraulic conductivity fields. Water Resour Res 39(3):WR001146

# Chapter 32
# The Origins of the Multiple-Point Statistics (MPS) Algorithm

**R. Mohan Srivastava**

**Abstract** First proposed in the early 1990s, the geostatistical algorithm known as multiple-point statistics (MPS) now enjoys widespread use, particularly in petroleum studies. It has become part of the toolkit that new practitioners are trained to use in several oil companies; it has been incorporated into commercial software; and research programs in many universities continue to tap into the central MPS idea of extracting statistical information directly from a training image. The inspiration for the development of a proof-of-concept MPS prototype code owes much to several different researchers and research programs in the late 1980s and early 1990s: the sequential algorithms pioneered at Stanford University, the work of Chris Farmer, then at UK Atomic Energy, and the growing use of outcrop studies by several oil companies. This largely accidental confluence of divergent theoretical perspectives, and of distinct practical workflows, serves as an example of how science often advances through the intersection of ideas that are not only disparate but even contradictory.

**Keywords** MPS · Multiple-point statistics · Conditional simulation
Training image

## 32.1 Introduction

Through the windows of the cottage, we watched the sun slip behind the trees on the ridge across the lake, turning the light dusting of snow from pink to red to crimson. As darkness settled outside, the windows became mirrors, lit by the flame from the logs in the fireplace, until all we could see was our two reflections, each resting comfortably in an armchair, wine glass in hand. We talked into the late evening, past the rising of the crescent moon, reminiscing about people, about

R. Mohan Srivastava (✉)
TriStar Gold Inc., #1100 – 120 Eglinton Avenue East, Toronto,
ON M4P 1E2, Canada
e-mail: MoSrivastava@tristargold.com



B. S. Daya Sagar et al. (eds.), *Handbook of Mathematical Geosciences*,
https://doi.org/10.1007/978-3-319-78999-6_32

ideas, about where it all began. We'd known each other for more than three decades, and were comfortable when talk lapsed into silence … and equally comfortable when silence gave way to a new thought, a different recollection, and the conversation flared up into a dispute about memory, about theory or about practice. Even when the wine bottle stood empty, and the embers in the fireplace seemed to be exhausted, the logs would sometimes adjust, one breaking and settling against another as sparks shot into the air. New fire from old.

It was December 2013, and I had succeeded in having my old advisor from Stanford, André Journel, visit me in Ontario to discuss a joint contribution to the volume on multiple-point statistics being compiled by Grégoire Mariethoz and Philippe Renard. Busy lives kept us from completing that task, but the conversation from that weekend by Lake Muskoka did become enough of an almost-paper that I was grateful for the opportunity of this 50th anniversary volume to complete what we began. Neither André nor I have much to contribute to modern MPS research; we are both "gray hairs" and now stand well back from the fire of leading-edge research. But our hair was once not so gray, and we were there at the beginning when we laid the kindling for what has become a remarkably rich idea. So our offering from that Lake Muskoka discussion is reflections on how the MPS came together. It is a tale familiar to science, with chance encounters, casual remarks that turn out to have great depth, cocktail napkins turned into whiteboards, heads shaking in disagreement: "that can't be right". As we yield the stage to the next generations of researchers, our hope is that others continue to recognize the value of cross-pollination, of interacting with others in the field, especially those who have ideas that contradict one's own beliefs. When one sturdy idea burns and breaks, settling against another, sparks fly and we have our best chance to ignite new understandings of both theory and practice.

## 32.2 1970s

### *32.2.1 A Hammer Without a Nail*

Although the theory of geostatistical simulation was firmly established by the early 1970s (Journel 1974), it had still not been widely accepted in practice by the end of the decade. The now-venerable turning bands algorithm was the only game in town when one wanted to create a conditional simulation. There were a handful of practical case study example of conditional simulation in the mining industry, but it remained a hard sell in an industry that prefers, even now, to report one single "best" estimate of mineral resources and reserves than to wrestle with a family of equi-probable outcomes.

## 32.3 1980s

### *32.3.1 Interest in Geostatistics Spreads to the Oil Industry*

Through the 1970s, the oil industry lagged behind the mining industry as an adopter of geostatistics. Many oil companies found value in some of the trend variants of kriging as additional tools in their contouring toolkit, especially when dealing with structural traps where trends are common in the elevation of the top of structure. Kriging with an external drift provided a good contouring solution in faulted reservoirs where seismic data provided strongly correlated indirect measurements of depth to the top of the reservoir (Maréchal 1984). But oil companies had many good contouring methods that worked well without any geostatistics, and it was not until the late 1980s when most of the major oil companies took notice of conditional simulation because it offered something new: the ability to do Monte Carlo analysis with 3D models of a reservoir's rock and fluid properties that honored data and that were geologically plausible.

### *32.3.2 New Simulation Tools and the Struggle for Visual Realism*

At Stanford, where I studied in the 1980s, research was supported by the Stanford Center for Reservoir Forecasting. The SCRF consortium's interest in risk analysis fueled a growing number of new geostatistical algorithms for creating realizations that honored continuous data (typically rock and fluid properties) and categorical data (typically lithologies): sequential indicator simulation (Alabert 1987), LU decomposition (Alabert 1987; Davis 1987), sequential gaussian simulation (Isaaks 1990; Gómez-Hernández 1991).

Despite having new algorithms for the conditional simulation of continuous variables, Stanford's toolkit still struggled to produce convincing simulations of categorical variables such as lithologies in a sand-shale sequence. Although indicator realizations could be made to honor indicator variogram models, the results usually were not convincing as artwork; they simply looked wrong. In Fig. 32.1, much of the (limited) success of the SIS simulation is due to the use of a trend model and to locally varying directions of maximum continuity, and not so much to the indicator kriging or to sequential simulation.

Boolean simulations that stochastically arranged prescribed geometries into a computer model usually won more approval for realism, but because these object-based algorithms were not pixel-based, they had difficulty with conditioning to well data, especially if there were lots of closely-spaced wells. In Fig. 32.1, the SIS realization is conditioned to 240 data points; but the object-based simulation, which produces a more satisfying result, is unconditional.

Through my time as a graduate student at Stanford, the Holy Grail of conditional simulation was a best-of-both-worlds algorithm that had the visual realism of

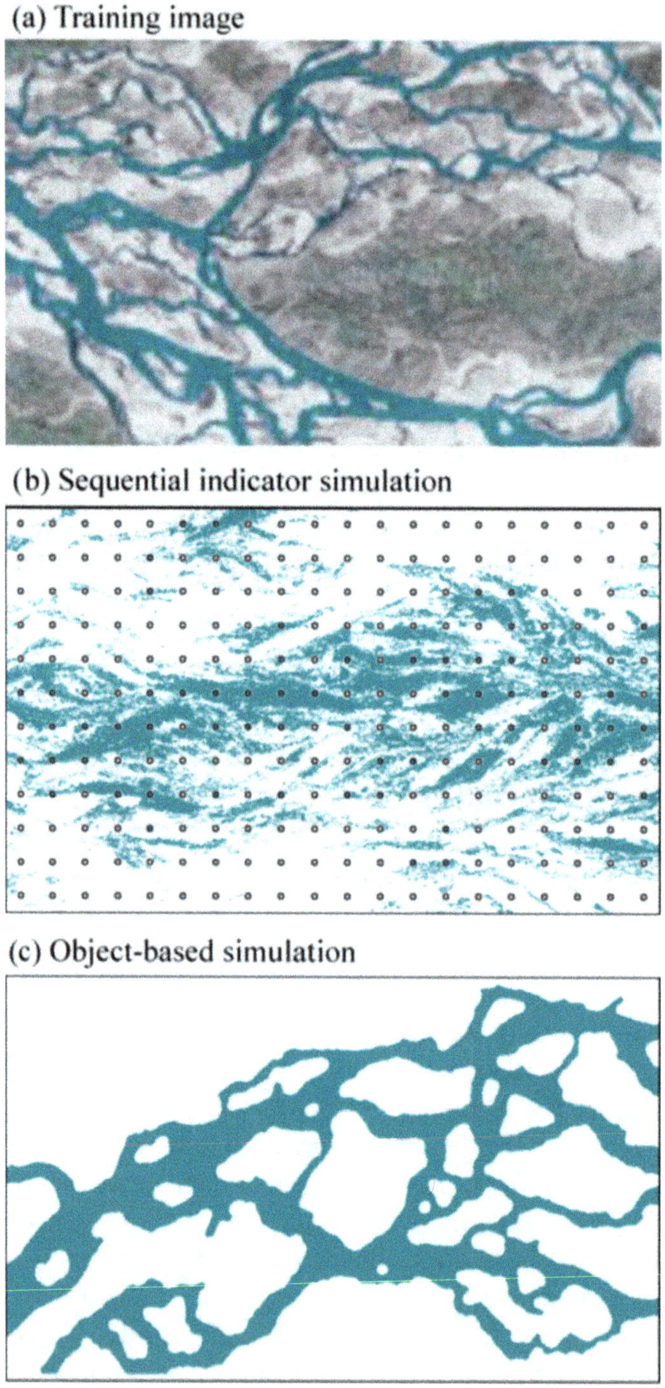


**Fig. 32.1** Examples of indicator simulation and object-based simulation of fluvial channels. The image at the top shows a training image (a satellite image of the Brahmaputra River) from which indicator variograms were calculated and used to create the SIS realization in the middle frame, conditioned to the data shown as circles. The same training image provides information on the distribution of parameters that describe object geometry; these were used as input to an object-based simulator, FLUVSIM (Deutsch and Tran 2002), to create the unconditional realization at the bottom. Although the object-based simulation succeeds in creating channels that are visually more coherent, it is difficult to condition to known lithologies at specific locations

object-based methods but that conditioned easily to hard data, no matter how dense. There were discussions at that time about the possibility that we might never achieve what we thought we wanted because of the fundamental difference between the statistical characteristics of an image and the meaning that knowledgeable experts extract from the image. In the example in Fig. 32.1, human vision allows us to see the entire set of meandering braided channels. Statistical summaries, especially variograms, do not "see" anything in its entirety; they see the image two points at a time. The analogy that André Journel often used was that it was like a blind person, trying to understand an object in front of him when he was allowed only to probe with the two forefingers. Limited to poking here and poking there, the blind person would struggle to tell the difference between an elephant and a rhinoceros.

The envy of the visual success of object-based realizations, and the desire to maintain the ease of conditioning with pixel-based methods, catalyzed a lot of discussion in the late 1980s about multi-point geostatistics. What would three-point or four-point or n-point variograms look like? How might they be calculated experimentally? How could they be modeled? How should they be used in an improved version of kriging?

### *32.3.3 Outcrops and Scanned Images as Analogs*

In the mining industry, where geostatistics was first embraced, drill hole spacing is typically on the order of tens of meters, close enough that the choice of a variogram model could be based on experimental variograms. In petroleum reservoirs, wells are typically spaced several hundreds of meters apart, sometimes thousands of meters. This practical reality of petroleum applications gave rise to an immediate practical problem when the oil industry took an interest in conditional simulation in the 1980s: where to get the closely-spaced information required to make experimental variograms?

The common advice in the 1980s was that outcrop studies could provide the data required to support statistical and geostatistical parameter choices, such as the length, anisotropy and orientation distributions required for object-based methods, or the variograms required for geostatistical methods. Outcrop studies did not begin in the 1980s; but this was the decade when they flourished. Many of the major oil companies, either individually or in consortiums, funded detailed quantitative studies of outcrops that could serve as good geological analogs for producing fields. And outcrop studies from earlier decades were dusted off and re-purposed as sources for data that could support parameterization of computer models.

Figure 32.2 shows an example of data from a 1960s outcrop study that was re-discovered by several researchers in the 1980s. It was created by digitizing shale streaks from a photograph of a cliff face of an outcrop of the Assakao Formation in the Tassili region of the central Sahara (Dupuy and du Prey 1968). Fifteen years after the data was first presented, Helge Haldorsen used the Assakao outcrop study as the basis for choosing the shale length distribution for object-based simulations

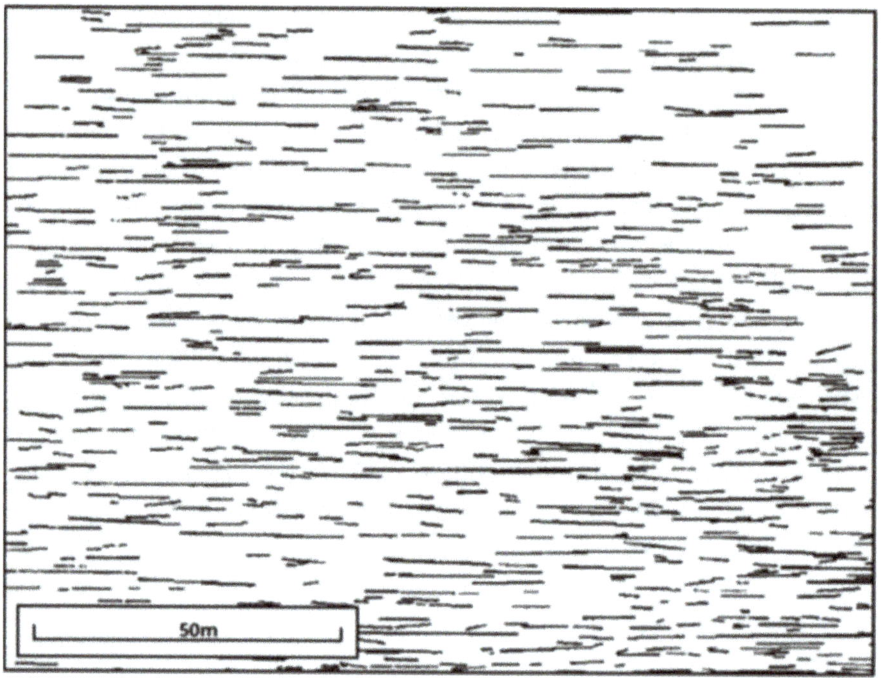


**Fig. 32.2** The Assakao Sandstone data set (from Desbarats 1987). The formation is generally sandstone (white) with occasional shale streaks (black)

of sand-shale sequences for his Ph.D. research (Haldorsen 1983; Haldorsen and Chang 1985). During the time when I studied at Stanford, I shared an office with Alec Desbarats to whom Helge had given the Assakao data for Alec's research on stochastic modeling of flow in sand-shale sequences (Desbarats 1987).

If a good outcrop analog was not available, one could (with fingers crossed and a prayer for absolution of sin) invoke a fractal argument and choose as an analog something with an entirely different scale. At a much larger scale than most reservoirs, satellite imagery, which started to become widely available in the 1980s, could serve as the source of information on spatial statistics. At the regional scale, or even at the scale of very large reservoirs, images like the top frame in Fig. 32.1 could help in sorting out statistical parameters for numerical simulation. And at a much smaller scale, there were scanned images of slabs of sedimentary rock at the scale of hand specimens, such as the example shown in Fig. 32.3.

Digitized images, whether of outcrops or of similar phenomena at different scales provide a basis for calculating not only experimental variograms but also multi-point statistics. When calculated from a rasterized image, the length distribution of shale streaks can be seen as a multi-point statistic. In the Assakao outcrop example shown in Fig. 32.2, where the individuals pixels are 20 × 20 cm, the probability of encountering a shale streak that is 20 m long can be calculated by scanning the image across each row, counting up the number of times we get a white pixel followed by 100 black pixels, then followed by a white pixel … then dividing this by the total number of shales of any length. Alec Desbarats did exactly this in his Ph.D. thesis when he wanted to test the fidelity of the synthetic

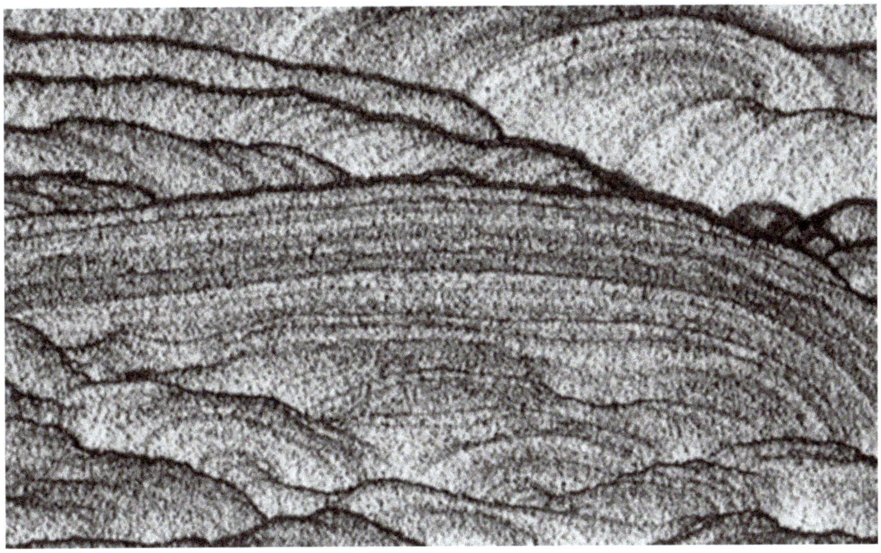

**Fig. 32.3** Digital image of a slab of cross-bedded sandstone from Utah

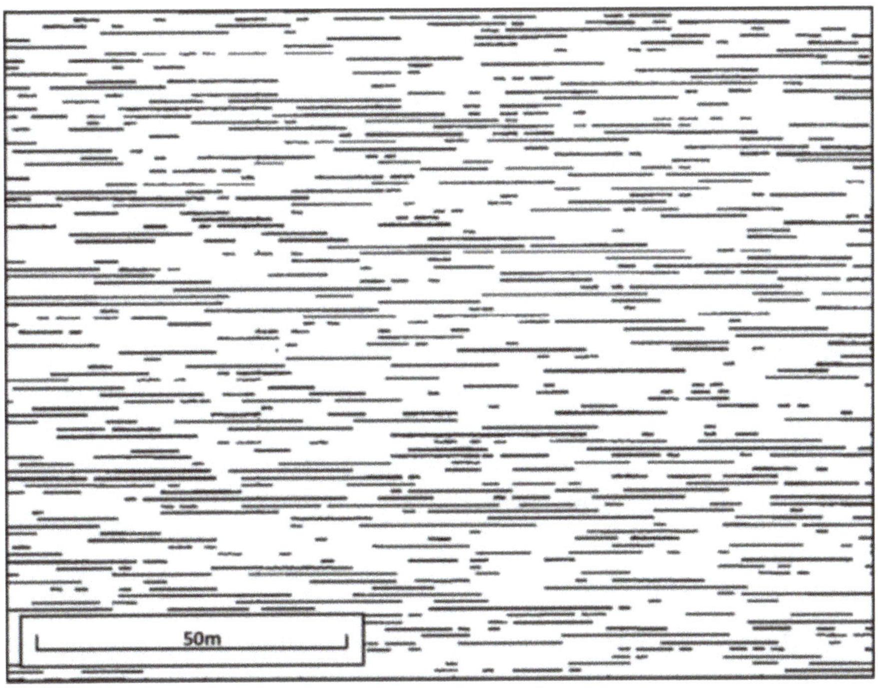


**Fig. 32.4** Indicator simulation of the Assakao outcrop image in Fig. 32.2 (from Desbarats 1987)

sand-shale sequences he had created using indicator simulation (Fig. 32.4). He knew he had the correct proportion of shales and that he had matched the indicator variogram; but he was curious about how well he had done on the multi-point statistic that Helge Haldorsen controlled directly in his simulations. Figure 32.5 shows the histograms of the shale length distributions from an indicator simulation of the Assakao outcrop, and from the original image; the indicator simulation shows

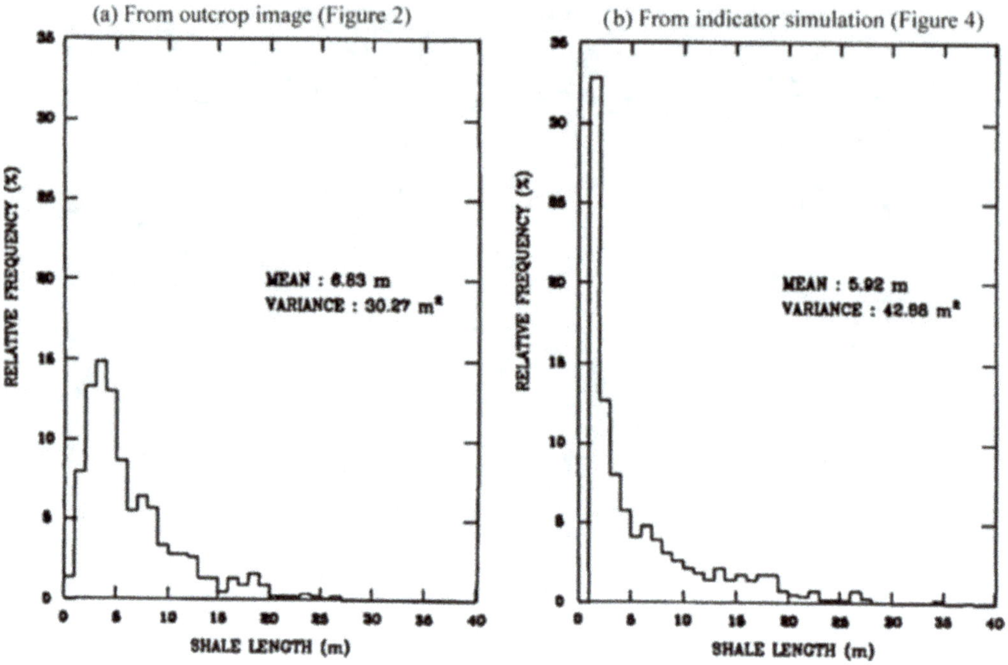


**Fig. 32.5** Histograms of shale lengths from Fig. 32.2 (left) and Fig. 32.4 (right) (from Desbarats 1987)

more very short shales than does the original image, with a lower mean length and higher variance.

Other similar studies at the same time by François Alabert showed the same result: indicator simulation produces realizations that show more short features and too few long features. The over-representation of short features is also obvious from a visual comparison of indicator simulations to the reality they try to mimic, e.g. the top two frames in Fig. 32.1, or the realization in Fig. 32.4 with the outcrop image in Fig. 32.2. The common explanation given at the time was that when an algorithm controls only the first and second-order moments (histogram, or indicator proportion, and the variogram) then the uncontrolled higher-order moments drift in the direction of disorder or maximum entropy.

### *32.3.4 Leaving the Ivory Tower and Getting on with Adult Life*

My years as a student at Stanford ended in 1988. Sold my bicycle, the one that hadn't been stolen. Gave up the wonderful room I had in a camping trailer behind a house in Palo Alto. Headed off into the world of consulting, with Neil Schofield and Roland Froidevaux as my partners in FSS International Consultants. The notion was simple: Neil and I were familiar with student poverty and didn't mind another year of living with little money. After a year, if we failed as consultants then we could get real jobs.

We managed not to fail, and each of the FSS partners found ourselves busy with clients who wanted advice and assistance with geostatistical studies. My workload was split between mining studies, where simulation was rarely discussed, and petroleum studies, where kriging was rarely discussed.

Even though my mining studies had little to do with stochastic modeling, there was one mining project that, in hindsight, probably planted some useful seeds for what later became the MPS prototype algorithm. It was a project in which some of the useful geological and numerical data were available only from paper records written by hand decades ago: drill logs with assay values transcribed manually. In the late 1980s, software for optical character recognition (OCR) struggled with handwriting; it still does today, but it was worse back then. Even though commercial OCR software could make no sense of the handwritten logs, my sense was that it should be possible to extract much of it automatically, instead of going through a time-consuming and error-prone process of manual data entry. The drill logs were neat and legible, and all of the key numerical values were written in boxes on a form. With only 11 possible characters in use, the ten digits and the decimal point, it seemed possible to me that the handwriting could be recognized by an algorithm that trained itself from actual images. I wrote a program that would search the scanned image (an eight-level grayscale raster), looking for islands of non-white in the appropriate boxes on the form. It would then show what it had found to the user, who would identify the symbol by typing in one of the 11 choices. After a few dozen examples of each of the 11 possibilities, the software was able to estimate the probability that a new small patch corresponded to each of the possibilities. It did this simply by direct pixel-to-pixel matching of grayscale levels, without any clever rescaling or rotation. If it could not establish a sufficiently high probability for one particular choice, it would drop pixels from the comparison and try again. The user would correct it when it made mistakes, and the software would store its acquired collection of confirmed examples in a growing database. As with most of my Mo-code, it took a bit of tinkering to get it to work well; but it ended up being used, and saved weeks of data entry from hundreds of old drill logs. We ended up calling the program "Am-I-Right" because that's how the program worked: by making guesses based on pixel-to-pixel pattern matching, and then checking with the user to see if that guess was correct.

### *32.3.5 Chris Farmer's Unexpected Claim*

1988 was also the year when I first met Chris Farmer, at the SPE Forum on reservoir characterization in Grindelwald, Switzerland. He was working on methods for numerically simulating reservoir rocks, recognized the benefits of a pixel-based approach, and had developed new ideas about what information to extract from outcrop studies and scanned images of analogs (Farmer, 1989). During my early years as a consultant, I managed to visit Chris at the UK Atomic Energy Agency's research centre at Winfrith. During this visit, he made a claim that seemed implausible … no, it actually seemed flat out wrong; but I was raised well by my parents, and knew that it was rude for a guest to precipitate an argument.

We had been talking about extracting indicator variogram and cross-variogram information from scanned images and Chris remarked that you have to be careful when you do this because if you try to make a realization exactly match all of the indicator variograms and cross-variograms from a scanned image, then you'll just get back the scanned image; and the purpose of creating realizations is not to exactly match one "true" image, but instead to sample a space of uncertainty that shares something in common with the original image. I checked if I understood him correctly: did he really mean that you can exactly … exactly … match an image just by reproducing its indicator variograms and cross-variograms? I knew (or thought I knew) that this wasn't true. Even with multiple indicators, all of the variograms and cross-variograms are still two-point statistics; you're still a blind person, feebly prodding either an elephant or a rhino.

Chris clarified that he did mean exactly, with one minor caveat: that you actually get two possible images which are 180° rotations of each other; you might end up with an upside-down elephant, but you'd easily be able to figure out that it wasn't a rhino. And he also explained that he meant that you match to the complete experimental indicator variograms for every possible separation distance and direction on the rasterized image. Even with these caveats, I still found his claim implausible; but kept thinking about why he would be so sure about this.

The other reason it was not worth getting into the details of why Chris was confused was that I agreed with the basic point he was trying to make: the purpose of what we have now taken to calling a "training image" is not to match it, but instead to use it as a guide for selected spatial statistical characteristics. You want to match the statistics, while conditioning to data, not replicate one training image.

## 32.4 1990s

### *32.4.1 Why Chris Farmer Was Right*

In 1991, the SPE Forum on reservoir characterization was held in Crested Butte, Colorado, and I had a chance to continue the discussion with Chris Farmer about indicator variograms and training images. When I explained, as diplomatically as I could, that I didn't think his claim was correct, he grabbed a nearby napkin, sketched a small grid, and colored in some pixels as black, white and gray. He agreed that I was right if we lived in a world of variogram models for random functions that are infinite in all directions. "But in the real world, things have edges," he explained patiently, "and this means there's only one pair of pixels in the original image that completely span the diagonal". He went on to show how you can actually deduce the grayscale levels for the two corner pixels (up to the

180° rotation) and then work inwards from the corners. The Appendix to this paper shows a small worked-out example of the trick that Chris explained.

As soon as he explained it, and I realized that I was the one who was wrong, Chris dismissed it as an algorithmic oddity, a cute and clever trick that has no practical value for simulating reservoir rocks, especially because the goal is never to exactly replicate the original image.

Even though I understood the principle behind the procedure of attacking the corners first and then working inwards, the algorithm still wasn't clear in my head, and I spent some time that year trying to write code for doing what Chris had described. I never did manage to work out all the special cases, and it ended up on the back burner as one more unfinished project.

### 32.4.2 *Back to the Ivory Tower: A Brief Escape from Adult Life*

In late 1991, my consulting business was thriving and growing; I had a small staff in Vancouver, and plenty of project work. But I was spending more time as an administrator and manager, neither of which I am good at, and less time doing the technical work that I enjoy.

My old advisor convinced me that I could let the staff run the show while I spent a year at Stanford, back in the ebb and flow of new ideas with his new crop of graduate students. Twenty five years later, I find it remarkable what was accomplished during that year: P-field simulation, co-located cokriging, and a proof-of-concept algorithm for multiple-point statistics. All of these new geostatistical methods that we investigated in 1992 began with a piece of Mo-code that did something useful, and not with theory; that came later. André comes at research from the side of theory that leads to equations that can be coded and tested. I tend to come at it the opposite way, with a piece of code that achieves a desired result and that then leads to the question "I wonder why that works?".

In the early part of 1992, with the luxury of time to do research again, I dusted off some of my back-burner projects, and came back to my attempt to code Chris Farmer's trick for replicating an image from its indicator variograms and cross-variograms. The details of the algorithm were still a mess, but I realized that I could get very close to a satisfactory result using simulated annealing, a possibility that came to the forefront because Clayton Deutsch was finishing his Ph.D. thesis on simulated annealing that year. I wrote a program that would start with a grid that had exactly the correct proportions of the gray levels, randomly scattered, and that would use simulated annealing to iteratively adjust the image by swapping pixels in order to push the experimental indicator variograms and cross-variograms of the evolving grid in the direction of a target values established by the complete indicator variograms of the original image. No variogram models were used; everything was done using look-up tables of variogram values. I used a photo of André, exhausted after a climb on Mount Whitney, as the training image, converted it to an eight-level grayscale image with seven indicators. 200 columns and rows, seven direct indicator variograms, 21 indicator cross-variograms, all calculated for every one of approximately 80,000 lags on the image. It took four days of run-time and hundreds of millions of swaps before the difference between the indicator variograms of the simulation and the image could not be reduced. It was hopelessly inefficient, but it confirmed for me that Chris Farmer had been right.

For me, the recognition that you can exactly match an image from a very complete and specific statistical summary of specific patterns was an eye-opener. Although it now probably seems fairly obvious, in the early 1990s, the wealth of information contained in an image's statistical summaries was not immediately apparent. Then, the normal workflow was to assemble statistical parameters by fitting models to experimental statistics. The experimental variogram, for example, was an important stepping-stone to a variogram model; but it was only a means to an end. We did not think of the massive look-up table of summary statistics for thousands of grouped pairs of data as something that could serve directly as an input parameter. But why not? Why in an age of computer power did we continue to create simplified mathematical models of statistical characteristics? Was it really necessary to boil the parameterization down to a few numbers, a nugget effect and a range, rather than leave the statistical summary in its original form as a massive look-up table? For me, this was the "aha" moment catalyzed by my belief, years earlier, that Chris Farmer's claim about indicator variograms was not correct. The reason I was wrong was that massive look-up tables of indicator variograms are a rich source of very detailed information. The mistake we were making was that we moved past this wealth of information and replaced it with a simple model.

The idea for the first prototype of an MPS simulation algorithm came from the accidental meeting of thoughts about the role of training images in reservoir simulation and the experience of having coded the Am-I-Right procedure for optical character recognition for a mining project. The principal difference between Am-I-Right and the MPS prototype is that, after scanning the image to build a probability distribution, the Am-I-Right procedure always took the most likely value while MPS used the distribution as a basis for random sampling.

The first tests of the MPS prototype were done on a digital image of a cross-bedded sandstone, like the one shown in Fig. 32.3. This was chosen because it presents curved structures that are difficult to capture with most geostatistical simulations, which tend to show straight features in the direction of maximum continuity unless an explicit attempt is made to use locally varying directions of anisotropy. Figure 32.6 shows the first published results of an MPS simulation (Guardiano and Srivastava 1992). That Tróia '92 paper used a two-level black-and-white training image because the first tests on an eight-level grayscale image were very slow; it would be several years before Sebastien Strebelle's Ph.D. research (Strebelle 2000) produced the first efficient and practical implementation of the original clumsy prototype.

Even though the first results were not brilliant, certainly not by today's standards, they did show that it was possible to impart to a simulation higher-order

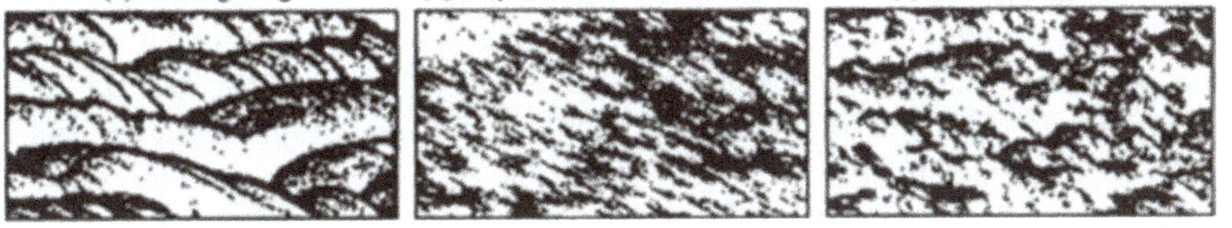


**Fig. 32.6** The first published example of results of an MPS simulation (from Guardiano and Srivastava 1992). The frame on the left shows the training image, a black-white image obtained from a digital photograph of a slab of cross-bedded sandstone. The middle frame shows a realization from sequential indicator simulation. The right frame shows a realization from the MPS prototype algorithm

connectivities and patterns that are not explicitly summarized in variograms. In the right frame of Fig. 32.6, it is the black pixels that make the thin curved arcs, while the contiguous regions of white pixels tend to be larger and blockier. The middle frame of Fig. 32.6 shows that these features are hard to capture in an indicator simulation, which tends to symmetrize the black and white geometries when the proportion is near 50%.

## 32.5 Concluding Thoughts

Where do ideas come from? Is it possible to create fertile conditions for innovation? Of the many who have studied these questions, my favorite is Steve Johnson, who wrote *Where Good Ideas Come From: The Natural History of Innovation*; he has presented his thoughts in a 2010 TED Talk and also in a short YouTube video (https://www.youtube.com/watch?v=NugRZGDbPFU). Much of what Johnson identifies as key elements of innovation are in evidence in the origins of the MPS simulation algorithm: the slow incubation of hunches, the borrowing and combining of ideas from other people with related hunches, the catalytic effect of recognizing error, and of finding the missing piece.

The one piece of Johnson's message that resonates most strongly with my experience is the importance of staying connected to others; he often concludes his presentations with the observation that innovation comes by chance, but chance favors the connected mind. By "connected mind" he means a mind that is connected to what others are doing, how they are thinking about similar problems. It is the hunches and cast-off ideas of those people that you'll end up borrowing and adapting to improve a hunch of your own that has still not reached fruition.

Of the many different ideas that ended up being woven together into the MPS prototype, there may be a dropped thread, something that might be research worth pursuing. It is the fact that complete indicator variograms and cross-variograms provide extremely rich and detailed information about an image, so rich and detailed that they can, in fact, be used to replicate the original image.

While replication of a training image should never be a goal, it's intriguing to think about what we might be able to do if we matched a small sub-set of the complete look-up table of all indicator variograms. We know that we get a "perfect" realization if we use 100% of the look-up table. Would the realization look "fairly good" or "completely ugly" if we decimated the complete look-up table and used only 10% of it, or only 1%? My own tests with the annealing version of this procedure, and the example in Appendix A, indicate that the indicator cross-variograms are sometimes not necessary, i.e. that you can achieve nearly perfect reproduction of the original image without them. Dropping all the indicator cross-covariances would considerably reduce the size of the look-up table, or any subset of it. Something worth trying?

My final reflection is on the beneficial tug-of-war between theory and practice. Throughout my career as a consultant, and tourist in academia, I have enjoyed discovering that the path to a solution sometimes starts when you enter the maze from the theory side, and sometimes starts from an entrance on the practical side. When theory leads you to the point of a set of equations, that need not be the end because there may be something useful to be learned in attempting to implement those equations in practice, in writing a piece of computer code that produces an answer in a reasonable amount of time. And, coming from the other end, having developed an algorithm that produces an intriguing result that seems "good" or "right", it's useful to try to work out why it works. Even if the answer came heuristically, the theory that explains why it's an approximately correct answer might reveal a generalization that makes it possible to improve the answer.

## Appendix: Example of Reconstructing a Grid from Its Indicator Variograms and Cross-Variograms

Figure 32.7 shows a tiny image with three levels of gray on a 3 × 3 grid. If we give values of 1, 2 and 3 to white, gray and black, the three levels give rise to two indicators: $I_1$ with a threshold between 1 and 2 and $I_2$ with a threshold between 2 and 3. There are two direct indicator variograms, $\gamma_1$ and $\gamma_2$, and one cross-variogram, $\gamma_{12}$. The nine locations give rise to 36 paired locations (not including the pairs that have zero separation). These 36 pairs are shown in Fig. 32.8, grouped into the 12 possible lags.

For any lag, the experimental indicator variogram is calculated by taking half the average squared difference between the paired indicators:

**Fig. 32.7** Example used to show how complete experimental indicator variograms can be used to reconstruct an image

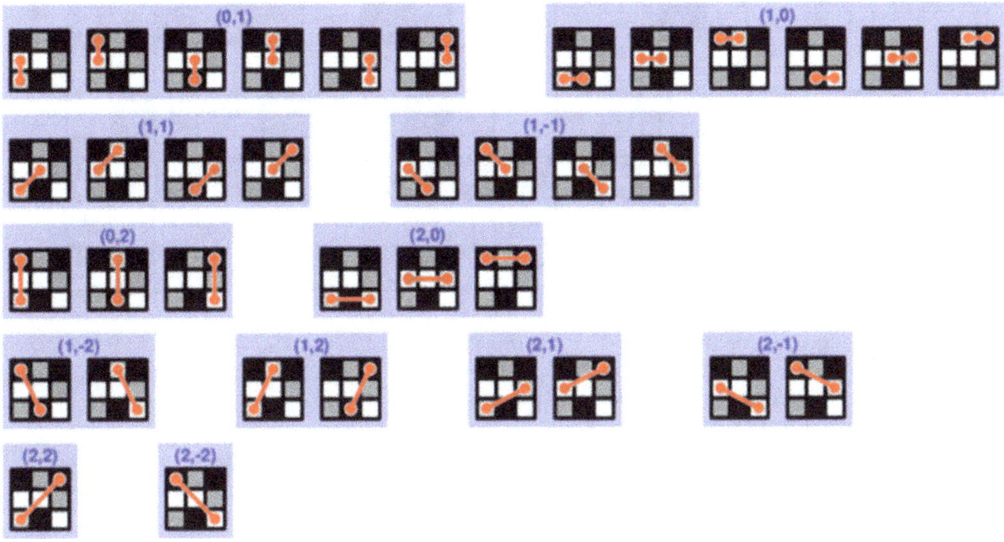


**Fig. 32.8** All 36 pairs in the image in Fig. 32.7, grouped into the 12 lags

$$\gamma(h) = \frac{1}{2N(h)} \sum [I(x) - I(x+h)]^2$$

Because the squared difference between 0 s and 1 s is always 0 or 1, all of the terms in the summation are either 0 or 1; the summation is simply a counting of the number of times that the indicators separated by $h$ are different.

**Table 32.1** Look-up table for the experimental indicator variograms and cross-variogram for every lag for the image in Fig. 32.7

| ΔX | ΔY | N | #Diff1 | #Diff2 | #Diff12 | $\gamma_{I1}$ | $\gamma_{I2}$ | $\gamma_{I12}$ |
|---|---|---|---|---|---|---|---|---|
| 0 | 1 | 6 | 5 | 3 | 3 | 0.42 | 0.25 | 0.25 |
| 1 | 0 | 6 | 2 | 4 | 2 | 0.17 | 0.33 | 0.17 |
| 1 | 1 | 4 | 3 | 2 | 3 | 0.38 | 0.25 | 0.38 |
| 1 | −1 | 4 | 2 | 2 | 3 | 0.25 | 0.25 | 0.38 |
| 0 | 2 | 3 | 1 | 3 | 2 | 0.17 | 0.50 | 0.33 |
| 2 | 0 | 3 | 2 | 0 | 1 | 0.33 | 0.00 | 0.17 |
| 1 | −2 | 2 | 1 | 0 | 1 | 0.25 | 0.00 | 0.25 |
| 1 | 2 | 2 | 0 | 0 | 1 | 0.00 | 0.00 | 0.25 |
| 2 | 1 | 2 | 1 | 1 | 2 | 0.25 | 0.25 | 0.50 |
| 2 | −1 | 2 | 0 | 1 | 1 | 0.00 | 0.25 | 0.25 |
| 2 | 2 | 1 | 0 | 1 | 0 | 0.00 | 0.50 | 0.00 |
| 2 | −2 | 1 | 1 | 1 | 1 | 0.50 | 0.50 | 0.50 |

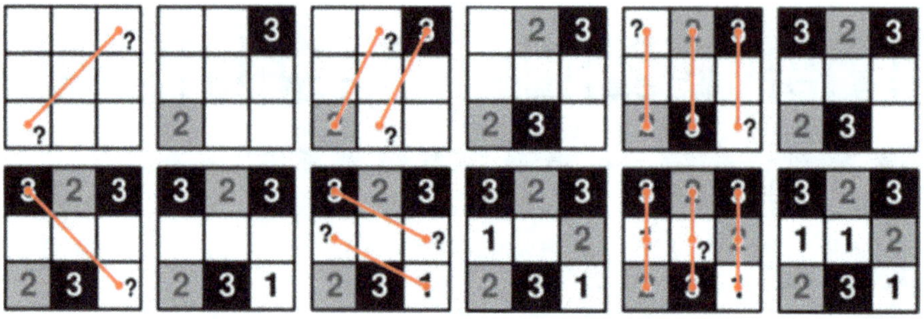

**Fig. 32.9** The sequence of steps used to interrogate Table 32.1 to deduce values in specific cells, the knowledge of which can then be used to fix the values of other cells by using other information from the look-up table. The sequence begins in the upper left where the look-up table is used for its information on the lag that spans the main diagonal. It then proceeds across the first row, down to the start of the second row, and across to the final solution at the lower left

For the image in Fig. 32.7, Table 32.1 gives the complete look-up table of the indicator variograms and cross-variograms in every lag, and includes for each lag the value of the summation term before the division by $2N(h)$, i.e. the number of pairs in each lag that have different indicators; these are in the columns headed #Diff1, #Diff2 and #Diff12.

Figure 32.9 shows a sequence of steps that can be used to interrogate Table 32.1 for the information that allows the values of specific cells to be deduced. It begins in the upper left with the (2, 2) lag that spans the main diagonal. There is only one pair that contributes to this lag and the (2, 2) row, (second from the bottom of Table 32.1) tells us that:

- the two $I_1$ indicators are the same, because of the 0 in the #Diff1 column
- the two $I_2$ indicators are different, because of the 1 in the #Diff2 column

The second of these facts says that the two values are either 2 and 3, or 1 and 3; but the second choice is contradicted by the first fact, so the only choice is a 2 in one cell and a 3 in the other. This gives us the next frame in Fig. 32.9, where a 2 has been fixed in the lower left and a 3 in the upper right. Note that this is exactly where the 180° rotation may occur because we can't tell which one is the 2 and which is the 3. But once we make a choice, everything else is fixed; so the worst that will happen is that the final solution will be rotated upside-down.

Proceeding across the first row of Fig. 32.9, the next thing we check is the (1, 2) lag, to which two pairs contribute. The look-up table entries for the (1, 2) lag, fifth row from the bottom, tell us that both pairs have the same $I_1$ and $I_2$ indicators, because of the 0s in the #Diff1 and #Diff2 columns. The only way that this can occur is if the value paired with the 2 in the lower left is also a 2, and the value paired with the 3 in the upper right is also a 3.

Continuing across the first row of Fig. 32.9, the next thing we check is the (0, 2) lag, to which three pairs contribute. The look-up table entries for the (0, 2) lag, fifth row from the top, tell us that all three pairs have different $I_2$ indicators, because of the 3 in the #Diff2 column. This tells us that the upper right corner must be a 3, and that the lower left is either a 1 or a 2.

The sequence continues on the second row, where we check the (2, −2) lag. There is only one pair that contributes to this lag, and this pair has different values for the I1 indicator, because of the 1 in the #Diff1 column. The only way this can happen is if the value in the lower left is a 1.

Continuing across the second row, the next thing we check is (2, −1) lag, to which two pairs contribute. The entries for the (2, −1) lag, third row from the bottom, tell us that both pairs have the same $I_1$ indicators, so the 1 in the bottom right must be paired with a 1, and the 3 in the upper left must be paired with either a 2 or a 3. For the same two pairs, one of the I2 indicators is the same and one is different; we know that the pair with the same I2 indicators is the pair of 1–1 values that we just fixed, so it's the other pair that must have different I2 indicators. We already know that the 3 must be paired with a 2 or a 3, so the only correct choice is a 2.

Moving along the second row, the last thing we check is the (0, 1) lag, to which there are six pairs that contribute. In the #Diff1 column, the top row in Table 32.1 tells us that five of the six pair have different $I_1$ indicators. With the eight values already fixed in previous steps, we can see three of those (0, 1) pairs: the 3–1 and 1–2 pairs in the first column and the 2–1 pair in the last column. But the only way we can get to five such pairs is if the middle column gives us two more. So the only correct choice for the middle cell is a 1 … which gives us the last value, and completely reconstructs the original image (Fig. 32.7) with <u>no</u> conditioning data, but with heavy use of the information in the complete table of indicator variograms.

Regardless of the size of the image, or of the number of levels in the grayscale (or number of colors in a color image), the approach of starting at the corners and working inwards will always work. There is enough information in the complete look-up table of experimental indicator variograms and cross-variograms that the corner pixels can be pinned down and then used to leverage the solution for the neighbors. In this particular example, the indicator cross-variogram was never needed for the final solution. It may be that the indicator cross-variograms are never needed, and that the image can always be exactly reconstructed (up to a 180° rotation) using only the indicator variograms.

## References

Alabert FG (1987) Stochastic imaging of spatial distributions using hard and soft data. MSc Dissertation Stanford University

Davis MW (1987) Production of conditional simulations via the LU triangular decomposition of the covariance matrix. Math Geol 19(2):91–98

Desbarats AJ (1987) Stochastic modeling of flow in sand-shale sequences. PhD Dissertation Stanford University

Deutsch CV, Tran TT (2002) FLUVSIM: a program for object-based stochastic modeling of fluvial depositional systems. Comput Geosci 28(4):525–535

Dupuy M, du Prey EL (1968) L'anisotropie d'écoulement en milieu poreux presentant des intercalations horizontals discontinues. 3ième colloque de la ARTFP, Pau, France, 23–26 Sept 1968

Farmer CL (1989) The mathematical generation of reservoir geology. In: Numerical rocks. Joint IMA/SPE European conference on the mathematics of oil recovery, Cambridge University, 25–27 July 1989

Gómez-Hernández JJ (1991) A stochastic approach to the simulation of block conductivity values conditioned upon data measured at a smaller scale. PhD Dissertation Stanford University

Guardiano FB, Srivastava RM (1992) Multivariate geostatistics: beyond bivariate moments. In: Soares A (ed) Geostatistics Tróia '92, pp 133–144

Haldorsen HH (1983) Reservoir characterization procedures for numerical simulation. PhD Dissertation University of Texas at Austin

Haldorsen HH, Chang DM (1985) Notes on stochastic shales: from outcrop to simulation model. In: Proceedings of the reservoir characterization technical conference sponsored by NIPER, Dallas, Texas, 29 Apr–1 May 1985

Isaaks EH (1990) The application of Monte Carlo methods to the analysis of spatially correlated data. PhD Dissertation Stanford University

Journel AG (1974) Geostatistics for conditional simulation of ore bodies. Econ Geol 69(5): 673–687

Maréchal A (1984) Kriging seismic data in presence of faults. In: Verly G, David M, Journel AG, Maréchal A (eds) Geostatistics for natural resources characterization, part I. Springer, pp 271–294

Strebelle S (2000) Sequential simulation drawing structures from training images. PhD Dissertation Stanford University

# Chapter 33
# Predictive Geometallurgy: An Interdisciplinary Key Challenge for Mathematical Geosciences

K. G. van den Boogaart and R. Tolosana-Delgado

**Abstract** Predictive geometallurgy tries to optimize the mineral value chain based on a precise and quantitative understanding of: the geology and mineralogy of the ores, the minerals processing, and the economics of mineral commodities. This chapter describes the state of the art and the mathematical building blocks of a possible solution to this problem. This solution heavily relies on all classical fields of mathematical geosciences and geoinformatics, but requires new mathematical and computational developments. Geometallurgy can thus become a new defining challenge for mathematical geosciences, in the same fashion as geostatistics has been in the first 50 years of the IAMG.

**Keywords** Geostatistics · Statistical scales · Microstructure · Computational geometry · Processing optimisation · Value of information · Mineral liberation analyser · QUEMSCAN

## 33.1 Introduction

*Geometallurgy*, from the Greek words for earth (geia), metal (metallo) and work (ergon), can be understood as the exploitation of a metallic ore based on a precise understanding of its geoscientific characteristics. Geometallurgy is hence a cooperation field for geoscientists and mineral processing engineers, something which has occurred in virtually all mining operations. A modern understanding of geometallurgy, what we could call *predictive geometallurgy*, proposes a quantitative approach to the subject. In rough terms, that requires optimizing the ore processing based on automated mineralogy and microstructure characterisation of the ore, coupled with geometallurgical tests. These are tests conducted at several scales (from lab to plant) along which the actual ore is processed in realistic conditions in order to study the differential behaviour of the several ore and waste mineral phases, and thus the enriching potential of the ore through the processes considered.

K. G. van den Boogaart · R. Tolosana-Delgado (✉)
Helmholtz Institute Freiberg for Resource Technology, Freiberg, Germany
e-mail: r.tolosana@hzdr.de



B. S. Daya Sagar et al. (eds.), *Handbook of Mathematical Geosciences*,
https://doi.org/10.1007/978-3-319-78999-6_33

As a subject, mathematical geosciences has always had a wide application in mining. Nowadays typical topics of the area are geostatistics, the analysis of data from special scales (such as compositional data or spherical data), numerical analysis of flow models, remote sensing, (mineral) potential modelling (for instance with weights of evidence), fractals, geodata standards, 3D geomodelling, or data integration techniques. The aim of this chapter is to show the deep link between geometallurgical problems and techniques from the main fields of mathematical geosciences.

Geometallurgy distinguishes primary and secondary properties of the ore (Coward et al. 2009). Primary properties are intrinsic to the ore and do not depend on the process. Secondary or response properties describe the behaviour of the ore during processing. The primary properties are observed by chemical assays, automated mineralogy (like with QUEMSCAN or Mineral Liberation Analyser—MLA—), X-ray methods, and other analytical instrumentation. Secondary properties are measured with geometallurgical tests, such as blasting tests, Bond mill test, flotation tests, magnetic separation, density separation and so on. These can even be conducted using the operation itself, that is, on the real plant. The secondary properties are used to predict the outcome and costs of the processing.

To the authors' knowledge, all studies conducted on predictive geometallurgy by mathematical geoscientists (Bye 2011; Boisvert et al. 2013; Rossi and Deutsch 2014; Hosseini and Asghari 2015; Tolosana-Delgado et al. 2015; Ortiz et al. 2015; Deutsch et al. 2016) consisted on appropriately predicting the secondary properties at each block of a mining block model, and proposing the mining and processing engineers to conduct their mine planning and plant scheduling based on those properties instead of on metal grades. The first step (Vann et al. 2011) is the geometallurgical analysis of the ore body with respect to its primary properties. Samples of similar primary properties or geology are often said to belong to the same *geometallurgical domain*. Conventional descriptive exploratory analysis like k-means clustering, PCA (Caciagli Warman 2015) or machine learning methods are nowadays used for this task. Moreover, primary properties are also interpolated to the block model, ideally with geostatistics.

The second step is a geometallurgical testwork, i.e. the characterisation of secondary properties of material from different geometallurgical domains. Often the goal of these tests is to define a mapping from the primary properties to the secondary properties, e.g. via more or less complex regression models (Keeney et al. 2011; Everett and Howard 2011; Sepulveda et al. 2017). Having it makes possible to populate the block model with estimated secondary geometallurgical properties and to infer the expected income and costs of each block. Such interpolation of secondary variables is often done on additive proxies (Ortiz et al. 2015; Deutsch et al. 2016). The result is typically called a *geometallurgical (block) model*.

This can be used in at least three different ways by an operation, to inform both in short- and long-term actions (McKay et al. 2016). First, the prediction of costs and recovery allows to assign monetary values to each block. These values can be used instead of grade as better proxy of cashflow in further calculations, like the mentioned ultimate pit or mine scheduling. Value is generated by minimizing capital costs, due to early exploitation of highly valuable parts of the deposit, and by an improved distinction between ore and waste (Bye 2011). Second, the predicted properties can be used as well to find matching ore partners in blending to reduce feed variability in the plant, and ensure constant plant operation conditions. Value is generated by lower risk of plant failure, optimal use capacity of all parts of the plant, and lower controlling efforts by the ability to find the optimal operation conditions empirically (Shaw et al. 2013). The third option is to use that knowledge to actively adapt the processing conditions to each portion of the varying feed. The value lies in higher recovery, lower operation costs, more extensive exploitation (Powell 2013; Tolosana-Delgado et al. 2015) and ultimately lower ecological footprint.

## 33.2 Process Modelling

With the exhaustion of simple-texture, single-commodity, easy-to-reach deposits, the mining industry has been confronted with the need to study a broad range of ore properties, beyond the classical grade. As mentioned in the introduction, predictive geometallurgy proposes to obtain a wealth of primary and secondary properties at each mining block in order to reproduce its behaviour through the processing chain and, ultimately, to predict its monetary value. This section focuses on such process modelling.

A couple of steps along the value chain after extraction and crushing, ores are treated with a variety of processes, mostly physical and physico-chemical, in order to liberate the several mineral grains and separate them in different streams. Later on, streams enriched in ore minerals are sent through metallurgical processes, mostly chemical and physical changes of state processes devised to break the crystal structure of the ore minerals and produce the final value metals. All these steps can be studied with two approaches. In the first one, each operation unit is considered as a black box, and data from both the conditions of operations and the properties of input and output streams are obtained in order to build empirical rules to predict the output streams (Matos Camacho et al. 2015). In the second strategy, these prediction laws are built in accordance with thermodynamical, chemical and physical first principles. These strategies are not mutually exclusive, as one can derive the form of a parametric predicting equation by first principles and fit the parameters with the empirical approach.

The first kind of processes mentioned, those mostly keeping the crystal structure of the minerals involved, include many different processes. Grinding and milling aim at splitting particles in order to produce single mineral, or *liberated*, particles. Sizing, magnetic separation, density separation and many other separation processes aim at splitting a *feed* stream into two or more streams with particles primarily classified according to one particular bulk volumetric property, like size, magnetic susceptibility or density. Finally, froth flotation aims at separating particles according to the hydrophobicity of its surface minerals as they fall through a bubble-rich 2- or 3-fluid medium (including water, gas, nonpolar liquids, oils). This is one of the most complex yet barely understood processes in minerals processing, including effects from fluid dynamics, surface physics, organic and anorganic chemistry. In processing plants, several of these processes might be combined so that the output streams of each processing unit is fed into other units, thus building serial or parallel chains, trees and even complex networks, with feed-back loops.

Particle based models (Lamberg 2011) are a particular simple and promising modelling strategy, primarily of use for such networks of minerals processing processes. Here, each particle of the general feed is given a probability of going to each one of the output streams of each processing unit, according to its singular properties and certain characteristics of the bulk material within the unit. As long as these probabilities can be considered constant in time, the transient behaviour of the system can be modelled with a simple system of first order differential equations with constant coefficients (Tolosana-Delgado et al. 2015). Other more complex settings, in particular, milling steps within loops, pose a much more complex challenge and remain yet unexplored to the authors' knowledge.

The second kind of processes typically destroy the ore mineral structure into a fluid state: a water solution (hydrometallurgy, electrometallurgy) or a melt (pyrometallurgy). All these processes can be modelled with relatively well-known thermo-electro-chemical reactions. Lack of space and a certain distance from the classical fields of mathematical geosciences made us leave the subject out of this contribution.

Whichever strategy of modelling is followed, it is necessary to characterise the frequency distribution of certain properties on the particle streams. The most obvious are the size and mineralogical composition of the particles, in exposed surface, mass and even in volume proportions. Derived from these, elemental deportment and liberation distribution are also relevant. Elemental deportment is the proportion of a given element mass apported by each mineral. The liberation distribution gives the volume (or mass) of particles containing a certain mineral in a (volume, mass or surface) proportion equal or larger than a threshold, as a function of that threshold. This is a cumulative distribution in the fashion of the better known recovery and tonnage curves in classical Geostatistics. Finally, more complex mineral association or paragenesis indicators do also matter, as often concentration processes do not target the value minerals themselves, but some accompanying, more abundant minerals. Next section discusses which instruments are used to measure these properties and which are the challenges brought with them to mathematical geoscientists.

## 33.3 Ore Characterisation

In the past, one-commodity grade was considered the sole and sufficient variable to characterize a mining block or a deposit. This variable could be more or less safely considered as a positive variable yet with an interval scale, according to the definition by Stevens (1946). This explains why Geostatistics was originally concerned with univariate properties following the properties of Gaussian or lognormal random fields (Journel and Huijbregts 1978).

However, the present and the future evaluation of a mining operation will require many more variables, kinds of scales and new geostatistical models. Multi-commodity grades, geochemistry and mineralogy, being vectors of positive or relative components (Pawlowsky-Glahn 2003; Boogaart and Tolosana-Delgado 2013), have already brought the need of considering multivariate ratio scales and compositional scales (Caciagli Warman 2015). The routine analysis of mineral and chemical properties by techniques like X-ray Fluorescence (XRF) or Instrumental Neutron Activation Analysis (INAA) for bulk geochemistry, X-ray Diffraction (XRD) for bulk mineralogy, or Electron Probe Microanalysis (EPMA), Proton-Induced X-ray emission (PIXE), Laser Ablation Inductively Coupled Mass Spectrometry (LA-ICP-MS) or Raman spectroscopy for single grain or locally resolved chemistry and mineralogy will ensure a continuous growth of compositional and multivariate positive data in predictive geometallurgy. The generalisation of microstructural analysis, with machines like QUEMSCAN, MLA or X-ray tomography (Bam et al. 2016; Becker et al. 2016), will make further primary properties easy to obtain: particle size curves (showing a distributional scale (Delicado 2008; Menafoglio et al. 2016a)), interphase mean contact length composition (a sort of two-way composition (Caracciolo et al. 2012)), grain size curves of each mineral phase (a discrete set of parallel distributions), deportment (a composition informing of the proportion of mass of a certain element contributed by each of its bearing minerals), and many more properties. Even the application of EBSD (electron backscatter diffraction) will make it possible to characterise the distribution of crystal orientations (spherical distributions) or its modal values (spherical directions). Spectral information is also produced by many instruments, and although spectra ar nowadays preferable interpreted in terms of chemical elements, minerals or paragenesis (Chlingaryan et al. 2015) before treatment, one might think of future applications in which core scanning or airborne spectral data are considered as informative on their own in a 3D geomodel. Consider that spectral information is easy and fast to obtain in the operation and thus could help to guide the extraction process and identify ore types during mining and further processing (Nguyen 2013).

Many of these characterisation techniques can be ordered in a chain of methods, where the more advanced methods provide more and more detail but at the price of lower precision, higher costs, and longer aquisition or turnaround times. For instance, XRD, though primarily measuring modal mineralogy, can be used to infer bulk geochemical composition, though with higher uncertainty than directly using XRF. Also, MLA, though primarily measuring grain and particle structures, can provide a modal mineralogy, but at higher costs than XRD. Finally, EBSD allows to characterize crystallites and defects, but can also be used to infer the mineralogical microfabric, albeit at longer measurement times than MLA for a fixed precision.

The other way around, inferring more advanced characteristics of the ore indirectly from cheaper measurements, is in general an inverse problem. Inverse problems are much more difficult to handle and often do not have a unique solution. For instance, inferring modal mineralogy from XRF is an *endmember problem*, and delivers at most equivalence classes of solutions (Tolosana-Delgado et al. 2011; Berry et al. 2011). Interpreting spectra into chemical and mineral compositions often requires as well unmixing the signal obtained as a linear mixture of known endmember spectra. Finally, inferring processing properties from primary properties might require statistical models or machine learning methods to approximate the inverse problem solution (e.g. Matos Camacho et al. (2015) for magnetic susceptibility from MLA data). In summary, each analytical method has a specific role to play, and several methods will be required to appropriately characterise all relevant aspects of the ores.

Another classical class of metrological problems appearing in ore characterisation is *instrumental calibration*, namely the inference of the composition of bulk samples or spots by comparing their signals with the signal obtained from a reference material or *standard* where the property is known, as well as the corresponding uncertainty. The specific challenges for geometallurgy are the high variability of natural materials, difficult to reflect in standards with comparable compositional and physical characteristics (called *matrix matched*), and to measure in a single method. This concerns many of the techniques mentioned before, like XRF, INAA, ICP-MS, PIXE and EPMA.

From the point of view of mathematical geosciences, these problems imply calibration problems, data fusion and consensus building. Data has often been collected during different periods with different instruments at different labs. Seldom all methods were applied to all locations. Different batches need to be made compatible and calibrated against each other. In the authors' opinion, solutions for such problems will require existing concepts and tools and new developments from geodata management, geo-ontology and geoinformatics.

Additionally, local analytics techniques (MLA, QUEMSCAN, X-ray tomography, PIXE, EPMA, Raman) bring their own problems to be solved with mathematical geosciences techniques. It is often very challenging or impossible to acquire standard material homogeneous at micron scale and matrix-matched to the ore samples. Geostatistical models have been proposed for supporting such local calibration efforts (Tolosana-Delgado et al. 2013).

Imaging techniques are also becoming more and more popular, at all spatial scales. More and more methods (hyperspectral satellite- and air-borne, drone-borne imaging, mine face imaging, core scanning, EBSD, MLA, X-Ray-CT, PIXE, ...) acquire images rather than only univariate or compositional information. On large scales, from the drill core to deposit scale, imaging gets a rising importance for the characterisation of the meso- to megastructure of the deposit, because selectivity of ore zones from barren zones during exploration, mining, extraction and waste pre-screening is highly dependent on such structures. If we focus on sub-millimeter scales, processing methods and processing costs react very sensitively to analogous microstructural properties: for instance, the type of intergrowth of minerals strongly conditions the necessary milling to achieve sufficient liberation (Perez-Barnuevo et al. 2013), and milling is one of the most cost intensive processing steps. Many of these methods measure spectral information at each pixel. Various supervised and unsupervised machine learning techniques have been used for mapping spectral information to geometallurgically relevant quantities (Decamp et al. 2015;

Harraden et al. 2016; Nguyen et al. 2016). Image processing analysing structure will thus become more and more relevant in geometallurgy.

Moreover surface imaging techniques like MLA or QUEMSCAN suffer of stereologic degradation: these instruments are devised to characterise geometric properties of 3D bodies, but only observe them on 2D sections. It is well-known that only some 3D properties can be estimated unbiasedly by averaging over their 2D counterparts. This allows e.g. to have certain confidence in properties like volumetric modal mineralogy (estimated from the proportions of pixels of the several minerals on the measured surface), mineral association as the proportion of surface of a mineral in contact with all other minerals (estimated from the proportion of contact lengths on the measured surface) or specific surfaces. But other highly relevant properties, like liberation distribution, grade curves, tonnage curves or particle and grain size distributions suffer significant stereological degradation (Perez-Barnuevo et al. 2012).

Open problems for the next generation of mathematical geoscientists will include, to mention a few, the development of widely accepted local analytics calibration procedures; the propagation of uncertainties through image analysis methods; or the integration of several analytical techniques through consensus-building, e.g. to deliver mutually consistent measurements of bulk mineral and chemical compositions as well as elemental deportment together with their uncertainties out of XRD, XRF, EPMA and MLA measurements of the same sample. Correcting stereological degradation is as well an open issue.

## 33.4 Orebody Modelling

The generation of large scale 3D models of the ore bodies is the classical key contribution of Mathematical Geosciences to the mining business. Nowadays, point and block kriging or simulation for grade variables and indicator-based techniques (indicator kriging, sequential indicator simulation, plurigaussian simulation) for categorical variables are accepted standard techniques. Beyond the framework of Gaussian random fields, cumulant based (Dimitrakopoulos et al. 2010; Minniakhmetov and Dimitrakopoulos 2017) and Copula based (Musafer et al. 2013, 2017) proposals, as well as multiple point geostatistics (MPS) can be found in scientific papers, though their penetration and acceptance in the industry is yet negligible. Multivariate issues are also seldom considered, though compositions (mineral or chemical) are geometallurgically relevant primary variables, and techniques do exist to predict or simulate them at both point (Pawlowsky 1989; Pawlowsky-Glahn and Burger 1992; Pawlowsky-Glahn and Olea 2004; Tolosana-Delgado 2006; Tolosana-Delgado et al. 2011; Mueller et al. 2014) and block support (Tolosana-Delgado et al. 2013) in a fashion consistent with their scale, namely delivering positive and constant-sum predictions/simulations abiding to a relative scale.

The geostatistical treatment of other geometallurgically relevant multivariate scales has received limited to no attention so far by the mathematical geosciences community. The challenges are multiple (Boogaart et al. 2013). Geometallurgical data from EBSD are known to exhibit spherical scales, for which a kriging approach is readily available (Boogaart and Schaeben 2002a, b). One-dimensional distributions are much more abundant, and methodological developments for kriging, cokriging and conditional simulation exist via functional analysis (Menafoglio et al. 2016a, b). Nevertheless, application to the many geometallurgical data with distributional scale still requires theoretical and practical developments. Upscaling of these geometallurgical properties present counter-intuitive characteristics: for instance, a categorical variable at point support gives rise to a compositional variable at block support, and while block kriging is generally thought to reduce uncertainty, block "estimates" of distributional and of categorical variables may very well exhibit higher entropy themselves. With a few exceptions based on geostatistical simulation (Deutsch et al. 2015), downscaling has not yet been systematically considered, but it may become a necessary tool to populate block models with smaller scale granularity, for instance for incorporating information from blast-hole analysis on the 3D models. Finally, the joint consistent modelling of several variables from different scales (for instance modal mineralogy, geochemistry, hardness and lithology) has received limited attention (see Maleki and Emery 2015 for a two-point case study with one continuous and one categorical variable), and only seminal ideas about the combination of Bayesian spaces (Boogaart et al. 2014), multigrid Markov Mesh Models (Stien and Kolbjornsen 2011; Kolbjornsen et al. 2014), generalized linear models and MPS have been presented for discussion (Boogaart et al. 2014).

It has been shown that the conditional distribution of the geostatistical simulation is highly relevant for optimal processing choices (Boogaart et al. 2013). Gaussian geostatistics only delivers that correctly in a Gaussian random field setting. Like with strategic mine planning (Dimitrakopoulos 2011; Goodfellow and Dimitrakopoulos 2017), non-linear simulation methods better reproducing the conditional distributions would thus be more appropriate for geometallurgical optimisation. However so far (April 2017), beyond single categorical variables, no case studies could show the added value of MPS methods in the context of geometallurgy. The fundamental difficulty appears to be producing sufficiently large, stochastically representative training images (Emery and Lantuejoul 2014), a problem made even worse by the many relevant variables, some with multivariate, compositional or distributional scales.

Besides the geometric modelling of the large-scale structure of a deposit, 3D Geomodelling offer also a tool for modelling and simulation of microstructure and texture of the ores. Stochastic simulation of such 3D geomodels of ores might be necessary to appropriately simulate breakage of microstructure by crushing, grinding and milling, as well as to offer an approach to stereological reconstruction. This is so because all these problems require an appropriate description of the geometric spatial relations between the mineral grains, and not just summaries of their composition. However, new concepts, models and techniques have to be developed to link the macroscale described by geostatistics and the microscale, possibly described by stochastic geometry.

Another challenge posed by such multi-scale (in the sense of spatial granularity), multi-scale (in the sense of statistical kinds of data), multi-step (data is added to the models at different times), multi-dimensional geometric modelling of ore bodies is the structuring, management and exploitation of the necessary data to appropriately provide input for the methods used. A more intimate link between geostatistical and geodatabases will be required for that, as flexible and sequential conditioning methods able to incorporate into the conditional distributions data on batches, as they become available. Sequential data assimilation techniques have been successfully used for this task in the assessment of univariate quantities (Wambeke and Benndorf 2017).

## 33.5 Decision Making

Geometallurgy touches on all levels of optimization of the mining operation, from exploration, investment, and strategic mine planning towards the daily operation. Each optimization task can be stated as a *w*-question, and delimits a certain scope of the decision to be taken.

Blending ores from different localities to ensure a stable feed properties for the plant presents the smallest decision scope, as it only changes *where* to mine and not *when* or *how to process*. Having the ability to predict mining and processing behaviour for different feed materials allows to better predict block values or machine time and maintenance requirements. Such better block values can be used in classical strategic mine planning tools for an optimal exploitation of the deposit, that is answering the *when* and *where* issues related to pushbacks and ultimate pit calculations. For this task a statistic model relating the primary geometallurgical properties, with secondary ones is typically enough (Vann et al. 2011). If the processing model is good enough to predict the value as a function of the processing choices, it can be used in conjunction with a geostatistical description of the geometallurgical ore properties to optimize the processing itself either for the whole deposit or each block (optimal adaptive processing) (Turner-Saad 2011; Tolosana-Delgado et al. 2015). Goodfellow and Dimitrakopoulos (2017) shows how blending, strategic mine planning and routing can be optimized together. The optimizability, i.e. the optimal achievable productivity, depends on very basic decisions like the size of selective mining units, available equipment and available data. The overall value of the mine and thus the decision to mine itself depends on all details. Boogaart et al. (2015) shows the relevance of the selective mining unit and the decision strategy for the value of the mine (*how to model*). Boogaart et al. (2016) shows how to quantify these values and the value of the available equipment, determining costs and available processing choices, before the actual mining operation starts. Such calculations are based on geostatistical simulation, and thus allow to optimize the geometallurgical approach (*how to optimize*) and the investment (*what to build*). Boogaart et al. (2016) show the substantial influence of the exploration plan and the data aquisition strategy (e.g. the influence of processing data) on the overall value of the operation and how quantifying the value of information can be used to optimize the geometallurgical exploration strategy. This offers a way to economically justify and timely plan extensive geometallurgical data aquisition campaigns (*what and when to measure*).

All these approaches rely on stochastic optimisation in a geostatistical framework for geometallurgical data combined with a geometallurgical processing model, both based on quantitative ore characterisations. That is, they rely on the mathematical tools described in the preceeding three sections. Applying these techniques is still a major geoinformational challenge including big data management, data fusion, massive parallel computing and real time data management (Jones and Moorhead 2013; Lopez et al. 2016).

## 33.6 Conclusions

Geometallurgy requires substantial geomathematical developments in all the classical fields of mathematical geosciences and geoinformatics. The challenges are beyond the classical solutions, e.g. a truly multivariate, multi-scale Geostatistics honoring non-Gaussian relationships is required; statistical analysis for various scales beyond positive data and compositions is required, in particular distributional data; a full space-time 3D data fusion and fast automated updating of models will be required; there are new challenges to the mathematical background of metrology including issues of local analytics, compositional calibration, and varying material matrices; structural characterisation on several scales from the ore body to the microfabric are needed on a quantitative level from limited 2D stereological data and supportive conditioning information (bulk mineralogy and geochemistry, accessory information on mineral stoichiometry, cristallographic defects, etc.); geostatistical models of the spatial variation of the microstructure throughout the deposits (i.e. a structure Geostatistics) needs to be developed; and so on.

The mathematical challenges of integrating characterisation, stochastic modelling, process simulation and optimisation, and data reconciliation, will extend to manmade and secondary resources (tailing dams, recycling, urban mining) and to the optimisation of other geosystems (water management, ecosystem management, urban ecosystems, the trisystem of energy-minerals-water), hence the lessons learnt from primary ores geometallurgy will be relevant for many fields beyond ore geology and mining. Beyond the classical fields of mathematical geosciences, geometallurgical questions will as well require solutions from mathematical disciplines uncommon at the IAMG, like optimisation, operations research and numerical process modelling. Thus, geometallurgy extends the scope of the IAMG towards these fields. In this way geometallurgy can become the scientific and economic driving force for the next generation of mathematical geosciences and geoinformatics.

## References

AusIMM (2011) First AusIMM international geometallurgical conference. AusIMM
AusIMM (2013) Second AusIMM international geometallurgical conference. AusIMM

AusIMM (2016) Third AusIMM international geometallurgical conference. AusIMM

Bam L, Miller J, Becker M, DeBeer F, Basson I (2016) X-ray computed tomography–determination of rapid scanning parameters for geometallurgical analysis of iron ore. See AusIMM (016), pp 209–219

Becker M, Jardine M, Miller J, Harris M (2016) X-ray computed tomography–a geometallurgical tool for 3d textural analysis of drill core? See AusIMM (2016), pp 231–240

Berry R, Hunt J, McKnight S (2011) Estimating mineralogy in bulk samples. See AusIMM (2011), pp 153–156

Boisvert JB, Rossi ME, Ehrig K, Deutsch CV (2013) Geometallurgical modeling at olympic dam mine, south australia. Math Geosci 45(8):901–925

Boogaart K, Konsulke S, Tolosana-Delgado R (2013) Non-linear geostatistics for geometallurgical optimisation. See AusIMM (2013), pp 253–257

Boogaart KGvd, Schaeben H (2002a) Kriging of regionalized directions, axes, and orientations I. Directions and axes. Math Geol 34(5):479–503

Boogaart K, Schaeben H (2002b) Kriging of regionalized directions, axes, and orientations II. Orientations. Math Geol 34(6):671–677

Boogaart K, Tolosana-Delgado R, Lehmann M, Mueller U (2014) On the joint multipoint simulation of discrete and continuous geometallurgical parameters. See Dimitrakopoulos (2014), pp 1–10

Boogaart K, Tolosana-Delgado, R, Matos Camacho S (2015) The effect of problem formulation on adaptive processing decisions

Boogaart K, Tolosana-Delgado R, Matos Camacho S (2016) Working with uncertainty in adaptive processing optimisation. Canadian Institute of Mining, Metallurgy and Petroleum, pp 1–9

Boogaart KGvd, Egozcue JJ, Pawlowsky-Glahn V (2014) Bayes hilbert spaces. Aust N Z J Stat 0–0. https://doi.org/10.1111/anzs.12074

Boogaart KGvd, Tolosana-Delgado R (2013) Analysing compositional data with R. Springer, Heidelberg, p 280

Boogaart Kvd, Tolosana-Delgado R, Mueller U, Matos Camacho S (2016) How details of the geometallurgical optimisation influence overall value. See AusIMM (2016), pp 303–311

Bye A (2011) Case studies demonstrating value from geometallurgical initiatives. See AusIMM (2011), pp 9–30

Caciagli Warman N (2015) Multi element geochemical modelling for mine planning: Case study from an epithermal gold deposit, pp 266–277

Caracciolo L, Tolosana-Delgado R, Le Pera E, von Eynatten H, Arribas J, Tarquini S (2012) Influence of granitoid textural parameters on sediment composition: implications for sediment generation. Sedimen Geol 280(SI):93–107

Chlingaryan A, Melkumyan A, Murphy R (2015) Classification of hyperspectral imagery using gaussian process with automated identification of the importance of information conveyed by different wavelengths

Coward S, Vann J, Dunham S, Stewart M (2009) The primary-response framework for geometallurgical variables. AusIMM, pp 109–113

Decamp X, Dislaire G, Barnabe P, Pirard E, Germay C (2015) A new approach to drill core scanning by combination of mechanical and optical tests: Preliminary results. In: Andre-Mayer A, Cathelineau M, Muchez P, Pirard E, Sindern S (eds) Mineral Resources in a Sustainable World, vols 1–5, 13th SGA Biennial Meeting on Mineral Resources in a Sustainable World, Nancy, FRANCE, 24–27 August 2015, pp 1395–1397

Delicado P (2008) Comparing methods for dimensionality reduction when data are density functions. In: Martín-Fernández JA, Daunis-i Estadella J (eds) Proceedings compositional data analysis workshop – CoDaWork'08, http://hdl.handle.net/10256/723.UniversitatdeGirona, http://ima.udg.es/Activitats/CoDaWork2008/

Deutsch J, Etsell T, Szymanski J, Deutsch C (2015) Downscaling and multiple imputation of metallurgical variables

Deutsch JL, Palmer K, Deutsch CV, Szymanski J, Etsell TH (2016) Spatial modeling of geometallurgical properties: techniques and a case study. Nat Resour Res 25(2):161–181

Dimitrakopoulos R (2011) Strategic mine planning under uncertainty. J Min Sci 47(2):138–150

Dimitrakopoulos R (ed) (2014) Advances in orebody modelling and strategic mine planning. AusIMM

Dimitrakopoulos R, Mustapha H, Gloaguen E (2010) High-order statistics of spatial random fields: exploring spatial cumulants for modeling complex non-gaussian and non-linear phenomena. Math Geosci 42(1):65–99

Emery X, Lantuejoul C (2014) Can a training image be a substitute for a random field model? Math Geosci 46(1, 2):133–147

Everett J, Howard T (2011) Predicting finished product properties in the mining industry from pre-extraction data. See AusIMM (2011), pp 205–215

Goodfellow R, Dimitrakopoulos R (2017) Simultaneous stochastic optimization of mining complexes and mineral value chains. Math Geosci 49(3, SI):341–360

Harraden C, Berry R, Lett J (2016) Proposed methodology for utilising automated core logging technology to extract geotechnical index parameters. See AusIMM (2016), pp 119–123

Hosseini SA, Asghari O (2015) Simulation of geometallurgical variables through stepwise conditional transformation in sungun copper deposit, Iran. Arab J Geosci 8(6):3821–3831

Jones J, Moorhead C (2013) Geometallurgical communication as a distributed information system. See AusIMM (2013), pp 133–137

Journel AG, Huijbregts CJ (1978) Mining geostatistics. Academic Press, London, p 600

Keeney L, Walters S, Kojovic T (2011) Geometallurgical mapping and modelling of comminution performance at the cadia east porphyry deposit. See AusIMM (2011), pp 73–83

Kolbjornsen O, Stien M, Kjonsberg H, Fjellvoll B, Abrahamsen P (2014) Using multiple grids in markov mesh facies modeling. Math Geosci 46(1, 2):205–225

Lamberg P (2011) Particles—the bridge between geology and metallurgy. In: Conference in minerals engineering, pp 1–16

Lopez A, Barberan A, Alarcon M, Vargas E, Ortiz J, Morales N, Emery X, Egana A, McFarlane A, Friedric C (2016) Progress towards data-driven mine planning via a virtual geometallurgical laboratory. See AusIMM (2016), pp 287–293

Maleki M, Emery X (2015) Joint simulation of grade and rock type in a stratabound copper deposit. Math Geosci 47(4):471–495

Matos Camacho S, Leißner I, Bachmann K, Boogaart K (2015) Inference of phase properties from sorting experiments and mla data

McKay N, Vann J, Ware W, Morley C, Hodkiewicz P (2016) Strategic and tactical geometallurgy—a systematic process to add and sustain resource value. See AusIMM (2016), pp 29–36

Menafoglio A, Secchi P, Guadagnini A (2016a) A class-kriging predictor for functional compositions with application to particle-size curves in heterogeneous aquifers. Math Geosci 48(4):463–485

Menafoglio A, Guadagnini A, Secchi P (2016b) Stochastic simulation of soil particle-size curves in heterogeneous aquifer systems through a bayes space approach. Water Resour Res 52(8):5708–5726

Minniakhmetov I, Dimitrakopoulos R (2017) Joint high-order simulation of spatially correlated variables using high-order spatial statistics. Math Geosci 49(1):39–66

Mueller U, Tolosana-Delgado R, van den Boogaart KG (2014) Simulation of compositional data: a nickel-laterite case study. See Dimitrakopoulos (2014)

Musafer G, Thompson M, Kozan E, Wolff R (2013) Copula-based spatial modelling of geometallurgical variables. See AusIMM (2013), pp 239–246

Musafer G, Thompson M, Kozan E, Wolff R (2017) Copula-based spatial modelling of geometallurgical variables. Nat Resour Res 26(2):223–236

Nguyen A, Jackson J, Nguyen K, Manlapig E (2016) A new semi-automated method to rapidly evaluate the processing variability of the orebody. See AusIMM (2016), pp 145–151

Nguyen K (2013) A new texture analysis technique for geometallurgy. See AusIMM (2013), pp 187–190

Ortiz J, Kracht W, Townley B, Lois P, Cardenas E, Miranda R, Alvarez M (2015) Workflows in geometallurgical prediction: challenges and outlook

Pawlowsky V (1989) Cokriging of regionalized compositions. Math Geol 21(5):513–521

Pawlowsky-Glahn V (2003) Statistical modelling on coordinates. In: Thió-Henestrosa S, Martín-Fernández JA (eds), Proceedings of CoDaWork'03, The 1st compositional data analysis workshop, Girona (E). Universitat de Girona, ISBN 84-8458-111-X, http://ima.udg.es/Activitats/CoDaWork2003/

Pawlowsky-Glahn V, Burger H (1992) Spatial structure analysis of regionalized compositions. Math Geol 24(6):675–691

Pawlowsky-Glahn V, Olea RA (2004) Geostatistical analysis of compositional data. In: DeGraffenreid JA (ed) Studies in mathematical geology, vol 7. Oxford University Press, Oxford

Perez-Barnuevo L, Pirard E, Castroviejo R (2012) Textural descriptors for multiphasic ore particles. Image Anal Stereol 31:175–184

Perez-Barnuevo L, Castroviejo R (2013) Automated characterisation of intergrowth textures in mineral particles. a case study. Miner Eng 52(SI):136–142

Powell M (2013) Utilising orebody knowledge to improve comminution circuit design and energy utilisation. See AusIMM (2013), pp 27–35

Rossi ME, Deutsch C (eds) (2014) Mineral resource estimation. Handbooks of modern statistical methods. Springer, Berlin

Sepulveda E, Dowd PA, Xu C, Addo E (2017) Multivariate modelling of geometallurgical variables by projection pursuit. Math Geosci 49(1):121–143

Shaw W, Khosrowshahi S, Weeks A (2013) Modelling geometallurgical variability-a case study in managing risks. See AusIMM (2013), pp 247–252

Stevens S (1946) On the theory of scales of measurement. Science 103:677–680

Stien M, Kolbjornsen O (2011) Facies modeling using a markov mesh model specification. Math Geosci 43(6):611–624

Tolosana-Delgado R (2006) Geostatistics for constrained variables: positive data, compositions and probabilities. Application to environmental hazard monitoring. Ph. D. Thesis, Universitat de Girona, Spain, p 198

Tolosana-Delgado R, van den Boogaart KG, Fiserova E, Hron K, Dunkl I (2015) Joint compositional calibration: a geochronological example. In: CoDaWork2015, the 6th International workshop on compositional data analysis, L'Escala Girona, Spain, 1–5 June 2015

Tolosana-Delgado R, Boogaart KGvd, Pawlowsky-Glahn V (2011) Geostatistics for compositions. In: Pawlowsky-Glahn V, Buccianti A (eds) Compositional data analysis: theory and applications. Wiley, New Jersey, pp 73–86

Tolosana-Delgado R, Boogaart Kvd, Konsulke S, Scholz A, Matos Camacho S, Christesen C, Rudolph M, Scharf C (2015) Optimizing a stepwise fractionation chain in mineral processing or metallurgy

Tolosana-Delgado R, Eynatten Hv, Karius V (2011) Constructing modal mineralogy from geochemical composition: a geometric-Bayesian approach. Comput Geosci 37(5, SI):677–691

Tolosana-Delgado R, Mueller U, Boogaart KGvd, Ward C, Gutzmer J (2015) Improving processing by adaption to conditional geostatistical simulation of block compositions. J South Afr Inst Min Metall 115(1):13–26

Tolosana-Delgado R, Mueller U, van den Boogaart KG, Ward C (2013) Block cokriging of a whole composition, pp 267–277

Turner-Saad G (2011) A cut-off of liberated adn selected ore minerals optimisation based on the geometallurgical concept. See AusIMM (2011), pp 263–272

Vann J, Jackson J, Coward S, Dunham S (2011) The geomet curve-a model for the implementation of geometallurgy. See AusIMM (2011), pp 35–43

Wambeke T, Benndorf J (2017) A simulation-based geostatistical approach to real-time reconciliation of the grade control model. Math Geosci 49(1):1–37

# Chapter 34
# Data Science for Geoscience: Leveraging Mathematical Geosciences with Semantics and Open Data

**Xiaogang Ma**

**Abstract** Mathematical geosciences are now in an intelligent stage. The freshly new data environment enabled by the Semantic Web and Open Data poses both new challenges and opportunities for the conduction of geomathematical research. As an interdisciplinary domain, mathematical geosciences share many topics in common with data science. Facing the new data environment, will data science inject new blood into mathematical geosciences, and can data science benefit from the achievements and experiences of mathematical geosciences? This chapter presents a perspective on these questions and introduces a few recent case studies on data management and data analysis in the geosciences.

## 34.1 Introduction

The global science community is facing a fresh data environment that never existed before. New generations of sensors, instruments and platforms extend the range of exploration and speed up the frequency of data collection. The quick updates in data storage facilities make it possible to archive and retrieve massive datasets in digital formats. The wide coverage of Internet and World Wide Web services allow researchers to share datasets and communicate with colleagues efficiently both in the office and from the field. As transparency, openness and reproducibility of research results and methods receive increasing attention, the science community is now promoting an open science culture (Nosek et al. 2015) and encouraging actions on open access, open data, open code and open samples (Easterbook 2014; Hey and Payne 2015; McNutt et al. 2016). In the domain of geoscience, significant progress has been achieved on open data, including those emanating from federal agencies such as data services of NASA, USGS, NOAA and community-built data portals such as OneGeology, EarthChem, RRUFF, PANGAEA, PaleoBioDB, and more.

X. Ma (✉)
Department of Computer Science, University of Idaho, 875 Perimeter Drive MS 1010, Moscow, ID 83844-1010, USA
e-mail: max@uidaho.edu

 
B. S. Daya Sagar et al. (eds.), *Handbook of Mathematical Geosciences*,
https://doi.org/10.1007/978-3-319-78999-6_34

A clear trend in open data actions is that the World Wide Web is used as the space for data storage, publication, discovery and access. Data resources on the Web provide convenience for geoscience researchers, and lay out the platform for cross-disciplinary collaboration and new scientific discoveries.

In addition to focused research topics within each discipline, geoscience researchers in the 21st century are now able to tackle more grand research questions (Fig. 34.1) that need broad perspectives, multidisciplinary collaboration and sustained data support. Studies on these questions will lead to the extension of our fundamental knowledge and understanding about the Earth system, which in turn will contribute to the application of geoscience in tackling social and economic issues that are relevant to human welfare. For example, the Future Earth, a ten-year initiative (2015–2025) coordinated by several international organizations, proposed eight key challenges to the global sustainability (Future Earth 2014): water-energy-food nexus, decarbonization, natural assets, cities, rural futures, human health, consumption and production, and social resilience. To grasp these tremendous opportunities and make innovative discoveries, geoscience researchers need the necessary data resources and skills. Although geoscience data are increasingly made available online, due to the heterogeneities inside them, many data are not ready for use by end users. The heterogeneities of geoscience data are reflected in the vast number of subjects, varied data structures and formats, and diverse terminologies (Berg-Cross et al. 2012; Ramachandran et al. 2006; Reitsma and Albrecht 2005). Methods and skills of both data management and data analysis are needed for conducting science within the inspiring and complex data environment of today.

Data management and data analysis are the two key concepts in data science (cf. Schutt and O'Neil 2013), which involves knowledge of library and information science, computer science, mathematics, statistics, and domain-specific disciplines. While the theoretical foundations of data science are still under development (Drineas and Huo 2016), there have already been many applications and

1. How did Earth and other planets form?
2. What happened during Earth's "dark age" (the first 500 million years)?
3. How did life begin?
4. How does Earth's interior work, and how does it affect the surface?
5. Why does Earth have plate tectonics and continents?
6. How are Earth's processes controlled by material properties?
7. What causes climate to change—and how much can it change?
8. How has life shaped Earth—and how has Earth shaped life?
9. Can earthquakes, volcanic eruptions, and their consequences be predicted?
10. How do fluid flow and transport affect the human environment?

**Fig. 34.1** The 10 grand research questions for the 21st century Earth sciences (National Research Council 2008)

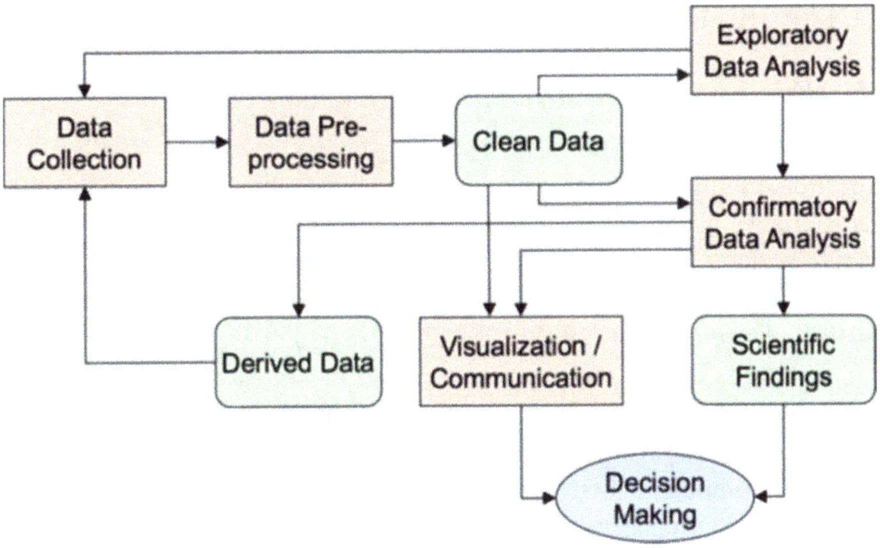


**Fig. 34.2** Primary steps in a data science process. From Schutt and O'Neil (2013) with changes

discussions of data science in recent years (Schutt and O'Neil 2013), and a general process of data science is emerging (Fig. 34.2). The steps and processes in Fig. 34.2 would be familiar to researchers in all disciplines mentioned above, as they are comparable to the widely-adopted hypothesis-driven research method in modern science. Nevertheless, there could remain many questions to be asked as we are now in the "inspiring and complex data environment": Do we have methods and techniques to improve the efficiency in each step? How to create a space and design an approach where researchers from the different disciplines can collaborate and leverage their individual capabilities to achieve a focused objective? What is the feature of data science in a domain-specific context, including geoscience?

Researchers of mathematical geosciences or geomathematics can have a lot to say about their experience and understanding of data science, because mathematical geosciences is a domain with a long history of incorporating knowledge from computer science, mathematics and statistics with geoscience (Agterberg 2014; Bonham-Carter 1994; Loudon 2000; Merriam 2004). Will the latest research progress of data science inject some new blood into the mathematical geosciences; and vice versa, can the methods and experiences in mathematical geosciences contribute to the theoretical developments of data science? The purpose of this chapter is to present a perspective on questions based on a review of the evolution of mathematical geosciences and a summary of the latest discussions of data science within the geoscience community. To support the presented perspective, a few recent case studies will be introduced in the second half of the chapter, with a focus on how data science can help leverage the existing capabilities in geoscience research and achieve new goals.

## 34.2 The Intelligent Stage of Mathematical Geosciences

### 34.2.1 Evolution of Mathematical Geosciences

Retrospection on the evolution of mathematical geosciences will help us understand the characteristics of this discipline as well as the opportunities it faces today. In an informative review, Merriam (2004) summarized the six stages in the development of quantitative geology: Origins (1650–1833), Formative (1833–1895), Exploration (1895–1941), Development (1941–1958), Automated (1958–1982), and Integration (1982–). The three earlier stages, over a period of almost 300 years, made use of various developments in both geoscience and mathematics, and more importantly the co-evolution between them. The latter three stages were characterized by the application of computers, first in geostatistics, simulation and modeling, and the organization of large datasets and later in all aspects of the geoscience workflow, including data capture, manipulation, analysis and documentation. Merriam (2004) also briefly mentioned the Internet and the potential challenges and opportunities in the connected virtual world, and he stated, "There is seemingly no limit to the information and communication revolution."

Indeed, coming to today, which is just about 12 years after Merriam's review paper, geomathematical researchers as well as the broad geoscience community already face the fresh data environment. We now have new instruments for measurement and observation, powerful facilities in data storage and transmission, improved interoperability of online datasets, and effective algorithms for data processing and analysis. New methods and technologies such as big data, open data, machine learning, data mining, data science, semantic web, natural language processing have been increasingly used in geoscience studies. The functionality of computers is being leveraged to a new level, where they are not only capable to represent "what is" known but can also show us "why" and help generate ideas on "how to" explore new findings. Ma (2015) proposed that the mathematical geosciences is now in an Intelligent stage (2014–). Besides these accelerated developments and applications of geomathematical methods within the geoscience disciplines, there are growing needs for using these methods in cross-disciplinary programs to address socio-economic issues that are of public concern (Freeden 2010).

In this intelligent stage, what we can do to leverage mathematical geosciences in various multidisciplinary studies? In this chapter, the author wants to address the need of refreshing our knowledge about the latest progress in open data and data science. For geoscience researchers, especially those who are not familiar with data science, knowing open data will be a key to understanding the general data science process and some featured works using datasets retrieved from the Web.

### 34.2.2 Characteristics of Open Data and Semantic Web

Most geoscience studies are driven by data. The term "open data" reflects people's desire of access to freely available datasets. Some open data are made accessible with specified licenses and copyrights, and others are without any limits or restrictions. The popularity of the Internet and the Web creates a wide space for the implementation of open data. For end users of open data, an issue of extreme concern is the data interoperability (Fig. 34.3). Researchers have discussed the levels of data interoperability from different aspects. The levels in the center of Fig. 34.3 (Brodaric 2007) are from a technical point of view. Systems level is fundamental, which means there should be the necessary protocols (e.g. TCP/IP for the Internet and HTTP for the Web) supporting data discovery and transmission. Syntax and Schematics levels are relevant to the data structures and models, for which an end user should be able to parse and analyze. Semantics level indicates that the meaning of data reflected in data model, terminology and encoding are made readable to machines and thus understandable to users. Pragmatics level means the data are suitable for the user's purpose and can contribute value in applications. The right part of Fig. 34.3 (Ma et al. 2011) explains these technical levels with layman's language, and it also adds that all the technologies and implementations at those levels should be legal and ethical from a point of view of social science.

The Semantic Web (Berners-Lee 2000) provides technological support to each level of data interoperability (Fig. 34.3). For geoscience researchers, the Semantic Web creates a space where datasets can be more efficiently annotated, published, discovered and accessed. The Semantic Web is an extension to the current World Wide Web (Berners-Lee et al. 2001). The Web is now in the transition from a Web of Documents to a Web of Data because of the embedded structures and meanings that did not exist before. Nevertheless, to add structure and meaning to the

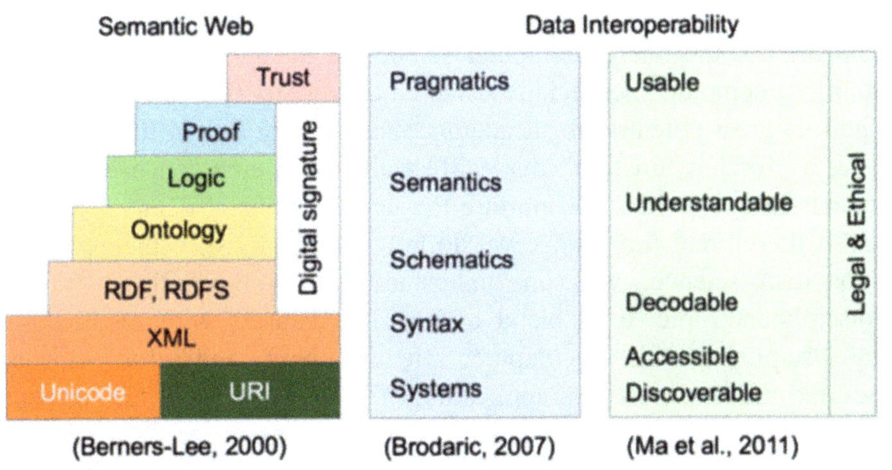


**Fig. 34.3** Levels of data interoperability and a comparison with the architecture of the Semantic Web. From Berners-Lee (2000), Brodaric (2007) and Ma et al. (2011)

information on the Web, definitions and representations of concepts and the interrelationships among concepts are needed (Berners-Lee. 2006). In the Semantic Web such definitions and representations are called ontologies. Each ontology is the formal specification of the shared conceptualization of a domain of study (Gruber 1995). In practice, ontologies can be of different forms, such as glossary, controlled vocabulary, conceptual schemas and detailed logic constraints, depending on the level of detail on conceptual specification. Semantic Web technologies provide the essential elements for modeling and encoding ontologies in machine-readable formats.

In the context of cross-disciplinary program with datasets from various resources and subjects and researchers from different knowledge domains, there could be a large number of ontologies addressing the various needs on knowledge engineering and concept representation. Those ontologies can be implemented to build innovative functions to support the discoverability, accessibility, understandability and usability of open data. For example, there can be projects on categorizing datasets and publications based on their subjects and keywords, recommending datasets or publications to a user based on his research interests, suggesting matches between datasets and scientific questions, and more. The data science domain recently also has proposed the topic "smart data" (Sheth 2014), which aims at using Semantic Web technologies to improve the efficiency in the transformation from massive datasets into actionable information.

### *34.2.3 Methodology of Deploying Data Science in Geoscience*

Although data science has already attracted significant attention in both academia and the industry, the theoretical foundations and technological systems of data science are still under development. In the summary report of a recent NSF-funded workshop (Drineas and Huo 2016), the emergence of data science as a discipline was compared to the rise of computer science in the 1950s along with the wide availability of computers, especially personal computers (PCs). The data deluge of today and its great potential for academia and industry are, in the report authors' language, a "forcing function" that will catalyze the emergence of data science departments in universities and nurture the development of data science as a discipline. At the current time, since we do not have established theoretical foundations for data science, we can understand the core of data science as a cross-disciplinary topic, or a blend of massive datasets with methodologies in existing disciplines, such as computer science, library and information science, statistics and mathematics. The application of data science will further extend the coverage of disciplines to other domains, such as geoscience.

In most scientific researches, including those in geoscience, a general research process includes the following steps: (1) Choose a general direction and do background research; (2) Generate a hypothesis; (3) Conduct experiments and collect data; (4) Analyze data and revise hypothesis; (5) Communicate results. We can compare those steps with the data science process in Fig. 34.2. Both processes follow a direction of data collection, data analysis and result communication, but there are also a few items worthy of further discussion. First, data science often faces a situation in which massive datasets are already in existence while we do not yet have a hypothesis. Second, the data science process addresses a step called data pre-processing, which detects the inconsistent, incomplete and incorrect parts in the datasets and takes actions to ensure the data quality before doing analysis. Data pre-processing is an essential step for large datasets collected from multiple sources. Third, the step of exploratory data analysis (EDA) offers clues for hypotheses in scientific research. EDA is a widely-used approach in statistics, and it covers many methods, such as scatterplot, box plot, residual plot, smoother, bag plot, and more (Brillinger 2011). The term "exploratory" explains the purpose of the method: it is flexible and can help look for things that we believe are not there or to be there (Tukey 1977). EDA helps address the shortage of research hypotheses for massive data that already exist. The functionality of EDA is comparable to the approach of data-driven abductive discovery (Hazen 2014). Abduction means the formation of a plausible explanation for an observation. Charles S. Pierce (1839–1914) viewed abduction as the first stage of scientific reasoning, i.e. to create a hypothesis. Then deduction will be carried out to determine the specific evidence needed to prove the hypothesis. After that, induction will be used to extrapolate a general rule or principle from the findings. Hazen (2014) summarized that abduction is to discover what we do not know we do not know, while deduction and induction are to discover what we know we do not know. This is comparable to Tukey's point of view on EDA (Tukey 1977).

One of the most significant challenges to deploy data science in geoscience is to create a space (physical and/or virtual) and establish an approach so that researchers from different disciplines can talk to each other. Science of today is highly compartmented into disciplines and there are considerable gaps between these, as reflected by differences in scientific subjects, research methods, terminologies used and even styles of working. The challenge of cross-disciplinary collaboration is like encouraging people to step out from their "comfort zones". Researchers in geoinformatics (Fox and McGuinness 2008; Ma et al. 2014b) have proposed a method called use case-driven iterative approach, and have successfully implemented it to facilitate the collaboration between data scientists and domain scientists in several projects. Each use case is a description of the process of a focused task. It can be used to identify scientific questions to ask, resources to be used to answer these questions and methods to be implemented to determine the answer. Through the documentation and analysis of a use case, data scientists and domain scientists (e.g. geologists) can understand the needs and aims of each other. As each use case is a focused small task, the collaborative team can spend a relatively short time to achieve the goal, and then can review, update and move on to the next use case. The process is iterative until the overall objective of a research program is realized.

## 34.3 Case Studies of Data Science in Geoscience

When applying data science to leverage current geoscience studies, the focus or highlight can consist of one or a few steps, depending on the target aimed at. For example, the target can be improving data discoverability and accessibility by updating building blocks and frameworks in the cyberinfrastructure. It can also be focused on finding patterns within massive datasets such as those from literature legacy or crow-sourcing databases. In this section, a few recent efforts and case studies will be introduced.

### *34.3.1 Coordinating Standards to Improve Data Interoperability*

In the domain of geoscience, a few recent achievements on data standards and their implementation were led by CGI-IUGS (http://www.cgi-iugs.org), the Commission for the Management and Application of Geoscience Information within the International Union for Geological Sciences. GeoSciML was proposed as a markup language for the exchange of general geoscience information on the Web (Sen and Duffy 2005). GeoSciML was built on top of the Geography Markup Language (GML) and the eXploration and Mining Markup Language (XMML). The first geoscience subjects covered in GeoSciML included boreholes and structural geology. Raw datasets such as those in geologic maps can be transformed into GeoSciML formats once the mapping between the original data structure and the GeoSciML schema is set up. This makes it easier for data exchange and sharing among organizations and nations. GeoSciML was successfully implemented in the OneGeology project (Jackson and Wyborn 2008). On the front end of the OneGeology data portal (http://portal.onegeology.org), users can access geologic map services in a standard data structure. At the back end of the portal, there are multiple data providers, distributed data servers and different data structures. GeoSciML acts as a mediator between those heterogeneous structures and improves the data interoperability. Another significant contribution from CGI-IUGS is the multi-lingual geoscience vocabularies. Initial projects on geologic time and rock type vocabularies were applied in the OneGeology-Europe project to harmonize geologic maps from around 20 European countries (Laxton et al. 2010). Standards derived from those vocabularies also became a part of INSPIRE, the Infrastructure for Spatial Information in Europe (http://inspire.jrc.ec.europa.eu).

Such efforts on data standards are an essential part of informatics, especially applied informatics that has a domain specific background. Comparing with the geoscience community at large, the number of people working on geoinformatics is low. The value and gains that data standard work can provide are often not fully understood within the geoscience community (Jackson and Wyborn 2008). The situation has been changing in recent years since the value of data science was recognized by increasingly more geoscience researchers. For instance, besides GeoSciML, CGI-IUGS also has developed EarthReousrceML for the exchange of information on mineral occurrences, mines and mining activity. CGI-IUGS's Terminology Working Group has published additional standardized vocabularies. The geoscience community has also collaborated with standard organizations to improve the visibility of data standard outputs. In 2017, GeoSciML was published as a standard of the Open Geospatial Consortium (OGC) (OGC 2017), making it one of the first domain-specific standards in OGC. Geoinformatics researchers also take the lead in coordinating data standards among different scientific disciplines. In 2016, CODATA, the International Council for Science's Committee on Data for Science and Technology, set up a task group on coordinating data standards amongst scientific unions (http://www.codata.org/task-groups/coordinating-data-standards). The aim of the group is to take stock of the progress on disciplinary data standards in different scientific unions, recognize the best practices and coordinate the development of future work. Data standards provide the basic-level technical support when we collect and analyze datasets in cross-disciplinary projects. They significantly reduce the workload on data pre-processing and data cleansing in a data science process (Fig. 34.2).

### 34.3.2 Openness, Provenance and Reproducibility of Research

Provenance and reproducibility are both regarded as important research topics in data science (Drineas and Hou 2016), and they are also essential parts of open science. The literal meaning of provenance is the origin of something. In data science, documenting provenance involves the annotation and interconnection of a network of research activities, people, organizations and resources involved in the production of scientific findings (Ma et al. 2014a). In 2013, the Semantic Web community released an ontology called PROV-O (Lebo et al. 2013). The three top classes Entity, Activity and Agent in PROV-O are easy to understand. The ontology also covers a list of subclasses and relationships that can be applied in domain specific applications. A recent successful implementation of PROV-O is the Global Change Information System (GCIS) (Tilmes et al. 2013), which is part of the U.S. Global Change Research Program (USGCRP, http://www.globalchange.gov). USGCRP is a multi-agency research program to "assist the Nation and the world to understand, assess, predict, and respond to human-induced and natural processes of global change." Every four or five years, USGCRP releases a National Climate Assessment Report with the latest scientific findings on different aspects global change. The most recent one was released in 2014. The initial aim of GCIS is to present the 2014 report and to incorporate integrated access to interlinked resources underpinning that report. The long-term goal of GCIS is to be a web-based source of authoritative, accessible, usable and timely information about global change.

Semantic Web technologies, including PROV-O, were applied in the design and development of GCIS. The project included four major parts: categorization, annotation, identification and linking (Ma et al. 2014a), which are coherent within the architecture of the Semantic Web (Berners-Lee 2000). With the well-documented provenance information on the GCIS website (https://data.globalchange.gov), users will be able to conduct innovative research on provenance tracing data mining. For example, they can seek answers for the question: What is NASA's contribution to the sea-level rise scenarios in the 2014 National Climate Assessment Report?

Reproducibility in data science and open science includes at least two levels of meaning. The first is replicability of a research output by using the datasets and methods in the research. The second is the derived value, which means the open datasets and methods from that research can be reused in new research and make substantial contributions (Beaulieu et al. 2017). To improve the reproducibility of scientific research, several technical frameworks can be applied and/or adapted, such as workflow platforms and provenance documentation. In a recent study about reproducible marine ecosystem assessment (Ma et al. 2017), the PROV-O ontology was extended and implemented in the Jupyter Notebook (http://jupyter.org) to capture and interconnect information from various resources in a scientific research project. Jupyter Notebook is an open-source web application that can be used to create workflow documents with codes, formulas, tables, diagrams, interactive visualizations and descriptive text. The developed ontology further enhanced the function of the platform in capturing and presenting scientific provenance information. The work was used in the Ecosystem Assessment Program of the U.S. NOAA Northeast Fisheries Science Center to support assessment reports of Large Marine Ecosystems. In the implementation, a user works within the Jupyter Notebook to write codes and text for data input, analysis, output and documentation. Once the notebook is completed, the provenance information is automatically captured using the structure defined in the ontology. The collected provenance information is machine-readable and can be archived for later use, such as verifying steps and outputs in the workflow or retrieving raw datasets used in any given step.

### *34.3.3 Leveraging Geoscience Data Legacy for New Discovery*

Geoscience is a domain with abundant literature resources, and much useful information can be extracted from the data legacy. A recent study, originally called PaleoDeepDive (Peters et al. 2014) and now GeoDeepDive (https://geodeepdive.org), has demonstrated the significant value of geoscience publication archives through the application of machine learning and data mining technologies. The domain of focus in GeoDeepDive is paleontology and its aim is to detect and extract fossil occurrence information from the massive scientific literature. The work leverages methods in natural language processing, entity recognition and extraction and knowledge graph construction to improve the efficiency of document processing and the quality of output datasets. In several complicated data extrication and reasoning tasks, the outputs of GeoDeepDive were comparable to the results collected by human experts of geologic history (Peters et al. 2014). Most recently, several publishers and research organizations have set up partnerships with GeoDeepDive and provided a huge number of publications for processing. By middle April 2017, the team has already processed more than 3.2 million documents. The extracted fossil records and their interrelationships can provide useful updates to existing databases, such as the Paleobiology Database (PBDB, https://paleobiodb.org/). PBDB, in turn, has set up interfaces and libraries such as those for Web-based data query and retrieval (Peters and McClennen 2015) and the R environment (Varela et al. 2015). These projects build up channels through which any geoscience researcher can easily access datasets of interest and integrate them with other datasets in their own projects.

A project ongoing in the author's group is about using an ontology to help integrate datasets from PBDB with geologic map services provided by USGS and, thus, to build an enriched data portal where users can discover and access more information. Previous works already have shown the functionality of ontology and data visualization in geoscience data services (Ma et al. 2012). In the ongoing project the focus is an ontology for the regional geologic time scale of North America, in addition to the established ontology for the global geologic time scale (Cox and Richard 2015). The geologic time scale of North America has unique classification and terminology for the time intervals at the Epoch and Age levels; for the levels of Eon, Era and Period it shares the architecture with the global standard. As the terminology in the regional standard has been used in geoscience research of the North American region, specific terms in the regional standard can now also be used as keywords in data search, such as in queries sent to PBDB. In the ontology for the regional geologic time scale of North America, detailed information on all time intervals and their relationships were captured and represented in a machine-readable format. A Web-based visualization was then developed for the ontology, and interactive functions were developed to deploy the visualization as a control panel for data search. When a user clicks a time term in the panel, a query will be sent to PBDB, and the retrieved fossil records from PBDB will be plotted in a map window. Our project also set up connections to the USGS data services, so the user can load geologic map layers onto the map window and browse the background geologic information of a location where a fossil was discovered. The multi-source information has the potential to stimulate discussion among users and help them propose new research questions.

### *34.3.4 Cross-Disciplinary Collaboration for Innovative Discoveries*

In early 2015, a research project focused on the co-evolution of geo- and biospheres was kicked off at the Carnegie Institution of Washington (http://dtdi.carnegiescience.edu).

The researchers in that project are from several universities and institutions and are with diverse knowledge backgrounds, making the research a real cross-disciplinary collaboration. The project proposed to deploy a data-driven abductive approach to discover patterns in the evolution of Earth's environment. A major task in the early stage of the project is to set up a Deep-Time Data Infrastructure (DTDI), which includes the enrichment of attributes (e.g. age information) in existing geo- and bio-databases, connections among geo-databases of petrology, mineralogy and geochemistry, the linkage between geo- and bio-databases, and open access and dissemination protocols for the built data infrastructure. Many open access data resources were considered for DTDI, including rruff.info (mineral species and properties), mindat.org (mineral species and localities), earthref.org (geochemistry and geomagnetism), geokem.com (igneous rock chemistry), metpetdb.rpi.edu (metamorphic petrology), earthchem.org (geochemistry, geochronology, petrology), vamps.mbl.edu (subsurface microbial ecosystem), pdb.org (protein structures), paleobiodb.org (paleobiology), and more. The user case-driven iterative method mentioned in Sect. 34.2.3 has been implemented to organize meetings and promote collaborations among researchers in the group. While the project is still ongoing, several interesting findings have already been achieved. One of them is the pattern of Large Number of Rare Events (LNRE) among the mineral species frequency distribution (Hystad et al. 2015). The work used the records of mineral species, localities and observations (species-locality pairs) from mindat.org and discovered the LNRE pattern. By extrapolating the domain of observation to be about four times the current size, the result in the LNRE model showed that there are about 1,500 new mineral species to be discovered. From that work, further studies on the population probabilities of all mineral species lead to the characterization of Earth-like planets, such as the Mars (Hystad et al. 2017).

As an affiliated scientist in the project mentioned above, the author led a project of using data visualization to study the co-relationships between mineral-forming elements and mineral species. The first study focused on a list of 30 key elements chosen by the research team (Ma et al. 2016). First, we built a 30 × 30 × 30 matrix and visualized it in a three-dimensional coordinate system, which made the matrix a fundamental framework to fill in records. Along each axis in this matrix we plotted the same arranged list of 30 elements as indices. Each cell in the matrix was first filled with the raw number of minerals in which elements X, Y, and Z coexist. A color spectrum was then applied to render each cell according to the value of the number in it. The process was intuitive, and the output in the three-dimensional matrix already showed interesting patterns in the co-relationships between elements and minerals. The visualized matrix was developed to be interactive in a web browser. Researchers can rotate the matrix and zoom into see details of a part, highlight a certain cell and see attributes in it, and slice one or more planes out from the matrix to see two-dimensional patterns. In another study, we extended the scale to all the 72 mineral-forming elements and constructed a 72 × 72 × 72 matrix. We then applied a chi-squared test to generate values to be filled and visualized in that matrix (Hummer et al. 2016). The mineralogical research question in that study was "Does the presence of element Z affect the correlation between elements X and Y in mineral species, and is the effect positive or negative?" Besides the completed case studies, many other interesting projects can be further developed with the three-dimensional matrix. For example, we can add data on electronegativity, ionic radius, atomic number, period, crustal abundance, etc. as associated parameters to each axis and test for different clustering of elements based those parameters.

## 34.4 Concluding Remarks

Mathematical geosciences are now in an intelligent stage. As a research domain, mathematical geosciences share many topics in common with the data science of today. A topic of great interest in deploying data science for geoscience is how to generate research questions or hypotheses when massive datasets are already in existence. In this chapter, the role of exploratory data analysis was analyzed for that purpose, and it was compared with the data-driven abductive approach. Semantic Web and Open Data create a freshly new data environment for conducting geo-mathematical studies. The Web is built as an open space where Anyone can say Anything on Any topic. The Semantic Web aims to facilitate data Interoperability on the Web, to improve Interactivity between humans and machines, and to inspire Intercreativity for exploring new things. For informatics, a major objective is to present the Right information to the Right person in the Right way. We can use the acronym AIR3 to represent those nine words with initial capital letters. AIR3 presents a broad vision of deploying data science for geoscience in the context of the Semantic Web and Open Data. To put this into practice, we need to create a physical and/or virtual space and implement an approach where researchers from different disciplines can step out from their 'comfort zones', talk to each other, and collaborate on focused research topics.

**Acknowledgements** This work was partly supported by W. M. Keck Foundation, the National Science Foundation (NSF) through the NSF Idaho EPSCoR Program (award number IIA-1301792) and by the University of Idaho ORED 2017 Seed Grant Program.

## References

Agterberg F (2014) Geomathematics: theoretical foundations, applications and future developments. Springer, Cham, Switzerland, 553 pp

Beaulieu SE, Fox PA, Di Stefano M, Maffei A, West P, Hare JA, Fogarty M (2017) Toward cyberinfrastructure to facilitate collaboration and reproducibility for marine integrated ecosystem assessments. Earth Sci Inf 10(1):85–97

Berg-Cross G, Cruz I, Dean M, Finin T, Gahegan M, Hitzler P, Hua H, Janowicz K, Li N, Murphy P, Nordgren B, Obrst L, Schildhauer M, Sheth A, Sinha K, Thessen A, Wiegand N, Zaslavsky I (2012) Semantics and ontologies for EarthCube. In: Janowicz K, Kessler C, Kauppinen T, Kolas D, Scheider S (eds) Proceedings of the workshop on GIScience in the big data age 2012, Columbus, OH, 5 pp

Berners-Lee T (2000) Semantic web on XML. Presentation at XML 2000 conference, Washington DC. http://www.w3.org/2000/Talks/1206-xml2k-tbl. Accessed 15 July 2013

Berners-Lee T (2006) Linked Data. http://www.w3.org/DesignIssues/LinkedData.html. Accessed on 02 July 2013

Berners-Lee T, Hendler J, Lassila O (2001) The semantic web. Sci Am 284(5):34–43

Bonham-Carter GF (1994) Geographic information systems for geoscientists: modeling with GIS. Pergamon, Elsevier Science Ltd., Kidlington, UK, 398 pp

Brillinger DR (2011) Exploratory data analysis. In: Badie B, Berg-Schlosser D, Morlino L (eds) International encyclopedia of political science. SAGE Publications, Thousand Oaks, CA, pp 530–537

Brodaric B (2007) Geo-pragmatics for the geospatial semantic web. Trans GIS 11(3):453–477

Cox SJ, Richard SM (2015) A geologic timescale ontology and service. Earth Sci Inf 8(1):5–19

Drineas P, Huo X (2016) NSF workshop report: theoretical foundations of data science (TFoDS), Arlington, VA, 20 pp. http://www.cs.rpi.edu/TFoDS/TFoDS_v5.pdf

Easterbrook SM (2014) Open code for open science? Nat Geosci 7(11):779–781

Fox P, McGuinness DL (2008) TWC semantic web technology. http://tw.rpi.edu/web/doc/TWC_SemanticWebMethodology. Accessed on 14 Apr 2017

Freeden W (2010) Geomathematics: its role, its aim, and its potential. In: Freeden W, Nashed MZ, Sonar T (eds) Handbook of geomathematics. Springer, Berlin, Heidelberg, pp 3–42

Future Earth (2014). Future earth 2025 vision. Future earth, Paris, France, 6 pp. http://www.futureearth.org/sites/default/files/future-earth_10-year-vision_web.pdf

Gruber TR (1995) Toward principles for the design of ontologies used for knowledge sharing. Int J Hum-Comput Stud 43(5–6):907–928

Hazen RM (2014) Data-driven abductive discovery in mineralogy. Am Miner 99(11–12):2165–2170

Hey T, Payne MC (2015) Open science decoded. Nat Phys 11(5):367–369

Hummer DR, Hazen RM, Ma X, Golden JJ, Downs RT, Liu C, Morrison SM, Meyer M (2016) Quantifying and visualizing earth's mineral chemistry through geologic time. GSA Annual Meeting, Denver, Colorado, USA

Hystad G, Downs RT, Hazen RM (2015) Mineral species frequency distribution conforms to a large number of rare events model: prediction of earth's "missing" minerals. Math Geosci 47:647–661

Hystad G, Downs RT, Hazen RM, Golden JJ (2017) Relative abundances of mineral species: a statistical measure to characterize earth-like planets based on earth's mineralogy. Math Geosci 49(2):179–194

Jackson I, Wyborn L (2008) One planet: OneGeology? The Google Earth revolution and the geological data deficit. Environ Geol 53(6):1377–1380

Laxton J, Serrano JJ, Tellez-Arenas A (2010) Geological applications using geospatial standards–an example from OneGeology-Europe and GeoSciML. Int J Digit Earth 3(S1):31–49

Lebo T, Sahoo S, McGuinness D (2013) PROV-O: the PROV ontology. http://www.w3.org/TR/prov-o/

Loudon TV (2000) Geoscience after IT: a view of the present and future impact of information technology on geoscience. Elsevier, Oxford, 142 pp

Ma X (2015) Geoinformatics in the semantic web. In: Schaeben H, Delgado RT, van den Boogaart KG, van den Boogaart R (eds) Proceedings of IAMG 2015, Freiberg, Germany, pp 18–26

Ma X, Asch K, Laxton JL, Richard SM, Asato CG, Carranza EJM, van der Meer FD, Wu C, Duclaux G, Wakita K (2011) Data exchange facilitated. Nat Geosci 4(12):814

Ma X, Beaulieu SE, Fu L, Fox P, Di Stefano M, West P (2017) Documenting provenance for reproducible marine ecosystem assessment in open science. In: Diviacoo P, Glaves HM, Leadbetter A (eds) Oceanographic and marine cross-domain data management for sustainable development. IGI Global, Hershey, PA, USA, pp 100–126

Ma X, Carranza EJM, Wu C, van der Meer FD (2012) Ontology-aided annotation, visualization and generalization of geological time scale information from online geological map services. Comput Geosci 40(3):107–119

Ma X, Fox P, Tilmes C, Jacobs K, Waple A (2014a) Capturing and presenting provenance of global change information. Nat Clim Change 4(6):409–413

Ma X, Zheng JG, Goldstein J, Zednik S, Fu L, Duggan B, Aulenbach S, West P, Tilmes C, Fox P (2014b) Ontology engineering in provenance enablement for the National Climate Assessment. Environ Model Softw 61:191–205

Ma X, Hummer D, Hazen RM, Golden JJ, Fox P, Meyer M (2016) Showing co-relationships between elements and minerals in a three-dimensional matrix. GSA Annual Meeting, Denver, Colorado, USA

McNutt M, Lehnert K, Hanson B, Nosek BA, Ellison AM, King JL (2016) Liberating field science samples and data. Science 351(6277):1024–1026

Merriam D (2004) The quantification of geology: from abacus to pentium: a chronicle of people, places, and phenomena. Earth Sci Rev 67(1–2):55–89

National Research Council (2008) Origin and evolution of Earth: research questions for a changing planet. National Academies Press, Washington, DC, 150 pp

Nosek BA, Alter G, Banks GC, Borsboom D, Bowman SD, Breckler SJ, Buck S, Chambers CD, Chin G, Christensen G, Contestabile M, Dafoe A, Eich E, Freese J, Glennerster R, Goroff D, Green DP, Hesse B, Humphreys M, Ishiyama J, Karlan D, Kraut A, Lupia A, Mabry P, Madon T, Malhotra N, Mayo-Wilson E, McNutt M, Miguel E, Paluck EL, Simonsohn U, Soderberg C, Spellman BA, Turitto J, VandenBos G, Vazire S, Wagenmakers EJ, Wilson R, Yarkoni T (2015) Promoting an open research culture. Science 348(6242):1422–1425

OGC (2017). OGC Geoscience Markup Language 4.1 (GeoSciML). Open Geospatial Consortium, 234 pp. http://www.opengeospatial.org/standards/geosciml

Peters SE, Zhang C, Livny M, Ré C (2014) A machine reading system for assembling synthetic paleontological data-bases. PLoS One 9:e113523. https://doi.org/10.1371/journal.pone.0113523

Peters SE, McClennen M (2015) The paleobiology database application programming interface. Paleobiology 42(1):1–7

Ramachandran R, Rushing J, Li X, Kamath C, Conover H, Graves S (2006) Bird's-eye view of data mining in geosciences. In: Sinha AK (ed) Geoinformatics: data to knowledge, geological society of America special papers, 397. The Geological Society of America, Boulder, CO, pp 235–247

Reitsma F, Albrecht J (2005) Modeling with the semantic web in the geosciences. IEEE Intell Syst 20(2):86–88

Schutt R, O'Neil C (2013) Doing data science: straight talk from the frontline. O'Reilly Media, Inc. Sebastopol, CA, 375 pp

Sen M, Duffy T (2005) GeoSciML: development of a generic geoscience markup language. Comput Geosci 31(9):1095–1103

Sheth A (2014) Transforming big data into smart data: deriving value via harnessing volume, variety, and velocity using semantic techniques and technologies. In: Proceedings of 2014 IEEE 30th international conference on data engineering (ICDE), Chicago, IL, pp 2–2

Tilmes C, Fox P, Ma X, McGuinness D, Privette AP, Smith A, Waple A, Zednik S, Zheng J (2013) Provenance representation for the national climate assessment in the global change information system. IEEE Trans Geosci Remote Sens 51(11):5160–5168

Tukey JW (1977) Exploratory data analysis. Addison-Wesley, Reading, PA, 688 pp

Varela S, González-Hernández J, Sgarbi LF, Marshall C, Uhen MD, Peters S, McClennen M (2015) paleobioDB: an R package for downloading, visualizing and processing data from the Paleobiology Database. Ecography 38(4):419–425

# Chapter 35
# Mathematical Morphology in Geosciences and GISci: An Illustrative Review

**B. S. Daya Sagar**

**Abstract** Georges Matheron and Jean Serra of the Centre of Mathematical Morphology, Fontainebleau founded Mathematical Morphology (MM). Since the birth of MM in the mid 1960s, its applications in a wide ranging disciplines have illustrated that intuitive researchers can find varied application-domains to extend the applications of MM. This chapter provides a concise review of application of Mathematical Morphology in Geosciences and Geographical Information Science (GISci). The motivation for this chapter stems from the fact that Mathematical Morphology is one of the better choices to deal with highly intertwined topics such as retrieval, analysis, reasoning, and simulation and modeling of terrestrial phenomena and processes. This chapter provides an illustrative review of various studies carried out by the author over a period of 25 years—related to applications of Mathematical Morphology and Fractal Geometry—in the contexts of Geosciences and Geographical Information Science (GISci). However, the reader is encouraged to refer to the cited publications to gather more details on the review provided in an abstract manner.

## 35.1 Introduction

A basic understanding of many geoscientific and geoengineering challenges across multiple spatial and/or temporal scales of terrestrial phenomena and processes is among the greatest of challenges facing contemporary sciences and engineering. Many space-time models explaining phenomena and processes of terrestrial relevance were descriptive in nature. Earlier, several toy models were developed via classical mathematics to explain possible phases in dynamical behaviors of complex systems. With the advent of computers with powerful graphics facilities, about three decades ago the interplay between numerical methods (generated via classical

B. S. Daya Sagar (✉)
Systems Science and Informatics Unit, Indian Statistical Institute-Bangalore Centre, 8th Mile, Mysore Road, RVCE PO, Bengaluru 560059, India
e-mail: bsdsagar@isibang.ac.in



B. S. Daya Sagar et al. (eds.), *Handbook of Mathematical Geosciences*,
https://doi.org/10.1007/978-3-319-78999-6_35

equations explaining the behaviors of dynamical systems) and graphics was shown to exist. That progress provided the initial impetus to visualize the systems' spatial and/or temporal behaviors that exhibit simple to complex patterns on graphical screens. One of the efficient ways of understanding the dynamical behavior of many complex systems of nature, society and science is possible through data acquired at multiple spatial and temporal scales. Data related to terrestrial (geophysical) phenomena at spatial and temporal intervals are available in numerous formats. The utility and application of such data could be substantially enhanced through related technologies documented in edited volumes and monographs of the recent past (Sagar 2001a, b, c, d, 2005a, b, 2009, 2013; Sagar and Rao 2003; Sagar et al. 2004; Sagar and Bruce 2005; Sagar and Serra 2010; Najman et al. 2012).

To understand the dynamical behavior of a phenomenon or a process, development of a good spatiotemporal model is essential. To develop a good spatiotemporal model, well-analyzed and well-reasoned information that could be extracted/retrieved from spatial and/or temporal data are important ingredients. Figure 35.1 shows a schematic illustrating the key links between the various phases where the involvement of Mathematical Morphology becomes obvious from the studies to be shown later in the chapter.

Mathematical Morphology—founded by Georges Matheron (1975) and Jean Serra (1982) has shown great impact in various fields including Geosciences and GISci—is one of the better choices to deal with all these key aspects mentioned. Mathematical morphology was founded by Georges Matheron (Agterberg 2001, 2004; Serra 1982, 1988). There are numerous representative publications related to mathematical morphology, to name a few: Serra (1982, 1988), Sternberg (1986), Beucher (1990, 1999), Soille (2003), Najman and Talbot (2010), Sagar (2013). Most notably, the comment on the issue of "What do Mathematical Geoscientists

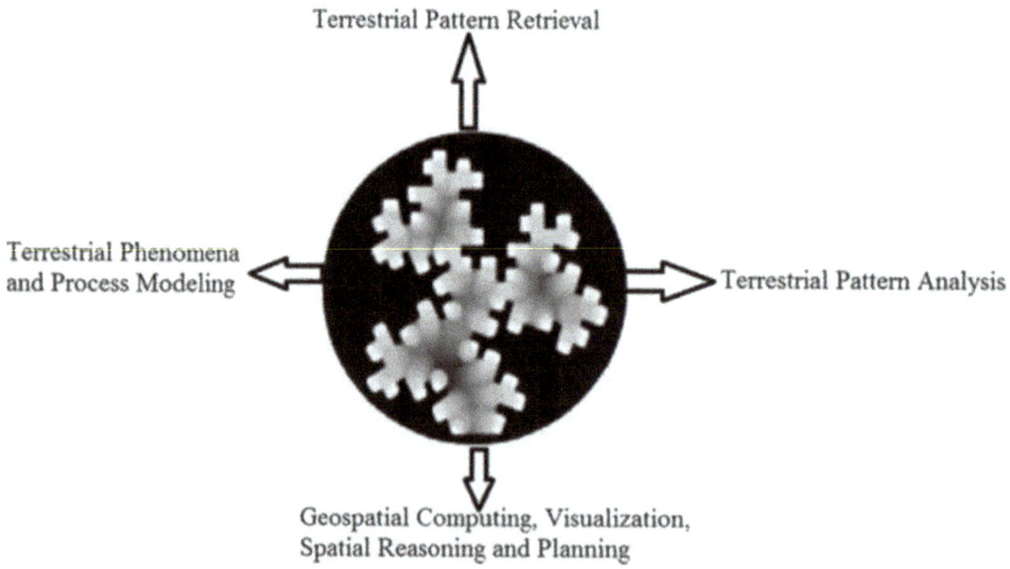


**Fig. 35.1** Mathematical morphology applications in several phases of studies of relevance to geosciences and geographical information science

**Table 35.1** Successful applications of MM transformations in geosciences, geomorphology, GISci-major references

| Morphological operator | Application domain | Major references |
|---|---|---|
| Binary and grayscale morphological erosion, dilation, opening, closing, multiscale morphological operations | Petrology, GISci, geosciences, remote sensing | Serra (1982), Sagar (2013), Brunet and Sills (2017), Beucher (1990, 1999) |
| Geodesic morphological operations | Remote sensing, GISci, geography, petrology | Lantuejoul (1978), Lantuejoul and Beucher (1981), Sagar and Lim (2008a, b), Challa et al. (2016) |
| Hit-or-miss transformation | Geomorphology, hydrology | Serra (1982), Tay et al. (2005a, b, c) |
| Morphological thinning, thickening, pruning | Hydrology, cartography | Soille (2003), Sagar (2013) |
| Morphological skeletonization | Cartography, hydrology, geomorphology | Sagar et al. (2000, 2003a, b), Soille (2003) |
| Skeletonization by zones of influence and weighted skeletonization by zones of Influence | Cartography, hydrology, geomorphology | Beucher (1990), Rajashekara et al. (2012), Sagar (2014a, b) |
| Granulometries and anti-granulometries | Petrology, geomorphology, hydrology | Serra (1982), Maragos (1989), Sagar (2013), Tay et al. (2005a, b, c 2007), Vardhan et al. (2013) |
| Morphological distances, hausdorff dilation (erosion) distances | GISci, limnology, biogeography, spatial planning | Serra (1988), Sagar (2010, 2013), Sagar and Lim (2015a, b) |
| Morphological interpolations and extrapolations | Geophysics, atmospheric science, geology, remote sensing, cartography | Sagar (2010) Brunet and Sills (2017), Rajashekara et al. (2012), Sagar (2014a, b), Sagar and Lim (2015a, b) |
| Watershed transformation | Hydrology, remote sensing, mapping, borehole studies, seismic data processing | Meyer (1980), Beucher and Meyer (1992), Rivest et al. (1992), Sagar (2007) |

Do?" made by Harbaugh (2014) includes the importance of mathematical morphology of geological features in making predictions. In this chapter we outline the successful applications of the most important concepts of mathematical morphology (Table 35.1) in the context of geosciences and Geographical Information Science (GISci).

While perceiving the terrestrial surfaces including geophysical and geomorphic basins (e.g. using Digital Elevation Models, Digital Bathymetric Models, cloud fields, microscale rock porous media etc.) as functions, planar forms (e.g. topographic depressions, water bodies, and threshold elevation regions, hillslopes) as sets, and abstract structures (e.g. networks and watershed boundaries) as skeletons, we make attempts to unravel key links for better understanding spatiotemporal behaviors of several terrestrial and/or spatial phenomena and processes between the following coherent aspects: (i) terrestrial pattern retrieval, (Sect. 35.2) (ii) terrestrial pattern analysis, (Sect. 35.3) (iii) simulation and modeling, (Sect. 35.4) and (iv) geocomputing, visualization, spatial reasoning and planning (Sect. 35.5).

## 35.2 Terrestrial Pattern Retrieval

Retrieving relevant information from precisely acquired spatial-temporal data of varied types about a specific complex system is a basic prerequisite to understand the spatial-temporal behavior of a system. Retrieval of information from a available spatiotemporal data acquired from a wide range of sources and a variety of formats, opens new horizons to the spatial statistical and geoscience communities. We have developed original spatial algorithms based on non-linear morphological transformations for retrieval of unique geophysical networks, mountain objects, segmentation of various geophysical objects, and pairing the geophysical spatial fields based on certain similarities (Sagar et al. 2000, 2003a, b; Sagar and Chockalingam 2004; Sathymoorthy et al. 2007; Chockalingam and Sagar 2003; Lim and Sagar 2008a, b; Lim et al. 2009, 2011; Sagar and Lim 2015a, b; Danda et al. 2016).

### *35.2.1 Mathematical Morphology in Extraction of Unique Topological Networks*

In contrast to other recent works, which have focused on extraction of channel networks via algorithms that fail to precisely extract networks from non-hilly regions (e.g. tidal regions), the algorithms we proposed can be generalized for application to both hilly (e.g. fluvial) and non-hilly (e.g. tidal) terrains, and also pore connectivity networks. These algorithms concerning the framework to extract multiscale geomorphologic networks via systematically decomposing elevation surfaces and/or decomposed threshold elevation regions into their abstract structures lead to valley and ridge connectivity networks. We proposed a framework to first decompose a binary fractal basin into fractal DEM from which two unique topological connectivity networks are extracted. These networks facilitate to segment Fractal-DEM (Fig. 35.2a) into sub-basins ranging from first to highest order (Fig. 35.2c). Results derived from a synthetic DEM (Fig. 35.2a) by applying one of these algorithms include unique topological connectivity networks similar to valley and ridge connectivity networks (Fig. 35.2b) and the hierarchically partitioned watersheds (Fig. 35.2c). We demonstrated the superiority of these stable algorithms which can be generalized to terrestrial surfaces of both fluvial and tidal types. This

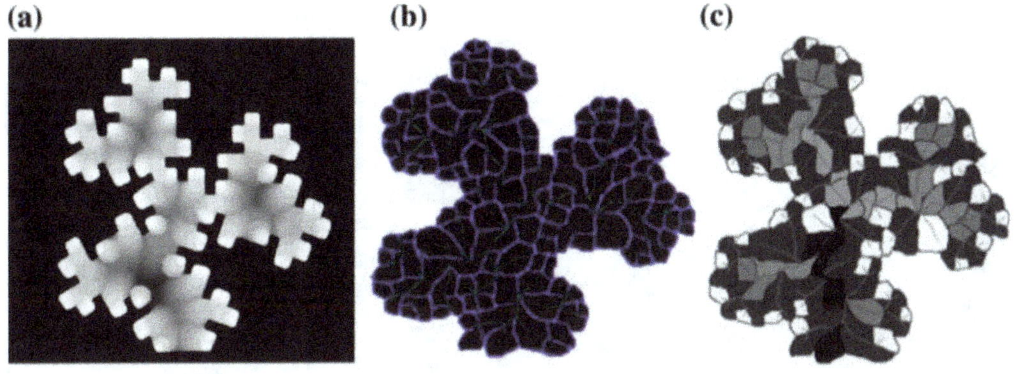


**Fig. 35.2** **a** simulated fractal DEM achieved through morphological decomposition procedure, **b** loop-like ridge connectivity and loopless channel connectivity networks, and **c** subbasins

work helps to solve basic problems that algorithms meant for extraction of unique terrestrial connectivity networks have faced for over three decades.

### *35.2.2 Retrieval of Morphologically Significant Regions*

Algorithms meant for morphological segmentation were demonstrated on a DEM, and mapped the physiographic features such as mountains, basins, and piedmont slopes from DEM (Fig. 35.3a); and the results are compared with that of other popular approaches (Fig. 35.3b).

Further, multiscale morphological opening was employed to segment binary fractal basins (Fig. 35.4a–c) that mimic geophysical basins, and cloud fields

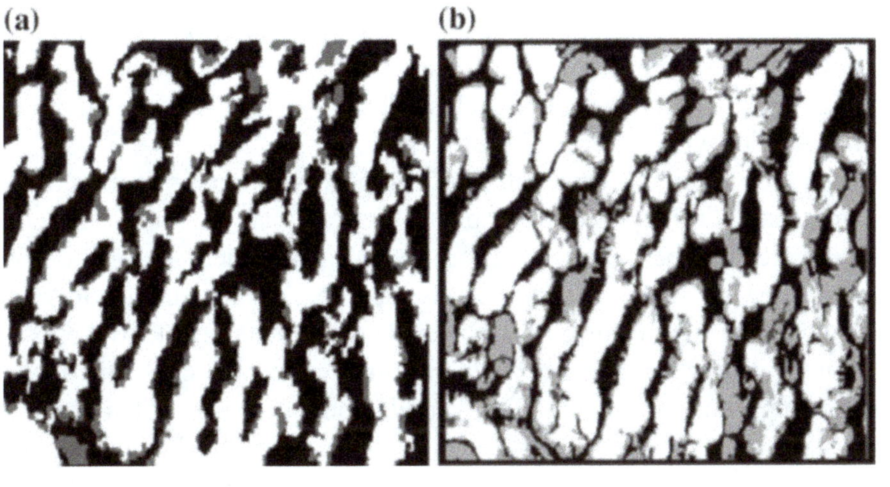


**Fig. 35.3** Mountain pixels are the pixels in white, the piedmont pixels are the pixels in gray, and the basin pixels are the pixels in black. **a** The results obtained using the newly developed algorithm. **b** The results obtained in Miliaresis and Argialas (1999). (From Sathymoorthy et al. 2007)

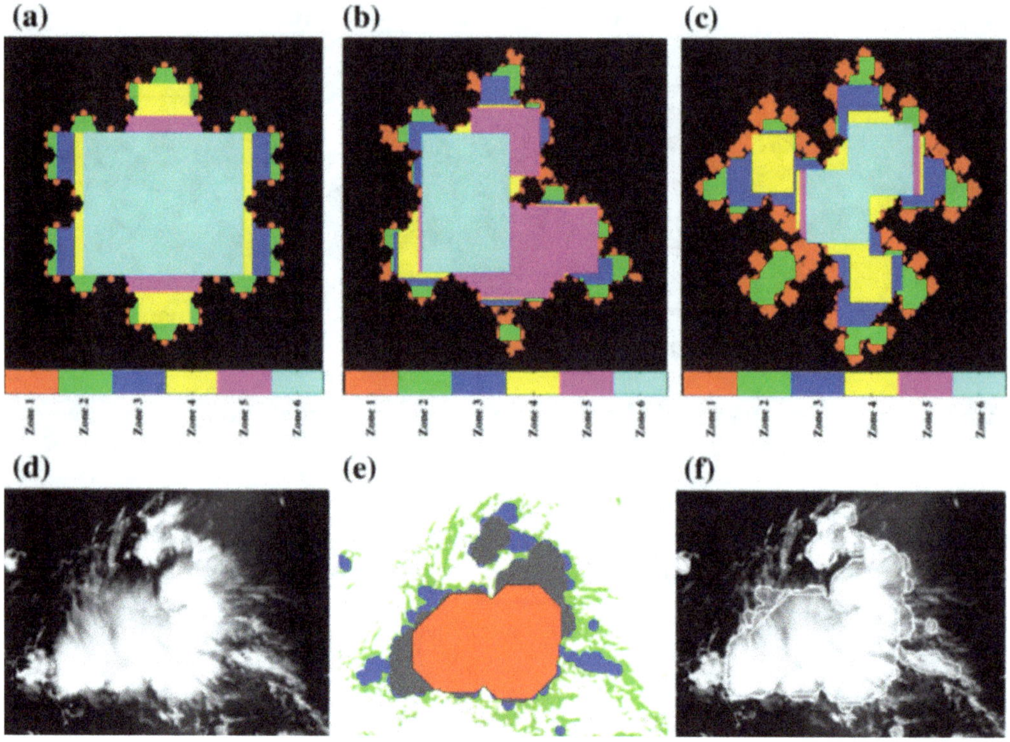


**Fig. 35.4** Morphologically significant zones decomposed from **a** Koch triadic fractal island, **b** Random Koch triadic fractal island, **c** Random Koch quadric fractal island, **d** Isolated Moderate Resolution Imaging Spectroradiometer (MODIS) cloud (cloud-1), **e** Color-coded binarized (by choosing threshold gray level value 128) cloud-1 images at three threshold-opening cycles superimposed on binarized original cloud-1 color-coded with green, and **f** boundaries of 12th, 32nd, and 100th opened cloud-1 images and thresholded original cloud-1 superimposed on the original cloud image

isolated from MODIS data into topologically prominent regions (Fig. 35.4d–f). We proposed granulometry-based segmentation of geophysical fields (e.g. DEMs, clouds, etc.) with demonstration on binary fractals of deterministic and random types (Fig. 35.4a–c), and on cloud fields (Fig. 35.4d–f) that have different compaction properties with varied cloud properties.

The approach based on computation of complexity measures of morphologically significant zones decomposed from binary fractal sets via multiscale convexity analysis—which can be implemented on several geophysical and geomorphologic fields (e.g. DEMs, clouds, binary fractals etc.) to segment them into regions of varied topological significance—has been demonstrated on cloud fields derived from MODIS data to better segment the regions within the cloud fields that have different compaction properties with varied cloud properties. This approach of fundamental importance can be extended to several geophysical and geomorphologic fields to segment them into regions of varied topological significance.

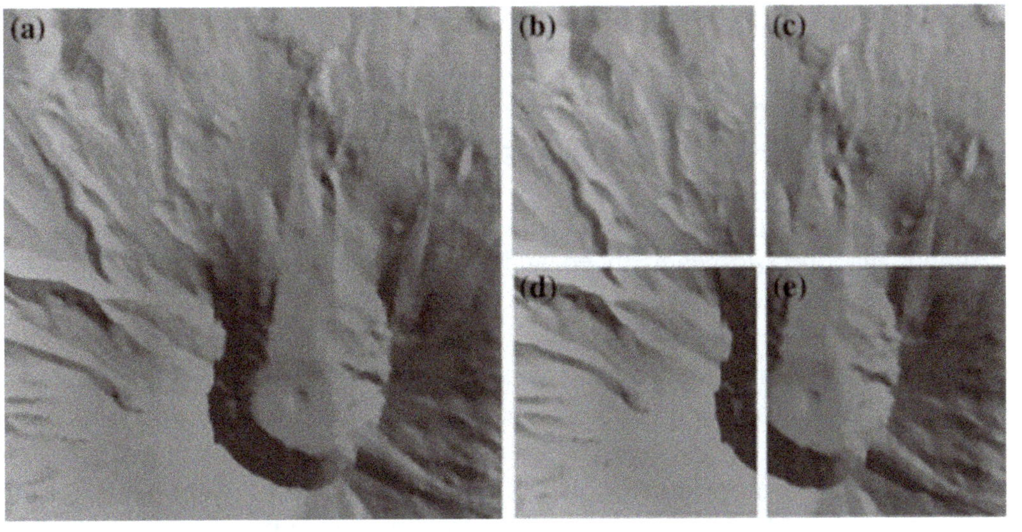


**Fig. 35.5** **a** Digital Elevation Model of size 256 × 256 pixels depicting Mount St Helens, **b–e** four quadrants of size 128 × 128 pixels partitioned from DEM (Fig. 35.5a) include top-left ($f^1$), top-right ($f^2$), bottom-left ($f^3$), and bottom-right ($f^4$) portions

### *35.2.3 Ranking of Best Pairs of Spatial Fields*

A new metric to quantify the degree of similarity between any two given spatial fields is proposed (Sagar and Lim 2015a, b). This metric based on morphological operations can be used for image classification, in particular hyperspectral image classification, to derive best pair(s) of spatial fields from among a large number of spatial fields available in a database. In this proposed approach to compute the ranks for every possible pair of spatial fields (grayscale images) in a database, the two major computations involved include (i) estimation of grayscale morphological distance between the source and target spatial fields, and (ii) the ratios between the areas of infima and suprema of source and target spatial fields. Using this approach, four spatial elevation fields (Fig. 35.5b–e), in other words four quadrants partitioned from Fig. 35.5a could be paired into best pair (Fig. 35.6a), medium best pair (Fig. 35.6b), and the least best pair (Fig. 35.6c).

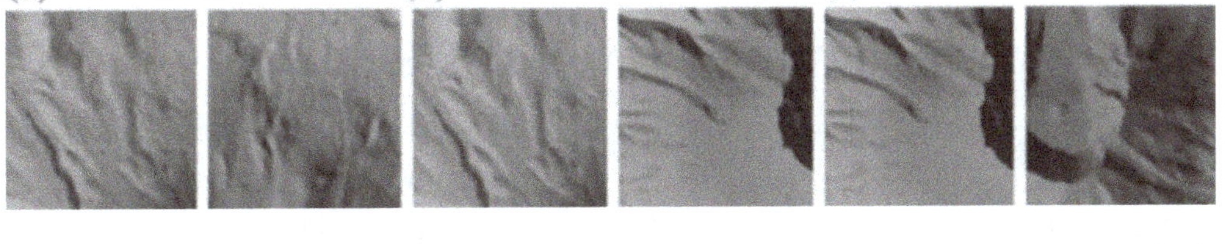


**Fig. 35.6** Three best ranked pairs of spatial elevation fields shown in Fig. 35.5b–e **a** ($f^1$,$f^2$), **b** ($f^1$,$f^3$), and **c** ($f^3$,$f^4$)

## 35.3 Terrestrial Pattern Analysis

Quantitative analyses of terrestrial phenomena and processes is one of the innovative new directions of geoscientific research. Analysis of terrestrial patterns—that include water bodies, valley and ridge connectivity networks, watersheds, hillslopes, mountain objects, elevation fields—at various spatial and temporal scales is an important aspect to better understand the dynamical behaviors of various terrestrial processes and surfaces. Over the decades, various quantitative approaches have been developed and successfully demonstrated. Some of these approaches include morphometric analysis of river networks, hypsometry, allometry, and granulometric analyses, and geodesic spectrum based analysis. In this section, we show some results through illustrations arrived at via applications of mathematical morphology in (i) morphometric and allometric analyses of river networks and water bodies and their corresponding zones of influence, (ii) deriving scale-invariant but shape-dependant power laws, (iii) deriving basin-specific geodesic spectrum, and (iv) DEM analysis.

### *35.3.1 Morphometry and Allometry of Networks*

Towards analyzing terrestrial surfaces we have shown unique ways to quantitatively characterize the spatiotemporal terrestrial complexity via scale-invariant measures that explain the commonly sharing physical mechanisms involved in terrestrial phenomena and processes. These contributions (Sagar and Rao 1995a, b, c, d; Sagar 1996, 1999a 2000a, b, 2001a, b, c, d 2007; Sagar et al. 1998a, b, 1999; Sagar and Tien 2004; Chockalingam and Sagar 2005; Tay et al. 2005a, b, c) highlighted the evidence of self-organization via scaling laws—in networks, hierarchically decomposed subwatersheds, and water bodies and their zones of influence, which evidently belong to different universality classes—which possess excellent agreement with geomorphologic laws such as Horton's Laws, Hurst exponents, Hack's exponent, and other power-laws given in non-geoscientific contexts. A host of allometric power-law relationships were derived that were in good accord with other established network models and real networks (Figs. 35.7, 35.8 and 35.9).

### *35.3.2 Allometry of Water Bodies and Their Zones of Influence*

Topologically, water bodies (Fig. 35.10a) are the first level topographic regions that get flooded, and as the flood level gets higher, adjacent water bodies merge. The looplike network that forms along all these merging points represents zones of influence (Fig. 35.10b) of each water body. The geometric organizations of these

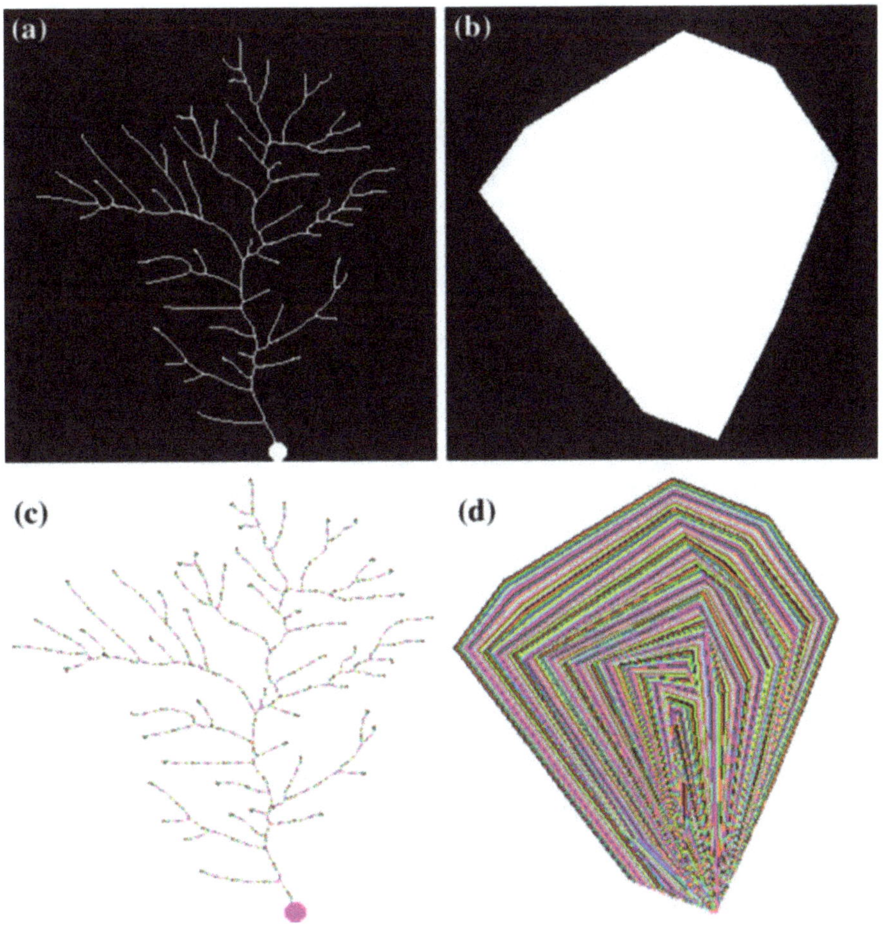


**Fig. 35.7** **a** An example of fourth-order channel network (nonconvex set) and **b** its convex hull. A stationary outlet is shown as a round dot in Fig 35.7a. **c** color-coded traveltime network pruned iteratively until it reaches the outlet and **d** color-coded union of convex hulls of networks pruned to different degrees

two phenomena are respectively sensitive and insensitive to perturbation due to exogenic processes. To demonstrate the allometric relationships of water bodies and their zones of influence, a large number of surface water bodies (irrigation tanks), situated in the floodplain region of certain rivers of India, which are retrieved from multi-date remotely sensed data were analyzed in 2-D space (Sagar et al. 1995a, b). Basic measures of these water bodies obtained by morphological analysis were employed to show fractal-length-area-perimeter relationships.

We found that these phenomena follow the universal scaling laws (Sagar et al. 2002; Sagar 2005a, b) found in other geophysical and biological contexts. In this work, universal scaling relationships among basic measures such as area, length, diameter, volume, and information about networks are exhibited by several natural phenomena to further retrieve and understand the common principles underlying organization of these phenomena. Some of the recent findings on universal scaling relations include relationships between brain and body, length and area (or volume),

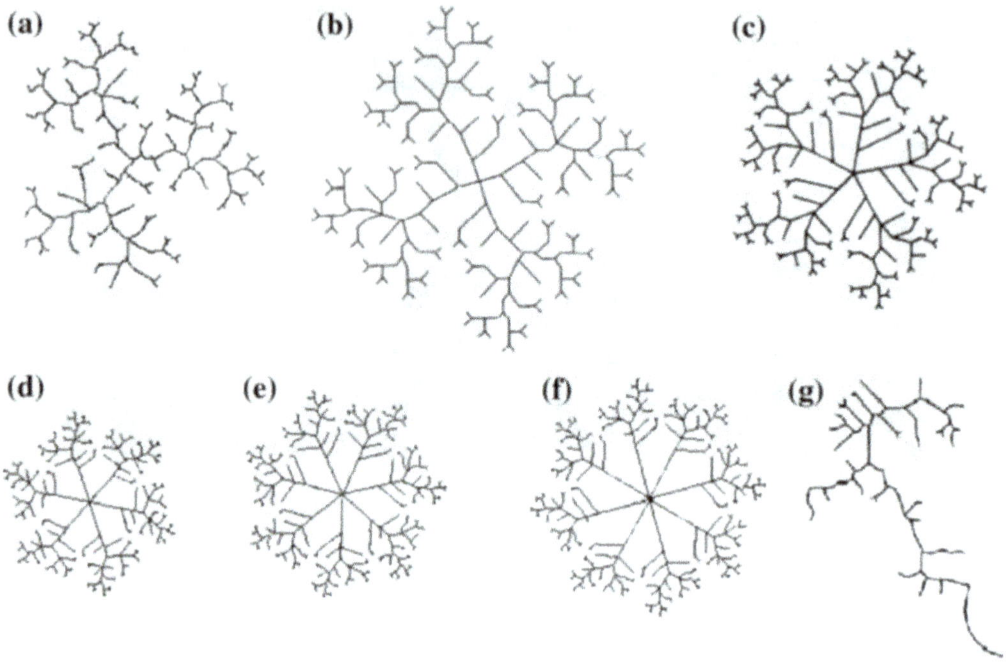


**Fig. 35.8** Networks in **a** three-sided fractal basin, **b** four-sided fractal basin, **c** five-sided fractal basin, **d** six-sided fractal basin, **e** seven-sided fractal basin, **f** eight-sided fractal basin, and **g** Nizamsagar reservoir. (From Sagar et al. 1998a, b, 1999, 2001)

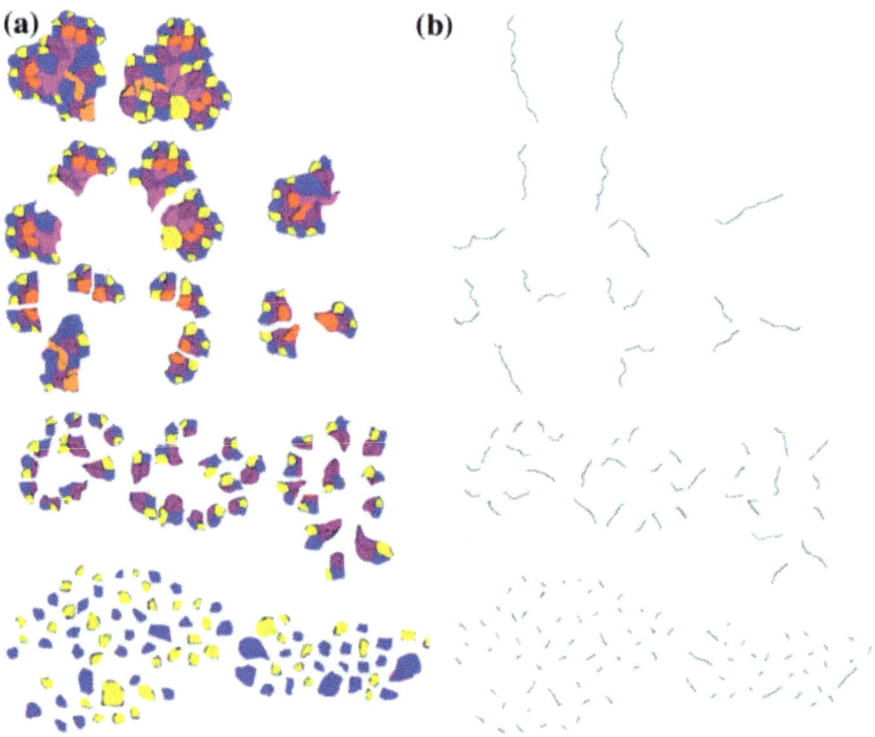


**Fig. 35.9** **a** sub-basins decomposed from a Hortonian F-DEM areas, and **b** corresponding main lengths

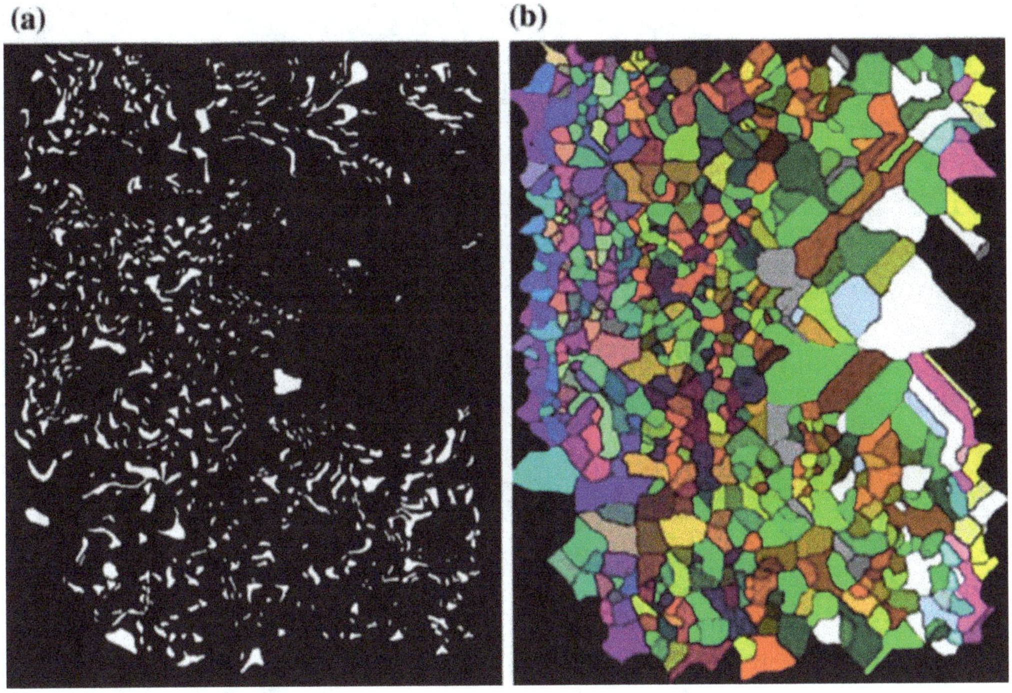


**Fig. 35.10** **a** A section consisting of a large number of small water bodies traced from the floodplain region of Gosthani River and **b** zones of influence of water bodies shown in Fig. 35.10a. Different colors are used to distinguish the adjacent influence zones

size and number, size and metabolic rate. In this study, we have shown a host of universal scaling laws in surface water bodies (Fig. 35.10a) and their zones of influence (Fig. 35.10b) that have similarities with several of these relationships encountered in various fields are shown.

### 35.3.3 *Morphometry of Non-network Space: Scale Invariant but Shape-Dependent Dimension*

In sequel works on terrestrial analysis, we argued that the universal scaling laws shown as examples in earlier section possess limited utility in exploring possibilities to relate them with geomorphologic processes. These arguments formed the basis for alternative methods (Radhakrishna et al. 2004; Teo et al. 2004; Sagar and Chockalingam 2004; Chockalingam and Sagar 2005; Tay et al. 2005a, b, 2007). Shape and scale based indexes provided to analyze and classify non-network space (hillslopes) (Sagar and Chockalingam 2004; Chockalingam and Sagar 2005), and terrestrial surfaces (Tay et al. 2005a, b, 2007) received wide attention. These methods that preserve the spatial and morphological variability yield quantitative results that are scale invariant but shape dependent, and are sensitive to terrestrial surface variations. “Fractal dimension of non-network space of a catchment basin”,

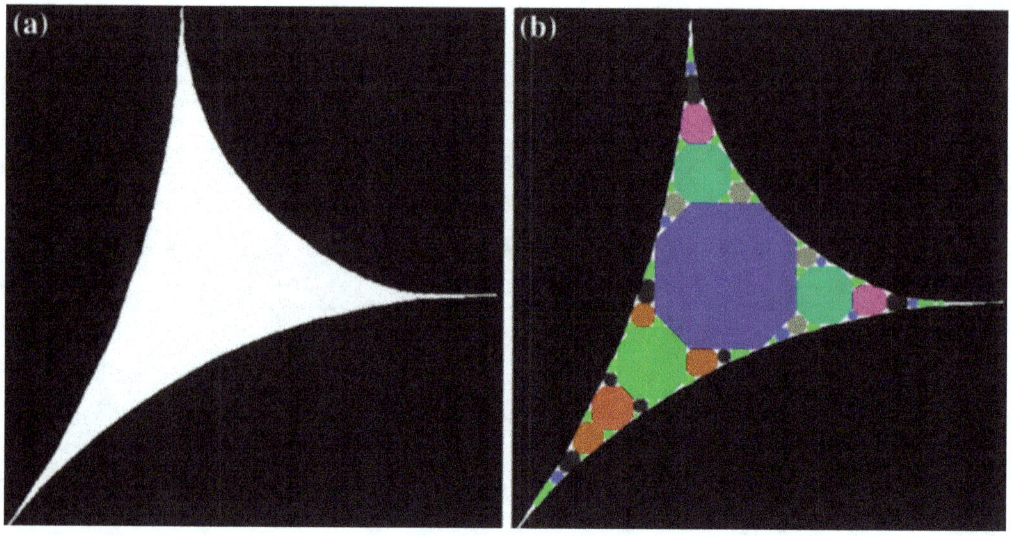


**Fig. 35.11** **a** Apollonian space, and **b** after decomposition by means of octagon

provides an approach to show basic distinction between the topologically invariant geomorphologic basins. It introduced morphological technique for hillslope decomposition that yields a scale invariant, but shape dependent, power-laws (Fig. 35.11a, b).

Varied degrees of topographically convex regions within a catchment basin represent varied degrees of hill-slopes. The non-network space, the characterization of which we focused on in our investigations, is akin to the space that is achieved by subtracting channelized portions contributed due to concave regions from the watershed space. This non-network space is akin to non-channelized convex region within a catchment basin. We proposed an alternative shape-dependent quantity akin to fractal dimension to characterize this non-network space (e.g.: Fig. 35.12a). Towards this goal, non-network space is decomposed, in two- dimensional discrete space, into simple non-overlapping disks (NODs) of various sizes by employing mathematical morphological transformations and certain logical operations (Fig. 35.12b). Furthermore, number of NODs of lesser than threshold radius is plotted against the radius, and computed the shape-dependent fractal dimension of non-network space. This study was extended to derive shape dependent scaling laws as the laws derived from network measurements are shape independent for realistic basins (Fig. 35.12). The relationship between number of NODs and the radius of the disk provides an alternative fractal-like dimension that is shape dependent. This was done with the aim to relate shape dependent power laws with geomorphic processes such as hill-slope processes and erosion.

Applications of mathematical morphology transformations are shown to decompose fractal basins (e.g.: Fig. 35.11a) into non-overlapping disks of various sizes (Fig. 35.11b) further to derive fractal power-laws based on number-radius relationships.

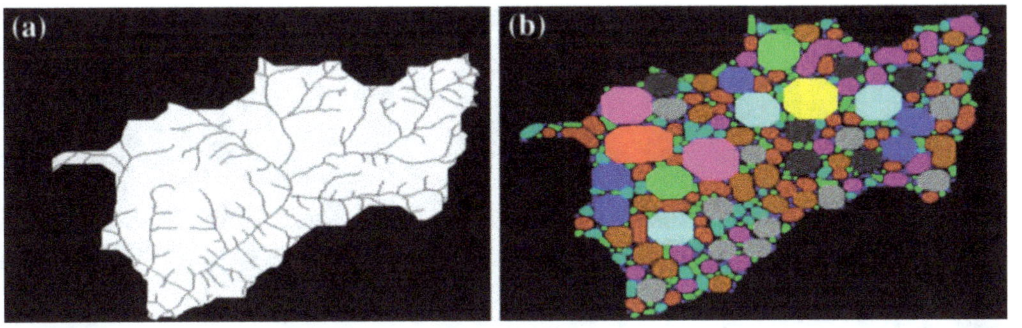


**Fig. 35.12** **a** 5th order channel network **c** of Durian Tungal catchment basin, basin X is reconstructed from this channel network via multiscale morphological closing transformation, **b** M = X\C

### *35.3.4 Geodesic Spectrum*

We have provided a novel geomorphologic indicator by simulating geodesic flow fields (Fig. 35.13d–f) within basins (Fig. 35.13a–c) consisting of spatially distributed elevation regions (Lim and Sagar 2008a, b), further to compute a geodesic spectrum that provides a unique one-dimensional geometric support.

This one-dimensional geometric support, in other words geodesic spectrum, outperforms the conventional width–function based approach which is usually derived from planar forms of basin and its networks–construction involves basin as

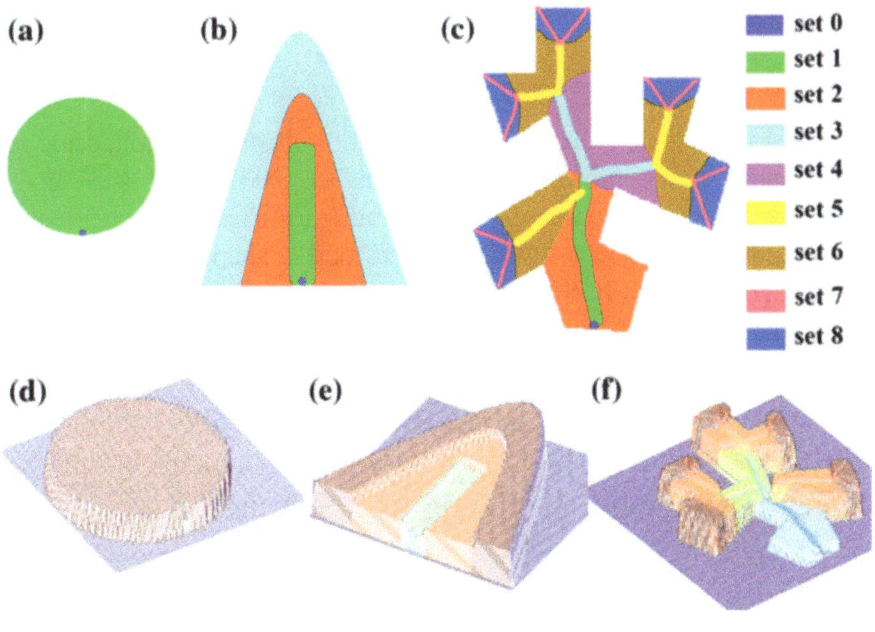


**Fig. 35.13** **a** a flat circular basin, **b** a basin with three spatially distributed elevation regions, **c** a fractal basin with channelised and non-channeled regions **d** flow fields with isotropic propagation in **a**, **e** isotropic flow fields within **b**, and **f** flow fields within **c** and orthogonality between the flow fields of channelized and non-channelized zones is obvious. (From Lim and Sagar 2008a, b.)

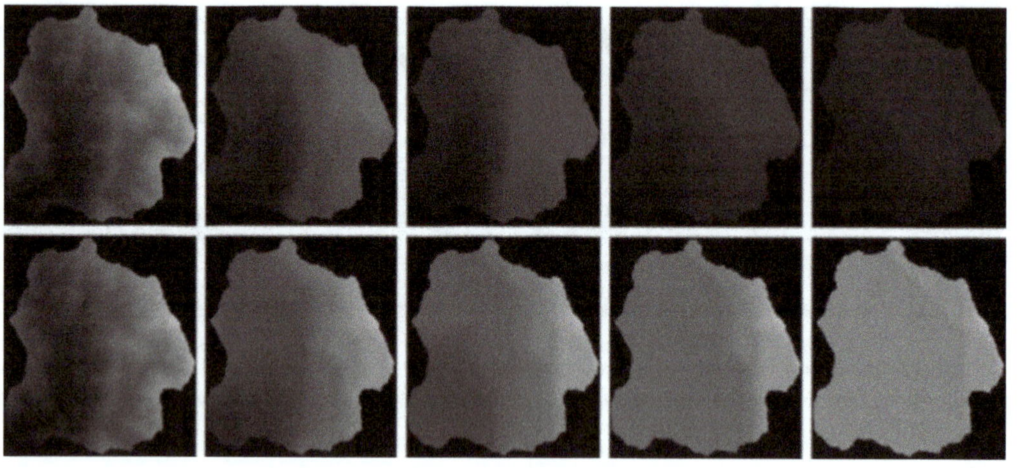

**Fig. 35.14** Basin 1 of Cameron Highlands is taken as an example to show the basin images at multiple scales generated via closing and opening. Basin 1 is located at the northern part of Cameron Highlands region, with a size of 3.1 km (east to west) 63.4 km (north to south). (Upper sequence) DEM at multiple scales generated via opening, and (Lower Panel) multiscale DEMs generated via closing

a random elevation field (e.g. Digital Elevation Model, DEM) and all threshold elevation regions decomposed from DEM for understanding the shape-function relationship much better than that of width function.

### *35.3.5 Granulometric and Anti-granulometric Analysis of Basin-DEMs*

Granulometric indexes derived for spatial elevation fields also yield scale invariant but shape-dependent measures (Tay et al. 2005a, b, c, 2007). DEMs are analyzed by following granulometries via multiscale opening (Fig. 35.14 upper panel), and antigranulometries (Fig. 35.14 lower panel) to derive shape-size complexity measures of foreground and background respectively that provide new indices to understand the terrestrial surfaces further to relate with several geomorphic processes.

## 35.4 Geomorphologic Modeling and Simulation

Simulations allow us to gain a significantly good understanding of complex geomorphologic systems in a way that is not possible with lab experiments. Effectively attaining these goals presents many computational challenges, which include the development of frameworks. The robustness of mathematical morphological operators combined with concepts of fractal geometry (Mandelbrot 1982) in modeling and simulations of certain geoscientific phenomena and processes is shown briefly with illustrative examples in this section. The phenomena and processes given emphasis in this section include geomorphologic features, basins and channel networks, landscapes, water bodies, symmetrical folds and ideal sand dunes. Besides providing approaches to simulate fractal-skeletal based channel network model and fractal landscapes, we have shown via the discrete simulations the varied dynamical behavioral phases of certain geoscientific processes (e.g. water bodies, ductile symmetric folds, sand dunes, landscapes) under nonlinear perturbations due to *endogenic* and *exogenic* nature of forces. For these simulations we employed nonlinear first order difference equations, bifurcation theory, fractal geometry, and nonlinear morphological transformations as the bases. The three complex systems that we focus on include the channelization process, surface water bodies, and elevation structures.

### 35.4.1 Geomorphologic Modeling: Concept of Discrete Force

Concept of discrete force was proposed from theoretical standpoint to model certain geomorphic phenomena, where geomorphologically realistic expansion and contractions, and cascades of these two transformations were proposed, and five laws of geomorphologic structures are proposed (Sagar et al. 1998a, b). A possibility to derive a discrete rule from a geomorphic feature (e.g. lake) undergoing morphological changes that can be retrieved from temporal satellite data was also proposed in this work, and explained (Fig. 35.15). Laws of geomorphic structures under the perturbations are provided and shown, through interplay between numerical simulations and graphic analysis as to how systems traverse through various behavioral phases.

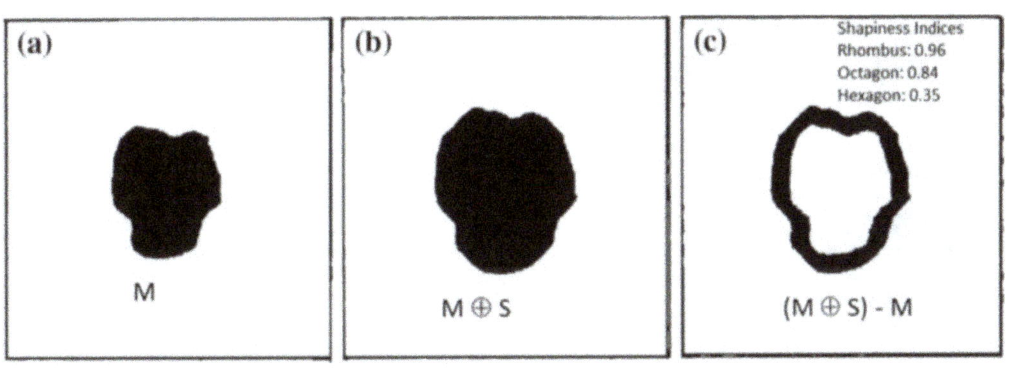


**Fig. 35.15** **a** Hypothetical geomorphic feature at time t, **b** geomorphic feature at time $t + 1$, and **c** difference in geomorphic feature from time $t$ to $t + 1$

### 35.4.2 Fractal-Skeletal Based Channel Network Model

Our work on channel network modelling Gastner and Newman (2004) and Sagar (2001c) represents unique contributions to the literature, which until recently were dominated by the classic random model. Fractal-skeletal based channel network model (F-SCN) was proposed by following certain postulates. We developed the Fractal-Skeletal Channel Network (F-SCN) model by employing morphological skeletonization to construct other classes of network models, which can exhibit various empirical features that the random model cannot. In the F-SCN model that gives rise to Horton laws, the generating mechanism plays an important role. Homogeneous and heterogeneous channel networks can be constructed by symmetric generator with non-random rules, and symmetric or asymmetric generators with random rules. Subsequently, F-SCNs (Fig. 35.16d–f) in different shapes of fractal basins (Fig. 35.16a–c) are generated and their generalized Hortonian laws (Fig. 35.16g, h) are computed which are found to be in good accord with other established network models such as Optimal Channel Networks (OCNs), and realistic rivers. F-SCN model is extended to generate more realistic dendritic branched networks.

### 35.4.3 Fractal Landscape via Morphological Decomposition

By applying morphological transformations on fractals of varied types are decomposed into topologically prominent regions (TPRs) (Fig. 35.17a) and each TPR is coded and a fractal landscape organization that is geomorphologically realistic is simulated (Fig. 35.17b) (Sagar and Murthy 2000).

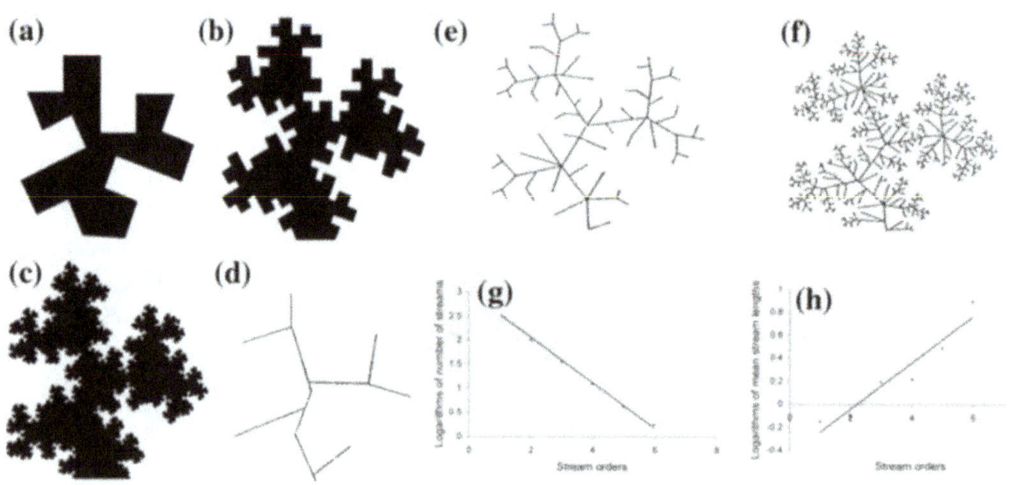


**Fig. 35.16** **a**, **b** and **c** Fractal basins after respective iterations. **d**, **e** and **f** An evolutionary sequence of F-SCNs after respective iterations, **g** Horton's law of number, and **h** Horton's law of mean length

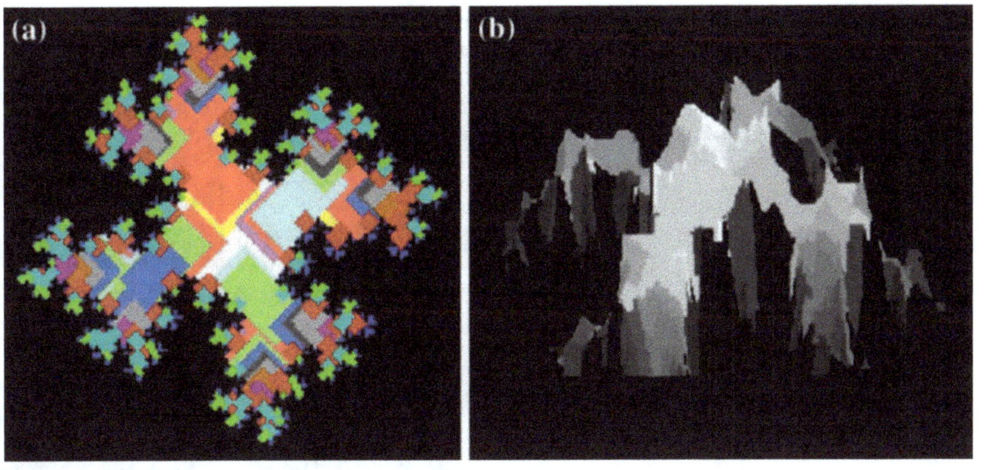


**Fig. 35.17** **a** A binary fractal basin after decomposition into TPRs **b** A fractal landscape generated from Fig. 35.17a. Light and dark regions of DEM are visualized as high and low elevations (vertical exaggeration: 7)

### *35.4.4 Discrete Simulations and Modeling the Dynamics of Small Water Bodies, Symmetrical Folds, and Sand Dunes*

In this subsection we show the fusion of computer simulations and modeling techniques in order to better understand certain terrestrial phenomena and processes with the ultimate goal of developing cogent models in discrete space further to gain a significantly good understanding of complex terrestrial systems in a way that is not possible with lab experiments. The three synthetic phenomena that are explained by generating attractors considered include water bodies (Sagar and Rao 1995a, b, c), symmetrical folds (Sagar 1998), and sand dunes (Sagar 1999b, 2000a, b, 2001a, 2005a, b; Sagar and Venu 2001; Sagar et al. 2003a, b).

#### 35.4.4.1 Discrete Simulations and Modeling the Dynamics of Small Water Bodies

Spatio-temporal patterns of small water bodies (SWBs) under the influence of temporally varied streamflow discharge behaviors are simulated in discrete space by employing geomorphologically realistic expansion and contraction transformations (Fig. 35.18). Expansions and contractions of SWBs to various degrees (e.g. Fig. 35.18B g–l), which are obvious due to fluctuations in streamflow discharge pattern (Fig. 35.18A, a–f), simulate the effects respectively owing to streamflow discharge that is greater or less than mean streamflow discharge. The cascades of expansion-contraction are systematically performed by synchronizing the streamflow discharge (Fig. 35.18A, a–f), which is represented as a template with definite

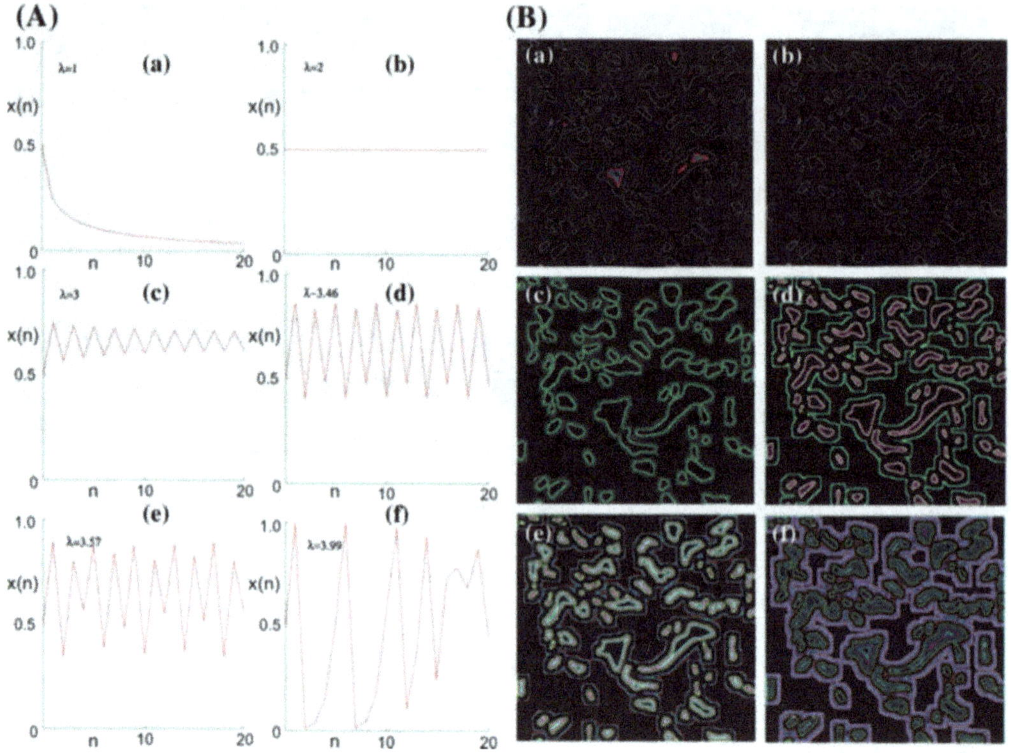

Fig. 35.18 **A** Streamflow discharge behavioral pattern at different environmental parameters. **a–f** $\lambda = 1, 2, 3, 3.46, 3.57$ and 3.99, and **B** Spatio-temporal organization of the surface water bodies under the influence of various streamflow discharge behavioral patterns at the environmental parameters at **a–f** $\lambda = 1, 2, 3, 3.46, 3.57$, and 3.99 are shown up to 20 time steps. In all the cases, the considered initial MSD, A0 = 0.5 (in normalized scale) is considered under the assumption that the water bodies attain their full capacity. It is illustrated only for the overlaid outlines of water bodies at respective time-steps with various λs

characteristic information, as the basis to model the spatio-temporal organization of randomly situated surface water bodies of various sizes and shapes.

We have shown the varied dynamical behavioral phases of certain geoscientific processes (e.g. water bodies) under nonlinear perturbations via the discrete simulations.

#### 35.4.4.2 Ductile Symmetrical Fold Dynamics

Under various possible time-dependent and time-independent strength of control parameter, in other words nonlinear perturbations, the three-limb symmetrical folds are transformed in a time sequential mode to simulate various possible fold dynamical behaviors (Fig. 35.19a, b) synchronizing trajectory behavior simulated via logistic equation with strength nonlinearity parameters 3.9 and 2.8 (Fig. 35.20a, b). We employed normalized fractal dimension values, and interlimb angles (IAs) as parameters along with strength of nonlinear parameters in this study. Bifurcation

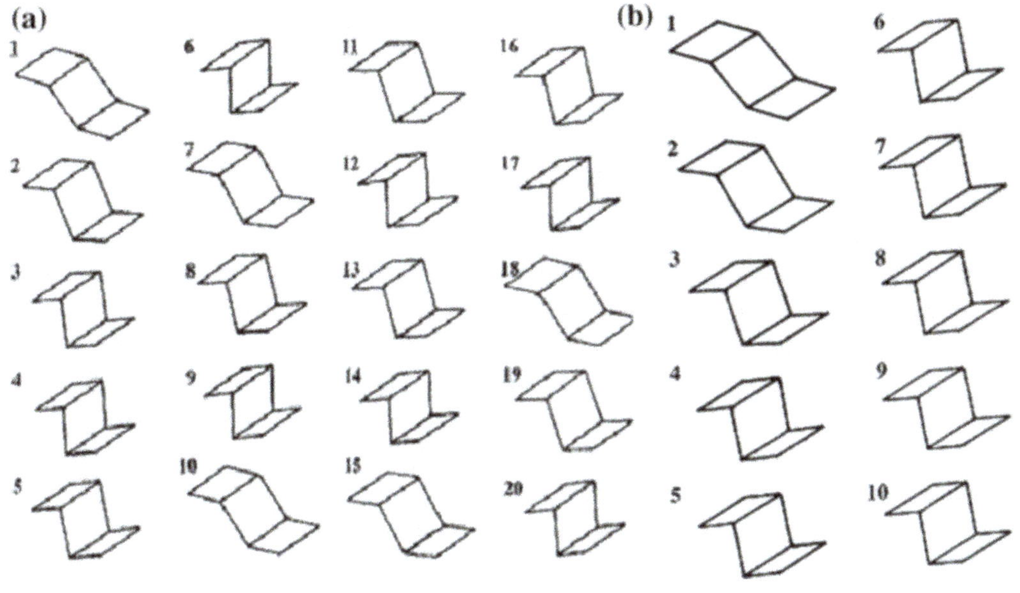

**Fig. 35.19** Evolution of a fold type with the strength of nonlinearities: **a** $\lambda = 3.9$ and **b** $\lambda = 2.8$. The numbers represent the discrete times. (From Sagar 1998)

diagrams are constructed for both time-dependent and time-independent fold dynamical behaviors, and the equations to compute metric universality by considering the interlimb angles computed at threshold strengths of nonlinearity parameters are proposed (Sagar 1998).

#### 35.4.4.3 Symmetrical Sand Dune Dynamics

Certain possible morphological behaviors with respective critical states represented by inter-slip face angles of a sand dune under the influence of non systematic processes are qualitatively illustrated by considering the first order difference equation that has the physical relevance to model the morphological dynamics of the sand dune evolution as the basis. It is deduced that the critical state of a sand dune under dynamics depends on the regulatory parameter that encompasses exodyanmic processes of random nature and the morphological configuration of sand dune. With the aid of the regulatory parameter, and the specifications of initial state of sand dune, morphological history of the sand dune evolution can be investigated. As an attempt to furnish the interplay between numerical experiments and theory of morphological evolution, the process of dynamical changes (Fig. 35.21) in the sand dune with a change in threshold regulatory parameter (e.g. Fig. 35.22) is modeled qualitatively for a better understanding. An equation to compute metric universality by considering attracting interslipface angles is also proposed. Avalanche size distribution in such a numerically simulated sand dune dynamics have also been studied.

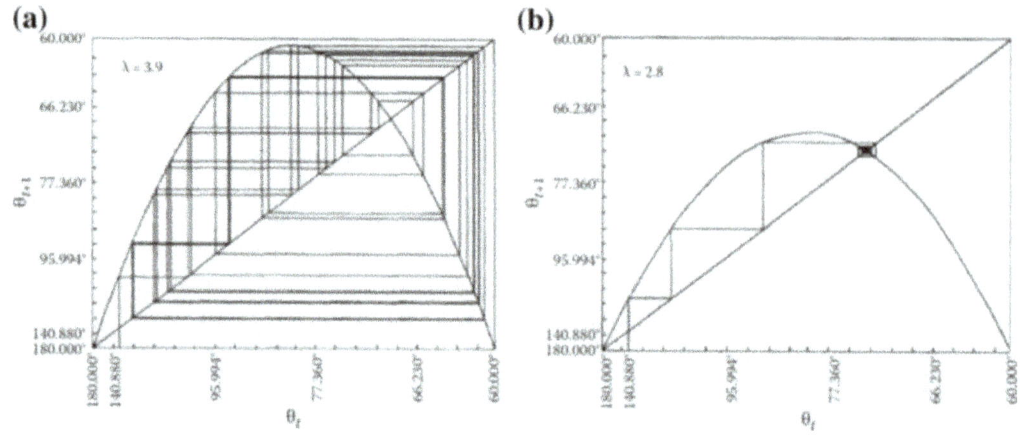


**Fig. 35.20** Logistic maps for the qualitative dynamical behavior of symmetric folds under evolution shown in Fig. 35.19a, b. It may be seen that the values mentioned on the abscissa are IAs in degrees for the symmetric fold with three limbs. (From Sagar 1998)

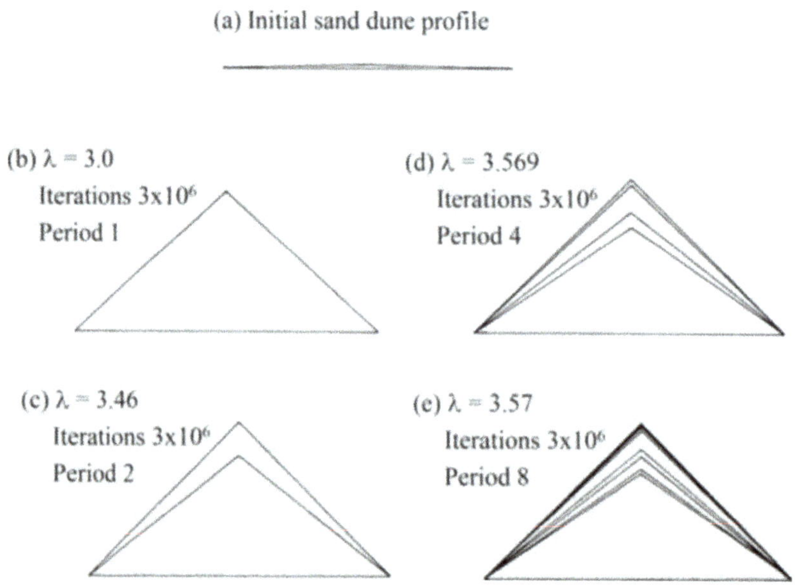


**Fig. 35.21** **a** Initial sand dune profile with $\alpha = 0.00001$ or $\theta = 179.57334$. The attractor sand dune profiles at various threshold regulatory parameters: **b** $\lambda = 3$, fixed point attractor sand dune; **c** $\lambda = 3.46$, period 2 attractor sand dunes; **d** $\lambda = 3.569$, period 4 attractor sand dunes; and **e** $\lambda = 3.57$, period 8 attractor sand dunes. The attractor sand dune profiles shown in **b–e** are by iterating $3 \times 104$ time steps. (From Sagar 1999a, b)

## 35.5 Geospatial Computing and Visualization

Mathematical morphology not only provides robust solutions in terrestrial pattern retrieval, analysis, and modeling and simulations but also provides numerous insights worth exploring to find solutions for the challenges encountered in GISci. In recent works—that include (i) binary and grayscale morphological interpolations,

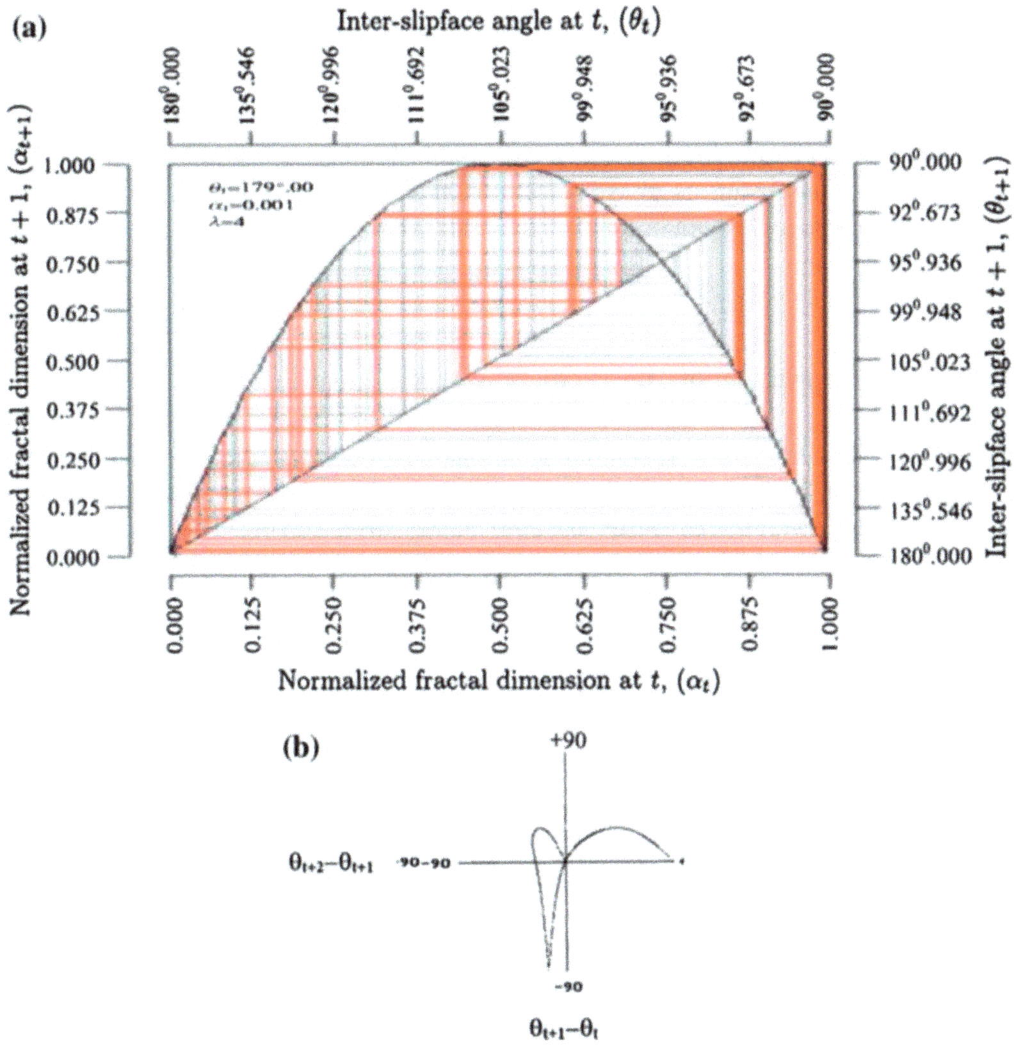


**Fig. 35.22** **a** A 1-D map plotted between $\theta_{t+1}$ versus $\theta_t$ for sand dune case $\lambda = 4$ and **b** return map plotted between $\theta_{t+1} - \theta_t$ versus $\theta_{t+2} - \theta_{t+1}$ for sand dune case with $\lambda = 4$. (From Sagar et al. 2003a, b; Sagar and Venu 2001)

SKIZ, WSKIZ and applications in spatiaotemporal visualizations, conversion of point-specific variable data into contiguous zonal maps (Rajashekara et al. 2012), morphing (Sagar and Lim 2015a, b) and variable-specific cartogram generation (Sagar 2014a, b), (ii) volumetric visualization of topologically significant components such as pore-bodies, pore-throats, and pore-channels (Teo and Sagar 2005, 2006), and (iii) spatial reasoning, planning, and interactions (Sagar et al. 2013; Vardhan et al. 2013; Sagar 2018)—one can realize on how robust approaches could be developed by considering mathematical morphological transformations.

### *35.5.1 Morphological Interpolations*

This subsection provides the applications of binary and grayscale morphological interpolations in hierarchical computation of morphological medians and in morphing, and the applications of SKIZ and WSKIZ in conversion of point-specific variable data into contiguous zonal map, and generation of variable-specific contiguous cartograms.

#### 35.5.1.1 Computation of Hierarchical Morphological Medians

Hausdorff-distance based (i) spatial relationships between the maps possessing bijection for categorization and (ii) nonlinear spatial interpolation in visualization of spatiotemporal behavior are proposed and demonstrated. This work (Sagar 2010, 2014a, b; Challa et al. 2016) concerns the development of frameworks with a goal to understand spatial and/or temporal behaviors of certain evolving and dynamic geomorphic phenomena. In (Sagar 2010), we have shown (i) how Hausdorff-Dilation and Hausdorff-Erosion metrics could be employed to categorize the time-varying spatial phenomena, and (ii) how thematic maps in time-sequential mode (Fig. 35.23a) can be used to visualize the spatiotemporal behaviour of a phenomenon, by recursive generation of median elements (Fig. 35.23b). Spatial interpolation, that was earlier seen as a global transform, is extended in Lim and Sagar (2008) by introducing *bijection* to deal with even connected components. This aspect solves problems of global nature in spatial-temporal GIS. Spatial Interpolation technique is found useful for spatial-temporal GIS and is demonstrated with validation on epidemic spread maps collected for eleven years between 1896 and 1906 (Fig. 35.23a–k, upper left panel). Morphological medians are computed between the epidemic spread maps staggered at two-year interval (Fig. 35.23a–k, upper right panel). Further morphological medians are computed in a hierarchical manner between every two epidemic spread maps of successive years (Fig. 35.23a, b in the lower panel).

#### 35.5.1.2 Grayscale Morphological Interpolation and Morphing

The computation of morphological medians between the thematic maps (binary images) demonstrated in the earlier subsection could be extended to the spatial fields (functions, e.g.: DEMs). This extended version is termed as grayscale morphological interpolation. We have demonstrated the application of grayscale morphological interpolations, computed hierarchically between the spatial fields (Fig. 35.24), to metamorphose a source-spatial field into a target-spatial field. Grayscale morphological interpolations are computed in a hierarchical manner

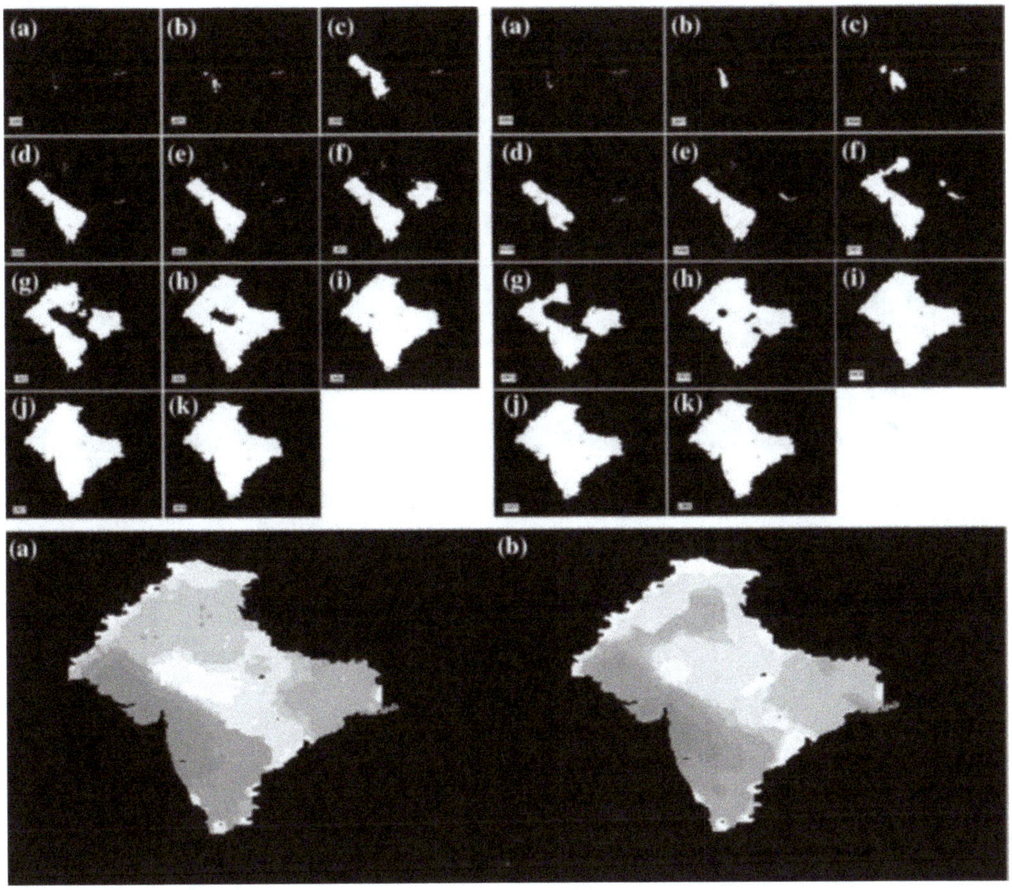


**Fig. 35.23** (Upper-Left Panel) **a–k** Spatial temporal maps that represent the geographic spread of bubonic plague in India between 1896 and 1906 at intervals of one year Maragos and Schafer (1986). The 11 spatial maps depicting the spread of plague were sequentially used to generate the maximum possible number of interpolated maps; (Upper right panel) **a** Original spatial map of the bubonic plague during 1896. **b–j** The first level median sets computed for $M(X^t, X^{t+2})$ for all "*t*" ranging from 1896 to 1905. **k** Original spatial map during 1906. For validation, the maps of Fig. b–j of upper left panel obtained as first-level median sets are, $M(X^t, X^{t+2})$ respectively, compared for all "*t*" with those *t* of Fig. 35.23b–j of upper left panel. These first-level median sets show a reasonable matching with the actual sets (Fig. 35.23b–j of upper left panel); (Lower Panel) Superimposed gray-coded **a** original spatial maps and **b** spatial maps generated via median set computations

(Fig. 35.25) with respect to non-flat structuring element, and found that the morphing, shown for transform source-spatial field into target-spatial field, created with respect to non-flat structuring element is more appropriate as the transition of source-spatial field into the target-spatial field across discrete time steps is smoother than that of the morphing shown with respect to flat structuring element (Sagar and Lim 2015a, b). This morphing shown via nonlinear grayscale morphological interpolations is of immense value in geographical information science, and in particular spatiotemporal geo-visualization.

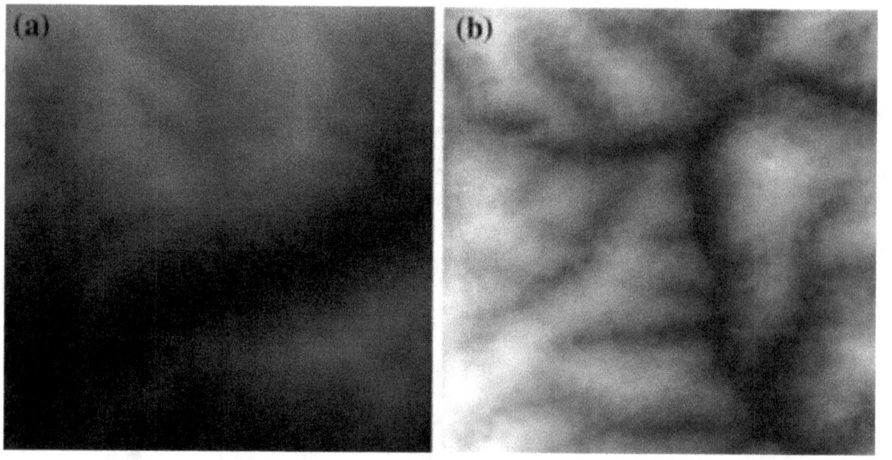


**Fig. 35.24** Smaller regions of DEMs: **a** Cameron Highlands, and **b** Petaling region

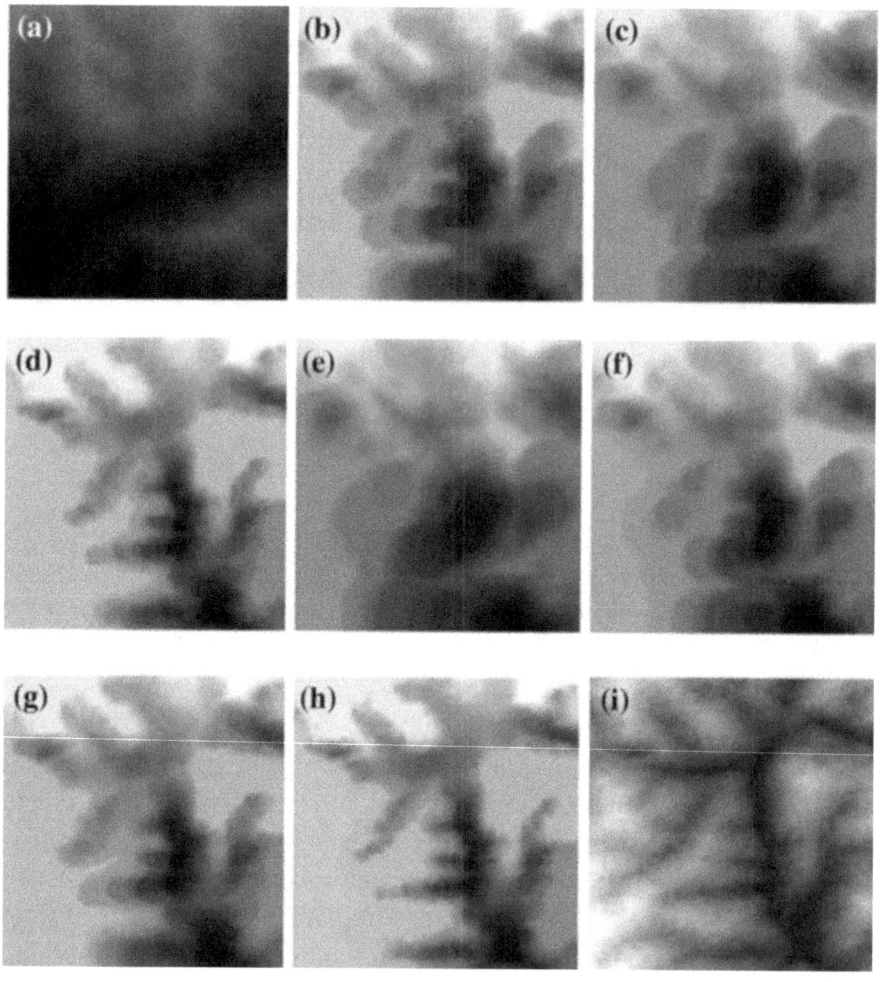


**Fig. 35.25** Generation of morphological medians generated by non-flat structuring element, between the DEMs shown in (**a**) and (**i**), at **b** zeroth level, **c**, **d** first level, and **e**–**h** second level

#### 35.5.1.3 Point-to-Polygon Conversion via WSKIZ

Data about many variables are available as numerical values at specific geographical locations in a noncontiguous form. We develop a methodology based on mathematical morphology to convert point-specific data into polygonal data. This methodology relies on weighted skeletonization by zone of influence (WSKIZ). This WSKIZ determines the points of contact of multiple frontlines propagating, from various points (e.g.: gauge stations) spread over the space, at the travelling rates depending upon the variable's strength. We demonstrate this approach for converting rainfall data available at specific rain gauge locations (points) (Fig. 35.26a) into a polygonal map (Fig. 35.26b) that shows spatially distributed zones of equal rainfall in a contiguous form (Rajashekara et al. 2012).

#### 35.5.1.4 Cartograms via WSKIZ

Visualization of geographic variables as spatial objects of size proportional to variable strength is possible via generating variable-specific cartograms. We developed a methodology based on mathematical morphology to generate contiguous cartograms. This approach determines the points of contact of multiple frontlines propagating, from centroids of various planar sets (states), at the travelling rates depending upon the variable's strength (Fig. 35.27a–d).

The contiguous cartogram generated via this algorithm preserves the global shape, and local shapes, and yields minimal area-errors. It is inferred from the comparative error analysis that this approach could be further extended by

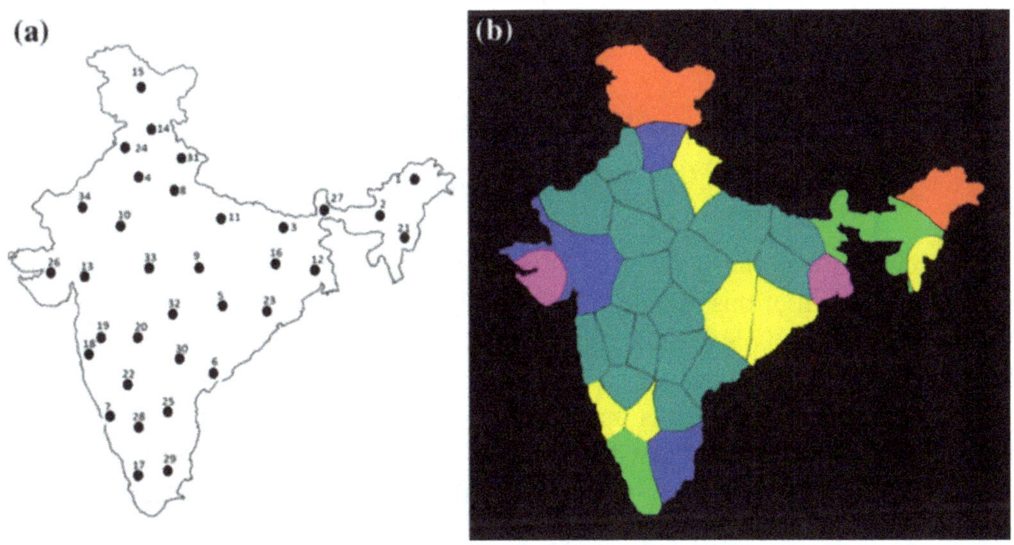


**Fig. 35.26** **a** 34 points (locations) of rain-gauge stations spread over India indexed ($A_1$–$A_{34}$), **b** Rainfall zonal map generated by having various possible propagation speeds, and the variable strengths in terms of propagation speeds

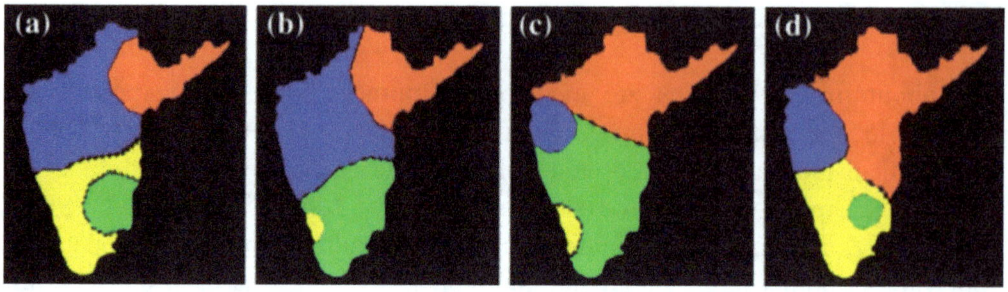


**Fig. 35.27** The variable strengths (in terms of propagation speeds are given as **a** $A_2 > A_4 > A_1 > A_3$, **b** $A_2 > A_1 > A_3 > A_4$, **c** $A_1 > A_3 > A_2 > A_4$, and **d** $A_1 > A_4 > A_2 > A_3$

exploring the applicability of additional characteristics of structuring element, which controls the dilation propagation speed and direction of dilation while generating variable-specific cartograms, to minimize the local shape errors, and area-errors. This algorithm addresses a decade-long problem of preservation of global and local shapes of cartograms. This approach was extended to generate a cartogram for a variable population to demonstrate the proposed approach. Further, the population cartograms for the USA generated via four other approaches (Kocmoud 1997; Keim et al. 2004; Gastner and Newman 2004; Gusein-Zade and Tikunov 1993) are compared with the morphology-based cartogram (Fig. 35.28a–f) in terms of errors with respect to area, local shape, and global shape. This approach for generating cartograms preserves the global shape at the expense of compromising with area-errors. It is inferred from the comparative error analysis that the proposed morphology-based approach could be further extended by exploring the applicability of additional characteristics of probing rule, which controls the dilation propagation speed and direction of dilation while performing WSKIZ, to minimize the local shape errors, and area-errors.

### *35.5.2 Visualization of Topological Components in a Volumetric Space*

Heterogeneous material is one that is composed of domains of different materials (phases). The aim of this module is to show how geometric descriptors derived via mathematical morphology and fractal analysis vary between the porous phases isolated from varied types of rocks at various spatial and spectral scales. It is evident from the recent works on Fontainebleau sandstone that the characteristics derived through computer assisted mapping and computer tomographic analysis were well correlated with the physical properties such as porosity, permeability, and conductance. Whatever the physical processes involved in altering the porous phase of material, we propose to emphasise quantifying the complexity of porous phase in both 2-D and 3-D domains. From a petrologic study perspective, such a quantitative characterization in both two- and three-dimensional spaces is of current interest.

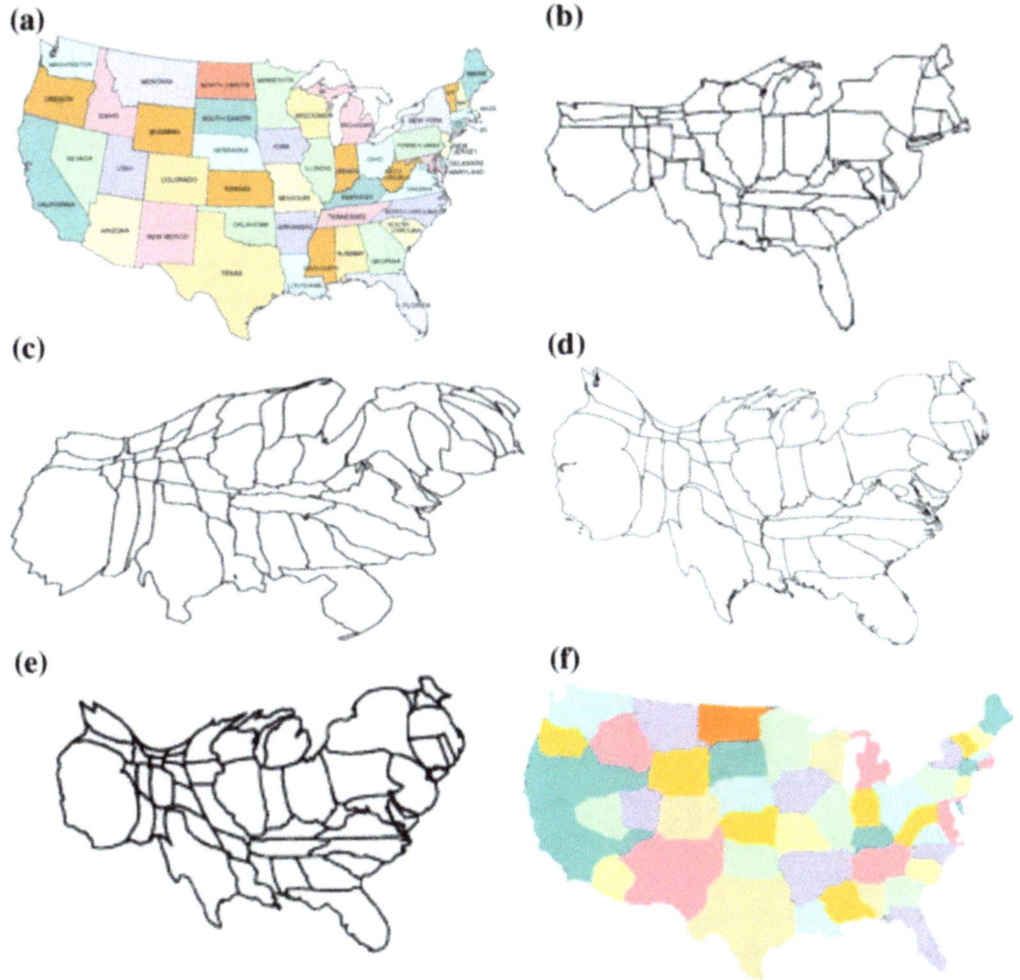

**Fig. 35.28** **a** Equal-area-projection map of USA. **b–e** Population cartograms generated for USA based on **b** Continuous cartogram (Kocmoud 1997), **c** cartodraw (Keim et al. 2004), **d** Gastner-Newman cartogram (Gastner and Newman 2004), **e** Area cartogram of the United States, with each county rescaled in proportion to its population (Gusein-Zade and Tikunov 1993), and **f** morphology-based cartogram (Sagar 2014a, b). U.S. population cartogram by Gusein-Zade and Tikunov (e: Reproduced with permission from Gusein-Zade and Tikunov 1993, page 172, Fig. 35.1, © 1993 American Congress on Surveying and Mapping). The color coding given in Fig. 35.28a is similar to that of Fig. 35.28f

Just like how CT scanning mechanism is employed to scan the brain to study several neurophysiologic processes, one can also employ such a CT-scanning mechanism, besides already existing scanning methods, to scan the rock bodies and store the scanned information in layered forms. Each layer depicts rock's cross sectional information at specific depth. Retrieval of three significant geometric and/or topologic components, describing organisation of porous medium, that include (a) pore channel, (b) pore throat, and (c) pore body in both 2-D and 3-D spaces is an important task. A 3-D fractal pore (Fig. 35.29a, b) simulated in such a way that it mimics the stacked layers of pore sections is converted into 3-D pore channel network (Fig. 35.29c, d), 3-D pore throats (Fig. 35.29e, f) and 3-D pore bodies (Fig. 35.29g, h). These decomposed pore features that are of topological importance would shed the light to derive geometric relations which further can be related with that of physical properties of porous structure.

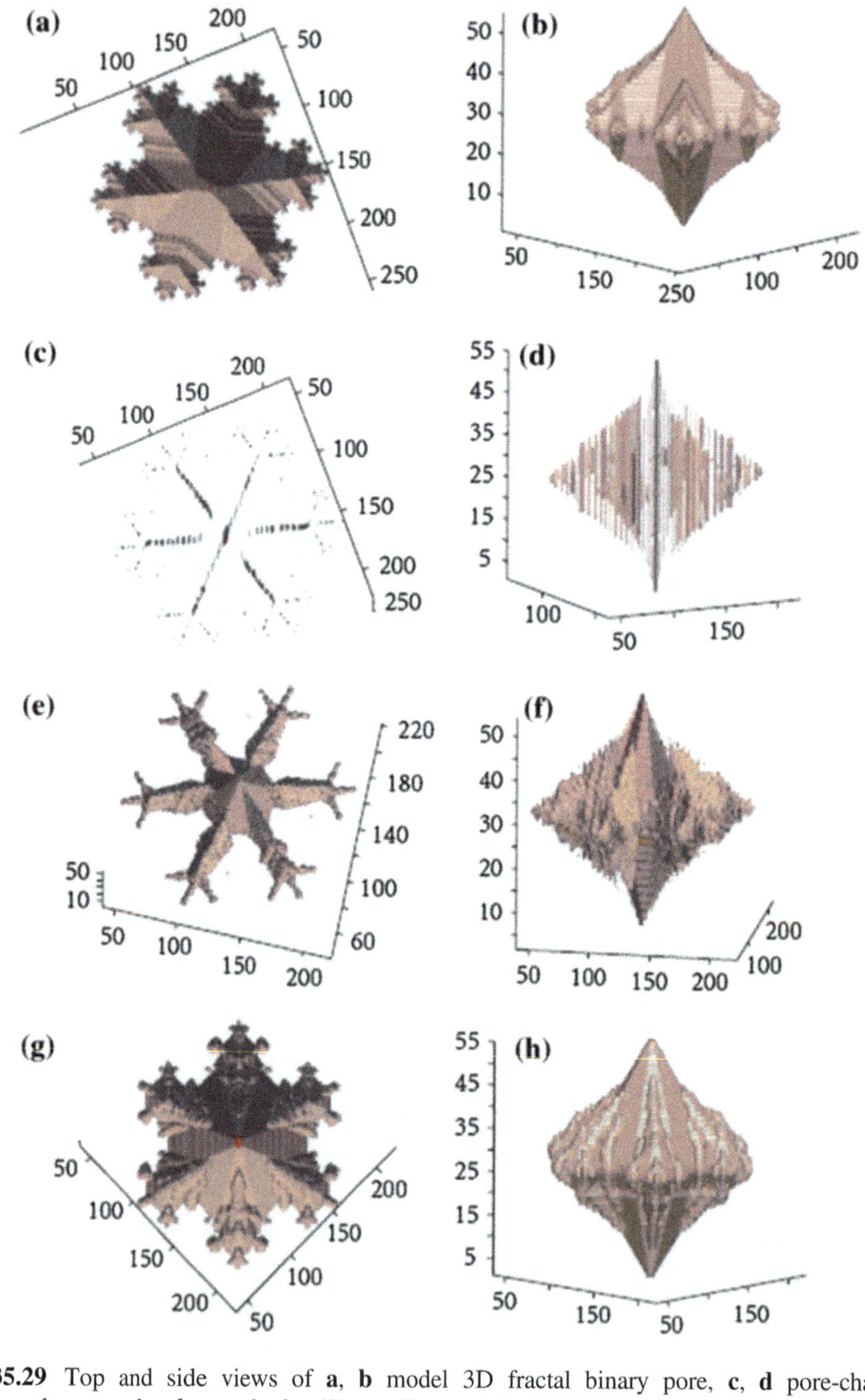


**Fig. 35.29** Top and side views of **a**, **b** model 3D fractal binary pore, **c**, **d** pore-channel, **e**, **f** pore-throat, and **g**, **h** pore-body. (*Source* Teo and Sagar 2006)

### 35.5.3 Spatial Reasoning and Planning

Mathematical morphology based algorithms developed and demonstrated shown in this subsection include to determine (i) strategically significant set(s) for spatial reasoning and planning, (ii) directional spatial relationship between areal objects (e.g.: lakes, states, sets) via origin-specific dilations, and (iii) spatial interactions via modified gravity model.

#### 35.5.3.1 Strategically Significant State (S)

Identification of a strategically significant set from a cluster of adjacent and/or non-adjacent sets depends upon the parameters that include size, shape, degrees of adjacency and contextuality, and distance between the sets. An example of cluster of sets includes continents, countries, states, cities, etc. The spatial relationships, deciphered via the parameters cited above, between such sets possess varied spatial complexities. Hausdorff dilation distance between such sets is considered to derive automatically the strategic set among the cluster of sets. The (i) dilation distances, (ii) length of boundary being shared, and (iii) degrees of contextuality and adjacency between origin-set and destination sets, which together provide solutions to derive strategically significant sets with respect to distance, degree of contextuality, degree of adjacency and length of boundary being shared. Simple mathematical morphologic operators and certain logical operations are employed in this study. Results drawn (Fig. 35.30)—by applying the proposed framework on a case study that involves spatial sets (states) decomposed from a spatial map depicting the country of India—are shown in Fig. 35.30.

This approach has been applied on data depicting randomly spread surface water bodies (Fig. 35.31a, b) and their corresponding zones of influence (Fig. 35.31c, d) within a subbasin to detect the strategically significant water body and zone of influence (Fig. 35.32a, b).

#### 35.5.3.2 Directional Spatial Relationship

We provide an approach to compute origin-specific morphological dilation distances between planar sets (e.g.: areal objects, spatially represented countries, states, cities, lakes) to further determine the directional spatial relationship between sets. Origin chosen for a structuring element that yields shorter dilation distance than that of the other possible origins of structuring element determines the directional spatial relationship between $A_i$ (origin-set) and $A_j$ (destination set). We demonstrate this approach on a cluster of spatial sets (states) decomposed from a spatial map depicting country India (Fig. 35.33a). This approach has potential to extend to any number (type) of sets on Euclidean space.

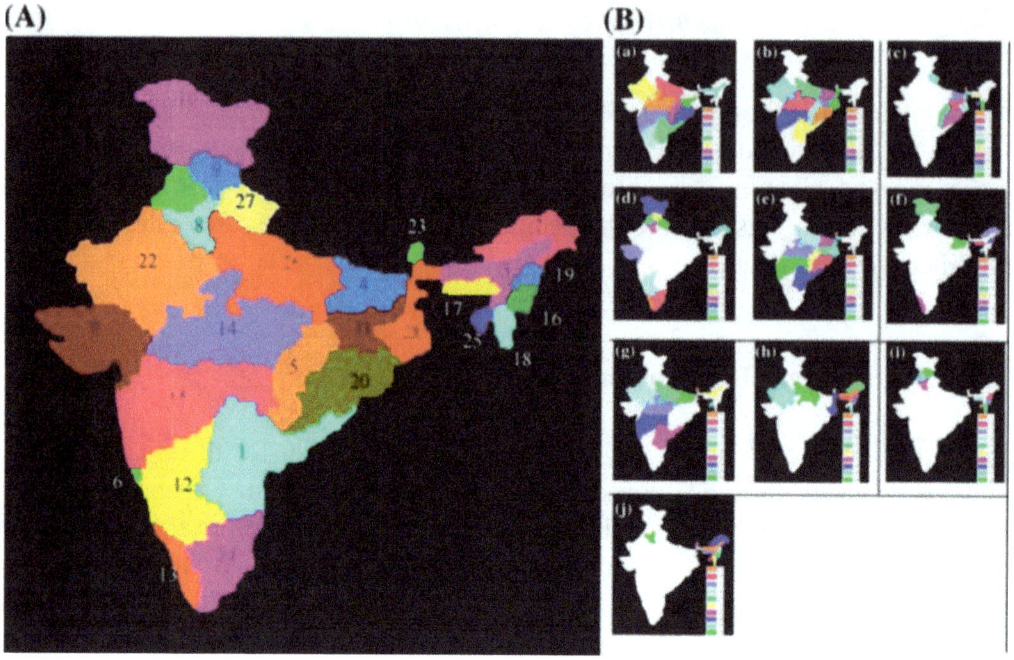


**Fig. 35.30** **A** Map of India (spatial system) with its constituent 28 states (subsets)—indexed according to alphabetical order are shown—Andhra Pradesh ($A_1$), Arunachal Pradesh ($A_2$), Assam ($A_3$), Bihar ($A_4$), Chhattisgarh ($A_5$), Goa ($A_6$), Gujarat ($A_7$), Haryana ($A_8$), Himachal Pradesh ($A_9$), Jammu & Kashmir ($A_{10}$), Jharkhand ($A_{11}$), Karnataka ($A_{12}$), Kerala ($A_{13}$), Madhya Pradesh ($A_{14}$), Maharashtra ($A_{15}$), Manipur ($A_{16}$), Meghalaya ($A_{17}$), Mizoram ($A_{18}$), Nagaland ($A_{19}$), Orissa ($A_{20}$), Punjab ($A_{21}$), Rajasthan ($A_{22}$), Sikkim ($A_{23}$), Tamilnadu ($A_{24}$), Tripura ($A_{25}$), Uttarapradesh ($A_{26}$), Uttarakhand ($A_{27}$), West Bengal ($A_{28}$), Union territories and Himalayan hill range that are parts Indian peninsular are not included in the figure. **B** Spatial representation of strategically important states in the order from 1 to 10 are shown in terms of twelve different parameters shown in Fig. 35.7. In each panel of this Figure, first 10 strategically significant states (please refer to the legend on each panel) are shown in different colors. These strategically significant sets with respect to **a** boundary being shared, **b** shortest distance from origin to destination states, **c** shortest total distance from destination states to origin state, **d** contextuality, **e** Hausdorff dilation distance, **f** spatial complexity involved in length of the boundary being shared, **g** spatial complexity in terms of contextuality, **h** spatial complexity in terms of distance from origin to destination states, **i** spatial complexity in terms of distance from destination states to origin state, **j** spatial complexity in terms of Hausdorff dilation distance from origin state to destination states. States with color-codes denote first ten strategically significant states, and the region with white space represents the states that are strategically non-significant with ranks starting from eleven to twenty eight

#### 35.5.3.3 Spatial Interactions

Hierarchical structures include spatial system (e.g. river basin), clusters of a spatial system (e.g. watersheds of a river basin), zones of a cluster (e.g. subwatersheds of a watershed), and so on. Variable-specific classification of the zones of a cluster of zones within a spatial system is the main focus of this work on spatial interactions. Variable-specific (e.g. resources) classification of zones is done by computing the levels of interaction between the *i*th and *j*th zones. Based on a heuristic argument,

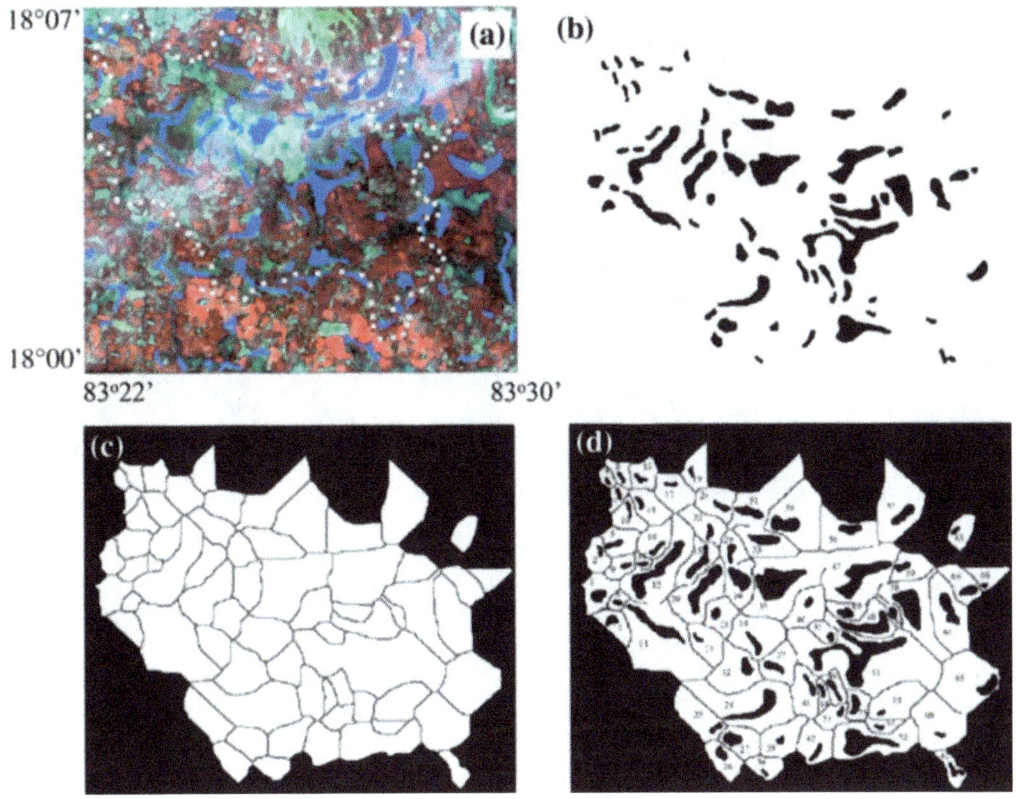


**Fig. 35.31** **a** Indian Remote Sensing satellite (IRS LISS-III) multispectral image of the study area, and the blue objects are water bodies traced from IRS LISS-III image with topographic map reference superposed on IRS LISS-III image, and white dots indicate the boundary of the considered cluster, **b** small water bodies, **c** zones of influence of corresponding water bodies, and **d** water bodies and zones of influence with labeling

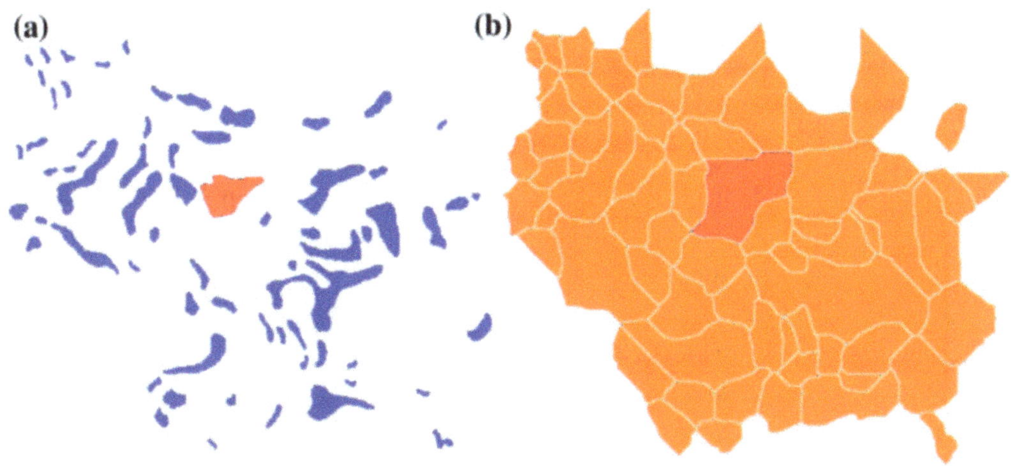


**Fig. 35.32** Spatially significant **a** water body with label 35 (Red Color), and **b** zone of water body influence labeled with 35 (Red Color)

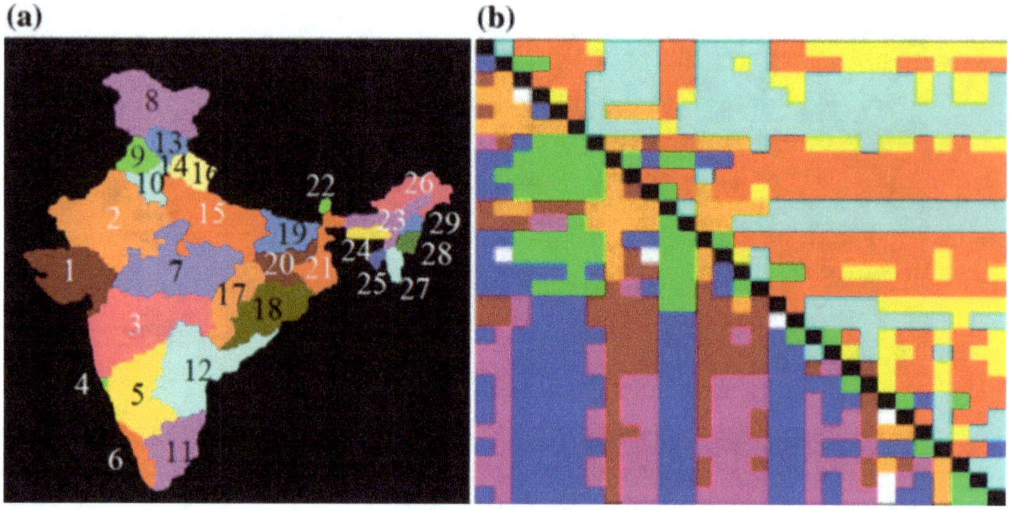


**Fig. 35.33** **a** Twenty nine sets (states of India) indexed according to alphabetical order are shown —Gujarat ($A_1$), Rajasthan ($A_2$), Maharashtra ($A_3$), Goa ($A_4$), Karnataka ($A_5$), Kerala ($A_6$), Madhya Pradesh ($A_7$), Jammu and Kashmir ($A_8$), Punjab ($A_9$), Haryana ($A_{10}$), Tamilnadu ($A_{11}$), Andhra Pradesh ($A_{12}$), Himachal Pradesh ($A_{13}$), Delhi ($A_{14}$), Uttar Pradesh ($A_{15}$), Uttaranchal ($A_{16}$), Chhattisgarh ($A_{17}$), Orissa ($A_{18}$), Bihar ($A_{19}$), Jharkhand ($A_{20}$), West Bengal ($A_{21}$), Sikkim ($A_{22}$), Assam ($A_{23}$), Meghalaya ($A_{24}$), Tripura ($A_{25}$), Arunachal Pradesh ($A_{26}$), Mizoram ($A_{27}$), Manipur ($A_{28}$), Nagaland ($A_{29}$). Union Territories are not considered. **b** Directional spatial relationship shown in colored matrix form in which there are 29 rows and 29 columns and a color in each grid cell explaining directional relationship between each state to other 28 states

we proposed a modified gravity model for the computations of levels of interaction between the zones. This argument is based on the following two facts: (i) the level of interaction between the *i*th and *j*th zones, with masses $m_i$ and $m_j$ is direction-dependent, and (ii) the level of interactions between the *i*th and *j*th zones with corresponding masses, situated at strategically insignificant locations would be much different (lesser) from that of the *i*th and *j*th zones with similar masses but situated at strategically highly significant locations. With the support of this argument, we provide a modified gravity model by incorporating the asymmetrical distances, and the product of location significance indexes of the corresponding zones. This modified gravity model yields level of interaction between the two zones that satisfies the realistic characteristic that is level of interaction between the zones is direction-dependent.

Each state of India is designated with ranks in terms of its (i) location significance index, (ii) strengths of interaction of all states with a specific state, (iii) strengths of interaction with other states, and (iv) strength out of (ii) and (iii) (Fig. 35.34a–d). Further by employing a modified gravity model, 28 states ($X_1$ to $X_{28}$) of India (Fig. 35.30A) are paired into best interacting to least interacting pairs with respect to areal extents of states as a variable (Fig. 35.35a–j).

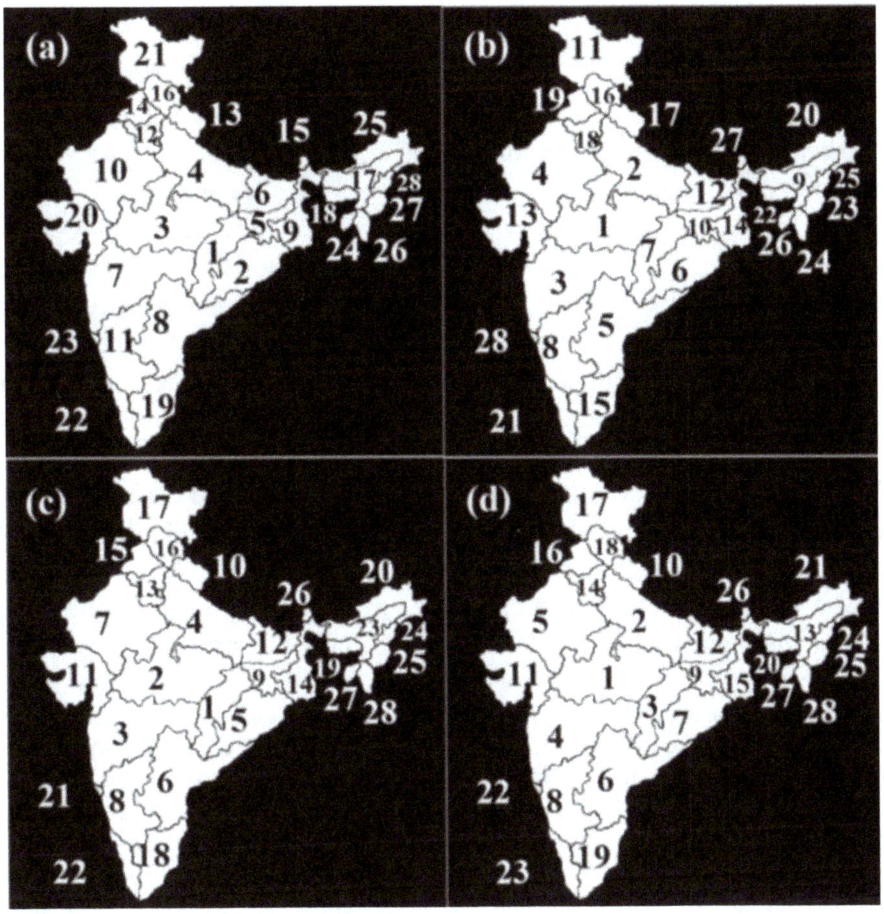


**Fig. 35.34** India map with each state designated with a rank with respect to four different parameters. **a** $\varphi X_i$, **b** $\max_i \left( \sum_j FX_{ij} \right)$, **c** $\max_j \left( \sum_i FX_{ji} \right)$, and **d** $\max \left( \max_i \left( \sum_j FX_{ij} \right), \max_j \left( \sum_i FX_{ji} \right) \right)$

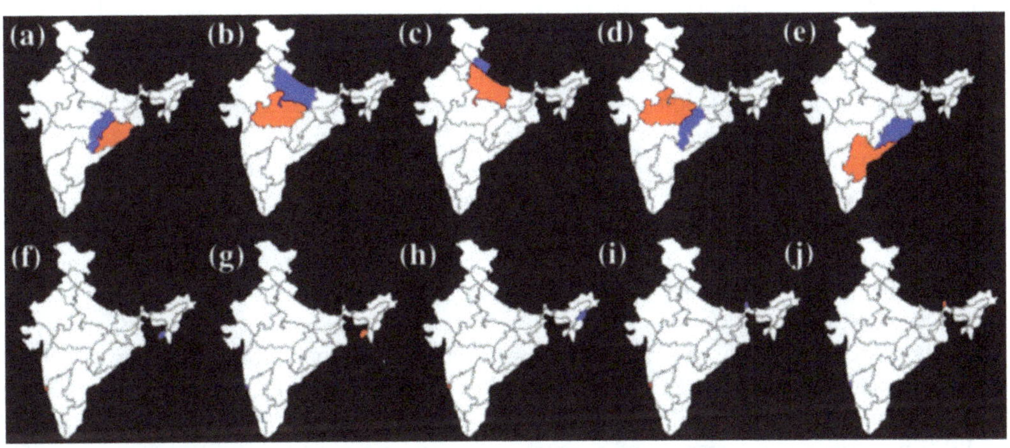


**Fig. 35.35** Five best pairs exhibited the high levels of interactions **a** $X_{20,5}$, **b** $X_{14,26}$, **c** $X_{26,27}$, **d** $X_{14,5}$, and **e** $X_{1,20}$. Five pairs exhibited the least levels of interactions **f** $X_{6,25}$, **g** $X_{25,6}$, **h** $X_{6,19}$, **i** $X_{6,23}$, and **j** $X_{23,6}$

## 35.6 Conclusions

From our attempts since early 1990s, we could clearly see a great potential for mathematical morphological transformations in the three aspects (retrieval, analysis and reasoning, and modeling) of relevance to geosciences and GISci. This chapter provided a brief illustrative review on how mathematical morphology could be applied to deal with varied topics of relevance to mathematical geosciences and geographical information science communities. Reader is encouraged to dig cited references for more details. Our studies show that there exist several open problems of relevance to the mathematical geosciences community. These open problems could be well-handled by mathematical morphology. Some of the recent advances of mathematical morphology and their applications in spatial data segmentation and morphological clustering were discussed. Applications of both classical and modern mathematical morphological transformations in geosciences and GISci are yet to be seen in full-length. It is our hope that most visible and highly distinguished scientists who are active in the IAMG activities would spread a word wide across and would spur the interest of young researchers to take the strides forward.

**Acknowledgements** I would like to gratefully acknowledge Jean Serra and collaborators, mentors, reviewers, examiners, friends, employers, well-wishers, and doctoral students—S.V.L.N. Rao, B.S.P. Rao, M. Venu, K.S.R. Murthy, Gandhi, Srinivas, Radhakrishnan, Lea Tien Tay, Chockalingam, Lim Sin Liang, Teo Lay Lian, Dinesh, Jean Serra, Gabor Korvin, Arthur Cracknell, Deekshatulu, Philippos Pomonis, Peter Atkinson, Hien-Teik Chuah, Laurent Najman, Jean Cousty, Christian Lantuejoul, Alan Tan, VC Koo, Rajesh, Ashok, Pratap, Rajashekhara, Sravan, Aditya, Sampriti, and several others. I am grateful to Frits Agterberg for his phenomenal support and suggestions that made this review chapter readable and understandable.

## References

Agterberg FP (2001) Appreciation of contributions by Georges Matheron and John Tukey to a mineral-resources research project. Nat Resour Res 10(4):287–295

Agterberg FP (2004) Georges matheron: founder of spatial statistics. Earth Sci Hist 23(2):325–334

Beucher S (1990) Segmentation d-images et morphologic mathematique. These Docteur en Morphologic Mathematique. Ecole des Mines de Paris

Beucher S (1999) Mathematical morphology and geology: when image analysis uses the vocabulary of earth science: a review of some applications. In: Proceedings of geovision—99. University of Liege, Belgium, 6–7 May 1999, pp 13–16

Beucher S, Meyer F (1992) The morphological approach to segmentation: the watershed transformation. In: Dougherty E (ed) Mathematical morphology in image processing. Marcel Dekker, NY, pp 433–481

Brunet D, Sills D (2017) A generalized distance transform: Theory and applications to weather analysis and forecasting. IEEE Trans Geosci Remote Sens 55(3):1752–1764

Challa A, Danda S, Sagar BSD (2016) Morphological interpolation for temporal changes. In: 2016 IEEE international geoscience and remote sensing symposium (IGARSS), Beijing, pp 3358–3361. https://doi.org/10.1109/igarss.2016.7729868

Chockalingam L, Sagar BSD (2003) Automatic generation of sub-watershed map from Digital Elevation Model: a morphological approach. Int J Pattern Recognit Artif Intell 17(2):269–274

Chockalingam L, Sagar BSD (2005) Morphometry of networks and non-network spaces. J Geophys Res-Solid Earth (American Geophysical Union) 110(B08203). https://doi.org/10.1029/2005jb003641

Danda S, Challa A, Sagar BSD (2016) A morphology-based approach for cloud detection. In: 2016 IEEE international geoscience and remote sensing symposium (IGARSS), Beijing, pp 80–83. https://doi.org/10.1109/igarss.2016.7729011

Gastner MT, Newman M (2004) Diffusion-based method for producing density-equalizing maps. Proc Natl Acad Sci 101(20):7499–7504

Gusein-Zade S, Tikunov V (1993) A new technique for constructing continuous cartograms. Cartogr Geogr Inf Syst 20(3):66–85

Harbaugh JW (2014) Letter to the editor: what do mathematical geoscientists do? IAMG Newsl 89:5

Keim D, North S, Panse C (2004) Cartodraw: a fast algorithm for generating contiguous cartograms. IEEE Trans Vis Comput Graph 10:95–110

Kocmoud C (1997) Constructing continuous cartograms: a constraint-based approach, MA thesis, Texas A & M University

Lantuejoul C (1978) La sequelettisation et son application aux mesures topologiques des mosaiques polycristallines. These de Docteur-Ingnieur. School of Mines, Paris, France

Lantuejoul C, Beucher S (1981) On the use of the geodesic metric in image analysis. J Microsc 121(1):39–49

Lim SL, Sagar BSD (2008a) Cloud field segmentation via multiscale convexity analysis. J Geophys Res-Atmos 113:D13208. https://doi.org/10.1029/2007jd009369. (17 pages)

Lim SL, Sagar BSD (2008b) Derivation of geodesic flow fields and spectrum in digital topographic basins, Discrete Dyn Nat Soc (Article ID 312870, 26 pages). https://doi.org/10.1155/2008/312870

Lim SL, Koo VC, Sagar BSD (2009) Computation of complexity measures of morphologically significant zones decomposed from binary fractal sets via multiscale convexity analysis. Chaos Solitons Fractals 41(3):1253–1262

Lim SL, Sagar BSD, Koo VC, Tay LT (2011) Morphological convexity measures for terrestrial basins derived from digital elevation models. Comput Geosci 37(9):1285–1294

Mandelbrot BB (1982) Fractal geometry of nature. WH Freeman & Co, San Francisco, p 468

Maragos PA (1989) Pattern spectrum and multiscale shape representation. IEEE Trans Pattern Anal Mach Intell 11:701–716

Maragos PA, Schafer RW (1986) Morphological skeleton representation and coding of binary images. In: IEEE transactions on acoustics, speech and signal processing, ASSP-34(5)

Matheron G (1975) Random sets and integral geometry. Wiley, New York

Meyer F (1980) Feature extraction by mathematical morphology in the field of quantitative cytology. Technical report of Ecole nationale superiere des mines de Paris, Fountainbleau, France

Miliaresis GC, Argialas DP (1999) Segmentation of physiographic features from global digital elevation model/GTOPO30. Comput Geosci 25:715–728

Najman L, Talbot H (2010) Mathematical morphology. Wiley, p 507

Najman L, Barrera J, Sagar BSD, Maragos P, Schonfeld D (2012) Preface on filtering and segmentation with mathematical morphology. IEEE J Sel Top Signal Process 6(7):736–886

Radhakrishnan P, Sagar BSD, Teo LL (2004) Estimation of fractal dimension through morphological decomposition. Chaos Solitons Fractals 21(3):563–572

Rajashekara HM, Vardhan P, Sagar BSD (2012) Generation of zonal map from point data via weighted skeletonization by influence zone. IEEE Geosci Remote Sens Lett 9(3):403–407

Rivest JF, Beucher S, Delhomme JP (1992) Marker-controlled segmentation: an application to electrical borehole. J Electron Imaging 1(2):136–142

Sagar BSD (1996) Fractal relations of a morphological skeleton. Chaos Solitons Fractals 7 (11):1871–1879

Sagar BSD (1998) Numerical simulations through first order nonlinear difference equation to study highly ductile symmetric fold (HDSF) dynamics: a conceptual study. Discret Dyn Nat Soc 2 (4):281–298

Sagar BSD (1999a) Estimation of number-area-frequency dimensions of surface water bodies. Int J Remote Sens 20(13):2491–2496

Sagar BSD (1999b) Morphological evolution of a pyramidal sandpile through bifurcation theory: a qualitative model. Chaos Solitons Fractals 10(9):1559–1566

Sagar BSD (2000a) Fractal relation of medial axis length to the water body area. Discret Dyn Nat Soc 4(1):97

Sagar BSD (2000b) Multi-fractal-interslipface angle curves of a morphologically simulated sand dune. Discret Dyn Nat Soc 5(2):71–74

Sagar BSD (2001a) Generation of self organized critical connectivity network map (SOCCNM) of randomly situated surface water bodies, Letters to Editor. Discret Dyn Nat Soc 6(3):225–228

Sagar BSD (2001b) Hypothetical laws while dealing with effect by cause in discrete space, Letter to the Editor. Discret Dyn Nat Soc 6(1):67–68

Sagar BSD (2001c) Quantitative spatial analysis of randomly situated surface water bodies through f-α spectra. Discret Dyn Nat Soc 6(3):213–217

Sagar BSD (ed) (2001) For a special issue of Journal of Mathematical Geosciences. In: Memory of the Late Professor SVLN Rao, vol 33, no 3. Kluwer Academic Publishers pp 245–396

Sagar BSD (2005a) Discrete simulations of spatio-temporal dynamics of small water bodies under varied streamflow discharges, (invited paper). Nonlinear Process Geophys (American Geophysical Union) 12:31–40

Sagar BSD (2005b) Qualitative models of certain discrete natural features of drainage environment. Allied Publishers Limited, New Delhi, p 232

Sagar BSD (2007) Universal scaling laws in surface water bodies and their zones of influence. Water Resour Res 43(2):W02416

Sagar BSD (ed) (2009) In: Proceedings on "spatial information retrieval, analysis, reasoning and modelling" seminar held during 18–20 Mar 2009, p 231

Sagar BSD (2010) Visualization of spatiotemporal behavior of discrete maps via generation of recursive median elements. IEEE Trans Pattern Anal Mach Intell 32(2):378–384

Sagar BSD (2013) Mathematical morphology in geomorphology and GISci. CRC Press, Boca Raton. p 546

Sagar BSD (2014a) Cartograms via mathematical morphology. Inf Vis 13(1):42–58

Sagar BSD (2014b) Erratum to visualization of spatiotemporal behavior of discrete maps via generation of recursive median elements. In: IEEE Trans Pattern Anal Mach Intell vol 36(3)

Sagar BSD (2018) Variable-specific classification of zones, pairs of zones, and clusters of a spatial system via modified gravity model. In: IEEE Trans Emerging Topics Comput (In Press). https://doi.org/10.1109/tetc.2016.2633436

Sagar BSD, Bruce LM (Eds.) (2005) Surficial mapping. IEEE Geosci Remote Sens Lett 2(4): 375–408

Sagar BSD, Chockalingam L (2004) Fractal dimension of non-network space of a catchment basin Geophys. Res. Lett (American Geophysical Union) 31(12):L12502

Sagar BSD, Lim SL (2015a) Ranks for pairs of spatial fields via metric based on grayscale morphological distances. IEEE Trans Image Process 24(3):908–918

Sagar BSD, Lim SL (2015b) Morphing of grayscale DEMs via morphological interpolations. IEEE J Sel Top Appl Earth Observ Remote Sens 8(11):5190–5198

Sagar BSD, Murthy KSR (2000) Generation of fractal landscape using nonlinear mathematical morphological transformations. Fractals 8(3):267–272

Sagar BSD, Rao BSP (1995a) Computation of strength of nonlinearity in Lakes, letter to the editor. Comput Geosci 21(3):445

Sagar BSD, Rao BSP (1995b) Fractal relation on perimeter to the water body area. Curr Sci 68(11):1129–1130

Sagar BSD, Rao BSP (1995c) Possibility on usage of return maps to study dynamics of lakes: hypothetical approach. Curr Sci 68(9):950–954

Sagar BSD, Rao BSP (1995d) Ranking of Lakes: Logistic maps. Int J Remote Sens 16(2):368–371
Sagar BSD, Rao CB (eds) (2003) Special issue on quantitative image morphology. Int J Pattern Recogn Artif Intell 17(2):163–330
Sagar BSD, Serra J (eds) (2010) Special Issue on "spatial information retrieval, analysis, reasoning and modelling". Int J Remote Sens 31(22):5747–6032
Sagar BSD, Tien TL (2004) Allometric power-law relationships in a Hortonian Fractal DEM. Geophys Res Lett (American Geophysical Union), 31(6):L06501
Sagar BSD, Venu M (2001) Phase space maps of a simulated sand dune: a scope. Discret Dyn Nat Soc 6(1):63–65
Sagar BSD, Gandhi G, Rao BSP (1995a) Applications of mathematical morphology on water body studies. Int J Remote Sens 16(8):1495–1502
Sagar BSD, Venu M, Rao BSP (1995b) Distributions of surface water bodies. Int J Remote Sens 16(16):3059–3067
Sagar BSD, Omoregie C, Rao BSP (1998a) Morphometric relations of fractal-skeletal based channel network model. Discret Dyn Nat Soc 2(2):77–92
Sagar BSD, Venu M, Gandhi G, Srinivas D (1998b) Morphological description and interrelationship between force and structure: a scope to geomorphic evolution process modelling. Int J Remote Sens 19(7):1341–1358
Sagar BSD, Venu M, Murthy KSR (1999) Do skeletal network derived from water bodies follow Horton's laws? J Math Geol 31(2):143–154
Sagar BSD, Venu M, Srinivas D (2000) Morphological operators to extract channel networks from digital elevation models. Int J Remote Sens 21(1):21–30
Sagar BSD, Srinivas D, Rao BSP (2001) Fractal skeletal based channel networks in a triangular initiator basin. Fractals 9(4):429–437
Sagar BSD, Rao CB, Raj B (2002) Is the spatial organization of larger water bodies heterogeneous? Int J Remote Sens 23(3):503–509
Sagar BSD, Murthy MBR, Radhakrishnan P (2003a) Avalanches in numerically simulated sand dune dynamics. Fractals 11(2):183–193
Sagar BSD, Murthy MBR, Rao CB, Raj B (2003b) Morphological approach to extract ridge-valley connectivity networks from digital elevation models (DEMs). Int J Remote Sens 24(3): 573–581
Sagar BSD, Rangarajan G, Veneziano D (eds) (2004) Special issue on fractals in geophysics. Chaos Solitons & Fractals 19(2):237–462
Sagar BSD, Rajesh N, Vardhan SA, Vardhan P (2013) Metric based on morphological dilation for the detection of spatially significant zones. IEEE Geosci Remote Sens Lett 10(3):500–504
Sathymoorthy D, Radhakrishnan P, Sagar BSD (2007) Morphological segmentation of physiographic features from DEM. Int J Remote Sens 28(15)
Serra J (1982) Image analysis and mathematical morphology. Academic Press, London, p 610
Serra J (1988) Image analysis and mathematical morphology: theoretical advances. Academic Press, p 411
Soille P (2003) Morphological image analysis, principles and applications, 2nd edn. Springer, Berlin
Sternberg SR (1986) Greyscale morphology. Comput Vis Graph Image Process 35:333–355
Tay LT, Sagar BSD, Chua HT (2005a) Derivation of terrain roughness indicators via Granulometries. Int J Remote Sens 26(18):3901–3910
Tay LT, Sagar BSD, Chuah HT (2005b) Allometric relationships between travel-time channel networks, convex hulls, and convexity measures. Water Resour Res (American Geophysical Union) 46(2):W06502. https://doi.org/10.1029/2005WR004092
Tay LT, Sagar BSD, Chuah HT (2005c) Analysis of geophysical networks derived from multiscale digital elevation models: a morphological approach. IEEE Geosci Remote Sens Lett 2(4): 399–403
Tay LT, Sagar BSD, Chuah HT (2007) Granulometric analysis of basin-wise DEMs: a comparative study. Int J Remote Sens 28(15):3363–3378

Teo LL, Sagar BSD (2005) Reconstruction of pore space from pore connectivity network via morphological transformations. J Microsc (Oxford) 219(Pt 2):76–85

Teo LL, Sagar BSD (2006) Modeling, characterization of pore-channel, throat and body. Discret Dyn Nat Soc (an International Journal) (Article ID 89280) 1–24

Teo LL, Radhakrishnan P, Sagar BSD (2004) Morphological decomposition of sandstone pore-space: fractal power-laws. Chaos Solitons & Fractals

Vardhan SA, Sagar BSD, Rajesh N, Rajashekara HM (2013) Automatic detection of orientation of mapped units via directional granulometric analysis. IEEE Geosci Remote Sens Lett 10(6):1449–1453

# Part V
# Reminiscences

# Chapter 36
# IAMG: Recollections from the Early Years

**John Cubitt and Stephen Henley**

John Cubitt and Stephen Henley, with contributions from T. Victor (Vic) Loudon, EHT (Tim) Whitten, John Gower, Daniel (Dan) Merriam, Thomas (Tom) Jones, and Hannes Thiergärtner

This chapter records some of the dramatic history of the first few years of the International Association for Mathematical Geology (IAMG, much later renamed the International Association for Mathematical Geosciences), and its subsequent development told mostly through recollections (both professional and personal) of some of its early members. It complements the paper by Václav Němec in this volume who discusses his own experiences leading up to and following the foundation of the Society.

The IAMG was formed on 22nd August 1968, in a meeting at the International Geological Congress in Prague, Czechoslovakia, attended by 20 scientists from around the world. This followed preparatory work by an ad hoc committee of 14 (not all of whom were able to attend the formation meeting) which formulated statutes and by-laws and proposed names of a first set of officers.

J. Cubitt (✉)
Newhaven, Church Street, Holt, Wrexham LL13 9JP, UK
e-mail: johnmcubitt@gmail.com

S. Henley
Resources Computing International Limited, 185 Starkholme Road, Matlock, Derbyshire DE4 5JA, UK





B. S. Daya Sagar et al. (eds.), *Handbook of Mathematical Geosciences*,
https://doi.org/10.1007/978-3-319-78999-6_36

## 36.1 The Birth of Mathematical Geology and the Origins of the IAMG

### *36.1.1 Vic Loudon*

The comprehensive framework for sharing geological knowledge developed over a long period, in the form of a shared network of scientific books and papers, maps, records, samples, specimens, reports, and guides—including the systematic output of regional and national geological surveys. Geological projects could contribute new information within a framework of existing knowledge and the requirements of publication. This framework, however, did not anticipate the arrival of the computer.

In the early 1960's some enthusiasts considered that computers could have an important role in creating new, widely shared mechanisms for analysing, exchanging and integrating numerical information. But to many geologists at that time, computers were a passing fad—surely the complexity of geological observation and thinking could not be reduced to mathematics, never mind its mechanical representation! Nevertheless, computer programs were shown to handle recurring statistical tasks, even if only in the detail of a geological study. They might also build on the work of others. But that requires communication, a shared objective, and in due course a shared framework.

At that early experimental stage, computer applications in geology were generally rather trivial, overlapping, uncoordinated and unpublishable. They were nevertheless essential to determine which possibilities might be fruitful, and which would be duplication. To help programmers to gain a broader view of similar work elsewhere, an informal 'Geologically Oriented Scheme for Sharing Information on Programming (GOSSIP)' was maintained at Reading University in England. Notes from various workers in geological computing were assembled and typed onto punched cards. These were sorted and revised by hand, the results printed on a typewriter connected to the keypunch, and mailed to the participants. The last of several editions was circulated in 1966 (GOSSIP 1966). It provides an insight into a fast-growing area where many individuals had been exploring possibilities independently, and beginning to develop an initial overview. Apart from one mention of information retrieval, the applications referred exclusively to numerical data.

Later, to quote Krumbein (1969): "…on the one hand we observe a growth in the complexity of programs, and on the other hand a spreading of essentially the same computer techniques through the many subfields of geology… the underlying methodology is so similar in all fields…that most speakers shifted emphasis from standard or conventional techniques to consideration of new and more analytical ways of setting up models applicable to their own fields."

### 36.1.2 John Cubitt and Stephen Henley

Merriam (1981) put Loudon's comments into a historical perspective by giving a helpful summary of the development of mathematical geology. This shows that the introduction of mathematical methods into the science of geology was very slow, until the advent of computer technology, despite the efforts of such notable scientists as Paolo Frisi, Charles Lyell, Paul Deshayes, Charles Babbage, and Lord Kelvin, as well as statisticians such as Karl Pearson and A.N. Kolmogorov, and others such as R. Everest, chief surveyor for India. It is well-known that the first edition of Lyell's Principles of Geology (1830) included statistical data that he used to justify his subdivision of the Tertiary; however, once the classification was accepted, this statistical scaffolding was not deemed important enough to be retained in subsequent editions.

The earliest consistent efforts towards routine application of quantitative methods in geology were made by A.B. Vistelius from 1941 onwards, while the use of computers was pioneered by W.C. Krumbein starting with a book in 1958 jointly written with L.L. Sloss (Krumbein and Sloss 1958). For the next ten years, there was a steadily increasing number and variety of publications on computational methods applied to geology mostly but not exclusively statistical.

### 36.1.3 Tim Whitten

Whitten noted that prior to 1968, different approaches to quantitative geology applied around the world. However, at the IAMG formation meeting, dissimilar approaches came together, having evolved principally in the Soviet Union, Western Europe, and U.S.A. Vistelius championed the concept that Mathematical Geology is a separate branch of science based on testing geological hypotheses mathematically, and that this should be IAMG's primary focus (Whitten 2003, 2004 pp. 384–5); for some years, he had contended it is not particularly important merely to manipulate geological data statistically. Dech and Henley (2003, p. 368) noted Vistelius (1991) considered that, if a science does not use mathematical modelling in constructing conclusions, "… it can be considered as belonging to the pre-Newtonian period, … behind the present-day level of research by approximately 300 years."

## 36.2 The Role of the Kansas Geological Survey in the Origins of the IAMG

### 36.2.1 Tom Jones

When I got to Northwestern University in 1967, I found several faculty members were quantitative, along with a few students. Krumbein was doing work in several areas at that time, notably geographic forms, Markov chains, and modifications of trend analyses. The Kansas Geological Survey Computer Contributions (KGSCC), spearheaded by Dan Merriam, provided an ongoing source of publications on mathematical geology and associated software.

### *36.2.2 Vic Loudon*

In the late 1960s, Dan Merriam led a pioneering group of geological programmers in the Kansas Geological Survey at the University of Kansas, describing the results of their computing activities in its own publication, the KGSCC. In 1967–8, Richard Reyment spent some time at the Kansas Geological Survey, on sabbatical leave from the University of Uppsala in Sweden. He was another of the prime movers in establishing the IAMG (its first General Secretary and subsequently its President, and in 2002 the recipient of that organization's Commendation). I was privileged to listen to one of their conversations, where they agreed that a formal body was desirable to assist and encourage documentation and communication of these developments.

### *36.2.3 Tim Whitten*

The momentum driving a founding meeting in 1968 really stemmed from the Kansas Survey folk—the main activist there was Dan Merriam, who was very keen on instituting an international society and I imagine it was he who got the meeting included in the IGC programme.

## 36.3 Name and Establishment of the Society

### *36.3.1 Vic Loudon*

Merriam (perhaps only in the wishful thinking of my biased mind), in the conversation referred to above, seemed to take the view that computer science, rather than mathematics, was the key issue. However, it seemed that the geological establishment at that time might find 'mathematics' more acceptable. Subsequently, Richard Reyment organised an ad hoc committee for the purpose of founding an association for the promotion of mathematical geology.

### 36.3.2 *John Gower*

I remember there being discussions on what name to give to the new Society and that somebody had suggested Geometrics echoing the names of the Biometrics and Psychometric Societies. It was noticed that Geometry had forestalled that suggestion so it became Mathematical Geology succeeded by Geomathematics succeeded by Mathematical Geosciences but perhaps geometrics was not so bad an idea as it seemed because originally geometry was about Measuring the Earth. Indeed, the mathematical geologists had nomenclatural problems from the start when, because of the political climate at that time, they could not appoint D.G. Krige from South Africa who would have been the obvious choice, to the Presidency. They made him a Councillor.

### 36.3.3 *Dan Merriam*

The 1968 IAMG foundation meeting followed considerable correspondence and fact-finding by the ad hoc committee whose Members were:

| | |
|---|---|
| F.P. Agterberg (Canada) | D.F. Merriam (USA) |
| C.J. Allègre (France) | V. Němec (Czechoslovakia) |
| F. Chayes (USA) | R.A. Reyment (Sweden) |
| J.C. Griffiths (USA) | E. Schlegel (DDR) |
| J.W. Harbaugh (USA) | A.B. Vistelius (USSR) |
| W.C. Krumbein (USA) | G.S. Watson (USA) |
| T.V. Loudon (UK) | E.H.T. Whitten (USA) |

This committee formulated a set of statutes and by-laws (largely written by R.A. Reyment in compliance with IUGS and ISI guidelines), made provision for establishing a journal, and proposed a slate of officers.

## 36.4 Foundation of IAMG Publications

### 36.4.1 *Tom Jones*

As time went on, the IAMG formed the journal Computers & Geosciences (C&G). The Kansas Geological Survey Computer Contributions (KGSCC) series was discontinued in 1970, probably in part due to C&G and as a result of Dan Merriam moving to Syracuse University to become Chairman of the Geology Department. The American Association of Petroleum Geologists (AAPG) formed a committee on Computer Applications, but I do not recall that it had much influence. A North American group formed MGUS (Mathematical Geologists of the United States) around the mid 70's with the goal that MGUS would eventually become a regional group tied to IAMG. Much later (I believe 1985) AAPG sponsored a computer-oriented magazine, GEOBYTE.

### *36.4.2 Vic Loudon*

To quote the IAMG website: 'The mission of the International Association for Mathematical Geosciences is to promote, worldwide, the advancement of mathematics, statistics and informatics in the geosciences'. It established a journal and a newsletter. From its inception in 1968, an important role of the IAMG has been publication—initially in its journal Mathematical Geology (now Mathematical Geosciences) which 'publishes original, high-quality, interdisciplinary papers focusing on quantitative methods and studies of the Earth, its natural resources and the environment.'

In 1975, Computers & Geosciences was established as a journal devoted to all aspects of computing in the geosciences. It was published by Elsevier with Merriam as its first Editor-in-Chief, and in due course became another IAMG publication. It publishes research papers on computer methods in the geosciences, such as spatial analysis, geomathematics, modelling, simulation, statistical and artificial intelligence methods, e-geoscience, geoinformatics, geomatics, geocomputation, image analysis, remote sensing and geographical information science.

These journals (including the later IAMG publication Natural Resources Research) filled a growing gap in the maturing area of computer applications, and became an essential part of geological computing. The earlier ad hoc sharing of results and many individually trivial, and therefore unpublishable, exploratory studies had helped to create the basis for their development and their integration. This is relevant now, as communication heads towards another looming gap, described later.

## 36.5 Prague

### *36.5.1 Dan Merriam*

The organizational meeting of the IAMG took place at the XXIII International Geological Congress (IGC) in Prague's New Technical University, Czechoslovakia, on the 22nd of August 1968. It was attended by 20 representatives from 10 different countries:

| | |
|---|---|
| F.P. Agterberg (Canada) | R.B. McCammon (USA) |
| F. Benkö (Hungary) | D.F. Merriam (USA) |
| D.J. Burdon (FAO of United Nations) | V. Němec (Czechoslovakia) |
| C.J. Dixon (UK) | R.A. Reyment (Sweden) |
| J.W. Harbaugh (USA) | D.A. Rodionov (USSR) |
| R. Hesse (FRG) | H. Thiergärtner (DDR) |
| R. Ivanov (Bulgaria) | A.B. Vistelius (USSR) |
| H. Knape (DDR) | G.S. Watson (USA) |
| V. Kutolin (USSR) | E.H.T. Whitten (USA) |
| T.V. Loudon (UK) | P. Wilkinson (UK) |

### 36.5.2 Tom Jones

Several Northwestern University faculty members went to the Prague IUGS meeting, but Krumbein and Whitten were the only ones who were associated with the founding of IAMG. Of course, when word came of the Soviet army moving into Prague during the IUGS, everyone at Northwestern University was concerned about safety issues, but no news was available to us. All went well, and they had lots of stories to tell upon their return, along with photos of tanks driving down the street in front of their hotel.

### 36.5.3 Tim Whitten

I was a founding IAMG Member in Prague in 1968 and, in several papers (Whitten 2003, 2004, 2005), I've alluded to that experience and to Vistelius' participation in the founding.

In many ways, 1968 was an extraordinary year that rocked the world (cf. Kurlansky 2004). Some enthusiasts gathered to create the IAMG in exciting, but tragic, times. Soviet troops had occupied the city on August 21st; guns of encircling Soviet tanks pointed at the University, which was the centre for printing and disseminating news. Vistelius was elected IAMG President and Krumbein 'Past President' (a designation he appreciated and found amusing!); both are fathers of geological modelling methodology.

Opening of the IGC itself was fine but it was immediately followed by the invasion. The founding meeting was therefore brief, hurried, and somewhat stressful because the Americans present were anxious to get away to complete and execute their evacuation plans (being organised by the US Embassy); they soon left Prague. However, despite the fact that I was an official delegate of Northwestern University, the organisers of the US evacuation wouldn't have anything to do with me, because I was on a UK passport. With most other delegates, I continued supporting and attending IGC sessions until, after a couple more days, the Czechs felt it necessary to terminate the Congress (at a very emotional hastily arranged closing ceremony). My friends in the Finnish contingent immediately promised I could evacuate with their party but, in the end, I learned the British Embassy was organising two coaches to drive out to Nuremberg in Bavaria—the route went through Pilsen and the passengers, being British, made the Czech drivers (against their concerns and protests) stop at the Pilsen brewery to have a last tankard apiece —thence to Nuremberg and a special BA plane via Amsterdam to Heathrow.

I had been on an excellent 14-day field trip right through Czechoslovakia before the Congress, mainly organised and led by Václav Němec—there always seemed to be an orchestra at dinner, stridently playing Dr. Zhivago, much to the consternation of the several Russian delegates.

Dr. Václav Němec from Prague deserves a word. He played a large role in the Prague IGC. In addition to his Prague home, my wife and I visited his attractive rustic cottage (in the forest someway up to the north) once whilst the country was still Russian occupied. He contributed quite a lot to one part of mathematical geology by regularly organising well-attended conferences at Príbram (not too far SW of Prague) through the 1970s—as appropriate to a mining town, there was quite a focus on mining issues and latterly on geo-ethics; these were loosely connected with IAMG. After truly awful food available during the conferences, he always organised a magnificent closing banquet (always pronounced 'basket')—don't know where he wrestled up the fine food and drink!

### 36.5.4 Dan Merriam

Modified from an interview with the Lawrence Journal World August 21 2008 with permission of the Merriam family–

In August 1968, the Soviet Union's Warsaw Pact allies rolled into the Eastern European country with tanks and planes to squash the movement known as the "Prague Spring," which sought more political and social freedoms during the Cold War years. Dan Merriam, who sadly died in 2017 after retiring from the Kansas Geological Survey, escaped the country safely on a train to Austria. He recorded his notes in Prague and mailed them back to Lawrence. Merriam lived through a tense time when more than 100 people were killed and Czechoslovakia's Communist Party leader, Alexander Dubcek, was arrested. Dubcek didn't return to Prague until 1989. Just before the invasion, geologists from around the world, including the Soviet Union, were there in August attending a session for the IGC to form a new organization, the International Association for Mathematical Geology.

British colleagues had driven Merriam and Stanford University geologist, John Harbaugh, into Prague for the conference. They were at a hotel in the eastern part of the city when at 2 a.m. on Aug. 21, low-flying airplanes suddenly woke Merriam.

"For some reason in my mind, I thought the Russians were coming, but it didn't occur to me that's what was happening," he said.

**Fig. 36.1** Dan Merriam and Trevor Ford (Leicester University, UK) searching for a way out of Prague August 1968. Copied with permission of the Merriam family

The invasion also shocked the native Czechs and even the Soviet delegates who attended the geology conference. On the eastern side of the city, Merriam didn't witness much destruction. His notes from those few days mention an eerie sense of calm in the eastern part of the city, apart from airplanes sweeping in and tanks rolling around. He noted "the tears in the eyes of the waitresses and the little knots of grim" in the neighbourhood along with several protests. Much of their news came from rumours on the street because radio stations had been bombed and the spread of information was spotty.

"There wasn't anything they could do. There wasn't anything we could do, either, but just watch and hope nothing happened," Merriam said.

The US Embassy had advised Merriam and his colleagues to stay in the hotel because transport from the city was impossible. Even though several members fled the city, the geological conference continued to meet for one day after the invasion.

"The new group, the International Association for Mathematical Geology, even elected its leadership, including President Andrei B. Vistelius, a geologist from the Soviet Union, while the tanks occupied the city," Merriam said.

"It had nothing to do with it, but it was kind of an interesting coincidence anyway," he said.

During that week back in Lawrence, Annie Merriam was on edge. She frequently called Harbaugh's wife, Josephine, to see whether there was any word. But she heard nothing. Finally, Dan Merriam and John Harbaugh had a chance to leave Prague (Fig. 36.1) on a train. It left the city even with tanks nearby, he said. As it approached the Austrian border, the lights went out, and soldiers came to check passports. The train eventually stopped in Vienna, where Merriam sent the telegram to his wife. He also mailed home his letter, which didn't arrive in Lawrence until after he returned home the next week.

It was only a few words but the short telegram Annie Merriam received at her home on Aug. 24, 1968, gave her a huge sense of relief.

"ARRIVED VIENNA OK = DAN = ."

"When that came, we were thrilled," said Annie Merriam. When he did return to Lawrence, it ended a tense chapter for his family. "Don't you ever go anywhere again," Annie Merriam said about her thoughts upon her husband's return. But he did continue his travels. He even returned to Prague in 1993 for the IAMG's 25th anniversary.

### *36.5.5 Vic Loudon*

A few months after our marriage, my wife and I set out from Reading in southern England in our Morris Minor, heading for the inauguration of IAMG. Apart from a stone-strike on the windscreen and its replacement before we reached the English Channel, the journey seemed uneventful. But odd things happened. Travelling through the beautiful Czech countryside, we were forcibly stopped at a secluded spot by a long, shiny black Mercedes. The driver came menacingly to our window: "Exchange foreign currency now, very good price!" The distraction of a passing truck let us escape. As we approached Prague, we noticed more and more heaps of cobblestones that had been lifted from the road and neatly arranged—road-works so tidy they looked like walls. We had booked a room at the Zlata Husa, now a luxury hotel, but then more mundane. The friendly receptionist, carrying our room key, showed us into a small alcove in the reception area, pressed a button, and the entire alcove, still open to the world, moved gently upwards through the ceiling, becoming an alcove (with us still in it) in the room above. She showed us to our bedroom overlooking the beautiful Wenceslas Square. But why did our door look as though it was cased in sheets of steel?

No matter. The hotel was in the centre of town, convenient for exploring the neighbourhood, which we eagerly proceeded to do. It was a long time ago, and I forget the precise order of events, but well remember enjoying walks through alleys and shops of the Old Town; the impromptu puppet show for our benefit in the back room of a tiny shop; and the crossing of the Vltava River by the ornate Charles Bridge, where the youth of the city were chatting in cheerful groups. On the other side was St Nicolas Church, with Prague Castle beyond.

In our bedroom at the Zlata Husa, about 4 a.m. on the 21st of August, we were wakened by planes flying at rooftop level. Did this happen often? But then it was followed by gunfire outside our window, and explosions nearby. Before dawn broke, the sound of tanks moving into position came from below. The armed forces of the Soviet Union and the Warsaw Pact countries had made their point, and the city was now under their control. A Google search for 'images for Prague spring photography' gives a good impression of the results.

The *Report of the 23rd Session of the International Geological Congress* records on page 20 that "On August 21st, 1968 and in the following days, the work of the Congress was interfered by the entry of foreign armies into Czechoslovakia. In result of the overall uncertainty, the blockage of bridges, tanks around the Congress Headquarters, shooting in the streets and other disturbances, a considerable part of the attending members was prevented to come to the Congress Headquarters or had to leave prematurely." Visiting geologists housed in the suburbs lacked any means of reaching the meeting. Merriam records that the US embassy negotiated a train to the border, by which they were evacuated (Lawrence Journal World 2008).

The IGC Report records on pages 200–201: "International Association for Mathematical Geology (IAMG). This Association was officially founded at Prague on August 22nd, when it held its General Assembly. The following officers were elected: President: A.B. Vistelius (USSR). Vice President: W.C. Krumbein (USA), G.S. Watson (USA), General Secretary: R.A. Reyment (Sweden), Treasurers: V. Němec (CSSR), T.V. Loudon (UK), Ordinary members: E.H.T. Whitten (USA), D. A. Rodionov (USSR), D.G. Krige (S. Africa), G. Matheron (France), F.P. Agterberg (Canada), S.N. Sengupta (India), Editor-in-Chief: D.F. Merriam (USA). The application for affiliation to the IUGS (International Union of Geological Sciences) of this Association was unanimously approved by the Council." And so, the IAMG was created, before a reduced but still quite substantial audience.

While I was attending the meetings, my wife took the opportunity to photograph the interesting happenings in the Old Town. A group of soldiers objected, and indicated that she should hand over her camera. They opened it to spoil the film, and returned it. A round of applause came from the on-lookers, perhaps realising that the film in the Instamatic camera would be unaffected.

A day or two later, when our business and sight-seeing were eventually complete, we felt that we should head for home and our anxious relatives. Getting out of Prague was no problem, returning on the same route as our arrival. But half-way to the border, a bridge across a river was blocked by the military, and the road closed to all. Despondently, we slowly retreated for about a mile, when an old man outside his cottage waved us down. We had no language in common, but looking about nervously he gesticulated towards a farm road a hundred yards away, making rippling movements with his hands, and repeating what sounded like the German word 'wasser'.

Not understanding, but with little to lose, we followed his directions and came to a wide stretch of water. It was the same river, and this might be a ford. While preparing to wade in and find out, a truck came the other way, water just reaching its axles. The ford and the road led us back to our intended route, now beyond the blocked bridge. So that kind man, at considerable risk to himself, had made possible our continued journey. Our fuel was running low, and all garages had been closed. Downhill coasting and gentle use of the accelerator brought us eventually to the border at Rosvadov, with the needle firmly set on Empty. A careful passport inspection, and we were through, greeted by US soldiers—pleasant, friendly and helpful. "There's a gas station just up there. Or ask that guy [pointing to their encampment], he'll fill you up, no charge." We took the first option, and soon were in open countryside. We stopped, got out the car and for a while just stood there together—still, silent, and subdued.

### *36.5.6 Hannes Thiergärtner*

Founding member of the IAMG -

I remember rather well the founding procedure of the IAMG. This event occurred for me as a drama in three unusual acts.

The prelude was to empower me to participate at the congress at all. Let me explain for our younger colleagues that the European world at that time was split into the western and the eastern blocks characterized by extremely different political-social systems and ways of life. I grew up, lived and worked in the former German Democratic Republic (GDR) that belonged to the "eastern world". Here, many, not to say most, things were centralized and provided from the top. Thus, it was nearly impossible to participate individually in an international congress. Participants have been selected, nominated and merged to so-called delegations. I worked in the Central Geological Institute in Berlin, the geological survey of the country, as young graduate in the field of mathematical geology and electronic data processing without international reputation. I never had a chance to be nominated for the IGC. On the other hand, I felt the opportunity to go there because it was the first IGC after World War II held in the Eastern Bloc and restrictions to visit the congress were still distinctly lower than in later years. So, I successfully requested the Director of our institute for vacation and paid the fee and all other requirements out-of-pocket. It was a unique courageous decision, for both the Director and myself. I travelled to the congress, was integrated into the official "delegation" and found accommodation in a student's hostel.

The main act played out in Prague. It was and is a wonderful and pulsating place. The townscape in the late 1960s was still characterized by the post war years, predominantly in greyish colours but nevertheless imposing and unique. A metro net did not yet exist at that time but the town centre was well developed by a dense tramway system. The organizers of the XXIII IGC had chosen for the opening ceremony the auditorium of the Charles University, the Carolinum, in the Prague historic centre—an amazing and venerable baroque hall with a super interior. The ceremony was impressive, indeed, and all participants hoped for a fruitful scientific exchange of ideas within the following days.

All attendees knew about the critical political situation because of the Czechoslovakian trends to reform their political system. My journey to join the congress session "Mathematical geology" passed the ministry of defence during these days. When I started to go in for the lectures on Wednesday and Tuesday (August 21–22), I had to walk in front of the Ministry between many tanks which had come from the Warsaw pact states and occupied the town. It was shocking! I do like to take photographs but I did not in this moment—it was too serious. The situation was ghastly and the agile Prague was silent.

I reached the session rooms without personal impairment. There I met so many colleagues I never had seen before but knew from the scientific literature, such as Frederik Agterberg, John W. Harbaugh, Vyachelav Kutolin, Victor T. Loudon, Richard B. McCammon, Václav Němec, Richard Reyment, Dmitri Alekseyevich Rodionov, Andrey Borisovich Vistelius or Eric Harold Timothy Whitten. Altogether 20 persons were present. It was simply great for the young fresh geologist from Germany! Regardless of the stressful situation, we founded the International Association for Mathematical Geology. The organisation was well prepared by Richard Reyment and it proceeded to elect its leading officers. I remember that the participants from the Eastern Bloc during a break agreed to vote for Andrei B. Vistelius as first president to ensure parity within the top of the association.

On the same day, all members of the GDR delegation got orders to meet at a very small railway station in the western part of Prague to "enter" one of the now rarely running trains to the German boundary. We left the hosting country in a night and fog action.

## 36.6 Subsequent Events Following Prague

### *36.6.1 John Cubitt*

As a second-year undergraduate at Leicester University at the time, I was almost unaware of the events of the IAMG foundation. All I can recollect is my tutor, Trevor Ford, and our Department Chairman, Professor Peter Sylvester-Bradley, returning from Prague with tall tales of the various lucky escapes. It must, however, have made some form of subconscious impression on my mind because less than a year later I mentioned to Trevor that I would like to go on to undertake postgraduate work in computer applications in geology. In that case he said, you need to meet someone and marched me out of his office and down the corridors of the Department of Geology. In a minute, we found the mystery person he wanted to introduce to me. He was striding down the corridor in cowboy boots, string tie and cowboy hat in his typical dynamic intimidating style, Dan Merriam. After brief introductions from Trevor, Dan talked about the Research Group at the Kansas Geological Survey and how I should undertake a Ph.D. at Leicester University but with the first year paid for and spent at the KGS. "That will be OK with the Department, won't it?" Dan said to Trevor and whether it was or not, the decision had been taken. Within a few months of whirlwind arrangements, I was on my way to Kansas and my career was underway (Dan subsequently took me to Syracuse University as well so I have much to be grateful to this amazing dynamic organiser for). This frenetic activity was typical of the rapid growth in the subject of mathematical geology and the IAMG at the time.

### *36.6.2 Hannes Thiergärtner*

The after-play led me back to the reality of those times. The founding of a new seminal association within the international geological community was ignored [in the Eastern Block] especially in the governmentally organized surveys. A policy of restriction was introduced step by step. Any contact—to say nothing of an IAMG membership—outside of the Eastern Bloc proved to be impossible and was strictly forbidden. I would however meet the majority of founding members of the IAMG again in 1984 during the XXVII IGC in Moscow where I could take an active part only on special request made by D.A. Rodionov at the GDR ministry of geology. But that is another story.

With the exception of my colleagues in Prague, Leningrad and Moscow, I was unable to renew my contacts to other founding members until after the German reunion (1990). Frits Agterberg was the first colleague I met in Potsdam (Germany). It was also 1990 and I could then renew my membership in the International Association for Mathematical Geosciences. I think we all have utilized this late time as well as possible to solve some common questions in our interesting field of science.

### *36.6.3 Stephen Henley*

As a humble Nottingham University postgraduate student in 1968, I wasn't at the IGC or the Prague launch of IAMG. However, I was deeply involved in computer applications and statistical analysis, processing what then seemed like huge volumes of data from the X-ray fluorescence spectrometer, and then making sense of the data using esoteric methods such as factor analysis, cluster analysis, and trend surface analysis. Under the mentoring eye of Peter Harvey, I joined IAMG as soon as I heard of its existence, in 1969—and have remained a member without a break since then. It is fair to say that mathematical geology shaped my entire career. As my Ph.D. studies came to an end in 1970, an opportunity arose in Australia.

The Bureau of Mineral Resources (now Geoscience Australia) suffered a mass resignation of several dozen geologists who left to join one of the periodic mining booms—this one in Western Australia, sparked by discoveries of major nickel deposits. Among those who left was their one computing 'expert', so my meagre computing experience was sufficient to gain me a position in Canberra, where I gained a broad experience of mathematical modelling and statistics in fields that included hydrogeology, exploration geochemistry, earth tides, and global scale geochemical modelling of Archaean evolution of the Earth (this last with Andrew Glikson, based on studies of some of the world's oldest rocks). After my return to the UK, I finally accepted my type-casting as a computer geologist and in 1973 joined the Computer Unit of the Institute of Geological Sciences (now the British Geological Survey). This small specialist unit occupied two rooms on the top floor of the Geological Museum in London, and had an IBM 1130 computer—which even then was of very limited capacity. However, we also had access to the much more powerful mainframe IBM 360/195 at the Atlas Computer Laboratory (ACL) in Oxfordshire.

The head of the Computer Unit was Dr T. Victor (Vic) Loudon who had pioneered generalised software development in his previous academic work at Reading University (the Rokdoc package) and was one of the founding members of IAMG. Rokdoc was the inspiration for a colleague Keith Jeffery to start the development of a general-purpose geological data handling system 'G-EXEC' which was built around the recently published ideas of IBM researcher Edgar Codd for relational database management. When I first met them, Keith and his co-worker Elizabeth Gill at ACL, were preparing an early version of G-EXEC: I walked into the office they were using to see the floor strewn with many piles of punched cards and reams of fan-folded lineprinter listings of the software. The whiteboard displayed a beautifully simple diagram of the system structure, and I was hooked.

**Table 36.1** Officers and Council of IAMG

| | 1968–72 | 1972–76 | 1976–80 |
|---|---|---|---|
| President | A.B. Vistelius (USSR) | R.A. Reyment (Sweden) | D.F. Merriam (USA) |
| Past president | W.C. Krumbein (USA)[a] | A.B. Vistelius (USSR) | R.A. Reyment (Sweden) |
| Vice president | G.S. Watson (USA) | A.T. Bharucha-Reid (USA) | G. Hill (Australia) |
| Treasurers<br>Western<br>Eastern | T.V. Loudon (UK)<br>V. Němec (Czech.) | J.C. Davis (USA)<br>V. Němec (Czech.) | J.C. Davis (USA)<br>V. Němec (Czech.) |
| Secretary General | R.A. Reyment (Sweden) | D.F. Merriam (USA) | E.H.T. Whitten (USA) |
| Council members | F.P. Agterberg (Canada)<br>D.G. Krige (S. Africa)<br>G. Matheron (France)<br>S.C. Robinson (Canada)<br>D.A. Rodionov (USSR)<br>S. Sengupta (India)<br>E.H.T. Whitten (USA) | H.A.F. Chaves (Brazil)<br>A.C. Cook (Australia)<br>J.E. Klovan (Canada)<br>P. Laffite (France)<br>G. Lea (UK)<br>D. Marsal (W. Germany)<br>E.H.T. Whitten (USA) | F.P. Agterberg (Canada)<br>K.L. Burns (USA)<br>G. de Marsily (France)<br>D. Gill (Israel)<br>D.M. Hawkins (S. Africa)<br>R.J. Howarth (UK)<br>W. Schwarzacher (UK) |
| Editors-in-chief Jour. Math. Geology | D.F. Merriam (USA) | D.F. Merriam (USA) | R.B. McCammon (USA) |
| Computers & Geosciences | — | D.F. Merriam (USA)<br>J.C. Davis (USA) | D.F. Merriam (USA)<br>J.C. Davis (USA) |
| Newsletter | G. Lea (UK) | | |

[a]served as Vice President

**Fig. 36.2** Official logo of IAMG

Soon after that, John Cubitt joined the team, and we formed a "gang of four" providing computing services to a wide range of users within IGS as well as supplying the software to other institutes in the Natural Environment Research Council and worldwide. The IGS Computer Unit itself was a research centre in its own right: John and I both worked together on the potential use of catastrophe theory as a geoscience modelling tool, though we were ahead of the times, and it was only when catastrophe theory was superseded by chaos theory that the potential became reality, in such fields as climatology and oceanography. Working with Jeff O'Leary, then at Leicester University, I also used the relatively new field of geostatistics in developing a 3D model of the Jwaneng diamond pipe in Botswana, but misgivings about the method, arising from that and other projects, led to development of more robust 'nonparametric' methods which formed the basis of a book (and led to my receiving the 1982 President's Award of IAMG).

The underlying G-EXEC concepts (and much of the software itself) were subsequently incorporated into other products including, in my case, the 'Datamine' mining software system. The rest, as they say, is history.

### 36.6.4 Dan Merriam

(From Merriam 1978, copied by permission of the Merriam family)–

A list of Officers and Council members of the Association is given in Table 36.1.

During the first year a call for members was made. A logo was designed according to specifications of D.F. Merriam by Charles Barksdale of the Kansas Geological Survey for use in connection with official Association business (Fig. 36.2). This logo was used on a certificate received by all charter members (those who joined during the first year). negotiations were complete with Plenum Press for a new journal, Journal of Mathematical Geology (JMG), which appeared first in 1969. It was made a quarterly in 1970 and a bimonthly in 1975. Also in 1975 the quarterly journal, Computers & Geosciences (C&G) was established with Pergamon Press the publisher.

The JMG focusses on geomathematics and mathematical geology, which includes geological arguments supported by numerical observations to purely mathematical models implemented with geological data. C&G is devoted to the rapid publication of computer programs of interest to earth scientists in widely used languages and their applications. A quarterly Newsletter contains general information of interest to members.

Each year the Association sponsors meetings, many in cooperation with other organisations. For example, IAMG cohosts the Geochautauqua held each year at Syracuse University and every other year a session in mathematical geology at the Pribram Mining Congress. At each IGC since Prague, we have sponsored or cosponsored several sessions of interest to our members. In addition, we have cohosted sessions at meetings of the American Association of Petroleum Geologists, and the Geological Information Society of the Geological Society of London. Proceedings for many of these meetings have been published either as special issues of the Journals or as hard-back books.

Seven national groups have been created and are functional. They are in the United States, Canada, Brazil, Great Britain, Czechoslovakia, Hungary, and Russia; others are in the formation stages. These national groups are active in disseminating information on geomathematics on a national level. Although national groups are autonomous, they are expected to coordinate their activities with the Association.

Operation of the Association is mainly through committees. The Project Committee is responsible for preparing the meetings at the next IGC which is held every four years. The Membership Committee is concerned with soliciting new members; the Finance Committee with soliciting money; and the Educational Committee with organizing material and activities to promote geomathematics. Each year a

**Table 36.2** IAMG committee chairmen

| | | 1968-72 | 1972-76 | 1976-80 |
|---|---|---|---|---|
| Standing committees | Projects | J.E. Klovan (Canada) | A.C. Cook (Australia) W.B. Hempkins (USA) | G. de Marsily (France) W.B. Hempkins (USA) |
| | Membership | G. Lea (UK) | M.K. Horn (USA) | J. Hefner (USA) |
| | Finance | H.A.F. Chaves (Brazil) | G.S. Koch (USA) | R. Till (UK) |
| | Education | G.F. Bonham-Carter (USA) | | |
| Ad hoc committee | CAI | | | F. Mutschler (USA) |

**Fig. 36.3** Design for Krumbein medal

committee (chaired by the President) selects the William Christian Krumbein medallist and another special committee selects the Best Paper for an award.

A special committee has undertaken the task of compiling a list of all computer-aided instruction (CAI) programs available and of interest to geologists and it will be distributed in the near future. There are also plans for compiling a list of computer software, the list will contain information on programs and their availability and limitations. Chairmen of the various committees are given in Table 36.2.

The Association maintains close contact with other organizations which share similar interests. For example, several members of the Association serve on the IUGS-sponsored COGEODATA Committee. Others are working on special projects for CODATA. The Association has a member on Scientific Committee 4 which evaluates quantitative aspects of projects for the IGCP. Liaison is maintained with the International Paleontological Association.

The William Christian Krumbein Medal is presented each year by the Association to an outstanding geomathematician. The first recipient was Professor John C. Griffiths of Pennsylvania State University, the second, Professor Walther Schwarzacher of Queen's University, Belfast, Northern Ireland, and the third, Dr. Frederik P. Agterberg of the Geological Survey of Canada, Ottawa. The recipient receives a medal with the likeness of William C. Krumbein on one side and the Association's logo on the other. The Medal was designed in 1977 by A. Pattison, sculptor of Florence, Italy and Winnetka, Illinois (Fig. 36.3).

The IAMG, in its short period of existence, has participated in and contributed to changes in the earth sciences. In the future the Association should play an even larger role in development of the science.

## 36.7 The Looming Gap

### *36.7.1 Vic Loudon*

The methodology of geological investigation and communication was initially formalised within the constraints imposed by the traditional mechanisms of pen, paper, typewriter, printing press, bookshops and libraries. It has been extended by computer techniques, formalised in a framework set by the manufacturers and providers of computer equipment and software, but is still based on and restricted by geological traditions, conventions and precedents. Geological surveys continue to provide geological maps world-wide, with defined scales of presentation, uniform stratigraphical classifications, and separate volumes of text, with cross-references to locations on the map.

These products provide a stable underlying shared basis for subsequent geological investigations, essential for accurate communication, including a consistent and coherent structure within which new investigations can build. This is achieved by results being confined within the rigid framework and slow-moving processes of conventional publication. Geological knowledge can potentially build on a wider framework, going far beyond its current traditions, conventions, limitations and precedents.

The global information structure is being remodelled, based on new technology with unfamiliar implications. Current developments in computer translation, voice recognition and speech synthesis point to a much more flexible future.

As in the mid-1960s, a significant gap may be developing between the future of geological communication and its current implementation of published papers and maps. Experimental initiatives might be a good starting point. Their results might be inappropriate for traditional patterns of communication, but information on their development could usefully be exchanged in an open and flexible forum, for which IAMG might be a suitable host.

## Appendix

A readable account in the Economist (2017) describes the power of deep learning: ‘an artificial intelligence technique in which a software system is trained using millions of examples, usually culled from the internet… Computers are, in short, getting much better at handling natural language in all its forms.’ But (p. 11): ‘Scientists do not know how the human brain draws on so many different kinds of knowledge at the same time. Programming a machine to replicate that feat is very much a work in progress.’

The conventional forms of scientific papers and the fixed scales of geological maps reflect the limitations and conventions of earlier technologies. Future development of our understanding of global geology can only be achieved through a multitude of investigations and experimental studies. Many geological developments will be based on local knowledge and requirements. Many will be too trivial for conventional publication but valuable in their own local context. Already, the computer technology for sharing detailed studies and strategies is well established. It could help to provide the essential background for a more comprehensive framework. It could lead to deeper evaluation and integration of data, text, graphic and cartographic information at all relevant levels of detail; rapid and appropriate response to input of new information; the routine calculation, depiction and quantitative assessment of multiple geological hypotheses; and the emergence of a never-ending dialogue between human input and computer implementation, supported by a multi-media interface for input and output.

This calls for developments that go far beyond the precedents and traditions of our established conventions, into an environment for geological information where users are motivated to carry forward an accessible shared understanding. Maps, data, illustrations, simulations, text explanations and scientific papers need not be separate entities nor restricted to a single scale. Input of new information can be rapid, with continual assessment and reassessment of its validity and relevance, and examination of its consistency with previous work.

## References

Dech VN, Henley S (2003) On the scientific heritage of Prof AB Vistelius. Math Geol 35(4): 363–379

Economist (2017) Technology quarterly: language. 7th Jan 2017, pp 1–12

GOSSIP (1966). http://nora.nerc.ac.uk/20280/1/GOSSIP_1966.pdf or http://nora.nerc.ac.uk/20280/

Krumbein WC (1969) The computer in geologic perspective. In: Merriam DF (ed) Computer applications in the earth sciences. Plenum Press, New York, pp 251–275

Krumbein WC, Sloss LL (1958) Stratigraphy and sedimentation. W H Freeman, San Francisco

Kurlansky M (2004) The year that rocked the World. Jonathon Cape Ltd, 448 pp

Lawrence Journal World (2008). http://www2.ljworld.com/documents/2008/aug/20/letter_and_telegram_daniel_merriam_his_wife/

Lyell C (1833)1830: Principles of geology. John Murray. (vol 1, Jan. 1830—vol 2, Jan. 1832—vol 3, May 1833)

Merriam DF (1978) The international association for mathematical geology—a brief history and record of accomplishments. Syracuse Univ Geol Contrib 5:1–6

Merriam DF (1981) Roots of quantitative geology; in down-to-earth statistics: solutions looking for geological problems. Syracuse Univ Geol Contrib 8:1–15

Vistelius AB (1991) Principles of Mathematical Geology. Kluwer Academic Publishers, 477 pp

Whitten EHT (2003) Mathematical geology in perspective: has objective hypothesis testing become overlooked? Math Geol 35(1):1–8

Whitten EHT (2004) AB Vistelius: first IAMG president and visionary mathematical geologist: earth science. History 23(2):384–389

Whitten EHT (2005) A half century of mathematical geology—the new threshold. Zeitschrift der Deutschen Geologischen Gesellschaft (Z dt geol Ges—Journal of the German Geological Society, Stuttgart) 155(2–4): 203–9

# Chapter 37
# Forward and Inverse Models Over 70 Years

**E. H. Timothy Whitten**

**Abstract** The transition over 70 years from qualitative rock description to attempted quantitative description of rocks and rock bodies (inverse modelling) and testing of process models with observation data (forward models) are outlined. Dramatic increases of readily measured variables, combined with almost unlimited computing power, yielded a plethora of varied inverse models, but limited attention has been given to critical sampling, variance, closure, 'black swan', and nonlinear issues; recent approaches to closure problems hold promise. Especially for plutonic rocks, paucity of quantitative process modelling left exciting forward-modelling opportunities neglected. Resulting challenges ahead are anticipated.

**Keywords** Sampling · Variance · Composition variability · Black swans
Granite composition

## 37.1 Birth of IAMG in 1968

In many different ways, 1968 was an extraordinary year that rocked the world (cf., Kurlansky 2004). Some 20 enthusiasts gathered at the XXIII International Geological Congress in Prague's New Technical University, Czechoslovakia, to create the International Association for Mathematical Geology in exciting, but tragic, times. Soviet troops had occupied the city a couple of days previously; guns of encircling Soviet tanks pointed at the university, which was the centre for printing and disseminating news. Vistelius was elected first IAMG President and Krumbein 'Past President' (a designation he appreciated and found amusing!); both are fathers of geological models.

---

E. H. T. Whitten (✉)
Riverside, Widecombe-in-the-Moor, Devon TQ13 7TF, UK
e-mail: timwhitten@btinternet.com



B. S. Daya Sagar et al. (eds.), *Handbook of Mathematical Geosciences*,
https://doi.org/10.1007/978-3-319-78999-6_37

At that meeting, dissimilar approaches came together, having evolved principally in the Soviet Union, Western Europe, and U.S.A. Vistelius championed the concept that Mathematical Geology is a separate branch of science based on testing geological hypotheses mathematically, and that this should be IAMG's primary focus (Whitten 2003; 2004, p. 384–5); for some years, he had contended it is not particularly important merely to manipulate geological data statistically. Dech and Henley (2003, p. 368) noted Vistelius (1991) considered that, if a science does not use mathematical modelling in constructing conclusions, "... it can be considered as belonging to the pre-Newtonian period, .... behind the present-day level of research by approximately 300 years."

## 37.2 In the Beginning (One Pre-1968 Experience)

Specializing in petrology in 1948, Hatch and Wells (1937) was my 'bible'. That descriptive, natural-history type, foundation meant it was thrilling in 1950 to visit Jacupiranga, the Brazilian jacupirangite type locality. For a Ph.D. project in 1948, it was recommended I look at 260 km$^2$ of coastal NW Ireland to see what is there; seventy years later, an unlikely method of identifying a thesis project. The area is red (granite) on the Geological Survey of Ireland 1:63,360 map (Hull et al. 1889).

A plan to record variability of granite across the area (including numerous islands in the Atlantic Ocean) was needed. Immediate problems in 1949 were devising (i) a scheme to collect representative samples, and (ii) realistic measurements (measurable in the field or laboratory) to reflect variability.

Unscientifically, a one-mile grid was oriented to maximize (by eye) grid nodes over outcrops (i.e., islands in the ocean and less peat-bog and drift-covered mainland areas). It was planned to collect samples (with hammer and chisel) at all nodes if possible. In the field, two compromises became necessary—using the nearest outcrop to nodes and accepting any hand-sample that could be hammered off.

Wet chemical analysis of numerous samples was beyond available resources; X-ray fluorescence analysis was then undeveloped. Point counting thin sections to determine mineral volume percentages with a Dollar (1937) mechanical stage was feasible, provided larger thin Sections. (3.3 × 2.3 cm) could be hand ground and stained with sodium cobaltinitrite—both challenging in 1949; this staining technique was described by Chayes (1952). Using a Chayes (1949) electrically-controlled stage improved point-counting accuracy. Studies of spacing and required number of counts (Chayes and Fairbairn 1951; Chayes 1954) suggested sufficiently large thin sections were being used. Manual contours for modal variables (e.g., K-feldspar volume percentage, colour index) at 44 grid nodes reflected considerable areal variation (Whitten 1957). Such contours were very controversial because they crossed ocean between islands and superficial deposits on land; also, no exposures occur in numerous grid squares. A senior reviewer deemed it impossible to draw contours across ocean (despite greater outcrop density with off-shore islands than on land with peat bogs, farming, etc.).

In 1958, I became a colleague at Northwestern University of W. C. Krumbein, who was pioneering quantitative description of sedimentary rocks. The University acquired an IBM360 mainframe computer; we used punch cards and wrote FORTRAN programs for statistical descriptors and surface-fitting algorithms for areally-distributed data (e.g., Whitten 1960, 1961). Analogous approaches began thriving at Kansas Geological Survey, Pennsylvania State University, etc. Krumbein developed the concept of descriptive, conceptual, and predictive models (Krumbein 1963; Krumbein and Graybill 1965, p. 13, *et seq.*; Whitten 1964). Driving to Leningrad to spend time at Vistelius' Institute for Mathematical Geology was a privilege in 1971.

## 37.3 Inverse and Forward Geology Problems

Vistelius (e.g., Vistelius 1977) differentiated *inverse* from *forward* problems. The objective with the former was describing the nature and variability of specified rocks, etc.; that is, with statistical or other techniques, formulating descriptive and/or genetic models for essentially arbitrary data for arbitrary variables. With forward problems, the objective was testing validity of genetic models (based on currently available information) for rocks, fold belts, etc. That is, testing whether a genetic model is supported or rejected by data for variables dictated by that model; many commonly measured variables are likely to be irrelevant for such testing (cf., Whitten 2005).

For sedimentary and metamorphic rocks inverse and forward problems present fewer difficulties. Thus, 'marine beach' can be defined descriptively by physical, chemical, and biological features that commonly enable marine-beach deposits to be recognised (e.g., in the stratigraphic column), or genetically by environmental conditions that result in beach formation (waves, currents, sediment transport, etc.). Similarly, as Bayly (1968) pointed out, metamorphic facies can be defined by presumed temperature and pressure during genesis (Eskola 1915, p. 114; Turner and Verhoogen 1951) or descriptively by diagnostic mineral assemblages (Fyfe et al. 1958). With igneous rocks (especially plutonic assemblages), geotectonics, etc., inter-relationships between the descriptive and genetic are commonly very debateable (Whitten et al. 1987a, p. 334).

## 37.4 Forward Models in Earth Sciences

Forward modelling is in its infancy and rare because, in most cases, little objective quantitative information is available about genetic factors, especially for plutonic rocks. Unlike many scientific fields, most earth-science domains do not permit *reproducible* experiment and testing. Vistelius (1972) used Tuttle and Bowen's (1958) experimental petrology to illustrate forward modelling of 'ideal granite', extending his method[1] to Omsukchan Granite, SE Asia (Vistelius and Romanova 1972), Malsburg Granite, Germany (Choubert and Vistelius 1972), etc.

Over the past decade, numerous "forward models" appeared in geophysical studies (petroleum, mining, water, volcanic activity) for prediction and extrapolation based on measured variables (e.g., Geol Soc Amer Symposium 2002; Sui et al. 2012; Butler and Zhang 2016). Butler and Sinha (2012, p. 168) stated such forward modelling is useful for interpreting data. McInerney et al. (2007) compared gravity data computed for a 3D geological model with new Bouguer data to iteratively improve their geological model, calling this forward modelling. Comparable usage occurs in biology (e.g., Tolwinski-Ward 2012). In such studies, inverse models have been honed with new data for sundry variables, producing improved inverse models (cf., iterative forward modelling, Schlumberger Limited 2016). However, such "forward modelling", albeit useful, is wholly different from testing genetic models with new variables prescribed by those models. Different distinctive terminology would prevent confusion.

Vistelius' forward-model definition is retained in this paper.

## 37.5 Inverse Models in Earth Sciences

Inverse–models reach into many earth-science domains. Manual contours for variability of Donegal granite modes (Whitten 1957) represented an inverse-problem approach; more-sophisticated inverse models followed as computing power facilitated trend-surface map preparation (e.g., Whitten 1960). Computing power soon resulted in every available data set being processed by every available statistical artifice, to explore whether anything interesting (and publishable) emerged. Such research provoked Vistelius' strident remarks at the IAMG founding meeting.

Inverse problems fall into two categories:

(a) analysis and description of available (or readily measured) data for geological entities (e.g., colour index in granite plutons; grain-size skewness in silt samples), and
(b) use of data to predict
    (i) useful features (e.g., gold content and location; subsurface sedimentary rock permeability variation) as with kriging and so-called 'geostatistics'

[1]Numerous papers by Vistelius and coworkers used the important and challenging discovery that grain transitions along linear traverses of many granitic rocks possess the Markov property, to suggest testing or erecting genetic crystallization models can be based on grain-transition probabilities. However, Whitten and Dacey (1975) and Whitten et al. (1975) demonstrated Markov chains in actual mineral sequences in varied rocks (including a calc-silicate granulite) is insufficient for establishing validity of the granite crystallization model.

(cf., Krige 1964; David 1977; Journel and Huijbregts 1978), or flooding or other risks (e.g., Burke et al. 2016), or

(ii) petrogenetic processes (e.g., infra-crustal origins of *I*- and *S*-type granites within orogenic belts (e.g., White and Chappell 1983; Chappell 1984; Chappell and Stephens 1988).

Speculation about petrogenetic processes that produced described rock assemblages has always been common. Over a thousand high-quality chemical analyses of major and many trace elements for southeast Australian granites led to partitioning samples into *I*-type or *S*-type granitoids with dissimilar sub-crustal origins, and to the restite genetic model (e.g., Chappell et al. 1988, 1987; Chappell and Stephens 1988). Analogous methods were used elsewhere (e.g., North American Peninsula Ranges, Silver and Chappell 1987). Such inverse models could afford excellent forward-modelling bases, if prescribing new variables with which to support or negate the supposed genetic model/s.

However, such inverse models are fraught with difficulties (Whitten 1991, p. 121). Use of different variable sets from Chappell and colleagues' chemical analyses can partition samples into an almost infinite set of descriptive suites. It is unrealistic to enunciate genetic scenarios for one set of descriptive suites, without concomitantly embracing all other coexisting sets defined by using different variables, sets of variables, variable weightings, etc. (Whitten et al. 1987a, p. 341; 1987b). Again, if techniques like cluster analysis were used to partition hundreds of samples on the basis of 36 chemical variables, normalization (to give each variable equal weight) would commonly be used, despite no a priori reason for each element being equally important. Different clusters emerge if one (or more) variable receives different weighting, and when more or less variables are included (Whitten et al. 1987b, p. 69; Whitten 1991, p. 121). Also, standard cluster analysis (and similar partitioning techniques) yield questionable results when percentage and/or parts-per-million data are used (cf., Aitchison 1986, p. 300).

However, where components are conserved throughout crystallisation within certain basic igneous rocks, molar ratios with a common constant denominator were shown to display, accurately and unequivocally, the actual chemical variability (e.g., Nicholls 1988; Stanley and Russell 1989). Molar-ratio diagrams for some Australian *I*- and *S*-suites seem to show chemical variations accurately, permitting quantitative objective testing of, say, the restite model (Whitten 1996). This technique for avoiding daunting closed-data problems deserves further examination, although, for many granites, lack of component conservation during crystallization may introduce difficulties.

## 37.6 The Samples Analysed

Statistical or mathematical analyses of available data are the relatively easy part. Statistical manipulation (inverse modelling) describes characteristics and variation of particular data, but *not* necessarily characteristics and variation of those variables in the rock samples from which the data were derived (or necessarily of variables of *petrogenetic* significance for forward modelling, or of direct *economic* importance).

Data come from samples (or geophysically-sampled rocks, etc.). It is important to assess how well available samples represent the *sampled population* of interest, and whether that sampled population permits realistic extrapolation to the *target population* of primary interest (cf. Whitten 1961). For example, where the objective is determining compositional variation of a pluton, the exposed surface is an arbitrary 2D section (or modestly 3D in mountainous terrain) through the original 3D mass, much of which is eroded away. Soil, vegetation, etc. always obscure major parts of 2D exposures; actual outcrops are disposed arbitrarily or preferentially, but not randomly. Analyses of those samples actually examined (samples collected from sampled outcrops) are necessarily used to estimate composition and variability of the sampled population, and subsequently the target population.

The significance of actual observed dependent data was reviewed by Whitten (2000, pp. 4 *et seq.*) who asserted that, in favourable circumstances, rigorous statistical inferences can be drawn about the sampled population on the basis of samples examined, and subsequently geologists can only use such inferences to make *subject-matter inferences* about the target population on the basis of previous geological experience (cf., Cochran et al. 1954, p. 19).

Unusually, such issues can be obvious. For example, road cuttings might expose significantly banded or layered rocks, but only some of those bands may be exposed in outcrops across neighbouring areas.

Serial thin sections from coarse-grained granite samples commonly yield modal values with considerable variance. Exposed igneous rocks may be porphyritic making collectable, representative, samples difficult to obtain. Commonly, samples of dissimilar size are required to estimate composition and variability of each variable. For variables measurable only by laboratory analyses (e.g., modal zircon percentage, trace-element weight percentages), an adequate sampling plan can be devised only following estimating the level of variance of each variable from analytical results. The classical example is Krumbein and Slack's (1956) determination that variance of their variable of interest within a black shale over many square kilometres of Illinois, USA, is greatest at their smallest level of sampling (thin-section level). Different rock types require dissimilar strategies (e.g., determining calcite volume percentage throughout a cratonic limestone requires a less-dense sampling plan than, say, assaying gold weight percentage within subsurface Witwatersrand conglomerates or apatite volume percentage in a granite).

For Rattlesnake Mountain Pluton, California (USA), Baird and Welday (1967) showed that, when variance of attributes is large at their smallest sampling level (hand-specimen level), adjacent samples yield dissimilar values and thus dissimilar areal-variability maps. For their monumental studies of Lachlan fold belt granitoids, Australia, Chappell and colleagues powdered very large samples (over a kilogram) from the mainly visually-homogeneous outcrops, with the intention of minimising major and trace-element variance at the sample level (e.g., White et al. 1977; Chappell 1978). Their sample size and reproducibility of their chemical analyses yielded reliable data. In many regions, they collected a sample from virtually every outcrop protruding through arid rolling pasture. Areas between widely scattered outcrops (sometimes a kilometre apart) were necessarily un-sampled and unknown; it is appropriate to question whether extant outcrops exist because composed of rocks less susceptible to weathering (compositionally dissimilar to the majority).

Generalising, each variable commonly has dissimilar variance in samples of a specified size. Variance tends to be large between small samples, especially when grain size is large, and, as sample size increases, variance between samples decreases to a minimum, before increasing again for extremely large samples (cf, Whitten 1968; 2000, p. 6).

Such issues have long been recognized in mining exploration. Moving-average methods, developed by Krige (e.g., 1964) for South African gold-bearing conglomerates were extended and explicitly controlled (in what is known as 'geostatistics') by levels of variance of variable/s, as expressed by semi-variograms (e.g., David 1977; Journel and Huijbregts 1978); observed large outlier values are accommodated within the 'nugget' effect. 'Nugget' aptly reflects very sparse, larger gold particles within the conglomerates, which affect predicted profitability of subsequent mining; nuggets are represented only occasionally in actual samples and resulting assay values (Whitten 2010, p. 250).

It is not uncommon for it to be assumed that, provided sampling has been 'adequate', variables of interest follow standard frequency distributions (normal, lognormal, etc.). Many common statistical algorithms assume input data are normally distributed; frequently, packaged computer programs normalise input data automatically (often with unspecified algorithms) prior to effecting statistical analyses. However, different normalisation algorithms can produce dissimilar resulting analyses.

## 37.7 The Black Swan Effect

Throughout the earth sciences, sporadic sample measurements are wholly dissimilar to those for the majority of samples. Not infrequently, analyses lying on the extreme wings of distribution curves (normal, lognormal, etc.), or beyond the tails, are discarded; although such analyses might be attributable to analytical error, many are likely to be real and *very* meaningful. In studying the influence of the improbable in the earth sciences, Whitten (2010) demonstrated that real, localised, anomalous data can reflect features of significant genetic and/or economic importance; the 'black swan' effect (cf., Taleb 2007). That is, such data can reflect important factors not previously considered in models and theories—factors that, after recognition, are likely to be found highly significant.

Throughout geological time, all manner of events occurred that appear to be wholly arbitrary with respect to formation of lithology, structure, palaeontology, etc., of rock units. Impact of a meteor with the Earth is a good example, because it can apparently affect substantially both current organic evolutionary patterns and ongoing physical processes (e.g., sedimentation). Consequently, some, but not necessarily all, dependent variables (with respect to space and time) might show anomalies reflected as outliers on a distribution curve (a nugget-like effect). Such phenomena reflect the operation of customary physico-chemical laws and the effects of irreducible elements of chance and indeterminism (Whitten 2010, pp. 250–1).

The traditional search for order and simplified description commonly deflects attention from important real black swans that require inclusion for realistic understanding of geological phenomena and natural hazards. Mandelbrot (1982) provided a beautiful introduction to fractal geometry in nature; more recently, fractal, chaos, and nonlinear approaches have helped expose basic characteristics of the physical world, whose fundamental significance throughout the earth sciences is rapidly becoming more clear. A report (Lovejoy et al. 2009) on 'geocomplexity' summarized the importance of nonlinear geophysical methods in elucidating rational bases for statistics and models of natural systems (including hazards), which previously were treated by ad hoc methods. That report reflected 15 authors' research ranging from earthquake dynamics, river-flood prediction, basalt columnar-joint formation, coastline topography, meteorological cloud models, and interaction of greenhouse gases and global warming. It concluded with a warning against (a) reliance on traditional state-of-the-art statistical techniques (and theories based on them) and (b) ignoring nonlinear methods which are often helpful for more-complete understanding of the natural world.

## 37.8 Concluding Thoughts

Throughout most geological domains, the qualitative-to-quantitative revolution via mathematical geology over the past half century has been awesome, made possible by numerical models and readily available data for greatly increased numbers of variables; all facilitated by hugely increased computing power. Investigations extend to variables whose variance cannot be estimated by eye (e.g., isotope ratios; electrical resistivity). The research is manifest in both IAMG Journals and other new approaches (e.g., 3-D visual digital models and virtual presentation of rocks and geological formations, De Paor 2016). Cataloguing, classifying, description, and presentation are often the useful goals, especially for economic geologists (e.g., oil-field research; kriging and 'geostatistics').

Pragmatic review emphasises that many basic (but apparently unexciting) problems enumerated five decades ago (e.g, variance; sampling), critical in inverse models for correctly portraying rock formations (rather than merely assembling data obtained from the rocks), have continued to receive little attention (Whitten 2003).

Birth, maturity, and old age characterise phases of all human endeavour. The past 50 years witnessed birth of IAMG and spreading of its influence throughout the earth sciences using inverse methods, but only initial recognition of the compelling importance of modelling forward problems (in Vistelius' meaning).

Inverse-problem studies will move into maturity as variance, sampling, and non-linear models underpin on-going research.

The challenging needs and goals of forward problems are reasonably obvious, but the complex issues involved have been addressed only occasionally (e.g., Vistelius and Romanova 1972; Maslov 2003). Commonly, forward problems will require non-linear process models (i.e., quantitative genetic models) that specify those variables required to test the hypothesis. The next 50 years await research towards that maturity in forward modelling. So-called forward models of recent geophysical studies must not obscure this challenge.

## References

Aitchison J (1986) The statistical analysis of compositional data. Chapman and Hall, London

Baird AK, Welday EE (1967) Geochemical and structural studies in batholithic rocks in southern California. Part II. Sampling of the Rattlesnake Mountain Pluton for chemical composition. Geol Soc Amer Bull 78:191–222

Bayly MB (1968) Introduction to petrology. Prentice-Hall, Englewood Cliffs, N.J

Burke H, Farrell R, Morgan D (2016) Models and flooding. Geoscientist 26:10–15

Butler SL, Sinha G (2012) Forward modeling of applied geophysics methods using Comsol and comparison with analytical and laboratory analog models. Comput Geosci 42:168–176

Butler SL, Sinha G, Zhan Z (2016) Forward modelling of geophysical electromagnetic methods using Comsol. Comput Geosci 87:1–10

Chappell BW (1978) Granitoids from the Moonbi district, New England batholith, eastern Australia. J Geol Sci Aust 25:267–283

Chappell BW (1984) Source rocks of *I*- and *S*-type granites in the Lachlan Fold Belt, southeastern Australia. Philos Trans R Soc A 310:693–707

Chappell BW, Stephens WE (1988) Origin of infracrustal (*I*-type) granite magmas. Trans R Soc Edinb Earth Sci 79:71–86

Chappell BW, White AJR, Hine R (1988) Granite provinces and basement terranes in the Lachlan Fold Belt, southeastern Australia. Aust J Earth Sci 35:505–521

Chappell BW, White AJR, Wyborn D (1987) The importance of residual source material (restite) in granite petrogenesis. J Petrol 28:1111–1138

Chayes F (1949) A simple point-counter for thin section analysis. Am Min 34:1–11

Chayes F (1952) Notes on the staining of potash feldspar with sodium cobaltinitrite in thin section. Am Min 37:337–340

Chayes F (1954) The theory of thin-section analysis. J Geol 62:92–101

Chayes F, Fairbairn HW (1951) A test of the precision of thin-section analysis by point counter. Am Min 36:704–712

Choubert B, Vistelius AB (1972) Pluton Malsburg (Schwarzwald, FRG): 1:25,000 map. Academy of Sciences, Nauka Press, Leningrad Laboratory of Mathematical Geology, Leningrad

Cochran WG, Mosteller E, Tukey JW, Jenkins WO (1954) Statistical problems of the Kinsey Report on sexual behaviour in the human male. American Statistical Association, Washington, DC

David MLR (1977) Geostatistical ore reserve estimation. Elsevier Scientific, New York

Dech VN, Henley S (2003) On the scientific heritage of Prof A. B. Vistelius. Math Geol 35: 363–379

De Paor DG (2016) Virtual rocks. GSA Today 26:4–11

Dollar ATJ (1937) An integrating micrometer for the geometric analysis of rocks. Minerol Mag 24:577–594

Eskola P (1915) On the relations between the chemical and mineralogical composition in the metamorphic rocks of the Orijärvi region. Comm Geol Finlande Bull 44:109–145

Fyfe WS, Turner FJ, Verhoogen J (1958) Metamorphic reactions and metamorphic facies. Geol Soc Am Mem 73

Geol Soc Amer Symposium (2002) Forward modelling in tectonics and structural geology. https://gsa.confex.com/gsa/2002AM/finalprogram/session_3025.htm. Accessed 10 Dec 2016

Hatch FH, Wells AK (1937) Petrology of the igneous rocks (Ninth Edit). Murby, London

Hull E et al (1889) Sheet 9 (Gweedore). 1″/mile Geological Survey of Ireland, Dublin

Journel A, Huijbregts CJ (1978) Mining geostatistics. Academic Press, New York

Krige D (1964) Recent developments in South Africa in the applications of trend surface and multiple regression techniques to gold ore valuations. Colo Sch Mines Quart 59:795–809

Krumbein WC (1963) A geological process-response model for analysis of beach phenomena. Bull Beach Eros Board 17:1–15

Krumbein WC, Graybill FA (1965) An introduction to statistical models in geology. McGraw-Hill, New York

Krumbein WC, Slack HA (1956) Statistical analysis of the low-level radioactivity of Pennsylvanian black fissile shale in Illinois. Geol Soc Am Bull 67:739–762

Kurlansky M (2004) 1968: the year that rocked the world. Cape, London

Lovejoy S, Agterberg F, Carsteanu A, Cheng Q, Davidsen J, Gaonac'h H, Gupta V, L'Heureux I, Liu W, Morris SW, Sharma S, Shcherbakov R, Tarquis A, Turcotte D, Uritsky V (2009) Nonlinear geophysics: why we need it. EOS, Trans Am Geophys Union 90(48):455–456

Mandelbrot BB (1982) The fractal geometry of nature. Freeman, San Francisco

Maslov LA (2003) Basic logical principles for analysis and synthesis of geological models. Georeasoning archives, Portsmouth, England. http://www.jiscmail.ac.uk/files/GEO-REASONING/maslov.pdf. Accessed 10 Dec 2016

McInerney P, Goldberg, A, Calcagno P, Courrioux G, Guillen R, Seikel R (2007) Improved 3D geology modelling using an implicit function interpolator and forward modelling of potential field data. In: Proceedings of the exploration 07: fifth decennial internat conference on mineral explore, pp 919–922

Nicholls J (1988) The statistics of Pearce element diagrams and the Chayes closure problem. Contrib Miner Pet 99:25–35

Schlumberger Limited (2016) Iterative forward modeling. http://www.glossary.oilfield.slb.com/Terms/i/iterative_forward_modeling.aspx. Accessed 10 Dec 2016

Silver LT, Chappell BW (1988) The Peninsula Ranges Batholith: an insight into the evolution of the Cordilleran batholiths of southwestern North America. Trans R Soc Edinb Earth Sci 79:105–121

Stanley CR, Russell JK (1989) Petrologic hypothesis testing with Pearce element ratio diagrams: derivation of diagram axes. Contrib Miner Pet 103:78–89

Sui W, Ehling-Economides C, Zhu D, Hill AD (2012) Determining multilayer formation profiles from transient temperature and pressure measurements. Pet Sci Technol 30:672–684

Taleb NN (2007) The black swan: the impact of the highly improbable. Random House, New York

Tolwinski-Ward S (2012) Forward and inverse modeling the nonlinear relationship between tree-ring width and climate—IMAGe seminar. https://www2.ucar.edu/for-staff/daily/announcements/2012-08-24/forward-and-inverse-modeling-nonlinear-relationship-between. Accessed 10 Dec 2016

Turner FJ, Verhoogen J (1951) Igneous and metamorphic petrology. McGraw-Hill, New York

Tuttle OF, Bowen NL (1958) Origin of granite in the light of experimental studies in the system $NaAlSi_3O_8$-$KAlSi_3O_8$-$SiO_2$-$HO_2$. Geol Soc Amer Mem 74, 153 pp

Vistelius AB (1972) Ideal granite and its properties I. The stochastic model. Math Geol 4:89–102

Vistelius AB (1977) Appendix. In Whitten EHT stochastic models in geology. J Geol 85:326–328

Vistelius AB (1991) Sovremenna li geologicheskaya nauka? (Does geology represent an up-to-date science?). Byulletin Akademii Nauk, USSR, Moskva 12:82–89

Vistelius AB, Romanova MA (1972) The concept of ideal granites and its use in petrographical survey and search tasks (on materials from the Mesozoic granitoids of northeast Asia) [in

Russian]. In: Vistelius AB (ed) Ideal granites – issue I. Acad Sci, Nauka Press, Leningrad Lab Math Geol, pp 4–47

White AJR, Chappell BW (1983) Granitoid types and their distribution in the Lachlan Fold Belt, southeastern Australia. Geol Soc Am Bull 159:21–34

White AJR, Williams IS, Chappell BW (1977) Geology of the Berridale 1:100000 Sheet 8625. Geol Survey New South Wales, Depart Mines 138 pp

Whitten EHT (1957) The Gola Granite (Co. Donegal) and its regional setting. Proc R Ir Acad 58B:245–292

Whitten EHT (1960) Quantitative evidence of palimpsestic ghost-stratigraphy from modal analysis of a granite complex. In: Reports international geological congress XXI session (Norden), Part 14, pp 182–193

Whitten EHT (1961) Quantitative areal modal analysis of granite complexes. Geol Soc Amer Bull 73:1331–1360

Whitten EHT (1964) Process-response models in geology. Geol Soc Am Bull 75:455–464

Whitten EHT (1968) Variance of some selected attributes in granitic rocks (in Russian): In Questions of mathematical geology (Anniv Vol for AB Vistelius), Leningrad Math Inst, Leningrad, USSR, pp 240–252. (English trans Topics in Mathematical Geology. Consultants' Bureau, NY 1970 pp 231–242)

Whitten EHT (1991) Granitoid suites. Geol J 26:117–122

Whitten EHT (1996) Molar-ratio and Harker diagrams in portraying the actual chemical variability of granitoid suites. J Geol Soc Lond 153:121–125

Whitten EHT (2000) Variability of igneous rocks and its significance. Proc Geol Assoc 111:1–15

Whitten EHT (2003) Mathematical geology in perspective: Has objective hypothesis testing become overlooked? Math Geol 35:1–8

Whitten EHT (2004) A. B. Vistelius: first IAMG President and visionary mathematical geologist. Earth Sci Hist 23(2):384–389

Whitten EHT (2005) A half century of mathematical geology—the new threshold. Zeits Deuts Geol Ges 155:203–209

Whitten EHT (2010) Influence of the improbable in earth sciences. Proc Geol Assoc 121:249–251

Whitten EHT, Dacey MF (1975) On the significance of certain Markovian features of granitic rocks. J Pet 16:429–453

Whitten EHT, Bornhorst TJ, Li G, Hicks DL, Beckwith JP (1987a) Suites, subdivision of batholiths, and igneous-rock classification: geological and mathematical conceptualization. Am J Sci 287:332–352

Whitten EHT, Dacey MF, Thompson KD (1975) Markovian grain relationships of a Grenville granulite. Am J Sci 275:1164–1182

Whitten EHT, Li G, Bornhorst TJ, Christenson P, Hicks DL (1987b) Quantitative recognition of granitoid suites within batholiths and other igneous assemblages. In: Size WB (ed) Use and abuse of statistical methods in the earth sciences. Oxford Univ Press, N Y, pp 55–73

# Chapter 38
# From Individual Personal Contacts 1962–1968 to My 50 Years of Service

**Václav Němec**

**Abstract** The author's initial personal random contacts with pioneers in introducing mathematics and computers to geology in Russia, USA and France evolved thanks to the 23rd International Geological Congress and the foundation of the IAMG in Prague 1968. An incredibly large set of colleagues from all over the world have continuously contributed to a long series of regular international sessions at the Mining Příbram Symposia—a unique East–West gateway for the IAMG during the period 1968–1989. Very intensive work has been continuing until 2000 with several new peaks. The author has used many positive international organizational experiences from the work for the IAMG in developing geoethics, where many experts of mathematical geology have brought a considerable contribution to this new field.

**Keywords** Mathematical geology · IAMG history · East–West contacts
Mining Příbram Symposia · Geoethics

## 38.1 Introduction

My way into geology did not follow an easy direct path. In 1951 my studies of economics (including courses of mathematics and statistics) were stopped because of political reasons I was not admitted to the final 4th year. Instead of my studies I spent the following 26 months in special army units for politically unreliable persons working in the coal mines of the Ostrava region. At the end of 1953 I started to work in a state enterprise for geology of industrial minerals. At that time, this was a Cinderella among the other sectors of uranium, coal or metals deposits. My chief

Founding and life IAMG member, Eastern Treasurer (1968–1980, 1984–1996), Krumbein medallist 1991, Prague, Czech Republic.

V. Němec (✉)
K rybníčkům 17, 100 00 Praha 10 - Strašnice, Czech Republic
e-mail: nemec.geo@seznam.cz


B. S. Daya Sagar et al. (eds.), *Handbook of Mathematical Geosciences*,
https://doi.org/10.1007/978-3-319-78999-6_38

appointed me as an assistant to two associate professors of the Charles University who were engaged by our enterprise because of a lack of our own graduated geologists. Both these men later became known as very famous professors: Zdeněk Pouba in economic geology and Zdeněk Špinar in palaeontology. I remained in friendly contact with both of them for the rest of their lives. In 1954 I was able to start distance university studies in applied geophysics (in order to study geology there was a condition of having several years of practice, whereas for my specialization it was only necessary to have finished military service—such was the life in those days). Mathematics was among the key disciplines of my studies and my regular work in my enterprise became more and more focussed on evaluating the results of geological projects concerned with computing ore reserves. I graduated in 1959.

At that time, our Cinderella was incorporated into a new enterprise covering exploration of all sorts of deposits except uranium. Despite some renewed political problems in 1960, I was appointed as chief of a special division for controlling the final reports of the company, being the only trained specialist when two of my new bosses arriving from other sectors preferred employment outside our company. On my own initiative I took my job as a consultant service discussing with my colleagues responsible for individual projects the appropriate methods for computing the reserves. Already in 1961, we started processes with the mechanization of work using punch cards. During a tourist trip to the USSR in 1962 I had my first occasional contacts with several colleagues in Moscow at the State Commission of Ore Reserves—I. D. Kogan was one of the top personalities (his son Robert later became my close friend). After a new reorganization in late 1962 I got a position in which it was possible to realize along with trained computer specialists new ways of applying computers for our specific professional needs.

In 1964—during my first trip behind the Iron Curtain after the Prague coup in February 1948—on a private family visit in the USA I had the chance to contact several colleagues in Colorado and Arizona working in the field of mathematical geology. The existence of the Tucson centre active in this field was discovered from literature by my colleague—economist and statistician *Blahomil Soukup*. My contacts with the organizers of the APCOM Symposia at that time held in Tucson and other US universities resulted in further interesting contacts. At the Colorado School of Mines *R. F. Hewlett* gave me the address of *Ivan P. Sharapov*. The following year (1965) this Russian scientist took a more than 2000 km long flight from Perm to Sochi in order to meet me in person for one weekend during my vacation in that famous Black Sea resort. Ivan was a man who despite incredible personal political problems (several years of arrest and concentration camps) continued to introduce mathematical statistics for applications in geology. He was extremely pleased to meet a colleague from abroad for the first time in his life in his 58th year. He had already established his own written contacts abroad and I obtained from him the addresses of such famous personalities as *Danie G. Krige* and *Georges Matheron*.

In 1965 I was among three Czech authors who published their papers at the APCOM Symposium in Tucson which in 1966 gave an impulse to *Dan Merriam* to contact us in the course of his visit to Europe including the Eastern territory (Krakow and Prague). Further progress in establishing new international contacts became extremely rapid and the approaching 23rd International Geological Congress in Prague (1968) brought me several special engagements among the organizers of the Congress as well as membership of the International Preparatory Committee (headed by *R. A. Reyment*) for the foundation of an international association for the application of mathematical methods and computers in geology (the exact name was under discussion).

In September 1967 during a private tourist trip to France I established personal contacts with Professor *Georges Matheron* and with several other French colleagues (*A. Carlier, Jean Serra*). In November 1967 I defended my doctoral thesis (RNDr.) at the Charles University in Prague in the field of economic and mathematical geology based on my first computerized model of three deposits for a cement factory in the suburbs of Prague.

In December 1967 I was the only foreign guest at the Second Siberian Symposium on Mathematical Methods in Geology and Geophysics in Novosibirsk (480 participants) where *Ivan Sharapov* and the local chief organizer *Yuri Voronin* helped me to contact many VIPs in this field from all parts of the USSR (including *Dmitry Rodionov*). When addressing the plenary meeting I invited people to attend the Prague Congress with a specialized session on mathematical geology and informed them about our plans to found a new international association. (*A. B. Vistelius* was the only member from the USSR on the international committee but he did not attend this Symposium).

## 38.2 IAMG Foundation (Prague 1968)

In 1968 an incredible optimism characterized both the hopeful political development of the Prague Spring as well as the preparations of the International Geological Congress and of the founding meeting of the IAMG. I already had the pleasure to describe more details of these events in the book for the IAMG Silver Jubilee (Němec 1993a).

The euphoric start of the 23rd International Geological Congress gave me the opportunity to meet in person for the first time many new colleagues already well known in the field of mathematical geology (*Frits Agterberg, R. B. McCammon, J. W. Harbaugh, R. A. Reyment, A. B. Vistelius, G. S. Watson,* and *E. H. T. Whitten*). Professor *W. C. Krumbein* informed me that his arrival would be delayed. But very early in the morning of Wednesday August 21 all plans were changed with the entry of five armies under the Warsaw Treaty. Because it was impossible to visit the Congress centre I spent part of that day with Professor Reyment, who was staying in a hotel near my home. He made several telephone calls with the Swedish Embassy. It appeared that the current situation prevented any prediction about the future of the Congress.

On the morning of Thursday August 22nd, 1968 special transportation was set up again for Congress participants and some of the Congress program was re-activated. It became possible to use the room reserved for the preliminary discussions planned prior to founding the new Association. The new situation only permitted essential formal administrative steps including the election of the first IAMG Council. Professor *R. A. Reyment* as the Chair of the meeting refused the suggestion of *John Harbaugh* to be elected as President (preferring the position of Secretary General) and asked to elect for this top position *A. B. Vistelius.* Both key functions were unanimously approved. *G. S. Watson* was elected as the Vice-President representing a liaison with the International Statistical Institute. My suggestion to elect the absent *Prof. W. C. Krumbein* to the post of the "Past President" was accepted as well. *T. V. Loudon* was elected as the Western Treasurer. Prof. Watson suggested me for the post of the Eastern Treasurer. After my election I started my official activity for the new Association by suggesting *D. Krige* and *G. Matheron* (in their absence) as IAMG Council members. *F. P. Agterberg, D. A. Rodionov* and *E. H. T. Whitten* as well as the absent *S. C. Robinson* and *S. Sengupta* were elected as further members of the Council while *D. F. Merriam* and *Graham Lea* (absent) were chosen as the first editors of intended IAMG publications. The first IAMG Council had a very good geographical distribution. The election of two Russian scientists to the Council on that day was a testimony in favour of absolute priority being given to personal professional quality avoiding any political concerns.

After a very emotional premature closing ceremony of the Congress on Friday August 23 afternoon I had the honour to represent the IAMG together with *A. B. Vistelius* and *E. H. T. Whitten* at a working meeting of the International Union of Geological Sciences where, in an accelerated process, our Association was officially approved as a new affiliated member. At that time I had no idea how many opportunities were to be awaiting me to work in the IAMG for so many years ahead including my service as the Eastern Treasurer altogether for six terms (1968–1980 and 1984–1996)!

## 38.3 Activities for the IAMG 1968–1993

Various activities of the new Association had to be negotiated, mostly using normal mail. Today it is already difficult to imagine the modest technical means of that time (without any fax or e-mail). However, some personal contacts helped me to make a start with my duties. At that time, my employer—the geo-exploration state enterprise under a new name of *Geoindustria* became the sole collective IAMG member in Czechoslovakia supporting my official activities abroad by financing a lot of my travel expenses.

In January 1969 I visited a conference of mining geodesy in Moscow and paid a visit to *A. B. Vistelius* in Leningrad. The possibility of visiting Western countries continued until the autumn of 1969 and I therefore had no problem to meet with many IAMG Council members at the Congress of the International Statistical Institute in London in August 1969, to spend three weeks in September in France attending a special course on geostatistics at Fontainebleau, and to accept with the consent of my employer the invitation of the Kansas Geological Survey in Lawrence (initiated by *Dan Merriam*) to work there from November 1969 until August 1970. This was an excellent opportunity for establishing many further (already global) useful contacts for my activities for the IAMG and for the international development of mathematical geology. I hold deep memories of my experiences from that time (Bonham-Carter et al. 2008), especially the colloquium on Geostatistics (Nemec 1970) held on the campus in Lawrence and the APCOM Symposium in Montreal (both in June 1970).

In addition to my stay in America I also had to work hard to fulfil my professional duties for Geoindustria. The following text will hopefully disclose how useful working at this cosmic speed during this starting period turned out to be for all the hyper-activities carried out during the remaining almost five decades of my further life.

## 38.4 Příbram—East–West Gate Near the Iron Curtain

As explained elsewhere (Němec 1993b) a symposium “The Mining Příbram in Science and Technique” was organised for the first time in 1962. The city of Příbram —located 60 km SW from Prague—had a long mining tradition going back to the thirteenth century. In November 1968 several Czech colleagues—mostly geophysicists from the Czechoslovakian Uranium Industry—organised a special session on *Mathematical Methods in Geology and Geophysics* for the first time. They also agreed to organise a special seminar on Geostatistics in Prague and I had the honour —in the course of my visit to France in September 1968—to invite *G. Matheron* and *J. Serra* to take part in that two-day seminar as well as in the new session in Příbram. Both guests were deeply impressed by both the Czech audience and hospitality and Prof. Matheron himself suggested continuing the Příbram meetings with co-sponsorship of the IAMG. I immediately started to promote that idea.

From 1969 I acted as the main convenor of that specialised international session, which actually came about as early as October 1969. We had guests from six countries, but it seemed impossible for *A. B. Vistelius* or *I. P. Sharapov* to attend the meeting (they sent in their written articles). Shortly after the meeting I left Prague to start my temporary work in Kansas. Through contact with the secretariat of the Symposium and with several Czech colleagues (*B. Soukup, M. Škubal*) it was possible for me to continue on from Lawrence with preparations for the next session at Příbram in 1970. Using my new contacts, I was able to successfully promote the idea of also holding these rendezvous at the above-mentioned meetings in Lawrence and Montreal. My work in Kansas terminated in August 1970 and in October there were already 26 foreign colleagues from 11 countries who participated in the Příbram session, together with about 55 participants from Czechoslovakia. We had several guests from America (*Michel David, Dan Merriam,* and *Tim Whitten*), one from India, and also *Dmitry Rodionov* appeared from Russia. Simultaneous translation was used for the first time. This was a very good start for further promotion of this kind of meeting which later took place regularly in October every year until 1973. The 1970 Příbram meeting can be classified as an important milestone of progress.

Since about 1965 the promotion of mathematical methods and computers in the Earth sciences became included in official activities within the framework of the Eastern bloc organization *COMECON (Countries of the Mutual Economic Aid)* and just in 1973 a regular meeting of specialists was planned and organized in Czechoslovakia. Many participants of previous regular meetings on this subject already knew Příbram. It became possible to find a way how to join the official meeting for COMECON delegates (it took place in a locality not far from Příbram) with the regular Symposium (all scientific papers presented in Příbram).

This arrangement made it possible to intensify the already existing East–West contacts. After 1973 the section on Mathematical Methods in Geology was regularly organized every second year—in 1983 again in conjunction with a special COMECON meeting. Many IAMG members from both the West and East were taking regular part in the meetings, e.g. *Tim Whitten* visited Příbram as IAMG Secretary General in 1977 and again as IAMG President in 1983. Also, representatives of COGEODATA were among the visitors and thanks to the initiative of *Jiří Hruška* on several occasions official meetings of that organization were arranged in Prague making it possible for their participants to also take part in the Mining Příbram Symposium. In 1989 and 1991 specific problems of geoinformatics were included in a separate parallel section of the Symposium.

Regular meetings of the specialized COMECON groups were organized in different COMECON countries according to their usual format which involved excluding visitors from other countries. However, both their meetings at Příbram in 1973 and 1983 were unique exceptions lifting scientific programs to a level accessible to all scientists from around the world. I was very lucky that this idea was adopted not only by top representatives of the Czechoslovak geological community but also by the representatives of the COMECON Secretariat in Moscow and by the authorities responsible for that sector especially in the USSR, Hungary, Poland and Yugoslavia.

From 1983 onwards the meetings of Příbram were regularly attended by participants of special courses on geochemistry organized regularly in Czechoslovakia by UNESCO with the School of Mines at Ostrava. At that time, I also had some written contact with UNESCO top representatives (see Fig. 38.1).

In 1987 the section was organized jointly with the GEOCHATAUQUA—held for the first time outside North America (unfortunately, without visitors from that part of the world).

The rapidly changing political situation in the Eastern bloc permitted in October 1989 (6 weeks prior to the November *velvet revolution*) the visit to the geo-mathematical section at Příbram Symposium of many people from the East (especially about 65 guests from the USSR). Altogether 125 visitors from 23

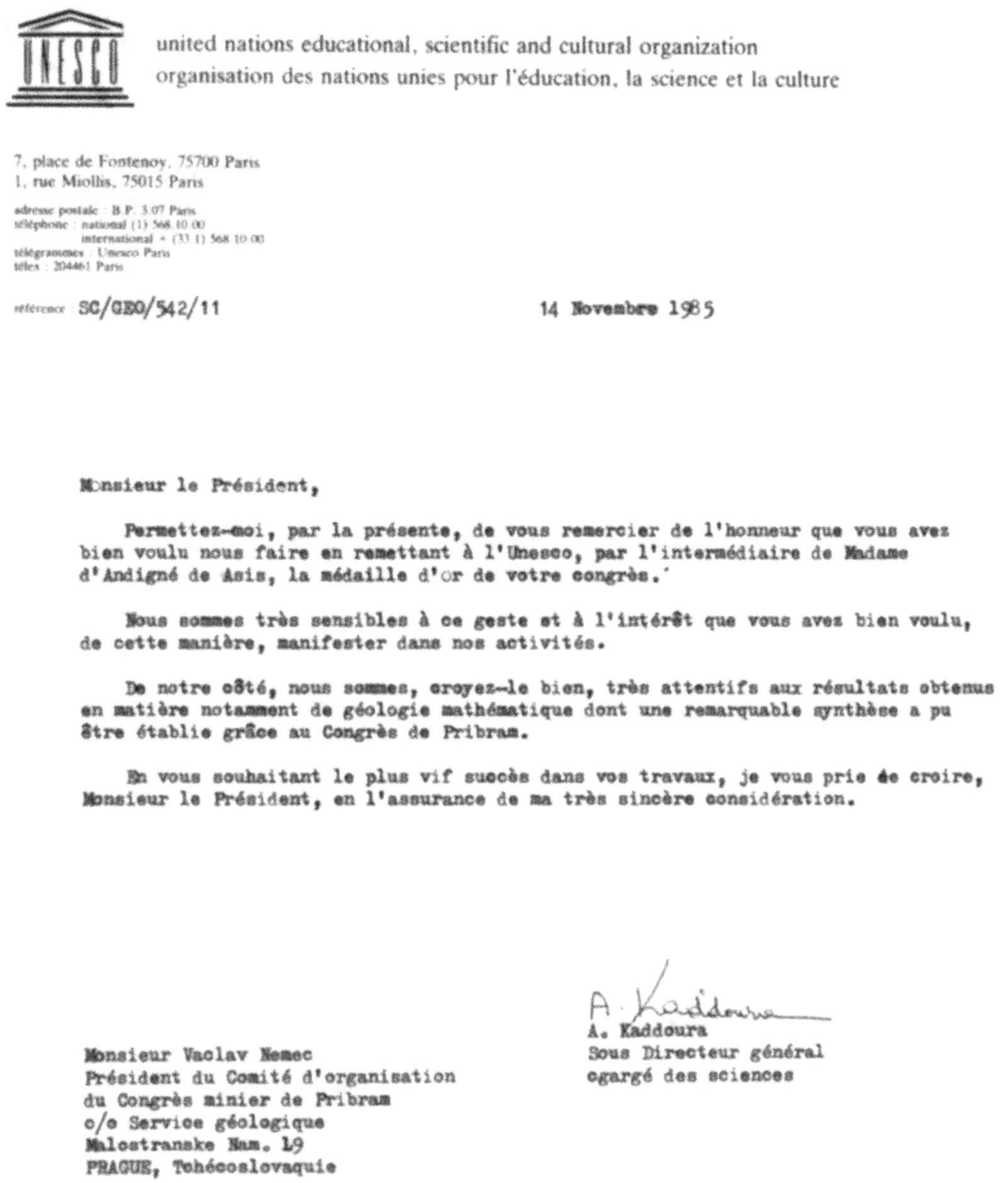
UNESCO

united nations educational, scientific and cultural organization
organisation des nations unies pour l'éducation, la science et la culture

7, place de Fontenoy, 75700 Paris
1, rue Miollis, 75015 Paris

adresse postale : B.P. 3.07 Paris
téléphone : national (1) 568.10.00
international + (33.1) 568.10.00
télégrammes : Unesco Paris
télex : 204461 Paris

référence : SC/GEO/542/11

14 Novembre 1985

Monsieur le Président,

Permettez-moi, par la présente, de vous remercier de l'honneur que vous avez bien voulu nous faire en remettant à l'Unesco, par l'intermédiaire de Madame d'Andigné de Asis, la médaille d'or de votre congrès.

Nous sommes très sensibles à ce geste et à l'intérêt que vous avez bien voulu, de cette manière, manifester dans nos activités.

De notre côté, nous sommes, croyez-le bien, très attentifs aux résultats obtenus en matière notamment de géologie mathématique dont une remarquable synthèse a pu être établie grâce au Congrès de Pribram.

En vous souhaitant le plus vif succès dans vos travaux, je vous prie de croire, Monsieur le Président, en l'assurance de ma très sincère considération.

A. Kaddoura
A. Kaddoura
Sous Directeur général
chargé des sciences

Monsieur Vaclav Nemec
Président du Comité d'organisation
du Congrès minier de Pribram
c/o Service géologique
Malostranske Nam. 19
PRAGUE, Tchécoslovaquie

**Fig. 38.1** Letter of the UNESCO Deputy Director General A. Kaddoura to Vaclav Nemec. The French text is a warm expression of thanks for the golden medal of the Mining Příbram Symposium appreciating regular co-operation of the international section of mathematical geology with UNESCO courses on geochemistry organized at the School of Mines in Ostrava

foreign countries (both East and West) with also about 125 colleagues from Czechoslovakia represented a new record of participation.

In 1991 the section was already organized in a new political and economic climate. Members of a new ad hoc *committee* of the IAMG appointed by the IAMG President *R. B. McCammon* for preparing the Silver Anniversary Meeting of the IAMG were present among the participants: *Dan Merriam, Frits Agterberg, Peter Dowd, Mike Hohn* (IAMG Secretary General), and *V. Němec*. Intensive talks were held in my home in Prague prior to the Symposium and everybody seemed to agree with my suggestion to prepare a joint Silver Anniversary Meeting of both IAMG and the Mining Příbram as a festive gathering of Western and Eastern colleagues in Prague following the format of the meetings of the Mining Příbram Symposium in 1993. The resulting information was communicated to all participants at Příbram.

At the IGC 1992 in Kyoto in my paper discussing the 15 geomathematical sessions held regularly at the Mining Příbram Symposia from 1968 until 1991 I had the pleasure to present the following impressive results:

- altogether 970 written contributions (total volume of 8696 printed pages),
- altogether 441 oral contributions (posters were used only marginally, mostly adding data for oral contributions),
- altogether 925 individual authors,
- altogether 30 countries with representatives from the whole world.

Only 45% of the published full texts or abstracts that were given were represented orally, because of the fact that not every author was allowed to come to Příbram. The State authorities, especially in the USSR and in Eastern Germany were watching and controlling the situation and more freedom for individual visitors only became evident in 1989 when the combination of both political and economic situations had become optimal for the possibility of travel to Příbram.

## 38.5 My Own Professional Work

In 1972 I was asked by the Central Geological Institute in Prague for a peer review of a book prepared by the Czech authors *Vladimír Sattran* and *Blahomil Soukup* about the application of mathematical methods in geology. It was published in the Czech language in 580 copies (Sattran and Soukup 1973). A large list of publications from prominent authors, both Western and Eastern, represented a very good review and the whole book reflected the actual situation and some promising future development trends.

In my own work at Geoindustria in Prague I had the possibility to continue developing new space and time models for various deposits as well as arranging the agendas for the Mining Příbram symposia. My continuing position in the IAMG Council was accepted by top representatives not only from my employer but also of the Czechoslovak Bureau of Geology. I had the chance to visit at least partially all the International Geological Congresses since 1980 (see Fig. 38.2 from Moscow 1984), and International Stratigraphic Congresses in Heidelberg (1971) and in Nice (1975), APCOM Symposia in Clausthal (1975) and London (1983). Every year I was a regular guest at geomathematical meetings organized in Krakow by Professor *Janusz Kotlarczyk*, in Freiberg (Saxony—Eastern Germany as section of large events), and at many meetings in various parts of the USSR as well as several meetings in Hungary (*Istvan Dienes, Endre Dudich*).

**Fig. 38.2** Václav Němec attending a session on mathematical geology at the International Geological Congress in Moscow (1984). The neighbour of Václav Němec is the highly respected French expert in geomorphology and petrography André Cailleux

I spent one month on a lecturing tour in Italy (1971—see Fig. 38.3) and another lecturing tour in Canada and the USA (1986); also, several working visits in Vietnam and in Mongolia should be mentioned because of the possibilities of making some special contacts with local academic circles (Professor *Ochir Gerel* in Ulaan Baatar).

In all these meetings I was presenting my own (sometimes co-authored) scientific papers, mostly in the domain of space and time models for various kinds of deposits. I always emphasized that special attention should be given to achieving a geologically correct solution by avoiding inappropriate mathematical processes (interpolation) leading to erroneous geological interpretations. My speciality also covered so-called inserted subsystems (Němec 1988).

Every opportunity was used for spreading information about the IAMG and about the possibility of visiting Příbram as the only relatively easily accessible East–West meeting point. The success was partially achieved thanks to my ability to communicate in different local languages.

I had also the possibility to officially invite several specialists to give individual courses or lectures in Prague (*Frits Agterberg, Tim Whitten,* and *Jan Harff*).

In the early 1970s I was already a guest lecturer at the Charles University in Prague, then in the late 80s at the Technical University of Košice and in 1991/92 at the Comenius University in Bratislava, providing special courses about applying

**ASSOCIAZIONE MINERARIA SUBALPINA**

presso Istituto di Arte Mineraria del Politecnico

Corso Duca degli Abruzzi 24 - TORINO - (Tel. 51 12 77)

Il Presidente

10129 Torino, 3 maggio 1971

Circolare 71/16

PROSSIMA CONFERENZA

Facendo seguito a quanto preannunciato con lettera circolare del 26/4 preciso che lunedì 10 maggio corrente, con inizio alle ore 18,15, nel= la consueta aula dell'Istituto di Geologia e Giacimenti Minerari, il Dott. Václaw NĚMEC, noto studioso di Geologia Matematica, appartenente alla "Geoindustria" di Praga e membro della International Association for Mathe= matical Geology, terrà - in lingua italiana - una conferenza su

APPLICAZIONI DELLA MATEMATICA
A PROBLEMI GIACIMENTOLOGICI E MINERARI

soffermandosi in particolare sulle questioni concernenti la valutazione dei giacimenti minerari con l'impiego di calcolatori elettronici.

Il Dott. Němec svolgerà poi - sempre nella stessa sede del Poli= tecnico - secondo il programma riportato in calce, delle lezioni di semina= rio comprendenti un'ampia discussione degli argomenti esposti, con l'esame di casi pratici.

I seminari sono specificamente riservati ai docenti ed agli stu= denti della Sezione di Ingegneria mineraria del Politecnico. Quei Soci però che avessero interesse agli argomenti, potranno intervenire dandone preven= tiva notizia alla Segreteria dell'Associazione. A tal riguardo, anzi, se= gnalo ancora che il Dott. Němec sarà particolarmente lieto di incontrare tecnici e studiosi del campo industriale minerario per discutere con essi applicazioni concrete della matematica per la soluzione di problemi di va= lutazione e gestione dei giacimenti.

(Prof.Dott.Ing. Lelio Stragiotti)

c. c. p. 2/42124

**Fig. 38.3** Announcement of a presentation of Dr. Němec and of his following seminar in Italy The Italian announcement signed by the Rector of the Polytechnic Institute in Torino Prof. Dr. Ing. Lelio Stragiotti informs about a special conference on the application of mathematics to problems of mineral deposits from the point of their exploration and mining exploitation followed by three days of seminars about the computerized evaluation of reserves of deposits. Seminars were reserved for teaching staff and students of the Institute but also accessible for specialists and members of the Sub Alps Mining Association (May 1971, all events were held in the Italian language) mathematical methods and models in the Earth sciences (including mining processes).

In 1987 I defended my higher scientific degree C.Sc. (candidate in sciences as a Ph.D. equivalent) at the Technical University in Košice (in the mining sciences). The work, without any supervisor, was based on summarizing my development of space and time models for optimizing the long-term mining processes at various kinds of deposits.

## 38.6 Two Separate Silver Anniversary Meetings of Mathematical Geologists in Prague (1993)

The idea to select Příbram 1993 for a broad international meeting in close co-operation with the IAMG had been discussed originally in 1986 during my trip to North America on the occasion of the Geochautauqua in Calgary and also when visiting *Dan Merriam* in Wichita. These talks continued in Washington DC at the International Geological Congress 1989, when the process of considerable change in the Eastern block was already starting. A few months later the velvet revolution in Czechoslovakia opened the door for fulfilling the idea in a more impressive way. The IAMG President *R. B. McCammon* in particular was emphasising his vision of a broad historical meeting of colleagues from both the West and East. All my activities at that period were oriented toward this goal and all authorities responsible for the Mining Příbram Symposium also agreed with such a vision.

With the help of my wife, Lidmila Němcová I arranged for contacts with the centre Krystal in Prague—working for three main Prague universities and administrated by the University of Economics (where my wife was teaching). This centre seemed to be the optimal place for holding the Silver Anniversary Meeting (technical equipment, advantage of relatively low prices in comparison with other possible centres, hotel capacity, very good access from the airport as well as from the down-town area, good personal contacts with administrators). We had also found several other possibilities of accommodation (some of them in the neighbourhood of Krystal)—at that time allowing people accommodation for only about 10 US$ per night. The members of the already aforementioned ad hoc committee were able to verify the situation as well as the IAMG President *R. McCammon* who paid his personal visit to Prague in November 1991. We also started to prepare a special "silver" medal for the Silver Anniversary meeting: *Antonín Ryčl*, secretary of the Příbram Symposium, introduced us to the famous Czech medallist *Lumír Šindelář* who after several discussions designed both marvellous sides of it. In April 1992 *John Davis* and *Jan Harff* visited Prague which, in addition to our intensive talks included a visit of the artist. We all expressed strong enthusiasm for the design of the medal and only a few small corrections seemed to be necessary. *John Davis* prepared on his PC a Memorandum of Discussions and later I also received this document from *Dan Merriam* with an accompanying letter giving full approval to all the results achieved. It was possible to arrange for the final production of the medal in the mint house of Kremnica and to continue with the standard preparations for the Jubilee meeting.

In the meantime, I was also very pleased when the IAMG President *Dick McCammon* announced to me by phone that I was elected as W. C. Krumbein medallist for 1991. This highest IAMG award primarily reflected my long-term service to the profession by organizing and keeping uninterrupted contact between East–West between mathematical geologists through the gateway of Příbram.

Unfortunately, some misunderstandings arose: one of them was connected with the side of the medal commemorating the liaison of Prague and Příbram with the IAMG (use of some religious symbols from the Saint Hill—a famous pilgrim locality at the border of Příbram). At that time the renewal of religious freedom was highly appreciated in Czechoslovakia and in other countries of the Eastern bloc. However, the American colleagues entertained different points of view for the standards of international contacts. After my arrival at the IGC in Kyoto (1992) I was asked to arrange with the artist to replace that side of the medal just by the official IAMG logo. Another idea consisted in separating the IAMG Silver Jubilee from the same jubilee of our meetings at Příbram. In my role as the IAMG officer I continued my loyal service to the Association, arranging for contacts with the *Carolina* agency as needed (enabling preparations for the IAMG Silver Anniversary meeting in the Krystal centre). On the other hand, I also had to prepare the Silver Anniversary meeting for the international section of the Mining Příbram Symposium. The respective authorities approved the use of the Krystal centre for that purpose for the days following the IAMG meeting. All potential participants of the "Příbram" Symposium (about 400!) were informed in time by me about the IAMG meeting as well. A special advertisement was published in the Czechoslovak monthly geological magazine.

The final solution resulted in two separate Silver Anniversary meetings taking place in Prague at the same Krystal centre. The IAMG sessions were visited by 152 (mostly Western) people, the Příbram sessions by 140 (mostly Eastern) people. Only about 40 persons attended both meetings. Just one compromise had been finally reached: a common half-a-day meeting accessible to both IAMG and Příbram participants focussed on the history of mathematical geology.

In the end I think that the various misunderstandings and misconceptions connected with the IAMG Silver Anniversary Meeting in Prague also had some positive consequences: more freedom was given to all local organizers of subsequent annual IAMG conferences and the IAMG Councils in the years following until 1999 continued to provide some financial and moral support for the geomathematical sessions organized by the Mining Příbram Symposium.

## 38.7 From The Silver to the Golden IAMG Jubilee

In 1994 I received a diploma of "engineer" from the University of Economics in Prague as restitution of the violation of my rights when I was not permitted to complete my studies of economics in 1951 in spite of good results in my studies.

I continued to organize the international meetings as part of the Mining Příbram Symposia in the years 1995, 1997 and 1999. These sessions were held again at the Krystal centre in Prague without any help from any official congress agency, and always with the moral and some financial support of the IAMG. *Mike Hohn*—the IAMG President—honoured the session in 1995 by his presence and was able to contact many Eastern participants. Financial support from the IAMG made it possible to pay local expenses and registration fees for about 15 foreign colleagues (for each session). We always had about 80–100 participants from abroad and the scientific level of presentations was good. The new economic situation in the Czech Republic led to decreasing participation from Czech colleagues who were represented by only a small minority.

Czech colleagues who helped me in my organization work until 1989 were not available anymore (being completely absorbed by other activities, retired or deceased). Western colleagues preferred to attend the official IAMG Annual Conferences. For some Eastern colleagues (especially from the countries of the former USSR) a new visa policy demanded lots of extra work for me as a volunteer organizer of the Příbram meetings. Therefore, I decided to stop further activities for the traditional session of "Mathematical Methods in Geology" organized 19 times between 1968 and 1999. I only revived this old tradition in 2011 on the occasion of the Mining Příbram Golden Jubilee Symposium, already reported in connection with my new field of interest in the following text of this article. Very positive remarks were published by Vera Pawlowsky in the Presidential Forum in the IAMG Newsletter (December 2011).

## 38.8 The IAMG Experiences Applied to Develop a New Discipline of Geoethics

With the inspiration and support of my wife Lidmila Němcová (expert on business ethics) I have worked since 1991 to establish a new discipline in the family of earth sciences—*geoethics*. Originally, the main reason was focussed on ethical problems connected with the non-renewability of mineral resources.

The relatively good start of the new discipline and its rapid development became possible thanks to our extensive contacts established especially in the former Eastern bloc where many colleagues had first-hand knowledge of and personal experience with the Mining Příbram Symposia and with their traditional sessions on mathematical methods in geology.

It is beyond the scope of this contribution to describe the proper development of this new field of interest. On the other hand, I feel it as my duty to express thanks to the IAMG representatives who supported these activities when the development was not yet covered by another association (AGID since 2004).

## 38.9 Conclusion

I started my final preparation of this article during the days following the death of the famous IAMG promoter Professor *Dan Merriam* as well as at the time of his funeral service in Lawrence. I have never changed my very positive evaluation of himself and of his merits for the IAMG as expressed in my Introduction to the *"Festschrift"* (Němec 1993a). I was deeply moved when reading in the official obituary about the Gold medal of the Mining Příbram Symposium 1970 which was the first place among a lot of other awards for his activities. His personality and his spirit will accompany the readers of this contribution at every page. It is impossible for me to put across his image on this occasion to anybody of the many very happy, pleasant and unforgettable events connected with Dan and other old fellows I had the privilege to meet during my long service to the IAMG.

Let me emphasize my personal conviction that just a trans-generational solidarity is the "secret" explaining the otherwise unbelievable success of the half-a-century IAMG history. A recipe for the further 50 years of the IAMG: Enthusiasm of the young generation should be always accompanied by life experiences and the know-how of the old pioneers.

Vivat IAMG!

## References

Bonham-Carter GF, Němec V, Merriam DF, Pengda Z, Cheng Q (2008) The role of Frederik Pieter Agterberg in the development of geomathematics. In: Bonham-Carter G, Cheng Q (eds) Progress in geomathematics. Springer, Berlin, Heidelberg, pp 5–21

Nemec V (1970) The law of regular structural pattern: its applications with special regard to mathematical geology. In: Merriam DF (ed) Geostatistics. Premium Press, New York, London, pp 63–78

Němec V (1988) Geomathematical models of ore deposits for exploration purposes. In: Namyslowska-Wilczynska B, Royer JJ (eds) Sciences de la Terre, Nr. 28, vol 2, pp 121–132

Němec V (1993a) Introduction. In: Davis JC, Herzfeld UC (eds) Computers in geology—25 years of progress. Oxford University Press, New York, Oxford, pp 3–11

Němec V (1993b) Fifteen international meetings of mathematical geologists at the Mining Příbram Symposia in Czechoslovakia (1968–1991). In: Nishiwaki N (Guest Editor) Recent advances in geomathematics and geoinformatics, pp 361–364. Geoinformatics 4(3). Japan Society of Geoinformatics, Osaka

Sattran V, Soukup B (1973) Použití matematických metod v geologii. ÚÚG Praha, 156 pp. In Czech

Numerous personal reports of V. Němec as well as reports of others about his activities can be found in the IAMG Newsletters at the IAMG website

# Chapter 39
# Andrey Borisovich VISTELIUS

**Stephen Henley**

**Abstract** This chapter provides a glimpse of the legacy of Professor Andrey Borisovich Vistelius, who served as the first President of the International Association for Mathematical Geoscientists (IAMG) during 1968–1972.

Professor Andrey Borisovich Vistelius (1915–1995) was arguably the founder of the field of mathematical geology, and he was the first President of the International Association for Mathematical Geology. As a 1982 recipient of the President's Prize (later renamed the Andrey Borisovich Vistelius Research Award) I consider it a great privilege to have been invited to contribute this chapter in his honour. The scientific heritage of Professor Vistelius is extremely rich. His active work on fundamental and applied problems of geology, and especially mathematical geology, continued to the last days of his life. He was responsible for more than 200 published works, each representing a significant contribution to science. His works

S. Henley (✉)
Resources Computing International Limited, 185 Starkholmes Road, Matlock, Derbyshire DE4 5JA, UK
e-mail: steve@vmine.net



B. S. Daya Sagar et al. (eds.), *Handbook of Mathematical Geosciences*,
https://doi.org/10.1007/978-3-319-78999-6_39

cover a wide range of subjects, with contributions to the development of stratigraphy, mineralogy, petrography, petrology and geochemistry. The mathematical approach to geoscientific research, pioneered by Vistelius, has gained recognition worldwide. As applied in practice, these works also represent building blocks to more effective methods of search for minerals. There have been a number of publications about Vistelius, and in attempting to present a rounded view of his life and works, this chapter quotes from them extensively: particularly Dvali et al. (1970), Romanova and Sarmanov (1970), Dech and Glebovitsky (2000), Merriam (2001), Henley (2003), Dech and Henley (2003), and Whitten (2004). I also wish to acknowledge unpublished sources including Whitten, the late Merriam, Pshenichny, and Dech.

## 39.1 Background

Andrey Borisovich Vistelius was born on 7th December 1915 into the family of a Russian nobleman. His father Boris Vistelius was a lawyer in St. Petersburg before the October Revolution of 1917. Boris's father (Andrey Borisovich's grandfather) occupied a senior position in the civil service of the Russian Empire. The relatives of Andrey's mother (the Bogaevsky family) included some distinguished academics. Thus, his maternal grandfather was a professor at the Imperial St. Petersburg Institute of Technology, and his uncle was rector of the Imperial St. Petersburg Academy of Art.

There is no published information on Vistelius' early childhood and how he and his family fared during the turbulent years of revolution and civil war. However, it is known that in 1935, after the assassination of Sergei Kirov, the communist leader of Leningrad (as St. Petersburg was renamed in 1924), Boris Vistelius with his wife and son Andrey (at that time a student aged 20) were exiled from Leningrad like many other intellectuals and noblemen. First the Vistelius family found themselves in a remote village in middle Russia, though later the family was allowed to settle in the city of Samara. Because of this forced deportation, A. B. Vistelius had to interrupt his education at the Leningrad State University (which he had entered in 1933).

His studies were resumed only by good luck. Stalin issued an edict with the slogan "sons are not responsible for their fathers' deeds", and Boris Vistelius sent a letter to Stalin which clearly received a positive reply. This allowed Andrey Vistelius to resume his studies in Leningrad and in 1939 he graduated brilliantly from the Department of Mineralogy which was headed at that time by Prof. S. M. Kurbatov, a pupil of Academician V. I. Vernadsky, the great mineralogist and geochemist who is considered one of the founders of geochemistry, biogeochemistry, and radiogeology.

A. B. Vistelius was a vivid and gifted personality. He had a very extensive knowledge of history and literature (both Russian and foreign), appreciated poetry and read English authors in the original. But geology and mathematics were his overwhelming passions. The research topics he investigated were always of great practical importance and at the same time lent themselves to the innovative and elegantly developed solutions which became a hallmark of Vistelius' work.

He was very sensitive to any dishonesty in science—and especially to political lies. He was known as a sharp-tongued man among his colleagues. Especially under Stalin's rule, officials did not like such people, and it was very hard for Andrey Vistelius to further his career. His scientific honesty, frankness and his manner of open and explicit expression of his viewpoint prevented his elevation to Academician of the Academy of Sciences, the highest scientific institution of the USSR. For the political appointees who, as a rule, were heads of all scientific establishments, he was an irritant, indeed an extreme nonconformist.

Thus, he never denied his aristocratic heritage, at a time when most descendants of noblemen in Russia were trying to obscure their origins, some even changing their surnames during the period of communist rule. In curricula vitae for job applications he repeatedly wrote that he was a nobleman by birth. Of course, copies of all these documents were compulsorily held by the KGB (Committee for State Security of the USSR), and his noble descent was an embarrassment for the scientific authorities, his employers.

During World War II, A. B. Vistelius was trapped in besieged Leningrad. He underwent all the sufferings of Leningradians. He was not enlisted into the army because of poor eyesight. However, despite the war, his studies continued, with award of his 'Candidacy' (roughly equivalent to a western Ph.D.) in 1941, and subsequently his Doctor of Science degree in 1948. After working as a senior scientist in several state organisations, and serving as a director of several geological 'expeditions' (the organisations in the USSR, and later the Russian Federation, responsible for regional geological mapping), he became the director of the newly created Laboratory of Mathematical Geology at the Steklov Mathematical Institute of the USSR Academy of Sciences in Leningrad.

In 1968, Vistelius was instrumental, with others, in founding the International Association for Mathematical Geology, and was elected its first president.

Although his circumstances meant that he was unable to participate in many of IAMG's activities, he continued work as a prolific researcher in Leningrad (subsequently St. Petersburg) with extensive publications in both English and Russian. Whitten (pers.comm.), during a visit to Leningrad in 1971, invited him to Northwestern University (Illinois) which Andrey Vistelius was finally able to accept for the Spring Quarter 1975, and his publication list reflects the results of research projects which he was able to undertake in the US during his time there.

He continued to work in St. Petersburg during the 1970s and 1980s, with a steady stream of research publications, in Russian and in English.

Professor Andrey Borisovich Vistelius died on 12 September, 1995. He continued to work until his last days, with lucidity and inventiveness of thought even in spite of serious illness. In 1992, not long before his death, Kluwer Academic Publishers printed an English translation of his life's work "Principles of

Mathematical Geology" (Vistelius 1992). This is a considerably reworked and enlarged English edition of his Russian monograph with the same title (Vistelius 1980).

## 39.2 Scientific Achievements and Insights

The scientific heritage of Prof. A. B. Vistelius is extremely rich. His active work on both fundamental and applied geology, and especially mathematical geology, continued to his last days. He was responsible for more than 200 published works, each of them presenting a very significant contribution to science. References to many of these are supplied below.

Reflecting the breadth of his knowledge and fields of interest, his works cover a wide range of subjects, dealing with research in the fields of stratigraphy, mineralogy, petrography, petrology and geochemistry. The application of mathematical methods, pioneered by Prof. Vistelius, has gained recognition worldwide. As applied in practice, these works represent a building block to more effective methods of search for minerals.

From his earliest post-graduate studies, Vistelius carved out a career which defined a whole new branch of science—mathematical geology.

The ideas of this newly created field of science were first vigorously supported by Academician Vernadsky and then by Academician Kolmogorov. The high value and prospects of Prof. Vistelius's ideas were emphasized in a review of his works, published by Nature, the international science journal, in 1947. Nevertheless, the ideological regime that reigned in the USSR forced mathematical geology to follow a most difficult path. At that time the Ideological Department of the Central Committee of the Communist Party of the USSR was concerned with purging various branches of science in any way connected with cybernetics, genetics and other newly developed fields which they proclaimed as contradicting Marxist-Leninist ideas. It is sufficient to remember the ill-starred session of the Academy of Agriculture of the USSR in 1948, with Academician Lysenko in the chair, whose actions contributed to the tragic death of Academician Vavilov, a botanist and geneticist of international fame.

For minds narrowed by ideology, mathematical geology was nothing but another suspicious field close to cybernetics. Prof. Vistelius and his group could not avoid this political minefield. Scientific life in the country was totally governed by communist administrators who, on the one hand, did not understand the ideas of Vistelius and sought to deny him the opportunity to work, and on the other hand wished to please higher party authorities. Prof. Vistelius with his unusual mathematical ideas appeared an ideal target. But the ideological attacks on him, fortunately, were not strong enough, and he was defending himself fiercely. This is why the ideological persecution did not bring tragic results. Nevertheless, the damage to his scientific career was considerable. He had to leave the All-Union Oil Geology

Research Institute (VNIGRI, Leningrad) where he had been developing the concept of phase differentiation of Paleozoic sedimentary carbonate rocks based on the theory of random functions (nevertheless, brilliantly defended by him in the same year, 1948, as his dissertation for the degree of Doctor of Science).

It is noteworthy that the academic summary "Introduction into the theory of random stationary processes" (the basis for studying phase differentiation of sedimentary carbonate rock), well-known today to mathematicians and specialists in applied science, was first presented only in 1952 by mathematician A. M. Yaglom. This shows that geological phenomena can become a principal material for creation and development of formal mathematical schemes also, as was repeatedly stated by Vistelius. At that period he closely collaborated with the distinguished mathematician, Academician A. N. Kolmogorov, and worked with him on a very important problem of sedimentology relating to the formation of sedimentary strata. As a result, Kolmogorov wrote a paper "Solution of one problem of the theory of probability, related to the problem of mechanism of bed formation" published in "Doklady AN SSSR" (Kolmogorov 1949). The methods of solving this problem were further discussed by M. F. Dacey in his paper "Models of bed formation" (Dacey 1979). There are other examples of such development of formal mathematical structures, for instance, mathematical investigations developing the formalisms of finite Markov chains and processes along with their geological applications, by mathematicians B. P. Harlamov and A. V. Faas in close collaboration with Vistelius.

In 1952 Prof. A. B. Vistelius was invited to join the Laboratory of Airborne Methods of the Academy of Sciences of the USSR (AS USSR). There, with the support of N. G. Kell, the director of the laboratory and a Corresponding Member of the Academy, he organized a group to carry out investigations not just in the field of airborne methods, but mainly in the field of mathematical geology. At this time (before 1960) his group researched several approaches to the problem of comparison of geological sections and reconstruction of the processes of bed formation using the theory of random processes. A. B. Vistelius was actively involved in development of methods of statistical evaluation and examination of hypotheses able to provide the necessary validity for comparison of a model with geological observations.

Despite the obvious importance of the results of Vistelius' work, and the support given by Academicians Kolmogorov, Korzhinsky, Belyankin, Linnik and later Artsimovich, the academic Department for Geology and Geography was too closely connected with the Ideological Department of the Central Committee of the Communist Party and impeded the development of mathematical geology whenever possible. In response, in 1961 the mathematical academicians transferred the group headed by Prof. Vistelius to the Leningrad Branch of the Steklov Institute of Mathematics (LOMI) of the USSR Academy of Sciences. The branch was headed by Prof. Petroshen, a well-known mathematician who specialized in seismic fields, and who encouraged the work of Vistelius' group. There it was set up formally as the Laboratory of Mathematical Geology. It is noteworthy that such a decision was an indication of the fact that the structure of the Academy of Sciences was like "a state within a state". Sometimes it was able to take actions which ran counter to the wishes of the Central Committee of the Communist Party.

The Academy of Sciences was precisely the right environment for initiating thorough field investigation, allowing disinterested scientific research, to develop the fundamental principles of mathematical geology. A. B. Vistelius, with broad experience in different fields of geology, developed ideas for the introduction of mathematics into geology systematically and with clarity of purpose.

By the end of the 1970s he demonstrated the advantages of using the methods of mathematical geology that he had developed to a range of questions in mineralogy, petrography, lithology, petrology and more general problems of regional geology in the fields of paleogeography, lithostratigraphy, and geochemistry. The results of his studies showed that mathematical methods were not to be confined to summarisation of geological information, or to identification of geological events and phenomena on the basis of numerical calculations, but could provide a means of expressing geological concepts in mathematical language. The line of inquiry that was defended by A. B. Vistelius and determined by that time as "mathematical geology" leads geology to a higher level, demanding more concrete and accurate notions about objects or processes under consideration than is possible without the application of mathematics.

His group's scientific work in LOMI, an outstanding internationally recognised mathematical research centre, however, entailed some specific problems. The mere principles of solving tasks of mathematical geology did not raise any objection in the institute, but the choice of propositions for each geological mathematical model remained hard to understand for mathematicians, including the hierarchy of the institute. The institute's administration consisted of theoretical mathematicians who needed only a sheet of paper and a pen for their work. It was hard to persuade them that geology needs field work and an experimental basis to obtain the data necessary to construct and verify models.

This is why Prof. Vistelius had to look for another more suitable host organisation for the Laboratory of Mathematical Geology. This difficulty, as well as the importance of mathematical geology, were met with understanding by A. P. Aleksandrov, the President of USSR Academy of Sciences, in 1986, and in the following year he moved the Laboratory of Mathematical Geology from the Department of Mathematics to the Department of Geology, Geochemistry, Geophysics and Mining of the Academy by attaching it to the Institute of Precambrian Geology and Geochronology (IGGD, AS USSR).

Then, however, it became immediately apparent that a traditional geologist and a mathematical geologist spoke different languages and the majority of geologists did not understand the mathematical approach to modelling geological phenomena despite the fact that mathematical geology had existed for more than forty years.

It seemed that transformation of the Laboratory of Mathematical Geology into an institute was overdue. The necessity of such a decision was repeatedly stressed by a number of senior scientists such as Academicians Sokolov and Laverov (who was an acting Vice-President of the Russian Academy of Sciences). But this idea was achieved only in 1991 when the Russian Academy of Natural Sciences (RANS) was founded. Prof. Andrey Vistelius was named an Honorary Member of this Academy at the first elections and charged with organization of an Institute of Mathematical Geology.

Vistelius' Laboratory of Mathematical Geology together with the Laboratory of Petrophysics and Mathematical Geology of the Earth's Crust Institute of St. Petersburg State University, constituted the basis of the institute. However, RANS is not a government institution and it had no support from the federal budget. For this reason RANS could not supply the Institute of Mathematical Geology with appropriate financing. The Ministry of Science and Technology of the Russian Federation agreed to subsidize the institute after difficult negotiations. The institute, for its part, took on large obligations in solving some practical geological problems by means of mathematical geology.

Dech and Glebovitsky (2000) give a detailed account of the many fields in which the work of Vistelius advanced geological knowledge through his deep understanding of underlying geological processes and innovative application of mathematical methods.

To understand fully Vistelius' immense contribution to the geosciences, it is necessary first to identify the different and complementary approaches to the subject. The two principal approaches can be summarised thus:

(1) development of genetic geological models and quantitative hypothesis testing of them: this is very close to standard scientific method, but because of the complexity of the subject, may not always be practicable
(2) the use of data to develop a numerical model which will often (indeed, usually) have no genetic significance: this is the statistical or data processing approach, where the emphasis is on finding patterns or structure in the data rather than understanding the underlying geological processes

Andrey Borisovich Vistelius, with a firm grounding in scientific method, was a strong advocate for genetic models and hypothesis testing. Not only was this theoretically more fulfilling, but also it did not generally require the massive computer power that was not available to him in the Soviet Union.

Vistelius' beliefs as expressed in 1968, were confirmed recently in a brief historical review (Dech and Henley 2003, p 368) of his 'scientific heritage', where it was noted that he

> *... supposed, and for good reason, that if a science does not use mathematical modelling in constructing its conclusions, "then it can be considered as belonging to the pre-Newtonian period, in other words such a science lags behind the present-day level of research by approximately 300 years" (Vistelius 1991). He understands that the new scientific paradigm of conceptual modelling of geological processes and objects will not be adopted by conservative geologists, the majority of whom continue to use old methods. And he writes that such a situation must be essentially changed, as to enter the twenty-first century with such a considerable time-delay is simply dangerous, not least for economic development.*

## 39.3 The International Association for Mathematical Geology

Vistelius' participation in the IGC in Prague in 1968 was fortuitous from several standpoints. Prior to the Congress, Reyment had been the first Visiting Research Scientist at the Kansas Geological Survey (1966–67) where the idea of an International Association for Mathematical Geology (IAMG) was conceived. The first hint of mathematical geology as a subject in its own right had actually come to Reyment's attention in the late 1940s from some of Vistelius' work. Reyment then visited Vistelius in Leningrad in the early 1960s while in the USSR as a research associate at Moscow University on exchange from the University of Stockholm. From his contact with Vistelius and his experience in Kansas, Reyment had the idea of sending a questionnaire to possible interested participants in such an organisation; he received an overwhelming positive response, and an especially enthusiastic one from Vistelius. Later, at an ISI (International Statistical Institute) meeting in Australia, Reyment conferred with a group of international scientists, including Chester Bliss, founder of the journal *Geometrics*, and the IAMG concept was nurtured (Reyment pers. comm., 1993). On April 9th, 1968, Reyment asked for approval of a proposed set of statutes in a letter "To all Committee members": "(1) *I am in agreement with the draft statutes of Professor Whitten, amended by Prof. Vistelius and Dr. Marsal and including suggestions from Dr. Agterberg, Mr. Schlegel, and Professor van Leckwijk, …*". The founding IAMG committee adopted these statutes, and the IAMG then applied for affiliation with the International Union of Geological Sciences (IUGS) and the International Statistical Institute (ISI). The proposal for affiliation with the IUGS was supported by S. Van der Heide, Secretary General of IUGS, and accepted at the Prague meeting as a result of prodding and cajoling by Reyment, and thus the IAMG was officially born.

Vistelius had served on an ad hoc exploratory committee and then was member of the Organizing Committee and attended, along with 19 other members, the first meeting of the committee in Prague. Eight of the attendees were from the Eastern Bloc; their attendance in Prague was allowed as being relatively 'safe.' It was the understanding of the other attendees that the 'Warsaw Pact' attendees were there on military visas (for reasons which were obvious later). The events during the Congress substantiated that understanding. Vistelius' participation in the IGC gave him visibility to Western scientists and those contacts (with Frits Agterberg, John Harbaugh, Tim Whitten, and Dan Merriam) were invaluable to him later.

Reyment had prepared a slate of officers to be ratified by the representatives, and it was no surprise he nominated Vistelius for president. Reyment was aware of and impressed by Vistelius' work (through his Russian publications and personal contact). He was an obvious choice for the position with Reyment's backing, and because Bill Krumbein, another possible choice for the office, was not interested, Vistelius was in but, Krumbein was elected the first past president! Reyment was elected Secretary General.

There was considerable discussion about the designation and focus for the new organisation. Proposed for the name of the Association's newly created journal were such adjectives as geometrics, geomathematics, mathematical geology, numerical, quantitative, etc. Vistelius championed 'mathematical geology' and, for a variety of reasons, that name was agreed on. The new Journal of Mathematical Geology was contracted to be published by Plenum Press. In 1969 in the first issue of the fledgling journal, Vistelius, as President of IAMG, wrote a Preface on the 'mathematization of geology' and contributed a short note.

At the inaugural meeting of IAMG, Andrey Vistelius championed the concept that Mathematical Geology is a separate branch of science (like Mathematical Physics) based on testing geological hypotheses mathematically, and that this science should be accepted as the primary focus of IAMG. He suggested it is not particularly important or interesting merely to manipulate geological data statistically. These had been his contentions for many years, though few of those present in 1968 appreciated the fact—and their primary objective was solely to initiate IAMG. It was not until several years later that their full significance and the historical importance of his earlier publications became clear to those outside the Soviet Union. Although it can be argued that Vistelius was largely correct, process modelling combined with objective hypothesis testing has received little attention among IAMG members over the ensuing years (Whitten 2003).

Because of the restrictions on travel and communication placed on Vistelius, most of the IAMG work load fell on Reyment as Secretary General and Merriam as editor of the new journal. Vistelius' direct contribution to the IAMG was minimal through no fault of his own, and later he served a 4-year stint on the Council helping prepare the IAMG sessions at the IGC in Moscow. Reyment succeeded Vistelius as president and by that time in 1972 the organisation was firmly established.

Vistelius attended few 'official' IAMG meetings. Because of his circumstances, it was difficult for him to make much direct contribution, except in name, to the activities of IAMG. Vistelius' unique and important scientific contributions, however, were recognized by the IAMG by awarding him the Krumbein Medal (the IAMG's highest honour) in 1980 (unfortunately he was unable to attend the IGC in Paris and collect his medal personally) and naming one of their awards in his honour. After IAMG created the Krumbein Medal in 1976, Merriam proposed another annual award for an outstanding young scientist, to be named in honour of Vistelius. The proposal was rejected by the Russian authorities on the grounds that such an honour could not be conferred upon a living person. Thus, the award was designated the President's Award in 1980 and subsequently changed to the Vistelius Award, as originally intended, after his death in 1995.

## 39.4 The “Father of Mathematical Geology”?

Andrey Vistelius has often been referred to as the “father of mathematical geology”. He was indeed the first president of IAMG, but there are many other pioneers in the field who could also be acknowledged by the title of “father” (including among others Krumbein, Griffiths, Matheron, Chayes, Krige, and Schwarzacher). Merriam (2001) names W. C. Krumbein as the “father of *computer* geology”, but of course this is not quite the same thing. Vistelius, himself, as noted above, was ambivalent towards the use of computers.

The history of development of mathematical geology [in the broad sense] is essentially two stories (East and West) with little connection or interaction until near the end of the 20th Century. The two schools developed independently and partly in parallel in response to changes in the science. The quantification of geology began in earnest from modest beginnings of a few quantitatively oriented researchers, such as Vistelius, Krumbein, and Griffiths among others.

Vistelius’ death in 1995 (Krumbein had died in 1979 and Griffiths in 1992), ended an extraordinary era in the growth of quantitative (mathematical) geology. Along with the rapid development of quantitative techniques and their adaptation to computers, these advances spread throughout the science and allowed rapid strides and changes to be made in the earth sciences.

Never before in the past, and probably never again in the future, will such rapid progress be made in such a short time, fostered by such a small group of dedicated, forwarding-thinking geo-giants.

## 39.5 Legacy

It is traditional to discuss the legacy of outgoing political leaders, to assess their place in history and to estimate the quality and quantity of their achievements in the light of effects on subsequent developments. Similar discussions take place over the legacy of our foremost scientists, among whose number Andrey Vistelius must surely be counted.

His rigorous scientific training led him to develop his ideas of applying mathematical methods in modelling geological processes, to allow statistical testing of hypotheses against real data. This contrasted starkly with the approach of many western geoscientists, of using data processing capabilities of computers to fit the data using standardised methods. The latter approach allowed the identification of patterns in data, but rarely provided scientific insight into the underlying geological processes. In the English-language literature, perhaps the outstanding example of Vistelius’ approach is the book *Computer Simulation in Geology* by Harbaugh and Bonham-Carter (1970) which identifies a wide range of geological process models which can be defined mathematically and implemented in computer code.

The process modelling approach pioneered by Vistelius is now making serious contributions to the geosciences. For example, in the work of Alison Ord, Bruce Hobbs, and colleagues in Australia and elsewhere, mathematical models from a number of hitherto separate fields have been combined into complex models with their recognition that the interactions of rock deformation, fluid flow, thermal transport, and chemical reaction are integral to geology. Prediction requires quantification of the processes and their interactions. What is observed is demonstrably multifractal so that we must explore and apply all that nonlinear dynamics has to offer (Ord and Henley 1997; Ord et al. 2002, 2007, 2012, 2016; Hobbs et al. 2010; Hobbs and Ord 2015, 2016).

The other approach is best typified by the field that is generally known as "geostatistics". Originating in the work of Matheron and many others, this uses purely mathematical concepts to fit models to the data. These models bear little or no relation to underlying geological processes, and the results are purely descriptive. In attempts to improve the quality of fit to the observed data sets, over the past 40 years progressively more complex mathematics has been developed, using assumptions about the statistical properties of data sets which have steadily less justification in the underlying geological processes. The history of development of geostatistics is reminiscent of the iterative refinement of the Ptolemaic astronomical model when circular planetary orbits were found to be incompatible with observations, and epicycles were added in an attempt to improve the fit. The problem, of course, was that the model was itself a mathematical fiction bearing no relation to the laws underlying planetary motions. Similarly, geostatistics is purely descriptive and bears no relationship to actual geological processes.

While geostatistics itself continues to be widely used, the more scientific approach espoused by Vistelius remains very much alive. Even though many of its practitioners are unaware of the debt of gratitude they owe to this pioneer, their work nonetheless is tribute enough.

A special issue of the Journal of Mathematical Geology (volume 35, number 4) dedicated to the memory of Vistelius was published in 2003 and contains papers by many of his former colleagues, as well as one previously unpublished paper by Vistelius himself (Dech et al. 2003; Vistelius and Pavlov 2003; Azimov and Shtukenberg 2003; Harlamov 2003; Voytekhovsky and Fishman 2003; Podkovyrov et al. 2003; Kotov 2003). The breadth of geoscientific subject matter and mathematical approaches shown by this collection of papers is ample illustration of the scientific legacy of Andrey Borisovich Vistelius.

## References

Azimov P, Shtukenberg A (2003) Numerical modeling of growth zoning at nonstationary crustallization of solid solutions: metamorphic garnets. Math Geol 35:405–430

Dacey MF (1979) Models of bed formation. Math Geol 11(6):655–668

Dech VN, Glebovitsky VA (2000) Establishment of mathematical geology in Russia: some brush strokes in the portrait of the founder of mathematical geology, Prof. A. B. Vistelius. Math Geol 32(8):897–918

Dech VN, Henley S (2003) On the scientific heritage of Prof. A. B. Vistelius. Math Geol 35 (4):363–379

Dech VN, Kiselev BV, Pisakin BN, Roshchinenko OM (2003) Math Geol 35:381–398

Dvali MF, Korzhinskii DS, Linnik YV, Romanova MA, Sarmanov OV (1970) The Life and work of Andrey Borisovich. Math geol 1–5. Chapter 1, Springer, New York

Harbaugh JW, Bonham-Carter G (1970) Computer simulation in geology. Wiley, p 575

Harlamov BP (2003) Formation of concentric rings around sources. Math Geol 35:431–450

Henley S (2003) Special issue: in honour of professor A. B. Vistelius: introduction. Math Geol 35 (4):361–362

Hobbs BE, Ord A, Spalla I, Gosso G, Zucali M (2010) The interaction of deformation and metamorphic reactions. In: Spalla MI, Marotta AM, Gosso G (eds) Advances in Interpretation of geological processes. Geological Society, London, Special Publications, vol 332, pp 189–222

Hobbs BE, Ord A (2015) Structural geology: the mechanics of deforming metamorphic rocks. Elsevier Inc., Netherlands, pp 665

Hobbs BE, Ord A (2016) Does non-hydrostatic stress influence the equilibrium of metamorphic reactions? Earth Sci Rev 103:190–233

Kolmogorov AN (1949) The decision of one task from the theory of probabilities connected to a question on the mechanism a layer of formation. Doklady USSR (Reports of Academy of Sciences of the USSR) 65(6):793–796

Kotov SR (2003) Dynamics of global climatic changes and possibility of their prediction using chemical data from Greenland ice. Math Geol 35:477–492

Merriam DF (2001) Andrey Borisovich Vistelius: a dominant figure in 20th Century mathematical geology. Nat Resour Res 10(4):297–304

Ord A, Henley S (1997) Fluid pumping: some exploratory numerical models. Phys Chem Earth 22:49–56

Ord A, Hobbs BE, Zhang Y, Broadbent GC, Brown M, Willetts G, Sorjonen-Ward P, Walshe JL, Zhao C (2002) Geodynamic modelling of the Century deposit, Mt Isa Province, Queensland. Aust J Earth Sci 49:1011–1039

Ord A, Hobbs BE, Regenauer-Lieb K (2007) Shear band emergence in granular materials—a numerical study. Int J Numer Anal Meth Geomech 31:373–393. https://doi.org/10.1002/nag.590

Ord A, Hobbs BE, Lester DR (2012) The mechanics of hydrothermal systems: I. Ore systems as chemical reactors. Ore Geol Rev 49:1–44

Ord A, Munro M, Hobbs B (2016) Hydrothermal mineralising systems as chemical reactors: wavelet analysis, multifractals and correlations. Ore Geol Rev 79. http://dx.doi.org/10.1016/j.oregeorev.2016.03.026

Podkovyrov VN, Graunov OV, Cullers RL (2003) A linear programming approach to determine the normative composition of sedimentary rocks. Math Geol 35:459–476

Romanova MA, Sarmanov OV (1970) The published works of A. B. Vistelius. Math Geol 6–12, chapter 2. Springer, New York

Vistelius AB (1980) Principles of mathematical geology. Nauka, Leningrad, p 389

Vistelius AB (1991) Whether is the geological science of modern?. Reports of Academy of Sciences of the USSR, no 12, pp 82–90

Vistelius AB (1992) Principles of mathematical geology. Kluwer Academic Publishers, Dordrecht–Boston–London, p 477

Vistelius AB, Pavlov VM (2003a) On rhythmical layering of rocks formed from basaltic magma. Math Geol 35:399–404

Voytekhovsky YL, Fishman MA (2003) Rock kriging with the microscope. Math Geol 35:451–458

Whitten EHT (2003) Mathematical geology in perspective: has objective hypothesis testing become overlooked? Math Geol 35:1–8

Whitten EHT (2004) A. B. Vistelius—first IAMG president and visionary mathematical geologist. Earth Sci History 23(2):384–389

# Publications of A. B. Vistelius

This note contains details of many of his published works (where he is sole or first named author), including many in the well known journal Doklady Akademii Nauk USSR (Papers of the Academy of Sciences of the USSR). After break-up of the USSR the journal is called Doklady Rossiyskoy Akademii Nauk (Papers of the Russian Academy of Sciences). These are supplemented by papers in many other journals, and monographs by A. B. Vistelius, some published in Russian, others in English.

Vistelius AB (1939) The Antarctic Continent. Vestn Znaniya 9:34–37

Vistelius AB (1939) Tourmaline in carbonate veins in the vicinity of Chupa Bay (Northern Karelia). Uch Zap Leningr Gos Univ (34):60–70

Vistelius AB (1939) Wulfenite in the Kyzyl-Kan mine (Northern Tadzhikistan. Uch Zap Leningr Gos Univ (49):56–62

Vistelius AB (1940) Geological history of Antarctica. Vestn Znaniya 2:36–38

Vistelius AB (1940) Mineralogy of the Andreevskii Mine (Khakassia), Uch Zap Leningr Gos Univ (44):148–157

Vistelius AB (1941) A composite dike from the area of the Gulshad lead-zinc deposit (Balkhash region). Doklady Akad Nauk USSR 31(6):559–601

Vistelius AB (1942) Mineral assemblages from the region of the Gulshad lead-zinc deposit (Balkhash region). Summary of dissertation for a scientific degree of Candidates of the geological and mineralogical sciences, Leningrad State University

Vistelius AB (1943) Notes on garnets from the vicinity of Lake Balkhash. Zap Vseross Mineral Obshch 72(3–4):167–173

Vistelius AB (1944) Notes on analytical geology. Doklady Akad Nauk USSR 44(1):27–31

Vistelius AB (1944) Physical properties of oil-bearing strata in the Permian system, author's abstract in scientific papers of petroleum specialists, Geol (1):81–82

Vistelius AB (1945a) Frequency distribution of the porosity coefficients and energy processes in *Spirifer* beds of the Buguruslan oil-bearing region. Doklady Akad Nauk USSR 49(1):44–47

Vistelius AB (1945b) On expression of results of fossilization of oscillation of the Earth's crust by means of the series $\sum_{i=0}^{k} e^{a_i x + b_i} \cos(\omega_i x + y_i)$. Doklady Akad Nauk USSR 49(7):531–535

Vistelius AB (1946) Porosity cycles and the phenomenon of phase differentiation of sedimentary strata. Doklady Akad Nauk USSR 54(6):519–521

Vistelius AB (1947a) On correlation of mesorhythms in the early Permian deposits of trans-Kama Tatariya and their stratigraphic significance. Doklady Akad Nauk USSR 55(3):241–244

Vistelius AB (1947b) Application of the correlation coefficient in the investigation of mineral paragenesis in clastic deposits. Doklady Akad Nauk USSR 55(4):343–345

Vistelius AB (1947c) Correlation of apatite and nepheline in the Kukisvumchorr-Jukspor sphene deposit (Khibiny tundra). Doklady Akad Nauk USSR 56(2):185–188

Vistelius AB (1947d) New confirmation of Goldschmidt's observations on the place of germanium in hard coals. Doklady Akad Nauk USSR 58(7):1455–1457

Vistelius AB, Sarmanov OV (1947) The stochastic basis of a geologically important probability distribution. Doklady Akad Nauk USSR 58(4):631–634

Vistelius AB (1948a) Geology of the Lower Kazanian deposits of the Buguruslan region. Sov Geol Sb (28):48–63

Vistelius AB (1948b) Some analytical methods of investigating rhythmicity. Sov Geol Sb (28):174–182

Vistelius AB (1948c) The measure of correlation between paragenetic members and methods for studying it. Zap Vses Mineralog Obschestva 77(2):147–158

Vistelius AB (1948d) The simplest types of problems in mathematical treatment of lithological observations. Litolog Sb Vses Nauchno-Issled Geologorazved Inst 1:125–131

Vistelius AB (1948e) The distribution of magnesite in Palaeozoic rocks on the eastern part of the Russian platform. Litolog Sb Vses Nauchno-Issled Geologorazved Inst 2:42–49

Vistelius AB (1948f) On the roundness of quartz sand grains on the Belinsky bank (Volga delta). Doklady Akad Nauk USSR 63(1):69–72

Vistelius AB (1949a) The mechanism of formation of sedimentary beds. Doklady Akad Nauk USSR 65(2):191–194

Vistelius AB (1949b) The mechanism of correlation of strata. Doklady Akad Nauk USSR 65 (4):535–538

Vistelius AB (1949c) Calcium sulphates in Palaeozoic rocks of the eastern part of the Russian platform. Geokhim Sb Vses Nauchno-Issled Geologorazved Inst 1:142–158

Vistelius AB (1950a) Palaeogeographic significance of correlation between bed thicknesses (according to data on the productive sequence of the Apsheron peninsula). Sb Vses Nauchno-Issled Geologorazved Inst 3:61–73

Vistelius AB (1950b) Porosity and chemical composition of the carbonate strata of the eastern part of the Russian platform. Trans Lab Gidrogeol Probl im F. P. Savaresnkogo 2:194–202

Vistelius AB (1950c) Distribution of enantiomorphic types of quartz. Zap Vsesoyuzn Mineral Obshch 79(3):191–195

Vistelius AB (1950d) On mineral composition of the heavy part of sands of the lower part of the productive sequence of the Apsheron peninsula, the Chokrak strata of southern Dagestan, and alluvium of the Volga. Doklady Akad Nauk USSR 71(2):367–370

Vistelius AB (1950e) Correlation between copper content in borax waters of Azerbaijan and the degree of mineralisation. Doklady Akad Nauk AzerbSSR 6(1):34–36

Vistelius AB, Miklukho-Maklay AD (1950) Lower Permian pebbles and cobbles from the productive sequence of the Apsheron peninsula. Doklady Akad Nauk USSR 72(2):369–372

Vistelius AB (1951a) Porosity rhythms in Kazanian deposits of southern Tataria. Trans Leningr Obshch Estestvoispyt 68(2):150–167

Vistelius AB (1951b) Correlation once again (a reply to S. V. Konstantov). Zap Vsesouyzn Mineral Obshch 80(1):79–80

Vistelius AB (1951c) The required number of grains for computation in immersions. Zap Vsesoyuzn Mineral Obshch 80(3):188–190

Vistelius AB (1951d) Probability of the effect of an aeolian field in the diagram of L. V. Rukhin on the field of residual types of sand. Izv Akad Nauk USSR Geol (1):155–156

Vistelius AB (1951e) The status of interpreting lithological observations, and methods of improvement. Izv Aakad Nauk USSR Ser Geol (3):90–104

Vistelius AB (1951f) On the origin of disthene in the middle Miocene of Dagestan. Doklady Akad Nauk USSR 79(1):133–136

Vistelius AB, Miklukho-Maklay AD (1951) Palaeozoic pebbles from the productive sequence of the Apsheron peninsula. Doklady Akad Nauk USSR 79(3):499–502

Vistelius AB (1952a) The Kirmankinskaya series of eastern Azerbaijan, 1. Principal lithologic features. Doklady Akad Nauk AzerbSSR 8(1):17–23

Vistelius AB (1952b) Probability distributions (in reply to V. S. Dmitrievskii). Izv Akad Nauk USSR Ser Geol (1):155–156

Vistelius AB (1952c) The mineralogy of the Miocene sandy-argillaceous sediments from southern Azerbaijan. Doklady Akad Nauk USSR 85(5):1155–1158

Vistelius AB, Zulfugarly DD (1952) Natural paragenetic associations of some components of oil in Azerbaijan. Izv Akad Nauk AzerbSSR (2):17–31

Vistelius AB, Sarsadskii NN (1952) Nature of changes in mineral content of concentrates during successive washing of sands. Zap Vsesoyuzn Mineral Obshch 80(2):143–150

Vistelius AB (1953) Rock salt. Bolshaya Sov Entsik, 2nd edn, vol 19, p 490

Vistelius AB (1953) Treatment of microstructural diagrams. Zap Vsesoyuzn Mineral Obshch 82 (4):271–280

Vistelius AB, Miklukho-Maklay AD, Ryabinin VN (1953) Devonian limestones from the red bed sequence of Tuarkyr. Doklady Akad Nauk USSR 90(2):231–234

Vistelius AB, Korobkov IA (1953) A new discovery of the Konka horizon on the Krasnovodsk plateau. Doklady Akad Nauk USSR 90(3):445–448

Vistelius AB (1954a) Sands of the middle and lower Volga, in a collection of papers of the Laboratory of Aerial Methods (Akad Nauk USSR) in memory of N. G. Kellyu

Vistelius AB (1954b) Mineralogical associations and characteristic paragenetic relations of the Aptian-Cenomanian clastic strata of the trans-Caspian region. Doklady Akad Nauk USSR 97 (3):503–506

Vistelius AB, Yaroslavskaya NN (1954) General features of the colour characteristics of Cretaceous clastic sand-silt strata of the trans-Caspian region. Doklady Akad Nauk USSR 95 (2):367–370

Vistelius AB, Sarmanov OV (1954) Remarks on a paper by P. A. Ryzhov: determining the accuracy of calculating reserves of mineral deposits, in investigations of problems of mine surveying, Sb no 29, pp 200–201

Vistelius AB (1955a) The Kirmakinskaya series of the East of Azerbaijan. II. Mineral associations. Doklady Akad Nauk USSR 103(1):117–120

Vistelius AB (1955b) Age of the lower part of the red bed sequence of the Cheleken Peninsula. Doklady Akad Nauk USSR 105(4):786–789

Vistelius AB (1956a) Subdividing the recent deposits of the eastern Caucasus and the northern Caspian region into districts according to mineral content. Doklady Akad Nauk USSR 111 (5):1067–1071

Vistelius AB (1956b) Problems of studying correlation in mineralogy and petrography. Zap Vsesoyuzn Mineral Obshch 85(1):58–73

Vistelius AB, Miklucho-Maklay AD (1956) The middle division of the productive sequence of the Apsheron Peninsula and the problem of its origin. Izv Akad Nauk USSR Ser Geol (4):77–94

Vistelius AB (1957a) Subdivision of unfossiliferous strata by quantitative-mineralogical, petrographic, and chemical features. Zap Vsesoyuzn Mineral Obshch 86(1):99–115

Vistelius AB (1957b) The statistics of microstructural diagrams. Zap Vsesoyuzn Mineral Obshch 86(6):691–703

Vistelius AB (1957c) Regional lithostratigraphy and conditions under which the productive sequence of the south-eastern Caucasus formed. Trans Leningr Obshch Estestvoisp 69(2):126–150

Vistelius AB (1957d) The nature of pre-Permian volcanism in Western Turkmenia. Doklady Akad Nauk USSR 117(5):867–869

Vistelius AB (1958a) A scheme of alluvial deposit zonation of the Pamirs according to their mineralogical associations. 118(6):1158–1161

Vistelius AB (1958b) Structural diagrams. Izd AN USSR 157, Moscow, Leningrad

Vistelius AB (1958c) Spectral brightness of sand-silt rocks of Aptian, Albiam, and Cenomanian ages in the trans-Caspian region, in geology of the Trans-Caspian (1):31–67

Vistelius AB (1958d) Dictionary of petroleum geology. The terms: autocorrelation, probability, probable error, dispersion, disperson analysis, phase differentiation, mathematical expectation, Gostoptekhizdat, Moscow

Vistelius AB (1958e) Volume-frequency analysis of sediments from thin-section data: a discussion. J Geol 66(2):224–226

Vistelius AB (1958f) Paragenesis of sodium, potassium, and uranium in volcanic rocks of Lassen Volcanic National Park, California. Geochim Cosmochim Acta 14(1-2):29–34
Vistelius AB, Korobkov IA (1958) Some questions on the geology of western Turkmenia. Izv Akad Nauk TurkmSSR (6):115–119
Vistelius AB (1959a) Geology at the University of Chicago (translation of a report by E. K. Olsen), Vestn Lenigr Gos Univ Ser geol i Geogr 3(18):137–138
Vistelius AB (1959b) Address at session of technical council of the ministry of the petroleum industry and the academy of sciences of turkmenia at Ashkhabad, 1956. Objectives and potentials of prospecting and exploratory work on oil and gas in the western parts of Central Asia. Izd Akad Nauk 352–355, TurkmSSR, Askhabad
Vistelius AB (1959c) Origin of the red bed sequence of the Cheleken peninsula. Experience of using absolute age of clastic mineral for solving problems of lithology and paleogeography. Doklady Akad Nauk USSR 125(6):1307–1310
Vistelius AB, Sarmanov OV (1959) Correlation between percentage values. Doklady Akad Nauk USSR 126(1):22–25
Vist elius AB (1960a) The morphometry of clastic particles. Trans Lab Aerometodov Akad Nauk USSR 9:135–202
Vistelius AB (1960b) Peculiarities of germanium concentration in coal (in reference to the review of V. M. Ershov). Izv Akad Nauk USSR Ser Geol (8):100
Vistelius AB (1960c) The skew frequency distributions and the fundamental law of the geochemical processes. J Geol 68(1):1–22
Vistelius AB, Korobkov LA, Romanova MA (1960d) Age of the red-beds sequence on the north-western Krasnovodsk plateau. Izv Akad Nauk TurkmenSSR Ser Fiz-Tekh, Khimik i Geol Nauk 3:108–111
Vistelius AB (1961a) Data on the lithostratigraphy of the productive sequence of Azerbaijan. Izd AN USSR 157. Leningrad
Vistelius AB (1961b) The middle Caspian land mass (reference to papers by Tamrazyan and Rikhter). Bull Mosk Obshchestva Ispitatelyei Prirody. Ser Geol 36(1):148–151
Vistelius AB (1961c) Sedimentation time trend functions and their application for correlation of sedimentary deposits. J Geol 69(6):703–728
Vistelius AB (1961d) Discussion of paper by F. Chayes On correlation between variables of constant sum. J Geophys Res 66(5):1601
Vistelius AB, Krylov AJ (1961) The absolute age of the clastic part of the sand-silt sediments of south-western Central Asia. Doklady Akad Nauk USSR 138(2):422–425
Vistelius AB, Sarmanov OV (1961) On the correlation between percentage values: major component correlation in ferro-magnesium micas. J Geol 69(2):145–153
Vistelius AB (1962a) Phosphorus in the granitoidal rocks of Tien Shan, Geokhimiya (2):116–135
Vistelius AB (1962b) Problems of mathematical geology. I On the history of the question. Geol i Geofiz Sibirsk Otd Akad Nauk USSR (12):3–9
Vistelius AB, Romanova MA (1962) Red beds deposits of the Cheleken peninsula (Lithostratigraphy and geological structure). Izd AN USSR, Leningrad, p 227p
Vistelius AB, Demina ME (1963a) Dispersion of clastic material in the Aptian-Cenomanian basin of south-eastern USSR. Doklady Akad Nauk USSR 150(6):1319–1322
Vistelius, AB, Demina ME (1963b) Rounding of quartz grains, in Geochemistry, Petrography, and Mineralogy of Sedimentary Rocks. In honour of L. V. Pustuvalov, pp 233–253
Vistelius AB, Yanovskaya TB (1963) Programming problems of geology and geochemistry for use with all-purpose electronic computers. Geol Rudn Mestorozhd (3):34–48
Vistelius AB (1963a) Phase differentiation of Palaeozoic sediments of the middle Volga and trans-Volga regions. Izd Akad Nauk USSR 203, Moscow/Leningrad
Vistelius AB (1963b) Functions of probability distributions of accessory element concentrations in rocks and minerals. Teor Veroyatnostei i ee Primeneniya 8(2):232–233

Vistelius AB (1963c) Problems of mathematical geology. II. Models of processes and paragenetic analysis. Geol i Geofiz Sibirsk Otd Akad Nauk USSR (7):3–17

Vistelius AB (1963d) Problems of mathematical geology. III. The random process. Geol i Geofiz Sibirsk Otd Akad Nauk USSR (12):3–10

Vistelius AB (1963e) Foreword to the book of F. Chayes: quantitative mineral analysis of rocks in thin Section under the microscope (Russian translation), IL, Moscow (in English: Petrographic modal analysis, Wiley, New York, 1956)

Vistelius AB (1963f) Functions of probability distributions of phosphorus concentrations in granitoids of Switzerland, the Guianas, and Equatorial Africa. Doklady Akad Nauk USSR 152 (6):1449–1452

Vistelius AB (1964a) Principal types of mathematical solutions of problems in modern geology. Razvedka i Okhrana Nedr (6):18–25

Vistelius AB (1964b) Problems of geochemistry and information measures. Sov Geol (12):5–26

Vistelius AB (1964c) The problem of formation of sedimentary beds (author's abstract of a report), Bull Mosk Obshchestva Ispitatyelei Prirody, Ser Geol, vol 39, no 3, p 148

Vistelius AB (1964d) Probability and statistical problems in geology. Trans. In: 4th mathematical conference, 1861, vol 2, Reports of sections, pp 329–335

Vistelius AB (1964e) Palaeogeographic reconstruction by absolute age determinations of sand particles. J Geol 72(4):483–486

Vistelius AB (1964f) Informational characteristics of frequency distribution in geochemistry. Nature 202(4938):1206

Vistelius AB (1964g) Stochastic model for the generation of the bedding of red-beds from the Cheleken peninsula (Caspian Sea). Bull Geol Soc Am. In: Program of Annual Meeting, p 213

Vistelius AB (1964h) A discussion on the statistical analysis of fabric diagrams. J Geol 4(1):224–228

Vistelius AB (1964i) Problems of geochemistry and information measures. Sov Geol (12):5–26

Vistelius AB, Romanova MA (1964) Distribution of the heavy fraction in sand deposits of Central Karakums. Doklady Akad Nauk USSR 158(4):860–863

Vistelius AB, Hurst VJ (1964) Phosphorus in granitic rocks of North America. Bull Geol Soc Am 75:1055–1092

Vistelius AB, Feygelson TS (1965) Theory of formation of sedimentary beds. Doklady Akad Nauk USSR 164(1):158–160

Vistelius AB, Faas AV (1965a) The nature of bed alternation in sections of some sedimentary sequences. Doklady Akad Nauk USSR 164(3):629–632

Vistelius AB, Faas AV (1965b) Variation of bed thicknesses in a section of the Palaeozoic flysch of the Southern Urals. Doklady Akad Nauk USSR 164(5):1115–1118

Vistelius AB (1966a) Probleme der mathematischen geologie, I Zur geschichte. Z Angew Geol 11 (5):265–268

Vistelius AB (1966b) Probleme der mathematischen geologie, II, Modelle von vorgängen und die analyse von paragenesen. Z Angew Geol 11(6):306–313

Vistelius AB (1966c) Probleme der mathematischen geologie, III, Der zufallsprozess. Z Angew Geol 11(7):356–359

Vistelius AB (1966d) Principaux types des solutions mathematiques des problemes geologiques actuels, BRGM, Service d'inform. Geol Centre Sci et Techn, Orleans-La Source, no 4701

Vistelius AB (1966e) Problemes de geologie mathematique, BRGM, Service d'inform. Geol Centre Sci et Techn, Orleans-La Source, no 4703

Vistelius AB (1966f) Structural diagrams. Pergamon Press, Oxford, p 178

Vistelius AB (1966g) About formation of the Belaya granodiorites on the Kamchatka Peninsula (an experiment in stochastic modeling). Doklady Akad Nauk USSR 167(5):1115–1118

Vistelius AB (1966h) A stochastic model of alaskite c0rystallization, and corresponding transition probabilities. Doklady Akad Nauk USSR 170(3):653–655

Vistelius AB (1966i) The red beds deposits of the Cheleken peninsula. Lithology. Experiment in the stochastic modeling of processes of formation of sedimentary beds. Izd Nauka (Publ. House Science), 1966, Leningrad, p 304

Vistelius AB (1966j) Review of book by V. F. Morkovkina. Chemical analyses of volcanic rocks and rock-forming minerals, Geokhimiyaa, no 5, pp 617–620

Vistelius AB (1966k) Trend surfaces. J S Afr Inst Min Met. In: Symposium on mathematical statistics and its computer applications in ore valuation, Johannesburg, pp 66–72

Vistelius AB (1967a) Crystallization of alaskites from the Karakuldgur river (Central Tyan Shan). Doklady Akad Nauk USSR 172(1):165–168

Vistelius AB (1967b) On the stochastic matrix of quasi-eutectic granites (granites from the area of Dharvar city in Northern Croatia as the example). Doklady Akad Nauk USSR 175(6):1363–1367

Vistelius AB (1967c) About the modes of crystallization and secondary minerals in some granites of the area near Khankay region (Primorje). Doklady Akad Nauk USSR 177(6):1411–1414

Vistelius AB (1967d) Studies in mathematical geology. Consultants Bureau, New York, p 294p

Vistelius AB (1968) Mathematical geology: a report of progress. Geocom Bull 1(8):229–269

vistelius AB (1968) Stochastic models of processes of sediment formation and their role in sedimentology. Physical and chemical processes and facies. Nauka, Moscow, pp 7–14

Vistelius AB (1969) On the Shap granites (Westmorland, England). Doklady Akad Nauk USSR 187(2):391–394

Vistelius AB, Romanova MA (1969) On the basic factors controlling the composition of present day sands of the Zaunguzsky Karakums. Doklady Akad Nauk USSR 188(1):173–176

Vistelius AB, Ruiz Fuller C (1969) On the origin of variations in the composition of granitic rocks of Chile and Bolivia. Math Geol 1(1):113–114

Vistelius AB (1969) Mathematical geology (the state and perspectives). Mathematical geology, a systematic reference index. Leningrad. lzd Bibl Akad Nauk USSR 41–56

Vistelius AB (1970) Certain factors controlling concentration of mercury in deposits of Severnaya Plavikovaya mountain and Vostochnaya Vershina of Khaydarkan group. Intern Geol Rev (12):691–702

Vistelius AB, Faas AV (1971) About the probability properties of sequences of grains of quartz, potassic feldspar and plagioclase in magmatic granites. Doklady Akad Nauk USSR 198 (4):925–928

Vistelius AB (1971) Some lessons of G-1-W-1 investigations. Math Geol 3(3):323–326

Vistelius AB, Faas AV (1972a) On the statistical identification of ideal granites. Doklady Akad Nauk USSR 203(3):670–673

Vistelius AB, Faas AV (1972b) About transformation of grain successions of quartz, potassic feldspar and plagioclase in ideal granites under the action of slight metasomatism. Doklady Akad Nauk USSR 203(6):1386–1389

Vistelius AB (1972) Ideal granite and its properties. I. Stochastic model. Math Geol 4(2):89–102

Vistelius AB, Lel'chuk GC, Talmud GA, Faas AV (1972) Statistical identification of ideal granites and products of their transformation. Ideal granites, no 2, Leningrad, Nauka, p 3–47

Vistelius AB (1973) Phosphorus in granitic rocks of Colombia. Math Geol 5(2):127–148

Vistelius AB (1974) Proper function and role of the international association for mathematical geology in the revolution in geological sciences, IAMG Newsletter, no 3, p l

Vistelius AB, Romanova MA (1976) On a degenerate case of a model of crystallization of ideal granites. Doklady Akad Nauk USSR 228(1):170–l73

Vistelius AB, Demina ME, Harlamov BP (1976) Random fields in some problems of Paleogeography and geochemical prospecting. In: 25th international geological congress, Sydney, CABS, 414–8/20, p 912

Vistelius AB, Demina ME, Harlamov BP (1976) The fundamental problem of exploration geochemistry and paleogeography of terrigenous components as a problem on the structure of random fields, MGC (Congress). XXV Session Dokl Sov Geol, Moscow, Nedra, pp 37–47

Vistelius AB (1976) Mathematical geology and development of geological sciences. Math Geol 8:3–8

Vistelius AB (1976) Mathematical Geology and the Progress of Geological Sciences. J Geol 84 (6):629–653

Vistelius AB (1977) Mathematical geology: its basic trends and problems. Sov Geol (1):11–34

Vistelius AB, Ivanov DN, Romanova MA, Talmud GA (1978) On the chemical composition of the Cretaceous and Paleogene volcanic rocks of the abyssal structure of the northern part of the Europe-Asia continent frontal zone. Doklady Akad Nauk USSR 242(2):386–389

Vistelius AB (1978) Platinoids and gold in silicate, troilite and metallic phases of chondrites. Zap VMO (3):257–270

Vistelius AB, Harlamov BP (1979) On three-dimensional packing with linear Markov sections. Dokl Akad Nauk USSR 245(2):431–434

Vistelius AB (1980) Principles of mathematical geology. Akad Nauk USSR Izd 389. Nauka

Vistelius AB, Ivanov DN, Romanova MA (1980) The experience of the use of two-dimensional regressions and application of frequency presentations of densities, 1980. Nauka, Leningrad

Vistelius AB, Faas AV (1981) A model of released albite with intergranular silicification of potassic feldspar. Dokl Akad Nauk USSR 260(2):410–414

Vistelius AB (1981) Gravitational stratification. In: Craig RG, Labovitz ML (eds) Future trends in geomathematics. Pion Press, London, pp 141–178

Vistelius AB (1986) About interlayer washout: in The theory of probability and mathematical statistics. Academician RAN Yu. V. Prochorov (ed) Nauka, Moscow, p 534 (The most fundamental works of the Academician A. N. Kolmogorov and subsequent development of their ideas)

Vistelius AB, Ivanov DN, Romanova MA (1982a) About main types of the Mesozoic granitoids of the North-East of Asia. Dokl Akad Nauk USSR 265(5):1213–1216

Vistelius AB, Ivanov DN, Romanova MA (1982b) About trends of potassium content variations in Mesozoic granitoids in the North-East of Asia and adjacent territories. Dokl Akad Nauk USSR 266(6):1444–1448

Vistelius AB, Ivanov DN, Romanova MA (1983a) On the regional oxidation-reduction regime of crystallization of the Mesozoic granitoids from the northern areas of East Asia. Dokl Akad Nauk USSR 268(1):147–151

Vistelius AB, Ivanov DN, Romanova MA (1983b) Principal tectonic elements of the structure of North East Asia and chemical composition of the Mesozoic granites. Dokl Akad Nauk USSR 271(1):141–145

Vistelius AB, Agterberg FP, Divi SR, Hogarth DD (1983) A stochastic model for the crystallization and textural analysis of a fine grained granitic stock near Meach Lake, Gatineau Park, Quebec. In: Geological survey, Canada, paper 81–21, p 62

Vistelius AB (1983) Commingling in the formation of zonal intrusions (with example of the zonal massif located in the vicinity of Gulshad settlement, North-West Balkhash). Dokl Akad Nauk USSR 272(2):457–461

Vistelius AB (1984) A model of commingling and its criterion (on the data of the zonal intrusion near the village of Gul'shad (near Balkhash). Moscow, Nauka, 27 MGK, Doklady, vol 20, pp 18–28

Vistelius AB, Faas AV (1984) On the location of secondary grains of albite and quartz between the primary grains of magmatic granites. Dokl Akad Nauk USSR 279:1461–1464

Vistelius AB (1985) On the correlation between chemical composition of granites and position of their intrusions in regional tectonic pattern of the Northern Part of Eastern Asia. In: Tectonics of Siberia, vol XV, Novosibirsk, Nauka Press, pp 36–41

Vistelius AB (1986) The model of spontaneous cleaning in zonal massif formation (with example of the Gulshad massif in Balkhash region). Dokl Akad Nauk USSR 289(1):180–184

Vistelius AB, Graunov O, Pavlov V (1989) Main stages of zonal massifs fomation: mathematical models of commingling, self-cleaning and pre-eutectic stabilization, Abstracts. In: 28th international geological congress, vol 3,Washington, D.C., USA, pp 304–305
Vistelius AB (1989) The oldest laboratory of mathematical geology, USSR. Math Geol 21: 921–932
Vistelius AB, Faas AV (1991) On the precision of age measurements by RB-Sr isochrons in nontrivial cases. Math Geol 23(8):999–1044
Vistelius AB (1991) Principles of mathematical geology. Kluwer Academic Publishers, pp 477
Vistelius AB, Pavlov VM (2003) On rhythmical layering of rocks formed from basaltic magma. Math Geol 35:399–404

# Chapter 40
# Fifty Years' Experience with Hidden Errors in Applying Classical Mathematical Geology

**Hannes Thiergärtner**

**Abstract** Classical mathematical geology is a branch of mathematical geosciences in which mathematical methods and models—not specifically developed for and not exclusive to specific geosciences—are applied to describe, to model and to analyse quantitatively geoscientific subjects and processes. It was the dominant approach in the 1960s to 1980s and it is still used today to solve numerous, mostly limited and less complex problems. The methods have been implemented in the form of algorithms in commercial software packages that are widely used in geological practice. Their application frequently assumes specific pre-conditions, which are often difficult, if not impossible, to verify. This situation can result in significantly spurious output and errors that are often not recognised (hidden errors). In this paper five case studies are used to demonstrate these errors. In particular, they demonstrate that small mistakes can lead to serious, but often unrecognised, misinterpretations. The main conclusion is that there is a need to improve education and training in classical mathematical geology especially for engineering sections of consulting firms, governmental agencies and individual consultants.

**Keywords** Mathematical geology · Application · Case studies
Error · History of the IAMG

## 40.1 Introduction and Definitions

The application of mathematical formulae and methods to solve geological problems started decades before the International Association for Mathematical Geosciences (IAMG) was founded. Initially, simple methods were used to compute derived parameters such as petrochemical mineral norms or grain size distributions and grain shapes. W. C. Krumbein in Chicago and A. B. Vistelius in the former

H. Thiergärtner (✉)
Department of Geosciences, Free University of Berlin, c/o Kohlisstraße 65,
12623 Berlin, Germany
e-mail: thiergartner@aol.com



B. S. Daya Sagar et al. (eds.), *Handbook of Mathematical Geosciences*,
https://doi.org/10.1007/978-3-319-78999-6_40

Leningrad (now again St. Petersburg) were the first to introduce probability-based statistical methods into geoscientific applications. In the 1960s sophisticated mathematical methods were increasingly developed and applied simultaneously with the development of electronic data processing. Numerous monographs were published to introduce these new tools to geologists (Table 40.1). Step by step, a new sub-discipline—termed "mathematical geology"—was established. It was within this context that the IAMG was established as an association within the International Union of Geosciences at the International Geological Congress in Prague 1968.

The majority of the methods introduced into the geosciences between the 1960s and 1980s were based on probability-statistical or heuristic models. Due to their high level of abstraction, these methods are equally applicable to the solution of analogous problems in other natural or social sciences provided the required data are available. Table 40.2 summarises some essential methods belonging to this group. For the purposes of this paper, these methods and models are classified as classical mathematical geology. (The term "geology" has recently been replaced by "geosciences" but the latter includes the former).

Classical mathematical geology applies mathematical methods and models, which comprise procedures that are not developed specifically for geosciences and which do not bear any direct relation to geological subjects or geological processes. They are extensively implemented in software packages such as Statistical Package for the Social Sciences (SPSS), and have been described in detail in the literature (e.g., Bühl 2016).

Over recent decades the development of mathematical geosciences has resulted in many new advanced models. These models have mostly been developed for specific geoscience applications such as basin modelling, groundwater flow models, contaminant transport models, heat flow models, and so they differ from the classical mathematical geology. This contribution does not cover these specific methods and models.

Classical mathematical geology models retain their applicability and practical advantages. They are helpful tools when other (specific) approaches are not available, when the development of a new model is disproportionately, when the geological problem does not require specific solutions or when limited questions are to be answered on the basis of few data. To date, this area of mathematical geology has not been replaced by later developments and it remains a useful component of the complete set of methodologies.

## 40.2 Hidden Errors and Case Study Examples

In the course of the past 50 years many correct, useful results have been generated by the application of classical mathematical geology. Whilst the application of classical mathematical geology does not necessarily result in incorrect or inaccurate solutions of geoscientific problems, it does have the potential to do so. Incorrect and

**Table 40.1** Early monographs of mathematical geology

| Author(s) | Year | Monograph | Field of application | Model(s) |
|---|---|---|---|---|
| Krige DG | 1962 | Statistical Applications in Mine Valuation. Journ. Inst. Mine Surveyors South Africa, vol 12 no 2 | South-African gold deposits | Geostatistics |
| Miller RL, Kahn JS | 1962 | Statistical analysis in the geological sciences. John Wiley & Sons, New York | Geology | Mathematical statistics |
| Matheron G | 1962 | Traité de géostatistique appliquée: tome 1. Éditions Technip, Paris | Mineral deposits | Regionalized variables |
| Matheron G | 1963 | Traité de géostatistique appliquée: tome 2 – Le Krigeage. Éditions Technip, Paris | Mineral deposits | Regionalized variables (kriging) |
| Whitten EHT | 1963 | A surface-fitting program suitable for testing geological models which involve areally distributed data. Tech. Rept. of ONR task no 389-135, Evanston (Ill.), no 2 | Mapping | Trend analysis |
| Whitten EHT | 1963 | Application of quantitative methods in the geochemical study of granite massifs. Royal Soc. Canada Spec. Publ. 6:76–123 | Geochemistry, igneous rock geology | Mathematical statistics |
| Formery P | 1964 | Course de géostatistique. École Polytechnique, Univ. de Montréal | Geosciences | Geostatistics |
| Родионов ДА [Rodionov DA] | 1964 | Funkcii raspredeleniya soderzhanii elementov i mineralov v izverzhennykh gornyx porodov [Distribution functions of the element and mineral content in eruptive rocks]. Izd. Nedra, Moskva | Geochemistry and mineralogy of eruptive rocks | Mathematical statistics (distribution functions) |
| Krumbein WC, Graybill FA | 1965 | An introduction to statistical models in geology. McGraw-Hill Book Co., New York | Geology | Modelling, mathematical statistics, mapping |
| Matheron G | 1965 | Thèses a la Faculté des Sciences de l'Université de Paris – Les Variables Régionalisées et Leur Estimation. Masson et C[ie] Éditeurs, Paris | Mathematical basics | Regionalized variables |
| Шарапов ИП [Šarapov IP] | 1965 | Primenenie matematicheskoy statistiki v geologii [Application of mathematical statistics in geology]. Izd. Nedra, Moskva | Geology, geochemistry | Mathematical statistics |
| Agterberg FP | 1966 | Markov schemes for multivariate well data. Miner. Industr. Experim. Station, Pennsylvania State Univ, Spec. Publ. 2–65 | Drill hole data | Semi-Markov processes |
| Smith FG | 1966 | Geological data processing using FORTRAN IV. Harper & Row, New York | Geosciences | FORTRAN procedures and application |
| Griffiths JC | 1967 | | Sedimentology | Sampling of sedimentary rock |

(continued)

**Table 40.1** (continued)

| Author(s) | Year | Monograph | Field of application | Model(s) |
|---|---|---|---|---|
| | | Scientific Method in the Analysis of Sediments. McGraw-Hill Book Co., New York | | |
| Gy P | 1967 | L'Échantillonnage des Minerais en Vrac: Tome 1 – Théorie Générale. Mémoires Bureau Recherche Géologiques et Minières, no 56, Édition B.R.G. M, Paris | Ore sampling | Probability theory of sampling and minimization of estimation errors |
| Serra J | 1967 | Échantillonnage et estimation locale de phénomènes de transition minière. Thèse de docteur, Fontainebleau | Sampling in mining | Geostatistics |
| Vistelius AB | 1967 | Studies in Mathematical Geology. Consultants Bureau, New York | Genesis of granites; sedimentary sequences | Semi-Markov processes |
| Marsal D | 1967 | Statistische Methoden für Erdwissenschaftler. E. Schweizerbart'sche Verlagsbuchhandlung, Stuttgart | Geology | Mathematical statistics and application |
| Thiergärtner H | 1968 | Grundprobleme der statistischen Behandlung geochemischer Daten. Freiberger Forschungsh. vol C237, Leipzig | Geochemistry | Mathematical statistics and application |
| Harbaugh JW, Merriam DF | 1968 | Computer Applications in Stratigraphic Analysis. John Wiley & Sons Inc., New York | Geology of sedimentary rocks | Mapping, trend-analysis, classification, simulation |
| Whitten EHT | 1968 | FORTRAN IV CDC 6400 Computer program to analyze subsurface fold geometry. Kansas Geol. Surv. Computer Contrib., vol 25 | Structure geology | FORTRAN procedures |
| Родионов ДА [Rodionov DA] | 1968 | Statisticheskie metody razgranicheniya geologicheskikh obyektov po kompleksu priznakov [Statistical methods for classification of geological objects based on a complex of attributes]. Izd. Nedra, Moskva | Bore hole data | Innovative approach of multivariate classification |
| Krumbein WC, Kauffman ME, McCammon RB | 1969 | Models in Geological Processes. An Introduction to Mathematical Geology. Amer. Geological Inst., Washington D.C. | Geology | Univariate und multivariate mathematical statistics |
| Бондаренко В.Н. [Bondarenko VN] | 1970 | Statisticheskie resheniya nekotorykh zadach geologii [Statistical solution of some geological problems]. Izd. Nedra, Moskva | Geochemistry | Mathematical statistics |
| Harbaugh JW, Bonham-Carter G | 1970 | Computer Simulation in Geology. Wiley-Interscience Publishers, New York | Geology | Computer-aided Modelling of |

(continued)

**Table 40.1** (continued)

| Author(s) | Year | Monograph | Field of application | Model(s) |
|---|---|---|---|---|
| | | | | geological processes, FORTRAN procedures |
| Reyment RA | 1971 | Introduction to Quantitative Paleoecology. Elsevier Publ. Amsterdam | Paleontology | Mathematical statistics |
| Blackith RE, Reyment RA | 1971 | Multivariate Morphometrics, 1st ed. Academic press Inc., London | Paleontology | Biometrics |
| Абрамович, ИИ, Груза ВВ [Abramovič II, Gruza VV] | 1972 | Facialno-formacionnyj analiz magmaticheskikh kompleksov [Facies analysis of magmatic rock complexes]. Izd. Nedra, Leningrad | Petrochemistry of eruptive rocks | Mathematical statistics |
| Davis JC | 1973 | Statistics and Data Analysis in Geology. John Wiley & Sons, New York | Geology | Statistical techniques and application |
| Masset J | 1973 | Un système de visualisation des variations géographiques d'un paramètre géologique. Sciences de la Terre, sér. Informat. Géol., no 1 | Geochemistry, stereography | Computer-aided mapping |
| Agterberg FP | 1974 | Geomathematics. Elsevier Publ., Amsterdam | Geology | Basics of mathematical geology |
| Davis JC, McCullagh MJ [eds.] | 1975 | Display and Analysis of Spatial Data. John Wiley–Blackwell, London/New York | Point, line, polygon and field information | Sampling, contouring, software, relief modelling, interpolation |
| Schwarzacher W | 1975 | Sedimentation models and quantitative stratigraphy. Elsevier Publ., Amsterdam | Sedimentary rocks | Stochastic models, semi-Markov process, time series, spectral analysis |
| Вистелиус АБ [Vistelius AB] | 1980 | Osnovy matematiheskoj geologii [Basics of mathematical geology]. Izd. Nauka, Leningrad | Genesis of granites; sedimentary sequences | Semi-Markov processes; theoretical foundation |

**Table 40.2** Selected models of classical mathematical geology

| Group of methods | Scope |
|---|---|
| Frequency distribution and descriptive statistics | Statistical description of random samples and estimation of expected values, deviation |
| Statistical tests | Analysis of the probability or significance of empirical values with respect to probability-theoretical functions or parameters; comparison between statistical parameter estimations of random samples |
| Correlation and regression analysis | Analysis of existence and kind of mutual association between attributes if they do not form a closed system |
| Cluster analysis R-mode | Non-supervised heuristic classification of variables describing geological objects and visualization of their mutual associations |
| Factor and principal component analysis | Reduction of the multivariate parameter space |
| Non linear mapping | Visualization of multivariate data structures in 2D plots |
| Analysis of variance | Analysis of the influence of extern factors onto measured values |
| Discriminant analysis | Supervised multivariate-statistical classification of geological objects and assignment of objects to given classes |
| Cluster analysis Q-mode | Non-supervised heuristic classification of multivariate described objects and visualization of their mutual associations |
| Octree modelling | Partition a 3D space into homogeneous octants regarding multivariate attributes |
| Markov chain analysis | Stochastic modelling of a sequence of (verbal) described multivariate observations and prediction of the probability to occur at not observed sectors |
| Time series analysis | Analysis of existence and kind of temporal or spatial trends in temporal or linear ordered sequences of measured values and forecasting |
| Trend surface analysis | Modelling of trends and anomalies in mapped data and interpolation |
| Spatial trend analysis | Modelling of trends and anomalies in 3D distributed measurements and interpolation |
| Geostatistical analysis | Modelling of main features of 1D, 2D, or 3D distributed data with minimized estimation error; interpolation and forecasting |
| Structure equation analysis | Multivariate-statistical modelling to estimate and to test correlative associations between dependent and independent observed and not observed variables |

frequently undetected errors can occur if the user is not sufficiently experienced with mathematical methods, with data processing problems or with the application of computer software. These problems occur mostly in geoscientific practice, especially when time and/or finance are restricted and the work is subject to pressure to produce positive results.

Spurious results or undetected errors result from apparently negligible inaccuracies such as unavailable or insufficient knowledge of data accuracy and precision, the uncritical use of values below the threshold of measurement, merging of different input data types, statistical parameter estimations without preliminary tests of the underlying type of frequency distribution, the restriction of correlation analyses to linear models, an unsuitable selection of non-supervised classification models and strategies, inclusion of non-informative attribute sets into the data file, missing information about the significance of statistical results, the acceptance of meaningless correlations, uncritical spatial or temporal extrapolation of trend-analytical results.

The application of classical mathematical geology methods and models requires frequent consideration of specific (mathematical) conditions such as the existence of a certain probability distribution, the independence of variables, a minimum number of observations, the proper treatment of missing values, a suitable choice of cluster model and strategy. Usually, long-term experience of the correct interpretation of results is necessary to avoid errors. All these fundamental conditions appear to be rarely included in training programmes and apparently insufficiently taught in courses. Commercial software is easy to handle but no signal alerts the user to the absence of essential pre-conditions and consequent occurrence of an inherent error in the results. Must computer-generated results be accepted as unbiased and reliable simply because they are produced by electronic equipment? Five selected cases derived from earlier projects will be used to demonstrate the problem in detail.

### 40.2.1 Bathymetric Map of the Azores

The archipelago of the Azores (Ilhas dos Açores) consists of nine islands and a reef area in the North Atlantic Ocean and is the result of partially active volcanoes. It covers an ocean surface between 31°30′ and 24°30′ W and 36°30′ and 40°00′ N (Fig. 40.1).

The Azores are situated on the Azores plateau, an area of thickened oceanic crust due to submarine volcanism caused by a hot spot at the Azores triple junction. The NE-SW striking Mid-Atlantic Ridge crosses the plateau between the Graciosa Island and Terceira and continues over São Jorge and Pico. Along this tectonic element, the North American plate and the Eurasian Plate drift to the west and the east respectively. The Corvo and the Flores islands belong to the American plate. The NW-SE striking Terceira rift runs from the island Graciosa over the São Miguel island to the southeast. This is the tectonic line along which the African plate is subducting under the Eurasian plate. The volcanic and seismic activity started in the Miocene epoch and the formation of the islands continued during the Neogene period.

This entire part of the Atlantic Ocean is of great geological and economic interest and is the target of numerous geoscientific expeditions. The sea floor

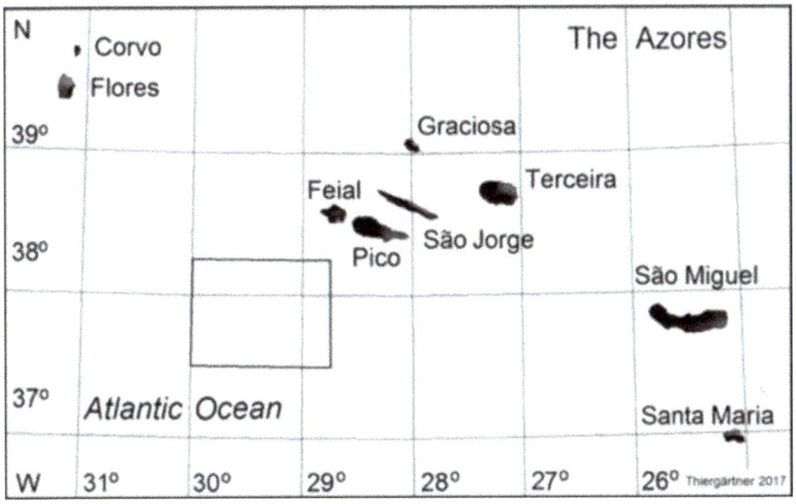


**Fig. 40.1** The Azores. Area of investigation

consists of basaltic rock and young volcanic glasses covered by abyssal clay and biogenous and clastic sediments (cf. Hübscher 2015). The close proximity to the crustal magmatic events causes the formation of important raw materials such as manganese nodules.

A fundamental component of marine survey expeditions is to make depth soundings of the locality. The depths measured in the early 1980s were interpolated by specialists at a computer centre to construct bathymetric contour lines. They used kriging interpolation, the results of which are shown in Fig. 40.2. These results do not reflect the expected predominant NW-SE striking structures described

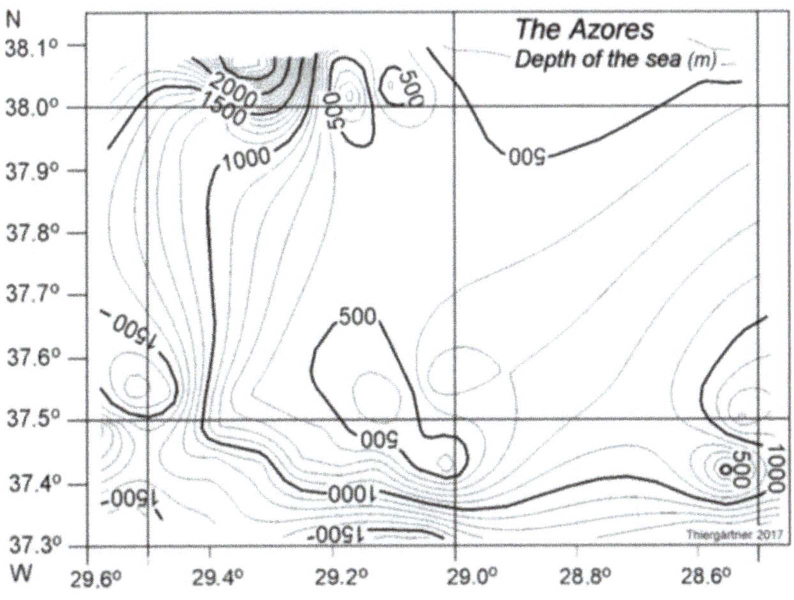


**Fig. 40.2** The Azores. Bathymetric contour lines based on inaccurate input data

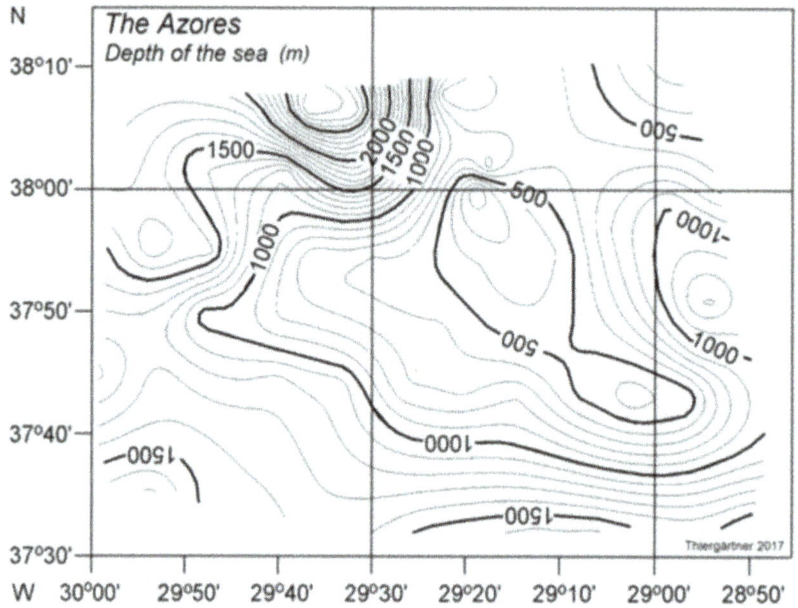


**Fig. 40.3** The Azores. Bathymetric contour lines based on corrected input data

above and give a distorted representation of the main morphological structures. Investigation showed that a suitable mathematical model was applied but the wrong input data were used: geodetic coordinates were used but the minutes were recorded as decimal places. This error was not detected in the computer centre. The result obtained after correcting the data is shown in Fig. 40.3 in which the map more closely reflects the main morphological structures of the investigated area (OpenSeaMap 2016).

### 40.2.2 *Granulometric Analysis of Coastal Sediments of the Southern Baltic Sea*

The Bay of Greifswald (Greifswalder Bodden) in Germany occupies the south-central part of the Baltic Sea. Holocene sand, gravel and boulder cover late Pleistocene till and basin sand. The recent material originated from an active cliff and from an abrasion platform (for details, see Niedermeyer et al. 2011). The fine, medium- and coarse-grained sediments show a lithological differentiation more or less parallel to the erosional shore line. The grain size is specified using the European standard DIN EN ISO 14688-1 (2013).

Knowledge of the characteristics of the sediment is important for designing measures to protect the coast and is necessary if the raw material is to be exploited for building purposes (cf. Börner 2011). One of the relevant parameters is the grain size. A principal component analysis was conducted to reduce the dimensions, or the number of manifest attributes, to a smaller number of latent components which

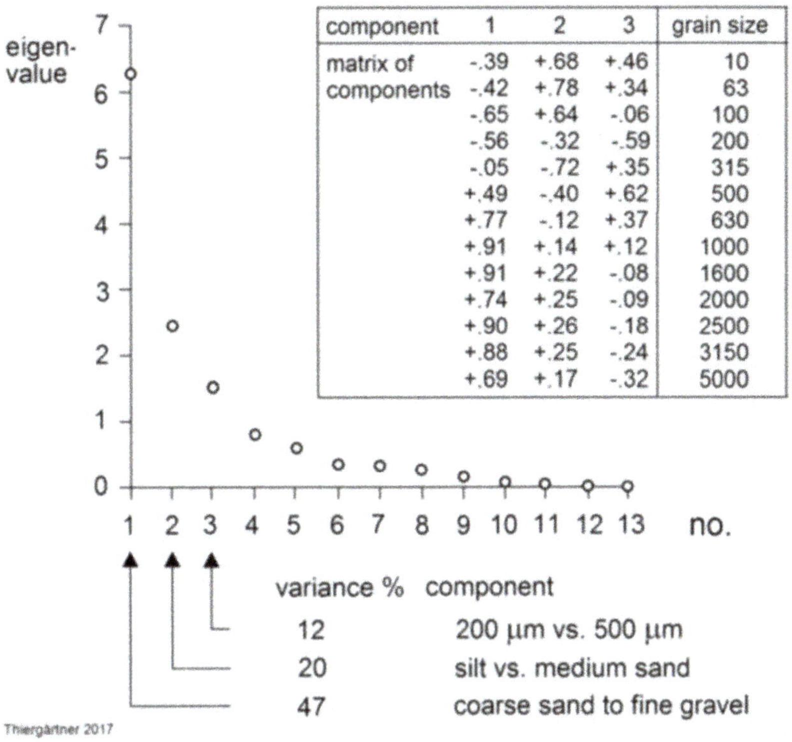


| component | 1 | 2 | 3 | grain size |
|---|---|---|---|---|
| matrix of | -.39 | +.68 | +.46 | 10 |
| components | -.42 | +.78 | +.34 | 63 |
| | -.65 | +.64 | -.06 | 100 |
| | -.56 | -.32 | -.59 | 200 |
| | -.05 | -.72 | +.35 | 315 |
| | +.49 | -.40 | +.62 | 500 |
| | +.77 | -.12 | +.37 | 630 |
| | +.91 | +.14 | +.12 | 1000 |
| | +.91 | +.22 | -.08 | 1600 |
| | +.74 | +.25 | -.09 | 2000 |
| | +.90 | +.26 | -.18 | 2500 |
| | +.88 | +.25 | -.24 | 3150 |
| | +.69 | +.17 | -.32 | 5000 |

**Fig. 40.4** Baltic Sea. Clastic sediments of the south coast. Principal component analysis of grain size data

largely explain the variance of the input data, and to avoid an undesirable multi-collinearity, i.e. to obtain a set of essential information (Fig. 40.4). A cluster (R) analysis should explain the relationships between the original grain size classes (Fig. 40.5). The result reflects only a trivial fact: the coastal sediments are mainly composed of silt and fine sand if they are not coarse-grained, and vice versa.

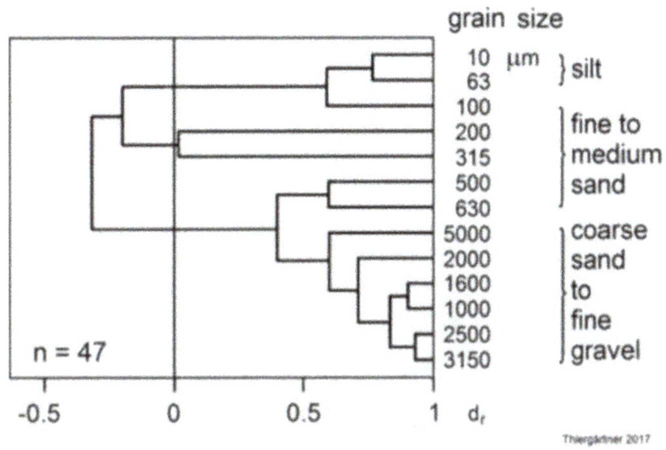


**Fig. 40.5** Baltic Sea. Clastic sediments of the south coast. Cluster (R) analysis of grain size data

The input information scaled in mass% was correctly recorded. However, the fact that the sum of all sieve fractions amounts to the constant sum of nearly 100% was ignored. This closed system means that mathematical results based on correlation among the attributes must be faulty. Chayes (1960a, b, 1971) and Vistelius and Sarmanov (1961) showed that the so-called percent correlation leads to unfeasible results. The modern approach to processing data that form a closed system was developed only later and therefore could not be applied (e.g., Pawlowsky-Glahn 2005; Pawlowsky-Glahn and Buccianti 2011).

### 40.2.3 Areal Distribution of Polycyclic Aromatics in an Abandoned Industrial Site

Until its abandonment an extensive industrial site in Germany was used for machine manufacture. During later assessment for redevelopment the site was investigated for possible ecological contamination. The disused, unsealed enterprise is located on near-surface Holocene sand and gravel. The consultants sampled and analysed twenty-five soil specimens and detected an appreciable concentration of polycyclic aromatic hydrocarbons (PAHs) at two locations. PAHs belong to a group of extremely carcinogenic substances. The 16 most important and persistent constituents are on the National Priority Pollutant List of the US-EPA. An occurrence of these hydrocarbons in subsoil typically requires appropriate remediation measures.

A map of the distribution of the pollutant within the site was constructed by means of kriging (Fig. 40.6) and an expensive soil excavation at well no. T20 over

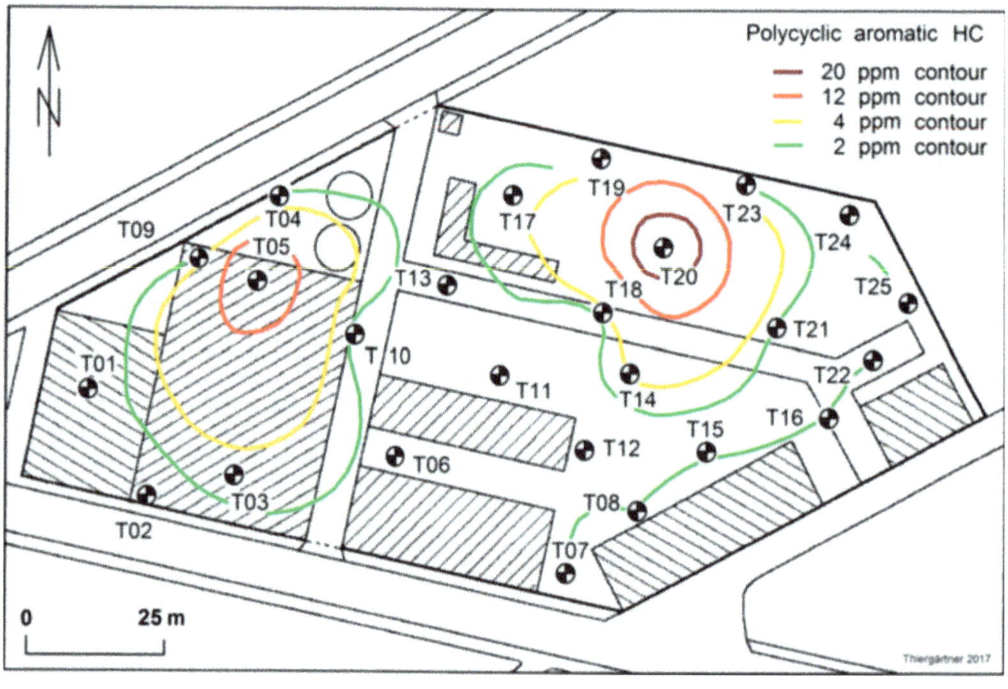


**Fig. 40.6** Contaminated industrial site. Contour map of the apparent distribution of polycyclic aromatics in subsoil

75 m$^2$ and at well no. T05 over 46 m$^2$ was proposed. The application of this geostatistical model to create a contour map is a widely used technique in mathematical geosciences. Similar cases have been repeatedly observed. Less well-known is that the analysed attributes must be interpolable between adjacent points.

PAHs include a wide spectrum of organic substances with relatively low solubility in water, e.g. naphthalene (32 mg l$^{-1}$), acenaphthylene (3.4 mg l$^{-1}$), acenaphthene (4 mg l$^{-1}$), fluorine (1.8 mg l$^{-1}$) and pyrene (0.134 mg l$^{-1}$). Their solubility in water and tendency to migrate into aquifers rises if solvents such as mineral oil, halogenated organic compounds or phenols are present. PAHs will be generated during coking processes or coal-gas generation but they never occur as waste products of machine manufacturing. A study by a project controller showed that a gas generation facility had been operational on the site until 1898. Coal tar was an unprofitable by-product at that time and it was frequently deposited near the factory. Thus, PAH bearing waste was also deposited locally, at distinct locations. Originally included fluids are removed by natural weathering processes over decades. At present, the remaining solid PAH components are persistent and relatively immobile (Stupp and Püttmann 2001). A result of this man-made impact is a spatially limited, although not tolerable, area of contamination. Any extension of these spatially limited occurrences caused by mathematical interpolation methods is meaningless.

The groundwater flow direction must be included in the risk evaluation if contaminants in unconsolidated subsoil are water-soluble and if they are able to migrate. Contour maps generated by standard kriging cannot consider this factor and its application would also result in an incorrect result.

The resulting insolubility of the pollutants under natural conditions causes their inability to migrate. Due to this property of the contamination, it is not correct to interpolate the detected PAH concentration values between observed locations. An isoline map predicts an area-wide contamination whereas only local and isolated pollution actually occurs. Later, it was recommended that the survey data be presented in the form of a point map (Fig. 40.7) and to focus future remediation on the observed hot spots.

A similar case study was discussed by Thiergärtner (1995).

### *40.2.4 Ore Grade Estimation in a Cassiterite Mine*

Tin ore has been mined for centuries in Altenberg (Saxony, Germany). Monzo-, aplite- and albite-granite intruded during the Cisuralian epoch (Permian) into Precambrian paragneiss and were followed by acid, fluorine and silica rich overcritical auras. Feldspar was mainly altered to quartz; lithium bearing mica, topaz, fluorite, and ore minerals such as cassiterite, wolframite, and molybdenite crystallised in the form of small grains. For details, see Weinhold (2002).

Thirty samples were taken from an exploration gallery to calculate the mean grade of the deposit yield in the investigated direction. The range of the metal

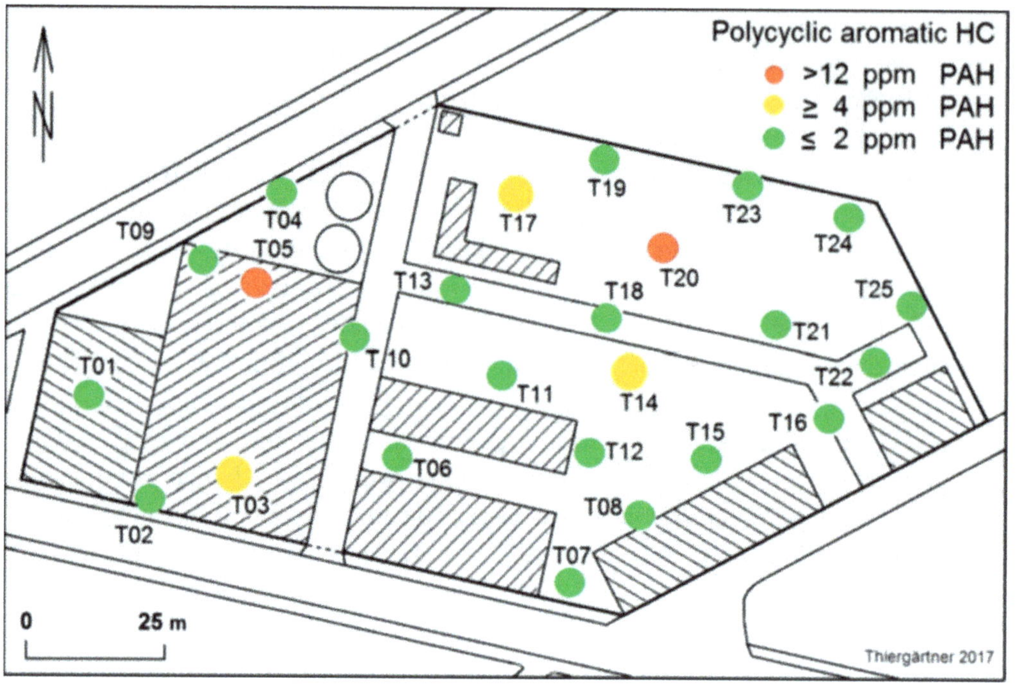


**Fig. 40.7** Contaminated industrial site. Hot spot map of polycyclic aromatics in subsoil

content was 0.07–4.22% tin. The arithmetic mean of all analysed sample values was computed "as usual" to be $m_{arith} = 0.755\%$ Sn. Inspection of the empirical histogram and the fitted normal distribution curve (Fig. 40.8) showed that the metal grade was extremely skewed to the right. A lognormal distribution was fitted to the input data (Fig. 40.9) and the arithmetic mean of the (decimal) logarithms of tin grade was calculated and the corresponding antilogarithm ($m_{geom} = 0.512\%$ Sn) was obtained. This value is less than the arithmetic mean. The geometric mean is a location parameter, such as the median or mode, and scarcely suitable to estimate the expected value of a population.

**Fig. 40.8** Tin grade. Arithmetical mean value estimation

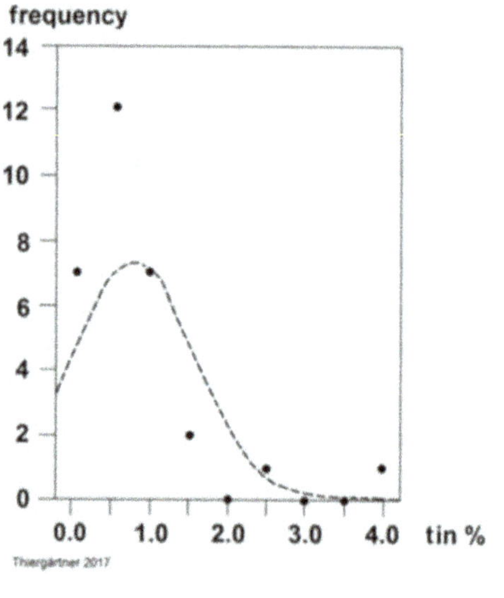

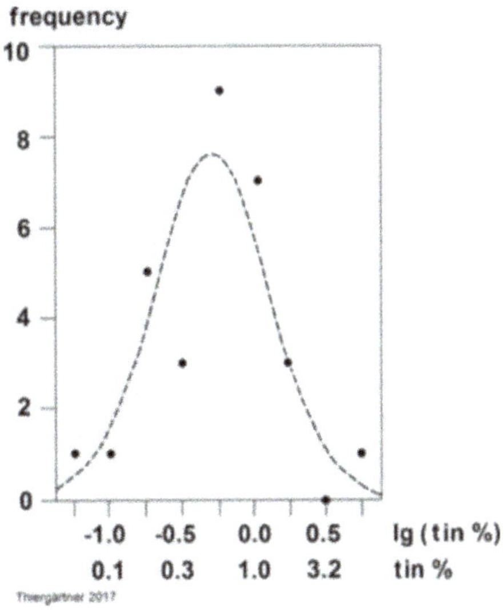


**Fig. 40.9** Tin grade. Lognormal mean value estimation

Which estimator should be applied? The statistical "best" estimator $\hat{E}(m_{lg})$ of the expected value for lognormally distributed data is calculated using Eq. 40.1 developed by Aitchison and Brown (1957) and applied e.g. by Dowd (1984):

$$\hat{E}(m_{lg}) = 10^{mean(lg\,X)} e^{k} \left(1 - \frac{k(k-1)}{n} + \cdots\right) \tag{40.1}$$

where k = 2.65095 var (lg X) and $n$ = number of observations.

This estimator is rather poorly known in geoscientific practice. The estimation results in $m_{lg}$ = 0.765% Sn for the given example. Only this value can be applied to estimate the mean tin grade of the investigated gallery in an unbiased way. The true ore grades of samples in an operating underground mine can be used to estimate the mean grade of un-mined volumes of ground and this is one of the most important parameters in determining economic feasibility.

### *40.2.5 Classification of a Doleritic Sill Using Trace Elements*

Tholeiitic basalt occurs in the Thuringian Forest (Germany) as Sakmarian doleritic sill (Lower Permian). It is intruded into a sandstone–siltstone formation. The contacts are metamorphosed. This sill was extensively described recently by Andreas and Voland (2010).

The matrix of the dolerite consists of pyroxene, plagioclase, olivine, alkali feldspar and some magnetite. The drill core was partitioned into seven sections by petrographical analysis (Table 40.3; Fig. 40.10a). Megascopic and mineralogical indications differ negligibly.

**Table 40.3** Vertical sections of doleritic sill, Thuringian Forest

| Zone | Depth (m below surface) | Description | Geological position | Symbol in Fig. 40.10 |
|---|---|---|---|---|
| 1 | 22.3–27.8 | Basalt, compact, greyish black | Marginal facies | B |
| 2 | 27.8–37.0 | Mainly olivine bearing dolerite with streaked intercalations of quartz dolerite | Transition facies | Doq |
| 3 | 37.0–120.0 | Quartz dolerite, light greyish green to greenish gray, with pale lathlike plagioclase crystals | Gravity fractionation | Dq |
| 4 | 120.0–230.0 | Olivine bearing dolerite | Gravity fractionation | Do |
| 5 | 230.0–348.0 | Olivine bearing dolerite with some biotite | Gravity fractionation | Dob |
| 6 | 348.0–376.0 | Basaltic compact rock, partly including xenolithic material | Transition facies | Bx |
| 7 | 376.0–377.5 | Basalt, compact | Marginal facies | B |

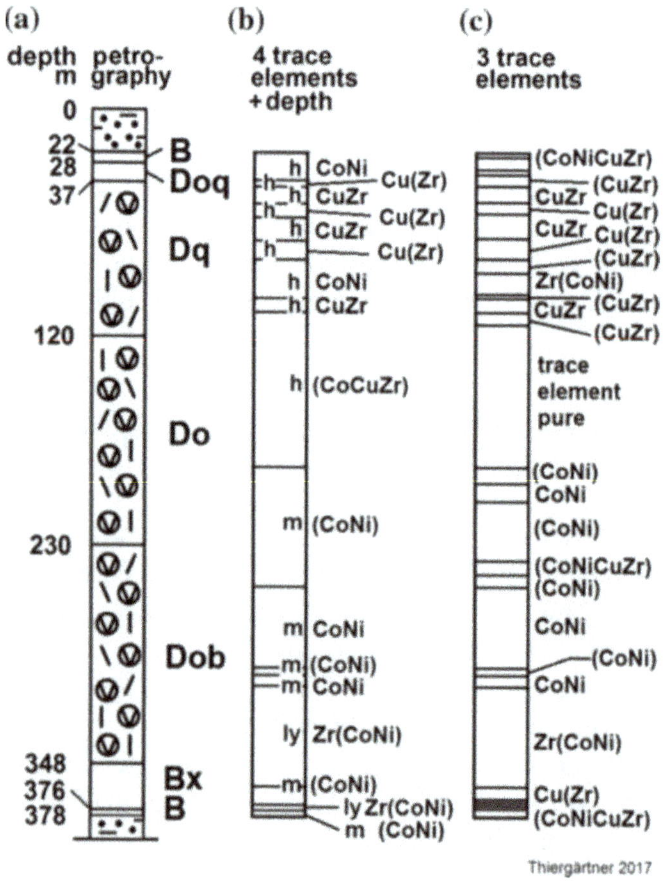


**Fig. 40.10** Doleritic sill. Geochemical classification (for explanation see Table 40.3)

The interesting question was to correlate this sequence with its chemical composition. Four trace elements were selected: cobalt and nickel which are fixed in the olivine mineral replacing magnesium, zircon as a constituent of feldspar or pyroxene, and copper which can occur in the form of microscopically small chalcopyrite crystals. These four chemical elements were analysed in 79 samples covering the whole sequence.

A hierarchical cluster analysis was carried out. It was based on z-scaled (normalized) input data to avoid an overestimation of attributes with large values, using the squared Euclidean distance measure and the Ward method. The cluster dendrogram shows clearly distinguishable classes and can be interpreted without difficulty. All resulting and interpretable classes have been assigned to the cross-section (Fig. 40.10b). Chemical symbols without brackets refer to high concentrations of an element in Fig. 40.10. Where medium concentrations occur they are enclosed in brackets and missing chemical symbols indicate low contents in the sample. The expressions High, Medium, and Low refer to the overall mean values. The figure shows first that the geochemical composition differs noticeably from the petrographical structure. The number of clearly distinguishable geochemical classes is low. Thick parts of the sill seem to be characterised by a similar micro-chemical composition. It is obvious that samples with lower depth (hanging-wall samples, marked by *h*) dominate the hanging-wall of the sill, and samples taken from the footwall (marked by *ly*) mainly occur at deeper levels. The middle section comprises samples that were collected at depths between 200 and 300 m (marked by *m*).

The results gave sufficient reason to review the methodological approach. First, it was noted that the depth was included as one of the input parameters and the parameter "depth" significantly influences the classification. Such procedures are not faulty in a mathematical sense but they accentuate the effect of neighbouring samples within a common class due to the similar value of the parameter "depth". Within the drill core neighbouring samples have a higher chance of falling into this common class than do the more distant samples. This effect should be avoided if not explicitly requested by the researcher. The relatively long sections of the profile with little or no geochemical variation can be explained by this effect. Secondly, the critical test showed that the inclusion of both cobalt and nickel into the analysis caused an overestimation of the olivine component. The linear pairwise correlation coefficient (Pearson) between Co and Ni was calculated as $r = +0.915$. Cobalt is not significantly correlated with copper or zircon, and copper and zircon are uncorrelated, too. This result could be expected from the relationships of the geochemical bonds.

A repeated cluster analysis based on the attributes Ni, Cu and Zr resulted in classes which were drawn into the rock sequence as shown in Fig. 40.10c. The influence of the depth is eliminated and the double effect of the trace elements Co and Ni—reflecting the olivine content—is reduced to only one factor. Much more detail is visible; i.e. a clear vertical geochemical differentiation can be recognised. In addition, the resulting geochemical classification of the rock profile does not simply correspond to the mineralogical and petrographical structure and displays more essential details than the first result. Although the mathematical model was chosen correctly, an incorrect set of input data was applied to solve the problem.

## 40.3 Conclusion and Suggestions

Classical mathematical geology is the application of non-specific mathematical methods and models to solve geoscientific problems. Its proper application frequently results in useful solutions but its misapplication can generate spurious results that may not be recognised. These hidden errors are not caused by the algorithms but by insufficient knowledge of their application and deficient experience with their use. They are avoidable.

A significant proportion of the methodological contributions to classic mathematical geology are written in an academic environment. New developments are mainly published in journals specialising in mathematical geosciences. However, only in rare cases are they evaluated by engineers and geoscientists working in engineering practices, mining companies, environmental bureaus, governmental agencies or individual consultants.

As a conclusion the following suggestions are offered to developers of mathematical-geoscientific methods, models, algorithms and software and to all academic teachers in the field of mathematical geosciences:

1. Instructive tutorials on the applicability and modes of application should be developed, introduced and delivered. This problem can be solved best by experienced geoscientists who are experienced in the "traps" of applying mathematical geology.
2. Informative, methodologically sound case studies should be published in widely-used geoscientific and eco-scientific journals as a means of improving the "daily" application of mathematical geosciences in practice.
3. A more critical view of users is required when assessing outputs from computer-generated mathematical-geological methods and models. In particular, belief in the absolute correctness of such outputs should be discouraged.
4. University studies and post-graduate education in the correct application of mathematical-geological methods should be combined with the development of new application fields (application research).

**Acknowledgements** I am grateful to F. P. Agterberg, Ottawa, and B. S. D. Sagar, Bangalore, for the invitation and encouragement to contribute on the occasion of the Golden Anniversary of the International Association for Mathematical Geosciences. Many thanks go to P. Dowd, Leeds (UK)/Adelaide (AU) and D. Little, Swavesey (UK), for an inspiring discussion of the draft and significant comments. I also would like to acknowledge the reviewers for their interesting suggestions to improve the article.

## References

Aitchison J, Brown JAC (1957) The lognormal distribution. University Press, Cambridge

Andreas D, Voland B (2010) Der Dolerit der Höhenberge –Teil eines eigenständigen Höhenberg-Intrusionsintervalls – sein Gesamtprofil in der Bohrung Schnellbach 1/62 und die Einordnung der Intrusion in den Ablauf der Rotliegendentwicklung des Thüringer Waldes [The dolerite of Höhenberge—discrete part of the Höhenberg intrusion intervall—profile of the drilling Schnellbach 1/62 and stratigraphic position of the intrusion in the Rotliegend]. Beitr Geol Thüringen NF 10:23–82

Börner A (2011) Geologie und Rohstoffgewinnung auf und um Rügen [Geology and mineral deposits on and around Rügen Island]. Proceedings Arbeitskreis Bergbaufolgelandschaften EDGG 245:9–19

Bühl A (2016) SPSS 23 – Einführung in die moderne Datenanalyse [SPSS 23—introduction into the modern data analysis], 15th edn. Pearson Studium, Hallbergmoos

Chayes F (1960a) On correlation between variables of constant sum. J Geophys Res 65:4185–4193

Chayes F (1960b) Correlation in closed tables. Annual report of the geophysical laboratory, vol 59. Carnegie Institution of Washington, pp 165–168

Chayes F (1971) Ratio correlation. University Press, Chicago

DIN EN ISO 14688-1 (2013) Geotechnical investigation and testing—identification and classification of soil—part 1

Dowd P (1984) Lognormal geostatistics. Sci Terre, sér Inf Géol 18:47–68

Hübscher C (2015) FS Meteor—Journey M 113 Azores Plateau. http://www.gzn.fau.de/krustendynamik/arbeitsgruppen/endogene-geodynamik/expeditionen/azoren-plateau-i-2015

Niedermeyer R, Lampe R, Janke W et al (2011) Die deutsche Ostseeküste [The German Baltic Sea Coast]. Schweizerbart Science Publishers, Stuttgart

OpenSeaMap (2016) OpenSeaMap—the free hydrographical chart. http://www.map.OpenSeaMap.org/depthcontour/Azores

Pawlowsky-Glahn V (2005) Forward—advances in compositional data. Math Geol 37:671–672

Pawlowsky-Glahn V, Buccianti A (eds) (2011) Compositional data analysis—theory and applications. Wiley, Hoboken

Stupp HD, Püttmann W (2001) Migrationsverhalten von PAK in Grundwasserleitern [Migration of PAH in aquifers]. altlasten spektrum 10:128–136

Thiergärtner H (1995) Correct modelling of groundwater contaminations. Hornická Příbram ve vědě a technice MD7:1–8

Vistelius AB, Sarmanov OV (1961) On the correlation between percentage values—major component correlation in ferro-magnesium micas. J Geol 69:145–153

Weinhold G (2002) Die Zinnerzlagerstätte Altenberg/Osterzgebirge [The tin ore deposit Altenberg/Eastern Ore Range, Saxony]. Sächs. Landesamt Umwelt Geologie ser. Bergbau in Sachsen, vol 9

# Chapter 41
# Mathematical Geology by Example: Teaching and Learning Perspectives

**James R. Carr**

**Abstract** Numerical examples and visualizations are presented herein as teaching aids for multivariate data analysis, spatial estimation using kriging and inverse distance methods, and the variogram as a standalone data analytical tool. Attention is focused on the practical application of these methods.

## 41.1 Introduction

An oxymoron. Mathematical geology has been characterized as such. Saying so, though, betrays ignorance, not of mathematics, but of geology. The science is inherently numerical. Minerals, for example, are quantifiable based on specific gravity, hardness, Miller index, and abundance. Rock classification in petrology and petrography is inherently dependent upon mineral frequency, determined in a manner identical to that which is used by the hematologist when classifying specimens of blood. Geologic structures are quantified by strike and dip, even abundance when characterizing the integrity of rock masses. Economic geologists and geochemists develop complex databases of samples, each associated with many elements, the analysis of which provides clues to ore genesis, water origin, environmental stresses, and rock classification, to name but a few applications. Geophysics and remote sensing provide enormous sets of numbers visualized as digital images. Far from being an oxymoron, mathematical geology is broadly defined as the application of theoretical and applied mathematics to the assessment of geologic data to aid in the interpretation of earth evolution.

The word, aid, is not chosen carelessly. No equation, no calculator, no computer, can substitute for the human ability to infer and interpret. Where equations, calculators, and computers can help with geologic interpretation is in the conversion of numbers to pictures, such as the case when converting numbers comprising a digital

J. R. Carr (✉)
Department of Geological Sciences and Engineering, University of Nevada, Reno, Reno, NV 89557, USA
e-mail: carr@unr.edu



B. S. Daya Sagar et al. (eds.), *Handbook of Mathematical Geosciences*,
https://doi.org/10.1007/978-3-319-78999-6_41

image into what mimics a photograph on a computer screen. Scientific visualization is the process of converting numerical information of any kind into a picture, hopefully improving its interpretation. The responsibility of interpretation remains, always, with the human analyst.

There tends to be an element of distrust of numbers. The quote, attributed to Benjamin Disraeli, is well known, "There are three kinds of lies: lies, damned lies, and statistics." Apparently, there is uncertainty regarding whether Disraeli actually made this quote. This saying was, however, widely used by the end of the 19th Century. Mark Twain, for example, writing in 1906: "Figures often beguile me, particularly when I have the arranging of them myself; in which case the remark attributed to Disraeli would often apply with justice and force: 'There are three kinds of lies: lies, damned lies, and statistics.'"—*Autobiography of Mark Twain.*

Perhaps mistrust of numbers is not as accurate as saying that there exists a reverence of numbers due to a fundamental insecurity about mathematical understanding. The presentation of a statistical analysis can be quite intimidating to those whose confidence in understanding the analytical methods is weak. Of course, the weak confidence can be taken advantage of by those less scrupulous, stating interpretations of numbers for which there is no clear justification. Thus the skepticism surrounding statistics—lies worse than damned lies.

Despite this ignorance, statistical analysis of data is the most widely applied mathematical method in the geological sciences. Geologists draw maps, with geostatistics, geographic information systems (GIS), and remote sensing fundamentally contributing to the process. Mine *geologists* are increasingly charged with ore reserve estimation and ore control using geostatistics. Other examples of applied statistics included bivariate and multivariate methods important for understanding the correlation between two or more variables. Other numerical methods of importance to geologic understanding are finite difference modeling for understanding ground water flow, geostatistical simulation for modeling uncertainty of spatial data, time-series (Fourier) analysis for identifying cycles in data strings over time or space, linear algebra for modeling landform and geologic structure morphology, fractal geometry for understanding scaling in geologic processes, and the application of neural networks to the modeling of geologic processes.

Some of these applications have proven less interesting to students of mathematical geology than others. Three and a half decades of teaching applied mathematics to earth scientists and engineers at a hardrock mining school provide the backdrop for the following observations. One, *graduate* students of economic geology, moreover economic geology professionals eagerly seek instruction and advice in multivariate methods applied to rock geochemistry data, with an emphasis on interpretation for a better geologic understanding of ore deposits. These students and professionals typically want a pure course on multivariate data analysis. Two, teaching kriging theory in an *undergraduate* course is a waste of time when the heavily parametric practice of spatial estimation is considered; industry often views universities as workshops for training mine geologists and engineers on the use of a particular choice of mine planning software, such as SURPAC, and teaching how to use the software and what choices to make for parameter definition is more than one course can provide. These students are less interested in kriging theory than they are in how to interpret a variogram, how to design a grid, what type of estimator to use, and more. The challenge in this case is how to answer these and other questions while not overstaying one's welcome when explaining theory. Thirdly, the variogram is popular among students and professionals as an analytical tool when kriging is not the primary goal; students and practitioners of remote sensing, in particular, use the variogram for various types of digital image processing.

Multivariate data analysis, the *practice* of kriging, and the variogram as a stand-alone data analytical tool are presented in this chapter with an emphasis on their teaching. Both teacher and student perspectives are presented to balance the discussion between tips for learning and advice for teaching.

## 41.2 Multivariate Analysis of Geochemical Data

During the 1930s decade, psychologists began to apply principal components methods to help with the interpretation of their data (e.g., Hotelling 1933; Young 1937). Many psychologists collect data on patients characterizing their behavioral traits. Principal components methods allow psychologists to group patients of similar behaviors resulting in a better understanding of them. Three decades later, sedimentologists (e.g., Imbrie and Purdy 1962; Klovan 1966) used principal components analysis to group samples of sediment based on sedimentological characteristics. In this case, the sediment sample is analogous to the patient and the sample characteristics are analogous to behavioral traits. How sediment samples group can be an indication of sediment source, depositional environment, composition, or some other condition of importance to geologic interpretation. A collection of papers published in 1983 (Howarth 1983) reviewed the application of multivariate analysis to geochemical prospecting. Tomes written on geochemistry (e.g. Albarède 1995) often discuss multivariate analysis applied to the interpretation of geochemical data.

Many mathematical methods have been developed to help with the analysis of multivariate data. An important goal of each of these methods is a reduction in the number of variables to enable a more efficient understanding. If there are M original variables, in other words, a smaller number of variables, B, is sought that define a lower multivariate sub-space. Then, the original M dimensional data are projected onto the lower sub-space to yield a plot (graph) that is visually inspected to appreciate data similarities and differences. For students and teachers alike, the ultimate goal of multivariate analysis is the creation of these plots, the study of which motivates *subjective* conclusions about data associations (Greenacre 1984).

In order to develop the plots, some mutually orthogonal coordinate system is needed. Many of the mathematical methods used to analyze multivariate data involve the reduction of the original data information into some matrix that is eigendecomposed to obtain eigenvalues, each associated with a unique eigenvector. The eigenvectors are mutually orthogonal. Moreover, these eigenvectors define the lower dimensional sub-space. They are the principal components of data intercorrelation information.

Can multivariate data analysis be taught without explaining, or at least reviewing, eigendecomposition of data? Answering yes leads to a teaching approach that treats the multivariate analytical algorithm as a black box. This approach relieves teachers of the chore of explaining a method that many students abhor. Undergraduate students, and even graduate students in some cases, are likely to skip class when eigenvalues and eigenvectors are to be discussed. Modern students are quick to dismiss that which they do not like, or find boring. For example, based on the experience of co-teaching mineralogy for the past five years, a discussion of crystallography often results in a rather empty classroom. A mental laziness is betrayed by students' behavior in this regard. It frustrates teachers wanting students to achieve an understanding of analytical methods deeper than the data in–data out black box.

Of course, answering no to the foregoing question and teaching multivariate data analysis outside the black box is confounded by the same student attitudes. Their learning cannot be forced. It can, however, be enticed by numerical examples that are straight-forward, explained in class, and reinforced by extracurricular calculations. Students can be shown that an understanding deeper than black box mysticism is relatively easy to achieve. What follows is a demonstration of this concept and is intended as an aid to instruction. Student understanding can be assessed by substituting the starting data table with alternative data.

### 41.2.1 Numerical Insight to Multivariate Data Analysis

Geochemical data from seven rock samples, each characterized by five elements (variables), are presented in the following table:

| Sample | Gold (Au) | Silver (Ag) | Copper (Cu) | Lead (Pb) | Zinc (Zn) |
|---|---|---|---|---|---|
| 1 | 15 | 0.4 | 21 | 21 | 15 |
| 2 | 25 | 0.3 | 14 | 15 | 3 |
| 3 | 19 | 0.5 | 12 | 19 | 4 |
| 4 | 37 | 0.5 | 24 | 17 | 7 |
| 5 | 33 | 0.3 | 14 | 13 | 5 |
| 6 | 12 | 0.4 | 21 | 29 | 5 |
| 7 | 12 | 0.4 | 13 | 19 | 5 |

*Note* values for Au, Ag, Cu, Pb, and Zn are in ppm

These data represent a five-dimensional variable space. The goal is to determine the eigenvectors for these data, the sub-space that will be used for plotting. A theorem presented by Eckart and Young (1936) holds that any real valued data matrix can be represented as the following product, [data] = [R-mode eigenvectors] [eigenvalues][transposed Q-mode eigenvectors]. In this case, Q-mode multivariate analysis is focused on the relationships among samples. R-mode multivariate analysis is one that focuses on the relationships among the variables.

To obtain the Q-mode result, the data matrix is multiplied by its transpose in the following order, [transposed data matrix][data matrix], to yield a square matrix as follows: $[\text{data}]^T[\text{data}] = [\text{Q-mode}]$:

$$\begin{pmatrix} 15 & 25 & 19 & 37 & 33 & 12 & 12 \\ 0.4 & 0.3 & 0.5 & 0.5 & 0.3 & 0.4 & 0.4 \\ 21 & 14 & 12 & 24 & 14 & 21 & 13 \\ 21 & 15 & 19 & 17 & 13 & 29 & 19 \\ 15 & 3 & 4 & 7 & 5 & 5 & 5 \end{pmatrix}$$

$$\begin{pmatrix} 15 & 0.4 & 21 & 21 & 15 \\ 25 & 0.3 & 14 & 15 & 3 \\ 19 & 0.5 & 12 & 19 & 4 \\ 37 & 0.5 & 24 & 17 & 7 \\ 33 & 0.3 & 14 & 13 & 5 \\ 12 & 0.4 & 21 & 29 & 5 \\ 12 & 0.4 & 13 & 19 & 5 \end{pmatrix} = \begin{pmatrix} 3957 & 61 & 2651 & 2685 & 920 \\ 61 & 1.16 & 48.4 & 54 & 17.9 \\ 2651 & 48.4 & 2163 & 2325 & 813 \\ 2685 & 54 & 2325 & 2687 & 860 \\ 920 & 17.9 & 813 & 860 & 374 \end{pmatrix}$$

The result of this multiplication is a square matrix, M × M in size, and M is the number of original variables, 5 in this case. This square matrix is the one from which eigenvalues and Q-mode eigenvectors are obtained [software is necessary for eigendecomposition, ironically rendering this step as a black box]:

$$\text{Eigenvalues:} \begin{pmatrix} 91.8 & 0 & 0 & 0 & 0 \\ 0 & 25 & 0 & 0 & 0 \\ 0 & 0 & 10.1 & 0 & 0 \\ 0 & 0 & 0 & 4.8 & 0 \\ 0 & 0 & 0 & 0 & 0.2 \end{pmatrix}$$

$$\text{Eigenvectors:} \begin{pmatrix} 0.655 & -0.73 & -0.10 & -0.17 & -0.007 \\ 0.011 & 0.005 & -0.01 & -0.17 & 1.0000 \\ 0.500 & 0.228 & 0.336 & 0.765 & -0.008 \\ 0.536 & 0.618 & -0.47 & -0.33 & -0.017 \\ 0.183 & 0.181 & 0.809 & -0.53 & -0.007 \end{pmatrix}$$

The eigenvectors are loaded column-wise. The eigenvalues are loaded into a matrix along the diagonal. All off-diagonal entries in the eigenvalue matrix are zero. These eigenvalues are actually the square roots of those computed directly from the R-mode matrix because the original data matrix is squared when multiplied by its transpose. By performing this multiplication to yield a square, symmetrical matrix, the eigenvectors from which are guaranteed to be orthogonal to one another. For example, if the first two eigenvectors are multiplied together, the result should be precisely zero:

$$3.59 \times -4.04 + 0.063 \times 0.028 + 2.74 \times 1.26 + 2.94 \times 3.42 + 1 \times 1 = 0.005 \approx 0$$

and the result would be precise if not for round-off error.

Working toward the goal of plotting samples 1 through 7, a first step involves multiplying the eigenvector matrix to the eigenvalue matrix:

$$\begin{pmatrix} 0.655 & -0.73 & -0.10 & -0.17 & -0.007 \\ 0.011 & 0.005 & -0.01 & -0.17 & 1.0000 \\ 0.500 & 0.228 & 0.336 & 0.765 & -0.008 \\ 0.536 & 0.618 & -0.47 & -0.33 & -0.017 \\ 0.183 & 0.181 & 0.809 & -0.53 & -0.007 \end{pmatrix}$$

$$\begin{pmatrix} 91.8 & 0 & 0 & 0 & 0 \\ 0 & 25 & 0 & 0 & 0 \\ 0 & 0 & 10.1 & 0 & 0 \\ 0 & 0 & 0 & 4.8 & 0 \\ 0 & 0 & 0 & 0 & 0.2 \end{pmatrix} = \begin{pmatrix} 60.13 & -18.25 & -1.01 & -0.82 & 0.00 \\ 1.01 & 0.13 & -0.10 & -0.82 & 0.20 \\ 45.90 & 5.70 & 3.39 & 3.67 & 0.00 \\ 49.20 & 15.45 & -4.75 & -1.58 & 0.00 \\ 16.80 & 4.53 & 8.17 & -2.54 & 0.00 \end{pmatrix}$$

The next step involves multiplying this resultant matrix by the original data matrix. Because the fifth column of this matrix represents values of practically zero, only the first four columns are used to obtain four factors:

$$\begin{pmatrix} 15 & 0.4 & 21 & 21 & 15 \\ 25 & 0.3 & 14 & 15 & 3 \\ 19 & 0.5 & 12 & 19 & 4 \\ 37 & 0.5 & 24 & 17 & 7 \\ 33 & 0.3 & 14 & 13 & 5 \\ 12 & 0.4 & 21 & 29 & 5 \\ 12 & 0.4 & 13 & 19 & 5 \end{pmatrix} \begin{pmatrix} 60.13 & -18.25 & -1.01 & -0.82 \\ 1.01 & 0.13 & -0.10 & -0.82 \\ 45.90 & 5.70 & 3.39 & 3.67 \\ 49.20 & 15.45 & -4.75 & -1.58 \\ 16.80 & 4.53 & 8.17 & -2.54 \end{pmatrix} =$$

| | Factor 1 | Factor 2 | Factor 3 | Factor 4 |
|---|---|---|---|---|
| 1 | 3151.00 | 238.40 | 78.80 | −6.84 |
| 2 | 2935.00 | −131.10 | −24.56 | −0.69 |
| 3 | 2696.00 | 33.38 | −36.13 | −12.13 |
| 4 | 4281.00 | −244.00 | 20.38 | 12.69 |
| 5 | 3351.00 | −298.90 | −6.80 | −9.17 |
| 6 | 3197.00 | 371.50 | −37.87 | 8.38 |
| 7 | 2337.00 | 171.40 | −17.49 | −5.18 |

The word, factor, heading each column is one of the new variables within the sub-space of the original data matrix. The numbers to the left are the sample numbers used in the original data able. The factors represent an orthogonal coordinate system to enable plotting these seven samples to determine their relationship to one another.

Relative significance of each factor with respect to the total data information is determined by summing the eigenvalues, then dividing each eigenvalue by this sum to obtain a proportion. The five eigenvalues sum to 131.9. Factor 1, for instance, represents 100 × (91.8/131.9) = 70% of the original data information content. The second factor associated with an eigenvalue equal to 25, incorporates 20% of the original data information content. If the seven samples are plotted using the first two factors, then the resultant plot represents 90% of the original data information content. This plot is shown in Fig. 41.1.

Notice that samples 2, 4, and 5 plot in the negative region with respect to Factor 2. These three samples are associated with the highest gold values. But, these samples are among the lowest for silver, lead, and zinc. Samples 1 and 6 are much higher in lead and zinc, but much lower for gold. Each factor is a function of all five of the data variables, Au, Ag, Cu, Pb, and Zn. For example, in the foregoing matrix multiplication involving the original data matrix, the Factor 1 "coordinate" for sample 1 is equal to:

$$15 \times 60.13 + 0.4 \times 1.01 + 21 \times 45.9 + 15 \times 49.2 + 15 \times 16.8 = 3151.0.$$

In reviewing this calculation, notice that it is:

$$\text{Au} - \text{value} \times 60.13 + \text{Ag} - \text{value} \times 1.01 + \text{Cu} - \text{value} \times 45.9 + \text{Pb} - \text{value} \times 49.2 + \text{Zn} - \text{value} \times 16.8.$$

Literally, the coordinate of a sample in any of the factors is a function of all the original variables, not just any one, or two. Because of this, the way samples plot in

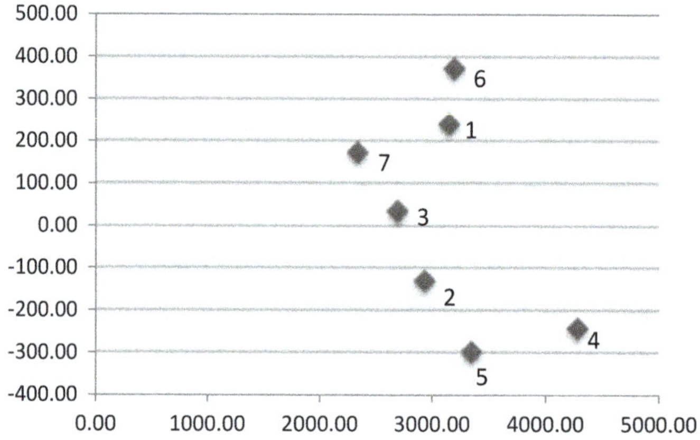


**Fig. 41.1** A plot of the seven data samples with respect to factor 1 (horizontal axis) and factor 2 (vertical axis). Sample numbers are shown near each plotting symbol

Fig. 41.1 reflects their similarity over all variables. A grouping of samples, then, suggests a rock-chemistry similarity that likely has importance to the interpretation of ore genesis.

To obtain the R-mode result, the original data matrix is once again squared, but in a different order of multiplication: $[\text{data}][\text{data}]^T = [\text{R-Mode}]$. Following the same steps above for the Q-mode result, resulting factors for plotting of the five variables are:

| | Factor1 | Factor 2 | Factor 3 | Factor 4 | Factor 5 |
|---|---|---|---|---|---|
| Au | 5542.00 | −463.80 | −7.27 | 2.41 | −0.28 |
| Ag | 96.58 | 2.98 | −0.54 | 0.37 | −0.03 |
| Cu | 4220.00 | 136.50 | 35.76 | −18.29 | −0.15 |
| Pb | 4527.00 | 378.80 | −46.61 | 6.99 | −0.13 |
| Zn | 1538.00 | 111.00 | 82.77 | 11.84 | −0.09 |

The relative importance of each factor with respect to original data information content is the same as for the Q-mode result because the eigenvalues are identical. Figure 41.2 presents a plot based on the first two factors.

Figure 41.2 suggests that gold (Au) is not closely associated with any one of the four other variables. Focusing only on factor 1, gold (Au) and silver (Ag) are on opposite sides of the horizontal axis. Often, variables plotting as such are inversely related; when one is higher in value, the other is lower in value. Further with respect to factor 1, zinc (Zn) is closer to silver and lead (Pb) and copper (Cu) are closer to gold. If, however, the focus is solely on factor 2, then gold and lead appear to be inversely related.

Software is necessary for larger data sets. Using this example and challenging students to follow it for data sets other than that which is used will not necessarily guarantee a deep understanding. But, when reviewing the output from multivariate software, students will have a general understanding of what happens to the input data and the jargon inherent to the method. Knowing why eigenvectors (factors) are

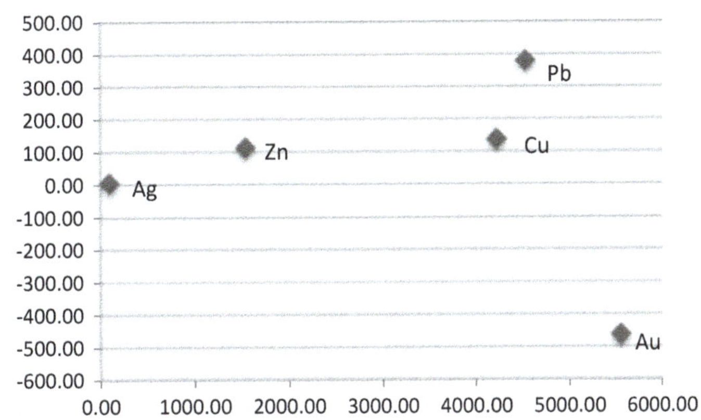


**Fig. 41.2** A plot of the variables based on factor 1 (horizontal axis) and factor 2 (vertical axis). Variable labels are shown next to each plotting symbol

used for developing plots gives students greater confidence when interpreting these plots.

An actual multivariate data set consisting of about 1000 samples, each associated with 50 elemental variables, is analyzed using the multivariate method known as correspondence analysis (Benzecri 1973). This multivariate method is the one preferred by the author for actual data analysis, but its mathematical presentation is not as straightforward as that presented above for principal components analysis. It is the opinion of the author that correspondence analysis yields plots that separate the data better than other methods. The result is shown in Fig. 41.3 for variables only (to reduce the clutter of the plot).

How elements are related is interpreted from Fig. 41.3. Manganese (Mn) is polarizing with all other elements plotting away from it along factor 1. Rocks higher in manganese are inferred to be much lower in the other elements. Given that the likely manganese mineral in this deposit is MnO (wad), moreover knowing that this mineral is black and sooty, could be useful knowledge in the field of where ore is, or is not, present. Factor 2 separates barium (Ba) from the precious metals. The likely barium mineral is barite, an easily recognized mineral in the field if crystalline. This element, too, may be useful for the approximate delineation of the ore zone in the field based on visual inspection.

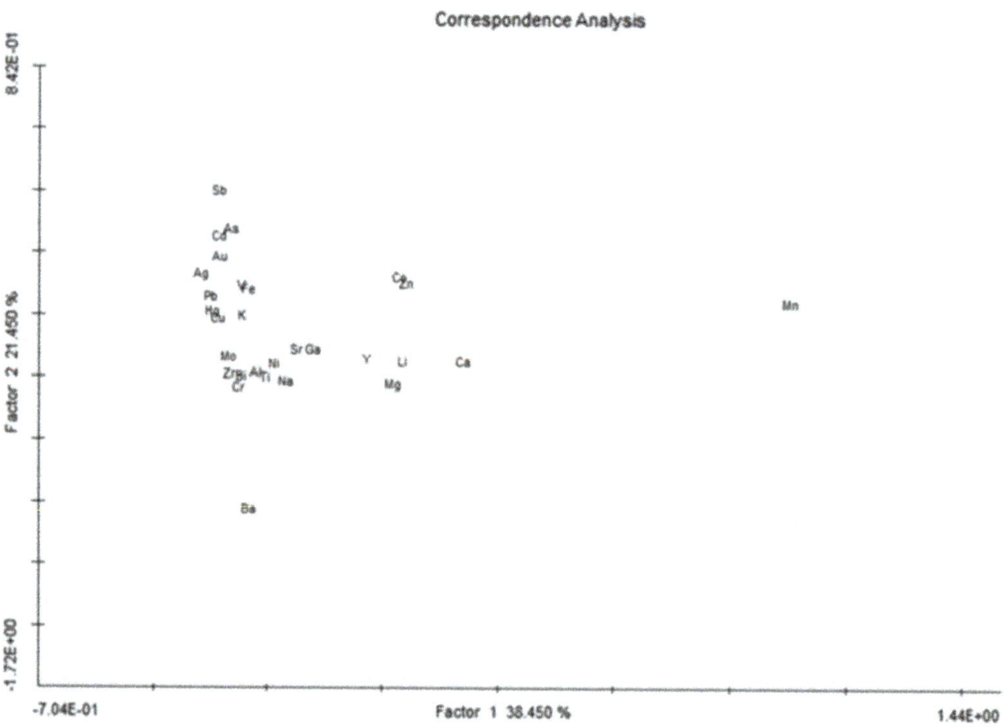


**Fig. 41.3** An R-mode plot of a multivariate geochemical set of data characterizing an ore deposit. The relative importance of each factor is indicated in the axis label. This plot was created by software that is presented in Carr (2002)

## 41.3 Geostatistics and Its Myriad Parameters

Decisions. A teacher of geostatistical estimation can spend weeks teaching geostatistical theory, broadly so by including polygonal and inverse distance strategies in addition to kriging. Weeks! Then faced with teaching the *practice* of estimation. The theory is complex, particularly in the case of kriging. And, yet, the outcome is highly vulnerable to the parameters selected for implementing theory. Figure 41.4, whereas not intended to be comprehensive, presents many of the decisions that must be made by a geostatistician when practicing the gridding of data.

A teacher can spend more or less time on geostatistical theory, lesser for undergraduates perhaps. Time, however, must still be devoted to explaining about and advising on the parameters that are necessary to estimation. Moreover, the

**A Noncomprehensive Table of Parameters**

Method of Gridding:

- Polygonal
- Inverse Distance Methods (ID)
- Kriging

| Kriging: | No Data Transform | With Data Transform |
|---|---|---|
| | Simple | Log-normal; with or without bias adjustment |
| | Ordinary | Indicator (true indicator using quantiles as cutoffs) |
| | | Cutoff |

Kriging (continued):

- Support, Block or Punctual
- Variogram Parameters:
  - Type of model
  - Nugget, Sill and Range

ID:

- Power term

Data:

- Transform options
  - None
  - Indicator
  - Natural logarithms
  - Hermitian
- Capping

Gridding:

- Grid dimensions
  - Number of rows
  - Number of columns
  - Number of levels (if three dimensional)
  - Spacing distance between rows, columns, and levels
  - Number, N, of nearest neighboring sample locations used for estimation
  - Search window radius for locating N nearest neighbors
    - Search window strategy: general, quadrant, octant
    - Anisotropy considerations: Major axis direction, plunge direction, dip, ratio of major axis/minor axis

**Fig. 41.4** Gridding a set of spatial data requires selecting the estimation algorithm for gridding, then defining parameters unique to the estimation algorithm. How to treat the data, raw or transformed, is another important decision. Likewise, the geometry of the grid must be designed influence of one or more of these parameters is best appreciated by visualizing estimation outcomes.

Estimation outcomes are visualized as color contour maps in the following demonstration. A collection of 2,500 mercury values were collected to assess the severity of site contamination after a flood. The variogram for these data is shown in Fig. 41.5.

The shape of the variogram in Fig. 41.5 is modeled using a spherical variogram model. This variogram shape is the most commonly observed for spatial data, regardless of the spatial phenomenon under study. The model is explicitly defined by setting the parameters for the nugget (found by extrapolating the calculated variogram backwards to intersect the y-axis at h = 0), the sill, that value of the variogram that is more or less constant once the range (of spatial correlation) is reached. In this example, these parameters are: nugget = 20 (rounded), sill = 117 (rounded), and the range is 90 m. Other parameters used in the following data visualization demonstration are as follows: (1) no data transform; (2) block support; (3) ordinary kriging; (4) general, isotropic search strategy with a radius equal to one-half the variogram range; (5) up to N = 10 nearest neighboring samples used for estimation; (6) inverse distance (power term = 1) and inverse distance squared (power term = 2) weighting presented for comparison to the kriging outcome; (7) grid parameters: 50 rows, each with 50 columns, spacing between rows and columns is 10 m. Outcomes are presented in Figs. 41.6 and 41.7.

A lower nugget value is seen to yield lesser smoothing during estimation (Fig. 41.6). A larger nugget yields more smoothing. With inverse distance methods, the larger the power term is, the less smoothing that results during estimation. The aesthetic appeal of a map is a subjective assessment. The amount of smoothing controls the complexity of the map. If larger scale aspects of a spatial region are of more interest than smaller scale aspects, then more smoothing should be used during estimation to downplay the smaller scales. On the other hand, if the desire is to visualize spatial variability down to the smallest possible scale that is allowed by the data, no to minimal smoothing should be used during estimation.

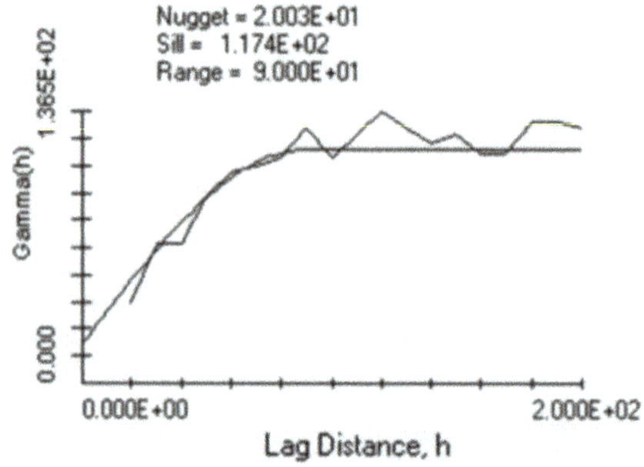


**Fig. 41.5** Variogram for 2,500 mercury values. The jagged line is the actual calculation outcome. The smooth, continuous line is a model fit to the calculation outcome. The model, in this case, is the spherical variogram model and its parameters, nugget, sill, and range, are listed above the variogram

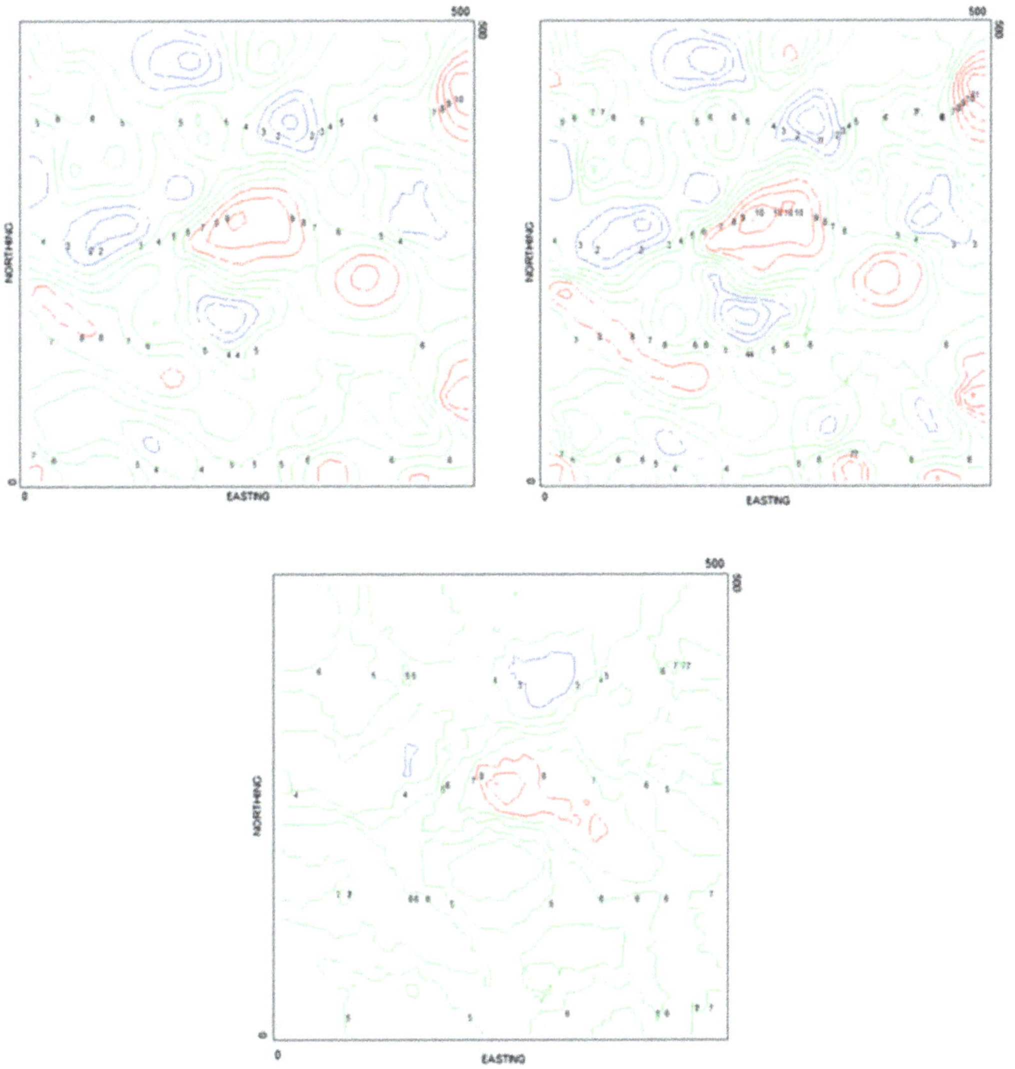


**Fig. 41.6** Three visualizations from kriging. Top, left map is based on the variogram parameters, nugget, sill, and range, that are listed in Fig. 41.5. Top, right map is based on the same variogram parameters, except the nugget value is set equal to zero. The bottom, left map is based on the nugget value set equal to the sill value; in this case, the outcome is a simple average estimation. Integer labels are used for the contour lines to indicate relative value from smaller, 1, to larger, 10. Color also indicates relative value from lower, blue, to higher, red

Indeed, there are similarities among these maps. Each map shows a zone of higher mercury values in the center, and two low zones at the left-center and top-center. These regions are associated with a higher density of spatial samples. Regions of the map that change appreciably when estimation parameters are changed are more sparsely sampled. The spatial distribution of mercury samples is shown in Fig. 41.8.

Differences among the contour map outcomes are noteworthy for spatial locations associated with sparser sampling. Moreover, these differences are more easily observed when increased smoothing is used during estimation.

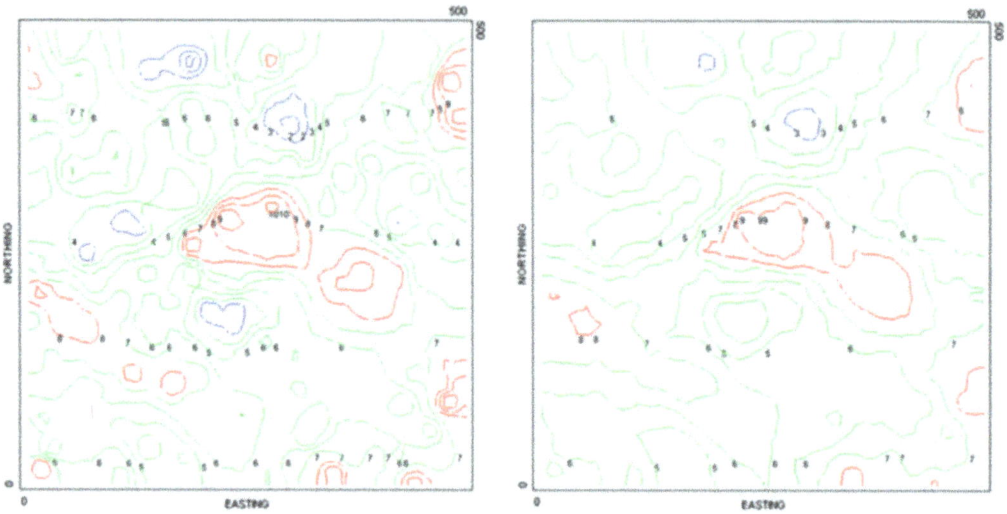


**Fig. 41.7** Outcomes from inverse distance squared weighting (left map) and inverse distance weighting (right map). The higher the power term is the lesser is the smoothing. This outcome is similar to decreasing the nugget value in kriging

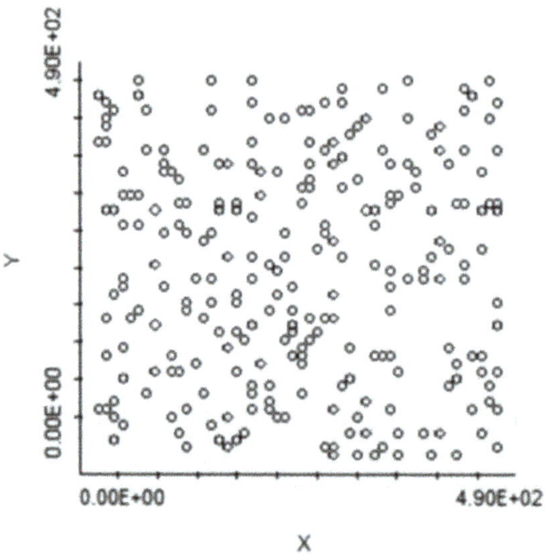


**Fig. 41.8** A map of the spatial locations of 2500 mercury samples within a 250,000 m$^2$ region

## 41.4 The Variogram as a Stand-Alone Data Analytical Tool

Kriging is not necessarily the ultimate goal of geostatistical analysis. The variogram as a stand-alone data analytical tool has a variety of uses that are independent of estimation. Examples are many and include noise isolation, texture classification of digital images, and self-affine fractal analysis and modeling. The concept of digital image texture is chosen for demonstration.

Four textures are illustrated in Fig. 41.9: water, playa, alluvium, and sedimentary rock outcrops.

Water and playa textures are similar, differing only in reflectivity. Variograms of these textures are likewise similar and indicate a predominant spatial randomness

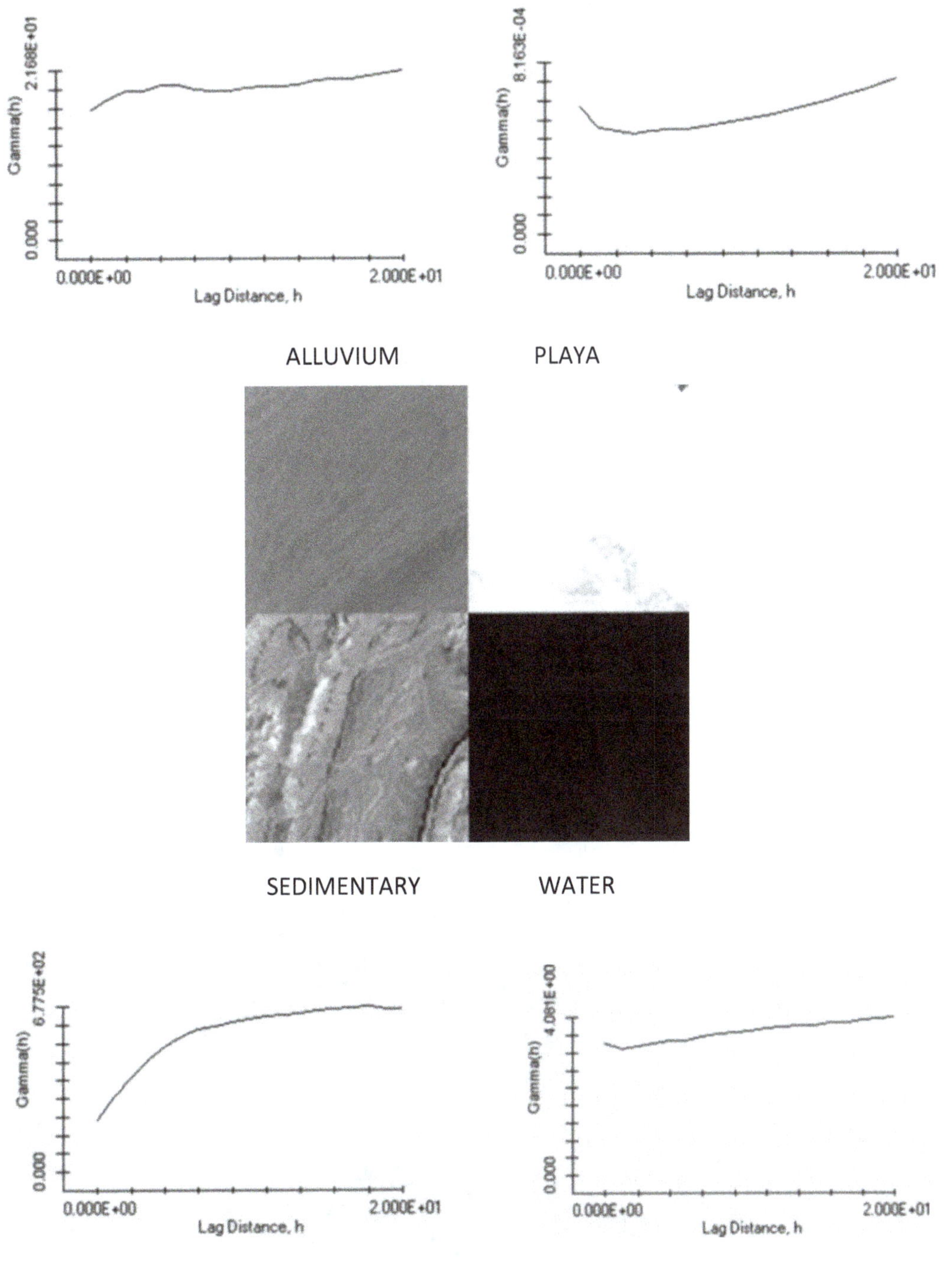


**Fig. 41.9** The center mosaic shows four textures extracted from a Landsat TM image, clockwise from top, left: alluvium, playa, water, and sedimentary rock outcrops. Variograms for these textures are likewise arranged

with little underlying signal. The variogram of alluvium texture indicates an underlying spatial structure that is heavily masked by noise (randomness). Unlike these other textures, sedimentary rock layer texture is predominantly signal. The variogram of this texture reveals the strong spatial structure and very low noise.

Digital image classification is a process that depends on automatic identification of classes, features on the ground, based on some form of signature, or characteristic for each class. The histogram of pixel values is one such signature that is often used when basing classification on pixel value. The variogram is a signature that is useful for classifying the texture of ground classes. The foregoing demonstration shows that variograms do differ for ground classes, but in ways that are not directly relatable to pixel values. Playa and water, for instance, are distinctly different in brightness, yet their variograms are similar in shape. The variogram has been used with considerable success for the classification of microwave images (e.g., Carr and Miranda 1998; Miranda et al. 1998). These images are inherently noisy due to microwave frequency additions and cancelations that impart what is known as speckle. The classification of texture using variogram signatures applied to less noisy images, such as those from the Landsat satellite, has not been extensively tested.

In the foregoing example, the images of alluvium, playa, water, and sedimentary rock are 100 × 100 pixel extracts from a band 3 (visible red) Landsat 7 image of

**Fig. 41.10** A band 3 (visible red) extract from a complete Landsat 7 scene, Path 39, Row 35, acquired on September 25, 2000

southern Nevada, U.S.A (Fig. 41.10). This image was selected for its varied textures.

The variogram signatures shown in Fig. 41.9 were applied to this image for the classification of texture. The understanding of what constitutes texture in a digital image takes some time to develop. Texture is not brightness, per se, but rather the unique patterns exhibited by groups of pixels. The outcome of textural classification is shown in Fig. 41.11.

The predominant texture seen in Fig. 41.11 is that of alluvium. The texture of water is not unique and is confused with the texture of alluvium. The texture of the shoreline of the lake is identified as playa. This lake (Lake Mead, Clark County, Nevada, U.S.A.) is an artificial reservoir that has a fluctuating water level that leaves an almost pure white calcium carbonate staining on the shoreline. Like playa sediments, the reflectivity of this material often saturates the satellite sensor resulting in identical textures. Sedimentary outcrop texture was often confused with alluvium, and shadows (northwest facing slopes) were often confused with water. Given that this image is of a harsh, arid environment (precipitation is less than 7 cm per year), the predominant alluvium texture makes sense.

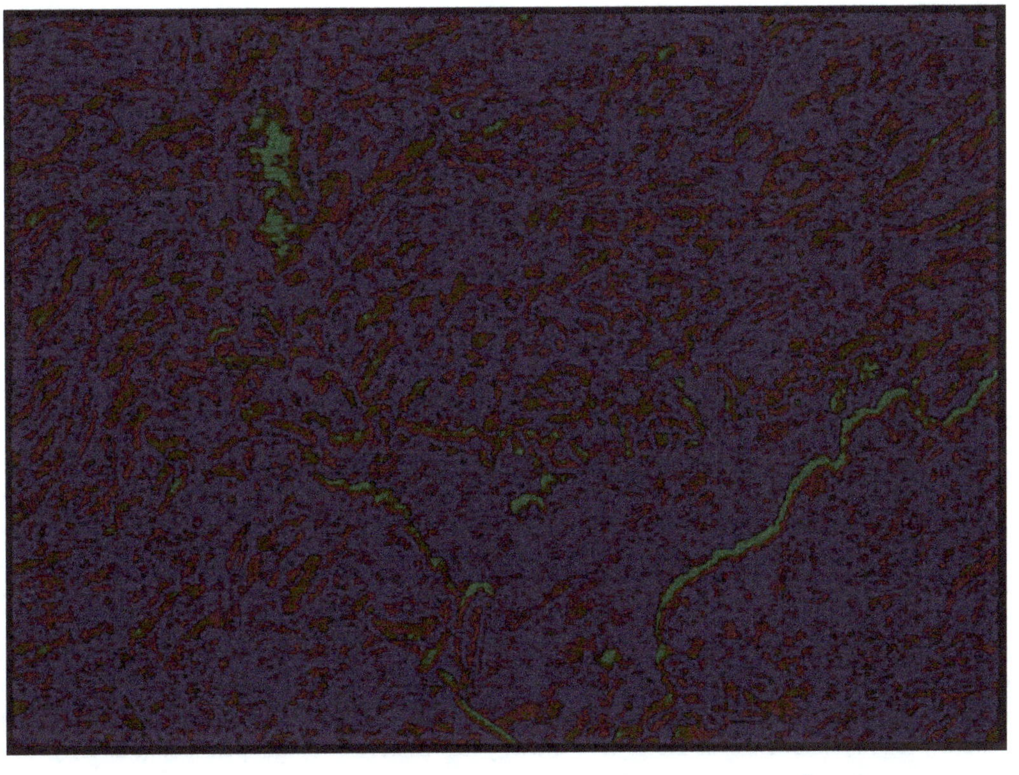

**Fig. 41.11** Outcome of texture classification based on variograms applied to the Landsat image shown in Fig. 41.10. Colors represent: water (red), playa (green), alluvium (blue), and sedimentary outcrops (yellow)

## References

Albarède F (1995) Introduction to geochemical modeling. Cambridge University Press

Benzècri JP (1973) L'analyse des donnèes. Tome 1: La taxinomie. Tome 2: L'analyse des correspondances. Dunod, Paris

Carr JR (2002) Data visualization in the geosciences. Prentice-Hall

Carr JR, Miranda FP (1998) The semivariogram in comparison to the co-occurrence matrix for classification of image texture. IEEE Trans Geosci Rem Sens 36(6):1945–1952

Eckart C, Young G (1936) The approximation of one matrix by another of lower rank. Psychometrika 1:211–218

Greenacre MJ (1984) Theory and applications of correspondence analysis. Academic Press

Hotelling H (1933) Analysis of a complex of statistical variables into principal components. J Educ Psychol 24(417–441):498–520

Howarth RJ (ed) (1983) Statistics and data analysis in geochemical prospecting. Elsevier

Imbrie J, Purdy EG (1962) Classification of modern Bahamian carbonate sediments. In: AAPG classification of carbonate rocks, a symposium, Mem 1, pp 253–272

Klovan JE (1966) The use of factor analysis in determining depositional environments from grain-size distributions. J Sediment Pet 36:115–125

Miranda FP, Fonseca L, Carr JR (1998) Semivariogram textural classification of JERS-1 (Fuyo-1) SAR data obtained over a flooded area of the Amazon rainforest. Int J Remote Sens 19(3): 549–556

Young G (1937) Matrix approximations and sub space fitting. Psychometrika 2:21–25

# Chapter 42
# Linear Unmixing in the Geologic Sciences: More Than A Half-Century of Progress

**William E. Full**

**Abstract** For more than a half-century, scientists have been developing a tool for linear unmixing utilizing collections of algorithms and computer programs that is appropriate for many types of data commonly encountered in the geologic and other science disciplines. Applications include the analysis of particle size data, Fourier shape coefficients and related spectrum, biologic morphology and fossil assemblage information, environmental data, petrographic image analysis, unmixing igneous and metamorphic petrographic variable and the unmixing and determination of oil sources, to name a few. Each of these studies used algorithms that were designed to use data whose row sums are constant. Non-constant sum data comprise what is a larger set of data that permeates many of our sciences. Many times, these data can be modeled as mixtures even though the row sums do not sum to the same value for all samples in the data. This occurs when different quantities of one or more end-member are present in the data. Use of the constant sum approach for these data can produce confusing and inaccurate results especially when the end-members need to be defined away from the data cloud. The approach to deal with these non-constant sum data is defined and called Hyperplanar Vector Analysis (HVA). Without abandoning over 50 years of experience, HVA merges the concepts developed over this time and extends the linear unmixing approach to more types of data. The basis for this development involves a translation and rotation of the raw data that conserves information (variability). It will also be shown that HVA is a more appropriate name for both the previous constant sum algorithms and future programs algorithms as well.

W. E. Full (✉)
GXStat, LLC, 1321 Farmstead, Wichita, KS 67208, USA
e-mail: bill@GXStat.com; BillFull@cox.net



B. S. Daya Sagar et al. (eds.), *Handbook of Mathematical Geosciences*,
https://doi.org/10.1007/978-3-319-78999-6_42

## 42.1 Introduction

Unmixing algorithms and programs have been used to solve many different types of geologic problems for more than 50 years. This approach has been developed by geologists for geologists and has been recently 'borrowed' by professionals in other fields. For the most part, the International Association for Mathematical Geosciences' publications Journal of Mathematical Geology (later renamed Journal of Mathematical Geosciences) and Computers & Geosciences have been the venue for the papers describing the developments and computer codes associated with the approaches described in this report. The history of linear unmixing tied to these papers is the topic of this manuscript along with extending the mathematics to make this approach more appropriate for more common types of geologic and petroleum data. The most recent name for these algorithms is Hyperplanar Vector Analysis (HVA)—a name that will be shown to be more appropriate than the other algorithms/program names that have been used in the past.

## 42.2 History of Constant Sum HVA

### *42.2.1 Determination of the Number of End-Members*

The rudiments of HVA started with a report to the Office of Naval Research by Imbrie (1963). In this report, the application of the cosine-theta similarity matrix was defined for the Q-mode factor analysis portions of HVA that were to follow. The cosine is used as a similarity index between two samples (Fig. 42.1a). When the angle between two samples approaches 0.0 (cosine approaches 1.0), the ratio of the two variables are assumed to nearly the same. Conversely, when a cosine approaches 0.0 ($\Theta = \pi/2$ radians), the two samples are considered very different from each other. In statistics, a cosine value of 0.0 would consider the two samples to be independent of each other. While the Imbrie (1963) approach never calculated a cosine function, it did accomplish the same thing by working with the unit vectors of each sample and with the unit sphere defined by these vectors which was subsequently rotated via an eigenvector rotation. The resulting matrix is the cosine-theta matrix defined for all the samples. Figure 42.1b shows the case where two vectors of differing length would produce a cosine $\Theta$ that would indicate that the two vectors would be the same as two vectors of exactly the same length. The constant sum approach assumes that the raw data represents vectors of equal length.

Working with vectors on the unit sphere is one of the fundamental differences between what we have been calling vector analysis and traditional factor analysis. Figure 42.2a illustrates the concept of a unit vector while Fig. 42.2b shows a cross-section of the unit sphere in two dimensions. In traditional factor analysis, in simplified terms, before the eigenvector rotation is performed, the mean of either the raw data or transformed data (usually the z-transform) is subtracted from the

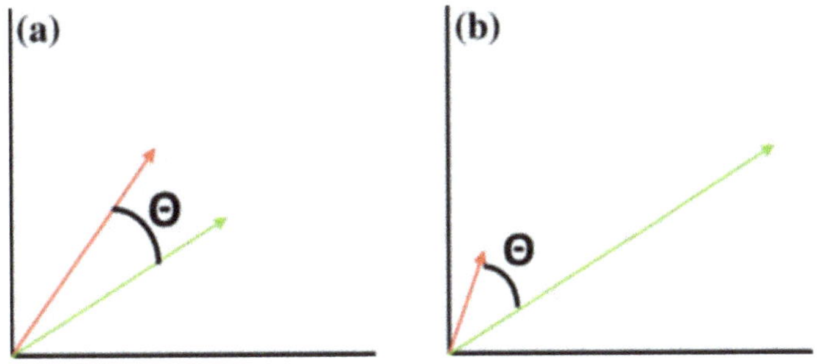


**Fig. 42.1** Example of the cosine as a measure of similarity where two samples are very similar to each other in terms of the ratio of the defining variables (**a**), and where the two samples are more dissimilar than the previous two samples (**b**). With constant sum models, both set of vectors would be considered as essentially the same

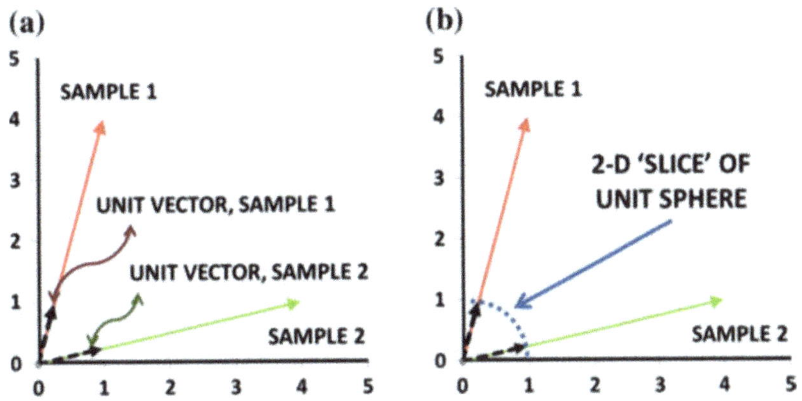


**Fig. 42.2** Every sample (row of data) can be considered a vector. The unit vector is the direction of this vector where the length of the unit vector is exactly 1.0 (**a**). The collections of the sample unit vectors are located on the unit sphere whose radius is 1.0 (**b**)

variance (or covariance matrix). This step in the procedure is a translation of the axes defining the system (Fig. 42.3). Figure 42.3 also shows in 2-dimensions that the use of the cosine-theta similarity approach does ultimately define eigenvectors and eigenvalues relative to the center of the unit sphere. It should be pointed out that using the approach of Imbrie (1963), the total variability (sum of squares of each coordinate in the space defined by the unit sphere) before and after the eigenvector rotation is simply the number of samples (N). If we have 45 samples, we will have variability in the unit sphere of 45.0. A FORTRAN-IV computer program to perform this procedure was published by Klovan and Imbrie (1971) and was named CABFAC (Columbia and Brown Factor Analysis). Unfortunately for a generation of students and practitioners, the terminology used in this and several of the subsequent programs was rooted in factor analysis.

The next step in the evolution of HVA was taken by Miesch (1976a, b). Miesch realized that the CABFAC program was really a combination of linear algebra and geometry. The eigenvector rotation defined by the previous authors was actually

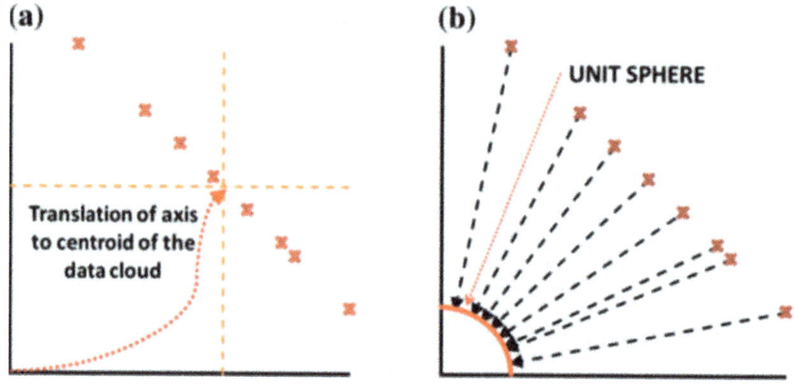


**Fig. 42.3** In traditional PCA or factor analysis, the subtraction of the mean is performed before the eigenvector rotation and is a translation of the axes to the center of the data (**a**). Of course, in a standard PCA or factor analysis, we would divide each value by the standard deviation of the corresponding variable. In contrast, the Q-Mode analysis defined by Imbrie (1963) defines the center of the unit sphere as the point of reference for the eigenvalue rotation (**b**)

capturing the geometry of the data on the unit sphere. This fact, in conjunction with the observation that with constant sum data the raw samples must fall on either a line (2-D), plane (3-D) or hyperplane (n-D), was a fundamental concept for Miesch. This was a different viewpoint about constant sum data than that reported by Chayes (1971). Miesch concluded that CABFAC can be used to tell us the real dimensionality of the data (must be less than or equal to the number of variables) and that with some additional programming, the end-members and relationships between these end-members and each sample (proportions) can be defined. Programs were created and published by Klovan and Miesch (1976) called EXTENDED CABFAC and QMODEL. These two programs, while still using the standard terminology of factor analysis, represented the foundation of the vector analysis unmixing approach that is used to this day. As a matter of fact, rotation procedures such as the orthogonal VARIMAX rotation (Kaiser 1958) are still performed in the programs.

Before we continue with the QMODEL evolution, a discussion of the ways that EXTENDED CABFAC helps us determine the number of appropriate dimensions to choose which is, in reality, the number of end-members present in the data. CABFAC presents us with several ways of defining the exact number or range of end-members that may be present in the data. Note that CABFAC does not tell us anything about what they look like—or the proportions relating these end-members to each sample. For the sake of this discussion, a data set was created wherein four end-members were mixed in known proportions. While the end-members were not constant sum (the sum of each end-member was not the same value), the collection of these data can still be informative, especially when we discuss non-constant sum analysis. The four end-members were taken from NURE stream sediment geochemical samples (Smith 1997) and this data set. For this section on constant sum algorithms, each sample in the data was transformed to a constant value of 1.0 before being submitted to CABFAC/SAWVEC/VECTOR/PVA routines.

The traditional approach used in the past is the scree plot (Fig. 42.4a). In this plot, the user looks for a break in the slope and then interprets this point as the maximum number of end-members present in the data. Note that like real data, Fig. 42.4a shows a case where the scree plot need not behave in an ideal sense. Miesch (1976a, b) recognized that since we are looking at how well the constant sum plane or hyperplane ‘fits’ the original data, back-calculated values from a **reduced** space defined by fewer than n eigenvectors can be directly compared to the variables defined in the raw data or **real** space. This back-calculation simply reverses the mathematics using a reduced number of eigenvectors ‘back’ into the raw data metric via matrix algebra. The comparison is made via the coefficient of determination (CD) function (Draper and Smith 2014) and the CD for each back-calculated variable to the original raw data for a given number of retained eigenvectors is plotted (Fig. 42.4b). Similarly, for each sample, total amount of original variability retained for a given number of eigenvectors is also calculated. This ratio is called the communality for a given sample and is the amount of variability retained by the reduced space divided by the total variability represented by that sample in real space. Figure 42.4c presents a few communality trends for arbitrary samples picked from the test data set. The collection of communalities for a given number of retained eigenvectors can be scanned to look for anomalies that may represent problematic data or the collection can be binned and plotted to assess the range of problems. In the past, a general ‘rule of thumb’ was that, scanning the columns of orthogonal coordinates (loadings) from the fewest to the highest number of end-members, the first time that approximately 5% or less of the data had communalities less than 0.99 and the coordinates had values less than 0.5, then that number of end-members was near the upper range for the maximum number of end-members. The reality was that lower communalities might be due to noise, measurement error, recording error, or it might be the hint of an additional end-member(s) which generally meant it could be more difficult for the modeling programs to define. Johnson (1997a, b) used the insight that by looking at plots of the back-calculated variables to the raw variables, further insights can be gleaned especially by those that want to visualize the ‘pile’ of numbers described earlier. Figure 42.4d displays some of those plots for a single variable. These plots have been called Johnson plots in the programs described later in this report.

Finally, if the assumption is that what is not included is in fact noise, there might not be enough information available that can be used to define any additional end-members. In such a case, the distribution of the variability relative to each ‘removed’ eigenvector can be examined. This is usually done by looking either at the ‘coordinates’ of the removed eigenvectors (similar to looking at the principal component loadings in Principal Components Analysis) and using external tools such as JMP Pro (1989–2017). The latest programs create appropriate data tables for this step, and for all of the previous steps with key information, that can be used in ancillary programs that have many more statistical functions and better graphics. One such example might be to examine the behavior of the ‘removed’ eigenvector coordinates to verify that the ‘removed’ eigenvectors do not contain meaningful information (i.e. whether they can be considered noise and not pertinent to the

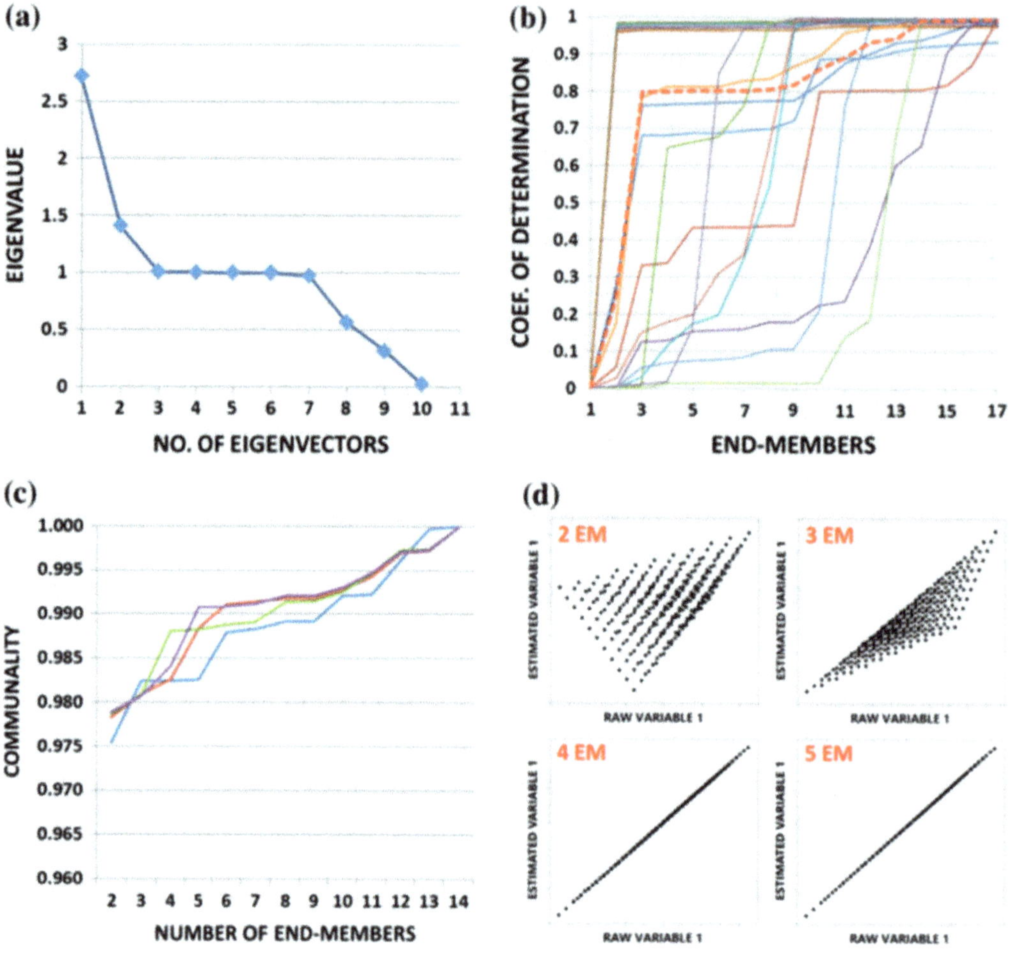


**Fig. 42.4** An example of the scree plot from the test data where the number of eigenvectors retained are plotted against individual eigenvalues (**a**). A plot of the CD's for the test data shows how each variable contributes to the overall choice of the number of end-members (**b**). Communalities for four samples are presented for the range of eigenvectors retained (**c**). Collection of Johnson plots showing the visual fit relative to a single variable as the number of end-members (EM) has increased (**d**)

overall model). The user would have to a priori establish criteria that defined noise in terms of the individual data used and/or by some distribution parameters such defined by mean and standard deviation, for instance.

### *42.2.2 Determination of the Composition of the End-Members and Proportions*

Klovan and Miesch (1976) developed the program QMODEL based on Miesch (1976a) in order to define the composition of the end-members and calculate the proportions relating each individual sample to this set of end-members. Given the choice of the number of end-members normally based on EXTENDED CABFAC, the procedure to define the compositions and proportions (oblique coordinates of the space defined by the end-member axes) is strictly linear algebra. The mathematics used up to this point is well defined in Miesch (1976a). QMODEL was designed to be a data modeling program that required interaction with the user. A discussion of these approaches and other alternatives can be found in Clarke (1978). There were several ways for this program to define end-members:

(1) Use the retained eigenvectors as end-members (principal factors)
(2) Use the VARIMAX axes as the end-members (VARIMAX factors)
(3) Use Imbrie's oblique end-members (the extremes in the reduced space—EXNORC routine)
(4) Use the extremes as defined by the back-calculated extremes in the raw space—the EXRAWC routine
(5) Define the end-members by the row indices of the set of samples (e.g. use the 5th and 12th sample as end-members)
(6) Define the actual composition of each of the end-members (these would normally be a set of end-members defined in the raw metric that the user would want to test)
(7) Externally define the end-members by defining the VARIMAX coordinates (loadings)—this would normally be done when the user has made multiple plots of the data in VARIMAX space

For each of the choices in the original QMODEL program, correct choices produced end-members that were realistic (defined by acceptable variables in the raw data space) and by proportions that were between 0.0 and 1.0. Problems arose with many data sets when the raw end-member compositions were unrealistic and/or the proportions were out of range. This problem is commonly encountered when there are many variables and samples which makes visualization of the location of the potential end-members difficult at best. To that end, new modeling approaches were devised that gave some automation toward the definition of proper end-members and proportions.

Full et al. (1981, 1982) devised two alternative methods that involves an iterative scheme that started with one of the original QMODEL choices above or with fuzzy cluster centers (Bezdek et al. 1984), and then allowed the program to define end-members external to the data, check their proportions for viability, change if needed the set of end-member compositions to the nearest viable location, and repeat the process until either the program shows no convergence or an acceptable solution is reached. The goal was to determine appropriate sets of end-members closest to the data cloud defined by the samples. This may be likened to trying to minimize the area or hyper-area that represents the planar/hyperplanar convex hull defined by the end-members. The computer code, along with some bug fixes to the EXRAWC and EXNORC subroutines, can be found in the appendix of Full (1981). A general discussion of these methods and their applications at the time can be found in Ehrlich and Full (1988). Alternatives to the aforementioned approaches can be found in Leinen and Pisias (1984) and Weltje (1997). Insights into the appropriate applications of these algorithms and recognizing how to detect problems with the underlying model were discussed in Williams et al. (1988a, b, Chaps. 15 and 19). Optimized data binning for continuous distributions that improved the results of these algorithms were presented in Full et al. (1984).

### *42.2.3 The Renaming to Polytopic Vector Analysis*

In the early 1980s, given the changes to the original CABFAC and QMODEL programs, the approach was renamed SAWVEC (South Carolina and Wichita Vector Analysis) and sometimes simply VECTOR. It was the recognition that the algorithms were dominated by vector algebra that prompted the name change. Circa 1990, the exact same approach was further renamed Polytopic Vector Analysis and applied under that name in Evans et al. (1992) and in many of the references mentioned in later in this report. Around this time, Sterling James Crabtree, then at the University of South Carolina, translated the FORTRAN IV code of Full (1981) into the C programming language and developed a Windows interface and ultimately called the program PVA. This program can be recognized by the fact that the first step after starting the program was to resize the introductory window.

The use of the term polytope has been problematic for this author even though the term was used in the original Full et al. (1981) algorithm. The field of polytopic mathematics has been around for over a century and was generally formulized by Coxeter (1948, 1973). Coxeter assumed that a polytope was a geometric construct in 4 or more dimensions with the degenerate cases being the point in 0 dimensions, the line segment in 1 dimension, the polygon in 2 dimensions and polyhedron in 3-dimensions representing polytopes of dimension 0, 1, 2 and 3 respectively. A search of the literature on polytopes shows that this field of mathematics is rich in various definitions of a polytope, depending for instance on whether you are talking about a convex hull in n-dimensions or more complex surfaces as in star-type polytopes. It is clear that for the geologist this can be a confusing landscape to travel through. A simplistic definition would be that a polytope is an n-dimensional geometric figure ($n > 3$) whose sides are planes or hyperplanes. The implicit assumption is that a polytope has some kind of volume or hypervolume. Henk et al. (1997) even developed equations for calculating this volume or hypervolume for many types of regular polytopes.

If a polytope can be considered as a region of n-dimensional space that is enclosed by hyperplanes (Coxeter 1973), then that causes problems for linear unmixing. If we consider a vector emanating from a point outside that region and look at the potential intersections of that vector with the polytope, the only possibilities for unique points would be if the vector intersected the vertices of the polytope. If the vector intersected a side, there could possibly be two or more points of intersections which would cause havoc with the uniqueness aspects of the unmixing model. The reality is that in the non-constant sum model, regardless of the number of dimensions (end-members), the data fall on a hyperplane when the number of dimensions is greater than 3. As we will see later, it is this fact that the extension of all of the previous algorithms to non-constant sum data can be realized. Because of the confusion associated with the term 'polytope' relative to the understanding of the previously described algorithms, they have been renamed Hyperplanar Vector Analysis (HVA).

### 42.2.4 Review of the Applications of Constant Sum Unmixing

The CABFAC, EXTENDED CABFAC-EXTENDED QMODEL, SAWVEC, VECTOR, PVA algorithms and programs (henceforth referred to as HVA family of algorithms) have found application in many geologic disciplines. Some of the earliest studies have involved the analysis of size data in both nearshore and lacustrine environments. These include the work of Klovan (1966) and Solohub and Klovan (1970) using traditional sieved size data. Fillon and Full (1984) used specialized equipment to define the size of particles on an individual basis and defined 5 different sources of deep sea sediment. As pointed out in Fillon and Full (1984) and Full et al. (1984), the success or failure of size analysis depends on the optimization of the size data using transforms such as the maximum entropy method.

In the field of grain shape analysis, the heart of the analytic scheme was the constant sum unmixing algorithms described above. The studies included sediment from Monterey Bay, CA (Porter et al. 1979). Brown et al. (1980), Reister et al. (1982), Mazzullo et al. (1982, 1984), Hudson and Ehrlich (1980), Smith et al. (1985), Tortora et al. (1986) and Evangelista et al. (1986, 1994, 1996) looked at sediment distributions along beaches, barrier islands, shelf and abyssal plains. Murillo-Jiménez et al. (2007) examined the sediment from a relatively large region along the southern coast of Baha California, MX. Material from more lithified material was studied by Mazzullo and Ehrlich (1980, 1983) and Civitelli et al. (1992). El-Awawdeh and Full (1996) looked at changes in key morphology in Florida Bay over time. The methods used in those studies were reviewed in Ehrlich and Full (1984a, b) and Zhao et al. (2004).

The biologic morphology and fossil assemblage scientists were early adapters of the HVA family of algorithms. Healy-Williams (1983, 1984) and Healy-Williams et al. (1997) worked with forams, Burke et al. (1986) with ostracodes and Kensington and Full (1994) with scallops. Williams et al. (1988a, b) looked at correlations of foram shapes with isotopic signatures. Assemblages of microfossils were unmixed in Gary et al. (2005) and Zellers and Gary (2007).

A major area of investigation using the HVA family of algorithms deals with environmental science. Detecting contaminates in soils and identifying their sources was reported by Ehrlich et al. (1994), Wenning and Erickson (1994), Doré et al.

(1996), Jarman et al. (1997), Johnson (1997a, b), Huntley et al. (1998), Bright et al. (1999), Johnson et al. (2000, 2001), Johnson and Quensen (2000), Nash and Johnson (2002), Nash et al. (2004), Barabas et al. (2004a, b), Magar et al. 2005, DeCaprio et al. (2005), Towey et al. (2012), Leather et al. (2012) and Megson et al. (2014). The Battelle Memorial Institute (2012) has listed PVA in their handbook for determining the sources of PCB in sediments.

The HVA family of algorithms is critical for the field of PIA (Petrographic Image Analysis). The literature includes Ehrlich and Horkowitz (1984), Ehrlich et al. (1984, 1991a, b, 1996, 1997), Ross et al. (1986), Scheffe and Full (1986), Full (1987), Etris et al. (1988), McCreesh et al. (1991), Ross and Ehrlich (1991), Ferm et al. (1993), Bowers et al. (1994, 1995), James (1995), Carr et al. (1996), Yannick et al. (1996), Anguy et al. (1999, 2002) and Sophie et al. (1999).

Igneous rock researchers have also been an adapter of these unmixing algorithms. These include Horkowitz et al. (1989), Stattegger and Morton (1992), Tefend et al. (2007), Vogel et al. (2008), Deering et al. (2008), Barclay et al. (2010), Szymanski et al. (2013), Lisowiec et al. (2015) and most recently by Blum-Oeste and Wörner (2016).

The unmixing of sources of oil using the HVA algorithms has been reported by Collister et al. (2004), Van de Wetering et al. (2015), Abrams et al. (2016) and Mudge (2016). The correlation between stratigraphy and chemical stratigraphic data was explored by McKenna et al. (1988). "Quasigeostopic potential vorticity" was explored in Evans et al. (1992). Mason and Ehrlich (1995) looked at aspects of well logs for basin exploration (1995). Full and James (2015) used the HVA (non-constant sum version) to decompose a large data set consisting of exploration data in order to better assess exploration and exploitation risk. At least two patents have mentioned using the HVA family of algorithms for analysis of the data derived from their process (Shafer and Ehrlich 1986; Nelson et al. 2013).

The above literature is by no-means the entire community of users of the unmixing approach began by Imbrie (1963). There have been verbal reports of researchers doing work with Shakespeare's plays, classifying business reports, analyzing social data and even applying these approached to marketing data. The success or failure of these studies cannot be directly ascertained, but represent some interesting applications.

## 42.3 Non-constant Sum Data and Algorithms

The previous sections, for the most part, dealt with rows of data whose row sum was the same or very similar for each sample (vector). This type of data is merely a subset of the data commonly encountered in the geologic sciences and, if you want to use the previous algorithms, you have to potentially degrade your data by transforming it to percentages or some other appropriate singular value. Oftentimes, this involves removing the absolute quantity involved with each sample. For example, if you have six glasses and pour into each glass a variable amount of three

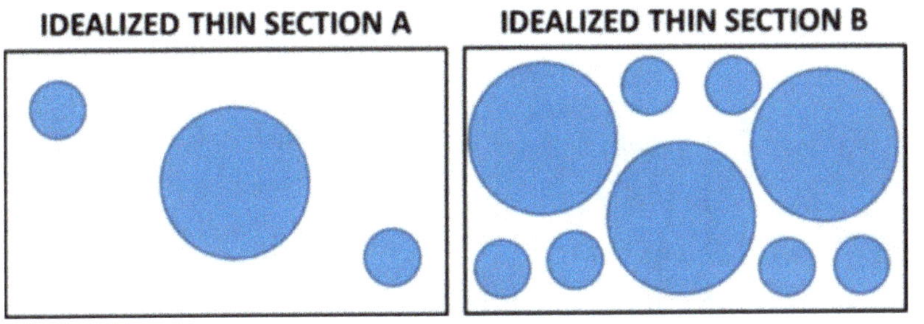


**Fig. 42.5** An example of two idealized images that would produce the same smooth-rough distributions in the petrographic image analysis system described in Ehrlich (1991a, b). Note that in image **a**, the porosity would be much greater than image **b** which would greatly affect the calculation of permeability and other petrophysical variables

solutions, some glasses might contain a greater volume and some a lesser volume—here the quantity of each solution might be important. The concept of unmixing might still be appropriate but would only be accurately defined in terms of end-member compositions and sample proportions in very special cases that will be discussed below. With petrographic image analysis which heavily uses the unmixing algorithms, two collections of imaged thin sections with vastly different porosities would ultimately have equal constant sum smooth-rough distributions (Fig. 42.5). Petrophysical logs, formation depths, seismic parameters and other petroleum related data are mostly non-constant sum in nature. There are many other types of data where the concept of mixtures and unmixing can be validly applied.

What happens when you try to apply the constant sum programs to inherently non-constant sum data? This topic was partially addressed by Klovan (1981) without addressing the application of determining end-members and proportions using the techniques described by Full et al. (1981, 1984). In his paper, Klovan notes that, if the data can be approximated by a plane or hyperplane parallel to the constant sum plane, then the aforementioned algorithms can be appropriately applied. However, Klovan (1981) acknowledges problems when the surface defined by the non-constant sum data is not parallel to the unit constant sum plane. Some of the problems can be demonstrated by a simple diagram in two dimensions (Fig. 42.6). Note that the midpoint of the non-constant sum segment does not correspond to the midpoint of the constant sum plane which would be the proportions reported for this point by the computer codes. Using some of the usual functions to create constant sum data that are available in the program would not help matters. A more complex series of transformations using trigonometry could be easily developed for 2 or 3 dimensions but would be difficult to visualize and cannot be easily generalized to n dimensions. Also note that Fig. 42.6 represents an example in two dimensions which intersects the two axes making the determination of end-member compositions a bit easier; they would be represented by the end-points of each line and whose compositions would be the raw data points defining these end-points. If end-members needed to be defined beyond the data

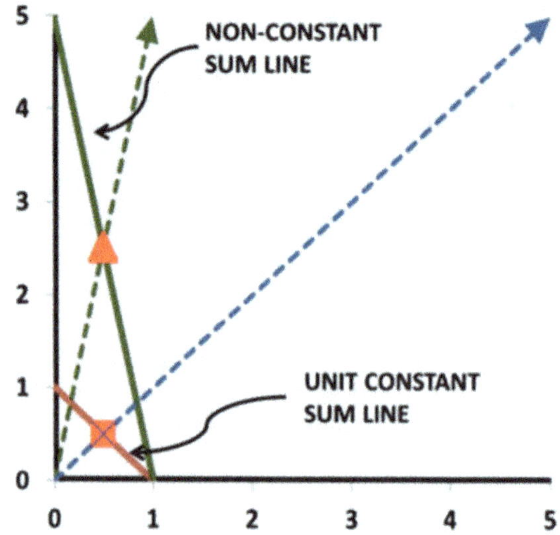


**Fig. 42.6** A simplistic example of some of the issues associated with using constant sum algorithms with non-constant sum data. The unit constant sum line is represented by the solid line passing through the points (1, 0) and (0, 1). The non-constant sum data is represented by the solid line at an oblique angle to the constant sum plane. The mid-points (0.5, 0.5) proportion of each line is represented with a symbol. Note that the extended unit vector (represented by the dashed line) that represents the midpoint of the constant sum system is divergent from the same unit vector that passes through the mid-point of the non-constant sum line segment

cloud, the definition of the end-member compositions would be very difficult when there are more than 3 dimensions.

How to deal with the non-constant sum problem was solved in the mid-1980s and has been used in petroleum industry projects and for research projects for the Department of Defense. The code was initially run on a 386-processor with 387-co-processor as well as IBM mainframes. It is only recently that the computer code has been written for Windows operating system with a Windows GUI. The abstract concept behind the approach to dealing with this type of data is to recognize that ultimately any mixing problem deals with data on either a line segment (in 2-d), a plane (2 or 3-d) or hyperplane in more than 3 dimensions. The goal then is to define that hyperplane and translate/rotate the data to a plane/hyperplane that is parallel to the unit constant sum plane where we can apply the usual constant sum approaches. Afterward, any time we want to know what the raw compositions are, we reverse the translation/rotation to bring us back into the original metric. In this way, the earlier approaches are not abandoned but can be efficiently extended to almost any other data that can be modeled as a mixture.

The procedure for this translation/rotation is the following:

(1) Remove the mean from the data. This is equivalent to the first step of principal components (Davis 2002; Draper and Smith 2014). The visualization for this step is that the axes defining the raw data are translated to the mean of the data with no loss of information.

(2) An eigenvector rotation is performed on this data. If we were to divide the variable standard deviation by each corresponding row of the matrix defined from the previous step above before this eigenvector rotation, we would have a standard principal component analysis. Since we have not done so, we have not altered the absolute position of the raw data in the data cloud nor the variance associated with the raw data—no loss of information. It should be noted that this step of the analysis is performed by the SVD computer algorithm (Golub and Reinsch 1970) programmed to use quad precision (128 bit) to minimize any information loss and to be able to run large raw data matrices. The rest of the HVA program currently runs in double precision.

(3) Create a new matrix G with the following definition:
Letting ANV = 1/NV where NV is the number of variables and ANX = SQRT (1 − ANV), then G is defined as an NV × NV matrix with every element −ANV/ ANX except along the main diagonal where the element is (1 − ANV)/ANX. Note that the sum of squares of each row element is 1 and each of the elements is orthogonal and represents spanning vectors for the constant sum plane.

(4) Using the Gram-Smith orthogonalization procedure (Cheney and Kincaid 2008), orthogonalize the matrix defined in the previous step. Call this matrix $G^0$.

(5) Create a new matrix G* where $G^* = G^0 * B$ where B is the set of previously defined eigenvectors in step 2. Note that since G* is an orthogonal matrix, then $G^{*-1} = G^{*T}$ where T is the notation for transpose (this fact is well known in mathematics: see for example Schwartz 2011). G* and $G^{*T}$ gives us the mechanism to go from the raw data space to a plane parallel to a constant sum plane. However, since this new reference system also contains the origin, the addition of a constant value will translate the plane/hyperplane away from this origin by a constant value to a position parallel to the constant sum plane/ hyperplane. In the program, this constant value is called AVAR and, based on experience, has been set to 2 * NV * (smallest value of the G* rotated coordinates) or 1.0 if this number is lower than 1.0.

In more simplistic terms, what we have done is to create an NV x NV matrix (NV = the number of variables) that will be used to rotate the raw data in order to create a one-to-one correspondence with a set of points in a plane/hyperplane parallel to a constant sum plane/hyperplane. This matrix was orthogonalized and the application of this rotation and translation results in the loss of no information. Since this is an orthogonal matrix, the transpose of this matrix is the inverse of the matrix and gives us the function to go from the constant sum hyperplane to the raw data. These functions allow for properly defined proportions and end-member compositions whether the end-members are contained in the data or not. Figure 42.7 illustrates what the procedure is doing in general.

The constant sum routines can then be applied as they were before only using the G* and $G^{*T}$ matrix defined above to move from the raw data hyperplane to the constant sum hyperplane with no (or minimum loss due to computational error) loss of information. This approach capitalizes on more than a half-century of previous

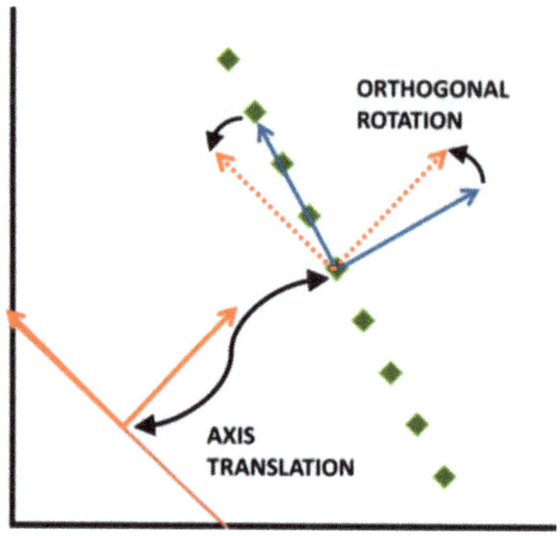


**Fig. 42.7** A 2-dimensional representation of the procedure to define the G* matrix procedure described in the text. Note that in 2-dimensions, the first eigenvector defines the direction of the line segment and the second the normal to this segment. The red axes represent the first eigenvector and the normal to the constant sum line. These axes are then translated to the mean of the non-constant sum data cloud defined by the green diamonds. The blue axes represent the first eigenvector and the normal to the non-constant sum line. This set of axes will be orthogonally rotated to the position of the constant sum axes (dotted axes), (i.e., the raw data will be defined by a new set of coordinates). Mathematically, this procedure will not result in information loss

algorithmic and programming experience. Furthermore, the appropriateness of the unmixing model in non-constant sum space can be checked by looking at the set of eigenvalues—data that do not fall on the mixing hyperplane will have a value other than 0.0 for the last eigenvalue. Additionally, by checking the raw data on a sample-to-sample basis with its equivalent location in the constant sum hyperplane via a similar function to the communality will allow the user to examine potentially aberrant data.

As a demonstration sample, using the previously defined test data set, we can compare the end-members and proportions when they are subjected to a constant sum approach (data was transformed to 100%) and the non-constant sum approach. The set of end-members are shown in Table 42.1 and randomly selected proportions for 10 of the original 296 samples are tabulated in Table 42.2. This data set will be made available from the GXStat website (www.GXSTat.com). Note that these data contained the end-members as samples and therefore no iterative schemes such as those described in Full et al. (1981, 1984) were used. It should be noted that, for the most part, the end-members are not that extreme compared to potential test end-members that could have been chosen. Mathematically, this is saying that, with the test data used in this example, most of the variables in the mixing hyperplane lie in portions of that hyperplane which can be modeled as constant sum (i.e. take away the handful of variables that lie in a section of the hyperplane that is most oblique to the constant sum plane, and the data might be able to be modeled using the constant sum algorithm). In the more common case

**Table 42.1** Test data set end-members (TEST EM) with constant sum end-members (CS EM) and the HVA non-constant sum end-members (HVA EM). Note the disparity between the constant sum end-members (central gray area) and the actual model end-members (white area on right of table). Also note how close these actual model end-members and the HVA end-members reflect each other. The variables represent parts per million data of 20 different elements

| VAR. # | TEST EM 1 | TEST EM 2 | TEST EM 3 | TEST EM 4 | CS EM 1 | CS EM 2 | CS EM 3 | CS EM 4 | HVA EM 1 | HVA EM 2 | HVA EM 3 | HVA EM 4 |
|---|---|---|---|---|---|---|---|---|---|---|---|---|
| 1 | 184.0 | 171.0 | 204.0 | 197.0 | 17.1 | 27.8 | 28.8 | 26.1 | 184.0 | 171.0 | 204.0 | 197.0 |
| 2 | 0.6 | 0.3 | 0.5 | 0.5 | 0.1 | 0.1 | 0.1 | 0.0 | 0.6 | 0.3 | 0.5 | 0.5 |
| 3 | 1.9 | 6.4 | 11.0 | 4.0 | 0.3 | 0.1 | 1.9 | 1.0 | 1.9 | 6.4 | 11.0 | 4.0 |
| 4 | 710.0 | 320.0 | 280.0 | 340.0 | 66.1 | 47.5 | 39.6 | 49.0 | 710.2 | 320.5 | 280.2 | 340.1 |
| 5 | 4.0 | 5.0 | 2.0 | 16.0 | 0.2 | 3.5 | -0.5 | 0.6 | 4.0 | 5.0 | 2.0 | 16.0 |
| 6 | 29.0 | 29.0 | 32.0 | 27.0 | 2.7 | 3.3 | 4.6 | 4.8 | 29.0 | 29.0 | 32.0 | 27.0 |
| 7 | 2.2 | 9.0 | 8.0 | 5.8 | 0.2 | 0.6 | 1.0 | 1.8 | 2.2 | 9.0 | 8.0 | 5.8 |
| 8 | 17.0 | 22.0 | 35.0 | 26.0 | 1.7 | 3.3 | 5.4 | 3.0 | 17.0 | 22.0 | 35.0 | 26.0 |
| 9 | 1.4 | 0.7 | 1.9 | 1.0 | 0.2 | 0.1 | 0.3 | 0.0 | 1.4 | 0.7 | 1.9 | 1.0 |
| 10 | 3.0 | 7.0 | 14.0 | 8.0 | 0.4 | 0.9 | 2.3 | 0.9 | 3.0 | 7.0 | 14.0 | 8.0 |
| 11 | 0.5 | 0.4 | 0.6 | 0.5 | 0.0 | 0.1 | 0.1 | 0.1 | 0.5 | 0.4 | 0.6 | 0.5 |
| 12 | 6.1 | 3.0 | 6.2 | 7.0 | 0.6 | 1.2 | 0.9 | 0.2 | 6.1 | 3.0 | 6.2 | 7.0 |
| 13 | 9.0 | 2.6 | 2.8 | 3.7 | 0.9 | 0.6 | 0.4 | 0.3 | 9.0 | 2.6 | 2.8 | 3.7 |
| 14 | 0.1 | 0.1 | 0.0 | 0.0 | 0.0 | 0.0 | 0.0 | 0.0 | 0.1 | 0.1 | 0.0 | 0.0 |
| 15 | 11.0 | 12.0 | 19.0 | 14.0 | 1.1 | 1.8 | 3.0 | 1.6 | 11.0 | 12.0 | 19.0 | 14.0 |
| 16 | 0.2 | 0.2 | 0.2 | 0.1 | 0.0 | 0.0 | 0.0 | 0.0 | 0.2 | 0.2 | 0.2 | 0.1 |
| 17 | 4.0 | 12.0 | 11.0 | 11.0 | 0.3 | 1.5 | 1.3 | 2.2 | 4.0 | 12.0 | 11.0 | 11.0 |
| 18 | 4.0 | 24.0 | 14.0 | 12.0 | 0.1 | 1.1 | 1.4 | 5.3 | 4.0 | 24.0 | 14.0 | 12.0 |
| 19 | 80.0 | 30.0 | 50.0 | 45.0 | 7.8 | 6.2 | 8.3 | 2.8 | 80.0 | 30.0 | 50.0 | 45.0 |
| 20 | 2.8 | 2.6 | 6.2 | 4.0 | 0.3 | 0.5 | 1.0 | 0.2 | 2.8 | 2.6 | 6.2 | 4.0 |

**Table 42.2** Ten randomly selected samples were picked to show the trends of the proportions from results of the application of the constant sum algorithms and the non-constant sum programs. The a prior proportions for each end-member used to create each test sample is given by the columns ORIG. PROP., the constant sum derived proportions by the grayed columns labeled CS PROP in the center of the table and the results of the non-constant sum application is given by the columns on the right of the table labeled HVA PROP. The average error for the proportions of the HVA results was found to be ±0.0004 which was largely attributed to the fact that the raw data was defined using two decimal-point accuracy. The sample numbers represent the sequence number of the row of the test data set

| SAMP. # | ORIG. PROP. EM 1 | ORIG. PROP. EM 2 | ORIG. PROP. EM 3 | ORIG. PROP. EM 4 | CS PROP. EM 1 | CS PROP. EM 2 | CS PROP. EM 3 | CS PROP. EM 4 | HVA PROP. EM 1 | HVA PROP. EM 2 | HVA PROP. EM 3 | HVA PROP. EM 4 |
|---|---|---|---|---|---|---|---|---|---|---|---|---|
| 14 | 0.70 | 0.00 | 0.00 | 0.30 | 0.78 | 0.00 | 0.00 | 0.22 | 0.70 | 0.00 | 0.00 | 0.30 |
| 32 | 0.60 | 0.10 | 0.20 | 0.10 | 0.70 | 0.07 | 0.15 | 0.08 | 0.60 | 0.10 | 0.20 | 0.10 |
| 81 | 0.40 | 0.10 | 0.40 | 0.10 | 0.51 | 0.08 | 0.33 | 0.08 | 0.40 | 0.10 | 0.40 | 0.10 |
| 95 | 0.30 | 0.60 | 0.10 | 0.00 | 0.41 | 0.50 | 0.09 | 0.00 | 0.30 | 0.60 | 0.10 | 0.00 |
| 107 | 0.30 | 0.10 | 0.30 | 0.30 | 0.40 | 0.08 | 0.26 | 0.26 | 0.30 | 0.10 | 0.30 | 0.30 |
| 121 | 0.20 | 0.40 | 0.30 | 0.10 | 0.28 | 0.34 | 0.28 | 0.09 | 0.20 | 0.40 | 0.30 | 0.10 |
| 154 | 0.20 | 0.20 | 0.00 | 0.60 | 0.28 | 0.17 | 0.00 | 0.55 | 0.20 | 0.20 | 0.00 | 0.60 |
| 166 | 0.20 | 0.10 | 0.40 | 0.30 | 0.28 | 0.08 | 0.36 | 0.28 | 0.20 | 0.10 | 0.40 | 0.30 |
| 208 | 0.10 | 0.20 | 0.00 | 0.70 | 0.15 | 0.18 | 0.67 | 0.00 | 0.10 | 0.20 | 0.00 | 0.70 |
| 252 | 0.00 | 0.50 | 0.30 | 0.20 | 0.00 | 0.48 | 0.31 | 0.21 | 0.00 | 0.50 | 0.30 | 0.20 |

where end-members need be defined external to the data cloud, the results would have been potentially far off and confused if the constant sum algorithm was applied. Also note that if the user did use the constant sum routines to define the composition of the end-members and either manually extracted the raw data of an internal end-member or the 'nearest' actual point (defined by the raw data) to the external end-member, it would be difficult to know how these points relate to all of the other data samples—the user would simply not know if all the data truly fall on a mixing plane or hyperplane. Finally, because HVA rotates the data to a plane parallel to the constant sum plane, when the data are inherently constant sum, no new program is needed.

Finally, it should be noted that this non-constant sum model will work for any mixing system that can be modeled as a plane or hyperplane. The dimensionality of the hyperplane must be less or equal to than the number of variables otherwise there will not be a unique solution to the end-member and proportions problem. This does bring up the case where a three end-member solution (defined by a triangle) in two dimensions can be solved using these algorithms. The G* rotation described above can potentially produce a plane or hyperplane that intersects with the origin defining an end-member consisting of the origin with (0, 0, …) as its composition. The interpretation of the origin as an end-member has been successful in previous studies when this situation has been encountered. It can be, however, a tricky proposition depending on the type of data being analyzed. It might be useful to substitute a value close to the origin for the definition of that end-member instead of using the origin as an end-member composition.

Areas of application of this approach have included chemo-stratigraphic data, correlation and mapping of wireline well logs, unmixing of oil compositions preserving volume of source material, determination of various forms of risk in exploration schema, correlating biologic assemblages to seismic stratigraphy, and determination of 'sweet spot' locations for oil exploitation, to name a few. Unfortunately, the results of these reports remain confidential. It is anticipated that these and new applications will be reported in the future in various literature.

## 42.4 Summary

Fifty years of research and development have given the geologic community a useful tool for the analysis of mixtures. It is anticipated at this time that this approach will last well into the future, especially since the program will be made available to anyone in any field they want. It should be noted however, that there are still untested areas of research in this field. The most appropriate approach for the definition of extreme end-members is still an open discussion. Generally, researchers have been looking at the extremes of the data and not looking so much at the bulk of the data. While much of the variable density of the raw data may be due to localized over-sampling problems (usually, we geologists sometimes just analyze the data we have!), there are other methods such as FUZZY clustering (Full et al. 1984; Bezdek et al. 1984) and algorithms that use FUZZY variables to define data density in terms of sets of point, lines, planes, hyperplanes and various n-dimensional spaces (Bezdek 1981).

Another area that needs some additional work is the definition of new criteria that will allow the various iterative schemes to know when the 'best' solution is achieved, when there might not be a complete convergence. In terms of computer programming, what would be beneficial is to be able to define one or more 'fixed' end-member(s) (the number being less than the original number of chosen end-members) and let the program determine other potentially viable end-members using the DENEG iteration scheme (i.e. one or more end-members want to be fixed in the analysis—the programs have always had ways of externally defining all of the end-members). Additionally, defining how the end-members interact with the modeled environment (such as when a geochemical component reaches a given level and precipitates out of the system) would also be of great use. This has been accomplished in the past by making alterations to the program, recompiling the code and proceeding with the newly built custom program. Being able to run this option without having to recompile would be quite useful. Another item on the wish list would be to convert the program out of FORTRAN IV, although the current program is very fast and FORTRAN has become a versatile programming language. This author acknowledges that there are fewer and fewer people who can program in this language, especially in the Windows environment. A language that has a 'better' future would be of great advantage, especially since the programs and algorithms may be used by a wider audience. Additionally, all of the mathematics needs to be described in one place along with a user manual that describes in detail not only all the options but also the whys and wherefores of particular options. It should be noted that the program has a built-in user manual but does not go into details of the more subtle nuances associated with the algorithms. These missing discussions will be the topic of various discussions available on the GXStat website (www.GXSTat.com). There is even some progress in producing an R version of the program for those who want to incorporate this approach into their projects. This flexibility will be of benefit to a large community of potential practitioners.

Finally, there is something that can be gleaned from the list of references. The access of researchers to the HVA family of algorithms has been somewhat limited by both changes in the computer industry (computer languages and graphic user's interfaces in addition to hardware) and by research association (i.e. who you know). It is for this reason that the complete source code and compiled code for the past algorithms and the HVA code discussed in this report will be made freely available from the GXStat website (www.GXSTat.com) or directly from the author. This, in addition to the test data set and additional research programs such as FUZZY n-Varieties written by this author, will also be made available (in FORTRAN, of course) through this outlet. This open access will allow others to contribute to the mathematics and algorithms, making them even more useful for the next 50 years.

**Acknowledgements** Many people have contributed to the development of the HVA family of algorithms and programs. It was one of the intents of this report to give them the due credit. Also credit goes to those that have spent a great deal of their career to the application and dissemination of the unmixing approach. To that end, the grand prize should go to Professor Robert Ehrlich, without whom many would not have known about the diverse applications of this approach. I would like to thank Drs. Magdalena and Nils Blum-Oeste for their comments and improvements on this manuscript and overall support along with pushing the subject material. Dr. Lucinda Brothers-Full also help with editing this manuscript. Finally, I would like to apologize profusely to anyone who felt I purposely left them off the list of references.

## References

Abrams MA, Greb MD, Collister JW, Thompson M (2016) Egypt far Western Desert basins petroleum charge system as defined by oil chemistry and unmixing analysis. Mar Pet Geol 77:54–74

Anguy Y, Belin S, Bernard D, Fritz B, Ferm JB (1999) Modeling physical properties of sandstone reservoirs by blending 2D image analysis data with 3D capillary pressure data. Phys Chem Earth (A) 24:581–586

Anguy Y, Belin S, Ehrlich R, Ahmadi A (2002) Interpretation of mercury injection experiments using a minimum set of porous descriptors derived by quantitative image analysis. Image Anal Stereol 21:127–132

Barabás N, Goovaerts P, Adriaens P (2004a) Modified polytopic vector analysis to identify and quantify a dioxin dechlorination signature in sediments: 1. Theory. Environ Sci Technol 38:1813–1820

Barabás N, Goovaerts P, Adriaens P (2004b) Modified polytopic vector analysis to identify and quantify a dioxin dechlorination signature in sediments. 2. Application to the Passaic River. Environ Sci Technol 38:1821–1827

Barclay J, Herd RA, Edwards BR, Christopher T, Kiddle EJ, Plail M, Donovan A (2010) Caught in the act: implications for the increasing abundance of mafic enclaves during the recent eruptive episodes of the Soufrière Hills Volcano, Montserrat. Geophys Res Lett 37:L00E09:1–L00E09:5. https://doi.org/10.1029/2010gl042509

Battelle Memorial Institute et al (2012) A handbook for determining the sources of PCB contamination in sediments: Technical report TR-NAVFAC EXWC-EV-1302, Technical report TR-NAVFAC EXWC-EV-1302, NAVFAC Engineering and Expeditionary Warfare Center, Port Hueneme, CA, 164 pp

Bezdek JC (1981) Pattern recognition with fuzzy objective function algorithms. In: Advanced applications in pattern recognition. Springer Science & Business Media, 272 pp

Bezdek JC, Ehrlich R, Full WE (1984) FCM: the fuzzy c-means clustering algorithm. Comput Geosci 10:191–203

Blum-Oeste M, Wörner G (2016) Central Andean magmatism can be constrained by three ubiquitous end-members. Terra Nova 28:434–440

Bowers MC, Ehrlich R, Howard JJ, Kenyon WE (1995) Determination of porosity types from NMR data and their relationship to porosity types derived from thin section. J Pet Sci Eng 13:1–14

Bowers MC, Ehrlich R, Clark R (1994) Determination of petrographic factors controlling permeability using petrographic image analysis and core data, Satun Field, Pattani Basin, Gulf of Thailand. Mar Pet Geol 11:148–156

Bright DA, Cretney WJ, MacDonald RW, Ikonomou MG, Grundy SL (1999) Differentiation of polychlorinated dibenzo-*p*-dioxin and dibenzofuran sources in coastal British-Columbia, Canada. Environ Toxicol Chem 18:1097–1108

Brown PJ, Ehrlich R, Colquhoun DJ (1980) Origin of patterns of quartz sand types on the southeastern United States continental shelf and implications on contemporary shelf sedimentation: Fourier grain shape analysis. J Sediment Res 50:1095–1100

Burke CD, Full WE, Gernant RE (1986) Recognition of fossil fresh water ostracodes: Fourier shape analysis. Lethaia 20:307–314

Carr MB, Ehrlich R, Bowers MC, Howard JJ (1996) Correlation of porosity types derived from NMR data and thin section image analysis in a carbonate reservoir. J Petrol Sci Eng 14:115–131

Chayes F (1971) Ratio correlation: a manual for students of petrology and geochemistry, Chicago and London. University of Chicago Press, 99 pp

Cheney W, Kincaid DR (2008) Linear algebra: theory and applications, 1st edn. Jones & Bartlett Learning, 740 pp

Civitelli G, Corda L, Evangelista S, Full WE (1992) Fourier quartz shape analysis: application to terrigenous sediments of Laga and Cellino Formations (Central Italy). Bollettino della Societa Geologica Italiana 111:355–366

Clarke TL (1978) An oblique factor analysis solution for the analysis of mixture. Math Geol 10:225–241

Collister J, Ehrlich R, Mango F, Johnson GW (2004) Modification of the petroleum system concept: origins of alkanes and isoprenoids in crude oils. Am Assoc Pet Geol Bull 88:587–611

Coxeter HSM (1948) Regular polytopes, London, Methuen, 341 pp

Coxeter HSM (1973) Regular polytopes, 3rd edn. Dover Publications, 368 pp

Davis JC (2002) Statistics and data analysis in geology, 3rd edn. Wiley, 638 pp

DeCaprio AP, Johnson GW, Tarbell AM, Carpenter DO, Chiarenzelli JR, Morse GS, Santiago-Rivera AL, Schymura MJ, Akwesasne Task Force on the Environment (2005) PCB exposure assessment by multivariate statistical analysis of serum congener profiles in an adult Native American population. Environ Res 98:284–302

Deering CD, Cole JW, Vogel TA (2008) A rhyolite compositional continuum governed by lower crustal source conditions in the Taupo Volcanic Zone, New Zealand. J Petrol 49:2245–2276

Doré TJ, Bailey AM, McCoy JW, Johnson GW (1996) An examination of organic/carbonate-bound metals in bottom sediments of Bayou Trepagnier, Louisiana. Trans Gulf Coast Assoc Geol Soc 46:109–116

Draper NR, Smith H (2014) Applied regression analysis. Wiley, 736 pp

Ehrlich R, Kennedy SK, Crabtree SJ, Cannon RL (1984) Petrographic image analysis, I. Analysis of reservoir pore complexes. J Sediment Petrol 54:1365–1376

Ehrlich R, Full WE (1984a) Optimal definition of class intervals of histograms or frequency plots. In: Beddow JK (ed) Particle characterization in technology. CRC Press, Inc., Boca Raton, Florida, pp 135–148

Ehrlich R, Full WE (1984b) Fourier shape analysis. In: Brown RM (ed) Pattern analysis in the marine environment. PAME Proceedings, pp 47–68

Ehrlich R, Cobaleda G, Ferm JB (1997) Relationship between petrographic pore types and core measurements in sandstones of the Monserrate Formation, Upper Magdalena Valley, Colombia, C.T.F Cienc. Tecnol. Futuro, pp 5–17

Ehrlich R, Horkowitz KO (1984) A strong transfer function links thin section data to reservoir physics. In: Proceedings of the 59th annual technical conference and exhibition, Society of Petroleum Engineers of AIMESPE-13263-MS, 7 pp

Ehrlich R, Horkowitz KO, Horkowitz JP, Crabtree SJ Jr (1991a) Petrography and reservoir physics I: objective classification of reservoir porosity. Am Assoc Pet Geol Bull 75:1547–1562

Ehrlich R, Etris EL, Brumfield D, Yuan LP, Crabtree SJ Jr (1991b) Petrography and reservoir physics III: physical models for permeability and formation factor. Am Assoc Pet Geol Bull 75:1579–1592

Ehrlich R, Wenning RJ, Johnson GW, Su SH, Paustenbach DJ (1994) A mixing model for polychlorinated dibenzo-*p*-dioxins and dibenzofurans in surface sediments from Newark Bay, New Jersey using polytopic vector analysis. Arch Environ Contam Toxicol 27:486–500

Ehrlich R, Etris EL, Crabtree SJ (1996) Physical relevance of pore types derived from thin section by petrographic image analysis. In: SCA international symposium on core analysis. Dallas, Texas, vol 1, no 90001, 28 pp

Ehrlich R, Full WE (1988) Sorting out geology-unmixing mixtures. In: Size W (ed) IAMG special publication. Oxford University Press, New York, pp 33–46

El-Awawdeh RT, Full WE (1996) Quantification of coastline and key morphology changes over time in Northern Florida Bay, using high resolution shape analysis and aerial photographs. In: Proceedings of the second international airborne remote sensing conference, Environmental Research Institute of Michigan (ERIM), Ann Arbor, MI, Part I, pp 419–428

Etris EL, Brumfield DS, Ehrlich R, Crabtree SJ Jr (1988) Relations between pores, throats and permeability: a petrographic/physical analysis of some carbonate grainstones and packstones. Carbonates Evaporites 3:17–32

Evangelista S, Full WE, Tortora P (1986) Analisi Della Forma Particelle Sedimentae: Una Applicazione ai Clasti Sabbiosi Di Ambrente Costiero. Mem Soc Geol It 35:823–826

Evangelista S, Full WE, Tortora P (1994) Fourier grain shape analysis as tool to quantify the contribution of the fluvial input to the coastal sedimentary budget: an example from the Port Stephen's area, New South Wales, Australia. Bollettino della Societa Geologica Italiana 113:729–747

Evangelista S, Full WE, Tortora P (1996) Contribution and dispersion of fluvial, beach, and shelf sands in the Bassa Maremma coastal system (Central Italy, Tyrrhenian Margin). Bollettino della Societa Geologica Italiana 116:195–217

Evans JC, Ehrlich R, Krantz D, Full WE (1992) A comparison between polytopic vector analysis and empirical orthogonal function analysis for analyzing quasi-geostrophic potential vorticity. J Geophys Res Green 97:2365–2378

Ferm JB, Ehrlich R, Crawford GA (1993) Petrographic image analysis and petrophysics: analysis of crystalline carbonates from the Permian basin, west Texas. Carbonates Evaporites 8:90–108

Fillon RH, Full WE (1984) Grain size variations of North Atlantic non-carbonate sediments and sources of terrigenous components. Mar Geol 59:13–15

Full WE, Ehrlich R, Kennedy SK (1984) Optimal configuration and information content of sets of frequency distributions. J Sediment Petrol 54:117–126

Full WE, Ehrlich R, Klovan JE (1981) EXTENTED QMODEL-Objective definition of external end members in the analysis of mixtures. Math Geol 13:331–344

Full WE (1981) Analysis of quartz detritus of complex provenance via analysis of shape. PhD dissertation, Department of Geological Sciences, University of South Carolina, Columbia, SC, 206 pp

Full WE (1987) The use of image analysis in petrography. In: Proceedings of electronic imaging '87, Boston, MA, pp 597–602

Full WE, James S (2015) Rapid exploration in a mature area in northwest Kansas: improving the definition of key reservoir characteristics using big data, search and discovery article # 41685, 30 pp

Full WE, Ehrlich R, Bezdek JC (1982) FUZZY QMODEL—a new approach for linear unmixing. Math Geol 14:259–270

Gary AC, Johnson GW, Ekart DD, Platon E, Wakefield MI (2005) A method for two-well correlation using multivariate biostratigraphical data. In: Powell AJ, Riding JB (eds) Recent developments in applied biostratigraphy, pp 205–218

Golub GH, Reinsch C (1970) Singular value decomposition and least squares solutions. Numer Math 14:403–420

Healy-Williams N, Ehrlich R, Full W (1997) Closed-form Fourier analysis: a procedure for extracting ecological information from foraminiferal test morphology. In: Fourier descriptors and the applications in biology. Cambridge University Press, Cambridge, pp 129–156

Healy-Williams N (1983) Fourier shape analysis of Globorotalia truncatulinoides from late quaternary sediments in the southern Indian Ocean. Mar Micropaleontol 8:1–15

Healy-Williams N (1984) Quantitative image analysys: application to planktonic foraminiferal paleoecology and evolution. Geobios Mémoire Spécial 8:425–432

Henk M, Richter-Gebert J, Ziegler GM (1997) Basic properties of convex polytopes. In: Goodman JE, O'Rourke J (eds) Handbook of discrete and computational geometry. CRC Press, Boca Raton, Chapter 19, pp 243–270

Horkowitz J, Stakes D, Ehrlich R (1989) Unmixing mid-ocean ridge basalts with EXTENDED QMODEL. Tectonophysics 165:42754

Hudson CB, Ehrlich R (1980) Determination of relative provenance contributions in samples of quartz sand using Q-mode factor analysis of Fourier grain shape data. J Sediment Petrol 50:1101–1110

Huntley SL, Carlson-Lynch H, Johnson GW, Paustenbach DJ, Finley BL (1998) Identification of historical PCDD/F sources in Newark Bay Estuary subsurface sediments using polytopic vector analysis and radioactive dating techniques. Chemosphere 36:1167–1185

Imbrie J (1963) Factor and vector analysis programs for analyzing geologic data, Office of Naval Research, Technical Report 6, Geography Branch, 83 pp

James RA (1995) Application of petrographic image analysis to the characterization of fluid-flow pathways in a highly-cemented reservoir: Kane Field, Pennsylvania, USA. J Pet Sci Eng 13:141–154

Jarman WM, Johnson GW, Bacon CE, Davis JA, Risebrough RW, Ramer R (1997) Levels and patterns of polychlorinated biphenyls in water collected from the San Francisco Bay and Estuary, 1993–1995. Fresenius' J Anal Chem 359:254–260

JMP Pro (1989–2017) JMP Pro software package. SAS Institute Inc., Cary, NC

Johnson GW, Ehrlich R, Full W (2001) Principal component analysis and receptor models in environmental forensics. In: Murphy BL, Morrison RD (eds) Introduction to environmental forensics. Academic Press, pp 461–515

Johnson GW (1997a) Application of polytopic vector analysis to environmental geochemical investigation. PhD dissertation, University of South Carolina, 134 pp

Johnson GW (1997b) Application of polytopic vector analysis to environmental geochemistry investigations. PhD dissertation, Department of Geological Sciences, University of South Carolina, Columbia, SC, 244 pp

Johnson GW, Quensen JF III (2000) Implications of PCB dechlorination on linear mixing models. Organohalogen Compd 45:280–283

Johnson GW, Jarman WM, Bacon CE, Davis JA, Ehrlich R, Risebrough R (2000) Resolving polychlorinated biphenyl source fingerprints in suspended particulate matter of San Francisco Bay. Environ Sci Technol 34:552–559

Kaiser HF (1958) The varimax criterion for analytic rotation in factor analysis. Psychometrika 23:187–200

Kensington E, Full WE (1994) Fourier analysis of scallop shells (Placopecten magellancius) in determining population structure. Can J Fish Aquat Sci 51:348–356

Klovan IE, Meisch AT (1976) EXTENDED CABFAC and QMODEL computer programs for Q-mode factor analysis of compositional data. Comput Geosci 1:161–178

Klovan JE (1966) The use of factor analysis in determining depositional environments from grain-size distributions. J Sediment Res 36:115–125

Klovan JE (1981) A generalization of extended 'Q'-mode factor analysis to data matrices with variable row sums. J Int Assoc Math Geol 13:217–224

Klovan JE, Imbrie J (1971) An algorithm and Fortran-IV program for large-scale Q-mode factor analysis and calculation of factor scores. Math Geol 3:61–77

Leather J, Durell D, Johnson G, Mills M (2012) Integrated Forensics approach to fingerprint PCB sources in sediments using rapid sediment characterization (RSC) and advanced chemical fingerprinting (ACF). Technical Document 3262, Final Report for the Environmental Security Technology Certification Program (ESTCP), 226 pp

Leinen M, Pisias N (1984) An objective technique for determining end-member compositions and for partitioning sediments according to their sources. Geochemica et Cosmochimica Acta 48:47–62

Lisowiec K, Słaby E, Förster HJ (2015) Polytopic vector analysis (PVA) modelling of whole-rock and apatite chemistry from the Karkonosze composite pluton (Poland, Czech Republic). Lithos 230:105–120

Magar VS, Johnson GW, Brenner R, Durell G, Quensen JF III, Foote E, Ickes JA, Peven-McCarthy C (2005) Long-term recovery of PCB-contaminated sediments at the Lake Hartwell Superfund Site: PCB Dechlorination. I. End-member characterization. Environ Sci Technol 39:3538–3547

Mason EW, Ehrlich R (1995) Automated, quantitative assessment of basin history from a multiwell analysis. Am Assoc Pet Geol Bull 79:711–724

Mazzullo J, Ehrlich R (1980) A vertical pattern of variation in the St. Peter Sandstone Fourier grain shape analysis. J Sediment Res 50:63–70

Mazzullo J, Ehrlich R (1983) Grain-shape variation in the St. Peter sandstone: a record of eolian and fluvial sedimentation of an early Paleozoic cratonic sheet sand. J Sediment Res 53:105–119

Mazzullo J, Ehrlich R, Hemming MA (1984) Provenance and areal distribution of late Pleistocene and Holocene quartz sand on the southern New England continental shelf. J Sediment Res 54:1335–1348

Mazzullo J, Ehrlich R, Pilkey OH (1982) Local and distal origin of sands in the Hatteras Abyssal Plain. Mar Geol 48:75–88

McCreesh CA, Ehrlich R, Crabtree SJ (1991) Petrography and reservoir physics II: relating thin section porosity to capillary pressure, the association between pore types and throat size. Am Assoc Pet Geol Bull 75:1563–1578

McKenna TE, Lerche I, Williams DF, Full WE (1988) Quantitative approaches to stratigraphic correlations and chemical stratigraphy. Transactions, GCAGS/GCS-SEPM, 38:128–135

Megson D, Brown TA, Johnson GW, O'Sullivan G, Bicknell AWJ, Votier SC, Lohan MC, Comber S, Kalin R, Worsfold PJ (2014) Identifying the provenance of Leach's storm petrels in the North Atlantic using polychlorinated biphenyl signatures derived from comprehensive two-dimensional gas chromatography with time-of-flight mass spectrometry. Chemosphere 114:195–202

Miesch AT (1976a) Q-mode factor analysis of geochemical and petrologic data matrices with constant row-sums. U.S. Geological Survey Professional Paper 574-G, 47 pp

Miesch AT (1976b) Q-mode factor analysis of compositional data. Comput Geosci 1:147–159

Mudge SM (2016) Statistical analysis of oil spill chemical composition data, 2nd edn. In: Stout S, Wang Z (eds) Standard handbook oil spill environmental forensics, pp 849–867

Murillo-Jiménez JM, Full W, Nava-Sánchez E, Vera-Camacho V, León-Manilla A (2007) Sediment sources of beach sand from the southern coast of the Baja California Peninsula, México—Fourier grain shape analysis. GSA Special Paper 420:297–318

Nash GD, Johnson GW (2002) Soil mineralogy anomaly detection in Dixie Valley, Nevada using hyperspectral data. In: Proceedings of the twenty-seventh workshop on geothermal reservoir engineering, Stanford University, Stanford, California, 28–30 Jan 2002 SGP-TR-171, pp 249–256

Nash GD, Johnson GW, Johnson S (2004) Hyperspectral detection of geothermal system related soil mineralogy anomalies in Dixie Valley, Nevada: a tool for exploration. Geothermics 33:695–711

Nelson RH Jr, Roberts JH, Siebert DJ, Massell WF, LeRoy SD, Denham LR, Ehrlich R, Coons RL (2013) Method for locating sub-surface natural resources. US Patent US 8344721 B2

Porter GA, Ehrlich R, Osborne RH, Combellich RA (1979) Sources and nonsources of beach sand along southern Monterey Bay, California: Fourier shape analysis. J Sediment Res Petrol 49:727–732

Reister DD, Shipp RC, Ehrlich R (1982) Patterns of quartz sand shape variation, Long Island littoral and shelf. J Sediment Res 52:1307–1314

Ross CM, Ehrlich R (1991) Objective measurement and classification of microfabrics and their relationship to physical properties. In: Bennett RH et al (eds) Microstructure of fine-grained sediments: from mud to shale. Springer, pp 353–358

Ross CM, Ehrlich R, Gellici JA, Richardson BO (1986) Relationship between wireline log response and quantitative pore geometry data in Shannon sandstone, Hartzog Draw Field, Wyoming. Am Assoc Pet Geol Bull 70:614–642

Scheffe G, Full WE (1986) Petrographic image analysis of a sandstone reservoir in Oklahoma. In: Proceedings of the Geotech symposium, Denver, pp 196–202

Schwartz JT (2011) Introduction to matrices and vectors. Dover Publications: Dover ed. Edition, 192 pp

Shafer JL, Ehrlich R (1986) Method and apparatus for determining conformity of a predetermined shape related characteristics of an object or stream of objects by shape analysis. US Patent 4,624,367

Smith MM, Ehrlich R, Ramirez R (1985) Quartz provenance changes through time: examples from two South Carolina barrier islands. J Sediment Res 55:483–494

Smith SM (1997) National geochemical database reformatted data from the national uranium resource evaluation (NURE) hydrogeochemical and stream sediment reconnaissance (HSSR) program (No. 97-492). US Geological Survey, US Dept. of the Interior

Solohub JT, Klovan JE (1970) Evaluation of grain-size parameters in lacustrine environments. J Sediment Petrol 40:81–101

Sophie B, Yannick A, Dominique D, Bertrand F, Ferm JB (1999) Characterization by image analysis of the micro-geometry of the porous network of an illitized sandstone reservoir, Alwyn, North Sea. Bulletin de la Societe Geologique de France 170:367–377

Stattegger K, Morton AC (1992) Enhancing tectonic and provenance information from detrital zircon studies: assessing terrane-scale sampling and grain-scale characterization. J Geol Soc 168:309–318

Szymanski DW, Patino LC, Vogel TA, Alvarado GE (2013) Evaluating complex magma mixing via polytopic vector analysis (PVA) in the Papagayo Tuff, Northern Costa Rica: processes that form continental crust. Geosciences 3:585–615

Tefend KS, Vogel TA, Flood TP, Ehrlich R (2007) Identifying relationships among silicic magma batches by polytopic vector analysis: a study of the Topopah Spring and Pah Canyon ash-flow sheets of the southwest Nevada volcanic field. J Volcanol Geoth Res 167:198–211

Tortora P, Evangelista S, Full WE (1986) Fourier grain shape analysis and its application to the southeastern coastline of Australia. In: National Geological Congress of Central Italy Proceedings, Rome, Italy, pp 1–6

Towey TP, Barabás N, Demond A, Franzblau A, Garabrant DH, Gillespie BW, Lepkowski J, Adriaens P (2012) Polytopic vector analysis of soil, dust, and serum samples to evaluate exposure sources of PCDD/FS. Environ Toxicol Chem 31:2191–2200

Van de Wetering N, Mayer B, Sanei H (2015) Chemostratigraphic associations between trace elements and organic parameters within the Duvernay formation, Western Canadian Sedimentary Basin. In: Unconventional Resources Technology Conference C: 2153955, 13 pp

Vogel TA, Hidalgo PJ, Patino L, Tefend KS, Ehrlich R (2008) Evaluation of magma mixing and fractional crystallization using whole-rock chemical analyses: polytopic vector analyses. Geochem Geophys Geosyst, vol 9. https://doi.org/10.1029/2007gc001790

Weltje GJ (1997) End-member modeling of compositional data: numerical-statistical algorithms for solving the explicit mixing problem. Math Geol 29:503–549

Wenning RJ, Erickson GA (1994) Interpretation and analysis of complex environmental data using chemometric methods. TrAC Trends Anal Chem 13:446–457

Williams DF, Lerche I, Full WE (1988a) Isotope chronostratigraphy: theory and methods. Academic Press, 345 pp

Williams DF, Ehrlich R, Spero HJ, Healy-Williams N, Gary AC (1988b) Shape and isotopic differences between conspecific foraminiferal morphotypes and resolution of paleoceanographic events. Palaeogeogr Palaeoclimatol Palaeoecol 64:153–162

Yannick A, Bernard D, Ehrlich R (1996) Towards realistic flow modeling. Creation and evaluation of two-dimensional simulated porous media: an image analysis approach. Surv Geophys 17:265–287

Zellers SD, Gary AC (2007) Unmixing foraminiferal assemblages: polytopic vector analysis applied to Yakataga formation sequences in the offshore Gulf of Alaska. Palaios 22:1443–1467

Zhao GT, Wei Z, Full WE, Chen Q, Lin YS (2004) Fourier shape analysis and its application in geology. Periodical of Ocean, University of China, vol 34, pp 429–436

# Chapter 43
# Pearce Element Ratio Diagrams and Cumulate Rocks

**J. Nicholls**

**Abstract** While this chapter is about Pearce element ratios, I've included some personal reflections as this book is a 50th Anniversary project of the IAMG. Pearce element ratios, Felix Chayes and the Chayes medal, came together on September 11, 2001. As the recipient of the Chayes Medal, I was in Cancún, Mexico on that fateful date to deliver a talk on Pearce element ratios. Pearce element ratios are designed to model processes of fractionation and accumulation in igneous systems. They are frequently used to extract information from analyses of rocks formed from melts produced by fractionation—volcanic suites. Rock bodies formed from the fractionated crystals—the cumulate rocks—have received practically no attention. From the standard paradigm describing the formation of cumulate rocks, based on studies of the Skaergaard Intrusion, one expects a predicted pattern of data points on a Pearce element ratio diagram. Points derived from the mean compositions of the units in the cumulate body should fall up-slope from the point representing the initial melt composition on a diagram that accounts for the cumulate assemblage. Points derived from the compositions of the inferred residual melts present at the beginning of crystallization of a unit in the rock body should fall down-slope from the point representing the initial magma. The distance between a point on the line of a Pearce element ratio diagram and the point representing the initial magma composition depends on (1) the size of the aliquot that crystallized to form the rock unit and (2) the ratio of crystals to melt in the mush that solidified to form the rock unit. Patterns extracted from computer simulations compared to analogous data points from units of the Skaergaard Intrusion indicate that the crystal mushes that formed the units of the Marginal Border Series had a smaller ratio of trapped melt to crystals than did coeval mushes forming the Upper Border Series. Simulation patterns further indicate that the LZa and UZa units of the Layered Series formed from assemblages with larger ratios of melt to crystals than did the respective coeval units, LZa* and UZa*, of the Marginal Border Series.

J. Nicholls (✉)
Department of Geoscience, University of Calgary, Calgary, AB T2N1N4, Canada
e-mail: jim.nicholls@shaw.ca; nichollj@ucalgary.ca



B. S. Daya Sagar et al. (eds.), *Handbook of Mathematical Geosciences*,
https://doi.org/10.1007/978-3-319-78999-6_43

**Keywords** Pearce element ratios • Cumulate rocks • Computer simulation Skaergaard intrusion

## 43.1 Introduction

Blue skies and balmy temperatures graced a tranquil world when I entered the lecture room of the hotel-conference center in Cancún, Mexico, venue of the 2001 International Association for Mathematical Geosciences (IAMG) meeting. It was an early Tuesday morning and I was on my way to ensure the equipment worked for the talk I was soon to deliver. I was looking forward to the day and feeling honoured as the recipient of the IAMG Felix Chayes Prize for Excellence in Research in Mathematical Petrology.

When the talk was over and people were thinking ahead to the coffee break and upcoming lectures, we left the lecture room. Up until that moment, we were unaware that the world had changed: hijackers had crashed murder-suicide planes into the World Trade Center in New York City. Those attending the meeting gathered around a TV and watched the horror of the south tower collapse; smoke and dust billowed down the streets of New York, chasing people as they ran for their lives. The north tower collapsed a few minutes later. Hijackers crashed another plane into the Pentagon, and a fourth had been brought down in a field in near Shanksville, Pennsylvania just minutes away from its target in Washington, D.C. It was September 11, 2001, referred to by nearly all as 9/11.

My talk on Pearce element ratio diagrams and their utility in evaluating petrologic hypotheses was largely forgotten, understandably, in the turmoil following the events of that morning. Pearce element ratios and the events of 9/11 have been inextricably linked in my mind since that terrible morning, which is why they come together in this chapter.

Pearce element ratios were conceived in the last century (Pearce 1968), as were the concepts and techniques needed to implement their application. Their defining characteristic is a denominator formed from concentrations of elements that enter the minerals crystallizing from igneous melts in negligible amounts. Pearce element ratios have been used to model the evolution of melts in volcanic systems (see Nicholls and Russell 2016 for recent applications and explanations of the concepts) but they have not seen much service in modeling changes in the concomitant rocks formed from the separated solids and the enclosed interstitial melts: the cumulate rocks. Pearce element ratios can provide insight into the evolution of such assemblages.

## 43.2 Outline of a Cumulate Rock Paradigm

Petrologists have developed a paradigm for the crystallization of a single magma body in a crustal magma chamber that explains many of the features of layered and cumulate igneous rocks. This paradigm originated in features found in the Skaergaard Intrusion of East Greenland (Wager and Deer 1939; Carmichael et al. 1974; McBirney 1989a, 1996).

Cumulate bodies are often huge; the Bushveld Complex in South Africa has an estimated volume between 370,000 and 1,000,000 km$^3$ (Cawthorn and Walvraven 1998). Each unit forms by crystallization of a portion or aliquot of the melt in the magma chamber at the time. The larger the unit, the larger the aliquot from which it formed.

A cumulate body can be enclosed by a shell of finer grained rock petrologists are wont to call a chilled margin. The standard inference is that the chilled margin represents the initial magma and that the composition of the chilled margin closely approximates the composition of the initial magma. However, the chilled margin of a large body can be a boundary layer formed by the reaction of the corrosive magma with the country rocks. If so, the composition of the chilled margin can differ from that of the initial magma in a way that depends on the composition of the country rock and on the extent of reaction between magma and country rock. Nevertheless, chilled margins need to be considered as possible samples of the initial magma.

### *43.2.1 The Skaergaard Intrusion*

The Skaergaard Intrusion in East Greenland is one of the most studied rock bodies on the face of the Earth. L.R. Wager discovered the intrusion in 1931 on a scientific expedition. He returned in 1932 on another expedition and again in 1935–36 when he organized and led the third expedition to map and study the intrusion. On this trip, W.A. Deer accompanied him. Publications on the petrology of the Skaergaard began with the report by Wager and Deer (1939). A facsimile of the report was issued in 1952 with a new preface and a list of papers published since the 1939 publication. The list contains 46 references. One can find several hundred references that target the Skaergaard in the literature published after 1952.

I never met Wager but I did meet Deer when he visited the University of California, Berkeley during my time there as a graduate student. On a field trip, he spoke briefly about working with Wager on the Skaergaard. Wager was a mountaineer and climber. In 1933, as a member of the British Expedition to Mount Everest, he climbed to more than 8595 m, setting a record for a climb without oxygen, a record that wasn't bested until 1978. It the preface to the original report, Wager and Deer wrote that the terrain was so demanding that the two-man mapping parties had to traverse roped together, which lends credence to the story in which

Deer was reputed to have said he woke scared nearly every day on the Skaergaard because Wager took them up and down cliffs and slopes where Deer would never go himself.

Significant contributions to Skaergaard petrology since the original Wager and Deer report in 1939 have come from Wager (1960), Wager and Brown (1968), Hoover (1989a, b), McBirney (1989a, b, 1996), Ariskin (2002), and Nielsen (2004) among many others. As a result, the Skaergaard Intrusion has become a standard of comparison against which the evolutionary paths of basaltic magmas are measured.

The major units of the intrusion are the Layered Series (LS), composed of relatively horizontal layers, the Marginal Border Series (MBS) composed of relatively steeply dipping layered rocks, and the Upper Border Series (UBS), composed, again, of relatively horizontal layers of rock (Fig. 43.1). The layers in the Layered Series and the Upper Border Series become approximately horizontal after removal of a post-intrusion tilting (McBirney 1989a). The smaller units, Lower Zone a, Lower Zone b (LZa, LZb), etc. (Fig. 43.1) are defined by mineralogical changes. For example, the coeval units of the Middle Zone (MZ, MZ* and $\beta$) are characterized by the absence of large, primary crystals of olivine (primocrysts) (McBirney 1989a). Olivine primocrysts occur throughout the rest of the intrusion.

The stratigraphic nomenclature has slightly changed with time. Earlier workers, for example, Wager and Deer (1939), Chayes (1970), Carmichael et al. (1974), and

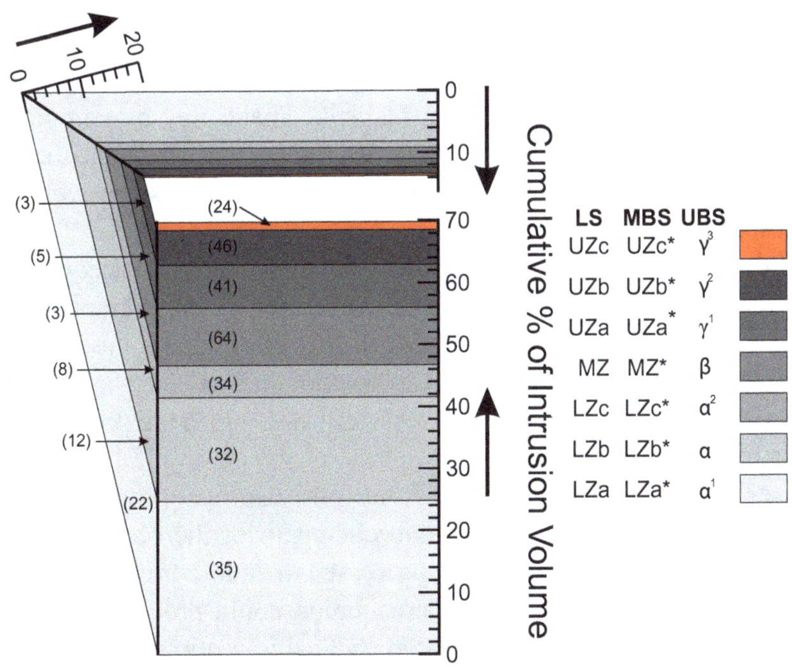


**Fig. 43.1** Rock units of the Skaergaard Intrusion (modified from Nielsen 2004). The Layered Series is interpreted to have formed by sedimentation of the crystallizing minerals onto the floor of the magma chamber. The Marginal Border Series (Hoover 1989a, b) and the Upper Border Series (Naslund 1984) are thought to have formed by plating of the crystallizing minerals on the walls and roof of the magma chamber. Labels in parentheses are number of analyses used to calculate the mean compositions of the rock units (McBirney 1989a)

Naslund (1984) called the Marginal Border Series and Upper Border Series (UBS) the Marginal Border Group and Upper Border Group. In addition, an asterisk has been attached to the names of the units of the Marginal Border Series to distinguish them from the units of the Layered Series.

The original floor of the magma is not exposed so the first rocks that formed on the floor are not available nor are samples from UZc* in the Marginal Border Series because of lack of outcrop (Hoover 1989a). The last or nearly last melt in the chamber is believed to have been caught between the crystal mush of solids and trapped melt that solidified as UZc* and the bottom of the UBS where the youngest unit of the UBS, the $\gamma^3$ unit, crystallized.

According to the paradigm, the rocks making up the intrusion formed by sedimentation of the crystallized minerals on the floor of the chamber and by plating minerals on the roof and walls. The solid assemblages that formed as sediments and as layers of plated minerals change with crystallization stage as do the mineral compositions. These mineral assemblages and mineral compositions found in the bottom, sides and top of the solidified magma chamber can be correlated and a stratigraphy of mineral assemblages and compositions provide coeval markers of crystallization stage. The rocks making up the intrusion consist of the mineral sediments and plated crystals plus melt trapped between the minerals; the trapped liquids later crystallize, creating intercumulus assemblages that, with the primocrysts, make up the rock units that fill the magma chamber.

Properties not emphasized but usually implicit in this paradigm are the ideas that the initial magma filling the magma chamber is uniform in composition and that the compositions of successive melts in the shrinking chamber maintain uniform compositions. These ideas may not be realistic. There may be compositional gradients as well as temperature and pressure gradients in the melt that induce the density currents that develop sedimentary structures, such as cross bedding, in the crystal mush.

In addition, the sedimentation-plating paradigm fails to account for several features of cumulate rocks, for example, repetition of stratigraphic units in the sedimentary layers (Bons et al. 2015). Mush formation above the magma interface (Bons et al. 2015) and double-diffusive convection in boundary layers (Huppert and Turner 1981; McBirney and Noyes 1979; McBirney 1985) are processes postulated to account for the repetition.

Processes behind the magma-mush front (post-cumulus processes, Sparks et al. 1985) can also affect the mineralogy and chemistry of the phases involved in the evolution of the magma body. These processes include convection in the trapped melt, compaction, and cementation. Cementation could produce significant chemical changes in the cumulate rock. Large, optically continuous crystals (poikilitic crystals) can be found enclosing previously formed primocrysts in both lava flows and in cumulate rocks. In the Skaergaard Intrusion and larger cumulate bodies, an interconnected crystal of pyroxene or plagioclase often fill the interstices between the primocrysts.

One infers the primocrysts were originally enclosed in a melt with the same composition as the melt that filled the magma chamber at the time and that melt was trapped between the primocrysts on the boundaries of the magma chamber. On crystallization of the interstitial melt, a single crystal can grow, fill the interconnected spaces, and displace the trapped melt. The melt, after undergoing post-cumulus processes, will differ in composition from the initially trapped melt. This modified melt could be expelled from the crystal pile and mix with the magma in the chamber, changing its composition. The large poikilitic crystals left behind would be part of the cement that holds the rock together.

Granted that processes in front of and behind the crystallization boundary can affect the resulting cumulate rock, the questions are: how effective are they in changing the rock composition and do they have a detectable influence on the composition of the melt in the chamber?

When the trapped melt crystallizes, permeability decreases, flow of melt from the cumulate mush slows, and its potential to change the composition of the melt in the magma chamber is lowered.

The difference in composition between the trapped melt and the melt in the magma chamber affects the composition of a mix of the two. If the trapped melt differs only slightly from the composition of the melt in the magma chamber, then the composition of a mix will differ from that of the composition of the melt in the magma chamber by a small amount, especially if the amount of trapped melt added to the mix is small.

Melt trapped in the crystal mush close to the crystallization boundary will be close in composition to the melt in the magma chamber. Farther from the boundary, the compositional differences will be larger. However, post-cumulus processes will act to decrease the volume of the trapped melt farther from the boundary. Processes like compaction, adcumulus growth (crystal growth on the surfaces of the primocrysts exposed to the interstitial melt), and cementation.

Expulsion of the trapped melt from the crystal mush could change the chemical composition of the melt in the magma chamber; however, the physical setting and processes could work in concert to keep the changes small.

Magma mixing, magma recharge, and magma mingling are labels for similar if not nearly identical processes. Simply put, the terms label the incorporation of one magma into another. If the invasive magma has a different composition than the original, the final body will have a different composition from the original (Anderson 1976; Carmichael 2004). Again, the effect of mixing on the chemistry of the combined magmas depends on how different the compositions are. The greater the differences, the greater the effect.

## 43.3 Pearce Element Ratio Patterns for Cumulate Rocks

The data to test any model of cumulate rock formation, Pearce element ratio or otherwise, comes from geologic maps, mineralogy, rock and mineral compositions, and rock textures. The more features of the data a model can predict, the stronger the model. If the model conforms to the data, the model is accepted as a description of the implied process that formed the rocks. If the model does not conform to the data, the model is rejected as an explanation (Nicholls and Russell 2016).

The numerators of the ratios plotted on the rectilinear axes of a Pearce element ratio diagram reflect the chemical changes in the melt-solid system caused by segregation and accumulation (sorting) of a specified mineral assemblage. Specification of the mineral assemblage allows us to create a model such that the compositions of melts and solid assemblages will fall on a line with the model slope. Only one rock composition is needed to locate the model line in Pearce element ratio space. The other analyses in the set of rock analyses can then be used to test the model. The specifics of the model dictate the slope of the line. Usually, the slope is one by design. Consequently, we can talk about up-slope and down-slope directions from a fixed point on the model line. If we select the point representing the chemistry of the melt present when the rock unit begins to form as the fixed point, then a point representing the chemistry of the derivative or residual melt will fall down-slope from the fixed point. Points representing the chemistry of crystal-melt mixtures (crystal mushes) will fall up-slope from the fixed point.

The general pattern expected for data points representing melts from a system undergoing sorting are known (Pearce 1968; Russell and Nicholls 1988; Nicholls and Russell 2016). The details of patterns expected in the data collected from cumulate bodies have not been explicitly investigated. A simple computer simulation of accumulation processes can delineate at least some of the expected patterns. Details of the simulation are described in the appendix.

The results of a simulation for a system with the composition listed in Table 43.1 are shown on Fig. 43.2. The Pearce element ratios plotted on Fig. 43.2 are:

$$(0.8\,\mathrm{Al} + 0.5\,\mathrm{Mg} + 0.4\,\mathrm{Ca})/\mathrm{K} \text{ versus } \mathrm{Si}/\mathrm{K}$$

The diagram was designed to describe the Pearce element ratios in the melts generated by fractionation (loss) of anorthite ($CaAl_2Si_2O_8$) and forsterite ($Mg_2SiO_4$) from the initial melt. The Pearce element ratio coordinates of the initial melt are shown with a black star on Fig. 43.2. The ratios derived from the compositions of the solids plus trapped melt are shown by filled circles.

**Table 43.1** Composition of the melts in the simulated crystallization processes

| Element | Melt 1 | Melt 2 |
|---|---|---|
| Si | 50% | 52% |
| Al | 20% | 19% |
| Mg | 15% | 14% |
| Ca | 10% | 10% |
| K | 3% | 2.5% |
| P | 2% | 2.5% |
| Size | 10000 m units | |
| Aliquot | 25% | |

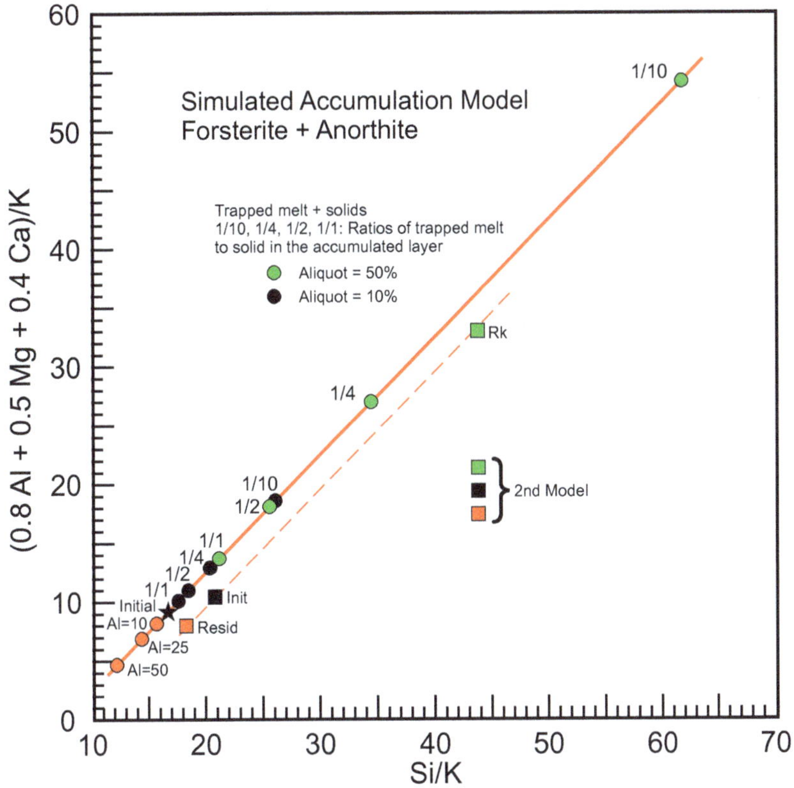


**Fig. 43.2** Pearce element ratio diagram for crystallization of a simulated system containing Si, Al, Mg, Ca, K, and P. Forsterite and anorthite are subtracted from the initial melt, leaving a residual melt that is trapped in the solid assemblage. Rocks formed by the simulated process would be composed of forsterite, anorthite, and solidified trapped melt (see appendix)

As expected, all the data points generated by the simulation fall on a line with a slope of one. The residual melts do produce points on the line that fall down-slope from the point representing the initial melt. Points representing the compositions of the accumulated solids and trapped melt and do plot up-slope from the point representing the initial melt (Fig. 43.2). These relationships are simply examples of the lever rule of phase diagrams (see Bloss 1994, pp. 304–306).

A second model is shown with a dashed line on Fig. 43.2. If the magma chamber undergoes recharge by a similar but not identical magma, we would expect the same ratio pair to describe the variation produced by crystallization of the second melt. The composition of the second simulated melt that produced the data points shown by squares is listed in Table 43.1. Mixing and crystallization of the mixed melts would produce data points falling between the two model-lines.

If the coordinates of the fixed point on a Pearce element ratio diagram are $(x_i, y_i)$, then the distance between the fixed point and another point on the model line with coordinates equal to $(x_j, y_j)$ will be given by:

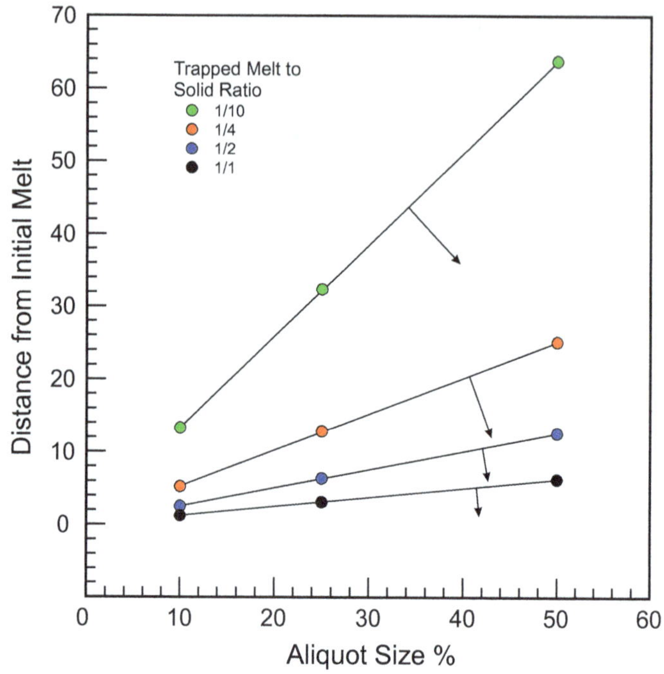


**Fig. 43.3** Plot of distance along the model line with a slope of one (red line in Fig. 43.2) from the point representing the initial melt (star in Fig. 43.2) against aliquot size. The arrows indicate direction of increasing ratio of trapped melt to cumulate solids

$$d = (x_i - x_j)\sqrt{2}$$

if the slope is equal to one and if the points representing cumulate assemblages fall exactly on the model line.

Two quantities determine the distance of a point from the fixed point: the size of the quantity of melt (aliquot) that crystallized to form the unit of crystals plus trapped melt and the amount of melt trapped in the crystal mush. Figure 43.3 shows how distance along the model line, aliquot size, and ratio of trapped melt to solid are related in the simulated system.

The two variables, distance along the model line and aliquot size, work in concert. The two are also quantities that can be extracted from sets of rock analyses and from geologic maps. The relationship between the two can be described by treating the ratio of the amount of trapped melt to the amount of accumulated crystals in a single unit of the cumulate rock body as a parameter. On a plot of aliquot size versus distance from the point representing the melt along the model line, lines of constant ratio of trapped melt to solid in the mush fan across the diagram. The smaller the ratio, the farther the line of constant ratio falls from the x-axis (Fig. 43.3).

Approximations of the amount of trapped melt could be made from estimates of petrographic modes (Chayes 1956; Nicholls and Stout 1986) of intercumulus assemblages versus primocrysts in thin section. However, distinguishing adcumulus growth from original growth material of the primocrysts is sometimes difficult. In addition, modal variations must underlie the large chemical variations found in the units of cumulate rocks (see below, Sect. 43.4). Consequently, petrographic assessment of the ratio of the volumes of trapped melt to primocrysts would require looking at many samples to get a precise value for a unit in the intrusion. At the present time, data to make a quantitative assessment of the agreement between model values and precise estimates of the petrographic modes are not available.

## 43.4 Compositions of Units of the Skaergaard Intrusion

A challenge to the construction of viable Pearce element models of cumulate rock formation arises from the chemical and mineralogical heterogeneity in the map units. The compositions of the constituent units must be determined as the mean of analyses from different locations in the unit. Mean values of the compositions and standard deviations for each constituent were published by McBirney (1989a, 1996) with data from Naslund (1984) for the Upper Border Series units. Dividing the standard deviations by the square root of the number of samples gives the standard errors of the means; the accepted measure of the uncertainty in a mean value. Standard errors of the means are large compared to analytical uncertainty (compare McBirney 1989a; Wright et al. 1975, p. 117). Analytical uncertainties are often two orders of magnitude smaller than the standard errors of the means. To make the two measures of uncertainty approximately equal, on the order of 10,000 samples would have to be analyzed for each unit.

When evaluating a model by comparing values from the model with the data, we expect certain criteria to be met if the model is successful. When testing models treating volcanic rocks, we expect model values to agree with the analytical data to within analytical uncertainty (Nicholls and Russell 2016; Nicholls and Stout 1988). Implicit in this expectation is the assumption that a sample from a lava flow is representative of the flow itself.

Estimates of the proportional volumes (Nielsen 2004) are shown on Fig. 43.1. The proportions, expressed as percentages of the volume of the intrusion were derived from the geologic maps of the body. It is worth explicitly noting that the quantitative entity plotted on Fig. 43.1 is volume, not thickness as has been traditionally plotted on similar looking graphs. Distances along the parallel lines have no real-world significance. The proportional volumes shown on Fig. 43.1 are not all independent (Nielsen 2004, p. 519). This dependence is revealed on Fig. 43.1 by the straight lines separating Layered Series volumes from the Marginal Border Series volumes and the Marginal Border Series volumes from the Upper Border Series volumes.

The abundant primocrysts in the intrusion are plagioclase, olivine, pyroxene (high-Ca augite and low-Ca pigeonite since inverted to orthopyroxene), and Fe-Ti oxides. The Middle Zone of the Layered Series, the Middle Zone of the Marginal Border Series, and the Upper Border Series $\beta$-zone lack olivine primocrysts, their place taken by low-Ca pyroxene.

### 43.4.1 Pearce Element Ratios and the Skaergaard Intrusion

We would like a Pearce element ratio design such that the products of sorting of all the mineral-melt assemblages in the intrusion would have compositions that generate points along a straight line with a slope of one. Unfortunately, nature prevents construction of such a diagram. The stoichiometry of olivine, $(Mg, Fe)_2SiO_4$, and low-Ca pyroxene, $(Mg, Fe)_2Si_2O_6$, with their different ratios of (Mg, Fe) to Si lead to an inconsistent set of algebraic equations in the design matrix (Nicholls and Russell 2016; Nicholls and Gordon 1994). We can, however, design two diagrams, one that accounts for sorting of olivine, plagioclase, augite, and Fe-Ti oxide and another that accounts for sorting of low-Ca pyroxene, plagioclase, augite, and Fe-Ti oxide.

Two ratio pairs that account for the abundant phases and their different compositions are:

[0.25 Al + 0.5(Fe + Mg) + 1.5 Ca + 2.75 Na]/K versus
(Si + 1.5 Ti)/K
(Olivine in the sorted assemblage)

and

[0.5 Al + Fe + Mg + Ca + 2.5 Na]/K versus
(Si + 3 Ti)/K
(Low-Ca pyroxene in the sorted assemblage)

Pearce element ratio diagrams for the two ratio pairs appear on Figs. 43.4 and 43.5. Figure 43.4 shows the diagram for olivine in the sorted assemblage whereas Fig. 43.5 shows a diagram for low-Ca pyroxene in the sorted assemblage.

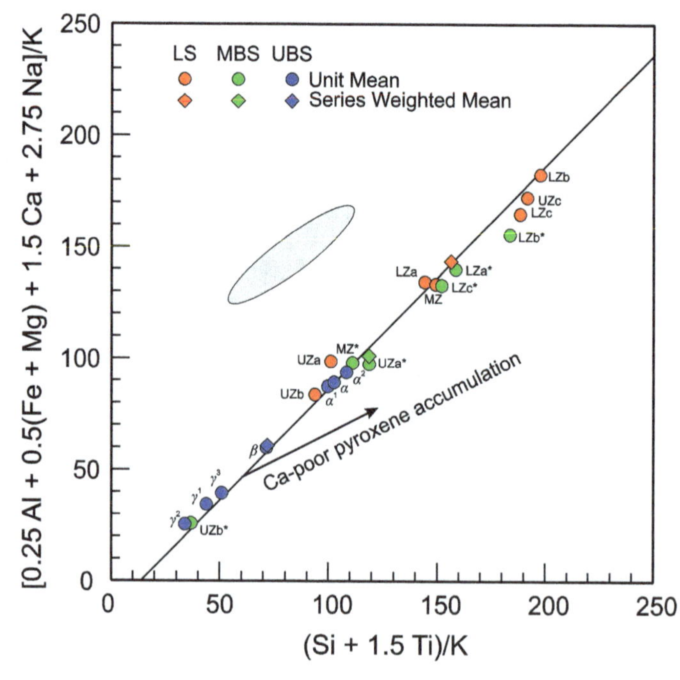


**Fig. 43.4** Pearce element ratio diagram designed to show the effects of sorting plagioclase, augite, olivine, and Fe-Ti oxide ($Usp_{75}$). Accumulation of Ca-poor pyroxene in addition to the listed minerals would cause data points to fall away from the model line along trends parallel to the arrow. The grey ellipse represents the size of the 1σ uncertainty in the location of the data point for UZb*

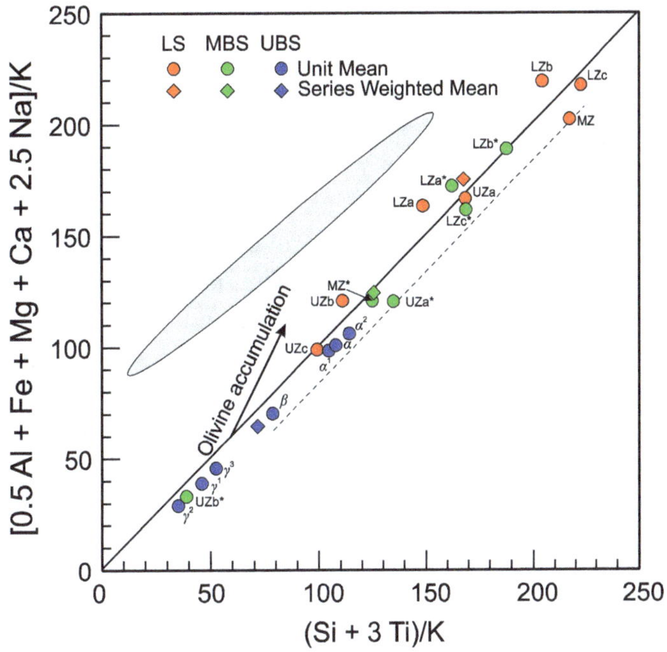


**Fig. 43.5** Pearce element ratio diagram designed to show the effects of sorting plagioclase, augite, Ca-poor pyroxene, and Fe-Ti oxide ($Usp_{75}$). Accumulation of olivine in addition to the listed minerals would cause data points to fall away from the model line along trends parallel to the arrow. The grey ellipse represents the size of the 1σ uncertainty in the location of the data point for UZb*

The points on the diagrams were calculated from the mean values of the compositions (McBirney 1989a, 1996). On both diagrams, the points are distributed along a trend with a slope of one but with considerable scatter; more scatter than found in trends calculated for suites of cogenetic volcanic rocks (compare Figs. 43.4 and 43.5 with diagrams in Nicholls and Russell 1991, 2016). The Skaergaard data span a larger range of values than do data from volcanic suites when plotted on similar Pearce element ratio diagrams. Data collected from basaltic volcanic suites, when plotted on comparable diagrams, span approximately 50 units (see Nicholls and Russell 1991). The Skaergaard data span approximately 250 units.

Although the number of analyses for several of the units in the Skaergaard Intrusion is large enough to make the mean values relatively stable in the sense that one more analysis would have a small effect on the mean, especially if the one analysis were for a rock like the ones analyzed. However, the large standard errors attached to the mean values opens the possibility that analyses of another set of samples of the same size collected from the same unit could result in a different set of means for the constituent oxide values.

Propagating the standard error of the means through the procedure for calculating the uncertainty in the location of a data point (Nicholls 1990b) produces large ellipses of 1σ analytical uncertainty in the location of the data point. The smallest ellipses for the data points shown on Figs. 43.4 and 43.5, belong to the points representing the mean of the UZb* unit of the Marginal Border Series.

The sizes of the uncertainty ellipses render them useless for testing the model. Almost any line with a slope of one will intercept the uncertainty ellipses. The model cannot be rejected because of the scatter of the data points off almost any line with a slope of one that we can pick.

Although the data points on Figs. 43.4 and 43.5 fall along a trend with a slope of one, the scatter about the trend precludes there being an obvious choice for a point through which to draw a model line. We could draw lines with unit slopes through every one of the data points but could not justify picking any one line over the others.

We can, however, calculate the mean compositions of each series (LS, MBS, UBS) by weighting the mean compositions of the units in the series by their respective relative volumes. The points derived from the weighted means are plotted as diamonds on Figs. 43.4 and 43.5. The points representing the weighted means do fall on a trend with a unit slope and with less scatter than do the full set of data points. It is a straight-forward procedure to find a line with a slope of one that falls closest, in the least-squares sense, to the three points representing the weighted mean compositions of the three series that make up the intrusion. The best fit lines for the weighted means fall close to the respective points (Figs. 43.4 and 43.5), well within any 1$\sigma$ error ellipse. These lines we will use as our model lines.

The inclusion of olivine or low-Ca pyroxene in the model assemblages produces no statistically significant difference in the efficacy of testing the models that I can see. If the lines defined by the weighted mean compositions for the three Series (LS, MBS, UBS) are the best models, then one would expect the points representing the Middle Zone rocks (MZ, MZ*, $\beta$) on Fig. 43.4 to deviate by falling below the line. They don't fall farther from the line than do points for the other units. Rather, they often fall closer to the line. Possibly, low-Ca pyroxene accumulated in the Middle Zone units in insufficient amounts to be detected with the olivine-sorting model.

On Fig. 43.5, one would expect the points representing the units outside the Middle Zone units to fall above a model line through points representing the Middle Zone rocks. The dashed line on Fig. 43.5 is a best fit line with a slope of one and is defined by the three Middle Zone values (MZ, MZ*, $\beta$). The data points for the other units displayed on Fig. 43.5 are displaced as expected if olivine sorting happened; they fall above the line.

The points representing the units (filled circles) fall in overlapping clusters along a trend with a slope of one with the larger units of the Layered Series generally falling up-slope from the points representing the Marginal Border Series units and with the Upper Border Series points falling farthest down-slope. This distribution is consistent with predictions from the computer simulations. The points representing Series compositions (filled diamonds) are also distributed as predicted by the computer simulation; the larger aliquot plots up-slope and the smaller aliquot down-slope.

The trends followed by the data points on Figs. 43.4 and 43.5 are consistent with the predictions of the models. Given the size of the uncertainties in the locations of the data points, there is no evidence that more than one magma was involved in the formation of the Skaergaard Intrusion.

## 43.5 Melts of the Skaergaard Intrusion

Three categories of melt crystallized to form the Skaergaard Intrusion: the melt that initially filled the magma chamber, the subsequent melts residual to each crystallization stage, and the melts trapped between the primocrysts. Melts trapped in the oldest part of a unit would have a different composition from melts trapped in the youngest part of a unit. Melt trapped in the youngest part of the unit would have the composition of the residual melt at the time of entrapment. Between crystallization of the oldest and youngest crystals in the units, the trapped melt would have compositions gradational between the two.

Any melt that existed in the Skaergaard crystallized long ago. Perforce, estimates of their compositions and their nature must be inferred. Melts whose compositions we can infer are those for the initial melt and the residual melts filling the magma chamber at the end of the formation of each rock unit and the beginning of the next.

### 43.5.1 *The Initial Melt*

Pearce element ratios for estimated compositions of the initial melt are plotted on Fig. 43.6. The initial melt composition should plot down slope from the point representing the mean composition of the Layered Series. Estimates of the initial Skaergaard magma have been made by Wager (1960), Hoover (1989a), McBirney (1996), Ariskin (1999), and Nielsen (2004). Wager (1960) used a composition from a sample from the chilled margin of the intrusion. Hoover (1989a) also used an analysis from a sample of the chilled margin but complimented it with melting experiments. Ariskin (1999) used thermodynamic modeling to make his estimates. AA1 (Fig. 43.6) is his preferred value. Nielsen (2004) based his estimate on volumes and average compositions complimented by comparison with chilled margin compositions and compositions of Tertiary basalts found near the intrusion. McBirney (1996) based his estimate on the mean composition of three samples from the chilled margin.

The estimates made by Wager (1960) and Ariskin (1999) do not fit the pattern we expect. A point representing an initial melt on a Pearce element ratio diagram should plot down-slope from the point representing our best estimate of the bulk composition of the intrusion (grey diamond, Fig. 43.6). I think it a tribute to the acumen of the estimators that all the preferred values fall close to the model line defined by the points representing the compositions of the weighted means of the major units of the intrusion.

### 43.5.2 *Residual Melts*

In addition to values for the mean compositions of the rock units of the Skaergaard Intrusion and estimates of the compositions of the initial melts, there are at least two

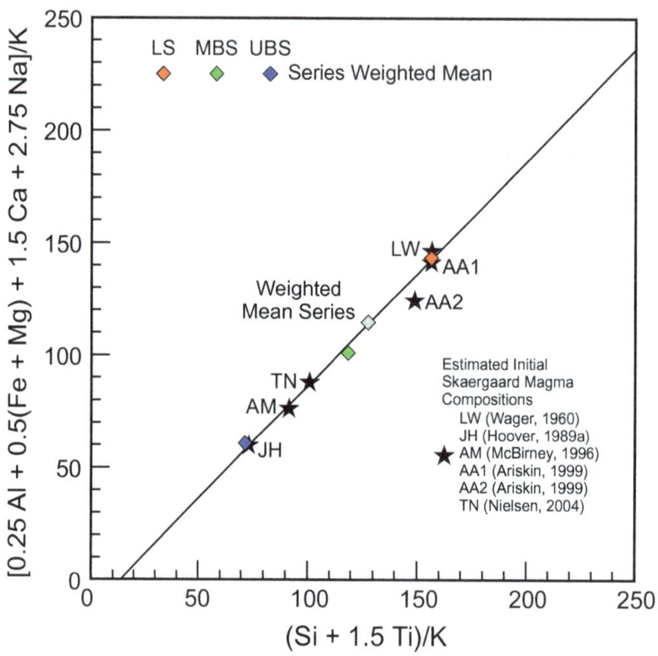


**Fig. 43.6** Pearce element ratio diagram showing the points derived from the mean compositions for Skaergaard rocks (McBirney 1996), estimates of the composition of the original Skaergaard magma (Wager 1960; Hoover 1989a, b; McBirney 1996; Nielsen 2004). The ratios plotted on the axes of the diagrams are designed such that melt and rock compositions should fall on a line with a slope of one if potassium (K) was conserved in the melts during crystallization of olivine, calcic pyroxene, plagioclase and an Fe-Ti oxide ($Usp_{75}$)

estimates of the compositions of the melt that filled the magma chamber at the time the particular crystal mush was in place: (1) experimentally determined compositions (McBirney 1996, red circles on Fig. 43.7) and (2) compositions derived through thermodynamic modeling (Ariskin 2002, green triangles on Fig. 43.7).

Felix Chayes was a petrologist who used mathematics in innovative ways to understand petrologic processes at a time when most petrologists knew little about mathematics. Among his many contributions was a small text that enhanced our understanding of the roles ratios can play in inferring petrologic processes (Chayes 1971). I met him but once at the 1967 meeting of the Geological Society of America in New Orleans. I was one of a number grad students and academics gathered in a night club. I later corresponded with him in the late 1980's about the efficacy of the correlation coefficient as a statistic for testing Pearce element ratio models. That correspondence caused me to use the designed slope of the line on a Pearce element ratio diagram as a characteristic of the model rather than a line fit to the data by least-squares methods. The designed line can then be compared to the data. Hence, one doesn't need the correlation coefficient to evaluate Pearce element ratio models. I think the same realization came independently to several others, notably Kelly Russell and Cliff Stanley, at about the same time.

In 1970 Chayes published a scheme for calculating residual melt compositions in the magma chamber and trapped in the mush during crystallization. His equation is:

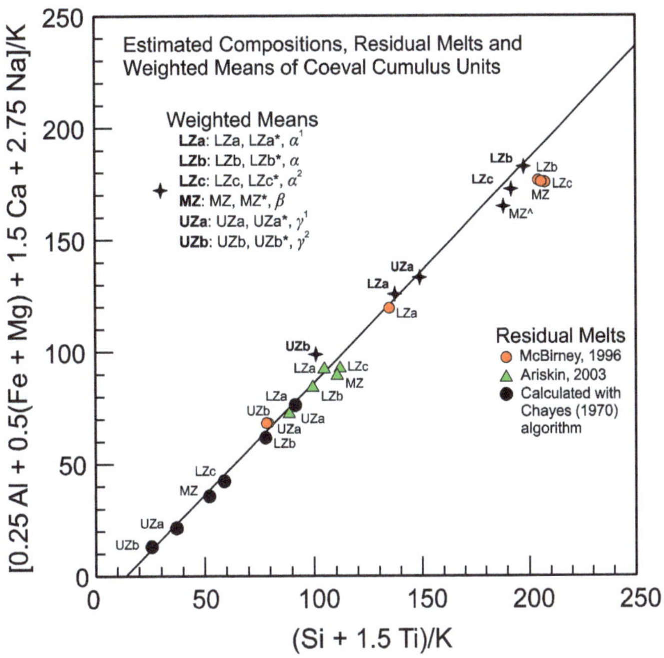


**Fig. 43.7** Pearce element ratio diagram showing the locations of points representing residual melt compositions at the end of the crystallization of the coeval units of the Skaergaard Intrusion. residual melt compositions estimated by McBirney (1996) and Ariskin (2003), and points calculated with Chayes (1970) algorithm

$$\mathbf{M}_{k+1} = \left[\mathbf{M}_1 - \sum_{j=1}^{k} (\mathrm{P}_j X_j)\right] \Big/ \left(1 - \sum_{j=1}^{k} \mathrm{P}_j\right), 0 < k < n$$

where the $\mathbf{M}_i$ are vectors whose elements are a set of oxide values in the residual melt and $n$ is the number of units in the intrusion. $\mathbf{M}_1$ is the vector containing the oxide values for the initial melt. The $P_j$ are the volumes or proportional volumes of the units in the intrusion. The $X_j$ are the mean values of the oxides in the units of the intrusion.

The values contained in the $\mathbf{M}_i$, $i > 1$, depend of the values contained in $\mathbf{M}_1$. Change the values in $\mathbf{M}_1$ and the values in $\mathbf{M}_i$ change.

All values for the initial melt, the $\mathbf{M}_1$, except those estimated by McBirney (1996) generate negative values for some of the oxides in the $\mathbf{M}_k$ at later stages in the evolution of the residual melts ($k > 3$). The Pearce element ratios for residual melts generated with Chayes' (1970) equation using McBirney's (1996) estimate for the values in the initial melt are shown with solid black circles on Fig. 43.7.

At any stage in the evolution of the Skaergaard Intrusion, the residual melt is simultaneously depositing crystals on the floor, walls and roof of the magma chamber, at least according to the simplest paradigm. The points to be compared, then, to the simulated patterns are the weighted means of the coeval units. Pearce element ratios for the three sets of residual melts: (Chayes 1970 algorithm; McBirney 1996; Ariskin 2003) can be compared on Fig. 43.7. McBirney's (1996) estimates for the compositions of the residual melts at the end of LZa, LZc, MZ, UZa, and UZb do not fit the expected pattern in that they plot up-slope from their respective cumulate compositions. All of the points representing the residual melt compositions estimated by Ariskin (2003) plot down-slope from the points representing their respective cumulate compositions as do residual melt compositions calculated with Chayes' (1970) algorithm. Only the latter however, fall in sequential order, a pattern expected for a series of melts formed by fractionation of a single initial magma.

### 43.5.3 Relative Amounts of Trapped Melt

We can make qualitative assessments of the amount of melt trapped in the cumulates by plotting the relative volumes of the units in the intrusion against the position of the Pearce element values along the model line (see Fig. 43.2).

Data points on Pearce element ratio diagrams need not fall exactly on model lines, which makes calculating distance along the model line less straight-forward than given in the formula above (see Sect. 43.3). To calculate distance from a point representing a melt composition to a point representing a cumulate composition we measure the distance along the model line between two points that are the closest to each of the two points in question. The point of closest approach will be along a line through the point and normal to the model line. An example is shown on Fig. 43.8 for the coeval Lower Zone units (LZa, LZa*, and $\alpha^1$). The points on Fig. 43.8 represent the initial melt composition (McBirney 1996, black star) and the mean compositions of the units (McBirney 1996 coloured circles).

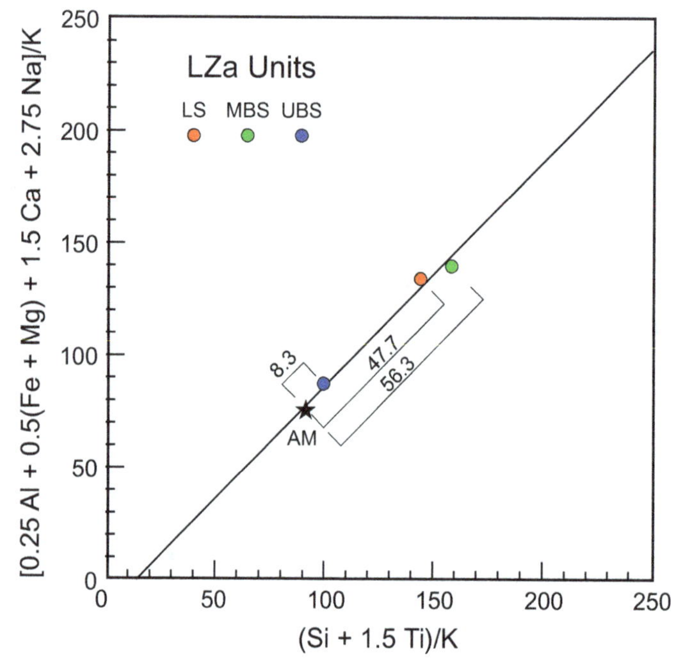


**Fig. 43.8** Pearce element ratio diagram showing distances along the model line between points representing weighted mean compositions of the LZa units and the point representing the estimate of the composition of the initial Skaergaard magma (McBirney 1996)

Figure 43.9 shows the distances along the model line versus unit size expressed as a percentage of the volume of the intrusion (Nielsen 2004). The left hand sides of the triangles defined by points representing coeval units are approximately vertical; in other words, the units whose points define the left hand sides of the triangles are approximately the same size. On Fig. 43.3, the ratio of trapped melt to primocryst in the crystal mush decreases upwards along a vertical line. If the same pattern carries over to real-world data, then the amount of trapped melt, relative to primocryst amount, is smaller in the UBS units than in the MBS units.

The lack of independence in the estimates of the volumes of the units does not in itself invalidate these conclusions. The estimates of the relative volumes may be correct; we just have less confidence that they may be. Because we are using the estimates in a qualitative fashion, the chances that our conclusions are reasonable improve.

Contours of equal trapped melt to primocryst ratio have a positive slope on Fig. 43.3, which illustrates the pattern of points in the simulation model. If the pattern applies to the real world, the upper boundaries, with negative slopes, of the triangles representing the coeval LZa and UZa units (red and yellow triangles) cannot be parallel to contours of equal ratio. We infer, then, that for these two sets of coeval units, the ratio of trapped melt to primocryst amount was smaller in the MBS units than in the LS units.

It is unlikely coeval units of the LS and the UBS would have the same ratios for trapped melt to crystals. Consequently, the lines drawn between points representing LS and UBS units are probably not lines of constant ratio (compare Fig. 43.3).

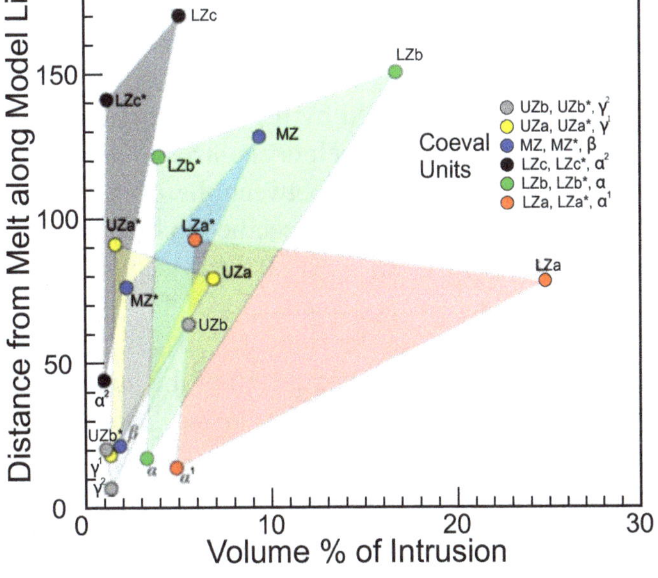


**Fig. 43.9** Distances of the locations of the points representing the compositions of coeval Skaergaard units from points representing the compositions of the residual melts (calculated with Chayes 1970, algorithm) along the model line versus the volumes of the units expressed as percentages of the volume of the intrusion (Nielsen 2004). The triangles connect vertices whose points represent properties of coeval units

## 43.6 Pearce Element Ratios, Cumulate Rocks, and September 11

Wager's discovery of the Skaergaard Intrusion and his recognition of its significance to igneous petrology and Pearce's insight that led to Pearce element ratios opened ways to decipher how cumulus rocks came to be. Understanding how these rocks came to be can affect our lives. They host ore deposits of chromium, nickel, and the platinum group elements (ruthenium, rhodium, palladium, osmium, iridium, and platinum), elements required by our civilization. To know more about how they came to be adds to our understanding of the Earth.

For nearly a decade I did little to extend the range of application of Pearce element ratios. An invitation to contribute to a review paper by *Geoscience Canada* led me to look at cumulate rocks through the lens of Pearce element ratios. Perhaps the perspective articulated by Stephen J. Gould in a piece he wrote for Canada's newspaper, the Toronto *Globe and Mail*, shortly after the events of 9/11 (Gould 2001) is apposite. His point: evil events, like 9/11, can cause big changes in our lives whereas many good events come in small packages. The good, however, by their number, eventually outweigh the evil. Maybe application of Pearce element ratios to the study of cumulate rocks can count as one of the small packages.

**Acknowledgements** Discussions with many people helped me learn about Pearce element ratios, in particular, Kelly Russell, Cliff Stanley, Terry Gordon, and Alex Wilson. Thanks to the late Tom Pearce for inventing Pearce element ratios.

## Appendix: Computer Simulation

The simulation will be for a single step or stage in the processes that lead to the development of a layered intrusion. The simulated system contains Si, Al, Mg, Ca, K, and P. Crystallization produces forsterite and anorthite with proportions of the two minerals constrained by the concentrations of Si, Al, Mg, and Ca in the melt. A fraction of the initial melt crystallizes to produce a melt modified in composition, some of which is trapped between the primocrysts.

Numbers that have to be specified to run the simulation are a composition for the initial melt (*im*[0], *im*[1], *im*[2], *im*[3], *im*[4]) where the items in the initial melt vector represent molar percentages of the elements: Si, Al, Mg, Ca, K, and P. The size (*S*) of the melt in the simulated magma chamber is entered into the simulation procedure, as is the percentage (*P*) of the initial melt, or aliquot that will supply the forsterite and anorthite crystals in the layer. The size is equal to the number of moles of the elements in the initial melt. The numbers of the different elements in the aliquot will designated as (*aq*[0], *aq*[1], *aq*[2], *aq*[3], *aq*[4], *aq*[5]).

One could assume the simulated magma chamber was initially uniformly mixed and filled with a homogeneous melt. If the composition of the system is known, the simulation could be made deterministic to within two adjustable parameters if a thermodynamic component were included in the model. This is a consequence of Duhem's theorem (see Nicholls 1990a, 2000, 2013). One could also make it deterministic by extracting the maximum amounts of forsterite and anorthite from the aliquot. To add some variability into the simulation, we will sample the initial melt to create the aliquot by following a constrained random number procedure. $P \times S/100$ random integers, $rn$, $n = 1 \ldots P \times S/100$, are generated from a uniform distribution between 0 and $P \times S/100$.

$$
\begin{aligned}
\text{If: } & 0 < rn < im[0] \times P \times S/100, \\
& aq[0] = aq[0] + 1 \\
\text{else if: } & im[0] \times P \times S/100 < rn < im[0] \times P \times S/100 + im[1] \times P \times S/100, \\
& aq[1] = aq[1] + 1 \\
\text{else if: } & im[0] \times P \times S/100 + im[1]\, P \times S/100 < rn < \\
& im[0] \times P \times S/100 + im[1] \times P \times S/100 + im[2] \times P \times S/100, \\
& aq[2] = aq[2] + 1 \\
\text{else: } & im[0] \times P \times S/100 + im[1] \times P \times S/100 + im[2] \times P \times S/100 < rn < \\
& im[0] \times P \times S/100 + im[1] \times P \times S/100 + im[2] \times P \times S/100 + im[3] \times P \times S/100, \\
& aq[3] = aq[3] + 1 \\
\text{and } & aq[4] = im[4] \times P \times S/100 \\
& aq[5] = im[5] \times P \times S/100.
\end{aligned}
$$

The last two equalities ensure that the two conserved elements, K and P, enter the aliquot in the same proportions as they are found in the initial melt.

From this new melt, forsterite and anorthite crystallize. The amounts of the two phases that can be extracted from the new melt are constrained by the composition of the aliquot. The amount of anorthite that can be extracted depends on the numbers of Ca and Al elements in the melt:

$$
\begin{aligned}
\text{if: } & aq[3] < aq[1]/2, \\
& \mathrm{An} = aq[3] \\
\text{else } & \mathrm{An} = aq[1]/2
\end{aligned}
$$

The amount of forsterite depends on the number of Mg elements in the melt.

$$\mathrm{Fo} = aq[2]/2$$

Using the amounts of anorthite and forsterite extracted from the aliquot, the numbers of elements in a new melt ($nm[0]$, $nm[1]$, $nm[2]$, $nm[3]$, $nm[4]$, $nm[5]$) are calculated by:

$$
\begin{aligned}
nm[0] &= im[0] - \mathrm{Fo} - 2\ \mathrm{An} \\
nm[1] &= im[1] - 2\ \mathrm{An} \\
nm[2] &= im[2] - 2\ \mathrm{Fo} \\
nm[3] &= im[3] - \mathrm{An} \\
nm[4] &= im[4] \\
nm[5] &= im[5]
\end{aligned}
$$

A melt with the new composition is then trapped between crystals to form the crystal mush. Solidification of the mush produces a layer in the cumulate rock body.

## References


Anderson AT (1976) Magma mixing: petrological process and volcanological tool. J Volcanol Geotherm Res 1:3–33

Ariskin AA (1999) Phase equilibria modeling in igneous petrology: use of COMAGMAT model for simulating fractionation of ferrobasaltic magmas and the genesis of high-alumina basalt. J Volcanol Geoth Res 90:115–162

Ariskin AA (2002) Geochemical thermometry of the Layered Series rocks of the Skaergaard Intrusion. Petrol 10(6):495–518

Ariskin AA (2003) The compositional evolution of differentiated liquids from the Skaergaard Layered Series as determined by geochemical thermometry. Russ J Earth Sci 5(1):1–29

Bloss FD (1994) Crystallography and crystal chemistry. Mineralogical Society of America, Washington, DC, USA

Bons PD, Baur A, Elburg MA, Lindhuber MJ, Marks MAW, Soesoo A, van Milligen BP, Walte NP (2015) Layered intrusions and traffic jams. Geology 43(1):71–74. https://doi.org/10.1130/g36276.1

Carmichael ISE (2004) Landmark papers. Volcanic petrology. In: Wood BJ (ed) Landmark papers, vol 1. Mineralogical Society of Great Britian & Ireland, London, England, p 319

Carmichael ISE, Turner FJ, Verhoogen J (1974) Igneous petrology. McGraw-Hill Book Company, New York, USA

Cawthorn RG, Walraven F (1998) Emplacement and crystallization time for the bushveld complex. J Petrol 39(9):1669–1687

Chayes F (1956) Petrographic modal analysis an elementary statistical appraisal. Wiley, New York, USA

Chayes F (1970) On estimating the magnitude of the Hidden Zone and the compositions of the residual liquids of the Skaergaard Layered Series. J Petrol 11(1):1–14

Chayes F (1971) Ratio correlation: a manual for students of petrology and geochemistry University of Chicago Press. Illinois, USA, Chicago

Gould SJ (2001) The good people of Halifax. In: Globe and mail, Toronto, Ontario, Canada

Hoover JD (1989a) The chilled marginal gabbro and other contact rocks of the Skaegaard Intrusion. J Petrol 30(2):441–476

Hoover JD (1989b) Petrology of the marginal border series of the Skaergaard Intrusion. J Petrol 30 (2):399–439

Huppert HE, Turner JS (1981) Double-diffusive convection. J Fluid Mech 106:299–329

McBirney AR (1985) Further considerations of double-diffusive stratification and layering in the Skaergaard Intrusion. J Petrol 26(4):993–1001. https://doi.org/10.1093/petrology/26.4.993

McBirney AR (1989a) The Skaergaard Layered Series: I. Structure and average compositions. J Petrol 30:363–397

McBirney AR (1989) Geological map of the Skaergaard intrusion, east Greenland. University of Oregon, Eugene, Oregon

McBirney AR (1996) The Skaergaard Intrusion. In: Cawthorn RG (ed) Layered intrusions. Elsevier, Amsterdam, Netherlands, pp 147–180

McBirney AR, Noyes RM (1979) Crystallization and layering of the Skaergaard Intrusion. J Petrol 20(3):487–554. https://doi.org/10.1093/petrology/20.3.487

Naslund HR (1984) Petrology of the Upper Border Series of the Skaergaard Intrusion. J Petrol 25:185–212

Nicholls J (1990a) Principles of thermodynamic modeling of igneous processes. In: Nicholls J, Russell JK (eds) Modern methods of igneous petrology—understanding magmatic processes, vol Mineralogical. Society of America. Washington, D.C., pp 1–23

Nicholls J (1990b) Stochiometric constraints on variations in Pearce element ratios and analytical uncertainty in hypothesis testing. In: Russell JK, Stanley CR (eds) Theory and application of pearce element ratios to geochemical data analysis, vol 8. Geological association of Canada. Vancouver, British Columbia, pp 73–98

Nicholls J (2000) Thermodynamics of a magmatic gas phase 50 years later: comments on a paper by John Verhoogen (1949). Can Mineral 38:1313–1328

Nicholls J (2013) Perfect fractionation in multicomponent, multiphase systems. Contrib Mineral Petrol 166(3):691–701. https://doi.org/10.1007/s00410-013-0897-y

Nicholls J, Gordon TM (1994) Procedures for the calculation of axial ratios on Pearce element-ratio diagrams. Can Mineral 32:969–977

Nicholls J, Russell JK (1991) Major-element chemical discrimination of magma batches in lavas from Kilauea volcano, Hawaii, 1954–1971 eruptions. Can Mineral 29:981–993

Nicholls J, Russell JK (2016) Igneous rock associations 20. Pearce element ratio diagrams: linking geochemical data to magmatic processes. Geosci Can 43(2):133–146. http://dx.doi.org/10.12789/geocanj.2016.43.095

Nicholls J, Stout MZ (1986) Electron beam analytical instruments and the determination of modes, spatial variations of minerals and textural features of rocks in polished section. Contrib Mineral Petrol 94:395–404

Nicholls J, Stout MZ (1997) Epitactic overgrowths and intergrowths of clinopyroxene on orthopyroxene: Implications for paths of crystallization, 1881 lava flow, Mauna Loa Volcano. Hawaii Can Mineral 35:909–922

Nielsen TFD (2004) The shape and volume of the Skaergaard Intrusion, Greenland: Implications for mass balance and bulk composition. J Petrol 45(3):507–530. https://doi.org/10.1093/petrology/egg092

Pearce TH (1968) A contribution to the theory of variation diagrams. Contrib Mineral Petrol 19:142–157

Russell JK, Nicholls J (1988) Analysis of petrologic hypotheses with Pearce element ratios. Contrib Mineral Petrol 99:25–35

Sparks RSJ, Huppert HE, Kerr RC, McKenzie DP, Tait SR (1985) Postcumulus processes in layered intrusions. Geol Mag 122(5):555–568

Wager LR (1960) The major element variation of the Layered Series of the Skaergaard Intrusion and a re-estimation of the average composition of the Hidden Layered Series and of the successive residual magmas. J Petrol 1(3):364–398

Wager LR, Brown GM (1968) Layered Igneous Rocks. San Francisco, W.H, Freeman and Company

Wager LR, Deer WA (1939) Geological investigations in east Greenland. Part III. The petrology of the Skaergaard intrusion. Kangerdlugssuaq East Greenland Meddeleser om Gronland 105 (4):1–352

Wright TL, Swanson DA, Duffield WA (1975) Chemical compositions of Kilauea east-rift lava, 1968–1971. J Petrol 16:110–133

# Chapter 44
# Reflections on the Name of IAMG and of the Journal

**Donald E. Myers**

**Abstract** This note is to highlight the transformation of the names of *International Association for Mathematical Geologists* and its flagship journal *Mathematical Geology* respectively into *International Association for Mathematical Geoscientists* and *Mathematical Geosciences*.

When first approached about submitting something for the special volume I thought the idea was a good one but was not sure what I might have to say that would be relevant and of interest. Initially I planned to simply reflect on my year as Distinguished Lecturer (2008) but somehow it didn't seem sufficient. Instead I want to reflect on three words in the name of the organization and also on the current title of the journal, i.e. *International, Association Mathematical, Geologists* and *Geosciences*. As anyone familiar with IAMG knows it was born in Prague in 1968 in the midst of what turned out to be a momentous event but it also returned to Prague to celebrate its 25th anniversary in 1993. I wasn't one of that moderately small but very influential group but I subsequently knew or still know many of them. I didn't really start working in the field until the early 1970s.

Prior to the 1970s I was only a mathematician but accidentally came in contact with two other faculty at the University of Arizona, Y. C. Kim (Mining Engineering) and De Verle Harris (Mineral Economics) as well as Art Warrick (Soils, Water and Engineering). Hence I was beginning to "Associate". Through those I learned about G. Matheron's work, met Frits Agterberg, André Journel and Shlomo Neuman (Hydrology), developed some collaboration with USGS in Denver and made plans to spend a sabbatical at the Centre de Géostatistique (Fontainebleau) in the spring of 1981. Ghislain de Marsily spent the academic year 1979–1980 at the University of Arizona in the Department of Hydrology. Through Art Warrick I knew of the work of Richard Webster, I was fortunate to be invited to participate in the NATO ASI at Lake Tahoe in 1983 and met many of the others in the very important group in mathematical geosciences.

D. E. Myers (✉)
Department of Mathematics, University of Arizona, Tucson, AZ 85721, USA
e-mail: myers@math.arizona.edu



B. S. Daya Sagar et al. (eds.), *Handbook of Mathematical Geosciences*,
https://doi.org/10.1007/978-3-319-78999-6_44

At this point it is important to note the change(s) that have taken place in the name of the journal. Initially most of the membership would have been geologists or mining engineers but clearly hydrology and soil science are a part of the geosciences so that the interests and membership were expanding in scope. In fairly short order geosciences grew to encompass "environmental sciences", "geography", "ecology", "image analysis", "remote sensing", "epidemiology", "atmospheric sciences" because the stress was on "geo" and not on "ology". Papers in the various soil science journals cited papers in the IAMG journal (and conversely), papers in the various American Geophysical Union cited papers in the IAMG journal (and conversely) and of course the petroleum industry was involved early with the collaboration between Fontainebleau and Shell Oil. It is likely that a list of referees for Mathematical *Geosciences* (and all the previous titles) would cross an ever increasing list of countries and institutions as well as areas of interest.

Except perhaps in France the work of G. Matheron was not really known in the mathematical/statistics community even though his signal paper appeared in the *J. of Applied Probability* in 1973. *Mathematical Reviews* still doesn't really have a category for mathematical geosciences other than geophysics. The statistics community likewise was slow to recognize mathematical geosciences. Most of the interest in Radial Basis functions either relates to solutions for partial differential equations or approximation theory.

The various editors (and publishers) of *Mathematical Geosciences* have been very interested in the impact ratings of the journal but it would be even more interesting to tabulate the number of different journals not closely related to mathematical geology that publish papers citing papers appearing in *Mathematical Geosciences* (including those that might have appeared twenty or thirty years ago. In many fields of science it is not uncommon for the significance or usefulness of a paper to appear many years later. This is especially true of pure mathematics.

As I have tried to point out that geosciences is a more encompassing term than geology (many university departments have changed their names to reflect this), the "mathematical" part of mathematical geosciences has also grown and expanded. In some ways statistics is an outgrowth of mathematics but it is also an outgrowth of agriculture (think of the work at Rothamstead Experimental Station and the many land grant universities in the US) but also the social sciences and economics/business. Statistics by its very nature is a very cross disciplinary applied area of interest. Another part of "mathematical" pertains to computing. The VAX computer and the software package BluePack were very much a part of the real growth of geostatistics, the desktop computer has created an even greater explosion. I first started teaching a class on geostatistics in 1982 and my students had to use a mainframe CDC 6400 with punch card input, it was terribly inconvenient but without that access the class would have had no practical value. The advances in computing and in access to computing have revolutionized the teaching of statistics in all its very forms.

Clearly IAMG was international from its original founding and that perspective has only grown with time. I can speak to that from a personal perspective both from my experience as the Distinguished Lecturer in 2008 but also as a referee/reviewer for the journal and attendance at various international meetings. I would also note the level of interest evident in the Questions appearing on the ResearchGate.net forum. It is truly international.

Sometimes old ideas come back in a different form. The Design of Experiments originated in applications to agriculture and often emphasized various forms of "plot design" but now it may be important in the design of aircraft wings and may incorporate kriging and/or cokriging. Google tells me that my paper on cokriging (J. of the International Assn of Mathematical Geologists, 1982) is being cited for applications very far afield from the problem I thought I was addressing when I wrote the paper. I am sure other authors of papers that appeared in this journal may have had a similar experience. It is a tribute to the vision of the founders of IAMG back in 1968. "Mining Geostatistics" was a classic when it appeared (the English version) and I am sure that many readers had no interest in mining but there were ideas and concepts in it that were useful for other kinds of problems. The proceedings of the NATO ASI (*Advanced Geostatistics in the Mining Industry*) became "Geostatistics for Natural Resources Characterization" in 1984. Who knows what the future will bring but IAMG and *Mathematical Geosciences* have made a significant contribution. They have influenced the development of mathematics, statistics, computing as well as the various fields that might be grouped under the heading "GEO-sciences".

# Chapter 45
# Origin and Early Development of the IAMG

**Frits Agterberg**

**Abstract** This chapter is primarily concerned with the first 15 years of our existence (I was a member of the IAMG Founding Committee, and on the 1968–1972 and 1996–1980 IAMG Councils). Daniel Merriam and Richard Reyment are the principal fathers of the IAMG, and many other scientists have contributed significantly to its origin and early development. Personal contacts with them are briefly described. These comments are supplementary to those already provided in earlier chapters by Founding Members and others who have made significant contributions to the IAMG originally. Special attention is paid to inputs by prominent mathematical statisticians with an interest in geology. I am grateful to all pioneers who have helped to establish the IAMG and provided a climate encouraging younger scientists, including myself, to pursue careers in their field of interest.

**Keywords** IAMG history · Richard Reyment · Daniel Merriam
Early mathematical geologists

## 45.1 Introduction

Perspectives on the origin and early development of the IAMG have already been provided in earlier chapters. Most of the following remarks are complementary to these other reminiscences. They are based on documents in the IAMG Archive, private information and what is publicly available on the IAMG Website including Newsletters from 1970 onward.

Richard Reyment had the original vision of establishing our organization as offspring from two parents: the International Union of Geological Sciences and the International Statistical Institute. As a successful example to follow for geologists, he took the biometrical society which was already in existence for quantitative

F. Agterberg (✉)
Geological Survey of Canada, 601 Booth Street, Ottawa, ON
K1A 0E8, Canada
e-mail: frits.agterberg@canada.ca; frits@rogers.com



B. S. Daya Sagar et al. (eds.), *Handbook of Mathematical Geosciences*,
https://doi.org/10.1007/978-3-319-78999-6_45

biologists and other life scientists, with its strong component of mathematical statistics. During 1966 and 1967, Reyment sought international support for the formation of our society. Especially mathematical statisticians were very supportive of his idea. He then organized the Founding Committee of the IAMG, although our name was to be chosen later. He invited me to be a member of his committee and chaired our inaugural meeting during the 23rd IGC in Prague where he became the IAMG's first Secretary General.

Daniel Merriam provided us with the essential publication and organizational background support for more than 30 years. In 1969 Dan was the founding Editor-in-Chief of the *Journal of the International Association for Mathematical Geology* (currently: *Mathematical Geosciences*), and in 1975 of *Computers & Geosciences*. Additionally, he was the chief organizer of numerous international meetings in our field, and editor of the proceedings for these meetings, as well as several other multi-author books. Later, in 2001, he took over as Editor-in-Chief of *Natural Resources Research*, our third international scientific journal that had originally been founded by Dick McCammon in 1992 under the name *Non-Renewable Resources*. In 1966, as Head of the Mathematical Geology Section, Kansas Geological Survey, Dan established the Distinguished Visiting Research Scientists program inviting mathematical geologists to work with him and his colleagues for successive one-year periods in Lawrence, Kansas. I was happy to accept Dan's invitation to occupy this position in 1969/70. During this fruitful year, my family and I were housed in the Sunflower apartments on the campus of Kansas University and received great hospitality. Merriam left Lawrence in 1976 to become Chair of the Geology Department, Syracuse University, where he commenced a new school for quantitative geoscientists. John Davis succeeded him at the Kansas Geological Survey.

Although originally educated in classical geology and geophysics at the University of Utrecht, I developed an interest in probability and statistics as a graduate student and published some papers on statistics applied in geology. Because of this, I was in 1962 invited to become "petrological statistician" at the Geological Survey of Canada (GSC) in Ottawa, initially to work within the framework of the Canadian Contribution to the International Upper Mantle Project and later to form their Geomathematics Section. The word "geomathematics" was used in analogy with "geophysics" and "geochemistry", but as a term it was never widely accepted. In 1982, engineers in photogrammetry had the idea of abbreviating the same word to "geomatics", which became widely accepted as a new discipline but is quite different from "mathematical geosciences".

GSC management allowed me to participate in the inaugural IAMG meeting on August 22nd, 1968, during the 23rd International Geological Congress in Prague. As described in earlier chapters, this event was disrupted and aborted because of the Russian-led occupation of Czechoslovakia. A list of participants in the inaugural meeting was included in its Minutes (see Appendix for final version of Minutes copied from the IAMG Archive) but several mathematical geologists including Bill Krumbein and Graeme Bonham-Carter, who had been planning to come to our first meeting, were prevented from coming to Prague to participate in the event.

Fortunately, my hotel was within walking distance of the Congress Centre and I also had been able to see several Founding Members before our meeting. Soon afterwards I was forced to leave Prague by car in a convoy of Dutch nationals led by the Dutch ambassador in the first car. Reyment had asked me to prepare minutes for our inaugural meeting and I handed him my first draft in Amsterdam where he, Geof Watson and I presented review papers at the Geostatistics Session organized during the 1968 meeting of the International Association of Statistics in the Physical Sciences (Section of the International Statistical Institute). This event helped to consolidate our affiliation with ISI. Formal affiliation with the IUGS had already been achieved in Prague.

## 45.2 Pioneers of Mathematical Geology

At its annual meetings the IAMG continues to honor five most eminent, pioneering scientists in our field: William Christian Krumbein, Andrey Borisovich Vistelius, John Cedric Griffiths, Felix Chayes and Georges Matheron. I was fortunate to know all five of them. Other leading scientists with strong IAMG involvements included John Tukey, Geof Watson, Danie Krige, Tim Whitten, Jean Serra and Walther Schwarzacher. Merriam and Howarth (2004) arranged for the publication of biographical articles on Matheron, Griffiths, Chayes, Reyment, Krumbein and Vistelius in a special edition of *Earth Sciences History*.

Krumbein (1936, 1939) already was developing important statistical techniques for geologists in the 1930s. My initial contact with him took place in the fall of 1961 when I was a postdoctorate fellow at the University of Wisconsin in Madison. My first assignment there was to perform statistical analysis of thousands of measurements on directional features taken by Ph.D. student Garrett Briggs in the Arkoma Basin of east-central Oklahoma (Agterberg and Briggs 1963). My report was reviewed by Krumbein before publication. His helpful comments included the suggestion to expand what initially was a brief footnote into a full section. It said that the circular normal (Von Mises) distribution for vectorial data converges to normal (Gaussian) form when dispersion around the vector mean approaches zero, so that standard (non-directional) statistical techniques including analysis of variance remain approximately applicable. Krumbein said that this remark solved a long-standing problem for him. Later, two of his Ph.D. students working with orientation data made use of this approach publishing their results in the first issue of our first IAMG journal (Jones and James 1969). I did not know at the time that Watson (1960) already had developed better approximations for statistical analysis of directional data. During his career, Krumbein continually sought the advice of mathematical statisticians including Franklin Graybill and John Tukey in order to stay on the right track. In 1963 the GSC invited him to Ottawa as a consultant, and I visited him at Northwestern University in a follow-up visit. Later I saw him regularly at scientific meetings, especially at those organized by Merriam in Lawrence, Kansas.

As a graduate student I gave an economic geology seminar on the skew frequency distribution of ore assays. In preparation I had read Krige's MSc thesis on microfilm in the library of the University of Utrecht. Its published version (Agterberg 1961) drew the attention of Danie Krige who wrote to me about it and became a good friend and esteemed colleague for more than 50 years. In 1963 he came to Ottawa on his way to the 3rd APCOM Symposium held at Stanford University. APCOM stands for "Applications of Computers and Operations Research in the Mineral Industries". With his wife Ansie and a colleague we went to Niagara Falls on a touristic outing. Danie persuaded GSC management that I should attend the 4th APCOM to be hosted by the Colorado School of Mines in 1964. Originally, APCOM meetings provided an important forum for mathematical geologists. I first met Dan Merriam, John Harbaugh, Tim Whitten and many others at early APCOMs.

In 1965 the GSC allowed me two months of travel abroad provided that I paid for my own travel expenses. First I went to the Netherlands where Codien Zwaardemaker invited me to dinner (we got married later that year; from 1993 onward she accompanied me to all IAMG annual meetings except one). From Amsterdam I went on to visit Krige in Johannesburg who took his family and me to the Kruger Park. Next there was the 8th Commonwealth Mining Congress in Australia, and finally the 5th APCOM at the University of Arizona, where I presented statistical analysis results for chemical analyses from the Muskox Layered Intrusion in northern Canada that was considered to be a sample of the upper mantle (Agterberg 1965). After this presentation John Griffiths came forward to congratulate me, also inviting me to present two papers instead of one at the next (1966) APCOM he would be hosting at the Pennsylvania State University. In those days, politicians in public paid more attention to oil and ore than today. The U.S. Secretary then in charge of mineral resources and mining gave the post-Symposium dinner speech. One of my two papers (Agterberg 1966) was entitled "Markov schemes for multivariate well data" and the Secretary singled this one out for a Cold War joke. Griffiths became one of my principal mentors. In 1968 Elsevier invited me to write a geomathematical textbook (Agterberg 1974). Griffiths and Merriam read all chapters and offered numerous helpful comments. Later I was honored to be invited to write the first chapter in the Griffiths commemorative book "Future Trends in Geomathematics" (Craig and Labovitz 1981).

Andrey Vistelius was the first IAMG President and his Laboratory of Mathematical Geology was used for our IAMG name. Tim Whitten, who was with Krumbein at Northwestern University, Evanston, Illinois, had invited him to come to North America in 1975 and for the last two weeks of this visit he was in the Geomathematics Section at the GSC in Ottawa. Before arrival, Vistelius had expressed the desire to sample a Canadian granite intrusion, preferably one with associated tin mineralization. There exists such a granite body in Nova Scotia but logistically we could not mount an expedition to sample it. Instead, with the help of other geologists we sampled the Meach Lake aplite body close to Ottawa. Aplite is fine-grained granite and this turned out to be a practical advantage, because thin sections of rock samples that could be cut in Ottawa were much smaller than the very large thin sections Vistelius had produced in Leningrad for counting frequencies of transitions between different minerals in granites. In total 104 thin sections were transition-counted and statistically analyzed. The rock body was interpreted to be "ideal granite" in which sequences of mineral grains are Markov chains (Vistelius et al. 1983). Later Xu et al. (2007) provided an alternative multifractal explanation of the Meach Lake aplite textures.

While Vistelius was in Ottawa, a preliminary itinerary was set up for my 6-week visit to the Soviet Union that took place two years later. It commenced with a 10-day stay in Novosibirsk where I participated in the Siberian Seminar on "Application of Mathematical Methods and Computers for Mineral Search and Prospecting" organized by Yuri Voronin. Václav Němec, IAMG Treasurer (East) was participating as well. Neither Vistelius nor Founding Member Dmitry Rodionov attended. Němec was our IAMG ambassador to the Soviet bloc countries (cf. Agterberg 1994). My Siberian Seminar contribution (Agterberg 1977) was the only presentation with slides. Initially, the organizers told that I could only show three slides, because other participants were not allowed to display more than three posters but they relented. A slide projector was brought in from another institute and all my slides were shown. Before I was leaving for Moscow on the next stop, Němec had warned me that during my upcoming visit to Rodionov and his colleagues I would be asked for an opinion on the work of Voronin and his team; he explained that a negative opinion could be detrimental because Moscow controlled funding of the Novosibirsk projects. I was careful in what I said. It was understood in the Soviet Union that the farther east you went, the more philosophical the mathematical approach to geology became. I learned at the Siberian Seminar that rocks are subject to the basic philosophical principle that the "whole is more than the sum of the parts".

The last two weeks of my visit to the Soviet Union were spent in Leningrad. Every day I arrived at the Laboratory of Mathematical Geology 2 h before Vistelius, who did most of his work at home where we went in the afternoon for discussions and a meal. As explained by Steve Henley, Vistelius was given a hard time under the communist regime because of his aristocratic roots. In order to accept an invitation for a lecture tour he had just received from Japan, he needed numerous approvals. The process, which involved various unpleasant interviews with officials plus extensive form-filling, took more than two weeks. On the day of my departure Vistelius received a phone call from somebody he referred to as a "foxtail" who communicated indirectly to him what could be interpreted as final travel approval. The foxtail did not communicate this in so many words but said that an official in Moscow had remarked that the Laboratory of Mathematical Geology in Leningrad did good work. This implied approval and Vistelius went indeed to Japan shortly afterwards. During our many discussions we were not always in total agreement. Vistelius held very strong opinions and was not at all impressed by geostatistics or geostatisticians. He felt that mathematical geology had to be "pure" and not contaminated with economic motivations. Even much later, after he had invited me to participate in a mathematical geology meeting, he pointed out that in his session there would be no room for statistics applied to ore deposits, but he suggested other topics on which I could report.

My recognition of the validity of French geostatistics took place in 1964 because of a curious incident. Our library had obtained a copy of the first book by Matheron (1963) but there had been a complaint from the public that this volume contained absolute nonsense and should be removed from the shelf. The head of the Library Committee approached me and asked for an evaluation because: "We don't want bad books on our shelves". My report was favorable and the book could stay. Although this is not universally known, Georges Matheron commenced his career at the French Geological Survey (BRGM) in 1954. One of his first publications (Matheron 1955a) concerns the Gara Gjebilet oolithic iron deposit in Algeria. It is a standard geological publication with detailed descriptions of the stratigraphy, structure and genesis of this deposit of Early Devonian age plus a folded geological map in the back. It seems that Matheron started out as a classical geologist but shortly afterwards he published a paper (Matheron 1955b) on applications of statistical methods for ore reserve estimation. This first paper foreshadowed the revolutionary approach to spatial statistics he was to bring about during the last 40 years of the 20th century. Like Vistelius, Matheron had strong opinions on topics that would be suitable for research. His first two Ph.D. students (Michel David and André Journel) ran into significant problems later on, when in some of their projects they deviated from what Matheron felt was appropriate for them. In 1968 Michel David had come to the École Polytechnique in Montreal and we collaborated on several projects. One of these involved correspondence analysis (Agterberg and David 1979). But one day David showed me a letter from Matheron stating that this work should be stopped immediately and that he should return to working full-time on geostatistics.

In 1968 Georges Matheron established the Centre de Morphologie Mathématique in Fontainebleau, as a research institute of the École des Mines de Paris. Jean Serra was his close collaborator. Matheron's preferred mode of work was to be in his office in Fontainebleau during the day. He would document his findings in limited-edition geostatistical notes. Fully concentrating on his research, he did not like to speak English nor extensive traveling. I visited him three times. Although for about 10 years my position at the GSC was classified as "bilingual", I never spoke French in Ottawa because all French Canadian colleagues spoke English. However, speaking French was a requirement for personal (and telephone) contact with Matheron. An extra benefit of making the geostatistical pilgrimage to Fontainebleau was that I could consult the numerous geostatistical notes in their library and could bring back to Ottawa any copies of particular interest. Today all these notes are freely available on a website maintained by the École des Mines de Paris. I am sure they continue to contain valuable information that is relatively unknown. During the late 1970s I programmed in FORTRAN some of the methods developed by Matheron and Serra. Twice, I received a *Computers & Geosciences* best-paper award for these efforts. I was pleased to be asked in 1975 to chair a session at the first Geostatistical World Conference held in Frascati, Italy, at which Georges Matheron presented a philosophical paper (Matheron 1976). At the 53rd Session of the International Statistical Institute in Seoul, August 2001, Georges Matheron was honoured as one of the greatest mathematical statisticians during the second half of the 20th century (cf. Baddeley 2001). After obtaining approval from Mrs. Matheron, the IAMG established its annual Georges Matheron lecture in 2005, delivered for the first time by Jean Serra at IAMG2006 in Liège. Our Matheron Lecture was modeled after the Fisher Memorial Lecture initiated by the International Statistical Institute in 1966.

Felix Chayes was a member of the IAMG Founding Committee and participated in many IAMG events. His numerous contributions have been documented by Howarth (2004). Upon his death in 1993 he left the IAMG a significant legacy in order to fund the biennial Felix Chayes Prize for Excellence in Research in Mathematical Petrology. For many years Chayes was involved in compiling large databases with worldwide data on Cenozoic volcanic rocks. This effort included directing International Geological Correlation Programme (IGCP) Project 163 (1977–1984) IGBA (Igneous data Base) which had supportive software as well. Close IAMG involvement with IGCP had been promoted by Merriam who also helped initiate IGCP Project 148 (1976–1983) "Quantitative Stratigraphy".

John Cubitt was the original leader of IGCP Project 148 but he left Syracuse University where he was with Merriam in 1977 to become a private consultant in the U.K. and I took over from him. We created a group of lecturers to present one-week short courses on the subject that eventually were held in as many as nine different countries. The strategy was to attract staff from oil companies in "developed" countries willing to pay registration fees that were later used to give the course in "developing" nations. Walther Schwarzacher and I were part of this "traveling circus". Originally, I had met Schwarzacher in Lawrence, Kansas, where we were both associated with Merriam's quantitative geology group. He was the IAMG's second Krumbein Medallist in 1977 (John Griffiths was the first a year earlier). In the IGCP Project 148 short course Schwarzacher lectured on lithostratigraphic correlation. Later he published a book that explained the Milankovitch theory (Schwarzacher 1993) according to which very small periodic variations in solar radiation create major climate changes on Earth. This idea had been anticipated by Croll (1875) as an explanation of the ice ages. Currently, the entire post-Cretaceous international geologic time scale is based on Milankovitch theory.

Walther and I had several things in common. In Europe we had attended similar high schools called "gymnasium" in both Austria and the Netherlands, at which the emphasis was on Latin and Greek. We still could recite some of the Odyssey to each other. Later I tried some of my ancient Greek on Roussos Dimitrakopoulos who smiled benevolently. The supervisor of Schwarzacher's Ph.D. project had been Bruno Sander at the University of Innsbruck. Later (in 1957) I took a short course at this university in order to learn micro-tectonics in preparation of my fieldwork during four successive summers in northern Italy (Agterberg 1961). The most important results of this doctoral thesis were included in Whitten (1966)'s textbook on structural geology. Later, Hannes Thiergärtner and Heinz Burger invited me to contribute further articles on this subject on two occasions. Original Alpine deformation patterns for the basement of the Italian Dolomites had to be re-interpreted in terms of rapid movements of the Adria microplate that presently keep on creating earthquakes in the Apennines (cf. Agterberg 2014).

## 45.3 Inputs from Mathematical Statisticians

Most important among the first mathematical statisticians was Ronald Fisher (1954) who suggested that geology with Lyell (1833) had been evolving as a more quantitative science but, rapidly, opposition against this development grew to the extent that Lyell's elaborate tables and statistical arguments (60 pages long) for his subdivision of the Tertiary were omitted from later editions of his *Principles of Geology*. In 1952 Fisher commenced giving regular talks on continental drift (cf. Fisher Box 1978. p. 440) lamenting that geophysicists and geologists were failing to take seriously Alfred Wegener's ideas on continental drift proposed in 1912. Plate tectonics only became generally accepted as a theory in the mid-1960s.

My Moscow stay in 1977 would have included visiting Andrey Nikolayevich Kolmogorov (1956) who originally formulated the axioms of probability calculus in his famous paper of 1931. Unfortunately, this visit had to be canceled for medical reasons. Like Krumbein in North America, Vistelius regularly consulted with mathematical statisticians and Kolmogorov was a major source of inspiration to him.

In 1983 the traveling circus of IGCP Project 148 was at the Indian Institute of Technology in Kharagpur. The lecturers included Geof Watson, 1968–1972 IAMG Vice President, who within 2 h filled an extra wide blackboard entirely with equations on the relationship between kriging and interpolation splines. It is doubtful that anybody in the audience (including me) could understand what he was talking about. Later I spent significant time understanding his subsequent paper on the subject (Watson 1984). I used smoothing splines extensively for estimating the ages of stage boundaries (with 95% error bars) in the International Geological Time Scale (Gradstein et al. 2004). Watson has done much to make Matheron's work in the fields of geostatistics and mathematical morphology better known in the English-speaking world. He persuaded Matheron (1975) to write his book on random sets and integral geometry. At the time Watson told me that there would be only three people in world able to understand this book from beginning to end.

Originally, Watson (1960) had developed statistical methods for directional features that were similar to methods for ordinary data originally developed by Fisher who was the world's most outstanding mathematical statistician during the first half of the 20th century. Fisher was from before my time. Some of our earliest IAMG members including Griffiths and Schwarzacher knew him personally. When I attended the 1963 congress of the International Statistical Institute in Ottawa, he had already left for Adelaide, Australia where he spent his last years in retirement. Fisher's life is described in detail by his daughter Joan Fisher Box (1978). During the latter part of the 19th century, Karl Pearson had introduced many basic statistical concepts including the Pearson correlation coefficient and goodness-of-fit tests for contingency tables, basing his approach on normal (Gaussian) distribution models. Fisher derived the mathematical equation for the frequency distribution of the Pearson correlation coefficient and introduced numbers of degrees of freedom for various statistical methods that became widely used, also by the early mathematical geologists. In these methods extensive use was made of independent identically distributed (*iid*) random variables, contrary to geostatistical applications in which the emphasis was on "regionalized" variables that generate observed values that are not stochastically independent but spatially correlated.

In 1966 the GSC allowed me to participate in the Advanced Statistical Seminar at the University of Wisconsin organized by Fisher's son-in-law Box. During the Icebreaker I was introduced to John Tukey who told me about his interest in geology. At this seminar he presented "The Fast Fourier Transform, for fun and profit" (cf. Cooley and Tukey 1965). Back in Ottawa, I received a box filled with about 2000 IBM cards for running the FFT in 1, 2, or 3 dimensions on our mainframe computer. During the next 25 years, Tukey commented on my projects at the GSC in three of the approximately 800 publications he authored or co-authored (cf. Agterberg 2001; Tukey 1984). Like Matheron, he was recognized at the 2001 ISI Congress in Seoul as one of the greatest mathematical statisticians alive during the second part of the 20th Century. With Watson who had become Chair of the Princeton University Statistics Department, where Tukey was a professor, he attended the 1969 Geostatistics Colloquium organized by Dan Merriam in Lawrence, Kansas, that also had Matheron, Krumbein and Serra as participants.

Watson owned a cottage on Blood Hill near Elizabethville in the Adirondacks, New York State, not too far from Ottawa. In those days, the GSC maintained a pool of cars with the words "Geological Survey of Canada" in big letters on the sides. I could use one if these cars to visit Watson during weekends. Once I drove Geof and some of his family members to Princeton where Tukey spotted us on the campus. He started laughing and pointing his finger at Watson suggesting that Geof had become a "geologist". Watson stimulated me to improve my mathematical skills. Pointing out some errors in a review of Agterberg (1974) he had, somewhat sarcastically, remarked that one could see I was not trained as a mathematical statistician. However, he would have granted me an MSc degree in this discipline. Subsequently I worked hard on my mathematics. In 1983 I organized a geomathematical workshop at the GSC in Ottawa with Geof Watson, Jean Serra and Benoit Mandelbrot among the presenters. Mandelbrot who had coined the word "fractal" like Matheron had been a student of Paul Lévy at the École Polytechnique in Paris. Other participants in our workshop included the directors of Carleton University's Centre of Mathematical Statistics who shortly afterwards invited me to become an Adjunct Professor in their Mathematics Department. I felt this was almost as good as a Ph.D. in mathematical statistics. Personally, I have always felt that this discipline offered me more challenges than conventional geology although this remains a scientific discipline in its own right.

## 45.4 Concluding Remarks

The preceding remarks are to a large extent personal like several reminiscences in earlier chapters. I have tried to add to these other contributions, above all attempting to bring out the generosity our pioneers extended to younger colleagues. By their research and contributions to the IAMG they insured a healthy organization that should continue to exist and expand for many years to come.

## Appendix: Minutes of the First Meeting of the International Association for Mathematical Geology, Prague, August 22, 1968

The meeting was attended by 20 representatives from 10 different countries (see attached list of participants).

After a general introduction by the acting chairman, R. A. Reyment, the following two problems were discussed:

1. Statutes and By-laws
2. Journal

The relatively short name of "International Association for Mathematical Geology (I.A.M.G.)" was adopted for the Society.

A. B. Vistelius proposed discussion of possible classes of membership and also which categories of members should be entitled to vote in the General Assembly. It was pointed out that the Association should consider the options of (a) voting by country (each country one vote) or (b) as individual scientists. However, membership should be open to all scientists. The possibility of having a fixed number of voting members was also discussed. It was felt that the latter procedure may be unfair to the larger countries.

Article 7 of the proposed Statutes (each member of I.A.M.G. one vote) was adopted. However, this discussion resulted in the following change in Article 10 of the proposed statutes:

1. There shall be two treasurers (East and West) instead of one, in order to meet the problem of non-convertible currencies.
2. There shall be only one representative on the Council appointed by the geologists of the host country for the next International Geological Congress.
3. The sentence "Not more than two ordinary members shall be from the same country" shall be replaced by "Representation on the Council shall reflect regional distribution of membership as stated in the by-laws."

The following by-law was adopted:

"By-law 7: Not more than two ordinary members, and/or four members of the Council shall be from the same country. This by-law shall be reviewed every four years by the General Assembly."

The matter of introducing a journal was discussed. First, the following by-law was accepted:

"By-law 8: The editor-in-chief, in consultation with the Council, shall be empowered to appoint up to four associate editors."

The Assembly adopted a motion initiated by G. S. Watson "that the Society shall have a journal".

After the acceptance of the statutes and by-laws had been reached and general agreement there shall be a journal, the chairman proposed to the Assembly the electing of the officers of the Council.

The following 13 members of the Council were elected:

A. B. Vistelius—President
G. S. Watson—Vice President (also president elect)
R. A. Reyment—Secretary General
V. Němec—Treasurer (east)
T. V. Loudon—Treasurer (west)
W. C. Krumbein—Past President (instead of Immediate Vice President, see by-law 9)
D. F. Merriam—Editor-in-Chief
D. F. Rodionov, S. P. Sen Gupta, F. P. Agterberg, G. Matheron, D. G. Krige, E. H. T. Whitten—Ordinary members.

The following by-law was accepted:

"By-law 9: For the first four years of the Society's life, instead of an immediate past president, there shall be an additional vice president."

Since some of the elected members were not present at this meeting, the following motion initiated by J. W. Harbaugh, was adopted:

"If an elected member should not wish to serve on the council, Professor Vistelius shall nominate the next member on the list." Prof. Vistelius has a list of persons eligible as ordinary members and the number of votes they received at the election.

P. Wilkinson moved that: "The Association encourages, in principle, the formation of national groups in mathematical geology and that the question of affiliation should be discussed at the next General Assembly in Montreal." This motion was adopted.

Finally, the policy and objectives for the journal were discussed. It was suggested that there should be a broad editorial program. Similar to that of the biometrical journal Biometrics. The editor-in-chief should prepare guidelines for the journal. The first issues should also contain educational papers.

The official languages of the organization are French, English, German and Russian. It is appreciated that the editing of papers in Russian may present a problem to the editor-in-chief, and in practice only two or three languages will be used for publication. All articles shall have an abstract in English.

List of participants, First meeting of International Association for Mathematical Geology, Prague, August 22, 1968.

---

R. A. Reyment (Sweden)
D. A. Rodionov (U.S.S.R.)
A. B. Vistelius (U.S.S.R.)
F. P. Agterberg (Canada)
H. Knape (G.D.R.)
H. Thiergärtner (G.D.R.)
G. S. Watson (U.S.A.)
V. Němec (Czechoslovakia)
D. J. Burdon (FAO of United Nations)
C. J. Dixon (U.K.)
P. Wilkinson (U.K.)
T. V. Loudon (U.K.)
R. Ivanov (Bulgaria)
V. Kutolin (U.S.S.R.)
F. Benkö (Hungary)
E. H. T. Whitten (U.S.A.)
R. B. McCammon (U.S.A.)
J. W. Harbaugh (U.S.A.)
R. Hesse (F.R.G.)
D. F. Merriam (U.S.A.)

## References


Agterberg FP (1961) Tectonics of the crystalline basement of the Dolomites in North Italy. Kemink, Utrecht (Open Access via Utrecht University Repository)

Agterberg FP (1965) Frequency distribution of trace elements in the Muskox layered intrusion. In: Dotson JC, Peters WC (eds) Short course and symposium on computers and computer applications in mining and exploration. University of Arizona, Tucson, pp G1–G33

Agterberg FP (1966) Markov schemes for multivariate well data. In: Proceedings, symposium on applications of computers and operations research in the mineral industries, Pennsylvania State University, State College, Pennsylvania, vol II, pp X1–X18

Agterberg FP (1974) Geomathematics. Elsevier, Amsterdam

Agterberg FP (1977) Probabilistic approach to mineral resource evaluation problems. In: Yu V (ed) Proceedings, Siberian seminar on "application of mathematical methods and computers for mineral search and prospecting", held in Novosibirsk, 23–26 May 1977, Computing Centre, Novosibirsk, U.S.S.R (in Russian)

Agterberg FP (1994) Václav Němec: Krumbein medallist. Math Geol 26(6):753–754

Agterberg FP (2001) Appreciation of contributions by Georges Matheron and John Tukey to a mineral-resources research project. Nat Resour Res 10(4):287–295

Agterberg FP (2014) Geomathematics: theoretical foundations, applications and future developments. Springer, Heidelberg, pp 290–301

Agterberg FP, Briggs G (1963) Statistical analysis of ripple marks in Atokan and Desmoinesian rocks in the Arkoma Basin of east-central Oklahoma. J Sediment Petrol 33:393–410

Agterberg FP, David M (1979) Statistical exploration. In: Weiss A (ed) Computer methods for the 80's. Society of Mining Engineers, New York, pp 90–115

Baddeley A (2001) Georges Matheron (1930–2000). Bull Intern. Statistical Institute, 53rd Session Proceedings. Tome LIX. Book 1:529–532

Cooley JW, Tukey JW (1965) An algorithm for the machine calculation of complex Fourier series. Math Comput 19:297–307

Craig RG, Labovitz (eds) (1981) Future trends of geomathematics. Pion, London

Croll J (1875) Climate and time in their geological relations. Appleton, New York

Fisher RA (1954) Expansion of statistics. Am Sci 42:275–282

Fisher Box J (1978) R. A. Fisher, the life of a scientist. Wiley, New York

Gradstein FM, Ogg J, Smith A, eds (2004) A geologic time scale. Cambridge University Press, Cambridge

Howarth RJ (2004) Not "just a petrographer": the life and work of Felix Chayes (1916–1993). Earth Sci Hist 23(2):343–364

Jones TA, James WR (1969) Statistical analysis of orientation data. J Int Assoc Math Geol 1: 129–136

Kolmogorov AN (1956) Foundations of the theory of probability (2nd English translation). Chelsea, New York

Krumbein WC (1936) Application of logarithmic moments to size frequency distributions of sediments. J Sediment Pet 6:35–47

Krumbein WC (1939) Preferred orientation of pebbles in sedimentary deposits. J Geol 47:673–706

Lyell C (1833) Principles of geology, 3rd edn. Murray, London

Matheron G (1955a) Les gisements de Gara Djebilet. Bulletin Scientifique et Économique du B.R. M.A., No 2, May 1955, published by Bureau de Recherches Minières de l'Algérie, pp 53–63

Matheron G (1955b) Application des methods statistiques à l'évaluation des gisements. Annales des Mines, décembre 1955, pp 50–75

Matheron G (1963) Traité de Géostatistique Appliquée. Mémoire BRGM 14. Éditions Technip, Paris

Matheron G (1975) Random sets and integral geometry. Wiley, New York

Matheron G (1976) Les concepts de base et l'évolution de la géostatistique minière. In: Guarascio M, David M, Huijbregts C (eds) Advanced geostatistics in the mining industry. Reidel, Dordrecht, pp 3–10

Merriam DF, Howarth RJ (2004) Pioneers in mathematical geology. Earth Sci Hist 23(2):314–432

Schwarzacher W (1993) Cyclostratigraphy and the Milankovitch theory. Elsevier, Amsterdam

Tukey JW (1984) Comments on a paper by F. P. Agterberg. Math Geol 16(6):590–594

Vistelius AB, Agterberg FP, Divi SR, Hogarth DD (1983) A stochastic model for the crystallization and textural analysis of a fine grained granitic stock near Meech Lake, Québec. Geological Survey of Canada Paper 81–21

Watson GS (1960) More significance tests on the sphere. Biometrika 47:87–91

Watson GS (1984) Smoothing and interpolating by kriging and with splines. Math Geol 16(6):601–615

Whitten EHT (1966) Structural geology of folded rocks. Rand McNally, Skokie, Illinois

Xu D, Cheng Q, Agterberg FP (2007) Scaling properties of ideal granitic sequences. Nonlinear Proc Geophys 14:237–246

www.ingramcontent.com/pod-product-compliance
Lightning Source LLC
LaVergne TN
LVHW062151200926
848484LV00007B/1896
*9781013276293*